Handbook
of
Biochemistry
and
Molecular Biology

CRC Handbook of Biochemistry and Molecular Biology

3rd Edition

Proteins
Volume II

Editor

Gerald D. Fasman, Ph. D.
Rosenfield Professor of Biochemistry
Graduate Department of Biochemistry
Brandeis University
Waltham, Massachusetts

CRC Press, Inc.
Boca Raton, Florida

Library of Congress Cataloging in Publication Data
Main entry under title:

Handbook of biochemistry and molecular biology.

    Previous editions published under title: Handbook of biochemistry.
    Includes bibliographies and indexes.
    CONTENT: A. Proteins. 2. —B. Nucleic acids.
2. v.—C. Liquids. carbohydrates, steroids.
    1. Biological chemistry — Handbooks, manuals, etc.
2. Molecular biology — Handbooks, manuals, etc.
1. Fasman. Gerald. D. II. Sober, Herbert Alexander,
1918-   [DNLM: 1. Nucleic acids — Tables. 2. Alkaloids — Tables. 3. Carbohydrates — Tables. 4. Lipids — Tables. 5. Steroids — Tables. QU16]
QP514.2.H34   1975   574.1'92   75-29514
ISBN 0-8493-0505-5

This book represents information obtained from authentic and highly regarded sources. Reprinted material is quoted with permission, and sources are indicated. A wide variety of references are listed. Every reasonable effort has been made to give reliable data and information, but the author and the publisher cannot assume responsibility for the validity of all materials or for the consequences of their use.

All rights reserved. This book, or any parts thereof, may not be reproduced in any form without written consent from the publisher.

Direct all inquiries to CRC Press, Inc., 2000 Corporate Blvd., N.W., Boca Raton, Florida, 33431.

© 1976 by CRC Press, Inc.
©1970, 1968 by The Chemical Rubber Co.
Second Printing, 1982
Third Printing, 1986
Fourth Printing, 1989

International Standard Book Number 0-8493-0503-9 (Complete Set)
International Standard Book Number 0-8493-0505-5
Former International Standard Book Number 0-87819-503-3 (Complete Set)
Former International Standard Book Number 0-87819-505-X (Proteins, Volume II)

Library of Congress Card No. 75-29514
Printed in the United States

Handbook of Biochemistry and Molecular Biology

3rd Edition

Proteins

Volume II

Editor
Gerald D. Fasman, Ph. D.
Rosenfield Professor of Biochemistry
Graduate Department of Biochemistry
Brandeis University
Waltham, Massachusetts

The following is a list of the four major sections of the *Handbook*, each consisting of one or more volumes

Proteins — Amino Acids, Peptides, Polypeptides, and Proteins

Nucleic, Acids — Purines, Pyrimidines, Nucleotides, Oligonucleotides, tRNA, DNA, RNA

Lipids, Carbohydrates, Steroids

Physical and Chemical Data, Miscellaneous — Ion Exchange, Chromatography, Buffers, Miscellaneous, e.g., Vitamins

ADVISORY BOARD

Gerald D. Fasman
Editor

Herbert A. Sober (deceased)
Consulting Editor

MEMBERS

Bruce Ames
  Professor, Department of Biochemistry
  University of California
  Berkeley, California 94720

Sherman Beychok
  Professor, Department of Biological
    Sciences
  Columbia University
  New York, New York 10027

Waldo E. Cohn
  Senior Biochemist, Biology Division
  Oak Ridge National Laboratory
  Oak Ridge, Tennessee 37830

Harold Edelhoch
  National Institute Arthritis, Metabolism
    and Digestive Diseases
  Department of Health, Education, and
    Welfare
  National Institutes of Health
  Bethesda, Maryland 20014

John Edsall
  Professor Emeritus, Biological Laboratories
  Harvard University
  Cambridge, Massachusetts 02138

Gary Felsenfeld
  Chief, Physical Chemistry Laboratory
  Laboratory of Molecular Biology
  National Institute of Arthritis,
    Metabolism, and Digestive Diseases
  National Institutes Of Health
  Bethesda, Maryland 20014

Edmond H. Fischer
  Professor, Department of Biochemistry
  University of Washington
  Seattle, Washington 98195

Victor Ginsburg
  Chief, Biochemistry Section, National
    Institute of Arthritis, Metabolism and
    Digestive Diseases
  Department of Health, Education, and
    Welfare
  National Institutes of Health
  Bethesda, Maryland 20014

Walter Gratzer
  MRC Neurobiology Unit
  Department of Biophysics
  Kings College
  University of London
  London
  England

Lawrence Grossman
  Professor, Department of Biochemical and
    Biophysical Sciences
  School of Hygiene and Public Health
  The Johns Hopkins University
  Baltimore, Maryland 21205

Frank Gurd
  Professor, Department of Chemistry
  Indiana University
  Bloomington, Indiana 47401

William Harrington
  Professor, Department of Biology
  The Johns Hopkins University
  Baltimore, Maryland 21218

William P. Jencks
  Professor, Graduate Department of
    Biochemistry
  Brandeis University
  Waltham, Massachusetts 02154

ADVISORY BOARD (continued)

O. L. Kline
Executive Officer
American Institute of Nutrition
9650 Rockville Pike
Bethesda, Maryland 20014

I. M. Klotz
Professor, Department of Chemistry
Northwestern University
Evanston, Illinois 60201

Robert Langridge
Professor, Department of Biochemistry
Princeton University
Princeton, New Jersey 08540

Philip Leder
Chief, Laboratory of Molecular Genetics
National Institute of Child Health
and Human Development
National Institutes of Health
Bethesda, Maryland 20014

I. Robert Lehman
Professor, Department Biochemistry
School of Medicine
Stanford University
Stanford, California 94305

Lawrence Levine
Professor, Graduate Department of
Biochemistry
Brandeis University
Waltham, Massachusetts 02154

John Lowenstein
Professor, Graduate Department of
Biochemistry
Brandeis University
Waltham, Massachusetts 02154

Emanuel Margoliash
Professor, Department of Biological
Sciences
Northwestern University
Evanston, Illinois 60201

Julius Marmur
Professor, Department of Biochemistry
and Genetics
Albert Einstein College of Medicine
New York, New York 10461

Alton Meister
Professor, Department of Biochemistry
Cornell University Medical College
New York, New York 10021

Kivie Moldave
Professor, Department of Biochemistry
California College of Medicine
University of California
Irvine, California 92664

D. C. Phillips
Professor, Laboratory of Molecular
Biophysics
Department of Zoology
Oxford University
Oxford
England

William D. Phillips
The Lord Rank Research Centre
Ranks Hove, McDougall Ltd.
Lincoln Road, High Wycombe
Bucks
England

G. N. Ramachandran
Professor, Molecular Biophysics Unit
Indian Institute of Science
Bangalore
India

Michael Sela
Professor, Department of Chemical
Immunology
The Weizmann Institute of Science
Rehovot
Israel

ADVISORY BOARD (continued)

Waclaw Szybalski
  Professor, McArdle Laboratory for
    Cancer Research
  The University of Wisconsin
  Madison, Wisconsin, 53706

Serge N. Timasheff
  Professor, Graduate Department of
    Biochemistry
  Brandeis University
  Waltham, Massachusetts 02154

Ignacio Tinoco, Jr.
  Professor, Department of Chemistry
  University of California
  Berkeley, California 94720

Bert L. Vallee
  Professor, Biophysics Research
    Laboratory
  Peter Bent Brigham Hospital
  Harvard Medical School
  Boston, Massachusetts 02115

CONTRIBUTORS

V. S. Ananthanarayahan
    Molecular Biophysics Unit
    Indian Institute of Science
    Bangalore 560012
    India

Lynne H. Botelho
    Department of Chemistry
    Indiana University
    Bloomington, Indiana 47401

Colin F. Chignell
    Section of Molecular Pharmacology
    Pulmonary Branch
    National Heart and Lung Institute
    National Institutes of Health
    Bethesda, Maryland 20014

Waldo E. Cohn
    Biology Division
    Oak Ridge National Laboratory
    Oak Ridge, Tennessee 37830

Dennis A. Darnall
    Department of Chemistry
    New Mexico State University
    Las Cruses, New Mexico 88003

Earl W. Davie
    Department of Biochemistry
    University of Washington
    Seattle, Washington 98195

Robert DeLange
    Department of Biological Chemistry
    School of Medicine
    The Center for the Health Sciences
    Los Angeles, California 90024

Leon Goldstein
    Biochemistry Department
    The George S. Wise Center for
        Life Sciences
    Tel-Aviv University
    Tel-Aviv, Israel

Frank R. N. Gurd
    Department of Chemistry
    Indiana University
    Bloomington, Indiana 47401

C. H. W. Hirs
    Division of Biological Sciences
    Indiana University
    Bloomington, Indiana 47401

Andrew H. Kang
    University of Tennessee Center for
        Health Sciences
    Veterans Administration Hospital
    Memphis, Tennessee 38104

Beatrice Kassell
    Department of Biochemistry
    The Medical College of Wisconsin
    Milwaukee, Wisconsin 53233

Donald M. Kirschenbaum
    Department of Biochemistry
    State University of New York
    Brooklyn, New York 11203

Walter Kisiel
    Department of Biochemistry
    University of Washington
    Seattle, Washington 98195

Irving M. Klotz
    Biochemistry Division
    Department of Chemistry
    Northwestern University
    Evanston, Illinois 60201

Halina Lis
    Department of Biophysics
    The Weizmann Institute of Science
    Rehovot, Israel

Laszlo Lorand
    Division of Biochemistry
    Department of Chemistry
    Northwestern University
    Evanston, Illinois 60201

Susan Lowey
    Rosenstiel Basic Medical Research Center
    Brandeis University
    Waltham, Massachusetts 02154

Pierre L. Masson
    International Institute of Cellular
        and Molecular Pathology
    Catholic University of Louvain
    B-1200, Brussels
    Belgium

Brian W. Matthews
    Institute of Molecular Biology
    University of Oregon
    Eugene, Oregon 97403

CONTRIBUTORS (continued)

Amadeo J. Pesce
Division of Nephrology
College of Medicine
University of Cincinnati
Cincinnati, Ohio 45267

Thomas D. Pollard
Department of Anatomy
Harvard Medical School
Boston, Massachusetts 02115

James R. Riordan
Biophysics Research Laboratory
Peter Bent Brigham Hospital
Boston, Massachusetts 02115

Lennart Roden
The University of Alabama
Birmingham, Alabama 35294

Walter C. Schneider
Laboratory of Physiology
National Cancer Institute
National Institutes of Health
Bethesda, Maryland 20014

Jerome M. Seyer
University of Tennessee Center for Health Sciences
Veterans Administration Hospital
Memphis, Tennessee 38104

Nathan Sharon
Department of Biophysics
The Weizmann Institute of Science
Rehovot, Israel

Malcolm M. Smith (deceased)

Morris Soodak
Graduate Department of Biochemistry
Brandeis University
Waltham, Massachusetts 02154

Thomas F. Spande
Laboratory of Chemistry
National Institute of Arthritis and Metabolic Diseases
National Institutes of Health
Bethesda, Maryland 20014

Pal Stenberg
Upplandsgatan
6 IV
S-11123 Stockholm
Sweden

Serge N. Timasheff
Graduate Department of Biochemistry
Brandeis University
Waltham, Massachusetts 02154

Bert L. Vallee
Harvard Medical School
Biophysics Research Laboratory
Peter Bent Brigham Hospital
Boston, Massachusetts 02115

M. Vijayan
Molecular Biophysics Unit
Indian Institute of Science
Bangalore 560012
India

Warren E. C. Wacker
Harvard University Health Services
Cambridge, Massachusetts 02138

Monica June Williams
Department of Biochemistry
The Medical College of Wisconsin
Milwaukee, Wisconsin 53233

B. Witkop
Laboratory of Chemistry
National Institute of Arthritis and Metabolic Diseases
National Institutes of Health
Bethesda, Maryland 20014

PREFACE

The rapid pace at which new data is currently accumulated in science presents one of the significant problems of today — the problem of rapid retrieval of information. The fields of biochemistry and molecular biology are two areas in which the information explosion is manifest. Such data is of interest in the disciplines of medicine, modern biology, genetics, immunology, biophysics, etc., to name but a few related areas. It was this need which first prompted CRC Press, with Dr. Herbert A. Sober as Editor, to publish the first two editions of a modern *Handbook of Biochemistry,* which made available unique, in depth compilations of critically evaluated data to graduate students, post-doctoral fellows, and research workers in selected areas of biochemistry.

This third edition of the *Handbook* demonstrates the wealth of new information which has become available since 1970. The title has been changed to include molecular biology; as the fields of biochemistry and molecular biology exist today, it becomes more difficult to differentiate between them. As a result of this philosophy, this edition has been greatly expanded. Also, previous data has been revised and obsolete material has been eliminated. As before, however, all areas of interest have not been covered in this edition. Elementary data, readily available elsewhere, has not been included. We have attempted to stress the areas of today's principal research frontiers and consequently certain areas of important biochemical interest are relatively neglected, but hopefully not totally ignored.

This third edition is over double the size of the second edition. Tables used from the second edition without change are so marked, but their number is small. Most of the tables from the second edition have been extensively revised, and over half of the data is new material. In addition, a far more extensive index has been compiled to facilitate the use of the Handbook. To make more facile use of the Handbook because of the increased size, it has been divided into four sections. Each section will have one or more volumes. The four sections are titled:

Proteins — Amino Acids, Peptides, Polypeptides, and Proteins
Nucleic Acids — Purines, Pyrimidines, Nucleotides, Oligonucleotides, tRNA, DNA, RNA
Lipids, Carbohydrates, Steroids
Physical and Chemical Data, Miscellaneous — Ion Exchange, Chromatography, Buffers, Miscellaneous, e.g., Vitamins

By means of this division of the data, we can continuously update the *Handbook* by publishing new data as they become available.

The Editor wishes to thank the numerous contributors, Dr. Herbert A. Sober, who assisted the Editor generously, and the Advisory Board for their counsel and cooperation. Without their efforts this edition would not have been possible. Special acknowledgments are due to the editorial staff of CRC Press, Inc., particularly Ms. Susan Cubar Benovich, Ms. Sandy Pearlman, and Mrs. Gayle Tavens, for their perspicacity and invaluable assistance in the editing of the manuscript. The editor alone, however, is responsible for the scope and the organization of the tables.

We invite comments and criticisms regarding format and selection of subject matter, as well as specific suggestions for new data (and their sources) which might be included in subsequent editions. We hope that errors and omissions in the data that appear in the Handbook will be brought to the attention of the Editor and the publisher.

Gerald D. Fasman
Editor
August 1975

PREFACE TO AMINO ACIDS, PEPTIDES, POLYPEPTIDES, AND PROTEINS, VOLUME II

The section of the *Handbook of Biochemistry and Molecular Biology* on Amino Acids, Peptides, Polypeptides, and Proteins is divided into three volumes. The second volume contains information mainly on Proteins.

Data on cleavage, chemical modification and hydrolysis of proteins are contained herein. Physical-chemical data on proteins, such as refractive index increments, molecular extinction coefficients, pK values, viscosity data, dimensions of the amide group, and protein conformation are listed. Molecular parameters of plasma proteins, glycoproteins, histones, muscle proteins, and many other proteins are tabulated. Properties of lectins, protein ligands, proteinase inhibitors, and proteoglycans are listed.

The third volume will contain additional material on proteins and the first volume contains information on amino acids, amino acid derivatives, etc.

Although the data, for which the editor alone is responsible is far from complete, it is hoped that these volumes will be of assistance to those working in the field of biochemistry and molecular biology.

Gerald D. Fasman
Editor
January 1976

THE EDITOR

Gerald D. Fasman, Ph.D., is the Rosenfield Professor of Biochemistry, Graduate Department of Chemistry, Brandeis University, Waltham, Massachusetts.

Dr. Fasman graduated from the University of Alberta in 1948 with a B.S. Honors Degree in Chemistry, and he received his Ph.D. in Organic Chemistry in 1952 from the California Institute of Technology, Pasadena, California. Dr. Fasman did postdoctoral studies at Cambridge University, England, Eidg. Technische Hochschule, Zurich, Switzerland, and the Weizmann Institute of Science, Rehovoth, Israel. Prior to moving to Brandeis University, he spent several years at the Children's Cancer Research Foundation at the Harvard Medical School. He has been an Established Investigator of the American Heart Association, a National Science Foundation Senior Postdoctoral Fellow in Japan, and recently was a John Simon Guggenheim Fellow.

Dr. Fasman is a member of the American Chemical Society, a Fellow of the American Association for the Advancement of Science, Sigma Xi, The Biophysical Society, American Society of Biological Chemists, The Chemical Society (London), the New York Academy of Science, and a Fellow of the American Institute of Chemists. He has published 180 research papers.

The Editor and CRC Press, Inc. would like to dedicate this third edition to the memory of Eva K. and Herbert A. Sober. Their pioneering work on the development of the Handbook is acknowledged with sincere appreciation.

TABLE OF CONTENTS

NOMENCLATURE

Biochemical Nomenclature 3
Nomenclature of Labeled Compounds 16
The Citation of Bibliographic References in Biochemical Journals 17
IUPAC Tentative Rules for the Nomenclature of Organic Chemistry   Section E.
  Fundamental Stereochemistry 21
A One-letter Notation for Amino Acid Sequences 59
Abbreviations and Symbols for the Description of the Conformation of Polypeptide Chains . . 63
Rules for Naming Synthetic Modification of Natural Peptides 79
The Nomenclature of Multiple Forms of Enzymes 84
Nomenclature of Iron-sulfur proteins 89
Enzyme Nomenclature 91
Recommendations for the Nomenclature of Human Immunoglobulins 173
The Nomenclature of Peptide Hormones 175
Nomenclature for Human Immunoglobulins 179
Notation for Human Immunoglobulin Subclasses 184
Notation for Genetic Factors of Human Immunoglobulins 186
An Extension of the Nomenclature for Immunoglobulins 191
Tentative Nomenclature for Blood Coagulation Factors 195

PROTEINS

Cyanogen Bromide Cleavage of Peptides and Proteins 199
Cyanogen Bromide Cleavage of Peptides and Proteins – Collagen 202
Specificity of Reagents Commonly Used to Chemically Modify Proteins 203
Hydrolysis of Proteins 206
Acid Hydrolysis of Proteins 208
Index to Physical-chemical Data of Proteins 222
Molecular Parameters of Purified Human Plasma Proteins 242
The Proteins of Blood Coagulation 254
Glycoproteins 257
Metalloproteins and Metalloenzymes 276
Characterization of Histones 293
Table of Histone Sequences 295
Enzymes Found in Normal Human Urine 301
Properties of Urokinase 302
Amino Acid Composition of Urokinase 302
Carbohydrate and Protein Composition of T-H Glycoprotein 303
Amino Acid Composition of T-H Glycoprotein 303
Amino Acid Composition of Human Retinol-binding Protein 303
Retinol-binding Protein 304
Tamm-Horsfall Mucoprotein 304
β_2 Microglobulin 305
Amino Acid Sequence of β_2 Microglobulin 305
Molecular Parameters of the Contractile Proteins 306
Proteins in Nonmuscle Cells 307
Subunit Constitution of Proteins 325
Refractive Index Increments of Proteins 372
Molar Absorptivity and $A_{1cm}^{1\%}$ Values for Proteins at Selected Wavelengths of the
  Ultraviolet and Visible Region 383

Properties of Purified Lectins 546
Ligand Binding to Plasma Albumin 554
Introduction to Proteinase Inhibitors 583
Specificities and Some Properties of Plant Proteinase Inhibitors 586
Amino Acid Composition of Plant Proteinase Inhibitors 605
Amino Acid Sequence of Plant Proteinase Inhibitors 611
Specificities and Some Properties of Animal Proteinase Inhibitors 618
Amino Acid Composition of Animal Proteinase Inhibitors 649
Amino Acid Sequences of Animal Proteinase Inhibitors 656
Specificities and Some Properties of Microbial Proteinase Inhibitors 661
Amino Acid Composition of Microbial Proteinase Inhibitors 664
Amino Acid Sequence of a Microbial Proteinase Inhibitor 665
Carbohydrate Composition of Selected Proteinase Inhibitors 666
Endo-γ-glutamin: ε-Lysine Transferases, Enzymes which Cross-link Proteins 669
Connective Tissue Polysaccharides (Glycosaminoglycans, Mucopolysaccharides) . . 686
Protein pK Values 689
Enzymatic Activities of Subcellular Fractions 697
Intrinsic Viscosity of Proteins in Native and Denatured States 721
Methods for the Immobilization of Enzymes 722
Dimensions of the Amino Acid Group, Amino Acid Side Chains, and the Peptide Linkage . . . 742
Protein Structures 760

INDEX 769

Nomenclature

BIOCHEMICAL NOMENCLATURE

This synopsis of the recommendations of the IUPAC-IUB Commission on Biochemical Nomenclature (CBN) was prepared by Waldo E. Cohn, Director, NAS-NRC Office of Biochemical Nomenclature (OBN, located at Biology Division, Oak Ridge National Laboratory, Oak Ridge, TN 37830), from whom reprints of the CBN publications listed below and on which the synopsis is based are available.

The synopsis is divided into three sections: Abbreviations, symbols, and trivial names. Each section contains material drawn from the documents (A1 to C1, inclusive) listed below, which deal with the subjects named.

Additions consonant with the CBN Recommendations have been made by OBN throughout the synopsis.

RULES AND RECOMMENDATIONS AFFECTING BIOCHEMICAL NOMENCLATURE AND PLACES OF PUBLICATION (AS OF FEBRUARY 1975)

I. IUPAC-IUB Commission on Biochemical Nomenclature
    A1. Abbreviations and Symbols [General; Section 5 replaced by A6]
    A2. Abbreviated Designation of Amino-acid Derivatives and Peptides (1965) [Revised 1971; Expands Section 2 of A1]
    A3. Synthetic Modifications of Natural Peptides (1966) [Revised 1972]
    A4. Synthetic Polypeptides (Polymerized Amino Acids) (1967) [Revised 1971]
    A5. A One-letter Notation for Amino-acid Sequences (1968)
    A6. Nucleic Acids, Polynucleotides, and their Constituents (1970)

    B1. (Nomenclature of Vitamins, Coenzymes, and Related Compounds)
        a. Miscellaneous [A, B's, C, D's, tocols, niacins; see B2 and B3]
        b. Quinones with Isoprenoid Side-chains: E, K, Q [Revised 1973]
        c. Folic Acid and Related Compounds
        d. Corrinoids: B-12's [Revised 1973]
    B2. Vitamins B-6 and Related Compounds [Revised 1973]
    B3. Tocopherols (1973)

    C1. Nomenclature of Lipids (1967) [Amended 1970; see also II, 2]
    C2. Nomenclature of α-Amino Acids (1974) [See also II, 5]

    D1. Conformation of Polypeptide Chains (1970) [See also III, 2]

    E1. Enzyme Nomenclature (1972)[a] [Elsevier (in paperback); Replaces 1965 edition.]
    E2. Multiple Forms of Enzymes (1971) [Chapter 3 of E1]
    E3. Nomenclature of Iron-sulfur Proteins (1973) [Chapter 6.5 of E1]
    E4. Nomenclature of Peptide Hormones (1974)

II. Documents Jointly Authored by CBN and CNOC [See III]
    1. Nomenclature of Cyclitols (1968) [Revised 1973]
    2. Nomenclature of Steroids (1968) [Amended 1971; Revised 1972]
    3. Nomenclature of Carbohydrates-I (1969)
    4. Nomenclature of Carotenoids (1972) [Revised 1975]
    5. Nomenclature of α-Amino Acids (1974) [Listed under I, C2 in the following table]

III. IUPAC Commission on the Nomenclature of Organic Chemistry (CNOC)
    1. Section A (Hydrocarbons), Section B (Heterocyclics): *J. Am. Chem. Soc.*, 82, 5545;[a] Section C (Groups containing N, Hal, S, Se/Te): *Pure Appl. Chem.*, 11, Nos. 1–2[a] [A, B, and C Revised 1969:[a] Butterworth's, London (1971)]
    2. Section E (Stereochemistry):[b] *J. Org. Chem.*, 35, 2489 (1970); *Biochim. Biophys. Acta*, 208, 1 (1970); *Eur. J. Biochem.*, 18, 151 (1970) [See also I, D1]

[a] No reprints available from OBN; order from publisher.
[b] Reprints available from OBN (in addition to all in IA to ID and II).

RULES AND RECOMMENDATIONS AFFECTING BIOCHEMICAL NOMENCLATURE AND PLACES OF PUBLICATION (AS OF FEBRUARY 1975)(continued)

IV. Physiochemical Quantities and Units (IUPAC)[a] *J. Am. Chem. Soc.,* 82, 5517 (1960) [Revised 1970: *Pure Appl. Chem.,* 21, 1 (1970)]

V. Nomenclature of Inorganic Chemistry (IUPAC) *J. Am. Chem. Soc.,* 82, 5523[a] [Revised 1971: *Pure Appl. Chem.,* 28, No. 1 (1971)][a]

VI. Drugs and Related Compounds or Preparations
    1. U.S. Adopted Names (USAN) No. 10 (1972) and Supplement [U.S. Pharmacopeial Convention, Inc., 12601 Twinbrook Parkway, Rockville, Md.]
    2. International Nonproprietary Names (INN) [WHO, Geneva]

CBN RECOMMENDATIONS APPEAR IN THE FOLLOWING PLACES[a]

	Arch. Biochem. Biophys.	Biochem. J.	Biochemistry	Biochim. Biophys. Acta	Eur. J. Biochem.	J. Biol. Chem.	Pure Appl. Chem.[b]	Biochimie (Bull. Soc.)[c]	Molek. Biol.[d]	Z. Phys. Chem.[e]
A1[f]	136,1	101,1	5,1445		1,259	241,527	40,(R)	50,3	1,872	348,245
A2(Revised)	150,1(R)	126,773(R)	11,1726(R)	263,205(R)	27,201(R)	247,977(R)	31,649(R)	49,121*	2,282*	348,256*
A3(Revised)	121,6*	104,17*	6,362*	133,1*	1,379*	242,555*	33,439(R)	49,325*	2,466*	348,262*
A4(Revised)[g]	151,597(R)	127,753(R)	11,9422(R)	278,211(R)	26,301(R)	247,323(R)	31,641	51,205(R)	5,492(R)	349,1013*
A5	125(3),i	113,1	7,2703	168,6	5,151	243,3557	40,	50,1577	3,473	350,793
A6[h]	145,425	120,449	9,4022	247,1	15,203	245,5171			6,167	351,1055
B1*	118,505	102,15		107,1(a–c)	2,1	241,2987		49,331		348,266
B1b(Revised)	165,1(R)	147,15(R)		387,397(R)	53,15(R)		38,439			
B1d(Revised)	161(2),iii(R)	147,1(R)	13,1555(R)		45,7(R)					
B2(Revised)	162,1(R)	137,417(R)	13,1056(R)	354,155(R)	40,325(R)	245,4229*	33,447(R)			351,1165*
B3(Revised)	165,6(R)	147,11(R)			46,217(R)					
C1[f]	123,409	105,897	6,3287	152,1	2,127	242,4845		50,1363	2,784	350,279
Amendments		116(5)		202,404	12,1	245,1511				
C2			14,449		53,1					
D1[i]	145,405	121,577	9,3471	229,1	17,193	245,6489			7,289	
E2	147,1	126,769	10,4825	258,1	24,1	246,6127				
E3	160,355	135,5	12,3582	310,295	35,1	248,5907		54,123		353,852
E4		151,1	14,2559			250,3215				
II,1(Revised)	128,269*	112,17*	8,2227	165,1*	5,1*	243,5809*	37,285(R)	51,3*		350,523*
II,2[f]	136,13	113,5	10,4994	164,453	10,1		31,285(R)	51,819		351,663
Amendments	147,4	127,613	10,3983	248,387	25,2					
II,3		125,673	10,4827	244,223	21,455	247,613				
II,4		127,741	14,1803	286,217	25,397	247,2633				
Amendments		151,507								

[a] Reprints available from OBN.
[b] No reprints available from OBN; order from publisher.
[c] In French.
[d] In Russian.
[e] In German.
[f] Also in other journals.
[g] Also in Biopolymers, 11, 321.
[h] J. Mol. Biol., 55, 299.
[i] J. Mol. Biol., 52, 1.

* First, unrevised version.
(R) = revised version.

ABBREVIATIONS

Abbreviations are distinguished from **symbols** as follows (taken from Reference A1):

    a. **Symbols,** for monomeric units in macromolecules, are used to make up abbreviated structural formulas (e.g., Gly-Val-Thr for the tripeptide glycylvalylthreonine) and can be made fairly systematic.

    b. **Abbreviations** for semi-systematic or trivial names (e.g., ATP for adenosine triphosphate; FAD for flavinadenine dinucleotide) are generally formed of three or four capital letters, chosen for brevity rather than for system. It is the indiscriminate coining and use of such abbreviations that has aroused objections to the use of abbreviations in general.

[Abbreviations are thus distinguished from symbols in that they (a) are for semi-systematic or trivial names, (b) are brief rather than systematic, (c) are usually formed from three or four capital letters, and (d) are not used — as are symbols — as units of larger structures. ATP, FAD, etc., are abbreviations. Gly, Ser, Ado, Glc, etc., are symbols (as are Na, K, Ca, O, S, etc.); they are sometimes useful as abbreviations in figures, tables, etc., where space is limited, but are usually not permitted in text. The use of abbreviations is permitted when necessary but is never required.]

1. Nucleotides (N = A, C, G, I, O, T, U, X, ψ — see Symbols)

NMP	Nucleoside 5'-phosphate
NDP	Nucleoside 5'-di(or pyro)phosphate
NTP	Nucleoside 5'-triphosphate

Prefix d indicates deoxy.

2. Coenzymes, vitamins

CoA(or CoASH)	Coenzyme A
CoASAc	Acetyl Coenzyme A
DPN[a]	Diphosphopyridine nucleotide
FAD	Flavin-adenine dinucleotide
FMN	Riboflavin 5'-phosphate
GSH	Glutathione
GSSG	Oxidized glutathione
NAD[b]	Nicotinamide-adenine dinucleotide (cozymase, Coenzyme I, diphosphopyridine nucleotide)
NADP[b]	Nicotinamide-adenine dinucleotide phosphate (Coenzyme II, triphosphopyridine nucleotide)
NMN	Nicotinamide mononucleotide
TPN[c]	Triphosphopyridine nucleotide

3. Miscellaneous

ACTH	Adrenocorticotropin, adrenocorticotropic hormone, or corticotropin
CM-cellulose	*O*-(Carboxymethyl)cellulose
DEAE-cellulose	*O*-(Diethylaminoethyl)cellulose
DDT	1,1,1-Trichloro-2,2-bis(*p*-chlorophenyl)ethane
EDTA	Ethylenediaminetetraacetate
Hb,HbCO,HbO$_2$	Hemoglobin, carbon monoxide hemoglobin, oxyhemoglobin
P_i	Inorganic orthophosphate

[a]Replaced by NAD (also DPN$^+$ by NAD$^+$, DPNH by NADH).
[b]Generic term; oxidized and reduced forms are NAD$^+$, NADH (NADP$^+$, NADPH).
[c]Replaced by NADP (also TPN$^+$ by NADP$^+$, TPNH by NADPH).

PP$_i$	Inorganic pyrophosphate
TEAE-cellulose	O-(Triethylaminoethyl)cellulose
Tris	Tris(hydroxymethyl)aminomethan (2-amino-2-hydroxymethylpropane-1,3-diol)

4. Nucleic Acids

DNA, RNA	Deoxyribonucleic acid, ribonucleic acid (or -nucleate)
hnRNA	Heterogeneous RNA
mtDNA	Mitochondrial DNA
cRNA	Complementary RNA
mRNA	Messenger RNA
nRNA	Nuclear RNA
rRNA	Ribosomal RNA
tRNA	Transfer RNA (generic term; sRNA should not be used for this or any other purpose)
tRNA^{Ala}	Alanine tRNA; tRNA$_1^{Ala}$, tRNA$_2^{Ala}$: isoacceptor alanine tRNA's
AA-tRNA	Aminoacyl-tRNA; aminoacylated tRNA; "charged" tRNA (generic term)
Ala-tRNA or Ala-tRNA^{Ala}	Alanyl-tRNA
tRNA^{Met}	Methionine tRNA (not enzymatically formylatable)
tRNA^{fMet} or tRNA$_f^{Met}$	Methionine tRNA, enzymatically formylatable to ...
fMet-tRNA	Formylmethionyl-tRNA (small f, to distinguish from fluorine F)

SYMBOLS

Symbols are distinguished from abbreviations in that they are designed to represent specific parts of larger molecules, just as the symbols for the elements are used in depicting molecules, and are thus rather systematic in construction and use. Symbols are not designed to be used as abbreviations and should not be used as such in text, but they may often serve this purpose when space is limited (as in a figure or table). Symbols are always written with a single capital letter, all subsequent letters being lower-case (e.g., Ca, Cl, Me, Ac, Gly, Rib, Ado), regardless of their position in a sequence, a sentence, or as a superscript or subscript.

Some abbreviations expressed in symbols (see also Section II F below), as examples of the use of symbols:

Dimethylsulfoxide	Me$_2$SO [a]
Tetranitromethane	(NO$_2$)$_4$C [b]
Guanidine hydrochloride	Gdn · HCl [c]
Guanidinium chloride	GdmCl
Cetyltrimethylammonium bromide	CtMe$_3$NBr [d]
Ethyl methanesulfonate	MeSO$_3$Et
Methylnitronitrosoguanidine	MeN$_2$O$_3$Gdn
-nitrosourea	-Nur [e]
-nitrosamine	-Nam [f]
-fluorene	-Fln
Aminofluorene	NH$_2$Fln
Acetylaminofluorene	AcNHFln [g]
Acetoxyacetylaminofluorene	Ac(AcO)NFln
N-Acetylneuraminic acid	AcNeu [h]

[a] Replaces DMSO.
[b] Replaces TNM.
[c] Replaces Gu, Gd, and G.
[d] Replaces CTAB (similarly for other ammonium compounds).
[e] Replaces NU.
[f] Replaces NA.
[g] Replaces AAF.
[h] Not NANA.

I. Phosphorylated Compounds (Reference A1)

-PO_3H_2 (or its ions)
-PO_2H-(or its ion)
-PO_2H-PO_3H_2 (or ions)

-*P*(or *P*-) ("p" in Nucleic Acids; see IV)
-*P*-(hyphen in Nucleic Acids; see IV)
-*P-P* or -*PP* or *PP*- (cf. PP_i in Abbreviations)

Examples:[a]

Glucose 6-phosphate
Phosphenolpyruvate (pyruvenol phosphate)
Fructose 1,6-bisphosphate (not di)
Creatine phosphate
Phosphocreatine

Glucose-6-*P* (or Glc-6-*p*; see II below).
*P-enol*Pyruvate or *enol*Pyruvate-*P* or *e* Prv-*P*[b]
Fructose-1,6-P_2 (or Fru-1,6-P_2; see II below).
Creatine-*P*
P-Creatine

[a] Note that symbols are hyphenated even where names are not.
[b] Recommended by OBN.

II. Peptides and Proteins (References A1–A5)
A. Symbols (Reference A2–A5)

1. Common amino acids

Name	Three-letter[a]	One-letter[b]	Name	Three-letter[a]	One-letter[b]
Alanine	Ala	A	Lysine[e]	Lys	K
Arginine	Arg	R	Methionine	Met	M
Asparagine	Asn [c,d,e]	N	Phenylalanine	Phe	F
Aspartic acid	Asp [d,e]	D	Proline	Pro	P
Cysteine	Cys [e]	C	Serine[e]	Ser	S
Glutamic acid	Glu [f]	E	Threonine[e]	Thr	T
Glutamine	Gln [e,f,g]	Q	Tryptophan[e]	Trp	W
Glycine	Gly	G	Tyrosine[e]	Tyr	Y
Histidine	His [e]	H	Valine	Val	V
Isoleucine	Ile	I	Unknown or "other"	AA [h]	X
Leucine	Leu	L			

[a] One capital, two small letters, at all times.
[b] For special uses and with special conventions; see III, *I* following.
[c] Or Asp (NH_2); see Footnotes e and g.
[d] Uncertainty as between Asp and Asn may be designated by Asx (or B).
[e] Substitution on a functional group may be indicated, as shown in Footnotes c and g, by parenthesis following the symbol, e.g., Cys (Cme), Ser (*P*); see C 2 below.
[f] Uncertainty, as between Glu and Gln, may be designated by Glx (or Z); pyroglutamate is pGlu or <Glu, not PCA.
[g] Or Glu (NH_2); see Footnotes c and e.
[h] See Abbreviations, Part 4 (AA-tRNA).

2. Less common amino acids

Name	Symbol
β-Alanine	βAla
Alloisoleucine	aIle
2-Aminoadipic acid	Aad
3-Aminoadipic acid	βAad
2-Aminobutyric acid	Abu
6-Aminocaproic acid[a]	εAhx [a]
2-Aminopimelic acid	Apm
2,4-Diaminobutyric acid	A_2bu [b]
2,2'-Diaminopimelic	A_2pm [b]
2,3-Diaminopropionic acid	A_2pr [b]
N-Ethylglycine, etc.	EtGly, etc.
Hydroxylysine	Hyl
allo-Hydroxylysine	aHyl
3-Hydroxyproline	3Hyp [c]
4-Hydroxyproline	4Hyp [c]
N-Methylglycine (sarcosine)	MeGly or Sar
N-Methylisoleucine	MeIle
N-Methylvaline, etc.	MeVal, etc.
Norleucine	Nle
Norvaline	Nva
Ornithine	Orn

[a] 6-Aminohexanoic acid is preferred; see Reference C2.
[b] The use of D (for di), T (for tri or tetra), etc., is undesirable. Hence, in this context, A_2 for diamino is recommended (cf. II F below).
[c] Or Pro(PH) for hydroxyproline.

B. Sequence, Direction, and Bonding (Reference A2)

→ peptide bond, originating in peptide -CO-
– peptide bond, originating in peptide -CO- of residue at left.
, separates symbols in unknown sequence (the entire unknown sequence is enclosed in parentheses).
| bond originating in first letter of symbol of a residue having a substituted functional group (-SH, 3- or 4-COOH, -OH, 6-NH_2, etc., or the remaining H of a peptide bond; see C2 below).

C. Substitution (Reference A2)

Groups substituted for hydrogen or for hydroxyl may be indicated either by their structural formulae, or by symbols, or by combinations of both, e.g.,

Benzoylglycine (hippuric acid)	PhCo-Gly or C_6H_5CO-Gly or Bz-Gly
Glycine methyl ester	Gly-OCH_3 or Gly-OMe
Trifluoroacetylglycine	CF_3CO-Gly

1. In α-NH_2 or α-COOH groups: horizontal dash (hyphen) to left or right, respectively.

N-Acetylglycine	Ac-Gly	N-Tosylphenylalanyl	
Glycine ethyl ester	Gly-OEt	chloromethyl ketone (TPCK)	Tos-Phe CH_2Cl
N^2-Acetyllysine	Ac-Lys		
Serine methyl ester	Ser-OMe		
O^1-Ethyl N-acetylglutamate	Ac-Glu-OEt	N-Ethyl-N-methylglycine	Et-(Me-)Gly or Et\>Gly / Me
Isoglutamine	Glu-NH_2		

2. On functional group: Vertical bond (see B above) or parentheses (see II A1 above, Footnote e).

4-Methyl aspartate	Asp or Asp(OMe) with OMe on vertical bonds above and below	O-Methyltyrosine	Tyr or Tyr or Tyr(Me) with Me on vertical bonds
N^6-Acetyllysine	Lys or Lys or Lys(Ac) with Ac above and below	S-Ethylcysteine	Cys or Cys or Cys(Et) with Et above and below
O-Phosphoserine	Ser or Ser or Ser( ) with P above and below	N^τ-Methylhistidine* (see 3.3) (telemethylhistidine)	His or His or His(τMe) with Me

similarly for N^π substitution (**prosmethylhistidine**)

N-Glycylsarcosine	Gly——Gly with Me below	or	Gly-(Me-)Gly	or	Gly-Sar
N-Glycyl-N-acetylglycine	Gly——Gly with Ac above	or	Gly-(Ac-)Gly	or	Ac-Gly>Gly
N,N-Diglycylglycine	Gly——Gly with Gly above	or	Gly$_2$ > Gly	or	Gly, Gly > Gly

*The prolonged and well-entrenched ambiguity in the nomenclature of the N-1 being the biochemist's N-3 and *vice versa*) led to a new trivial system for designating these substances: The imidazole N *nearer* the alanine residue is designated *pros* (symbol π) and the one *far*ther *tele* (symbol τ), to give the following names and symbols: *pros*methylhistidine or π-methylhistidine, His(πMe); *tele*methylhistidine or τ-methylhistidine, His(τMe).

D. Polypeptides: Follow Rules for Substitution (C above) (Reference Az)

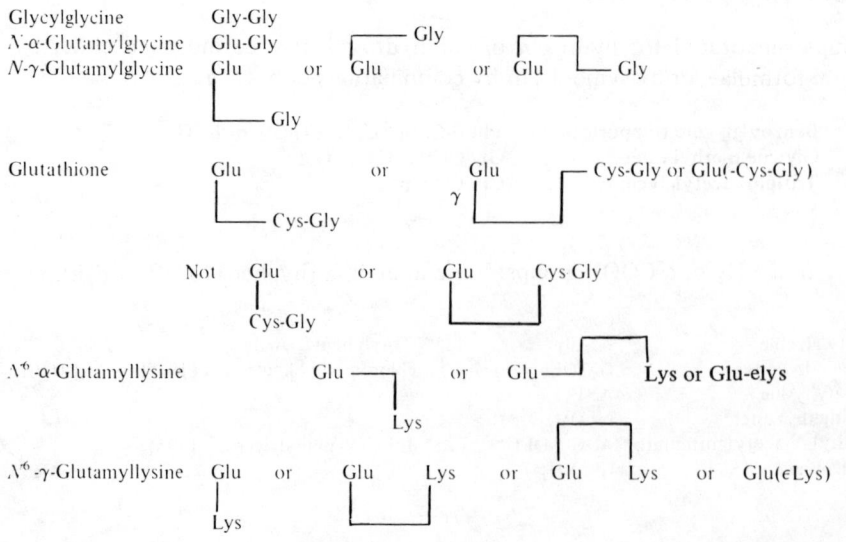

Glycyllysylglycine dihydrochloride $+H_2$ — Gly-Lys-Gly — OH · 2HCl

Its N^6-formylderivative   Gly-Lys-Gly - or   Gly-Lys(CHO)-Gly
                                  |
                                  CHO

E. Cyclic Polypeptides (Reference A2)

1. Homodetic: Gramicidin S

cyclo-Val-Orn-Leu-DPhe-Pro-Val-Orn-Leu-DPhe-Pro
Val-Orn-Leu-DPhe-Pro-Val-Orn-Leu-DPhe-Pro

```
  ┌→ Val → Orn → Leu — DPhe → Pro ┐
  └  Pro ← DPhe ← Leu — Orn ← Val ┘
```

2. Heterodetic:

Oxytocin   Cys-Tyr-Ile-Asn-Gln-Cys-Pro-Leu-Gly-NH$_2$
             └────────────┘

Cyclic ester of threonylglycylglycylglycine

┌─────────────────┐
Thr-Gly-Gly-Gly ┘   or   H – Thr-Gly-Gly-Gly ┐

F. Substituents (Reference A2)

1. NH$_2$ protecting groups of the urethan type (partial list)

Benzyloxycarbonyl-	Z- or Cbz-[a]	*p*-Methoxyphenylazobenzyloxycarbonyl-	Mz-
p-Nitrobenzyloxycarbonyl-	Z(NO$_2$)-	*p*-Phenylazobenzyloxycarbonyl-	Pz-
p-Bromobenzyloxycarbonyl-	Z(Br)-	*t*-Butoxycarbonyl-	Boc-[b] or Bu^{t}OCO-
p-Methoxybenzyloxycarbonyl-	Z(OMe)-	Cyclopentyloxycarbonyl-	Poc- or cPeOCO-

2. Other N- protecting groups (partial list)

Acetyl-	Ac-	Maleoyl- (–OC–CH=CH–CO–)	Mal-[e] or Mal<
Benzoyl-(C$_6$H$_5$CO-)	PhCO- or Bz-	Maleyl- (HOOC–CH=CH–CO–)	Mal-
Benzyl- (C$_6$H$_5$CH$_2$-)	PhCH$_2$- or Bzl	Methylthiocarbamoyl-	MeNHCS-[a] or Mtc-[f]
Benzylthiomethyl-	PhSCH$_2$- or Btm-	*o*-Nitrophenylthio-	Nps-
Carbamoyl-	NH$_2$CO- (preferred to Cbm)	Phenylthiocarbamoyl-	PhNHCS-[a] or Ptc-[f]
		Phthaloyl-	–Pht- or Pht<
1-Carboxy-2-nitrophenyl-5-thio-	Nbs-	Phthalyl-	Pht-
3-Carboxypropionyl- (HOOC–CH$_2$–CH$_2$–CO–)	Suc-	Succinyl- (–OC–CH$_2$–CH$_2$–CO–)	–Suc- or Suc<
		Tetrahydropyranyl-	H$_4$pyran-[c,g]

Dansyl-(5-dimethylamino-naphthalene-inonaphthalene-1-sulfonyl)	Dns-[c] or dansyl[a]	Tosyl- (p-tolylsulfonyl)	Tos- or tosyl
Dinitrophenyl-	N_2ph-[a,c] or Dnp	Trifluoroacetyl-	CF_3CO- or F_3Ac-[a,h]
Formyl-	HCO-[d] or CHO-	Trityl- (triphenylmethyl)	Ph_3C-[a,i] or Trt-
p-Iodophenylsulfonyl (pipsyl)	Ips or pipsyl		

3. Substituents at carboxyl group

Benzyloxy- (benzyl ester)	$-OCH_2$Ph or $-$OBzl	p-Nitrophenoxy- (p-nitrophenyl ester)	$-$ONph
Cyanomethoxy- (cyanomethyl ester)	$-OCH_2$CN or $-$OMeCN	p-Nitrophenylthio-	$-$SNph
Diphenylmethoxy- (benzhydryl ester)	$-OCHPh_2$ or $-$OBzh	Phenylthio- (phenylthiolester)	$-$SPh
Ethoxy- (ethyl ester)	$-$OEt	1-Piperidino-oxy-	$-$OPip
Methoxy- (methyl ester)	$-$OMe	8-Quinolyloxy-	$-$OQu
		Succinimido-oxy-	$-$ONSuc
		Tertiary butoxy- (t-butyl ester)	$-OBu^t$

[a] Preferred.
[b] Not BOC or tBOC.
[c] The use of D for di and T for tri (or tetra) is discouraged. Recognized symbols with numerical subscripts are recommended.
[d] fMet is approved for formylmethionine.
[e] MalNEt is recommended for N-ethylmaleimide (not NEM).
[f] Mtc and Ptc have been used to denote methyl- and phenylthiohydantoins (e.g., Ptc-Leu). Since this incorrectly implies the substitution of an amino acid by a "phenyl (or methyl) thiohydantoyl" group, the correct representation, CS-Leu-NPh or PhNCS-Leu, or, in text, Leu>PhNCS, is recommended.
[g] Not THP or Thp (see Footnote c).
[h] Not TFA (see Footnote c).
[i] Or trityl.

4. Other substituents (and reagents)

2-Aminoethyl-[a]	$-(CH_2)_2NH_2$ or Aet-[a,b]	Chloroethylamine	$Cl(CH_2)_2NH_2$ or AetCl
Carbamoylmethyl-	$-CH_2CONH_2$ or Ncm-	Chloroacetamide	$ClCH_2CONH_2$ or NcmCl
Carboxymethyl-	$-CH_2CO_2H$ or Cxm-	Chloroacetic acid	$ClCH_2CO_2H$ or CxmCl
p-Carboxyphenylmercuri-	-HgBzOH	p-Chloromercuribenzoate	$ClHgBzO^-$ [c]
1-Carboxy-2-mitrophenyl-5-thio-nitrophenyl-5-thio-	Nbs-	5,5'-Dithiobis(2-nitrobenzoic acid) (2-nitrobenzoic acid)	Nbs_2 [d]
Diazoacetyl-	N_2CHCO- or N_2<Ac-		
-Diisopropylphosphor	$iPr_2 P$-[e]	Diisopropylfluorophosphate	$iPr_2 P$-F [f]
Dinitrophenyl-	N_2ph [g]	Fluorodinitrobenzene	N_2ph-F [h]
Hydroxyethyl-	$-(CH_2)_2$OH or HOEt-	Ethylene oxide	$(CH_2)_2$O or Et>O
		N-Ethylmaleimide	MalNEt [i]
		Tetrahydrofuran	H_4furan [j]
		Tosyllysyl chloromethyl ketone	Tos-LysCH$_2$Cl [k]

		Tosylarginine methyl ester	Tos-ArgOMe [l]
Trifluoroacetyl-	F_3Ac-[m]	Trifluoroacetic acid	F_3AcOH
Trimethylsilyl-	Me_3Si-[n]	Tetramethylsilane	Me_4Si [n]

[a] For -ethylamine, -Etn; for -ethanolamine (see Lipids), -OEtn.
[b] Not AET.
[c] Replaces PCMB, pCMB, and CMB.
[d] Replaces DTNB.
[e] Replaces, DIP and Dip.
[f] Replaces DPF, DFP, DIPF, etc.
[g] Replaces DNP and Dnp.
[h] Replaces FDNB.
[i] Replaces NEM.
[j] Replaces THF. Similarly, H_4 folate.
[k] Replaces TLCK (similarly for TPCK, etc.).
[l] Replaces TAME (similarly for other N-substituted amino-acid esters. See C1 above).
[m] Replaces TFA.
[n] Not TMS- or TMS. Similarly, Me_2SO, not DMSO; NAc_3, not NTA.

G. Polymerized Amino Acids (Synthetic Polypeptides) (Reference A4)

1. Linear polymers (only normal peptide links are involved).

   a. Homopolymer: polylysine; poly(Lys) or $(Lys)_n$ (n may be replaced by a number).

   b. Copolymer, alternating sequence: poly(alanine-lysine); poly(Ala-Lys) or $(Ala-Lys)_n$.

   c. Copolymer, random sequence, composition unspecified: poly(alanine, lysine); poly(Ala,Lys) or $(Ala,Lys)_n$.

   d. Copolymer, random sequence, molar percentages ($\Sigma = 100\%$) known: poly($DLGlu^{56}Lys^{38}DTyr^6$) or $(DLGlu^{56}Lys^{38}DTyr^6)_n$ (only lysine is L).

   e. Block polymer of poly(Glu) linked via α-COOH to α-NH_2 of poly(Lys): poly(Glu^{56})-poly(Lys^{44}) or $(Glu^{56})_n$-$(Lys^{44})_n$.

   f. Block polymer, a repeating series of the known sequence Glu-Lys-Lys-Tyr: poly(Glu-Lys_2-Tyr) or $(Glu$-Lys_2-$Tyr)_n$.

   g. Block polymer of two repeating series: poly$(Glu$-$Lys)^{25}$-poly(Ala-Tyr_2-$Glu)^{12.5}$ or $(Glu$-$Lys)_n^{25}$-$(Ala$-Tyr_2-$Glu)_n^{12.5}$ (molar percentages = 100).

2. Branched graft polymers (functional groups are involved).

   a. Main chain is a repeating sequence (see 1f above), sidechain is of random sequence, connection is from ϵ-NH_2 of a lysine to an unknown group in the sidechain

```
        poly(Asp³⁰Glu⁵⁰)         or    (Asp³⁰Glu⁵⁰)ₙ
              |                             |
        poly(Glu-Lys₂-Tyr)⁵           (Glu-Lys₂-Tyr)ₙ⁵
```

or

```
  ┌─────────────────────────┐       ┌─────────────────────────┐
poly(Glu-Lys₂-Tyr)⁵   poly(Asp³⁰Glu⁵⁰)  or  (Glu-Lys₂-Tyr)ₙ⁵   (Asp³⁰Glu⁵⁰)ₙ
```

b. Main chain of unknown sequence, linked via ϵ-NH_2 of a lysine to the α-COOH of an L-tyrosine in the sidechain, which is a block polymer (no analytical data):

$$\text{poly(Ala)-poly(Tyr)} \quad\text{or}\quad [(Ala)_n\text{-}(Tyr)_n]$$
$$\text{poly(DLAla,Lys)} \qquad\qquad (DLAla,Lys)_n$$

or

$$\text{poly(DLAla,Lys)} \quad (\text{poly(Ala)-poly(Tyr)}) \quad\text{or}\quad \text{poly(DLAla,Lys)--poly(Ala)-poly(Tyr)}$$

Note: The points of attachment of Lys and Tyr cannot be specified in the last example. This system, depending on double hyphens to express functional group involvement, is not recommended.

c. Two linear copolymers of unknown sequence, triply linked between ϵ-NH_2 groups of lysines and γ-COOH residues of glutamates:

$$\begin{array}{c} \text{poly}(Lys^{1\,6}Ala^{2\,0}) \\ (3)| \\ \text{poly}(Glu^{3\,5}Tyr^{2\,9}) \end{array} \quad\text{or}\quad \begin{array}{c} (Lys^{1\,6}Ala^{2\,0})_n \\ (3)| \\ (Glu^{3\,5}Tyr^{2\,9})_n \end{array}$$

or

$$\text{poly}(Glu^{3\,5}Tyr^{2\,9}) \quad \text{poly}(Lys^{1\,6}Ala^{2\,0}) \quad\text{or}\quad \text{poly}(Tyr^{2\,9}Glu^{3\,5}) \;\frac{\gamma\epsilon}{(3)}\; \text{poly}(Lys^{1\,6}Ala^{2\,0})$$
$$(3)$$

(See comment under b with respect to last example).

d. Linear, random-sequence chain attached via terminal α-COOH group (of either Tyr or Glu) to NH_2 terminal of poly(DLalanine) in turn connected, via terminal COOH, to ϵ-NH_2 groups(s) of poly(L-lysine) (no analytical data):

$$(Tyr,Glu)_n\text{-}(DLAla)_n \quad Lys_n$$

$$\text{or poly(Tyr, Glu)-poly(DLAla)} \quad\text{poly(Lys)}$$

H. Synthetic Modifications of Natural Peptides (Reference A3)

Modification and Name	Abbreviation
1. Replacement:	
a. of 8th residue in vasopressin by citrulline:	
[8-Citrulline] vasopressin	[Cit^8] vasopressin
b. at 5 and 7 positions in hypertensin II:	
[5-Isoleucine, 7-alanine] hypertensin II	[Ile^5, Ala^7] hypertensin II
2. Extension of X by valyl residue, at N and C terminals:	
valyl-X	Val-X
X(yl)-valine	X(yl)-Val
3. Insertion of tyrosine residue between 4th and 5th residue	
4a-endo-tyrosine-hypertensin II	endo-$Tyr^{4\,a}$-hypertensin II
4. Removal of proline from position 7 in oxytocin:	
des-7-proline-oxytocin	des-Pro^7-oxytocin

5. Substitution of valine on ε-nitrogen of a lysine at position 2 in peptide X: $N^{\epsilon 2}$-valyl-X $N^{\epsilon 2}$-Val-X
6. Substitution of valine on γ-carboxyl of glutamate at position 3 in peptide X: $C^{\gamma 3}$-X(yl)-valine $C^{\gamma 3}$-X(yl)-Val
7. Fragments, or partial sequences: fragments from α-MSH.

    Ac-Ser-Tyr-Ser-Met-Glu-His-Phe-Arg-Trp-Gly-Lys-Pro-Val-NH$_2$ α-MSH
        1   2   3   4   5   6   7   8   9 10 11 12 13

    Met-Glu-His-Phe-Arg-Trp-Gly   α-MSH-(4–10)-heptapeptide
     6   10

    His-Phe-Arg-Lys-Pro-Val-NH$_2$ α-MSH-(6–8)-(11–13)-hexapeptide amide
     6   8   11   13

I. One-letter Notation[a] (Reference A5)

1. Symbols: see II.A.1 above (NH$_2$ terminal at left, COOH terminal at right).
2. Known sequence: space[b] between symbols.
3. Unknown sequence: comma[c] between symbols, parentheses[d] enclosing.
4. Adjacent unknown sequences: = replaces)
5. Uncertainty as to sequence or terminus: / (see examples b and c).

Examples: a.   (Ala, Cys, Asp) (Arg, Ser) (Gly, His, Ile) Lys-Leu-Met-Asn-Pro-Gln
becomes   (A,   C, D = R, S = G, H, I) K L M N P Q

         b. A   C D E F G H I K L M N P Q
         c. (A.   C. D = R, S = G. H. I) K L/ M N/   P Q/

In c., the tripeptides A . C . D and G . H . I are not of known sequence, but are inferred by analogy with the known peptide b.; the inference is expressed by periods instead of commas. The comma between R and S indicates that no inference as to sequence can be drawn for this dipeptide. The internal slashes indicate that no connection between L and M, and N and P, has been proven, although KL, MN and PQ are each of known internal sequence. The final slash indicates that Q has not been proven to be the COOH terminal residue of the entire peptide, although it is the terminus of the PQ dipeptide.

[a]For display of very long sequences or computer use only.
[b]In place of hyphen in three-letter system. Spaces must be equal to characters, as in typing. So must commas, dots, and all other symbols.
[c]As in three-letter system; becomes a dot (period) when sequence is inferred but not demonstrated (see example c).
[d]The double symbol,)(, is replaced by = (see 4) to preserve equal spacing.

NOMENCLATURE OF LABELED COMPOUNDS

The statement below was adopted by the IUB Commission of Editors of Biochemical Journals* (CEBJ) and appears, in the same or in similar form, in the Instructions to Authors of their journals. This system originated with the Chemical Society (London) and was subsequently adopted by the American Chemical Society (Handbook for Authors, 1967). It was adopted by CEBJ in 1971 and is the only system currently permitted in the pages of their journals.

ISOTOPICALLY LABELED COMPOUNDS

The symbol for the isotope introduced is placed in *square* brackets directly attached to the front of the name (word), as in [^{14}C]urea. When more than one position in a substance is labeled by means of the same isotope and the positions are not indicated (as below), the number of labeled positions is added as a right-hand subscript, as in [^{14}C$_2$]glycollic acid. The symbol "U" indicates uniform and "G" general labeling, e.g., [U-^{14}C]glucose (where the ^{14}C is uniformly distributed among all six positions) and [G-^{14}C]glucose (where the ^{14}C is distributed among all six positions, but not necessarily uniformly); in the latter case it is often sufficient to write simply "[^{14}C]glucose."

The isotopic prefix precedes that part of the name to which it refers, as in sodium [^{14}C]formate, iodo[^{14}C$_2$]acetic acid, 1-amino[^{14}C]methylcyclopentanol (H$_2$N–^{14}CH$_2$–C$_5$H$_8$–OH), α-naphth[^{14}C]oic acid (C$_{10}$H–^{14}CO$_2$H), 2-acetamido-7-[^{131}I]iodofluorene, fructose 1,6-[1-^{32}P]diphosphate, D-[^{14}C]glucose, 2H-[2-^{2}H]pyran, S-[8-^{14}C]adenosyl[^{35}S]methionine. Terms such as "^{131}I-labeled albumin" should not be contracted to "[^{131}I]albumin" (since native albumin does not contain iodine), and "^{14}C-labeled amino acids" should similarly not be written as "[^{14}C]amino acids" (since there is no carbon in the amino group).

When isotopes of more than one element are introduced, their symbols are arranged in alphabetical order, including ^{2}H and ^{3}H for deuterium and tritium, respectively.

When not sufficiently distinguished by the foregoing means, the positions of isotopic labeling are indicated by Arabic numerals, Greek letters, or prefixes (as appropriate), placed within the square brackets and before the symbol of the element concerned, to which they are attached by a hyphen; examples are [1-^{2}H]ethanol (CH$_3$–C^{2}H$_2$–OH), [1-^{14}C]aniline, L-[2-^{14}C]leucine (or L-[α-^{14}C]-leucine), [*carboxy*-^{14}C]leucine, [*Me*-^{14}C]isoleucine, [2,3-^{14}C]maleic anhydride, [6,7-^{14}C]xanthopterin, [3,4-^{13}C,^{35}S]methionine, [2-^{13}C; 1-^{14}C]acetaldehyde, [3-^{14}C; 2,3-^{2}H; ^{15}N]serine.

The same rules apply when the labeled compound is designated by a standard abbreviation or symbol, other than the atomic symbol, e.g. [γ-^{32}P]ATP.

For simple molecules, however, it is often sufficient to indicate the labeling by writing the chemical formulae, e.g. ^{14}CO$_2$, H$_2$^{18}O, ^{2}H$_2$O (not D$_2$O), H$_2$^{35}SO$_4$, with the prefix superscripts attached to the proper atomic symbols in the formulae. The square brackets are not to be used in these circumstances, nor when the isotopic symbol is attached to a word that is not a chemical name, abbreviation or symbol (e.g. ^{131}I-labeled).

---

*CEBJ consists of the Editors-in-Chief of the following journals: *Archives of Biochemistry and Biophysics, Biochemical Journal, Biochemistry, Biochimica et Biophysica Acta, Biochimie, European Journal of Biochemistry, Hoppe-Seyler's Zeitschrift für Physiologische Chemie, Journal of Biochemistry, Journal of Biological Chemistry, Journal of Molecular Biology,* and *Molekulyarnaya Biologiya;* corresponding members include *Proceedings of the National Academy of Sciences* (U.S.A.) and approximately 40 others.

THE CITATION OF BIBLIOGRAPHIC REFERENCES IN BIOCHEMICAL JOURNALS RECOMMENDATIONS (1971)*

IUB Commission of Editors of Biochemical Journals (CEBJ)

These Recommendations were reviewed by the Commission in August 1972, when it was decided to publish them.

PREAMBLE

Two basic systems for the citation of references are used at present. The so-called Harvard System (where names of authors and the date are cited in the text, and the reference list is in alphabetical order) and the Numbering System (where numbers, but not necessarily names of authors, are cited in the text, and the reference list is in order of citation in the text). Several ways of quoting references in the list are in current use.

The Commission is of the opinion, arrived at as a result of much consultation between many senior editors, that it is unlikely that all journals would accept a recommendation to use either the Harvard or the Numbering System to the exclusion of the other. It believes, however, that most biochemists will accept the need for, and indeed welcome, a substantial degree of unification of practices, there being no strong case for the individuality of each journal on this issue. Accordingly, the Commission makes the following Recommendations to all biochemical journals; the reasons for some of them are given. The Recommendations deal first with the way in which references should be cited in the list; the proposal is suitable for journals adopting either the Harvard or the Numbering System. Secondly, there are Recommendations about the way in which each of these systems is used. Thirdly, abbreviations for titles of journals and a few other points are considered. Implementation of the Recommendations would mean that any very small differences between journals in their practices would be of the type that can be attended to at the redactory stage of preparation for press. The Commission recognizes that it cannot deal with a number of smaller problems concerning citations that arise from time to time.

RECOMMENDATIONS

1. Citations of References in the List of References Should Be as Follows

Braun, A., Brown, B. & LeBrun, C. (1971) *Journal*, 11, 111–113.

Notes: (a) This form can be used by both systems.

(b) Journals using the Numbering System should arrange the references in numerical order beside the number (which can be italicized or in brackets according to the house custom of the journal).

(c) Journals using the Harvard System should arrange the references in alphabetical order, whatever the language, except in certain situations (see Recommendation 4a below).

(d) This recommendation incorporates the following points:
  i. Initials after surnames (full first names are not given in the list).
  ii. The use of the symbol "&" is recommended if at all possible because of its widespread usage and the fact that it is independent of the language. No comma before "&."

---

*From IUB Commission of Editors of Biochemical Journals (CEBJ), *J. Biol. Chem.*, 248(21), 7279–7280 (1973). With permission.

iii. Year in parentheses (this follows immediately after the authors' names because it is essential to the Harvard System).

iv. Journal title (abbreviated). This can be in italics according to house practice (see Recommendation 7 below concerning journal title abbreviations).

v. Volume number. This can be in heavy type or italics according to house practice.

vi. A few journals do not have volume numbers in which case the page numbers should follow immediately after the abbreviated journal title.

If it is necessary to quote both a volume and a part number, the reference should read: Brown, B. (1971) *Journal,* 11, pt 1, 121–123.

vii. First and last pages should be given. The Commission decided to make this Recommendation mainly on the basis of evidence that the additional information provided by quoting the last page was being required increasingly in many types of library and information retrieval services. Citation of the last page (as well as the first) has been requested for some time by the secondary and abstracting journals. Citation of both first and last pages is also an aid in the prevention of errors.

viii. The number of stops and commas is kept as small as possible.

(e) Authors' names and the abbreviated name of the journal when repeated in the next reference should be spelled out in full; ibid. and similar terms should not be used.

(f) Recommendations of the IUPAC-IUB Commission on Biochemical Nomenclature (CBN) and similar documents should be referred to as: Commission on Biochemical Nomenclature (1970) followed by a journal reference.

(g) Junior should be abbreviated to "Jr," not "jun."

2. Numbering System in the Text

The use of authors' names is permissible as authors wish; only the initial letter of the name should be in capital type. Numbers can be inserted in parentheses or as superscripts according to house custom. The printing of references at the foot of the page on which they are first quoted is considered to be helpful with the Numbering System but is not part of the Recommendation because the extra cost is generally considered to be prohibitive.

3. Harvard System in the Text

For multi-author papers, it is recommended that:

a. Not more than two authors to be named either on the first or any subsequent occasion;

b. et al. should be used for three or more authors on every occasion;

c. Each name to have the initial letter in capital type only.

Examples (Harvard System style):

Braun et al. (1969) did some work that was confirmed by LeBrun (1970).

These results (Braun et al., 1969; LeBrun, 1970) have been discussed by Brown & Braun (1971).

The same Recommendation (without the year) applies when authors are quoted in the text in the Numbering System.

4. Harvard System in the List of References

a. A special problem arises in the list when there are several papers by, e.g., Green et al. in the same or over several years. While the list could be in strict alphabetical order of the full reference, the reader will find no clue in the text to the alphabetical status of the names of the second and subsequent authors (see Recommendations 3a and 3b). It is therefore recommended that all the papers by Green et al. (that is by Green and more than one co-author) should be arranged, irrespective of the names of the other authors, in chronological order (over many years if necessary) and designate tham a, b, c, etc.

Examples:

Green, G. (1970) etc.
Green, G. & Brown, B. (1971) etc.
Green, G. & White, W. (1969) etc.
Green, G., White, W. & Black, B. (1968a) etc.   sequence governed by order or date of
                                                 publication, as far as can be ascertained.
Green, G., Brown, B. & Black, B. (1968b) etc.
Green, G., White, W., Black, B. & Brown, B. (1969) etc.
Green, G., Black, B. & Brown, B. (1970) etc.

  b. Names beginning with "Mc" should be listed under "Mc" and not under "Mac," to decide alphabetical order.

  c. Names beginning with "De," "Van," or "von," etc. should be arranged under D or V/v, etc.

5. Reference to Books

These should appear in text like any reference to a journal paper. The reference in the list should read: Brown, B. & Braun, A. (1971) in *Book Title* (LeBrun, C., ed.), pp. 1–20, Publisher, Town.

Notes:

  a. If a volume number has to be quoted, this would appear before the pp. as, e.g., "vol. 2," with the number in Arabic numerals (even when Roman numerals are printed on the cover of the book).

  b. Where an author wishes to refer to a specific page within a book reference, this should be given in the text.

Example (in text): ". . . discussed on p. 21 of Braun et al.(1971)."

6. Other Forms of References

  a. *In the press.* It is recommended that (i) this should mean that the paper has been finally accepted by a journal, (ii) it is quoted in the text (both systems) just as any other paper, (iii) the year quoted should be the best estimate revised if necessary at proof stage, and (iv) the full citation in the list to read: Braun, A. & Brown, B. (1971) *Journal*, in the press.

  b. *Submitted for publication* should be used in a typescript only when it is reasonable to expect that it will be possible to alter the quotation to a final form at a stage before publication; if such alteration cannot be made then the name of the journal involved should be stated.

  c. The use of *in preparation* and *private communication* should not be allowed because they have no real value.

  d. *Personal communication* and *unpublished work* should be permitted in the text only, i.e., not in the list of references. Editors may require to see written evidence of the former.

7. Abbreviations for Journal Titles

Most biochemical journals use the *Chemical Abstract** system but a few use the World List, 4th Edition. The Commission noted that the latest information available (International List of Periodical Title Word Abbreviations prepared for the UNISIST/ICSU-AB Working Group on Bibliographical Descriptions) suggests that the abbreviations that will be recommended finally by ICSU will be very similar to those now used by *Chemical Abstracts*.

Believing that complete uniformity on this issue is highly desirable now and estimating that it may be a few more years before ICSU finally reports, the Commission recommends that all biochemical journals should now use the *Chemical Abstracts* (American Chemical Society) system. The Commission believes that any changes that will be required when ICSU eventually issues recommendations on this point will be comparatively minor ones.

8. Implementation of these Recommendations

The Commission at its meeting in Menton, May 7 to 8, 1971, has taken the view that the degree of uniformity envisaged in the Recommendations is highly desirable and therefore further recommends to all biochemical journals that the changes required should be made as soon as possible. The Commission recognizes that all journals will have to make some changes (in most cases these are minor) from their present established practices to implement these Recommendations in full. It considers that the possible objections of difficulties even for a commercial publisher with an established "house style" are outweighed by the advantage that conformity of style in the citation of references will prove to the authors, editors, and readers upon whom all journals depend for their existence.

---

*The journal-title abbreviations in *Biological Abstracts* are essentially the same in *Chemical Abstracts*. A *List of Serials with Title Abbreviations* is available from BioSciences Information Service of Biological Abstracts, 2100 Arch Street, Philadelphia, PA 19103.

IUPAC TENTATIVE RULES FOR THE NOMENCLATURE OF ORGANIC CHEMISTRY SECTION E. FUNDAMENTAL STEREOCHEMISTRY*

International Union of Pure and Applied Chemistry

INTRODUCTION

This Section of the IUPAC Rules for Nomenclature of Organic Chemistry differs from previous Sections in that it is here necessary to legislate for words that describe concepts as well as for names of compounds.

At the present time, concepts in stereochemistry (that is, chemistry in three-dimensional space) are in the process of rapid expansion, not merely in organic chemistry, but also in biochemistry, inorganic chemistry, and macromolecular chemistry. The aspects of interest for one area of chemistry often differ from those for another, even in respect to the same phenomenon. This rapid evolution and the variety of interests have led to development of specialized vocabularies and definitions that sometimes differ from one group of specialists to another, sometimes even within one area of chemistry.

The Commission on the Nomenclature of Organic Chemistry does not, however, consider it practical to cover all aspects of stereochemistry in this Section E. Instead, it has two objects in view: To prescribe, for basic concepts, terms that may provide a common language in all areas of stereochemistry; and to define the ways in which these terms may, so far as necessary, be incorporated into the names of individual compounds. The Commission recognizes that specialized nomenclatures are required for local fields; in some cases, such as carbohydrates, amino acids, peptides and proteins, and steroids, international rules already exist; for other fields, study is in progress by specialists in Commissions or Subcommittees; and further problems doubtless await identification. The Commission believes that consultations will be needed in many cases between different groups within IUPAC and IUB if the needs of the specialists are to be met without confusion and contradiction between the various groups.

The Rules in this Section deal only with Fundamental Stereochemistry, that is, the main principles. Many of these Rules do little more than codify existing practice, often of long standing; however, others extend old principles to wider fields, and yet others deal with nomenclature that is still subject to controversy.

Rule E-0

The stereochemistry of a compound is denoted by an affix or affixes to the name that does not prescribe the stereochemistry; such affixes, being additional, do not change the name or the numbering of the compound. Thus, enantiomers, diastereoisomers, and *cis–trans* isomers receive names that are distinguished only by means of different stereochemical affixes. The only exceptions are those trivial names that have stereochemical implications (for example, fumaric acid, cholesterol).

Note: In some cases (see Rules E-2.23 and E-3.1) stereochemical relations may be used to decide between alternative numberings that are otherwise permissible.

E-1. Types of Isomerism

E-1.1. The following nonstereochemical terms are relevant to the stereochemical nomenclature given in the Rules that follow.

---

*From *IUPAC Inf. Bull. Append. Tentative Nomencl. Sym. Units Stand.*, No. 35, August 1974, pp. 36–80. With permission.

(a) The term structure may be used in connection with any aspect of the organization of matter.

Hence: structural (adjectival)

(b) Compounds that have identical molecular formulas but differ in the nature or sequence of bonding of their atoms or in arrangement of their atoms in space are termed isomers.

Hence: isomeric (adjectival)
　　　　isomerism (phenomenological)

Examples:

$H_3C-O-CH_3$ is an isomer of H_3C-CH_2-OH

$$\underset{H}{\overset{H_3C}{>}}C=C\underset{H}{\overset{CH_3}{<}} \text{ is an isomer of } \underset{H_3C}{\overset{H}{>}}C=C\underset{H}{\overset{CH_3}{<}}$$

(In this and other Rules a broken line denotes a bond projecting behind the plane of the paper, and a thickened line denotes a bond projecting in front of the plane of the paper. In such cases a line of normal thickness denotes a bond lying in the plane of the paper.)

(c) The constitution of a compound of given molecular formula defines the nature and sequence of bonding of the atoms. Isomers differing in constitution are termed constitutional isomers.

Hence: constitutionally isomeric (adjectival)
　　　　constitutional isomerism (phenomenological)

Example:

$H_3C-O-CH_3$ is a constitutional isomer of H_3C-CH_2-OH.

Note: Use of the term "structural" with the above connotation is abandoned as insufficiently specific.

E-1.2. Isomers are termed stereoisomers when they differ only in the arrangement of their atoms in space.

Hence: stereoisomeric (adjectival)
　　　　stereoisomerism (phenomenological)

Examples:

$$\underset{H}{\overset{H_3C}{>}}C=C\underset{H}{\overset{CH_3}{<}} \text{ is a stereoisomer of } \underset{H}{\overset{H_3C}{>}}C=C\underset{CH_3}{\overset{H}{<}}$$

$$\underset{HO}{\overset{H}{\diagdown}}\overset{CHO}{\underset{|}{C}}\diagup^{CH_2OH} \text{ is a stereoisomer of } \underset{H}{\overset{HO}{\diagdown}}\overset{CHO}{\underset{|}{C}}\diagup^{CH_2OH}$$

E-1.3. Stereoisomers are termed *cis–trans* isomers when they differ only in the positions of atoms relative to a specified plane in cases where these atoms are, or are considered as if they were, parts of a rigid structure.

Hence: *cis–trans* isomeric (adjectival)
cis–trans isomerism (phenomenological)

Examples:

E-1.4. Various views are current regarding the precise definition of the term "configuration." (a) Classical interpretation: The configuration of a molecule of defined constitution is the arrangement of its atoms in space without regard to arrangements that differ only as after rotation about one or more single bonds. (b) This definition is now usually limited so that no regard is paid also to rotation about π bonds or bonds of partial order between one and two. (c) A third view limits the definition further so that no regard is paid to rotation about bonds of any order, including double bonds.

Molecules differing in configuration are termed configurational isomers.

Hence: configurational isomerism

Notes: (1) Contrast conformation (Rule E-1.5). (2) The phrase "differ only as after rotation" is intended to make the definition independent of any difficulty of rotation, in particular independent of steric hindrance to rotation. (3) For a brief discussion of views (a) to (c), see Appendix 1. It is hoped that a definite consensus of opinion will be established before these Rules are made "Definitive."

Examples: The following pairs of compounds differ in configuration:

(ii) [structures of cyclohexane-1,2-diol isomers]

(iii) [structures of decalin isomers]

(iv) [structures of 2,3-dimethyl-2-butene related dienes]

These isomers (iv) are configurational in view (a) or (b) but are conformational (see Rule E-1.5) in view (c)

E-1.5. Various views are current regarding the precise definition of the term "conformation." (a) Classical interpretation: The conformations of a molecule of defined configuration are the various arrangements of its atoms in space that differ only as after rotation about single bonds. (b) This is usually now extended to include rotation about π bonds or bonds of partial order between one and two. (c) A third view extends the definition further to include also, rotation about bonds of any order, including double bonds.

Molecules differing in conformation are termed conformational isomers.

Hence: conformational isomerism

Notes: All the Notes to Rule E-1.4 apply also to E-1.5.

Examples: Each of the following pairs of formulas represents a compound in the same configuration but in different conformations.

(a, b, c) [Newman projections of ethane, staggered and eclipsed]

(a, b, c) [Newman projections of butane conformers]

(a, b, c)

(b, c)

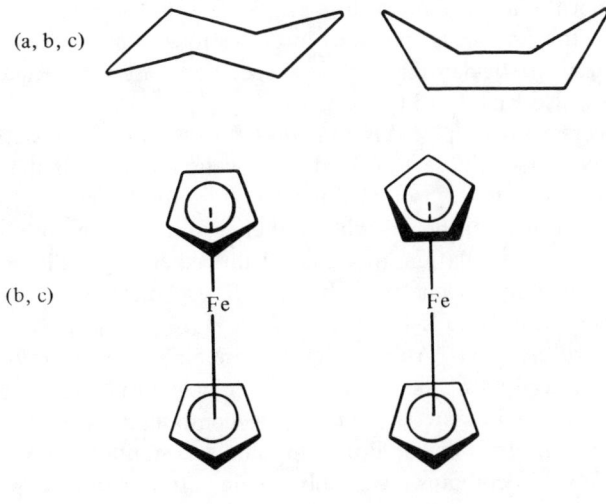

(c) See Example (iv) to Rule E-1.4.

E-1.6. The terms relative stereochemistry and relative configuration are used with reference to the positions of various atoms in a compound relative to one another, especially, but not only, when the actual positions in space (absolute configuration) are unknown.

E-1.7. The terms absolute stereochemistry and absolute configuration are used with reference to the known actual positions of the atoms of a molecule in space.*

E-2. cis–trans Isomerism†

Preamble. The prefixes *cis* and *trans* have long been used for describing the relative positions of atoms or groups attached to nonterminal doubly bonded atoms of a chain or attached to a ring that is considered as planar. This practice has been codified for hydrocarbons by IUPAC.** There has, however, not been agreement on how to assign *cis* or *trans* at terminal double bonds of chains or at double bonds joining a chain to a ring. An obvious solution was to use *cis* and *trans* where doubly bonded atoms formed the backbone and were nonterminal and to enlist the sequence-rule preferences to decide other cases; however, since the two methods, when generally applied, do not always produce analogous results, it would then be necessary to use different symbols for the two procedures. A study of this combination showed that both types of symbols would often be required in one name and, moreover, it seemed wrong in principle to use two symbolisms for essentially the same phenomenon. Thus it seemed to the Commission wise to use only the sequence-rule system, since this alone was applicable to all cases. The same decision was taken independently by Chemical Abstracts Service who introduced *Z* and *E* to correspond more conveniently to *seqcis* and *seqtrans* of the sequence rule.

It is recommended in the Rules below that these designations *Z* and *E* based on the sequences rule shall be used in names of compounds, but *Z* and *E* do not always correspond to the classical *cis* and *trans* which show the steric relations of like or similar

---

*Determination of absolute configuration became possible through work by Bijvoet, J. M., Peerdeman, A. F., and van Bommel, A. J., *Nature,* 168, 271 (1951); cf. Bijvoet, J. M., *Proc. Kon. Ned. Akad. Wetensch.,* 52, 313 (1949).

†These Rules supersede the Tentative Rules for olefinic hydrocarbons published in the Comptes rendus of the 16th IUPAC Conference, New York, N.Y., 1951, pp. 102–103.

**Blackwood, J. E., Gladys, C. L., Loening, K. L., Petrarca, A. E., and Rush, J. E., *J. Amer. Chem. Soc.,* 90, 509 (1968); Blackwood, J. E., Gladys, C. L., Petrarca, A. E., Powell, W. H., and Rush, J. E., *J. Chem. Doc.,* 8, 30 (1968).

groups that are often the main point of interest. So the use of *Z* and *E* in names is not intended to hamper the use of *cis* and *trans* in discussions of steric relations of a generic type or of groups of particular interest in a specified case (see Rule E-2.1 and its Examples and Notes, also Rule E-5.11).

It is also not necessary to replace *cis* and *trans* for describing the stereochemistry of substituted monocycles (see Subsection E-3). For cyclic compounds the main problems are usually different from those around double bonds; for instance, steric relations of substitutents on rings can often be described either in terms of chirality (see Subsection E-5) or in terms of *cis–trans* relationships, and, further, there is usually no single relevant plane of reference in a hydrogenated polycycle. These matters are discussed in the Preambles to Subsections E-3 and E-4.

E-2.1. *Definition of cis–trans*. Atoms or groups are termed *cis* or *trans* to one another when they lie respectively on the same or on opposite sides of a reference plane identifiable as common among stereoisomers. The compounds in which such relations occur are termed *cis–trans* isomers. For compounds containing only doubly bonded atoms, the reference plane contains the doubly bonded atoms and is perpendicular to the plane containing these atoms and those directly attached to them. For cyclic compounds, the reference plane is that in which the ring skeleton lies or to which it approximates. When qualifying another word or a locant, *cis* or *trans* is followed by a hyphen. When added to a structural formula, *cis* may be abbreviated to *c*, and *trans* to *t* (see also Rule E-3.3).

Examples: (Rectangles here denote the reference planes and are considered to lie in the plane of the paper.)

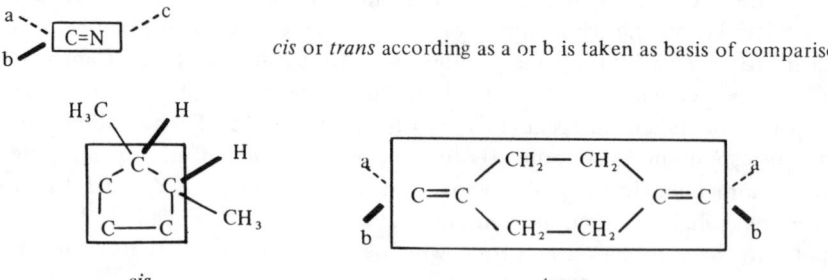

The groups or atoms a,a are the pair selected for designation but are not necessarily identical; b,b are also not necessarily identical but must be different from a,a.

cis or *trans* according as a or b is taken as basis of comparison

Notes: The formulas above are drawn with the reference plane in the plane of the paper, but for doubly bonded compounds it is customary to draw the formulas so that this plane is perpendicular to that of the paper; atoms attached directly to the doubly bonded atoms then lie in the plane of the paper and the formulas appear as, for instance

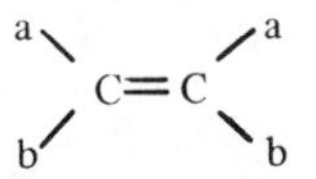

$$\begin{array}{c} a \diagdown \qquad \diagup a \\ C{=}C \\ b \diagup \qquad \diagdown b \end{array}$$

cis

Cyclic structures, however, are customarily drawn with the ring atoms in the plane of the paper, as above. However, care is needed for complex cases, such as

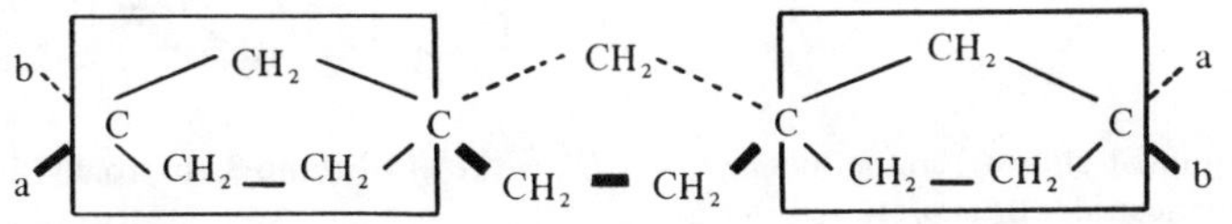

[structure showing bicyclic system with labels a and b on opposite sides]

The central five-membered ring lies (approximately) in a plane perpendicular to the plane of the paper. The two a groups are *trans* to one another; so are the b groups; the outer cyclopentane rings are *cis* to one another with respect to the plane of the central ring. *cis* or *trans* (or Z or E; see Rule E-2.21) may also be used in cases involving a partial bond order when a limiting structure is of sufficient importance to impose rigidity around the bond of partial order. An example is

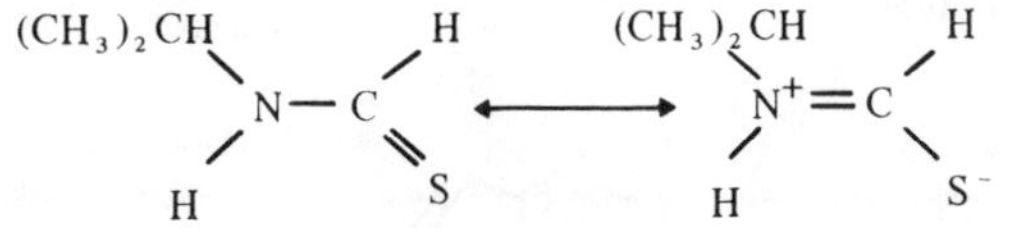

[resonance structures of a thioamide / thioiminium zwitterion with (CH$_3$)$_2$CH and H on N, and H and S (or S$^-$) on C]

trans (or E)

E-2.2. *cis-trans* Isomerism around Double Bonds.

E-2.21. In names of compounds steric relations around one or more double bonds are designated by affixes Z and/or E, assigned as follows. The sequence-rule-preferred* atom or group attached to one of a doubly bonded pair of atoms is compared with the sequence-rule-preferred atom or group attached to the other of that doubly bonded pair of atoms; if the selected pair are on the same side of the reference plane (see Rule 2.1) an italic capital letter Z prefix is used; if the selected pair are on opposite sides an italic capital letter E prefix is used.† These prefixes, placed in parentheses and followed by a hyphen, normally precede the whole name; if the molecule contains several double bonds, then each prefix is immediately preceded by the lower or less primed locant of the relevant double bond.

Examples:

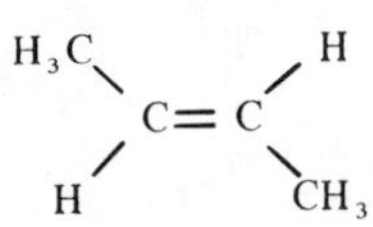

(E)-2-Butene	(Z)-2-Methyl-2-butenoic acid** or (Z)-2-methylisocrotonic acid (see Exceptions below)	(E)-2-Methyl-2-butenoic acid†† or (E)-2-Methylcrotonic acid (see Exceptions below)

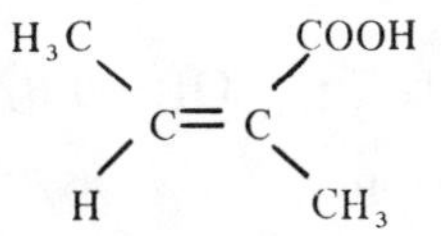

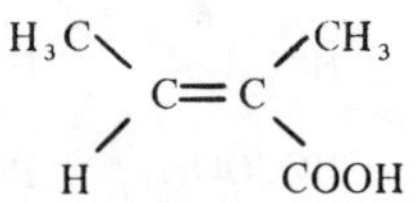

*For sequence-rule preferences see Appendix 2.
†These prefixes may be rationalized as from the German *zusammen* (together) and *entgegen* (opposite).
**The name angelic acid is abandoned because it has been associated with the designation *trans* with reference to the methyl groups.
††The name tiglic acid is abandoned because it has been associated with the designation *cis* with reference to the methyl groups.

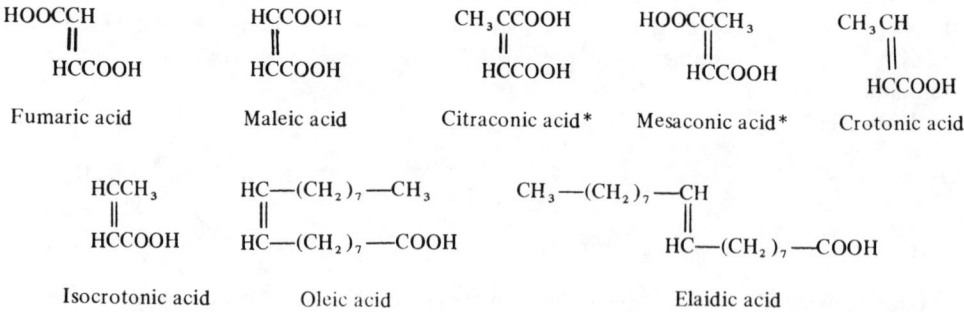

Exceptions to Rule E-2.21. The following are examples of accepted trivial names in which the stereochemistry is prescribed by the name and is not cited by a prefix.

E-2.22 (*Alternative to Part of E-2.21*). (a) When more than one series of locants starting from unity is required to designate the double bonds in a molecule, or when the name consists of two words, the Z and E prefixes together with their appropriate locants may be placed before that part of the name where ambiguity is most effectively removed.

(b) [Alternative to (a)] When several Z or E prefixes are required they are arranged in

---

*Systematic names are recommended for derivatives of these compounds formed by substitution on carbon.

order as follows: Of the four atoms or groups attached to each doubly bonded pair of atoms, that one preferred by the sequence rule is selected; the single atoms or groups thus selected are then arranged in their sequence rule order (determined in respect of their position in the whole molecule), and the prefixes Z and/or E for the respective double bonds are placed in that order, but *without* their locants.

Note: In method (a) the final choice is left to an author or editor because of the variety of cases met and because the problems are not always the same in different languages. The presence of the locants usually eases translation from the name to a formula, but this method (a) may involve the logical difficulty explained for the third example below. Method (b) always gives a single unambiguous order and is not subject to the logical difficulty just mentioned, but translation from the name to the formula is harder than for method (a). Method (a) may be more suitable for cursive text, and method (b) for compendia. If method (b) is used it should be used whenever more than one double bond is involved, but method (a) is to be used only under the special conditions detailed in the rule.

Examples:

(a) (2E,4Z)-2,4-Hexadienoic acid
(b) (E,Z)-2,4-Hexadienoic acid

(a) (2E,4Z)-5-Chloro-2,4-hexadienoic acid
(b) (Z,E)5-Chloro-2,4-hexadienoic acid

(a) 3-[(E)-1-Chloropropenyl]-(3Z,5E)-3,5-heptadienoic acid
(b) (E,Z,E)-3-(1-Chloropropenyl)-3,5-heptadienoic acid

[The last example shows the disadvantages of both methods. In method (a) there is a fault of logic, namely, the 3Z,5E are not the property of the unsubstituted heptadienoic acid chain, but the 3Z arises only because of the side chain that is cited before the 3Z,5E. In method (b) it is some trouble to assign the E,Z,E to the correct double bonds.]

(a) (1Z,3E)-1,3-Cyclododecadiene
(b) (Z,E)-1,3-Cyclododecadiene

[The lower locant is assigned to the Z double bond.]

(a) 5-Chloro-4-(E-sulfomethylene)-(2E,5Z)-2,5-heptadienoic acid
(b) (Z,E,E)-5-Chloro-4-(sulfomethylene)-2,5-heptadienoic acid

[In application of the sequence rule, the relation of the SO_3H to CCl (rather than to C-3), and of the CH_3 to Cl, are decisive.]

(a) Butanone (E)-oxime*
(b) (E)-Butanone oxime

(a) 2-Chlorobenzophenone (Z)-hydrazone
(b) (Z)-2-Chlorobenzophenone hydrazone

(a) (E)-2-Pentenal (Z)-semicarbazone
(b) (Z,E)-2-Pentenal semicarbazone

(a) Benzil (Z,E)-dioxime
(b) (Z,E)-Benzil dioxime

E-2.23. When Rule C-13.1 or E-2.22(b) permits alternatives, preference for lower locants and for inclusion in the principal chain is allotted as follows, in the order stated, so far as necessary: Z over E groups; cis over trans cyclic groups; R over S groups (also r over s, etc., as in the sequence rule); if the nature of these groups is not decisive, then the lower locant for such a preferred group at the first point of difference.

Examples:

(a) (2Z,5E)-2,5-Heptadienedioic acid
(b) (E,Z)-2,5-Heptadienedioic acid

[The lower numbers are assigned to the Z double bond.]

*The terms syn, anti, and amphi are abandoned for such compounds.

$$\underset{Cl}{\overset{H}{\diagup}}\underset{Z}{\overset{1}{C}=\overset{2}{C}}\underset{\diagup}{\overset{H}{\diagdown}}\overset{3}{C}=\overset{2}{C}\underset{H}{\overset{5}{\diagdown}CH_3}$$

$$\underset{Cl}{\overset{H}{\diagup}}\underset{E}{C=C}\underset{H}{\diagdown}$$

(a) 1-Chloro-3-[2-chloro-(E)-vinyl]-(1Z,3Z)-1,3-pentadiene
(b) (E,Z,Z)-1-Chloro-3-(2-chlorovinyl)-1,3-pentadiene

[According to Rule C-13.1 the principal chain must include the C=C–CH₃ group because this gives lower numbers to the double bonds (1,3 rather than 1,4); then the Cl-containing Z group is chosen for the remainder of the principal chain in accord with Rule E-2.23.]

$$\underset{(S)\text{-}CH_3CH_2CH(CH_3)}{\overset{(R)\text{-}\overset{6}{C}H_2\overset{5}{C}H_2\overset{4}{C}H(CH_3)}{}}\overset{3}{C}=\overset{2}{C}\underset{H}{\overset{\overset{1}{COOH}}{}}$$

(a,b) (Z)-(4R)-3-[(S)-sec-Butyl]-4-methyl-2-hexenoic acid

[The principal chain is chosen to include the (R)-group, and the prefix Z refers to the (R)-group.]

E-3. Relative Stereochemistry of Substituents in Monocyclic Compounds[†]

Preamble. The prefixes *cis* and *trans* are commonly used to designate the positions of substituents on rings relative to one another; when the ring is, or is considered to be, rigidly planar or approximately so and is placed horizontally, these prefixes define which groups are above and which below the (approximate) plane of the ring. This differentiation is often important, so this classical terminology is retained in Subsection E-3; since the difficulties inherent in end groups do not arise for cyclic compounds, it is unnecessary to resort to the less immediately informative E/Z symbolism.

When the *cis–trans* designation of substituents is applied, rings are considered in their most extended form; reentrant angles are not permitted; for example

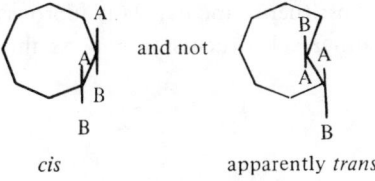

cis   apparently *trans*

The absolute stereochemistry of optically active or racemic derivatives of monocyclic compounds is described by the sequence-rule procedure (see Rule E-5.9 and Appendix 2). The relative stereochemistry may be described by a modification of sequence-rule symbolism as set out in Rule E-5.10. If either of these procedures is adopted, it is then superfluous to use also *cis* or *trans* in the names of individual compounds.

[†]Formulas in Examples to this Rule denote relative (not absolute) configurations.

E-3.1. When alternative numberings of the ring are permissible according to the Rules of Section C, that numbering is chosen which gives a *cis* attachment at the first point of difference; if that is not decisive, the criteria of Rule E-2.23 are applied. The prefixes *cis* and *trans* may be abbreviated to *c* and *t*, respectively, in names of compounds when more than one such designation is required.
Examples:

1,*c*-2,*t*-3-Trichlorocyclohexane

1-(Z)-Propenyl-*trans*-3-(*E*)-propenylcyclohexane

E-3.2. When one substituent and one hydrogen atom are attached at each of two positions of a monocycle, the steric relations of the two substituents are expressed as *cis* or *trans*, followed by a hyphen and placed before the name of the compound.
Examples:

cis-1,2-Dichlorocyclopentane

trans-2-Chloro-1-cyclopentanecarboxylic acid

trans-2-Chloro-4-nitro-1,
1-cyclohexanedicarboxylic acid

E-3.3. When one substituent and one hydrogen atom are attached at each of more than two positions of a monocycle, the steric relations of the substituents are expressed by adding *r* (for *reference* substituent), followed by a hyphen, before the locant of the lowest numbered of these substituents and *c* or *t* (as appropriate), followed by a hyphen, before the locants of the other substituents to express their relation to the reference substituent.
Examples:

r-1,*t*-2,*c*-4-Trichlorocyclopentant
(not *r*-1, *t*-2, *t*-4, which would follow from the alternative direction of numbering; see Rule E-3.1)

t-5-Chloro-*r*-1, *c*-3-cyclohexanedicarboxylic acid

E-3.4. When two different substituents are attached at the same position of a monocycle, then the lowest numbered substituent named as suffix is selected for designation as reference group in accordance with Rule E-3.2 or E-3.3; or, if none of the substituents is named as suffix, then of the lowest numbered pair that one preferred by the sequence rule is selected as reference group; and the relation of the sequence-rule preferred group at each other position, relative to the reference group, is cited as *c* or *t* (as appropriate).

Examples:

1,*t*-2-Dichloro-*r*-1-cyclopentanecarboxylic acid

r-1-Bromo-1-chloro-*t*-3-ethyl-3-methylcyclohexane
(alphabetical order of prefixes)

c-3-Bromo-3-chloro-*r*-1-cyclopentanecarboxylic acid

2-Crotonoyl-*t*-2-isocrotonoyl-*r*-1-cyclopentane-carboxylic acid

E-4. Fused Rings

Preamble. In simple cases the relative stereochemistry of substituted fused-ring systems can be designated by the methods used for monocycles. For the absolute stereochemistry of optically active and racemic compounds the sequence-rule procedure can be used in all cases (see Rule E-5.9 and Appendix 2), and for related relative stereochemistry the procedure of Rule E-5.10 can be applied. Sequence-rule methods are, however, not descriptive of geometrical shape for other than quite simple cases. There is as yet no generally acceptable system for designating in an immediately interpretable manner the stereochemistry of polycyclic bridged ring compounds (for instance, the *endo–exo* nomenclature, which should solve one set of problems, has been used in different ways). These and related problems (e.g., cyclophanes, catenanes) will be considered in a later document.

E-4.1. Steric relations at saturated bridgeheads common to two rings are denoted by *cis* or *trans*, followed by a hyphen and placed before the name of the ring system, according to the relative positions of the exocyclic atoms or groups attached to the bridgeheads. Such rings are said to be *cis* fused or *trans* fused.

Examples:

cis-Decalin

1-Methyl-*trans*-bicyclo[8.3.1]tetradecane

E-4.2. Steric relations at more than one pair of saturated bridgeheads in a polycyclic compound are denoted by *cis* or *trans*, each followed by a hyphen and, when necessary, the corresponding locant of the lower numbered bridgehead and a second hyphen, all placed before the name of the ring system. Steric relations between the nearest atoms* of *cis*- or *trans*-bridgehead pairs may be described by affixes *cisoid* or *transoid*, followed by a hyphen and, when necessary, the corresponding locants and a second hyphen, the whole placed between the designations of the *cis*- or *trans*-ring junctions concerned. When a choice remains among nearest atoms, the pair containing the lower numbered atom is selected; *cis* and *trans* are not abbreviated in such cases. In complex cases, however, designation may be more simply effected by the sequence-rule procedure (see Appendix 2).

Examples:

cis-cisoid-trans-Perhydrophenanthrene

cis-cisoid-4a, 10a-*trans*-Perhydroanthracene
or *rel*-(4a*R*, 8a*S*, 9a*S*, 10a*S*)-Perhydroanthracene†

trans-3a-*cisoid*-3a, 4a-*cis*-4a-Perhydrobenz[*f*]indene
or *rel*-(3a*R*, 4a*S*, 8a*R*, 9a*R*)-Perhydrobenz[*f*]indene

E-5. Chirality

E-5.1. The property of nonidentity of an object with its mirror image is termed chirality. An object, such as a molecule in a given configuration or conformation, is termed chiral when it is not identical with its mirror image; it is termed achiral when it is identical with its mirror image.

Notes: (1) Chirality is equivalent to handedness, the term being derived from the Greek Χειρ = hand.

(2) All chiral molecules are molecules of optically active compounds, and molecules of all optically active compounds are chiral. There is a 1:1 correspondence between chirality and optical activity.

(3) In organic chemistry the discussion of chirality usually concerns the individual molecule or, more strictly, a model of the individual molecule. The chirality of an assembly of molecules may differ from that of the component molecules, as in a chiral quartz crystal or in an achiral crystal containing equal numbers of dextrorotatory and levorotatory tartaric acid molecules.

(4) The chirality of a molecule can be discussed only if the configuration or conformation of the molecule is specifically defined or is considered as defined by

---

*The term "nearest atoms" denotes those linked together through the smallest number of atoms, irrespective of actual separation in space. For instance, in the second Example to this Rule, the atom 4a is "nearer" to 10a than to 8a.

†For the designation *rel*, see Rule E-5.10.

common usage. In such discussions structures are treated as if they were (at least temporarily) rigid. For instance, ethane is configurationally achiral although many of its conformations, such as (A), are chiral; in fact, a configuration of a mobile molecule is chiral only if all its possible conformations are chiral; and conformations of ethane such as (B) and (C) are achiral.

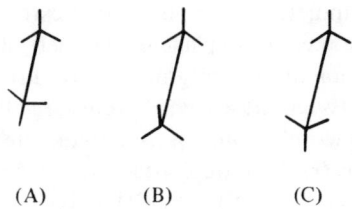

(A)   (B)   (C)

Examples:

```
      CHO              CHO                        CH₂OH
       |                |                          |
       C                C                          C
   H ⟋ | ⟍ OH     HO ⟋ | ⟍ H              H ⟋ | ⟍ OH
       |                |                          |
      CH₂OH            CH₂OH                      CH₂OH
       (D)              (E)                        (F)
```

(D) and (E) are mirror images and are not identical, not being superposable. They represent chiral molecules. They represent (D) dextrorotatory and (E) levorotatory glyceraldehyde.

(F) is identical with its mirror image. It represents an achiral molecule, namely, a molecule of *1,2,3*-propanetriol (glycerol).

E-5.2. The term asymmetry denotes absence of any symmetry. An object, such as a molecule in a given configuration or conformation, is termed asymmetric if it has no element of symmetry.

Notes: (1) All asymmetric molecules are chiral, and all compounds composed of them are therefore optically active; however, not all chiral molecules are asymmetric since some molecules having axes of rotation are chiral.

(2) Notes (3) and (4) to Rule E-5.1 apply also in discussions of asymmetry.

Examples:

```
      CHO
       |
       C
   H ⟋ | ⟍ OH
       |
      CH₂OH
```

has no element of symmetry and represents a molecule of an optically active compound.

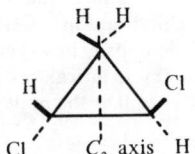

has a C_2 axis of rotation; it is chiral although not asymmetric, and is therefore a molecule of an optically active compound.

E-5.3. (a) An asymmetric atom is one that is tetrahedrally bonded to four different atoms or groups, none of the groups being the mirror image of any of the others.

(b) An asymmetric atom may be said to be at a chiral center since it lies at the center of a chiral tetrahedral structure. In a general sense, the term "chiral center" is not restricted to tetrahedral structures; the structure may, for instance, be based on an octahedron or tetragonal pyramid.

(c) When the atom by which a group is attached to the remainder of a molecule lies at a chiral center, the group may be termed a chiral group.

Notes: (1) The term "asymmetric," as applied to a carbon atom in rule E-5.3 (a), was chosen by van't Hoff because there is no plane of symmetry through a tetrahedron whose corners are occupied by four atoms or groups that differ in scalar properties. For differences of vector sense between the attached groups, see Rule E-5.8.

(2) In Subsection E-5 the word "group" is used to denote the series of atoms attached to one bond. For instance, in (i) the groups attached to C* are $-CH_3$, $-OH$, $-CH_2CH_3$, and $-COOH$; in (ii) they are $-CH_3$, $-OH$, $-COCH_2CH_2CH_2$, and $-CH_2CH_2CH_2CO$.

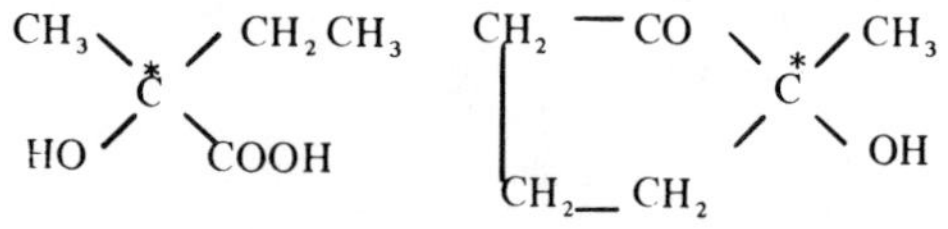

(3) For the chiral axis and chiral plane (which are less common than the chiral center), see Appendix 2.

(4) There may be more than one chiral center in a molecule and these centers may be identical, or structurally different, or structurally identical but of opposite chirality; however, the presence of an equal number of structurally identical chiral groups of opposite chirality, and no other chiral group, leads to an achiral molecule. These statements apply also to chiral axes and chiral planes. Identification of the sites and natures of the various factors involved is essential if the overall chirality of a molecule is to be understood.

(5) Although the term "chiral group" is convenient for use in discussions it should be remembered that chirality attaches to molecules and not to groups or atoms. For instance, although the *sec*-butyl group may be termed chiral in dextrorotatory 2-*sec*-butyl-naphthalene, it is not chiral in the achiral compound $(CH_3CH_2)(CH_3)CH-CH_3$.

Examples:

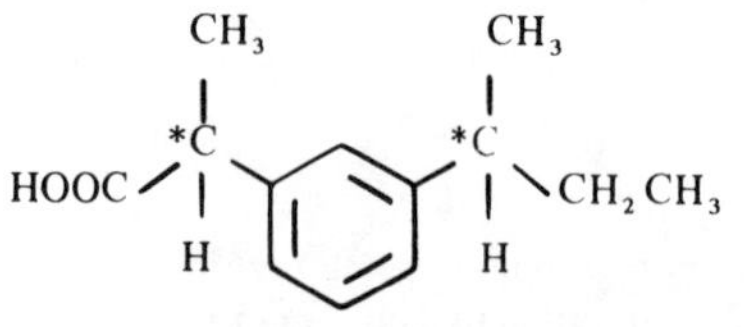

In this chiral compound there are two asymmetric carbon atoms, marked C*, each lying at a chiral center. These atoms form part of different chiral groups, namely, $-CH(CH_3)-COOH$ and $-CH(CH_3)CH_2CH_3$.

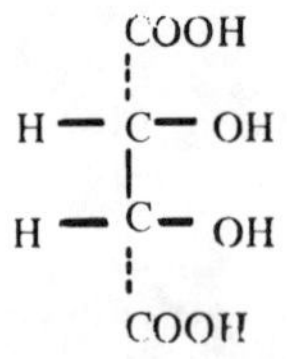

In this molecule (*meso*-tartaric acid) the two central carbon atoms are asymmetric atoms and each is part of a chiral group $-CH(OH)COOH$. These groups, however, although structurally identical, are of opposite chirality, so that the molecule is achiral.

E-5.4. Molecules that are mirror images of one another are termed enantiomers and may be said to be enantiomeric. Chiral groups that are mirror images of one another are termed enantiomeric groups.

Hence: enantiomerism (phenomenological)

Note: Although the adjective enantiomeric may be applied to groups, enantiomerism strictly applies only to molecules [see Note (5) to Rule E-5.3].

Examples: The following pairs of molecules are enantiomeric.

(i) [Fischer projections of glyceraldehyde enantiomers: CHO/C/H/OH/CH₂OH and CHO/C/HO/H/CH₂OH]

(ii) [Fischer projections of tartaric acid-like structures: COOH/H–C–OH/HO–C–H/COOH and COOH/HO–C–H/H–C–OH/COOH]

(iii) [Biphenyl atropisomers with CH₃ and COOH substituents]

(iv) [Two cyclooctene structures] Cyclooctene

(v) [Two cyclobutane structures with Cl, Br, NH₂, CH₂ substituents]

(vi) [Two structures with Cl-substituted phenyl rings bonded to sec-butyl groups]

The *sec*-butyl groups in (vi) are enantiomeric.

E-5.5. When equal amounts of enantiomeric molecules are present together, the product is termed racemic, independently of whether it is crystalline, liquid, or gaseous. A homogeneous solid phase composed of equimolar amounts of enantiomeric molecules is termed a racemic compound. A mixture of equimolar amounts of enantiomeric molecules present as separate solid phases is termed a racemic mixture. Any homogeneous solid containing equimolar amounts of enantiomeric molecules is termed a racemate.

Examples: The mixture of two kinds of crystal (mirror-image forms) that separate below 28° from an aqueous solution containing equal amounts of dextrorotatory and levorotatory sodium ammonium tartrate is a racemic mixture.

The symmetrical crystals that separate from such a solution above 28°, each containing equal amounts of the two salts, provide a racemic compound.

E-5.6. Stereoisomers that are not enantiomeric are termed diastereoisomers.

Hence: diastereoisomeric (adjectival)
diastereoisomerism (phenomenological)

Note: Diastereoisomers may be chiral or achiral.
Examples:

```
      COOH              COOH
       |                 |
  H — C — OH        H — C — OH
       |       and       |            are diastereoisomers; the former is achiral,
  H — C — OH       HO — C — H         and the latter is chiral.
       |                 |
      COOH              COOH


      COOH              COOH
       |                 |
  H — C — OH        H — C — OH
       |       and       |            are diastereoisomers; both are chiral.
  H — C — OH       HO — C — H
       |                 |
      CH₃               CH₃
```

E-5.7. A compound whose individual molecules contain equal numbers of enantiomeric groups, identically linked, but no other chiral group, is termed a *meso* compound.
Example:

```
                              COOH
                               |
      COOH                H — C — OH
       |                       |
  H — C — OH             HO — C — H
       |                       |
  H — C — OH             HO — C — H
       |                       |
      COOH                H — C — OH
                               |
                              COOH

   meso-Tartaric acid      Galactaric acid
```

E-5.8. An atom is termed pseudoasymmetric when bonded tetrahedrally to one pair of enantiomeric groups (+)-a and (−)-a and also to two atoms or groups b and c that are different from group a, different from each other, and not enantiomeric with each other.
Examples:

```
         H   H   H                    H   H   H
         |   |   |                    |   |   |
   HOOC—C — C*— C—COOH          HOOC—C — C*— C—COOH
         |   |   |                    |   |   |
         HO  OH  OH                   HO  CH₂ OH
                                          |
           (A)                        CH₃—C — CH₂CH₃      (B)
                                          |
   C* are pseudoasymmetric               H
```

Notes: (1) The orientation, in space, of the atoms around a pseudoasymmetric atoms is not reversed on reflection; for a chiral atom (see Note to Rule E-5.3) this orientation is always reversed.

(2) Molecules containing pseudoasymmetric atoms may be achiral or chiral. If ligands b and c are both achiral, the molecule is achiral as in the first example to this Rule. If either or both of the nonenantiomeric ligands b and c are chiral, the molecule is chiral, as in the second example to this Rule, that is the molecule is not identical with its mirror image. A molecule (i) is also chiral if b and c are enantiomeric, that is, if the molecule can be symbolized as (ii), but then, by definition, it does not contain a pseudoasymmetric atom.

(3) Compounds differing at a pseudoasymmetric atom belong to the larger class of diastereoisomers.

(4) In example (A), interchange of H and OH on C* gives a different achiral compound, which is an achiral diastereoisomer of (A) (see Rule E-5.6). In example (B), diastereoisomers are produced by inversion at C* or °C, giving in all four diastereoisomers, all chiral because of the $-CH(CH_3)CH_2CH_3$ group.

E-5.9. Names of chiral compounds whose absolute configuration is known are differentiated by prefixes *R, S*, etc., assigned by the sequence-rule procedure (see Appendix 2), preceded when necessary by the appropriate locants.

Examples:

(*R*)-Glyceraldehyde   (*S*)-Glyceraldehyde

(6a*S*,12*S*,5'*R*)-Rotenone   Methyl phenyl (*R*)-sulfoxide

E-5.10. (a) Names of compounds containing chiral centers, of which the relative but not the absolute configuration is known, are differentiated by prefixes *R*, S** (spoken R star, S star), preceded when necessary by the appropriate locants, these prefixes being assigned by the sequence-rule procedure (see Appendix 2) on the arbitrary assumption that the prefix first cited is *R*.

(b) In complex cases the stars may be omitted and, instead, the whole name is prefixed by *rel* (for *relative*).

(c) When only relative configuration is known, enantiomers are distinguished by a prefix (+) or (−), referring to the direction of rotation of plane-polarized light passing through them (wavelength, temperature, solvent, and/or concentration should also be specified, particularly when known to affect the sign).

(d) When a substituent of known absolute chirality is introduced into a compound of which only the relative configuration is known, then starred symbols R^*, S^* are used and not the prefix *rel*.

Note: This Rule does not form part of the procedure formulated in the sequence-rule papers by Cahn, Ingold, and Prelog (see Appendix 2).

Examples:

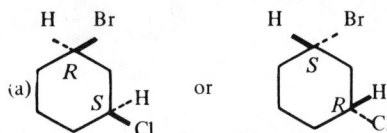

(1R^*, 3S^*,)-1-Bromo-3-chlorocyclohexane

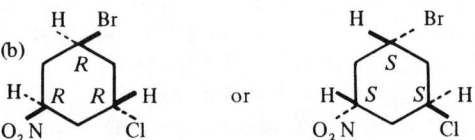

rel-(1R,3R,5R)-1-Bromo-3-chloro-5-nitrocyclohexane

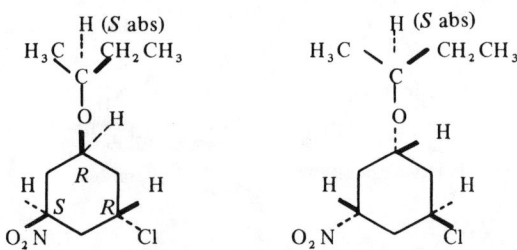

(1R^*,3R^*,5S^*)-[(1S)-*sec*-Butoxy]-3-chloro-5-nitrocyclohexane

E-5.11. When it is desired to express relative or absolute configuration with respect to a class of compounds, specialized local systems may be used. The sequence rule may, however, be used additionally for positions not amenable to treatment by the local system.

Examples:

gluco, arabino, etc., combined when necessary with D or L, for carbohydrates and their derivatives [see IUPAC-IUB Tentative Rules for Carbohydrate Nomenclature; see also *J. Org. Chem.*, 28, 281 (1963)].

D, L for amino acids and peptides [see Comptes rendus of the 16th IUPAC Conference, New York, N.Y., 1951., pp. 107–108; also published in *Chem. Eng. News*, 30, 4522 (1952)].

D, L, and a series of other prefixes and trivial names for cyclitols and their derivatives [see IUPAC-IUB Tentative Rules for the Nomenclature of Cyclitols, 1967, *IUPAC Inf. Bull.*, No. 32, 51 (1968); also published in *J. Biol. Chem.*, 243, 5809 (1968)].

α, β, and a series of trivial names for steroids and related compounds [see IUPAC-IUB Revised Tentative Rules for the Nomenclature of Steroids, 1967, *IUPAC Inf. Bull.*, No. 33, 23 (1968); also published in *J. Org. Chem.*, 34, 1517 (1969)].

The α, β system for steroids can be extended to other classes of compounds such as terpenes and alkaloids when their absolute configurations are known; it can also be combined with stars or the use of the prefix *rel* when only the relative configurations are known.

In spite of the Rules of Subsection E-2, *cis* and *trans* are used when the arrangement of the atoms constituting an unsaturated backbone is the most important factor, as, for instance, in polymer chemistry and for carotenoids. When a series of double bonds of the same stereochemistry occurs in a backbone, the prefix all-*cis* or all-*trans* may be used.

E-5.12. (a) An achiral object having at least one pair of features that can be distinguished only by reference to a chiral object or to a chiral reference frame is said to be prochiral, and the property of having such a pair of features is termed prochirality. A consequence is that, if one of the paired features of a prochiral object is considered to differ from the other, the resultant object is chiral.

(b) In a molecule an achiral center or atom is said to be prochiral if it would be held to be chiral when two attached atoms or groups, that taken in isolation are indistinguishable, are considered to differ.

Notes: (1) For a tetrahedrally bonded atom this requires a structure Xaabc (where none of the groups a, b, or c is the enantiomer of another).

(2) For a fuller exploration of this concept, which is of particular importance to biochemists and spectroscopists, and for its extension to axes, planes, and unsaturated compounds, see Hanson, K. R., *J. Am. Chem. Soc.*, 88, 2731 (1966).

Examples:

```
                         CHO
                          |
         CH₃          H — C — OH
          |               |
     H — C — H        H — C — H
          |               |
         OH              OH

         (A)             (B)
```

In both examples (A) and (B), the methylene carbon atom is prochiral; in both cases it would be held to be at a chiral center if one of the methylene hydrogen atoms were considered to differ from the other. An actual replacement of one of these protium atoms by, say, deuterium would produce an actual chiral center at the methylene carbon atom; as a result, compound (A) would become chiral, and compound (B) would be converted into one of two diastereoisomers.

E-5.13. Of the identical pair of atoms or groups in a prochiral compound, that one which leads to an (R) compound when considered to be preferred to the other by the sequence rule (without change in priority with respect to other ligands) is termed *pro-R*, and the other is termed *pro-S*.

Example:

$$\begin{array}{c} CHO \\ | \\ H^1 \text{---} C \text{---} OH \\ / \\ H^2 \end{array}$$

H^1 is *pro-R*.
H^2 is *pro-S*.

E-6. Conformations

E-6.1. A molecule in a conformation into which its atoms return spontaneously after small displacements is termed a conformer.

Examples:

are different conformers.

E-6.2. (a) When, in a six-membered saturated ring compound, atoms in relative positions 1, 2, 4, and 5 lie in one plane, the molecule is described as in the chair or boat conformation according as the other two atoms lie, respectively, on opposite sides or on the same side of that plane.

Examples:

Chair   Boat

Note: These and similar representations are idealized, minor divergences being neglected.

(b) A molecule of a monounsaturated six-membered ring compound is described as being in the half-chair or half-boat conformation according as the atoms not directly bound to the doubly bonded atoms lie, respectively, on opposite sides or on the same side of the plane containing the other four (adjacent) atoms.

Examples:

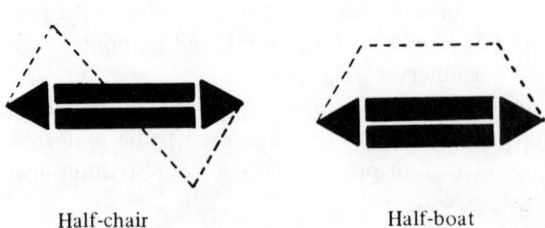

Half-chair   Half-boat

(c) A median conformation through which one boat form passes during conversion into the other boat form is termed a twist conformation. Similar twist conformations are involved in conversion of a chair into a boat form or vice versa.

Examples:

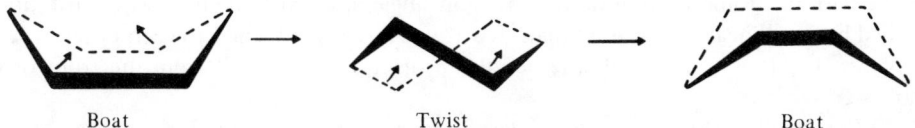

Boat   Twist   Boat

E-6.3. (a) Bonds to a tetrahedral atom in a six-membered ring are termed equatorial or axial according as they or their projections make a small or a large angle, respectively, with the plane containing a majority of the ring atoms.* Atoms or groups attached to such bonds are also said to be equatorial or axial, respectively.

Notes: (1) See, however, pseudoequatorial and pseudoaxial [Rule E-6.3(b)]. (2) The terms equatorial and axial may be abbreviated to e and a when attached to formulas; these abbreviations may also be used in names of compounds and are there placed in parentheses after the appropriate locants, for example, 1(e)-bromo-4(a)-chlorocyclohexane.

Examples:

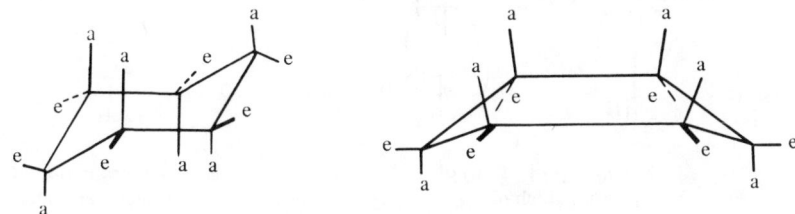

(b) Bonds from atoms directly attached to the doubly bonded atoms in a monounsaturated six-membered ring are termed pseudoequatorial or pseudoaxial according as the angles that they make with the plane containing the majority of the ring atoms approximate those made by, respectively, equatorial or axial bonds from a saturated six-membered ring. Pseudoequatorial and pseudoaxial may be abbreviated to e' and a', respectively, when attached to formulas; these abbreviations may also be used in names, then being placed in parentheses after the appropriate locants.

Example:

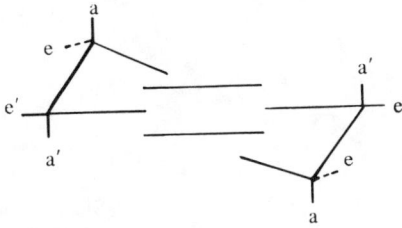

E-6.4. Torsion angle: In an assembly of attached atoms X–A–B–Y, where neither X nor Y is collinear with A and B, the smaller angle subtended by the bonds X–A and Y–B in a plane projection obtained by viewing the assembly along the axis A–B is termed the

---

*The terms axial, equatorial, pseudoaxial, and pseudoequatorial [see Rule E-6.3(b)] may be used also in connection with other than six-membered rings if, but only if, their interpretation is then still beyond dispute.

torsion angle (denoted by the Greek lower case letter theta θ or omega ω). The torsion angle is considered positive or negative according as the bond to the front atom X or Y requires rotation to the right or left, respectively, in order that its direction may coincide with that of the bond to the rear selected atom Y or X. The multiplicity of the bonding of the various atoms is irrelevant. A torsion angle also exists if the axis for rotation is formed by a collinear set of more than two atoms directly attached to each other.

Notes: (1) It is immaterial whether the projection be viewed from the front or the rear.

(2) For the use of torsion angles in describing molecules see Rule E-6.6.

Examples: (For construction of Newman projections, as here, see Rule E-7.2.)

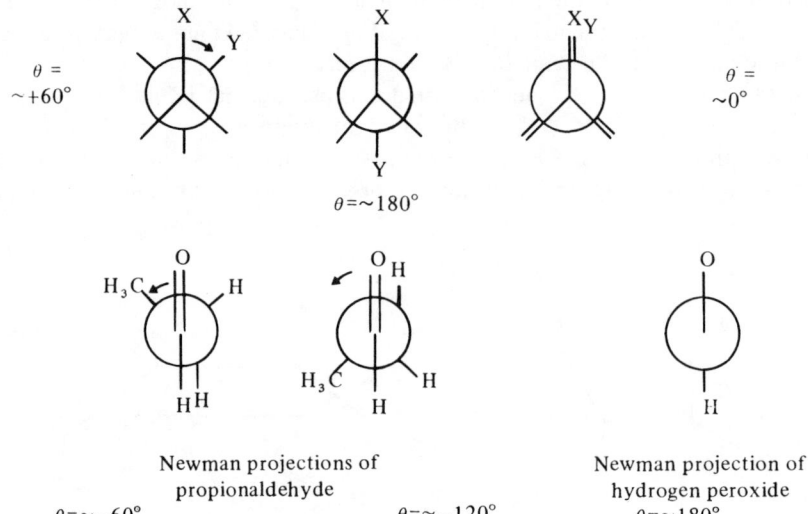

Newman projections of propionaldehyde

Newman projection of hydrogen peroxide
$\theta = \sim 180°$

E-6.5. If two atoms or groups attached at opposite ends of a bond appear one directly behind the other when the molecule is viewed along this bond, these atoms or groups are described as eclipsed, and that portion of the molecule is described as being in the eclipsed conformation. If not eclipsed, the atoms or groups and the conformation may be described as staggered.

Examples:

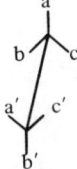

Eclipsed conformation.
The pairs a/a′, b/b′, and c/c′ are eclipsed.

Staggered conformation.
All the attached groups are staggered.

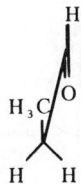

Projection of CH_3CH_2CHO.
The CH_3 and the H of the CHO are eclipsed.
The O and H's of CH_2 in CH_2CH_3 are staggered.

E-6.6. Conformations are described as synperiplanar (*sp*), synclinal (*sc*), anticlinal (*ac*), or antiperiplanar (*ap*) according as the torsion angle is within ±30° of 0°, ±60°, ±120°, or ±180°, respectively; the letters in parentheses are the corresponding abbreviations. Atoms or groups are selected from each set to define the torsion angle according to the following criteria: (1) if all the atoms or groups of a set are different, that one of each set that is preferred by the sequence rule; (2) if one of a set is unique, that one; or (3) if all of a set are identical, that one which provides the smallest torsion angle.
Examples:

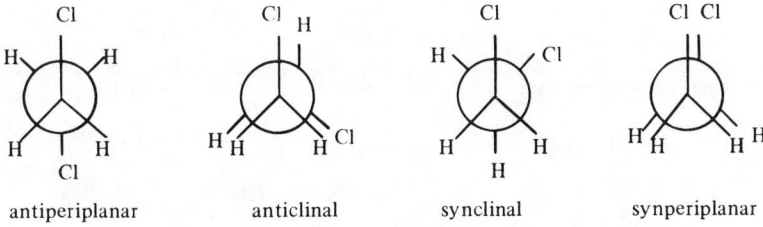

In the above conformations, all CH_2Cl-CH_2Cl, the two Cl atoms decide the torsion angle.

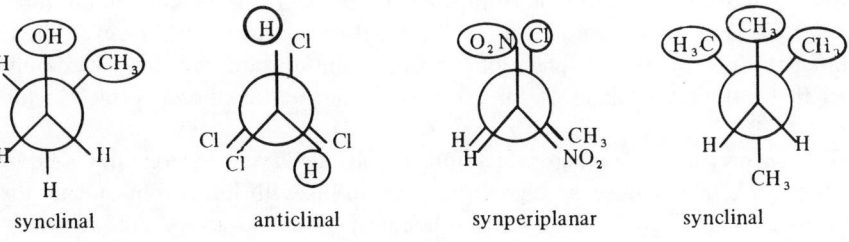

Criterion for:				
rear atom	2	2	1	3
front atom	2	2	1	2

	(CH$_3$)$_2$N–NH$_2$ synclinal*	CH$_3$CH$_2$–COCl anticlinal	(CH$_3$)$_2$CH–CONH$_2$ antiperiplanar
Criterion for:			
rear atom	2	2	2
front atom	2	1	1

E-7. Stereoformulas

E-7.1. In a Fischer projection the atoms or groups attached to a tetrahedral center are projected on to the plane of the paper from such an orientation that atoms or groups appearing above or below the central atom lie behind the plane of the paper and those appearing to left and right of the central atom lie in front of the plane of the paper, and that the principal chain appears vertical with the lowest numbered chain member at the top.

Examples:

Orientation   Fischer projection

Notes: (1) The first of the two types of Fischer projection should be used whenever convenient.

(2) If a Fischer projection formula is rotated through 180° in the plane of the paper, the upward and downward bonds from the central atom still project behind the plane of the paper, and the sideways bonds project in front of that plane. If, however, the formula is rotated through 90° in the plane of the paper, the upward and downward bonds now project in front of the plane of the paper and the sideways bonds project behind that plane.

E-7.2. To prepare a Newman projection, a molecule is viewed along the bond between two atoms; a circle is used to represent these atoms with lines from outside the circle toward its center to represent bonds to other atoms; the lines that represent bonds to the nearer and the further atom end at, respectively, the center and the circumference of the circle. When two such bonds would be coincident in the projection, they are drawn at a small angle to each other.[†]

---

*The lone pair of electrons (represented by two dots) on the nitrogen atoms are the unique substituents that decide the description of the conformation (these are the "phantom atoms" of the sequence-rule symbolism).

[†]Cf. Newman, M. S., *Rec. Chem. Progr.*, 13, 111 (1952); *J. Chem. Educ.*, 33, 344 (1955); *Steric Effects in Organic Chemistry*, John Wiley & Sons, New York, 1956, 5.

Examples:

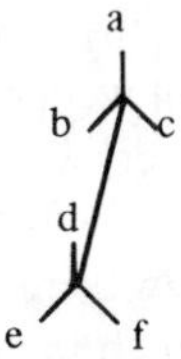

Perspective	Newman projection	Perspective	Newman projection

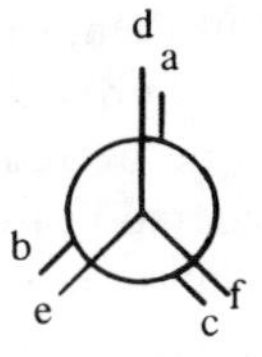

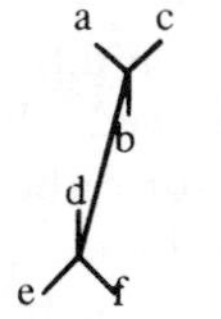

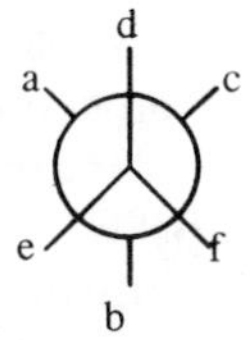

E-7.3. *General Note*; Formulas that display stereochemistry should be prepared with extra care so as to be unambiguous and, whenever possible, self-explanatory. It is inadvisable to try to lay down rules that will cover every case, but the following points should be borne in mind.

A thickened line (━) denotes a bond projecting from the plane of the paper toward an observer, a broken line (- - -) denotes a bond projecting away from an observer, and, when this convention is used, a full line of normal thickness (———) denotes a bond lying in the plane of the paper. A wavy line (∼∼) may be used to denote a bond whose direction cannot be specified or, if it is explained in the text, a bond whose direction it is not desired to specify in the formula. Dotted lines (· · · · · ·) should preferably not be used to denote stereochemistry, and never when they are used in the same paper to denote mesomerism, intermediate states, etc. Wedges should not be used as complement to broken lines (but see below). Single large dots have sometimes been used to denote atoms or groups attached at bridgehead positions and lying above the plane of the paper, with open circles to denote them lying below the plane of the paper, but this practice is strongly deprecated.

Hydrogen or other atoms or groups attached at sterically designated positions should never be omitted.

In chemical formulas, rings are usually drawn with lines of normal thickness, that is, as if they lay wholly in the plane of the paper even though this may be known not to be the case. In a formula such as (I) it is then clear that the H atoms attached at the A/B ring junction lie further from the observer than these bridgehead atoms, that the H atoms attached at the B/C ring junction lie nearer to the observer than those bridgehead atoms, and that X lies nearer to the observer than the neighboring atom of ring C.

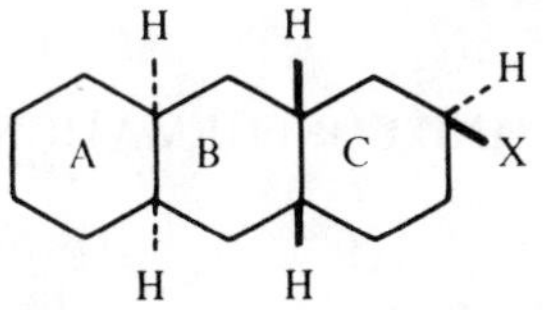

(I)

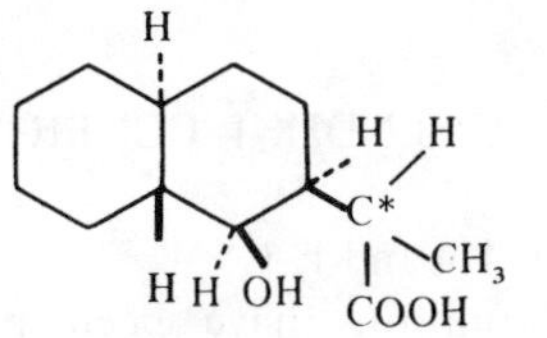

(II)

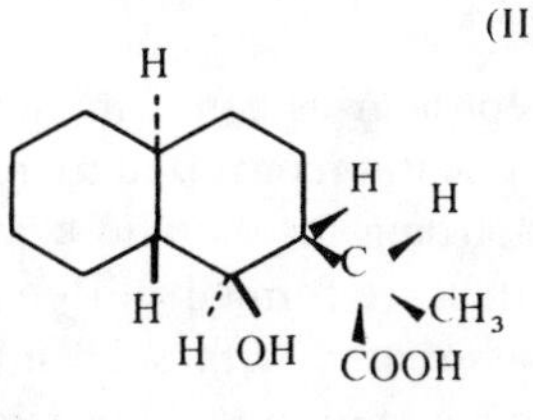

(III)

However, ambiguity can then sometimes arise, particularly when it is necessary to show stereochemistry within a group such as X attached to the rings that are drawn planar. For instance, in formula (II), the atoms O and C*, lying above the plane of the paper, are attached to ring B by thick bonds, but then, when showing the stereochemistry at C*, one finds that the bond *from* C* *to* ring B projects away from the observer and so should be a broken line. Such difficulties can be overcome by using wedges in place of lines, the broader end of the wedge being considered nearer to the observer, as in (III).

In some fields, notably for carbohydrates, rings are conveniently drawn as if they lay perpendicular to the plane of the paper, as represented in (IV); however, conventional formulas such as (V), with the lower bonds considered as the nearer to the observer, are so well established that is is rarely necessary to elaborate this to form (IV).

(IV)   (V)

By a similar convention, in drawings such as (VI) and (VII), the lower sets of bonds are considered to be nearer than the upper to the observer. In (VII), note the gaps in the rear lines to indicate that the bonds crossing them pass in front (and thus obscure sections of the rear bonds). In some cases, when atoms have to be shown as lying in several planes, the various conventions may be combined, as in (VIII). In all cases the overriding aim should be clarity.

(VI)   (VII)   (VIII)

APPENDIX 1. CONFIGURATION AND CONFORMATION

See Rules E-1.4 and E-1.5.

Various definitions have been propounded to differentiate configurations from conformations.

The original usage was to consider as conformations those arrangements of the atoms of a molecule in space that can be interconverted by rotation(s) around a single bond, and as configurations those other arrangements whose interconversion by rotation requires bonds to be broken and then re-formed differently. Interconversion of different configurations will then be associated with substantial energies of activation, and the various species will be separable, but interconversion of different conformations will normally be associated with less activation energy, and the various species, if separable, will normally be more readily interconvertible. These differences in activation energy and stability are often large.

Nevertheless, rigid differentiation on such grounds meets formidable difficulties. Differentiation by energy criteria would require an arbitrary cut in a continuous series of values. Differentiation by stability of isolated species requires arbitrary assumptions about conditions and halflives. Differentiation on the basis of rotation around single bonds meets difficulties connected both with the concept of rotation and with the selection of single bonds as requisites, and these need more detailed discussion here.

Enantiomeric biaryls are nowadays usually considered to differ in conformation, any difficulty in rotation about the 1,1' bond due to steric hindrance between the neighboring groups being considered to be overcome by bond bending and/or bond stretching, even though the movements required must closely approach bond breaking if these substituents are very large. Similar doubts about the possibility of rotation occur with a molecule such as (A), where rotation of the benzene ring around the oxygen-to-ring single bonds affords easy interconversion if x is large but appears to be physically impossible if x is small; and no critical size of x can be reasonably established. For reasons such as this, Rules E-1.4 and E-1.5 are so worded as to be independent of whether rotation appears physically feasible or not (see Note 2 to those Rules).

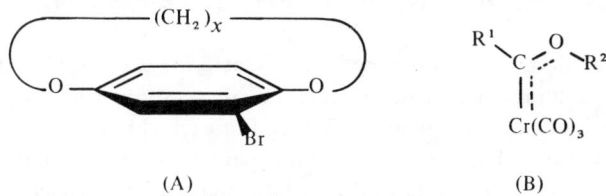

(A)   (B)

The second difficulty arises in the many cases where rotation is around a bond of fractional order between one and two, as in the helicenes, crowded aromatic molecules, metallocenes, amides, thioamides, and carbene-metal coordination compounds (such as B). The term conformation is customarily used in these cases and that appears a reasonable extension of the original conception, though it will be wise to specify the usage if the reader might be in doubt.

When interpreted in these ways, Rules E-1.4 and E-1.5 reflect the most frequent usage of the present day and provide clear distinctions in most situations. Nevertheless, difficulties remain and a number of other usages have been introduced.

It appears to some workers that once it is admitted that change of conformation may involve rotation about bonds of fractional order between one and two, it is then illogical to exclude rotation about classical double bonds because interconversion of open-chain *cis-trans* isomers depends on no fundamentally new principle and is often relatively easy, as for certain alkene derivatives such as stilbenes and for *azo* compounds, by irradiation. This extension is indeed not excluded by Rules E-1.4 and E-1.5, but if it is applied that fact should be explicitly stated.

A further interpretation is to regard a stereoisomer possessing some degree of stability (that is, one associated with an energy hollow, however shallow) as a configurational isomer, the other arrangements in space being termed conformational isomers; the term conformer (Rule E-6.1) is then superfluous. This definition, however requires a knowledge of stability (energy relations) that is not always available.

In another view, a configurational isomer is any stereoisomer that can be isolated or (for some workers) whose existence can be established (for example, by physical methods); all other arrangements then represent conformational isomers; but it is then impossible to differentiate configuration from conformation without involving experimental efficiency or conditions of observation.

Yet another definition is to regard a conformation as a precise description of a configuration in terms of bond distances, bond angles, and dihedral angles.

In none of the above views except the last is attention paid to extension or contraction of the bond to an atom that is attached to only one other atom, such as —H or =O. Yet such changes in interatomic distance due to nonbonded interactions may be important, for instance, in hydrogen bonding, in differences due to crystal form, in association in solution, and in transition states. This area may repay further consideration.

Owing to the circumstances outlined above, the Rules E-1.4 and E-1.5 have been deliberately made imprecise, so as to permit some alternative interpretations, but they are not compatible with all the definitions mentioned above. The time does not seem ripe to legislate for other than the commoner usages or to choose finally between these. It is, however, encouraging that no definition in this field has (yet) involved atomic vibrations for which, in all cases, only time-average positions are considered.

Finally it should be noted that an important school of thought uses conformation with the connotation of "a particular geometry of the molecule, i.e., a description of atoms in space in terms of bond distances, bond angles, and dihedral angles," a definition much wider than any discussed above.

APPENDIX 2. OUTLINE OF THE SEQUENCE-RULE PROCEDURE

The sequence-rule procedure is a method of specifying the absolute molecular chirality (handedness) of a compound, that is, a method of specifying which of two enantiomeric forms each chiral element of a molecule exists. For each chiral element in the molecule it provides a symbol, usually R or S, which is independent of nomenclature and numbering. These symbols define the chirality of the specific compound considered; they may not be the same for a compound and some of its derivatives; and they are not necessarily constant for chemically similar situations within a chemical or a biogenetic class. The procedure is applied directly to a three-dimensional model of the structure, and not to any two-dimensional projection thereof.

The method has been developed to cover all compounds with ligancy up to four and with ligancy six,[*] and for all configurations and conformations of such compounds. The following is an outline confined to the most common situations; it is essential to study the original papers, especially the 1966 paper,[†] before using the sequence rule for other than fairly simple cases.

General Basis

The sequence rule itself is a method of arranging atoms or groups (including chains and rings) in an order of precedence, often referred to as an order of preference; for discussion this order can conveniently be generalized as $a > b > c > d$, where $>$ denotes "is preferred to."

The first step, however, in considering a model is to identify the nature and position of each chiral element that it contains. There are three types of the chiral element, namely, the chiral center, the chiral axis, and the chiral plane. The chiral center, which is very much the most commonly met, is exemplified by an asymmetric carbon atom with the tetrahedral arrangement of ligands, as in (1). A chiral axis is present in, for instance, the chiral allenes such as (2) or the chiral biaryl derivatives. A chiral plane is exemplified by the plane containing the benzene ring and the bromine and oxygen atoms in the chiral compound (3), or by the underlined atoms in the cycloalkene (4). Clearly, more than one

---

[*]Ligancy refers to the number of bonds from an atom, independently of the nature of the bonds.
[†]Cahn, R. S., Ingold, C., and Prelog, V., *Angew. Chem. Int. Ed.,* 5, 385 (1966); errata, 5, 511 (1966); *Angew. Chem.,* 78, 413 (1966). Earlier papers: Cahn, R. S. and Ingold, C. K., *J. Chem. Soc.* (Lond.), 612 (1951); Cahn, R. S., Ingold, C., and Prelog, V., *Experientia,* 12, 81 (1956). For a partial, simplified account see Cahn, R. S., *J. Chem. Educ.,* 41, 116 (1964); errata, 41, 503 (1964).

type of chiral element may be present in one compound; for instance, group "a" in (2) migh be a *sec*-butyl group which contains a chiral center.

The Chiral Center

Let us consider first the simplest case, namely, a chiral center (such as carbon) with four ligands, a, b, c, and d, which are all different atoms tetrahedrally arranged as in CHFClBr. The four ligands are arranged in order of preference by means of the sequence rule; this contains five subrules, which are applied in succession so far as necessary to obtain a decision. The first subrule is all that is required in a great majority of actual cases; it states that ligands are arranged in order of decreasing atomic number, in the above case (a) Br > (b) Cl > (c) F > (d) H. There would be two (enantiomeric) forms of the compound and we can write these as (5) and (6). In the sequence-rule procedure the model is viewed from the side remote from the least-preferred ligand (d), as illustrated. Then, tracing a path from a to b to c in (5) gives a clockwise course, which is symbolized by (*R*) (Latin *rectus*, right; for right hand); in (6) it gives an anticlockwise course, symbolized as (*S*) (Latin *sinister*, left). Thus (5) would be named (*R*)-bromochlorofluoromethane, and (6) would be named (*S*)-bromochlorofluoromethane. Here already it may be noted that converting one enantiomer into another changes each *R* to *S*, and each *S* to *R*, always. It will be seen also that the chirality prefix is the same whether the alphabetical order is used, as above, for naming the substituents or whether this is done by the order of complexity (giving fluorochlorobromomethane).

Next, suppose we have $H_3C-CHClF$. We deal first with the atoms directly attached to the chiral center; so the four ligands to be considered are Cl > F > C (of CH_3) > H. Here the H's of the CH_3 are not concerned, because we do not need them in order to assign our symbol.

However, atoms directly attached to a center are often identical, as, for example, the underlined C's in $H_3\underline{C}-CHCl-\underline{C}H_2OH$. For such a compound we at once establish a preference (a) Cl > (b, c) $\underline{C},\underline{C}$ > (d) H. Then to decide between the two $\underline{C}$'s we work outward, to the atoms to which they in turn are directly attached and we then find which we can conveniently write as C(H,H,H) and C(O,H,H). We have to compare H,H,H with O,H,H, and since oxygen has a higher atomic number than hydrogen we have O > H and thence the complete order $Cl > C$ (of CH_2OH) $> C$ (of CH_3) $> H$, so that the chirality symbol can then be determined from the three-dimensional model.

$$-\overset{H}{\underset{H}{C}}-H \quad \text{and} \quad -\overset{O}{\underset{H}{C}}-H$$

We must next meet the first complication. Suppose that we have a molecule (7).

$$\text{(b) } H_3C-\underline{C}HCl-\overset{Cl\ (a)}{\underset{H\ (d)}{C}}-\underline{C}HF-OH \text{ (c)}$$

(7) (S)

To decide between the two C's we first arrange the atoms attached to them in *their* order of preference, which gives $\underline{C}$(Cl,C,H) on the left and $\underline{C}$(F,O,H) on the right. Then we compare the preferred atom of one set (namely, Cl) with the preferred atom (F) of the other set, and as $Cl > F$ we arrive at the preferences $a > b > c > d$ shown in (7) and chirality (S). If, however, we had a compound (8) we should have met $\underline{C}$(Cl,C,H) and C(Cl,O,H) and, since the atoms of first preference are identical (Cl), we should have had to make the comparisons with the atoms of second preference, namely, $O > C$, which to the different chirality (R) as shown in (8).

$$\text{(c) } H_3C-\underline{C}HCl-\overset{Cl\ (a)}{\underset{H\ (d)}{C}}-\underline{C}HCl-OH \text{ (b)}$$

(8) (R)

Branched ligands are treated similarly. Setting them out in full gives a picture that at first sight looks complex but the treatment is in fact simple. For instance, in compound (9) a first quick glance again shows (a) $Cl > $ (b, c) $\underline{C},\underline{C} > $ (d) H: When we expand the two C's we find they are both $\underline{C}$(C,C,H), so we continue exploration. Considering first the left-hand ligand we arrange the branches and their sets of atoms in order thus: C(Cl,H,H) $> $ C(H,H,H). On the right-hand side we have C(O,$\underline{C}$,H) $> $ C(O,$\underline{\underline{H}}$,H) (because $\underline{C} > \underline{\underline{H}}$). We compare first the preferred of these branches from each side and we find C(Cl,H,H) $> $ C(O,C,H) because $Cl > O$, and that gives the left-hand branch preference over the right-hand branch. That is all we need to do to establish chirality (S) for this highly branched compound (9). Note that it is immaterial here that, for the lower branches, the right-hand C(O,H,H) would have been preferred to the left-hand C(H,H,H); we did not need to reach that point in our comparisons and so we are not concerned with it; but we should have reached it if the two top (preferred) branches had both been the same CH_2Cl.

Rings, when we met during outward exploration, are treated in the same way as branched chains.

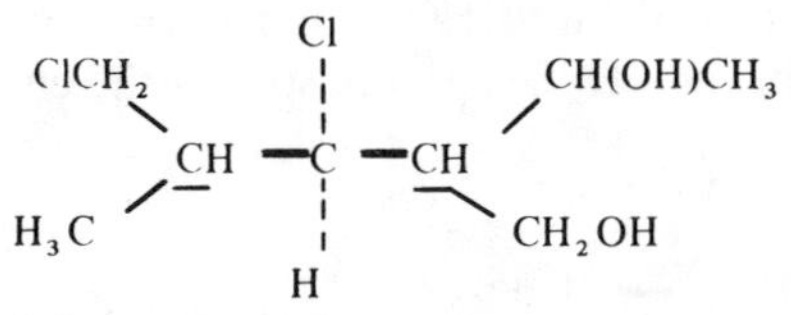

(9)

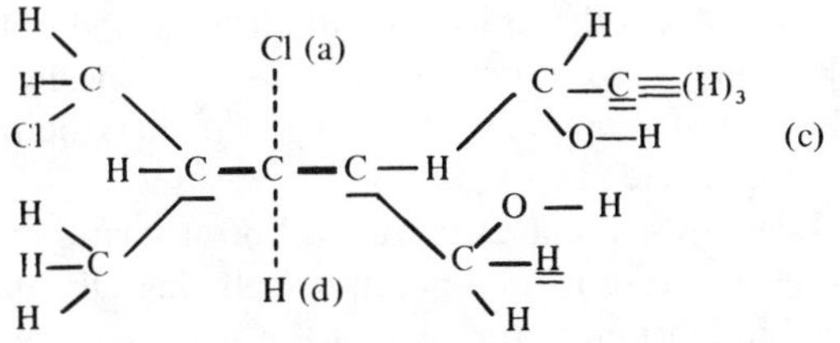

(9) (S)

With these simple procedures alone, quite complex structures can be handled; for instance, the analysis alongside Formula (10) for natural morphine explains why the specification is as shown. The reason for considering C-12 as C(C,C,C) is set out in the next paragraphs.

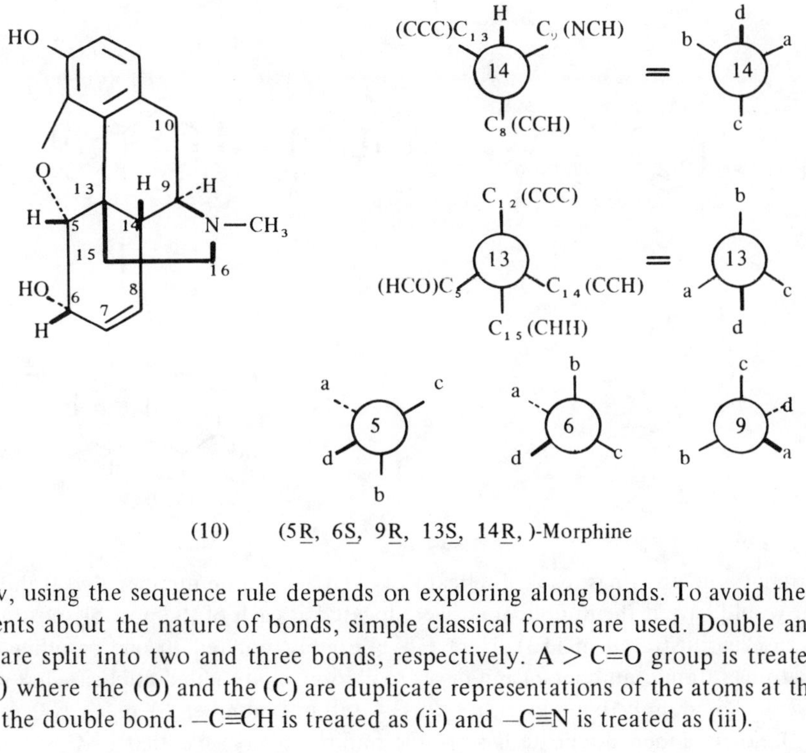

(10)   (5R, 6S, 9R, 13S, 14R,)-Morphine

Now, using the sequence rule depends on exploring along bonds. To avoid theoretical arguments about the nature of bonds, simple classical forms are used. Double and triple bonds are split into two and three bonds, respectively. A >C=O group is treated as (i) (below) where the (O) and the (C) are duplicate representations of the atoms at the other end of the double bond. —C≡CH is treated as (ii) and —C≡N is treated as (iii).

(i)   (ii)   (iii)

Thus in D-glyceraldehyde (11) the CHO group is treated as C(O,(O),H) and is thus preferred to the C(O,H,H) of the CH$_2$OH group, so that the chirality symbol is (R).

$$\text{(d) H} - \underset{\underset{\text{CH}_2\text{OH (c)}}{|}}{\overset{\overset{\text{CHO (b)}}{|}}{\text{C}}} - \text{OH (a)}$$

D-Glyceraldehyde
(11) (R)

Only the doubly bonded atoms themselves are duplicated, and not the atoms or groups attached to them; the duplicated atoms may thus be considered as carrying three phantom atoms (see below) of atomic number zero. This may be important in deciding preferences in certain complicated cases.

Aromatic rings are treated as Kekulé structures. For aromatic hydrocarbon rings it is immaterial which Kekulé structure is used because "splitting" the double bonds gives the same result in all cases; for instance, for phenyl the result can be represented as (12a) where "(6)" denotes the atomic number of the duplicate representations of carbon.

For aromatic hetero rings, each duplicate is given an atomic number that is the mean of what it would have if the double bond were located at each of the possible positions. A complex case is illustrated in (13). Here C-1 is doubly bonded to one or other of the nitrogen atoms (atomic number 7) and never to carbon, so its added duplicate has atomic number 7; C-3 is doubly bonded either to C-4 (atomic number 6) or to N-2 (atomic number 7), so its added duplicate has atomic number 6½; so has that of C-8; but C-4a may be doubly bonded to C-4, C-5, or N-9, so its added duplicate has atomic number 6.33.

One last point about the chiral center may be added here. Except for hydrogen, ligancy, if not already four, is made up to four by adding "phantom atoms" which have atomic number zero and are thus always last in order of preference. This has various uses but perhaps the most interesting is where nitrogen occurs in a rigid skeleton, as, for example, in α-isosparteine (14). Here the phantom atom can be placed where the nitrogen

SOME COMMON GROUPS IN ORDER OF SEQUENCE-RULE PREFERENCE[a]

A. Alphabetical Order (Higher Number Denotes Greater Preference)

64 Acetoxy	38 Carboxyl	9 Isobutyl	55 Nitroso
36 Acetyl	74 Chloro	8 Isopentyl	6 n-Pentyl
48 Acetylamino	17 Cyclohexyl	20 Isopropenyl	61 Phenoxy
21 Acetylenyl	52 Diethylamino	14 Isopropyl	22 Phenyl
10 Allyl	51 Dimethylamino	69 Mercapto	47 Phenylamino
43 Amino	34 2,4-Dinitrophenyl	58 Methoxy	54 Phenylazo
44 Ammonio $^+H_3N-$	28 3,5-Dinitrophenyl	39 Methoxycarbonyl	18 Propenyl
37 Benzoyl	59 Ethoxy	2 Methyl	4 n-Propyl
49 Benzoylamino	40 Ethoxycarbonyl	45 Methylamino	29 1-Propynyl
65 Benzoyloxy	3 Ethyl	71 Methylsulfinyl	12 2-Propynyl
50 Benzyloxycarbonylamino	46 Ethylamino	66 Methylsulfinyloxy	73 Sulfo
13 Benzyl	68 Fluoro	72 Methylsulfonyl	25 m-Tolyl
60 Benzyloxy	35 Formyl	67 Methylsulfonyloxy	30 o-Tolyl
41 Benzyloxycarbonyl	63 Formyloxy	70 Methylthio	23 p-Tolyl
75 Bromo	62 Glycosyloxy	11 Neopentyl	53 Trimethylammonio
42 ter-Butoxycarbonyl	7 n-Hexyl	56 Nitro	32 Trityl
5 n-Butyl	1 Hydrogen	27 m-Nitrophenyl	15 Vinyl
16 sec-Butyl	57 Hydroxy	33 o-Nitrophenyl	31 2,6-Xylyl
19 tert-Butyl	76 Iodo	24 p-Nitrophenyl	26 3,5-Xylyl

B. Increasing Order of Sequence Rule Preference

1 Hydrogen	20 Isopropenyl	39 Methoxycarbonyl[b]	58 Methoxy
2 Methyl	21 Acetylenyl	40 Ethoxycarbonyl[b]	59 Ethoxy
3 Ethyl	22 Phenyl	41 Benzyloxycarbonyl[b]	60 Benzyloxy
4 n-Propyl	23 p-Tolyl	42 tert-Butoxycarbonyl[b]	61 Phenoxy
5 n-Butyl	24 p-Nitrophenyl	43 Amino	62 Glycosyloxy
6 n-Pentyl	25 m-Tolyl	44 Ammonio $^+H_3N-$	63 Formyloxy
7 n-Hexyl	26 3,5-Xylyl	45 Methylamino	64 Acetoxy
8 Isopentyl	27 m-Nitrophenyl	46 Ethylamino	65 Benzoyloxy
9 Isobutyl	28 3,5-Dinitrophenyl	47 Phenylamino	66 Methylsulfinyloxy
10 Allyl	29 1-Propynyl	48 Acetylamino	67 Methylsulfonyloxy
11 Neopentyl	30 o-Tolyl	49 Benzoylamino	68 Fluoro
12 2-Propynyl	31 2,6-Xylyl	50 Benzyloxycarbonylamino	69 Mercapto HS—
13 Benzyl	32 Trityl	51 Dimethylamino	70 Methylthio CH_3S-
14 Isopropyl	33 o-Nitrophenyl	52 Diethylamino	71 Methylsulfinyl
15 Vinyl	34 2,4-Dinitrophenyl	53 Trimethylammonio	72 Methylsulfonyl
16 sec-Butyl	35 Formyl	54 Phenylazo	73 Sulfo HO_3S-
17 Cyclohexyl	36 Acetyl	55 Nitroso	74 Chloro
18 1-Propenyl	37 Benzoyl	56 Nitro	75 Bromo
19 tert-Butyl	38 Carboxyl	57 Hydroxy	76 Iodo

[a] ANY alteration to structure, or substitution, etc., may alter the order of preference.
[b] These groups are ROC(=O)—.

lone pair of electrons is; then N-1 appears as shown alongside the formula; and the chirality (R) is the consequence. The same applies to N-16. Phantom atoms are similarly used when assigning chirality symbols to chiral sulfoxides (see example to Rule E-5.9).

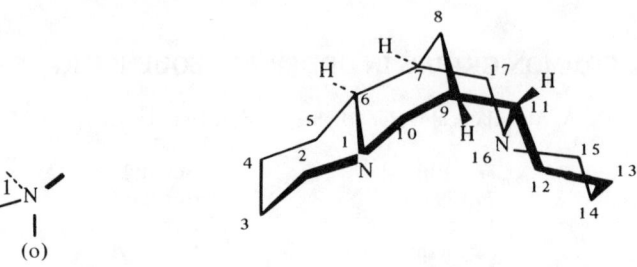

(14) (1*R*, 6*R*, 7*S*, 9*S*, 11*R*, 16*R*)-Sparteine

Symbolism

In names of compounds, the *R* and *S* symbols, together with their locants, are placed in parentheses, normally in front of the name, as shown for morphine (10) and sparteine (14), but this may be varied in indexes or in languages other than English. Positions within names are required, however, when more than a single series of numerals is used, as for esters and amines. When relative stereochemistry is more important than absolute stereochemistry, as for steroids or carbohydrates, a local system of stereochemical designation may be more useful and sequence-rule symbols need then be used only for any situations where the local system is insufficient.

Racemates containing a single center are labeled (*RS*). If there is more than one center the first is labeled (*RS*) and the others are (*RS*) or (*SR*) according to whether they are *R* or *S* when the first is *R*. For instance, the 2,4-pentanediols $CH_3-CH(OH)-CH_2-CH(OH)-CH_3$ are differentiated as

one chiral form (2*R*,4*R*)–
other chiral form (2*S*,4*S*)–
meso compound (2*R*,4*S*)–
racemic compound (2*RS*,4*RS*)–

Finally the principles by which some of the least rare of other situations are treated will be very briefly summarized.

Pseudoasymmetric Atoms

A subrule decrees that *R* groups have preference over *S* groups and this permits pseudoasymmetric atoms, as in abC(c-*R*)(c-*S*) to be treated in the same way as chiral centers, but as such a molecule is achiral (not optically active) it is given the lower case symbol *r* or *s*.

Chiral Axis

The structure is regarded as an elongated tetrahedron and viewed along the axis — it is immaterial from which end it is viewed; the nearer pair of ligands receives the first two positions in the order of preference, as shown in (15) and (16).

(16)

Chiral Plane

The sequence-rule-preferred atom directly attached to the plane is chosen as "pilot atom." In compound (3) this is the C of the left-hand CH$_2$ group. Now this is attached to the left-hand oxygen atom in the plane. The sequence-rule-preferred path from this oxygen atom is then explored in the plane until a rotation is traced which is clockwise (R) or anticlockwise (S) when viewed from the pilot atom. In (3) this path is O → C → C(Br) and it is clockwise (R).

Other Subrules

Other subrules cater for new chirality created by isotopic labeling (higher mass number preferred to lower) and for steric differences in the ligands. Isotopic labeling rarely changes symbols allotted to other centers.

Octahedral Structures

Extensions of the sequence rule enable ligands arranged octahedrally to be placed in an order of preference, including polydentate ligands, so that a chiral structure can then always be represented as one of the enantiomeric forms (17) and (18). The face 1—2—3 is observed from the side remote from the face 4—5—6 (as marked by arrows), and the path 1 → 2 → 3 is observed; in (17) this path is clockwise (R), and in (18) it is anticlockwise (S).

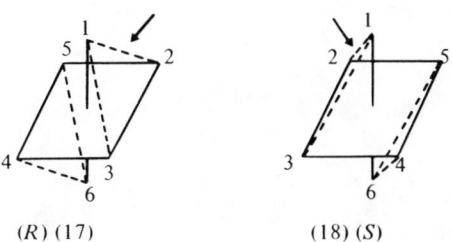

(R) (17)   (18) (S)

Conformations

The torsion angle between selected bonds from two singly bonded atoms is considered. The selected bond from each of these two atoms is that to a unique ligand, or otherwise to the ligand preferred by the sequence rule. The smaller rotation needed to make the front ligand eclipsed with the rear one is noted (this is the rotatory characteristic of a helix); if this rotation is right-handed it leads to a symbol P (plus); if left-handed to M (minus). Examples are

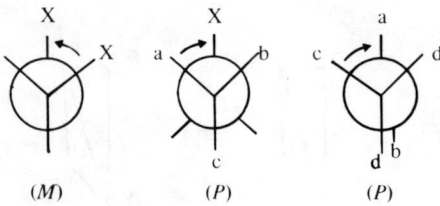

(M)   (P)   (P)

Details and Complications

For details and complicating factors the original papers should be consulted. They include treatment of compounds with high symmetry or containing repeating units (e.g., cyclitols), also π bonding (metallocenes, etc.), mesomeric compounds and mesomeric radicals, and helical and other secondary structures.

A ONE-LETTER NOTATION FOR AMINO ACID SEQUENCES[*][†]
TENTATIVE RULES

IUPAC-IUB Commission on Biochemical Nomenclature (CBN)

INTRODUCTION

1. General Considerations

Various difficulties are encountered when presenting the formulas of long protein sequences in the usual three-letter symbols.[1] Space is often at a premium. A one-letter code minimizes this difficulty and has other distinct advantages. In summarizing large amounts of data or in the alignment of homologous protein sequences, it is important that the patterns in the sequences be condensed and simplified as much as possible. Computer techniques are increasingly applied for the storage of sequences of hundreds of amino acid residues and for their evaluation. For this purpose, a one-letter code is the best solution. Finally, a one-letter code is useful in the labeling of individual amino acid side-chains in three-dimensional pictures of protein molecules.

The possibility of using one-letter symbols was mentioned by Gamow and Ycas[2] in 1958. The idea was systematized by Šorm et al.[3] in 1961. It was used by this group[4-10] and also by Fitch[11] in several papers on the structure of proteins. In extensive compilations of protein structures, Eck and Dayhoff[12-14] systematically used one-letter symbols derived partly from the code of Šorm and Keil. Independent proposals were made by Wiswesser[15] and by Braunstein.[16]

In view of the increasing number of different notations and the attending problems, the IUPAC-IUB Commission on Biochemical Nomenclature (CBN) has undertaken the task of drafting a single notation for one-letter symbols. The present proposal was evolved by a CBN subcommission (composed of B. Keil, R. V. Eck, M. O. Dayhoff, and W. E. Cohn); it is based principally on the most recent summary published by Dayhoff and Eck.[14]

2. Limits of Application

In publications, CBN recommends that one-letter symbols be used only in comparisons of long sequences in tables, lists, or figures, and for such special use as tagging three-dimensional models of proteins. They should not be used in simple text nor for original reports of experimental details of sequences. This system is not suitable for reporting the details of peptide synthesis, for example, where a fuller description of substituents is needed and where uncommon amino acids may occur. It should not be used in papers where the single-letter system for nucleoside sequences is employed (Reference 1a, sections 5.4 and 5.5), as in representing codons, etc.

RULES

3. Principles of the One-letter Code

3.1. The letter written at the left-hand end is that of the amino acid residue carrying the free amino group and the letter written at the right-hand end is that of the amino acid residue carrying the free carboxyl group. The absence of punctuation beyond either end

---

*Document of the IUPAC-IUB Commission on Biochemical Nomenclature (CBN), approved by IUPAC and IUB in March 1968 and published by permission of the International Union of Pure and Applied Chemistry, the International Union of Biochemistry, and the official publishers to the International Union of Pure and Applied Chemistry, Messrs. Butterworths Scientific Publications.
†From *Pure Appl. Chem.*, 31(4), 639–645 (1972). With permission.

of a sequence implies that it is known to be the amino or carboxyl end of the protein. A fragmentary sequence is to be preceded or followed by a slash (/) to indicate that it is not known to be the end of the complete protein (see comment in section 8.2).

3.2. Initial letters are used where there is no ambiguity. There are six such cases — cysteine, histidine, isoleucine, methionine, serine and valine. All the other amino acids share the initial letters A, G, L, P, or T and assignments of them must therefore be somewhat arbitrary. These letters are assigned to the most frequently occurring and structurally most simple amino acids. On the basis, the letters A, G, L, P, and T are assigned to alanine, glycine, leucine, proline, and threonine, respectively.

3.3. The assignment of the other abbreviations is more arbitrary. However, certain clues are helpful. Two are phonetically suggestive, F for *ph*enylalanine, and R for *ar*ginine. For tryptophan, the double ring in the molecule is associated with bulky letter W. The letters N and Q are assigned to asparagine and glutamine, respectively; D and E are assigned to aspartic acid and glutamic acid, respectively. This leaves lysine and tyrosine, to which K and Y are assigned. These are chosen rather than any of the few other remaining letters because they are alphabetically nearest the initial letters L and T. U and O are avoided because U is easily confused with V in handwritten work and O is confused with G, Q, C, and D in imperfect computer print-outs and also with zero. J is avoided for linguistic reasons.

3.4. Two other abbreviations are necessary in order to avoid ambiguity. B is assigned to aspartic acid or asparagine when this distinction has not been determined. Z is assigned when glutamic acid and glutamine have not been distinguished. X means that the identity of an amino acid is undetermined, or the amino acid is atypical.

4. Abbreviations in Alphabetical Order

These are listed in Tables 1 and 2.

5. Spacing

A very important use of the one-letter notation is in presenting alignments of many homologous sequences. In printing, it often happens that the alignment is not perfectly maintained because of the variable size of the letters and the variable amount of punctuation. This effect can be very troublesome in extensive comparisons. Therefore, a single typewriter space is left between letters, either as a blank or occupied by punctuation (see Sections 6, 7, 8). The alignment is preserved by allowing exactly the same spacing for each letter, each blank, and each punctuation mark, as in typewritten material or, if printed, as in "typewriter type font."

6. Known and Unknown Sequences*

A blank between letters indicates that the sequence is known. (See also comment in Section 8.2.) As in the three-letter notation, parentheses and commas are used to indicate regions in which the sequence is unknown or undetermined.

Example. The β-corticotropin releasing factor, using three-letter symbols:

Ser-Tyr-Cys-Phe-His(Asn, Gln)Cys(Pro, Val)Lys-Gly

or one-letter symbols:

S Y C F H(N,Q)C(P,V)K G

*The sequence quoted for β-corticotropin releasing factor has been withdrawn from the 1969 *Atlas* (see References 12–14). Hence the sequence used in Section 6 should be regarded only as a hypothetical example for purposes of illustration.

7. Juxtaposition of Unknown Sequences Known to be Connected

Consider the two sequences, one completely known, the other containing peptides of unknown internal sequence.

  a) Ala-Cys-Asp-Glu-Phe-Gly-His-Ile-Lys-Leu-Met-Asn- Pro-Gln
  b) (Ala,Cys,Asp)(Arg,Ser)(Gly,His,Ile)Lys-Leu-Met-Asn- Pro-Gln

In one-letter notation, these become:

  a) A C D   E   F   G H I   K L M N P Q
  b) (A , C , D) (R , S) (G , H , I) K L M N P Q
        ↑   ↑

In the second illustration, two punctuation marks have been crowded into each of two single spaces (indicated by the arrows). In a computer print-out, this would not be possible. A single one-space symbol must be used. Here = is used for)(to indicate the end of one unknown sequence and the beginning of another, as shown below.

  a) A C D   E   F   G H I   K L M N P Q
  b) (A , C , D = R , S = G , H , I) K L M N P Q
      ↑ ↑

8. Juxtaposition of Residues Inferred, but not Known, to be Connected

Consider the following case in which peptides from a second sequence (d) can be aligned with a known, related sequence (c).

  c) A C D   E   F   G H I   K L M N P Q
  d) (A . C . D = R , S = G . H . I) K L / M N / P Q /

8.1. In this illustration, the sequences of two of the fragments (A.C.D and G.H.I in d), while not determined, are inferred with good confidence, which is indicated by dots instead of commas between their residues. Where such inferences cannot be made with confidence, commas, which retain their original connotation of "unknown sequence" (Section 6), should be used, as in the R,S dipeptide.

8.2. The two internal slashes (/) separate adjacent amino acids that come from different peptides not proven experimentally to be connected. The third (end) slash indicates that Q is not experimentally proven to be at the carboxyl end of the protein, although it is at the carboxyl end of the P—Q dipeptidyl residue.

Comment. The absence of punctuation at the beginning or end of a complete polypeptide or protein sequence indicates the known amino or carboxyl terminal, respectively (see Section 3.1).

8.3. Depending on the experimental details and the nature of the inferences to be represented, even more elaborate punctuation may sometimes be required. It is essential, however, that only one character (or a blank space of similar size) appear between the single letters to preserve the spacing that is essential for comparisons (see Section 5).

Table 1

Amino acid	One-letter symbol
*A*lanine	A
*A*rginine	R
*A*sparagine	N } B[a]
*A*spartic acid	D
*C*ysteine	C
*G*lutamine	Q } Z[b]
*G*lutamic acid	E
*G*lycine	G
*H*istidine	H
*I*soleucine	I
*L*eucine	L
*L*ysine	K
*M*ethionine	M
*Ph*enylalanine	F
*P*roline	P
*S*erine	S
*T*hreonine	T
*T*ryptophan	W
*T*yrosine	Y
*V*aline	V
Unknown or "other".	X

[a]For aspartic acid *or* asparagine.
[b]For glutamic acid *or* glutamine.

Table 2

One-letter symbol	Three-letter symbol	Amino acid
A	Ala	*A*lanine
B	Asx	Aspartic acid *or* asparagine
C	Cys	*C*ysteine
D	Asp	*A*spartic acid
E	Glu	*G*lutamic acid
F	Phe	*Ph*enylalanine
G	Gly	*G*lycine
H	His	*H*istidine
I	Ile	*I*soleucine
K	Lys	*L*ysine
L	Leu	*L*eucine
M	Met	*M*ethionine
N	Asn	*A*sparagine
P	Pro	*P*roline
Q	Gln	*G*lutamine
R	Arg	*A*rginine
S	Ser	*S*erine
T	Thr	*T*hreonine
V	Val	*V*aline
W	Trp	*T*ryptophan
X	–	Unknown or "other"
Y	Tyr	*T*yrosine
Z	Glx	*G*lutamic acid *or* glutamine

REFERENCES

1a. IUPAC-IUB Tentative Rules, *J. Biol. Chem.,* 241, 527 (1966); see also *Eur. J. Biochem.,* 1, 259, (1967).
1b. IUPAC-IUB Tentative Rules, *J. Biol. Chem.,* 241, 2491 (1966); see also *Eur. J. Biochem.,* 1, 375 (1967). (Replaced in 1971. See "Symbols for Amino-acid Derivatives and Peptides.")
2. **Gamow and Yčas,** *Symp. on Information Theory in Biology,* Pergamon Press, New York, 1958.
3. **Šorm, Keil, Vaněček, Tomášek, Mikeš, Meloun, Kostka, and Holeyšovský,** *Collect. Czech. Chem. Commun.,* 26, 531 (1961).
4. **Mikeš, Holeyšovský, Tomášek, Keil, and Šorm,** *Collect. Czech. Chem. Commun.,* 27, 1964 (1962).
5. **Holeyšovský, Alexijev, Tomášek, Mikeš, and Šorm,** *Collect. Czech. Chem. Commun.,* 27, 2662 (1962).
6. **Šorm and Keil,** *Adv. Protein Chem.,* 17, 1967 (1962).
7. **Mikeš, Holeyšovský, Tomášek, and Šorm,** *6th Int. Congr. Biochem.,* New York, 1964, Abstr. II-136, p. 169.
8. **Keil, Prosík, and Šorm,** *Biochim. Biophys. Acta,* 78, 559 (1963).
9. **Mikeš, Prusik, and Svoboda,** *Collect. Czech. Chem. Commun.,* 29, 1193 (1964).
10. **Šorm, Holeyšovský, Mikeš, and Tomášek,** *Collect. Czech. Chem. Commun.,* 30, 2103 (1965).
11. **Fitch,** *J. Mol. Biol.,* 16, 1, 9, 17 (1966).
12. **Dayhoff, Eck, Chang, and Sochard,** *Atlas of Protein Sequence and Structure,* National
13. **Eck and Dayhoff,** *Atlas of Protein Sequence and Structure,* National Biomedical Research Foundation, Washington, D.C., 1968.
14. **Dayhoff and Eck,** *Atlas of Protein Sequence and Structure,* National Biomedical Research Foundation, Washington, D.C., 1968.
15. **Wiswesser,** *Chem. Eng. News,* 42, 4 (1964).
16. **Braunstein,** personal proposal to CBN.
17. **Schally and Bowers,** *Metabolism,* 13, 1190 (1964).

ABBREVIATIONS AND SYMBOLS FOR THE DESCRIPTION OF THE CONFORMATION OF POLYPEPTIDE CHAINS TENTATIVE RULES (1969)*

IUPAC-IUB Commission on Biochemical Nomenclature

Preamble

These Rules are based on "A Proposal of Standard Conventions and Nomenclature for the Description of Polypeptide Conformation" (Edsall et al.),[8] and have been prepared by a subcommission set up by the IUPAC-IUB Commission on Biochemical Nomenclature in 1966.[†] The original proposals have been modified so as to bring them as far as possible into line with the system of nomenclature current in the fields of organic and polymer chemistry.

Two Recommendations are appended to the Rules, the first dealing with the terms configuration and conformation, and the second with primary, secondary, and tertiary structure. These are formulated as recommendations rather than rules because there is at present no general agreement about their definition.

Note: Two alternative notations are recommended throughout. That with superscripts and subscripts may be used when it is unlikely to cause confusion, e.g., in printed or manuscript material; that without is to be used where superscripts or subscripts may cause confusion, or are technically difficult or impossible, e.g., in computer outputs. In the latter connection the following Roman equivalents of Greek letters are recommended:

α	A	ρ	R
β	B	τ	T
γ	G	ν	U
δ	D	ϕ	F
ϵ	E	χ	X
ξ	Z	ψ	Q
η	H	ω	W

Rule 1. General Principles of Notation

1.1. *Designation of Atoms*. The atoms of the main chain are denoted thus:

$$- \text{NH} - \overset{|}{\text{C}^\alpha \text{H}^\alpha} - \text{CO} -$$

Where confusion might arise the following additional symbolism may be used:

$$- \text{N'H'} - \overset{|}{\text{C}^\alpha \text{H}^\alpha} - \text{C'O'} -$$

1.2. Amino-acid residues, $-$NH-CHR-CO$-$, are numbered sequentially from the amino-terminal to the carboxyl-terminal end of the chain, the residue number being denoted i.
    Example:

$$C^\alpha \text{ of the } i\text{th residue is written } C_i^\alpha \text{ or } C\alpha(i)$$

[†]The members of the subcommission were J. C. Kendrew (Chairman), W. Klyne, S. Lifson, T. Miyazawa, G. Némethy, D. C. Phillips, G. N. Ramachandran, and H. A. Scheraga. In addition, the following assisted in the work of the subcommission: R. S. Cahn, R. Diamond, J. T. Edsall, P. J. Flory, C. K. Ingold, A. Liquori, V. Prelog, and J. A. Schellman.
*From IUPAC-IUB Commission on Biochemical Nomenclature, *Pure Appl. Chem.*, 40(3), in press. With permission.

1.3. For some purposes it is more convenient to group together the atoms –CHR-CO-NH–. These groups are described as "peptide units," and the peptide unit number, like the residue number, is denoted i. It will be noted that the two numbers are identical for all atoms except NH; generally there will be no confusion, because a single document will use either "residues" alone, or "peptide units" alone, but in the latter case explicit reference must be made to this usage at the beginning. If confusion might arise, the symbols N_i^* and H_i^* are to be used for these atoms in the ith peptide unit, N_i and H_i in the ith residue (so that $N_i^* = N_{i+1}$).

Example:

```
                        Peptide Unit No. 2
                       |                  |
NH₂ — CHR₁ — CO ┼ NH ┼ CHR₂ — CO ┼ NH ┼ CHR₃
                |                 |
                |   Residue No. 2  |
```

| Residue notation | N_2 | C_2^α | C_2 | N_3 |
| Peptide unit notation | N_1^* | C_2^α | C_2 | N_2^* |

Notes: (i) Residue notation is used throughout these Rules.

(ii) Whether "residues" or "peptide units" are being used, ϕ_i and ψ_i always refer to torsion angles about the same C_i^α.

1.4. *Bond Lengths*. If a bond A–B be denoted A_i–B_j or A_i (see Rules 3.1, 4.5), the bond length is written $b(A_i, B_j)$ [or $b(Ai, Bj)$, or b_i^A (or $bA(i)$)]. An abbreviated notation for use in side chains is indicated in Rule 4.5.

Note: The symbol previously recommended for bond length was l. This symbol is no longer recommended, partly because it is easily confused with 1 in many type fonts, and partly because it is also used for vibration amplitude in electron diffraction and spectroscopy.

1.5. *Bond Angles*. The bond angle included between three atoms $A_i\diagup^{B_j}\diagdown C_k$ is written $\tau(A_i, B_j, C_k)$, which may be abbreviated, if there is no ambiguity, to $\tau(B_j)$ or τ_j^B or $\tau B(j)$.

1.6. *Torsion Angles*. If a system of four atoms $\begin{smallmatrix}A & & & D\\ \diagdown & & \diagup\\ & B-C &\end{smallmatrix}$ is projected onto a plane normal to bond B–C, the angle between the projection of A–B and the projection of C–D is described as the *torsion angle** of A and D about bond B–C; this angle may also be described as the angle between the plane containing A, B, and C, and the plane containing B, C, and D. The torsion angle is written in full as $\theta(A_i, B_j, C_k, D_l)$ which may be abbreviated, if there is no ambiguity, to $\theta(B_j, C_k)$, $\theta(B_j)$, or θ_j^B, etc. In the eclipsed conformation in which the projections of A–B and C–D coincide, θ is given the value 0° (synplanar conformation). A torsion angle is considered positive ($+\theta$) or negative ($-\theta$) according as, when the system is viewed along the central bond in the direction B → C (or C → B), the bond to the front atom A (or D) must be rotated to the right or to the left, respectively, in order that it may eclipse the bond to the rear atom D (or A); note that it is immaterial whether the system be viewed from one end or the other. These relationships are illustrated in Figure 1.

---

*The terms *dihedral angle* and *internal rotation angle* are also used to describe this angle, and may be regarded as alternatives to *torsion angle* though the latter has been used throughout these Rules.

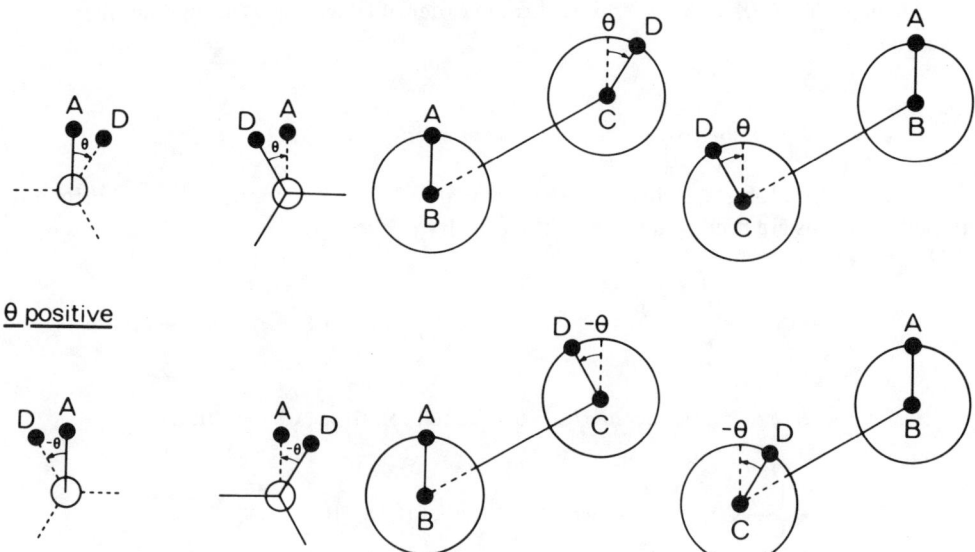

θ positive

FIGURE 1. Newman and perspective projections illustrating positive and negative torsion angles. Note that a right-handed turn of the bond to the front atom about the central bond gives a positive value of θ from whichever end the system is viewed.

Notes: (i) Angles are measured in the range $-180° < \theta \leq +180°$, rather than from 0° to 360°, so that the relationship between enantiomeric configurations or conformations can be readily appreciated.

(ii) The symbols actually used to describe the various torsion angles important in polypeptides are ϕ, ψ, ω, ν, and χ (see Rules 3.2, 4.5.2). In the above, θ is used simply as an illustrative generic symbol covering all these.

Rule 2. The Sequence Rule, and Choice of Torsion Angle

2.1. The Rules here enunciated for use in the field of synthetic polypeptides and proteins are in general harmony with the Sequence Rule of Cahn, Ingold, and Prelog,* with the exceptions of Rules 2.1.1 and 2.2.2 (cases II and III), and later Rules dependent upon these. The Sequence Rule was formulated as a universal and unambiguous means of designating the "handedness" or chirality of an element of asymmetry. It includes Subrules for the purpose of arranging atoms or groups in an order of precedence or preference, and this system may conveniently be used in the description of steric relationships across single bonds (see Klyne and Prelog).[11] Here its function is to determine the priority of precedence of different atoms or groups attached to the same atom. However, Rule 2.1.1 below overrides the precedences of the Sequence Subrules, providing a new "local" (specialist) system for use with the general Sequence Rule.† After application of Rule 2.1.1, the normal procedure of the Sequence Rule is applied, but modified by Rule 2.2.2; in this connection the only parts of the Sequence Rule required are given in Rules 2.1.2 to 2.1.5.

2.1.1. The main chain is given formal priority over branches, notwithstanding any conflict with the following rules. Thus the main chain has precedence at C^α over the side chain, and at C' over O'.

Note: This rule has not yet been formally accepted except in the present context.

---

*See Cahn, Ingold, and Prelog[7] IUPAC Tentative Rules for the Nomenclature of Organic Chemistry, Section E, in IUPAC Information Bulletin No. 35, pp. 36–80.[10] Earlier papers: Cahn and Ingold[5] Cahn, Ingold and Prelog.[6] For a partial, simplified account see Cahn[4] Eliel.[9]

†Other local systems are available analogously for steroids, carbohydrates and cyclitols, where the Sequence Rule is applied when the local system does not suffice.

2.1.2. The order of (decreasing) priority is the order of (decreasing) atomic number. Example:

$$\text{In } Br-\underset{\underset{H}{|}}{\overset{\overset{Cl}{|}}{C}}-CH_3 \text{ the order of priority is } Br, Cl, CH_3, H.$$

2.1.3. If two atoms attached to the central atom are the same, the ligands attached to these two atoms are used to determine the priority:

Examples:

(i) In $CH_3-CH_2-\underset{\underset{H}{|}}{\overset{\overset{Cl}{|}}{C}}-CH_3$ the order is Cl, (CH_3-CH_2), CH_3, H. ($C^xH_2-CH_3$ takes

precedence over C^yH_3 because C^x is bonded to C, H, H, and C^y to H, H, H).

(ii) In $HO-\underset{\underset{H}{|}}{\overset{\overset{CH_2Cl}{|}}{C}}-CH_2OH$ the order is OH, CH_2Cl, CH_2OH, H.

(iii) In $CH_3-CH_2-\underset{\underset{H}{|}}{\overset{\overset{OH}{|}}{C}}-CH(CH_3)_2$ the order is OH, $CH(CH_3)_2$, CH_2-CH_3, H.

2.1.4. A double bond is formally treated as though it were split. Thus $>C=O$ is treated as $>\underset{\underset{(O)}{|}}{C}-\underset{\underset{(C)}{|}}{O}.$

Example:

$$\text{In } CH_3-CO-OH \text{ the order is } =O, -OH, CH_3.$$

2.1.5. If two ligands are distinguished only by having different masses (e.g., deuterium and hydrogen), the heavier takes precedence.

Example:

$$\text{In } Br-\underset{\underset{H}{|}}{\overset{\overset{D}{|}}{C}}-CH_3 \text{ the order is } Br, CH_3, D, H.$$

Note: This rule is to be used only if the two previous rules do not give a decision.

2.2 Choice of Torsion Angle and Numbering of Branches (Tetrahedral Configurations)

2.2.1. If, in a compound P–B–C–E, the Sequence Rule gives the priorities $A>P, Q$ and $D>E>F$, then the Principal Torsion Angle, θ, is that measured by reference to the atoms A–B–C–D as in Rule 1.6 above.

The branches beginning at C are numbered $C_{\overline{1}}D$, $C_{\overline{2}}E$ and $C_{\overline{3}}F$.

2.2.2. If two branches are identical, and the third is different (or nonexistent), they are numbered in a clockwise sense when viewed in the direction B → C, as follows (see Figure 2).

2.2.3. If all three branches are identical, that giving the smallest positive or negative

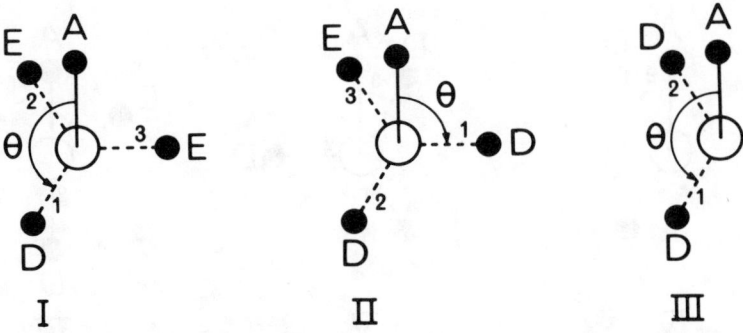

FIGURE 2. Tetrahedral configurations. Case I: D > E = E. D has the highest priority and is given the smallest number (1). Case II: D = D > E. E has the lowest priority and is given the largest number (3). Case III: D = D, numbered 1 and 2 (E is nonexistant). In each case the Principal Torsion Angle is measured between A–B and Branch 1.

Notes: (i) The rule given in Case II differs from Conformational Selection Rule (b) of the Sequence Rule (see Cahn, Ingold, and Prelog, p. 406)[7] according to which if an identity among the groups of a set leaves one group unique, the unique group is fiducial. The reason for the difference is that the Sequence Rule would define Principal Torsion Angle in terms of a hydrogen atom whenever a single such atom formed part of the set; in the X-ray technique, nearly always used to establish structures of the type under discussion, hydrogen atoms are usually unobservable, and even at best not accurately locatable, so that the position of one used to define a Principal Torsion Angle could only be established by calculation based on (perhaps unjustified) assumptions about the bond angles concerned. These considerations apply with even more force to Case III, where one branch is nonexistent; The "phantom atom" of zero atomic number would be given highest priority because it is unique.
(ii) In Case III the clockwise passage from CD^1 to CD^2 shall be by the shorter of the two possible routes.

value of the Principal Torsion Angle is normally* assigned the highest priority and the lowest number (1) (see Figure 3, IV, V); if two branches have torsion angles respectively +60° and −60°, the former is chosen (see Figure 3, VI). The others are numbered in a clockwise sense when viewed in the direction B → C.

Note: Rule 2.2.3 introduces a new principle, not invoked in 2.2.1 or 2.2.2, that the precedence depends on the conformation. This must necessarily be done since in this case the branches are distinguishable only in this respect. (The same applies to Rule 2.3.2 below).

2.3 *Choice of Torsion Angle and Numbering of Branches (Planar Trigonal Configurations)*

2.3.1. If, in a compound P–B–C such that B, C, D, and E are coplanar, or nearly so,

$$\begin{matrix} A & & D \\ \backslash & & / \\ & B-C & \\ / & & \backslash \\ Q & & E \end{matrix}$$

the Sequence Rule gives the priorities A > P, Q and D > E, then the Principal Torsion Angle is that measured by reference to atoms A–B–C–D as in Rule 1.6 above.

The branches beginning at C are numbered $C_{\overline{1}}D$, $C_{\overline{2}}E$.

2.3.2. If the branches are identical, that giving the smallest positive or negative value of

---

*The qualification "normally" is added to avoid the need to renumber the branches, if by chance the rule would demand this in consequence of a movement during refinement of a structure. In this or similar cases, the symbolism should remain unchanged.

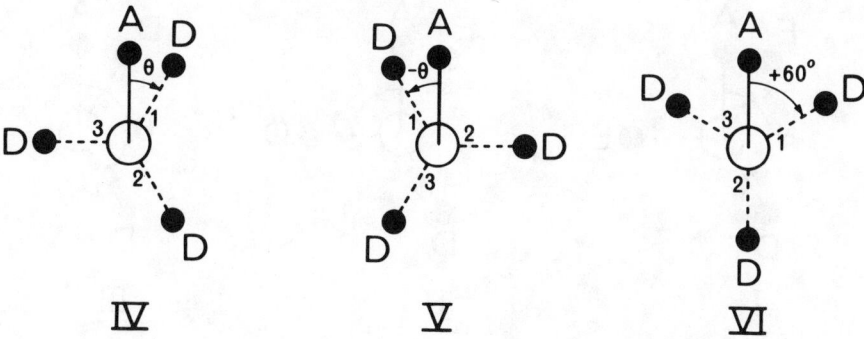

FIGURE 3. Tetrahedral configurations. Three identical branches: IV — general case, θ positive; V — general case, θ negative; VI — θ = +60°.

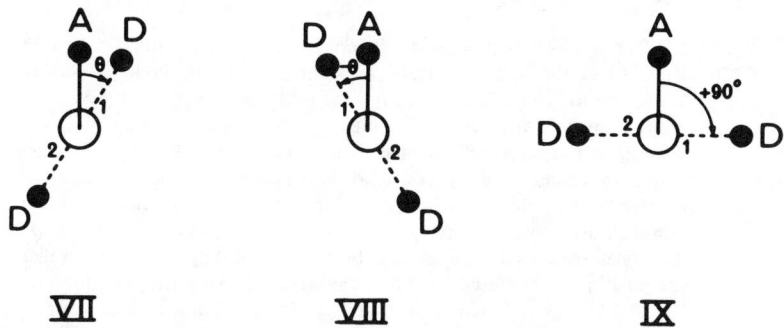

FIGURE 4. Planar trigonal configurations. Identical branches: VII — θ positive; VIII — θ negative; IX — θ = +90°.

the Principal Torsion Angle is normally assigned the highest priority and the lowest number (1); if the two branches have torsion angles respectively +90° and −90°, the former is chosen (see Figure 4).

Rule 3. The Main Chain (or Polypeptide Backbone)

3.1. *Designation of Bonds.* Bonds between main-chain atoms are denoted by the symbols of the two atoms terminating them, e.g., $N_i-C_i^\alpha$, $C_i^\alpha-C_i$, C_i-N_{i+1}, C_i-O_i, N_i-H_i. Abbreviated symbols should not be used. Bond lengths are written $b(C_i, N_{i+1})$, etc.

3.2. *Torsion Angles*

3.2.1. The Principal Torsion Angle describing rotation about $N-C^\alpha$ is denoted by ϕ, that describing rotation about $C^\alpha-C$ is denoted by ψ, and that describing rotation about $C-N$ is denoted by ω. The symbols ϕ_i, ψ_i, and ω_i are used to denote torsion angles of bonds within the ith residue in the case of ϕ and ψ, and between the ith and $(i+1)$th residues in the case of ω; specifically, ϕ_i refers to the torsion angle of the sequence of atoms C_{i-1}, N_i, C_i^α, C_i; ψ_i to the sequence N_i, C_i^α, C_i, N_{i+1}; and ω_i to the sequence C_i^α, C_i, N_{i+1}, C_{i+1}^α (see Figure 5). In accordance with Rules 1.6 and 2.1.1, these torsion angles are ascribed zero values for eclipsed conformation of the main-chain atoms N, C^α, and C, that is, for the so-called *cis*-conformations (see Table 1).

Notes: (i) This convention differs from that proposed by Edsall et al.[8] The new designation of angles may be derived from the old by adding 180° to, or subtracting 180°

Table 1
MAIN-CHAIN TORSION ANGLES FOR VARIOUS CONFORMATIONS IN PEPTIDES OF L-AMINO ACIDS

ϕ	Rotation about N–C$^\alpha$		ψ	Rotation about C$^\alpha$–C	
0°	C$^\alpha$–C trans	⎫	0°	C$^\alpha$–N trans	⎫
+60°	C$^\alpha$–H cis	⎪	+60°	C$^\alpha$–R cis	⎪
+120°	C$^\alpha$–R trans	⎬ to N–H	+120°	C$^\alpha$–H trans	⎬ to C–O
+180°	C$^\alpha$–C cis	⎪	+180°	C$^\alpha$–N cis	⎪
–120°	C$^\alpha$–H trans	⎪	–120°	C$^\alpha$–R trans	⎪
–60°	C$^\alpha$–R cis	⎭	–60°	C$^\alpha$–H cis	⎭

Notes: (i) *trans* to N_i–H_i is the same as *cis* to N_i–C_i^{-1}; *trans* to C_i–O_i is the same as *cis* to C_i–N_{i+1} (see Figure 5).

(ii) For the description of D-amino acids, interchange C$^\alpha$–H and C$^\alpha$–R in the Table.

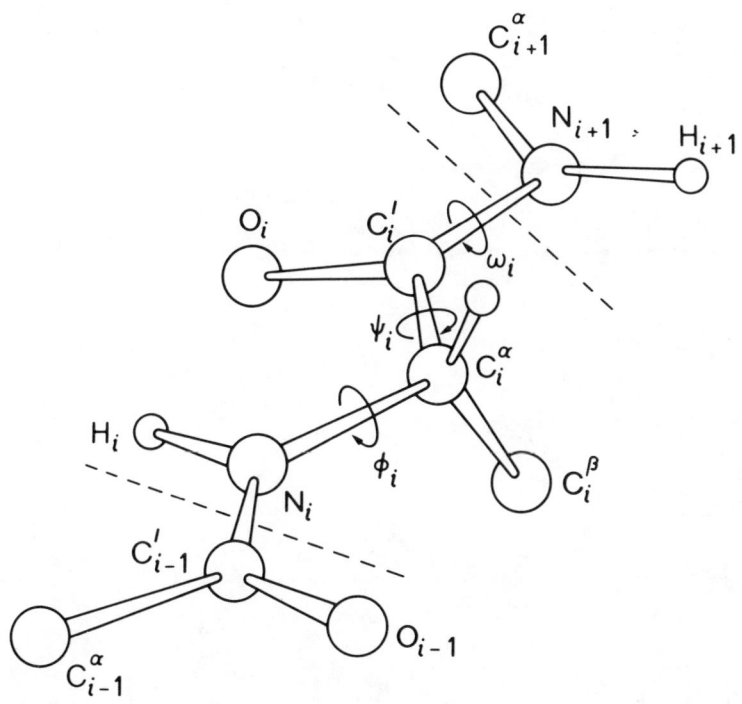

FIGURE 5. Perspective drawing of a section of polypeptide chain representing two peptide units. The limits of a *residue* are indicated by dashed lines, and recommended notations for atoms and torsion angles are indicated. The chain is shown in a fully extended conformation ($\phi_i = \psi_i = \omega_i = +180°$), and the residue illustrated is L-.

from, the latter. (This statement is precisely correct only if the peptide bond is exactly planar, which is not generally the case in experimentally determined structures.)

(ii) Owing to the partial double-bond character of CO=NH, it is normally possible for ω to assume values only in the neighborhood of 0° or 180°. $\omega \sim 180°$ is the value which is generally found (i.e., the *trans*-conformation).

(iii) A "fully-extended" polypeptide chain is characterized by $\phi = \psi = \omega = +180°$. The case of $\phi = \psi = 0°$ would involve the relations indicated in Table 1.

(iv) Table 2 gives values of ϕ and ψ for various well-known regular structures. It is noteworthy that a right-handed α-helix has negative torsion angles.

Table 2
APPROXIMATE TORSION ANGLES FOR SOME REGULAR STRUCTURES

(*Note*: For a fully extended chain $\phi = \psi = \omega = +180°$)

	ϕ	ψ	ω	Reference
Right-handed α-helix (α-poly-L-alanine)	−57°	−47°	+180°	1
Left-handed α-helix	+57°	+47°	+180°	1
Parallel-chain pleated sheet	−119°	+113°	+180°	16
Antiparallel-chain pleated sheet (β-poly-L-alanine)	−139°	+135°	−178°	3
Polyglycine II	−80°	+150°	+180°	15
Collagen	−51°, −76°, −45°	153°, 127°, 148°	+180°	19
Poly-L-proline I	−83°	+158°	0°	14, 17
Poly-L-proline II	−80°	+85°	+180°	2

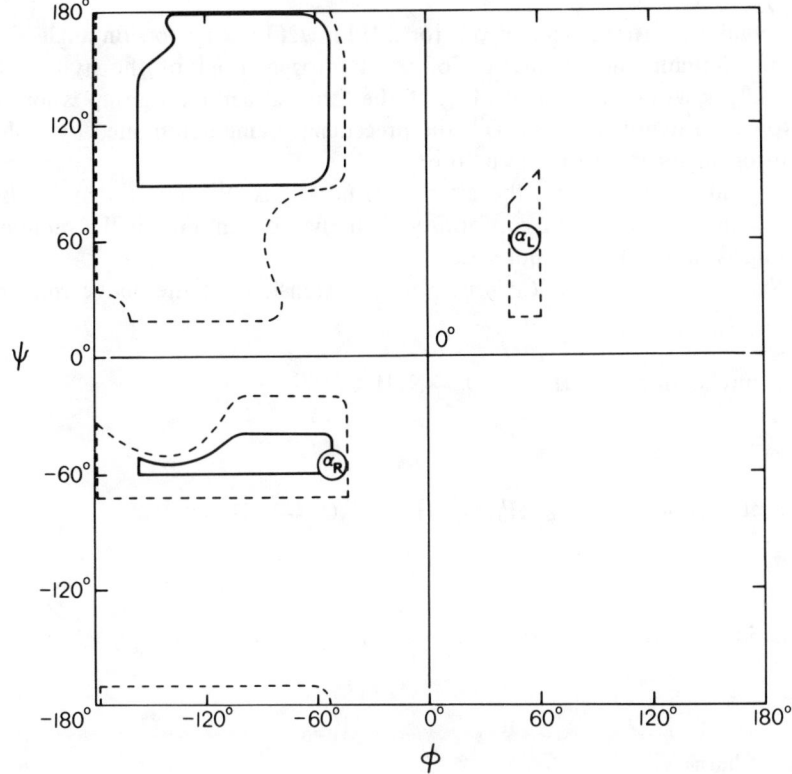

FIGURE 6. Typical conformational map (Ramachandran, Ramakrishnan, and Sasisekharan[13]) transposed into the standard conventions.

Note: This diagram is identical to that of Edsall et al.[8] except that the origin is now at the center, instead of at the lower left-hand corner. The solid lines enclose the freely allowed values of ϕ and ψ for an alanyl residue in a polypeptide; the dotted lines enclose "outer limit" values based on the shortest known van der Waals' radii in related structures. Analogous diagrams for other residues, and for slightly different assumptions, are given by Ramachandran and Sasisekharan;[14] note that these authors used the earlier convention with the origin at the corner.

(v) Figure 6 is a typical conformational map [$(\phi-\psi)$ plot] using the Rules enunciated above.

3.2.2. There may occasionally be a need to consider torsion angles differing from zero for the sequences of atoms O=C–N–C$^\alpha$ and C$^\alpha$–C–N–H, in cases where C=O or N–H lie out of the peptide plane. These angles may be represented ν^O and ν^H (Greek *upsilon*).

3.3. *Chain Terminations*

3.3.1. If the terminal amino group of the chain is protonated the three hydrogen atoms are denoted H_1^1, H_1^2, and H_1^3; the hydrogen atom giving the smallest (positive or negative) value of the Principal Torsion Angle H–N–C$^\alpha$–C is denoted H_1^1, and the others are numbered in a clockwise sense when viewed in the direction C$^\alpha \to$ N. The corresponding torsion angles are denoted ϕ_1^1, ϕ_1^2, and ϕ_1^3. If the terminal amino group is not protonated, the hydrogen atoms are denoted H_1^1 and H_1^2 in accordance with Rule 2.2.2 and the corresponding torsion angles ϕ_1^1 and ϕ_1^2.

3.3.2. At the carboxyl-terminus of the chain (i = T) the double bonded oxygen is written as O' and the other oxygen as O'', thus $C^\alpha-C\begin{smallmatrix}\diagup\!\diagup O' \\ \diagdown O''-H''\end{smallmatrix}$. The torsion angles about the C^α–C bond are written ψ_T^1 and ψ_T^2 [or $\psi 1(T)$, $\psi 2(T)$]; the torsion angle about the C–O″ bond, defining the orientation of the hydrogen atom of the hydroxyl group relative to C^α, is writen θ_T^C [or $\theta C(T)$]. If the terminal carboxyl group is ionized, the oxygen atoms are denoted O′ and O″, the precedence being determined by Rule 2.3.2, and the torsion angles are written as before.

Note: Instead of O′ and O″, the alternative notations O^1 and O^2 may be used. ψ_T may be used instead of ψ_T^1, in conformity with the convention for the middle of the chain, so long as confusion does not arise.

3.3.3. *Substituted Terminal Groups*. Natural extensions of the above rules may be devised, e.g.,

    i. *N*-formyl Group $H_0—C_0O_0—N_1H_1C_1^\alpha H_1^\alpha— \cdots$

    ii. *N*-acetyl Group $C_0(H_0^1, H_0^2, H_0^3)—C_0O_0—N_1H_1—C_1^\alpha H_1^\alpha— \cdots$

    iii. C-amido Group $C_T^\alpha H_T^\alpha—C_T\overset{O_T'}{\underset{N_{T+1}}{\|}}\begin{smallmatrix}H_{T+1}^1\\ \\H_{T+3}^2\end{smallmatrix}$

Rule 4. Side Chains

4.1. Atoms are lettered or lettered and numbered from C^α, and bonds are numbered from C^α, working outwards away from the main chain.

4.2. *Designation of Atoms other than Hydrogen*. Atoms other than hydrogen are designated in the usual way by Greek letters, β, γ, δ, etc., e.g., C_i^β [or $C\beta(i)$], N_i^ζ [or $N\zeta(i)$].

Note: The notations for the amino acids normally occurring in proteins are given in Table 3.

4.3. *Designation of Branches*. If a side chain is branched, the branches are numbered 1 and 2, the order being determined.

    i. in cases where the branches are different, by application of Rule 2.2.1 or 2.3.1,
    ii. in cases where two branches are identical (e.g., in valine, phenylalanine), by the application of Rule 2.2.2 (valine) or 2.3.2 (phenylalanine).

Nonhydrogen atoms in different branches are designated by the Greek letter indicating their degree of remoteness from C^α and by the number of their branch (see Rule 2.2 and 2.3); e.g., in valine $C_i^{\gamma 1}$ and $C_i^{\gamma 2}$ [or $C\gamma 1(i)$, $C\gamma 2(i)$]. The branch number need not be indicated where no ambiguity results, e.g., in threonine O^γ and C^γ instead of $O^{\gamma 1}$ and $C^{\gamma 2}$, in hydroxyproline O^δ, C^δ instead of $O^{\delta 1}$, $C^{\delta 2}$, and in histidine C^δ, N^ϵ etc. instead of $C^{\delta 2}$, $N^{\epsilon 2}$. For asparagine or glutamine, in cases where nitrogen and oxygen in the amide group have not yet been distinguished, these atoms may be written $(NO)^{\delta 1}$, $(NO)^{\delta 2}$, or $(NO)^{\epsilon 1}$, $(NO)^{\epsilon 2}$, the indices 1 and 2 being determined by Rule 2.3.2.

4.4. *Designation of Hydrogen Atoms*. Hydrogen atoms are designated by the Greek letter and/or number of the atom to which they are attached, e.g., in valine H_i^β [or $H\beta(i)$]. Where three hydrogen atoms are attached to a single nonhydrogen atom, they are designated 1, 2, and 3; in the situation $A—B—C\begin{smallmatrix}H\\ \\H\end{smallmatrix}$, the hydrogen atom giving the smallest

Table 3
SYMBOLS FOR ATOMS AND BONDS IN THE SIDE CHAINS OF THE COMMONLY OCCURRING L-AMINO ACIDS

(a) Unbranched Side Chains

Alanine: $C^\alpha \underset{1}{\longrightarrow} C^\beta$

Serine: $C^\alpha \underset{1}{\longrightarrow} C^\beta \underset{2}{\longrightarrow} O^\gamma$

Cysteine: $C^\alpha \underset{1}{\longrightarrow} C^\beta \underset{2}{\longrightarrow} S^\gamma$

Cystine: $C^\alpha_i \underset{1i}{\longrightarrow} C^\beta_i \underset{2i}{\longrightarrow} S^\gamma_i \underset{3i}{\longrightarrow} S^\gamma_k \underset{2k}{\longrightarrow} C^\beta_k \underset{1k}{\longrightarrow} C^\alpha_k$

Methionine: $C^\alpha \underset{1}{\longrightarrow} C^\beta \underset{2}{\longrightarrow} C^\gamma \underset{3}{\longrightarrow} S^\delta \underset{4}{\longrightarrow} C^\epsilon$

Lysine: $C^\alpha \underset{1}{\longrightarrow} C^\beta \underset{2}{\longrightarrow} C^\gamma \underset{3}{\longrightarrow} C^\delta \underset{4}{\longrightarrow} C^\epsilon \underset{5}{\longrightarrow} N^\zeta$

(b) Branched Side Chains

Valine: $C^\alpha \underset{1}{\longrightarrow} C^\beta$ with branches to $C^{\gamma 1}$ (2,1) and $C^{\gamma 2}$ (2,2)

Threonine: $C^\alpha \underset{1}{\longrightarrow} C^\beta$ with branches to $O^{\gamma 1}$ and $C^{\gamma 2}$ (2,2)

Isoleucine: $C^\alpha \underset{1}{\longrightarrow} C^\beta$ with branches to $C^{\gamma 1} \underset{3,1}{\longrightarrow} C^{\delta 1}$ (2,1) and $C^{\gamma 2}$ (2,2)

Leucine: $C^\alpha \underset{1}{\longrightarrow} C^\beta \underset{2}{\longrightarrow} C^\gamma$ with branches to $C^{\delta 1}$ (3,1) and $C^{\delta 2}$ (3,2)

Aspartic acid: $C^\alpha \underset{1}{\longrightarrow} C^\beta \underset{2}{\longrightarrow} C^\gamma$ with branches to $O^{\delta 1}$ (3,1) and $O^{\delta 2}$ (3,2) or $-C^\gamma \begin{smallmatrix} \nearrow O^{\delta 1} \\ \searrow O^{\delta 2}-H \end{smallmatrix}$ (with 3,1 double bond and 3,2 single bond)

Asparagine: $C^\alpha \underset{1}{\longrightarrow} C^\beta \underset{2}{\longrightarrow} C^\gamma$ with branches to $O^{\delta 1}$ (3,1) and $N^{\delta 2}$ (3,2)

Table 3 (continued)
SYMBOLS FOR ATOMS AND BONDS IN THE SIDE CHAINS OF THE COMMONLY OCCURRING L-AMINO ACIDS

(b) Branched Side Chains (continued)

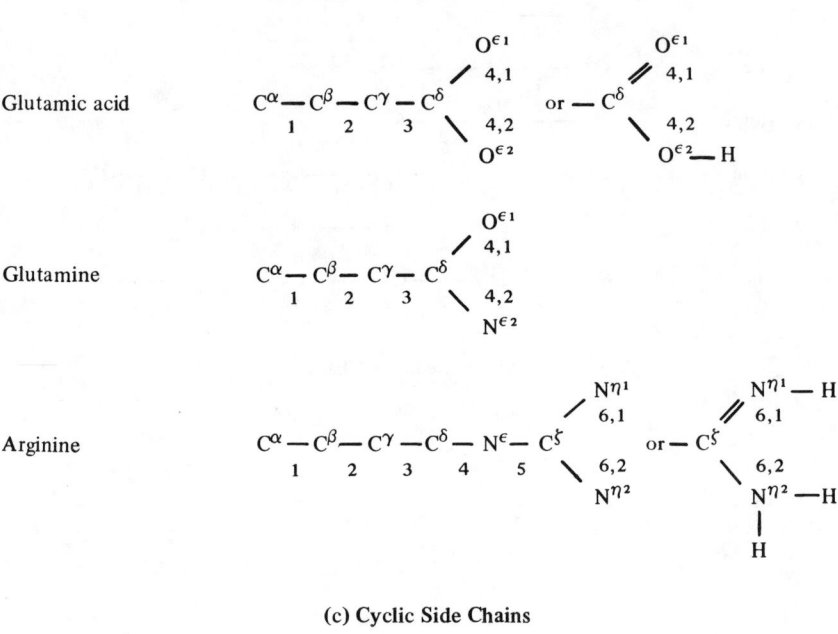

(c) Cyclic Side Chains

Proline

Hydroxyproline

Histidine

Phenylalanine

Table 3 (continued)
SYMBOLS FOR ATOMS AND BONDS IN THE SIDE CHAINS OF THE COMMONLY OCCURRING L-AMINO ACIDS

(c) Cyclic Side Chains (continued)

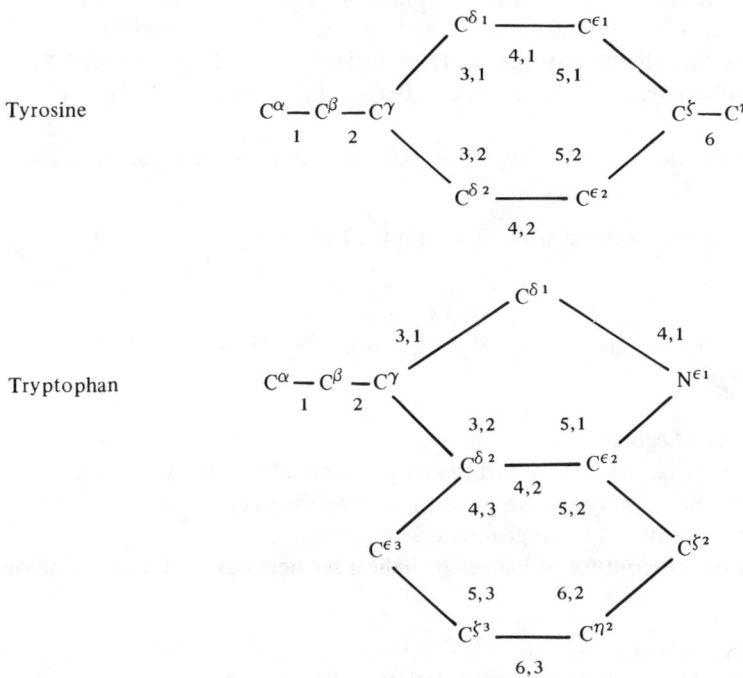

(positive or negative) value of the Principal Torsion Angle is designated 1, and the others are numbered in a clockwise sense when viewed in the direction B → C (see Rule 2.2.3, which also covers the case where $\theta = \pm 60°$); e.g., in valine $H_i^{\gamma 11}$, $H_i^{\gamma 12}$, $H_i^{\gamma 13}$ and $H_i^{\gamma 21}$, $H_i^{\gamma 22}$, $H_i^{\gamma 23}$ [or $H\gamma 11(i)$, etc.]. Where only two hydrogen atoms are present, they are designated in accordance with Rule 2.2.2, Case I for $-CH_2-R$ and Case III for $-NH_2$.

4.5. *Designation of Bonds and Torsion Angles*

4.5.1. Bonds are designated by means of the two atoms terminating them, e.g., $C_i^\alpha - C_i^\beta$, $N_i^\zeta - H_i^{\zeta 2}$, or, if no ambiguity results, by the symbol of the first atom of the bond, e.g., C_i^α, $C_i^{\gamma 1}$. In superscripts the bond may be denoted either by α; β; $\gamma 1$; $\gamma 2$ etc. or by 1; 2; 3,1; 3,2 etc. Bond lengths are denoted $b(C_i^\alpha, C_i^\beta)$, bC_i^α, b_1^1, $b_i^{3,1}$ etc.

4.5.2. Torsion angles are denoted by χ, and are specified by two (or three) superscripts, the first one (or two) (in the situation A–B–C–E with D above and F below) indicating the bond B–C about which the angle is measured, and the last indicating whether the angle is measured relative to D, E, or F. The Principal Torsion Angle is defined by Rule 2.2.1, and if there is no ambiguity the last superscript may be omitted in referring to it.

Thus, in valine, $\chi_i^{2,1}$ and $\chi_i^{2,2}$ refer to the torsion angles specifying atoms $C_i^{\gamma 1}$ and $C_i^{\gamma 2}$; in leucine, $\chi_i^{3,1,1}$, $\chi_i^{3,1,2}$, and $\chi_i^{3,1,3}$ refer to the torsion angles specifying the three hydrogen atoms attached to $C^{\delta 1}$. If there is no ambiguity, the Principal Torsion Angles may be referred to, in valine and leucine, as χ_i^2 and $\chi_i^{3,1}$, respectively. Corresponding notations without superscripts are $\chi 2,1(i)$, $\chi 2(i)$; $\chi 3,1,1(i)$, $\chi 3,1(i)$.

Note: By the Sequence Rule, when $\chi_1 = 0$, C^γ (or $C^{\gamma 1}$) is in the eclipsed position relative to N.

Rule 5.

5.1. *Polarity of Hydrogen Bonds*. In specifying a hydrogen bond as directed from residue i to residue k (or from atom X_i to atom Y_k), the direction X–H to :Y is implied; i.e., the atom covalently linked to the hydrogen atom is mentioned first.

Example: In the α-helix the N–H of residue i is hydrogen-bonded to the C=O of residue $(i-4)$. Therefore, the α-helix is described as having i to $(i-4)$, or $(5-1)$, hydrogen bonding.

5.2. *Dimensions of Hydrogen Bonds*. Dimensions may be denoted by natural extensions of the nomenclature given above. For example, in $\diagdown N_i - H_i \cdots O_k = C_k \diagup$, the following symbols might be used: $b(H_i, O_k)$, $\tau(N_i, H_i, O_k)$, $\tau(H_i, O_k, C_k)$, $\theta(H_i, O_k)$, $\theta_i(N, H)$, $\theta_k(C, O)$.

Rule 6. Helical Segments

A regular helix is strictly of infinite length, with all ϕ's identical and all ψ's identical. A helical *segment* of polypeptide chain may be defined *either* in terms of ϕ and ψ, *or* in terms of symmetry and hydrogen-bond arrangement.

6.1. In the description of helices or helical segments, the following symbols should be used:

n = number of residues per turn,
h = unit height (translation per residue along the helix axis),
t = $360°/n$ = unit twist (angle of rotation per residue about the helix axis).

6.2. *Definition in Terms of ϕ and ψ*. Under this definition, a helical segment is referred to as a (ϕ, ψ) helix; thus a right-handed α-helix would be a $(-57°, -47°)$ helix. The *first* and *last* residues of the helical segment are taken to be the first and last residues which have ϕ and ψ values equal to those defining the helix, within limits which should be defined in the context. No account is taken of hydrogen-bonding arrangements.

6.3. *Definition in Terms of Symmetry and Hydrogen-bond Arrangement*. A helix is referred to as an n_r helix, where

n = number of residues per turn,
r = number of atoms in ring formed by a hydrogen bond and the segment of main chain connecting its extremities.

Thus an α-helix would be 3.6_{13}. The *first* helical residue is taken as the first whose CO group is *regularly* bonded to NH along the helix (in the case of an α-helix, to the NH of the fifth residue); the *last* helical residue is the last whose NH is *regularly* hydrogen-bonded to CO along the helix (in the case of an α-helix, to the CO of the residue last but four). Irregular hydrogen-bonding arrangements are not considered to form part of the helix.

Notes: (i) A helical segment defined by Rule 6.2 may, but need not necessarily, be two residues shorter than the same segment defined by Rule 6.3.

(ii) These rules prescribe no definitions for irregular helical segments.

APPENDIX

Recommendation A. Conformation and Configuration

There is at present no agreed definition of these two terms for general stereochemical usage.

In polypeptide chemistry, the term "conformation" should be used, in conformity with current usage, to describe different spatial arrangements of atoms produced by rotation about covalent bonds; a change in conformation does not involve the breaking of chemical bonds (except hydrogen bonds) or changes in chirality (see Cahn, Ingold, and Prelog.[7])

On the other hand, in polypeptide chemistry the term "configuration" is currently used to describe spatial arrangements of atoms whose interconversion requires the formal breaking and making of covalent bonds (*note*: this usage takes no account of the breaking or making of hydrogen bonds). [Cf. a more extensive discussion in IUPAC Tentative Rules for the Nomenclature of Organic Chemistry, Section E, Fundamental Stereochemistry, IUPAC Information Bulletin No. 35, pp. 71–80 (*Biochim. Biophys. Acta,* 208, 1 (1970)).]

Recommendation B. Definitions of Primary, Secondary, Tertiary, and Quaternary Structure

These concepts, originally introduced by Linderstrøm-Lang,[1,2]* cannot be defined with precision, but the definitions given below may be helpful.

B.1. The *primary structure* of a segment of polypeptide chain or of a protein is the amino-acid sequence of the polypeptide chain(s), without regard to spatial arrangement (apart from configuration at the α-carbon atom).

Note: This definition does not include the positions of disulphide bonds, and is therefore not identical with "covalent structure."

B.2. The *secondary structure* of a segment of polypeptide chain is the local spatial arrangement of its main-chain atoms without regard to the conformation of its side chains or to its relationship with other segments.

B.3. The *tertiary structure* of a protein molecule, or of a subunit of a protein molecule, is the arrangement of all its atoms in space, without regard to its relationship with neighboring molecules or subunits.

B.4. The *quaternary structure* of a protein molecule is the arrangement of its subunits in space and the ensemble of its inter-subunit contacts and interactions, without regard to the internal geometry of the subunits.

Note: A protein molecule not made up of at least potentially separable subunits (not connected by covalent bonds) possesses no quaternary structure. Examples of proteins without quaternary structure are ribonuclease (1 chain) and chymotrypsin (3 chains).

REFERENCES

1. **Arnott and Dover,** *J. Mol. Biol.,* 30, 209 (1967).
2. **Arnott and Dover,** *Acta Cryst.,* B24, 599 (1968).
3. **Arnott, Dover, and Elliott,** *J. Mol. Biol.,* 30, 201 (1967).
4. **Cahn,** *J. Chem. Educ.,* 41, 116 (1964).
5. **Cahn and Ingold,** *J. Chem. Soc.,* 612 (1951).
6. **Cahn, Ingold, and Prelog,** *Experientia,* 12, 81 (1956).
7. **Cahn, Ingold, and Prelog,** *Angew. Chem. Int. Ed.,* 5, 385, 511 (1966); *Angew. Chem.,* 78, 413 (1966).

*The use of the terms "primary, secondary, tertiary, and quaternary structure" has been criticized as being imprecise by Wetlaufer.[18] He has proposed an alternative terminology.

8. **Edsall, Flory, Kendrew, Liquori, Némethy, Ramachandran, and Scheraga,** *J. Biol. Chem.,* 241, 1004 (1966); *Biopolymers,* 4, 130 (1966); *J. Mol. Biol.,* 15, 339 (1966).
9. **Eliel,** *Stereochemistry of Carbon Compounds,* McGraw Hill, New York, 1962, 92.
10. IUPAC Information Bulletin No. 35, 36 (1969).
11. **Klyne and Prelog,** *Experientia,* 16, 521 (1960).
12. **Linderstrøm-Lang,** *Proteins and Enzymes, Lane Memorial Lectures,* Stanford University Press, 1952, 54.
13. **Ramachandran, Ramakrishnan, and Sasisekharan,** *J. Mol. Biol.,* 7, 95 (1963).
14. **Ramachandran and Sasisekharan,** *Advan. Prot. Chem.,* 23, 283 (1968).
15. **Ramachandran, Sasisekharan, and Ramakrishnan,** *Biochim. Biophys. Acta,* 112, 168 (1966).
16. **Schellman and Schellman,** in *The Proteins,* Vol. 2, Neurath, Ed., Academic Press, New York, 1964, 1.
17. **Traub and Schmueli,** in *Aspects of Protein Structure,* Ramachandran, Ed., Academic Press, London, 1963, 81.
18. **Wetlaufer,** *Nature,* 190, 1113 (1961).
19. **Yonath and Traub,** *J. Mol. Biol.,* 43, 461 (1969).

RULES FOR NAMING SYNTHETIC MODIFICATIONS OF NATURAL PEPTIDES TENTATIVE RULES* †

IUPAC-IUB Commission on Biochemical Nomenclature

During the last few years, chemists have made many compounds that are variants of naturally occurring peptides (or proteins) having trivial names. Therefore, the need has arisen for "semitrivial" names to designate these variants without the necessity of designating every residue in the chain.

After discussion with active workers in the field, the following proposals are put forward; they are based on the names used by du Vigneaud and his collaborators [cf. Bodanszky and du Vigneaud, *J. Am. Chem. Soc.*, 81, 1258 (1959); Popenoe, Lawler, and du Vigneaud, *J. Am. Chem. Soc.*, 74, 3713 (1952)] and the symbols introduced by Schwyzer et al. [cf. Rittel, Iselin, Kappeler, Riniker, and Schwyzer, *Angew. Chem.*, 69, 179 (1957); Riniker and Schwyzer, *Helv. Chim. Acta*, 44, 685 (1961); see also *J. Biol. Chem.*, 241, 2491 (1966)].

This draft has been prepared by a subcommittee consisting of J. S. Fruton, W. Klyne, and R. Schwyzer. The subcommittee is greatly indebted to many colleagues for helpful suggestions, notably to V. du Vigneaud, J. Rudinger, H. B. F. Dixon, and P. E. Verkade, chairman of the IUPAC Commission on Organic Nomenclature.

These proposals are *not* suitable for application to "abnormal" links in a peptide sequence; e.g., to disulfide links or γ-peptide links. They are *only* suitable for modifications involving normal α-peptide links.

RULES

1. Replacement

In a polypeptide of trivial name X, if the qth amino acid residue (starting from the N-terminal end of the chain) is *replaced* by the amino acid residue Abc, the semitrivial name of the modified polypeptide is [q-amino acid] X and the abbreviated form, chiefly for use in tables, is [Abc^q] X.

Examples:

[8-Citrulline] vasopressin, [Cit^8] vasopressin [Bodanszky and Birkimer, *J. Am. Chem. Soc.*, 84, 4963 (1962)]. [5-Isoleucine, 7-alanine] hypertensin II, [Ile^5, Ala^7] hypertensin II [Seu, Smeby, and Bumpus, *J. Am. Chem. Soc.*, 84, 3883 (1962)].

Comments

(a) In the full name, the replacement amino acid is designated by its *own full name*, *not* the name of its radical (cf. 4 below). This name, and the position of replacement, are given in square brackets [], as for isotopic replacement.

(b) In the abbreviated form, the amino acid residues are designated by the standard three-letter symbols [*J. Biol. Chem.*, 241, 527, 2491 (1966); *Biochim. Biophys. Acta*, 121, 1 (1966)], the first letter *only* being a capital, in square brackets [].

(c) In the abbreviated form, the *position* of substitution is indicated in a special

---

*Document of the IUPAC-IUB Commission on Biochemical Nomenclature (CBN), approved by CBN in July 1966 and published by permission of the International Union of Pure and Applied Chemistry, the International Union of Biochemistry, and the official publishers to the International Union of Pure and Applied Chemistry, Messrs. Butterworths Scientific Publications. (Revised 1970).

†From *Pure Appl. Chem.*, 31(4), 647–653 (1972). With permission. (Revised version).

fashion, i.e., by a superior numeral[q], to indicate that it is a *residue,* not an individual atom, that is being replaced and also for the reason indicated in comment d.

(d) The nature of the residue replaced is *not* designated in either the full or the abbreviated name. This is contrary to a general principle of organic nomenclature requiring that an atom (or group) that is replaced should (unless it is hydrogen) be clearly designated, as in 2-amino-2-deoxy-D-glucose. It has been decided *not* to insist on the designation of the residue replaced in these semitrivial names in order to keep the names as short as possible, and because the form of nomenclature in Rule 1 clearly differs from ordinary substitution nomenclature.

(e) A partial analogy may be drawn with the form used for isotopic replacement, where the isotope symbol is indicated in square brackets before the name.

(f) The replacement of an amino-acid residue by its enantiomer may be shown logically by the application of this rule as follows: the replacement in X of L-alanine at position 7 by D-alanine results in [7-D-alanine] X with the abbreviation [D-Ala^7] X. An example may be found in Boissonnas, Guttman, and Pless [*Experientia,* 22 (1966)], dealing with the D-Ser^1 ......derivative of β-corticotropin; the natural compound has L-serine in position 1. Another example is the [α-D-Asp^1] hypertensin II of Riniker and Schwyzer [*Helv. Chim. Acta,* 47, 2357 (1964)].

2. Extension

The compounds obtained by the extension of polypeptide X at either (a) the N-terminal end or (b) the C-terminal end are designated by the kinds of names and abbreviations shown below; these are in accordance with the general principles of polypeptide nomenclature [*J. Biol. Chem.,* 241, 2491 (1966); *Biochim. Biophys. Acta,* 121, 1 (1966)]. (See "Symbols for Amino-acid Derivatives and Peptides.")

Examples:

(a)		Extension at N-terminal end:
		Aminoacyl-X Abc-X
e.g.,		Valyl-X Val-X
or		Valylglycyl-X Val-Gly-X (for extension by two residues)
(b)		Extension at C-terminal end:
		X-yl-amino acid X-yl-Abc
e.g.,		X-yl-leucine X-yl-Leu
		(where X-yl is the trivial name of polypeptide X with the ending -yl).

Comment

This rule is not applicable to the extension at the C-terminal of natural peptides having a terminal α-carboxamido group, as in the case of oxytocin or α-melanophore-stimulating hormone (α-MSH). It has been suggested that new names be given to the peptides having a free terminal α-carboxyl group (e.g., oxytocinoic acid) and that extension at the C-terminal end be denoted as in the example given above (e.g., oxytocinoyl-Abc).

3. Insertion

The compound obtained by the *insertion* of an additional amino acid residue Abc in the position between the qth and (q + 1)th residues of a polypeptide X is named qa-endo-amino acid-X (abbreviated form, endo-Abc^{qa}-X).

Example:

4a-endo-tyrosine-hypertensin II; endo-Tyr^{4a}-hypertensin II

Comments

(a) This form has analogies in other fields where endo implies the insertion of something into a structure (e.g., endo-methylene). The prefix or index qa is based on analogies with the steroids where the atoms inserted in a ring after atom number q are designated qa, qb, etc.

(b) The prefix homo is *not* suitable for designating the insertion of a whole residue, since it is commonly used to modify the names of *individual* amino acids, e.g., homoserine.

(c) Multiple insertions, and insertion of two or more residues together in the same place in the chain, are shown by a logical extension of this rule. For example, the insertion into the polypeptide X of threonine between residues 4 and 5, and of valine and glycine (*in that order*) between residues 6 and 7, is shown by the name "endo-4a-threonine,6a-valine,6b-glycine-X" and the abbreviation "endo-Thr^{4a}, (Val^{6a}-Gly^{6b})-X."

4. Removal

The compound obtained by the formal *removal* of an amino acid residue from a polypeptide X in position q is designated by the name des-q-amino acid-X, abbreviated des-Abc^q-X.

Example:

des-7-proline-oxytocin; des-Pro^7-oxytocin [Jacquenoud and Boissonas, *Helv. Chim. Acta*, 45, 1462 (1962)]

Comment

(a) Removal of a whole residue is indicated as is the removal of a ring in steroids, e.g., des-A-androstane.

(b) "de" is *not* suitable as a prefix because it is easily confused, in speaking, with D (for configuration).

5. Substitution Forming a Side Chain

The compound formed by the substitution of an additional amino acid residue as a side chain into a polypeptide X is named by applying the ordinary rules of nomenclature to the trivial name.

(a) If the substitution is on a side-chain *amino* group of polypeptide X, the name of the additional amino *residue* is written (with the termination "yl") and prefixed by symbols indicating the position of substitution (residue number and atom).

Example:

An imaginary compound (A)

$$\begin{array}{c} \text{Val} \\ | \\ \text{Ala-Lys-Ala}\ldots\ldots \\ 1\quad 2\quad 3 \\ \text{A} \end{array}$$

in which a valyl group is substituted at the ε-amino group of lysine at position 2 of the chain of a peptide X is named $N^{\epsilon 2}$-valyl-X (abbreviated $N^{\epsilon 2}$-Val-X).

(b) If the substitution is on a side-chain *carboxyl* group of polypeptide X, the additional amino acid having a free α-carboxyl group, the substituted derivative is named by specifying the position of substitution (residue number, and atom) and is given the designation "X-yl-amino acid."

Example:

An imaginary compound (B)

$$\begin{array}{c} \text{Val} \\ | \\ \text{Ala-Leu-Glu-Ala} \ldots \ldots \\ 1 \quad 2 \quad 3 \quad 4 \\ \text{B} \end{array}$$

in which a valine residue is substituted into the δ-carboxyl group of glutamic acid in position 3 of the chain of a peptide X would be named $C^{\delta 3}$-X-yl-valine (abbreviated $C^{\delta 3}$-X-yl-Val).

Comment

Note the importance of clear distinction from *replacement* as indicated in Rule 1.

6. Partial Sequences (Fragments)

Polypeptide sequences that form fragments of a longer sequence that already has a trivial name may be designated as follows. The *trivial name* is followed by numbers giving the positions of the first and last amino acids, and then usual *Greek* designation giving the number of amino acid units in the fragment; thus

Trivial name (-X–Y-) . . . . . . peptide.

Example: from α-MSH

$$\begin{array}{c} \text{Ac-Ser-Tyr-Ser-Met-Glu-His-Phe-Arg-Trp-Gly-Lys-Pro-} \\ 1 \quad 2 \quad 3 \quad 4 \quad 5 \quad 6 \quad 7 \quad 8 \quad 9 \quad 10 \quad 11 \quad 12 \end{array}$$

$$\begin{array}{c} \text{-Val-NH}_2 \quad \text{α-MSH} \\ 13 \end{array}$$

we may have

$$\begin{array}{c} \text{Met-Glu-His-Phe-Arg-Trp-Gly} \\ 4 \hspace{4cm} 10 \end{array}$$

α-MSH-(4−10)-heptapeptide

and

$$\begin{array}{c} \text{His-Phe-Arg-Lys-Pro-Val-NH}_2 \\ 6 \quad\quad 8 \quad 11 \quad\quad 13 \end{array}$$

α-MSH-(6−8)-(11−13)-hexapeptide amide

The last example illustrates the nomenclature for a composition sequence of two fragments, and also for an amide-terminal group.

SUMMARY WITH EXAMPLES

The Systematic Application of These Principles to the Name of an Imaginary Pentapeptide "Iupaciubin"[a] May Illustrate the Symbolism

Rule	Operation	Short name	Structure
	(Fundamental name)	Iupaciubin	1 2 3 4 5 Ala-Lys-Glu-Tyr-Leu
1.	Replacement	[Phe^4]iupaciubin[b]	4 Ala-Lys-Glu-Phe-Leu
2a.	Extension (N terminal)	Arginyl-iupaciubin, Arg-iupaciubin	1 5 Arg-Ala-Lys-Glu-Tyr-Leu
2b.	Extension (C terminal)	Iupaciubyl-methionine, iupaciubyl-Met	1 5 Ala-Lys-Glu-Tyr-Leu-Met
3.	Insertion	Endo-Thr$^{2\,a}$-iupaciubin	2 2a 3 Ala-Lys-Thr-Glu-Tyr-Leu
4.	Removal	Des-Glu^3-iupaciubin	2 4 Ala-Lys-Tyr-Leu
5a.	Side-chain substitution on amino group	$N^{\epsilon 2}$-Val-iupaciubin	Val |ϵ Ala-Lys-Glu-Tyr-Leu 2
5b.	Side-chain substitution on carboxyl group	C-$^{\delta 3}$-Iupaciubyl-valine	Val |δ Ala-Lys-Glu-Tyr-Leu 3
6.	Partial sequence	Iupaciubin-(2–4)-tripeptide	2 3 4 Lys-Glu-Tyr

[a] To symbolize the harmonious cooperation of IUPAC and IUB.
[b] Note that only for *replacement* are square brackets required.

THE NOMENCLATURE OF MULTIPLE FORMS OF ENZYMES RECOMMENDATIONS (1971)*

IUPAC-IUB Commission on Biochemical Nomenclature (CBN)

Recommendations on the nomenclature of multiple forms of enzymes were made by a subcommittee set up by the International Union of Biochemistry and were published in 1964 in a number of journals.[1] Since that time, there have appeared a large number of publications in this field; some hundreds of enzymes have been shown to exist as heteromorphs. It seemed appropriate at this time to review the earlier recommendations[1] in the light of current knowledge. For that purpose, the IUPAC-IUB Commission on Biochemical Nomenclature (CBN) set up a subcommittee consisting of G. Brewer, O. Hoffmann-Ostenhof, R. S. Holmes, P. Karlson (Convenor), B. Keil, G. B. Kitto, C. L. Markert, C. J. Masters, F. Moyer, J. Scandalios, C. R. Shaw, E. C. Slater, R. Tashian, and E. C. Webb. The subcommittee reported to CBN, the report was discussed, and the following recommendations were finally adopted.

I. Definition of Isozymes or Isoenzymes

The 1964 Committee recommended[1] that "multiple enzyme forms" in a single species should be known as isoenzymes (or isozymes). It is known that enzymes catalyzing essentially the same reaction may differ in various ways, as shown in Table 1. In this table, some prominent examples are also listed.

According to the original definition, which was meant to be purely operational, all of these multiple forms should be termed "isozymes" or "isoenzymes." However, most biochemists feel that the term "isoenzymes" should be restricted to those forms arising from genetic control of primary protein structure. Genetically determined differences in primary structure are the reason for the multiplicity in Groups 1 to 3 of Table 1, but not in Groups 4 to 7. Indeed, scientists working on conjugated or derived enzymes do not use the isoenzyme terminology for characterization of their multiple forms.

It is therefore recommended:

1. The term "multiple forms of the enzyme . . ." should be used as a broad term covering all proteins possessing the same enzyme activity and occurring naturally in a single species.

2. The term "isoenzyme" of "isozyme" should apply only to those multiple forms of enzymes arising from genetically determined differences in primary structure, and *not* to those derived by modification of the same primary sequence.

Nevertheless, the term "isozyme" or "isoenzyme" is being used as an operational term in dealing with enzyme proteins with the same catalytic activities but separable by suitable methods (e.g., electrophoresis) and where knowledge of the nature of multiplicity is lacking. These terms should not be used in cases in which other multiple forms, e.g., conjugated or derived, are primarily under study.

The present recommendations deal only with isoenzymes and genetic variants.

II. Nomenclature of Isozymes or Isoenzymes

The earlier committee[1] recommended that individual isoenzymes (isozymes) should be distinguished and numbered on the basis of electrophoretic mobility, with the number 1 being assigned to that form having the highest mobility toward the anode.

*From IUPAC-IUB Commission of Biochemical Nomenclature, *Pure Appl. Chem.,* 40(3), in press. With permission.

Table 1
MULTIPLE FORMS OF ENZYMES

Artifacts Occurring During Preparation are Outside
the Scope of This Document

Group	Reason of multiplicity	Example
1	Genetically independent proteins	Malate dehydrogenase in mitochondria and cytosol
2	Heteropolymers (hybrids) of two or more polypeptide chains, noncovalently bound	Lactate dehydrogenase
3	Genetic variants (allelic)	Glucose-6-phosphate dehydrogenase in man
4	Proteins conjugated with other groups	Phosphorylase a and b
5	Proteins derived from one polypeptide chain	The family of chymotrypsins arising from chymotrypsinogen
6	Polymers of a single subunit	Glutamate dehydrogenase of mol wt 1,000,000 and 250,000
7	Conformationally different forms	All allosteric modifications of enzymes

Of all of the means available to indicate the different properties of isoenzymes (electrophoresis, chromatography, kinetic criteria, chemical structure, etc.), electrophoresis is most widely used for the following reasons:

    a. It is still extremely important in the study of enzyme heterogeneity to avoid any artifactual consequences of the handling or "purification" of enzymes, and zone electrophoresis is one procedure in which the resolution of individual protein species is not unduly influenced by their application in the original state (i.e., as tissue homogenates, etc.).

    b. The degree of resolution is another factor of prime importance in this field, and resolution by electrophoretic procedures is generally more effective than other methods of protein separation.

    c. Electrophoresis offers the advantages of rapidity and broad applicability.

The recognition of these facts and the consequent wide utilization of electrophoretic procedures by workers in this field has built up a substantial literature on the subject and has facilitated communication and reference.

Other alternatives suggested for the distinction of isoenzyme forms include kinetic criteria and structural data. While kinetic criteria are extremely valuable as an adjunct to investigations of enzyme multiplicity, they cannot provide information on the extent of heterogeneity. With regard to structural criteria, it has long been the hope of many workers in the field that the interrelationships of multiple forms might soon be delineated in chemical terms. However, such information is available only for very few enzymes systems, so that it would not be useful at this stage to make general recommendations about isoenzyme nomenclature based on structural considerations. When structural details are available, a system similar to that used for hemoglobin, in which the polypeptide chains are represented by Greek letters, could be used. [The use of upper case Roman letters (as have been used for lactate dehydrogenase and aldolase) would be more acceptable.]

Isoenzymes (isozymes) or their subunits should not be labeled on the basis of tissue distribution (e.g., brain type, heart type) since confusion can arise on account of species variation; homologous forms may occur in altogether different tissues in other species.

It is therefore recommended:

3. In naming isozymes (isoenzymes), the normal enzyme name (either systematic or trivial)[2] should be used, followed by a number. The numbers should be allotted consecutively, preferably on the basis of electrophoretic mobility under defined conditions, with the lower numbers given to the forms with the higher mobility towards the anode. In photographs or diagrams of electrophoretic results, the anode should be oriented to the top or right-hand side of the page.

4. Where complex isoenzyme (isozyme) patterns occur, with major groups each composed of several different electrophoretic zones, the numbers may be used to designate the major groups, with subscript lower case letters used consecutively for the individual subzones (1_a, 1_b, 1_c, 2_a, 2_b, etc.).

5. For unambiguous identification of particular isozymes (isoenzymes), additional parameters such as molecular weight, stability, or subunit structure should be given where available. Subunits may be denoted by upper case Roman letters or lower case Greek letters, but not by terms based on tissue distribution.

III. Nomenclature of Genetically Variant Enzymes

In the case of genetic variants, a flexible and open system is desirable. The chief consideration is that the investigator recognize that the new variant that he describes will probably not be the last one discovered, and, in the case of some enzymes, a very large number may eventually be found.

A special committee on the nomenclature of glucose-6-phosphate dehydrogenase in man (of which more than 50 genetic variants are known) was convened by the World Health Organization, and the report of its recommendations was published in abbreviated form in several journals;[3] it forms the basis for certain of the recommendations presented here.

This report considers only the naming of the variant enzymes; it does not consider the designation of genotype and phenotype symbols, since these have been standardized by several appropriate genetic groups and differ somewhat among various organisms. The name of the enzyme variant can be readily adapted to the appropriate genetic terminologies.

It is therefore recommended:

6. In naming genetic variants, the normal enzyme name (either systematic or trivial) should be used, followed by a trivial name for the variant. This can be of any sort provided that it is sufficiently adaptable. A suitable system (as has been used for glucose-6-phosphate dehydrogenase) is to use the name of the town, university, country, etc., where the variant was discovered. The use of superscript letters or numbers is considered acceptable initially, but, once it is apparent that more than a few variants are being found, the trivial nomenclature should be initiated, to avoid using the same letter for two different variants.

7. When two variants originally considered as different are subsequently discovered to be identical, the name first used for that variant should then be accepted, the other one being discontinued.

REFERENCES

1. Webb, E. C., *Lancet,* 1, 1110 (1964); *Nature,* 203, 821 (1964); *Experientia,* 20, 592 (1964); *Z. Klin. Chem. Klin. Biochem.,* 2, 160 (1964); *Postepy Biochem.,* 10, 525 (1964); *Enzymol. Biol. Clin.,* 5, 124 (1965).
2. *Enzyme Nomenclature: Recommendations (1964) of the the International Union of Biochemistry on the Nomenclature and Classification of Enzymes, together with their Units and the Symbols of Enzyme Kinetics,* Elsevier Publishing Company, Amsterdam, 1965.
3. *Biochem. Genet.,* 1, 198 (1967).

The following Tentative Rules of the IUPAC Commission on the Nomenclature of Organic Chemistry (CNOC) and the IUPAC-IUB Commission on Biochemical Nomenclature (CBN), some of which are issued jointly with the IUPAC Commission on the Nomenclature of Organic Chemistry (CNOC) are available from Waldo E. Cohn, Director, NAS-NRC Office of Biochemical Nomenclature, Oak Ridge National Laboratory, P.O. Box Y, Oak Ridge, TN 37830, U.S.A. Joint Rules are marked with an asterisk.

NOMENCLATURE

Nomenclature of vitamins, coenzymes and related compounds: Trivial names of miscellaneous compounds of importance in biochemistry; Nomenclature of quinones with isoprenoid side chains; Nomenclature and symbols for folic acid and related compounds; Nomenclature of corrinoids [see *Eur. J. Biochem.,* 2, 1 (1967)].

The nomenclature of lipids. A document for discussion [see *Eur J. Biochem.,* 2, 127 (1967)]; admendments [see *Eur. J. Biochem.,* 12, 1 (1970)].

The nomenclature of cyclitols* [see *Eur. J. Biochem.,* 5, 1 (1968)].

The nomenclature of steroids* [see *Eur. J. Biochem.,* 10, 1 (1969)]; corrections [see *Eur. J. Biochem.,* 12, 1 (1970)].

Nomenclature for vitamins B_6 and related compounds [see *Eur. J. Biochem.,* 17, 1 (1970)].

IUPAC Tentative rules for the nomenclature of organic chemistry. Section E. Fundamental stereochemistry [see *Eur. J. Biochem.,* 18, 151 (1971)].

Tentative rules for carbohydrate nomenclature. Part 1* [see *Eur. J. Biochem.,* 21, 455 (1971)]; correction [see *Eur. J. Biochem.,* 22, 592 (1971)].

The nomenclature of multiple forms of enzymes [this document].

ABBREVIATIONS AND SYMBOLS

Abbreviations and symbols for chemical names of special interest in biological chemistry [see *Eur. J. Biochem.,* 1, 259 (1967)].

Abbreviated designation of amino acid derivatives and peptides [see *Eur. J. Biochem.,* 1, 375 (1967)].

Rules for naming synthetic modifications of natural peptides [see *Eur. J. Biochem.,* 1, 379 (1967)].

Abbreviated nomenclature of synthetic polypeptides (polymerized amino acid) [see *Eur. J. Biochem.,* 3, 129 (1967)]; correction [see *Eur. J. Biochem.,* 12, 2 (1970)].

A one-letter notation for amino-acid sequences [see *Eur. J. Biochem.,* 5, 151 (1968)].

Abbreviations and symbols for nucleic acids, polynucleotides, and their constituents [see *Eur. J. Biochem.,* 15, 203 (1970)]; corrections [see *Eur. J. Biochem.,* 18, 588 (1971)].

Abbreviations and symbols for the description of the conformation of polypeptide chains [see *Eur. J. Biochem.,* 17, 193 (1970)].

A document, OBN-5, describing the (American) NAS-NRC Office of Biochemical Nomenclature, and listing other rules affecting biochemical nomenclature is available from its Director, Dr. Waldo E. Cohn [see also *J. Chem. Doc.*, 7, 72 (1967)].

NOMENCLATURE OF IRON-SULFUR PROTEINS RECOMMENDATIONS (1973)*

IUPAC-IUB Commission on Biochemical Nomenclature (CBN)

On August 19, 1967, an informal meeting organized by T. Kimura was held in Tokyo, Japan, to discuss the needs and desires of establishing a systematic nomenclature for the so-called "non-heme iron proteins". The 18 scientists attending that meeting — all actively investigating the chemistry or biological function of this unique class of proteins — agreed that the time was propitious to stem the proliferation of trivial names that had developed in the last few years, and that the IUPAC-IUB Commission on Biochemical Nomenclature (CBN) be requested to form a subcommittee to establish tentative rules for nomenclature. This recommendation was accepted by CBN at their meeting in Bellagio, Italy, in July 1968 and a subcommission was established.

A formal meeting of the subcommission was held on October 17, 1968, and a provisional system of nomenclature was discussed. Of primary importance was the unanimous agreement that the term "non-heme iron proteins" be abandoned. It was proposed that the general category of iron-containing proteins should have a subdivision composed of "iron-sulfur proteins." Further it was agreed that the terms "ferredoxin" and "rubredoxin" be retained and their usage expanded.

Following this meeting, opinions were obtained by correspondence with members of the subcommission and a meeting of the subcommittee on June 7, 1971, approved the following recommendations.

RECOMMENDATIONS

1. Proteins containing iron may be divided into three groups: Hemoproteins, iron-sulfur proteins, and other iron proteins. The last group includes ferritin, transferrin, and the oxygenases. The term "iron-sulfur proteins" refers only to those proteins where the iron is shown to be liganded with inorganic sulfur or cysteine sulfur. When the heme-iron atom in hemoproteins is also liganded with inorganic sulfur or cysteine sulfur, the protein is classified as a hemoprotein.

2. The iron-sulfur proteins may be subdivided into four catergories:

2.1. *Ferredoxin*. This group comprises those iron-sulfur proteins with an equal number of iron and labile sulfur atoms, and a negative midpoint redox potential at pH 7. They are characterized by an EPR (electron-paramagnetic resonance) signal with $g < 2$ for the reduced protein. Ferredoxins are present in plants, animals, and bacteria. The source should always be stated. Examples: Chloroplast ferredoxin, adrenal ferredoxin (formerly called adrenodoxin), *Pseudomonas putida* ferredoxin (formerly called putidaredoxin), *Clostridium acidi-urici* ferredoxin.

Ferredoxin may be abbreviated Fd.

2.2. *High-potential Iron-sulfur Proteins*. Certain microorganisms contain a unique class of iron-sulfur proteins containing acid-labile sulfur, but differing from the ferredoxins in their physical properties. No EPR signal has been detected with the reduced form of this type of protein. The oxidized form is paramagnetic with an EPR signal with a **g**-value of about 2. At pH 7, the midpoint potential is positive. Until further characterized, the descriptive but cumbersome name "high-potential iron-sulfur protein" should be retained with the source indicated as a prefix, e.g., chromatium high-potential iron-sulfur protein.

2.3. *Rubredoxins*. This group comprises those iron-sulfur proteins without acid-labile

*From IUPAC-IUB Commission on Biochemical Nomenclature, *IUPAC Inf. Bull. Append. Tentative Nomencl. Sym. Units Stand.*, No. 32, August 1973. With permission.

sulfur characterized by having iron in a typical mercaptide coordination, i.e., one center surrounded by 4 cysteine or equivalent sulfur ligands. Oxidized rubredoxin has a distinctive EPR spectrum with a line at $g = 4.3$ whereas the reduced pigment gives no discernible EPR signal. The redox potential for those rubredoxins now characterized are negative at pH 7.0. The full name should be listed as (source) rubredoxin (function), e.g., *Pseudomonas oleovorans* rubredoxin, alkane ω-hydroxylation.

2.4. *Conjugated Iron-sulfur Proteins*. This group comprises those proteins containing iron and labile sulfur or iron in a typical mercaptide coordination, but also containing additional prosthetic groups. Many of the iron-containing flavoproteins, molybdenum iron proteins, or molybdenum-iron flavoproteins are included. Frequently these proteins may contain, as a component part of the enzyme complex, characteristics (EPR, optical spectra, or redox properties) similar to a protein classified in 2.1 to 2.3. However, since they are now considered in other nomenclature systems, no specific system of naming is now recommended. If desired, a cross-reference to this category of proteins may be included in addition to the present name in order to avoid ambiguity.

The committee has developed the above system of nomenclature fully cognizant that considerably more information will be required before a more definitive nomenclature can be developed. The basis for the above system of nomenclature rests strongly on the chemical properties of the proteins with a number of physical criteria used for further differentiation. It is hoped that the suggested nomenclature is sufficiently flexible to encompass new proteins discovered without the need to generate further trivial names.

The recommended classification is illustrated schematically in Figure 1.

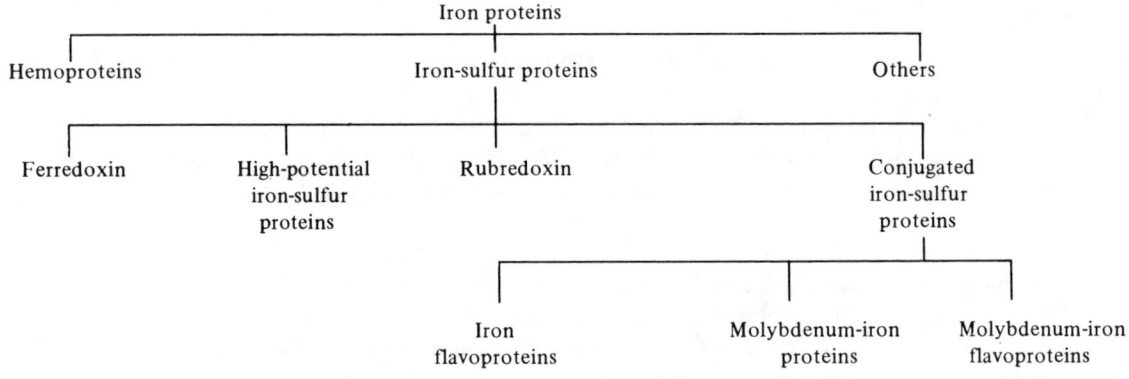

FIGURE 1. Classification of iron proteins.

ENZYME NOMENCLATURE
RECOMMENDATIONS (1972)*

IUPAC-IUB Commission on Biochemical Nomenclature

HISTORICAL INTRODUCTION

The rapid growth in the science of enzymology and the great increase in the number of enzymes known have given rise to many difficulties of terminology in recent years. By about 1955 it had become evident that the nomenclature of the subject, in the absence of any guiding authority, was getting out of hand. The naming of enzymes by individual workers had proved far from satisfactory in practice. In many cases the same enzymes became known by several different names, while conversely there were cases in which the same name was given to different enzymes. Many of the names conveyed little or no idea of the nature of the reactions catalyzed, and similar names were sometimes given to enzymes of quite different types. To meet this situation, various attempts to bring order into the general nomenclature of enzymes, or into that of particular groups of enzymes, were made by individuals or small groups of specialists. But none of the resulting nomenclatures met with general approval.

Furthermore, no general agreement had been reached on the nomenclature of the coenzymes, on which so many names of enzymes inevitably depend; in the equations of enzyme kinetics different systems of mathematical symbols were used by different workers; and the standardization of enzymes was in a chaotic state, owing to the multiplicity of arbitrarily defined units of enzyme activity.

In view of this state of affairs, the General Assembly of the International Union of Biochemistry decided, during the third International Congress of Biochemistry in Brussels in August 1955, to set up an International Commission on Enzymes. This step was taken in consultation with the International Union of Pure and Applied Chemistry.

The International Commission on Enzymes was established in 1956 by the President of the International Union of Biochemistry, Professor M. Florkin, with the advice of an *ad hoc* Committee. The following members were appointed by the Bureau of the International Union of Biochemistry:

A. E. Braunstein, U.S.S.R.; S. P. Colowick, U.S.A.; P. A. E. Desnuelle, France; M. Dixon, U. K. (Chairman); W. A. Engelhardt, U.S.S.R.; E. F. Gale, U. K.; O. Hoffmann-Ostenhof, Austria; A. L. Lehninger, U.S.A.; K. Linderstrøm-Lang, Denmark; F. Lynen, Germany.

Corresponding Members: F. Egami, Japan; L. F. Leloir, Argentina.

In 1959, on the death of K. Linderstrøm-Lang, E. C. Webb (United Kingdom, later Australia) joined the commission.

The terms of reference of the Enzyme Commission, as laid down by the *ad hoc* Committee, were as follows

To consider the classification and nomenclature of enzymes and coenzymes, their units of activity and standard methods of assay, together with the symbols used in the description of enzyme kinetics.

The enzyme Commission faced many difficulties arising from the uncontrolled naming of the rapidly increasing number of known enzymes. Some of the names in use were

---

*From IUPAC-IUB Commission on Biochemical Nomenclature, Enzyme Nomenclature, Recommendations (1972), Elsevier, Amsterdam, 1973.

definitely misleading; others conveyed little or nothing about the nature of the reaction catalyzed, as for example, **diaphorase, Zwischenferment, catalase**. Enzymes catalyzing essentially similar reactions sometimes had names implying that they belong to different groups, while some enzymes of different types had been placed in the same group, for example, the *pyrophosphorylases* had included both glycosyl-transferases and phospho-transferases. In some cases a name, which had been well established for many years with a definite meaning, such as the term *synthetase,* had been employed later with different meanings, causing confusion.

One of the main tasks given to the Commission was therefore to see how the nomenclature of enzymes could best be brought into a satisfactory state, and whether a code of systematic rules could be devised that would serve as a guide for the consistent naming of new enzymes in the future. At the same time, the Commission realized the difficulties that would be caused by a large number of changes of well-known enzyme names, and the desirability of retaining the existing names wherever there was no good reason for making an alteration. Nevertheless, the overriding consideration was to reduce the confusion and prevent further confusion from arising. Its task could not have been accomplished without causing some inconvenience, for this was the inevitable result of having allowed the problem to drift for a considerable time.

Throughout its work, the Enzyme Commission was in close touch with the Biological Chemistry Nomenclature Commission of the International Union of Pure and Applied Chemistry. In addition, it considered many comments and suggestions from various experts in the field; 52 formal documents were circulated and discussed in several meetings. Finally, the Commission prepared a report, which was presented to the General Assembly of the International Union of Biochemistry at their meeting in Moscow, 1961, and was adopted. The nomenclature set out in that report has been widely used in scientific journals, textbooks, etc. since 1961.

Subsequently, the Council of I.U.B. dissolved the Commission on Enzymes and set up a Standing Committee on Enzymes consisting of S. P. Colowick, O. Hoffmann-Ostenhof, A. L. Lehninger, and E. C. Webb (Secretary). This Standing Committee discussed the comments and criticisms received on the published report of the Enzyme Commission and prepared in 1964 a second version, the *Recommendations (1964) of the International Union of Biochemistry on the Nomenclature and Classification of Enzymes.*

The function of the Standing Committee on Enzymes was then taken over by the IUPAC-IUB Commission on Biochemical Nomenclature (CBN). This Commission was orginally set up to deal with the nomenclature of various compounds of biochemical interest. At a meeting in September 1969, it was decided that the *Recommendations on Enzyme Nomenclature* should be revised, mainly to include the many enzymes discovered in recent years, and an Expert Committee on Enzymes was formed, consisting of A. E. Braunstein, J. S. Fruton, O. Hoffmann-Ostenhof, B. L. Horecker, W. B. Jakoby, P. Karlson, B. Keil, E. C. Slater, E. C. Webb (convenor), and W. J. Whelan. The first question considered by the Expert Committee was whether the nomenclature and classification systems should be completely revised in the light of new knowledge. It was decided that the principles formulated by the Enzyme Commission and the Standing Committee on Enzymes should still be followed and that the new enzymes should be incorporated into the Enzyme List and code numbers should be assigned to them. On the other hand, it was agreed that the introductory chapters of *Recommendations (1964) on the Nomenclature and Classification of Enzymes* should be completely revised, and some changes made in the format of the Enzyme List. The forthcoming revision of that list was announced in scientific journals, and all workers in the field were invited to submit comments, suggestions, and experimental data on any new enzymes.

As the result of a careful search of the literature and many helpful suggestions from individual biochemists, more than 900 new enzymes have been included, and 51 old entries have been deleted. The enzyme list now contains 1,770 entries, compared with 874 in the 1964 list. It was generally fairly easy to assign a proper place and code number to the new enzymes; this proves that the present classification is, at least, a good working system. For some special fields, expert groups were formed; for example, expert groups reported on cytochromes, oxygenases, nucleases, carbohydrases, peptidases, and lyases.*

The Commission on Biochemical Nomenclature is grateful for the assistance given by members of these expert groups, by individual biochemists who sent suggestions, and by Margaret Nugent, who carried out the intensive literature search. The Commission is also indebted to the University of Queensland and its officers for assistance in the preparation of the report, and especially to Miriam Armstrong who typed all drafts as well as the material now published.

As a result of this work, this new Report on Enzyme Nomenclature was prepared and adopted by the International Union of Biochemistry and the International Union of Pure and Applied Chemistry.

THE CLASSIFICATION AND NOMENCLATURE OF ENZYMES

1. General Principles

Because of their close interdependence, it is convenient to deal with the classification and nomenclature together in one chapter.

The first general principle of these "Recommendations" is that names purporting to be names of enzymes, especially those ending in -ase, should only be used for single enzymes, i.e., single catalytic entities. They should not be applied to systems containing more than one enzyme. When it is desired to name such a system on the basis of the overall reaction catalyzed by it, the word **system** should be included in the name. For example, the system catalyzing the oxidation of succinate by *molecular oxygen*, consisting of succinate dehydrogenase, cytochrome oxidase, and several intermediate carriers, should not be named **succinate oxidase**, but it may be called the **succinate oxidase system**. Other systems consisting of several structurally and functionally linked enzymes (and cofactors) are the pyruvate dehydrogenase system, the similar 2-oxoglutarate dehydrogenase system, and the fatty acid synthetase system.

In this context it is appropriate to express disapproval of a loose and misleading practice that is currently rather frequent in biological literature. It consists in designation of a natural substance (or even of a hypothetical active principle), responsible for a physiological or biophysical phenomenon that cannot be described in terms of a definite chemical reaction, by the name of the phenomenon in conjugation with the suffix -ase, which implies an individual enzyme. Some recent examples of such **phenomenase** nomenclature, which should be discouraged even if there are reasons to suppose that the particular agent may have enzymic properties, are **permease, translocase, reparase, joinase, replicase, codase**, etc.

---

*Membership of the expert groups was as follows

Cytochromes: W. D. Bonner, Jr., D. V. DerVartanian, J. Le Gall, H. A. Harbury, E. F. Hartee, M. D. Kamen, M. R. Lemberg, E. Margoliash, E. C. Slater (convenor), Lucille Smith, Joan Whiteley, and R. J. P. Williams.
Oxygenases: O. Hayaishi (convenor), M. Katagiri, R. Sato, S. Senoh, B. Tamaoki, N. Tamiya, and I. Yamazaki.
Nucleases: W. E. Cohn, Halinka Sierakowska, and D. Shugar (convenor).
Carbohydrases: E. C. Heath, Elizabeth Neufeld, E. T. Reese, and W. J. Whelan (convenor).
Peptidases: S. Buchs, M. Gruber, B. S. Hartley, B. Keil (convenor), V. N. Orekhovich, G. Pfleiderer, R. Stambaugh, D. R. Whitaker, and W. L. Williams.
Lyases: A. E. Braunstein, B. L. Horecker (convenor), H. L. Kornberg, P. Srere, E. E. Snell, and W. A. Wood.

The second general principle is that enzymes are classified and named according to the reaction they catalyze. The chemical reaction catalyzed is the specific property which distinguishes one enzyme from another, and it is logical to use it as the basis for the classification and naming of enzymes.

Several alternative bases for classification and naming had been considered, e.g., chemical nature of the enzyme (whether it is a flavoprotein, a haemoprotein, a pyridoxal-phosphate-protein, a cuproprotein, and so on), or chemical nature of the substrate (nucleotides, carbohydrates, proteins, etc.). The first cannot serve as a general basis, for only a minority of enzymes have such identifiable prosthetic groups. The chemical nature of the enzyme has, however, been used exceptionally in this revised report in certain cases where classification based on specificity is difficult, for example, with the proteinases (sub-subgroups 3.4.21 to 3.4.24). The second basis for classification is hardly practicable, owing to the great variety of substances acted upon and because it is not sufficiently informative unless the type of reaction is also given. It is the overall reaction, as expressed by the formal equation, that should be taken as the basis. Thus, the intimate mechanism of the reaction, and the formation of intermediate complexes of the reactants with the enzyme, is not taken into account, but only the observed chemical change produced by the complete enzyme reaction. For example, in those cases in which the enzyme contains a prosthetic group which serves to catalyze transfer from a donor to an acceptor (e.g., flavin, biotin, or pyridoxal-phosphate enzymes) the name of the prosthetic group is not included in the name of the enzyme. Nevertheless, where alternative names are possible, the mechanism may be taken into account in choosing between them.

A consequence of the adoption of the chemical reaction as the basis for naming enzymes is that a systematic name cannot be given to an enzyme until it is known what chemical reaction it catalyzes. This applies, for example, to a few enzymes which have so far not been shown to catalyze any chemical reaction, but only isotopic exchanges; the isotopic exchange gives some idea of one step in the overall chemical reaction, but the reaction as a whole remains unknown.

A second consequence of this concept is that a certain name designates not a single enzyme protein but a group of proteins with the same catalytic property. Enzymes from different sources (various bacterial, plant, or animal species) are classified as one entry. The same applies to isoenzymes (see below). However, there are exceptions to this general rule. Some are justified because the mechanism of the reaction or the substrate specificity is so different as to warrant different entries in the enzyme list. This applies, for example, to the two cholinesterases, EC 3.1.1.7 and 3.1.1.8, the two citrate hydro-lyases, EC 4.2.1.3 and 4.2.1.4, and the two amine oxidases, EC 1.4.3.4 and 1.4.3.6. Others are mainly historical, e.g., acid and alkaline phosphatase.

A third general principle adopted is that the enzymes are divided into groups on the basis of the *type of reaction* catalyzed, and this together with the name(s) of the substrate(s) provides a basis for naming individual enzymes. It is also the basis for classification and code numbers.

Special problems attend the classification and naming of enzymes which catalyze complicated transformations that can be resolved into several sequential or coupled intermediary reactions of different types, all catalyzed by a single enzyme (not an enzyme system). Some of the steps may be spontaneous noncatalytic reactions, while one or more intermediate steps depend on catalysis by the enzyme. Wherever the nature and sequence of intermediary reactions is known or can be presumed with confidence, classification and naming of the enzyme should be based on the *first enzyme-catalyzed* step, which is essential to the subsequent transformation; the latter can be indicated by a supplementary term in parentheses, e.g., L-malate glyoxylate-lyase (CoA-acetylating) (EC 4.1.3.2, cf. Section 3).

For the classification according to the type of reaction catalyzed, it is occasionally necessary to choose between alternative ways of regarding a given reaction. Some considerations of this type are outlined in Section 3. In general, that alternative should be selected which fits in best with the general system of classification and reduces the number of exceptions.

One important extension of this principle is the question of the direction in which the reaction is written for the purposes of classification. To simplify the classification, the direction chosen should be the same for all enzymes in a given class, even if this direction has not been demonstrated for all. Thus the systematic names, on which the classification and code numbers are based, may be derived from a written reaction, even though only the reverse of this has been actually demonstrated experimentally.

2. Systematic and Trivial Names

The first Enzyme Commission gave much thought to the question of a systematic and logical nomenclature for enzymes and finally recommended that there should be two nomenclatures for enzymes, one systematic, and one working or trivial. The systematic name of an enzyme, formed in accordance with definite rules, showed the action of an enzyme as exactly as possible, thus identifying the enzyme precisely. The trivial name was sufficiently short for general use, but not necessarily very systematic; in a great many cases it was a name already in current use. The introduction of (often cumbersome) systematic names has been strongly criticized. It has been pointed out that in many cases the reaction catalyzed, given in parentheses, is not much longer than the systematic name and can serve just as well for identification, especially in conjunction with the code number.

The Commission for Revision of Enzyme Nomenclature has discussed this problem at length. It was decided to give the *trivial* names more prominence in the enzyme list; they now follow immediately after the code number, and are described as *Recommended Name*. Also, in the index the recommended names appear in bold roman. Nevertheless, it was decided to retain the systematic names as the basis for classification as well as for identification (to be given only once in the paper) for the following reasons:

  i. the code number alone is only useful for identification of an enzyme when a copy of the Enzyme List is at hand, while the systematic name is self-explanatory;

  ii. the systematic name stresses the type of reaction, the reaction equation does not;

  iii. systematic names can be formed for new enzymes by the discoverer, by application of the rules, while **code numbers should not be assigned by individuals**;

  iv. recommended names for new enzymes are generally formed as a condensed version of the systematic name; therefore, the systematic names are helpful in finding recommended names that are in accordance with the general pattern.

It is recommended that for enzymes, which are not the main subject of a paper or abstract, the recommended names should be used, but they should be identified at their first mention by their code numbers and source. Where an enzyme is the main subject of a paper or abstract, its code number, systematic name, or, alternatively, the reaction equation, and source should be given at its first mention; thereafter the recommended name may be used. In the light of the fact that enzyme names and code numbers refer to reactions catalyzed rather than to discrete proteins, *it is of special importance to give also the source of the enzyme* for full identification; in cases where multiple forms are known to exist, knowledge of this should be included where available.

When a paper deals with an enzyme that is not yet in the enzyme list, the author may introduce a new name and, if desired, a new systematic name, both formed according to the recommended rules. A number should be assigned only by CBN.

The enzyme list at the end of this volume, in contrast to the earlier versions, contains one or more references for each enzyme. It should be stressed that no attempt has been made to provide a complete bibliography, or to refer to the first description of an enzyme. The references are intended to provide sufficient evidence for the existence of an enzyme catalyzing the reaction as set out. In those cases where there is a major paper describing the purification and specificity of an enzyme, this has been quoted to the exclusion of earlier and later papers. In some cases separate references are given for animal, plant, and bacterial enzymes.

3. Scheme of Classification and Numbering of Enzymes

The first Enzyme Commission, in its report in 1961, devised a system for classification of enzymes which also serves as a basis for assigning code numbers to them. These code numbers, which are now widely in use, contain four elements separated by points, with the following meaning:

    i. the first number shows to which of the six main divisions (classes) the enzyme belongs,
    ii. the second figure indicates the subclass,
    iii. the third figure gives the sub-subclass,
    iv. the fourth figure is the serial number of the enzyme in its sub-subclass.

The subclasses and sub-subclasses are formed according to principles indicated below; the full key to the classification is given in Section 5.

The main divisions and suclasses are

1. Oxidoreductases — To this class belong all enzymes catalyzing oxido-reduction reactions. The substrate which is oxidized is regarded as hydrogen donor. The systematic name is based on *donor:acceptor oxidoreductase*. The recommended name will be *dehydrogenase*, wherever this is possible; as an alternative, *reductase* can be used. *Oxidase* is only used in cases where O_2 is the acceptor.

The second figure in the code number of the oxidoreductases indicates the group in the hydrogen donor which undergoes oxidation: 1 denotes a —CHOH—group, 2 an aldehyde- or keto-group, and so on, as listed in Section 5.

The third figure indicates the type of acceptor involved: 1 denotes NAD(P), 2 a cytochrome, 3 molecular oxygen, 4 a disulphide, 5 a quinone or related compound, etc.

It should be noted that in reactions with a nicotinamide coenzyme, this is always regarded as acceptor, even if this direction of the reaction is not readily demonstrated. The only exception is the subclass 1.6, in which NAD(P)H is the donor; some other redox catalyst is the acceptor.

2. Transferases — Transferases are enzymes transferring a group, e.g., the methyl group or a glycosyl group, from one compound (generally regarded as donor) to another compound (generally regarded as acceptor). The systematic names are formed according to the scheme *donor:acceptor grouptransferase*. The recommended names are normally formed according to *acceptor grouptransferase* or *donor grouptransferase*. In many cases, the donor is a cofactor (coenzyme) charged with the group to be transferred. A special case is that of the aminotransferases (see below).

Some transferase reactions can be viewed in different ways. For example, the enzyme-catalyzed reaction

$$X - Y + Z = X + Z - Y$$

may be regarded either as a transfer of the group Y from X to Z, or as a breaking of the X – Y bond by the introduction of Z. Where Z represents phosphate or arsenate, the process is often spoken of as "phosphorolysis" or "arsenolysis," respectively, and a number of enzyme names based on the pattern of *phosphorylase* have come into use. These names are not suitable for a systematic nomenclature, because there is no reason to single out these particular enzymes from the other transferases, and it is better to regard them simply as *Y-transferases*.

Another problem is posed in the enzyme-catalyzed transamination reactions. They involve the transfer of one electron pair and a proton, together with the NH_2 group, from a primary amine to an oxo compound, according to the general equation

$$R^1 - CHNH_2 - R^2 + R^3 - CO - R^4 \xrightarrow{enzyme} R^1 - CO - R^2 + R^3 - CHNH_2 - R^4.$$

The reaction can formally be considered as oxidative deamination of the donor (e.g., amino acid) linked with reductive amination of the acceptor (e.g., oxo acid), and the transaminating enzymes (pyridoxal-phosphate-proteins) might be classified as oxidoreductases. However, the unique, distinctive feature of the reaction is the transfer of the amino group (by a well-established mechanism involving covalent substrate-coenzyme intermediates), which justifies allocation of these enzymes among the transferases as a special subgroup (2.6.1, aminotransferases).

The second figure in the code number of transferases indicates the group transferred: a one-carbon group in 2.1, an aldehydic or ketonic group in 2.2, a glycosyl group in 2.3, and so on.

The third figure gives further information on the group transferred: e.g., subclass 2.1 is subdivided into methyltransferases (2.1.1), hydroxymethyl- and formyltransferases (2.1.2), and so on; only in subclass 2.7, the third figure indicates the nature of the acceptor group.

3. Hydrolases – These enzymes catalyze the hydrolytic cleavage of C–O, C–N, C–C and some other bonds, including phosphoric anhydride bonds. While the systematic name always includes *hydrolase,* the recommended name is, in many cases, formed by the name of the substrate with the suffix ...*ase*. It is understood that the name of the substrate with this suffix means a hydrolytic enzyme.

A number of hydrolases acting on ester, glycosyl, peptide, amide, or other bonds are known to catalyze not only hydrolytic removal of a particular group from their substrates, but likewise the transfer of this group to suitable acceptor molecules. In principle, all hydrolytic enzymes might be classified as transferases, since hydrolysis itself can be regarded as transfer of a specific group to water as the acceptor. Yet, in most cases, the reaction with water as the *acceptor* was discovered earlier and is considered as the main physiological function of the enzyme. This is why such enzymes are classified as hydrolases rather than as transferases.

Some hydrolases (especially among the esterases and glycosidases pose problems because they have a very wide specificity and it is not easy to decide if two preparations described by different authors (may be from different sources) have the same catalytic properties, or if they should be listed under separate entries. An example is vitamin A esterase (formerly EC 3.1.1.12, now believed to be identical with EC 3.1.1.1). To some extent the choice must be arbitrary; however, separate entries should be given only when the specificities are sufficiently different.

Another problem is the so-called "esterolytic" proteases, which hydrolyse ester bonds in appropriate substrates even more rapidly than natural peptide bonds. In this case, classification among the peptide hydrolases was based on historical priority and presumed physiological function.

The second figure in the code number of the hydrolases indicates the nature of the bond hydrolyzed: 3.1 are the esterases, 3.2 the glycosidases, and so on (cf. Section 5).

The third figure normally specifies the nature of the substrate, e.g., in the esterases the carboxylic ester hydrolases (3.1.1), thiol ester hydrolases (3.1.2), phosphoric monoesterases (3.1.3); in the glycosidases the O-glycosidases (3.2.1), N-glycosidases (3.2.2), etc. Exceptionally, in the case of the peptidyl-peptide hydrolases, the third figure is based on the catalytic mechanism as shown by active center studies or the effect of pH (cf. Section 5 for the full key).

4. Lyases – Lyases are enzymes cleaving C–C, C–O, C–N, and other bonds by elimination, leaving double bonds, or conversely adding groups to double bonds. The systematic name is formed according to the pattern *substrate group-lyase*. The hyphen is an important part of the name, and to avoid confusion should not be omitted, e.g., *hydro-lyase* not "hydrolyase." In the recommended names, expressions like *decarboxylase, aldolase,* and *dehydratase* (in case of elimination of water) are used. In cases where the reverse reaction is much more important, or the only one demonstrated, *synthase* (not synthetase) may be used in the name. Various subclasses of the lyases include pyridoxal-phosphate enzymes that catalyze the elimination of β- or γ-substituent from an α-amino acid, followed by a replacement of this substituent by some other group. In the overall replacement reaction, no unsaturated end product is formed; therefore, these enzymes might formally be classified as alkyl-transferases (EC 2.5.1. . . .). However, there is ample evidence that the replacement is a two-step reaction involving the transient formation of enzyme-bound α,β (or β,γ)-unsaturated amino acids. According to the rule that the first reaction is indicative for classification, these enzymes are correctly classified as lyases. Examples are tryptophan synthase (EC 4.2.1.20) and cystathionine β-synthase (EC 4.2.1.22).

The second figure in the code number indicates the bond broken: 4.1 are carbon-carbon-lyases, 4.2 carbon-oxygen-lyases, and so on.

The third figure gives further information on the group eliminated (e.g., CO_2 in 4.1.1, H_2O in 4.2.1).

5. Isomerases – These enzymes catalyze geometric or structural changes within one molecule. According to the type of isomerism, they may be called *racemases, epimerases,* cis-trans-*isomerases, isomerases, tautomerases, mutases,* or *cyclo-isomerases.*

In some cases, the interconversion in the substrate is brought about by intramolecular oxidoreduction (5.3); since hydrogen donor and acceptor are the same molecule, and no oxidized product appears, they are not classified as oxidoreductases even if they may contain firmly bound $NAD(P)^+$.

The subclasses are formed according to the type of isomerism, the sub-subclasses to the type of substrates.

6. Ligases (synthetases) – Ligases are enzymes catalyzing the joining together of two molecules coupled with the hydrolysis of a pyrophosphate bond in ATP or a similar triphosphate. The bonds formed are often *high energy* bonds. The systematic names are formed on the system *X:Y ligase (ADP-forming)*. In the recommended nomenclature the term *synthetase* may be used, if no other short term (e.g., carboxylase) is available. Names of the type "X-activating enzyme" should not be used.

The second figure in the code number indicates the bond formed: 6.1 for C–O bonds (enzymes acylating tRNA), 6.2 for C–S bonds (acyl-CoA derivatives) etc. Sub-subclasses are only in use in the C–N ligases (cf. Section 5).

In a few cases it is necessary to use the word **other** in the description of subclasses and sub-subclasses. They have been provisionally given the figure 99 in order to leave space for new subdivisions. Actually, in the revised Enzyme List presented here, a number of new subclasses and sub-subclasses have been introduced.

Some enzymes have been deleted from the List, some others have been renumbered. However, the old numbers have *not* been allotted to new enzymes; rather the place has been left vacant and cross reference is made according to the following scheme:

[n.m.o.p.   *Deleted entry:old name]*

or

[n.m.o.p.   *Transferred entry:now EC q.r.s.t. – recommended name]*

Entries for reclassified enzymes transferred from one position in the List to another are followed, for reference, by a comment indicating the former number.

It is regarded as important that the same policy be followed in future revisions and extensions of the Enzyme List, which will become necessary from time to time, and will have to be carried out by future Enzyme Commissions or Expert Committees of CBN.

4. Rules for Classification and Nomenclature

(a) GENERAL RULES AND GUIDELINES

Guidelines for recommended names	Rules for systematic nomenclature on which the classification is based
1. Generally accepted trivial names of substrates may be used in enzyme names. The prefix D- should be omitted for common sugars and L- for individual amino acids, unless ambiguity would be causes. If desired, α, β, γ may be used instead of numbers to indicate positions where such usage is widely established at present; in general, it is not necessary to indicate positions of substituents in recommended names unless it is necessary to prevent two different enzymes having the same name.	1. To produce usable systematic names, accepted trivial names of substrates forming part of the enzyme names should be used. Where no accepted and convenient trivial names exist, the official IUPAC rules of nomenclature should be applied to the substrate name. The 1, 2, 3 system of locating substituents should be used instead of the α,β,γ system; α, β should be used for indicating configuration, although group names such as β-aspartyl-, γ-glutamyl, and also β-alanine, γ-lactone, are permissible. For nucleotide radicals, *adenylyl* (not *adenyl*), etc. should be the form used. The name oxo acids (not keto acids) may be used as a class name, and for individual compounds in which $-CH_2-$ has been replaced by $-CO-$, *oxo* should be used. The prefix *keto* should be used for derivatives of sugars in which $-CHOH-$ has been replaced by $-CO-$.

2. Where the substrate is normally in the form of an anion, its name should end in *-ate* rather than *-ic*; e.g., *lactate dehydrogenase,* not "lactic dehydrogenase" or "lactic acid dehydrogenase."

3. Commonly used abbreviations for substrates, e.g., ATP, may be used in names of enzymes, but the use of new abbreviations (*not listed in recommendations of the IUPAC-IUB Commission on Biochemical Nomenclature*) should be discouraged. Chemical formulas should not be used instead of names of substrates. Abbreviations for names of enzymes, e.g., GDH should not be used.

4. Names of substrates such as glucose phosphate, which are normally written with a space, should be

(a) GENERAL RULES AND GUIDELINES (continued)

Guidelines for recommended names

Rules for systematic nomenclature on which the classification is based

hyphenated when they form part of the enzyme names, e.g., glucose-6-phosphate dehydrogenase (EC 1.1.1.49*).

5. The use as enzyme names of *descriptions* such as *condensing enzyme, Zwischenferment, acetate-activating enzyme,* and *pH 5 enzyme* should be discontinued as soon as the catalyzed reaction is known. The word *activating* should not be used in the sense of converting the substrate into a substance which reacts further; all enzymes act by activating their substrates, and the use of the word in this sense may lead to confusion.

6. If it can be avoided, a recommended name should not be based on a substance which is not a true substrate, e.g., enzyme EC 4.2.1.17 should not be called *crotonase*, since it does not act on crotonate.

7. Where a name in common use gives some indication of the reaction and is not incorrect or ambiguous, its continued use is recommended. In other cases a recommended name is based on the same general principles as the systematic name (see opposite), but with a minimum of detail, to produce a name which is short enough for convenient use. A few names of proteolytic enzymes ending in *-in* are retained; all other enzyme names should end in *-ase*. (*The termination -ese should be not used.*)

7. Systematic names consist of two parts. The first contains the name of the substrate or, in the case of a bimolecular reaction, of the two substrates separated by a colon, with small and equal spaces before and after the colon. The second part, ending in *-ase,* indicates the nature of the reaction. Where additional information is needed to make the reaction clear, a phrase indicating the reaction or a product should be added in parentheses after the second part of the name e.g., (*ADP-forming*), (*dimerizing*), (*CoA-acylating*).

8. A number of generic words indicating a type of reaction may be used in either recommended or systematic names: *oxidoreductase, oxygenase, transferase* (with a prefix indicating the nature of the group transferred), *hydrolase, lyase, racemase, epimerase, isomerase, mutase, ligase,* and *synthetase*.

9. A number of additional generic words indicating reaction types are used in recommended names, but not in the systematic nomenclature, e.g., *dehydrogenase, reductase, oxidase, peroxidase, kinase, tautomerase, deaminase, dehydratase,* etc.

10. The direct attachment of *-ase* to the name of the substrate will indicate that the enzyme brings about hydrolysis.

10. The suffix *-ase* will never be attached directly to the name of the substrate.

11. The name *dehydrase* which has been used for both dehydrogenating and dehydrating enzymes, should not be used. *Dehydrogenase* will be used for the former and *dehydratase* for the latter.

12. Where possible, recommended names should normally be based on a reaction direction which has been demonstrated, e.g., *dehydrogenase* or *reductase, decarboxylase* or *carboxylase*.

12. In the case of reversible reactions, the direction chosen for naming should be the same for all the enzymes in a given class, even if this direction has not been demonstrated for all. Thus systematic names may be based on a

---

*This follows standard practice in phrases where two nouns qualify a third.

(a) GENERAL RULES AND GUIDELINES (continued)

Guidelines for recommended names	Rules for systematic nomenclature on which the classification is based
	written reaction, even though only the reverse of this has been actually demonstrated experimentally.
	13. When the overall reaction includes two different charges, e.g., an oxidative demethylation, the classification and systematic name should be based whenever possible, on the one (or the first one) catalyzed by the enzyme; the other function(s) should be indicated by adding a suitable participle in parentheses, e.g., sarcosine:oxygen oxidoreductase (*demethylating*) (EC 1.5.3.1); D-aspartate:oxygen oxidoreductase (*deaminating*) (EC 1.4.3.1); L-serine hydro-lyase (*adding indole*) (EC 4.2.1.20).
	Other examples of such additions are (*decarboxylating*), (*cyclizing*), (*acceptor-acylating*), and (*isomerizing*).

14. When an enzyme catalyzes more than one type of reaction, the name should normally refer to one reaction only. Each case must be considered on its merit, and the choice must be, to some extent, arbitrary. Other important activities of the enzyme may be indicated in the List under "Reaction" or "Comments."

Similarly, when an enzyme acts on more than one substrate (or pair of substrates), the name should normally refer only to one substrate (or pair of substrates), although in certain cases it may be possible to use a term which covers a whole group of substrates, or an alternative substrate may be given in parentheses.

15. A group of enzymes with rather similar specificities should normally be described by a single entry. However, when the specificity of two enzymes catalyzing the same reactions is sufficiently different (the degree of difference being a matter of arbitrary choice) two separate entries may be made, e.g., EC 1.2.1.4 and EC 1.2.1.7. Separate entries are also appropriate for enzymes having similar catalytic functions, but known to differ basically with regard to reaction mechanism or to the nature of the catalytic groups, e.g., amine oxidase (flavin-containing) (EC 1.4.3.4) and amine oxidase (pyridoxal-containing) (EC 1.4.3.6).

(b) RULES AND GUIDELINES FOR PARTICULAR CLASSES OF ENZYMES

Guidelines for recommended names	Rules for systematic nomenclature on which the classification is based
Class 1	
16. The terms *dehydrogenase* or *reductase* will be used much as hitherto. The latter term is appropriate when hydrogen transfer from the substance mentioned as donor in the systematic name is not readily demonstrated. *Transhydrogenase* may be retained for a few well-established cases. *Oxidase* will by used only for cases where O_2 acts as an acceptor, and *oxygenase* only for those cases where the O_2 molecule is directly incorporated into the substrate. *Peroxidase* is used for enzymes using H_2O_2 as acceptor. *Catalase* must be regarded as exceptional. Where no ambiguity is caused, the	16. All enzymes catalyzing oxidoreductions should be named *oxidoreductases* in the systematic nomenclature, and the names formed on the pattern "donor:acceptor oxidoreductase."

(b) RULES AND GUIDELINES FOR PARTICULAR CLASSES OF ENZYMES (continued)

| Guidelines for recommended names | Rules for systematic nomenclature on which the classification is based |

Class 1 (continued)

second reactant is not usually named; but where required to prevent ambiguity, it may be given in parentheses, e.g., EC 1.1.1.1, *alcohol dehydrogenase* and EC 1.1.1.2, *alcohol dehydrogenase* (*NADP⁺*).

17. For oxidoreductases using NAD^+ or $NADP^+$, the coenzyme should always be named as the acceptor except for the special case of Section 1.6 (*enzymes whose normal physiological function is regarded to be reoxidation of the reduced coenzyme*). Where the enzyme can use either coenzyme, this should be indicated by writing $NAD(P)^+$.

18. Where the true acceptor is unknown and the oxidoreductase has only been shown to react with artificial acceptors, the word *acceptor* should be written in parentheses, e.g., EC 1.3.99.1, succinate:(*acceptor*) oxidoreductase.

19. Oxidoreductases which bring about the incorporation of molecular oxygen into one donor or into either or both of a pair of donors will be named *oxygenase*. If only one atom of oxygen is incorporated the term *monooxygenase* will be used; if both atoms of O_2 are incorporated, the term *dioxygenase* will be used.

19. Oxidoreductases bringing about the incorporation of oxygen into one of paired donors should be named on the pattern "donor, donor:oxygen oxidoreductase (hydroxylating)."

Class 2

20. Only one specific substrate or reaction product is generally indicated in the recommended names, together with the group donated or accepted.
   The forms, *aminotransferase,* etc. may be replaced if desired by the corresponding forms *transaminase,* etc.
   A number of special words are used to indicate reaction types, e.g., *kinase* to indicate a phosphate transfer from ATP, to the named substrate (not *phosphokinase*), *pyrophosphokinase* for a similar transfer of pyrophosphate, and *phosphomutase* for an apparent intramolecular phosphate transfer.

20. Enzymes catalyzing group-transfer reactions should be named *transferase* and the names formed on the pattern "donor:acceptor *group-transferred-transferase*," e.g., ATP:acetate phosphotransferase (EC 2.7.2.1). A figure may be prefixed to show the position to which the group is transferred, e.g., ATP:D-fructose 1-phosphotransferase (EC 2.7.1.3). The spelling *transphorase* should not be used. In the case of the phosphotransferases, ATP should always be named as the donor. In the case of the aminotransferases involving 2-oxoglutarate, the latter should always be named as the acceptor.

21. The prefix denoting the group transferred should as far as possible, be noncommittal with respect to the mechanism of the transfer, e.g., *phospho-*.

(b) RULES AND GUIDELINES FOR PARTICULAR CLASSES OF ENZYMES (continued)

Guidelines for recommended names	Rules for systematic nomenclature on which the classification is based

Class 3

22. The direct addition of *-ase* to the name of the substrate generally denotes a hydrolase. Where this is difficult, e.g., for EC 3.1.2.1, the word *hydrolase* may be used. Enzymes should not normally be given separate names merely on the basis of optimal conditions for activity. The acid and alkaline phosphatases (EC 3.1.3.1 to 3.1.3.2) should be regarded as special cases and not as examples to be followed. The recommended name *lysozyme* is also exceptional.

22. Hydrolyzing enzymes should be systematically named on the pattern *substrate hydrolase*. Where the enzyme is specific for the removal of a particular group, the group may be named as a prefix, e.g., adenosine *amino*hydrolase (EC 3.5.4.4). In a number of cases this group can also be transferred by the enzyme to other molecules, and the hydrolysis itself might be regarded as a transfer of the group to water.

Class 4

23. The old names *decarboxylase, aldolase*, etc., are retained; and *dehydratase* (not *dehydrase*) is used for the hydro-lyases. *Synthetase* should not be used for any enzymes in this class. The term *synthase* may be used instead for any enzyme in this class (or any class other than Class 6) when it is desired to emphasize the synthetic aspect of the reaction.

23. Enzymes removing groups from substrates nonhydrolytically, leaving double bonds (or adding groups to double bonds) should be called *lyases* in the systematic nomenclature. Prefixes such as *hydro-* and *ammonia-* should be used to denote the type of reaction, e.g., L-malate *hydro*-lyase (EC 4.2.1.2). Decarboxylases should be regarded as carboxy-lyases. A hyphen should always be written before *lyase* to avoid confusion with hydrolases, carboxylases, etc.

24. Where the equilibrium warrants it, or where the enzyme has long been named after a particular substrate, the reverse reaction may be taken as the basis of the name, using *hydratase, carboxylase*, etc., e.g., *fumarate hydratase* for EC 4.2.1.2 (in preference to *fumarase*, which suggests an enzyme hydrolyzing fumarate).

24. The complete molecule, not either of the parts into which it is separated, should be named as the substrate.
The part indicated as a prefix to *-lyase* is the more characteristic and usually, but not always, the smaller of the two reaction products. This may either be the removed (saturated) fragment of the substrate molecule, as in *ammonia-, hydro-, thiol*-lyases, etc. or the remaining unsaturated fragment, e.g., in the case of *carboxy-, aldehyde-,* or *oxo acid*-lyases.

25. Various subclasses of the lyases include a number of strictly specific or group-specific pyridoxal-5-phosphate enzymes that catalyze *elimination* reactions of β- or γ-substituted α-amino acids. Some closely related pyridoxal-5-phosphate-containing enzymes, e.g, tryptophan synthase (EC 4.2.1.20) and cystathionine β-synthase (EC 4.2.1.22) catalyze *replacement* reactions in which a β- or γ-substituent is replaced by a second reactant without creating a double bond. Formally, these enzymes appear to be transferases rather than lyases. However, there is evidence that in these cases the elimination of the β- or γ-substituent and the formation of an unsaturated intermediate is the first step in the reaction. Thus, applying Rule 13, these enzymes are correctly classified as lyases.

Class 5

26. In this class, the recommended names are, in general, similar to the systematic names which indicate the basis of classification.

27. *Isomerase* will be used as a general name for enzymes in this class. The types of isomerization will be indicated by prefixes, e.g., maleate *cis-trans*-isomerase (EC 5.2.1.1), phenylpyruvate keto-enol-

(b) RULES AND GUIDELINES FOR PARTICULAR CLASSES OF ENZYMES (continued)

Guidelines for recommended names	Rules for systematic nomenclature on which the classification is based

Class 5 (continued)

isomerase (EC 5.3.2.1), 3-oxo-steroid Δ^4-Δ^5-isomerase (EC 5.3.3.1). Enzymes catalyzing an aldose-ketose interconversion will be known as *ketol-isomerases*, e.g., L-arabinose ketol-isomerase (EC 5.3.1.4). When the isomerization consists of an intramolecular transfer of a group, the enzyme will be named a *mutase*, e.g., EC 5.4.1.1, and when it consists of an intramolecular lyase-type reaction, e.g., EC 5.5.1.1, it will be named a *lyase (decyclizing)*.

28. Isomerases catalyzing inversions at asymmetric centers should be termed *racemases* or *epimerases*, according to whether the substrate contains one, or more than one, center of asymmetry; compare, for example, EC 5.1.1.5 with EC 5.1.1.7. A numerical prefix to the word *epimerase* should be used to show the position of the inversion.

Class 6

29. Since the enzymes of this class are still generally known as *synthetases*, this designation is retained, in order to avoid extensive changes in the existing nomenclature. *Synthetase* should not be used for enzymes which do not involve nucleoside triphosphates (see Rule 23).

29. The class of enzymes catalyzing the linking together of two molecules, coupled with the breaking of a pyrophosphate link in ATP, etc. should be known as *ligases*. These enzymes have previously been known as *synthetases;* however, this terminology differs from all other systematic enzyme names in that it is based on the product and not on the substrate. For these reasons, a new systematic class name was necessary.

30. The recommended names should be formed on the pattern *X-Y synthetase,* where X-Y is the substance formed by linking X and Y. Exceptionally where Y is CO_2 the name *X carboxylase* may be used. Names of the type *X-activating enzyme* will not be used (see Rule 5).

30. The systematic names should be formed on the pattern *X:Y ligase (ADP-forming)*, where X and Y are the two molecules to be joined together. The phrase shown in parentheses indicates both that ATP is the triphosphate involved, and also that the terminal pyrophosphate link is broken. Thus, the reaction is X + Y + ATP = X–Y + ADP + P.

31. In the special case where glutamine acts as an ammonia donor, this is indicated by adding in parentheses (*glutamine-hydrolyzing*).

31. In this case, the name *amido-ligase* should be used in the systematic nomenclature.

5. Key to Numbering and Classification of Enzymes

1. Oxidoreductases

 1.1 *Acting on the CH-OH group of donors*
   1.1.1 With NAD^+ or $NADP^+$ as acceptor
   1.1.2 With a cytochrome as acceptor
   1.1.3 With oxygen as acceptor
   1.1.99 With other acceptors

 1.2 *Acting on the aldehyde or keto group of donors*
   1.2.1 With NAD^+ or $NADP^+$ as acceptor
   1.2.2 With a cytochrome as acceptor
   1.2.3 With oxygen as acceptor
   1.2.4 With a disulphide compound as acceptor
   1.2.7 With an iron-sulphur protein as acceptor
   1.2.99 With other acceptors

1.3   Acting on the CH-CH group of donors
        1.3.1   With NAD⁺ or NADP⁺ as acceptor
        1.3.2   With a cytochrome as acceptor
        1.3.3   With oxygen as acceptor
        1.3.7   With an iron-sulphur protein as acceptor
        1.3.99 With other acceptors

1.4   Acting on the $CH-NH_2$ group of donors
        1.4.1   With NAD⁺ or NADP⁺ as acceptor
        1.4.3   With oxygen as acceptor
        1.4.4   With a disulphide compound as acceptor
        1.4.99 With other acceptors

1.5   Acting on the CH-NH group of donors
        1.5.1   With NAD⁺ or NADP⁺ as acceptor
        1.5.3   With oxygen as acceptor
        1.5.99 With other acceptors

1.6   Acting on NADH or NADPH
        1.6.1   With NAD⁺ or NADP⁺ as acceptor
        1.6.2   With a cytochrome as acceptor
        1.6.4   With a disulphide compound as acceptor
        1.6.5   With a quinone or related compound as acceptor
        1.6.6   With a nitrogenous group as acceptor
        1.6.7   With an iron-sulphur protein as acceptor
        1.6.99 With other acceptors

1.7   Acting on other nitrogenous compounds as donors
        1.7.2   With a cytochrome as acceptor
        1.7.3   With oxygen as acceptor
        1.7.7   With an iron-sulphur protein as acceptor
        1.7.99 With other acceptors

1.8   Acting on a sulphur group of donors
        1.8.1   With NAD⁺ or NADP⁺ as acceptor
        1.8.2   With a cytochrome as acceptor
        1.8.3   With oxygen as acceptor
        1.8.4   With a disulphide compound as acceptor
        1.8.5   With a quinone or related compound as acceptor
        1.8.6   With a nitrogenous group as acceptor
        1.8.7   With an iron-sulphur protein as acceptor
        1.8.99 With other acceptors

1.9   Acting on a haem group of donors
        1.9.3   With oxygen as acceptor
        1.9.6   With a nitrogenous group as acceptor
        1.9.99 With other acceptors

1.10   Acting on diphenols and related substances as donors
        1.10.2 With a cytochrome as acceptor
        1.10.3 With oxygen as acceptor

1.11   Acting on hydrogen peroxide as acceptor

1.12   Acting on hydrogen as donor
        1.12.1 With NAD⁺ or NADP⁺ as acceptor
        1.12.2 With a cytochrome as acceptor
        1.12.7 With an iron-sulphur protein as acceptor

1.13   Acting on single donors with incorporation of molecular oxygen (oxygenases)
        1.13.11 With incorporation of two atoms of oxygen

1.13.12 With incorporation of one atom of oxygen (internal monooxygenases or internal mixed function oxidases)
1.13.99 Miscellaneous (requires further characterization)

1.14 *Acting on paired donors with incorporation of molecular oxygen*
1.14.11 With 2-oxoglutarate as one donor and incorporation of one atom each of oxygen into both donors
1.14.12 With NADH or NADPH as one donor and incorporation of two atoms of oxygen into one donor
1.14.13 With NADH or NADPH as one donor and incorporation of one atom of oxygen
1.14.14 With reduced flavin or flavoprotein as one donor and incorporation of one atom of oxygen
1.14.15 With a reduced iron-sulphur protein as one donor and incorporation of one atom of oxygen
1.14.16 With reduced pteridine as one donor and incorporation of one atom of oxygen
1.14.17 With ascorbate as one donor and incorporation of one atom of oxygen
1.14.18 With another compound as one donor and incorporation of one atom of oxygen
1.14.99 Miscellaneous (requires further characterization)

1.15 *Acting on superoxide radicals as acceptor*

1.16 *Oxidizing metal ions*
1.16.3 With oxygen as acceptor

1.17 *Acting on $-CH_2-$ groups*
1.17.1 With NAD^+ or $NADP^+$ as acceptor
1.17.4 With a disulphide compound as acceptor

2. Transferases

2.1 *Transferring one-carbon groups*
2.1.1 Methyltransferases
2.1.2 Hydroxymethyl-, formyl-, and related transferases
2.1.3 Carboxyl- and carbamoyltransferases
2.1.4 Amidinotransferases

2.2 *Transferring aldehyde or ketonic residues*

2.3 *Acyltransferases*
2.3.1 Acyltransferases
2.3.2 Aminoacyltransferases

2.4 *Glycosyltransferases*
2.4.1 Hexosyltransferases
2.4.2 Pentosyltransferases
2.4.99 Transferring other glycosyl groups

2.5 *Transferring alkyl or aryl groups, other than methyl groups*

2.6 *Transferring nitrogenous groups*
2.6.1 Aminotransferases
2.6.3 Oximinotransferases

2.7 *Transferring phosphorus-containing groups*
2.7.1 Phosphotransferases with an alcohol group as acceptor
2.7.2 Phosphotransferases with a carboxyl group as acceptor
2.7.3 Phosphotransferases with a nitrogenous group as acceptor
2.7.4 Phosphotransferases with a phospho-group as acceptor
2.7.5 Phosphotransferases with regeneration of donors (apparently catalyzing intramolecular transfers)

        2.7.6   Diphosphotransferases
        2.7.7   Nucleotidyltransferases
        2.7.8   Transferases for other substituted phospho-groups
        2.7.9   Phosphotransferases with paired acceptors

   2.8 *Transferring sulphur-containing groups*
        2.8.1   Sulphurtransferases
        2.8.2   Sulphotransferases
        2.8.3   CoA-transferases

3. **Hydrolases**

   3.1 *Acting on ester bonds*
        3.1.1   Carboxylic ester hydrolases
        3.1.2   Thiolester hydrolases
        3.1.3   Phosphoric monoester hydrolases
        3.1.4   Phosphoric diester hydrolases
        3.1.5   Triphosphoric monoester hydrolases
        3.1.6   Sulphuric ester hydrolases
        3.1.7   Diphosphoric monoester hydrolases

   3.2 *Acting on glycosyl compounds*
        3.2.1   Hydrolysing O-glycosyl compounds
        3.2.2   Hydrolysing N-glycosyl compounds
        3.2.3   Hydrolysing S-glycosyl compounds

   3.3 *Acting on ether bonds*
        3.3.1   Thioether hydrolases
        3.3.2   Ether hydrolases

   3.4 *Acting on peptide bonds (peptide hydrolases)*
        3.4.11   α-Aminoacylpeptide hydrolases
        3.4.12   Peptidylamino-acid or acylamino-acid hydrolases
        3.4.13   Dipeptide hydrolases
        3.4.14   Dipeptidylpeptide hydrolases
        3.4.15   Peptidyldipeptide hydrolases
        3.4.21   Serine proteinases
        3.4.22   SH-proteinases
        3.4.23   Acid proteinases
        3.4.24   Metalloproteinases
        3.4.99   Proteinases of unknown catalytic mechanism

   3.5 *Acting on carbon-nitrogen bonds, other than peptide bonds*
        3.5.1   In linear amides
        3.5.2   In cyclic amides
        3.5.3   In linear amidines
        3.5.4   In cyclic amidines
        3.5.5   In nitriles
        3.5.99   In other compounds

   3.6 *Acting on acid anhydrides*
        3.6.1   In phosphoryl-containing anhydrides
        3.6.2   In sulphonyl-containing anhydrides

   3.7 *Acting on carbon-carbon bonds*
        3.7.1   In ketonic substances

   3.8 *Acting on halide bonds*
        3.8.1   In C-halide compounds
        3.8.2   In P-halide compounds

   3.9 *Acting on phosphorus-nitrogen bonds*

3.10   Acting on sulphur-nitrogen bonds

3.11   Acting on carbon-phosphorus bonds

4. **Lyases**

   4.1   Carbon-carbon lyases
       4.1.1   Carboxy-lyases
       4.1.2   Aldehyde-lyases
       4.1.3   Oxo-acid-lyases
       4.1.99   Other carbon-carbon lyases

   4.2   Carbon-oxygen lyases
       4.2.1   Hydro-lyases
       4.2.2   Acting on polysaccharides
       4.2.99   Other carbon-oxygen lyases

   4.3   Carbon-nitrogen lyases
       4.3.1   Ammonia-lyases
       4.3.2   Amidine-lyases

   4.4   Carbon-sulphur lyases

   4.5   Carbon-halide lyases

   4.6   Phosphorus-oxygen lyases

   4.99   Other lyases

5. **Isomerases**

   5.1   Racemases and epimerases
       5.1.1   Acting on amino acids and derivatives
       5.1.2   Acting on hydroxy acids and derivatives
       5.1.3   Acting on carbohydrates and derivatives
       5.1.99   Acting on other compounds

   5.2   Cis-trans isomerases

   5.3   Intramolecular oxidoreductases
       5.3.1   Interconverting aldoses and ketoses
       5.3.2   Interconverting keto- and enol-groups
       5.3.3   Transposing C=C bonds
       5.3.4   Transposing S–S bonds
       5.3.99   Other intramolecular oxidoreductases

   5.4   Intramolecular transferases
       5.4.1   Transferring acyl groups
       5.4.2   Transferring phosphoryl groups
       5.4.3   Transferring amino groups
       5.4.99   Transferring other groups

   5.5   Intramolecular lyases

   5.99   Other isomerases

6. **Ligases (synthetases)**

   6.1   Forming carbon-oxygen bonds
       6.1.1   Ligases forming aminoacyl-tRNA and related compounds

   6.2   Forming carbon-sulphur bonds
       6.2.1   Acid-thiol ligases

6.3   Forming carbon-nitrogen bonds
    6.3.1   Acid-ammonia ligases (amide synthetases)
    6.3.2   Acid-amino-acid ligases (peptide synthetases)
    6.3.3   Cyclo-ligases
    6.3.4   Other carbon-nitrogen ligases
    6.3.5   Carbon-nitrogen ligases with glutamine as amido-N-donor

6.4   Forming carbon-carbon bonds

6.5   Forming phosphate ester bonds

UNITS OF ENZYMIC ACTIVITY

The presence of an enzyme is generally recognized by the occurrence of the chemical reaction that it catalyzes, and the amount of the enzyme present may be determined by measuring the rate of the reaction, which under suitable conditions, is proportional to this quantity.

In the older literature, the rate of reaction was expressed in a variety of ways, often directly related to measured quantities such as change in absorbance, volume of gas absorbed or evolved, or volume of a titration reagent used.

The Enzyme Commission, in 1961, proposed that the reaction rate should be expressed as micromoles of substrate converted per minute, and the enzyme unit was defined as "that amount (of enzyme) which will catalyze the transformation of one micromole of the substrate per minute under standard conditions."

In the above definition, the precise meaning of the term "amount (of enzyme)" is not specified. Since the definition is meant to apply to enzymes that have never been isolated and weighed, "amount" cannot be identified with mass. It certainly does not correspond to the chemical concept of "amount of substance," which can only be defined if the chemical identity and the molecular weight have been established. "Amount" is then, in this context, a completely abstract concept that is useful for certain computational purposes; it is *not* a determinable physicochemical quantity.

On the other hand, many enzymes have now been isolated, weighed, and characterized chemically. It has therefore become desirable to regard enzymic activity in the same manner as other physicochemical properties; it should be defined operationally and bear a definite relationship to the mass and the amount of substance.

It is accordingly recommended that:

    a. the concept of "enzyme unit" as a physically undefined "amount of enzyme" be abandoned;

    b. *enzymic activity* be defined as the rate of reaction of substrate that may be attributed to catalysis by an enzyme;

    c. the phenomenological coefficient that relates the activity *under specified conditions* to the mass, be called *specific activity;*

    d. the phenomenological coefficient that relates the activity *under specified conditions* to the amount of enzyme substance (usually expressed in moles) be called the *molar activity.*

The activity of enzymes that have not been isolated can be measured, but not resolved into its constituent factors, mass and specific activity.

1. The Katal as the Unit of Enzymic Activity

It is recommended that the unit in which enzymic activity is expressed be the amount of activity that converts one mole* of substrate per second. The new unit of enzymic activity is named the *katal* (symbol: *kat*).

The Commission realizes that an activity of one katal, corresponding to the conversion of one mole of substrate per second, will very often be too great for practical use. In most cases, activities will be expressed in *microkatals* (μkat), *nanokatals* (nkat), or *picokatals* (pkat) corresponding to reaction rates of micromoles, nanomoles, or picomoles per second, respectively.

Enzymic activities expressed in the former *enzyme units* (U) and in *katals* may be interconverted by one of the following relationships:

$$1 \text{ kat} = 1 \text{ mol/s} = 60 \text{ mol/min} = 60 \times 10^6 \text{ } \mu\text{mol/min} \simeq 6 \times 10^7 \text{ U}$$

or

$$1 \text{ U} \simeq 1 \text{ } \mu\text{mol/min} = \frac{1}{60} \text{ } \mu\text{mol/s} = \frac{1}{60} \text{ } \mu\text{kat} = 16.67 \text{ nkat}.$$

It is recognized that the old enzyme unit will remain in use during a transition period. However, its use should be discouraged progressively so that it may eventually be abandoned.

The *katal*, being based on the second as unit of time, fits much better with the rate constants used in chemical kinetics than does the old enzyme unit, since reaction rates are commonly expressed in moles (or appropriate submultiples of the mole) per second.

Ambiguities might arise in some cases from the expression *conversion of one mole of substrate* . . . . . , for instance in a bimolecular reaction of the type 2A = B + C, or with polymers like starch from which monomers are removed consecutively. In such cases, the rate of reaction may be expressed as the *number of cycles of a stated reaction per second*, the conversion of one mole of substrate being normally equivalent to *as many reaction cycles as there are carbon atoms in 0.012 kg of the pure nuclide*[12]*C*. In other words, in the case of a reaction 2A = B + C, two moles of A converted per second represent an activity of 1 katal. In the second example, one mole of reducing group formed from starch, corresponding to the number of glycosidic bonds broken, forms the logical basis for calculating activity.

It is recommended that enzyme assays be based, wherever possible, on measurements of initial rates and not on amounts of substrate converted by the end of a given period of time, unless it is known that the reaction rate remains constant throughout this period. If the reaction rate falls off appreciably during the assay, for example, because of the formation of inhibitory products, or, in reversible reactions, because the reverse reaction is no longer negligible, the amount of substrate converted is not proportional to the activity of the catalyst.

2. Derived Quantities

A number of other quantities are related to the enzymic activity, as follows:

The *specific activity* of an enzyme preparation that has been defined above, will normally be expressed as katals per kilogram of protein, or suitable multiples thereof, e.g., microkatals per kilogram (μkat/kg).

The *molar activity* of an enzyme is expressed as katals per mole of enzyme.

The *concentration of enzymic activity* in a solution is defined as activity divided by

---

*The *mole* has been defined as the *amount of substance* of a system that contains as many elementary entities as there are carbon atoms in 0.012 kilograms of the nuclide [12]C.

volume of solution and is expressed as katals per liter, or suitable multiples thereof, e.g., microkatals per liter (μkat/l).

SYMBOLS OF ENZYME KINETICS

Obviously, it would be of great advantage if all authors used the same system of symbols in their mathematical equations. Therefore, the Commission on Enzyme Nomenclature (1961) made recommendations for symbols representing often used quantities, such as rates of reaction, rate constants, etc.

These recommendations have been widely followed. They are repeated here in abbreviated form.

1. Rates of Reaction

The observed *rate* of an enzyme reaction will be represented as v.* The rate of reaction referred to in the Michaelis Equation (see below) is the initial rate, which is denoted v_0, although where no confusion will be caused, the subscript may be omitted. v_0 may be referred to as *the initial steady-state rate*.

The maximum value of v_0 at infinite substrate concentration (i.e., the maximum value as defined by the Michaelis Equation) will be represented by V.

2. Combination of Enzymes with Substrate: Michaelis Constant

Since the days of Michaelis and Menten, it has been more or less standard practice to describe equilibria between enzyme and substrate by use of dissociation constants according to the general equation

$$K = \frac{[E][S]}{[ES]}$$

where [E], [S] and [ES] represent the concentrations at equilibrium of free enzyme, free substrate, and active complex, respectively. Under normal conditions, [S] represents the initial total concentration of substrate.

The general use of dissociation constants does not conform with the convention used by physical chemists and adopted (although not specifically in connection with enzymology) by the International Union of Pure and Applied Chemistry. However, for reasons stated *in extenso* in the Recommendations of 1964, the consistent use of dissociation constants in enzymology is to be preferred over the alternation of association and dissociation constants as required by the systems of physical chemistry. The rule that equilibria should be expressed as dissociation constants also applies to the combination of enzyme with inhibitor and enzyme with product or vice versa, dissociation of enzyme-product complex to enzyme plus product.

One prolific source of misunderstanding has been the indiscriminate use of the name *Michaelis constant* and the two symbols K_s and K_m for two quite different quantities; one of these quantities is the equilibrium (dissociation) constant of the reaction ES = E + S; the other is the substrate concentration at which $v = V/2$. The two quantities are not equal unless the breakdown of ES is sufficiently slow to enable a true equilibrium of ES with free E and S to be set up.

In order to avoid such confusion, the following recommendations were made by the 1961 Commission and are endorsed by the Commission on Biochemical Nomenclature:

---

*In physical chemical usage, the symbol ξ is used for the rate of reaction; however since v is widely used by biochemists the Commission on Biochemical Nomenclature does not recommend any change in established practice.

The name Michaelis constant and the symbol K_m are used for the substrate concentration at which $v = V/2$.

The name substrate constant and the symbol K_s are used for the equilibrium (dissociation) constant of the reaction ES = E + S.

The name inhibition constant and the symbol K_i are used for the equilibrium (dissociation) constant of the reaction EI = E + I.

3. Rate Constants

In accordance with established convention, small k will be used to represent a rate constant, capital K an equilibrium constant.

Unfortunately, a number of different systems of numbering rate constants for the sequential stages of enzyme reactions have been used; three systems were described in Enzyme Nomenclature (1964). In the light of the recommendation given in *The Manual of Symbols and Terminology for Physicochemical Quantities and Units* (1970), produced by the International Union of Pure and Applied Chemistry, that "elementary processes should be labeled in such a manner that reversed processes are immediately recognizable," the following system is now recommended:

$$E + S \underset{k_{-1}}{\overset{k_{+1}}{\rightleftharpoons}} ES \underset{k_{-2}}{\overset{k_{+2}}{\rightleftharpoons}} EP \underset{k_{-3}}{\overset{k_{+3}}{\rightleftharpoons}} E + P$$

This system is symmetrical, and cannot be confused with any of the systems used, which do not include both + and − signs. However, where no ambiguity would be caused, the + signs may be omitted.

ES and EP formally represent forms of the enzyme that have arisen from the free enzyme by reaction with the substrate. They may retain all or only a part of the substrate molecule. The fraction of the enzyme in the form of ES or EP is represented by the saturation function Y_s. During the initial steady state (i.e., [P] = 0),

$$\overline{Y}_s = \frac{\alpha}{1 + \alpha}, \text{ where } \alpha = s/K_m.$$

An enzyme may exist in two conformations, in equilibrium with one another.

$$R \rightleftharpoons T$$

Both conformations may be enzymically active but with different rate constants. The symbol c is used for the ratio $K_m^R : K_m^T$ and L for the ratio [T]:[R], in the absence of substrate. If, even in presence of substrate, the equilibrium between free enzyme in the R and T forms is maintained, then, during the initial steady state

$$\overline{Y}_s = \frac{\alpha(1 + cL)}{(1 + L) + \alpha(1 + cL)}$$

The fraction of the enzyme in the R state is represented by the state function, $\overline{R}$ which is equal to

$$\frac{1 + \alpha}{(1 + L) + \alpha(1 + cL)}$$

4. Cooperative Effects

When an enzyme molecule contains more than one substrate-binding site, cooperative (both positive and negative) binding may occur. This will not be the case when an enzyme molecule is made up of independently reacting subunits, but can occur when an oligomeric protein can exist in two conformations. For the special case, where the free enzyme in the R and T states are in equilibrium even in the presence of substrate

$$\overline{Y}_s = \frac{\alpha(1+\alpha)^{n-1} + Lc\alpha(1+c\alpha)^{n-1}}{(1+\alpha)^n + L(1+c\alpha)^n}$$

and

$$\overline{R} = \frac{(1+\alpha)^n}{(1+\alpha)^n + L(1+c\alpha)^n}$$

where n is the number of identical substrate-binding sites in the oligomer.

If c is small and L large, or vice versa, positive cooperativity, characterized by a sigmoidal $\overline{Y}_s$ against α plot, will be obtained. The degree of cooperativity will be larger the greater the value of n.

The degree of cooperativity is often calculated from the so-called Hill plot of log $\overline{Y}_s/(1-\overline{Y}_s)$ against log α. The slope of the line thus obtained is equal to 1 at low and high values of α and reaches a maximum value at intermediate values of α. The maximum value of the slope is called the Hill coefficient, but this can be mistaken for the number of subunits. In general, the two values are not identical. It is recommended then, that the symbol h be used for the Hill coefficient.

Cooperativity, both positive and negative, is also found when the reaction of one molecule of substrate with the enzyme induces a conformation change leading to greater or less reactivity with a second molecule. Substrate inhibition is a special case of negative cooperativity.

The general equation for sigmoidal v_o against s curves is

$$v_o^{-1} = a + bs^{-2} + cs^{-3} \quad \ldots\ldots\ldots\ldots$$

A rate equation of this form may also be obtained when the enzyme catalyzes the reaction between two different substrates, for each of which there is only one binding site in the enzyme, when alternative pathways for the enzyme-catalyzed reaction exist.

5. List of Recommended Symbols for Enzyme Kinetics

E	=	Free enzyme
S	=	Free substrate
ES	=	Enzyme-substrate complex
EP	=	Enzyme-product complex
R,T	=	Two conformational forms of an enzyme
e	=	Total enzyme concentration
s	=	Total substrate concentration
i	=	Total inhibitor concentration
v	=	Rate of reaction catalyzed by an enzyme
v_o	=	Initial steady-state reaction rate
V	=	Value of v when the enzyme is saturated with substrate, as given by the Michaelis equation
K_m	=	*Michaelis constant*, i.e., concentration of substrate at which $v = V/2$
K_s	=	*Substrate constant*, i.e., equilibrium (dissociation) constant of the reaction ES = E + S
K_i	=	*Inhibitor constant*, i.e., equilibrium (dissociation) constant of the reaction EI = E + I
k_{+n}, k_{-n}	=	Rate constants of the forward and backward reactions in the nth step of an enzyme reaction

c	=	$K_m^R : K_m^T$, where R and T refer to two conformations of the enzyme
L	=	[T]:[R] in absence of substrate
α	=	s/K_m
$\bar{Y}_s$	=	Saturation function
$\bar{R}$	=	Fraction of enzyme in the R state
n	=	Number of identical substrate-binding sites in an oligomeric enzyme
h	=	Hill coefficient

NOMENCLATURE OF ELECTRON-TRANSFER PROTEINS

1. General Introduction

The processes of oxidation in living cells are catalyzed by enzyme systems that transfer hydrogen atoms or electrons in successive steps from an initial donor to a final acceptor. Examples of such systems are

1. Oxidation of intermediary metabolites by molecular oxygen in the mitochondria of animal, plant, and protist cells, and also in the protoplasmic membranes of those protists whose cells do not contain mitochondria. This enzyme system is commonly referred to as the respiratory chain.

2. Light-driven oxidation of water in chloroplasts of green plants and in membranes of blue-green protists, and the light-driven oxidation of water, or other suitable reductants, in the chromatophores of certain protists. These enzyme systems are commonly referred to as the photosynthetic chain.

3. The oxygenation of compounds by the introduction of one or both of the atoms of molecular oxygen.

4. The reduction of cytidinediphosphate to deoxycytidinediphosphate by NADH.

The initial and final steps of electron transfer in these chains of enzymes are catalyzed by discrete enzymes that can be separated from the other components and studied by conventional methods of enzymology. The initial dehydrogenation of a substrate such as succinate, or a reduced coenzyme such as NADH, is catalyzed by a dehydrogenase that is relatively specific for the hydrogen donor but not for the acceptor. Thus, artificial acceptors such as ferricyanide, phenazine methosulphate, and methylene blue are often used in the study of dehydrogenases. Similarly, ferricytochrome c and quinones find application as electron acceptors even with enzymes that do not catalyze transfer of electrons to such acceptors in vivo. For example, in the mitochondrial respiratory chain, electrons from NADH are transferred to ubiquinone by an enzyme complex that could be given the systematic name NADH:ubiquinone oxidoreductase. However, although this enzyme complex (Complex I) can be separated from the rest of the respiratory chain, the isolated complex reacts sluggishly with added Q-10. Preparations of high particle weight, isolated by digestion of mitochondrial membranes with phospholipase at 30°C, actively catalyze the oxidation of NADH by ferricyanide but react very slowly with Q-10 or even with the more soluble homologues of lower molecular weight. Since the transfer of electrons from NADH to cytochrome c requires the participation of a second complex (Complex III), it is not surprising that the isolated enzyme catalyzing the oxidation of NADH by ferricyanide does not react with ferricytochrome c. However, it acquires this property after certain treatments (e.g., with acid ethanol) that lead to the splitting off from the complex of high molecular weight of a smaller protein molecule. Although this smaller molecule retains its ability to react with NADH, EPR spectrometry shows that the iron-sulphur centers are modified by the extraction. It is possible then that the ability to react with cytochrome c is an artifact. In any case, it is unlikely that a single protein of low molecular weight catalyzes the oxidation of NADH by cytochrome c in the intact mitochondria, since this span of the respiratory chain is linked with two phosphorylation steps and contains a number of other electron carriers. It is, then, misleading to name the enzyme NADH:cytochrome c oxidoreductase, although there is no objection to referring to the NADH-(ferri)cytochrome c (oxido)reductase activity.

The protein of low molecular weight also catalyzes the reduction by NADH of ubiquinone homologues of low molecular weight. Although this would appear to provide a basis for naming NADH:ubiquinone oxidoreductase, this reaction, unlike that between NADH and Q-10 in the mitochondrial membrane, is not inhibited by rotenone. Thus, it is likely that a step involved in the reaction between NADH and ubiquinone in the intact membrane is by-passed in the reaction catalyzed by the isolated protein of low molecular weight. Naming it NADH:ubiquinone oxidoreductase would, therefore, be misleading, although once again there is no objection to referring to NADH-Q (oxido)reductase activity.

For these reasons, the enzyme responsible for the primary dehydrogenation of NADH in the mitochondrial respiratory chain is named simply NADH dehydrogenase (systematic name, NADH:(acceptor) oxidoreductase, EC 1.6.99.3).

The subsequent steps of electron transfer in the mitochondrial respiratory chain are catalyzed by electron-transfer carriers, both nonprotein (ubiquinone) and protein (flavoproteins, cytochromes, copper, and iron-sulphur). Although the protein carriers, being catalytically active proteins, satisfy the most all-embracing definition of an enzyme, many do not readily fit in with the scheme of enzyme nomenclature, since they catalyze hydrogen or electron transfer from another enzyme to yet a third enzyme. Moreover, since much is known about the electron-carrying center of these enzymes, it is more appropriate to classify them on the basis of chemical structure of the prosthetic groups and the manner of their attachment to the protein.

Some electron-transfer carriers can be readily classified both on the basis of chemical structure and as an enzyme. Those that catalyze the initial dehydrogenation of substrate or coenzyme in the respiratory chain, such as succinate dehydrogenase (EC 1.3.99.1) or NADH dehydrogenase (EC 1.6.99.3), or the final reduction of $NADP^+$ in the photosynthetic chain (ferredoxin-$NADP^+$ reductase, EC 1.6.7.1) are clear examples. The enzyme catalyzing the final electron transfer to oxygen (cytochrome c oxidase, EC 1.9.3.1) has also long been classified as an enzyme, even though it contains four electron-carrying centers per catalytic unit (two distinguishable copper atoms, two distinguishable haem a groups), and there are insufficient reasons for discontinuing current practice in this respect. The analogous Complex III (see above) has not, however, been listed as QH_2:ferricytochrome c reductase, since it is usual to consider cytochromes b and c_1, which are present in this complex, as more discrete identities than cytochromes a and a_3 in cytochrome c oxidase (also called Complex IV). This is, however, clearly a borderline case, and increasing knowledge may well necessitate a change in this standpoint. For the time being, it is recommended that the terms QH_2-cytochrome reductase system or activity should be used. The same terminology can be applied to larger segments of the chain, for example, the NADH-cytochrome c reductase system or even the succinate oxidase system to refer to those segments lying between NADH and cytochrome c, and between succinate and oxygen, respectively.

Six types of electron-transfer proteins have been identified:

1. Flavoproteins
2. Proteins containing reducible disulphide groups
3. Cytochromes
4. Iron-sulphur proteins
5. Cuproteins
6. Molybdoproteins

2. Flavoproteins

Flavoproteins contain two types of prosthetic groups, FMN (e.g., NADH dehydrogenase, EC 1.6.99.3) and FAD. The FMN is noncovalently bound. FAD may be noncovalently bound [e.g., in lipoamide dehydrogenase (NADH), EC 1.6.4.3] or covalently bound via a methylene bridge between the benzene ring of the isoallooxazine and an amino acid residue in the protein (e.g., succinate dehydrogenase, EC 1.3.99.1).

Those flavoproteins that are known to catalyze well-defined chemical reactions are classified in the appropriate subgroup of the list of enzymes, with one exception. This is the so-called electron-transfer flavoprotein that catalyzes the transfer of electrons from butyryl-CoA dehydrogenase (EC 1.3.99.2), acyl-CoA dehydrogenase (EC 1.3.99.3), sarcosine dehydrogenase (EC 1.5.99.1), and dimethylglycine dehydrogenase (EC 1.5.99.2) to the respiratory chain. Its function is quite analogous to that of the cytochromes. Flavoproteins that form part of electron-transfer systems are non-autoxidizable. In addition, flavoprotein oxidases, catalyzing the direct oxidation of substrates by oxygen, are known (e.g., D-amino-acid oxidase, EC 1.4.3.3).

3. Proteins Containing Reducible Disulphide

Lipoylproteins containing lipoic acid covalently bound by an acid amide link between its carboxyl group and the 6-amino group of a lysine residue in the protein are involved in the oxidation of both pyruvate and 2-oxoglutarate. The disulphide group of the lipoic acid is both reduced and acylated by pyruvate dehydrogenase (lipoate) (EC 1.2.4.1) and oxoglutarate dehydrogenase (EC 1.2.4.2.). Thus, lipoylproteins act as both hydrogen and acyl acceptors.

Proteins of low molecular weight containing reducible cystine bridges are involved in the transfer of hydrogen atoms from reduced NADPH dehydrogenase to enzymes catalyzing the reduction of "active sulphate" to sulphite, methionine sulphoxide to methionine, and cytidinediphosphate to deoxycytidinediphosphate. In the latter case, the cystine-containing carrier has been called thioredoxin.

4. Cytochromes

A cytochrome is defined as a haemoprotein whose characteristic mode of action involves transfer of reducing equivalents, associated with a reversible change in oxidation state of the prosthetic group. Four major groups of cytochromes have been established:

1. Cytochromes *a*. Cytochromes in which the haem prosthetic group contains a formyl side-chain, e.g., haem *a*.

2. Cytochromes *b*. Cytochromes with protohaem or a related haem (without formyl group) as prosthetic group, not covalently bound to protein.

3. Cytochromes *c*. Cytochromes with covalent linkages between the haem side-chains and protein. This group includes all cytochromes with prosthetic groups linked in this way, not only those with thioether linkages.

4. Cytochromes *d*. Cytochromes with a tetrapyrollic chelate of iron as prosthetic group in which the degree of conjugation of double bonds in less than is porphyrin, e.g., dihydroporphyrin(chlorin).

Use of the small unprimed italicized letter implies that the haem prosthetic group is in a haemochrome* linkage. To indicate that in both the oxidized and reduced forms the

---

*In the Recommendations (1964) a haemoprotein is defined as a compound of haem in which the fifth and sixth coordination places are each occupied by a nitrogen atom. Since it is now clear that this definition is too restrictive e.g., in cytochrome *c* one of these two places is occupied by a methionine sulphur atom, it becomes necessary to extend this definition. A haemochrome is now defined as a low-spin compound of haem in which the fifth and sixth coordination places are occupied by strong-field ligands. The ferrohaemochrome has a characteristic spectrum in which the absorbance at the α-band maximum is greater than at the β-band maximum.

haem prosthetic group is not in a haemochrome linkage, a primed small letter, e.g., c' is used (see *Enzyme Nomenclature* (1964) for discussion of the "RHP" problem).

In the case of a cytochrome having two or more different haem groups attached to a specific protein, each different haem should be indicated, e.g., cytochrome cd (present in *Pseudomonas aeruginosa*). In the case of a cytochrome having two or more of the same haem groups attached to a specific protein but in different environments, so that one or more is in a haemochrome linkage and one or more in a nonhaemochrome linkage, both types of linkage should be indicated by using both the unprimed and primed small italicized letter, appropriate for the haem in question.† The classification of a haemoprotein as a cytochrome takes precedence over a possible classification on the basis of an additional prosthetic group (e.g., cytochrome b_2 contains FMN, cytochrome aa_3 contains Cu).

Although ever since their discovery by Keilin, it has been customary to assign the cytochromes to the groups a, b, and c (group d was introduced by the IUB Enzyme Commission in 1961), according to the nature and mode of binding of the prosthetic haem moieties, the further classification of cytochromes within the four groups has given rise to difficulties. Thus individual cytochromes were identified by means of consecutive numerical subscripts (e.g., b, b_1, b_2 etc.), an arbitrary convention subject to the vicissitudes of chronology and the whims of investigators. In their 1961 Recommendations, the Enzyme Commission established two principles as a basis for discussion: First, the uncontrolled proliferation of numerical subscripts should be halted; second, the basis of the classification should be chemical rather than functional. In the second edition of these Recommendations (1964) the chapter on cytochromes was again prepared with these principles in mind. At the same time an attempt was made to plan a rational use of subscripts for describing bacterial cytochromes. For example, it was proposed that the two c-type cytochromes isolated from *Azotobacter vinelandii* be no longer referred to as cytochromes c_4 and c_5, respectively, but be included under cytochrome c_2 which, like c_4 and c_5 has an isoelectric point below pH 7, does not react with mammalian cytochrome c oxidase, has an α-band at 550 to 552 nm and a redox potential (E'_o) at pH 7 of 300 to 340 mV.

The experience of the past ten years is that the first aim — to halt the uncontrolled proliferation of numerical subscripts — has been achieved. However, the proposal to include all the bacterial c-type cytochromes of high redox potential under the term cytochrome c_2 has not been followed. The chapter on cytochromes in the 1964 Recommendations included a procedure for allocating names to newly discovered cytochromes. The proposal was that a first, temporary name should specify the wavelength of the α-band of the reduced form together with the source, e.g., haemoprotein 560 (*Bacterium X*). Following elucidation of the cytochrome nature of the new protein, and its group, a final systematic name would be allotted, e.g., cytochrome b (560 *Bacterium X*). For the selection of a trivial name based upon the italic small letter and subscript number, a standing committee was to be set up; it was felt that only in this way could confusion be avoided. This naming procedure has, however, been virtually disregarded by investigators in the cytochrome field who have preferred to use the α-band

---

†The name of a cytochrome normally gives no indication of the number of identical haem molecules contained in a single molecule of the cytochrome. Cytochrome c oxidase (EC 1.9.3.1) contains only one kind of haem (haem a), but in the presence of certain ligands one half of these haem molecules behaves as if bound in haemochrome linkage and the other half in nonhaemochrome linkage. This could be indicated by the name cytochrome aa'. However, it has been suggested that in the absence of any ligand or cytochrome c, all haem groups are in haemochrome linkage. If this is substantiated, cytochrome a would be the correct name. While these questions are still under discussion, the name cytochrome aa_3 should be retained.

wavelength as a subscript, e.g., cytochrome b_{560} or, where the group was still uncertain, cytochrome$_{560}$. Moreover, the standing committee was not set up.

In retrospect, the disadvantages of the recommended procedure have become clear. First, the need for an abbreviated or trivial name arises as soon as a new cytochrome is described. Secondly, the use of consecutive numerical subscripts based upon chronology was useful when comparatively few subscripts were current. However, were the number of subscripts to increase, their uninformative nature would be a growing handicap to comprehension among biochemists generally. Thirdly, the use of wavelengths provides a natural basis for a system of nomenclature that is largely based upon spectrophotometry.

It is now recommended that:

1. At first mention in a publication, the name should be expanded to the systematic name that will include the source in parentheses, e.g., cytochrome b_1 (*Bacterium X*). Since the need has arisen to refer to different b and c type cytochromes present in the same cell or in the same organelle of a given cell type, it will often be necessary to add distinguishing characteristics, such as microsomal, tetrahaem, high-potential, etc.

2. The names of the already well-established cytochromes with consecutive subscript numbering, listed below, are retained. All cytochromes not fitting in this category should be given a name based upon the α-band wavelength (in nm) and written thus: cytochrome c-554.

3. The α-band wavelengths used should be determined at room temperature, not liquid-air temperature, and should, if possible be obtained from absolute absorption spectra of the purified protein under carefully defined conditions. Since an error of 1 nm in assigning the position of the band alters the name, care should be taken to standardize the spectrophotometer with standard lines, e.g., with those given by Nd (III).

Cytochrome P-450

The naming of cytochrome *P*-450, the autoxidizable cytochrome involved in the hydroxylation electron-transport chain, does not conform to these or to earlier recommendations. The 450 refers to the Soret peak of the compound of the reduced cytochrome with CO. Since its prosthetic group is probably protohaem, it presumably belongs to the b group, but until this is settled it is undesirable to make a radical change in the name. Moreover, it is difficult to use the weak α-band as a basis for nomenclature.

Cytochrome o

Cytochrome o is the name given to a group of autoxidizable cytochromes possibly functioning as terminal oxidases in certain bacteria. Many, if not all, belong to the b group.

It is recommended that:

4. The name cytochrome o be discarded and that those now bearing this name be classified in the appropriate main group.

Helicorubin and Cytochrome h

Helicorubin belongs to the cytochrome b group. Whether or not cytochrome h belongs to this group or to the cytochromes c depends upon whether or not the haem prosthetic group is bound by covalent ester linkage.

Future Developments

Information on the chemical structure of the c cytochromes including the amino acid sequences and tertiary structures is accumulating so rapidly that this group may provide a good opportunity for attempting a rational chemical classification, which may turn out to have a phylogenetic basis.

LIST OF CYTOCHROMES

Cytochrome *a* Group

Cytochrome aa_3 — Identical with cytochrome *c* oxidase of eucaryotes. The α-band of the reduced cytochromes is at 605 nm, the γ (Soret) at 444 nm. The reduced cytochrome combines with CO with a shift of the α-band to 590 nm and the γ-band to 430 nm; it also combines with cyanide with a shift of the α-band to 590 nm with little effect on the γ-band. The reduced form is autoxidizable. In the presence of cyanide, one half of the haem molecule is autooxidizable.

Cytochrome a_1 — Present in certain bacteria (e.g., *Acetobacter pasteurianum*). Its ferro form is autoxidizable and combines with cyanide without appreciable change in the position of the α-band (590 nm).

Cytochrome *c* Group

Cytochrome *b* — Present in mitochondria of eucaryotes and in chloroplasts. Different *b* species are indicated by the position of the α-band (in nm) of the reduced species, e.g., *b*-562, *b*-563, *b*-566. Chloroplast cytochrome *b*-563 is also known as cytochrome b_6.

Cytochrome b_1 — Present in certain bacteria (e.g., *Escherichia coli*). Its characteristic absorption band is the α-band of the ferro form at 559 nm.

Cytochrome b_2 — Present in yeast. It contains FMN as a second prosthetic group and acts as a lactate dehydrogenase (EC 1.1.2.3).

Cytochrome b_3 — Present in microsomal material from nonphotosynthetic plant tissues. Its characteristic absorption band is the α-band of the ferro form at 559 nm.

Cytochrome b_5 — Present in animal microsomes. It is reduced by NADH in the presence of cytochrome b_5 reductase (EC 1.6.2.2). Its characteristic band is the α-band at 556 nm.

Cytochrome b_7 — Present in spadices of various *Arum* species. Its characteristic absorption is the α-band of the ferro form at 560 nm. It is autoxidizable.

Cytochrome *b* Group *

Cytochrome *c* — Present in mitochondria of eucaryotes. It possesses an isoelectric point above pH 7 and the oxidation of the ferro form of O_2 is catalyzed by cytochrome *c* oxidase (EC 1.9.3.1). The midpoint redox potential over most of the physiological pH range is close to 250 mV. The characteristic absorption bands of the ferro form are at 550 nm (α), 520 nm (β), and 415 nm (γ).

Cytochrome c_1 — Present in mitochondria of eucaryotes. The characteristic absorption band of the ferro form is at 553 nm (α).

Cytochrome c_2 — Soluble, monohaem cytochromes with characteristic binding of the haem *c* prosthetic group through thio-ether linkage, found in nonsulphur photosynthetic bacteria. Equivalent forms may also be present in other bacteria (often termed cytochrome c_4, cytochrome c_5, etc.). Reaction with mitochondrial cytochrome *c* oxidase is sluggish or absent.

Cytochrome c_3 — A class of low-potential cytochromes (midpoint redox potential negative at pH 7) including both mono- and multi-haem forms, found so far in many sulphate-reducing bacteria, some photosynthetic bacteria, and in one blue-green alga.

Cytochrome c_6 — Also known as cytochrome *f*. A class of high-potential *c* cytochromes found in chloroplasts and other photosynthetic membranes.

---

*A variety of soluble *c*-type cytochromes encountered in bacteria remain officially uncategorized. These include Azotobacter cytochromes c_4 and c_5, flavocytochromes *c*, cytochromes c_{551}, etc., which appear in multi- as well as mono-haem forms, and cytochromes *c*-555, *c*-553, 556 of Chromatium. As insufficient structural information is available, no attempt has been made to assign official names to these cytochromes.

LIST OF CYTOCHROMES (continued)

Cytochrome c Group (continued)

Cytochrome c' A class of variant cytochromes c (sometimes known as "RHP" or cytochromoid c) with haem c prosthetic groups bound most probably in all cases through thio-ether linkage, as in cytochrome c, but lacking haemochrome linkages to extraplanar ligands of the protein at physiological pH. Found in purple photosynthetic bacteria, as well as in at least one nitrate-reducing pseudomonad (*Alculigenes* sp.). In the soluble forms usually isolated, NO and CO are the only ligands that react readily with the reduced cytochromes, and NO is the only ligand that reacts readily with the oxidized cytochrome.

Cytochrome d Group

Cytochrome d Also known as cytochrome a_2. Present in many bacteria. Its characteristic absorption bands are at about 645–650 nm (ferri) and 625–630 nm (ferro). The ferrocytochrome is autoxidizable and combines with CO with a shift of the absorption band to about 635 nm.

5. Iron-sulphur Proteins

The iron-sulphur proteins may be grouped into four categories:

  i. Ferredoxin
  ii. High-potential iron-sulphur protein
  iii. Rubredoxin
  iv. Conjugated iron-sulphur protein

Ferredoxin

This group comprises those iron-sulphur proteins with an equal number of iron and labile sulphur atoms and a negative midpoint redox potential at pH 7. They are characterized by an EPR signal with a line at $g < 2$ for the reduced protein. The source should always be stated.

Examples:

  Chloroplast ferredoxin
  Adrenal ferredoxin (formerly called adrenodoxin)
  Pseudonomas putida ferredoxin (formerly called putidaredoxin)
  Clostridium acidi-urici ferredoxin

Ferredoxin may be abbreviated Fd.

High-potential Iron-sulphur Protein

Certain microorganisms contain a unique class of iron-sulphur proteins containing acid-labile sulphur, but differing from the ferredoxins by their physical properties. No EPR signal has been detected with the reduced form of this type of protein. The oxidized form is paramagnetic with an EPR signal with a g value of about 2. At pH 7, the midpoint potential is positive. Until further characterized, the descriptive but cumbersome name "high-potential iron-sulphur protein" should be retained with the source indicated as a prefix, e.g., Chromatium high-potential iron-sulphur protein.

Rubredoxin

This group comprises those iron-sulphur proteins without acid-labile sulphur characterized by having iron in a typical mercaptide coordination, i.e., one center, surrounded by four cysteine or equivalent sulphur ligands. Oxidized rubredoxin has a distinctive EPR spectrum with a line at $g = 4.3$, while the reduced pigment gives no discernible EPR signal. The redox potential for those rubredoxins now characterized is negative at pH 7. The full name should be listed as (source) rubredoxin (function), e.g., *Pseudomonas oleovorans* rubredoxin, alkane ω-hydroxylation.

Conjugated Iron-sulphur Protein
This group comprises those iron-sulphur proteins also containing additional prosthetic groups. Many of the iron-containing flavoproteins, molybdo-iron proteins, or molybdo-iron flavoproteins are included in this group.

6. Other Metalloproteins
Cuproproteins
Oxidases containing copper appear in the List of Enzymes (e.g., monophenol monooxygenase, EC 1.14.18.1). Cytochrome *c* oxidase, classified both as an enzyme (EC 1.9.3.1) and a cytochrome (cytochrome aa_3), also contains copper. An electron-transfer protein, plastocyanine, is part of the photosynthetic chain in chloroplasts. Azurin and stellacyanin are also electron-transfer proteins.

Molybdoproteins
Molybdenum is present in xanthine oxidase (EC 1.2.3.2), which also contains FAD and iron-sulphur centers. Aldehyde oxidase (EC 1.2.3.1) also contains molybdenum as well as iron and flavin. Nitric-oxide reductase (EC 1.7.99.2) contains molybdenum as well as iron and inorganic sulphur, but not flavin. Nitrate reductase (NADPH) (EC 1.6.6.3) contains molybdenum and iron.

INDEX TO THE ENZYME LIST

Recommended names are printed in bold roman letters; other names are printed in ordinary roman. The *systematic names,* used as a basis of classification, are in italics.
Names of deleted entries are shown in small roman letters.

Abequosyltransferase	2.4.1.60	α-*(Acetamidomethylene) succinate amidohydrolase (deaminating, decarboxylating)*	3.5.1.29 ǂ
2-Acetamido-1-N-(4-L-aspartyl)-2-deoxy-β-D-glucosylamine glucosyl-amidohydrolase	3.5.1.37	α-**(Acetamidomethylene)succinate hydrolase**	3.5.1.29
2-Acetamido-2-deoxy-α-D-galactoside acetamidodeoxygalactohydrolase	3.2.1.49	*Acetate: CoA ligase (AMP-forming)*	6.2.1.1
		Acetate CoA-transferase	2.8.3.8
2-Acetamido-2-deoxy-β-D-galactoside acetamidodeoxygalactohydrolase	3.2.1.53	**Acetate kinase**	2.7.2.1
		Acetate thiokinase	6.2.1.1
2-Acetamido-2-deoxy-D-glucose amidohydrolase	3.5.1.33	*Acetic-ester hydrolase*	3.1.1.6
		Acetoacetate carboxy-lyase	4.1.1.4
2-Acetamido-2-deoxy-D-glucose-1,6-bisphosphate:2-acetamido-2-deoxy-D-glucose-1-phosphate phosphotransferase	2.7.5.2	**Acetoacetate decarboxylase**	4.1.1.4
		Acetoacetyl-CoA hydrolase	3.1.2.11
		Acetoacetyl-CoA hydrolase	3.1.2.11
2-Acetamido-2-deoxy-D-glucose-6-phosphate amidohydrolase	3.5.1.25	**Acetoacetyl-CoA reductase**	1.1.1.36
		Acetoacetyl-CoA thiolase	2.3.1.9
2-Acetamido-2-deoxy-D-glucose-6-phosphate ketol-isomerase (deaminating)	5.3.1.11	**Acetoacetylglutathione hydrolase**	3.1.2.8
		S-Acetoacetylglutathione hydrolase	3.1.2.8
2-Acetamido-2-deoxy-α-D-glucoside acetamidodeoxyglucohydrolase	3.2.1.50	S-Acetoacetylhydrolipoate hydrolase	[3.1.2.9]
		Acetoin dehydrogenase	1.1.1.5
2-Acetamido-2-deoxy-β-D-glucoside acetamidodeoxyglucohydrolase	3.2.1.30	*Acetoin:NAD$^+$ oxidoreductase*	1.1.1.5
		Acetoin racemase	5.1.2.4
2-Acetamido-2-deoxy-β-hexoside acetamidodeoxyhexohydrolase	3.2.1.52	*Acetoin racemase*	5.1.2.4
		Acetokinase	2.7.2.1

INDEX TO THE ENZYME LIST (continued)

Acetolactate decarboxylase	4.1.1.5
2-Acetolactate methylmutase	5.4.99.3
2-Acetolactate mutase	5.4.99.3
Acetolactate pyruvate-lyase (carboxylating)	4.1.3.18
Acetolactate synthase	4.1.3.18
Acetol kinase	2.7.1.29
Acetyl activating enzyme	6.2.1.1
N-Acetyl-β-alanine amidohydrolase	3.5.1.21
N-Acetyl-β-alanine deacetylase	3.5.1.21
Acetylaminodeoxyglucose kinase	[2.7.1.9]
Acetylcholine hydrolase	3.1.1.7
Acetylcholinesterase	3.1.1.7
Acetyl-CoA:acetyl-CoA C-acetyltransferase	2.3.1.9
Acetyl-CoA:N-acetylneuraminate 4-O-acetyltransferase	2.3.1.44
Acetyl-CoA:N-acetylneuraminate 7(or 8)-O-acetyltransferase	2.3.1.45
Acetyl-CoA acetyltransferase	2.3.1.9
Acetyl-CoA acylase	3.1.2.1
Acetyl-CoA:[acyl-carrier-protein] S-acetyltransferase	2.3.1.38
Acetyl-CoA acyltransferase	2.3.1.16
Acetyl-CoA:D-amino-acid N-acetyltransferase	2.3.1.36
Acetyl-CoA:2-amino-2-deoxy-D-glucose N-acetyltransferase	2.3.1.3
Acetyl-CoA:2-amino-2-deoxy-D-glucose-6-phosphate N-acetyltransferase	2.3.1.4
Acetyl-CoA:arylamine N-acetyltransferase	2.3.1.5
Acetyl-CoA→arylamine transacetylase	2.3.1.5
Acetyl-CoA:L-aspartate N-acetyltransferase	2.3.1.17
Acetyl-CoA:carbon-dioxide ligase (ADP-forming)	6.4.1.2
Acetyl-CoA carboxylase	6.4.1.2
Acetyl-CoA:carnitine O-acetyltransferase	2.3.1.7
Acetyl-CoA:chloramphenicol 3-O-acetyltransferase	2.3.1.28
Acetyl-CoA:choline O-acetyltransferase	2.3.1.6
Acetyl-CoA→choline transacetylase	2.3.1.6
Acetyl-CoA:cortisol O-acetyltransferase	2.3.1.27
Acetyl-CoA deacylase	3.1.2.1
Acetyl-CoA:dihydrolipoate S-acetyltransferase	2.3.1.12
Acetyl-CoA:galactoside 6-O-acetyltransferase	2.3.1.18
Acetyl-CoA→D-glucosamine transacetylase	2.3.1.4
Acetyl-CoA:L-glutamate N-acetyltransferase	2.3.1.1
Acetyl-CoA:glycine C-acetyltransferase	2.3.1.29
Acetyl-CoA:L-histidine N-acetyl-transferase	2.3.1.33
Acetyl-CoA:L-homoserine O-acetyltransferase	2.3.1.31
Acetyl-CoA:hydrogen-sulphide S-acetyltransferase	2.3.1.10
Acetyl-CoA hydrolase	3.1.2.1
Acetyl-CoA hydrolase	3.1.2.1
Acetyl-CoA:imidazole N-acetyltransferase	2.3.1.2
Acetyl-CoA:malonate CoA-transferase	2.8.3.3
Acetyl-CoA:orthophosphate acetyltransferase	2.3.1.8
Acetyl-CoA:propionate CoA-transferase	2.8.3.1
Acetyl-CoA:L-serine O-acetyltransferase	2.3.1.30
Acetyl-CoA synthetase	6.2.1.1
Acetyl-CoA:thioethanolamine S-acetyltransferase	2.3.1.11
Acetyl-CoA:D-tryptophan N-acetyltransferase	2.3.1.34
Acetylesterase	3.1.1.6
α-N-Acetylgalactosaminidase	3.2.1.49
β-N-Acetylgalactosaminidase	3.2.1.53
N-Acetylglucosamine deacetylase	3.5.1.33
N-Acetyl-D-glucosamine kinase	2.7.1.59
N-Acetylglucosamine-6-phosphate deacetylase	3.5.1.25
Acetylglucosaminephosphate isomerase	5.3.1.11
Acetylglucosamine phosphomutase	2.7.5.2
α-N-Acetylglucosaminidase	3.2.1.50
β-N-Acetylglucosaminidase	3.2.1.30
trans-N-Acetylglucosaminosylase	2.4.1.16
β-N-Acetylglucosaminylsaccharide fucosyltransferase	2.4.1.65
6-O-Acetylglucose deacetylase	3.1.1.33
6-O-Acetyl-D-glucose hydrolase	3.1.1.33
Acetylglutamate kinase	2.7.2.8
N-Acetyl-L-glutamate-5-semialdehyde:NADP$^+$ oxidoreductase (phosphorylating)	1.2.1.38
N-Acetyl-γ-glutamyl-phosphate reductase	1.2.1.38
β-N-Acetylhexosaminidase	3.2.1.52
β-N-Acetylhexosaminidase	3.2.1.52
N-Acetyl-L-histidine amidohydrolase	3.5.1.34
Acetylhistidine deacetylase	3.5.1.34
O-Acetyl-L-homoserine acetate-lyase (adding methanethiol)	4.2.99.10
Acetylindoxyl oxidase	1.7.3.2
N-Acetylindoxy:oxygen oxidoreductase	1.7.3.2
Acetylmethylcarbinol racemase	5.1.2.4
N-Acetylmuramoyl-L-alanine amidase	3.5.1.28
N-Acetylneuraminate 4-O-acetyltransferase	2.3.1.44
N-Acetylneuraminate 7(or 8)-O-acetyltransferase	2.3.1.45
N-Acetylneuraminate lyase	4.1.3.3
N-Acetylneuraminate pyruvate-lyase	4.1.3.3

INDEX TO THE ENZYME LIST (continued)

N-Acetylneuraminate pyruvate-lyase (pyruvate-phosphorylating)	4.1.3.19
N-Acetylneuraminate synthase	4.1.3.19
N-Acetylneuraminic acid aldolase	4.1.3.3
N^2-Acetyl-L-ornithine amidohydrolase	3.5.1.16
Acetylornithine aminotransferase	2.6.1.11
Acetylornithine deacetylase	3.5.1.16
N^2-Acetyl-L-ornithine:L-glutamate N-acetyltransferase	2.3.1.35
N^2-Acetyl-L-ornithine:2-oxoglutarate aminotransferase	2.6.1.11
Acetylphosphate→ADP transphosphatase	2.7.2.1
Acetyl-phosphate:L-lysine N^6-acetyl-transferase	2.3.1.32
O-Acetyl-L-serine acetate-lyase (adding hydrogen-sulphide)	4.2.99.8
Acetylserotonin methyltransferase	2.1.1.4
Acid:CoA ligase (AMP-forming)	6.2.1.3
Acid:CoA ligase (GDP-forming)	6.2.1.10
Acid maltase	3.2.1.3
Acid phosphatase	3.1.3.2
Acid phosphomonoesterase	3.1.3.2
Acid proteinase, Aspergillus	3.4.23.6
Acid proteinase, Endothia	3.4.23.10
Acid proteinase, Penicillium janthinellum	3.4.23.7
Acid proteinase, Rhizopus	3.4.23.9
Acid proteinase, Trametes	3.4.99.25
Acid RNase	3.1.4.23
Aconitase	4.2.1.3
cis-Aconitate carboxy-lyase	4.1.1.6
Aconitate decarboxylase	4.1.1.6
Aconitate hydratase	4.2.1.3
Aconitate Δ-isomerase	5.3.3.7
Aconitate $\Delta^2-\Delta^3$-isomerase	5.3.3.7
Acrocylindricum proteinase	3.4.99.1
Acrosin	3.4.21.10
Acrosomal proteinase	3.4.21.10
Actinomycin lactonase	3.1.1.39
Actinomycin lactone-hydrolase	3.1.1.39
Acyl-activating enzyme	6.2.1.1, 6.2.1.2, 6.2.1.3
Acyl-[acyl-carrier-protein]:O-(2-acyl-sn-glycero-3-phospho)-ethanolamine O-acyltransferase	2.3.1.40
Acyl-[acyl-carrier-protein] desaturase	1.14.99.6
Acyl-[acyl-carrier-protein], hydrogen-donor:oxygen oxidoreductase	1.14.99.6
Acyl-[acyl-carrier-protein]:malonyl-[acyl-carrier-protein] C-acyltransferase (decarboxylating)	2.3.1.41
Acyl-[acyl-carrier-protein]:NAD^+ oxidoreductase	1.3.1.9
Acyl-[acyl-carrier-protein]:$NADP^+$ oxidoreductase	1.3.1.10
Acyl-[acyl-carrier-protein]—phospholipid acyltransferase	2.3.1.40
Acylamidase	3.5.1.4
Acylamide amidohydrolase	3.5.1.4
2-Acylamido-2-deoxy-D-glucose 2-epimerase	5.1.3.8
2-Acylamido-2-deoxy-D-glucose-6-phosphate 2-epimerase	5.1.3.9
N-Acylamino-acid amidohydrolase	3.5.1.14
Acylase	3.5.1.4
N-Acyl-L-aspartate amidohydrolase	3.5.1.15
Acylcarnitine hydrolase	3.1.1.28
O-Acylcarnitine hydrolase	3.1.1.28
[Acyl-carrier-protein] acetyltransferase	2.3.1.38
[Acyl-carrier-protein] malonyltransferase	2.3.1.39
[Acyl-carrier-protein] 4'-pantetheine-phosphohydrolase	3.1.4.14
[Acyl-carrier-protein] phosphodiesterase	
Acylcholine acyl-hydrolase	3.1.1.8
Acyl-CoA:(acceptor) oxidoreductase	1.3.99.3
Acyl-CoA:acetate CoA-transferase	2.8.3.8
Acyl-CoA:acetyl-CoA C-acyltransferase	2.3.1.16
Acyl-CoA:1-acylglycero-3-phosphocholine O-acyltransferase	2.3.1.23
Acyl-CoA:1-O-(1'-alk-1'-enyl)-glycero-3-phosphocholine O-acyltransferase	2.3.1.25
Acyl-CoA:cholesterol O-acyltransferase	2.3.1.26
Acyl-CoA dehydrogenase	1.3.99.3
Acyl-CoA dehydrogenase ($NADP^+$)	1.3.1.8
Acyl-CoA desaturase	1.14.99.5
Acyl-CoA:1,2-diacylglycerol O-acyltransferase	2.3.1.20
Acyl-CoA:dihydroxyacetone-phosphate O-acyltransferase	2.3.1.42
Acyl-CoA:sn-glycerol-3-phosphate O-acyltransferase	2.3.1.15
Acyl-CoA:glycine N-acyltransferase	2.3.1.13
Acyl-CoA, hydrogen-donor: oxygen oxidoreductase	1.14.99.5
Acyl-CoA:$NADP^+$ oxidoreductase	1.3.1.8
Acyl-CoA:sphingosine N-acyltransferase	2.3.1.24
Acyl-CoA synthetase	6.2.1.3
Acyl-CoA synthetase (GDP-forming)	6.2.1.10
Acyl dehydrogenase	1.3.99.3
Acylglucosamine 2-epimerase	5.1.3.8
Acylglucosamine-6-phosphate 2-epimerase	5.1.3.9
S-Acylglutathione hydrolase	3.1.2.7
Acylglycerol palmitoyltransferase	2.3.1.22
N^6-Acyl-L-lysine amidohydrolase	3.5.1.17
Acyl-lysine deacylase	3.5.1.17
N-Acyl-D-mannosamine kinase	2.7.1.60
Acylmuramoylalaninase	3.4.12.5
Acylmuramoyl-alanine carboxypeptidase	3.4.12.5
N-Acylmuramoyl-L-alanine hydrolase	3.4.12.5
Acylneuraminate cytidylyltransferase	2.7.7.43
N-Acylneuraminate-9-phosphatase	3.1.3.29
N-Acylneuraminate-9-phosphate phosphohydrolase	3.1.3.29

INDEX TO THE ENZYME LIST (continued)

N-Acylneuraminate-9-phosphate pyruvate-lyase (pyruvate-phosphorylating)	4.1.3.20
N-Acylneuraminate-9-phosphate synthase	4.1.3.20
Acylneuraminyl hydrolase	3.2.1.18
Acylphosphatase	3.6.1.7
Acyl-phosphate—hexose phosphotransferase	2.7.1.61
Acyl-phosphate:D-hexose phosphotransferase	2.7.1.61
Acylphosphate phosphohydrolase	3.6.1.7
5'-Acylphosphoadenosine acylhydrolase	3.6.1.20
5'-Acylphosphoadenosine hydrolase	3.6.1.20
N-Acylsphingosine amidohydrolase	3.5.1.23
Acylsphingosine deacylase	3.5.1.23
Adenase	3.5.4.2
Adenine aminase	3.5.4.2
Adenine aminohydrolase	3.5.4.2
Adenine deaminase	3.5.4.2
Adenine phosphoribosyltransferase	2.4.2.7
Adenosinase	3.2.2.7
Adenosine aminohydrolase	3.5.4.4
Adenosine-3',5'-bisphosphate 3'-phosphohydrolase	3.1.3.7
Adenosine deaminase	3.5.4.4
Adenosine diphosphatase	3.6.1.5
Adenosine kinase	2.7.1.20
Adenosine nucleosidase	3.2.2.7
Adenosine(phosphate) aminohydrolase	3.5.4.17
Adenosine(phosphate) deaminase	3.5.4.17
Adenosine ribohydrolase	3.2.2.7
Adenosinetetraphosphatase	3.6.1.14
Adenosinetetraphosphate phosphohydrolase	3.6.1.14
Adenosinetriphosphatase	3.6.1.3
Adenosinetriphosphatase (Mg-activated)	[3.6.1.4]
Adenosylhomocysteinase	3.3.1.1
S-Adenosyl-L-homocysteine homocysteinylribohydrolase	3.2.2.9
S-Adenosyl-L-homocysteine hydrolase	3.3.1.1
S-Adenosylhomocysteine nucleosidase	3.2.2.9
S-Adenosyl-L-methionine: N-acetylserotonin O-methyltransferase	2.1.1.4
S-Adenosyl-L-methionine alkyltransferase (cyclizing)	2.5.1.4
S-Adenosyl-L-methionine carboxylyase	4.1.1.50
S-Adenosyl-L-methionine:carnosine N-methyltransferase	2.1.1.22
S-Adenosyl-L-methionine:catechol O-methyltransferase	2.1.1.6
Adenosylmethionine cyclotransferase	2.5.1.4
S-Adenosylmethionine decarboxylase	4.1.1.50
S-Adenosyl-L-methionine:O-demethylpuromycin O-methyltransferase	2.1.1.38
S-Adenosyl-L-methionine:DNA(cytosine-5'-)-methyltransferase	2.1.1.37
S-Adenosyl-L-methionine:fatty-acid O-methyltransferase	2.1.1.15
S-Adenosyl-L-methionine:1,4-α-D-glucan 6-O-methyltransferase	2.1.1.18
S-Adenosyl-L-methionine:glycine methyltransferase	2.1.1.20
S-Adenosyl-L-methionine:guanidinoacetate N-methyltransferase	2.1.1.2
S-Adenosyl-L-methionine:histamine N-methyltransferase	2.1.1.8
S-Adenosyl-L-methionine:L-homocysteine S-methyltransferase	2.1.1.10
S-Adenosyl-L-methionine:myo-inositol 1-methyltransferase	2.1.1.39
S-Adenosyl-L-methionine:myo-inositol 3-methyltransferase	2.1.1.40
S-Adenosyl-L-methionine:2-iodophenol methyltransferase	2.1.1.26
S-Adenosyl-L-methionine:magnesium-protoporphyrin O-methyltransferase	2.1.1.11
S-Andenosyl-L-methionine:L-methionine S-methyltransferase	2.1.1.12
S-Adenosyl-L-methionine:nicotinamide N-methyltransferase	2.1.1.1
S-Adenosyl-L-methionine:nicotinate N-methyltransferase	2.1.1.7
S-Adenosyl-L-methionine:phenol O-methyltransferase	2.1.1.25
S-Adenosyl-L-methionine:phenylethanolamine N-methyltransferase	2.1.1.28
S-Adenosyl-L-methionine:phosphatidylethanolamine N-methyltransferase	2.1.1.17
S-Adenosyl-L-methionine:protein N-methyltransferase	2.1.1.23
S-Adenosyl-L-methionine:protein O-methyltransferase	2.1.1.24
S-Adenosyl-L-methionine:thiol S-methyltransferase	2.1.1.9
S-Adenosyl-L-methionine:tRNA (adenine-1-)-methyltransferase	2.1.1.36
S-Adenosyl-L-methionine:tRNA (cytosine-5'-)-methyltransferase	2.1.1.29
S-Adenosyl-L-methionine:tRNA (guanine-1'-)-methyltransferase	2.1.1.31
S-Adenosyl-L-methionine:tRNA (guanine-2-)-methyltransferase	2.1.1.32
S-Adenosyl-L-methionine:tRNA (guanine-7-)-methyltransferase	2.1.1.33
S-Andenosyl-L-methionine:tRNA (guanosine-2'-)-methyltransferase	2.1.1.34
S-Adenosyl-L-methionine:tRNA (purine-2 or -6-)-methyltransferase	2.1.1.30
S-Adenosyl-L-methionine:tRNA (uracil-5-)-methyltransferase	2.1.1.35
S-Adenosyl-L-methionine:tyramine N-methyltransferase	2.1.1.27

INDEX TO THE ENZYME LIST (continued)

S-*Adenosyl*-L-*methionine:unsaturated-phospholipid methyltransferase*	2.1.1.16
S-*Adenosyl*-L-*methionine:zymosterol methyltransferase*	2.1.1.41
Adenylate cyclase	4.6.1.1
Adenylate kinase	2.7.4.3
Adenyl cyclase	4.6.1.1
Adenylic acid deaminase	3.5.4.6
Adenylosuccinase	4.3.2.2
Adenylosuccinate AMP-lyase	4.3.2.2
Adenylosuccinate lyase	4.3.2.2
Adenylosuccinate synthase	6.3.4.4
Adenylosuccinate synthetase	6.3.4.4
Adenylpyrophosphatase	3.6.1.3
Adenylyl cyclase	4.6.1.1
Adenylyl-[L-glutamate:ammonia ligase (ADP-forming)] adenylylhydrolase	3.1.4.15
Adenylyl-[glutamine-synthetase] hydrolase	3.1.4.15
Adenylylsulphatase	3.6.2.1
Adenylylsulphate kinase	2.7.1.25
Adenylylsulphate reductase	1.8.99.2
Adenylylsulphate sulphohydrolase	3.6.2.1
ADP → ADP transphosphatase	2.7.4.3
ADP aminohydrolase	3.5.4.7
ADPase	3.6.1.5
ADP deaminase	3.5.4.7
ADPglucose:1,4-α-D-glucan 4-α-glucosyltransferase	2.4.1.21
ADPglucose pyrophosphorylase	2.7.7.27
ADPglucose-starch glucosyltransferase	2.4.1.21
ADPphosphoglycerate phosphatase	3.1.3.28
ADPphosphoglycerate phosphohydrolase	3.1.3.28
ADP:D-ribose-5-phosphate adenylyltransferase	2.7.7.35
ADPribose phosphorylase	2.7.7.35
ADPribose pyrophosphatase	3.6.1.13
ADPribose ribophosphohydrolase	3.6.1.13
ADP:sugar-1-phosphate adenylyltransferase	2.7.7.36
ADPsugar phosphorylase	2.7.7.36
ADPsugar pyrophosphatase	3.6.1.21
ADPsugar sugarphosphohydrolase	3.6.1.21
ADP-sulfurylase	2.7.7.5
ADP:sulphate adenylyltransferase	2.7.7.5
Adrenalin oxidase	1.4.3.4
Adrenodoxin reductase	1.6.7.1
Aero-glucose dehydrogenase	1.1.3.4
Aeromonas proteolytica **aminopeptidase**	3.4.11.10
Aeromonas proteolytica endopeptidase	3.4.24.4
Aero-oxalo dehydrogenase	1.2.3.4
A-esterase	3.1.1.2
Agarase	3.2.1.81
Agaritine γ-glutamyltransferase	2.3.2.9
Agarose 3-glycanohydrolase	3.2.1.81
Agavain	3.4.99.2
Agmatinase	3.5.3.11
Agmatine amidinohydrolase	3.5.3.11
Agmatine deiminase	3.5.3.12
Agmatine iminohydrolase	3.5.3.12
D-Alanine:D-alanine ligase (ADP-forming)	6.3.2.4
Alanine aminotransferase	2.6.1.2
D-Alanine aminotransferase	2.6.1.21
Alanine carboxypeptidase	3.4.12.11
Alanine dehydrogenase	1.4.1.1
L-Alanine:4,5-dioxovalerate aminotransferase	2.6.1.43
Alanine–glyoxylate aminotransferase	2.6.1.44
L-Alanine:glyoxylate aminotransferase	2.6.1.44
L-Alanine:malonate-semialdehyde aminotransferase	2.6.1.18
L-Alanine:NAD+ oxidoreductase (deaminating)	1.4.1.1
Alanine–oxo-acid aminotransferase	2.6.1.12
L-Alanine:2-oxo-acid aminotransferase	2.6.1.12
β-Alanine–oxoglutarate aminotransferase	2.6.1.19
D-Alanine:2-oxoglutarate aminotransferase	2.6.1.21
L-Alanine:2-oxoglutarate aminotransferase	2.6.1.2
Alanine–oxomalonate aminotransferase	2.6.1.47
L-Alanine:oxomalonate aminotransferase	2.6.1.47
D-Alanine:poly(phosphoribitol) ligase (AMP-forming)	6.1.1.13
β-Alanine–pyruvate aminotransferase	2.6.1.18
Alanine racemase	5.1.1.1
Alanine racemase	5.1.1.1
L-Alanine:tRNA^{Ala} ligase(AMP-forming)	6.1.1.7
D-Alanylalanine synthetase	6.3.2.4
β-Alanyl-CoA ammonia-lyase	4.3.1.6
β-Alanyl-CoA ammonia-lyase	4.3.1.6
O-Alanylphosphatidylglycerol synthase	2.3.2.11
D-Alanyl-poly(phosphoribitol) synthetase	6.1.1.13
D-Alanyl-sRNA synthetase	[6.1.1.8]
Alanyl-tRNA:phosphatidylglycerol alanyltransferase	2.3.2.11
Alanyl-tRNA synthetase	6.1.1.7
L-Alanyl-tRNA:UDP-N-acetylmuramoyl-L-alanyl-D-glutamyl-L-lysyl-D-alanyl-D-alanine N^6-alanyltransferase	2.3.2.10
Alcalase Novo	3.4.21.14
Alcohol:(acceptor) oxidoreductase	1.1.99.8
Alcohol dehydrogenase	1.1.1.1
Alcohol dehydrogenase (acceptor)	1.1.99.8
Alcohol dehydrogenase (NADP+)	1.1.1.2
Alcohol dehydrogenase (NAD(P)+)	1.1.1.71
Alcohol:NAD+ oxidoreductase	1.1.1.1
Alcohol:NADP+ oxidoreductase	1.1.1.2
Alcohol:NAD(P)+ oxidoreductase	1.1.1.71
Alcohol oxidase	1.1.3.13
Alcohol:oxygen oxidoreductase	1.1.3.13
Aldehyde dehydrogenase	1.2.1.3
Aldehyde dehydrogenase (acylating)	1.2.1.10
Aldehyde dehydrogenase (NADP+)	1.2.1.4

INDEX TO THE ENZYME LIST (continued)

Aldehyde dehydrogenase (NAD(P)+)	1.2.1.5
Aldehyde → DPN transhydrogenase	1.2.1.3
Aldehyde:NAD+ oxidoreductase	1.2.1.3
Aldehyde:NAD+ oxidoreductase (CoA-acylating)	1.2.1.10
Aldehyde:NADP+ oxidoreductase	1.2.1.4
Aldehyde:NAD(P)+ oxidoreductase	1.2.1.5
Aldehyde → O$_2$ transhydrogenase	1.2.3.1, 1.2.3.2
Aldehyde oxidase	1.2.3.1
Aldehyde:oxygen oxidoreductase	1.2.3.1
Aldehyde reductase	1.1.1.1
Aldehyde →TPN transhydrogenase	1.2.1.4
Aldehydrase	1.2.3.2
Alditol:NADP+ 1-oxidoreductase	1.1.1.21
D-*Aldohexoside:(acceptor) 5-oxidoreductase*	1.1.99.13
Aldoketomutase	4.4.1.5
Aldolase	4.1.2.13
L-3-Aldonate dehydrogenase	1.1.1.45
Aldonolactonase	3.1.1.18
D-Aldopantoate dehydrogenase	1.2.1.33
Aldose dehydrogenase	1.1.1.121
Aldose 1-epimerase	5.1.3.3
Aldose 1-epimerase	5.1.3.3
Aldose mutarotase	5.1.3.3
D-*Aldose:NAD+ 1-oxidoreductase*	1.1.1.121
Aldose reductase	1.1.1.21
Alginase	[3.2.1.16]
Alginate lyase	4.2.2.3
Alginate synthase	2.4.1.33
Ali-esterase	3.1.1.1
Alkaline phosphatase	3.1.3.1
Alkaline phosphomonoesterase	3.1.3.1
Alkaline proteinase, Aspergillus	3.4.21.15
Alkaline proteinase, Streptomyces	3.4.21.4
Alkaline RNase	3.1.4.22
Alkane 1-hydroxylase	1.14.15.3
Alkane 1-monooxygenase	1.14.15.3
Alkane, reduced-rubredoxin:oxygen 1-oxidoreductase	1.14.15.3
1-(1′-alk-1′-enyl)-glycero-3-phosphinicocholine aldehydohydrolase	3.3.2.2
Alkenyl-glycerophosphinicocholine hydrolase	3.3.2.2
3-Alkylcatechol 2,3-dioxygenase	1.13.11.25
S-Alkylcysteine lyase	[4.4.1.6]
3-Alkyl-2,3-dihydroxy-acid:NADP+ oxidoreductase (isomerizing)	1.1.1.86
Alkylhalidase	3.8.1.1
Alkyl-halide:glutathione S-alkyltransferase	2.5.1.12
Alkyl-halide halidohydrolase	3.8.1.1
Allantoate amidinohydrolase	3.5.3.4
Allantoate amidinohydrolase (decarbonylating)	3.5.3.9
Allantoate deiminase	3.5.3.9
Allantoicase	3.5.3.4
Allantoin amidohydrolase	3.5.2.5
Allantoinase	3.5.2.5
Alliin alkyl-sulphenate-lyase	4.4.1.4
Alliin lyase	4.4.1.4
Alliinase	4.4.1.4
Allokinase	2.7.1.55
Allose kinase	2.7.1.55
Allothreonine aldolase	[4.1.2.6]
Allyl-alcohol dehydrogenase	1.1.1.54
Allyl-alcohol:NADP+ oxidoreductase	1.1.1.54
Allylic-terpene-diphosphate:isopentenyldiphosphate terpenoid-allyltransferase	2.5.1.11
Alternaria endopeptidase	3.4.21.16
Altronate dehydratase	[4.2.1.7]
D-*Altronate:NAD+ 3-oxidoreductase*	1.1.1.58
Amidase	3.5.1.4
ω-**Amidase**	3.5.1.3
ω-Amidodicarboxylate amidohydrolase	3.5.1.3
Amidophosphoribosyltransferase	2.4.2.14
Amide oxidase	1.4.3.4
Amine oxidase (flavin-containing)	1.4.3.4
Amine oxidase (pyridoxal-containing)	1.4.3.6
Amine:oxygen oxidoreductase (deaminating) (flavin-containing)	1.4.3.4
Amine:oxygen oxidoreductase (deaminating) (pyridoxal-containing)	1.4.3.6
D-*Amino-acid:(acceptor) oxidoreductase (deaminating)*	1.4.99.1
Amino-acid acetyltransferase	2.3.1.1
D-*Amino-acid acetyltransferase*	
D-**Amino-acid acetyltransferase**	2.3.1.36
D-**Amino-acid dehydrogenase**	1.4.99.1
L-**Amino-acid dehydrogenase**	1.4.1.5
L-*Amino-acid:NAD+ oxidoreductase (deaminating)*	1.4.1.5
D-Amino-acid → O$_2$ transhydrogenase	1.4.3.3
L-Amino-acid → O$_2$ transhydrogenase	1.4.3.2
Amino-acid oxhydrase	1.4.3.2, 1.4.3.3
D-**Amino-acid oxidase**	1.4.3.3
L-**Amino-acid oxidase**	1.4.3.2
D-*Amino-acid:oxygen oxidoreductase (deaminating)*	1.4.3.3
L-*Amino-acid:oxygen oxidoreductase (deaminating)*	1.4.3.2
Amino-acid racemase	5.1.1.10
Amino-acid racemase	5.1.1.10
Aminoacylase	3.5.1.14
Aminoacylase II	3.5.1.15
α-*Aminoacyl-dipeptide hydrolase*	3.4.11.4
Aminoacyl-histidine dipeptidase	3.4.13.3
Aminoacyl-L-histidine hydrolase	3.4.13.3

INDEX TO THE ENZYME LIST (continued)

Aminoacyl-L-lysine (-L-arginine) hydrolase	3.4.13.4
Aminoacyl-lysine dipeptidase	3.4.13.4
Aminoacyl-methylhistidine dipeptidase	3.4.13.5
Aminoacyl-prosmethyl-L-histidine hydroiase	3.4.13.5
α-Aminoacyl-peptide hydrolase	3.4.11.3
α-Aminoacyl-peptide hydrolase (Aeromonas proteolytica)	3.4.11.10
α-Aminoacyl-peptide hydrolase (cytosol)	3.4.11.1
α-Aminoacyl-peptide hydrolase (microsomal)	3.4.11.2
Aminoacylproline aminopeptidase	3.4.11.9
Aminoacyl-L-proline hydrolase	3.4.13.9
Aminoacylprolyl-peptide hydrolase	3.4.11.9
Aminoacyl-tRNA hydrolase	3.1.1.29
Aminoacyl-tRNA hydrolase	3.1.1.29
2-Aminoadipate aminotransferase	2.6.1.39
L-2-Aminoadipate:2-oxoglutarate aminotransferase	2.6.1.39
Aminoadipate-semialdehyde dehydrogenase	1.2.1.31
L-2-Aminoadipate-6-semialdehyde:NAD(P)$^+$ oxidoreductase	1.2.1.31
Aminobenzoate carboxy-lyase	4.1.1.24
Aminobenzoate decarboxylase	4.1.1.24
Aminobutyraldehyde dehydrogenase	1.2.1.19
4-Aminobutyraldehyde:NAD$^+$ oxidoreductase	1.2.1.19
Aminobutyrate aminotransferase	2.6.1.19
4-Aminobutyrate:2-oxoglutarate aminotransferase	2.6.1.19
N^2-(4-Aminobutyryl)-L-lysine hydrolase	3.4.13.4
Aminocarboxymuconate-semialdehyde decarboxylase	4.1.1.45
2-Amino-2-deoxy-D-gluconate ammonia-lyase (isomerizing)	4.3.1.9
Aminodeoxygluconate dehydratase	4.2.1.26
2-Amino-2-deoxy-D-gluconate hydro-lyase (deaminating)	4.2.1.26
2-Amino-2-deoxy-D-glucose-6-phosphate ketol-isomerase (amino-transferring)	5.3.1.19
2-Amino-2-deoxy-D-glucose-6-phosphate ketol-isomerase (deaminating)	5.3.1.10
(2-Aminoethyl)phosphonate aminotransferase	2.6.1.37
(2-Aminoethyl)phosphonate:pyruvate aminotransferase	2.6.1.37
2-Amino-4-hydroxy-6-hydroxymethyl-7,8-dihydropteridine-diphosphate:4-aminobenzoate 2-amino-4-hydroxydihydropteridine-6-methenyltransferase	2.5.1.15
2-Amino-4-hydroxy-6-hydroxymethyldihydropteridine pyrophosphokinase	2.7.6.3
2-Amino-4-hydroxypteridine aminohydrolase	3.5.4.11
2-Amino-4-hydroxy-6-(D-erythro-1′,2′,3′-trihydroxypropyl)-7,8-dihydropteridine glycolaldehyde-lyase	4.1.2.25
Aminoimidazolase	3.5.4.8
4-Aminoimidazole aminohydrolase	3.5.4.8
L-3-Aminoisobutyrate aminotransferase	2.6.1.22
L-3-Aminoisobutyrate:2-oxoglutarate aminotransferase	2.6.1.22
D-3-Aminoisobutyrate—pyruvate aminotransferase	2.6.1.40
Aminolaevulinate aminotransferase	2.6.1.43
Aminolaevulinate dehydratase	4.2.1.24
5-Aminolaevulinate hydro-lyase (adding 5-aminolaevulinate and cyclizing)	4.2.1.24
δ-Aminolaevulinate synthase	2.3.1.37
Aminomalonate decarboxylase	[4.1.1.10]
(R)-3-Amino-2-methylpropionate—pyruvate aminotransferase	2.6.1.40
(R)-3-Amino-2-methylpropionate:pyruvate aminotransferase	2.6.1.40
1-(4-Amino-2-methylpyrimid-5-ylmethyl)-3-(β-hydroxyethyl)-2-methylpyridinium-bromide aminohydrolase	3.5.4.20
Aminomuconate-semialdehyde dehydrogenase	1.2.1.32
2-Aminomuconate-6-semialdehyde:NAD$^+$ oxidoreductase	1.2.1.32
5-Aminonorvaleramidase	3.5.1.30
5-Aminonorvaleramide amidohydrolase	3.5.1.30
Aminopeptidase, Aeromonas proteolytica	3.4.11.10
Aminopeptidase (cytosol)	3.4.11.1
Aminopeptidase (microsomal)	3.4.11.2
Aminopeptidase, particle-bound	3.4.11.2
o-Aminophenol oxidase	1.10.3.4
o-Aminophenol:oxygen oxidoreductase	1.10.3.4
D-Aminopropanol dehydrogenase	1.1.1.74
L-Aminopropanol dehydrogenase	1.1.1.75
D-1-Aminopropan-2-ol:NAD$^+$ oxidoreductase	1.1.1.74
L-1-Aminopropan-2-ol:NAD$^+$ oxidoreductase	1.1.1.75
Aminopropyltransferase	2.5.1.16
5-Aminovalerate aminotransferase	2.6.1.48
5-Aminovalerate:lipoate oxidoreductase (cyclizing)	1.4.4.1
5-Aminovalerate:NAD$^+$ oxidoreductase (cyclizing)	1.4.1.6
5-Aminovalerate:2-oxoglutarate aminotransferase	2.6.1.48
Ammonia:(acceptor) oxidoreductase	1.7.99.1
Ammonia:ferredoxin oxidoreductase	1.7.7.1
Ammonia kinase	2.7.3.8
AMP aminase	3.5.4.6
AMP aminohydrolase	3.5.4.6
AMP deaminase	3.5.4.6
AMP nucleosidase	3.2.2.4

INDEX TO THE ENZYME LIST (continued)

AMP phosphoribohydrolase	3.2.2.4	L-Arabinitol:NAD⁺ 2-oxidoreductase	
AMP:pyrophosphate phosphoribosyl-transferase	2.4.2.7	(L-ribulose-forming)	1.1.1.13
AMP pyrophosphorylase	2.4.2.7	L-Arabinitol:NAD⁺ 4-oxidoreductase (L-xylulose-forming)	1.1.1.12
AMP sulphite:(acceptor) oxidoreductase	1.8.99.2	α-L-**Arabinofuranosidase**	3.2.1.55
Amygdalase	3.2.1.21	α-L-Arabinofuranoside arabinofurano-hydrolase	3.2.1.79
α-**Amylase**	3.2.1.1		
β-**Amylase**	3.2.1.2	α-L-Arabinofuranoside arabinohydrolase	3.2.1.55
γ-Amylase	3.2.1.3	α-L-**Arabinofuranoside hydrolase**	3.2.1.79
Amyloglucosidase	3.2.1.3	D-**Arabinokinase**	2.7.1.54
Amylo-1,6-glucosidase	3.2.1.33	L-**Arabinokinase**	2.7.1.46
Amylopectin branching enzyme	2.4.1.18	**Arabinonate dehydratase**	4.2.1.5
Amylopectin 6-glucanohydrolase	3.2.1.69	L-**Arabinonate dehydratase**	4.2.1.25
Amylopectin 6-glucanohydrolase	3.2.1.69	D-Arabinonate hydro-lyase	4.2.1.5
Amylopectin-1,6-glucosidase	[3.2.1.9]	L-Arabinonate hydro-lyase	4.2.1.25
Amylophosphorylase	2.4.1.1	**Arabinonolactonase**	3.1.1.15
Amylosucrase	2.4.1.4	D-**Arabinonolactonase**	3.1.1.30
Ancistrodom rhodastoma proteinase	3.4.21.5	D-Arabinono-γ-lactone hydrolase	3.1.1.30
Androst-4-ene-3,17-dione, hydrogen-donor:oxygen oxidoreductase (lactonizing)	1.14.99.12	L-Arabinono-γ-lactone hydrolase	3.1.1.15
		D-**Arabinose dehydrogenase**	1.1.1.116
Androstene-3,17-dione hydroxylase	1.14.99.12	L-**Arabinose dehydrogenase**	1.1.1.46
4-Androstene-3,17-dione monooxygenase	1.14.99.12	D-**Arabinose dehydrogenase (NAD(P)⁺)**	1.1.1.117
		Arabinose isomerase	5.3.1.3
Angiotensinase	3.4.99.3	L-**Arabinose isomerase**	5.3.1.4
Aniline, reduced-flavoprotein:oxygen oxidoreductase (4-hydroxylating)	1.14.14.1	D-Arabinose ketol-isomerase	5.3.1.3
		L-Arabinose ketol-isomerase	5.3.1.4
Anserinase	3.4.13.5	D-Arabinose:NAD⁺ 1-oxidoreductase	1.1.1.116
Anthranilate 2,3-dioxygenase (deaminating)	1.14.12.2	L-Arabinose:NAD⁺ 1-oxidoreductase	1.1.1.46
		D-Arabinose:NAD(P)⁺ 1-oxidoreductase	1.1.1.117
Anthranilate 1,2-dioxygenase (deaminating, decarboxylating)	1.14.12.1	**Arabinosephosphate isomerase**	5.3.1.13
Anthranilate hydroxylase	1.14.12.1, 1.14.12.2, 1.14.16.3	D-Arabinose-5-phosphate ketol-isomerase	5.3.1.13
		Arabinosidase	3.2.1.55
		Aralkyl-halide:glutathione S-aralkyl-transferase	2.5.1.14
Anthranilate 3-hydroxylase	1.14.16.3	**Arene monooxygenase (epoxidizing)**	1.14.99.8
Anthranilate 3-monooxygenase	1.14.16.3	**Arene-oxide hydratase**	4.2.1.64
Anthranilate,NADPH:oxygen oxidoreductase (2,3-hydroxylating, deaminating)	1.14.12.2	**Arginase**	3.5.3.1
		D-**Arginase**	3.5.3.10
		Arginine amidinase	3.5.3.1
Anthranilate,NAD(P)H:oxygen oxidoreductase (1,2-hydroxylating, deaminating, decarboxylating)	1.14.12.1	D-Arginine amidinohydrolase	3.5.3.10
		L-Arginine amidinohydrolase	3.5.3.1
		Arginine aminopeptidase	3.4.11.6
Anthranilate phosphoribosyltransferase	2.4.2.18	L-Arginine carboxy-lyase	4.1.1.19
Anthranilate synthase	4.1.3.27	**Arginine carboxypeptidase**	3.4.12.7
Anthranilate, tetrahydropteridine:oxygen oxidoreductase (3-hydroxylating)	1.14.16.3	**Arginine decarboxylase**	4.1.1.19
		Arginine deiminase	3.5.3.6
D-Apiitol:NAD⁺ 1-oxidoreductase	1.1.1.114	**Arginine dihydrolase**	3.5.3.6
D-**Apiose reductase**	1.1.1.114	L-Arginine:glycine amidinotransferase	2.1.4.1
APS-kinase	2.7.1.25	L-Arginine iminohydrolase	3.5.3.6
Apyrase	3.6.1.5	**Arginine kinase**	2.7.3.3
Aquocob(I)alamin adenosyltransferase	2.5.1.17	**Arginine 2-monooxygenase**	1.13.12.1
Aquocobalamin reductase	1.6.99.8	L-Arginine:oxygen 2-oxidoreductase (decarboxylating)	1.13.12.1
D-**Arabinitol dehydrogenase**	1.1.1.11		
L-**Arabinitol dehydrogenase**	1.1.1.12	**Arginine racemase**	5.1.1.9
L-**Arabinitol dehydrogenase (ribulose-forming)**	1.1.1.13	Arginine racemase	5.1.1.9
		L-Arginine:tRNA^Arg ligase (AMP-forming)	6.1.1.19
D-Arabinitol:NAD⁺ 4-oxidoreductase	1.1.1.11	Argininosuccinase	4.3.2.1
		L-Argininosuccinate arginine-lyase	4.3.2.1
		Argininosuccinate lyase	4.3.2.1

INDEX TO THE ENZYME LIST (continued)

Argininosuccinate synthetase	6.3.4.5
L-Arginyl(L-lysyl)-peptide hydrolase	3.4.11.6
Arginyltransferase	2.3.2.8
L-Arginyl-tRNA:protein arginyltransferase	2.3.2.8
Arginyl-tRNA synthetase	6.1.1.19
Aromatic-L-amino-acid carboxy-lyase	4.1.1.28
Aromatic L-amino-acid decarboxylase	4.1.1.28
Arthrobacter serine proteinase	3.4.21.17
Aryl acylamidase	3.5.1.13
Aryl-acylamide amidohydrolase	3.5.1.13
Aryl-alcohol dehydrogenase	1.1.1.90
Aryl-alcohol dehydrogenase ($NADP^+$)	1.1.1.91
Aryl-alcohol:NAD^+ oxidoreductase	1.1.1.90
Aryl-alcohol:$NADP^+$ oxidoreductase	1.1.1.91
Aryl-alcohol oxidase	1.1.3.7
Aryl-alcohol:oxygen oxidoreductase	1.1.3.7
Aryl-aldehyde dehydrogenase	1.2.1.29
Aryl-aldehyde dehydrogenase ($NADP^+$)	1.2.1.30
Aryl-aldehyde:NAD^+ oxidoreductase	1.2.1.29
Aryl-aldehyde:$NADP^+$ oxidoreductase (ATP-forming)	1.2.1.30
Arylamine acetylase	2.3.1.5
Arylamine acetyltransferase	2.3.1.5
Arylamine sulphotransferase	2.8.2.3
Aryl-chloride:glutathione S-aryltransferase	2.5.1.13
Arylesterase	3.1.1.2
Aryl-ester hydrolase	3.1.1.2
Aryl-formylamine amidohydrolase	3.5.1.9
Aryl 4-hydroxylase	1.14.14.1
Aryl 4-monooxygenase	1.14.14.1
Arylsulphatase	3.1.6.1
Aryl-sulphate sulphohydrolase	3.1.6.1
Aryl sulphotransferase	2.8.2.1
Asclepain	3.4.22.7
Ascorbase	1.10.3.3
L-Ascorbate–cytochrome b_5 reductase	1.10.2.1
Ascorbate 2,3-dioxygenase	1.13.11.13
L-Ascorbate:ferricytochrome b_5 oxidoreductase	1.10.2.1
Ascorbate oxidase	1.10.3.3
L-Ascorbate:oxygen oxidoreductase	1.10.3.3
Ascorbate:oxygen 2,3-oxidoreductase (bond-cleaving)	1.13.11.13
Ascorbino dehydrogenase	1.10.3.3
Ascorbino oxhydrase	1.10.3.3
Asparaginase	3.5.1.1
Asparaginase II	3.5.1.1
L-Asparagine amidohydrolase	3.5.1.1
L-Asparagine:hydroxylamine γ-aspartyltransferase	2.3.2.7
Asparagine→hydroxylamine transaspartase	2.3.2.7
Asparagine→α-ketoacid transamidase	2.6.1.14
Asparagine–oxo-acid aminotransferase	2.6.1.14
L-Asparagine:2-oxo-acid aminotransferase	2.6.1.14
Asparagine synthetase	6.3.1.1
Asparagine synthetase (ADP-forming)	6.3.1.4
Asparagine synthetase (glutamine-hydrolysing)	6.3.5.4
1-L-β-Aspartamido-2-acetamido-1,2-dideoxy-β-D-glucose amidohydrolase	3.5.1.26
Aspartase	4.3.1.1
Aspartate acetyltransferase	2.3.1.17
Aspartate aminopeptidase	3.4.11.7
Aspartate aminotransferase	2.6.1.1
D-Aspartate aminotransferase	2.6.1.21
L-Aspartate:ammonia ligase (ADP-forming)	6.3.1.4
L-Aspartate:ammonia ligase (AMP-forming)	6.3.1.1
Aspartate ammonia-lyase	4.3.1.1
L-Aspartate ammonia-lyase	4.3.1.1
Aspartate carbamoyltransferase	2.1.3.2
L-Aspartate 4-carboxy-lyase	4.1.1.12
Aspartate carboxypeptidase	3.4.12.9
Aspartate 1-decarboxylase	[4.1.1.11]
Aspartate 4-decarboxylase	4.1.1.12
L-Aspartate:L-glutamine amido-ligase (AMP-forming)	6.3.5.4
Aspartate kinase	2.7.2.4
D-Aspartate→O_2 transhydrogenase	1.4.3.1
D-Aspartate oxidase	1.4.3.1
L-Aspartate:2-oxoglutarate aminotransferase	2.6.1.1
D-Aspartate:oxygen oxidoreductase (deaminating)	1.4.3.1
Aspartate-semialdehyde dehydrogenase	1.2.1.11
L-Aspartate-β-semialdehyde hydro-lyase (adding pyruvate and cyclizing)	4.2.1.52
L-Aspartate-β-semialdehyde:$NADP^+$ oxidoreductase (phosphorylating)	1.2.1.11
Aspartate transcarbamylase	2.1.3.2
L-Aspartate:tRNA^{Asp} ligase (AMP-forming)	6.1.1.12
Aspartic oxidase	1.4.3.1
Aspartoacylase	3.5.1.15
1-β-Aspartyl-2-acetamido-1,2-dideoxy-D-glucosylamine L-asparaginohydrolase	3.2.2.11
β-Aspartylacetylglucosaminidase	3.2.2.11
β-L-Aspartyl-amino-acid hydrolase	3.4.13.10
β-Aspartyldipeptidase	3.4.13.10
Aspartylendopeptidase	3.4.99.4
Aspartylglucosylaminase	3.5.1.26
Aspartylglucosylamine deaspartylase	3.5.1.26
L-α-Aspartyl(L-α-glutamyl)-peptide hydrolase	3.4.11.7
4-L-Aspartylglycosylamine amidohydrolase	3.5.1.37
β-Aspartyl peptidase	3.4.13.10
Aspartyltransferase	2.3.2.7
Aspartyl-tRNA synthetase	6.1.1.12
Aspergillopeptidase A	3.4.23.6
Aspergillopeptidase B	3.4.21.15
Aspergillus acid proteinase	3.4.23.6
Aspergillus alkaline proteinase	3.4.21.15
Aspergillus oryzae neutral proteinase	3.4.24.4
Aspergillus oryzae ribonuclease	3.1.4.8
Aspergillus proteinase B	3.4.21.15

INDEX TO THE ENZYME LIST (continued)

Assimilatory nitrate reductase	1.6.61, 1.6.6.2	ATP:deoxythymidinemonophosphate phosphotransferase	2.7.4.9
ATP:2-acetamido-2-deoxy-D-glucose 6-phosphotransferase	2.7.1.59	ATP:dephospho-CoA 3'-phosphotransferase	2.7.1.24
ATP:acetate phosphotransferase	2.7.2.1	ATP:5'-dephosphopolynucleotide 5'-phosphotransferase	2.7.1.78
ATP:N-acetyl-L-glutamate 5'-phosphotransferase	2.7.2.8	ATP-diphosphatase	3.6.1.5
ATP:2-acylamino-2-deoxy-D-mannose 6-phosphotransferase	2.7.1.60	ATP diphosphohydrolase	3.6.1.5
ATP:adenosine 5'-phosphotransferase	2.7.1.20	ATP:diphosphoinositide 5-phosphotransferase	2.7.1.68
ATP→adenosine transphosphatase	2.7.1.20	ATP:5-diphosphomevalonate carboxy-lyase (dehydrating)	4.1.1.33
ATP:adenylylsulphate 3'-phosphotransferase	2.7.1.25	ATP→DPN transphosphatase	2.7.1.23
ATP:D-allose 6-phosphotransferase	2.7.1.55	ATP:erythritol 4-phosphotransferase	2.7.1.27
ATP:2-amino-2-deoxy-D-glucose phosphotransferase	2.7.1.8	ATP:FMN adenylyltransferase	2.7.7.2
ATP:2-amino-2-deoxy-scyllo-inositol phosphotransferase	2.7.1.65	ATP→FMN transadenylase	2.7.7.2
ATP aminohydrolase	3.5.4.18	ATP:formate phosphotransferase	2.7.2.6
ATP:2-amino-4-hydroxy-6-hydroxymethyl-7,8-dihydropteridine 6'-pyrophosphotransferase	2.7.6.3	ATP:D-fructose-1-phosphate 6-phosphotransferase	2.7.1.56
ATP:ammonia phosphotransferase	2.7.3.8	ATP:D-fructose-6-phosphate 1-phosphotransferase	2.7.1.11
ATP:AMP phosphotransferase	2.7.4.3	ATP→fructose-6-phosphate transphosphatase	2.7.1.4
ATP:(d)AMP phosphotransferase	2.7.4.11	ATP:D-fructose 1-phosphotransferase	2.7.1.3
ATP:aquocob(I)alamin Co-adenosyl-transferase	2.5.1.17	ATP:D-fructose 6-phosphotransferase	2.7.1.4
ATP:L-arabinose 1-phosphotransferase	2.7.1.46	ATP→fructose transphosphatase	2.7.1.3
ATP:D-arabinose 5-phosphotransferase	2.7.1.54	ATP:L-fuculose 1-phosphotransferase	2.7.1.51
ATP→D-arabinose transphosphatase	2.7.1.54	ATP:D-galactose 1-phosphotransferase	2.7.1.6
ATP:L-arginine N$^\omega$-phosphotransferase	2.7.3.3	ATP→galactose transphosphatase	2.7.1.6
ATP→L-arginine transphosphatase	2.7.3.3	ATP:D-galacturonate 1-phosphotransferase	2.7.1.44
ATPase	3.6.1.3 3.6.1.8	ATP:D-gluconate 6-phosphotransferase	2.7.1.12
ATP:L-aspartate 4-phosphotransferase	2.7.2.4	ATP→gluconate transphosphatase	2.7.1.12
ATP→aspartate transphosphatase	2.7.2.4	ATP:α-D-glucose-1-phosphate adenylyltransferase	2.7.7.27
ATP:butyrate phosphotransferase	2.7.2.7	ATP:D-glucose-1-phosphate 6-phosphotransferase	2.7.1.10
ATP:carbamate phosphotransferase	2.7.2.2	ATP→glucose-1-phosphate transphosphatase	2.7.1.10
ATP:carbamate phosphotransferase (dephosphorylating)	2.7.2.5	ATP:D-glucose 6-phosphotransferase	2.7.1.2
ATP:carbamate phosphotransferase (dephosphorylating, amido-transferring)	2.7.2.9	ATP→glucose transphosphatase	2.7.1.1, 2.7.1.2
ATP:choline phosphotransferase	2.7.1.32	ATP:D-glucuronate 1-phosphotransferase	2.7.1.43
ATP→choline transphosphatase	2.7.1.32	ATP:[L-glutamate:ammonialigase (ADP-forming)] adenylyltransferase	2.7.7.42
ATP citrate (pro-3S)-lyase	4.1.3.8		
ATP:citrate oxaloacetate-lyase (pro-3S-CH$_2$·COO$^-$→acetyl-CoA; ATP-dephosphorylating)	4.1.3.8	ATP:D-glyceraldehyde 3-phosphotransferase	2.7.1.28
ATP:CMP phosphotransferase	2.7.4.14	ATP→glyceraldehyde transphosphatase	2.7.1.28
ATP:creatine N-phosphotransferase	2.7.3.2	ATP:D-glycerate 3-phosphotransferase	2.7.1.31
ATP→creatine transphosphatase	2.7.3.2	ATP:glycerol 3-phosphotransferase	2.7.1.30
ATP deaminase	3.5.4.18	ATP:(d)GMP phosphotransferase	2.7.4.8
ATP:deoxyadenosine 5'-phosphotransferase	2.7.1.76	ATP:guanidinoacetate N-phosphotransferase	2.7.3.1
ATP:6-deoxy-L-galactose 1-phosphotransferase	2.7.1.52	ATP:guanidinoethyl-methyl-phosphate phosphotransferase	2.7.3.7
ATP:deoxynucleosidemonophosphate phosphotransferase	2.7.4.13	ATP:D-hexose 6-phosphotransferase	2.7.1.1
		ATP→hexose transphosphatase	2.7.1.1

INDEX TO THE ENZYME LIST (continued)

ATP:L-homoserine O-phosphotransferase	2.7.1.39
ATP:hydroxyacetone phosphotransferase	2.7.1.29
ATP:hypotaurocyamine N^{ω}-phosphotransferase	2.7.3.6
ATP:inosine 5′-phosphotransferase	2.7.1.73
ATP:myo-inositol 1-phosphotransferase	2.7.1.64
ATP:C-55-isoprenoid-alcohol phosphotransferase	2.7.1.66
ATP:2-keto-3-deoxy-D-galactonate phosphotransferase	2.7.1.58
ATP:2-keto-3-deoxy-D-gluconate 6-phosphotransferase	2.7.1.45
ATP:2-keto-D-gluconate 6-phosphotransferase	2.7.1.13
ATP:lombricine N^{ω}-phosphotransferase	2.7.3.5
ATP:mannitol 1-phosphotransferase	2.7.1.57
ATP:D-mannose 6-phosphotransferase	2.7.1.7
ATP:L-methionine S-adenosyltransferase	2.5.1.6
ATP→methionine transadenosylase	2.5.1.6
ATP:2-methyl-4-amino-5-hydroxymethyl-pyrimidine 5-phosphotransferase	2.7.1.49
ATP:2-methyl-4-amino-5-phosphomethyl-pyrimidine phosphotransferase	2.7.4.7
ATP:4-methyl-5-(2′-hydroxyethyl)-thiazole 2′-phosphotransferase	2.7.1.50
ATP:mevalonate 5-phosphotransferase	2.7.1.36
ATP monophosphatase	3.6.1.3
ATP:NAD^{+} 2′-phosphotransferase	2.7.1.23
ATP→nicotinamide mononucleotide transadenylase	2.7.7.1
ATP:nicotinatemononucleotide adenylyltransferase	2.7.7.18
ATP:NMN adenylyltransferase	2.7.7.1
ATP:(d)NMP phosphotransferase	2.7.4.12
ATP:nucleosidediphosphate phosphotransferase	2.7.4.6
ATP→nucleosidediphosphate transphosphatase	2.7.4.6
ATP:nucleosidemonophosphate phosphotransferase	2.7.4.4
ATP:oxaloacetate carboxy-lyase (transphosphorylating)	4.1.1.49
ATP:pantetheine-4′-phosphate adenylyltransferase	2.7.7.3
ATP:pantetheine 4′-phosphotransferase	2.7.1.34
ATP:pantothenate 4′-phosphotransferase	2.7.1.33
ATP:phosphatidylinositol 4-phosphotransferase	2.7.1.67
ATP:3-phospho-D-glycerate 1-phosphotransferase	2.7.2.3
ATP phosphohydrolase	3.6.1.3
ATP:5-phosphomevalonate phosphotransferase	2.7.4.2
ATP phosphoribosyltransferase	2.4.2.17
ATP:phosphorylase-b phosphotransferase	2.7.1.38
ATP:polynucleotide adenylyltransferase	2.7.7.19
ATP:polyphosphate phosphotransferase	2.7.4.1
ATP:protamine O-phosphotransferase	2.7.1.70
ATP:protein phosphotransferase	2.7.1.37
ATP:pyridoxal 5-phosphotransferase	2.7.1.35
ATP→pyridoxal transphosphatase	2.7.1.35
ATP pyrophosphatase	3.6.1.8
ATP pyrophosphate-lyase (cyclizing)	4.6.1.1
ATP pyrophosphohydrolase	3.6.1.8
ATP:pyruvate,orthophosphate phosphotransferase	2.7.9.1
	2.7.9.1
ATP:pyruvate 2-O-phosphotransferase	2.7.1.40
ATP:L-rhamnulose 1-phosphotransferase	2.7.1.5
ATP:riboflavin 5′-phosphotransferase	2.7.1.26
ATP→riboflavin transphosphatase	2.7.1.26
ATP:D-ribose-5-phosphate 1-phosphotransferase	2.7.1.18
ATP:D-ribose-5-phosphate pyrophosphotransferase	2.7.6.1
ATP→ribose-5-phosphate transphosphatase	2.7.1.18
ATP:D-ribose 5-phosphotransferase	2.7.1.15
ATP→ribose transphosphatase	2.7.1.15
ATP:N-ribosylnicotinamide 5′-phosphotransferase	2.7.1.22
ATP:D-ribulose-5-phosphate 1-phosphotransferase	2.7.1.19
ATP→ribulose-5-phosphate transphosphatase	2.7.1.19
ATP:D-ribulose 5-phosphotransferase	2.7.1.47
ATP:L (or D)-ribulose 5-phosphotransferase	2.7.1.16
ATP:sedoheptulose 7-phosphotransferase	2.7.1.14
ATP:shikimate 5-phosphotransferase	2.7.1.71
ATP:streptidine 5-phosphotransferase	2.7.1.72
ATP-sulfurylase	2.7.7.4
ATP:sulphate adenylyltransferase	2.7.7.4
ATP:taurocyamine N^{ω}-phosphotransferase	2.7.3.4
ATP:thiamine-diphosphate phosphotransferase	2.7.4.15
ATP:thiamine pyrophosphotransferase	2.7.6.2
ATP→thiamine transphosphatase	2.7.6.2
ATP:thymidine 5′-phosphotransferase	2.7.1.75
ATP:tRNA adenylyltransferase	2.7.7.25
ATP:uridine 5′-phosphotransferase	2.7.1.48
ATP:D-xylulose 5-phosphotransferase	2.7.1.17
ATP:L-xylulose 5-phosphotransferase	2.7.1.53
Atropin esterase	3.1.1.1
Azobenzene reductase	1.6.6.7
Azotobacter nuclease	3.1.4.9
Bacillus amyloliquefaciens serine proteinase	3.4.21.15
Bacillus licheniformis proteinase	3.4.21.14
Bacillus macerans amylase	2.4.1.19

INDEX TO THE ENZYME LIST (continued)

Bacillus pumilis proteinase	3.4.21.14	Biotin—[propionyl-CoA-carboxylase (ATP-hydrolysing)] synthetase	6.3.4.10
Bacillus subtilis neutral proteinase	3.4.24.4	Biotin—[methylmalonyl-CoA-carboxyltransferase] synthetase	6.3.4.9
Bacillus thermoproteolyticus neutral proteinase	3.4.24.4	Biotinyl-CoA synthetase	6.2.1.11
Bacterial proteinase *Novo*	3.4.21.14	Bisphosphoglycerate phosphatase	3.1.3.13
Bacteroides collagenase	3.4.24.3	*2,3-Bisphospho-D-glycerate:2-phospho-D-glycerate phosphotransferase*	2.7.5.3
Baker's yeast proteinase	3.4.22.9		
Baker's yeast proteinase A	3.4.23.8	*2,3-Bisphospho-D-glycerate 2-phosphohydrolase*	3.1.3.13
Baker's yeast proteinase B	3.4.22.9		
Baker's yeast proteinase C	3.4.22.9	**Bisphosphoglyceromutase**	2.7.5.4
Barbiturase	3.5.2.1	Bisphosphoglyceromutase	5.4.2.1
Barbiturate amidohydrolase	3.5.2.1	Blood-group-substance α-galactosyltransferase	2.4.1.37
Batyl-alcohol, hydrogen-donor:oxygen oxidoreductase	1.14.99.17	BPN′	3.4.21.14
Bee proteinase	3.4.21.4	Branched-chain-amino-acid aminotransferase	2.6.1.42
Benzaldehyde dehydrogenase (NAD⁺)	1.2.1.28		
Benzaldehyde dehydrogenase (NADP⁺)	1.2.1.7	*Branched-chain-amino-acid:2-oxoglutarate aminotransferase*	2.6.1.42
Benzaldehyde:NAD⁺ oxidoreductase	1.2.1.28		
Benzaldehyde:NADP⁺ oxidoreductase	1.2.1.7	Branching enzyme	2.4.1.18
Benzaldehyde→PN transhydrogenase	1.2.1.7	Brewer's yeast proteinase	3.4.22.9
Benzamidase	3.5.1.14	Brewer's yeast proteinase β	3.4.22.9
Benzene 1,2-dioxygenase	1.14.12.3	**Bromelain (juice)**	3.4.22.5
Benzene hydroxylase	1.14.12.3	**Bromelain (stem)**	3.4.22.4
Benzene,NADH:oxygen 1,2-oxidoreductase	1.14.12.3	**Bromelin**	3.4.22.5
		D(−)-Butanediol dehydrogenase	1.1.1.4
Benzo[a]pyrene,reduced-flavoprotein:oxygen oxidoreductase (3-hydroxylating)	1.14.14.2	**L(+)-Butanediol dehydrogenase**	1.1.1.76
		D(−)-2,3,-Butanediol:NAD⁺ reductase	1.1.1.4
Benzoate 1,2-dioxygenase	1.13.99.2	*L(+)-2,3-Butanediol:NAD⁺ oxidoreductase*	1.1.1.76
Benzoate hydroxylase	1.13.99.2		
Benzoate:oxygen oxidoreductase	1.13.99.2	Butyleneglycol dehydrogenase	1.1.1.4
Benzopyrene hydroxylase	1.14.14.2	*Butyrate:CoA ligase (AMP-forming)*	6.2.1.2
Benzopyrene 3-monooxygenase	1.14.14.2	Butyrate CoA-transferase	[2.8.3.4]
N-Benzoylamino-acid aminohydrolase	3.5.1.32	**Butyrate kinase**	2.7.2.7
Benzoylcholinesterase	3.1.1.8	γ-**Butyrobetaine,2-oxoglutarate dioxygenase**	1.14.11.1
Benzoylformate carboxy-lyase	4.1.1.7		
Benzoylformate decarboxylase	4.1.1.7	Butyrylcholine esterase	3.1.1.8
Benzyl-thiocyanate isomerase	5.99.1.1	*Butyryl-CoA:(acceptor) oxidoreductase*	1.3.99.2
B-esterase	3.1.1.1	**Butyryl-CoA dehydrogenase**	1.3.99.2
Betainaldehyde→DPN transhydrogenase	1.2.1.8	*Butyryl-CoA:orthophosphate butyryltransferase*	2.3.1.19
Betaine-aldehyde dehydrogenase	1.2.1.8		
Betaine-aldehyde:NAD⁺ oxidoreductase	1.2.1.8	**Butyryl-CoA synthetase**	6.2.1.2
Betaine—homocysteine methyltransferase	2.1.1.5	Butyryl dehydrogenase	1.3.99.2
Betaine:L-homocysteine S-methyltransferase	2.1.1.5	Caffeate 3,4-dioxygenase	1.13.11.22
Bilirubin:NAD(P)⁺ oxidoreductase	1.3.1.24	Callose synthetase	[2.4.1.34]
Biliverdin reductase	1.3.1.24	**Camphor 1,2-monooxygenase**	1.14.15.2
Biotin-amide amidohydrolase	3.5.1.12	**Camphor 5-monooxygenase**	1.14.15.1
Biotin:apo-[3-methylcrotonoyl-CoA:carbondioxide ligase (ADP-forming)] ligase (AMP-forming)	6.3.4.11	*Camphor,reduced-putida-ferredoxin:oxygen oxidoreductase (5-hydroxylating)*	1.14.15.1
Biotin:apo-[methylmalonyl-CoA:pyruvate carboxyltransferase] ligase (AMP-forming)	6.3.4.9	*Camphor,reduced-rubredoxin:oxygen oxidoreductase (1,2-lactonizing)*	1.14.15.2
Biotin:apo-[propionyl-CoA:carbon-dioxide ligase (ADP-forming)] ligase (AMP-forming)	6.3.4.10	Canavanase	3.5.3.1
Biotin:CoA ligase (AMP-forming)	6.2.1.11	Carbamate kinase	2.7.2.2
Biotinidase	3.5.1.12	*N-Carbamoyl-β-alanine amidohydrolase*	3.5.1.6
Biotin—[methylcrotonoyl-CoA-carboxylase] synthetase	6.3.4.11	*N-Carbamoyl-L-aspartate amidohydrolase*	3.5.1.7

INDEX TO THE ENZYME LIST (continued)

Carbamoylaspartate decarboxylase	[4.1.1.13]
Carbamoylaspartic dehydrase	3.5.2.3
Carbamoylphosphate:L-aspartate carbamoyltransferase	2.1.3.2
Carbamoylphosphate:L-ornithine carbamoyltransferase	2.1.3.3
Carbamoyl-phosphate synthase (ammonia)	2.7.2.5
Carbamoyl-phosphate synthase (glutamine)	2.7.2.9
Carbamylaspartotranskinase	2.1.3.2
Carbonate dehydratase	4.2.1.1
Carbonate hydro-lyase	4.2.1.1
Carbonic anhydrase	4.2.1.1
Carobxycathepsin	3.4.15.1
Carboxydismutase	4.1.1.39
N^2-*(1-Carboxyethyl)-L-arginine:NAD$^+$ oxidoreductase(L-arginine-forming)*	1.5.1.11
3-Carboxyethylcatechol 2,3-dioxygenase	1.13.11.16
α-Carboxylase	4.1.1.1
Carboxylesterase	3.1.1.1
Carboxylic-ester hydrolase	3.1.1.1
4-Carboxymethylbut-3-enolide(1,4) enol-lactone-hydrolase	3.1.1.24
L-5-Carboxymethylhydantoin amidohydrolase	3.5.2.4
Carboxymethylhydantoinase	3.5.2.4
4-Carboxymethyl-4-hydroxyisocrotonolactonase	[3.1.1.16]
4-Carboxymethyl-4-hydroxyisocrotonolactone lyase (decyclizing)	5.5.1.1
Carboxy-*cis-cis*-muconate cyclase	5.5.1.5
3-Carboxy-*cis-cis*-muconate cycloisomerase	5.5.1.2
4-Carboxymuconolactone carboxy-lyase	4.1.1.44
4-Carboxymuconolactone decarboxylase	4.1.1.44
3-Carboxymuconolactone lyase (decyclizing)	5.5.1.5
4-Carboxymuconolactone lyase (decyclizing)	5.5.1.2
Carboxypeptidase A	3.4.12.2
Carboxypeptidase B	3.4.12.3
Carboxypeptidase C	3.4.12.1
Carboxypeptidase, yeast	3.4.12.8
1-(2'-Carboxyphenylamino)-1-deoxyribulose-5-phosphate carboxy-lyase (cyclizing)	4.1.1.48
Carboxypolypeptidase	3.4.12.2
Carnitine acetyltransferase	2.3.1.7
Carnitine carboxy-lyase	4.1.1.42
Carnitine decarboxylase	4.1.1.42
Carnitine dehydrogenase	1.1.1.108
Carnitine:NAD$^+$ oxidoreductase	1.1.1.108
Carnitine palmitoyltransferase	2.3.1.21
Carnosinase	3.4.13.3
Carnosine N-methyltransferase	2.1.1.22
Carnosine synthetase	6.3.2.11
β-**Carotene 15,15'-dioxygenase**	1.13.11.21
Carotene oxidase	1.13.11.12
β-Carotene:oxygen 15,15'-oxidoreductase (bond-cleaving)	1.13.11.21
κ-**Carrageenanase**	3.2.1.83
κ-*Carrageenan 4-β-glycanohydrolase*	3.2.1.83
Catalase	1.11.1.6
Catechol 1,2-dioxygenase	1.13.11.1
Catechol 2,3-dioxygenase	1.13.11.2
Catechol methyltransferase	2.1.1.6
Catechol oxidase	1.14.18.1
Catechol oxidase (dimerizing)	1.1.3.14
Catechol oxygen 1,2-oxidoreductase (decyclizing)	1.13.11.1
Catechol:oxygen 2,3-oxidoreductase (decyclizing)	1.13.11.2
Catechol:oxygen oxidoreductase (dimerizing)	1.1.3.14
Cathepsin B	3.4.22.1
Cathepsin B_1	3.4.22.1
Cathepsin B'	3.4.22.1
Cathepsin C	3.4.14.1
Cathepsin D	3.4.23.5
Cathepsin E	3.4.23.5
CDPabequose epimerase	5.1.3.10
CDPabequose 2-epimerase	5.1.3.10
CDPabequose:D-mannosyl-rhamnosyl-galactose-1-diphospholipid abequosyltransferase	2.4.1.60
CDPcholine:N-acylsphingosine cholinephosphotransferase	2.7.8.3
CDPcholine:1,2-diacylglycerol cholinephosphotransferase	
CDPcholine:1,2-diacylglycerol choline phosphotranspherase	2.7.8.10
CDPdiglyceride:sn-glycerol-3-phosphate phosphatidyltransferase	2.7.8.5
CDPdiglyceride−inositol phosphatidyltransferase	2.7.8.11
CDPdiglyceride:myo-inositol phosphatidyltransferase	2.7.8.11
CDPdiglyceride pyrophosphorylase	2.7.7.41
CDPdiglyceride−serine O-phosphatidyltransferase	2.7.8.8
CDPdiglyceride:L-serine O-phosphatidyltransferase	2.7.8.8
CDPethanolamine:1,2-diacylglycerol ethanolaminephosphotransferase	2.7.8.1
CDPethanolamine:L-serine ethanolaminephosphotransferase	2.7.8.4
CDPglucose 4,6-dehydratase	4.2.1.45
CDPglucose 4,6-hydro-lyase	4.2.1.45
CDPglucose pyrophosphorylase	2.7.7.33
CDPglycerol glycerophosphotransferase	2.7.8.12
CDPglycerol phosphoglycerylhydrolase	3.6.1.16
CDPglycerol:poly(glycerophosphate) glycerophosphotransferase	2.7.8.12
CDPglycerol pyrophosphatase	3.6.1.16
CDPglycerol pyrophosphorylase	2.7.7.39

INDEX TO THE ENZYME LIST (continued)

CDP-4-keto-6-deoxy-D-glucose reductase	1.17.1.1
CDP-4-keto-3,6-dideoxy-D-glucose:NAD(P)⁺ 3-oxidoreductase	1.17.1.1
CDPparatose epimerase	5.1.3.10
CDPribitol pyrophosphorylase	2.7.7.40
CDPribitol:teichoic-acid phosphoribitoltransferase	2.4.1.55
Cellobiase	3.2.1.21
Cellobiose epimerase	5.1.3.11
Cellobiose 2-epimerase	5.1.3.11
Cellobiose:orthophosphate α-glucosyltransferase	2.4.1.20
Cellobiose phosphorylase	2.4.1.20
Cellodextrin phosphorylase	2.4.1.49
Cellulase	3.2.1.4
Cellulose polysulphatase	3.1.6.7
Cellulose-sulphate sulphohydrolase	3.1.6.7
Cellulose synthase (GDP-forming)	2.4.1.29
Cellulose synthase (UDP-forming)	2.4.1.12
Cephalosporin amido-β-lactam-hydrolase	3.5.2.8
Cephalosporinase	3.5.2.8
Ceramide cholinephosphotransferase	2.7.8.3
Cerebroside-sulphatase	3.1.6.8
Cerebroside-3-sulphate 3-sulphohydrolase	3.1.6.8
C-esterase	3.1.1.6
Chalcone isomerase	5.5.1.6
Chitinase	3.2.1.14
Chitin synthase	2.4.1.16
Chitin–UDP acetylglucosaminyltransferase	2.4.1.16
Chitobiase	[3.2.1.29]
Chitodextrinase	3.2.1.14
Chloramphenicol acetyltransferase	2.3.1.28
Chloride:hydrogen-peroxide oxidoreductase	1.11.1.10
Chloride peroxidase	1.11.1.10
Chlorophyllase	3.1.1.14
Chlorophyll chlorophyllido-hydrolase	3.1.1.14
Cholate:CoA ligase (AMP-forming)	6.2.1.7
Cholate thiokinase	6.2.1.7
Cholestenol Δ-isomerase	5.3.3.5
Δ⁷-Cholestenol Δ⁷–Δ⁸-isomerase	5.3.3.5
5α-Cholest-7-en-3β-ol:oxygen Δ⁵-oxidoreductase	1.3.3.2
Cholestenone 5α-reductase	1.3.1.22
Cholestenone 5β-reductase	1.3.1.23
Cholesterol acyltransferase	2.3.1.26
Cholesterol esterase	3.1.1.13
Cholesterol 20-hydroxylase	[1.14.1.9]
Cholesterol:NADP⁺ Δ⁷-oxidoreductase	1.3.1.21
Cholesterol oxidase	1.1.3.6
Cholesterol:oxygen oxidoreductase	1.1.3.6
Choline:(acceptor) oxidoreductase	1.1.99.1
Choline acetylase	2.3.1.6
Choline acetyltransferase	2.3.1.6
Choline dehydrogenase	1.1.99.1
Choline esterase I	3.1.1.7
Choline esterase II (unspecific)	3.1.1.8
Choline kinase	2.7.1.32
Choline phosphatase	3.1.4.4
Cholinephosphate cytidylyltransferase	2.7.7.15
Cholinephosphotransferase	2.7.8.2
Cholinesterase	3.1.1.8
Cholinesterase	3.1.1.7
Cholinesulphatase	3.1.6.6
Choline-sulphate sulphohydrolase	3.1.6.6
Choline sulphotransferase	2.8.2.6
Choloyl-CoA synthetase	6.2.1.7
Choloylglycine hydrolase	3.5.1.24
Chondroitin ABC eliminase	4.2.2.4
Chondroitin ABC lyase	4.2.2.4
Chondroitin ABC lyase	4.2.2.4
Chondroitin AC eliminase	4.2.2.5
Chondroitin AC lyase	4.2.2.5
Chondroitin AC lyase	4.2.2.5
Chondroitinase	3.1.6.4, 4.2.2.4, 4.2.2.5
Chondroitinsulphatase	3.1.6.4
Chondroitin sulphate lyase	4.2.2.5
Chondroitin-sulphate sulphohydrolase	3.1.6.4
Chondroitin sulphotransferase	2.8.2.5
Chondro-4-sulphatase	3.1.6.9
Chondro-6-sulphatase	3.1.6.10
Chorismate mutase	5.4.99.5
Chorismate pyruvate-lyase (amino-accepting)	4.1.3.27
Chorismate pyruvatemutase	5.4.99.5
Chymase	3.4.23.4
Chymopapain	3.4.22.6
Chymopapain A	3.4.22.6
Chymopapain B	3.4.22.6
Chymosin	3.4.23.4
Chymotrypsin	3.4.21.1
Chymotrypsin A	3.4.21.1
Chymotrypsin B	3.4.21.1
Chymotrypsin C	3.4.21.2
Citraconate hydratase	4.2.1.35
Citramalate CoA-transferase	2.8.3.7
(+)-Citramalate hydro-lyase	4.2.1.34
(−)-Citramalate hydro-lyase	4.2.1.35
Citramalate lyase	4.1.3.22
Citramalate pyruvate-lyase	4.1.3.22
Citramalyl-CoA hydro-lyase	4.2.1.56
Citramalyl-CoA lyase	4.1.3.25
Citramalyl-CoA pyruvate-lyase	4.1.3.25
Citrase	4.1.3.6
Citratase	4.1.3.6
Citrate aldolase	4.1.3.6
Citrate cleavage enzyme	4.1.3.8
Citrate condensing enzyme	4.1.3.7
Citrate dehydratase	4.2.1.4
Citrate hydro-lyase	4.2.1.4
Citrate (isocitrate) hydro-lyase	4.2.1.3
Citrate (pro-3S)-lyase	4.1.3.6
Citrate oxaloacetone-lyase (pro-3S-$CH_2 \cdot COO^-\rightarrow$acetate)	4.1.3.6

INDEX TO THE ENZYME LIST (continued)

Citrate oxalacetate-lyase (pro-*3S*-$CH_2 \cdot COO^- \to$ *acetyl-CoA*)	4.1.3.28
Citrate oxaloacetate-lyase (pro-*3S*-$CH_2 \cdot COO^- \to$ *acetyl-CoA*)	4.1.3.7
Citrate (*re*)-synthase	4.1.3.28
Citrate (*si*)-synthase	4.1.3.7
Citridesmolase	4.1.3.6
Citritase	4.1.3.6
Citrogenase	4.1.3.7
Citrullinase	3.5.1.20
L-Citrulline:L-aspartate ligase (AMP-forming)	6.3.4.5
L-Citrulline N^5-*carbamoyldihydrolase*	3.5.1.20
Citrulline phosphorylase	2.1.3.3
Clearing factor lipase	3.1.1.34
Clostridiopeptidase A	3.4.24.3
Clostridiopeptidase B	3.4.22.8
Clostridium histolyticum collagenase	3.4.24.3
Clostridium histolyticum **collagenase 2**	3.4.99.5
Clostridium histolyticum **proteinase B**	3.4.22.8
Clostridium oedematiens β- and γ-toxins	3.1.4.3
Clostridium welchii α-toxin	3.1.4.3
Clostripain	3.4.22.8
CMP-N-acetylneuraminate: D-galactosylglycoprotein N-acetylneuraminyltransferase	2.4.99.1
dCMP aminohydrolase	3.5.4.12
dCMP deaminase	3.5.4.12
CMP-3-deoxy-D-*manno*-octulosonate pyrophosphorylase	2.7.7.38
CMPsialate pyrophosphorylase	2.7.7.43
CMPsialate synthase	2.7.7.43
CoA:apo-[acyl-carrier-protein] pantetheinephosphotransferase	2.7.8.7
CoAS—S glutathione reductase (NADPH)	1.6.4.6
Cocain esterase	3.1.1.1
Cocoonase	3.4.21.4
Coenzyme A:oxidized-glutathione oxidoreductase	1.8.4.3
Coeruloplasmin	1.16.3.1
Collagenase	3.4.24.3
Collagenase A	3.4.24.3
Collagenase 2, *Clostridium histolyticum*	3.4.99.5
Condensing enzyme	4.1.3.7
Coproporphyrinogenase	1.3.3.3
Coproporphrinogen oxidase	1.3.3.3
Coproporphrinogen:oxygen oxidoreductase (decarboxylating)	1.3.3.3
Corticosterone 18-hydroxylase	1.14.15.5
Corticosterone 18-monooxygenase	1.14.15.5
Corticosterone, reduced-adrenal-ferredoxin:oxygen oxidoreductase (18-hydroxylating)	1.14.15.5
Cortisol acetyltransferase	2.3.1.27
Cortisone reductase	1.1.1.53
Cortisone α-reductase	1.3.1.4
Cortisone β-reductase	1.3.1.3
p-Coumarate hydroxylase	1.14.17.2
p-Coumarate 3-monooxygenase	1.14.17.2
Crayfish low molecular weight proteinase	3.4.99.6
Crayfish proteinase	3.4.21.4
Creatinase	3.5.3.3
Creatine amidinohydrolase	3.5.3.3
Creatine kinase	2.7.3.2
Creatinine deiminase	3.5.4.21
Creatinine iminohydrolase	3.5.4.21
Crotalus atrox proteinase	3.4.24.1
Crotonase	4.2.1.17
Crotonoyl- [acyl-carrier-protein] hydratase	4.2.1.58
CTP:N-acylneuraminate cytidylyltransferase	2.7.7.43
dCTP aminohydrolase	3.5.4.13
dCTPase	3.6.1.12
CTP:cholinephosphate cytidylyltransferase	2.7.7.15
dCTP deaminase	3.5.4.13
CTP:3-deoxy-D-manno-octulosonate cytidylyltransferase	2.7.7.38
CTP:ethanolaminephosphate cytidylyltransferase	2.7.7.14
CTP:D-glucose-1-phosphate cytidylyltransferase	2.7.7.33
CTP:sn-glycerol-3-phosphate cytidylyltransferase	2.7.7.39
dCTP nucleotidohydrolase	3.6.1.12
CTP:phosphatidate cytidylyltransferase	2.7.7.41
CTP:D-ribitol-5-phosphate cytidylyltransferase	2.7.7.40
CTP synthetase	6.3.4.2
CTP:tRNA cytidylyltransferase	2.7.7.21
Cucurbitacin Δ^{23}**-reductase**	1.3.1.5
Cyanase	3.5.5.3
Cyanate aminohydrolase	3.5.5.3
Cyanate hydrolase	3.5.5.3
β-Cyanoalanine synthase	4.4.1.9
3':5'-Cyclic-AMP 5'-nucleotidohydrolase	3.1.4.17
3':5'-Cyclic-AMP phosphodiesterase	3.1.4.17
Cyclic-dextrin dextrin-hydrolase (decyclizing)	3.2.1.54
2':3'-Cyclic-nucleosidemonophosphate phosphodiesterase	3.1.4.16
Cyclodextrinase	3.2.1.54
Cyclodextrin glucanotransferase	2.4.1.19
Cycloheptaglucanase	3.2.1.12
Cyclohepta-D-glucan 4-glucanohydrolase	3.2.1.12
Cyclohexaglucanase	3.2.1.13
Cyclohexa-D-glucan 4-glucanohydrolase	3.2.1.13
Cystathionase	4.4.1.1
β-Cystathionase	4.4.1.8
γ-Cystathionase	4.4.1.1
L-Cystathionine cysteine-lyase (deaminating)	4.4.1.1
Cystathionine L-homocysteine-lyase (deaminating)	4.4.1.8
Cystathionine β-lyase	4.4.1.8
Cystathionine γ-lyase	4.4.1.1

INDEX TO THE ENZYME LIST (continued)

Cystathionine β-synthase	4.2.1.22
Cystathionine γ-synthase	4.2.99.9
Cysteamine dehydrogenase	[1.8.1.1]
Cysteamine dioxygenase	1.13.11.19
Cysteamine:oxygen oxidoreductase	1.13.11.19
Cysteine aminotransferase	2.6.1.3
Cysteine desulphhydrase	4.4.1.1
Cysteine dioxygenase	1.13.11.20
L-Cysteine hydrogen-sulphide-lyase (adding HCN)	4.4.1.9
L-Cysteine hydrogen-sulphide-lyase (adding sulphite)	4.4.1.10
Cysteine lyase	4.4.1.10
L-Cysteine:2-oxyglutarate aminotransferase	2.6.1.3
L-Cysteine:oxygen oxidoreductase	1.13.11.20
Cysteinesulphinate decarboxylase	[4.1.1.29]
Cysteine synthase	4.2.99.8
L-Cysteine:tRNA^{Cys} ligase (AMP-forming)	6.1.1.16
Cysteinyl-glycine dipeptidase	3.4.13.6
L-Cysteinyl-glycine hydrolase	3.4.13.6
Cysteinyl-tRNA synthetase	6.1.1.16
Cystine desulphhydrase	4.4.1.1
Cystine reductase (NADH)	1.6.4.1
Cystyl-aminopeptidase	3.4.11.3
Cytidine aminohydrolase	3.5.4.5
Cytidine deaminase	3.5.4.5
Cytidylate kinase	2.7.4.14
Cytochrome a_3	1.9.3.1
Cytochrome b_5 reductase	1.6.2.2
Cytochrome *cd*	1.9.3.2
Cytochrome c_3 hydrogenase	1.12.2.1
Cytochrome *c* oxidase	1.9.3.1
Cytochrome c^Δ reductase	1.6.99.3
Cytochrome *c* reductase (NADPH)	[1.6.2.3]
Cytochrome oxidase	1.9.3.1
Cytochrome oxidase, Pseudomonas	1.9.3.2
Cytochrome peroxidase	1.11.1.5
Cytosine aminohydrolase	3.5.4.1
Cytosine deaminase	3.5.4.1
Cytosol aminopeptidase	3.4.11.1
DDT-dehydrochlorinase	4.5.1.1
Deamido-NAD$^+$:ammonia ligase (AMP-forming)	6.3.1.5
Deamido-NAD$^+$:L-glutamine amido-ligase (AMP-forming)	6.3.5.1
Deamido-NAD$^+$ pyrophosphorylase	2.7.7.18
Debranching enzyme	3.2.1.41, 3.2.1.68, 3.2.1.69
Decarboxylating isocitrate→PN transhydrogenase	1.1.1.41, 1.1.1.42
Decarboxylating L-malate→PN transhydrogenase	1.1.1.38, 1.1.1.39, 1.1.1.40
Decarboxylating 6-phosphogluconate→TPN transhydrogenase	1.1.1.44
Decylcitrate oxaloacetate-lyase (CoA-acylating)	4.1.3.23
Decylcitrate synthase	4.1.3.23
7-Dehydrocholesterol reductase	1.3.1.21
Dehydropeptidase II	3.5.1.14
5-Dehydroquinate dehydratase	4.2.1.10
5-Dehydroquinate hydro-lyase	4.2.1.10
O-Demethylpuromycin methyltransferase	2.1.1.38
D-enzyme	2.4.1.25
Deoxyadenosine kinase	2.7.1.76
5'-Deoxyadenosyl-(5'),3-aminopropyl-(1),methylsulphonium-salt:putrescine 3-aminopropyltransferase	2.5.1.16
(Deoxy)adenylate kinase	2.7.4.11
Deoxy-CTPase	3.6.1.12
Deoxycytidine aminohydrolase	3.5.4.14
Deoxycytidine deaminase	3.5.4.14
Deoxycytidine kinase	2.7.1.74
Deoxycytidinetriphosphatase	3.6.1.12
Deoxycytidylate hydroxymethyltransferase	2.1.2.8
Deoxycytidylate kinase	2.7.4.14
6-Deoxy-L-galactose:NAD$^+$ 1-oxidoreductase	1.1.1.122
2-Deoxy-D-gluconate dehydrogenase	1.1.1.125
2-Deoxy-D-gluconate:NAD$^+$ 3-oxidoreductase	1.1.1.125
Deoxy-GTPase	3.1.5.1
Deoxygunanylate kinase	2.7.4.8
4-Deoxy-L-*threo*-5-hexulose-uronate ketol-isomerase	5.3.1.17
4-Deoxy-L-threo-5-hexosulose-uronate ketol-isomerase	5.3.1.17
Deoxynucleosidemonophosphate kinase	2.7.4.13
Deoxynucleosidetriphosphate:DNA deoxynucleotidyltransferase	2.7.7.7
3'-Deoxynucleotidase	3.1.3.34
Deoxynucleotide 3'-phosphatase	3.1.3.34
3-Deoxy-D-*manno*-octulosonate aldolase	4.1.2.23
3-Deoxy-D-manno-octulosonate D-arabinose-lyase	4.1.2.23
3-Deoxy-*manno*-octulosonate cytidylytransferase	2.7.7.38
Deoxyriboaldolase	4.1.2.4
Deoxyribocyclobutadipyrimidine pyrimidine-lyase	4.1.99.3
Deoxyribodipyrimidine photolyase	4.1.99.3
Deoxyribonuclease I	3.1.4.5
Deoxyribonuclease II	3.1.4.6
Deoxyribonucleate 5'-dinucleotidohydrolase	3.1.4.31
Deoxyribonucleate (double-stranded) 5'-nucleotidohydrolase	3.1.4.27

INDEX TO THE ENZYME LIST (continued)

Deoxyribonucleate 5'-nucleotidohydrolase	3.1.4.26
Deoxyribonucleate oligonucleotidohydrolase	3.1.4.30
Deoxyribonucleate 3'-oligonucleotidohydrolase	3.1.4.6
Deoxyribonucleate 5'-oligonucleotidohydrolase	3.1.4.5
Deoxyribonucleate 5'-oligonucleotidohydrolase (ATP-hydrolyzing)	3.1.4.33
Deoxyribonucleate polynucleotidohydrolase (ATP- and S-adenosyl-L-methionine-dependent)	3.1.4.32
Deoxyribonucleate (single-stranded) 5'-oligonucleotidohydrolase	3.1.4.25
Deoxyribonucleodepolymerase	3.1.4.5
2'-Deoxyribonucleoside-diphosphate:oxidized thioredoxin 2'-oxidoreductase	1.17.4.1
2'-Deoxyribonucleoside-triphosphate:oxidized-thioredoxin 2'-oxidoreductase	1.17.4.2
Deoxyribonucleotide 3'-phosphohydrolase	3.1.3.34
2-Deoxy-D-ribose-5-phosphate acetaldehyde-lyase	4.1.2.4
Deoxyribose-phosphate aldolase	4.1.2.4
Deoxyuridine:orthophosphate deoxyribosyltransferase	2.4.2.23
Deoxyuridine phosphorylase	2.4.2.23
Deoxyuridinetriphosphatase	3.6.1.23
Dephospho-CoA kinase	2.7.1.24
Dephospho-CoA pyrophosphorylase	2.7.7.3
Dephosphophosphorylase kinase	2.7.1.38
Desulphoheparin sulphotransferase	2.8.2.8
Dextranase	3.2.1.11
Dextransucrase	2.4.1.5
Dextrin dextranase	2.4.1.2
Dextrin→dextran transglucosidase	2.4.1.2
Dextrin 6-α-glucanohydrolase	3.2.1.10
Dextrin 6-α-glucosidase	3.2.1.33
Dextrin 6-glucosyltransferase	2.4.1.2
Dextrin glycosyltransferase	2.4.1.25
DFPase	3.8.2.1
DHAP synthase	4.1.2.14
2,3-Di-O-acyl-1-O-(β-D-galactosyl)-D-glycerol acylhydrolase	3.1.1.26
Diacylglycerol acylhydrolase	3.1.1.34
Diacylglycerol acyltransferase	2.3.1.20
Diacylglycerol lipase	3.1.1.34
2,2-Dialkyl-L-amino-acid carboxy-lyase (amino-transferring)	4.1.1.64
Dialkylamino-acid decarboxylase (pyruvate)	4.1.1.64
Diamine aminotransferase	2.6.1.29
Diamine oxidase	1.4.3.6
Diamine: 2-oxoglutarate aminotransferase	2.6.1.29
Diamino-acid aminotransferase	2.6.1.8
Diaminobutyrate−pyruvate aminotransferase	2.6.1.46
L-2,4-Diaminobutyrate:pyruvate aminotransferase	2.6.1.46
D-2,6-Diaminohexanoate aminomutase	5.4.3.4
L-3,6-Diaminohexanoate aminomutase	5.4.3.3
Diamino→O$_2$ transhydrogenase	1.4.3.6
Diamino oxhydrase	1.4.3.6
meso-2,6-Diaminopimelate carboxy-lyase	4.1.1.20
Diaminopimelate decarboxylase	4.1.1.20
Diaminopimelate epimerase	5.1.1.7
2,6-LL-Diaminopimelate 2-epimerase	5.1.1.7
2,5-Diaminovalerate: 2-oxoglutarate aminotransferase	2.6.1.8
Diaphorase	1.6.4.3
Diastase	3.2.1.1, 3.2.1.2
N^5-(1,3-Dicarboxypropyl)-L-lysine:NAD$^+$ oxidoreductase (L-glutamate-forming)	1.5.1.9
N^5-(1,3-Dicarboxypropyl)-L-lysine:NAD$^+$ oxidoreductase (L-lysine-forming)	1.5.1.7
N^5-(1,3-Dicarboxypropyl)-L-lysine:NADP$^+$ oxidoreductase (L-glutamate-forming)	1.5.1.10
N^5-(1,3-Dicarboxypropyl)-L-lysine:NADP$^+$ oxidoreductase (L-lysine-forming)	1.5.1.8
Diglyceride acyltransferase	2.3.1.20
Diglyceride lipase	3.1.1.34
Diguanosinetetraphosphatase	3.6.1.17
Diguanosine-tetraphosphate guanylohydrolase	3.6.1.17
cis-1,2-Dihydrobenzene-1,2-diol dehydrogenase	1.3.1.19
trans-1,2-Dihydrobenzene-1,2-diol dehydrogenase	1.3.1.20
cis-1,2-Dihydrobenzene-1,2-diol:NAD$^+$ oxidoreductase	1.3.1.19
trans-1,2-Dihydrobenzene-1,2-diol:NADP$^+$ oxidoreductase	1.3.1.20
7,8-Dihydrobiopterin:NADP$^+$ oxidoreductase	1.1.1.153
4,5α-Dihydrocortisone:NADP$^+$ Δ^4-oxidoreductase	1.3.1.4
4,5β-Dihydrocortisone:NADP$^+$ Δ^4-oxidoreductase	1.3.1.3
Dihydrocoumarin hydrolase	3.1.1.35
Dihydrocoumarin hydrolase	3.1.1.35
23,24-Dihydrocucurbitacin:NAD(P)$^+$ Δ$^{2\,3}$-oxidoreductase	1.3.1.5
2,3-Dihydro-2,3-dihydroxybenzoate dehydrogenase	1.1.1.109
2,3-Dihydro-2,3-dihydroxybenzoate:NAD$^+$ oxidoreductase	1.1.1.109
2,3-Dihydro-2,3-dihydroxybenzoate synthase	3.3.2.1
7,8-Dihydro-7,8-dihydroxykynurenate:NAD$^+$ oxidoreductase	1.3.1.18
Dihydrodiol hydro-lyase (arene-oxide-forming)	4.2.1.64
Dihydrofolate dehydrogenase	1.5.1.4

INDEX TO THE ENZYME LIST (continued)

7,8-Dihydrofolate:NADP⁺ oxidoreductase	1.5.1.4
Dihydrofolate synthetase	6.3.2.12
Dihydro-α-lipoic acid→DPN transhydrogenase	1.6.4.3
Dihydroneopterin aldolase	4.1.2.25
Dihydro-orotase	3.5.2.3
L-5,6-Dihydro-orotate amidohydrolase	3.5.2.3
L-5,6-Dihydro-orotate:NAD⁺ oxidoreductase	1.3.1.14
L-5,6-Dihydro-orotate:NADP⁺ oxidoreductase	1.3.1.15
Dihydro-orotate oxidase	1.3.3.1
L-5,6-Dihydro-orotate:oxygen oxidoreductase	1.3.3.1
Dihydropicolinate synthase	4.2.1.52
Dihydropteridine reductase	1.6.99.7
Dihydropteroate:L-glutamate ligase (ADP-forming)	6.3.2.12
Dihydropteroate pyrophosphorylase	2.5.1.15
Dihydropteroate synthase	2.5.1.15
Dihydropyrimidinase	3.5.2.2
5,6-Dihydropyrimidine amidohydrolase	3.5.2.2
D-erythro-Dihydrosphingosine:NADP⁺ 3-oxidoreductase	1.1.1.102
Dihydrosphingosine-1-phosphate aldolase	4.1.2.27
Dihydrosphingosine-1-phosphate palmitaldehyde-lyase	4.1.2.27
Dihydrouracil dehydrogenase	1.3.1.1
Dihydrouracil dehydrogenase (NADP⁺)	1.3.1.2
5,6-Dihydrouracil:NAD⁺ oxidoreductase	1.3.1.1
5,6-Dihydrouracil:NADP⁺ oxidoreductase	1.3.1.2
Dihydroxyacetone-phosphate acyltransferase	2.3.1.42
Dihydroxyacetone-phosphate phospho-lyase	4.2.99.11
Dihydroxyacetonetransferase	2.2.1.2
Dihydroxyacid dehydratase	4.2.1.9
2,3-Dihydroxyacid hydro-lyase	4.2.1.9
2,3-Dihydroxybenzoate carboxy-lyase	4.1.1.46
2,3-Dihydroxybenzoate 1,2-dioxygenase	1.13.11.14
2,3-Dihydroxybenzoate:oxygen 1,2-oxidoreductase (decyclizing, decarboxylating)	1.13.11.14
2,3-Dihydroxybenzoate:L-serine ligase	6.3.2.14
2,3-Dihydroxybenzoylserine synthetase	6.3.2.14
3,4-Dihydroxy-trans-cinnamate:oxygen 3,4-oxidoreductase (decyclizing)	1.13.11.22
Dihydroxyfumarate carboxy-lyase	4.1.1.54
Dihydroxyfumarate decarboxylase	4.1.1.54
2,3-Dihydroxyindole 2,3-dioxygenase	1.13.11.23
2,3-Dihydroxyindole:oxygen 2,3-oxidoreductase (decyclizing)	1.13.11.23
Dihydroxyisovalerate dehydrogenase (isomerizing)	1.1.1.89
2,3-Dihydroxyisovalerate:NADP⁺ oxidoreductase (acetolactate-forming)	1.1.1.89
7,8-Dihydroxykynurenate 8,8a-dioxygenase	1.13.11.10
7,8-Dihydroxykynurenate oxygenase	1.13.11.10
7,8-Dihydroxykynurenate:oxygen 8,8a-oxidoreductase (decyclizing)	1.13.11.10
11α,15-Dihydroxy-9-oxoprost-13-enoate:NAD⁺ 15-oxidoreductase	1.1.1.141
3,4-Dihydroxyphenylacetate 2,3-dioxygenase	1.13.11.15
3,4-Dihydroxyphenylacetate 3,4-dioxygenase	1.13.11.7
3,4-Dihydroxyphenylacetate:oxygen 2,3-oxidoreductase (decyclizing)	1.13.11.15
3,4-Dihydroxyphenylacetate:oxygen 3,4-oxidoreductase (decyclizing)	1.13.11.7
Dihydroxyphenylalanine aminotransferase	2.6.1.49
Dihydroxyphenylalanine ammonia-lyase	4.3.1.11
3,4-Dihydroxy-L-phenylalanine ammonia-lyase	4.3.1.11
3,4-Dihydroxy-L-phenylalanine:2-oxoglutarate aminotransferase	2.6.1.49
3,4-Dihydroxyphenylethylamine, ascorbate:oxygen oxidoreductase (β-hydroxylating)	1.14.17.1
β(2,3-Dihydroxyphenyl)propionate:oxygen 1,2-oxidoreductase (decyclizing)	1.13.11.16
4,5-Dihydroxyphthalate carboxy-lyase	4.1.1.55
4,5-Dihydroxyphthalate decarboxylase	4.1.1.55
2,5-Dihydroxypyridine 5,6-dioxygenase	1.13.11.9
2,5-Dihydroxypyridine oxygenase	1.13.11.9
2,5-Dihydroxypyridine:oxygen 5,6-oxidoreductase (decyclizing)	1.13.11.9
3,4-Dihydroxy-9,10-secoandrosta-1,3,5(10)-triene-9,17-dione 4,5-dioxygenase	1.13.11.25
3,4-Dihydroxy-9,10-secoandrosta-1,3,5(10)-triene-9,17-dione:oxygen 4,5-oxidoreductase (decyclizing)	1.13.11.25
β-(3,5-Diiodo-4-hydroxyphenyl)-lactate:NAD⁺ oxidoreductase	1.1.1.96
Diiodophenylpyruvate reductase	1.1.1.96
Diiodotyrosine aminotransferase	
Diiodotyrosine aminotransferase	2.6.1.24
3,5-Diiodo-L-tyrosine:2-oxoglutarate aminotransferase	2.6.1.24
Diidopropylfluorophosphonate halogenase	3.8.2.1
Di-isopropyl phosphorofluoridase	3.8.2.1
Di-isopropyl-phosphorofluoridate fluorohydrolase	3.8.2.1
2,3-Diketocamphane lactonizing enzyme	1.14.15.2

INDEX TO THE ENZYME LIST (continued)

β-Diketonase	3.7.1.2
Dimethylallyldiphosphate:isopentenyldiphosphate dimethylallyltransferase	2.5.1.1
Dimethylallyltransferase	2.5.1.1
Dimethylaniline monooxygenase (*N*-oxide-forming)	1.14.13.8
N,N-*Dimethylaniline,NADPH:oxygen oxidoreductase (N-oxide-forming)*	1.14.13.8
Dimethylaniline oxidase	1.14.13.8
Dimethylaniline-*N*-oxide aldolase	4.1.2.24
N,N-*Dimethylaniline*-N-*oxide formaldehyde-lyase*	4.1.2.24
4,4-Dimethyl-5α-cholest-7-en-3β-ol, hydrogen-donor:oxygen oxidoreductase	1.14.99.16
N,N-*Dimethylglycine:(acceptor) oxidoreductase (demethylating)*	1.5.99.2
Dimethylglycine dehydrogenase	1.5.99.2
Dimethylmalate dehydrogenase	1.1.1.84
3,3-Dimethyl-D-malate:NAD⁺ oxidoreductase (decarboxylating)	1.1.1.84
Dimethylpropiothetin dethiomethylase	4.4.1.3
S-*Dimethyl-β-propiothetin dimethylsulphide-lyase*	4.4.1.3
6,7-Dimethyl-8-(1′-D-ribityl)lumazine:6,7-dimethyl-8-(1′-D-ribityl)lumazine 2,3-butanediyltransferase	2.5.1.9
Dimethylthetin–homocysteine methyltransferase	2.1.1.3
Dimethylthetin:L-homocysteine S-methyltransferase	2.1.1.3
3,5-Dinitrotyrosine aminotransferase	2.6.1.26
Dinucleotide nucleotidohydrolase	3.6.1.9
2,4-Dioxotetrahydropyrimidinenucleotide:pyrophosphate phosphoribosyltransferase	2.4.2.20
Dioxotetrahydropyrimidine phosphoribosyltransferase	2.4.2.20
Dioxotetrahydropyrimidine-ribonucleotide pyrophosphorylase	2.4.2.20
2,5-Dioxovalerate dehydrogenase	1.2.1.26
2,5-Dioxovalerate:NADP⁺ oxidoreductase	1.2.1.26
Dipeptidase	3.4.13.11
Dipeptide hydrolase	3.4.13.11
Dipeptidyl peptidase	3.4.14.1
Dipeptidylpeptide hydrolase	3.4.14.1
Dipeptidyl-transferase	3.4.14.1
o-Diphenol oxidase	1.14.18.1
p-Diphenol oxidase	1.14.18.1
1,3-Diphosphoglycerate→ADP transphosphatase	2.7.2.3
Diphosphoinositide kinase	2.7.1.68
Disproportionating enzyme	2.4.1.25
Disulphoglucosamine-6-sulphatase	3.1.6.11
DNAase	3.1.4.5
DNA (cytosine-5-)-methyltransferase	2.1.1.37
DNA 5′-dinucleotidohydrolase	3.1.4.31
DNA joinase	6.5.1.1, 6.5.1.2
DNA nucleotidylexotransferase	2.7.7.31
DNA nucleotidyltransferase	2.7.7.7
DNA phosphatase-exonuclease	3.1.4.27
DNA polymerase	2.7.7.7
DNA repair enzyme	6.5.1.1, 6.5.1.2
DNA restriction enzyme	3.1.4.32
DNase	3.1.4.5
DNase I	3.1.4.5
DNase II	3.1.4.6
DNase R-K	3.1.4.32
Donor:hydrogen-peroxide oxidoreductase	1.11.1.7
DOPA decarboxylase	4.1.1.28
Dopamine β-hydroxylase	1.14.17.1
Dopamine β-monooxygenase	1.14.17.1
Double-stranded-deoxyribonucleate 5′-nucleotidohydrolase	3.1.4.28
Double-stranded-ribonucleate 3′-oligonucleotidohydrolase	3.1.4.24
DPNase	3.2.2.5
DPN·H₂→aldehyde transhydrogenase	1.1.1.1
DPN·H₂→cystine transhydrogenase	1.6.4.1
DPN·H₂→cytochrome *c* transelectronase	1.6.99.3
DPN·H₂→dihydroxyacetone-phosphate transhydrogenase	1.1.1.8
DPN·H₂→glyoxylate transhydrogenase	1.1.1.26
DPN·H₂→nitrate transhydrogenase	1.6.6.1, 1.6.6.2
DPN·H₂→nitrite transhydrogenase	1.6.6.4
DPN·H₂→pyruvate transhydrogenase	1.1.1.27
DPN hydrolase	3.2.2.5
DPN kinase	2.7.1.23
DT-diaphorase	1.6.99.2
8,11,14-Eicosatrienoate, hydrogen-donor:oxygen oxidoreductase	1.14.99.1
Elastase	3.4.21.11
Endodeoxyribonuclease	3.1.4.30
Endodeoxyribonuclease (ATP- and *S*-adenosyl-methionine-dependent)	3.1.4.32
Endodeoxyribonuclease (ATP-hydrolyzing)	3.1.4.33
Endo-1,2-β-glucanase	3.2.1.71
Endo-1,3-α-glucanase	3.2.1.59
Endo-1,3-ζ-glucanase	3.2.1.39
Endo-1,3-β-glucanase	3.2.1.6
Endo-1,4-β-glucanase	3.2.1.4
Endo-1,6-β-glucanase	3.2.1.75
Endo-1,4-β-mannanase	3.2.1.78
Endonuclease, spleen	3.1.4.7
Endopeptidase, *Aeromonas proteolytica*	3.4.24.4
Endopeptidase, Alternaria	3.4.21.16
Endopolyphosphatase	3.6.1.10
Endoribonuclease III	3.1.4.24
Endothia acid proteinase	3.4.23.10
Endo-1,3-β-xylanase	3.2.1.32
Endo-1,4-β-xylanase	3.2.1.8
Enolase	4.2.1.11
Enoyl-[acyl-carrier-protein] reductase	1.3.1.9
Enoyl-[acyl-carrier-protein] reductase (NADPH)	1.3.1.10

INDEX TO THE ENZYME LIST (continued)

Enoyl-CoA hydratase	4.2.1.17	Exo-1,3-β-xylosidase	3.2.1.72
Enoyl-CoA reductase	1.3.1.8	Exo-1,4-β-xylosidase	3.2.1.37
Enoyl hydrase	4.2.1.17		
Enoylpyruvate transferase	2.5.1.7	Factor XA	3.4.21.6
Enterokinase	3.4.21.9	*FAD nucleotidohydrolase*	3.6.1.18
Enteropeptidase	3.4.21.9	FAD pyrophosphatase	3.6.1.18
Epoxide hydratase	4.2.1.63	FAD pyrophosphorylase	2.7.7.2
trans-Epoxysuccinate hydratase	4.2.1.37	Farnesylpyrophosphate synthetase	2.5.1.1
5α-Ergosta-7,22-diene-3β, 5-diol		Fatty acid desaturase	1.14.99.5
5,6-hydro-lyase	4.2.1.62	*Fatty-acid-hydroperoxide isomerase*	5.3.99.1
Erythritol kinase	2.7.1.27	Fatty acid ω-hydroxylase	1.14.15.3
Erythrocuprein	1.15.1.1	Fatty-acid methyltransferase	2.1.1.15
Erythrose isomerase	5.3.1.2	Fatty-acid peroxidase	1.11.1.3
D-Erythrose ketol-isomerase	5.3.1.2	Fatty acid thiokinase (long chain)	6.2.1.3
Erythrulose-1-phosphate formaldehyde-		Fatty acid thiokinase (medium chain)	6.2.1.2
lyase	4.1.2.2	Ferredoxin hydrogenase	1.12.7.1
Erythrulose-1-phosphate synthease	4.1.2.2	Ferredoxin–NADP⁺ reductase	1.6.7.1
Escherichia coli exo-RNase II	3.1.4.20	Ferredoxin–nitrite reductase	1.7.7.1
Escherichia coli RNase I	3.1.4.23	Ferrihaemoprotein P_{450} reductase	1.6.2.4
Ethanolamine ammonia-lyase	4.3.1.7	Ferrochelatase	4.99.1.1
Ethanolamine ammonia-lyase	4.3.1.7		
Ethanolamine oxidase	1.4.3.8	*Ferrocytochrome c:hydrogen-peroxide*	
Ethanolamine:oxygen oxidoreductase		*oxidoreductase*	1.11.1.5
(deaminating)	1.4.3.8	*Ferrocytochrome c:iron oxidoreductase*	1.9.99.1
Ethanolaminephosphate cytidylyl-		*Ferrocytochrome c:oxygen oxidoreductase*	1.9.3.1
transferase	2.7.7.14	*Ferrocytochrome c_2:oxygen oxidoreductase*	1.9.3.2
Ethanolaminephosphate phospho-		*Ferrocytochrome:nitrite oxidoreductase*	1.9.6.1
lyase	4.2.99.7	Ferroxidase	1.16.3.1
Ethanolaminephosphate phospho-		Fibrinase	3.4.21.7
lyase (deaminating)	4.2.99.7	Fibrinogenase	3.4.21.5
Ethanolaminephosphotransferase	2.7.8.1	Fibrinolysin	3.4.21.7
Ethylene reductase	1.3.99.2	Ficin	3.4.22.3
2-Ethylmalate glyoxylate-lyase		*Flavanone lyase (decyclizing)*	5.5.1.6
(CoA-butyrylating)	4.1.3.10	Flavokinase	2.7.1.26
2-Ethylmalate synthase	4.1.3.10	FMN adenylyltransferase	2.7.7.2
Etiocholanolone 3α-dehydrogenase	1.1.1.152	Folic acid reductase	1.5.1.3
Euphorbain	3.4.99.7	Formaldehyde dehydrogenase	1.2.1.1
Exodeoxyribonuclease I	3.1.4.25	*Formaldehyde:NAD⁺ oxidoreductase*	1.2.1.1
Exodeoxyribonuclease II	3.1.4.26	Formamidase	3.5.1.9
Exodeoxyribonuclease III	3.1.4.27	Formate dehydrogenase	1.2.1.2
Exodeoxyribonuclease IV	3.1.4.28	Formate dehydrogenase (cytochrome)	1.2.2.1
Exo-β-fructosidase	3.2.1.80	Formate→DPN transhydrogenase	1.2.1.1
Exo-1,3-α-glucanase	3.2.1.84	*Formate:ferricytochrome b_1 oxido-*	
Exo-1,3-β-glucosidase	3.2.1.58	*reductase*	1.2.2.1
Exo-1,4-α-glucosidase	3.2.1.3	Formate hydrogenlyase	1.2.1.2
Exo-1,4-β-glucosidase	3.2.1.74	Formate kinase	2.7.2.6
Exo-1,6-α-glucosidase	3.2.1.70	*Formate:NAD⁺ oxidoreductase*	1.2.1.2
Exo-maltotetraohydrolase	3.2.1.60	*Formate:tetrahydrofolate ligase*	
Exo-1,2-1,3-α-mannosidase	3.2.1.77	*(ADP-forming)*	6.3.4.3
Exonuclease III	3.1.4.27	Formic dehydrogenase	1.2.1.1
3'-Exonuclease	3.1.4.18	Formicohydrogenlyase	1.2.1.2
5'-Exonuclease	3.1.4.1	Formiminoaspartate deiminase	3.5.3.5
λ-Exonuclease	3.1.4.28	*N-Formimino-L-aspartate imino-*	
Exonuclease, spleen	3.1.4.18	*hydrolase*	3.5.3.5
Exopolygalacturonase	3.2.1.67	Formiminoglutamase	3.5.3.8
Exopolygalacturonate lyase	4.2.2.9	*N-Formimino-L-glutamate*	
Exo-poly-α-galacturonosidase	3.2.1.82	*formiminohydrolase*	3.5.3.8
Exopolyphosphatase	3.6.1.11	*5-Formiminotetrahydrofolate*	
Exoribonuclease	3.1.4.20	*ammonia-lyase (cyclizing)*	4.3.1.4

INDEX TO THE ENZYME LIST (continued)

Formiminotetrahydrofolate cyclodeaminase	4.3.1.4	β-h-Fructosidase	3.2.1.26
5-*Formiminotetrahydrofolate:L-glutamate N-formiminotransferase*	2.1.2.5	Fructuronate reductase	1.1.1.57
		Fucoidanase	3.2.1.44
5-*Formiminotetrahydrofolate:glycine N-formiminotransferase*	2.1.2.4	Fucokinase	2.7.1.52
Formylase	3.5.1.9	2-O-α-L-*Fucopyranosyl-β-D-galactoside fucohydrolase*	3.2.1.63
N-*Formyl-L-aspartate amidohydrolase*	3.5.1.8	L-Fucose dehydrogenase	1.1.1.122
Formylaspartate deformylase	3.5.1.8	Fucose-1-phosphate guanylyltransferase	2.7.7.30
Formyl-CoA hydrolase	3.1.2.10	β-D-Fucosidase	[3.2.1.38]
Formyl-CoA hydrolase	3.1.2.10	α-L-Fucosidase	3.2.1.51
Formylkynureninase	3.5.1.9	1,2-α-L-Fucosidase	3.2.1.63
N-*Formyl-L-methionine amidohydrolase*	3.5.1.31	α-L-*Fucoside fucohydrolase*	3.2.1.51
		Fucosyl-galactose acetylgalactosaminyl-transferase	2.4.1.40
Formylmethionine deformylase	3.5.1.31	L-Fuculokinase	2.7.1.51
N-*Formyl-L methionylaminoacyl-tRNA amidohydrolase*	3.5.1.27	L-Fuculosephosphate aldolase	4.1.2.17
N-Formylmethionylaminoacyl-tRNA deformylase	3.5.1.27	L-*Fuculose-1-phosphate L-lactaldehyde-lyase*	4.1.2.17
		Fumarase	4.2.1.2
10-*Formyltetrahydrofolate amidohydrolase*	3.5.1.10	Fumarate hydratase	4.2.1.2
		Fumarate reductase	1.3.99.1
5-Formyltetrahydrofolate cyclo-ligase	6.3.3.2	Fumarate reductase (NADH)	1.3.1.6
5-*Formyltetrahydrofolate cyclo-ligase (ADP-forming)*	6.3.3.2	Fumarhydrogenase	1.3.99.1
		Fumaric aminase	4.3.1.1
Formyltetrahydrofolate deformylase	3.5.1.10	Fumaric hydrogenase	1.3.99.1
Formyltetrahydrofolate dehydrogenase	1.5.1.6	Fumarylacetoacetase	3.7.1.2
5-*Formyltetrahydrofolate:L-glutamate N-formyltransferase*	2.1.2.6	4-*Fumarylacetoacetate fumarylhydrolase*	3.7.1.2
		Funduspepsin	3.4.23.1
10-*Formyltetrahydrofolate:L-methionyl-tRNA N-formyltransferase*	2.1.2.9		
10-*Formyltetrahydrofolate:NADP+ oxidoreductase*	1.5.1.6	Galactarate dehydratase	4.2.1.42
		D-*Galactarate hydro-lyase*	4.2.1.42
10-*Formyltetrahydrofolate:5'-phosphoribosyl-5-amino-4-imidazolecarboxamide formyltransferase*	2.1.2.3	Galactinol–raffinose galactosyl-transferase	2.4.1.67
		Galactitol dehydrogenase	1.1.1.16
Formyltetrahydrofolate synthetase	6.3.4.3	*Galactitol:NAD+ 3-oxidoreductase*	1.1.1.16
2,1-β-D-*Fructan fructanohydrolase*	3.2.1.7	Galactocerebroside sulphotransferase	2.8.2.11
2,6-β-D-*Fructan fructanohydrolase*	3.2.1.65	Galactokinase	2.7.1.6
2,6-β-D-*Fructan 6-β-fructofuranosyl fructohydrolase*	3.2.1.64	Galactolipase	3.1.1.26
		Galactonate dehydratase	4.2.1.6
β-D-*Fructan fructohydrolase*	3.2.1.80	D-*Galactonate hydro-lyase*	4.2.1.6
2,6-β-Fructan 6-levanbiohydrolase	3.2.1.64	Galactonolactone dehydrogenase	1.3.2.3
β-Fructofuranosidase	3.2.1.26	L-*Galactono-γ-lactone:ferricytochrome c oxidoreductase*	1.3.2.3
β-D-*Fructofuranoside fructohydrolase*	3.2.1.26	Galactose dehydrogenase	1.1.1.48
Fructokinase	2.7.1.4	Galactose dehydrogenase (NADP+)	1.1.1.120
D-*Fructose:(acceptor)5-oxidoreductase*	1.1.99.11	D-*Galactose:NAD+ 1-oxidoreductase*	1.1.1.48
Fructose-bisphosphate aldolase	4.1.2.13	D-*Galactose:NADP+ 1-oxidoreductase*	1.1.1.120
D-*Fructose-1,6-bisphosphate D-glyceraldehyde-3-phosphate-lyase*	4.1.2.13	Galactose oxidase	1.1.3.9
		D-*Galactose:oxygen 6-oxidoreductase*	1.1.3.9
D-*Fructose-1,6-bisphosphate 1-phosphohydrolase*	3.1.3.11	Galactose-1-phosphate thymidylyltransferase	2.7.7.32
Fructose-1,6-bisphosphate triosephosphate-lyase	4.1.2.13	Galactose-1-phosphate uridylyltransferase	2.7.7.10
D-Fructose 5-dehydrogenase	1.1.99.11	Galactose-6-sulphatase	2.5.1.5
D-*Fructose:NADP+ 5-oxidoreductase*	1.1.1.124	*Galactose-6-sulphate alkyltransferase (cyclizing)*	2.5.1.5
D-*Fructose-6-phosphate D-erythrose-4-phosphate-lyase (phosphate-acetylating)*	4.1.2.22	Galactose-6-sulphurylase	2.5.1.5
Fructose-6-phosphate phosphoketolase	4.1.2.22	α-Galactosidase	3.2.1.22

INDEX TO THE ENZYME LIST (continued)

β-Galactosidase	3.2.1.23	GDPmannose:undecaprenyl-phosphate	
Galactoside acetyltransferase	2.3.1.18	mannosyltransferase	2.4.1.54
α-D-Galactoside galactohydrolase	3.2.1.22	GDP mannuronate:alginateD-mann-	
β-D-Galactoside galactohydrolase	3.2.1.23	uronyltransferase	2.4.1.33
D-Galactosyl-N-acylsphingosine		Gelatinase	3.4.23.2
galactohydrolase	3.2.1.46	Gentiobiase	3.2.1.21
Galactosylceramidase	3.2.1.46	Gentisate carboxy-lyase	4.1.1.62
Galactosylgalactosylglucosylceramidase	3.2.1.47	Gentisate decarboxylase	4.1.1.62
D-Galactosyl-D-galactosyl-D-glucosyl-		Gentisate 1,2-dioxygenase	1.13.11.4
ceramide galactohydrolase	3.2.1.47	Gentisate oxygenase	1.13.11.4
1-O-α-D-galactosyl-myo-inositol:raffinose		Gentisate:oxygen 1,2-oxidoreductase	
galactosyltransferase	2.4.1.67	(decyclizing)	1.13.11.4
Galactowaldenase	5.1.3.2	Geranoyl-CoA:carbon-dioxide ligase	
Galacturonokinase	2.7.1.44	(ADP-forming)	6.4.1.5
Gallate carboxy-lyase	4.1.1.59	Geranoyl-CoA carboxylase	6.4.1.5
Gallate decarboxylase	4.1.1.59	Geranyldiphosphate:isopentenyidi-	
Gastricsin	3.4.23.3	phosphate geranyltransferase	2.5.1.10
GDP-6-deoxy-D-talose dehydrogenase	1.1.1.135	Geranyltransferase	2.5.1.10
GDP-6-deoxy-D-talose:NADP⁺		Gliocladium proteinase	3.4.99.8
4-oxidoreductase	1.1.1.135	1,4-α-Glucan branching enzyme	2.4.1.18
GDP fucose: β-2-acetamido-2-deoxy-D-		1,4-α-D-Glucan: 1,4-α-D-glucan	
glucosaccharide 4-α-L-fucosyltransferase	2.4.1.65	6-α-(1,4-α-glucano)-transferase	2.4.1.18
GDP fucose–glycoprotein fucosyltrans-		1,4-α-D-Glucan:1,4-α-D-glucan	
ferase	2.4.1.68	(D-glucose) 6-α-glucosyltransferase	2.4.1.24
GDPfucose:glycoprotein fucosyl-		1,4-α-D-Glucan:1,6-α-D-glucan	
transferase	2.4.1.68	6-α-glucosyltransferase	2.4.1.2
GDPfucose–lactose fucosyltransferase	2.4.1.69	1,4-α-D-Glucan:1,4-α-D-glucan	
GDPfucose:lactose fucosyltransferase	2.4.1.69	4-α-glycosyltransferase	2.4.1.25
GDPfucose pyrophosphorylase	2.7.7.30	1,2-β-D-Glucan glucanohydrolase	3.2.1.71
GDPglucose:1.4-β-D-glucan 4-β-glucosyl-		1,3-β-D-Glucan glucanohydrolase	3.2.1.39
transferase	2.4.1.29	1,3-(1,3,;1,4)-α-D-Glucan 3-glucano-	
GDPglucose glucohydrolase	3.2.1.42	hydrolase	3.2.1.59
GDPglucose–glucosephosphate		1,3-(1,3;1,4)-β-D-Glucan 3(4)-glucano-	
glucosyltransferase	2.4.1.36	hydrolase	3.2.1.6
GDPglucose:D-glucose-6-phosphate		1,3-1,4-α-D-Glucan 4-glucanohydrolase	3.2.1.61
α-glucosyltransferase	2.4.1.36	1,3,-1,4-β-D-Glucan 4-glucanohydrolase	3.2.1.73
GDPglucose pyrophosphorylase	2.7.7.34	1,4-α-D-Glucan glucanohydrolase	3.2.1.1
GDPglucosidase	3.2.1.42	1,4-(1,3;1,4)-β-D-glucan 4-glucano-	
GDPhexose pyrophosphorylase	2.7.7.29	hydrolase	3.2.1.4
GDPmannose 4,6-dehydratase	4.2.1.47	1,6-α-D-Glucan 6-glucanohydrolase	3.2.1.11
GDPmannose dehydrogenase	1.1.1.132	1,6-β-D-Glucan glucanohydrolase	3.2.1.75
GDPmannose:glucomannan 1,4-β-mannosyl-		1,4-α-D-Glucan 4-α-(1,4-α-glucano)-	
transferase	2.4.1.32	transferase (cyclizing)	2.4.1.19
GDPmannose:heteroglycan 2,3-α-mannosyl-		1,3-β-D-Glucan glucohydrolase	3.2.1.58
transferase	2.4.1.48	1,3-α-D-Glucan 3-glucohydrolase	3.2.1.84
GDPmannose 4,6-hydro-lyase	4.2.1.47	1,4-α-D-Glucan glucohydrolase	3.2.1.3
GDPmannose α-mannosyltransferase	2.4.1.48	1,4-β-D-Glucan glucohydrolase	3.2.1.74
GDPmannose:NAD⁺ 6-oxidoreductase	1.1.1.132	1,6-α-D-Glucan glucohydrolase	3.2.1.70
GDP:D-mannose-1-phosphate		1,4-α-Glucan 6-α-glucosyltransferase	2.4.1.24
guanylyltransferase	2.7.7.22	1,4-α-D-Glucan maltohydrolase	3.2.1.2
GDPmannose–phosphatidyl-myo-inositol		1,4-α-D-Glucan maltotetraohydrolase	3.2.1.60
α-mannosyltransferase	2.4.1.57	1,4-α-D-Glucan:orthophosphate	
GDPmannose:1-phosphatidyl-myo-inositol		α-glucosyltransferase	2.4.1.1
α-mannosyltransferase	2.4.1.57	4-α-Glucanotransferase	2.4.1.25
GDPmannose:phosphomannan		1,3-β-Glucan synthase	2.4.1.34
mannosephosphotransferase	2.7.8.9	1,3-β-Glucan–UDP glucosyltransferase	2.4.1.34
GDPmannose phosphorylase	2.7.7.22	Glucarate dehydratase	4.2.1.40
GDPmannose–undecaprenyl-		D-Glucarate hydro-lyase	4.2.1.40
phosphate mannosyltransferase	2.4.1.54	Glucoamylase	3.2.1.3

INDEX TO THE ENZYME LIST (continued)

Glucoinvertase	3.2.1.20
Glucokinase	2.7.1.2
Glucomannan 4-β-mannosyltransferase	2.4.1.32
D-Gluconate:(acceptor) 2-oxidoreductase	1.1.99.3
Gluconate dehydratase	4.2.1.39
Gluconate dehydrogenase	1.1.99.3
D-Gluconate hydro-lyase	4.2.1.39
D-Gluconate:NAD(P)$^+$ 5-oxidoreductase	1.1.1.69
Gluconokinase	2.7.1.12
Gluconolactonase	3.1.1.17
D-Glucono-δ-lactone hydrolase	3.1.1.17
Glucosaminate ammonia-lyase	4.3.1.9
Glucosamine acetylase	2.3.1.3
Glucosamine acetyltransferase	2.3.1.3
Glucosamine kinase	2.7.1.8
Glucosaminephosphate acetyltransferase	2.3.1.4
Glucosaminephosphate isomerase	5.3.1.10
Glucosaminephosphate isomerase (glutamine-forming)	5.3.1.19
D-Glucose:(acceptor) 1-oxidoreductase	1.1.99.10
Glucose aerodehydrogenase	1.1.3.4
α-D-Glucose-1,6-bisphosphate:deoxy-D-ribose-1-phosphate phosphotransferase	2.7.5.6
α-D-Glucose-1,6-bisphosphate:α-D-glucose-1-phosphate phosphotransferase	2.7.5.1
Glucose dehydrogenase	1.1.1.47
D-Glucose dehydrogenase	1.1.1.118
Glucose dehydrogenase (Aspergillus)	1.1.99.10
Glucose dehydrogenase (NADP$^+$)	1.1.1.119
Glucose→DPN transhydrogenase	1.1.1.47
Glucose isomerase	5.3.1.18
D-Glucose ketol-isomerase	5.3.1.18
D-Glucose:NAD$^+$ 1-oxidoreductase	1.1.1.118
D-Glucose:NADP$^+$ 1-oxidoreductase	1.1.1.119
β-D-Glucose:NAD(P)$^+$ 1-oxidoreductase	1.1.1.47
β-Glucose→O$_2$ transhydrogenase	1.1.3.4
Glucose oxhydrase	1.1.3.4
Glucose oxidase	1.1.3.4
β-D-Glucose:oxygen 1-oxidoreductase	1.1.3.4
Glucose-1-phosphatase	3.1.3.10
Glucose-6-phosphatase	3.1.3.9
Glucose-1-phosphate adenylyltransferase	2.7.7.27
Glucose-1-phosphate cytidylyltransferase	2.7.7.33
Glucose-6-phosphate dehydrogenase	1.1.1.49
D-Glucose1-phosphate:D-glucose-1-phosphate 6-phosphotransferase	2.7.1.41
D-Glucose-1-phosphate:D-glucose 6-phosphotransferase	2.7.5.5
Glucose-1-phosphate guanylyltransferase	2.7.7.34
Glucosephosphate isomerase	5.3.1.9
Glucose-6-phosphate isomerase	5.3.1.9
D-Glucose-6-phosphate ketol-isomerase	5.3.1.9
D-Glucose-6-phosphate:NADP$^+$ 1-oxidoreductase	1.1.1.49
Glucose-1-phosphate phosphodismutase	2.7.1.41
D-Glucose-1-phosphate phosphohydrolase	3.1.3.10
D-Glucose-6-phosphate phosphohydrolase	3.1.3.9
D-Glucose-1-phosphate:riboflavin 5'-phosphotransferase	2.7.1.42
Glucose-1-phosphate thymidylyltransferase	2.7.7.24
Glucose-1-phosphate uridylyltransferase	2.7.7.9
Glucose phosphomutase	2.7.5.1, 2.7.5.5
Glucose (1→6) phosphomutase	2.7.5.1
α-Glucosidase	3.2.1.20
β-Glucosidase	3.2.1.21
α-1,3-Glucosidase	[3.2.1.27]
trans-N-Glucosidase	2.4.2.6
D-Glucoside 3-dehydrogenase	1.1.99.13
α-D-Glucoside glucohydrolase	3.2.1.20
β-D-Glucoside glucohydrolase	3.2.1.21
Glucosidosucrase	3.2.1.20
Glucosulphatase	3.1.6.3
D-Glucosyl-N-acylsphingosine glucohydrolase	3.2.1.45
Glucosylceramidase	3.2.1.45
3-O-β-D-Glucosylglucose:orthophosphate glucosyltransferase	2.4.1.31
Glucuronate isomerase	5.3.1.12
D-Glucuronate ketol-isomerase	5.3.1.12
Glucuronate reductase	1.1.1.19
β-Glucuronidase	3.2.1.31
β-D-Glucuronide glucuronosohydrolase	3.2.1.31
Glucuronokinase	2.7.1.43
D-Glucuronolactone dehydrogenase	1.1.1.70
D-Glucurono-δ-lactone hydrolase	3.1.1.19
D-Glucurono-γ-lactone:NAD$^+$ 1-oxidoreductase	1.1.1.70
Glucuronolactone reductase	1.1.1.20
Δ4,5-β-D-Glucuronosyl-(1,4)-2-acetamido-2-deoxy-D-galactose-4-sulphate 4-sulphohydrolase	3.1.6.9
Δ4,5-β-D-Glucuronosyl-(1,4)-2-acetamido-2-deoxy-D-galactose-6-sulphate 6-sulphohydrolase	3.1.6.10
Glucuronosyl-disulphoglucosamine glucuronidase	3.2.1.56
1,3-D-Glucuronosyl-2-sulphamido-2-deoxy-6-O-sulpho-β-D-glucose glucuronohydrolase	3.2.1.56
Glutamate acetyltransferase	2.3.1.35
L-Glutamate:ammonia ligase (ADP-forming)	6.3.1.2
L-Glutamate 1-carboxy-lyase	4.1.1.15
Glutamate carboxypeptidase	3.4.12.10
D-Glutamate cyclase	4.2.1.48
L-Glutamate:L-cysteine γ-ligase (ADP-forming)	6.3.2.2
Glutamate decarboxylase	4.1.1.15
Glutamate dehydrogenase	1.4.1.2
Glutamate dehydrogenase (NADP$^+$)	1.4.1.4
Glutamate dehydrogenase (NAD(P)$^+$)	1.4.1.3
L-Glutamate→DPN tranhsydrogenase	1.4.1.2
Glutamate formiminotransferase	2.1.2.5

INDEX TO THE ENZYME LIST (continued)

Glutamate formyltransferase	2.1.2.6
D-*Glutamate hydro-lyase (cyclizing)*	4.2.1.48
L-*Glutamate:methylamine ligase (ADP-forming)*	6.3.4.12
Glutamate mutase	5.4.99.1
L-*Glutamate:NAD$^+$ oxidoreductase (deaminating)*	1.4.1.2
L-*Glutamate:NADP$^+$ oxidoreductase (deaminating)*	1.4.1.4
L-*Glutamate:NAD(P)$^+$ oxidoreductase (deaminating)*	1.4.1.3
D-Glutamate→O_2 transhydrogenase	1.4.3.7
D-**Glutamate oxidase**	1.4.3.7
D-*Glutamate:oxygen oxidoreductase (deaminating)*	1.4.3.7
Glutamate racemase	5.1.1.3
Glutamate racemase	5.1.1.3
Glutamate synthase	2.6.1.53
L-Glutamate→TPN transhydrogenase	1.4.1.4
L-*Glutamate:tRNA^{Glu} ligase (AMP-forming)*	6.1.1.17
Glutamic-alanine transaminase	2.6.1.2
Glutamic-aspartic transaminase	2.6.1.1
Glutamic dehydrogenase	1.4.1.2–1.4.1.4
Glutamic-oxaloacetic transaminase	2.6.1.1
D-Glutamic oxidase	1.4.3.7
Glutamic-pyruvic transaminase	2.6.1.2
Glutaminase	3.5.1.2
D-**Glutaminase**	3.5.1.35
Glutaminase II	2.6.1.15
D-*Glutamine amidohydrolase*	3.5.1.35
L-*Glutamine amidohydrolase*	3.5.1.2
Glutamine–fructose-6-phosphate aminotransferase	5.3.1.19
Glutamine:D-glutamyl-peptide glutamyltransferase	2.3.2.1
Glutamine–*scyllo*-inosose aminotransferase	2.6.1.50
Glutamine→β-ketoacid transamidase	2.6.1.15
Glutamine–oxo-acid aminotransferase	2.6.1.15
L-*Glutamine:2-oxo-acid aminotransferase*	2.6.1.15
L-*Glutamine:2-oxoglutarate aminotransferase (NADPH-oxidizing)*	2.6.1.53
L-*Glutamine:2,4,6/3,5-pentahydroxycyclohexanone aminotransferase*	2.6.1.50
Glutamine phenylacetyltransferase	2.3.1.14
Glutamine synthetase	6.3.1.2
Glutamine-synthetase adenylyltransferase	2.7.7.42
L-*Glutamine:tRNA^{Gln} ligase (AMP-forming)*	6.1.1.18
Glutaminyl-tRNA cyclotransferase	2.3.2.5
L-*Glutaminyl-tRNA γ-glutamyltransferase (cyclizing)*	2.3.2.5
Glutaminyl-tRNA synthetase	6.1.1.18
(γ-L-Glutamyl)-L-amino-acid γ-glutamyltransferase (cyclizing)	2.3.2.4
γ-Glutamylcyclotransferase	2.3.2.4
γ-L-*Glutamyl-L-cysteine:glycine ligase (ADP-forming)*	6.3.2.3
γ-*Glutamyl-cysteine synthetase*	6.3.2.2
α-**Glutamyl-glutamate dipeptidase**	3.4.13.7
α-L-*Glutamyl-L-glutamate hydrolase*	3.4.13.7
N^3-*(γ-L-glutamyl)-4-hydroxymethylphenylhydrazine:(acceptor) γ-glutamyltransferase*	2.3.2.9
γ-**Glutamylmethylamide synthetase**	6.3.4.12
(γ-Glutamyl)-peptide:amino-acid γ-glutamyltransferase	2.3.2.2
γ-Glutamyltransferase	2.3.2.2
D-Glutamyltransferase	2.3.2.1
Glutamyl-tRNA synthetase	6.1.1.17
Glutarate:CoA ligase (ADP-forming)	6.2.1.6
Glutarate-semialdehyde dehydrogenase	1.2.1.20
Glutarate-semialdehyde:NAD$^+$ oxidoreductase	1.2.1.20
Glutaryl-CoA:(acceptor) oxidoreductase (decarboxylating)	1.3.99.7
Glutaryl-CoA dehydrogenase	1.3.99.7
Glutaryl-CoA synthetase	6.2.1.6
Glutathione *S*-alkyltransferase	2.5.1.12
Glutathione *S*-aralkyltransferase	2.5.1.14
Glutathione *S*-aryltransferase	2.5.1.13
Glutathione–CoAS-SG transhydrogenase	1.8.4.3
Glutathione:cystine oxidoreductase	1.8.4.4
Glutathione–cystine transhydrogenase	1.8.4.4
Glutathione:dehydroascorbate oxidoreductase	1.8.5.1
Glutathione dehydrogenase (ascorbate)	1.8.5.1
Glutathione:homocystine oxidoreductase	1.8.4.1
Glutathione–homocystine transhydrogenase	1.8.4.1
Glutathione:hydrogen-peroxide oxidoreductase	1.11.1.9
Glutathione peroxidase	1.11.1.9
Glutathione:polyolnitrate oxidoreductase	1.8.6.1
Glutathione:protein-disulphide oxidoreductase	1.8.4.2
Glutathione reductase (NAD(P)H)	1.6.4.2
Glutathione synthetase	6.3.2.3
Glutathione thiolesterase	3.1.2.7
Glyceraldehyde-phosphate dehydrogenase	1.2.1.12
Glyceraldehyde-phosphate dehydrogenase (NADP$^+$)	1.2.1.9
Glyceraldehyde-phosphate dehydrogenase (NADP$^+$) (phosphorylating)	1.2.1.13
D-*Glyceraldehyde-3-phosphate ketol-isomerase*	5.3.1.1
D-*Glyceraldehyde-3-phosphate:NAD$^+$ oxidoreductase (phosphorylating)*	1.2.1.12
D-*Glyceraldehyde-3-phosphate:NADP$^+$ oxidoreductase*	1.2.1.9

INDEX TO THE ENZYME LIST (continued)

D-*Glyceraldehyde-3-phosphate:NADP⁺ oxidoreductase (phosphorylating)*	1.2.1.13
Glycerate dehydrogenase	1.1.1.29
Glycerate kinase	2.7.1.31
D-*Glycerate:NAD⁺ oxidoreductase*	1.1.1.29
D-*Glycerate:NAD(P)⁺ oxidoreductase*	1.1.1.60
D-*Glycerate:NAD(P)⁺ 2-oxidoreductase*	1.1.1.81
D-*Glycerate:NAD(P)⁺ oxidoreductase (carboxylating)*	1.1.1.92
D-*Glycerate-2-phosphate phosphohydrolase*	3.1.3.20
Glycerate phosphomutase	2.7.5.3, 2.7.5.4
Glycerate(3→2)-phosphomutase	2.7.5.3, 5.4.2.1
Glycerol dehydratase	4.2.1.30
Glycerol dehydrogenase	1.1.1.6
Glycerol dehydrogenase (NADP⁺)	1.1.1.72
Glycerol→DPN transhydrogenase	1.1.1.6
Glycerol hydro-lyase	4.2.1.30
Glycerol kinase	2.7.1.30
Glycerol-monoester hydrolase	3.1.1.23
Glycerol:NAD⁺ 2-oxidoreductase	1.1.1.6
Glycerol:NADP⁺ oxidoreductase	1.1.1.72
Glycerol-1-phosphatase	3.1.3.21
Glycerol-2-phosphatase	3.1.3.19
sn-*Glycerol-3-phosphate:(acceptor) oxidoreductase*	1.1.99.5
Glycerolphosphate acyltransferase	2.3.1.15
Glycerol-3-phosphate cytidylyltransferase	2.7.7.39
Glycerol-3-phosphate dehydrogenase	1.1.99.5
Glycerol-3-phosphate dehydrogenase (NAD⁺)	1.1.1.8
sn-*Glycerol-3-phosphate dehydrogenase (NAD(P)⁺)*	1.1.1.94
sn-*Glycerol-3-phosphate:NAD⁺ 2-oxidoreductase*	1.1.1.8
sn-*Glycerol-3-phosphate:NAD(P)⁺ 2-oxidoreductase*	1.1.1.94
Glycerolphosphate phosphatidyltransferase	2.7.8.5
Glycerol-1-phosphate phosphohydrolase	3.1.3.21
Glycerol-2-phosphate phosphohydrolase	3.1.3.19
Glycerophosphatase	3.1.3.1, 3.1.3.2
Glycerophosphinicocholine diesterase	3.1.4.2
Glyceryl ether cleaving enzyme	1.14.99.17
Glyceryl-ether monooxygenase	1.14.99.17
L-*3-Glycerylphosphinicocholine glycerophosphohydrolase*	3.1.4.2
Glycinamide ribonucleotide synthetase	6.3.4.13
Glycine acetyltransferase	2.3.1.29
Glycine acyltransferase	2.3.1.13
Glycine amidinotransferase	2.1.4.1
Glycine aminotransferase	2.6.1.4
Glycine carboxypeptidase	3.4.12.8
Glycine dehydrogenase	1.4.1.10
Glycine formiminotransferase	2.1.2.4
Glycine methyltransferase	2.1.1.20
Glycine:NAD⁺ oxidoreductase (deaminating)	1.4.1.10
Glycine–oxaloacetate aminotransferase	2.6.1.35
Glycine:oxaloacetate aminotransferase	2.6.1.35
Glycine:2-oxoglutarate aminotransferase	2.6.1.4
Glycine synthase	2.1.2.10
Glycine:tRNA^{Gly} ligase (AMP-forming)	6.1.1.14
Glycocholase	3.5.1.24
Glycocyaminase	3.5.3.2
Glycogenase	3.2.1.1, 3.2.1.2
Glycogen branching enzyme	2.4.1.18
Glycogen 6-glucanohydrolase	3.2.1.68
Glycogen phosphorylase	2.4.1.1
Glycogen (starch) synthase	2.4.1.11
Glycogen synthetase *a* kinase	2.7.1.37
Glycolaldehyde dehydrogenase	1.2.1.21
Glycolaldehyde:NAD⁺ oxidoreductase	1.2.1.21
Glycolaldehydetransferase	2.2.1.1
Glycol hydro-lyase (epoxide-forming)	4.2.1.63
Glycollate:NAD⁺ oxidoreductase	1.1.1.26
Glycollate:NADP⁺ oxidoreductase	1.1.1.79
Glycollate→O₂ transhydrogenase	1.1.3.1
Glycollate oxidase	1.1.3.1
Glycollate:oxygen oxidoreductase	1.1.3.1
Glycoprotein β-galactosyltransferase	2.4.1.38
Glycosulphatase	3.1.6.3
Glycyl-glycine dipeptidase	3.4.13.1
Glycyl-glycine hydrolase	3.4.13.1
Glycyl-leucine dipeptidase	3.4.13.2
Glycyl-L-leucine hydrolase	3.4.13.2
Glycyl-tRNA synthetase	6.1.1.14
Glyoxalase I	4.4.1.5
Glyoxalase II	3.1.2.6
Glyoxalate carbo-ligase	4.1.1.47
Glyoxylate carboxy-lyase (dimerizing)	4.1.1.47
Glyoxylate dehydrogenase	1.2.1.17
Glyoxylate:NADP⁺ oxidoreductase (CoA-oxalylating)	1.2.1.17
Glyoxylate oxidase	1.2.3.5
Glyoxylate:oxygen oxidoreductase	1.2.3.5
Glyoxylate reductase	1.1.1.26
Glyoxylate reductase (NADP⁺)	1.1.1.79
Glyoxylate transacetase	4.1.3.2
GMP-N-acetylneuraminate:galactosylglycoprotein N-acetylneuraminyltransferase	2.4.99.1
GMP reductase	1.6.6.8
GMP synthetase	6.3.4.1
GMP synthetase (glutamine-hydrolyzing)	6.3.5.2
dGTPase	3.1.5.1
GTP cyclohydrolase	3.5.4.16
GTP:6-deoxy-L-galactose-1-phosphate guanylyltransferase	2.7.7.30
GTP 7,8-8,9-dihydrolase	3.5.4.16
GTP:α-D-glucose-1-phosphate guanylyltransferase	2.7.7.34
GTP:α-D-hexose-1-phosphate guanylyltransferase	2.7.7.29
GTP:5-hydroxy-L-lysine O-phosphotransferase	2.7.1.81

INDEX TO THE ENZYME LIST (continued)

GTP–mannose-1-phosphate guanylyl-transferase	2.7.7.13
GTP: α-D-mannose-1-phosphate guanylyltransferase	2.7.7.13
GTP:oxaloacetate carboxy-lyase (transphosphorylating)	4.1.1.32
GTP pyrophosphate-lyase (cyclizing)	4.6.1.2
dGTP triphosphohydrolase	3.1.5.1
Guanase	3.5.4.3
Guanidinoacetate amidinohydrolase	3.5.3.2
Guanidinoacetate kinase	2.7.3.1
Guanidinoacetate methyltransferase	2.1.1.2
Guanidinobutyrase	3.5.3.7
4-Guanidinobutyrate amidinohydrolase	3.5.3.7
Guanine aminase	3.5.4.3
Guanine aminohydrolase	3.5.4.3
Guanine deaminase	3.5.4.3
Guanosine aminase	3.5.4.15
Guanosine aminohydrolase	3.5.4.15
Guanosine deaminase	3.5.4.15
Guanosine:orthophosphate ribosyltransferase	2.4.2.15
Guanosine phosphorylase	2.4.2.15
Guanylate cyclase	4.6.1.2
Guanylate kinase	2.7.4.8
Guanyl cyclase	4.6.1.2
Guanyloribonuclease	3.1.4.8
Guanylyl cyclase	4.6.1.2
L-Gulonate dehydrogenase	1.1.1.45
L-Gulonate:NAD$^+$ 3-oxidoreductase	1.1.1.45
L-Gulonate:NADP$^+$ 1-oxidoreductase	1.1.1.19
L-Gulono-γ-lactone hydrolase	3.1.1.18
L-Gulono-γ-lactone:NADP$^+$ 1-oxidoreductase	1.1.1.20
L-Gulonolactone oxidase	1.1.3.8
L-Gulono-γ-lactone:oxygen 2-oxidoreductase	1.1.3.8
Haem, hydrogen-donor:oxygen oxidoreductase (α-methene-oxidizing, hydroxylating)	1.14.99.3
Haemocuprein	1.15.1.1
Haem oxygenase (decyclizing)	1.14.99.3
Haloacetate dehalogenase	3.8.1.3
Haloacetate halidohydrolase	3.8.1.3
2-Haloacid dehalogenase	3.8.1.2
2-Haloacid halidohydrolase	3.8.1.2
Halogenase	3.8.1.1
Heparinase	4.2.2.7
Heparinase	[3.2.1.19]
Heparin eliminase	4.2.2.7
Heparin lyase	4.2.2.7
Heparin lyase	4.2.2.7
Heparin-sulphate eliminase	4.2.2.8
Heparin-sulphate lyase	4.2.2.8
Heparitinsulphate lyase	4.2.2.8
Heparitin sulphotransferase	2.8.2.12
Heptulokinase	2.7.1.14
Heterophosphatase	2.7.1.1

Hexokinase	2.7.1.1
Hexosediphosphatase	3.1.3.11
Hexose oxidase	1.1.3.5
D-Hexose:oxygen 1-oxidoreductase	1.1.3.5
Hexosephosphate aminotransferase	5.3.1.19
Hexose-1-phosphate guanylyl-transferase	2.7.7.29
Hexosephosphate isomerase	5.3.1.9
Hexose-1-phosphate nucleotidyl-transferase	2.7.7.28
Hexose-1-phosphate uridylyl-transferase	2.7.7.12
Hippurate hydrolase	3.5.1.32
Hippuricase	3.5.1.14
Histaminase	1.4.3.6
Histamine methyltransferase	2.1.1.8
Histidase	4.3.1.3
Histidinase	4.3.1.3
Histidine acetyltransferase	2.3.1.33
L-Histidine-β-alanine ligase (AMP-forming)	6.3.2.11
Histidine aminotransferase	2.6.1.38
Histidine ammonia-lyase	4.3.1.3
L-Histidine ammonia-lyase	4.3.1.3
L-Histidine carboxy-lyase	4.1.1.22
Histidine α-deaminase	4.3.1.3
Histidine decarboxylase	4.1.1.22
L-Histidine:2-oxoglutarate aminotransferase	2.6.1.38
L-Histidine:tRNA^{His} ligase (AMP-forming)	6.1.1.21
Histidinol dehydrogenase	1.1.1.23
L-Histidinol:NAD$^+$ oxidoreductase	1.1.1.23
Histidinolphosphatase	3.1.3.15
Histidinol-phosphate aminotransferase	2.6.1.9
L-Histidinol-phosphate:2-oxoglutarate aminotransferase	2.6.1.9
L-Histidinol-phosphate phosphohydrolase	3.1.3.15
Histidyl-tRNA synthetase	6.1.1.21
Histozyme	3.5.1.14
Holo-[acyl-carrier-protein] synthase	2.7.8.7
Homoaconitate hydratase	4.2.1.36
Homocitrate synthase	4.1.3.21
Homocysteine desulphhydrase	4.4.1.2
L-Homocysteine hydrogen-sulphide-lyase (deaminating)	4.4.1.2
Homocysteine methyltransferase	2.1.1.10
Homogentisate 1,2-dioxygenase	1.13.11.5
Homogentisate oxygenase	1.13.11.5
Homogentisate:oxygen 1,2-oxidoreductase (decyclizing)	1.13.11.5
Homogentisicase	1.13.11.5
Homoprotocatechuate oxygenase	1.13.11.7
Homoserine acetyltransferase	2.3.1.31
Homoserine deaminase	4.4.1.1
Homoserine dehydratase	4.4.1.1
Homoserine dehydrogenase	1.1.1.3
Homoserine kinase	2.7.1.39
L-Homoserine:NAD$^+$ oxidoreductase	1.1.1.3

INDEX TO THE ENZYME LIST (continued)

Hurain	3.4.99.9
Hyaluronate 3-glycanohydrolase	3.2.1.36
Hyaluronate 4-glycanohydrolase	3.2.1.3.5
Hyaluronate lyase	4.2.2.1
Hyaluronate lyase	4.2.2.1
Hyaluronidase	3.2.1.35, 3.2.1.36, 4.2.2.1
Hyaluronoglucosidase	3.2.1.35
Hyaluronoglucuronidase	3.2.1.36
Hybrid nuclease	3.1.4.34
Hydantoinase	3.5.2.2
Hydrogenase	1.12.1.2, 1.12.2.1, 1.12.7.1
Hydrogen dehydrogenase	1.12.1.2
Hydrogen:ferredoxin oxidoreductase	1.12.7.1
Hydrogen:ferricytochrome c_3 oxidoreductase	1.12.2.1
Hydrogenlyase	1.12.7.1
Hydrogen:NAD⁺ oxidoreductase	1.12.1.2
Hydrogen-peroxide:hydrogen-peroxide oxidoreductase	1.11.1.6
Hydrogen-sulphide:(acceptor) oxidoreductase	1.8.99.1
Hydrogen-sulphide acetyltransferase	2.3.1.10
Hydrogen-sulphide:ferredoxin oxidoreductase	1.8.7.1
Hydrogen-sulphide:NADP⁺ oxidoreductase	1.8.1.2
Hydroperoxide isomerase	5.3.99.1
D-2-Hydroxyacid:(acceptor) oxidoreductase	1.1.99.6
D-**2-Hydroxyacid dehydrogenase**	1.1.99.6
L-**2-Hydroxyacid oxidase**	1.1.3.15
L-2-Hydroxyacid: oxygen oxidoreductase	1.1.3.15
Hydroxyacid racemase	5.1.2.1
D-3-Hydroxyacyl-[acyl-carrier-protein]:NADP⁺ oxidoreductase	1.1.1.100
3-Hydroxyacyl-CoA dehydrogenase	1.1.1.35
L-3-Hydroxyacyl-CoA hydro-lyase	4.2.1.17
L-3-Hydroxyacyl-CoA:NAD⁺ oxidoreductase	1.1.1.35
D-3-Hydroxyacyl-CoA:NADP⁺ oxidoreductase	1.1.1.36
β-Hydroxyacyl dehydrogenase	1.1.1.35
Hydroxyacylglutathione hydrolase	3.1.2.6
S-2-Hydroxyacylglutathione hydrolase	3.1.2.6
S-(2-Hydroxyalkyl)glutathione alkyl-epoxide-lyase	4.4.1.7
***S*-(Hydroxyalkyl)glutathione lyase**	4.4.1.7
D-Hydroxyaminoacid dehydratase	4.2.1.14
L-Hydroxyaminoacid dehydratase	4.2.1.13
3-Hydroxyanthranilate 3,4-dioxygenase	1.13.11.6
3-Hydroxyanthranilate oxidase	1.10.3.5
3-Hydroxyanthranilate:oxygenase	1.13.11.6
3-Hydroxyanthranilate:oxygen oxidoreductase	1.10.3.5
3-Hydroxyanthranilate:oxygen 3,4-oxidoreductase (decyclizing)	1.13.11.6
3-Hydroxyaspartate aldolase	4.1.3.14
erythro-3-Hydroxyaspartate dehydratase	4.2.1.38
erythro-3-Hydroxy-L-aspartate glyoxylate-lyase	4.1.3.14
erythro-3-Hydroxy-L-aspartate hydro-lyase (deaminating)	4.2.1.38
3-Hydroxybenzoate carboxy-lyase	4.1.1.61
p-**Hydroxybenzoate decarboxylase**	4.1.1.61
3-Hydroxybenzoate, hydrogen-donor:oxygen oxidoreductase (4-hydroxylating)	1.14.99.13
p-**Hydroxybenzoate hydroxylase**	1.14.13.2
3-Hydroxybenzoate 4-hydroxylase	1.14.99.13
3-Hydroxybenzoate 4-monooxygenase	1.14.99.13
4-Hydroxybenzoate 3-monooxygenase	1.14.13.2
4-Hydroxybenzoate,NADPH: oxygen oxidoreductase (3-hydroxylating)	1.14.13.2
3-Hydroxybenzyl-alcohol dehydrogenase	1.1.1.97
3-Hydroxybenzyl-alcohol:NADP⁺ oxidoreductase	1.1.1.97
3-Hydroxybutyrate dehydrogenase	1.1.1.30
4-Hydroxybutyrate dehydrogenase	1.1.1.61
Hydroxybutyrate-dimer hydrolase	3.1.1.22
β-Hydroxybutyrate→DPN transhydrogenase	1.1.1.30
D-3-Hydroxybutrate:NAD⁺ oxidoreductase	1.1.1.30
4-Hydroxybutyrate:NAD⁺ oxidoreductase	1.1.1.61
D-3-Hydroxybutyryl-[acyl-carrier-protein] hydro-lyase	4.2.1.58
D-3-Hydroxybutyryl-CoA dehydratase	4.2.1.55
3-Hydroxybutyryl-CoA epimerase	5.1.2.3
3-Hydroxybutyryl-CoA 3-epimerase	5.1.2.3
D-3-Hydroxybutyryl-CoA hydro-lyase	4.2.1.55
3-D-(3-D-Hydroxybutyryloxy)-butyrate hydrolase	3.1.1.22
2-Hydroxy-3-carboxyadipate dehydrogenase	1.1.1.87
2-Hydroxy-3-carboxyadipate hydro-lyase	4.2.1.36
2-Hydroxy-3-carboxyadipate:NAD⁺ oxidoreductase (decarboxylating)	1.1.1.87
3-Hydroxy-3-carboxyadipate 2-oxo-glutarate-lyase (CoA-acetylating)	4.1.3.21
(+)-L-Hydroxy-4-carboxymethyl-isocrotonolactone $\Delta^2-\Delta^3$-isomerase	5.3.3.4
α-**Hydroxycholanate dehydrogenase**	1.1.1.52
3α-Hydroxy-5β-cholanate:NAD⁺ oxidoreductase	1.1.1.52
4-Hydroxycinnamate, ascorbate:oxygen oxidoreductase(3-hydroxylating)	1.14.17.2
ω-**Hydroxydecanoate dehydrogenase**	1.1.1.66

INDEX TO THE ENZYME LIST (continued)

10-Hydroxydecanoate:NAD$^+$ oxidoreductase	1.1.1.66	3-Hydroxy-3-methylglutaryl-CoA hydrolase	3.1.2.5
D-3-Hydroxydecanoyl-[acyl-carrier-protein] dehydratase	4.2.1.60	3-Hydroxy-3-methylglutaryl-CoA hydro-lyase	4.2.1.18
D-3-Hydroxydecanoyl-[acyl-carrier-protein] hydro-lyase	4.2.1.60	Hydroxymethylglutaryl-CoA lyase	4.1.3.4
		Hydroxymethylglutaryl-CoA reductase	1.1.1.88
D-2-Hydroxy-3,3-dimethyl-3-formyl-propionate:NAD$^+$ 4-oxidoreductase	1.2.1.33	Hydroxymethylglutaryl-CoA reductase (NADPH)	1.1.1.34
Hydroxyethylthiazole kinase	2.7.1.50	Hydroxymethylglutaryl-CoA synthase	4.1.3.5
D-2-Hydroxy-fatty acid dehydrogenase	1.1.1.98	2-Hydroxy-2-methyl-3-oxobutyrate carboxy-lyase	4.1.1.5
L-2-Hydroxy-fatty acid dehydrogenase	1.1.1.99	4-Hydroxy-4-methyl-2-oxoglutarate aldolase	4.1.3.17
3-Hydroxy-L-glutamate 1-carboxy-lyase	4.1.1.16		
Hydroxyglutamate decarboxylase	4.1.1.16	4-Hydroxy-4-methyl-2-oxoglutarate pyruvate-lyase	4.1.3.17
4-Hydroxy-L-glutamate:2-oxoglutarate aminotransferase	2.6.1.23	3-Hydroxy-2-methyl-pyridine-4,5-dicarboxylate 4-carboxy-lyase	4.1.1.51
4-Hydroxyglutamate transaminase	2.6.1.23	3-Hydroxy-2-methyl-pyridine-4,5-dicarboxylate 4-decarboxylase	4.1.1.51
L-2-Hydroxyglutarate:(acceptor) oxidoreductase	1.1.99.2	Hydroxymethylpyrimidine kinase	2.7.1.49
2-Hydroxyglutarate dehydrogenase	1.1.99.2	5-Hydroxy-N-methylpyroglutamate synthase	3.5.1.36
2-Hydroxyglutarate glyoxylate-lyase (CoA-propionylating)	4.1.3.9	5-Hydroxymethyluracil,2-oxoglutarate dioxygenase	1.14.11.5
2-Hydroxyglutarate synthase	4.1.3.9		
3-Hydroxyisobutyrate dehydrogenase	1.1.1.31	5-Hydroxymethyluracil,2-oxo-glutarate:oxygen oxidoreductase	1.14.11.5
3-Hydroxyisobutyrate:NAD$^+$ oxidoreductase	1.1.1.31	5-Hydroxymethyluracil oxygenase	1.14.11.5
3-Hydroxyisobutyryl-CoA hydrolase	3.1.2.4	6-Hydroxynicotinate reductase	1.3.7.1
3-Hydroxyisobutyryl-CoA hydrolase	3.1.2.4	6-Hydroxy-D-nicotine oxidase	1.5.3.6
3-Hydroxy-3-isohexenylglutaryl-CoA hydro-lyase	4.2.1.57	6-Hydroxy-L-nicotine oxidase	1.5.3.5
3-Hydroxy-3-isohex-3-enylglutaryl-CoA isopentenylacetoacetyl-CoA-lyase	4.1.3.26	6-Hydroxy-D-nicotine:oxygen oxidoreductase	1.5.3.6
3-Hydroxy-3-isohexenylglutaryl-CoA lyase	4.1.3.26	6-Hydroxy-L-nicotine:oxygen oxidoreductase	1.5.3.5
Hydroxylamine oxidase	1.7.3.4	3-Hydroxyoctanoyl-[acyl-carrier-protein] dehydratase	4.2.1.59
Hydroxylamine:oxygen oxidoreductase	1.7.3.4	3-Hydroxyoctanoyl-[acyl-carrier-protein] hydro-lyase	4.2.1.59
Hydroxylamine reductase	1.7.99.1	2-Hydroxy-3-oxoadipate carboxylase	4.1.3.15
Hydroxylamine reductase (NADH)	1.6.6.11	2-Hydroxy-3-oxoadipate glyoxylate-lyase (carboxylating)	4.1.3.15
ω-Hydroxylase	1.14.15.3	Hydroxyoxobutyrate aldolase	[4.1.2.1]
Hydroxylysine kinase	2.7.1.81	4-Hydroxy-2-oxoflutarate glyoxylate-lyase	4.1.3.16
4-Hydroxymandelonitrile hydroxybenzaldehyde-lyase	4.1.2.11	4-Hydroxy-2-oxoglutarate glyoxylate-lyase	4.1.3.16
Hydroxymandelonitrile lyase	4.1.2.11	3-Hydroxypalmitoyl-[acyl-carrier-protein] dehydratase	4.2.1.61
2-Hydroxy-4-methyl-3-carboxyvalerate hydro-lyase	4.2.1.33		
2-Hydroxy-4-methyl-3-carboxyvalerate:NAD$^+$ oxidoreductase	1.1.1.85	3-Hydroxypalmitoyl-[acyl-carrier-protein] hydro-lyase	4.2.1.61
3-Hydroxy-4-methyl-3-carboxyvalerate 2-oxo-3-methylbutyrate-lyase (CoA-acetylating)	4.1.3.12	p-Hydroxyphenylacetate 3-hydroxylase	1.14.13.3
		4-Hydroxyphenylacetate 3-monooxygenase	1.14.13.3
3-Hydroxy-3-methylglutaryl-CoA acetoacetate-lyase	4.1.3.4	4-Hydroxyphenylacetate,NADH:oxygen oxidoreductase(3-hydroxylating)	1.14.13.3
3-Hydroxy-3-methylglutaryl-CoA acetoacetyl-CoA-lyase (CoA-acetylating)	4.1.3.5	2-Hydroxyphenylpropionate hydroxylase	1.14.13.4
		3-(2-Hydroxyphenyl)-propionate,NADH:oxygen oxidoreductase (3-hydroxylating)	1.14.13.4
Hydroxymethylglutaryl-CoA hydrolase	3.1.2.5	2-Hydroxyphenylpropionate:NAD$^+$ oxidoreductase	1.3.1.11
		4-Hydroxyphenylpyruvate dioxygenase	1.13.11.27

INDEX TO THE ENZYME LIST (continued)

4-Hydroxyphenylpyruvate:oxygen oxidoreductase (hydroxylating, decarboxylating)	1.13.11.27
Hydroxyproline epimerase	5.1.1.8
Hydroxyproline 2-epimerase	5.1.1.8
4-Hydroxy-L-proline:NAD$^+$ oxidoreductase	1.1.1.104
3-Hydroxypropionate dehydrogenase	1.1.1.59
3-Hydroxypropionate:NAD$^+$ oxidoreductase	1.1.1.59
15-Hydroxyprostaglandin dehydrogenase	1.1.1.141
Hydroxypyruvate carboxy-lyase	4.1.1.40
Hydroxypyruvate decarboxylase	4.1.1.40
Hydroxypyruvate reductase	1.1.1.81
10-D-Hydroxystearate 10-hydro-lyase	4.2.1.53
D-2-Hydroxystearate:NAD$^+$ oxidoreductase	1.1.1.98
L-2-Hydroxystearate:NAD$^+$ oxidoreductase	1.1.1.99
β-Hydroxysteroid dehydrogenase	1.1.1.51
3α-Hydroxysteroid dehydrogenase	1.1.1.50
3β-Hydroxy-Δ^5-steroid dehydrogenase	1.1.1.145
11β-Hydroxysteroid dehydrogenase	1.1.1.146
16α-Hydroxysteroid dehydrogenase	1.1.1.147
20α-Hydroxysteroid dehydrogenase	1.1.1.149
20β-Hydroxysteroid dehydrogenase	1.1.1.53
21-Hydroxysteroid dehydrogenase	1.1.1.150
21-Hydroxysteroid dehydrogenase (NADP$^+$)	1.1.1.151
16-Hydroxysteroid epimerase	5.1.99.2
16-Hydroxysteroid 16-epimerase	5.1.99.2
3α-Hydroxy-5β-steroid:NAD$^+$ 3-oxidoreductase	1.1.1.152
3β-Hydroxy-Δ^5-steroid:NAD$^+$ 3-oxidoreductase	1.1.1.145
17β-Hydroxysteroid:NAD$^+$ 17-oxidoreductase	1.1.1.63
21-Hydroxysteroid:NAD$^+$ 21-oxidoreductase	1.1.1.150
3α-Hydroxysteroid:NAD(P)$^+$ oxidoreductase	1.1.1.50
3(or 17)β-Hydroxysteroid:NAD(P)$^+$ oxidoreductase	1.1.1.51
11β-Hydroxysteroid:NADP$^+$ 11-oxidoreductase	1.1.1.146
16α-Hydroxysteroid:NAD(P)$^+$ 16 oxidoreductase	1.1.1.147
17α-Hydroxysteroid:NAD(P)$^+$ 17-oxidoreductase	1.1.1.148
17β-Hydroxysteroid:NADP$^+$ 17-oxidoreductase	1.1.1.64
20α-Hydroxysteroid:NAD(P)$^+$ 20-oxidoreductase	1.1.1.149
21-Hydroxysteroid:NADP$^+$ 21-oxidoreductase	1.1.1.151
3β-Hydroxysteroid sulphotransferase	2.8.2.2
5α-Hydroxysterol dehydratase	4.2.1.62
Hydroxytryptophan decarboxylase	4.1.1.28
Hyponitrite reductase	1.6.6.6
Hypotaurine dehydrogenase	1.8.1.3
Hypotaurine:NAD$^+$ oxidoreductase	1.8.1.3
Hypotaurocyamine kinase	2.7.3.6
Hypoxanthine oxidase	1.2.3.2
Hypoxanthine phosphoribosyltransferase	2.4.2.8
D-Iditol dehydrogenase	1.1.1.15
L-Iditol dehydrogenase	1.1.1.14
D-Iditol:NAD$^+$ 5-oxidoreductase	1.1.1.15
L-Iditol:NAD$^+$ 5-oxidoreductase	1.1.1.14
L-Idonate:NADP$^+$ 2-oxidoreductase	1.1.1.128
L-Iduronidase	3.2.1.76
Imidazoleacetate 4-monooxygenase	1.14.13.5
Imidazoleacetate, NADH:oxygen oxidoreductase (hydroxylating)	1.14.13.5
Imidazoleacetate:5-phosphoribosyldiphosphate ligase (ATP- and pyrophosphate-forming)	6.3.4.8
Imidazole acetylase	2.3.1.2
Imidazole acetyltransferase	2.3.1.2
Imidazoleglycerolphosphate dehydratase	4.2.1.19
D-erythro-Imidazoleglycerolphosphate hydro-lyase	4.2.1.19
Imidazolonepropionase	3.5.2.7
4-Imidazolone-5-propionate amidohydrolase	3.5.2.7
4-Imidazolone-5-propionate hydro-lyase	4.2.1.49
Imidazol-5-yl-lactate dehydrogenase	1.1.1.111
Imidazol-5-yl-lactate:NAD(P)$^+$ oxidoreductase	1.1.1.111
Imidodipeptidase	3.4.13.9
Iminodipeptidase	3.4.13.8
Iminopeptidase	3.4.11.5
IMP:L-aspartate ligase (GDP-forming)	6.3.4.4
IMP cyclohydrolase	3.5.4.10
IMP dehydrogenase	1.2.1.14
IMP 1,2-hydrolase (decyclizing)	3.5.4.10
IMP:NAD$^+$ oxidoreductase	1.2.1.14
IMP:pyrophosphate phosphoribosyltransferase	2.4.2.8
IMP pyrophosphorylase	2.4.2.8
Indanol dehydrogenase	1.1.1.112
Indanol:NAD(P)$^+$ oxidoreductase	1.1.1.112
Indoleacetaldoxime dehydratase	4.2.1.29
3-Indoleacetaldoxime hydro-lyase	4.2.1.29
Indole 2,3-dioxygenase	1.13.11.17
Indoleglycerolphosphate aldolase	[4.1.2.8]
Indole-3-glycerol-phosphate synthase	4.1.1.48
Indolelactate dehydrogenase	1.1.1.110
Indolelactate:NAD$^+$ oxidoreductase	1.1.1.110
Indole:oxygen 2,3-oxidoreductase (decyclizing)	1.13.11.17
Indophenolase	1.9.3.1
Indophenol oxidase	1.9.3.1
Inorganic pyrophosphatase	3.6.1.1
scyllo-**Inosamine kinase**	2.7.1.65
Inosinase	3.2.2.2
Inosinate nucleosidase	3.2.2.12
5'-Inosinate phosphoribohydrolase	3.2.2.12
Inosine kinase	2.7.1.73
Inosine nucleosidase	3.2.2.2
Inosine phosphorylase	2.4.2.1
Inosine ribohydrolase	3.2.2.2
Inositol 2-dehydrogenase	1.1.1.18

INDEX TO THE ENZYME LIST (continued)

myo-*Inositol-hexakisphosphate 1-phosphohydrolase*	3.1.3.8
myo-*Inositol-hexakisphosphate 6-phosphohydrolase*	3.1.3.26
myo-Inositol 1-kinase	2.7.1.64
myo-Inositol 1-methyltransferase	2.1.1.39
myo-Inositol 3-methyltransferase	2.1.1.40
myo-*Inositol:NAD$^+$ 2-oxidoreductase*	1.1.1.18
myo-Inositol oxygenase	1.13.99.1
myo-*Inositol:oxygen oxidoreductase*	1.13.99.1
1L-myo-Inositol-1-phosphatase	3.1.3.25
1L-myo-*Inositol-1-phosphate lyase (isomerizing)*	5.5.1.4
1L-myo-*Inositol-1-phosphate phosphohydrolase*	3.1.3.25
myo-*Inositol-1-phosphate synthase*	5.5.1.4
myo-*Inosose-2-dehydratase*	4.2.1.44
Insulinase	3.4.99.10
Inulase	3.2.1.7
Inulinase	3.2.1.7
Inulosucrase	2.4.1.9
Invertase	3.2.1.26
Invertin	3.2.1.26
Iodide:hydrogen-peroxide oxidoreductase	1.11.1.8
Iodide peroxidase	1.11.1.8
Iodinase	1.11.1.8
Iodophenol methyltransferase	2.1.1.26
Iodotyrosine deiodase	1.11.1.8
Iron–cytochrome *c* reductase	1.9.99.1
Iron(II):oxygen oxidoreductase	1.16.3.1
Isoamylase	3.2.1.68
Isochorismatase	3.3.2.1
Isochorismate hydroxymutase	5.4.99.6
Isochorismate pyruvate-hydrolase	3.3.2.1
Isochorismate synthase	5.4.99.6
Isocitrase	4.1.3.1
Isocitratase	4.1.3.1
Isocitrate dehydrogenase (NAD$^+$)	1.1.1.41
Isocitrate dehydrogenase (NADP$^+$)	1.1.1.42
threo-*D-Isocitrate glyoxylate-lyase*	4.1.3.1
Isocitrate lyase	4.1.3.1
threo-*D-Isocitrate:NAD$^+$ oxidoreductase (decarboxylating)*	1.1.1.41
threo-*D-Isocitrate:NADP$^+$ oxidoreductase (decarboxylating)*	1.1.1.42
Isocitric dehydrogenase	1.1.1.41
Isocitritase	4.1.3.1
Isohexenylglutaconyl-CoA hydratase	4.2.1.57
L-*Isoleucine:tRNA^{Ile}ligase (AMP-forming)*	6.1.1.5
Isoleucyl-tRNA synthetase	6.1.1.5
Isomaltase	3.2.1.10
Isopentenyldiphosphate Δ-isomerase	5.3.3.2
Isopentenyldiphosphate Δ^3 – Δ^2-isomerase	5.3.3.2
2-*Isopentenyldiphosphate:tRNA 2-isopentenyl transferase*	2.5.1.8
Isophenoxazine synthase	1.10.3.4
Isoprenoid-alcohol kinase	2.7.1.66
Isopropanol dehydrogenase (NADP$^+$)	1.1.1.80
2-Isopropylmalate dehydratase	4.2.1.33
2-Isopropylmalate dehydrogenase	1.1.1.85
2-Isopropylmalate synthase	4.1.3.12
Isopullulanase	3.2.1.57
Itaconyl-CoA hydratase	4.2.1.56
Joinase	6.5.1.1, 6.5.1.2
Kallikrein	3.4.21.8
Keratinase, Streptomyces	3.4.99.11
Keratinase, Trichophyton	3.4.99.12
α-Ketoacid carboxylase	4.1.1.1
3-Ketoacid CoA-transferase	2.8.3.5
β-Ketoacyl-CoA→CoA transacylase	2.3.1.16
3-Ketoacyl-CoA thiolase	2.3.1.16
2-Keto-3-deoxy-L-arabonate aldolase	4.1.2.18
2-Keto-3-deoxy-L-arabonate dehydratase	4.2.1.43
2-*Keto-3-deoxy-L-arabonate glycolaldehyde-lyase*	4.1.2.18
2-*Keto-3-deoxy-L-arabonate hydrolyase*	4.2.1.43
2-Keto-3-deoxygalactonate kinase	2.7.1.58
2-*Keto-3-deoxy-D-galactonate-6-phosphate D-glyceraldehyde-3-phosphate-lyase*	4.1.2.21
2-Keto-3-deoxy-D-glucarate aldolase	4.1.2.20
5-Keto-4-deoxy-D-glucarate dehydratase	4.2.1.41
5-*Keto-4-deoxy-D-glucarate hydro-lyase (decarboxylating)*	4.2.1.41
2-*Keto-3-deoxy-D-glucarate tartronate-semialdehyde-lyase*	4.1.2.20
2-Keto-3-deoxy-D-gluconate dehydrogenase	1.1.1.126
2-Keto-3-deoxy-D-gluconate dehydrogenase (NAD(P)$^+$)	1.1.1.127
2-*Keto-3-deoxy-D-gluconate:NAD(P)$^+$ 5-oxidoreductase*	1.1.1.127
2-*Keto-3-deoxy-D-gluconate:NADP$^+$ 6-oxidoreductase*	1.1.1.126
Ketodeoxygluconokinase	2.7.1.45
5-Keto-D-fructose reductase	1.1.1.123
5-Keto-D-fructose reductase (NADP$^+$)	1.1.1.124
2-*Keto-D-gluconate:(acceptor) 5-oxidoreductase*	1.1.99.4
Ketogluconate dehydrogenase	1.1.99.4
5-Keto-D-gluconate 2-reductase	1.1.1.128
5-Keto-D-gluconate 5-reductase	1.1.1.69
Ketogluconokinase	2.7.1.13
β-Ketoglutaric dehydrogenase	1.2.4.2
β-Ketoglutaric-isocitric carboxylase	1.1.1.41
3-Keto-L-gulonate carboxy-lyase	4.1.1.34
Keto-L-gulonate decarboxylase	4.1.1.34
3-Keto-L-gulonate dehydrogenase	1.1.1.130
3-*Keto-L-gulonate:NAD(P)$^+$ 2-oxidoreductase*	1.1.1.130
Ketohexokinase	2.7.1.3
Ketol-acid reductoisomerase	1.1.1.86
Ketone—aldehyde mutase	4.4.1.5
Ketopantoaldolase	4.1.2.12
β-Keto-reductase	1.1.1.35
Ketose-1-phosphate aldolase	[4.1.2.7]

INDEX TO THE ENZYME LIST (continued)

Ketotetrose-phosphate aldolase	4.1.2.2
β-Ketothiolase	2.3.1.16
Kininase II	3.4.15.1
Kininogenase	3.4.21.8
Kininogenin	3.4.21.8
Kynurenate-7,8-dihydrodiol dehydrogenase	1.3.1.18
Kynurenate, hydrogen-donor:oxygen oxidoreductase (hydroxylating)	1.14.99.2
Kynurenate 7,8-hydroxylase	1.14.99.2
Kynureninase	3.7.1.3
Kynurenine aminotransferase	2.6.1.7
Kynurenine formamidase	3.5.1.9
L-Kynurenine hydrolase	3.7.1.3
Kynurenine 3-hydroxylase	1.14.13.9
Kynurenine 3-monooxygenase	1.14.13.9
L-Kynurenine, NADPH:oxygen oxidoreductase (3-hydroxylating)	1.14.13.9
L-Kynurenine:2-oxoglutarate aminotransferase (cyclizing)	2.6.1.7
Laccase	1.14.18.1
Lactaldehyde dehydrogenase	1.2.1.22
D-Lactaldehyde dehydrogenase	1.1.1.78
D-Lactaldehyde:NAD$^+$ oxidoreductase	1.1.1.78
L-Lactaldehyde:NAD$^+$ oxidoreductase	1.2.1.22
Lactaldehyde reductase	1.1.1.77
Lactaldehyde reductase (NADPH)	1.1.1.55
β-Lactamase I	3.5.2.6
β-Lactamase II	3.5.2.8
Lactase	3.2.1.23
Lactate dehydrogenase	1.1.1.27
D-Lactate dehydrogenase	1.1.1.28
Lactate dehydrogenase (cytochrome)	1.1.2.3
D-Lactate dehydrogenase (cytochrome)	1.1.2.4
D-Lactate:ferricytochrome c oxidoreductase	1.1.2.4
L-Lactate:ferricytochrome c oxidoreductase	1.1.2.3
Lactate–malate transhydrogenase	1.1.99.7
Lactate 2-monooxygenase	1.13.12.4
D-Lactate:NAD$^+$ oxidoreductase	1.1.1.28
L-Lactate:NAD$^+$ oxidoreductase	1.1.1.27
Lactate → O$_2$ transhydrogenase	1.13.12.4
Lactate:oxaloacetate oxidoreductase	1.1.99.7
Lactate oxidase	[1.1.3.2]
Lactate oxidative decarboxylase	1.13.12.4
L-Lactate:oxygen 2-oxidoreductase (decarboxylating)	1.13.12.4
Lactate racemase	5.1.2.1
Lactate racemase	5.1.2.1
Lactic acid dehydrogenase	1.1.1.27, 1.1.1.28, 1.1.2.3, 1.1.2.4
Lacticoracemase	5.1.2.1
Lactonase	3.1.1.17
γ-Lactonase	3.1.1.25
γ-Lactone hydrolase	3.1.1.25
Lactose synthase	2.4.1.22
Lactoyl-CoA dehydratase	4.2.1.54
Lactoyl-CoA hydro-lyase	4.2.1.54
Lactoyl-glutathione lyase	4.4.1.5
S-Lactoyl-glutathione methylglyoxallyase (isomerizing)	4.4.1.5
Laminaribiose phosphorylase	2.4.1.31
Laminarinase	3.2.1.6
Laminarinase	3.2.1.39
Lathosterol oxidase	1.3.3.2
Lecithin acyltransferase	2.3.1.43
Lecithinase A	3.1.1.4
Lecithinase B	3.1.1.5
Lecithinase C	3.1.4.3
Lecithinase D	3.1.4.4
Lecithin:cholesterol acytransferase	2.3.1.43
Leucine aminopeptidase	3.4.11.1
Leucine aminotransferase	2.6.1.6
Leucine dehydrogenase	1.4.1.9
L-Leucine:NAD$^+$ oxidoreductase (deaminating)	1.4.1.9
L-Leucine:2-oxoglutarate aminotransferase	2.6.1.6
L-Leucine:tRNA^{Leu} ligase (AMP-forming)	6.1.1.4
Leucyltransferase	2.3.2.6
L-Leucyl-tRNA:protein leucyltransferase	2.3.2.6
Leucyl-tRNA synthetase	6.1.1.4
Leyanase	3.2.1.65
Levansucrase	2.4.1.10
Lichenase	3.2.1.73
Lichenase	3.2.1.6
Licheninase	[3.2.1.5]
Limit dextrinase	3.2.1.10, 3.2.1.41
Limonin-D-ring-lactonase	3.1.1.36
Limonoate-D-ring-lactone lactone-hydrolase	3.1.1.36
Linamarin synthase	2.4.1.63
Linoleate isomerase	5.2.1.5
Linoleate Δ^{12}-cis-Δ^{11}-trans-isomerase	5.2.1.5
Linoleate:oxygen oxidoreductase	1.13.11.12
Lipase	3.1.1.3
Lipoamide dehydrogenase (NADH)	1.6.4.3
Lipoate acetyltransferase	2.3.1.12
Lipophosphodiesterase I	3.1.4.3
Lipophosphodiesterase II	3.1.4.4
Lipoprotein lipase	3.1.1.34
Lipoxidase	1.13.11.12
Lipoxygenase	1.13.11.12
Lipoyl dehydrogenase	1.6.4.3
Lohmann's enzyme	2.7.3.2
Lombricine kinase	2.7.3.5
Luciferin sulphotransferase	2.8.2.10
Lysine acetyltransferase	2.3.1.32
Lysine 2,3-aminomutase	5.4.3.2
L-Lysine 2,3-aminomutase	5.4.3.2
β-Lysine 5,6-aminomutase	5.4.3.3
D-Lysine 5,6-aminomutase	5.4.3.4
L-Lysine 6-aminotransferase	2.6.1.36
L-Lysine carboxy-lyase	4.1.1.18
Lysine decarboxylase	4.1.1.18
Lysine hydroxylase	1.14.11.4
Lysine 2-monooxygenase	1.13.12.2
D-α-Lysine mutase	5.4.3.4

151

INDEX TO THE ENZYME LIST (continued)

β-Lysine mutase	5.4.3.3	Maleate hydratase	4.2.1.31
L-Lysine:2-oxoglutarate 6-aminotransferase	2.6.1.36	Maleate isomerase	5.2.1.1
Lysine, 2-oxoglutarate dioxygenase	1.14.11.4	*Maleate cis-trans-isomerase*	5.2.1.1
L-Lysine:oxygen 2-oxidoreductase (decarboxylating)	1.13.12.2	**Maleylacetoacetate isomerase**	5.2.1.2
Lysine racemase	5.1.1.5	*4-Maleylacetoacetate cis-trans-isomerase*	5.2.1.2
Lysine racemase	5.1.1.5	**Maleylpyruvate isomerase**	5.2.1.4
L-Lysine:tRNA^{Lys}ligase (AMP-forming)	6.1.1.6	*3-Maleylpyruvate cis-trans-isomerase*	5.2.1.4
Lysolecithin acyl-hydrolase	3.1.1.5	Malic dehydrogenase	1.1.1.37
Lysolecithin acylmutase	5.4.1.1	"Malic" enzyme	1.1.1.38,
Lysolecithin 2,3-acylmutase	5.4.1.1		1.1.1.39,
Lysolecithin acyltransferase	2.3.1.23		1.1.1.40
Lysolecithinase	3.1.1.5	**Malonate CoA-transferase**	2.8.3.3
Lysolecithin migratase	5.4.1.1	**Malonate-semialdehyde dehydratase**	4.2.1.27
Lysophospholipase	3.1.1.5	**Malonate-semialdehyde dehydrogenase**	1.2.1.15
Lysosomal α-glucosidase	3.2.1.3	**Malonate-semialdehyde dehydrogenase (acetylating)**	1.2.1.18
Lysozyme	3.2.1.17	*Malonate-semialdehyde hydro-lyase*	4.2.1.27
Lysyltransferase	2.3.2.3	*Malonate-semialdehyde:NAD(P)$^+$ oxidoreductase*	1.2.1.15
L-Lysyl-tRNA:phosphatidylglycerol 3'-lysyltransferase	2.3.2.3	*Malonate-semialdehyde:NAD(P)$^+$ oxidoreductase (decarboxylating, CoA-acetylating)*	1.2.1.18
Lysyl-tRNA synthetase	6.1.1.6		
α-Lytic proteinase	3.4.21.12	*Malonyl-CoA:[acyl-carrier-protein] S-malonyltransferase*	2.3.1.39
β-Lytic proteinase *(Mycobacterium sorangium)*	3.4.99.13	Malony-CoA carboxyltransferase	[2.1.3.4]
D-Lyxose ketol-isomerase	5.3.1.15	Malonyl-CoA decarboxylase	[4.1.1.9]
D-Lyxose ketol-isomerase	5.3.1.15	Maltase	3.2.1.20
		Maltodextrin phosphorylase	2.4.1.1
Macerans amylase	2.4.1.19	Maltose→amylose transglucosidase	2.4.1.25
Magnesium-protoporphyrin methyltransferase	2.1.1.11	Maltose 3-glucosyltransferase	[2.4.1.6]
Malate:CoA ligase (ADP-forming)	6.2.1.9	Maltose 4-glucosyltransferase	[2.4.1.3]
Malate condensing enzyme	4.1.3.2	*Maltose:orthophosphate α-glucosyltransferase*	2.4.1.8
Malate dehydrogenase	1.1.1.37	Maltose→orthophosphate transglucosidase	2.4.1.8
D-Malate dehydrogenase	1.1.1.83		
Malate dehydrogenase (decarboxylating)	1.1.1.39	**Maltose phosphorylase**	2.4.1.8
Malate dehydrogenase (decarboxylating) (NADP$^+$)	1.1.1.40	*Malyl-CoA glyoxylate-lyase*	4.1.3.24
Malate dehydrogenase (NADP$^+$)	1.1.1.82	**Malyl-CoA lyase**	4.1.3.24
Malate dehydrogenase (oxaloacetate-decarboxylating)	1.1.1.38	**Malyl-CoA synthetase**	6.2.1.9
L-Malate glyoxylate-lyase (CoA-acetylating)	4.1.3.2	**Mandelate racemase**	5.1.2.2
D-Malate hydro-lyase	4.2.1.31	*Mandelate racemase*	5.1.2.2
L-Malate hydro-lyase	4.2.1.2	*Mandelonitrile benzaldehyde-lyase*	4.1.2.10
D-Malate:NAD$^+$ oxidoreductase	1.1.1.83	**Mandelonitrile lyase**	4.1.2.10
L-Malate:NAD$^+$ oxidoreductase	1.1.1.37	**Mannanase**	3.2.1.25
L-Malate:NAD$^+$ oxidoreductase (decarboxylating)	1.1.1.39	*1,4-β-D-Mannan mannohydrolase*	3.2.1.78
L-Malate:NAD$^+$ oxidoreductase (oxaloacetate-decarboxylating)	1.1.1.38	*1,2-1,3-α-D-Mannan mannohydrolase*	3.2.1.77
		Mannase	3.2.1.25
L-Malate:NADP$^+$ oxidoreductase	1.1.1.82	**Mannitol dehydrogenase**	1.1.1.67
L-Malate:NADP$^+$ oxidoreductase (oxaloacetate-decarboxylating)	1.1.1.40	**Mannitol dehydrogenase (cytochrome)**	1.1.2.2
		Mannitol dehydrogenase (NADP$^+$)	1.1.1.138
L-Malate-NAD transhydrogenase	1.1.1.37	*D-Mannitol:ferricytochrome 2-oxidoreductase*	1.1.2.2
Malate oxidase	1.1.3.3	**Mannitol kinase**	2.7.1.57
L-Malate:oxygen oxidoreductase	1.1.3.3	*D-Mannitol:NAD$^+$ 2-oxidoreductase*	1.1.1.67
Malate synthase	4.1.3.2	*D-Mannitol:NADP$^+$ 2-oxidoreductase*	1.1.1.138
Malate synthetase	4.1.3.2	**Mannitol-1-phosphatase**	3.1.3.22
Malayan pit viper proteinase	3.4.21.5	**Mannitol-1-phosphate dehydrogenase**	1.1.1.17
		D-Mannitol-1-phosphate:NAD$^+$ 2-oxidoreductase	1.1.1.17

INDEX TO THE ENZYME LIST (continued)

D-Mannitol-1-phosphate phosphohydrolase	3.1.3.22
Mannokinase	2.7.1.7
Mannonate dehydratase	4.2.1.8
D-Mannonate dehydrogenase (NAD(P)$^+$)	1.2.1.34
D-Mannonate hydro-lyase	4.2.1.8
D-Mannonate:NAD$^+$ 5-oxidoreductase	1.1.1.57
D-Mannonate:NAD(P)$^+$ 6-oxidoreductase	1.1.1.131
D-Mannonate:NAD(P)$^+$ 6-oxidoreductase (D-mannuronate-forming)	1.2.1.34
Mannose isomerase	5.3.1.7
D-Mannose ketol-isomerase	5.3.1.7
Mannose-1-phosphate guanylyltransferase	2.7.7.22
Mannosephosphate isomerase	5.3.1.8
Mannose-6-phosphate isomerase	5.3.1.8
D-Mannose-6-phosphate ketol-isomerase	5.3.1.8
α-Mannosidase	3.2.1.24
β-Mannosidase	3.2.1.25
α-D-Mannoside mannohydrolase	3.2.1.24
β-D-Mannoside mannohydrolase	3.2.1.25
Mannuronate reductase	1.1.1.131
Melibiase	3.2.1.22
Melilotate dehydrogenase	1.3.1.11
Melilotate hydroxylase	1.14.13.4
Melilotate 3-monooxygenase	1.14.13.4
Menadione reductase	1.6.99.2
3-Mercaptopyruvate:cyanide sulphurtransferase	2.8.1.2
3-Mercaptopyruvate sulphurtransferase	2.8.1.2
Mesaconate hydratase	4.2.1.34
Metalloenzymes, microbial	3.4.24.4
Metaphosphatase	3.6.1.10, 3.6.1.11
Metaphosphate→ADP transphosphatase	2.7.4.1
Metapyrocatechase	1.13.11.2
Methenyltetrahydrofolate cyclohydrolase	3.5.4.9
5,10-Methenyltetrahydrofolate 5-hydrolase (decyclizing)	3.5.4.9
5,10-Methenyltetrahydrofolate:5-phosphoribosylglycinamide formyltransferase	2.1.2.2
5,10-Methenyltetrahydrofolate synthetase	6.3.3.2
Methionine adenosyltransferase	2.5.1.6
D-Methionine aminotransferase	2.6.1.41
L-Methionine carboxy-lyase	4.1.1.57
Methionine decarboxylase	4.1.1.57
Methionine S-methyltransferase	2.1.1.12
D-Methionine:pyruvate aminotransferase	2.6.1.41
Methionine racemase	5.1.1.2
Methionine racemase	5.1.1.2
Methionine synthase	4.2.99.10
L-Methionine:tRNA^{Met} ligase (AMP-forming)	6.1.1.10
Methionyl-tRNA formyltransferase	2.1.2.9
Methionyl-tRNA synthetase	6.1.1.10
4-Methoxybenzoate,hydrogen-donor:oxygen oxidoreductase (O-demethylating)	1.14.99.15
4-Methoxybenzoate monooxygenase (O-demethylating)	1.14.99.15
Methylamine—glutamate methyltransferase	2.1.1.21
Methylamine:L-glutamate N-methyltransferase	2.1.1.21
N-Methylamino-acid oxidase	1.5.3.2
N-Methyl-L-amino-acid:oxygen oxidoreductase (demethylating)	1.5.3.2
2-Methyl-4-amino-5-hydroxymethylpyrimidinediphosphate:4-methyl-5-(2-phosphoethyl)-thiazole 2-methyl-4-aminopyrimidine-5-methenyltransferase	2.5.1.3
β-Methylaspartase	4.3.1.2
Methylaspartate ammonia-lyase	4.3.1.2
L-threo-3-Methylaspartate ammonia-lyase	4.3.1.2
L-threo-3-Methylaspartate carboxyaminomethylmutase	5.4.99.1
Methylaspartate mutase	5.4.99.1
Methylbutyrase	3.1.1.1
3-Methylcrotonoyl-CoA:carbon-dioxide ligase (ADP-forming)	6.4.1.4
Methylcrotonoyl-CoA carboxylase	6.4.1.4
Methylcysteine synthase	4.2.1.22
2-Methylene-glutarate carboxy-methylene-methylmutase	5.4.99.4
2-Methylene-glutarate mutase	5.4.99.4
Methylene hydroxylase	1.14.15.1
3-Methyleneoxindole reductase	1.3.1.17
5,10-Methylenetetrahydrofolate:D-alanine hydroxymethyltransferase	2.1.2.7
5,10-Methylenetetrahydrofolate:ammonia hydroxymethyltransferase (carboxylating, reducing	2.1.2.10
Methylenetetrahydrofolate dehydrogenase	1.5.1.5
5,10-Methylenetetrahydrofolate:deoxycytidylate 5-hydroxymethyltransferase	2.1.2.8
5,10-Methylenetetrahydrofolate:glycine hydroxymethyltransferase	2.1.2.1
5,10-Methylenetetrahydrofolate:NADP$^+$ oxidoreductase	1.5.1.5
5,10-Methylenetetrahydrofolate reductase	1.1.1.68
Methylglutaconyl-CoA hydratase	4.2.1.18
N-Methylglutamate synthase	2.1.1.21
Methylglyoxalase	4.4.1.5
Methylglyoxal synthase	4.2.99.11
Methylhydroxypyridine-carboxylate dioxygenase	1.14.12.4
2-Methyl-3-hydroxypyridine-5-carboxylate, NAD(P)H:oxygen oxidoreductase (decylizing)	1.14.12.4
Methylhydroxypyridine-carboxylate oxidase	1.14.12.4
5-OMethyl-myo-inositol:NAD$^+$ oxidoreductase	1.1.1.143

INDEX TO THE ENZYME LIST (continued)

5D-5-O-Methyl-chiro-inositol:NADP⁺ oxidoreductase	1.1.1.142
Methylitaconate Δ-isomerase	5.3.3.6
Methylitaconate Δ² – Δ³-isomerase	5.3.3.6
N⁶-Methyl-lysine oxidase	1.5.3.4
N⁶-Methyl-L-lysine:oxygen oxidoreductase (demethylating)	1.5.3.4
Methylmalonate-semialdehyde dehydrogenase (acylating)	1.2.1.27
Methylmalonate-semialdehyde:NAD⁺ oxidoreductase (CoA-propionylating)	1.2.1.27
Methylmalonyl-CoA carboxyltransferase	2.1.3.1
R-Methylmalonyl-CoA carboxy-lyase	4.1.1.41
Methylmalonyl-CoA CoA-carbonyl-mutase	5.4.99.2
Methylmalonyl-CoA decarboxylase	4.1.1.41
Methylmalonyl-CoA mutase	5.4.99.2
Methylmalonyl-CoA:pyruvate carboxyltransferase	2.1.3.1
Methylmalonyl-CoA racemase	5.1.99.1
Methylmalonyl-CoA racemase	5.1.99.1
Methylmethionine-sulphonium-salt hydrolase	3.3.1.2
S-Methyl-L-methionine-sulphonium-salt hydrolase	3.3.1.2
3-Methyloxindole:NADP⁺ oxidoreductase	1.3.1.17
N-Methyl-2-oxoglutaramate hydrolase	3.5.1.36
N-Methyl-2-oxoglutaramate methylamidohydrolase	3.5.1.36
6-Methylsalicylate carboxy-lyase	4.1.1.52
6-Methylsalicylate decarboxylase	4.1.1.52
2-Methylserine hydroxymethyltransferase	2.1.2.7
Methylsterol hydroxylase	1.14.99.16
Methylsterol monooxygenase	1.14.99.16
5-Methyltetrahydrofolate:NAD⁺ oxidoreductase	1.1.1.68
5-Methyltetrahydropteroyl-L-glutamate:L-homocysteine S-methyltransferase	2.1.1.13
5-Methyltetrahydropteroyl-tri-L-glutamate:L-homocysteine S-methyltransferase	2.1.1.14
Methylthiophosphoglycerate phosphate	3.1.3.14
1-Methylthio-3-phospho-D-glycerate phosphohydrolase	3.1.3.14
Metridium proteinase A	3.4.21.3
Mevaldate reductase	1.1.1.32
Mevaldate reductase (NADPH)	1.1.1.33
Mevalonate kinase	2.7.1.36
Mevalonate:NAD⁺ oxidoreductase	1.1.1.32
Mevalonate:NAD⁺ oxidoreductase (CoA-acylating)	1.1.1.88
Mevalonate:NADP⁺ oxidoreductase	1.1.1.33
Mevalonate:NADP⁺ oxidoreductase (CoA-acylating)	1.1.1.34

Mexicanain	3.4.99.14
Microbial metalloenzymes	3.4.24.4
Microbial neutral proteinase	3.4.24.4
Micrococcal nuclease	3.1.4.7
Microsomal aminopeptidase	3.4.11.2
Monoacylglycerol lipase	3.1.1.23
Monoamine→O₂ transhydrogenase	1.4.3.4
Monoamine oxidase	1.4.3.4
Monobutyrase	3.1.1.1
Monodehydroascorbate reductase (NADH)	1.6.5.4
Monoglyceride acyltransferase	2.3.1.22
Monomethylaminoacid→O₂ transhydrogenase	1.5.3.2
Monophenol,dihydroxyphenylalanine:oxygen oxidoreductase	1.14.18.1
Monophenol monooxygenase	1.14.18.1
Monophenol oxidase	1.14.18.1
Monophosphatidylinositol inositolphosphohydrolase	
Monophosphatidylinositol inositolphosphohydrolase	3.1.4.10
Monophosphatidylinositol phosphodiesterase	3.1.4.10
Mucinase	3.2.1.35, 3.2.1.36, 4.2.2.1
Muconate cycloisomerase	5.5.1.1
Muconolactone Δ-isomerase	5.3.3.4
Mucopeptide N-acetylmuramoylhydrolase	3.2.1.17
Mucopeptide amidohydrolase	3.5.1.28
Mucopeptide glycohydrolase	3.2.1.17
Mucopolysaccharide α-L-iduronohydrolase	3.2.1.76
Mucor proteinase	3.4.23.10
Mung bean nuclease	3.1.4.9
Muramidase	3.2.1.17
Muramoyl-pentapeptide carboxypeptidase	3.4.12.6
Muscle phosphorylase a and b	2.4.1.1
Mutarotase	5.1.3.3
Mycobacterium sorangium α-lytic proteinase	3.4.21.12
Mycobacterium sorangium β-lytic proteinase	3.4.99.13
Mycodextranase	3.2.1.61
Myokinase	2.7.4.3
Myrosinase	3.2.3.1
Myrosulphatase	[3.1.6.5]
NADase	3.2.2.5
NAD⁺ glycohydrolase	3.2.2.5
NADH:(acceptor) oxidoreductase	1.6.99.3
NADH:aquocobalamin oxidoreductase	1.6.99.8
NADH:cob(II)alamin oxidoreductase	1.6.99.9
NADH:L-cystine oxidoreductase	1.6.4.1
NADH dehydrogenase	1.6.99.3
NADH dehydrogenase (quinone)	1.6.99.5
NADH:ferricytochrome b₅ oxidoreductase	1.6.2.2
NADH:ferricytochrome oxidoreductase	1.6.2.4
NADH:hydrogen-peroxide oxidoreductase	1.11.1.1
NADH:hydroxylamine oxidoreductase	1.6.6.11

INDEX TO THE ENZYME LIST (continued)

NADH:hyponitrite oxidoreductase	1.6.6.6
NADH:lipoamide oxidoreductase	1.6.4.3
NADH:monodehydroascorbate oxidoreductase	1.6.5.4
NADH:nitrate oxidoreductase	1.6.6.1
NADH:(quinone-acceptor) oxidoreductase	1.6.99.5
NADH:rubredoxin oxidoreductase	1.6.7.2
NADH:trimethylamine-N-oxide oxidoreductase	1.6.6.9
NAD$^+$ kinase	2.7.1.23
NAD$^+$ nucleosidase	3.2.2.5
NAD$^+$ peroxidase	1.11.1.1
NAD(P)$^+$ glycohydrolase	3.2.2.6
NADPH:CoAS–Sglutathione oxidoreductase	1.6.4.6
NADPH–cytochrome c reductase	1.6.2.4
NADPH–cytochrome c_2 reductase	1.6.2.5
NADPH–cytochrome reductase	1.6.2.4
NADPH dehydrogenase	1.6.99.1
NADPH dehydrogenase (quinone)	1.6.99.6
NAD(P)H dehydrogenase (quinone)	1.6.99.2
NADPH:6,7-dihydropteridine oxidoreductase	1.6.99.7
NADPH:dimethylaminoazobenzene oxidoreductase	1.6.6.7
NADPH:ferredoxin oxidoreductase	1.6.7.1
NADPH:ferricytochrome c_2 oxidoreductase	1.6.2.5
NADPH:ferricytochrome oxidoreductase	1.6.2.4
NADPH:GMP oxidoreductase (deaminating)	1.6.6.8
NADPH:hydrogen-peroxide oxidoreductase	1.11.1.2
NADPH:NAD$^+$ oxidoreductase	1.6.1.1
NADPH:nitrate oxidoreductase	1.6.6.3
NAD(P)H:nitrate oxidoreductase	1.6.6.2
NAD(P)H:nitrite oxidoreductase	1.6.6.4
NAD(P)H:4-nitroquinoline-N-oxide oxidoreductase	1.6.6.10
NAD$^+$ phosphohydrolase	3.6.1.22
NAD(P)H:oxidized-glutathione oxidoreductase	1.6.4.2
NADPH:oxidized-thioredoxin oxidoreductase	1.6.4.5
NAD(P)H:protein-disulphide oxidoreductase	1.6.4.4
NADPH:(quinone-acceptor) oxidoreductase	1.6.99.6
NAD(P)H:(quinone-acceptor) oxidoreductase	1.6.99.2
NAD(P)$^+$ nucleosidase	3.2.2.6
NADP$^+$ peroxidase	1.11.1.2
NAD(P)$^+$ transhydrogenase	1.6.1.1
NAD$^+$ pyrophosphatase	3.6.1.22
NAD$^+$ pyrophosphorylase	2.7.7.1
NAD$^+$ synthetase	6.3.1.5
NAD$^+$ synthetase (glutamine-hydrolysing)	6.3.5.1
Nagarse proteinase	3.4.21.14
Naphthalene, hydrogen-donor:oxygen oxidoreductase (1,2-epoxidizing)	1.14.99.8
NDPhexose pyrophosphorylase	2.7.7.28
NDP:sugar-1-phosphate nucleotidyltransferase	2.7.7.37
NDPsugar phosphorylase	2.7.7.37
Neuraminidase	3.2.1.18
Neurospora crassa endonuclease	3.1.4.21
Neutral proteinase	3.4.24.4
Nicotinamidase	3.5.1.19
Nicotinamide amidohydrolase	3.5.1.19
Nicotinamide methyltransferase	2.1.1.1
Nicotinamidenucleotide:pyrophosphate phosphoribosyltransferase	2.4.2.12
Nicotinamide phosphoribosyltransferase	2.4.2.12
Nicotinate dehydrogenase	1.5.1.13
Nicotinate methyltransferase	2.1.1.7
Nicotinatemononucleotide adenylytransferase	2.7.7.18
Nicotinatemononucleotide pyrophosphorylase (carboxylating)	2.4.2.19
Nicotinate:NADP$^+$ 6-oxidoreductase (hydroxylating)	1.5.1.13
Nicotinatenucleotide–dimethylbenzimidazole phosphoribosyltransferase	2.4.2.21
Nicotinatenucleotide:dimethylbenzimidazole phosphoribosyltransferase	2.4.2.21
Nicotinatenucleotide:pyrophosphate phosphoribosyltransferase	2.4.2.11
Nicotinatenucleotide:pyrophosphate phosphoribosyltransferase (carboxylating)	2.4.2.19
Nicotinate phosphoribosyltransferase	2.4.2.11
Nicotine:(acceptor)6-oxidoreductase (hydroxylating)	1.5.99.4
Nicotine dehydrogenase	1.5.99.4
Nitrate-ester reductase	1.8.6.1
Nitrate reductase	1.7.99.4
Nitrate reductase (cytochrome)	1.9.6.1
Nitrate reductase (NADH)	1.6.6.1
Nitrate reductase (NADPH)	1.6.6.3
Nitrate reductase (NAD(P)H)	1.6.6.2
Nitric-oxide:(acceptor) oxidoreductase	1.7.99.3
Nitric-oxide:ferricytochrome c oxidoreductase	1.7.2.1
Nitric-oxide reductase	1.7.99.2
Nitrilase	3.5.5.1
Nitrile aminohydrolase	3.5.5.1
Nitrite:(acceptor)oxidoreductase	1.7.99.4
Nitrite reductase	1.7.99.3
Nitrite reductase (cytochrome)	1.7.2.1
Nitrite reductase (NAD(P)H)	1.6.6.4
β-**Nitroacrylate reductase**	1.3.1.16
Nitroethane oxidase	1.7.3.1
Nitroethane:oxygen oxidoreductase	1.7.3.1
Nitrogen:(acceptor) oxidoreductase	1.7.99.2
Nitrogenase	1.7.99.2
3-Nitropropionate:NADP$^+$ oxidoreductase	1.3.1.16

INDEX TO THE ENZYME LIST (continued)

Nitroquinoline-N-oxide reductase	1.6.6.10
NMN adenylyltransferase	2.7.7.1
NMN pyrophosphorylase	2.4.2.12
Noradrenalin N-methyltransferase	2.1.1.28
Notatin	1.1.3.4
NTP:deoxycytidine 5'-phosphotransferase	2.7.1.74
NTP:hexose-1-phosphate nucleotidyltransferase	2.7.7.28
NTP polymerase	2.7.7.19
Nuclease, Azotobacter	3.1.4.9
Nuclease, hybrid	3.1.4.34
Nuclease, Micrococcal	3.1.4.7
Nuclease, Mung bean	3.1.4.9
Nucleate endonuclease	3.1.4.9
Nucleate 3'-oligonucleotidohydrolase	3.1.4.7
Nucleate 5'-oligonucleotidohydrolase	3.1.4.9
Nucleosidase	3.2.2.1
Nucleoside-2':3'-cyclic-phosphate 3'-nucleotidohydrolase	3.1.4.16
Nucleoside deoxyribosyltransferase	2.4.2.6
Nucleosidediphosphatase	3.6.1.6
Nucleosidediphosphate kinase	2.7.4.6
Nucleosidediphosphate phosphohydrolase	3.6.1.6
Nucleosidemonophosphate kinase	2.7.4.4
Nucleoside-5'-phosphoacylate acylhydrolase	3.6.1.24
Nucleoside phosphoacylhydrolase	3.6.1.24
Nucleoside phosphotransferase	2.7.1.77
Nucleoside:purine (pyrimidine) deoxyribosyltransferase	2.4.2.6
Nucleoside:purine (pyrimidine) ribosyltransferase	2.4.2.5
Nucleoside ribohydrolase	3.2.2.8
Nucleoside ribosyltransferase	2.4.2.5
Nucleosidetriphosphate–adenylate kinase	2.7.4.10
Nucleosidetriphosphate:AMP phosphotransferase	2.7.4.10
Nucleosidetriphosphate:DNA deoxynucleotidylexotransferase	2.7.7.31
Nucleoside triphosphatase	3.6.1.15
Nucleosidetriphosphate pyrophosphatase	3.6.1.19
Nucleosidetriphosphate pyrophosphohydrolase	3.6.1.19
Nucleosidetriphosphate:RNA nucleotidyltransferase	2.7.7.6
Nucleotidase	3.1.3.31
3'-Nucleotidase	3.1.3.6
5'-Nucleotidase	3.1.3.5
Nucleotide:3'-deoxynucleoside 5'-phosphotransferase	2.7.1.77
Nucleotide phosphohydrolase	3.1.3.31
Nucleotide pyrophosphatase	3.6.1.9
Octanol dehydrogenase	1.1.1.73
Octanol:NAD$^+$ oxidoreductase	1.1.1.73
Octopine dehydrogenase	1.5.1.11
Oestradiol 17α-dehydrogeanse	1.1.1.148
Oestradiol 17β-dehydrogenase	1.1.1.62
α-Oestradiol→DPN transhydrogenase	1.1.1.62
Oestradiol-17β, hydrogen-donor:oxygen oxidoreductase (6β-hydroxylating)	1.14.99.11
Oestradiol 6β-hydroxylase	1.14.99.11
Oestradiol 6β-monooxygenase	1.14.99.11
Oestradiol-17β:NAD$^+$ 17-oxidoreductase	1.1.1.62
17β-Oestradiol–UDP glucuronyltransferase	2.4.1.59
Oestriol 2-hydroxylase	[1.14.1.11]
Oestriol–UDP 16α-glucuronyltransferase	2.4.1.61
Oestriol–UDP 17β-glucuronyltransferase	2.4.1.42
Oestrone sulphotransferase	2.8.2.4
Old yellow enzyme	1.6.99.1
Oelate hydratase	4.2.1.53
Oligodeoxyribonucleate exonuclease	3.1.4.29
Oligodeoxyribonucleate 5'-nucleotidohydrolase	3.1.4.29
Oligogalacturonide lyase	4.2.2.6
Oligogalacturonide lyase	4.2.2.6
Oligoglucan-branching glycosyltransferase	2.4.1.24
1,3-β-Oligoglucan:orthophosphate glucosyltransferase	2.4.1.30
1,4-β-Oligoglucan:orthophosphate α-glucosyltransferase	2.4.1.49
1,3-β-Oligoglucan phosphorylase	2.4.1.30
Oligo-1,3-glucosidase	3.2.1.39
Oligo-1,6-glucosidase	3.2.1.10
Oligonucleate 3'-nucleotidohydrolase	3.1.4.18
Oligonucleate 5'-nucleotidohydrolase	3.1.4.1
Oligonucleotidase	3.1.4.19
Oligonucleotide 5'-nucleotidohydrolase	3.1.4.19
Opheline kinase	2.7.3.7
Ophio-aminoacid oxidase	1.4.3.2
Orcinol hydroxylase	1.14.13.6
Orcinol 2-monooxygenase	1.14.13.6
Orcinol, NADH:oxygen oxidoreductase (2-hydroxylating)	1.14.13.6
Ornithine 4,5-aminomutase	5.4.3.1
L-Ornithine 4,5-aminomutase	5.4.3.1
Ornithine carbamoyltransferase	2.1.3.3
L-Ornithine carboxy-lyase	4.1.1.17
Ornithine decarboxylase	4.1.1.17
Ornithine–oxo-acid aminotransferase	2.6.1.13
L-Ornithine:2-oxo-acid aminotransferase	2.6.1.13
Ornithine transcarbamylase	2.1.3.3
Orotate phosphoribosyltransferase	2.4.2.10
Orotate reductase	1.3.1.14
Orotate reductase (NADPH)	1.3.1.15
Orotidine-5'-phosphate carboxy-lyase	4.1.1.23
Orotidine-5'-phosphate decarboxylase	4.1.1.23
Orotidine-5'-phosphate:pyrophosphate phosphoribosyltransferase	2.4.2.10
Orotidine-5'-phosphate pyrophosphorylase	2.4.2.10
Orotodylic acid phosphorylase	2.4.2.10
Orsellinate carboxy-lyase	4.1.1.58
Orsellinate decarboxylase	4.1.1.58
Orthophenolase	1.14.18.1

INDEX TO THE ENZYME LIST (continued)

Orthophosphate:oxaloacetate carboxy-lyase (phosphorylating)	4.1.1.31
Orthophosphoric-monoester phosphohydrolase (acid optimum)	3.1.3.2
Orthophosphoric-monoester phosphohydrolase (alkaline optimun)	3.1.3.1
Oxalacetate β-decarboxylase	4.1.1.3
Oxalate carboxy-lyase	4.1.1.2
Oxalate:CoA ligase (AMP-forming)	6.2.1.8
Oxalate CoA-transferase	2.8.3.2
Oxalate decarboxylase	4.1.1.2
Oxalate→O_2 transhydrogenase	1.2.3.4
Oxalate oxidase	1.2.3.4
Oxalate:oxygen oxidoreductase	1.2.3.4
Oxaloacetase	3.7.1.1
Oxaloacetate acetylhydrolase	3.7.1.1
Oxaloacetate carboxy-lyase	4.1.1.3
Oxaloacetate decarboxylase	4.1.1.3
Oxaloacetate keto–enol-isomerase	5.3.2.2
Oxaloacetate tautomerase	5.3.2.2
Oxaloacetate transacetase	4.1.3.7
Oxaloglycollate reductase (decarboxylating)	1.1.1.92
3-Oxalomalate glyoxylate-lyase	4.1.3.13
Oxalomalate lyase	4.1.3.13
Oxalooxyhydrase	1.2.3.4
Oxalyl-CoA carboxy-lyase	4.1.1.8
Oxalyl-CoA decarboxylase	4.1.1.8
Oxalyl-CoA synthetase	6.2.1.8
2,3-Oxidosqualene cycloartenol-cyclase	5.4.99.8
2,3-Oxidosqualene lanosterol-cyclase	5.4.99.7
2,3-Oxidosqualene mutase (cyclizing, cycloartenol-forming)	5.4.99.8
2,3-Oxidosqualene mutase (cyclizing, lanosterol-forming)	5.4.99.7
Oximinotransferase	2.6.3.1
2-Oxo-acid carboxy-lyase	4.1.1.1
3-Oxoacyl-[acyl-carrier-protein] reductase	1.1.1.100
3-Oxoacyl-[acyl-carrier-protein] synthase	2.3.1.41
3-Oxoadipate CoA-transferase	2.8.3.6
3-Oxoadipate enol-lactonase	3.1.1.24
2-Oxoaldehyde dehydrogenase	1.2.1.23
2-Oxoaldehyde:NAD(P)$^+$ oxidoreductase	1.2.1.23
2-Oxobutyrate:ferredoxin oxidoreductase (CoA-propionylating)	1.2.7.2
2-Oxobutyrate synthase	1.2.7.2
Oxoglutarate dehydrogenase	1.2.4.2
2-Oxoglutarate:ferredoxin oxidoreductase (CoA-succinylating)	1.2.7.3
2-Oxoglutarate:lipoate oxidoreductase (decarboxylating and acceptor-succinylating)	1.2.4.2
2-Oxoglutarate synthase	1.2.7.3
2-Oxoisocaproate dehydrogenase	1.2.4.3
2-Oxoisocaproate:lipoate oxidoreductase (decarboxylating and acceptor-isovalerylating)	1.2.4.3
Oxoisomerase	5.3.1.9
2-Oxoisovalerate dehydrogenase	1.2.1.25
2-Oxoisovalerate dehydrogenase (lipoate)	1.2.4.4
2-Oxoisovalerate:lipoate oxidoreductase (decarboxylating and acceptor-isobutyrylating)	1.2.4.4
2-Oxoisovalerate:NAD$^+$ oxidoreductase (CoA-isobutyrylating)	1.2.1.25
3-Oxolaurate carboxy-lyase	4.1.1.56
3-Oxolaurate decarboxylase	4.1.1.56
2-Oxopantoate formaldehyde-lyase	4.1.2.12
4-Oxoproline reductase	1.1.1.104
(3'-Oxo-prop-2'-enyl)-2-amino-but-2-ene-dioate carboxy-lyase	4.1.1.45
D-3-Oxosphinganine reductase	1.1.1.102
3-Oxosteroid:(acceptor) Δ^1-oxidoreductase	1.3.99.4
3-Oxo-5α-steroid:(acceptor) Δ^4-oxidoreductase	1.3.99.5
3-Oxo-5β-steroid:(acceptor) Δ^4-oxidoreductase	1.3.99.6
3-Oxosteroid Δ^1-dehydrogenase	1.3.99.4
3-Oxo-5-α-steroid Δ^4-dehydrogenase	1.3.99.5
3-Oxo-5β-steroid Δ^4-dehydrogenase	1.3.99.6
3-Oxosteroid Δ^4–Δ^5-isomerase	5.3.3.1
3-Oxo-5α-steroid:NADP$^+$ Δ^4-oxidoreductase	1.3.1.22
3-Oxo-5β-steroid:NADP$^+$ Δ^4-oxidoreductase	1.3.1.23
6-Oxotetrahydronicotinate dehydrogenase	1.3.7.1
Oxytocinase	3.4.11.3
PI	3.4.23.1, 3.4.23.3
PII	3.4.23.1
PIII	3.4.23.3
Paecilomyces proteinase	3.4.99.15
Palmitate:hydrogen-peroxide oxidoreductase	1.11.1.3
Palmitoyl-CoA:acylglycerol O-palmitoyltransferase	2.3.1.22
Palmitoyl-CoA:L-carnitine O-palmitoyltransferase	2.3.1.21
Palmitoyl-CoA hydrolase	3.1.2.2
Palmitoyl-CoA hydrolase	3.1.2.2
Palmitoyldihydroxyacetone-phosphate reductase	1.1.1.101
1-Palmitoylglycerol-3-phosphate:NADP$^+$ oxidoreductase	1.1.1.101
Pancreatic DNase	3.1.4.5
Pancreatic RNase	3.1.4.22
Pancreatopeptidase E	3.4.21.11
Pantetheine kinase	2.7.1.34
Pantetheinephosphate adenylyltransferase	2.7.7.3
Pantoate activating enzyme	6.3.2.1
L-Pantoate:β-alanine ligase (AMP-forming)	6.3.2.1
Pantoate dehydrogenase	1.1.1.106

INDEX TO THE ENZYME LIST (continued)

D-Pantoate:NAD⁺ oxidoreductase	1.1.1.106	*Peptidyllysine, 2-oxoglutarate:oxygen*	
Pantothenase	3.5.1.22	*5-oxidoreductase*	1.14.11.4
Pantothenate amidohydrolase	3.5.1.22	*Peptidylprolyl-amino-acid hydrolase*	3.4.12.4
Pantothenate kinase	2.7.1.33	**Peptidyltryptophan 2,3-dioxygenase**	1.13.11.26
Pantothenate synthetase	6.3.2.1	*Peptidyltryptophan:oxygen 2,3oxidoreductase*	
N-(L-Pantothenoyl)-L-cysteine		*(decyclizing)*	1.13.11.26
carboxy-lyase	4.1.1.30	*Peptidyl-L-tyrosine hydrolase*	3.4.12.12
Pantothenoylcysteine decarboxylase	4.1.1.30	**Perillyl-alcohol dehydrogenase**	1.1.1.144
Papain	3.4.22.2	*Perillyl-alcohol:NAD⁺ oxidoreductase*	1.1.1.144
Papainase	3.4.22.2	**Peroxidase**	1.11.1.7
Paraoxonase	3.1.1.2	Phage SP3 DNase	3.1.4.31
Parapepsin I	3.4.23.2	Phaseolin	3.4.21.13
Parapepsin II	3.4.23.3	**Phaseolus proteinase**	3.4.21.13
Particle-bound aminopeptidase	3.4.11.2	Phenolase	1.14.18.1
Pectate lyase	4.2.2.2	Phenol hydroxylase	1.14.13.7
Pectate transeliminase	4.2.2.2	**Phenol O-methyltransferase**	2.1.1.25
Pectinase	3.2.1.15 '	**Phenol 2-monooxygenase**	1.14.13.7
Pectin demethoxylase	3.1.1.11	*Phenyl,NADPH:oxygen oxidoreductase*	
Pectin depolymerase	3.2.1.15	*(2-hydroxylating)*	1.14.13.7
Pectinesterase	3.1.1.11	Phenol sulphotransferase	2.8.2.1
Pectin lyase	4.2.2.10	*Phenylacetyl-CoA:L-glutamine*	
Pectin methoxylase	3.1.1.11	*α-N-phenylacetyltransferase*	2.3.1.14
Pectin methylesterase	3.1.1.11	Phenylalaninase	1.14.16.1
Pectin pectyl-hydrolase	3.1.1.11	**Phenylalanine ammonia-lyase**	4.3.1.5
Penicillin amidase	3.5.1.11	*L-Phenylalanine ammonia-lyase*	4.3.1.5
Penicillin amidohydrolase	3.5.1.11	*L-Phenylalanine carboxy-lyase*	4.1.1.53
Penicillin amido-β-lactamhydrolase	3.5.2.6	**Phenylalanine decarboxylase**	4.1.1.53
Penicillinase	3.5.2.6	Phenylalanine 4-hydroxylase	1.14.16.1
Penicillium janthinellum acid		**Phenylalanine 4-monooxygenase**	1.14.16.1
proteinase	3.4.23.7	**Phenylalanine racemase (ATP-hydrolyzing)**	5.1.1.11
Penicillium notatum extracellular		*Phenylalanine racemase (ATP-hydrolyzing)*	5.1.1.11
proteinase	3.4.99.16	*L-Phenylalanine, tetrahydropteridine:oxygen*	
Penicillopepsin	3.4.23.7	*oxidoreductase (4-hydroxylating)*	1.14.16.1
24,6/3,5-Pentahydroxycyclohexanose		*L-Phenylalanine:tRNA^{Phe} ligase*	
hydro-lyase	4.2.1.44	*(AMP-forming)*	6.1.1.20
Pentosealdolase	[4.1.2.3]	**Phenylalanyl-tRNA synthetase**	6.1.1.20
P-enzyme	2.4.1.1	*Phenylpyruvate carboxy-lyase*	4.1.1.43
PEP carboxyphosphotransferase	4.1.1.38	**Phenylpyruvate decarboxylase**	4.1.1.43
Pepsin	3.4.23.1	*Phenylpyruvate keto–enol-isomerase*	5.3.2.1
Pepsin IV	3.4.23.2	**Phenylpyruvate tautomerase**	5.3.2.1
Pepsin A	3.4.23.1	**Phenylserine aldolase**	4.1.2.26
Pepsin B	3.4.23.2	*L-threo-3-Phenylserine benzaldehyde-*	
Pepsin C	3.4.23.3	*lyase*	4.1.2.26
Pepsin D	3.4.23.1	**Phloretin-glucosidase**	3.2.1.62
Peptidase β	3.4.22.9	**Phloretin hydrolase**	3.7.1.4
Peptidase A	3.4.23.7	*Phlorizin glucohydrolase*	3.2.1.62
Peptidase P	3.4.15.1	Phlorizin hydrolase	3.2.1.62
Peptidoglycan endopeptidase	3.4.99.17	**Phosphatase, acid**	3.1.3.2
Peptidyl-L-alanine hydrolase	3.4.12.11	**Phosphatase, alkaline**	3.1.3.1
Peptidyl-L-amino-acid hydrolase	3.4.12.2	**Phosphate acetyltransferase**	2.3.1.8
Peptidyl-L-amino-acid (-L-proline)		**Phosphate butyryltransferase**	2.3.1.19
hydrolase	3.4.12.1	Phosphatidase	3.1.1.4
Peptidyl-L-arginine hydrolase	3.4.12.7	*Phosphatidate 1-acylhydrolase*	3.1.1.32
Peptidyl-L-aspartate hydrolase	3.4.12.9	**Phosphatidate cytidylyltransferase**	2.7.7.41
Peptidyl dipeptidase	3.4.15.1	**Phosphatidate phosphatase**	3.1.3.4
Peptidyldipeptide hydrolase	3.4.15.1	*L-α-Phosphatidate phosphohydrolase*	3.1.3.4
Peptidyl-L-glutamate hydrolase	3.4.12.10	*Phosphatide 2-acyl-hydrolase*	3.1.1.4
Peptidyl-glycine hydrolase	3.4.12.8	**Phosphatide cytidylyltransferase**	2.7.7.41
Peptidyl-L-lysine(-L-arginine) hydrolase	3.4.12.3	Phosphatidolipase	3.1.1.4

INDEX TO THE ENZYME LIST (continued)

Phosphatidylcholine cholinephosphohydrolase	3.1.4.3
Phosphatidylcholine phosphatidohydrolase	3.1.4.4
Phosphatidylethanolamine methyltransferase	2.1.1.17
Phosphatidylglycerophosphate phosphatase	3.1.3.27
Phosphatidylglycerophosphate phosphohydrolase	3.1.3.27
Phosphatidyl-inositol-bisphosphate phosphatase	3.1.3.36
Phosphatidyl-myo-inositol-4,5-bisphosphate phosphohydrolase	3.1.3.36
Phosphatidylinositol kinase	2.7.1.67
Phosphoacetylglucosamine mutase	2.7.5.2
Phospho-N-acetylmuramoyl-pentapeptide-transferase	2.7.8.13
Phosphoacylase	2.3.1.8
Phosphoadenylate 3'-nucleotidase	3.1.3.7
Phosphoadenylylsulphatase	3.6.2.2
3'-Phosphoadenylylsulphate:arylamine sulphotranferase	2.8.2.3
3'-Phosphoadenylylsulphate:choline sulphotransferase	2.8.2.6
3'Phosphoadenylylsulphate:chondroitin 4'-sulphotransferase	2.8.2.5
3'-Phosphoadenylylsulphate:N-desulphoheparin N-sulphotransferase	2.8.2.8
3'-Phosphoadenylylsulphate:galactocerebroside 3'-sulphotransferase	2.8.2.11
3'-Phosphoadenylylsulphate:heparitin N-sulphotransferase	2.8.2.12
3'-Phosphoadenylylsulphate:3β-hydroxysteroid sulphotransferase	2.8.2.2
3'-Phosphoadenylylsulphate:luciferin sulphotransferase	2.8.2.10
3'-Phosphoadenylylsulphate:oestrone 3'-sulphotransferase	2.8.2.4
3'-Phosphoadenylylsulphate:phenol sulphotransferase	2.8.2.1
3'-Phosphoadenylylsulphate 3'-phosphatase	3.1.3.30
3'Phosphoadenylylsulphate 3'phosphohydrolase	3.1.3.30
3'-Phosphoadenylylsulphate sulphohydrolase	3.6.2.2
3'-Phosphoadenylylsulphate:L-tryosine-methyl-ester sulphotransferase	2.8.2.9
3'-Phosphoadenylylsulphate:UDP-2-acetamido-2-deoxy-D-galactose-4-sulphate 6-sulphotransferase	2.8.2.7
Phosphoamidase	3.9.1.1
Phosphoamide hydrolase	3.9.1.1
Phosphodeoxyriboaldolase	4.1.2.4
Phosphodeoxyribomutase	2.7.5.6
Phosphodiesterase	3.1.4.7, 3.1.4.18
Phosphodiesterase I	3.1.4.1
Phosphodiesterase II	3.1.4.18
Phospho*enol*pyruvate → ADP transphosphatase	2.7.1.40
Phospho*enol*pyruvate carboxykinase	4.1.1.49
Phospho*enol*pyruvate carboxykinase (ATP)	4.1.1.49
Phospho*enol*pyruvate carboxykinase (GTP)	4.1.1.32
Phospho*enol*pyruvate carboxykinase (pyrophosphate)	4.1.1.38
Phospho*enol*pyruvate carboxylase	4.1.1.31
Phospho*enol*pyruvate carboxylase	4.1.1.32, 4.1.1.38, 4.1.1.49
Phospho*enol*pyruvate kinase	2.7.1.40
Phospho*enol*pyruvate–protein phosphotransferase	2.7.3.9
*Phospho*enol*pyruvate:protein phosphotransferase*	2.7.3.9
*Phospho*enol*pyruvate:UDP-2-acetamido-2-deoxy-D-glucose 2-enoyl-1-carboxyethyltransferase*	2.5.1.7
Phosphoenol transphosphorylase	2.7.1.40
1-Phosphofructokinase	2.7.1.56
6-Phosphofructokinase	2.7.1.11
Phosphoglucokinase	2.7.1.10
Phosphoglucomutase	2.7.5.1
Phosphoglucomutase (glucose-cofactor)	2.7.5.5
Phosphogluconate dehydratase	4.2.1.12
Phosphogluconate dehydrogenase	1.1.1.43
Phosphogluconate dehydrogenase (decarboxylating)	1.1.1.44
6-Phospho-D-gluconate hydro-lyase	4.2.1.12
6-Phospho-D-gluconate-δ-lactone hydrolase	3.1.1.31
6-Phospho-D-gluconate:NADP⁺ 2-oxidoreductase (decarboxylating)	1.1.1.44
6-Phospho-D-gluconate:NAD(P)⁺ 2-oxidoreductase	1.1.1.43
6-Phosphogluconate → TPN transhydrogenase	1.1.1.44
Phosphogluconic(acid) dehydrogenase	1.1.1.43, 1.1.1.44
6-Phosphogluconic carboxylase	1.1.1.44
6-Phosphogluconic dehydrogenase	1.1.1.43, 1.1.1.44
6-Phosphogluconolactonase	3.1.1.31
Phosphoglucosamine acetylase	2.3.1.4
Phosphoglucosamine transacetylase	2.3.1.4
3-Phospho-D-glycerate carboxylyase (dimerizing)	4.1.1.39
2-Phosphoglycerate dehydratase	4.2.1.11
Phosphoglycerate dehydrogenase	1.1.1.95
2-Phospho-D-glycerate hydro-lyase	4.2.1.11
Phosphoglycerate kinase	2.7.2.3
3-Phosphoglycerate:NAD⁺ 2-oxidoreductase	1.1.1.95
Phosphoglycerate phosphatase	3.1.3.20
Phosphoglycerate phosphomutase	5.4.2.1
3-Phosphoglycerate (1 → 2)-phosphomutase	2.7.5.4
D-Phosphoglycerate 2,3-phosphomutase	5.4.2.1
Phosphoglyceromutase	2.7.5.3

INDEX TO THE ENZYME LIST (continued)

3-Phospho-D-glyceroyl-phosphate:3-phospho-D-glycerate phosphotransferase	2.7.5.4
Phosphoglycollate phosphatase	3.1.3.18
2-Phosphoglycollate phosphohydrolase	3.1.3.18
Phosphohexokinase	2.7.1.11
Phosphohexomutase	5.3.1.9
Phosphohexose isomerase	5.3.1.9
Phosphohistidinoprotein—hexose phosphotransferase	2.7.1.69
Phosphohistidinoprotein:hexose phosphotransferase	2.7.1.69
O-Phosphohomoserine phospholyase (adding water)	4.2.99.2
6-Phospho-2-keto-3-deoxy-galactonate aldolase	4.1.2.21
Phospho-2-keto-3-deoxy-gluconate aldolase	4.1.2.14
6-Phospho-2-keto-3-deoxy-D-gluconate D-glyceraldehyde-3-phosphate-lyase	4.1.2.14
Phospho-2-keto-3-deoxy-heptonate aldolase	4.1.2.15
7-Phospho-2-keto-3-deoxy-D-arabino-heptonate D-erythrose-4-phosphate-lyase (pyruvate-phosphorylating)	4.1.2.15
Phospho-2-keto-3-deoxyoctonate aldolase	4.1.2.16
8-Phospho-2-keto-3-deoxy-D-octonate D-arbinose-5-phosphate-lyase (pyruvate-phosphorylating)	4.1.2.16
Phosphoketolase	4.1.2.9
Phosphoketotetrose aldolase	4.1.2.2
Phospholipase A$_1$	3.1.1.32
Phospholipase A$_2$	3.1.1.4
Phospholipase B	3.1.1.5
Phospholipase C	3.1.4.3
Phospholipase D	3.1.4.4
Phosphomannan mannosephosphotransferase	2.7.8.9
Phosphomannose isomerase	5.3.1.8
Phosphomethylpyrimidine kinase	2.7.4.7
Phosphomevalonate kinase	2.7.4.2
Phosphomonoesterase	3.1.3.1, 3.1.3.2
Phosphonoacetaldehyde hydrolase	3.11.1.1
2-Phosphonoacetaldehyde phosphohydrolase	3.11.1.1
4'-Phospho-L-pantothenate:L-cysteine ligase	6.3.2.5
4'Phospho-N-(L-pantothenoyl)-L-cysteine carboxy-lyase	4.1.1.36
Phosphopantothenoyl-cysteine decarboxylase	4.1.1.36
Phosphopantothenoyl-cysteine synthetase	6.3.2.5
Phosphopentosisomerase	5.3.1.6
Phosphopentokinase	2.7.1.19
Phosphopentomutase	2.7.5.6
Phosphoprotein phosphatase	3.1.3.16
Phosphoprotein phosphohydrolase	3.1.3.16
Phosphopyruvate carboxylase	4.1.1.32, 4.1.1.38
Phosphopyruvate carboxylase (ATP)	4.1.1.49
Phosphopyruvate hydratase	4.2.1.11
Phosphoramidate—hexose phosphotransferase	2.7.1.62
Phosphoramidate:hexose 1-phosphotransferase	2.7.1.62
Phosphoriboisomerase	5.3.1.6
Phosphoribokinase	2.7.1.18
5-Phospho-α-D-ribose-1-diphosphate:xanthine phosphoribosyltransferase	2.4.2.22
5-Phosphoribosylamine:glycine ligase (ADP-forming)	6.3.4.13
5-Phosphoribosylamine:pyrophosphate phosphoribosyltransferase (glutamate-amidating)	2.4.2.14
5'-Phosphoribosylamine synthetase	6.3.4.7
Phosphoribosylaminoimidazolecarboxamide formyltransferase	2.1.2.3
Phosphoribosylaminoimidazole carboxylase	4.1.1.21
5'Phosphoribosyl-5'-amino-4-imidazole-carboxylate carboxy-lyase	4.1.1.21
Phosphoribosylaminoimidazole-succinocarboxamide synthetase	6.3.2.6
Phosphoribosylaminoimidazole synthetase	6.3.3.1
Phosphoribosyl-AMP cyclohydrolase	3.5.4.19
1-N-(5'-Phospho-D-ribosyl)-AMP 1,6-hydrolase	3.5.4.19
N-(5-Phosphoribosyl)-anthranilate:pyrophosphate phosphoribosyltransferase	2.4.2.18
Phosphoribosylanthranilate pyrophosphorylase	2.4.2.18
1-(5'-Phosphoribosyl)-ATP:pyrophosphate phosphoribosyltransferase	2.4.2.17
Phosphoribosyl-ATP pyrophosphorylase	2.4.2.17
5'-Phosphoribosyl-4-carboxy-5-aminoimidazole:L-aspartate ligase (ADP-forming)	6.3.2.6
Phosphoribosyldiphosphate amidotransferase	2.4.2.14
N-(5'-Phospho-D-ribosylformimino)-5-amino-1-(5'-phosphoribosyl)-4-imidazolecarboxamide isomerase	5.3.1.16
N-(5'-Phospho-D-ribosylformimino)-5-amino-1-(5'-phosphoribosyl)-4-imidazolecarboxamide ketol-isomerase	5.3.1.16
5'-Phosphoribosylformylglycinamide:L-glutamine amido-ligase (ADP-forming)	6.3.5.3
5'-Phosphoribosylformylglycinamidine cyclo-ligase (ADP-forming)	6.3.3.1
Phosphoribosylformylglycinamidine synthetase	6.3.5.3
Phosphoribosylglycinamide formyltransferase	2.1.2.2
Phosphoribosylglycinamide synthetase	6.3.4.13
5'-Phosphoribosylimidazoleacetate synthetase	6.3.4.8
Phosphoribulokinase	2.7.1.19

INDEX TO THE ENZYME LIST (continued)

Phosphoribulose epimerase	5.1.3.1
Phosphorylase	2.4.1.1
Phosphorylase *a*	2.4.1.1
Phosphorylase-a phosphohydrolase	3.1.3.17
Phosphorylase *b* kinase kinase	2.7.1.37
Phosphorylase kinase	2.7.1.38
Phosphorylase phosphatase	3.1.3.17
Phosphorylating D-3-phospho-glyceraldehyde→DPN transhydrogenase	1.2.1.12
Phosphorylcholine-glyceride transferase	2.7.8.2
Phosphorylcholine transferase	2.7.7.15
Phosphorylethanolamine transferase	2.7.7.14
Phosphosaccharomutase	5.3.1.9
Phosphoserine aminotransferase	2.6.1.52
O-Phospho-L-serine: 2-oxoglutarate aminotransferase	2.6.1.52
Phosphoserine phosphatase	3.1.3.3
O-Phosphoserine phosphohydrolase	3.1.3.3
Phosphotransacetylase	2.3.1.8
Phosphotriose isomerase	5.3.1.1
Photoreactivating enzyme	4.1.99.3
Phylloquinone reductase	1.6.99.2
Phytase	3.1.3.26
1-Phytase	3.1.3.8
6-Phytase	3.1.3.26
Phytate 6-phosphatase	3.1.3.26
Pinguinain	3.4.99.18
D-Pinitol dehydrogenase	1.1.1.142
L-Pipecolate:(acceptor) oxidoreductase	1.5.99.3
L-**Pipecolate dehydrogenase**	1.5.99.3
Plant RNase	3.1.4.23
Plasmalogen synthase	2.3.1.25
Plasmin	3.4.21.7
Poly(1,4-β-(2-acetamido-2-deoxy-D-glucoside))glycanohydrolase	3.2.1.14
1,4-β-Poly-*N*-acetylglucosaminidase	3.2.1.14
Poly(deoxyribonucleotide):poly(deoxyribonucleotide) ligase (AMP-forming)	6.5.1.1
Poly(deoxyribonucleotide):poly(deoxyribonucleotide) ligase (AMP-forming, NMN-forming)	6.5.1.2
Poly(1,2-α-fucoside-4-sulphate) glycanohydrolase	3.2.1.44
Poly(1,4-α-D-galactosiduronate) digalacturonohydrolase	3.2.1.82
Polygalacturonase	3.2.1.15
Poly(galacturonate) hydrolase	3.2.1.67
Poly(1,4-α-D-galacturonide) exo-lyase	4.2.2.9
Poly(1,4-α-D-galacturonide) galacturonohydrolase	3.2.1.67
Poly(1,4-α-D-galacturonide) glycanohydrolase	3.2.1.15
Poly(1,4-α-D-galacturonide) lyase	4.2.2.2
Poly-β-glucosaminidase	3.2.1.14
Polyglucuronide lyase	[4.2.99.5]
Poly(isoprenol)-phosphate galactosephotransferase	2.7.8.6
Poly(1,4-β-D-mannuronide) lyase	4.2.2.3
Poly(methoxygalacturonide) lyase	4.2.2.10
2′(3′)-Polynucleotidase	3.1.3.32
5′-Polynucleotidase	3.1.3.33
Polynucleotide adenylyltransferase	2.7.7.19
Polynucleotide 5′-hydroxyl-kinase	2.7.1.78
Polynucleotide ligase	6.5.1.1
Polynucleotide ligase (NAD⁺)	6.5.1.2
Polynucleotide 3′-phosphatase	3.1.3.32
Polynucleotide 5′-phosphatase	3.1.3.33
Polynucleotide 3′-phosphohydrolase	3.1.3.32
Polynucleotide 5′-phosphohydrolase	3.1.3.33
Polynucleotide phosphorylase	2.7.7.8
Polynucleotide synthetase (ATP)	6.5.1.1
Polynucleotide synthetase (NAD⁺)	6.5.1.2
Polyol dehydrogenase	1.1.1.14
Polyol dehydrogenase (NADP⁺)	1.1.1.139
Polyphenol oxidase	1.14.18.1
Polyphosphatase	3.6.1.10
Polyphosphate depolymerase	3.6.1.10
Polyphosphate–glucose phosphotransferase	2.7.1.63
Polyphosphate:D-glucose 6-phosphotransferase	2.7.1.63
Polyphosphate kinase	2.7.4.1
Polyphosphate phosphohydrolase	3.6.1.11
Polyphosphate polyphosphohydrolase	3.6.1.10
Polyphosphorylase	2.4.1.1
Polyribonucleotide nucleotidyltransferase	2.7.7.8
Polyribonucleotide:orthophosphate nucleotidyltransferase	2.7.7.8
Polysaccharide methyltransferase	2.1.1.18
Porphobilinogen ammonia-lyase (polymerizing)	4.3.1.8
Porphobilinogen synthase	4.2.1.24
Porphyran sulphatase	2.5.1.5
Prenol-diphosphate pyrophosphohydrolase	3.1.7.1
Prenol pyrophosphatase	3.1.7.1
Prenyltransferase	2.5.11
PR-enzyme	3.1.3.17, 4.1.99.3
Prephenate dehydratase	4.2.1.51
Prephenate dehydrogenase	1.3.1.12
Prephenate dehydrogenase (NADP⁺)	1.3.1.13
Prephenate hydro-lyase (decarboxylating)	4.2.1.51
Prephenate:NAD⁺ oxidoreductase (decarboxylating)	1.3.1.12
Prephenate:NADP⁺ oxidoreductase (decarboxylating)	1.3.1.13
Procaine esterase	3.1.1.1
Progesterone, hydrogen-donor:oxygen oxidoreductase (hydroxylating)	1.14.99.4
Progesterone,hydrogen-donor:oxygen oxidoreductase(11α-hydroxylating)	1.14.99.14
Progesterone hydroxylase	1.14.99.4
Progesterone 11α-hydroxylase	1.14.99.14
Progesterone monooxygenase	1.14.99.4
Progesterone 11α-monooxygenase	1.14.99.14

INDEX TO THE ENZYME LIST (continued)

Progesterone reductase	1.1.1.145
Prolidase	3.4.13.9
Prolinase	3.4.13.8
Proline aminopeptidase	3.4.11.5
Proline carboxypeptidase	3.4.12.4
Proline dipeptidase	3.4.13.9
Proline hydroxylase	1.14.11.2
L-Proline:NAD(P)⁺ 2-oxidoreductase	1.5.1.1
L-Proline:NAD(P)⁺ 5-oxidoreductase	1.5.1.2
Proline,2-oxoglutarate dioxygenase	1.14.11.2
Proline racemase	5.1.1.4
Proline racemase	5.1.1.4
D-**Proline reductase**	1.4.1.6
D-**Proline reductase (dithiol)**	1.4.4.1
L-Proline:tRNA^Pro ligase (AMP-forming)	6.1.1.15
L-Prolyl-amino-acid hydrolase	3.4.13.8
Prolyl dipeptidase	3.4.13.8
L-Prolylglycine dipeptidase	3.4.13.8
Prolyl-glycyl-peptide,2-oxoglutarate:oxygen oxidoreductase	1.14.11.2
L-Prolyl-peptide hydrolase	3.4.11.5
Prolyl-tRNA synthetase	6.1.1.15
Pronase	3.4.21.4, 3.4.24.4
Propanediol dehydratase	4.2.1.28
1,2-Propanediol hydro-lyase	4.2.1.28
D(or L)-1,2-Propanediol:NAD⁺ oxidoreductase	1.1.1.77
1,2-Propanediol:NADP⁺ oxidoreductase	1.1.1.55
Propanediol-phosphate dehydrogenase	1.1.1.7
1,2-Propanediol-1-phosphate:NAD⁺ oxidoreductase	1.1.1.7
2-Propanol:NADP⁺ oxidoreductase	1.1.1.80
Propionate CoA-transferase	2.8.3.1
Propionyl-CoA:carbon-dioxide ligase (ADP-forming)	6.4.1.3
Propionyl-CoA carboxylase	4.1.1.41
Propionyl-CoA carboxylase (ATP-hydrolyzing)	6.4.1.3
3-Propylmalate glyoxylate-lyase (CoA-valerylating)	4.1.3.11
3-Propylmalate synthase	4.1.3.11
Prostaglandin synthase	1.14.99.1
Protaminase	3.4.12.3
Protamine kinase	2.7.1.70
Protease A of sea anemone	3.4.21.3
Protein(arginine)methyltransferase	2.1.1.23
Proteinase A, baker's yeast	3.4.23.8
Proteinase, Acrocylindricum	3.4.99.1
Proteinase, acrosomal	3.4.21.10
Proteinase A, Metridium	3.4.21.3
Proteinase B, *Clostridium histolyticum*	3.4.22.8
Proteinase B, yeast	3.4.22.9
Proteinase, *Crotalus atrox*	3.4.24.1
Proteinase, Gliocladium	3.4.99.8
Proteinase, α-lytic	3.4.21.12
Proteinase, Paecilomyces	3.4.99.15
Proteinase, *Penicillium notatum* (extracellular)	3.4.99.16
Proteinase, Phaseolus	3.4.21.13
Proteinase, Scopulariopsis	3.4.99.20
Proteinase, Sepia	3.4.24.2
Proteinase, Streptococcal	3.4.22.10
α-Proteinase, Tenebrio	3.4.21.18
β-Proteinase, Tenebrio	3.4.99.24
Proteinase, viper	3.4.21.5
Proteinase β, yeast	3.4.22.9
Protein disulphide-isomerase	5.3.4.1
Protein disulphide-isomerase	5.3.4.1
Protein-disulphide reductase (glutathione)	1.8.4.2
Protein-disulphide reductase (NAD(P)H)	1.6.4.4
Protein kinase	2.7.1.37
Protein methylase I	2.1.1.23
Protein methylase II	2.1.1.24
Protein O-methyltransferase	2.1.1.24
Protein phosphatase	3.1.3.16
Protein–UDP acetylgalactosaminyltransferase	2.4.1.41
Proteus aeruginosa neutral proteinase	3.4.24.4
Prothrombinase	3.4.21.6
Protocatechuate carboxy-lyase	4.1.1.63
Protocatechuate decarboxylase	4.1.1.63
Protocatechuate 3,4-dioxygenase	1.13.11.3
Protocatechuate 4,5-dioxygenase	1.13.11.8
Protocatechuate oxygenase	1.13.11.3
Protocatechuate 4,5-oxygenase	1.13.11.8
Protocatechuate:oxygen 3,4-oxidoreductase (decyclizing)	1.13.11.3
Protocatechuate:oxygen 4,5-oxidoreductase (decyclizing)	1.13.11.8
Protocollagen hydroxylase	1.14.11.2
Protohaem ferro-lyase	4.99.1.1
Pseudocholinesterase	3.1.1.8
Pseudomonas cytochrome oxidase	1.9.3.2
Psychosine–UDP galactosyltransferase	2.4.1.23
Pterin deaminase	3.5.4.11
N-Pteroyl-L-glutamate hydrolase	3.4.12.10
Ptyalin	3.2.1.1
Pullulanase	3.2.1.41
Pullulan 4-glucanohydrolase	3.2.1.57
Pullulan 6-glucanohydrolase	3.2.1.41
Purinedeoxyriboside→pyrimidine transdeoxyribosidase	2.4.2.6
Purine nucleosidase	3.2.2.1
Purine-nucleoside:orthophosphate ribosyltransferase	2.4.2.1
Purine nucleoside phosphorylase	2.4.2.1
Pyloruspepsin	3.4.23.3
Pyranose oxidase	1.1.3.10
Pyranose:oxygen 2-oxidoreductase	1.1.3.10
Pyrazolylalanine synthase	4.2.1.50
Pyridine nucleotide transhydrogenase	1.6.1.1
Pyridoxal dehydrogenase	1.1.1.107
Pyridoxal kinase	2.7.1.35
Pyridoxal:NAD⁺ oxidoreductase	1.1.1.107
Pyridoxamine:oxaloacetate aminotransferase	2.6.1.31
Pyridoxamine–oxalacetate transaminase	2.6.1.31

INDEX TO THE ENZYME LIST (continued)

Pyridoxaminephosphate oxidase	1.4.3.5
Pyridoxaminephosphate:oxygen oxidoreductase (deaminating)	1.4.3.5
Pyridoxamine:pyruvate aminotransferase	2.6.1.30
Pyridoxamine–pyruvate transaminase	2.6.1.30
5-Pyridoxate dioxygenase	1.14.12.5
5-Pyridoxate,NADPH:oxygen oxidoreductase (decyclizing)	1.14.12.5
5-Pyridoxate oxidase	1.14.12.5
Pyridoxin dehydrogenase	1.1.1.65
Pyridoxin:NADP$^+$ oxidoreductase	1.1.1.65
Pyridoxin 4-oxidase	1.1.3.12
Pyridoxin:oxygen 4-oxidoreductase	1.1.3.12
Pyridoxol:(acceptor)5'-oxidoreductase	1.1.99.9
4-Pyridoxolactonase	3.1.1.27
4-Pyridoxolactone hydrolase	3.1.1.27
Pyridoxol 5'-dehydrogenase	1.1.99.9
Pyrimidine-nucleoside:orthophosphate ribosyltransferase	2.4.2.2
Pyrimidine-nucleoside phosphorylase	2.4.2.2
Pyrimidine 5'-nucleotide nucleosidase	3.2.2.10
Pyrimidine-5'-nucleotide phosphoribo-(deoxyribo)hydrolase	3.2.2.10
Pyrimidine phosphorylase	2.4.2.3, 2.4.2.4
Pyrimidine transferase	2.5.1.2
Pyrithiamine deaminase	3.5.4.20
Pyrocatechase	1.13.11.1
o-Pyrocatechuate decarboxylase	4.1.1.46
o-Pyrocatechuate oxygenase	1.13.11.14
Pyroglutamate aminopeptidase	3.4.11.8
L-Pyroglutamyl-peptide hydrolase	3.4.11.8
Pyrophosphatase, inorganic	3.6.1.1
Pyrophosphate–glycerol phosphotransferase	2.7.1.79
Pyrophosphate:glycerol 1-phosphotransferase	2.7.1.79
Pyrophosphate:oxaloacetate carboxylyase (transphosphorylating)	4.1.1.38
Pyrophosphate phosphohydrolase	3.6.1.1
Pyrophosphate–serine phosphotransferase	2.7.1.80
Pyrophosphate:L-serine O-phosphotransferase	2.7.1.80
Pyrophosphomevalonate decarboxylase	4.1.1.33
Pyrrolidonecarboxylate peptidase	3.4.11.8
1-Pyrroline-5-carboxylate:NAD$^+$ oxidoreductase	1.5.1.12
Pyrroline-2-carboxylate reductase	1.5.1.1
Pyrroline-5-carboxylate reductase	1.5.1.2
1-Pyrroline dehydrogenase	1.5.1.12
Pyrrolooxygenase	1.13.11.26
Pyruvate:carbon-dioxide ligase (ADP-forming)	6.4.1.1
Pyruvate carboxylase	6.4.1.1
Pyruvate decarboxylase	4.1.1.1
Pyruvate dehydrogenase	1.2.2.2, 1.2.4.1
Pyruvate dehydrogenase (cytochrome)	1.2.2.2
Pyruvate dehydrogenase (lipoate)	1.2.4.1
Pyruvate:ferredoxin oxidoreductase (CoA-acetylating)	1.2.7.1
Pyruvate:ferricytochrome b_1 oxidoreductase	1.2.2.2
Pyruvate kinase	2.7.1.40
Pyruvate:lipoate oxidoreductase decarboxylating and acceptor-acetylating)	1.2.4.1
Pyruvate, orthophosphate dikinase	2.7.9.1
Pyruvate oxidase	1.2.3.3
Pyruvateoxime;acetone oximinotransferase	2.6.3.1
Pyruvate:oxygen oxidoreductase (phosphorylating)	1.2.3.3
Pyruvate synthase	1.2.7.1
Pyruvic carboxylase	6.4.1.1
Pyruvic decarboxylase	4.1.1.1
Pyruvic dehydrogenase	1.2.2.2, 1.2.4.1
Pyruvic-malic carboxylase	1.1.1.38, 1.1.1.40
Pyruvic oxidase	1.2.3.3
Q-enzyme	2.4.1.18
Quercetin 2,3-dioxygenase	1.13.11.24
Quercetin:oxygen 2,3-oxidoreductase (decyclizing)	1.13.11.24
Quercitrinase	3.2.1.66
Quercitrin 3-rhamnohydrolase	3.2.1.66
Quinate dehydrogenase	1.1.1.24
Quinate:NAD$^+$ oxidoreductase	1.1.1.24
Quinone reductase	1.6.99.2
Quinone reductase	[1.6.5.1]
Rattlesnake venom proteinase	3.4.24.1
Renal dipeptidase	3.4.13.11
Renin	3.4.99.19
Rennin	3.4.23.4
R-enzyme	3.2.1.41
	6.5.1.1, 6.5.1.2
Respiratory nitrate reductase	1.7.99.4
Retinal dehydrogenase	1.2.1.36
all-trans-Retinal 11-cis-trans-isomerase	5.2.1.3
Retinal:NAD$^+$ oxidoreductase	1.2.1.36
Retinene isomerase	5.2.1.3
Retinol dehydrogenase	1.1.1.105
Retinol:NAD$^+$ oxidoreductase	1.1.1.105
Retinol-palmitate esterase	3.1.1.21
Retinol-palmitate hydrolase	3.1.1.21
L-Rhamnose isomerase	5.3.1.14
L-Rhamnose ketol-isomerase	5.3.1.14
α-L-Rhamnosidase	3.2.1.40
β-L-Rhamnosidase	3.2.1.43
α-L-Rhamnoside rhamnohydrolase	3.2.1.40
β-L-Rhamnoside rhamnohydrolase	3.2.1.43
Rhamnulokinase	2.7.1.5
Rhamnulosephosphate aldolase	4.1.2.19

INDEX TO THE ENZYME LIST (continued)

L-Rhamnulose-1-phosphate L-lact-aldehyde-lyase	4.1.2.19
Rhizopus acid proteinase	3.4.23.9
Rhodanese	2.8.1.1
Ribitol dehydrogenase	1.1.1.56
Ribitol:NAD$^+$ 2-oxidoreductase	1.1.1.56
D-Ribitol-5-phosphate cytidylyltransferase	2.7.7.40
Ribitol-5-phosphate dehydrogenase	1.1.1.137
D-Ribitol-5-phosphate:NAD(P)$^+$ 2-oxidoreductase	1.1.1.137
Riboflavinase	3.5.99.1
Riboflavin hydrolase	3.5.99.1
Riboflavin kinase	2.7.1.26
Riboflavin phosphotransferase	2.7.1.42
Riboflavin synthase	2.5.1.9
Ribokinase	2.7.1.15
Ribonuclease I	3.1.4.22
Ribonuclease II	3.1.4.23
Ribonucleate 3'-guanylooligonucleotidohydrolase	3.1.4.8
Ribonucleate 5'-nucleotidogydrolase	3.1.4.20
Ribonucleate 3'-oligonucleotidohydrolase	3.1.4.23
Ribonucleate 3'-pyrimidinooligonucleotidohydrolase	3.1.4.22
Ribonucleoside-diphosphate reductase	1.17.4.1
Ribonucleoside-triphosphate reductase	1.17.4.2
3'-Ribonucleotide phosphohydrolase	3.1.3.6
5'-Ribonucleotide phosphohydrolase	3.1.3.5
D-Ribose dehydrogenase (NADP$^+$)	1.1.1.115
D-Ribose:NADP$^+$ 1-oxidoreductase	1.1.1.115
Ribose-5-phosphate adenylyltransferase	2.7.7.35
Ribose-5-phosphate:ammonia ligase (ADP-forming)	6.3.4.7
Ribosephosphate isomerase	5.3.1.6
D-Ribose-5-phosphate ketol-isomerase	5.3.1.6
Ribose-1-phosphate→purine transribosidase	2.4.2.1
Ribose-1-phosphate→pyrimidine transribosidase	2.4.2.3, 2.4.2.4
Ribosephosphate pyrophosphokinase	2.7.6.1
Ribosylhomocysteinase	3.3.1.3
S-Ribosyl-L-homocysteine ribohydrolase	3.3.1.3
Ribosylnicotinamide kinase	2.7.1.22
N-Ribosyl-purine ribohydrolase	3.2.2.1
N-Ribosylpyrimidine nucleosidase	3.2.2.8
Ribulokinase	2.7.1.16
D-Ribulokinase	2.7.1.47
Ribulosebisphosphate carboxylase	4.1.1.39
Ribulosephosphate 3-epimerase	5.1.3.1
L-Ribulosephosphate 4-epimerase	5.1.3.4
D-Ribulose-5-phosphate 3-epimerase	5.1.3.1
L-Ribulose-5-phosphate 4-epimerase	5.1.3.4
Ricinine aminohydrolase	3.5.5.2
Ricinine nitrilase	3.5.5.2
RNA adenylating enzyme	2.7.7.19
RNAase I	3.1.4.22
RNA-DNA-hybrid ribonucleotidohydrolase	3.1.4.34
RNA nucleotidyltransferase	2.7.7.6
RNA polymerase	2.7.7.6
RNase	3.1.4.22
RNase H	3.1.4.34
RNase N$_1$	3.1.4.8
RNase T$_1$	3.1.4.8
Robison ester dehydrogenase	1.1.1.49
Rubredoxin–NAD$^+$ reductase	1.6.7.2
Rubredoxin reductase	1.6.7.2
Saccharase	3.2.1.26
Saccharogen amylase	3.2.1.2
Saccharopine dehydrogenase (NAD$^+$, L-glutamate-forming)	1.5.1.9
Saccharopine dehydrogenase (NAD$^+$, lysine-forming)	1.5.1.7
Saccharopine dehydrogenase (NADP$^+$, L-glutamate-forming)	1.5.1.10
Saccharopine dehydrogenase (NADP$^+$, lysine-forming)	1.5.1.8
Salicylate hydroxylase	1.14.13.1
Salicylate 1-monooxygenase	1.14.13.1
Salicylate,NADH:oxygen oxidoreductase (1-hydroxylating,decarboxylating)	1.14.13.1
Sarcosine: (acceptor) oxidoreduct-(demethylating)	1.5.99.1
Sarcosine dehydrogenase	1.5.99.1
Sarcosine oxidase	1.5.3.1
Sarcosine:oxygen oxidoreductase (demethylating)	1.5.3.1
Schardinger enzyme	1.2.3.2
Scopulariopis proteinase	3.4.99.20
Sea anemone protease A	3.4.21.3
Sealase	6.5.1.1
Sedoheptulokinase	2.7.1.14
Sedoheptulose-7-phosphate:D-glyceraldehyde-3-phosphate dihydroxyacetonetransferase	2.2.1.2
Sedoheptulose-7-phosphate:D-glyceraldehyde-3-phosphate glycolaldehydetransferase	2.2.1.1
Sepia proteinase	3.4.24.2
Sepiapterin reductase	1.1.1.153
Sequoyitol dehydrogenase	1.1.1.143
Serine acetyltransferase	2.3.1.30
Serine aldolase	2.1.2.1
Serine deaminase	4.2.1.16
D-Serine dehydratase	4.2.1.14
L-Serine dehydratase	4.2.1.13
L-Serine dehydratase	4.2.1.16
Serine dehydrogenase	1.4.1.7
Serine-ethanolaminephosphate phosphodiesterase	3.1.4.13
Serine–glyoxylate aminotransferase	2.6.1.45
L-Serine:glyoxylate aminotransferase	2.6.1.45

INDEX TO THE ENZYME LIST (continued)

L-Serine hydro-lyase(adding homocysteine)	4.2.1.22
L-Serine hydro-lyase(adding indole)	4.2.1.20
L-Serine hydro-lyase(adding pyrazole)	4.2.1.50
D-Serine hydro-lyase(deaminating)	4.2.1.14
L-Serine hydro-lyase(deaminating)	4.2.1.13
Serine hydroxymethlyase	2.1.2.1
Serine hydroxymethyltransferase	2.1.2.1
L-Serine:NAD⁺ oxidoreductase (deaminating)	1.4.1.7
Serine-phosphinico-ethanolamine ethanolaminephosphohydrolase	3.1.4.13
Serine-phosphinico-ethanolamine synthase	2.7.8.4
Serine proteinase, Arthrobacter	3.4.21.17
Serine proteinase, Bacillus amyloliquefaciens	3.4.21.15
Serine–pyruvate aminotransferase	2.6.1.51
L-Serine:pyruvate aminotransferase	2.6.1.51
Serinesulphate ammonia-lyase	4.3.1.10
L-Serine-O-sulphate ammonia-lyase (pyruvate-forming)	4.3.1.10
Serine sulphhydrase	4.2.1.22
L-Serine:tRNA^{Ser} ligase(AMP-forming)	6.1.1.11
Seryl-tRNA synthetase	6.1.1.11
Shikimate dehydrogenase	1.1.1.25
Shikimate kinase	2.7.1.71
Shikimate:NADP⁺ oxidoreductase	1.1.1.25
Sialidase	3.2.1.18
Sialyltransferase	2.4.99.1
Single-stranded-nucleate endonuclease	3.1.4.21
Single-stranded-nucleate 5'-oligonucleotidohydrolase	3.1.4.21
Sinigrase	3.2.3.1
Sinigrinase	3.2.3.1
Sinigrin sulphohydrolase	[3.1.6.5]
Snake venom zinc proteinase	3.4.24.1
Solanain	3.4.99.21
Sorbitol dehydrogenase	1.1.1.14
Sorbitol→DPN transhydrogenase	1.1.1.14
D-Sorbitol-6-phosphate dehydrogenase	1.1.1.140
D-Sorbitol-6-phosphate: NAD⁺ 2-oxidoreductase	1.1.1.140
L-Sorbose:(acceptor)5-oxidoreductase	1.1.99.12
Sorbose dehydrogenase	1.1.99.12
L-Sorbose:NADP⁺ 5-oxidoreductase	1.1.1.123
L-Sorbose oxidase	1.1.3.11
L-Sorbose:oxygen 5-oxidoreductase	1.1.3.11
Spermine oxidase	[1.5.3.3]
Sphingomyelin cholinephosphohydrolase	3.1.4.12
Sphingomyelin phosphodiesterase	3.1.4.12
Sphingosine acyltransferase	2.3.1.24
Sphingosine cholinephosphotransferase	2.7.8.10
Spleen endonuclease	3.1.4.7
Spleen exonuclease	3.1.4.18
Spleen phosphodiesterase	3.1.4.7, 3.1.4.18
Spreading factor	3.2.1.35, 3.2.1.36, 4.2.2.1
Squalene epoxidase	1.14.99.7
Squalene,hydrogen-donor:oxygen oxidoreductase(2,3-epoxidizing)	1.14.99.7
Squalene hydroxylase	[1.14.1.3]
Squalene monooxygenase(2,3-epoxidizing)	1.14.99.7
Squalene oxydocyclase	1.14.99.7, 5.4.99.7
sRNA adenylyltransferase	[2.7.7.20]
S-S rearrangase	5.3.4.1
Staphylokinase	3.4.99.22
Starch(bacterial glycogen)synthase	2.4.1.21
Starch phosphorylase	2.4.1.1
Steapsin	3.1.1.3
Stearyl-CoA→L-α-glycerophosphate transstearylase	2.3.1.15
Stem bromelain	3.4.22.4
Steroid 4,5-dioxygenase	1.13.11.25
Steroid,hydrogen-donor:oxygen oxidoreductase(17α-hydroxylating)	1.14.99.9
Steroid,hydrogen-donor:oxygen oxidoreductase(21-hydroxylating)	1.14.99.10
Steroid 11β-hydroxylase	1.14.15.4
Steroid 17α-hydroxylase	1.14.99.9
Steroid 21-hydroxylase	1.14.99.10
Steroid Δ-isomerase	5.3.3.1
Steroid-lactonase	3.1.1.37
Steroid 11β-monooxygenase	1.14.15.4
Steroid 17α-monooxygenase	1.14.99.9
Steroid 21-monooxygenase	1.14.99.10
Steroid,reduced-adrenal-ferredoxin:oxygen oxidoreductase(11β-hydroxylating)	1.14.15.4
Sterol-ester hydrolase	3.1.1.13
Δ^{24}-Sterol methyltransferase	2.1.1.41
Sterol-sulphatase	3.1.6.2
Sterol-sulphate sulphohydrolase	3.1.6.2
Stipiatonate carboxy-lyase (decyclizing)	4.1.1.60
Stipitatonate decarboxylase	4.1.1.60
Streptidine kinase	2.7.1.72
Streptococcal proteinase	3.4.22.10
Streptomyces alkaline proteinase	3.4.21.4
Streptomyces collagenase	3.4.24.3
Streptomyces griseus neutral proteinase	3.4.24.4
Streptomyces keratinase	3.4.99.11
Streptomyces naraensis neutral proteinase	3.4.24.4
Subtilisin	3.4.21.14
Subtilisin A	3.4.21.14
Subtilisin B	3.4.21.14
Subtilopeptidase A	3.4.21.14
Subtilopeptidase B	3.4.21.14
Subtilopeptidase C	3.4.21.14
Succinate:(acceptor)oxidoreductase	1.3.99.1
Succinate:CoA ligase(ADP-forming)	6.2.1.5

INDEX TO THE ENZYME LIST (continued)

Succinate:CoA ligase(GDP-forming)	6.2.1.4	Sucrose→orthophosphate transglucosidase	2.4.1.7
Succinate dehydrogenase	1.3.99.1	**Sucrose-phosphatase**	3.1.3.24
Succinate:NAD⁺ oxidoreductase	1.3.1.6	*Sucrose-6^F-phosphate phosphohydrolase*	3.1.3.24
Succinate-semialdehyde dehydrogenase	1.2.1.24	**Sucrose-phosphate synthase**	2.4.1.14
Succinate-semialdehyde dehydrogenase (NAD(P)⁺)	1.2.1.16	Sucrosephosphate–UDP glucosyltransferase	2.4.1.14
Succinate-semialdehyde:NAD⁺ oxidoreductase	1.2.1.24	Sucrose phosphorylase	2.4.1.7
Succinate-semialdehyde:NAD(P)⁺ oxidoreductase	1.2.1.16	**Sucrose synthase**	2.4.1.13
Succinic dehydrogenase	1.3.99.1	Sucrose–UDP glucosyltransferase	2.4.1.13
Succinic thiokinase	6.2.1.4, 6.2.1.5	Sugar-phosphatase	3.1.3.23
		Sugar-1-phosphate adenylyltransferase	2.7.7.36
Succinyl-CoA→acetoacetyl-CoA CoA-transferase	2.8.3.5	**Sugar-1-phosphate nucleotidyltransferase**	2.7.7.37
Succinyl-CoA acylase	3.1.2.3	*Sugar-phosphate phosphydrolase*	3.1.3.23
Succinyl-CoA:citramalate CoA-transferase	2.8.3.7	*Sugar-sulphate sulphohydrolase*	3.1.6.3
Succinyl-CoA:glycine C-succinyltransferase (decarboxylating)	2.3.1.37	Sulfokinase	2.8.2.1
Succinyl-CoA hydrolase	3.1.2.3	Sulfurylase	2.7.7.4
Succinyl-CoA hydrolase	3.1.2.3	*2-Sulphamido-2-deoxy-D-glucose sulphamidase*	3.10.1.1
Succinyl-CoA:oxalate CoA-transferase	2.8.3.2	*2-Sulphamido-2-deoxy-6-O-sulpho-D-glucose 6-sulphohydrolase*	3.1.6.11
Succinyl-CoA:3-oxo-acid CoA-transferase	2.8.3.5	Sulphatase	3.1.6.1
Succinyl-CoA:3-oxoadipate CoA-transferase	2.8.3.6	**Sulphate adenylyltransferase**	2.7.7.4
Succinyl-CoA synthetase(ADP-forming)	6.2.1.5	**Sulphate adenylyltransferase(ADP)**	2.7.7.5
Succinyl-CoA synthetase(GDP-forming)	6.2.1.4	**Sulphite dehydrogenase**	1.8.2.1
N-Succinyl-LL-2,6-diaminopimelate amindohydrolase	3.5.1.18	*Sulphite:ferricytochrome c oxidoreductase*	1.8.2.1
Succinyl-diaminopimelate aminotransferase	2.6.1.17	**Sulphite oxidase**	1.8.3.1
Succinyl-diaminopimelate desuccinylase	3.5.1.18	*Sulphite:oxygen oxidoreductase*	1.8.3.1
N-Succinyl-L-2,6-diaminopimelate:2-oxoglutarate aminotransferase	2.6.1.17	**Sulphite reductase**	1.8.99.1
O-Succinyl-L-homoserine succinate-lyase (adding cysteine)	4.2.99.9	**Sulphite reductase (ferredoxin)**	1.8.7.1
Succinyl–β-ketoacyl-CoA transferase	2.8.3.2	**Sulphite reductase (NADPH)**	1.8.1.2
Sucrase	3.2.1.26	**Sulphoglucosamine sulphamidase**	3.10.1.1
Sucrose→amylose transglucosidase	2.4.1.4	**Sulphur dioxygenase**	1.13.11.18
Sucrose→dextran transglucosidase	2.4.1.5	*Sulphur:oxygen oxidoreductase*	1.13.11.18
Sucrose:2,1-β-D-fructan 1-β-fructosyltransferase	2.4.1.9	**Superoxide dismutase**	1.15.1.1
Sucrose:2,6-β-D-fructan 6-β-fructosyltransferase	2.4.1.10	*Superoxide:superoxide oxidoreductase*	1.15.1.1
Sucrose 1-fructosyltransferase	2.4.1.9		
Sucrose 6-fructosyltransferase	2.4.1.10	Tabernamontanain	3.4.99.23
Sucrose-glucan glucosyltransferase	2.4.1.4	Tabunase	3.8.2.1
Sucrose:1,4-α-D-glucan 4-α-glucosyltransferase	2.4.1.4	**Tagaturonate reductase**	1.1.1.58
Sucrose:1,6-α-D-glucan 6-α-glucosyltransferase	2.4.1.5	Takadiastase	3.4.23.6
Sucrose α-glucohydrolase	3.2.1.48	**Tannase**	3.1.1.20
Sucrose α-glucohydrolase	3.2.1.48	*Tannin acyl-hydrolase*	3.1.1.20
Sucrose glucosyltransferase	2.4.1.7	**Tartrate dehydratase**	4.2.1.32
Sucrose 6-glucosyltransferase	2.4.1.5	**Tartrate dehydrogenase**	1.1.1.93
Sucrose→inulin transfructosidase	2.4.1.9	*meso*-**Tartrate dehydrogenase**	1.3.1.7
Sucrose→levan transfructosidase	2.4.1.10	**Tartrate epimerase**	5.1.2.5
Sucrose:orthophosphate α-glucosyltransferase	2.4.1.7	*Tartrate epimerase*	5.1.2.5
		L(+)-Tartrate hydro-lyase	4.2.1.32
		meso-Tartrate hydro-lyase	4.2.1.37
		Tartrate:NAD⁺ oxidoreductase	1.1.1.93
		meso-Tartrate:NAD⁺ oxidoreductase	1.3.1.7
		Tartronate-semialdehyde carboxylase	4.1.1.47
		Tartronate-semialdehyde reductase	1.1.1.60
		Tartronate-semialdehyde synthase	4.1.1.47
		Taurocyanine kinase	2.7.3.4
		dTDP-4-amino-4,6-dideoxy-D-glucose aminotransferase	2.6.1.33

INDEX TO THE ENZYME LIST (continued)

dTDP-4-amino-4,6-dideoxy-D-glucose:2-oxoglutarate aminotransferase	2.6.1.33
dTDP-6-deoxy-L-mannose:NADP⁺ 4-oxidoreductase	1.1.1.133
dTDP-6-deoxy-L-talose dehydrogenase	1.1.1.134
dTDP-6-deoxy-L-talose:NADP⁺ 4-oxidoreductase	1.1.1.134
dTDPgalactose pyrophosphorylase	2.7.7.32
dTDPglucose 4,6-dehydratase	4.2.1.46
dTDPglucose 4,6-hydro-lyase	4.2.1.46
dTDP-4-keto-6-deoxy-D-glucose 3,5-epimerase	5.1.3.13
dTDP-4-ketorhamnose 3,5-epimerase	5.1.3.13
dTDP-4-ketorhamnose reductase	1.1.1.133
Teichoic-acid synthase	2.4.1.55
Tenebrio α-proteinase	3.4.21.18
Tenebrio β-proteinase	3.4.99.24
T-enzyme	2.4.1.24
Terminal addition enzyme	2.7.7.31
Terminal deoxyribonucleotidyltransferase	2.7.7.31
Terpenoid-allyltransferase	2.5.1.11
Testololactone lactone-hydrolase	3.1.1.37
Testosterone 17β-dehydrogenase	1.1.1.63
Testosterone 17β-dehydrogenase (NADP⁺)	1.1.1.64
Testosterone→DPN transhydrogenase	1.1.1.63
Tetrahydrofolate dehydrogenase	1.5.1.3
5,6,7,8-Tetrahydrofolate:NADP⁺ oxidoreductase	1.5.1.3
1,4,5,6-Tetrahydro-6-oxo-nicotinate:ferredoxin oxidoreductase	1.3.7.1
Tetrahydropteroylglutamate methyltransferase	2.1.1.13
Tetrahydropteroyltriglutamate methyltransferase	2.1.1.14
2',4,4',6'-Tetrahydroxydehydrochalcone 1,3,5-trihydroxybenzenehydrolase	3.7.1.4
Tetrahydroxypteridine cycloisomerase	5.5.1.3
Tetrahydroxypteridine lyase (isomerizing)	5.5.1.3
T₂ exonuclease	3.1.4.29
Thermolysin	3.4.24.4
Thiaminase	3.5.99.2
Thiaminase I	2.5.1.2
Thiaminase II	3.5.99.2
Thiamine:base 2-methyl-4-aminopyrimidine-5-methenyltransferase	2.5.1.2
Thiamine-diphosphate kinase	2.7.4.15
Thiamine hydrolase	3.5.99.2
Thiamine kinase	2.7.6.2
Thiaminephosphate pyrophosphorylase	2.5.1.3
Thiamine pyridinylase	2.5.1.2
Thiamine pyrophosphokinase	2.7.6.2
Thiocyanate isomerase	5.99.1.1
Thioethanolamine acetyltransferase	2.3.1.11
Thiogalactoside acetyltransferase	2.3.1.18
Thioglucosidase	3.2.3.1
Thioglucoside glucohydrolase	3.2.3.1
Thiolase	2.3.1.9
Thiol methyltransferase	2.1.1.9
Thiol oxidase	1.8.3.2
Thiol:oxygen oxidoreductase	1.8.3.2
Thioltransacetylase A	2.3.1.12
Thioltransacetylase B	2.3.1.11
β-Thionase	4.2.1.22
Thioredoxin reductase (NADPH)	1.6.4.5
Thiosulphate:cyanide sulphurtransferase	2.8.1.1
Thiosulfate cyanide transsulfurase	2.8.1.1
Thiosulphate sulphurtransferase	2.8.1.1
L-Threonate dehydrogenase	1.1.1.129
L-Threonate:NAD⁺ oxidoreductase	1.1.1.129
Threonine aldolase	2.1.2.1
Threonine aldolase	[4.1.2.5]
Threonine deaminase	4.2.1.16
Threonine dehydratase	4.2.1.16
L-Threonine 3-dehydrogenase	1.1.1.103
L-Threonine hydro-lyase (deaminating)	4.2.1.16
L-Threonine:NAD⁺ oxidoreductase	1.1.1.103
Threonine racemase	5.1.1.6
Threonine racemase	5.1.1.6
Threonine synthase	4.2.99.2
L-Threonine:tRNA^Thr ligase (AMP-forming)	6.1.1.3
Threonyl-tRNA synthetase	6.1.1.3
Thrombase	3.4.21.5
Thrombin	3.4.21.5
Thymidine 2'-hydroxylase	1.14.11.3
Thymidine kinase	2.7.1.75
Thymidine:orthophosphate deoxyribosyltransferase	2.4.2.4
Thymidine,2-oxoglutarate dioxygenase	1.14.11.3
Thymidine,2-oxoglutarate:oxygen oxidoreductase	1.14.11.3
Thymidine phosphorylase	2.4.2.4
Thymidylate 5'-nucleotidase	3.1.3.35
Thymidylate 5'-phosphatase	3.1.3.35
Thymidylate 5'-phosphohydrolase	3.1.3.35
Thymine,2-oxoglutarate dioxygenase	1.14.11.6
Thymine,2-oxoglutarate:oxygen oxidoreductase (7-hydroxylating)	1.14.11.6
Thymonuclease	3.1.4.5
Thymonucleodepolymerase	3.1.4.5
Thyroid galactosyltransferase	2.4.1.38
Thyroid hormone aminotransferase	2.6.1.26
Thyroid peptidase	3.4.12.12
Thyroid peptide carboxypeptidase	3.4.12.12
Thyroxine aminotransferase	2.6.1.25
Thyroxine:2-oxoglutarate aminotransferase	2.6.1.25
T₂-induced deoxynucleotide kinase	2.7.4.12
dTMP kinase	2.7.4.9
TPN·H₂→β-aspartylphosphate transhydrogenase	1.2.1.11
TPNH–cytochrome c reductase	1.6.2.4
TPNH₂→cytochrome c transelectronase	1.6.2.4
TPN·H₂→4-dimethylaminoazobenzene transhydrogenase	1.6.6.7
TPN·H₂→DPN transhydrogenase	1.6.1.1
TPN·H₂→glutathione transhydrogenase	1.6.4.2

INDEX TO THE ENZYME LIST (continued)

TPN·H$_2$→nitrate transelectronase	1.6.6.2, 1.6.6.3
TPN · H$_2$→O$_2$ transhydrogenase	1.6.99.1
Trametes acid proteinase	3.4.99.25
Transaldolase	2.2.1.2
Transcarboxylase	2.1.3.1
Transhydrogenase	1.6.1.1
Transketolase	2.2.1.1
Transoximinase	2.6.3.1
Transphosphoribosidase	2.4.2.7, 2.4.2.8
α,α-**Trehalase**	3.2.1.28
α,α-Trehalose glucohydrolase	3.2.1.28
α,α-Trehalose:orthophosphate β-glucosyltransferase	2.4.1.64
Trehalose-phosphatase	3.1.3.12
Trehalose-6-phosphate phosphohydrolase	3.1.3.12
α,α-**Trehalose-phosphate synthase (GDP-forming)**	2.4.1.36
α,α-**Trehalose-phosphate synthase (UDP-forming)**	2.4.1.15
Trehalosephosphate–UDP glucosyltransferase	2.4.1.15
α,α-**Trehalose phosphorylase**	2.4.1.64
Triacetate-lactonase	3.1.1.38
Triacetolactone hydrolase	3.1.1.38
Triacylglycerol acyl-hydrolase	3.1.1.3
Triacylglycerol lipase	3.1.1.3
Tributyrase	3.1.1.3
1,1,1-Trichloro-2,2-bis(p-chlorophenyl)-ethane hydrogen-chloride-lyase	4.5.1.1
Trichophyton collagenase	3.4.24.3
Trichophyton keratinase	3.4.99.12
Triglyceride lipase	3.1.1.3
3α,7α,12α-Trihydroxy-5β-cholan-24-oyl-glycine amidohydrolase	3.5.1.24
17,20β,21-Trihydroxysteroid:NAD$^+$ oxidoreductase	1.1.1.53
L-3,5,3'-Triiodothyronine:2-oxoglutarate aminotransferase	2.6.1.26
Trimetaphosphatase	3.6.1.2
Trimetaphosphate hydrolase	3.6.1.2
Trimethylamine-N-oxide reductase	1.6.6.9
4-Trimethylaminobutyrate,2-oxoglutarate:oxygen oxidoreductase (3-hydroxylating)	1.14.11.1
Trimethylsulphonium-chloride:tetrahydrofolate N-methyltransferase	2.1.1.19
Trimethylsulphonium–tetrahydrofolate methyltransferase	2.1.1.19
Triokinase	2.7.1.28
Triosephosphate dehydrogenase	1.2.1.9, 1.2.1.12
Triosephosphate dehydrogenase (NADP$^+$)	1.2.1.13
Triosephosphate isomerase	5.3.1.1
Triosephosphate mutase	5.3.1.1
Tripeptide aminopeptidase	3.4.11.4
Triphosphatase	3.6.1.3
Triphosphoinositide inositol-trisphosphohydrolase	3.1.4.11
Triphosphoinositide phosphatase	3.1.3.36
Triphosphoinositide phosphodiesterase	3.1.4.11
tRNA (adenine-1-)-methyltransferase	2.1.1.36
tRNA adenylyltransferase	2.7.7.25
tRNA CCA-pyrophosphorylase	2.7.7.21, 2.7.7.25
tRNA cytidylyltransferase	2.7.7.21
tRNA (cytosine-5-)-methyltransferase	2.1.1.29
tRNA (guanine-1-)-methyltransferase	2.1.1.31
tRNA (guanine-2-)-methyltransferase	2.1.1.32
tRNA (guanine-7-)-methyltransferase	2.1.1.33
tRNA (guanosine-2'-)-methyltransferase	2.1.1.34
tRNA isopentenyltransferase	2.5.1.8
tRNA (purine-2- or -6-)-methyltransferase	2.1.1.30
tRNA (uracil-5-)-methyltransferase	2.1.1.35
Tropinesterase	[3.1.1.10]
True cholinesterase	3.1.1.7
Trypsin	3.4.21.4
α-Trypsin	3.4.21.4
β-Trypsin	3.4.21.4
ψ-Trypsin	3.4.21.4
Trypsinogen-kinase	3.4.23.6
D-**Tryptophan acetyltransferase**	2.3.1.34
Tryptophan aminotransferase	2.6.1.27
Tryptophanase	4.1.99.1
Tryptophanase	1.13.11.11
Tryptophan decarboxylase	4.1.1.28
Tryptophan 2,3-dioxygenase	1.13.11.11
Tryptophan 5-hydroxylase	1.14.16.4
L-Tryptophan indole-lyase (deaminating)	4.1.99.1
Tryptophan 2-monooxygenase	1.13.12.3
Tryptophan 5-monooxygenase	1.14.16.4
L-Tryptophan:2-oxoglutarate aminotransferase	2.6.1.27
Tryptophan oxygenase	1.13.11.11
L-Tryptophan:oxygen 2-oxidoreductase (decarboxylating)	1.13.12.3
L-Tryptophan:oxygen 2,3-oxidoreductase (decyclizing)	1.13.11.11
Tryptophan peroxidase	[1.11.1.4]
Tryptophan–phenylpyruvate aminotransferase	2.6.1.28
L-Tryptophan:phenylpyruvate aminotransferase	2.6.1.28
Tryptophan pyrrolase	1.13.11.11
Tryptophan synthase	4.2.1.20
L-Tryptophan,tetrahydropteridine:oxygen oxidoreductase (5-hydroxylating)	1.14.16.4
L-Tryptophan:tRNA^{Trp} ligase (AMP-forming)	6.1.1.2
Tryptophanyl-tRNA synthetase	6.1.1.2
dTTP:α-D-galactose-1-phosphate thymidylytrnasferase	2.7.7.32
dTTP:α-D-glucose-1-phosphate thymidylyltransferase	2.7.7.24
Tyraminase	1.4.3.4
Tyramine N-methyltransferase	2.1.1.27

INDEX TO THE ENZYME LIST (continued)

Tyramine oxidase	1.4.3.9
Tyramine oxidase	1.4.3.4
Tyramine:oxygen oxidoreductase (deaminating)	1.4.3.9
Tyrosinase	1.14.18.1
β-Tyrosinase	4.1.99.2
Tyrosine aminotransferase	2.6.1.5
L-Tyrosine carboxy-lyase	4.1.1.25
Tyrosine carboxypeptidase	3.4.12.12
Tyrosine decarboxylase	4.1.1.25
Tyrosine-ester sulphotransferase	2.8.2.9
Tyrosine 3-hydroxylase	1.14.16.2
Tyrosine 3-monooxygenase	1.14.16.2
L-Tyrosine:2-oxoglutarate aminotransferase	2.6.1.5
Tyrosine phenol-lyase	4.1.99.2
L-Tyrosine phenol-lyase (deaminating)	4.1.99.2
Tyrosine-pyruvate aminotransferase	[2.6.1.20]
L-Tyrosine,tetrahydropteridine:oxygen oxidoreductase (3-hydroxylating)	1.14.16.2
L-Tyrosine:tRNA^{Tyr} ligase (AMP-forming)	6.1.1.1
Tyrosyl-tRNA synthetase	6.1.1.1
Ubiquinone reductase	[1.6.5.3]
UDP-2-acetamido-2-deoxy-D-galactose:O-α-L-fucosyl-(1,2-D-galactose acetamidodeoxygalactosyltransferase	2.4.1.40
UDP-2-acetamido-2-deoxy-D-galactose:protein acetamidodeoxygalactosyltransferase	2.4.1.41
UDP-2-acetamido-2-deoxy-D-glucose:chitin 4-β-acetamidodeoxyglucosyltransferase	2.4.1.16
UDP-2-acetamido-2-deoxy-D-glucose 4-epimerase	5.1.3.7
UDP-2-acetamido-2-deoxy-D-glucose:glycoprotein 2-acetamido-2-deoxy-D-glucosyltransferase	2.4.1.51
UDP-2-acetamido-2-deoxy-D-glucose:17α-hydroxysteroid-3-D-glucuronoside 17α-acetamidodeoxyglucosyltransferase	2.4.1.39
UDP-2-acetamido-2-deoxy-D-glucose:lipopolysaccharide 2-acetamido-2-deoxy-D-glucosyltransferase	2.4.1.56
UDP-2-acetamido-2-deoxy-D-glucose:NAD$^+$ 6-oxidoreductase	1.1.1.136
UDP-2-acetamido-2-deoxy-D-glucose:poly-(ribitol phosphate) 2-acetamido-2-deoxy-glucosyltransferase	2.4.1.70
UDPacetylgalactosamine–protein acetylgalactosaminyltransferase	2.4.1.41
UDP-*N*-acetylgalactosamine-4-sulphate sulphotransferase	2.8.2.7
UDP-*N*-acetylglucosamine dehydrogenase	1.1.1.136
UDPacetylglucosamine 4-epimerase	5.1.3.7
UDP-*N*-acetylglucosamine–glycoprotein *N*-acetylglucosaminyltransferase	2.4.1.51
UDP-*N*-acetylglucosamine–lipopolysaccharide *N*-acetylglucosaminyltransferase	2.4.1.56
UDPacetylglucosamine–poly(ribitol phosphate) acetylglucosaminyltransferase	2.4.1.70
UDPacetylglucosamine pyrophosphorylase	2.7.7.23
UDPacetylglucosamine–steroid acetylglucosaminyltransferase	2.4.1.39
UDP-N-acetylmuramate:L-alanine ligase (ADP-forming)	6.3.2.8
UDP-N-acetylmuramoyl-L-alanine:D-glutamate ligase (ADP-forming)	6.3.2.9
UDP-*N*-acetylmuramoyl-alanine synthetase	6.3.2.8
UDP-N-acetylmuramoyl-L-alanyl-D-glutamate:meso-2,6-diaminopimelate ligase (ADP-forming)	6.3.2.13
UDP-N-acetylmuramoyl-L-alanyl-D-glutamate:L-lysine ligase (ADP-forming)	6.3.2.7
UDP-*N*-acetylmuramoyl-*L*-alanyl-*D*-glutamate synthetase	6.3.2.9
UDP-*N*-acetylmuramoyl-*L*-alanyl-*D*-glutamyl-meso-2,6-diaminopimelate synthetase	6.3.2.13
UDP-N-acetylmuramoyl-L-alanyl-D-glutamyl-L-lysine:D-alanyl-D-alanine ligase (ADP-forming)	6.3.2.10
UDP-*N*-acetylmuramoyl-*L*-alanyl-*D*-glutamyl-*L*-lysine synthetase	6.3.2.7
UDP-*N*-acetylmuramoyl-*L*-alanyl-*D*-glutamyl-*L*-lysyl-*D*-alanyl-*D*-alanine synthetase	6.3.2.10
UDP-N-acetylmuramoyl-L-alanyl-D-γ-glutamyl-L-lysyl-D-alanyl-D-alanine:undecaprenoid-alcohol-phosphate phospho-N-acetylmuramoyl-pentapeptide-transferase	2.7.8.13
UDPacetylmuramoylpentapeptide lysine N^6-alanyltransferase	2.3.2.10
UDP-N-acetylmuramoyl-tetrapeptidyl-D-alanine alanine-hydrolase	3.4.12.6
UDP-4-amino-2-acetamido-2,4,6-trideoxyglucose aminotransferase	2.6.1.34
UDP-4-amino-2-acetamido-2,4,6-trideoxyglucose:2-oxoglutarate aminotransferase	2.6.1.34
UDParabinose 4-epimerase	5.1.3.5
UDP-L-arabinose 4-epimerase	5.1.3.5
UDPgalactose:2-acetamido-2-deoxy-D-galactosyl-(N-acetylneuraminyl)-D-galactosyl-D-glucosyl-N-acylsphingosine galactosyltransferase	2.4.1.62
UDPgalactose:2-acetamido-2-deoxy-D-glucosyl-glycopeptide galactosyltransferase	2.4.1.38
UDPgalactose–*N*-acylsphingosine galactosyltransferase	2.4.1.47
UDPgalactose:N-acylsphingosine galactosyltransferase	2.4.1.47
UDPgalactose–ceramide galactosyltransferase	2.4.1.62

INDEX TO THE ENZYME LIST (continued)

UDPgalactose—collagen galactosyltransferase	2.4.1.50
UDPgalactose—1,2-diacylglycerol galactosyltransferase	2.4.1.46
UDPgalactose:1,2-diacylglycerol 3-O-galactosyltransferase	2.4.1.46
UDPgalactose:O-α-L-fucosyl-(1,2)-D galactose α-D-galactosyltransferase	2.4.1.37
UDPgalactose—glucose galactosyltransferase	2.4.1.22
UDPgalactose:D-glucose 4-β-galactosyltransferase	2.4.1.22
UDPgalactose—2-hydroxyacylsphingosine galactosyltransferase	2.4.1.45
UDPgalactose:2-(2-hydroxyacyl)-sphingosine galactosyltransferase	2.4.1.45
UDPgalactose:5-hydroxylysine-collagen galactosyltransferase	2.4.1.50
UDPgalactose—lipopolysaccharide galactosyltransferase	2.4.1.44
UDPgalactose:lipopolysaccharide galactosyltransferase	2.4.1.44
UDPgalactose—mucopolysaccharide galactosyltransferase	2.4.1.74
UDPgalactose:mucopolysaccharide galactosyltransferase	2.4.1.74
UDPgalactose:C-55-poly(isoprenol)-phosphate galactosephosphotransferase	2.7.8.6
UDPgalactose—sphingosine β-galactosyltransferase	2.4.1.23
UDPgalactose:sphingosine β-galactosyltransferase	2.4.1.23
UDPgalacturonate—polygalacturonate α-galacturonosyltransferase	2.4.1.43
UDPgalacturonate:1,4-α-poly-D-galacturonate 4-α-galacturonosyltransferase	2.4.1.43
UDPglucose—arylamine glucosyltransferase	2.4.1.71
UDPglucose:arylamine N-glucosyltransferase	2.4.1.71
UDPglucose—cellulose glucosyltransferase	2.4.1.12
UDPglucose—collagen glucosyltransferase	2.4.1.66
UDPglucose dehydrogenase	1.1.1.22
UDPglucose—DNA α-glucosyltransferase	2.4.1.26
UDPglucose:DNA α-glucosyltransferase	2.4.1.26
UDPglucose—DNAβ-glucosyltransferase	2.4.1.27
UDPglucose:DNA β-glucosyltransferase	2.4.1.27
UDPglucose 4-epimerase	5.1.3.2
UDPglucose 4-epimerase	5.1.3.2
UDPglucose—fructose glucosyltransferase	2.4.1.13
UDPglucose:D-fructose 2-α-glucosyltransferase	2.4.1.13
UDPglucose—fructosephosphate glucosyltransferase	2.4.1.14
UDPglucose:D-fructose-6-phosphate 2-α-glucosyltransferase	2.4.1.14
UDPglucose:α-D-galactose-1-phosphate uridylyltransferase	2.7.7.12
UDPglucose:galactosyl-lipopolysaccharide glucosyltransferase	2.4.1.73
UDPglucose—β-glucan glucosyltransferase	2.4.1.12
UDPglucose—1,3-β-glucan glucosyltransferase	2.4.1.34
UDPglucose:1,3-β-D-glucan 3-β-glucosyltransferase	2.4.1.34
UDPglucose:1,4-β-D-glucan 4-β-glucosyltransferase	2.4.1.12
UDPglucose—glucosephosphate glucosyltransferase	2.4.1.15
UDPglucose:D-glucose-6-phosphate 1-α-glucosyltransferase	2.4.1.15
UDPglucose—glucosyl-DNA β-glucosyltransferase	2.4.1.28
UDPglucose:D-glucosyl-DNA β-glucosyltransferase	2.4.1.28
UDPglucose—glycogen glucosyltransferase	2.4.1.11
UDPglucose:glycogen 4-α-glucosyltransferase	2.4.1.11
UDPglucose:2-hydroxyisobutyronitrile β-glucosyltransferase	2.4.1.63
UDPglucose:5-hydroxylysine-collagen glucosyltransferase	2.4.1.66
UDPglucose—lipopolysaccharide glucosyltransferase I	2.4.1.58
UDPglucose—lipopolysaccharide glucosyltransferase II	2.4.1.73
UDPglucose:lipopolysaccharide glucosyltransferase	2.4.1.58
UDPglucose:NAD$^+$ 6-oxidoreductase	1.1.1.22
UDPglucose:phenol β-glucosyltransferase	2.4.1.35
UDPglucose—poly(glycerol phosphate) α-glucosyltransferase	2.4.1.52
UDP-glucose:poly(glycerol phosphate) α-glucosyltransferase	2.4.1.52
UDPglucose—poly(ribitol phosphate) β-glucosyltransferase	2.4.1.53
UDPglucose:poly(ribitol phosphate) β-glucosyltransferase	2.4.1.53
UDPglucose pyrophosphorylase	2.7.7.9
UDPglucosyltransferase	2.4.1.35
UDPglucuronate carboxy-lyase	4.1.1.35
UDPglucuronate decarboxylase	4.1.1.35
UDPglucuronate 4-epimerase	5.1.3.6
UDPglucuronate 4-epimerase	5.1.3.6
UDPglucuronate 5′-epimerase	5.1.3.12
UDPglucuronate 5′-epimerase	5.1.3.12
UDPglucuronate β-glucuronosyltransferase (acceptor-unspecific)	2.4.1.17
UDPglucuronate:17β-hydroxysteroid 17β-glucuronosyltransferase	2.4.1.42
UDPglucuronate—oestradiol glucuronosyltransferase	2.4.1.59
UDPglucuronate:17β-oestradiol 3-glucuronosyltransferase	2.4.1.59
UDPglucuronate—oestriol 16α-glucuronosyltransferase	2.4.1.61
UDPglucuronate:oestriol 16α-glucuronosyltransferase	2.4.1.61
UDPglucuronate—oestriol 17β-glucuronosyltransferase	2.4.1.42

INDEX TO THE ENZYME LIST (continued)

UDPglucuronate → phenol transglucuronidase	2.4.1.17
UDPglucuronosyltransferase	2.4.1.17
UDPxylose:1,4-β-D-xylan 4-β-xylosyltransferase	2.4.1.72
UMP:pyrophosphate phosphoribosyltransferase	2.4.2.9
UMPpyrophosphorylase	2.4.2.9
Unsaturated acyl-CoA hydratase	4.2.1.17
Unsaturated acyl-CoA reductase	1.3.99.2
Unsaturated fatty acid→O_2 transhydrogenase	1.13.11.12
Unsaturated-phospholipid methyltransferase	2.1.1.16
Unspecific diphosphate phosphohydrolase	3.6.1.15
Uracil:(acceptor) oxidoreductase	1.2.99.1
Uracil dehydrogenase	1.2.99.1
Uracil phosphoribosyltransferase	2.4.2.9
Urate oxidase	1.7.3.3
Urate:oxygen oxidoreductase	1.7.3.3
Urateribonucleotide:orthophosphate ribosyltransferase	2.4.2.16
Urateribonucleotide phosphorylase	2.4.2.16
Urea amidohydrolase	3.5.1.5
Urea:carbon-dioxide ligase (ADP-forming) decarboxylating, deaminating)	6.3.4.6
Urea carboxylase (hydrolyzing)	6.3.4.6
Urease	3.5.1.5
Ureidoglycollate lyase	4.3.2.3
(−)-Ureidoglycollate urea-lyase	4.3.2.3
β-Ureidopropionase	3.5.1.6
Ureidosuccinase	3.5.1.7
Uricase	1.7.3.3
Uricate→O_2 transhydrogenase	1.7.3.3
Urico oxhydrase	1.7.3.3
Uridine kinase	2.7.1.48
Uridine nucleosidase	3.2.2.3
Uridine:orthophosphate ribosyltransferase	2.4.2.3
Uridine phosphorylase	2.4.2.3
Uridine ribohydrolase	3.2.2.3
Uridyl transferase	2.7.7.12
Urocanase	4.2.1.49
Urocanate hydratase	4.2.1.49
Urokinase	3.4.99.26
Uronate dehydrogenase	1.2.1.35
Uronate:NAD⁺ 1-oxidoreductase	1.2.1.35
Uronic isomerase	5.3.1.12
Uronolactonase	3.1.1.19
Uroporphyrinogen-III carboxy-lyase	4.1.1.37
Uroporphyrinogen decarboxylase	4.1.1.37
Uroporphyrinogen I synthase	4.3.1.8
Urushiol oxidase	1.14.18.1
UTP:2-acetamido-2-deoxy-α-D-glucose-1-phosphate uridylytransferase	2.7.7.23
UTP:ammonia ligase (ADP-forming)	6.3.4.2
dUTPase	3.6.1.23
UTP:α-D-galactose-1-phosphate uridylyltransferase	2.7.7.10
UTP:α-D-glucose-1-phosphate uridylyltransferase	2.7.7.9
dUTP nucleotidohydrolase	3.6.1.23
UTP: α-D-xylose-1-phosphate uridylyltransferase	2.7.7.11
L-Valine carboxy-lyase	4.1.1.14
Valine decarboxylase	4.1.1.14
Valine dehydrogenase (NADP⁺)	1.4.1.8
Valine−isoleucine aminotransferase	2.6.1.32
L-Valine:3-methyl-2-oxovalerate aminotransferase	2.6.1.32
L-Valine:NADP⁺ oxidoreductase (deaminating)	1.4.1.8
L-Valine:tRNA^{Val} ligase (AMP-forming)	6.1.1.9
Valyl-tRNA synthetase	6.1.1.9
Vinylacetyl-CoA Δ-isomerase	5.3.3.3
Vinylacetyl-CoA Δ³−Δ²-isomerase	5.3.3.3
Vitamin A esterase	[3.1.1.12]
Vitamin $B_{12\,r}$ reductase	1.6.99.9
Xanthine and aldehyde→O_2 transhydrogenase	1.2.3.2
Xanthine dehydrogenase	1.2.1.37
Xanthine:NAD⁺ oxidoreductase	1.2.1.37
Xanthine oxidase	1.2.3.2
Xanthine:oxygen oxidoreductase	1.2.3.2
Xanthine phosphoribosyltransferase	2.4.2.22
Xanthosine-5′-phosphate:ammonia ligase (AMP-forming)	6.3.4.1
Xanthosine-5′-phosphate:L-glutamine amido-ligase (AMP-forming)	6.3.5.2
Xylanase	3.2.1.32
1,4-β-Xylan synthase	2.4.1.72
1,3-β-D-Xylan xylanohydrolase	3.2.1.32
1,4-β-D-Xylan xylanohydrolase	3.2.1.8
1,3-β-D-Xylan xylohydrolase	3.2.1.72
1,4-β-D-Xylan xylohydrolase	3.2.1.37
Xylitol:NAD⁺ 2-oxidoreductase (D-xylulose-forming)	1.1.1.9
Xylitol:NADP⁺ 1-oxidoreductase (D-xylose-forming)	1.1.1.139
Xylitol:NADP⁺ 4-oxidoreductase (L-xylulose-forming)	1.1.1.10
Xylobiase	3.2.1.37
L-Xylose dehydrogenase	1.1.1.113
Xylose isomerase	5.3.1.5
D-Xylose ketol-isomerase	5.3.1.5
L-Xylose:NADP⁺ 1-oxidoreductase	1.1.1.113
Xylose-1-phosphate uridylyltransferase	2.7.7.11
β-Xylosidase	3.2.1.37
Xylulokinase	2.7.1.17
L-Xylulokinase	2.7.1.53
D-Xylulose-5-phosphate D-glyceraldehyde-3-phosphate-lyase (phosphate-acetylating)	4.1.2.9

INDEX TO THE ENZYME LIST (continued)

D-Xylulose reductase	1.1.1.9	Yeast proteinase A	3.4.23.8
L-Xylulose reductase	1.1.1.10	Yeast proteinase B	
Yeast carboxypeptidase	3.4.12.8	Zwischenferment	1.1.1.49
Yeast proteinase β	3.4.22.9	Zymohexase	4.1.2.13

RECOMMENDATIONS FOR THE NOMENCLATURE OF HUMAN IMMUNOGLOBULINS*†

International Union of Immunological Sciences

The following is a document prepared by a Subcommittee** for Human Immunoglobulins of the IUIS Nomenclature Committee, and reproduced by permission of its chairman.††

Preamble

The final draft was developed after receiving additional suggestions from colleagues who had assisted in reaching agreement on the earlier nomenclatures published in 1964 and 1969. The report of the Subcommittee has been reviewed and approved by the Nomenclature Committee of the IUIS.

Terminology for Immunoglobulin Molecules

Following a proposal made in 1964,[1] two symbols (Ig and γ) have been used interchangeably to designate human or animal immunoglobulins. Although it was pointed out that Ig is a logical symbol for immunoglobulins, the symbol γ was retained as an acceptable substitute, mainly in view of the tradition that had long associated it with the immunoglobulins.

In recent years there has been a trend among editors and authors to give increasing preference to the symbol Ig. A major reason for dissatisfaction with the existing dual terminology, Ig and γ, is that the symbol γ is also employed to designate the heavy polypeptide chains of a particular class of immunoglobulins.

▲ It is therefore proposed to discontinue the use of the symbol γ for the term immunoglobulin and to apply γ to designate exclusively the heavy chains of immunoglobulin G (IgG). Symbols such as γG1, γD, etc., should be replaced by IgG1, IgD, etc. The term γ-globulins should not be used as a synonym for immunoglobulins.

Use of the Symbols L and K

The symbol L is now being used in two senses, firstly as an abbreviation for light chains and secondly to designate that type of immunoglobulin molecules whose light chains are of the lambda variety. The intrinsic defect in this terminology becomes apparent in expressions such as "L chains of the L type" as opposed to "L chains of the K type."

---

*Reprinted with permission from International Union of Immunological Sciences, *Biochemistry*, 11(18), 3311–3312 (1972). Copyright by the American Chemical Society.
† ▲ indicates revisions of References 1 and 2.
**Composition of the Subcommittee: Dr. J. D. Capra (Mount Sinai School of Medicine, New York, NY 10029), Dr. M. O. Dayhoff (National Biomedical Research Foundation, Georgetown University Medical Center, Washington, DC 20007), Dr. H. H. Fudenberg (University of California Medical Center, San Francisco, CA 94122), Dr. J. F. Heremans (University of Louvain, B-1200 Brussels, Belgium), Dr. N. Hilschmann (Max-Planck-Institut, D-3400 Göttingen, Federal Republic of Germany), Dr. L. E. Hood (California Institute of Technology, Pasadena, CA 91109), Dr. C. Milstein (University Postgraduate Medical School, Cambridge, CB2 2QH, England), Dr. E. Osserman (Columbia University, College of Physicians and Surgeons, New York, NY 10032), Dr. F. W. Putnam [chairman] (Indiana University, Bloomington, IN 47401), Dr. D. S. Rowe (WHO International Reference Centre for Immunoglobulins, Institute of Biochemistry, CH-1005 Lausanne, Switzerland), Dr. A. Solomon (University of Tennessee Memorial Research Center and Hospital, Knoxville, TN 37920), Dr. W. D. Terry (National Institutes of Health, Bethesda, MD 20014), and Dr. G. Torrigiani (Immunology Unit, World Health Organization, CH-1211 Geneva 27, Switzerland).
††Reprints may be obtained from Dr. Waldo E. Cohn, Office of Biochemical Nomenclature, Biology Division, Oak Ridge National Laboratory, Post Office Box Y, Oak Ridge, TN 37830.

▲ It is therefore proposed to restrict the use of the symbol L to the designation of light chains as opposed to the symbol H for heavy chains and to discard the use of the symbol K. The terms kappa type and lambda type should be used to indicate the type of whole molecules or isolated light chains formerly described as belonging to the K type and L type, respectively.

▲ This amendment to the terminology proposed in 1964 also makes it necessary to discontinue the use of symbols such as IgGK or IgML, etc., which should be discarded in favor of notations such as IgG(κ), IgM(λ), etc.

Use of the Terms Classes, Subclasses, Types, Subtypes, Groups, and Subgroups

The previous proposals[1,2] for the use of the terms classes, subclasses, types, and subtypes are retained. Specifically, these terms may be applied both to the entire molecule and to its chains.

The terms type and subtype designate variants of nonallotypic nature defined by characteristics of the constant (C_L) regions of light chains. The terms class and subclass designate variants of nonallotypic nature defined by characteristics of the constant (C_H) regions of heavy chains.

All variable regions associated with κ chains, λ chains, or H chains should be defined as forming a group. Three such groups, to be called the V_κ group, the V_λ group, and the V_H group, have so far been characterized. The variable regions from the V_κ group and the V_λ group appear to be associated exclusively with constant regions from, respectively, κ-type and λ-type light chains. In contrast, the variable regions from the V_H group seem to occur in association with the constant regions from any of the heavy chain classes.

Within a group of variable regions it is possible to distinguish a number of subgroups.

▲ It is now clear that the nomenclature earlier proposed for subgroups[2] needs revision. Criteria for the differentiation of subgroups are being developed and will form the basis for future recommendations. Current information can be obtained from Dr. F. Putnam, Chairman of the Subcommittee on Human Immunoglobulins of the International Union of Immunological Societies.

Similarly to the proposal for the terms class and subclass, type and subtype, the terms group and subgroup may also be used to characterize the variable region of the immunoglobulin molecule.

Note: These recommendations have been prepared after consultation with the IUPAC-IUB Commission on Biochemical Nomenclature (CBN).

REFERENCES

1. *Bull. W.H.O.,* 30, 447 (1964).
2. *Bull. W.H.O.,* 41, 975 (1969).

THE NOMENCLATURE OF PEPTIDE HORMONES*†
RECOMMENDATIONS (1974)

IUPAC-IUB Commission on Biochemical Nomenclature

In the last two decades, the structures of many peptide hormones have been elucidated and other peptide hormones have been obtained in pure form. However, there are presently no accepted guidelines for nomenclature in this field. Therefore, the IUPAC-IUB Commission on Biochemical Nomenclature (CBN) appointed a subcommittee consisting of R. A. Acher, R. A. Boissonnas, H. B. F. Dixon, R. Guillemin, P. Karlson (Convenor), H. Rasmussen, J. Rudinger, and N. A. Yudayev to discuss the question of the nomenclature of peptide hormones and to report to CBN. The International Union of Physiological Sciences nominated, as consultants to the subcommittee, A. Brodish, S. M. McCann, and F. Ulrich. The present Recommendations are based on the report of this subcommittee.

1. General Principles

Naturally occurring oligo- and polypeptides are generally referred to by trivial names; their systematic names are so cumbersome that they are of little use. Most of the peptide hormones already have well established trivial names indicating either natural source (e.g., insulin) or physiological action (e.g., relaxin, prolactin). However, some of the trivial names are so long that these hormones are known mainly by abbreviations (e.g., FSH for follicle-stimulating hormone). This is unfortunate, and it was therefore considered advisable to create suitable names for those peptide hormones not already having well established short trivial names. Three principles have been observed:

   a. new names for hormones of the adenohypophysis bear the ending "-tropin;"
   b. hypothalamic releasing factors (hormones) bear the ending "-liberin;"
   c. hypothalamic release-inhibiting factors (hormones) bear the ending "-statin" (see below).

2. Trivial Names

The trivial names proposed for peptide hormones are given in the "Appendix."
Abbreviations of the new names are not proposed, and the use of currently fashionable abbreviations is discouraged.

3. Species Designation

Since peptide hormones show species variation in their amino-acid sequence, their names are essentially "generic names," and are insufficient to specify a single chemical compound. It is therefore recommended that authors add to the name of each hormone the species from which the hormone was isolated, or at least indicate the biological source(s) where appropriate in each paper.

4. Special Groups of Hormones

(a) *Hypothalamic Factors (Hormones)* — The hypothalamic "releasing factors" or "releasing hormones" have no well established trivial names. It is recommended that the

---

*Document of the Commission on Biochemical Nomenclature (CBN) of the International Union of Pure and Applied Chemistry (IUPAC) and the International Union of Biochemistry (IUB), approved by CBN, IUPAC, and IUB in 1974 and published by permission of IUPAC and IUB.

†From *IUPAC Inf. Bull. Append. Tentative Nomencl. Sym. Units Stand.*, No. 48, September 1975. With permission. (Also in *J. Biol. Chem.*, 250, 3215 (1975) and elsewhere.)

trivial names given in the "Appendix" be used for the releasing factors (hormones). They are based on the ending "-liberin" added to the prefix of the pituitary hormone released by the factor. Thus, "thyroliberin" indicates the hypothalamic peptide stimulating the release (and perhaps also the biosynthesis) of thyrotropin, the corresponding tropic hormone, from the pituitary gland. (Note that the ending "-tropin" is no longer retained in the name; it is implied in the definition of "-liberin.")

The names of those factors inhibiting the release (and perhaps the synthesis) of pituitary hormones are formed in a similar way with the suffix "-statin."

(b) *Pituitary Hormones* — Most of the hormones of the adenohypophysis have acceptable trivial names ending in -tropin.* The committee has created the missing names for follicle-stimulating hormone, "follitropin," and for luteinizing hormone, "lutropin." It is recommended that pituitary hormones discovered in the future also be named with the ending -tropin. This suffix should be restricted to pituitary and similar hormones and should not be used for, e.g., crustacean hormones acting on pigment cells.

Some placental hormones are physiologically very similar to pituitary hormones. They are named accordingly with the prefix "chorio-," e.g., choriogonadotropin for chorionic gonadotropin.

(c) *Invertebrate Peptide Hormones* — Though some of the invertebrate peptide hormones have been isolated in pure form and their amino-acid compositions have been determined, the field has not yet developed to a stage where a list of names seems warranted.

It is, however, recommended that the suffixes defined above for hypothalamic and pituitary hormones not be used in a different sense in invertebrates. Thus, a crustacean color change hormone acting on, e.g., erythrophores, should *not* be named "erythrotropin," a hormone causing release of eggs and/or sperm in sea urchins should *not* be called "gametoliberin."

---

* The committee has re-evaluated the arguments for and against the suffix "-trophin," still used by many anatomists and physiologists. Since the bioassay systems are based mostly on effects other than the trophic one, it was decided to recommend "-tropin" for general usage in biochemistry.

APPENDIX

LIST OF PEPTIDE HORMONES[a]

Trivial name	Other names	Current abbreviation[b]
1. Hypothalamic Factors		
Corticoliberin	Corticotropin-releasing factor	CRF
Folliberin	Follicle-stimulating-hormone-releasing factor	FSH-RF
Gonadoliberin[c]	Gonadotropin-releasing factor	(LH/FSH-RF)
Luliberin	Luteinizing hormone-releasing factor	LH-RF (LRF)
Melanoliberin	Melanotropin-releasing factor	MFR
Melanostatin	Melanotropin release-inhibiting factor	MIF
Prolactoliberin	Prolactin-releasing factor	PRF
Prolactostatin	Prolactin release-inhibiting factor	PIF
Somatoliberin	Somatotropin-releasing factor; growth hormone-releasing factor	SRF GH-RF
Somatostatin	Somatotropin release-inhibiting factor	
Thyroliberin	Thyrotropin-releasing factor	TRF
2. Pituitary and Related Hormones		
Choriogonadotropin[d]	Chorionic gonadotropin	CG
Choriomammotropin	Chorionic somatomammotropin	CS
Corticotropin	Adrenocorticotropic hormone	ACTH
Follitropin	Follicle-stimulating hormone	FSH
Gonadotropin[e]	Gonadotropin hormone	
Glumitocin[f]	[Ser^4, Gln^8]Ocytocin[f]	
Isotocin[g]	[Ser^4, Ile^8]Ocytocin[f]	
Lipotropin	Lipotropic hormone	LPH
Lutropin	Luteinizing hormone; (Interstitial cell-stimulating hormone)	LH (ICSH)
Melanotropin[h]	Melanocyte-stimulating hormone	MSH

[a]For convenience, some biologically active peptides that may not fulfill all criteria of a hormone are included.

[b]Abbreviations, old or new, are not recommended; they are given here for identification purposes only.

[c]This name indicates a hypothalamic substance releasing gonadotropin. It may also be used for the decapeptide isolated from pig hypothalami and known as luteinizing hormone/follicle-stimulating-hormone releasing factor, abbreviated LH/FSH-RF,[b] since the peptide induces the release of both lutropin and follitropin in constant proportions and thus carries the activity of both luliberin and folliberin (see also Footnote e).

[d]The chorionic gonadotropins have in most species (including man) the action of both follitropin and lutropin and are therefore termed "gonadotropins."

[e]Gonadotropin is to be used for hormones having the activity of both follitropin and lutropin, like the gonadotropins of cold-blooded vertebrates. It may also be used for impure preparations containing lutropin and follitropin.

[f]In elasmobranch fishes.

[g]In bony fishes.

[h]Two peptides have been sequenced and designated α-melanotropin and β-melanotropin.

APPENDIX (continued)

LIST OF PEPTIDE HORMONES

Trivial name	Other names	Current abbreviation[b]

2. Pituitary and Related Hormones (continued)

Trivial name	Other names	Current abbreviation[b]
Mesotocin[i]	[Ile^8]Ocytocin[f]	
Ocytocin[j] (oxytocin)		OXT
Prolactin	Mammatropic hormone; mammatropin; lactotropic hormone; lactotropin	PRL
Somatotropin	Somatropic hormone; growth hormone	STH GH
Thyrotropin	Thyrotropic hormone	TSH
Urogonadotropin[k]	(Human) Menopausal gonadotropin	HMG
Vasopressin	Adiuretin; antidiuretic hormone	VP, ADH
Vasotocin	[Arg^8]Ocytocin[f]	

3. Other Peptide Hormones

Trivial name	Other names	Current abbreviation[b]
Angiotensin	Angiotensin II	
Bradykinin	Kinin-9	
Calcitonin	Thyrocalcitonin	
Erythropoietin		
Gastrin		
Gastrin sulphate	Gastrin II	
Glucagon	Hyperglycemic factor	(HGF)
Insulin		
Kallidin	Kinin-10	
Pancreozymin	Cholecystokinin	
Parathyrin[d]	Parathyroid hormone; Parathormone	
Proangiotensin	Angiotensin I	
Relaxin		
Secretin		
Somatomedin[m]	Sulfation factor	
Thymopoietin[n]	Thymin	

[i]In birds and reptiles.

[j]The name of this hormone is derived from Greek ωκυτοκοσ (OKYTOKOS = fast birth, prompt delivery), *not* from the Greek οξυσ (oxys = acid; fast). The spelling ocytocin should therefore be preferred; moreover, it avoids confusion with oxy, meaning "related to oxygen." However, oxytocin is in wide use, especially in the English language. Therefore, both spellings are listed as optional.

[k]Most work has been done on the human hormone, known as Human Menopausal Gonadotropin (HMG); it is a pituitary hormone, chemically changed during passage through the kidney. Due to its occurrence in urine, it has been termed "urogonadotropin."

[l]Parathyrin is a new name suggested here. The synonym Parathormone is a proprietary name.

[m]The name "somatomedin" was suggested by a group working in the field [*Nature* 235, 107 (1972)].

[n]A polypeptide from the thymus. The name proposed was suggested in a letter to *Nature* [249, 863 (1974)] to avoid confusion with the earlier "thymine" from nucleic acids. "Thymin" should be abandoned.

NOMENCLATURE FOR HUMAN IMMUNOGLOBULINS*†

Definition of Immunoglobulins

Immunoglobulins are proteins of animal origin endowed with known antibody activity, and certain proteins related to them by chemical structure and hence antigenic specificity. Related proteins for which antibody activity has not been demonstrated are included — for example, myeloma proteins, Bence-Jones proteins, and naturally occurring subunits of the immunoglobulins.

Immunoglobulins are not restricted to the plasma but may be found in other body fluids or tissues, such as urine, spinal fluid, lymph-nodes, spleen, etc. Proteins may occur which fulfill the above requirements but which have widely differing physicochemical properties such as electrophoretic mobility, sedimentation coefficient, and diffusion coefficient, and different chemical properties such as carbohydrate content, amino-acid composition of polypeptide chains, etc. Immunoglobulins do not include the components of the complement system.

Terminology

Advances in biological and chemical studies of immunoglobulins have emphasized the need for a nomenclature and classification to meet new requirements. Terminology should be based on the identification of the polypeptide chains which by complementation make up the molecule. This principle should allow a precise description of different molecules. In addition, broader terms are needed for grouping classes of molecules with common physicochemical and biological characteristics. This can be achieved by minor modifications of current terminologies.

The following notation for immunoglobulins provides:

a. An abbreviated notation to designate the major classes of immunoglobulins;
b. Terms to designate the polypeptide chains of the immunoglobulin molecules.

Whenever possible terms already in common usage have been retained.

Abbreviated Notation

Two symbols for immunoglobulins are considered to be appropriate. The logical abbreviation for the term immunoglobulin would be the symbol Ig. Since the symbol Ig may present some difficulties in verbal communication and, furthermore, since the symbol γ has been commonly employed to designate the immunoglobulins, both symbols γ and Ig are recognized as appropriate: These symbols should be accompanied by a capital letter designating a specific class of immunoglobulin, for example, γG, γM ..., IgG, IgM ..., as noted below. The symbol should *not* be followed by arabic numbers, which in the past have been used to indicate electrophoretic mobilities.

The following symbols proposed for major classes of immunoglobulin molecules are based on differences associated with heavy chains.

Present usage	Proposed usage	Present usage	Proposed usage
γ, 7Sγ, 6.6Sγ, γ_2, γ_{SS} β_2A, γ_1A	γG or IgG γA or IgA	γ_1M, β_2M, 19Sγ,γ-macroglobulin	γM or IgM

*From *Bull. Org. mond Santé, Bull. W.H.O.*, 30, 447 (1964). Reprinted by permission of World Health Organization, Geneva, Switzerland.

†See revision: Recommendations for the Nomenclature of Human Immunoglobins.

The following symbols are proposed for types of molecules identified by the properties of light chains.

Present usage	Proposed usage	Present usage	Proposed usage
Type I, 1, B	Type K	Type II, 2, A	Type L

Notation for Polypeptide Chains

Two major groups of polypeptide chains are now recognized as occurring in immunoglobulin molecules:

a. The L or B chains (molecular weight about 20,000) are at this time the only subunits known to be common to the three major classes of immunoglobulins;

b. The H or A chains (molecular weight about 50,000) determine the distinctive properties of each class.

It is suggested that the terms L and H, and A and B to denote chains be discarded, in view of the more precise terminology proposed below. For simplicity, however, it is advisable to designate two major groups of polypeptide chains by the terms "light" and "heavy." The designation of these polypeptide chains may be revised as new evidence accumulates. For example, the possibility that the "heavy" chain is in fact several chains has been considered. Until definite proof of this hypothesis that the heavy chain is composed of two or more chains is provided, the heavy chain will be considered as a single chain in the following description.

a. *Heavy chains.* It is proposed that the heavy chains be designated by small Greek letters corresponding to the Roman capital letters used for the immunoglobulin classes.

Immunoglobulin class	Heavy chain	Immunoglobulin chain	Heavy chain
γG or IgG	γ(gamma)	γM or IgM	μ(mu)
γA or IgA	α(alpha)		

It is recognized that the symbol γ has been used in two different designations. No ambiguity should occur, however, since γ is not used alone in either case. The symbol is always used in association with either the word immunoglobulin or chain, for example:

"γG-immunoglobulin" for the class of immunoglobulin, and
"γ-chain" for the polypeptide chain.

Use of the term γ-chain (rather than another Greek letter) to designate the heavy chain of the IgG or γG-immunoglobulin molecule preserves an association with the distinctive properties of this major component within the immunoglobulin system.

b. *Light chains.* Two forms of light chain (Type I and Type II) are recognized as occurring in man. To conform with the use of Greek letters for distinctive polypeptide chains, the following symbols are proposed:

	Proposed usage	
Present usage	Abbreviated notation for immunoglobulins	Notation for light chains
Type I	Type K	κ(kappa)
Type II	Type L	λ(lambda)

c. *Molecular formulas.* The proposed system is similar to that in use for hemoglobin. Some typical formulas would be:

Present usage	Abbreviated notation for immunoglobulins	Molecular formulas
7Sγ – type I	γGK or IgGK	$\gamma_2 \kappa_2$
7Sγ – type II	γGL or IgGL	$\gamma_2 \lambda_2$
γ_1A – type I	γAK or IgAK	$\alpha_2 \kappa_2$
γ_1A – type II	γAL or IgAL	$\alpha_2 \lambda_2$
γ_1M – type I	γMK or IgMK	$(\mu_2 \kappa_2)_n$*
γ_1M – type II	γML or IgML	$(\mu_2 \lambda_2)_n$*
Urinary globulin composed of light chains		κ_2 and λ_2 (if in the dimer form)

If it is desirable to indicate that the protein is a myeloma protein, Bence-Jones protein, etc., this information can be added in parentheses after the formula.

Classes and Subclasses

There may be different degrees of molecular association within an immunoglobulin class, and this distinction may be made by the term dimers, trimers and polymers of a particular unit. In addition the sedimentation coefficient or the molecular weight may be cited.

It may become necessary to recognize a new class of immunoglobulins when molecules containing a novel heavy chain are discovered. This new heavy chain should differ from the presently known heavy chains to the same extent that the latter differ from one another. Similar principles should be applied to the differentiation of new light chains. When new polypeptide chains are characterized, they should be designated with a new small letter of the Greek alphabet. If new immunoglobulins contain chains which can be recognized as members of presently known classes even though they differ in some significant detail, then it may become necessary to define them as new *subclasses*. Until adequate characterization of such novel polypeptide chains is achieved, it is suggested that a noncommittal designation be used, and enclosed in parentheses as illustrated below. Terms indicative of origin, such as geographical names and the names of patients, could be used. For example, a newly found immunoglobulin with the characteristics of class G, thought to be a new subclass, if identified in Prague, could be designated IgG(Pr). If several such components are identified simultaneously, a combination of abbreviated words and numbers could be used – γG(Pr-1), γG(Pr-2), etc.

When new immunoglobulins, which can be recognized as subclasses, have been well characterized, they will be indicated by the symbol of the major class followed by a small Roman letter (IgAa, IgAb), to be read as A subclass a. Similarly, the new chain responsible for the subclass should be indicated as αa, αb, etc.

The WHO Reference Centre for Immunoglobulins is being established at the Institute of Biochemistry, Lausanne, Switzerland, and investigators are encouraged to consult the Centre concerning notation of new immunoglobulins.

Fragments

The term "fragment" is reserved for portions of the molecule obtained as a result of the cleavage of peptide bonds. Proteolytic enzymes produce heterogeneous groups of fragments which nevertheless can be distinguished by immunological and chemical properties. It is useful to name these fragments, but it is stressed that a single designation does not imply homogeneity.

*n in this case may be 6.

Fragments Produced by Digestion with Papain

Present usage	Proposed usage
A, C, S(I, II)*	Fab-fragment (antigen-binding)
B, F(III)*	Fc-fragment (crystallizable)
A piece	Fd-fragment

If it is necessary to specify the chain from which fragments are derived, the name of the chain may be subscripted — for example, Fcγ or Fdγ. For 5S fragments obtained by pepsin digestion and having two antibody sites, the notation would be $F(ab')_2$. The univalent pepsin fragments should be designated as Fab'. If possible details of molecular formulas may be appended. For fragments derived from immunoglobulins without known antibody activity, the same notation should be used.

Immunoglobulins in Proliferative Disorders of Lymphocytic or Plasmacytic Cells

It is suggested that the distinctive proteins related to the immunoglobulins be classified as "discrete components" or "pathological proteins," and that they be classified further in relation to the corresponding normal protein. It is recognized that at this time no decision can be made as to whether these are large amounts of normally present immunoglobulins or abnormal proteins related to them.

In the abbreviated notation, these proteins should not be classified precisely in the absence of an appropriate clinical diagnosis. In patients with multiple myeloma or macroglobulinaemia they should be termed (a) G myeloma or A myeloma globulin; (b) M-macroglobulin (Waldenström); (c) light-chain protein or Bence-Jones protein depending on the thermosolubility properties associated with them; (d) heavy-chain proteins with the appropriate class designation.

As is the case for normal immunoglobulins, the notation for polypeptide chains can be used regardless of the disease process with which they are associated. This provides for those instances where relatively homogeneous immunoglobulins are associated with pathological states other than those mentioned above. Under those circumstances the diagnosis can be listed separately.

Nomenclature of Immunoglobulins in Other Species

In the nomenclature of immunoglobulins of different species, it is recommended that the same principles used in devising a nomenclature for human immunoglobulins be applied. As many criteria as possible, for example, immunological cross-reactivity, estimation of molecular weight, carbohydrate content and biological properties, should be employed to establish whatever similarities in chemical structure may exist between animal and human immunoglobulin classes.

For example, in animal sera the use of γM or IgM is recommended for a class of immunoglobulins with molecular weight of the order of magnitude of 10^6. In addition, other properties of the human γM or IgM — such as high carbohydrate content, dissociation by mercaptoethanol, and antigenic relationship — should be considered.

*The parentheses enclose terms used for fragments in the rabbit.

Notation for Genes, Genotypes, and Allotypes

As a logical consequence of the criteria followed for outlining the above proposed terminology, the gene notation should be based on the specification of the molecular subunit controlled by the gene in question. For immunoglobulins, unusual genetic situations may occur, such as somatic mutation, repeats, etc., posing problems which at the moment cannot be precisely foreseen. Moreover, identification of individual variants under genetic control has been limited until now to variants detected by serological reagents (Gm groups, allotypes). The problem of genetic terms related to immunoglobulins, as well as terms for serological variants, would best be dealt with by a committee of investigators active in the field of immunoglobulins, and representing different disciplines (genetics, immunology, protein chemistry).

NOTATION FOR HUMAN IMMUNOGLOBULIN SUBCLASSES*†

Henry G. Kunkel, John L. Fahey, Edward C. Franklin, Elliott F. Osserman, and William D. Terry**

Human immunoglobulin molecules can be divided into at least four major groups or classes, referred to either as IgG, IgA, IgM, and IgD, or as γG, γA, γM, and γD, as suggested in 1964.

Recent studies in many laboratories have shown that several of these classes can be further divided into subclasses, and consequently, several temporary subclass designations have come into use.

Considerable information has been obtained concerning immunoglobulin G and A subclasses, and it now seems appropriate (1) to put forward a uniform notation for the immunoglobulin G subclasses, and (2) to propose a general scheme for the notation of additional immunoglobulin subclasses as they are described.

The general proposal is as follows

1. Subclasses should be indicated by an arabic numeral following the letter denoting the class. Arabic rather than Roman numerals are suggested because of their greater simplicity. Confusion with the genetic factors can be avoided by listing genetic factors in parentheses, preferably preceded by the locus.

2. Subclasses should be numbered on the basis of relative concentration in normal serum or on the basis of relative frequency of occurrence as myeloma proteins.

3. When potential new subclasses are first identified, the investigator should employ a temporary designation in accordance with the nomenclature proposed in 1964[1] (initial, city, etc.), but this designation should not resemble the numerical terminology herein proposed.

4. Final numerical subclass notations should be used only when several laboratories have exchanged reagents and agreed on the categorization.

The specific proposal for IgG (γG) is shown below:

Current	Occurrence as myeloma proteins (%)	Proposed	Polypeptide heavy chain (γ-chain)
We or γ2b or C	70–80	IgG1 or γG1	γ1
Ne or γ2a	13–18	IgG2 or γG2	γ2
Vi or γ2c or Z	6–8	IgG3 or γG3	γ3
Ge or γ2d	3	IgG4 or γG4	γ4

Thus a type K myeloma protein currently classed as Vi, Z-type or γ2c would be called IgG3-K or γG3-K; and the heavy chain would be designated γ3.

---

*From *Bull. Org. mond Santé, Bull. W.H.O.*, 35, 953 (1966). Reprinted by permission of World Health Organization, Geneva, Switzerland.
†See revision: Recommendations for the Nomenclature of Human Immunoglobins.
**H. G. Kunkel, Rockefeller University, New York, N.Y., J. L. Fahey, National Cancer Institute, National Institutes of Health, Bethesda, Md., E. C. Franklin, Department of Medicine, New York University School of Medicine, New York, N.Y., E. F. Osserman, Department of Medicine, College of Physicians and Surgeons of Columbia University, New York, N.Y., W. D. Terry, National Cancer Institute, National Institutes of Health, Bethesda, Md.

It is hoped that the proposed scheme will prove a useful aid in communicating current information about human immunoglobulin subclasses while retaining flexibility for designating new subclasses in the future.

Addendum

The three subclasses of human immunoglobulin designated as γA, γB, and γC by S. Dray correspond to subclasses γG2, γG1, and γG3, respectively, in the proposed notation.

REFERENCE

1. *Bull. W.H.O.,* 30, 447 (1964).

NOTATION FOR GENETIC FACTORS OF HUMAN IMMUNOGLOBULINS*

In a previous memorandum stating recommendations for a uniform notation for human immunoglobulins, it was considered advisable that analogous recommendations be made in the future for genes, genotypes, and allotypes. In that memorandum, classes of human immunoglobulins which are common to all normal individuals of a given species were considered. These classes of immunoglobulins can be distinguished from each other by a number of means including their antigenic specificity (isotypic specificity).

Another kind of classification, which is at present based on serological typing, distinguishes variants which occur among proteins of the above classes. The corresponding determinants, which are present in some groups of individuals within the species, define antigenic specificities which in animals are commonly termed as "allotypic specificities" and in man as "factors." Many of these antigenic specificities determine genetic polymorphisms in the different races of man and other animals.

The genetic factors of human immunoglobulins are currently identified by serological tests which measure the inhibition by the test material of the agglutination of erythrocytes precoated with incomplete antibody. The agglutinating antibodies occur in selected sera from patients with rheumatoid arthritis or from normal subjects. The incomplete coating antibodies are usually IgG immunoglobulin molecules and the agglutinating reagents are usually IgM molecules. The whole serum, an isolated immunoglobulin, or an immunoglobulin subunit may be used in such an inhibition test.

The term "factor" defines a particular property of certain immunoglobulins which enables them to be identified by a given set of reagents, but it should be noted that the characterization of factors by these means does not preclude the possibility of a crossreaction.

Family studies have shown that the factors involved are inherited according to Mendelian principles. Two series of factors controlled by independent genetic systems are known; one (the Gm system) is associated with the heavy polypeptide chain of some IgG molecules, and the other (the Inv system) is associated with light polypeptide chains of immunoglobulin molecules.

The nomenclature currently in use is based on serological tests, with genetical analysis providing linkage group information. In higher organisms, the term locus (and the corresponding terminology for genes and genotypes) may be ambiguously used for indicating either a complex chromosomal region or a structural gene, and owing to the nature of serological analysis a one-to-one correspondence between one factor and one gene or one region of a gene cannot be inferred at the present time. For these reasons, the term "system" is used here for a unit of closely linked genetic information which cannot be separated into subunits at the level of resolution allowed by formal human genetics.

A new notation should provide a symbol which suggests the relation of the system to the structure it influences, which does not go beyond established fact and which is sufficiently flexible to meet future needs. It should relate to prior terminology for the same system if at all possible, and provide for different systems a series of symbols which differ from each other by at least two letters when practicable. The following recommendations have been designed to meet these requirements as far as possible.

Notation for the Factors of the Gm and Inv Systems

At the present time, the accumulated information is insufficient to permit precise correlation of the genetic loci for the human immunoglobulins with the structures they influence. Consequently, it seems advisable at this time to retain the present symbols Gm

---

*From *Bull. Org. mond Santé, Bull. W.H.O.*, 30, 721 (1965). Reprinted by permission of World Health Organization, Geneva, Switzerland.

and Inv to describe the two currently recognized systems. For these systems, particularly at the Gm locus, an ever-increasing complexity has been developing in recent years. To avoid these complexities and the implication of possibly false relationships, we recommend a noncommittal, flexible numerical notation, similar to that suggested for the *Salmonella* antigens, Rh blood groups, and histocompatibility antigens in the mouse.

The assignment of numbers to the already known Gm and Inv factors has been done using the principles outlined below and considering also the order of their discovery and the known relationships of the Gm factors to the subgroups of the polypeptide chain of IgG.

Notation for the factors at
the Gm locus*

Original	New	Original	New
a	1	e	8
x	2	p	9
b^W and b^2	3	b^α	10
f	4	b^β	11
b and b^1	5	b^γ	12
c	6	b^3	13
r	7	b^4	14

Notation for the factors at
the Inv locus

Original	New
1 (lower-case L)	1
a	2
b	3

Either of two methods is suitable for recording the phenotype of an individual:

a. Record the result obtained with each set of reagents: e.g., Gm(1, − 2, 3, 4, 5); Inv(1, − 2, 3). Punctuation includes commas between the reactions with each set of reagents. This would mean that the sample was tested with reagents which correspond to the factors Gm(1), Gm(2), Gm(3), Gm(4), and Gm(5) and Inv(1), Inv(2), and Inv(3); and that it was positive with the test systems for Gm(1), Gm(3), Gm(4), and Gm(5), and negative for Gm(2); and positive for Inv(1) and (3), and negative for Inv(2).

b. Record only the positive results: e.g., Gm(1, 3, 4, 5); Inv(1, 3). In this case, the factors tested cannot be determined from the recorded phenotype and reference must be made to the list of reagents used for this information.

It is recognized that studies with several sets of reagents of Gm factors 7–14 inclusive have not yet been extensively published. Nor is the closeness of the relationship between Gm3 and Gm4 precisely known at the present time. These factors have nevertheless been assigned separate numbers here. Future decisions in the form of deletions of some numbers are foreseen.

The phenotypes of isolated discrete (e.g., myeloma) proteins may be recorded on the same basis as whole serum and by adding the symbol for the immunoglobulin subtype if known. For example: phenotypes

γ_2b: Gm(1, 2); or γWe: Gm(1, 2); κ: Inv(1); etc.

* A number has not been assigned to Gm (D) because it has been found to be an artifact and not a Gm factor.

Notation for Genes

Symbols for genes should be underlined when typed and italicized when printed. The factors determined by the gene should be indicated by superscript digits.

It is understood that the assignment of a genotype to a phenotype can be done only after genetic analysis, i.e., the observation of family segregation. Assignment of "the most probable" genotype in the absence of family data should be clearly indicated.

Examples:

Phenotype*	Population	Most probable genotype
Gm(1, 5)	Caucasoid	Gm^1/Gm^5
Gm(1, 5)	Negroid	$Gm^{1,5}/Gm^{1,5}$
Gm(1, 2)	Caucasoid or Mongoloid	$Gm^{1,2}/Gm^{1,2}$ or $Gm^{1,2}/Gm^1$

It should be emphasized that findings in other immunological systems indicate that serological specificities may appear in the phenotype which may not be represented in the genotype (that is, specificities resulting from interaction of different genes). Although such specificities have not yet been found in the human immunoglobulins, the proposed system of nomenclature may be used to record them.

The use of genetic symbols containing several numbers is a convenient notation to describe the inheritance of complex patterns of serological reactions (phenogroups). The symbols do not imply that regions corresponding to the *individual* numbers do or do not exist in the relevant genes. For example, a gene $Gm^{1,2}$ does not necessarily have part of its structure determining factor 1 and a different part determining factor 2.

Notation for New Systems

It is possible that new factors in addition to those currently known will be discovered and shown to be inherited independently of any of the known systems. If this occurs it seems advisable to relate the system symbol to the immunoglobulins and to the appropriate polypeptide chain and to rename the currently established Gm and Inv systems so that their symbols reflect the structural units they control. In this event, it is suggested that the symbol (I) be used in the description of all the loci for the immunoglobulins.

Notation for New Factors

There is no wish to restrict the freedom of investigators to publish data in any way they desire. On the other hand, the usefulness of the proposed system of notation will be seriously jeopardized if numerical symbols are added to the present list in an uncoordinated way. For these reasons, the following recommendations are made:

a. If the serological and family data are insufficient to meet the criteria for recognition of new factors set out below, either because of insufficient quantity of the reagent or because the frequency of the factor in the population precludes definitive family studies, it is recommended that the author give the new factor a provisional name using a two-letter abbreviation derived from town name, name of patient, etc. This should be preceded by the nonspecific generic term I, rather than a specific system symbol, e.g., I(pr) and I(−pr) for the phenotypes.

b. If the factor which has not yet been assigned a number is known to be part of

---

*All samples were tested for Gm(1), Gm(2), and Gm(5).

the Gm or Inv systems, it can be indicated in the parentheses after the numbered factors, for example: Gm(1, − 2, 3, 4, 5, pr) or Gm(1, 3, 4, 5, pr).

   c. If the author believes that the criteria are fulfilled and that the linkage relations have been established, the use of a number instead of the provisional name is warranted. To avoid duplicate use of the same number a request for a number should be sent to the WHO International Reference Center for Genetic Factors of Human Immunoglobulins, addressed to Dr. C. Ropartz, Directeur, Centre départemental de Transfusion sanguine et de Génétique humaine, 609, Chemin de la Bretèque, 76-Boi-Guillaume, France.

   d. If the factor is shown to belong to one of the already known linkage groups (systems), it would be assigned the next available number of that group, e.g., Inv(4).

   e. If the factor is shown to belong to a different system, it would be given the number 1. If it is located on a particular immunoglobulin class, the system symbol could be chosen according to the suggestions made above. In choosing the system symbol for a factor not localized on one particular class of immunoglobulins, transposition of the provisional letter into the system symbol might be useful. The phenotype I(pr) would in this instance be changed to IPr(1).

None of these suggestions should be taken to imply any formal control over the future use of numbers to describe factors inherited in the immunoglobulins. The sole intention is to avoid duplication of numbers or their use for poorly defined factors, and to assist in the rapid inclusion of reported factors into the proposed system of notation. Circulation of data relevant to their identification to other investigators would be helpful in this respect.

Criteria for Recognition of New Factors

To establish that a test system detects a "new" factor, it is recommended that the following conditions be fulfilled:

   a. The reaction system employed should show complete reproducibility. With the current hemaglutination-inhibition technique, the difference between "inhibiting" and "noninhibiting" normal sera from the same population should be three tubes or higher in a twofold dilution test system.

   b. Since antisera may contain mixtures of antibodies of different specificities, steps should be taken to show that the observed "new" reaction patterns are not due to an obvious mixture of antibodies of previously known specificities. (Absorption experiments using cells coated with selected antibodies, inhibition tests with purified myeloma proteins, and tests for consistency using cells coated with antibodies from different persons may give information concerning these matters.)

   c. It is essential to specify all reagents used in any study because,

   1. two reagents (antiglobulin and a cell coat) are necessary to define a factor;

   2. the same coated cells in some cases may be used for defining more than one factor; and

   3. the antiglobulin from one person may be used with different cell coats to define more than one factor. For example:

Factor	Antiglobulin	Dilution	Cell coat	Dilution
Gm(5)	Dr.*	1/8	Anti-D V.S.*	1/5
Gm(13)	Th.*	1/8	Anti-D V.S.*	1/5
Gm(6)	Bomb.*	1/32	Anti-D Warren*	1/10
Gm(5)	Bomb.*	1/32	Anti-D Berg*	1/10

*Identification of reagent.

d. Enough family data should be obtained to establish the pattern of inheritance of the new factor.

e. Tests with isolated immunoglobulins from positive and negative sera should be performed to show that the new factor is associated with immunoglobulins. Tests with agammaglobulinaemic sera might also be useful.

f. The reactions observed with the new set of reagents should be different in at least one race from those of any previously described set of reagents. It is suggested that a useful standard for deciding that the new factor differs from known factors is a χ^2 test of the comparisons of the reaction with different reagent sets, giving values with a $P \leqslant 0.02$. However, one clearly established and inherited difference, confirmed in several laboratories, may be acceptable as evidence for a new factor.

Reference Panel of Sera

In order to maintain uniformity in reagent sets used to detect established differences between the immunoglobulins of different individuals, the specificities of the reagent sets should be defined by their reactions with a reference panel of normal human sera. It is hoped that such a panel will be made available through the WHO International Reference Centre for Genetic Factors of Human Immunoglobulins.

This memorandum was drafted by the following signatories following discussions held during the meeting of the WHO Scientific Group on Genes, Genotypes and Allotypes of Immunoglobulins, that took place from 31 May to 5 June 1965 in Geneva. A French version will be published in a later issue.

R. Ceppellini, Institute of Medical Genetics, University of Turin, Turin, Italy; S. Dray, Department of Microbiology, University of Illinois, Chicago, Ill., USA; J. L. Fahey, National Cancer Institute, National Institutes of Health, Bethesda, Md., USA; E. C. Franklin, Department of Medicine, New York University School of Medicine, New York, N.Y., USA; H. Fudenberg, Department of Medicine, University of California, San Francisco Medical Center, San Francisco, Calif., USA; P. G. H. Gell, Department of Experimental Pathology, School of Medicine, University of Birmingham, Birmingham, England; H. C. Goodman, Immunology, World Health Organization, Geneva, Switzerland; R. Grubb, Institute of Bacteriology, Lund, Sweden; M. Harboe, Institutt for Eksperimentell Medisinsk Forskning, Oslo, Norway; R. L. Kirk, Human Genetics, World Health Organization, Geneva, Switzerland; J. Oudin, Institut Pasteur, Paris, France; C. Ropartz, Centre départemental de Transfusion sanguine et de Génétique humaine, 76-Bois-Guillaume, France; O. Smithies, Department of Medical Genetics, University of Wisconsin, Madison, Wis., USA; A. G. Steinberg, Case Western Reserve University, Cleveland, Ohio, USA; Z. Trnka, Immunology, World Health Organization, Geneva, Switzerland.

AN EXTENSION OF THE NOMENCLATURE FOR IMMUNOGLOBULINS*†

Earlier proposals for the nomenclature of human immunoglobulins[1-4] have been generally accepted. Recent studies of the remarkable heterogeneity of immunoglobulins and of antibodies have produced additional findings on human immunoglobulins which require the extension of the existing terminology. Certain new terms are therefore proposed for well-established properties of the structure and amino acid sequences of the heavy and light chains of these proteins. These additions to the terminology should help in defining those regions of antibody molecules which are responsible for their two major kinds of functions – i.e., the function of antigen binding and those other functions which are manifested by biological properties such as complement fixation.

Definition of the Regions of Immunoglobulin Molecules

It is now clear that the polypeptide chains of immunoglobulins consist of two well-defined regions which are here designated "the variable region" and "the constant region." The varabile region has been so designated because of the diversity of its amino acid sequences, whereas the constant region is relatively invariable in molecules of the same class and type. It is recommended that the variable and constant regions be termed the "V region" and the "C region," respectively. It is proposed that the symbols "V_L" and "C_L" be generic terms for the variable and the constant regions of light chains and that "V_H" and "C_H" be generic terms for the corresponding regions of heavy chains. If it is desired to specify a particular class or subclass of heavy chain, the symbol "H" could be replaced by the symbol of the chain. For example, the C_H region of a heavy chain of a molecule of the IgG1 subclass would be designated "$C_{\gamma 1}$ region." Similarly, if it is desired to specify a particular type of light chain, the symbol "L" could be replaced by the symbol for the chain: for example, "V_κ region," "C_γ region."**

The exact length of these regions cannot be specified. In human light chains the V_L and C_L regions have about the same number of amino acid residues. For example, in the κ chains so far studied the V_L regions consist of 107 to 113 residues beginning at the amino terminus and the C_L regions consist of the remaining 107 residues. Heavy chains have been less extensively studied than have light chains. For gamma chains the available data on complete sequences of human heavy chains show that the V_H region is of approximately the same length as the V_L region of the same molecule and that the C_H region consists of 3 linearly arranged adjacent regions which show homologies with C_L and with each other. Less extensive data on other human gamma chains and on rabbit gamma chains suggest that this will be a general finding in chains of the gamma class. It is

---

*This memorandum was drafted following discussions held during a WHO meeting on the nomenclature of immunoglobulins that took place on 9–11 June 1969 in Prague. The signatories are R. Asofski, National Institute of Allergy and Infectious Diseases, Bethesda, Md., R. A. Binaghi, Laboratoire de Médecine expérimentale, Collège de France, Paris, France, G. M. Edelman, The Rockefeller University, New York, N.Y., H. C. Goodman, Immunology, World Health Organization, Geneva, Switzerland, J. F. Heremans, Cliniques universitaires St. Pierre, Université catholique de Louvain, Louvain, Belgium, L. Hood, National Cancer Institute, Bethesda, Md., E. A. Kabat, Columbia-Presbyterian Medical Center, New York, N.Y., J. Rejnek, Department of Immunology, Institute of Microbiology, Prague, Czechoslovakia, D. S. Rowe, Director, WHO International Reference Centre for Immunoglobulins, Lausanne, Switzerland, P. A. Small, Jr., Department of Microbiology, College of Medicine, University of Florida, Gainesville, Fla., Z. Trnka, Immunology, World Health Organization, Geneva, Switzerland.
†See revision: Recommendations for the Nomenclature of Human Immunoglobins.
**The symbol "L" has been suggested previously for the L type of immunoglobulins [*Bull. W.H.O.*, 30, 447 (1964)]. The usage of the subscript "$_L$," as in "V_L" and "C_L" is distinct: Subscript "$_L$" specifies light chains irrespective of type.

therefore proposed that such regions in γ1 heavy chains, for example, be named "homology region $C_{\gamma 1}1$," "homology region $C_{\gamma 1}2$," "homology region $C_{\gamma 1}3$," and so on, beginning with the homology region which is closest to the amino terminus of the chain. If similar homology regions are found in other classes, the appropriate chain symbol would be employed, e.g., "homology region $C_\alpha 1$," "homology region $C_\mu 1$," etc.

Nomenclature of Half-cystinyl Residues

It is suggested that half-cystinyl residues be designated by Roman numerals, the first residue being that which is closest to the amino terminus. An Arabic numeral (to be placed as a subscript) corresponds to the number of the amino acid residue in the chain being described. This will satisfactorily permit the representation of intrachain and interchain disulfide bonds as well as of free sulfhydryl groups. Intrachain bonds can be designated, for example "disulfide bond $\kappa(I_{23}-II_{88})$." Interchain bonds can be designated as follows: "Disulfide bond $\kappa V_{214}-\gamma IV_{200}$" and "disulfide bond $\gamma 1VI_{226}-\gamma 1VI_{226}$." In all matters of chemical terminology the rules of IUPAC-IUB[5] should be observed.

Definition of Groups and Subgroups

All variable regions associated with a light chain of given type are defined as a "group." Subdivisions may be distinguished within a group. These subdivisions are called "subgroups" and are designated by Roman numerals. In human κ-chain groups at least 3 discrete sets of sequences have been recognized. These are now called "subgroup $V_{\kappa I}$," "subgroup $V_{\kappa II}$," and "subgroup $V_{\kappa III}$." Similarly, in human γ chains at least five sets of sequences have now been recognized: These are called "subgroup $V_{\lambda I}$," etc. The correspondence between the present usage and the proposed usage is shown in the accompanying table. The subgroups have been designated I, II, and III, etc., in order of their frequency of occurrence in the chains which have so far been studied. This follows the nomenclature of subclasses of heavy chains of IgG, which are also numbered in order of frequency of occurrence.[3] Prototype sequences of 20 residues of the subgroups beginning at the amino terminus are given in the Annex to this paper.

Definition of Classes and Subclasses and Types and Subtypes

C_H and C_L regions show heterogeneity. This is the basis of the recognition of classes of heavy chains and types of light chains. Subdivisions of classes have previously been designated subclasses. It is proposed that similar subdivisions of types be designated "subtypes" and that the principles for their nomenclature should be the same as those for subclasses. The terms "classes" and "subclasses," "types," and "subtypes" specifically exclude differences arising owing to allelism.

Formulas for Immunoglobulin Molecules

The over-all structure of immunoglobulins may be represented by formulas. Thus, an IgG1 molecule with κ light chains of subgroup I could be written

$$[(V_{\kappa I}C_\kappa)(V_\gamma C_{\gamma 1})]$$

If it is desired to include homology regions, this could be expanded to

$$[(V_{\kappa I}C_\kappa)(V_\gamma C_{\gamma 1}1 C_{\gamma 1}2 C_{\gamma 1}3)]_2$$

Presence inside the square brackets indicates association of light and heavy chains, the formula for each of which is in parentheses. The subscript 2 outside the brackets indicates that the association of 2 such units forms the molecule. An IgM molecule with κ light chains may be represented, by use of the brace, as

$$\{[(V_\kappa C_\kappa)(V_\mu C_\mu)]_2\}_5$$

Formulas could be especially useful in representing certain γA immunoglobulins in which the two light chains are linked as $(V_\kappa C_\kappa)_2$ and the heavy chains as $(V_\alpha C_\alpha)_2$ to give

$$[(V_\kappa C_\kappa)_2 (V_\alpha C_\alpha)_2]$$

Immunoglobulins of Secretions

The IgA in the external secretions of man and several other species has a molecular structure characterized by the presence of an additional component which is lacking in IgA derived from serum. This additional component is sometimes encountered not bound to IgA. It is suggested that it be designated "secretory component" or "S component" and that other terms such as "transport piece" should not be used. The term "secretory" in the definition is not meant to imply any function but rather to indicate the characteristic association with secretions of IgA possessing this additional component.

LIGHT CHAIN SUBGROUPS

Light chain	Present usage	Proposed usage
κ chains[6-8]	$S\kappa_I$, Tra, basic group I	$V_\kappa I$
	$S\kappa_{II}$,[a] Smi,[a] basic group III	$V_\kappa II$
	$S\kappa_{II}$,[a] Smi,[a] basic group	$V_\kappa III$
λ chains[9,10]	I, $S\lambda_{II}$	$V_\lambda I$
	II,[a] $S\lambda_I$	$V_\lambda II$
	IV, other	$V_\lambda III$
	II,[a] $S\lambda_{III}$	$V_\lambda IV$
	III, other	$V_\lambda V$

Contributed by D. S. Rowe, *Bull. W.H.O.*, (1970).

[a] These subgroups were not further separated at the time of publication.

REFERENCES

1. *Bull. W.H.O.*, 30, 447 (1964).
2. *Bull. W.H.O.*, 33, 721 (1965).
3. *Bull. W.H.O.*, 35, 953 (1966).
4. *Bull. W.H.O.*, 38, 151 (1968).
5. IUPAC-IUB Commission on Biochemical Nomenclature, *J. Biol. Chem.*, 241, 527, 2491 (1966).
6. **Hood, Gray, Sanders, and Dreyer,** *Cold Spring Harbor Symp. Quant. Biol.*, 32, 133 (1967).
7. **Niall and Edman,** *Nature*, 216, 263 (1967).
8. **Milstein,** *Nature*, 216, 330 (1967).
9. **Langer, Steinmetz-Kayne, and Hilschmann,** *Hoppe-Seyler's Z. Physiol. Chem.*, 349, 945 (1968).
10. **Hood and Ein,** *Nature*, 220, 764 (1968).

APPENDIX

PROTOTYPE AMINO-TERMINUS SEQUENCES OF THE SUBGROUPS OF HUMAN κ AND λ LIGHT CHAINS

Subgroups	Sequence[a]
	κ chains
$V_{\kappa I}$	Asp-Ile-Gln-Met-Thr-Gln-Ser-Pro-Ser-Ser-Leu-Ser-Ala-Ser-Val-Gly-Asp-Arg-Val-Thr-
$V_{\kappa II}$	Glu-Ile-Val-Leu-Thr-Gln-Ser-Pro-Gly-Thr-Leu-Ser-Leu-Ser-Pro-Gly-Glu-Arg-Ala-Thr-
$V_{\kappa III}$	Asp-Ile-Val-Met-Thr-Gln-Ser-Pro-Leu-Ser-Leu-Pro-Val-Thr-Pro-Gly-Glu-Pro-Ala-Ser-
	λ chains
$V_{\lambda I}$	pGlu-Ser-Val-Leu-Thr-Gln-Pro-Pro-[ ]-Ser-Val-Ser-Gly-Ala-Pro-Gly-Gln-Arg-Val-Thr-
$V_{\lambda II}$	pGlu-Ser-Ala-Leu-Thr-Gln-Pro-Ala-[ ]-Ser-Val-Ser-Gly-Ser-Pro-Gly-Gln-Ser-Ile-Thr-
$V_{\lambda III}$	Tyr-Val-Leu-Thr-Gln-Pro-Pro-[ ]-Ser-Val-Ser-Val-Ser-Pro-Gly-Gln-Thr-Ala-Ser-
$V_{\lambda IV}$	pGlu-Ser-Ala-Leu-Thr-Gln-Pro-Pro-[ ]-Ser-Ala-Ser-Gly-Ser-Pro-Gly-Gln-Ser-Val-Thr-
$V_{\lambda V}$	Ser-Glu-Leu-Thr-Gln-Pro-Pro-[ ]-Ala-Val-Ser-Val-Ala-Leu-Gly-Gln-Thr-Val-Arg-

[a] pGlu, the residue derived from pyrollid-2-one-5-carboxylic acid.

TENTATIVE NOMENCLATURE FOR BLOOD COAGULATION FACTORS

In the field of blood coagulation research a number of different names have often been used for what appears to be a single factor. The International Committee for the Nomenclature of Blood Clotting Factors (now the International Committee on Haemostasis and Thrombosis of the International Society on Thrombosis and Haemostasis) has adopted for some of the blood coagulation factors a tentative nomenclature which has been widely used. The terms that have been adopted are given with a partial list of synonyms arranged alphabetically. Details on the deliberations of the Committee and the Sub-committee reports were published in *Thrombosis et Diathesis Haemorrhagica* 3, 435 (1959); Suppl 1, 265 (1960); Suppl 5, 347 (1962); and Suppl 13, 411 and 419 (1964).

Committee Nomenclature	Synonyms and Symbols
Factor I	Fibrinogen
Factor II	Prothrombin
Factor III	Thromboplastin (Tissue Thromboplastin)
Factor IV	Calcium ions
Factor V	Ac-globulin (AcG) Labile factor Proaccelerin
(Factor VI not assigned)	
Factor VII	Autoprothrombin I Proconvertin Serum prothrombin conversion accelerator (Spca) Stable factor
Factor VIII	Antihemophilic factor (AHF) Antihemophilic globulin (AHG) Facteur antihémophilique A Plasma thromboplastic factor Platelet cofactor I Thromboplastinogen
Factor IX	Autoprothrombin II Christmas factor Facteur antihémophilique B Plasma thromboplastic factor B Plasma thromboplastin component (PTC) Platelet cofactor II
Factor X	Prower factor Stuart factor Stuart-Prower factor
Factor XI	Plasma thromboplastin antecedent (PTA)
Factor XII	Hageman factor
Factor XIII	Fibrin stabilizing factor (FSF) Fibrinase Laki-Lorand factor (L-L factor)

Contributed by Robert H. Wagner.

Proteins

CYANOGEN BROMIDE CLEAVAGE OF PEPTIDES AND PROTEINS

Peptide or protein	Mol wt	Methionine residues	Solvent	Reaction conditions[a] BrCN Met	Temperature (°C)	Time (hr)	Reference
Tryptophan synthetase, fragment	2,000	1	0.1 N HCl	30	R.T.	24	1
Gastrin	2,000	2	Aqueous HF	X's	R.T.	24	2
Ribonuclease, bovine							
S peptide of bovine	2,160	1	0.1 N HCl	30	R.T.	24	3
Aldolase, active site peptide, rabbit muscle	3,000	1	0.1 N HCl	X's	R.T.	24	4
Cholecystokinin-pancreozymin	3,500	3	0.1 N HCl	X's	21	24	5
Thyrocalcitonin, porcine	3,600	1	70% HCOOH	180	25	24	6
Thyrocalcitonin, bovine	3,600	1	70% HCOOH	50	25	24	7
Cytochrome-c 551, *Pseudomonas*	8,000	2	0.1 N HCl	<30	R.T.	24	8
Hormone, parathyroid	9,000	2	0.1 N HCl	X's	R.T.	24	9
Acyl carrier protein, *E. coli*	9,000	1	99% HCOOH	200	25	24	10
Ferredoxin, green alga	10,200	1	70% HCOOH	85	25	24	11
Histone, calf thymus (GAR)	10,600	1	—	—	—	—	12, 13
Cytochrome-c	12,000	2	0.1 N HCl	X's	40	24	14
Cytochrome-c, baker's yeast	12,000	2	0.1 N HCl	X's	36	24	15, 16
Cytochrome-c, horse-heart	12,000	2	—	—	—	—	17
Cytochrome-c, wheat germ	12,000	2	—	X's	—	—	18
Cytochrome-c_2, *Rhodospirillum rubrum*	12,840	2	70% HCOOH	50	30	30	19
Thioredoxin	12,000	1	(HCl) (F_3CCOOH) 70% HCOOH	300	25	24	20
Ribonuclease, bovine pancreatic	13,700	4	0.1 N HCl	30	25	24	21
Ribonuclease, cross-linked bovine pancreatic	13,700	4	0.1 N HCl	—	R.T.	24	22
Ribonuclease A, S-methylmethionine-29	13,700	4	0.1 N HCl	30	R.T.	24	23
Azurin	14,000	6	0.1 N HCl	X's	R.T.	24	24
Lysozyme, hen egg	14,400	2	70% HCOOH	X's	—	—	25
Lysozyme, Bacteriophage λ	—	3 (one N-term)	0.1 N HCl	90–100	30	30	26
α-Lactalbumin	16,000	1	60% HCOOH	X's	R.T.	—	27
Nuclease, extracellular *Staph. aureus*	16,500	4	70% HCOOH	30	25	20	28
Myoglobin	18,000	2	0.1 N HCl	40	R.T.	24	29
Growth hormone, human	21,500	3	70% HCOOH	X's	R.T.	—	30
Growth hormone, bovine	20,800	3	70% HCOOH	X's	25	24	31, 32
Trypsinogen, bovine pancreatic	24,000	2	0.2 N HCl	40	30	30	33
Chymotrypsin	25,000	2	0.1 N HCl	<30	R.T.	24	34
Carboxypeptidase A	34,600	3	70% HF	X's	R.T.	24	35
S-Sulfopepsin	35,000	4	80% HCOOH	60	37	20	36
Streptokinase, *Streptococcus*	48,000	4	70% HCOOH	250	23	24	37
Kininogen, human	50,000	2	0.25 N HCl	5	35	20	38
Collagen	100,000	6–8	0.1 N HCl	100	30	15	39, 40
Collagen, rat skin							
α 1-chain	93,000	7	0.1 N HCl	100	30	4	41
α 2-chain	95,000	5	0.1 N HOAc (HCl)	200	30	4	42
Collagen, chick skin							
α 1-chain	95,000	9	0.1 N HCl	100	30	4	43
Collagen, chick bone							
α 1-chain	95,000	9	0.1 N HCl	150	30	4	44
α 2-chain	95,000	5	0.1 N HCl	150	30	4	45
β-Galactosidase, *E. coli*	135,000	24	70% HCOOH	50	R.T.	16–20	46

[a] X's signifies excess reagent: R.T., room temperature.

CYANOGEN BROMIDE CLEAVAGE OF PEPTIDES AND PROTEINS (continued)

Peptide or protein	Mol wt	Methionine residues	Solvent	BrCN Met	Temperature (°C)	Time (hr)	Reference
Immunoglobulin IgG, rabbit, partial cleavage, active fragment	~150,000	10	0.3 N HCl	200	R.T.	4	47
complete cleavage, inactive fragment	~150,000	10	70% HCOOH	X's	R.T.	24	
Immunoglobulin IgG, human	150,000	20	70% HCOOH	X's	R.T.	4	48–52
Aldolase, rabbit muscle (4 units of 40,000)	160,000	4 × 3	70% HCOOH	30	R.T.	22	4
Myosin A	500,000	130	0.1 N HCl	>50	25	22	53
Thyroglobulin, native (2 subunits of 330,000)	660,000	46	70% HCOOH	150	25	20	54

Reaction conditions[a]

Compiled by T. F. Spande and B. Witkop.

REFERENCES

1. Guest and Yanofsky, *J. Biol. Chem.*, 241, 1 (1966).
2. Gregory, Hardy, Jones, Kennel, and Sheppard, *Nature*, 204, 931 (1964).
3. Gross and Witkop, *Biochemistry*, 6, 745 (1967).
4. Lai, C. Y., Hoffee, P., and Horecker, B. L., personal communication. 1969.
5. Mutt and Jorpes, *Biochem. Biophys. Res. Commun.*, 26, 392 (1967).
6. Potts, Niall, Keutmann, Brewer, and Debtos, *Proc. Natl. Acad. Sci. U.S.A.*, 59, 1321 (1968).
7. Brewer and Ronan, *Proc. Natl. Acad. Sci. U.S.A.*, 63, 940 (1969).
8. Ambler, *Biochem. J.*, 96, 32P (1965).
9. Potts and Aurbach, in *The Parathyroid Glands: Ultrastructure, Secretion and Function*, Gaillard, Talmadge, and Budy, Eds., University of Chicago Press, Chicago, 1965.
10. Vanaman, Wakil, and Hill, *J. Biol. Chem.*, 243, 6409 (1968).
11. Sugeno and Matsubara, *J. Biol. Chem.*, 244, 2979 (1969).
12. Sautiere, Starbuck, and Busch, *Fed. Proc.*, 27, 777 (1968).
13. DeLange and Fambrough, *Fed. Proc.*, 27, 392 (1968).
14. Chu and Yasunobu, *Biochim. Biophys. Acta*, 89, 148 (1964).
15. Narita, Titani, Yaoi, and Murakami, *Biochim. Biophys. Acta*, 77, 688 (1963).
16. Titani and Narita, *J. Biochem.* (Tokyo), 56, 241 (1964).
17. Blacj and Leaf, *Biochem. J.*, 96, 693 (1964).
18. Stevens, Glazer, and Smith, *J. Biol. Chem.*, 242, 2764 (1967).
19. Dus, Sletten, and Kamen, *J. Biol. Chem.*, 243, 5507 (1968).
20. Holmgren and Reichard, *Eur. J. Biochem.*, 2, 187 (1967).
21. Gross and Witkop, *J. Biol. Chem.*, 237, 1856 (1962).
22. Marfey, Uziel, and Little, *J. Biol. Chem.*, 240, 3270 (1965).
23. Link and Stark, *J. Biol. Chem.*, 243, 1082 (1968).
24. Ambler and Brown, *J. Mol. Biol.*, 9, 825 (1964).
25. Bonavida, Miller, and Sercarz, *Fed. Proc.*, 26, 339 (1967); *Biochemistry*, 8, 968 (1969).
26. Black and Hogness, *J. Biol. Chem.*, 244, 1982 (1969).
27. Brew, Vanaman, and Hill, *J. Biol. Chem.*, 242, 3747 (1967).
28. Taniuchi, Anfinsen, and Sodja, *J. Biol. Chem.*, 242, 4736 (1967).
29. Edmundson, *Nature*, 198, 354 (1963).
30. Li, Liu, and Dixon, *J. Am. Chem. Soc.*, 88, 2050 (1966).
31. Fellows, Nutting, Rogol, and Kostyo, *Fed. Proc.*, 27, 434 (1968).
32. Fellows and Rogol, *J. Biol. Chem.*, 244, 1567 (1969).
33. Hofmann, *Biochemistry*, 3, 356 (1965).
34. Koshland, Strumeyer, and Ray, *Brookhaven Symp. Biol.*, 15, 101 (1962).
35. Bargetzi, Thompson, Sampath, Kumar, Walsh, and Neurath, *J. Biol. Chem.*, 239, 3767 (1964).
36. Kostka, Moravek, Kluh, and Keil, *Biochim. Biophys. Acta*, 175, 459 (1969).
37. Morgan and Henschen, *Biochim. Biophys. Acta*, 181, 93 (1969).
38. Axen, Gross, Witkop, Pierce, and Webster, *Biochem. Biophys. Res. Commun.*, 23, 92 (1966).

CYANOGEN BROMIDE CLEAVAGE OF PEPTIDES AND PROTEINS (continued)

39. Bornstein and Piez, *Biochemistry,* 5, 3460 (1966).
40. Nordwig and Dick, *Biochim. Biophys. Acta,* 97, 179 (1965).
41. Butler, Piez, and Bornstein, *Biochemistry,* 6, 3771 (1967).
42. Fietzek and Piez, *Biochemistry,* 8, 2129 (1969).
43. Kang, Piez, and Gross, *Biochemistry,* 8, 1506 (1969).
44. Miller, Lane, and Piez, *Biochemistry,* 8, 30 (1969).
45. Lane and Miller, *Biochemistry,* 8, 2134 (1969).
46. Steers, Craven, Anfinsen, and Bethune, *J. Biol. Chem.,* 240, 2478 (1969).
47. Cahnmann, Arnon, and Sela, *J. Biol. Chem.,* 240, PC 2762 (1965).
48. Waxdal, Konigsberg, and Edelman, *Biochemistry,* 7, 1967 (1968).
49. Yoo and Pressman, *Fed. Proc.,* 27, 683 (1968).
50. Press, Piggott and Porter, *Biochem. J.,* 99, 356 (1966).
51. Wikler, Köhler, Sinoda, and Putnam, *Fed. Proc.,* 27, 559 (1968).
52. Edelman, Cunningham, Gall, Gottlieb, Rutishauser, and Waxdal, *Proc. Natl. Acad. Sci. U.S.A.,* 63, 78 (1969).
53. Young, Blanchard, and Brown, *Proc. Natl. Acad. Sci. U.S.A.,* 61, 1087 (1968).
54. Nissley, Cittanova, and Edelhoch, *Biochemistry,* 8, 443 (1969).

This table originally appeared in Sober, Ed., *Handbook of Biochemistry and selected data for Molecular Biology,* 2nd ed., Chemical Rubber Co., Cleveland, 1970.

CYANOGEN BROMIDE CLEAVAGE OF PEPTIDES AND PROTEINS – COLLAGEN

Peptide or protein	Molecular weight	Methenine residues	Solvent	BrCN Met	Temp °C	Time (hr)	Ref.
Collagen	100,000	6–8	0.1 N HCl	100	30	15	1,2
Rat skin							
α1(I) chain	93,000	7	0.1 N HCl	100	30	4	3
α2 chain	95,000	5	0.1 N HOAC (HCl)	200	30	4	4
Chick skin							
α1(I) chain	95,000	9	0.1 N HCl	100	30	4	5
Chick bone							
α1(I) chain	95,000	9	0.1 N HCl	150	30	4	6
α2 chain	95,000	5	0.1 N HCl	150	30	4	6
Human skin							
α1(I) chain	95,000	7	70% formic acid	150	30	4	7
α2 chain	95,000	5	70% formic acid	150	30	4	7
Human cartilage							
α1(II) chain	95,000	9	70% formic acid	150	30	4	8
Human skin							
α1(III) chain	95,000	8	70% formic acid	150	30	4	9

Compiled by Andrew H. Kang and Jerome M. Seyer.

REFERENCES

1. Bonnstein and Piez, *Biochemistry,* 5, 3460 (1966).
2. Nordwig and Dick, *Biochem. Biophys. Acta,* 97, 179 (1965).
3. Butler, Piez, and Bonnstein, *Biochemistry,* 6, 3771 (1967).
4. Kang, Piez, and Gross, *Biochemistry,* 8, 1506 (1969).
5. Miller, Lane, and Piez, *Biochemistry,* 8, 30 (1969).
6. Lane and Miller, *Biochemistry,* 8, 2134 (1969).
7. Epstein, Scott, Miller, and Piez, *J. Biol. Chem.,* 246, 1718 (1971).
8. Miller and Lunde, *Biochemistry,* 12, 3153 (1974).
9. Chung, Keele, and Miller, *Biochemistry,* 13, 3458 (1974).

SPECIFICITY OF REAGENTS COMMONLY USED TO CHEMICALLY MODIFY PROTEINS

	Amino	Imidazole	Guanidinyl	Indole	Thio Ether	Disulfide	Sulfhydryl	Hydroxyl	Phenol	Carboxyl	Reference
Acetic anhydride	+	±	−	−	−	−	+	±	+	−	1
2-Acetoxy-5-nitrobenzyl chloride	−	−	−	+	−	−	+	−	+	−	2
N-Acetylimidazole	±	−	−	−	−	−	+	−	+	−	3
Acrylonitrile	+	−	−	−	−	−	+	−	−	−	4
Aldehyde/NaBH$_4$	+	−	−	−	−	−	−	−	−	−	5
Azobenzene-2-sulfenyl bromide	−	−	−	+	−	−	+	−	−	−	6
Bromoacetamido-4-nitrophenol	−	−	−	−	+	−	−	−	−	−	7
D,L-α-Bromo-β-(5-imidazolyl) propionic acid	−	−	−	−	−	−	+	−	−	−	8
N-Bromosuccinimide	−	+	−	+	−	−	+	−	+	−	9
Butanedione	±	−	+	−	−	−	−	−	−	−	10
Carbodiimides, water soluble	−	−	−	−	−	−	−	−	−	+	11
N-Carboxyanhydrides	+	−	−	−	−	−	−	−	−	−	12
NBD chloride	+	−	−	−	−	−	+	−	−	−	13
Citraconic anhydride	+	−	−	−	−	−	+	±	−	−	14
Cyanate	+	−	−	−	−	−	+	+	−	−	15
Cyanogen bromide	−	−	−	−	+	−	+	−	−	−	16
1,2-Cyclohexanedione	+	−	+	−	−	−	−	−	−	−	17
Diacetyl (trimer, dimer)	+	−	+	−	−	−	−	−	−	−	18
Diazoacetates, diazomethane	−	−	−	−	−	−	+	−	−	+	19
Diazonium fluoroborates	+	+	−	−	−	−	−	−	+	−	20
Diazonium salts	+	+	−	±	−	−	−	−	+	−	21
Diazonium 1-H-tetrazole	+	+	−	−	−	−	+	−	+	−	22
1,5-Difluoro-2,4-dinitrobenzene	+	+	−	−	−	−	+	−	+	−	23
4,4′-Difluoro-3,3′-dinitrophenyl sulfone	+	−	−	−	−	−	−	−	+	−	24
Diketene	+	−	−	−	−	−	+	±	+	−	25
Dimethyl adipimidate dimethyl suberimidate	+	−	−	−	−	−	−	−	−	−	26
4,4′-Bis-dimethylaminodiphenyl carbinol	−	−	−	−	−	−	+	−	−	−	27
5-Dimethyaminonapthalenesulfonyl chloride	+	+	−	−	−	−	+	−	+	−	28
Dimethyl(2-methoxy-5-nitrobenzyl) sulfonium bromide	−	−	−	+	−	−	−	−	−	−	29
2,4-Dinitro-5-fluoroaniline	+	+	−	−	−	−	+	−	+	−	30
Dinitrofluorobenzene	+	+	−	−	−	−	+	−	+	−	31
5,5′-Dithiobis(2-nitrobenzoic) acid	−	−	−	−	−	−	+	−	−	−	32
Ethoxyformic anhydride	+	+	−	−	−	−	−	−	+	−	33
Ethyleneimine	−	−	−	−	−	−	+	−	−	−	34
N-Ethylmaleimide	+	−	−	−	−	−	+	−	−	−	35
Ethyl thiotrifluoroacetate	+	−	−	−	−	−	−	−	−	−	36
Fluorescein mercuric acetate	−	−	−	−	−	−	+	−	−	−	37
Formaldehyde	+	+	+	+	−	−	+	−	+	−	19
Glyoxal	+	−	+	−	−	−	−	−	−	−	38
Guanyl-3,5-dimethylpyrazole nitrate	+	−	−	−	−	−	−	−	−	−	39
Haloacetates	+	+	−	−	+	−	+	−	−	+	40
Hydrogen peroxide	−	−	−	+	+	+	+	−	−	−	41
2-Hydroxy-5-nitrobenzylbromide	−	−	−	+	−	−	+	−	−	−	42
N-Hydroxysuccinimide esters	+	−	−	−	−	−	−	−	+	−	43

SPECIFICITY OF REAGENTS COMMONLY USED TO CHEMICALLY MODIFY PROTEINS (continued)

	Amino	Imidazole	Guanidinyl	Indole	Thio Ether	Disulfide	Sulfhydryl	Hydroxyl	Phenol	Carboxyl	Reference
Iodine	−	+	−	+	−	−	+	−	+	−	19
Iodine monochloride	−	+	−	+	−	−	+	−	+	−	44
Iodoacetamide	−	+	−	−	−	−	+	−	−	+	45
O-Iodosobenzoate	−	−	−	−	−	−	+	−	−	−	46
Maleic anhydride	+	+	−	−	−	−	+	−	±	−	47
Mercurials, heavy metals	−	+	−	−	−	−	+	−	−	−	19
P-Mercuribenzoate	−	−	−	−	−	−	+	−	−	−	35
Methanol/HCl	−	−	−	−	−	−	−	−	−	+	48
2-Methoxy-5-nitrotropone	+	−	−	−	−	−	−	−	−	−	49
Methyl acetimidate	+	−	−	−	−	−	−	−	−	−	50
O-Methylisourea	+	−	−	−	−	−	−	−	−	−	51
Methyl 4-mercaptobutyrimidate HCl	−	−	−	−	−	−	+	−	−	−	52
Methyl p-nitrobenzene sulfonate	−	−	−	−	−	−	+	−	−	−	53
O-Nitrophenylsulfenyl chloride	−	−	−	+	−	−	+	−	−	−	54
BNPS-skatole	−	−	−	+	−	−	+	−	−	−	55
Nitrosyldisulfonate, K	−	−	−	−	−	−	−	−	+	−	56
Nitrous acid	+	−	−	−	−	−	+	−	+	−	19
Performic acid	−	−	−	+	+	+	+	−	−	−	57
Phenylglyoxal	+	−	+	−	−	−	−	−	−	−	58
Photooxidation	−	+	−	+	+	−	−	−	+	−	59
Salicylaldehyde	+	−	−	−	−	−	−	−	−	−	60
Sodium borohydride	−	−	−	−	−	+	−	−	−	−	61
Succinic anhydride	+	−	−	−	−	−	+	±	+	−	62
N-Succinimidyl 3-(4-hydroxyphenyl) propionate	+	−	−	−	−	−	−	−	−	−	63
Sulfenyl halides	−	−	−	+	−	−	+	−	−	−	64
Sulfite	−	−	−	−	−	+	+	−	−	−	19
Sulfonyl halides	+	+	−	−	−	−	+	−	+	−	65
Tetranitromethane	−	−	−	+	+	−	+	−	+	−	66
Tetrathionate	−	−	−	−	−	−	+	−	−	−	67
Thiols	−	−	−	−	−	+	−	−	−	−	68
2,4,6-Tribromo-4-methyl cyclohexanedione	−	−	−	+	−	−	−	−	−	−	69
Trinitrobenzenesulfonic acid	+	−	−	−	−	−	+	−	−	−	70

Compiled by James F. Riordan.

REFERENCES

1. Riordan and Vallee, *Methods Enzymol.*, 25, 494 (1972).
2. Horton and Koshland, *Methods Enzymol.*, 25, 468 (1972).
3. Riordan and Vallee, *Methods Enzymol.*, 25, 500 (1972).
4. Riehm and Scheraga, *Biochemistry*, 5, 93 (1966).
5. Means and Feeney, *Biochemistry*, 7, 2192 (1968).
6. Fontana and Scoffone, *Methods Enzymol.*, 25, 482 (1972).
7. Hille and Koshland, *J. Am. Chem. Soc.*, 89, 5945 (1967).
8. Yankeelov and Jolley, *Biochemistry*, 11, 159 (1972).
9. Witkop, *Adv. Protein Chem.*, 16, 261 (1961).
10. Riordan, *Biochemistry*, 12, 3915 (1973).
11. Carraway and Koshland, *Methods Enzymol.*, 25, 616 (1972).

SPECIFICITY OF REAGENTS COMMONLY USED TO CHEMICALLY MODIFY PROTEINS (continued)

12. Sela and Arnon, *Methods Enzymol.*, 25, 553 (1972).
13. Faber et al., *Anal. Biochem.*, 53, 290 (1973).
14. Atassi and Habeeb, *Methods Enzymol.*, 25, 546 (1972).
15. Stark, *Biochemistry*, 4, 588, 1030 (1965).
16. Gross, *Methods Enzymol.*, 11, 238 (1967).
17. Patthy and Smith, *J. Biol. Chem.*, 250, 557 (1975).
18. Yankeelov et al., *J. Am. Chem. Soc.*, 90, 1664 (1968).
19. Herriott, *Adv. Protein Chem.*, 3, 169 (1947).
20. Wofsy et al., *Biochemistry*, 1, 1031 (1962).
21. Riordan and Vallee, *Methods Enzymol.*, 25, 521 (1972).
22. Sokolovsky and Vallee, *Biochemistry*, 5, 3574 (1966).
23. Cuatrecasas et al., *J. Biol. Chem.*, 244, 406 (1968).
24. Zahn and Zuber, *Berufsdermatosen*, 86, 172 (1953).
25. Singhal and Atassi, *Biochemistry*, 10, 1756 (1971).
26. Hunter and Ludwig, *Methods Enzymol.*, 258, 585 (1972).
27. Rohrback, *Anal. Biochem.*, 52, 127 (1973).
28. Gray, *Methods Enzymol.*, 11, 139 (1967).
29. Horton and Tucker, *J. Biol. Chem.*, 245, 3397 (1970).
30. Bergman and Bentov, *J. Org. Chem.*, 26, 1480 (1961).
31. Sanger, *Biochem. J.*, 42, 287 (1948).
32. Ellman, *Arch. Biochem. Biophys.*, 82, 70 (1959).
33. Rosen and Fedoresak, *Biochim. Biophys. Acta*, 130, 401 (1966).
34. Cole, *Methods Enzymol.*, 11, 315 (1967).
35. Riordan and Vallee, *Methods Enzymol.*, 25, 449 (1972).
36. Goldberger, *Methods Enzymol.*, 11, 317 (1967).
37. Karush et al., *Anal. Biochem.*, 9, 100 (1964).
38. Nayaka et al., *Biochim. Biophys. Acta*, 194, 301 (1969).
39. Habeeb, *Biochim. Biophys. Acta*, 34, 294 (1959).
40. Gurd, *Methods Enzymol.*, 11, 532 (1967).
41. Neumann, *Methods Enzymol.*, 25, 393 (1972).
42. Koshland et al., *J. Am. Chem. Soc.*, 86, 1448 (1964).
43. Blumberg et al., *Isr. J. Chem.*, 12, 643 (1974).
44. Koshland et al., *J. Biol. Chem.*, 238, 1343 (1963).
45. Gurd, *Methods Enzymol.*, 25, 424 (1972).
46. Hellerman et al., *J. Am. Chem. Soc.*, 63, 2551 (1941).
47. Butler et al., *Biochem. J.*, 112, 679 (1969).
48. Means and Feeney, *Chemical Modification of Proteins*, Holden-Day, San Francisco, 1971, 139.
49. Tamaoki et al., *J. Biochem.* (Tokyo), 62, 7 (1967).
50. Hunter and Ludwig, *J. Am. Chem. Soc.*, 84, 3491 (1962).
51. Kimmel, *Methods Enzymol.*, 11, 584 (1967).
52. Traut et al., *Biochemistry*, 12, 3266 (1973).
53. Heinrikson, *Biochem. Biophys. Res. Commun.*, 41, 967 (1970).
54. Scoffone et al., *Biochemistry*, 7, 971 (1968).
55. Fontana, *Methods Enzymol.*, 25, 419 (1972).
56. Wiseman and Woodward, *Biochem. Soc. Trans.*, 2, 594 (1974).
57. Moore, *J. Biol. Chem.*, 238, 235 (1963).
58. Takahashi, *J. Biol. Chem.*, 243, 6171 (1968).
59. Westhead, *Methods Enzymol.*, 25, 401 (1972).
60. Williams and Jacobs, *Biochim. Biophys. Acta*, 154, 323 (1968).
61. Light and Sinha, *J. Biol. Chem.*, 242, 1358 (1967).
62. Klapper and Klotz, *Methods Enzymol.*, 25, 531 (1972).
63. Bolton and Hunter, *Biochem. J.*, 133, 529 (1973).
64. Fontana and Scoffone, *Methods Enzymol.*, 25, 482 (1972).
65. Means and Feeney, *Chemical Modification of Proteins*, Holden-Day, San Francisco, 1971, 97.
66. Riordan and Vallee, *Methods Enzymol.*, 25, 515 (1972).
67. Liu, *J. Biol. Chem.*, 242, 4029 (1967).
68. Anfinsen and Haber, *J. Biol. Chem.*, 236, 1361 (1961).
69. Burstein et al., *Isr. J. Chem.*, 5, 65 (1967).
70. Fields, *Methods Enzymol.*, 25, 464 (1972).

ACID HYDROLYSIS OF PROTEINS

The reactions that occur during acid hydrolysis of a protein are complex. They are influenced by the nature and composition of the protein itself, by the presence of impurities, particularly metal ions in the acid used for hydrolysis, and by oxygen. Few systematic studies have been made. Definitive data on the destructions of serine and threonine during acid hydrolysis of proteins are provided by Rees,[1] who studied the stabilities of these amino acids by themselves as well as in mixtures of other amino acids so constituted as to mimic hydrolyates of various proteins. Similar, though less complete data were obtained by Hirs et al.[2] for these amino acids and for cystine and tyrosine. Tryptophan is usually destroyed extensively during acid hydrolysis, particularly in the presence of oxygen. Addition of thioglycollic acid to the medium has been found to provide effective protection[3] — C. H. W. Hirs.

Table 1
DESTRUCTION OF SERINE AND THREONINE

Amino acid mixture	Type of N	Recovery after 24 hours (%)	Amino acid mixture	Type of N	Recovery after 24 hours (%)
Serine alone	Serine-N	89.5	(Insulin)	Serine-N	86.0
Threonine alone	Threonine-N	94.7		Threonine-N	93.8
(Edestin)	Serine-N	89.2	(β-Lactoglobulin)	Serine-N	89.4
	Threonine-N	93.3		Threonine-N	94.0
(Horse Globin)	Serine-N	87.6			
	Threonine-N	95.9			

(From Rees[1])

Table 2
DESTRUCTION OF SERINE, THREONINE, CYSTINE AND TYROSINE

Amino acid in mixture simulating bovine serum albumin hydrolysate	Recovery (%) after various times (hours) Time			
	22	70	23	71
Serine	83.7	61.0	84.9	60.3
Threonine	92.4	81.0	92.7	80.8
Cystine	96.2	83.1	98.0	84.6
Tyrosine	89.7	76.6	94.9	89.2

(From Hirs, et al.[2])

Table 3
RECOVERY OF TRYPTOPHAN FROM
VARIOUS PROTEINS AFTER ACID
HYDROLYSIS IN THE PRESENCE OF 4%
THIOGLYCOLLIC ACID

Protein	Recovery (%)
Cytochrome c, bovine	96
Ferredoxin, spinach	89
Chymotrypsin, bovine	89
TMV protein	93
Hg-papain	98
Tryptophanase	88

(From Matsubara and Sasaki[3])

REFERENCES

1. **Rees,** *Biochem. J.,* 40, 632 (1946).
2. **Hirs, Stein, and Moore,** *J. Biol. Chem.,* 211, 941 (1954).
3. **Matsubara and Sasaki,** *Biochem. Biophys. Res. Commun.,* 35, 175 (1969).

These tables originally appeared in Sober, Ed., *Handbook of Biochemistry and selected data for Molecular Biology,* 2nd. Ed., Chemical Rubber Co., Cleveland, 1970.

HYDROLYSIS OF PROTEINS

HYDROLYSIS OF PEPTIDES IN ACID SOLUTION[1]

	Relative rate of hydrolysis (Gly-Gly = 1)					
Peptide[a]	2 N HCl, 99°C (2)	10 N HCl-glacial acetic acid (50 : 50), 37°C (3)	2 N HCl, 104°C (4)	1 N HCl, 104°C (5)	Dowex-50 (5)	0.8 N HCl, 54.5°C (6)
DL-Ala-Gly	0.69	0.62	—	0.62	0.61	0.56
Ala-Leu[b]	0.32	—	—	—	—	—
DL-Ala-DL-Asp	—	—	—	2.15	0.60	—
Ala-Ser	1.1	—	—	—	—	—
Gly-L-Ala	0.37	—	—	—	—	—
Gly-D-Ala	0.40	—	—	—	—	—
Gly-DL-Ala	—	0.62	—	—	—	—
Gly-Asp	1.94	—	—	—	—	—
Gly-Gly	1.0	1.0	1.0	1.0	1.0	1.0
Gly-Leu	0.34	0.40	0.47	0.48	0.49	0.48
Gly-Ser	1.83	—	—	—	—	—
Gly-Tyr	0.43	—	—	—	—	0.52
Gly-Try	—	0.35	—	—	—	0.44
Gly-DL-Val	—	0.31	—	0.34	0.38	—
Gly-Leu-Gly	—	—	0.35 (Gly, Leu)	—	—	—
Gly-Leu-Gly	—	—	0.65 (Leu, Gly)	—	—	—
Leu-Asp	0.86	—	—	—	—	—
Leu-Glu	0.23	—	—	—	—	—
Leu-Gly	0.23	—	0.23	—	—	0.18
DL-Leu-Gly	—	0.23	—	0.22	0.22	—
Leu-Gly-Leu	—	—	0.22 (Leu- Gly)	—	—	—
Leu-Gly-Leu	—	—	1.55 (Gly, Leu)	—	—	—
Leu-D-Leu	0.06	—	—	—	—	—
DL-Leu-DL-Leu	—	0.045	—	—	—	—
Leu-Try	—	0.041	—	—	—	—
Pro-Phe	0.29	—	—	—	—	—
Pro-Tyr	—	—	—	0.116	0.04	—
Ser-Ala	0.74	—	—	—	—	—
Ser-Gly	0.40	—	—	—	—	—
Ser-Ser	0.40	—	—	—	—	—
DL-Val-Gly	—	0.015	—	—	—	—
DL-Val-DL-Ile	—	—	—	0.0086	0.0091	—

Compiled by C. H. W. Hirs.

[a]All optically active amino acids are L isomers unless otherwise stated.
[b]Optical configuration undetermined.

REFERENCES

1. Hill, *Advanc. Protein Chem.*, 20, 37 (1965).
2. Harris, Cole, and Pon, *Biochem. J.*, 62, 154 (1956).
3. Synge, *Biochem. J.*, 39, 351 (1945).
4. Long and Lillycrop, *Trans. Faraday Soc.*, 59, 907 (1963).
5. Whitaker and Deatherage, *J. Amer. Chem. Soc.*, 77, 3360 (1955).
6. Lawrence and Moore, *J. Amer. Chem. Soc.*, 73, 3973 (1951).

ENZYMATIC HYDROLYSIS OF PROTEINS AND POLYPEPTIDES

Enzyme	Native protein	Enzyme	Native protein
Trypsin	Human γ-globulin[1]	Pepsin	Horse diphtheria toxin[26]
	Horse diphtheria antitoxin[2]		Pepsin[27]
	Human serum albumin[3]		Botulinum toxin[28]
	Insulin[4]		Catalase[7]
	Trypsin[5]		ACTH[29]
	Enolase[6]		Ribonuclease[30]
	Catalase[7]		Human serum albumin[31]
	Adolase[5]	Chymotrypsin	Human γ-globulin[32]
	Ribonuclease[8]		Growth hormone[33]
	Ribonuclease S[9]		Oxytocin[34]
	Fibrinogen[10]		Insulin[35]
	Myosin[11]		Ribonuclease[36]
	Tropomyosin[12]		Human serum albumin[31]
	Bovine plasma albumin[13]		Chymotrypsin[37]
Papain	Rabbit γ-globulin[14]		Growth hormone[38]
	Human γ-globulin[15]		Ribonuclease[30]
	Human serum globulins[16]		Ribonuclease T$_1$[39]
	Lipovitellin[17]	Leucine amino-peptidase	Insulin[40]
	Thyroglobulin[18]		Oxytocin[34]
Carboxypeptidase A	Insulin[19]		Papain[41]
	Enolase[6]		ACTH[42]
	Hemoglobin[a][20]		Enolase[6]
	ACTH[21]	Subtilisin	Ribonuclease[43]
	Soybean trypsin inhibitor[22]		Ovalbumin[44]
	Tobacco mosaic virus[23]		Human hemoglobin[45]
	Aldolase[24]		Cytochrome c[46]
	Crotoxin[25]	S. griseus protease	Taka-amylase[47]
		Collagenase	Collagen[48]

Compiled by C. H. W. Hirs.

[a]Carboxypeptidase A and B.

REFERENCES

1. **Schrohenloher**, *Arch. Biochem. Biophys.*, 101, 456 (1963).
2. **Northrop**, *J. Gen. Physiol.*, 25, 465 (1941–1942); Rothen, *J. Gen. Physiol.*, 25, 487 (1941–1942).
3. **Lapresle, Kaminski, and Tanner**, *J. Immunol.*, 82, 94 (1959); Kaminski and Tanner, *Biochim. Biophys. Acta*, 33, 10 (1959).
4. **Nicol**, *Biochem. J.*, 75, 395 (1960); Carpenter and Baum, *J. Biol. Chem.*, 237, 409 (1962).
5. **Bresler, Glikina, and Fenkel**, *Dokl. Akad. Nauk. SSR*, 96, 565 (1954); Chernikov, *Biokimiya*, 21, 295 (1956); Hess and Wainfon, *J. Amer. Chem. Soc.*, 80, 501 (1958).
6. **Malmstrom**, in *Symposium on Protein Structure*, Neuberger, Ed., John Wiley & Sons, New York, 1958, 338.
7. **Anan**, *J. Biochem. (Tokyo)*, 45, 211, 227 (1958).
8. **Ooi, Rupley, and Scheraga**, *Biochemistry*, 1, 432 (1963).
9. **Allende and Richards**, *Biochemistry*, 1, 295 (1962).
10. **Mihalyi and Godfrey**, *Biochim. Biophys. Acta*, 67, 73 (1963).
11. **Mihalyi and Harrington**, *Biochim. Biophys. Acta*, 36, 447 (1959).
12. **de Milstein and Bailey**, *Biochim. Biophys. Acta*, 49, 412 (1961).
13. **Richard, Beck, and Hoch**, *Arch. Biochem. Biophys.*, 90, 309 (1960).
14. **Porter**, *Biochem. J.*, 46, 479 (1950); *Biochem. J.*, 73, 119 (1959); Putnam, Tan, Lynn, Easley, and Shunsuke, *J. Biol. Chem.*, 237, 717 (1962); Fleischman, Porter, and Press, *Biochem. J.*, 88, 220 (1963).
15. **Hsiao and Putnam**, *J. Biol. Chem.*, 236, 122 (1961).
16. **Deutsch, Steihm, and Morton**, *J. Biol. Chem.*, 236, 2216 (1961).
17. **Glick**, *Arch. Biochem. Biophys.*, 100, 192 (1963).
18. **O'Donnell, Baldwin, and Williams**, *Biochim. Biophys. Acta*, 28, 294 (1958).

ENZYMATIC HYDROLYSIS OF PROTEINS AND POLYPEPTIDES (continued)

19. **Lens,** *Biochim. Biophys. Acta,* 3, 367 (1949); Harris, *J. Amer. Chem. Soc.,* 74, 2944 (1952), Harris and Li, *J. Amer. Chem. Soc.,* 74, 2945 (1952); Slobin and Carpenter, *Biochemistry,* 2, 16 (1963).
20. **Antonini, Wyman, Zito, Rossi-Fanelli, and Caputo,** *J. Biol. Chem.,* 238, PC60 (1961).
21. **Harris and Li,** *J. Biol. Chem.,* 213, 499 (1955).
22. **Davie and Neurath,** *J. Biol. Chem.,* 212, 507 (1955).
23. **Harris and Knight,** *J. Biol. Chem.,* 214, 215 (1955).
24. **Drechsler, Boyer, and Kowolsky,** *J. Biol. Chem.,* 234, 2627 (1959).
25. **Fraenkel-Conrat and Singer,** *Arch. Biochem. Biophys.,* 60, 64 (1956).
26. **Pope,** *Brit. J. Expt. Pathol.,* 20, 132, 201 (1939); Peterman and Pappenheimer, *J. Phys. Chem.,* 45, 1 (1941).
27. **Perlmann,** *Nature,* 173, 406 (1954); Tokuyasu and Funatsu, *J. Biochem. (Tokyo),* 52, 103 (1962).
28. **Wagman,** *Arch. Biochem. Biophys.,* 100, 414 (1963).
29. **Li, Geschwind, Cole, Raacke, Harris, and Dixon,** *Nature,* 176, 687 (1955).
30. **Anfinsen,** *J. Biol. Chem.,* 221, 405 (1956); Ginsberg and Schachman, *J. Biol. Chem.,* 235, 108 (1960), *J. Biol. Chem.,* 235, 115 (1960).
31. **Kaminski and Tanner,** *Biochim. Biophys. Acta,* 33, 10 (1959).
32. **Hanson and Johansson,** *Nature,* 187, 600 (1960).
33. **Li, Papkoff, and Hayashida,** *Arch. Biochem. Biophys.,* 85, 97 (1959).
34. **Golubow and du Vigneaud,** *Proc. Soc. Expt. Biol. Med.,* 112, 218 (1963).
35. **Ginsberg and Schachman,** *J. Biol. Chem.,* 235, 108 (1960); *J. Biol. Chem.,* 235, 115 (1960); Butler, Phillips, Stephen, and Creeth, *Biochem. J.,* 46, 44 (1950).
36. **Rupley and Scheraga,** *Biochemistry,* 2, 421 (1963).
37. **Gladner and Neurath,** *J. Biol. Chem.,* 206, 911 (1954).
38. **Harris, Li, Condliffe, and Pon,** *J. Biol. Chem.,* 209, 133 (1954).
39. **Takahashi,** *J. Biochem. (Tokyo),* 52, 72 (1962).
40. **Hill and Smith,** *J. Biol. Chem.,* 228, 577 (1957); Smith, Hill, and Borman, *Biochim. Biophys. Acta,* 29, 207 (1958).
41. **Hill and Smith,** *J. Biol. Chem.,* 235, 2332 (1960).
42. **White,** *J. Amer. Chem. Soc.,* 77, 4691 (1955).
43. **Richards and Vithayathil,** *J. Biol. Chem.,* 234, 1459 (1959); Gordillo, Vithayathil, and Richards, *Yale J. Biol. Med.,* 34, 582 (1962).
44. **Ottsen,** *Compt. Rend. Trav. Lab. Carlsberg,* 30, 211 (1958).
45. **Ottesen and Schroeder,** *Acta Chem. Scand.,* 15, 926 (1961).
46. **Nozaki, Yamanaka, Horro, and Okunuki,** *J. Biochem. (Tokyo),* 44, 453 (1957).
47. **Toda and Akabori,** *J. Biochem. (Tokyo),* 53, 95 (1963).
48. **von Hippel and Harrington,** *Biochim. Biophys. Acta,* 36, 427 (1959).

ENZYMATIC HYDROLYSIS OF SEVERAL CONJUGATED PROTEINS

Protein	Enzymes	Products isolated
Azaserine labeled enzyme	Papain, pronase, aminopeptidase	C^{14}-labeled azaserinepeptides[1]
Rabbit γ-globulin	Papain	Glycopeptide[2]
Bovine globulin of colostrum	Papain	Glycopeptide[2]
Human γ-globulin	Papain	Glycopeptide[3]
Cytochrome c	Pepsin, trypsin	Heme peptides[4]
Ovalbumin	Pepsin, trypsin, chymotrypsin, mold protease	Glycopeptides[5]
Ovalbumin	Pancreatin	Glycopeptide[6]
Ovalbumin	Trypsin, chymotrypsin	Glycopeptides[7]
Chondroitin sulfate complex	Papain	Glycopeptide[8]
Fetuin	Papain, trypsin, chymotrypsin, pepsin, subtilisin	Glycopeptides[9]
Chromatium heme protein	Pepsin	Heme peptide[10]
Phosphorylase	Chymotrypsin	Pyridoxal peptide[11]

Compiled by C. H. W. Hirs.

REFERENCES

1. Dawid, French, and Buchanan, *J. Biol. Chem.*, 238, 2178 (1963).
2. Nolan and Smith, *J. Biol. Chem.*, 237, 453 (1962).
3. Rosevear and Smith, *J. Biol. Chem.*, 236, 425 (1961).
4. Tuppy, in *Symposium on Protein Structure,* Neuberger, Ed., John Wiley, New York, 1958, 66.
5. Johansen, Marshall, and Neuberger, *Biochem. J.*, 78, 518 (1961).
6. Jevons, *Nature*, 181, 1346 (1958).
7. Cunningham, Nuenke, and Nuenke, *Biochim. Biophys. Acta,* 26, 660 (1957).
8. Muir, *Biochem. J.,* 69, 195 (1958); Anderson, Hoffman, and Meyer, *Biochim. Biophys. Acta,* 74, 309 (1963).
9. Spiro, *J. Biol. Chem.*, 237, 382 (1962).
10. Dus, Bartsch, and Kamen, *J. Biol. Chem.*, 237, 3083 (1962).
11. Fischer, Kent, Snyder, and Krebs, *J. Amer. Chem. Soc.*, 80, 2906 (1958).

SPECIFICITY OF TRYPSIN TOWARD SYNTHETIC SUBSTRATES

Substrates[a]	Relative rate of hydrolysis	Substrates[a]	Relative rate of hydrolysis
Benzoyl-L-argininamide[1]	100	L-Arginyl-L-leucine[2]	0.036
Glycyl-L-lysinamide[1]	91	L-Arginylglycine[2]	0.029
L-Alanyl-L-lysinamide[1]	200	L-Arginylphenylalanine[2]	0.029
L-α-Aminobutyryl-L-lysinamide[1]	406	L-Arginylglutamic acid	0.017
L-Norleucyl-L-lysinamide[1]	424	Cbz-nitro-L-arginyl-L-leucine[2]	0
β-Alanyl-L-lysinamide[1]	82	Cbz-nitro-L-arginyl-L-phenylalanine	0
L-Valyl-L-lysinamide[1]	324	Cbz-nitro-L-argininamide[2]	0
L-Leucyl-L-lysinamide[1]	155	Benzoyl-L-homoargininamide[3]	0
L-Phenylalanyl-L-lysinamide[1]	124		

Compiled by C. H. W. Hirs.

[a] At 0.05 M substrate, 25°–40°C, pH 7.5–7.8. Cbz, carbobenzoxy.

REFERENCES

1. Izumiya, Yamashita, Uchio, and Kitagawa, *Arch. Biochem. Biophys.,* 90, 170 (1960).
2. Van Orden and Smith, *J. Biol. Chem.,* 208, 751 (1954).
3. Shields, Hill, and Smith, *J. Biol. Chem.,* 234, 1747 (1959).

SPECIFICITY OF CHYMOTRYPSIN FOR HYDROLYSIS OF PEPTIDE BONDS IN PROTEINS AND POLYPEPTIDES[a]

Type of bond	Type of bond	Type of bond	Type of bond
-Thr-Asn⋯Arg-Asn[d]	-Glu-Gln⋯Ala-Arg[h]	-Lys-His⋯Ile-Ile[e]	-Leu-Met⋯Glu-Tyr[c]
-Val-Asn⋯CMC-Ala[d]	-Val-Gln⋯Ala-Ser[j]		
	-Ala-Gln⋯Lys-His[e]	-Lys-His⋯Lys-Thr[c,f]	-Lys-Met⋯Ile-Phe[c]
	-Tyr-Gln⋯Lys-Met[j]		
-Ala-Asn⋯Lys-Asn[f]	-Ser-Gln⋯Val-Thr[h]	-Asp-Ile⋯Asn-Leu[h]	-Ala-Met⋯Lys-Arg[d]
		-Val-Lys⋯Ala-His[l]	
-Lys-Asn⋯Lys-Gly[c]	-Lys-Gln⋯[h]	-Glu-Lys⋯Gly-Gly[c]	-Thr-Met⋯Ser[d]
			-Lys-Met⋯Val-Thr[j]
-Lys-Asn⋯Val-Ala[e]	-Val-Gly⋯Lys-Lys[k]	-Val-Lys⋯Gly-His[b]	-Ala-Phe⋯Pro-Leu[k]
-Thr-Asn⋯Val-Lys[b]	-Leu-His⋯Ala-His[b]	-Gly-Lys⋯Lys-Arg[h]	-Phe-Ser⋯[j]
-Lys-Asn⋯[i]	-Val-His⋯Ala-Ser[b]	-Ile-Lys⋯Lys-Lys[f]	-Thr-Ser⋯Ala-Ala[e]
	-Ala-His⋯Gly-Lys[j]		
-Val-CySO$_3$H⋯Ser-Leu[g]	-Gly-His⋯Gly-Lys[b]	-Lys-Lys⋯Lys[f]	-Tyr-Thr⋯Ala-Ala[f]
-Thr-Glu⋯Gln-Ala[h]	-Leu-His⋯Gly-Leu[c,f]	-Gly-Met⋯Asn-Ala[d]	-Val-Thr⋯Ala-Leu[h]
		-Ile-Met⋯Gly-Asn[j]	

Compiled by C. H. W. Hirs.

[a]In the same substrates a total of 32 phenylalanyl bonds, 36 leucyl bonds, 24 tyrosyl bonds, and 6 tryptophanyl bonds were also hydrolyzed.
[b]Human hemoglobin α-chain.[1]
[c]Horse cytochrome c.[2]
[d]Egg-white lysozyme.[3]
[e]Ribonuclease.[4]
[f]Human cytochrome c.[5]
[g]Insulin.[6]
[h]Tobacco mosaic virus.[7]
[i]Papain.[8]
[j]Human hemoglobin, γ-chain.[9]
[k]Ovine corticotropin.[10]

SPECIFICITY OF CHYMOTRYPSIN FOR HYDROLYSIS OF PEPTIDE BONDS IN PROTEINS AND POLYPEPTIDES (continued)

REFERENCES

1. **Hill and Konigsberg**, *J. Biol. Chem.*, 237, 3151 (1962).
2. **Margoliash**, *J. Biol. Chem.*, 237, 2161 (1962).
3. **Canfield**, *J. Biol. Chem.*, 238, 2698 (1963).
4. **Hirs, Stein, and Moore**, *J. Biol. Chem.*, 221, 151 (1956); Hirs, Moore, and Stein, *J. Biol. Chem.*, 235, 633 (1960); Smyth, Stein, and Moore, *J. Biol. Chem.*, 237, 1845 (1962).
5. **Matsubara and Smith**, *J. Biol. Chem.*, 238, 2732 (1963).
6. **Sanger and Tuppy**, *Biochem. J.*, 49, 463, 481 (1951); Sanger and Thompson, *Biochem. J.*, 53, 353, 366 (1953).
7. **Anderer, Uhlig, Weber, and Schramm**, *Nature*, 186, 922 (1960).
8. **Light and Smith**, *J. Biol. Chem.*, 237, 2537 (1962).
9. **Schroeder, Shelton, Shelton, Cormick, and Jones**, *Biochemistry*, 2, 992 (1963).
10. **Leonis, Li, and Chung**, *J. Amer. Chem. Soc.*, 81, 419 (1959).

PRODUCTS FORMED ON CHYMOTRYPTIC HYDROLYSIS OF α-CHAINS OF HUMAN GLOBIN UNDER TWO DIFFERENT CONDITIONS OF HYDROLYSIS[a]

Peptide substrate, α-chain, human hemoglobin.
Bonds hydrolyzed are indicated by arrows

$\qquad$ 1 $\qquad$ ↓ 5 $\qquad$ ↓ 10 $\qquad$ ↓ 15
Val-Leu-Ser-Pro-Ala-Asp-Lys-Thr-Asn-Val-Lys-Ala-Ala-Try-Gly

Products on hydrolysis for 6 hr at pH 8, 25°C, with α-chymotrypsin : substrate molar ratio of 1 : 150

Cα14	Val-Leu-Ser-Pro-Ala-Asp-Lys-Thr-Asn-Val-Lys-Ala-Ala-Try	(75% yield)
Cα10	Ser-Pro-Ala-Asp-Lys-Thr-Asn-Val-Lys-Ala-Ala-Try	(20% yield)

Products on hydrolysis for 24 hr at pH 9, 30°C, with α-chymotrypsin : substrate molar ratio of 1 : 150

Cα14	Val-Leu-Ser-Pro-Ala-Asp-Lys-Thr-Asn-Val-Lys-Ala-Ala-Try	(15% yield)
Cα10	Ser-Pro-Ala-Asp-Lys-Thr-Asn-Val-Lys-Ala-Ala-Try	(10% yield)
Cα15	Val-Lys-Ala-Ala-Try	(15% yield)

Peptide, substrate, α-chain, human hemoglobin.
Bonds hydrolyzed are indicated by arrows

$\qquad$ ↓ 25 $\qquad$ ↓ 30 $\qquad$ 35
Tyr-Gly-Ala-Glu-Ala-Leu-Glu-Arg-Met-Phe-Leu-Ser

Products on hydrolysis for 6 hr at pH 8, 25°C, with α-chymotrypsin : substrate molar ratio of 1 : 150

Cα9	Gly-Ala-Glu-Ala-Leu-Glu-Arg-Met-Phe	(80% yield)

Products on hydrolysis for 24 hr at pH 9, 30°C, with α-chymotrypsin : substrate molar ratio of 1 : 150

Cα9	Gly-Ala-Glu-Ala-Leu-Glu-Arg-Met-Phe	(32% yield)
Cα16	Glu-Arg-Met-Phe	(50% yield)
Cα29	Gly-Ala-Glu-Ala-Leu	(15% yield)

Compiled by C. H. W. Hirs.

[a] Only portions of the α-chain (substrate) are shown (*1*).

REFERENCES

1. Hill and Konigsberg, *J. Biol. Chem.*, 237, 3151 (1962).

CONDITIONS FOR CHYMOTRYPTIC HYDROLYSIS OF CERTAIN PROTEINS

Substrate	Substrate (μmoles)	Chymotrypsin[a] (μmoles)	Volume of reaction (ml)	Enzyme substrate molar ratio	Temperature (°C)	pH	Time of hydrolysis (hr)
Performate oxidized ribonuclease[1]	72	0.20	100	1 : 360	25	7.0	24
Oxidized A-chain insulin[2]	22	0.1	5	1 : 220	37	7.5	24
Oxidized B-chain insulin[3]	17	0.1	5	1 : 170	25	7.5–8.0	24
Ovine corticotropin[4]	7.5	0.046	5	1 : 160	40	9.0	24
Human hemoglobin, α-chain[5]	121	0.80	200	1 : 150	30	9.0	24
Human hemoglobin α-chain[5]	60	0.40	90	1 : 150	25	8.0	6
Carboxymethyl lysozyme[6]	70	0.8	100	1 : 87	37	8.0	2
Equine cytochrome c[7]	100	2.4	120	1 : 42	22	7.8	29
Human cytochrome c[8]	70	2.3	90	1 : 33	Room temp.	7.85	26
Oxidized papain[9]	10	0.48	15	1 : 21	39	7.6	6

Compiled by C. H. W. Hirs.

[a] Assumed molecular weight 25,000.

REFERENCES

1. **Hirs, Moore, and Stein,** *J. Biol. Chem.,* 235, 633 (1960).
2. **Sanger and Thompson,** *Biochem. J.,* 53, 353, 366 (1953).
3. **Sanger and Tuppy,** *Biochem. J.,* 49, 463, 481 (1951).
4. **Léonis, Li, and Chung,** *J. Amer. Chem. Soc.,* 81, 419 (1959).
5. **Hill and Konigsberg,** *J. Biol. Chem.,* 237, 3151 (1962).
6. **Canfield,** *J. Biol. Chem.,* 238, 2698 (1963).
7. **Margoliash,** *J. Biol. Chem.,* 237, 2161 (1962).
8. **Matsubara and Smith,** *J. Biol. Chem.,* 238, 2732 (1963).
9. **Light and Smith,** *J. Biol. Chem.,* 237, 2537 (1962).

SPECIFICITY OF PEPSIN FOR HYDROLYSIS OF PEPTIDE BONDS IN PROTEINS AND POLYPEPTIDES[a]

Type of bond	Type of bond	Type of bond	Type of bond	Type of bond
-Asn-Arg···Ile-[c]	-Met-Glu···His-[j]	-Leu-Gly···Arg-COO-[f]	-Gly-Ser···Tyr-[c]	-Leu-Tyr···Leu-[g]
-Val-Asn···Phe-[b]	-Ile-Glu···Leu-[c]	-Leu-Gly···Asp-[h]	-Val-Thr···Ala-[c]	-Ser-Tyr···Ser-[c]
-Phe-Asn···Thr-[d]	-Glu-Glu···Lys-[c]	-Gly-Gly···Glu-[f]	-Val-Thr···Leu-[b]	-Pro-Tyr···Val-[e]
-Gly-Asn···Try-[d]	-Ala-Glu···Phe-[b]	-Asp-Gly···Leu-[f]	-Asn-Thr···Phe-[e]	-Ile-Val···Ala-[e]
-Val-Asp···Glu-[f]	-Lys-Glu···Phe-[f,h]	-Arg-Gly···Phe-[g]	-Ala-Thr···Val-[c]	-Ala-Val···Asp-[c]
-Ser-Asp···Phe-[c]	-Gly-Glu···Tyr-[b]	-Arg-Gly···Tyr-[d]	-Arg-Thr···Val-[c]	-Asn-Val···Asp-[f]
-Ala-Asp···Pro-[c]	-Asp-Glu···Val-[f]	-Val-His···Ala-[b]	-Val-Thr···Val-[c]	-Ser-Val···CySO$_3$H-[e,h]
-Val-Asp···Pro-[f,c]	-Val-Gln···Ala-[e,h]	-Glu-His···Phe-[j]	-Ala-Try···Gly-[b]	-CySH-Val···Leu-[f]
-Leu-CMC···Asn-[d]	-Asn-Gln···His-[g]	-His-His···Phe-[b]	-Arg-Try···Try-[d]	-Asn-Val···Lys-[b]
-Try-CMC···Asn-[d]	-Tyr-Gln···Leu-[g]	-His-Lys···Leu-[b]	-Ala-Try···Val-[d]	-Ala-Val···Thr-[f]
-Val-Glu···Ala-[g]	-Asn-Gln···Phe-[c]	-Phe-Lys···Leu-[b]	-Asn-Try···Val-[d]	-Gln-Val···Try-[c]
-Leu-Glu···Asn-[g]	-Arg-Gln···Phe-[c]	-Arg-Met···Phe-[b]	-Asn-Tyr···CySO$_3$H-[g]	-Val-Val···Tyr-[f]
		-Ser-Ser···Asp-[d]		
		-Leu-Ser···Ser-[h]		
-Val-Glu···Gln-[g]	-Ser-Gln···Val-[c]	-Val-Ser···Thr-[b]	-Leu-Tyr···Gln-[g]	

Compiled by C. H. W. Hirs.

[a] In the same substrates a total of 22 alanyl bonds, 34 leucyl bonds, and 23 phenylalanyl bonds were also hydrolyzed. Asn, Asparagine; Gln, glutamine; CMC, S-carboxymethyl-cystine.
[b] Human hemoglobin, α-chain.[1]
[c] Tobacco mosaic virus.[2]
[d] Egg-white lysozyme.[3]
[e] Ribonuclease.[4]
[f] Human hemoglobin, β-chain.[1]
[g] Insulin.[5]
[h] Human hemoglobin, γ-chain.[6]
[j] β-MSH.[7]

REFERENCES

1. **Hill and Konigsberg**, *J. Biol. Chem.*, 237, 351 (1962).
2. **Anderer, Uhlig, Weber, and Schramm**, *Nature*, 186, 922 (1960).
3. **Canfield**, *J. Biol. Chem.*, 238, 2698 (1963).
4. **Hirs, Moore, and Stein**, *J. Biol. Chem.*, 235, 633 (1960).
5. **Sanger and Tuppy**, *Biochem. J.*, 49, 463, 481 (1951); Sanger and Thompson, *Biochem. J.*, 53, 353, 366 (1953).
6. **Schroeder, Shelton, Shelton, Cormick, and Jones**, *Biochemistry*, 2, 992 (1963).
7. **Harris and Roos**, *Biochem. J.*, 71, 434 (1959).

PEPTIC HYDROLYSIS OF BONDS FORMED BY THE AROMATIC AMINO ACIDS IN HUMAN α- AND β-CHAINS

Substrate	Hydrolysis amino-terminal to aromatic residue	Hydrolysis carboxyl-terminal to aromatic residue	Bonds not hydrolyzed
α-Chain[a]	-Glu-Glu···Tyr-Gly- -Arg-Met···Phe-Leu- -Val-Asn···Phe-Lys- -Ala-Glu···Phe-Thr-	-Ala-Tyr-···Gly-Lys- -Met-Phe···Leu-Ser- -His-Phe···Asp-Leu- -Tyr-Phe···Pro-His- -Ser-Phe···Pro-Thr- -Asn-Phe···Val-Lys- -Lys-Phe···Leu-Ala-	-Thr-Tyr•••Phe-Pro-[c]
β-Chain[b]	-Ala-Leu···Tyr-Gly- -Val-Val···Tyr-Pro- -Arg-Phe···Phe-Glu- -Gly-Ala···Phe-Ser- -Glu-Asn···Phe-Arg- -His-His···Phe-Gly- -Lys-Glu···Phe-Thr- -Ala-Ala···Tyr-Gln-	-Arg-Phe···Phe-Glu- -Ser-Phe···Gly-Asp- -Thr-Phe···Ala-Thr-	-Pro-Try···Thr-Gln- -His-Lys···Tyr-His-COO$^-$ -Lys-Tyr···His-COO$^-$

Compiled by C. H. W. Hirs.

[a]Conditions of hydrolysis: 2 g α-chain in 200 ml at pH 2.8. 20 mg of pepsin, 16 hr hydrolysis at room temperature.[1]
[b]Conditions of hydrolysis: 0.8 g β-chain in 80 ml at pH 2, 8 mg of pepsin, 1 hr hydrolysis at 25°C.[2]
[c]Hydrolysis at -Phe-Pro- observed.

REFERENCES

1. **Konigsberg and Hill,** *J. Biol. Chem.*, 237, 3157 (1962).
2. **Konigsberg, Goldstein, and Hill,** *J. Biol. Chem.*, 238, 2028 (1963).

PEPTIC HYDROLYSIS OF BONDS FORMED BY LEUCINE IN HUMAN α- AND β-CHAINS

Substrate	Hydrolysis amino-terminal to leucine	Hydrolysis carboxyl-terminal to leucine	Bond not hydrolyzed
α-Chain[1]	-Glu-Ala···Leu-Glu-[a] -Met-Phe···Leu-Ser- -His-Lys···Leu-Arg- -Phe-Lys···Leu-Leu- -Val-Thr···Leu-Ala-[a] -Lys-Phe···Leu-Ala-[a]	-Ala-Leu···Glu-Arg-[a] -Ala-Leu···Thr-Asn- -Ala-Leu···Ser-Ala- -Ala-Leu···Ser-Asp- -Asp-Leu···His-Ala- -Leu-Leu···Val-Thr- -Thr-Leu···Ala-Ala-[a] -Ser-Leu···Asp-Lys- -Phe-Leu···Ala-Ser-[a] -Val-Leu···Thr-Ser-	^+H_3N-Val-Leu···Ser-Pro- -Asp-Leu···Ser-His- -Lys-Leu···Leu-Ser-[b] -$CySO_3H$-Leu···Leu-Val-[c] -Ala-His···Leu-Pro-
β-Chain[2]	-Arg-Leu···Leu-Val- -Arg-Leu···Leu-Gly- -Cys-Val···Leu-Ala- -Asn-Ala···Leu-Ala-	-Ala-Leu···Try-Gly- -Ala-Leu···Gly-Arg- -Arg-Leu···Leu-Val- -Asp-Leu···Ser-Thr- -His-Leu···Asp-Asn- -Asn-Leu···Lys-Gly- -Thr-Leu···Ser-Glu- -Glu-Ser···His-CMC- -Lys-Leu···His-Val- -Arg-Leu···Leu-Gly-	-His-Leu···Thr-Pro- -Val-Leu···Gly-Ala- -Gly-Leu···Ala-His- -Val-Leu···Val-CySH

Compiled by C. H. W. Hirs.

[a] Hydrolysis at bonds on either side of leucine was observed.
[b] Hydrolysis at -Lys-Leu- and -Leu-Ser- bonds was observed.
[c] Hydrolysis at -Leu-Val- bond was observed.

REFERENCES

1. **Konigsberg and Hill,** *J. Biol. Chem.,* 237, 3157 (1962).
2. **Konigsberg, Goldstein, and Hill,** *J. Biol. Chem.,* 238, 2028 (1963).

SPECIFICITY OF PAPAIN FOR HYDROLYSIS OF PEPTIDE BONDS IN PROTEINS AND POLYPEPTIDES[a]

Compiled by C. H. W. Hirs.

Type of bond	Type of bond	Type of bond	Type of bond
-Asp-Ala···Asn-[d]	-Pro-Asn···Ala-[d]	-Glu-Glu···Lys-COO-[c]	-Lys-Ile···Phe-[d]	-Gly-Lys···Lys-[d]
-Val-Ala···Asn-[b]	-Pro-Asn···Leu-[d]	-Val-Gly···Ala-[b]	-Ala-Leu···Glu-[b]	-Pro-Lys···Lys-[d]
-Pro-Ala···Glu-[b]	-Glu-Asn···Phe-[c]	+H$_3$N-Gly···Ala-[b]	+H$_3$N-Leu···Glu-[d]	-Lys-Lys···Tyr-[d]
-Val-Ala···Gly-[b]	-Val-Asp···Asp-[b]	-Lys-Gly···Ile-[d]	-Ala-Leu···Ser-[b]	+H$_3$N-Lys···Thr-[d]
+H$_3$N-Val-Ala···His-[b]	-Asp-Asp···Met-[b]	-His-Gly···Ser-[b]	-His-Leu···Thr-[c]	-Glu-Phe···Thr-[b,c]
-Asn-Ala···Leu-[b]	-Leu-Glu···Asn-[d]	-Ala-Gly···Val-[b]	-Asn-Lys···Asn-[d]	+H$_3$N-Ser···Ala-[b]
-Glu-Ala···Leu-[b]	-Pro-Glu···Asn-[c]	-Leu-His···Ala-[b]	-Glu-Lys···Gly-[d]	-Leu-Ser···Asp-[b]
+H$_3$N-Ser-Ala···Leu-[b]	-Pro-Glu···Asn-[c]	-Ala-His···Leu-[b]	+H$_3$N-Lys···Gly-[d]	-Leu-Ser···His-[b]
-Pro-Ala···Val-[b]	-Pro-Glu···Glu-[c]	-Val-His···Leu-[c]	-Lys-Lys···Ile-[d]	-Ile-Thr···Tyr-[d]
				-Thr-Tyr···Phe-[b]

[a] Of the peptides examined, none contained arginine. Bonds formed by this amino acid should be very susceptible to hydrolysis.
[b] Human hemoglobin, α-chain.[1]
[c] Human hemoglobin, β-chain.[2]
[d] Cytochrome c.[3]

REFERENCES

1. Konigsberg and Hill, *J. Biol. Chem.*, 237, 3157 (1962).
2. Konigsberg, Goldstein, and Hill, *J. Biol. Chem.*, 238, 2028 (1963).
3. Margoliash, *J. Biol. Chem.*, 237, 2161 (1962).

THE HYDROLYSIS OF SEVERAL PEPTIDES FROM α- AND β-CHAINS OF HUMAN HEMOGLOBIN BY PAPAIN[a]

Peptide	Peptide sequences	Yield (%)
A	Asp-Leu-Ser-His-Gly-Ser-Ala 　　　↑　　　↑	
Papain 1A	Asp-Leu-Ser	95
Papain 2A	Ser-Ala	70
Papain 3A	His-Gly	90
B	Val-Asp-Pro-Val-Asn-Phe-Lys 　　　　　　↑	
Papain 1B	Val-Asp-Pro-Val-Asn	75
Papain 2B	Phe-Lys	95
C	Ser-Ala-Leu-Ser-Asp-Leu-His-Ala-His-Lys 　↑　　↑　　　↑	
Papain 1C	Leu-Ser	25
Papain 2C	Ser-Ala	55
Papain 3C	Asp-Leu-His	50
Papain 4C	Leu-Ser-Asp-Leu-His	45
Papain 5C	Ala-His-Lys	95
D	Gly-Ala-Glu-Ala-Leu-Gly-Arg 　↑　↑　↑　↑　↑　↑	
Papain 1D	Arg	Not measured
Papain 2D	Ala-Glu-Ala	20
Papain 3D	Ala- or Ala	30
Papain 4D	Gly	20
Papain 5D	Leu-Glu-Arg	35
Papain 6D	Ala-Glu-Ala-Leu	45
E	Val-Ala-His-Val-Asp-Asp-Met-Pro-Asn-Ala-Leu 　↑　　　↑　↑　　　　↑　↑	
Papain 1E	Ala-Leu	20
Papain 2E	Val-Ala	75
Papain 3E	Asp-Met-Pro-Asn-Ala	70
Papain 4E	His-Val-Asp	45
Papain 5E	His-Val-Asp-Asp	55
F	Val-His-Leu-Thr-Pro-Glu-Glu-Lys 　↑　↑　　　↑　↑	
Papain 1F	Glu	36
Papain 2F	Leu	33
Papain 3F	Leu-Thr-Pro-Glu	34
Papain 4F	Thr-Pro-Glu	20
Papain 5F	Glu-Lys	40
Papain 6F	Lys	42
Papain 7F	Val-His	82
G	Val-Val-Ala-Gly-Val-Ala-Asn-Ala 　　↑　↑　　　↑	
Papain 1G	Gly	95
Papain 2G	Asn-Ala	75
Papain 3G	Val-Ala	65
Papain 4G	Val-Val-Ala	70

[a]Hydrolyses were performed at 37°–40°C for 15–18 hr at pH 5.5, with papain concentrations of 0.001–0.005%. The peptides have been numbered arbitrarily.[1]

THE HYDROLYSIS OF SEVERAL PEPTIDES FROM α- AND β–CHAINS OF HUMAN HEMOGLOBIN BY PAPAIN[a] (continued)

Peptide	Peptide sequences	Yield (%)
H	His-Val-Asp-Pro-Glu-Asn-Phe-Arg ↑ ↑	
Papain 1H	Asn	30
Papain 2H	His-Val-Asp-Pro-Glu	40
Papain 3H	His-Val-Asp-Pro-Glu-Asn	30
Papain 4H	Phe-Arg	80
I	Thr-Tyr-Phe-Pro-His-Phe ↑	
Papain 1I	Phe-Pro-His-Phe	95
Papain 2I	Thr-Tyr	80

Compiled by C. H. W. Hirs.

REFERENCE

1. **Konigsberg and Hill,** *J. Biol. Chem.,* 237, 3157 (1962); Konigsberg, Goldstein, and Hill, *J. Biol. Chem.,* 238, 2028 (1963).

SPECIFICITY OF SUBTILISIN FOR HYDROLYSIS OF PEPTIDE BONDS IN PROTEINS AND POLYPEPTIDES

Type of bond	Type of bond	Type of bond	Type of bond	Type of bond
-Ala-Ala···Lys-^{d}	-Val-CMC···Ala-^{d}	-Asn-Gln···His-^{f}	-Met-Lys···Arg-^{d}	-Ala-Thr···Asn-^{d}
-CyS-Ala···Ser-^{f}	-Arg-CMC···Asn-^{d}	-Arg-Gln···Phe-^{b}	-Leu-Met···Asp-^{c}	-Asp-Ser···Gly-^{e}
-Val-Ala···Try-^{d}	-Leu-CMC···Asn-^{d}	-Glu-Gly···Gly-^{e}	-Lys-Met···Glu-^{e}	-Arg-Thr···Val-^{b}
-Ser-Arg···Arg-^{c}		-Leu-CySO$_3$H···Gly-^{f}	-Ala-Met···Lys-^{d}	-Arg-Try···Gly-^{a}
-Glu-Arg···Gly-^{f}		-Ser-Gly···Lys-^{f}	$^+$H$_3$N-Phe···Val-^{f}	-Ala-Try···Ile-^{d}
-Arg-Arg···Val-^{b}		-Val-CySO$_3$H···Gly-^{f}	-His-Phe···Arg-^{a}	-Gln-Try···Leu-^{c}
-Met-Asn···Ala-^{d}	-CyS-CyS···Ala-^{f}	-Arg-Gly···Thr-^{b}	-Gly-Phe···Phe-^{f}	-Asn-Try···Val-^{d}
-CMC-Asn···Asp-^{d}	-Val-CyS···Ser-^{f}	-Gln-His···Leu-^{f}	-Thr-Phe···Thr-^{e}	-Arg-Tyr···Asn-^{b}
-Val-Asn···Gln-^{f}	-Val-Glu···Ala-^{f}	-His-Leu···CySO$_3$H-^{f}	-Phe-Phe···Tyr-^{f}	-Asn-Tyr···CyS-^{f}
-Ser-Asn···Phe-^{d}	-Leu-Glu···Asn-^{f}	-Gln-Leu···Glu-^{f}	-Asp-Phe···Val-^{c}	-Asp-Tyr···Gly-^{d}
-Ile-Asn···Ser-^{d}	-CySO$_3$H-Glu···Gly-^{e}	-Ile-Leu···Gln-^{d}	-Gly-Pro···Val-^{e}	-Leu-Tyr···Leu-^{f}
-Arg-Asn···Thr-^{d}	-Thr-Gln···Ala-^{d}	-Ser-Leu···Gly-^{d}	-CMC-Ser···Ala-^{d}	-Lys-Tyr···Leu-^{c}
-Met-Asn···Thr-COO-^{c}	-Val-Gln···Ala-^{d}	-Ala-Leu···Tyr-^{f}	-Thr-Ser···Asp-^{c}	-Pro-Tyr···Lys-^{a}
-Asn-Asp···Gly-^{d}	-Ala-Gln···Asp-^{c}	-Ser-Leu···Tyr-^{f}	-Gly-Ser···His-^{f}	-Gly-Tyr···Ser-^{d}
-Asp-Asp···Ala-^{b}	-Glu-Gln···Cys-S-^{f}	-His-Leu···Val-^{f}	-Tyr-Ser···Lys-^{c}	-Phe-Tyr···Thr-^{f}
-Val-Asp···Asp-^{b}	-Asn-Gln···Glu-^{f}	-Tyr-Leu···Val-^{f}	-Gly-Ser···Thr-^{d}	$^+$H$_3$N-Phe-Val···Asn-^{f}
-Gly-Asp···Gly-^{d}	-Ser-Gln···Glu-^{c}	-Pro-Lys···Ala-COO-^{f}	-Ile-Thr···Ala-^{d}	-Leu-Val···CySO$_3$H^f
				-Pro-Val···CySO$_3$H^e

Compiled by C. H. W. Hirs.

^{a}β-MSH.[1]
^{b}Tobacco mosaic virus.[2]
^{c}Glucagon.[3]
^{d}Egg-white lysozyme.[4]
^{e}Diisopropyltrypsin peptide.[5]
^{f}Insulin.[6]

REFERENCES

1. **Harris and Roos**, *Biochem. J.*, 71, 434 (1959).
2. **Tsugita, Gish, Young, Fraenkel-Conrat, Knight, and Stanley**, *Proc. Natl. Acad. Sci. (USA)*, 46, 1463 (1960).
3. **Bromer, Staub, Sinn, and Behrens**, *J. Amer. Chem. Soc.*, 79, 2801 (1957); Bromer, Sinn, and Behrens, *J. Amer. Chem. Soc.*, 79, 2807 (1957).
4. **Canfield**, *J. Biol. Chem.*, 238, 2698 (1963).
5. **Dixon, Kauffman, and Neurath**, *J. Amer. Chem. Soc.*, 80, 1260 (1958); Dixon, Kauffman, and Neurath, *J. Biol. Chem.*, 233, 1373 (1958).
6. **Tuppy**, in *Symposium on Protein Structure*, Neuberger, Ed., John Wiley & Sons, New York, 1958, 66; Haugaard and Haugaard, *Compt. Rend. Trav. Lab. Carlsberg*, 29, 350 (1955).

These tables originally appeared in Sober, Ed., *Handbook of Biochemistry and selected data for Molecular Biology*, 2nd ed., Chemical Rubber Co., Cleveland, 1970.

INDEX TO PHYSICAL–CHEMICAL DATA OF PROTEINS

Malcolm H. Smith (deceased)

This index follows earlier compilations, notably by Svedberg[235] and Edsall,[55] and are similarly concerned primarily with information relating to ultracentrifuge studies, particularly sedimentation-velocity studies. The references are taken from the article by Malcolm H. Smith in the 2nd Edition of the Handbook of Biochemistry.

Table 1
GLOBULAR PROTEINS

Protein	References	Protein	References
Acetylcholinesterase, *Electrophorus*	467, 519	D-Amino acid oxidase holo-enzyme, pig kidney	271
Acetyl-CoA synthetase, bovine heart	471	D-Amino acid oxidase, pig kidney	153
Acid proteinase, *Aspergillus*	413	D-Amino acid oxidase, Michaelis complex, pig kidney	270
Acylase I, (Hippuricase) hog kidney	485		
Acyl phosphatase, bovine brain	190	L-Amino acid oxidase, moccasin snake	55, 369
Adenine deaminase, calf intestinal mucosa	25	L-Amino acid oxidase, rat kidney	161, 541
S-Adenine deaminase	133	δ-Aminolevulinate dehydratase, mouse	511
Adrenocorticotropin, pig	55, 223, 320	Amylase, malt	42
Adrenocorticotropin, sheep	55, 319, 323	α-Amylase, *B. subtilis*	67
Adrenocorticotropin, sheep pituitary	137, 231	α-Amylase, pig pancreas	41
Adrenocorticotropin A, pig pituitary	137	α-Amylase, *Pseudomonas*	148
Aequorin, *Aquorea*	452	β-Amylase, sweet potato	55, 375
Alanine dehydrogenase, *B. subtilis*	503	α-Amylase (Zn free), *B. subtilis*	67
Albumin, bovine	185, 188, 354	Anthranilate synthetase, *E. coli*	542
Albumin, bovine serum	55, 188, 408	Anticatalase	72
Albumin, canine serum	253	Apoferritin, Guinea pig	73
Albumin, carp	96	Apoferritin, horse spleen	196
Albumin, cow serum	55, 311, 334, 348, 351, 352, 358, 405	D-Arabinose dehydrogenase, pseudomonad	476
Albumin, dog serum	4	Arachin, *Arachis hypogaea* (ground nut)	108, 111
Albumin esterase, mouse	470	Arginine phosphotransferase, crustacean	61
Albumin, horse serum	55, 235, 297, 336, 356, 402, 403, 404, 411	Ascorbate oxidase apoenzyme, *Cucurbita pepo*	491
Albumin, human serum	55, 177, 300, 351, 410	Ascorbate oxidase, *Cucurbita pepo condensa* (squash)	55, 368
Albumin, rat serum	280, 468	Ascorbate oxidase, *Cucurbita pepo* (inactivated)	491
Alcohol dehydrogenase, horse liver	58, 245		
Alcohol dehydrogenase, yeast	245	Ascorbate oxidase, *Cucurbita pepo* (native)	491
Aldolase	55, 373		
Aldolase, bovine liver, 25°C	184	Ascorbate oxidase, *Cucurbita pepo* (reduced)	491
Aldolase, rabbit liver	500		
Aldolase, rabbit skeletal muscle	243	Aspartate transaminase, ox heart	536
Aldose dehydrogenase, pseudomonad	476	Aspartate transaminase, ox heart (apo-enzyme)	536
Alkaline phosphatase, calf	64		
Alkaline phosphatase, *E. coli*	76	apo-Aspartate transaminase, yeast	490
Amandin	235, 289, 379	Autoprothrombin c, bovine	278
Amino acid oxidase, *Crotalus*	261	Autoprothrombin II, bovine	92
D-Amino acid oxidase apo-enzyme, pig kidney	271	Azurin, *Bordetella*	447

Table 1 (continued)
GLOBULAR PROTEINS

Protein	References	Protein	References
Bacillus phlei protein	235, 286	Cocosin	235, 377
Bence-Jones protein	235, 349	Cobalamin protein B, hog gastric mucosa	262
Bence-Jones protein, human urine	50		
Blue latex protein, *Rhus vernicifera*	175	Colbalamin protein A, hog gastric mucosa	262
Blue protein, *Pseudomonas*	234		
		Collagenase	212
		Conalbumin, chicken egg white	75
Canavalin, Jack bean	235, 295	Fe-Conalbumin, chicken egg white	75
Carbamate kinase	85	Conarachin	109
Carbamyl phosphate synthetase, frog liver	149	Concanavalin A, Jack bean	235, 295, 298
		Concanavalin B, Jack bean	235, 295
Carbonic anhydrase, human	172	Creatine kinase BB, chicken	535
Carbonic anhydrase, mammalian RBC, 25°C	55, 415	Creatine kinase BB, rabbit	535
		Creatine kinase MM, chicken	535
Carbonic anhydrase, ox blood	55, 328	Creatine kinase MM, rabbit	535
Carbonic anhydrase B	140	Crotonase	228
Carbonic anhydrase X_1, human RBC	194	Crotoxin, *Crotalus terrificus* (rattlesnake)	235, 294
Carbonic anhydrase Y, human RBC	194		
α-Carboxylase, wheat germ	218	Cryptocytochrome c, *Pseudomonas*	234
Carboxylesterase, pig kidney	557	Cytochrome, *Chlorobium*	78
Carboxypeptidase	55, 337, 397	Cytochrome, *Rhodopseudomonas*	162
Carboxypeptidase B, pig	69	Cytochrome a, mammalian heart	241
Catalase, bovine liver	100, 202, 235, 303	Cytochrome b_1, E. coli	516
		Cytochrome b_2, yeast	7, 274
Catalase, horse liver	3	Cytochrome c, beef heart	20
Catalase, human blood	34	Cytochrome c, bovine heart	55, 56, 321
Catalase, *Micrococcus*	34	Cytochrome c, *Chromatium*	9
Catalase, pig blood	165	Cytochrome c, horse heart	55, 321, 469
Cathepsin C	88	Cytochrome c, human	449
Cathepsin, cod spleen	458	Cytochrome c, pig heart	55, 321
Ceruloplasmin, human	185, 188, 350	Cytochrome c, rust fungus	167
Ceruloplasmin, human plasma	203	Cytochrome c, vertebrate	235
Ceruloplasmin, pig	185	Cytochrome c, vertebrate heart	235, 289, 407
Ceruloplasmin, pig blood	179	Cytochrome c_1, bovine heart	283
Chloroperoxidase, *Caldariomyces fumago*	466	Cytochrome c_1, *Rhodopseudomonas*	178
		Cytochrome f	44
Chlorocruorin, *Sabella*	6	Cytochrome h	121
Choline acetyltransferase, human placenta	473	Cytochrome oxidase, P. aeruginosa	99
Choline acetyltransferase, rabbit brain	473	DDT dehydrochlorinase	142
Chymopapain	53	Dehydropeptidase II, bovine kidney	214
Chymotrypsin, bovine pancreas	55, 297, 330, 338, 341, 400	Denitrifying enzyme, bacterial (*Pseudomonas*?)	409
		Dextran saccharose, *Leuconostoc*	502
Chymotrypsin β, bovine pancreas	55, 327	Diaphorase, pig heart	204
Chymotrypsin inhibitor, potato	461	Dihydrolipoic dehydrogenase, E. coli	126
Chymotrypsin α (dimer), bovine pancreas	55, 291, 296	Dihydrolipoic dehydrogenase, *Spinacea aleracea*	493
Chymotrypsin α (monomer), bovine pancreas	55, 291	Diphtheria antitoxin	55, 362, 412
		Diphtheria toxin	55, 299
Chymotrypsinogen, bovine pancreas	55, 293, 330		
Chymotrypsinogen α, bovine pancreas	55, 263, 327	Edestin	235, 289, 380
Chymotrypsinogen β, bovine pancreas	55, 327	Enolase	55
Citrate cleaving enzyme, rat liver	512	Enolase, rabbit muscle, 1°C	98
Citrate-oxalacetate lyase, *Aerobacter*	23, 514	Erythrocuprin, human	232
Clostridium toxin	55, 419, 420	Erythrocuprein, human RBC	124
Clupein sulphate	285	Excelsin	235, 289, 379

Table 1 (continued)
GLOBULAR PROTEINS

Protein	References	Protein	References
Ferredoxin, *Clostridium*	237	γ-Globulin, bovine	55, 406
Ferri-cytochrome *c*, bovine heart	171	γ-Globulin, cow	55, 358, 365, 370, 408
Ferro-cytochrome *c*, bovine heart	171		
Fibrinogen, cow	55, 358	γ-Globulin, cow serum	55, 372
Fibrinogen, human	55, 418	γ-Globulin, equine serum	188, 366
Ficin	37	γ-Globulin, horse	177, 180, 188
Flavin mononucleotide (FMN), (dimer) 25°C	544	γ-Globulin, horse serum	55, 336, 370
		γ-Globulin, human	185
Fluorokinase	258	γ-Globulin, human	26, 55, 177, 185, 359
Follicle-stimulating hormone, bovine pituitary	531		
		γ-Globulin, human serum	55, 307
Follicle-stimulating hormone, sheep	55, 301	γ-Globulin, pig serum	55, 371
Fucan, *Fucus vesiculosus*	13	γ-Globulin, rabbit	31
Fumarase, hog heart	55, 374	γ-Globulin, rabbit serum	55, 365, 394
		γ_1-Globulin, human plasma	240
Galactosamino-glycan, bovine cornea	555	Glucosamino-glycan, bovine cornea	555
Galactose dehydrogenase, pseudomonad	476	Glucose oxidase, *P. amagasakiense*	130
β-Galactosidase, *E. coli* ML35	284	Glucose oxidase-microside protein, *Penicillium*	501
β-Galactosidase, *E. coli* ML309	284	Glucose oxidase (Notatin), *Penicillium*	33
Gliadin	70, 235, 322, 398	Glucose-6-phosphate dehydrogenase, human R.B.C.	498
Globulin, *Acacia* seed	43	Glutamate dehydrogenase, bovine liver	55, 305
Globulin, *Arachis* seed	43	Glutamate dehydrogenase, chicken liver	74, 513
Globulin, *Astragalus* seed	43		
Globulin, *Avena* seed	43	Glutamic-aspartic transaminase	107
Globulin, bovine	235, 307, 382	Glyceraldehyde phosphomutase, yeast	243
Globulin, *Cytisus* seed	43	D-Glyceraldehyde-3-phosphate dehydrogenase	60
Globulin, *Dolichos* seed	43		
Globulin, *Ervum* seed	43	D-Glyceraldehyde-3-phosphate dehydrogenase, mammalian	60
Globulin, *Festuca* seed	43		
Globulin, *Genista* seed	43	D-Glyceraldehyde-3-phosphate dehydrogenase, rabbit muscle	60, 243
Globulin, *Glycine* seed	43		
Globulin, *Hordeum vulgare* seed	43	D-Glyceraldehyde-3-phosphate dehydrogenase (+ KCN), rabbit muscle	71
Globulin, horse serum	235, 289, 411		
Globulin, *Lathyrus* seed	43	Glycerokinase, *Candida*	11
Globulin, *Lotus* seed	43	Glycoprotein, bovine plasma	14
Globulin, *Lupinus angustifolius* seed	43, 112	α-Glycoprotein, pleural fluid	22
Globulin, *Medicago* seed	43	Glycoprotein, rat	68
Globulin, *Panicum* seed	43	α-Glycoprotein, fetal calf serum	465
Globulin, *Phaseolus* seed	43	Gonadotrophin, pregnant mare	135
Globulin, *Phleum* seed	43	Growth hormone, pituitary	55, 137, 183, 347, 399
Globulin, pig	235, 307, 382		
Globulin, *Pisum* seed	43		
Globulin, *Secale* seed	43	Haemopexin	209
Globulin, *Trifolium* seed	43	Haptoglobin type *1-1*, human	188, 417
Globulin, *Triticum vulgare* seed	43	Hemagglutinin, castor bean	239
Globulin, *Vicia* seed	43	Hemagglutinin, cold	260
Globulin, *Zea* seed	43	Hemagglutinin, soya bean	181, 256
α-Globulin, barley	43	Hemicellulose A, *Chlorella*	173
α-Globulin, human	188, 343	Hemocyanin, *Achatina*	235, 317
α-Globulin, human plasma	177	Hemocyanin, *Agriolimax*	235, 317
α_2-Globulin	177	Hemocyanin, *Arion*	235, 317
β_1-Globulin	177	Hemocyanin, *Astacus*	235, 364
β-Globulin, human plasma	177	Hemocyanin, *Buccinum*	235, 317, 364
γ-Globulin, barley	43	Hemocyanin, *Busycon*	235, 364

Table 1 (continued)
GLOBULAR PROTEINS

Protein	References	Protein	References
Hemocyanin, *Calocaris*	235, 317	Hemoglobin, *Eumenia*	235, 317
Hemocyanin, *Cancer*	235, 364	Hemoglobin, frog	464
Hemocyanin, *Carcinus*	235, 364	Hemoglobin, *Gasterophilus*	2
Hemocyanin, *Chirodothea*	235, 317	Hemoglobin, *Gasterosteus*	235, 317
Hemocyanin component, *Helix pomatia*	27	Hemoglobin, *Glycera*	235, 317
		Hemoglobin, *Haemopsis*	235, 317
Hemocyanin, *Eledone*	235, 289, 317, 364	Hemoglobin, hedgehog	235, 317
		Hemoglobin, hen	235, 317
Hemocyanin, *Eupagurus*	235, 317	Hemoglobin, *Hirudo*	235, 317
Hemocyanin, *Euscorpius*	235, 317	Hemoglobin, horse	235, 376, 464
Hemocyanin, *Helix pomatia*	27, 235, 289, 364	CO-hemoglobin, horse	55, 287
		Hemoglobin, human	49, 55, 235, 287, 289, 348, 360, 398, 464
Hemocyanin, *Homarus*	235, 289, 364		
Hemocyanin, *Hyas*	235, 317	Hemoglobin, *Lampetra*	5, 136
Hemocyanin, *Limax*	235, 364	Hemoglobin, *Lucioperca*	235, 317
Hemocyanin, *Limnea*	235, 317	Hemoglobin, *Lumbricus*	235, 317
Hemocyanin, *Limulus*	235, 364	Hemoglobin, *Lumbrinereis*	235, 317
Hemocyanin, *Littorina*	235, 317, 364	Hemoglobin, *Marphysa sanguinea*	518
Hemocyanin, *Loligo*	235, 364, 508	Hemoglobin, mouse	464
Hemocyanin, *Maia*	234, 317	Hemoglobin, *Myxine*	235, 317
Hemocyanin, *Nephrops*	235, 289, 364	Hemoglobin, *Neireis*	235, 317
Hemocyanin, *Neptunea*	235, 364	Hemoglobin, *Notomastus*	235, 317
Hemocyanin, octopus	235, 289, 364	Hemoglobin, *Opsanus*	235, 317
Hemocyanin, *Ommatostrephes sloani pacificus* (Squid)	176	Hemoglobin, ox	464
		Hemoglobin, *Paramecium*	225
Hemocyanin, *Pagurus*	235, 317	Hemoglobin, *Parus*	235, 317
Hemocyanin, *Palaemon*	235, 317	Hemoglobin, *Pectinaria*	235, 317
Hemocyanin, *Palinurus*	235, 289, 364	Hemoglobin, perienteric fluid, *Ascaris*	510
Hemocyanin, *Paludina vivipara*	27, 235, 317		
Hemocyanin, *Pandalus*	235, 364	Hemoglobin, *Picus*	235, 317
Hemocyanin, *Rossia*	235, 289, 364	Hemoglobin, pig	464
Hemocyanin, *Sepia*	235, 364	Hemoglobin, pigeon	235, 317, 464
Hemocyanin, *Sepiola*	235, 317	Hemoglobin, *Planorbis*	235, 314, 317
Hemocyanin, *Squilla*	235, 317	Hemoglobin, *Pleuronectes*	235, 317
Hemocyanin, *Tonicella*	235, 317	Hemoglobin, *Polymnia*	235, 317
Hemoglobin, *Aethelges*	235, 317	Hemoglobin, *Prionotus*	235, 317
Hemoglobin, *Andara (= Arca)*	235, 272, 317	Hemoglobin, *Protopterus*	235, 317
Hemoglobin, *Anguilla*	235, 317	Hemoglobin, rabbit	235, 317, 464
Hemoglobin, *Anguis*	235, 317	Hemglobin, *Raja*	235, 317
Hemoglobin, ape	235, 317	Hemoglobin, *Rana*	235, 317, 474
Hemoglobin, *Arenicola*	235, 381	Hemoglobin, *Salmo*	235, 317
Hemoglobin, cat	235, 317	Hemoglobin, sheep	464
Hemoglobin, *Chameleon*	235, 317	Hemoglobin, *Syrnium*	235, 317
Hemoglobin, chicken	464	Hemoglobin, *Tautoga*	235, 317
Hemoglobin, *Chironomous*	235, 317	Hemoglobin, *Thyone*	235, 317
Hemoglobin, *Coluber*	235, 317	Hemoglobin, *Tubifex*	207
Hemoglobin, *Corvus*	235, 317	Hemoglobin, turtle	464
Hemoglobin, cow	235, 317	Hemoglobin A, human	472
CO-hemoglobin, cow	55, 287, 410	Hemoglobin E, human	472
Hemoglobin, *Cyprinus*	235, 317	Hemoglobin F, human	472
Hemoglobin, *Daphnia*	235, 381	Hemoglobin H, human (i)	10
Hemoglobin, dog	235, 317, 464	Hemoglobin H, human (ii)	10
Hemoglobin, dogfish	464	Hemoglobin I, *Bufo*	235, 317
Hemoglobin, duck	235, 317, 464	Hemoglobin I, *Chrysemys*	235, 317
Hemoglobin, *Eisenia*	235, 317	Hemoglobin I, *Lacerta*	235, 317
Hemoglobin, *Esox*	235, 317		

Table 1 (continued)
GLOBULAR PROTEINS

Protein	References
Hemoglobin I, legume	59
Hemoglobin I, *Salamandra*	235, 317
Hemoglobin II, *Bufo*	235, 317
Hemoglobin II, *Chrysemys*	235, 317
Hemoglobin II, *Lacerta*	235, 317
Hemoglobin II, legume	59
Hemoglobin II, *Salamandra*	235, 317
Hemoglobin III, *Bufo*	235, 317
Hemoglobulin, *Lampetra*	235, 314
Hemoglobulin, *Lumbricus*	235, 289
Hemoglobulin, *Planorbis*	235, 376
δ-Hemolysin, staphylococcal	495
Homogentisate oxygenase, *Pseudomonas fluoresens*	509
Hordein	235, 318
Hyaluronic acid	255
p-Hydroxybenzoate hydroxylase, *Pseudomonas putida*	481
Hydroxyproline epimerase, *Pseudomonas*	448
20-β-Hydroxy steroid dehydrogenase *Streptomyces*	497
L-Iditol dehydrogenase	219
Insulin	39, 55, 287, 308–310, 311, 313, 344, 345, 346, 414, 421, 556
Interferon	29
Interstitial-cell stimulating hormone (ICSH), human pituitary	455
Iron-binding protein, cow milk	79
Isocitrate lyase, *Pseudomonas*	506
Isocitric enzyme	163
Iso-hemagglutinin	188, 300
Kallikrein inactivator, bovine parotid	128
Keratinase, *Streptomyces*	454
Laccase	166, 174
Laccase A, *Polyporus*	475
Laccase B, *Polyporus*	475
Lactalbumin, cow	55, 289
Lactalbumin, cow	55
Lactalbumin, human	144
Lactate dehydrogenase, ox heart	157, 168
Lactate dehydrogenase, pig heart	105
Lactic dehydrogenase H, beef heart	281, 282
Lactic dehydrogenase H, chicken	282
Lactic dehydrogenase M, beef heart	282
Lactic dehydrogenase M, chicken	282
Lactogenic hormone (prolactin), sheep	138
Lactoglobulin	235, 289, 339

Protein	References
β-Lactoglobulin	55, 334, 336, 340, 355, 430
β-Lactoglobulin, goat	8
$β_2$-Lactoglobulin, human	208
Lactoperoxidase	247
Lactotransferrin	16, 160
Legumin, *Pisum sativum* (pea)	43
Lipase, milk	277
Lipoamide dehydrogenase, human liver	539
α-Lipoprotein, human	188, 205
α-Lipoprotein, human serum	205
Lipoxidase, soya bean	246
α-Livetin	150
β-Livetin	150
Luciferase, *Cypridina*	66, 250
Luteinising hormone, pig	55, 361
Luteinising hormone, pig	55, 367
Lysozyme, chicken egg white	55, 227, 312, 315, 316, 324
Lysozyme, *Papaya*	224
α-Macroglobulin, rat serum	468
Malate dehydrogenase, beef heart	63
Malate dehydrogenase, bovine heart	217
Malate dehydrogenase, *B. subtilis*	496
Malate dehydrogenase, horse heart	248
Malate dehydrogenase, ox heart	248
Malate dehydrogenase, ox heart, 2°C	45
Malate dehydrogenase, pig heart	248
Malate-lactate transhydrogenase, *Micrococcus lactilyticus*	534
Malic dehydrogenase, pig heart	268
Megacin	97
Merino wool protein, SCMK B1	445
Merino wool protein, SCMK B2	445
Metallothionein, horse kidney	115
Metapyrocatechase, *Pseudomonas*	487
Met-Hemoglobin, horse	55, 357
Methylmalonyl racemase, *Propionibacterium*	459
Milk protein, mouse	199
Monoamine oxidase, ox plasma	273
Monoamine oxidase, plasma	505
Mucoprotein, urinary	154
Myeloperoxidase, infected dog uterus	57
Myogen A, rabbit skeletal muscle	81
Myoglobin, *Aplysia*	200
Myoglobin, guinea pig	94
Myoglobin, horse	200
Myoglobin, horse heart	235, 289, 290
Myoglobin, seal	201
Myoglobin, tortoise	101
Myoglobin, Tunny	192
CO-myoglobin I, mammalian	56
Myokinase	30
Myosin	55, 389, 416

Table 1 (continued)
GLOBULAR PROTEINS

Protein	References
NAD glycohydrolase	463
Nerve protein, lobster	155
Neuraminase, *Vibrio cholerae*	189
Nuclear protein, chicken RBC	440
Nuclear protein, RP2-L, rat carcinoma	443
Old yellow enzyme	244
Old yellow enzyme, brewer's yeast	235, 302
Ornithine transcarbamylase	197
Ovalbumin	55, 235, 348, 398, 408
Oxytocic hormone	55, 331
Papain	226
Paramyosin, *Venus mercenaria*	545
Penicillase, *Bacillus*	187
Penicillinase	89
Pepsin	55, 289, 335, 342, 396, 457, 515
Pepsinogen	457
Peptidase A, *Pencillium*	492
Peroxidase, wheat	215
Peroxidase, wheat germ	215
Peroxidase II, horse radish	35, 247
Phenol oxidase, *Calliphora* larva	116
Phosphoenolpyruvate carboxykinase, pig liver	537
Phosphoglyceric acid mutase, rabbit muscle	54
Phosphoglyceric acid mutase, yeast	54
Phosphoglycerate mutase, chicken muscle	462
Phosphoglucomutase	243
Phospholipase A-I, *Crotalus adamanteus*	488
Phospholipase A-II, *Crotalus adamanteus*	488
Phosphorylase a	84
Phosphorylase b	83
Phosphorylase, potato	134
Phosphorylase, rabbit heart	276
Phosphorylase, rabbit muscle	276
Phosvitin calcium complex	113
Phosvitin magnesium complex	113
Phycocyanin, *Ceramium*	55, 376
Phycoerythrin, *Ceramium*	235, 376, 378
Phytohemagglutinin (lectin), *Phaseolus vulgaris*	538
Plastocyamin, spinach	118
Polyphenol oxidase, *Camelia sinensis* (tea)	482
Pomelin	235, 326
Pre-albumin	188, 350
Procarboxypeptidase	123
Prolactin	55, 332

Protein	References
Prolactin, sheep	453
Protease, *Bacillus thermoproteolyticus*	533
Protease, bacterial	158
Protease, *Streptococcus*	170
Protein, *Lupinus* seed	112
Protein B, lobster nerve	155
Protein X, human blood	49
"A" Protein, *E. coli*	95
Proteinase, *B. subtilis*	446
Proteinase, *Pseudomonas*	102
Proteinase, *Tetrahymena*	51
Proteolytic inhibitor, blood	82
Prothrombin, bovine	92
Prothrombin, bovine	92, 188
Protochlorophyll-protein complex, bean leaf	18
ϵ-Protoxin, *Clostridium*	456
Pseudocholinesterase, horse	106
Pyridine nucleosidase, bull semen	1
Pyrophosphatase	55, 206, 401
Pyrophosphatase, yeast	206
Pyruvic acid oxidase	211
C-Reactive protein	269
Red protein, bovine milk	47
Red protein, cow milk	86
Rhodanese, bovine liver	229
RHP, *Chromatium*	9
RHP, *Rhodospirillum rubrum*	9
Ribonuclease, bovine pancreas	55, 288, 325, 451, 515
Ribonuclease, *B. subtilis*	444
Ribonuclease, pancreatic	444
Ribonuclease T_1	252
Ribonucleoprotein	90
Ricin	114
Ricin D	489
Scarlet fever toxin	55, 333
Secretin phosphate, hog intestine	91, 235
Somatotrophin, beef	137
Somatotropin, human	137
Somatotrophin, sheep	137
Somatotrophin, whale	137
Soya bean protein	267
Succinic dehydrogenase, bacterial	259
Sulfate-binding protein, *Salmonella typhimurium*	532
Taka-amylase A	103, 238
T.B. bacillus protein	235, 286
T Component, equine	188, 366
Thetin-homocysteinase	52
Thiogalactoside transacetylase, *E. coli*	479
Thrombin, bovine	92, 441
Thyroglobulin, pig	48, 235, 289, 383

Table 1 (continued)
GLOBULAR PROTEINS

Protein	References	Protein	References
Transamidinase, hog kidney	486	Triose-phosphate dehydrogenase, turkey	499
Transferrin, human	139, 188, 350, 477, 480, 483, 484, 494	Triose-phosphate dehydrogenase, yeast	499
		Tropomyosin, rabbit muscle	55, 422
Transferrin, human plasma	484	Trypsin	55, 297, 329
Transferrin, monkey	480	Trypsin inhibitor, bovine pancreas	442
Transferrin, pig	363	Trypsin inhibitor, human	188
Transferrin, porcine	188, 363	α-Trypsin inhibitor, human serum	28
Transferrin, rat	480	Trypsin inhibitor, soya bean	15, 193, 450
Transglutaminase, pig liver, 25°C	543	Trypsinogen, bovine pancreas	55, 292
Triose-phosphate dehydrogenase, bovine	499	Tryptophan synthetase B, *E. coli*	478
		Tryptophanase, *E. coli*	504, 507
Triose-phosphate dehydrogenase, chicken	499	Turnip yellow mosaic virus protein	147
		Tyrosinase, *Psalliota* (mushroom)	145
Triose-phosphate dehydrogenase, *E. coli*	499	Tyrosinase L, *Neurospora crassa*	469
		Tyrosinase S, *Neurospora crassa*	469
Triose-phosphate dehydrogenase, halibut	499	Tryosine transaminase, rat kidney	540
		Tween hydrolyzing enzyme	257
Triose-phosphate dehydrogenase, human	499	Urease, Jack bean	235, 304
Triose-phosphate dehydrogenase, lobster	499	Venom substrate, bovine plasma	65
		Vicilin, *Pisum sativum* (pea)	43
Triose-phosphate dehydrogenase, pheasant	499	Viper venom coagulant	265
Triose-phosphate dehydrogenase, rabbit	499	Wheat germ hemoprotein 550	279
Triose-phosphate dehydrogenase, sturgeon	499	Yeast protein	141
		Zein	70

Compiled by Malcolm H. Smith.

Table 2
FIBROUS PROTEINS

Protein	References
F_1-Actin, polymerized	529
F_2-Actin, polymerized	529
G-Actin, cod	38
G-ADP actin, rabbit skeletal muscle	156
α-Amylase, *B. stearothermophilus*	146
Casein, mouse milk	198
α-Casein, cow milk	236
β-Casein, cow milk	236
Ceruloplasmin, human	117
Apo-ceruloplasmin, human	117
Chloroplast protein, spinach leaf	36
Clostridium toxin	55, 419
Collagen, cod skin	275
Collagen, cod swim bladder	275
Collagen A, earth worm cuticle	152
Collagen B, earth worm cuticle	152
α-Crystallin, bovine lens	17, 182
β-Crystallin, bovine lens	17
α-Crystallin, bovine lens cortex	182
Cytochrome b_5, calf liver	233
Elastin (urea-graded), bovine ligamentum nuchae	24
Euglobulin, human pathological	185
Fetuin, bovine	188
Fetuin, calf	186
Fetuin, cow fetus	186, 230
Fibrinogen, bovine	62
Fibrinogen, cow	55, 390
Fibrinogen, horse	210
Fibrinogen, human	32, 55, 177
Globulin, horse antipneumococcus	235, 307
Globulin, human "cold-insoluble"	188, 393
β_1-Globulin, human plasma	177
Eu-γ-globulin, human plasma	210
Globulin, metal-binding, human plasma	210
Gollagen, cod swim bladder	275
Gonadotrophin, human urinary	143
Glycoprotein ("ovoglycoprotein"), egg white	521
M_2-Glycoprotein, bovine	523
α-Glycoprotein, fetal calf serum	465
α-Glycoprotein, human	188, 385
α_1-Glycoprotein	188, 386
Haptoglobin, type 1-1, human	188, 387
Haptoglobin I-(methemoglobin-Hb)[2]	87
Haptoglobin II	87
Haptoglobin II-methemoglobin complex	87
Hemocyanin component, *Paludina vivipara*	27
Heparin fraction	132
β-Histone	40
γ-Histone	40
β-Histone, aggregated	40

Protein	References
Histone I, calf thymus	251
Histone II, calf thymus	251
Iridine, *Salmo*	285
Lewis blood group substance	122
Lipoeuglobulin III, human	188, 391
Lipoprotein, human serum (HDL-3)	93
β-Lipoprotein, human	55, 306
α_1-Lipoprotein, human	177, 188, 338, 388
Lipoprotein, human serum (HDL-2)	93
Lipovitellin	254
β-Lipovitellin	129
β-Livetin, hen egg	264
α_2-Macroglobulin, human	188, 395
α_2-Macroglobulin, human plasma	210
Mucin, pig submaxillary gland	524
Mucoid, bovine oestrous cervical	77
Mucoid, bovine pregnancy cervical	77
Mucoprotein, cartilage	12
Mucoprotein, cow milk	104
Mucoprotein, human	188, 392
Mucoprotein, human plasma	222
Mucoprotein, human urinary	242
Mucoprotein, sheep submaxillary gland	80
Mucoprotein, ovine urinary	528
Mucoprotein, urinary	154, 213
Myosin	55, 110, 389
Myosin, bovine	525
Myosin, cod	38, 527
Myosin, dog heart	526
γ-Myosin	120
Myxomyosin	249
Peptomyosin B	21
Phosvitin	113, 159
Phosvitin, hen serum	159
Plasminogen, human plasma	46, 216
α_3-Protein, canine	253
Pseudo γ-globulin, human plasma	210
α_1-Seromucoid, human plasma	210
Sialoprotein, bovine cortical bone (EDTA extracted)	266
Sialoprotein (phosphate extracted), bovine cortical bone	266
Silk fibroin	520
Spore peptide, *B. megatherium*	191
Tropocollagen, calf skin	522
Tropomyosin, adult blowfly	127
Tropomyosin, larval blowfly	127
Tropomyosin, *Pinna nobilis*	119
Tropomyosin, rabbit	127
Zein	55, 384

Compiled by Malcolm H. Smith.

Table 3
VERTEBRATE SERUM PROTEINS

Vertebrate serum[a]	$S \times 10^{13}$	Vertebrate serum[a]	$S \times 10^{13}$
Ameriurus nebulosus (Bullhead)	3.1	Dromiceius novaehollandie (Emu)	3.8
	5.5		7.4
	9.0		16.4
	12.3	Electrophorus electricus (Electric eel)	3.8
Ancistrodon piscivorus (Cottonmouth moccasin)	3.3		6.7
	5.6		8.8
	14.0		12.9
Bufo marinus (Marine toad)	3.6	Erinaceus europaeus (European hedgehog)	3.5
	6.8		4.7
	13.3		14.6
Caiman sclerops	3.6	Eudocimus albus (White ibis)	4.0
	7.8		6.7
	13.3		18.9
Carcharhinus limbatus (Black-tipped shark)	2.6	Felis domesticus (Cat)	3.3
	7.9		5.9
	10.9		16.5
Chelydra serpentina (Snapping turtle)	3.7	Florida thula (Snowy egret)	3.8
	8.8		17.5
	12.5	Ginglymostoma cirratum (Nurse shark)	3.2
	15.2		6.0
Chrysemys picta (Painted turtle)	4.7		13.0
	6.7	Haliaetus leucoryphas (Fishing eagle)	3.8
	15.0		6.5
	22.1		16.7
Corvus brachyrhyncus (Crow)	3.3	Heterodontus francisci (Horned shark)	4.9
	6.3		7.1
	~13		11.7
	15.0		15.6
Crocodylus morleti (Crocodile)	4.3	Homo sapiens (Man)	3.9
	6.7		5.4
	15.7		15.0
Crotalus adamanteus (Eastern diamondback rattlesnake)	4.2	Ictalurus punctata (Catfish)	3.4
	7.1		7.1
	16.2		13.7
Crotalus viridis (Prairie rattlesnake)	3.7	Iguana iguana	3.5
	4.8		5.8
	15.3		14.5
Cyprinus carpio (Carp)	4.3		18.9
	6.5	Jabiru mycteria (Jabiru stork)	3.9
	10.9		7.1
	13.8		18.2
Dasyatus americana (Sting ray)	3.6	Lampropeltis getulus (King snake)	3.5
	8.8		6.3
	16.1		~8
Deirochelys reticularia (Chicken turtle)	3.6		14.9
	4.4	Lepisosteus osseus (Gar)	4.0
	13.8		13.1
	15.1	Lepidosiren paradoxa (Lungfish)	4.6
Didelphis marsupialis (Opossum)	3.9		18.0
	5.3	Macropus rufo (Red kangaroo)	3.9
	9.6		6.4
	15.4		16.9

[a]Unfractionated serum proteins. The S values were obtained from the sedimentation patterns of serum itself, during centrifugation at ambient room temperature, "very close to 20° and within the range 20–22°C" (530; and Seal, personal communication).

Table 3 (continued)
VERTEBRATE SERUM PROTEINS

Vertebrate serum[a]	$S \times 10^{13}$	Vertebrate serum[a]	$S \times 10^{13}$
Masticopus flagellum (Western coachwhip)	3.6	*Siren lacertina* (Salamander)	4.2
	6.2		~6
	9.6		14.0
	14.9	*Sphenicus humboldti* (Humboldt's penguin)	4.0
Natrix gramhi (Graham's watersnake)	3.7		7.5
	6.5		10.0
	8.1		15.1
	15.5	*Sphyrna tiburo* (Bonnet shark)	3.3
			6.1
Orycteropus afer (Aardvark)	4.5		10.4
	5.5		14.5
	12.7	*Sus scrofa* (Pig)	3.4
			5.8
Petromyzon marinus (Lamprey)	3.6		14.9
	5.6	*Sylvilagus floridanus* (Cottontail rabbit)	4.0
	11.6		4.8
			16.7
Rana catesbeiana (Bullfrog)	4.1	*Tachyglossus aculeatus* (Echidna)	3.6
	7.8		5.7
	14.5		15.5
	16.8	*Terrapene carolina* (Terrapin)	3.5
Rattus norvegicus (Hooded rat)	3.3		4.0
	6.4		7.6
	14.8		13.0
Rhea rhea	3.5	*Xenopus laevi* (Clawed toad)	4.2
	6.0		6.5
	10.3		9.2
	15.0		17.6

Compiled by Malcolm H. Smith.

Table 4
VIRUSES

Protein	References
B. megatherium G phage	164
Bromegrass mosaic virus	19
Chicken sarcoma virus	55, 434
Horse encephalitis virus	55, 432
Influenza virus PR 8	55, 435, 436
Phage fd	151
Phage fr	151
Polyhedral silkworm virus	55, 439
Potato latent mosaic virus	55, 425
Rabbit papilloma virus	55, 169, 433
Southern bean mosaic virus	55, 424
T_7 Bacteriophage	554
T2 phage	55, 437, 438
Tobacco mosaic virus	55, 420, 429
Tobacco mosaic virus TM 58	55, 431
Tobacco necrosis virus	55, 423
Tomato bushy stunt virus	55, 426, 427
Turnip yellow mosaic virus	147

Compiled by Malcolm H. Smith.

Table 5
PARTICLES AND ORGANELLES

Protein	References
Bacterial ribosome	550
Bacterial ribosome particle	550
Fatty acid synthetase particle, pigeon liver	549
Ganglioside micelles, ox-brain	548
α-Ketoglutarate dehydrogenase system, *E. coli*	125
Lysolecithin sol, egg	547
Lysosome, rat liver	353
Mitochondrion, rat liver	353
Pyruvate dehydrogenase system, *E. coli*	125
Sodium taurodeoxycholate micelles	546
Yeast ribosome	551

Compiled by Malcolm H. Smith.

Table 6
CARBOHYDRATES

Protein	References
Fucan, *Fucus vesiculosus*	13
Galactosamino-glycan, bovine cornea	555
Glucosamino-glycan, bovine cornea	555
Hemicellulose A, *Chlorella*	173
Hyaluronic acid	255
Insulin	556

Compiled by Malcolm H. Smith.

REFERENCES

1. Abdel-Latif and Alivisatos, *J. Biol. Chem.*, 237, 500 (1962).
2. Adair, Ogston, and Johnston, *Biochem. J.*, 40, 867 (1946).
3. Agner, in Wyman, *Advanc. Protein Chem.*, 4, 407 (1948).
4. Allerton, Elwyn, Edsall, and Spahr, *J. Biol. Chem.*, 237, 85 (1962).
5. Allison, Cecil, Charlwood, Gratzer, Jacobs, and Snow, *Biochim. Biophys. Acta*, 42, 43 (1960).
6. Antonini, Rossi-Fanelli, and Caputo, *Arch. Biochem. Biophys.*, 97, 343 (1962).
7. Armstrong, Coates, and Morton, *Biochem. J.*, 86, 136 (1963).
8. Askonas, *Biochem. J.*, 58, 332 (1954).
9. Bartsh and Kamen, *J. Biol. Chem.*, 235, 825 (1960).
10. Benesch, *Nature*, 194, 840 (1962).
11. Bergmeyer, Holz, Kauder, Mollering, and Wieland, *Biochem. Z.*, 333, 471 (1961).

12. Bernardi, *Nature,* 180, 93 (1957).
13. Bernardi and Springer, *J. Biol. Chem.,* 237, 75 (1962).
14. Bezkorovainy and Doherty, *Arch. Biochem. Biophys.,* 96, 491 (1962).
15. Birk, Gertler, and Khalef, *Biochem. J.,* 87, 281 (1963).
16. Blanc and Isliker, *Bull. Soc. Chim. Biol.,* 43, 929 (1961).
17. Bloemendal, Bont, Jongkind, and Wisse, *Nature,* 193, 437 (1962).
18. Boardman, *Biochim. Biophys. Acta,* 62, 63 (1962).
19. Bockstahler and Kaesberg, *Biophys. J.,* 2, 1 (1962).
20. Bomstein, Goldberger, and Tisdale, *Biochim. Biophys. Acta,* 50, 527 (1961).
21. Bourdillon, *J. Amer. Chem. Soc.,* 77, 5308 (1955).
22. Bourrillon, Michon, and Got, *Biochim. Biophys. Acta,* 47, 243 (1961).
23. Bowen and Rogers, *Biochem. J.,* 84, 46 P (1962).
24. Bowen, *Biochem. J.,* 55, 766 (1953).
25. Brady and O'Connell, *Biochim. Biophys. Acta,* 62, 216 (1962).
26. Bridgman, *J. Amer. Chem. Soc.,* 68, 857 (1946).
27. Brohult, *J. Phys. Chem.,* 51, 206 (1947).
28. Bundy and Maehl, *J. Biol. Chem.,* 234, 1124 (1959).
29. Burke, *Biochem. J.,* 78, 556 (1961).
30. Callaghan, *Biochem. J.,* 67, 651 (1957).
31. Cammack, *Nature,* 194, 745 (1962).
32. Caspary and Kekwick, *Biochem. J.,* 67, 41 (1957).
33. Cecil and Ogston, *Biochem. J.,* 42, 229 (1948).
34. Cecil and Ogston, *Biochem. J.,* 43, 205 (1948).
35. Cecil and Ogston, *Biochem. J.,* 49, 105 (1951).
36. Chiba, *Arch. Biochem. Biophys.,* 90, 294 (1960).
37. Cohen, *Nature,* 182, 659 (1958).
38. Connell, *Biochem. J.,* 70, 81 (1958).
39. Creeth, *Biochem. J.,* 53, 41 (1958).
40. Cruft, Mauritzen, and Stedman, *Proc. Roy. Soc. Ser. B,* 149, 21 (1958).
41. Danielson, *Nature,* 160, 899 (1947).
42. Danielson, *Nature,* 162, 525 (1948).
43. Danielson, *Biochem. J.,* 44, 387 (1949).
44. Davenport and Hill, *Proc. Roy. Soc. Ser. B,* 139, 327 (1951–52).
45. Davies and Kun, *Biochem. J.,* 66, 307 (1957).
46. Davies and Englert, *J. Biol. Chem.,* 235, 1011 (1960).
47. Derechin and Johnson, *Nature,* 194, 473 (1962).
48. Derrien, Michel, Pedersen, and Roche, *Biochim. Biophys. Acta,* 3, 436 (1949).
49. Derrien and Reynaud, *Compt. Rend. Acad. Sci.* (France), 252, 214 (1961).
50. Deutsch, *J. Biol. Chem.,* 216, 97 (1955).
51. Dickie and Liener, *Biochim. Biophys. Acta,* 64, 41 (1962).
52. Durell, Anderson, and Cantoni, *Biochim. Biophys. Acta,* 26, 270 (1957).
53. Ebata and Yasunobu, *J. Biol. Chem.,* 237, 1086 (1962).
54. Edelhoch, Rodwell, and Grisolia, *J. Biol. Chem.,* 228, 891 (1957).
55. Edsall, in *The Proteins,* 1st ed., Neurath and Bailey Eds., Academic Press, New York IB, 1953, 549.
56. Ehrenberg, *Acta Chem. Scand.,* 11, 1257 (1957).
57. Ehrenberg and Agner, *Acta Chem. Scand.,* 12, 95 (1958).
58. Ehrenberg and Dalziel, *Acta Chem. Scand.,* 12, 465 (1958).
59. Ellfolk, *Acta Chem. Scand.,* 14, 1819 (1960).
60. Elödi, *Acta Physiol. Acad. Scient. Hung.,* 13, 199 (1958).
61. Elödi and Szorenyi, *Acta Physiol. Acad. Scient. Hung.,* 9, 367 (1956).
62. Ende, Meyerhoff, and Schulz, *Z. Naturforsch B,* 13, 713 (1958).
63. Englard and Breiger, *Biochim. Biophys. Acta,* 56, 571 (1962).
64. Engström, *Biochim. Biophys. Acta,* 52, 36 (1961).
65. Esnouf and Williams, *Biochem.,* 84, 62 (1963).
66. Fedden and Chase, *Biochim. Biophys. Acta,* 32, 176 (1959).
67. Fischer, Sumerwell, Junge, and Stein, *Proc. IV Biochem. Congress VIII,* 124 (1958).
68. Fishkin and Berenson, *Arch. Biochem. Biophys.,* 95, 130 (1961).
69. Folk, Piez, Carroll, and Gladner, *J. Biol. Chem.,* 235, 2272 (1960).
70. Foster and French, *J. Amer. Chem. Soc.,* 67, 687 (1964).
71. Fox and Dandliker, *J. Biol. Chem.,* 218, 53 (1956).
72. Friedberg, *Experientia,* 18, 164 (1962).
73. Friedberg, *Can. J. Biochem.,* 40, 983 (1962).

74. Frieden, *Biochim. Biophys. Acta,* 62, 421 (1962).
75. Fuller and Briggs, *J. Amer. Chem. Soc.,* 78, 5253 (1956).
76. Garen and Levinthal, *Biochim. Biophys. Acta,* 38, 470 (1960).
77. Gibbons and Glover, *Biochem. J.,* 73, 217 (1959).
78. Gibson, *Biochem. J.,* 79, 151 (1961).
79. Gordon, Ziegler, and Basch, *Biochim. Biophys. Acta,* 60, 410 (1962).
80. Gottschalk and McKenzie, *Biochim. Biophys. Acta,* 54, 226 (1961).
81. Gralén, *Biochem. J.,* 33, 1342 (1939).
82. Gray, Priest, Blatt, Westphal, and Jensen, *J. Biol. Chem.,* 235, 57 (1960).
83. Green, *J. Biol. Chem.,* 158, 315 (1945).
84. Green and Cori, *J. Biol. Chem.,* 151, 21 (1943).
85. Grisolia, Harmon, and Raijman, *Biochim. Biophys. Acta,* 62, 293 (1962).
86. Groves, *J. Amer. Chem. Soc.,* 82, 3345 (1960).
87. Guinand, Tonnelat, Boussier, and Jayle, *Bull. Soc. Chim. Biol.,* 38, 329 (1956).
88. de la Haba, Cammarata, and Timasheff, *J. Biol. Chem.,* 234, 316 (1959).
89. Hall and Ogston, *Biochem. J.,* 62, 401 (1956).
90. Hamilton, Cavalieri, and Peterman, *J. Biol. Chem.,* 237, 1155 (1962).
91. Hammarsten, Agren, Hammarsten, and Wilander, *Biochem. Z.,* 264, 275 (1933).
92. Harmison and Seegers, *J. Biol. Chem.,* 237, 3074 (1962).
93. Hazelwood, *J. Amer. Chem. Soc.,* 80, 2152 (1958).
94. Helwig and Greenburg, *J. Biol. Chem.,* 198, 695 (1952).
95. Henning, Helinski, Chao, and Yanofsky, *J. Biol. Chem.,* 237, 1523 (1962).
96. Henrotte, *Arch. Int. Physiol. Biochim.,* 62, 294 (1954).
97. Holland, *Biochem. J.,* 78, 641 (1961).
98. Holt and Wold, *J. Biol. Chem.,* 236, 3227 (1961).
99. Horio, Higashi, Yamanaka, Matsubara, and Okunuki, *J. Biol. Chem.,* 236, 944 (1961).
100. Sumner and Gralén, *J. Biol. Chem.,* 125, 33 (1935).
101. Huys, *Arch. Int. Physiol. Biochim.,* 62, 296 (1954).
102. Inoue, Nakagawa, and Morihara, *Biochim. Biophys. Acta,* 73, 125 (1963).
103. Isemura and Fujita, *J. Biochem.* (Japan), 44, 443 (1957).
104. Jackson, Coulson, and Clark, *Arch. Biochem. Biophys.,* 97, 373 (1962).
105. Jaenicke and Pfleiderer, *Biochim. Biophys. Acta,* 60, 615 (1962).
106. Jansz and Cohen, *Biochim. Biophys. Acta,* 56, 531 (1962).
107. Jenkins, Yphantis, and Sizer, *J. Biol. Chem.,* 234, 51 (1959).
108. Johnson, *Trans. Faraday Soc.,* 42, 28 (1946).
109. Johnson and Naismith, *Discuss. Faraday Soc.,* 13, 98 (1953).
110. Johnson and Rowe, *Biochem. J.,* 74, 432 (1960).
111. Johnson and Shooter, *Biochim. Biophys. Acta,* 5, 361 (1950).
112. Joubert, *Biochim. Biophys. Acta,* 16, 370 (1955).
113. Joubert and Cook, *Can. J. Biochem.,* 36, 399 (1958).
114. Kabat, Heidelberger, and Bezer, *J. Biol. Chem.,* 168, 629 (1947).
115. Kägi and Vallee, *J. Biol. Chem.,* 236, 2435 (1961).
116. Karlson and Liebau, *Hoppe-Seyler's Z. Physiol. Chem.,* 326, 135 (1961).
117. Kasper and Deutsch, *J. Biol. Chem.,* 238, 2325 (1963).
118. Katoh, Shiratori, and Takamiya, *J. Biochem.* (Japan), 51, 32 (1962).
119. Kay, *Biochim. Biophys. Acta,* 27, 469 (1958).
120. Kay and Pabst, *J. Biol. Chem.,* 237, 727 (1962).
121. Keilin, *Biochem. J.,* 64, 663 (1956).
122. Kekwick, *Biochem. J.,* 50, 471 (1952).
123. Keller, Cohen, and Neurath, *J. Biol. Chem.,* 223, 457 (1956).
124. Kimmel, Markowitz, and Brown, *J. Biol. Chem.,* 234, 46 (1959).
125. Koike, Reed, and Carroll, *J. Biol. Chem.,* 235, 1924 (1960).
126. Koike, Shah, and Reed, *J. Biol. Chem.,* 235, 1939 (1960).
127. Kominz, Maruyama, Levenbrook, and Lewis, *Arch. Biochem. Biophys.,* 63, 106 (1962).
128. Kraut, Korbel, Scholtan, and Schultz, *Hoppe-Seyler's Z. Physiol. Chem.,* 321, 90 (1960).
129. Kratohvil, Martin, and Cook, *Can. J. Biochem.,* 40, 877 (1962).
130. Kusai, Bekuzu, Hagihara, Okunuki, Yamauchi, and Nakai, *Biochim. Biophys. Acta,* 40, 555 (1960).
131. Lauffer, *J. Amer. Chem. Soc.,* 66, 1188 (1944).
132. Laurent, *Arch. Biochem. Biophys.,* 92, 224 (1961).
133. Lee, *J. Biol. Chem.,* 227, 993 (1957).
134. Lee, *Biochim. Biophys. Acta,* 43, 18 (1960).
135. Legault-Démare, Clauser, and Jutisz, *Bull. Soc. Chim. Biol.,* 43, 897 (1961).

136. Lenhert, Love, and Carlson, *Biol. Bull.*, 111, 293 (1956).
137. Li, (1958) *Symposium on Protein Structure,* Neuberger, Ed., Methuen, London, 1958, 302.
138. Li, Cole, and Coval, *J. Biol. Chem.*, 229, 153 (1957).
139. Bezkorovainy and Rafelson, *Arch. Biochem. Biophys.*, 107, 302 (1964).
140. Lindskog, *Biochim. Biophys. Acta*, 39, 218 (1960).
141. Lindquist, *Biochim. Biophys. Acta*, 10, 443 (1953).
142. Lipke and Kearns, *J. Biol. Chem.*, 234, 2123 (1959).
143. Lundgren, Gurin, Bachman, and Wilson, *J. Biol. Chem.*, 142, 367 (1942).
144. Maeno and Kiyosawa, *Biochem. J.*, 83, 271 (1962).
145. Mallette and Dawson, *Arch. Biochem. Biophys.*, 23, 29 (1949).
146. Manning, Campbell, and Foster, *J. Biol. Chem.*, 236, 2958 (1961).
147. Markham, *Discuss. Faraday Soc.*, 11, 221 (1951).
148. Markovitz, Klein, and Fischer, *Biochim. Biophys. Acta*, 19, 267 (1956).
149. Marshall, Metzenberg, and Cohen, *J. Biol. Chem.*, 236, 2229 (1961).
150. Martin, Vandegaer, and Cook, *Can. J. Biochem.*, 35, 241 (1957).
151. Marvin and Hoffman-Berling, *Nature*, 197, 517 (1963).
152. Maser and Rice, *Biochim. Biophys. Acta*, 63, 255 (1962).
153. Massey, Palmer, and Bennett, *Biochim. Biophys. Acta*, 48, 1 (1961).
154. Maxfield, *Arch. Biochem. Biophys.*, 55, 382 (1955).
155. Maxfield and Hartley, *J. Biophys. Biol. Cytol.*, 1, 279 (1955).
156. Mihashi, *Arch. Biochem. Biophys.*, 107, 441 (1964).
157. Millar, *J. Biol. Chem.*, 237, 2135 (1962).
158. Mills and Wilkins, *Biochim. Biophys. Acta*, 30, 63 (1958).
159. Mok, Martin, and Common, *Can. J. Biochem.*, 39, 109 (1961).
160. Montreuil, Tonnelat, and Mullet, *Biochim. Biophys. Acta*, 45, 413 (1960).
161. Moore, *J. Biol. Chem.*, 161, 597 (1945).
162. Morita, *J. Biochem.* (Japan), 48, 870 (1960).
163. Moyle and Dixon, *Biochem. J.*, 63, 548 (1956).
164. Murphy and Philipson, *J. Gen. Physiol.*, 45, 155 (1962).
165. Nagahisa, *J. Biochem.* (Japan), 51, 216 (1962).
166. Nakamura, *Biochim. Biophys. Acta*, 30, 44 (1958).
167. Neilands, *J. Biol. Chem.*, 197, 701 (1952).
168. Neilands, *J. Biol. Chem.*, 208, 225 (1954).
169. Neurath, Cooper, Sharp, Taylor, Beard, and Beard, *J. Biol. Chem.*, 140, 293 (1940).
170. Nomoto and Narahashi, *J. Biochem.* (Japan), 46, 1645 (1959).
171. Nozaki, *J. Biochem.* (Japan), 47, 592 (1960).
172. Nyman, *Biochim. Biophys. Acta*, 52, 1 (1961).
173. Olaitan and Northcote, *Biochem. J.*, 82, 509 (1962).
174. Omura, *J. Biochem.* (Japan), 50, 264 (1961).
175. Omura, *J. Biochem.* (Japan), 50, 394 (1961).
176. Omura, Fujita, Yamada, and Yamato, *J. Biochem.* (Japan), 50, 400 (1961).
177. Oncley, Scatchard, and Brown, *J. Phys. Colloid. Chem.*, 51, 184 (1947).
178. Orlando, *Biochim. Biophys. Acta*, 57, 373 (1962).
179. Osaki, *J. Biochem.* (Japan), 48, 190 (1960).
180. Pain, *Biochem. J.*, 88, 234 (1963).
181. Pallansch and Liener, *Arch. Biochem. Biophys.*, 45, 366 (1953).
182. Papaconstantinou, Resnik, and Saito, *Biochim. Biophys. Acta*, 60, 205 (1962).
183. Papkoff, Li, and Liu, *Arch. Biochem. Biophys.*, 96, 216 (1962).
184. Peanasky and Lardy, *J. Biol. Chem.*, 233, 371 (1958).
185. Pedersen, in *Lés Protéines,* Stoops, Ed., Neuviéme Conseil de Chimie, Bruxelles, 1953.
186. Pedersen, *J. Phys. Chem.*, 51, 164 (1947).
187. Pollock and Torriani, *Compt. Rend. Acad. Sci.* (France), 237, 276 (1953).
188. Phelps and Putnam, in *Plasma Proteins,* Putnam, Ed., Academic Press, New York, 1960, 143.
189. Pye and Curtain, *J. Gen. Microbiol.*, 24, 423 (1961).
190. Raijman, Grisolia, and Edelhoch, *J. Biol. Chem.*, 235, 2340 (1960).
191. Record and Grinstead, *Biochem. J.*, 58, 85 (1954).
192. Renard, *Arch. Int. Physiol. Biochem.*, 61, 466 (1953).
193. Wu and Scheraga, *Biochemistry*, 1, 698 (1962).
194. Reynaud, Rametta, Savary, and Derrien, *Biochim. Biophys. Acta*, 77, 521 (1963).
195. Richterich, Temperli, and Aebi, *Biochim. Biophys. Acta*, 56, 240 (1962).
196. Rothen, *J. Biol. Chem.*, 152, 679 (1944).

197. Rogers and Novelli, *Arch. Biochem. Biophys.*, 96, 398 (1962).
198. Ross and Moore, *Biochim. Biophys. Acta*, 16, 293 (1955).
199. Ross and Moore, *Biochim. Biophys. Acta*, 15, 50 (1954).
200. Rossi-Fanelli, Antonini, and Poveledo, in *Symposium on Protein Structure*, Neuberger, Ed., Methuen, London, 1958, 144.
201. Rumen and Appella, *Arch. Biochem. Biophys.*, 97, 128 (1962).
202. Samejima, Kamata, and Shibata, *J. Biochem.* (Japan), 51, 181 (1962).
203. Sanders, Miller, and Richard, *Arch. Biochem. Biophys.*, 84, 60 (1959).
204. Savage, *Biochem. J.*, 67, 146 (1957).
205. Seanu, Lewis, and Bumpus, *Arch. Biochem. Biophys.*, 74, 390 (1958).
206. Schachman, *J. Gen. Physiol.*, 35, 451 (1952).
207. Scheler and Schneiderat, *Acta Biol. Med. Germ.*, 3, 588 (1959).
208. Schultze, Heide, and Haupte, *Naturwiss.*, 48, 719 (1961).
209. Schultze, Heide, and Haupte, *Naturwiss.*, 48, 696 (1961).
210. Schultze, Schmidtberger, and Haupte, *Biochem. Z.*, 329, 490 (1958).
211. Schweet, Katchman, Bolk, and Jagannathan, *J. Biol. Chem.*, 196, 563 (1952).
212. Seifter, Gallop, Klein, and Meilman, *J. Biol. Chem.*, 234, 285 (1959).
213. Seppala, *Scand. J. Clin. Lab. Invest.*, 13, 665 (1961).
214. Shack, *J. Biol. Chem.*, 180, 411 (1949).
215. Shin and Nakamura, *J. Biochem.* (Japan), 50, 500 (1961).
216. Shulman, Alkjaersig, and Sherry, *J. Biol. Chem.*, 233, 91 (1958).
217. Siegel and Englard, *Biochim. Biophys. Acta*, 54, 67 (1961).
218. Singer and Pensky, *J. Biol. Chem.*, 196, 375 (1952).
219. Smith, *Biochem. J.*, 83, 135 (1962).
220. Smith, *Biochem. J.*, 89, 45P (1963).
221. Smith, *J. Theoret. Biol.*, 13, 261 (1966).
222. Smith, Brown, Weimer, and Winzler, *J. Biol. Chem.*, 185, 569 (1950).
223. Smith, Brown, Ghosh, and Sayers, *J. Biol Chem.*, 187, 631 (1950).
224. Smith, Kimmel, Brown, and Thompson, *J. Biol. Chem.*, 215, 67 (1955).
225. Smith, George, and Preer, *Arch. Biochem. Biophys.*, 99, 313 (1962).
226. Smith, Hill, and Kimmel, in *Symposium on Protein Structure*, Neuberger Ed., Methuen, London, 1958, 182.
227. Sophianopoulos, Rhodes, Holcomb, and Van Holde, *J. Biol. Chem.*, 237, 1107 (1962).
228. Stern, del Campillo, and Raw, *J. Biol. Chem.*, 218, 971 (1956).
229. Sörbo, *Acta Chem. Scand.*, 7, 1129 (1953).
230. Spiro, *J. Biol. Chem.*, 235, 2860 (1960).
231. Squire and Li, *J. Amer. Chem. Soc.*, 83, 3521 (1961).
232. Stansell and Deutsch, *J. Biol. Chem.*, 240, 4306 (1965).
233. Strittmatter, *J. Biol. Chem.*, 235, 2492 (1960).
234. Suzuki and Iwasaki, *J. Biochem.* (Japan), 52, 193 (1962).
235. Svedberg and Pedersen, in *The Ultracentrifuge*, Oxford University Press, London, 1940.
236. Sullivan, Fitzpatrick, Stanton, Annino, Kissel, and Palermiti, *Arch. Biochem. Biophys.*, 55, 455 (1955).
237. Tagawa and Arnon, *Nature*, 195, 537 (1962).
238. Takagi and Toda, *J. Biochem.* (Japan), 52, 16 (1962).
239. Takashi, Funatsu, and Funatsu, *J. Biochem.* (Japan), 52, 50 (1962).
240. Takahashi and Schmid, *Biochim. Biophys. Acta*, 63, 343 (1962).
241. Takemori, Sekuzu, and Okunuki, *Biochim. Biophys. Acta*, 51, 464 (1961).
242. Tamm, Bugher, and Horsfall, *J. Biol. Chem.*, 212, 125 (1955).
243. Taylor and Lowry, *Biochim. Biophys. Acta*, 20, 109 (1956).
244. Theorell and Åkeson, *Arch. Biochem. Biophys.*, 65, 439 (1956).
245. Theorell and Bonnichsen, *Acta Chem. Scand.*, 5, 1105 (1951).
246. Theorell, Holman, and Akeson, *Acta Chem. Scand.*, 1, 571 (1947).
247. Theorell and Pedersen, in Wyman, *Advanc. Protein Chem.*, 4, 407 (1948).
248. Thorne, *Biochim. Biophys. Acta*, 59, 624 (1962).
249. Ts'o, Eggman, and Vinograd, *Biochim. Biophys. Acta*, 25, 532 (1957).
250. Tsuji and Sowinski, *J. Cell. Comp. Physiol.*, 58, 125 (1961).
251. Ui, *Biochim. Biophys. Acta*, 25, 493 (1957).
252. Ui and Tarutani, *J. Biochem.* (Japan), 49, 9 (1961).
253. Uspenskaya, Alekseenko, Rodionov, and Solov'eva, *Biokhimiya*, 26, 592 (1961).
254. Vandegaer, Reichmann, and Cook, *Arch. Biochem. Biophys.*, 62, 328 (1956).
255. Varga, Pietruszkiewicz, and Ryan, *Biochim. Biophys. Acta*, 32, 155 (1959).
256. Wada, Pallansch, and Leiner, *J. Biol. Chem.*, 233, 395 (1958).

257. Wallach, Ko, and Marshall, *Biochim. Biophys. Acta,* 59, 690 (1962).
258. Warner, *Arch. Biochem. Biophys.*, 78, 494 (1958).
259. Warringa, Smith, Guiditta, and Singer, *J. Biol. Chem.*, 230, 97 (1958).
260. Weber, *Vox Sang. N.S.*, 1, 37 (1956).
261. Wellner and Meister, *J. Biol. Chem.*, 235, 2013 (1960).
262. Wijmenga, Thompson, Stern, and O'Connell, *Biochim. Biophys. Acta,* 13, 144 (1954).
263. Wilcox, Kraut, Wade, and Neurath, *Biochim. Biophys. Acta,* 24, 72 (1957).
264. Williams, *Biochem. J.,* 83, 346 (1962).
265. Williams and Esnouf, *Biochem. J.,* 84, 52 (1963).
266. Williams and Peacocke, *Biochim. Biophys. Acta,* 101, 327 (1965).
267. Wolf and Briggs, *Arch. Biochem. Biophys.*, 85, 186 (1959).
268. Wolfe and Nielands, *J. Biol. Chem.*, 221, 61 (1956).
269. Wood, McCarty, and Slater, *J. Exp. Med.*, 100, 71 (1954).
270. Yagi and Ozawa, *Biochim. Biophys. Acta,* 62, 397 (1962).
271. Yagi, Ozawa, and Ooi, *Biochim. Biophys. Acta,* 54, 199 (1961).
272. Yagi, Mishima, Tsujimura, Sato, and Egami, *Compt. Rend. Sos. Biol.*, 149, 2283 (1955).
273. Yamada and Yasunobu, *J. Biol. Chem.*, 237, 1511 (1962).
274. Yamashita and Okunuki, *J. Biochem.* (Japan), 52, 117 (1962).
275. Young and Lorimer, *Arch. Biochem. Biophys.*, 92, 183 (1961).
276. Yunis, Fischer and Krebs, *J. Biol. Chem.*, 237, 2109 (1962).
277. Chandon, Shahani, Hill, and Scholz, *Enzymologia,* 26, 87 (1963).
278. Seegers, Cole, Harrison and Marciniak, *Can. J. Biochem.*, 41, 1047 (1963).
279. Wasserman and Burris, *Phytochemistry,* 4, 413 (1965).
280. Jungblut and Turba, *Biochem. Z.,* 337, 88 (1963).
281. Markert and Appella, *Ann. N.Y. Acad. Sci.*, 94, 678 (1961).
282. Pesce, McKay, Stolzenbach, Cahn and Kaplan, *J. Biol. Chem.*, 239, 1753 (1964).
283. Criddle, Bock, Green, and Tisdale, *Biochemistry,* 1, 827 (1962).
284. Sund and Weber, *Biochem. Z.,* 337, 24 (1963).
285. Gehatia and Hashimoto, *Biochim. Biophys. Acta,* 69, 212 (1963).
286. Seibert, Pedersen and Tiselius, *J. Exp. Med.*, 68, 413 (1938).
287. Pedersen, *Cold Spring Harbor Symp.*, 14, 140 (1949).
288. Rothen, *Gen. Physiol.*, 24, 203 (1940).
289. Polson, *Thesis,* University of Stellenbosch, 1937.
290. Theorell, *Biochem. Z.,* 268, 46 (1934).
291. Schwert and Kaufman, *J. Biol. Chem.*, 190, 807 (1951).
292. Tietze, *J. Biol. Chem.*, 204, 1 (1953).
293. Schwert, *J. Biol. Chem.*, 190, 799 (1951).
294. Gralén, and Svedberg, *Biochem. J.,* 32, 1375 (1938).
295. Sumner, Gralén, and Eriksson- Quensel, *J. Biol. Chem.*, 125, 45 (1938).
296. Schwert, *J. Biol. Chem.*, 179, 655, (1949).
297. Bergold, *Z. Naturforsch.*, 1, 100 (1946).
298. Agrawal and Goldstein, *Biochim. Biophys. Acta,* 133, 376 (1967).
299. Petermann and Pappenheimer, *J. Phys. Chem.*, 45, 1 (1941); Tiselius and Dahl, *Ark. Kemi.,* 14B No. 31, 7 (1941).
300. Pedersen, *Ultracentrifugal Studies on Serum and Serum Fractions,* Almqvist and Wiksell, Boktryckeri AB, Uppsala, 1945.
301. Li and Pedersen, *J. Gen. Physiol.*, 35, 629 (1952).
302. Kekwick and Pedersen, *Biochem. J.,* 30, 2201 (1936).
303. Sumner and Gralén, *Science,* 87, 284 (1938); Sumner and Gralén, *J. Biol. Chem.*, 125, 33 (1935).
304. Sumner, Gralén, and Eriksson-Quensel, *J. Biol. Chem.*, 125, 37 (1938).
305. Olson and Anfinsen, *J. Biol. Chem.*, 197, 67 (1952).
306. Pedersen, *J. Phys. Colloid Chem.*, 51, 156 (1947).
307. Kabat, *J. Exp. Med.*, 69, 103 (1939).
308. Fredericq and Neurath, *J. Amer. Chem. Soc.*, 72, 2684 (1950).
309. Moody, *Ph.D. Thesis,* University of Wisconsin, 1944; quoted in Williams, *Annu. Rev. Phys. Chem.*, 2, 412 (1951).
310. Gutfreund, *Biochem. J.,* 50, 564 (1952).
311. Taylor, *Arch. Biochem. Biophys.*, 36, 357 (1952).
312. Abraham, *Biochem. J.,* 33, 622 (1939).
313. Oncley, Ellenbogen, Gitlin, and Gurd, *J. Phys. Chem.*, 56, 85 (1952).
314. Svedberg and Eriksson-Quensel, *J. Amer. Chem. Soc.*, 56, 1700 (1934).
315. Alderton, Ward and Fevold, *J. Biol. Chem.*, 157, 43 (1945).
316. Passynskii and Plaskeyev, *C.R. Acad. Sci. URSS,* 48, 579 (1945).
317. Svedberg and Hedenius, *Biol. Bull.*, 66, 191 (1934).

318. Quensel and Svedberg, *Compt. Rend. Lab. Carlsberg,* 22, 441 (1938).
319. Li and Pedersen, *Arch. Biochem. Biophys.,* 36, 462 (1952).
320. Sayers, White and Long, *J. Biol. Chem.,* 149, 425 (1943).
321. Atlas and Farber, *J. Biol. Chem.,* 219, 31 (1956).
322. Krejci and Svedberg, *J. Amer. Chem. Soc.,* 57, 946 (1935).
323. Li, Evans, and Simpson, *J. Biol. Chem.,* 149, 413 (1943).
324. Wetter and Deutsch, *J. Biol. Chem.,* 192, 237 (1951).
325. Vilbrandt, Tennent, and Hakala, quoted in Bridgman and Williams, *Ann. N.Y. Acad. Sci.,* 43, 195 (1942).
326. Krejci and Svedberg, *J. Amer. Chem. Soc.,* 56, 1706 (1934).
327. Smith, Brown, and Laskowski, *J. Biol. Chem.,* 191, 639 (1951).
328. Peterman and Hakala, *J. Biol. Chem.,* 145, 701 (1942).
329. Cunningham, Tietze, Green, and Neurath, *Discuss. Faraday Soc.,* No. 13, 58 (1953).
330. Schwert and Kaufman, *J. Biol. Chem.,* 180, 517 (1949).
331. Van Dyke, Chow, Greep, and Rothen, *J. Pharmacol. Exp. Ther.,* 74, 190 (1942).
332. Li, Lyons, and Evans, *J. Biol. Chem.,* 140, 43 (1941).
333. Krejci, Stock, Sanigar, and Kraemer, *J. Biol. Chem.,* 142, 785 (1942).
334. Miller and Golder, *Arch. Biochem. Biophys.,* 36, 249 (1952).
335. Steinhardt, *J. Biol. Chem.,* 123, 543 (1938).
336. Johnston and Ogston, *Trans. Faraday Soc.,* 42, 789 (1946).
337. Smith, Brown, and Hanson, *J. Biol. Chem.,* 180, 33 (1949).
338. Hess and Williams, cited in Oncley, *Ann. N.Y. Acad. Sci.,* 41, 121 (1942).
339. Pederson, *Biochem. J.,* 30, 948 (1936).
340. Bain and Deutsch, *Arch. Biochem. Biophys.,* 16, 221 (1948).
341. Schwert, *J. Biol. Chem.,* 179, 655 (1949).
342. Philpot, *Biochem. J.,* 29, 2458 (1935).
343. Wallenius, Trautman, Kunkel, and Franklin, *J. Biol. Chem.,* 225, 253 (1957).
344. Gutfreund and Ogston, *Biochem. J.,* 40, 432 (1946).
345. Gutfreund, *Biochem. J.,* 42, 544 (1948).
346. Miller and Anderson, *J. Biol. Chem.,* 144, 459 (1942).
347. Smith, Brown, Fishman, and Wilhelmi, *J. Biol. Chem.,* 177, 305 (1949).
348. Kegeles and Gutter, *J. Amer. Chem. Soc.,* 73, 3770 (1951).
349. Svedberg and Sjogren, *J. Amer. Chem. Soc.,* 51, 3594 (1929).
350. Schultze and Schwick, *Clin. Chim. Acta,* 4, 15 (1959).
351. Charlwood, *Biochem. J.,* 51, 113 (1952).
352. Creeth, *Biochem. J.,* 51, 10 (1952).
353. Corbett, *Biochem. J.,* 102, 43P (1967).
354. Loeb and Scheraga, *J. Phys. Chem.,* 60, 1633 (1956); Wagner and Scheraga, *J. Phys. Chem.,* 60, 1066 (1956).
355. Ogston, *Proc. Roy. Soc. A* (London), 196, 272 (1949).
356. Kekwick, *Biochem. J.,* 32, 552 (1938).
357. Gutfreund, cited in Boyes-Watson, Davidson, and Perutz, *Proc. Roy. Soc. A* (London), 191, 83 (1947).
358. Koenig, *Arch. Biochem. Biophys.,* 25, 241 (1950); Koenig and Pedersen, *Arch. Biochem. Biophys.,* 25, 97 (1950).
359. Ikenaka, Gitlin, and Schmid, *J. Biol. Chem.,* 240, 2868 (1965).
360. Moon and Reiner, *J. Biol. Chem.,* 156, 411 (1944).
361. Meyer, Thompson, Palmer, and Khorazo, *J. Biol. Chem.,* 113, 303 (1936).
362. Northrop, *J. Gen. Physiol.,* 25, 465 (1941-42); Rothen, *J. Gen. Physiol.,* 25, 487 (1941-42).
363. Laurell and Ingleman, *Acta Chem. Scand.,* 1, 770 (1947).
364. Eriksson-Quensel and Svedberg, *Biol. Bull.,* 71, 498 (1936).
365. Cann, Brown, Stringer, Shumaker, and Kirkwood, *Science,* 114, 30 (1951).
366. Largier, *Arch. Biochem. Biophys.,* 77, 350 (1958).
367. Shedlovsky, Rothen, Greep, Van Dyke, and Chow, *Science,* 92, 178 (1940).
368. Dunn and Dawson, *J. Biol. Chem.,* 189, 485 (1951).
369. Singer and Kearney, *Arch. Biochem.,* 29, 190 (1950).
370. Smith and Brown, *J. Biol. Chem.,* 183, 241 (1950).
371. Koenig, *Arch. Biochem.,* 23, 229 (1949).
372. Hess and Deutsch, *J. Amer. Chem. Soc.,* 70, 84 (1948).
373. Glikina and Finogenov, *Biokhimiya,* 15, 457 (1950).
374. Massey, *Biochem. J.,* 51, 490 (1952).
375. Englard and Singer, *J. Biol. Chem.,* 187, 213 (1950).
376. Tiselius and Gross, *Kolloid Z.,* 66, 11 (1934).
377. Sjögren and Spychalski, *J. Amer. Chem. Soc.,* 52, 4400 (1930).
378. Eriksson-Quensel, *Biochem. J.,* 32, 585 (1938).
379. Svedberg and Sjögren, *J. Amer. Chem. Soc.,* 52, 279 (1930).

380. Svedberg and Stamm, *J. Amer. Chem. Soc.,* 51, 2170 (1929).
381. Svedberg and Eriksson, *J. Amer. Chem. Soc.,* 55, 2834 (1933).
382. Kabat and Pederson, *Science,* 87, 372 (1938).
383. Heidelberger and Pedersen, *J. Gen. Physiol.,* 19, 95 (1935).
384. Watson, Arrhenius and Williams, *Nature,* 137, 322 (1936).
385. Schmid, *Biochem. Biophys. Acta,* 21, 399 (1956).
386. Schultze, Göllner, Heide, Schönenberger and Schwick, *Z. Naturforsch.,* 10b, 463 (1955).
387. Jayle and Boussier, *Exposés Annu. Biochim. Méd.,* 17, 157 (1955).
388. Shore, *Arch. Biochem. Biophys.,* 71, 1 (1957).
389. Snellman and Erdös, *Biochim. Biophys. Acta,* 2, 650 (1948).
390. Ehrlich, Shulman, and Ferry, *J. Amer. Chem. Soc.,* 74, 2258 (1952).
391. Sandor and Slizewicz, *Bull. Soc. Chim. Biol.,* 39, 857 (1957).
392. Brown, Baker, Peterkofsky, and Kauffman, *J. Amer. Chem. Soc.,* 76, 4244 (1954).
393. Edsall, Gilbert, and Scheraga, *J. Amer. Chem. Soc.,* 77, 157 (1955).
394. Nichol and Deutsch, *J. Amer. Chem. Soc.,* 70, 80 (1948).
395. Schönenberger, Schmidtberger, and Schultze, *Z. Naturforsch.,* 13b, 761 (1958).
396. Northrop, *J. Gen. Physiol.,* 13, 739 (1930).
397. Putnam, Neurath, Elkins, and Segal, *J. Biol. Chem.,* 166, 603 (1946).
398. Lamm and Polson, *Biochem. J.,* 30, 528 (1936).
399. Li, *J. Phys. Colloid Chem.,* 51, 218 (1947).
400. Kunitz and Northrop, *J. Gen. Physiol.,* 18, 433 (1935).
401. Kunitz, cited by Schachman, *J. Gen. Physiol.,* 35, 451 (1952).
402. Champagne, *J. Chim. Phys.,* 48, 627 (1951).
403. Kojiro and Wanatabe, *Rep. Radiat. Chem. Res. Inst. Tokyo Univ.,* No. 4, 7-8 (1949).
404. Cooper and Neurath, cited in Neurath, *Chem. Rev.,* 30, 357 (1942).
405. Stern, Singer, and Davis, *J. Biol. Chem.,* 167, 321 (1947).
406. Kahn and Polson, quoted in Hess and Deutsch, *J. Amer. Chem. Soc.,* 70, 84 (1948).
407. Theorell, *Biochem. Z.,* 285, 207 (1936).
408. Dayhoff, Perlmann, and MacInnes, *J. Amer. Chem. Soc.,* 74, 2515 (1952).
409. Iwasaki, Shidara, Suzuki, and Nori, *J. Biochem.* (Japan), 53, 299 (1963).
410. Adair and Adair, *Proc. Roy. Soc. A, London,* 190, 341 (1947).
411. Svedberg and Sjögren, *J. Amer. Chem. Soc.,* 50, 3318 (1928).
412. Northrop, *J. Gen. Physiol.,* 25, 465 (1941-2); Polson, *J. Gen. Physiol.,* 25, 487 (1941-2).
413. Ichishima and Yoshida, *Nature,* 207, 525 (1965).
414. Sjögren and Svedberg, *J. Amer. Chem. Soc.,* 53, 2657 (1931).
415. Eirich and Rideal, *Nature,* 146, 541 (1940).
416. Johnson and Landolt, *Discuss. Faraday Soc.,* No. 11, 179 (1951).
417. Nyman, *Scand. J. Clin. Lab. Invest.,* 11, Suppl 39 (1959).
418. Armstrong, Budka, Morrison, and Hasson, *J. Amer. Chem. Soc.,* 69, 1747 (1947).
419. Putnam, Lamanna, and Sharp, *J. Biol. Chem.,* 176, 401 (1948).
420. Kegeles, *J. Amer. Chem. Soc.,* 68, 1670 (1946).
421. Tietze and Neurath, *J. Amer. Chem. Soc.,* 75, 1758 (1953).
422. Bailey, Gutfreund, and Ogston, *Biochem. J.,* 43, 279 (1948).
423. Ogston, *Br. J. Exp. Pathol.,* 26, 286 (1945).
424. Miller and Price, *Arch. Biochem.,* 10, 467 (1946).
425. Lauffer and Cartwright, *Arch. Biochem. Biophys.,* 38, 371 (1952).
426. Stanley and Anderson, *J. Biol. Chem.,* 139, 325 (1941).
427. Neurath and Cooper, *J. Biol. Chem.,* 135, 455 (1940).
428. McFarlane and Kekwick, *Biochem. J.,* 32, 1607 (1938).
429. Neurath and Saum, *J. Biol. Chem.,* 126, 435 (1938).
430. Cecil and Ogston, *Biochem. J.,* 43, 592 (1948).
431. Schramm and Bergold, *Z. Naturforsch.,* B2, 108 (1947).
432. Taylor, Sharp, Beard, and Beard, *J. Inf. Dis.,* 71, 110, 115 (1942).
433. Sharp, Taylor, and Beard, *J. Biol. Chem.,* 163, 289 (1940).
434. Claude, *Science,* 91, 77 (1940); Pollard, *Br. J. Exp. Pathol.,* 20, 429 (1939).
435. Gard and Magnus, *Arkiv. Kemi.,* 24B No. 8, 1 (1947).
436. Stanley and Lauffer, *J. Phys. Colloid Chem.,* 51, 148 (1947).
437. Loring, Morton, and Schwerdt, *Proc. Soc. Exp. Biol. Med.,* 62, 291 (1946).
438. Hook, Beard, Taylor, Sharp, and Beard, *J. Biol. Chem.,* 165, 241 (1946).
439. Bergold, *Z. Naturforsch.,* B2, 122 (1947).
440. Kuehl, *Biochim. Biophys. Acta,* 71, 531 (1962).
441. Schrier, Broomfield, and Scheraga, *Arch. Biochem. Biophys.,* Supp 1, 309 (1962).

442. Kassell, Radicevic, Barlow, Peanasky, and Laskowski, *J. Biol. Chem.*, 238, 3274 (1963).
443. Busch, Hnilica, Chien, Davis, and Taylor, *Cancer Res.*, 22, 637 (1962).
444. Hartley, Rushizky, Greco, and Sober, *Biochemistry*, 2, 794 (1963).
445. Gillespie, *Aust. J. Biol. Sci.*, 16, 241 (1963).
446. Ganno, *J. Biochem.* (Japan), 58, 556 (1965).
447. Sutherland and Wilkinson, *J. Gen. Microbiol.*, 30, 105 (1962).
448. Adams and Norton, *J. Biol. Chem.*, 239, 1525 (1964).
449. Matsubara, Chu, and Yasunobu, *Arch. Biochem. Biophys.*, 101, 209 (1963).
450. Rackis, Sasame, Mann, Anderson, and Smith, *Arch. Biochem. Biophys.*, 98, 471 (1962).
451. Kickhöfen and Burger, *Biochim. Biophys. Acta*, 65, 190 (1962).
452. Shimomura, Johnson, and Saiga, *J. Cell-Comp. Physiol.*, 62, 1 (1963).
453. Reisfeld, Williams, Cirillo, Tong, and Brink, *J. Biol. Chem.*, 239, 1777 (1964).
454. Nickerson and Durand, *Biochim. Biophys. Acta*, 77, 87 (1963).
455. Squire, Li, and Anderson, *Biochemistry*, 1, 412 (1962).
456. Habeeb, *Can. J. Biochem.*, 42, 545 (1964).
457. Williams and Rajagopalan, *J. Biol. Chem.*, 241, 4951 (1966).
458. Siebert, Schmitt, and Traxler, *Hoppe-Seyler's Z. Physiol. Chem.*, 332, 160 (1963).
459. Allen, Kellermeyer, Stjernholm, Jacobson, and Wood, *J. Biol. Chem.*, 238, 1637 (1963).
460. Gosselin-Rey, *Arch. Int. Physiol. Biochem.*, 73, 313 (1965).
461. Balls and Ryan, *J. Biol. Chem.*, 238, 2976 (1963).
462. Torralba and Grisolia, *J. Biol. Chem.*, 241, 1713 (1966).
463. Green and Bodansky, *J. Biol. Chem.*, 240, 2574 (1965).
464. Chiancone, Vecchim, Forlani, Antonini, and Wyman, *Biochim. Biophys. Acta*, 127, 549 (1966).
465. Turner, *Biochim. Biophys. Acta*, 69, 518 (1963).
466. Morris and Hagers, *J. Biol. Chem.*, 241, 1763 (1966).
467. Hargreaves, Wanderley, Hargreaves, and Gonçales, *Biochim. Biophys. Acta*, 67, 641 (1963).
468. Boffa, Jacquot-Armand, and Fine, *Biochim. Biophys. Acta*, 86, 514 (1964).
469. Fling, Horowitz, and Heinemann, *J. Biol. Chem.*, 238, 2045 (1963).
470. Popp, Heddle, Canning, and Allen, *Biochim. Biophys. Acta*, 115, 113 (1966).
471. Campagnari and Webster, *J. Biol. Chem.*, 238, 1628 (1963).
472. Ganguly, Gupta, and Chatterjea, *Nature*, 199, 919 (1963).
473. Bull, Feinstein, and Morris, *Nature*, 201, 1326 (1964).
474. Tantori, Vivaldi, Corta, Salvati, Sorcini, and Velani, *Arch. Biochem. Biophys.*, 109, 404 (1965).
475. Mosbach, *Biochim. Biophys. Acta*, 73, 204 (1963).
476. Cline and Hu, *J. Biol. Chem.*, 240, 4498 (1965).
477. Roberts, Makey, and Seal, *J. Biol. Chem.*, 241, 4907 (1966).
478. Wilson and Crawford, *J. Biol. Chem.*, 240, 4801 (1965).
479. Goldwasser, *J. Biol. Chem.*, 238, 3306 (1963).
480. Charlwood, *Biochem. J.*, 88, 394 (1963).
481. Hosokawa and Stanier, *J. Biol. Chem.*, 241, 2453 (1966).
482. Gregory and Bendall, *Biochem. J.*, 101, 569 (1966).
483. Nagler, Kochwa, and Wasserman, *Proc. Soc. Exp. Biol.*, 111, 746 (1962).
484. Mahling, *Z. Naturforsch.*, 18B, 1 (1963).
485. Bruns and Schultze, *Biochem. Z.*, 336, 162 (1962).
486. Conconi and Grazi, *J. Biol. Chem.*, 240, 2461 (1965).
487. Nozaki, Kagamiyama, and Hayaishi, *Biochem. Biophys. Res. Commun.*, 11, 65 (1963).
488. Saito and Hanahan, *Biochemistry*, 1, 521 (1962).
489. Ishiguro, Takahashi, Hayashi, and Funatsu, *J. Biochem.* (Tokyo), 56, 325 (1964).
490. Schreiber, Eckstein, Maas, and Hoker, *Biochem. Z.*, 340, 21 (1964).
491. Clark, Poillon, and Dawson, *Biochim. Biophys. Acta*, 118, 82 (1966).
492. Hofmann and Shaw, *Biochim. Biophys. Acta*, 92, 543 (1964).
493. Matthews and Reed, *J. Biol. Chem.*, 238, 1869 (1962).
494. Schultze, Schönenberger, and Schwick, *Biochem. Z.*, 328, 267 (1956).
495. Yoshida, *Biochim. Biophys. Acta*, 71, 544 (1963).
496. Yoshida, *J. Biol. Chem.*, 240, 1113 (1965).
497. Gehatia, *Z. Naturforsch.*, B 17, 432 (1962).
498. Chung and Langdon, *J. Biol. Chem.*, 238, 2309 (1963).
499. Allison and Kaplan, *J. Biol. Chem.*, 239, 2140 (1964).
500. Christen, Göschke, Lenthardt, and Schmid, *Helv. Chim. Acta*, 48, 1050 (1965).
501. Monakhov and Neifakh, *Biokhimia*, 27, 494 (1962).
502. Ebert and Schenk, *Z. Naturforsch.*, B 17, 732 (1962).

503. Yoshida and Freese, *Biochim. Biophys. Acta,* 92, 33 (1964).
504. Burns and Demoss, *Biochim. Biophys. Acta,* 65, 233 (1962).
505. Yamada, Gee, Ebata, and Yasunobu, *Biochim. Biophys. Acta,* 81, 165 (1964).
506. Shiio, Shiio, and McFadden, *Biochim. Biophys. Acta,* 96, 114 (1965).
507. Newton, Morino, and Snell, *J. Biol. Chem.,* 240, 1211 (1965).
508. Van Holde and Cohen, *Biochemistry,* 3, 1803 (1965).
509. Adachi, Iwayama, Tanioka, and Takeda, *Biochim. Biophys. Acta,* 118, 88 (1966).
510. Okazaki, Briehl, Wittenberg, and Wittenberg, *Biochim. Biophys. Acta,* 111, 496 (1965).
511. Coleman, *J. Biol. Chem.,* 241, 5511 (1966).
512. Inoue, Adachi, Suzuki, Fukinish, and Takada, *Biochem. Biophys. Res. Commun.,* 21, 432 (1965).
513. Rogers, Geiger, Thompson, and Hellerman, *J. Biol. Chem.,* 238, 481 (1963).
514. Bowen and Rogers, *Biochim. Biophys. Acta,* 67, 633 (1963).
515. McMeekin, Wilensky, and Groves, *Biochem. Biophys. Res. Commun.,* 7, 151 (1962).
516. Fujita, Itagak, and Sato, *J. Biochem.* (Japan), 53, 282 (1963).
517. Davison and Freifelder, *J. Mol. Biol.,* 5, 643 (1962).
518. Chew, Scutt, Oliver, and Lugg, *Biochem. J.,* 94, 378 (1965).
519. Lawler, *J. Biol. Chem.,* 238, 132 (1963).
520. Rao and Pandit, *Biochim. Biophys. Acta,* 94, 238 (1965).
521. Ketterer, *Life Sci.,* 1, 163 (1962).
522. Rice, Casassa, Kerwin, and Maser, *Arch. Biochem. Biophys.,* 105, 409 (1964).
523. Bezkorovainy, *Biochemistry,* 2, 10 (1963).
524. Hashimoto, Hashimoto, and Pigman, *Arch. Biochem. Biophys.,* 104, 282 (1964).
525. Fryar and Gibbs, *Experientia,* 19, 493 (1963).
526. Brahms and Kay, *J. Mol. Biol.,* 5, 132 (1962).
527. Connell, *Biochim. Biophys. Acta,* 74, 374 (1963).
528. Cornelius, Pangborn, and Heckly, *Arch. Biochem. Biophys.* 101, 403 (1963).
529. Johnson, Napper, and Rowe, *Biochim. Biophys. Acta,* 74, 365 (1963).
530. Roberts and Seal, *Comp. Biochem. Physiol.,* 16, 327 (1965).
531. Jutisz, Hermier, Colonge, and Courrier, *Ann. Endocrinol.* (Paris), 26, 670 (1965).
532. Pardee, *J. Biol. Chem.,* 241, 5886 (1966).
533. Ohta, Ogura, and Wada, *J. Biol. Chem.,* 241, 5919 (1966).
534. Allen, *J. Biol. Chem.,* 241, 5266 (1966).
535. Dawson, Eppenberger, and Kaplan, *J. Biol. Chem.,* 242, 210 (1967).
536. Marino, Greco, Scardi, and Zito, *Biochem. J.,* 99, 589 (1966).
537. Chang and Lane, *J. Biol. Chem.,* 241, 2413 (1966).
538. Takahashi, Ramachandramurthy, and Liener, *Biochem. Biophys. Acta,* 133, 123 (1967).
539. Ide, Huyakawa, Okabe, and Koike, *J. Biol. Chem.,* 242, 54 (1967).
540. Nakano, *J. Biol. Chem.,* 242, 73 (1967).
541. Nakano and Danowski, *J. Biol. Chem.,* 241, 2075 (1966).
542. Baker and Crawford, *J. Biol. Chem.,* 241, 5577 (1966).
543. Folk and Cole, *J. Biol. Chem.,* 241, 5518 (1966).
544. Gibson, Massey, and Atherton, *Biochem. J.,* 85, 369 (1962).
545. Lowey, Kucera, and Holtzer, *J. Mol. Biol.,* 7, 234 (1963).
546. Laurent and Person, *Biochim. Biophys. Acta,* 106, 616 (1965).
547. Perrin and Saunders, *Biochim. Biophys. Acta,* 84, 216 (1964).
548. Gammack, *Biochem. J.,* 88, 373 (1963).
549. Yang, Bock, Hsu, and Porter, *Biochim. Biophys. Acta,* 110, 608 (1965).
550. De Ley, *J. Gen. Microbiol.,* 34, 219 (1964).
551. De Ley, *J. Gen. Microbiol.,* 37, 153 (1964).
552. Bogdanova, Gavrilova, Dvorkin, Kiselev, and Spirin, *Biokhimiya,* 27, 387 (1962).
553. Mitra and Kaesberg, *J. Mol. Biol.,* 14, 558 (1965).
554. Davison and Freifelder, *J. Mol. Biol.,* 5, 635 (1962).
555. Laurent and Anseth, *Exp. Eye. Res.,* 1, 99 (1961).
556. Phelps, *Biochem. J.,* 95, 41 (1965).
557. Franz and Krisch, *Biochem. Biophys. Res. Commun.,* 23, 816 (1966).

These tables originally appeared in Sober, *Handbook of Biochemistry and selected data for Molecular Biology,* 2nd ed., Chemical Rubber Co., Cleveland, 1970.

MOLECULAR PARAMETERS OF PURIFIED HUMAN PLASMA PROTEINS[a]

No.	Names and synonyms	$E_{280\,nm}$	$S_{20,w} \cdot 10^{-13}$ cm/sec dyne	$D_{20,w} \cdot 10^{-7}$ cm²/sec	$\bar{v}_{20}$	Molecular wt	f/f_0	η	pI	Electrophoretic mobility pH 8.6 barbital I = 0.1	Amount in normal plasma (mg/ml)	References
1	Prealbumin (tryptophan rich)	13.2	4.2	—	—	61,000	—	—	4.7	7.6	0.28–0.35	1
	Thyroxin-binding prealbumin (TBPA)	14.1				62,500						2,3
2	Albumin Serum albumin	5.8	4.6	6.1	0.733	54,000 69,000	1.28	0.042	4.9	5.92	35–45	4,5 1
3	α₁-Lipoprotein High-density lipoproteins											
	HDL₂	—	5.5	—	1.093	435,000	—	—	—	—	0.37–1.17	1
	HDL₃	—	5.0	—	1.149	195,000	—	—	—	—	2.17–2.7	1
	HDL₁.₁₈₅	—	4.6	5.2	0.873	166,000–175,000	—	—	—	—	—	6
	VHDL₁	—	4.5	5.4	0.866	148,000–153,800	—	—	—	—	—	
4	Orosomucoid α₁-Acid glycoprotein	8.9	3.11	5.27	0.675	44,100	1.78	0.069	2.7	5.2	0.75–1.0	1
5	4.6 S-Postalbumin	8.0	4.6	—	—	—	—	—	—	—	—	1
6	α₁-T-Glycoprotein	6.0	3.3[b]	—	—	55,000–60,000	—	—	—	—	—	1
7	Tryptophan-poor α₁-glycoprotein	4.5	—	—	—	—	—	—	—	—	0.06–0.10	7
	Transcortin Corticosteroid binding globulin (CBG)	6.45	3.79	6.15	0.708	51,700	1.42	—	—	4.9	0.041 ± 0.004	8,9
8	α₁-Antitrypsin	5.3	3.41	5.2	0.646 0.728	45,000 50,000	—	0.068	4.0	5.42	2.1–4.0	1 10

[a] $S_{20,w} \cdot 10^{-13}$ sedimentation coefficient in Svedberg units, referred to water at 20°, and extrapolated to zero concentration; $D_{20,w} \cdot 10^{-7}$, diffusion coefficient, referred to water at 20°; $\bar{v}_{20}$, partial specific volume of the protein (generally at 20°); f/f_0, frictional ratio; η, intrinsic viscosity; pI, isoelectric point; $E_{280\,nm}$, absorption of 1% solution, 1 cm light path at 280 nm.

[b] Not extrapolated to zero concentration.

MOLECULAR PARAMETERS OF PURIFIED HUMAN PLASMA PROTEINS (continued)

No.	Names and synonyms	$E_{280\,nm}$	$S_{20,w} \cdot 10^{-13}$ cm/sec dyne	$D_{20,w} \cdot 10^{-7}$ cm²/sec	$\bar{v}_{20}$	Molecular wt	f/f_o	η	pI	Electrophoretic mobility pH 8.6 barbital $I = 0.1$	Amount in normal plasma (mg/ml)	References
9	α_1 X-Glycoprotein α_1-Antichymotrypsin	6.0	3.9[b]	—	—	68,000	—	—	—	—	0.4–0.6	1,7
10	α_1 B-Glycoprotein α_1-Easily precipitable glycoprotein	9.0	3.8[b]	—	—	50,000	—	—	—	—	0.19–0.25	1,7
11	9.5 S-α_1-Glycoprotein CM-Protein III	18.2	9.5	—	0.730	308,000	—	—	5.0	—	0.055	11
12	Activated factor X of blood coagulation Factor X_a Thrombokinase Autoprothrombin C Activated Stuart factor Activated Prower factor	—	2.1	—	—	25,000 ±2,000	—	—	—	—	0.005–0.01	12,13
13	Zn-α_2-Glycoprotein	18	3.2	—	0.706	41,000	—	—	3.8	4.2	0.042–0.054	1,7
14	Thyroxin-binding globulin (TBG)	7.25	3.0	7.39	—	36,500	1.31	—	3.8	—	—	14
15	Antithrombin III Heparin cofactor	6.9	3.4	—	—	64,000 65,000	—	—	—	—	0.015 0.25–0.35 in serum 0.33–0.48 in plasma	15 7
16	Gc-globulin Group-specific components	6.3	3.7[b]	—	—	50,000	—	—	—	—	0.63–0.87	1,16,17
17	Activated C_{1s} factor of complement C_{1s} C_1-esterase	16.7	4.0	—	—	79,000	—	—	—	—	0.033 ± 0.006	18,19
18	Inter-α-trypsin inhibitor π Protein	7.1	6.4[b]	—	—	180,000	—	—	—	—	0.28–0.30	1,20

243

MOLECULAR PARAMETERS OF PURIFIED HUMAN PLASMA PROTEINS (continued)

No.	Names and synonyms	$E_{280\,nm}$	$S_{20,w} \cdot 10^{-13}$ cm/sec dyne	$D_{20,w} \cdot 10^{-7}$ cm²/sec	$\bar{v}_{20}$	Molecular wt	f/f_0	η	pI	Electrophoretic mobility pH 8.6 barbital I = 0.1	Amount in normal plasma (mg/ml)	References
19	Anaphylatoxin inactivator 10 S-Inter-α-globulin Kininase Carboxypeptidase N	—	9.5	2.9	—	325,000	—	—	—	—	0.03–0.04	21
20	Retinol-binding protein	18.7 19.8	2.3	9.8	0.720	21,000	1.28	—	4.4–4.8	—	0.04–0.06	22,23 24
21	α₂ HS-Glycoprotein Ba-α₂ glycoprotein	5.6	3.3	—	—	49,000	—	—	4.1–4.3	4.2	0.30–0.90	1,25
22	Factor II of blood coagulation Prothrombin	8.84	5.11	5.14	0.719	70,200	1.54	—	3.8	—	0.110–0.230	26,28
23	Soluble fibroblast antigen SF antigen	12.7 —	— 13.5	— —	0.710 —	68,700 355,000	— —	— —	— —	— —	0.100 —	27 29
24	C_1 Inactivator C_1-Esterase inhibitor α₂-Neuraminoglycoprotein	4.5	3.67	—	0.667	104,000	—	—	—	2.7–2.8	0.170–0.300	7,30
25	Factor D of properdin C_3-Proactivator convertase C_3-PA-convertase C_3-PAse Pro-GBGase	—	3.0	—	—	25,000	—	—	—	—	—	31,32
26	Histidine-rich 3.8 S α₂-glycoprotein CM protein I	—	3.8	—	—	55,500	—	—	5.6–6.2	—	0.090	33
27	C_9-Factor of complement	—	4.5	—	—	79,000	—	—	—	—	0.001	32,34

MOLECULAR PARAMETERS OF PURIFIED HUMAN PLASMA PROTEINS (continued)

No.	Names and synonyms	$E_{280\,nm}$	$S_{20,w} \cdot 10^{-13}$ cm/sec dyne	$D_{20,w} \cdot 10^{-7}$ cm^2/sec	$\bar{v}_{20}$	Molecular wt	f/f_0	η	pI	Electrophoretic mobility pH 8.6 barbital I = 0.1	Amount in normal plasma (mg/ml)	References
28	Pregnancy-associated α_2-glycoprotein (PAG) Pregnancy-zone protein Serum factor XH α_2-Pregnoglobulin α_2 AP-glycoprotein SP3	—	12	—	0.724	359,000	—	—	4.7	—	0.003 in men 0.015 in women	35,36,37
29	Haptoglobin Type 1-1 2-1 2-2	12.0	4.4 4.3; 6.5[b] 7.5	4.7	0.766	100,000	—	—	4.1 4.5[c]	—	1.0—2.2 1.6—3.0 1.2—2.6	1
30	Ceruloplasmin	14.9	7.1	4.4	0.714	134,000	—	0.044	4.4	4.6	0.27—0.39	1,38,39
31	Cholinesterase Pseudocholinesterase	—	12	—	—	348,000	—	—	3.0	3.1	0.008—0.011	1
32	α_2-Macroglobulin	8.1	19.6 18.1	2.41	0.735	820,000 725,000	1.43 1.57	0.068	5.4	4.2	2.65 ± 0.55 in men 3.35 ± 0.57 in women	1 40,41
33	Kininogen	—	—	—	—	70,000	—	—	4.5—4.7	—	0.25—0.50	42,43
34	α_2-Lipoproteins Low-density lipoproteins (LDL)	—	Sf > 12	—	<.1019[e]	5 × 10^6 — 20 × 10^6	—	—	—	—	1.5—2.3	1
35	8 Sα_3-Glycoprotein	10	7.87	—	—	220,000	—	—	4.2—4.6	—	0.03—0.04	44
36	Cold insoluble globulin	12.8	12.3	—	—	—	—	—	—	—	2.6 × 10^{-3} 3.8 × 10^{-3}	45
37	Plasminogen Profibrinolysin	17	4.2	4.31	0.715	81,000	1.8	0.08	5.6 6.3—8.6	3.7	0.48 ± 0.09	1,46,47 48,42

[c] pH 8.5.
[e] For lipoproteins specific densities are given instead of partial specific volumes.

MOLECULAR PARAMETERS OF PURIFIED HUMAN PLASMA PROTEINS (continued)

No.	Names and synonyms	$E_{280\,nm}$	$S_{20,w} \cdot 10^{-13}$ cm/sec dyne	$D_{20,w} \cdot 10^{-7}$ cm^2/sec	$\bar{v}_{20}$	Molecular wt	f/f_0	η	pI	Electrophoretic mobility pH 8.6 barbital I = 0.1	Amount in normal plasma (mg/ml)	References
38	Plasmin Fibrinolysin	—	3.9	—	0.715	75,400	1.5	—	6.1–8.4	—	—	42,48
39	C_3c-Factor of complement β_1A-Globulin	6.7	4.0	—	—	151,000	—	—	—	—	—	49
40	Sialic-acid-free β_1-glycoprotein	11.2	2.9	7.5	0.702	30,700–31,500	1.37	—	4.4	3.3	2×10^{-3}	50
41	Hemopexin Heme-binding β-globulin	19.7 and 21.8 with heme	4.8	—	0.702	57,000	—	—	—	3.1	0.5–1.0	1,51,52
42	Transferrin Siderophilin	14.9 with Fe^{3+} 11.4 without Fe^{3+}	4.9	5.85	0.725	76,000	1.37	0.055	5.2 with Fe^{3+} 5.5 without Fe^{3+}	3.63 with Fe^{3+} 3.03 without Fe^{3+}	2–4	1,53–56
43	Sex steroid binding protein 17-hydroxysteroid binding β-globulin Pregnancy associated β_1-glycoprotein (SP2) Testosterone-estradiol-binding globulin Steroid-binding β-globulin	12.5	4.5 4.1	—	0.665	52,000 65,000	—	—	—	—	0.001–0.012 in men 0.003–0.015 in women	57–60 61
44	Pregnancy-specific β_1-glycoprotein (SP1)	11.2	4.6	—	—	90,000	—	—	—	—	0.05–0.2 during pregnancy	62,63

MOLECULAR PARAMETERS OF PURIFIED HUMAN PLASMA PROTEINS (continued)

No.	Names and synonyms	$E_{280\,nm}$	$S_{20,w} \cdot 10^{-13}$ cm/sec dyne	$D_{20,w} \cdot 10^{-7}$ cm²/sec	$\bar{v}_{20}$	Molecular wt	f/f_o	η	pI	Electrophoretic mobility pH 8.6 barbital I = 0.1	Amount in normal plasma (mg/ml)	References
45	β-Lipoproteins Low-density lipoproteins (LDL)	—	$Sf = 0.12$	—	1.019 1.063^e	3.2×10^6	—	—	—	3.1^d	2.8–4.4	1
46	C_1-Proesterase	—	—	—	—	104,000–118,000	—	—	—	—	—	64
47	C_{1s} C_3b-Inactivator C_3b-INA Conglutinogen-activating-factor (KAP)	—	8.4 4.9 5.5–6.0	5.0	—	180,000 — 100,000	—	—	—	—	0.025	65 66,67 68
48	C_2-Factor of complement	—	5.5	—	—	117,000	—	—	—	—	0.003	32,34
49	Factor B of properdin C_3-Proactivator Glycine-rich β-glycoprotein (GBG)	8.8	6.2	—	—	80,000	—	—	—	—	0.094–0.46	69,70
50	Factor VIII of blood coagulation Antihemophilic A globulin	—	23.7	—	—	1,120,000	—	—	—	—	0.01	71,72
51	C_{11}-Factor of complement	7.5	3.5	—	—	188,000	1.56	—	—	—	—	73
52	$β_2$-Microglobulin	16.8	1.6	13.3	0.727	11,500	1.06–1.16	—	—	—	1.3×10^{-3} 2.5×10^{-3}	74,75
53	C_5-Factor of complement $β_1$ F-Globulin	—	8.7	—	—	206,000	—	—	—	—	0.075	34,76
54	C_3-Factor of complement $β_1$ C-Globulin	10.0	9.5	3.6	—	185,000	—	—	—	—	1.2–1.5	1,49, 77–79

^{d} Phosphate buffer, pH 7.8.

MOLECULAR PARAMETERS OF PURIFIED HUMAN PLASMA PROTEINS (continued)

No.	Names and synonyms	$E_{280\,nm}$	$S_{20,w} \cdot 10^{-13}$ cm/sec dyne	$D_{20,w} \cdot 10^{-7}$ cm²/sec	$\bar{v}_{20}$	Molecular wt	f/f_0	η	pI	Electrophoretic mobility pH 8.6 barbital I = 0.1	Amount in normal plasma (mg/ml)	References
55	C_4-Factor of complement $\beta_1 E$-Globulin	8.8	10	—	—	209,000	—	—	—	—	0.43	34,80,81
56	C_6-Factor of complement	—	5.7	4.0	—	95,000–125,000	—	—	—	—	0.06	82
57	C_7-Factor of complement	—	5.7	—	—	120,000–140,000	—	—	—	—	0.06	32,83
58	Fibrinogen Factor I of blood coagulation	13.6	7.63	1.97	0.723	341,000	—	—	5.8	2.1	2–6	1
59	β_2-Glycoprotein III	13.2	2.14[b]	—	—	35,000	—	—	—	—	0.05–0.15	7,84
60	Activated factor B of properdin C_3-Activator Glycine-rich γ-Glycoprotein (GGG)	11.0	4.2	—	—	60,000	—	—	—	—	0.18	7,69,85,86
61	β_2-Glycoprotein II	9.4	2.9	—	0.710	40,000	—	—	6.2	1.6	0.15–0.30	87
62	Factor XI of blood coagulation Plasma thromboplastin antecedent (PTA)	—	6.9	—	—	158,000 170,000	—	—	— 9.1	—	0.007–0.009	88 42
63	Prekallikrein Fletcher factor	—	5.2	—	—	107,000 127,000	—	—	5.9 8.7	—	0.05–0.1	88,89 42
64	Factor XII of blood coagulation Hageman factor	—	4.5	—	—	90,000 110,000	—	—	6.1 5.8–7.5	—	0.01	88,90 42
65	C_8-Factor of complement	—	8.0	—	—	153,000	—	—	—	—	—	32,34

MOLECULAR PARAMETERS OF PURIFIED HUMAN PLASMA PROTEINS (continued)

No.	Names and synonyms	$E_{280\,nm}$	$S_{20,w} \cdot 10^{-13}$ cm/sec dyne	$D_{20,w} \cdot 10^{-7}$ cm^2/sec	$\bar{v}_{20}$	Molecular wt	f/f_o	η	pI	Electrophoretic mobility pH 8.6 barbital I = 0.1	Amount in normal plasma (mg/ml)	References
66	Factor XIII of blood coagulation Fibrin stabilizing factor (FSF) Laki or Lorand factor Precursor of plasma transglutaminase, of transamidase, of fibrinoligase, or of fibrinase Protein A (active part) Protein S (carrier)	13.8	9.9 7.4	2.5	0.730	320,000 150,000 176,000	1.8	—	—	—	— 0.013	91–93 91,94
67	Immunoglobulin E IgE γE	15.3	8.2	3.71	0.713	190,000	—	—	—	—	$6 \times 10^{-5} - 10^{-3}$	95
68	Immunoglobulin D IgD γD	17	6.14–7.04	—	0.717	172,000–184,000	—	—	—	—	0.003–0.4	96
69	Immunoglobulin A IgA γA (monomer)	13.4	6.3–6.9	3.0–3.6	0.725	162,000	—	—	—	2.1	3.28	97
70	Immunoglobulin M IgM 19Sγ-Globulin	12.5	18–20	1.71–1.75	0.723	1,000,000	2	0.106–0.162	5.1–7.8	2.1	0.8–0.9	98
71	Immunoglobulin G IgG 7 Sγ-Globulin	13.8	6.6–7.2	4.0	0.739	153,000	1.38	0.06	5.8–7.3	1.2	12–18	1,97
72	Plasminogen proactivator	—	—	—	—	100,000	—	—	8.9	—	—	42
73	0.6 S-γ$_2$-Globulin	—	0.6	1.1	0.757	5,100–5,700	1.49	—	—	—	0.55×10^{-3}	99

MOLECULAR PARAMETERS OF PURIFIED HUMAN PLASMA PROTEINS (continued)

No.	Names and synonyms	$E_{280\,nm}$	$S_{20,w} \cdot 10^{-13}$ cm/sec dyne	$D_{20,w} \cdot 10^{-7}$ cm^2/sec	$\bar{v}_{20}$	Molecular wt	f/f_0	η	pI	Electrophoretic mobility pH 8.6 barbital I = 0.1	Amount in normal plasma (mg/ml)	References
74	C-Reactive protein (CRP)	—	7.5	—	—	129,000	—	—	—	—	0.02×10^{-3} 13.5×10^{-3}	100–102
75	$C_{1}q$-Factor of complement	6.8	11.1 10.2	2.5	—	410,000	1.7	—	—	—	0.19	103,104 32
76	Properdin	—	5.2	2.15	0.700	184,000	—	—	—	—	0.015	32,105,106
77	Lactoferrin Lactotransferrin Lactosiderophilin Ekkrinosiderophilin Red milk protein	14.6 with Fe^{3+} 11.2 without Fe^{3+}	4.93	5.6	0.716	77,000	—	—	8.2–9.2	—	0.4×10^{-3} 1.6×10^{-3}	107–109
78	Lysozyme	25.65	2.19	11.92	0.721	15,000–16,000			10.5–11.0		5×10^{-3} 15×10^{-3}	110,111

Compiled by Pierre L. Masson.

MOLECULAR PARAMETERS OF PURIFIED HUMAN PLASMA PROTEINS (continued)

REFERENCES

1. Schultze and Heremans, in *Molecular Biology of Human Proteins with Special Reference to Plasma Proteins*, Vol. 1, Elsevier, Amsterdam, 1966, 150.
2. Raz and Goodman, *J. Biol. Chem.*, 244, 3230 (1969).
3. Rask, Per, and Nilsson, *J. Biol. Chem.*, 246, 6087 (1971).
4. Blake, Swan, Rerat, Berthou, Laurent, and Rerat, *J. Mol. Biol.*, 61, 217 (1971).
5. Branch, Robbins, and Edelhoch, *J. Biol. Chem.*, 246, 6011 (1971).
6. Alaupovic, Sanbar, Furman, Sullivan, and Walraven, *Biochemistry*, 5, 4044 (1966).
7. Becker, Schwick, and Störiko, *Clin. Chem.*, 15, 649 (1969).
8. Muldoon and Westphal, *J. Biol. Chem.*, 242, 5636 (1967).
9. Van Baelen and De Moor, *J. Clin. Endocrinol. Metab.*, 39, 160 (1974).
10. Crawford, *Arch. Biochem. Biophys.*, 156, 215 (1973).
11. Haupt, Heimburger, Kranz, and Baudner, *Hoppe-Seyler's Z. Physiol. Chem.*, 353, 1841 (1972).
12. Berre, Østerud, Christensen, Holm, and Prydz, *Biochem. J.*, 135, 791 (1973).
13. Dombrose and Seegers, *Thromb. Res.*, 3, 737 (1973).
14. Sterling, Hamada, Takemura, Brenner, Newman, and Inada, *J. Clin. Invest.*, 50, 1758 (1971).
15. Marshall and Pensky, *Arch. Biochem. Biophys.*, 146, 76 (1971).
16. Jungfer and Tremper, *Z. Immunitaetsforsch. Exp. Klin. Immunol.*, 143, 6 (1972).
17. Kitchin and Bearn, *Proc. Soc. Exp. Biol. Med.*, 118, 304 (1965).
18. Müller-Eberhard, *Annu. Rev. Biochem.*, 38, 389 (1969).
19. Nagaki and Stroud, *J. Immunol.*, 105, 170 (1970).
20. Steinbuch, *Rev. Fr. Transfus.*, 14, 61 (1971).
21. Bokisch and Müller-Eberhard, *J. Clin. Invest.*, 49, 2427 (1970).
22. Peterson and Berggård, *J. Biol. Chem.*, 246, 25 (1971).
23. Peterson, *J. Biol. Chem.*, 246, 34 (1971).
24. Haupt and Heide, *Blut*, 24, 94 (1972).
25. Weeke and Krasilnikoff, *Acta Med. Scand.*, 192, 149 (1972).
26. Tishkoff, Williams, and Brown, *Thromb. Diath. Haemorrh.*, 24, 325 (1970).
27. Lanchantin, Hart, Friedmann, Saavedra, and Mehl, *J. Biol. Chem.*, 243, 5479 (1968).
28. Takeda, *Thrombosis*, 27, 472 (1972).
29. Ruoslahti and Vaheri, *Nature*, 248, 789 (1974).
30. Haupt, Heimburger, Kranz, and Schwick, *Eur. J. Biochem.*, 17, 254 (1970).
31. Müller-Eberhard and Götze, *J. Exp. Med.*, 135, 1003 (1972).
32. Ruddy, *Transplant. Proc.*, VI, 1 (1974).
33. Heimburger, Haupt, Kranz, and Baudner, *Hoppe-Seyler's Z. Physiol. Chem.*, 353, 1133 (1972).
34. Müller-Eberhard, *Adv. Immunol.*, 8, 1 (1968).
35. Von Schoultz and Stigbrand, *Acta Obstet. Gynecol. Scand.*, 52, 51 (1973).
36. Bohn, *Behring Inst. Mitt.*, 54, 56 (1974).
37. Von Schoultz and Stigbrand, *Biochim. Biophys. Acta*, 359, 303 (1974).
38. Ryden, *Eur. J. Biochem.*, 26, 380 (1972).
39. Morell, *J. Biol. Chem.*, 244, 3494 (1969).
40. Jones, Creeth, and Kekwick, *Biochem. J.*, 127, 187 (1972).
41. James, Johnson, and Fudenberg, *Clin. Chim. Acta*, 14, 207 (1966).
42. Austen, *Transplant. Proc.*, VI, 39 (1974).
43. Habermann, in *Bradykinin, Kallidin and Kallikrein*, Erdös, Ed., Springer-Verlag, Berlin, 1970, 250.
44. Haupt, Baudner, Kranz, and Heimburger, *Eur. J. Biochem.*, 23, 242 (1971).
45. Mosesson and Umfleet, *J. Biol. Chem.*, 245, 5728 (1970).
46. Magoon, Austen, and Spragg, *Clin. Exp. Immunol.*, 17, 345 (1974).
47. Davies and Englert, *J. Biol. Chem.*, 235, 1011 (1960).
48. Barlow, Summaria, and Robbins, *J. Biol. Chem.*, 244, 1138 (1969).
49. Bokisch, Müller-Eberhard, and Cochrane, *J. Exp. Med.*, 129, 1109 (1969).
50. Labat, Ishiguro, Fujisaki, and Schmid, *J. Biol. Chem.*, 244, 4975 (1969).
51. Seery, Hathaway, and Müller-Eberhard, *Arch Biochem. Biophys.*, 150, 269, (1972).
52. Müller-Eberhard, *N. Engl. J. Med.*, 283, 1090 (1970).
53. Roberts, Makey, and Seal, *J. Biol. Chem.*, 241, 4907 (1966).

MOLECULAR PARAMETERS OF PURIFIED HUMAN PLASMA PROTEINS (continued)

54. Oncley, Scatchard, and Brown, *J. Phys. Colloid Chem.*, 51, 184 (1947).
55. Roop and Putnam, *J. Biol. Chem.*, 242, 2507 (1967).
56. Wenn and Williams, *Biochem. J.*, 108, 69 (1968).
57. Mercier-Bodard, Alfsen, and Baulieu, in *Karolinska Symp. Res. Meth. Reproductive Endocrinology, 2nd Symp., Steroid Assay by Protein Binding,* March 23 to 25, 1970, Diczfaluzy, Ed., Bogtrykkeriet Forum, Copenhagen, 1970, 202.
58. de Moor, *Ann. Endocrinol.*, 31, 429 (1970).
59. Bohn and Kranz, *Arch. Gynaekol.*, 215, 63 (1973).
60. Corvol, Chrambach, Rodbard, and Bardin, *J. Biol. Chem.*, 246, 3435 (1971).
61. Bohn, *Blut,* 29, 17 (1974).
62. Bohn, *Blut,* 24, 292 (1972).
63. Bohn, *Arch. Gynaekol.*, 216, 347 (1974).
64. Sakai and Stroud, *J. Immunol.*, 110, 1010 (1973).
65. Okamura, Muramatu, and Fujii, *Biochim. Biophys. Acta,* 295, 252 (1973).
66. Ruddy and Austen, *J. Immunol.*, 107, 742 (1971).
67. Ruddy and Austen, *J. Immunol.*, 102, 533 (1969).
68. Lachmann and Müller-Eberhard, *J. Immunol.*, 100, 691 (1968).
69. Götze and Müller-Eberhard, *J. Exp. Med.*, 134, 90s (1971).
70. Boenisch and Alper, *Biochim. Biophys. Acta,* 221, 529 (1970).
71. Legaz, Schmer, Counts, and Davie, *J. Biol. Chem.*, 248, 3946 (1973).
72. Zimmerman, Ratnoff, and Powell, *J. Clin. Invest.*, 50, 244 (1971).
73. Valet and Cooper, *J. Immunol.*, 112, 1667 (1974).
74. Berggård and Bearn, *J. Biol. Chem.*, 243, 4095 (1968).
75. Karlsson, *Immunochemistry,* 11, 111 (1974).
76. Nilsson, Tomar, and Taylor, *Immunochemistry,* 9, 709 (1972).
77. Budzko, Bokisch, and Müller-Eberhard, *Biochemistry,* 10, 1166 (1971).
78. Humair, *Helv. Med. Acta,* 34, 279 (1968).
79. Ogg, Cameron, and White, *Lancet,* II, 78 (1968).
80. Haupt, Heide, and Schwick, *Klin. Wochenschr.*, 48, 550 (1970).
81. Schreiber and Müller-Eberhard, *J. Exp. Med.*, 140, 1324 (1974).
82. Arroyave and Müller-Eberhard, *Immunochemistry,* 8, 995 (1971).
83. Thompson and Lachmann, *J. Exp. Med.*, 131, 629 (1970).
84. Schwick, Haupt, and Heide, *Klin. Wochenschr.*, 46, 981 (1968).
85. Haupt and Heide, *Clin. Chim. Acta,* 12, 419 (1965).
86. Boenisch and Alper, *Biochim. Biophys. Acta,* 214, 135 (1970).
87. Haupt, Schwick, and Störiko, *Humangenetik,* 5, 291 (1968).
88. Wuepper, in *Inflammation Mechanisms and Control,* Lepow and Ward, Eds., Academic Press, N.Y., 1972.
89. Colman, Mattler, and Sherry, *J. Clin. Invest.*, 48, 11 (1969).
90. Webster, *Fed. Proc.*, 27, 84 (1968).
91. Schwartz, Pizzo, Hill, and McKee, *J. Biol. Chem.*, 248, 1395 (1973).
92. Loewy, *Ann. N. Y. Acad. Sci.*, 202, 41 (1972).
93. Loewy, *Thromb. Diath. Haemorrh. Suppl.*, 13, 109 (1963).
94. Bohn, Haupt, and Kranz, *Blut,* 25, 235 (1972).
95. Bennich and Johansson, *Adv. Immunol.*, 13, 1 (1971).
96. Spiegelberg, *Contemp. Top. Immunochem.*, 1, 165 (1972).
97. Heremans, in *The Antigens,* Vol. 2, Sela, Ed., Academic Press, N.Y., 1974, 365.
98. Metzger, *Adv. Immunol.*, 12, 57 (1970).
99. Nimberg and Schmid, *J. Biol. Chem.*, 247, 5056 (1972).
100. Wood, McCarty, and Slater, *J. Exp. Med.*, 100, 71 (1954).
101. Gotschlich and Edelman, *Proc. Natl. Acad. Sci.*, 54, 558 (1965).
102. Kindmark, *Scand. J. Clin. Lab. Invest.*, 29, 407 (1972).
103. Calcott and Müller-Eberhard, *Biochemistry,* 11, 3443 (1972).
104. Reid, Lowe, and Porter, *Biochem. J.*, 130, 749 (1972).
105. Minta and Lepow, *Immunochemistry,* 11, 361 (1974).
106. Pensky, Hinz, Todd, Wedgwood, Boyer, and Lepow, *J. Immunol.*, 100, 142 (1968).

MOLECULAR PARAMETERS OF PURIFIED HUMAN PLASMA PROTEINS (continued)

107. **Querinjean, Masson, and Heremans,** *Eur. J. Biochem.,* 20, 420 (1971).
108. **Roberts, Masson, and Heremans,** in *VII Int. Symp. Chromatography and Electrophoresis,* Brussels, September 1972, Presses Académiques Européennes, Brussels, 1972, 280.
109. **Rümke, Visser, Kwa, and Hart,** *Folia Med. Neerl.,* 14, 156 (1971).
110. **Parry, Chandan, and Shahani,** *Arch. Biochem. Biophys.,* 103, 59 (1969).
111. **Osserman and Lawlor,** *J. Exp. Med.,* 124, 921 (1966).

THE PROTEINS OF BLOOD COAGULATION

Common name	Roman numeral[a]	Associated names	Conversion or activated product	Stage of participation in blood coagulation	Name of disease[b]	References
Fibrinogen	I	—	Fibrin	Late, intrinsic,[e] and extrinsic[f]	Afibrinogenemia or hypofibrinogenemia	1
Prothrombin[g]	II	—	II$_a$, thrombin, biothrombin	Late, intrinsic, and extrinsic	Prothrombin deficiency	2–4
Tissue thromboplastin	III	Extrinsic thromboplastin	—	Early, extrinsic	—	
Calcium	IV	—	—	Early, middle, late, intrinsic, and extrinsic	—	
Proaccelerin	V	Labile factor, plasma accelerator globulin, plasma Ac-globulin, thrombogene	Activated proaccelerin,[h] accelerin, Factor VI	Middle, intrinsic, and extrinsic	Parahemophilia	5, 6
Precursor of serum prothrombin conversion accelerator (ProSPCA)[g]	VII	Proconvertin, stabile factor, cofactor V, cothromboplastin, kappa factor, serozyme, autoprothrombin I	VII$_a$, serum prothrombin, conversion accelerator (SPCA), convertin[i]	Middle, intrinsic, and extrinsic	ProSPCA deficiency	7, 8
Antihemophilic factor (AHF)	VIII	Antihemophilic globulin (AHF), platelet cofactor I, antihemophilic globulin A, thromboplastinogen, plasma thromboplastin factor, plasmokinase	Activated AHF[h]	Middle, intrinsic	Classic hemophilia, hemophilia A	9–11
Christmas factor[g]	IX	Plasma thromboplastin component (PTC), autoprothrombin II, platelet cofactor II, antihemophilic factor B	IX$_a$, activated Christmas factor, prephase accelerator, PTC′	Middle, intrinsic	Christmas disease, hemophilia B	12
Stuart factor[g]	X	Prower factor, venom substrate, autoprothrombin III	X$_a$, activated Stuart factor, activated venom substrate, autoprothrombin C, thrombokinase	Middle, intrinsic, and extrinsic	Stuart factor deficiency	13, 14
Plasma thromboplastin antecedent (PTA)	XI	Antihemophilic factor C	XI$_a$, activated PTA, third prothromboplastic factor, activation product	Early, intrinsic	PTA deficiency	15, 16
Hageman factor	XII	Surface factor, clot-promoting factor, contact factor, the fifth plasma thromboplastin precursor	XII$_a$, activated Hageman factor	Early, intrinsic	Hageman trait	17, 18
Fibrin stabilizing factor (FSF)	XIII	Laki-Lorand factor, fibrinase	XIII$_a$, activated fibrin stabilizing factor, FSF′	Late, intrinsic, and extrinsic	—	19, 20
Prekallikrein	—	Fletcher factor, kallikreinogen	Kallikrein	Early, intrinsic	Fletcher factor disease	21
Plasminogen	—	Profibrinolysin	Plasmin, fibrinolysin	—	—	22

Note: Footnotes appear on page 256.

THE PROTEINS OF BLOOD COAGULATION (continued)

Common name	Mode of inheritance	Plasma concentration (μg/ml plasma)	Molecular weight (daltons)	Polypeptide chains	Molecular weight of chains (daltons)	Amino terminus	References
Fibrinogen	Disorder of both sexes, transmitted as autosomal recessive trait	1700–4000	340,000	2 α / 2 β / 2 γ	63,500[d] / 56,000[d] / 47,000[d]	Ala / Glu[j] / Tyr	1
Prothrombin[g]	Disorder of both sexes, transmitted as autosomal recessive trait	70	68,000–72,000	1	—	Ala	2–4
Tissue thromboplastin							
Calcium	—	90–115	—	—	—	—	
Proaccelerin	Disorder of both sexes, transmitted as autosomal recessive trait	Trace	310,000–400,000[c]	—	—	—	5, 6
Precursor of serum prothrombin conversion accelerator (ProSPCA)[g]	Disorder of both sexes, transmitted as autosomal recessive trait	0.13	45,500[c] / 60,000[d]	1	—	Ala	7, 8
Antihemophilic factor (AHF)	Disorder primarily of males, transmitted as sex-linked recessive trait	5–10	1.1×10^6 [c,d]	—	200,000[d]	—	9–11
Christmas factor[g]	Disorder primarily of males, transmitted as sex-linked recessive trait	3	54,000[c]	1	—	Tyr	12
Stuart factor[g]	Disorder of both sexes, transmitted as autosomal recessive trait	2–5	55,000[c]	2	39,000 / 16,000	Trp / Ala	13, 14
Plasma thromboplastin antecedent (PTA)	Disorder of both sexes, transmitted as autosomal recessive trait	Trace	160,000[c,d]	2	80,000	—	15, 16
Hageman factor	Disorder of both sexes, transmitted as autosomal recessive or dominant trait	15–47	82,000[c] / 80,000[d]	1	—	—	17, 18
Fibrin stabilizing factor (FSF)	Disorder of both sexes, transmitted as autosomal recessive trait	Trace	320,000[d]	2 a / 2 b	75,000 / 88,000	N-AcSer / Glu	19, 20
Prekallikrein	—	Trace	90,000[c]	1	—	—	21
Plasminogen	—	Trace	87,000[d]	1	—	Glu	22

Compiled by Walter Kisiel and Earl W. Davie.

THE PROTEINS OF BLOOD COAGULATION (continued)

[a] Roman numerals have been assigned to the well recognized clotting factors by an international nomenclature committee (Wright, *J. Am. Med. Assoc.*, 170, 325, 1959).
[b] Clinical manifestations vary from epitaxes and mild bruising to excessive bleeding following injury and surgery for afibrinogenemia, parahemophilia, ProSPCA deficiency, and Stuart deficiency. Classic hemophilia and Christmas disease are often characterized initially by bleeding in the joints of the knees, ankles, and elbows; spontaneous bleeding is unusual in PTA deficiency, while Hageman trait shows no bleeding tendency.
[c] Protein isolated from bovine plasma.
[d] Protein isolated from human plasma.
[e] Coagulation reactions yielding thrombin due to the interaction of constituents found only in plasma.
[f] Coagulation reactions yielding thrombin due to the interaction of constituents found in plasma and tissue extracts.
[g] Requires vitamin K for biosynthesis.
[h] Activated proaccelerin (accelerin, Factor VI) and activated AHF are thrombin-modified proteins with increased clotting activity. Originally, accelerin was thought to be a new clotting factor and was assigned roman numeral VI.
[i] Proconvertin and convertin have been applied collectively to Factor VII and Stuart factor and their activated forms, respectively.
[j] Pyrrolidone carboxylic acid.

GENERAL REFERENCES

Esnouf and Macfarlane, *Advances in Enzymology,* Nord, Ed., Interscience, N. Y., 1968, 255.
Heimburger and Trobisch, *Angew. Chem.*, 10, 85(1971).
Ratnoff, *Progress in Hemostasis and Thrombosis,* Spaet, Ed., Grune and Stratton, N. Y., 1972, 39.
Davie and Kirby, *Current Topics in Cellular Regulation,* Horecker and Stadtman, Eds., Academic Press, New York, 1972, 51.
Davie and Fujikawa, *Ann. Rev. Biochem.*, 44, 799 (1975).

REFERENCES

1. Doolittle, *Adv. Protein Chem.*, 27, 1 (1973).
2. Cox and Hanahan, *Biochim. Biophys. Acta,* 207, 49 (1970).
3. Pirkle, McIntosh, Theodor, and Vernon, *Thromb. Res.*, 2, 461 (1973).
4. Kisiel and Hanahan, *Biochim. Biophys. Acta,* 304, 103 (1973).
5. Philip, Moran, and Colman, *Biochemistry,* 9, 2212 (1970).
6. Hanahan, Rolfs, and Day, *Biochim. Biophys. Acta,* 286, 205 (1972).
7. Kisiel and Davie, *Biochemistry,* 14, in press (1975).
8. Laake and Ellingsen, *Thromb. Res.*, 5, 539 (1974).
9. Schmer, Kirby, Teller, and Davie, *J. Biol. Chem.*, 247, 2512 (1972).
10. Legaz, Schmer, Counts, and Davie, *J. Biol. Chem.*, 248, 3946 (1972).
11. Shapiro, Anderson, Pizzo, and McKee, *J. Clin. Invest.*, 52, 2198 (1973).
12. Fujikawa, Thompson, Legaz, Meyer, and Davie, *Biochemistry,* 12, 4938 (1973).
13. Jackson and Hanahan, *Biochemistry,* 7, 4506 (1968).
14. Fujikawa, Legaz, and Davie, *Biochemistry,* 11, 4882 (1972).
15. Kato, Legaz, and Davie, unpublished results.
16. Wuepper, in *Inflammation: Mechanism and Control,* Lepow and Ward, Eds., Academic Press, N.Y., 1972, 93.
17. Schoenmakers, Matze, Haanen, and Zilliken, *Biochem. Biophys. Acta,* 101, 166 (1965).
18. Revak, Cochrane, Johnston, and Hugli, *J. Clin. Invest.*, 54, 619 (1974).
19. Schwartz, Pizzo, Hill, and McKee, *J. Biol. Chem.*, 248, 1395 (1973).
20. Takagi and Doolittle, *Biochemistry,* 13, 750 (1974).
21. Takahashi, Nagasawa, and Suzuki, *J. Biochem.*, 71, 471 (1972).
22. Walther, Steinman, Hill, and McKee, *J. Biol. Chem.*, 249, 1173 (1974).

GLYCOPROTEINS

Morris Soodak

Glycoproteins may be defined "as conjugated proteins containing as prosthetic group(s) one or more heterosaccharide(s), usually branched, with a relatively low number of sugar residues, lacking a serially repeating unit and bound covalently to the polypeptide chain."

At the present time, 14 types of carbohydrate-protein linkages have been found in living matter:

	Amino acid	Monosaccharide
1.	Alanine	Muramic acid
2.	Asparagine	N-Acetylglucosamine
3.	Cysteine	Glucose
4.	Cysteine	Galactose
5.	Hydroxylysine	Galactose
6.	Hydroxyproline	Arabinose
7.	Serine	N-Acetylglucosamine
8.	Serine	Galactose
9.	Serine	Mannose
10.	Serine	Xylose
11.	Threonine	N-Acetylgalactosamine
12.	Threonine	Fucose
13.	Threonine	Galactose
14.	Threonine	Mannose

The not uncommon heterogeneity of essentially pure glycoprotein preparations is attributable in part to the variable carbohydrate composition, minor proteolysis in vivo or during resolutions, genetic polymorphism, or all three. This heterogeneity when combined with the difficulties involved in the analyses of the carbohydrate constituents of the glycoproteins results in a variability of the analytical data found in the literature.

The two-volume work, *Glycoproteins*, edited by A. Gottschalk, 1972, Elsevier Publishing Company, should be consulted for an understanding of these problems.

The tables compiled by Richard J. Winzler for the 2nd Edition are reproduced again. Richard Winzler, who was one of the Founding Fathers in this field, died September 28, 1972.

Additions and updating for some proteins are appended to his tables.

Table 1
GLYCOPROTEINS, HUMAN PLASMA*

Protein	Molecular weight	$S_{20,w}$	Carbohydrate content, g/100 g				Reference
			Hexose	Acetyl hexosamine	Sialic acid	Fucose	
1. α_1-Antitrypsin	44,100	3.11	14.7	13.9	12.1	0.7	1
2. Ceruloplasmin	160,000	7.08	3.0	2.4	2.4	0.2	1
3. α_1-Easily precipitable glycoprotein	50,000	3.8	4.8	4.4	3.7	0.4	1
4. Fibrinogen	341,000	7.63	1.0	0.9	0.6	–	1
5. β_{1A}- and β_{1C}-Globulin	–	6.9	1.8	0.6	0.45	0.16	1
6. G_c-Globulin	50,800	3.7	2.0	2.0	0	0.2	1
7. α_{1x}-Glycoprotein	–	3.9	8.0	8.0	6.0	0.7	1
8. α_{2HS}-Glycoprotein [Ba-α_2-glycoprotein]	49,000	3.3	5.2	3.9	4.1	0.2	1
9. β_2-Glycoprotein	–	2.9	6.7	5.8	4.4	0.2	1
10. Haptoglobin	100,000	4.4	7.8	5.3	5.3	0.2	1
11. Hemopexin	80,000	4.8	9.0	7.4	5.8	0.4	1
12. γA-Immunoglobulin	–	7.0	3.2	2.3	1.8	0.22	1
13. γG-Immunoglobulin	156,000	7.0	1.1	1.3	0.3	0.2	1
14. γM-Immunoglobulin	1,000,000	19.0	5.4	4.4	1.3	0.7	1
15. Myeloma globulin	–	–	3.7	3.0	1.0	–	2
16. Inter-α-trypsin inhibitor	–	6.4	3.4	3.5	2.1	0.1	1
17. α_2-Macroglobulin	820,000	19.6	3.6	2.9	1.8	0.1	1, 3
18. α_2-Neuramino glycoprotein	–	3.7	12.0	13.0	17.0	0.6	1
19. Orosomucoid [α_1-acid glycoprotein]	44,100	3.11	14.7	13.9	12.1	0.7	1
20. Plasminogen	143,000	4.28	6.7	5.8	4.4	0.2	1, 4
21. 4.6 S-Postalbumin	–	4.6	4.0	2.8	3.0	0.2	1
22. Prothrombin	–	–	4.6	2.9	4.2	–	5
23. Rheumatoid factor	–	19.0	5.4	3.6	1.8	–	6
24. Transcortin	58,500	3.0–4.1	5.4	4.7	3.2	0.8	7
25. Transferrin	90,000	5.5	2.4	2.0	1.4	0.07	1
26. Tryptophan-poor α_1-glycoprotein	55,000–60,000	3.3	5.5	4.5	3.4	0.3	1
27. Zn-α_2-Glycoprotein	41,000	3.2	7.0	4.0	7.0	0.2	1
28. α_2-β_1-Glycoprotein	60,000		10.7	10.8	7.5	0.5	8
29. Glycine-rich – γ protein		4.3	8.5	–	1.3	–	9
30. Coagulation factor VIII	1,100,000		2.2	3.2	0.9	–	10
31. Coagulation factor XIII	320,000		1.9	1.6	1.2	0.2	11, 12
32. Prothrombin	70,000		2.8	3.8	2.3	–	13, 14
33. C1q of complement	400,000		8.5	1.0	0.5	–	15, 16
34. α-Fetoprotein	70,000		2.2	1.4	1.2	–	17

Compiled by Morris Soodak.

*There are thought to be more than 100 blood plasma proteins and a great many of them are glycoproteins.

Table 1 (continued)
GLYCOPROTEINS, HUMAN PLASMA

REFERENCES

1. Schultze and Heremans, in *Molecular Biology of Human Proteins,* Elsevier, Amşterdam I, 1966, 182.
2. Dawson and Clamp, *Biochem. J.,* 107, 341 (1968).
3. Dunn and Spiro, *J. Biol. Chem.,* 242, 5549 (1967).
4. Slotta and Gonzalez, *Biochemistry,* 3, 285 (1964).
5. Schultze and Schwick, *Clin. Chim. Acta,* 4, 15 (1959).
6. Kunkel, Franklin, and Müller-Eberhard, *J. Clin. Invest.,* 38, 424 (1959).
7. Slaunwhite, Schneider, Wissler, and Sandberg, *Biochemistry,* 5, 3527 (1966).
8. Iwasaki and Schmid, *J. Biol. Chem.,* 245, 1814 (1970).
9. Boenisch and Alper, *Biochim. Biophys. Acta,* 214, 135 (1970).
10. Legaz, Schmer, Counts, and Davie, *J. Biol. Chem.,* 248, 3946 (1973).
11. Bohn, *Ann. N.Y. Acad. Sci.,* 202, 256 (1972).
12. Schwartz, Pizzo, Hill, and McKee, *J. Biol. Chem.,* 246, 5851 (1971).
13. Kisiel and Hanahan, *Biochim. Biophys. Acta,* 304, 103 (1973).
14. Saavedra and Mehl, *J. Biol. Chem.,* 243, 5479 (1968).
15. Calcott and Müller-Eberhard, *Biochemistry,* 11, 3443 (1972).
16. Müller-Eberhard, *Annu. Rev. Biochem.,* 44, 697 (1975).
17. Ruoslahti and Seppälä, *Int. J. Cancer,* 7, 218 (1971).

Table 2
PLASMA GLYCOPROTEINS (OTHER THAN HUMAN)

Protein	Source	Carbohydrate content, g/100 g				Reference
		Neutral sugar	Acetyl hexosamine	Sialic acid	Fucose	
1. Fetuin	Fetal calf	7.6	6.6	8.7	—	1, 2
2. M_2-Glycoprotein	Cow	6.1	7.7	10.3	—	1, 3, 4
3. Gonadotropin-transporting protein	Horse	14.0	13.3	12.0	—	1, 5
4. Corticosteroid-transporting protein	Rabbit	11.4	11.6	8.7	—	6
5. Immunoglobulin IgA	Horse	2.1	1.9	0.9	—	7
6. Immunoglobulin IgA	Cow	0.9	1.8	0.3	—	8
7. Immunoglobulin IgG	Horse	1.1	1.1	0.2	—	7
8. Immunoglobulin IgG	Rabbit	1.2	1.2	0.2	—	9
9. Immunoglobulin light chain	Mouse urine	7.8	7.4	1.8	—	10
10. α_2-Macroglobulin	Fetal calf	2.7	1.6	0.7	—	1, 11
11. α_2-Macroglobulin	Horse serum	4.3	4.3	3.6	—	12
12. α_2-Macroglobulin	Rabbit serum	6.5	4.0	1.6–3.3	—	13
13. α_2-Macroglobulin	Pig serum	8–10	—	2.5–4.0	—	14
14. α_2-Macroglobulin	Rat serum	10.3	—	—	—	15
15. γ-Inhibitor of viral hemagglutination	Horse serum	8.6	3.9	2.5	—	16
16. α_1-Macroglobulin	Rabbit	6.5	4.95	3.3	—	17
17. α-Acid glycoprotein	Cow	12.0	14.5	10–16.2	—	18–20
18. α-Glycoprotein	Dog	11.2	11.1	8.0	—	20, 21
19. α-Glycoprotein	Guinea pig	9.6	8.3	6.7	—	20, 22
20. α-Glycoprotein	Horse	12.3	8.6	6.8	—	20

Table 2 (continued)
PLASMA GLYCOPROTEINS (OTHER THAN HUMAN)

Protein	Source	Carbohydrate content, g/100 g				Reference
		Neutral sugar	Acetyl hexosamine	Sialic acid	Fucose	
21. α-Glycoprotein	Rabbit	12.3	11.2	6.9	—	20
22. α-Glycoprotein	Rat	9.9–15.7	7.9–10.2	5.1–10.0	—	20, 23
23. α_1-Glycoprotein	Sheep	11.1	11.2	13.5	0.5	24
24. Hageman factor	Cow	5.9	5.9	4.4	—	25
25. α_1-Acute phase protein	Rat	4.1	7.3	3.3	—	26
26. Transferrin	Hen	1.2	1.6	0.7	—	27
27. M_1-Glycoprotein	Cow	11.3–18.6	12.6	7.2	0.9	28
28. M_1-Glycoprotein	Pig	13.8–23.1	13.6	7.4	25.3	28
29. M_1-Glycoprotein	Hen	13.8–20.4	15.0	10.1	0.4	28
30. M_2-Glycoprotein	Cow	6.2–10.8	6.5	5.1	0	28
31. M_2-Glycoprotein	Pig	8.3–12.0	6.6	5.0	14.8	28
32. α-Glycoprotein	Sheep	6.2–7.1	6.0–8.1	4.4–4.8	1.1	29, 30
33. Prothrombin	Beef	3.7	2.5	4.2	—	31
34. Freezing-point depression glycoproteins[3]	Fish	28	29	—	—	32, 33
35. Progesterone binding globulin I	Pig	27.7	31.3	12.7	0.8	34
Progesterone binding globulin II		24.5	25.0	12.2	0.7	
36. Coagulation factor XII	Cow	5.9	6.0	4.4	0.5	35
37. Coagulation factor XI	Cow	4.2	4.9	3.6	—	36
38. Coagulation factor IX	Cow	10.6	6.5	8.7	—	37
39. Coagulation factor VIII	Cow	3.8	5.2	0.6	—	38
40. Coagulation factor X	Cow	2.9	4.4	3.8	—	39, 40
41. Prothrombin	Cow	3.5	2.0	3.7	—	41, 42

Compiled by Morris Soodak.

REFERENCES

1. Eylar, *J. Theor. Biol.*, 10, 89 (1965).
2. Spiro, *J. Biol. Chem.*, 235, 2860 (1960).
3. Bezkorovainy, *Biochemistry*, 2, 10 (1963).
4. Bezkorovainy and Doherty, *Arch. Biochem. Biophys.*, 96, 491 (1962).
5. Bourrillon, Got, and Marcy, *Bull. Soc. Chim. Biol.*, 40, 87 (1958).
6. Chader and Westphal, *Fed. Proc.*, 26, Abstracts-140 (1967).
7. Schultze, *Clin. Chim. Acta*, 4, 610 (1959).
8. Nolan and Smith, *J. Biol. Chem.*, 237, 453 (1962).
9. Fleischman, Porter, and Press, *Biochem. J.*, 88, 220 (1963).
10. Melchers, Lennox, and Facon, *Biochem. Biophys. Res. Commun.*, 24, 244 (1966).
11. Marr, Owen, and Wilson, *Biochim. Biophys. Acta*, 63, 276 (1962).
12. Boretti, di Marco, and Julita, *G. Microbiol.*, 12, 55 (1964).
13. Got, Cheftel, Font, and Moretti, *Biochim. Biophys. Acta*, 136, 320 (1967).
14. Jacquot-Armand and Guinand, *Biochim. Biophys. Acta*, 133, 289 (1967).
15. Fisher and Canning, *Natl. Cancer Inst. Monogr.*, 21, 403 (1966).
16. Pepper, *Biochim. Biophys. Acta*, 156, 327 (1968).
17. Got, Mouray, and Moretti, *Biochim. Biophys. Acta*, 107, 278 (1965).
18. Bezkorovainy, *Biochim. Biophys. Acta*, 101, 336 (1965).

Table 2 (continued)
PLASMA GLYCOPROTEINS (OTHER THAN HUMAN)

19. Bezkorovainy, *Arch. Biochem. Biophys.*, 110, 558 (1965).
20. Weimer and Winzler, *Proc. Soc. Exp. Biol. Med.*, 90, 458 (1955).
21. Athenios, Kukral, and Winzler, *Arch. Biochem. Biophys.*, 106, 338 (1964).
22. Simkin, Skinner, and Seshadri, *Biochem. J.*, 90, 316 (1964).
23. Kawasaki, Koyama, and Yamashina, *J. Biochem.*, 60, 554 (1966).
24. Campbell, Schneider, Howe, and Durand, *Biochim. Biophys. Acta*, 148, 137 (1967).
25. Schoenmakers, Matze, Haanen, and Zilliken, *Biochim. Biophys. Acta*, 93, 433 (1964); 101, 166 (1965).
26. Gordon and Louis, *Biochem. J.*, 113, 481 (1969).
27. Williams, *Biochem. J.*, 108, 57 (1968).
28. Grant, Martin, and Anastassiadis, *J. Biol. Chem.*, 242, 3912 (1967).
29. Samy, *Arch. Biochem. Biophys.*, 121, 703 (1967).
30. Das, *Biochim. Biophys. Acta*, 58, 52 (1962).
31. Magnusson, *Ark. Kemi.*, 23, 285 (1965).
32. DeVries, Vandenheede, and Feeney, *J. Biol. Chem.*, 246, 305 (1971).
33. Glochner, Newman, and Uhlenbruck, *Biochem. Biophys. Res. Commun.*, 66, 701 (1975).
34. Burton, Harding, Aboul-Hosn, MacLaughlin, and Westphal, *Biochemistry*, 13, 3554 (1974).
35. Schoenmakers, Matze, Haanen, and Zilliken, *Biochim. Biophys. Acta*, 101, 166 (1965).
36. Kato, Legaz, and Davie, unpublished results in *Annu. Rev. Biochem.*, 44, 799 (1975), by Davie and Fujikawa.
37. Fujikawa, Thompson, Leger, Meyer, and Davie, *Biochemistry*, 12, 4938 (1973).
38. Legaz, Weinstein, Heldebrandt, and Davie, *Ann. N.Y. Acad. Sci.*, 240, 43 (1975).
39. Fujikawa, Legaz, and Davie, *Biochemistry*, 11, 4882 (1972).
40. Jackson, *Biochemistry*, 11, 4873 (1972).
41. Fujikawa, Coan, Enfield, Titani, Erickson, and Davie, *Proc. Natl. Acad. Sci. U.S.A.*, 71, 427 (1974).
42. Nelsestuen and Suttie, *J. Biol. Chem.*, 247, 6096 (1972).

Table 3
GLYCOPROTEINS, BLOOD GROUP SUBSTANCES, AND MUCINS

Glycoprotein	Source	Carbohydrate content g/100 g				Reference
		Neutral sugar	Acetyl hexosamine	Sialic acid	Fucose	
1. A substance	Ovarian cysts	19.8–26.0	34.4–45.3	1.3–2.9	17.0–19.5	1
2. B substance	Ovarian cysts	33.0–38.1	29.5–34.0	1.9–5.1	16.2–20.8	1
3. AB substance	Ovarian cysts	26.6–31.8	32.0–35.7	1.0–1.7	17.3–17.9	1
4. H substance	Ovarian cysts	22.1–25.1	28.4–32.1	1.5–4.3	16.7–21.6	1
5. Le[a] substance	Ovarian cysts	27.5–31.8	32.6–37.7	3.0–18.0	8.6–13.1	1
6. M and N substances (influenza virus receptor)	Human erythrocytes	13.1–14.8	12.1–18.0	15.7–24.4	0.7–1.1	2–5
7. Infectious mononucleosis antigen	Ovine erythrocytes	—	13.5	7.0	—	6
8. Infectious mononucleosis antigen	Bovine erythrocytes	—	5.5	4.5	—	6
9. Fucose-containing glycoprotein	Human erythrocyte membranes	7.0	4.0	—	3.0	7
10. Cervical mucin	Bovine	27.5	32.9	13.8	5.1	8–10
11. Cervical mucin	Human	46.0	24.9	8.0	11.0	8, 11
12. Submaxillary mucin	Bovine	0.7–3.6	18.3–31.1	22.4–30.8	2.8	12
13. Submaxillary mucin	Ovine	0.45	13.8–15.0	23.8–26.0	0.4	12, 13
14. Submaxillary mucin	Canine	15.9	24.8	9.2	12.1	12
15. Submaxillary mucin	Porcine	11.5–18.4	18.5–25.8	15.3–16.1	6.0–7.6	12, 14, 15
16. Sublingual mucin	Bovine	22.0	27.0	20.9	7.1	12, 16
17. Salivary mucin	Collicallia secretions	18.0	17.3	18.0	—	17
18. Fish skin mucin	Loach	9.8	9.8	22.0	—	18
19. Fish skin mucins	Several species	1.5–10.2	0.7–5.1	0.25–2.0	—	19
20. Bile mucin	Human bile	24.0	17.2	—	—	20
21. Meconium glycoprotein	Human meconium	28.0	29.0	9.5	7.9	4, 5
22. Intestinal mucus	Human – cystic fibrosis patient	9.0	19.7	12.6	6.4	21
23. Mucin	Human colloid breast tumor	20.0	39.6	1.7	—	22

Table 3 (continued)
GLYCOPROTEINS BLOOD GROUP SUBSTANCES, AND MUCINS

Glycoprotein	Source	Carbohydrate content g/100 g				Reference
		Neutral sugar	Acetyl hexosamine	Sialic acid	Fucose	
24. Bronchial sulfated glycoprotein	Human cystic fibrosis patient	29.0	33.5	3.4	16.6	23
25. Jelly coat sulfated glycoprotein	Sea urchin eggs	3.7	3.0	70.0	9.2	24

Compiled by Morris Soodak.

REFERENCES

1. Watkins, in *Glycoproteins*, Gottschalk, Ed., Elsevier, Amsterdam, 1966, 462.
2. Kathan, Johnson, and Winzler, *J. Exp. Med.*, 113, 37 (1961).
3. Kathan and Adamany, *J. Biol. Chem.*, 242, 1716 (1967).
4. Bezkorovainy, Springer, and Hotta, *Biochim. Biophys. Acta*, 115, 501 (1966).
5. Springer, Nagai, and Tegtmeyer, *Biochemistry*, 5, 3254 (1966).
6. Springer and Fletcher, in *Organ Transplantation*, Haymer, Ricken, and Letterer, Eds., Schattauer Verlag, Stuttgart, 1969, 35.
7. Uhlenbruck, Hansen, and Pardoe, *Z. Physiol. Chem.*, 349, 737 (1968).
8. Buddecke, in *Glycoproteins*, Gottschalk, Ed., Elsevier, Amsterdam, 1966, 558.
9. Gibbons, *Biochem. J.*, 73, 209 (1959).
10. Gibbons, *Biochem. J.*, 89, 380 (1963).
11. Gibbons and Roberts, *Ann. N.Y. Acad. Sci.*, 106, 218 (1963).
12. Pigman and Gottschalk, in *Glycoproteins*, Gottschalk, Ed., Elsevier, Amsterdam, 1966, 434.
13. Bhargava and Gottschalk, *Biochim. Biophys. Acta*, 127, 223 (1966).
14. de Salegui and Plonska, *Arch. Biochem. Biophys.*, 129, 49 (1969).
15. Carlson, *J. Biol. Chem.*, 243, 616 (1968).
16. Katzman and Eylar, *Arch. Biochem. Biophys.*, 117, 623 (1966).
17. Howe, Lee, and Rose, *Arch. Biochem. Biophys.*, 95, 512 (1961).
18. Turumi and Saito, *Tohoku J. Exp. Med.*, 58, 247 (1953).
19. Wessler and Werner, *Acta Chem. Scand.*, 11, 1240 (1957).
20. Tiba, *Tohoku J. Exp. Med.*, 52, 103 (1950).
21. Johansen, *Biochem. J.*, 87, 63 (1963).
22. Adams, *Biochem. J.*, 94, 368 (1965).
23. Roussel, Lamblin, Degand, Walker-Nasir, and Jeanloz, *J. Biol. Chem.*, 250, 2114 (1975).
24. Hotta, Hamazaki, Kurokawa, and Isaka, *J. Biol. Chem.*, 245, 5434 (1970).

Table 4
GLYCOPROTEINS, TISSUES

Glycoprotein isolated from	Source	Carbohydrate content, g/100 g				Reference
		Neutral sugar	Acetyl hexosamine	Sialic acid	Fucose	
1. Aorta	Cow	5.5	5.8	7.1	0.8	1, 2, 3
2. Aorta	Human	4.0	4.2	2.2	–	4
3. Bone	Cow	11.3	10.7	20.7	0.7	5–8
4. Bone	Rabbit	7.8	10.5	5.8	1.0	9
5. Cornea	Cow	1.69	1.55	0.22	0–1.2	10
6. Fetal skin	Cow	5.5	5.2	5.9	–	1, 11
7. Vitreous body	Cow	4.3	2.6	0.4	–	1, 12
8. Sarcolemma	Frog	4.1	1.1	1.0	–	13
9. Spleen (apohemosiderin)	Horse	2.4	1.2	0.5	0.3	14
10. Granulation tissue	Rat	12.2	5.8	12.9	0.9	1, 15
11. Albuminoid	Human platelets	5.1	0.9	0.8	–	16
12. Sericin A	Bombyx mori cocoons	0.3	0.8	–	–	17
13. Sericin B	Bombyx mori cocoons	0.8	2.26	–	–	17
14. β-Parotin	Bovine paratid gland	2.5	0.6	–	–	18
15. Intrinsic factor	Hog stomach	18.6	–	–	–	19
16. Collagen	Earthworm	12.0	2.5	0	–	20, 21
17. Collagen	Calf	1.5	0.4	0	–	20, 22
18. Collagen	Dog basement membrane	0.42	0.03	–	–	23
19. Collagen	Bovine cornea	1.9	4.4	0.5	–	24
20. Sperm binding protein	Sea urchin eggs	3.3	3.4	–	–	25
21. Membrane glycoprotein	Pig platelets	17.7	18.4	9.6	–	26
22. Aorta (intima)	Pig	2.5	2.0	1.2	0.6	27
23. Brain G. P. –350 soluble and membrane-bound	Calf	9.4	6.6	2.4	1.0	28, 29
24. Brain	Pigeon					30
G. P. 10–BI		11.1	20.0	0.6	–	
G. P. 10–BII		16.7	22.0	1.0	–	
25. Brain-sulfated glycopeptide	Cow	31.2–37.5	17.5–27.7	13.5–25.0	3.3–5.8	31
26. Brain-tubulin	Pig	0.38	0.57	0.26	0.1	32, 33
27. Rhodopsin	Cow	1.7	2.2	–	–	34
28. Anterior lens capsule	Calf	9.6	2.0	0.3	0.25	35
29. Anterior lens capsule	Cow	11.3	0.9	0.1	0.25	35
30. Anterior lens capsule	Dog	10.4	1.2	0.4	0.6	36
31. Posterior lens capsule	Calf	7.6	2.8	0.6	0.24	35
32. Posterior lens capsule	Cow	10.0	1.2	0.2	0.3	35
33. Glomerular basement membrane	Cow	6.3	2.1	–	0.2	37
34. Glomerular basement membrane	Man	5.5	1.5	0.7	0.2	38
35. Glomerular basement membrane	Dog	6.2	1.8	2.0	0.75	36

Table 4 (continued)
GLYCOPROTEINS, TISSUES

Glycoprotein isolated from	Source	Carbohydrate content, g/100 g				Reference
		Neutral sugar	Acetyl hexosamine	Sialic acid	Fucose	
36. Glycoprotein component VII	Bovine, glomerular membrane	3.7	3.9	1.5	0.2	39
37. Collagen	Carp, swim bladder	0.5	0.1	0.02	0.01	40
38. Collagen	Shark, elastoidin	0.9	0.004	—	—	41
39. Collagen	Bovine Achilles tendon	0.8	0.6	0.06	0.04	40
40. Collagen	Metridium, body wall	7.4	0.5	—	0.3	42
41. Collagen	Lumbricus, whole cuticle	13.1	1.7	—	1.33	43
42. Collagen	Loligo, cephalic cartridge	2.7	0.2	—	0.05	43

Compiled by Morris Soodak.

REFERENCES

1. Buddecke, in *Glycoproteins,* Gottschalk, Ed., Elsevier, Amsterdam, London, New York, 1966, 558.
2. Berenson and Fishkin, *Arch. Biochem. Biophys.,* 97, 18 (1962).
3. Radhakrishnamurthy, Fishkin, Hubbell, and Berenson, *Arch. Biochem. Biophys.,* 104, 19 (1964).
4. Barnes and Partridge, *Biochem. J.,* 109, 883 (1968).
5. Williams and Peacocke, *Biochim. Biophys. Acta,* 101, 327 (1965).
6. Andrews and Herring, *Biochim. Biophys. Acta,* 101, 239 (1965).
7. Herring and Kent, *Biochem. J.,* 89, 405 (1963).
8. Andrews, Herring, and Kent, *Biochem. J.,* 104, 705 (1967).
9. Burckard, Havez, and Dautrevaux, *Bull. Soc. Chim. Biol.,* 48, 851 (1966).
10. Robert and Dische, *Biochem. Biophys. Res. Commun.,* 10, 209 (1963).
11. Bourrillon and Got, *Biochim. Biophys. Acta,* 58, 63 (1962).
12. Berman, *Biochim. Biophys. Acta,* 83, 27 (1964).
13. Kono, Kakuma, Homma, and Fukuda, *Biochim. Biophys. Acta,* 88, 155 (1964).
14. Ludewig and Glover, *Arch. Biochem. Biophys.,* 113, 654 (1966).
15. Fishkin and Berenson, *Arch. Biochem. Biophys.,* 95, 130 (1961).
16. Bezkorovainy and Rafelson, *J. Lab. Clin. Med.,* 64, 212 (1964).
17. Sinohara and Asano, *J. Biochem.* (Tokyo), 62, 129 (1967).
18. Ito, Okabe, and Namba, *Endocrinol. Jap.,* 12, 249 (1965).
19. Williams, Ellenbogen, and Esposito, *Proc. Soc. Exp. Biol. Med.,* 87, 400 (1954).
20. Eylar, *J. Theor. Biol.,* 10, 89 (1965).
21. Maser and Rice, *Biochim. Biophys. Acta,* 63, 255 (1962).
22. Gross, Dumsha, and Glazer, *Biochim. Biophys. Acta,* 30, 293 (1958).
23. Kefalides, *Biochem. Biophys. Res. Commun.,* 22, 26 (1966).
24. Bosmann and Jackson, *Biochim. Biophys. Acta,* 170, 6 (1968).
25. Aketa, Tsuzuki, and Onitake, *Exp. Cell Res.,* 50, 676 (1968).
26. Mullinger and Manley, *Biochim. Biophys. Acta,* 170, 282 (1968).
27. Wagh and Roberts, *Biochemistry,* 11, 4222 (1972).
28. van Nieuw Amerongen, van den Eijnden, Heijlman, and Roukema, *J. Neurochem.,* 19, 2195 (1972).
29. van Nieuw Amerongen and Roukema, *J. Neurochem.,* 23, 85 (1947).
30. Bogoch, in *Protein Metabolism of the Nervous System,* Plenum Press, New York, 1970, 555.

Table 4 (continued)
GLYCOPROTEINS, TISSUES

31. Arima, Muramatsu, Saigo, and Egami, *Jap. J. Exp. Med.*, 39, 301 (1969).
32. Margolis, Margolis, and Shelanski, *Biochem. Biophys. Res. Commun.*, 47, 432 (1972).
33. Feit and Shelanski, *Biochem. Biophys. Res. Commun.*, 66, 920 (1975).
34. Heller, *Biochemistry*, 7, 2907 (1968).
35. Fukushi and Spiro, *J. Biol. Chem.*, 244, 2041 (1969).
36. Kefalides and Denduchis, *Biochemistry*, 8, 4613 (1969).
37. Spiro, *J. Biol. Chem.*, 242, 1915 (1967).
38. Beisswenger and Spiro, *Science*, 168, 596 (1970).
39. Ohno, Riquetti, and Hudson, *J. Biol. Chem.*, 250, 7780 (1975).
40. Spiro, *J. Biol. Chem.*, 244, 602 (1969).
41. Sastry and Ramachandran, *Biochim. Biophys. Acta*, 97, 281 (1965).
42. Katzman and Jeanloz, *Fed. Proc.*, 29, 599 (1970).
43. Spiro and Bhoyroo, in *Glycoproteins, Part B*, Gottschalk, Ed., Elsevier, 1972, 986.

Table 4A
GLYCOPROTEINS, CELL MEMBRANES

Protein	Source	Carbohydrate content g/100 g				Reference
		Neutral sugar	Acetyl hexosamine	Sialic acid	Fucose	
1. Glycophorin	Human erythrocytes	14.9	17.7	25.4	1.0	1, 2
2. Anion exchange protein	Human erythrocytes	3.8	1.7	0.6	1.4	3
3. GP-II	Human erythrocytes	4.9	5.4	7.8	–	4
4. GP-III		7.6	7.2	8.1	–	4
5. Major glycoprotein	Horse erythrocytes	9.4	12.4	33.7	0.6	5
6. Major glycoprotein	Swine erythrocytes	18.0	23.4	27.2	1.7	5
7. Major glycoprotein	Swine erythrocytes	6.4	8.6	36.3	–	5
8. Lipopolysaccharide receptor	Human erythrocytes	7.8	12.7	16.3	0.9	6
9. H-2^b Alloantigen fragment	Mouse spleen cells	5.4	4.3	1.3	–	7, 8
10. H-2^d Alloantigen fragment	Mouse spleen cells	3.5	3.5	1.0	–	7, 8
11. Glycopeptide of (Na$^+$ + K$^+$) adenosine-triphosphatase	Canine renal medulla	2.2–3.7	4.8	1.7	–	9

Compiled by Morris Soodak.

REFERENCES

1. Javaid and Winzler, *Biochemistry*, 13, 3635 (1974).
2. Tomita and Marchesi, *Proc. Natl. Acad. Sci. U.S.A.*, 72, 2964 (1975).
3. Ho and Guidotti, *J. Biol. Chem.*, 250, 675 (1975).
4. Fujita and Cleve, *Biochim. Biophys. Acta*, 382, 172 (1975).
5. Fujita and Cleve, *Biochim. Biophys. Acta*, 406, 206 (1975).
6. Springer, Adye, Bezkorovainy, and Jirgensons, *Biochemistry*, 13, 1379 (1974).
7. Shimada and Nathanson, *Biochemistry*, 8, 4048 (1969).
8. Schwartz, Kato, Cullen, and Nathanson, *Biochemistry*, 12, 2157 (1973).
9. Kyte, *J. Biol. Chem.*, 247, 7642 (1972).

Table 4B
GLYCOPROTEINS, ENVELOPED VIRUSES

Protein	Source	Carbohydrate content g/100 g				Reference
		Neutral sugar	Acetyl hexosamine	Sialic acid	Fucose	
1. Membrane protein	Sindbis virus	5.1	8.7	0.7–1.2	0.5	1
2. Membrane protein E_1	Semliki Forest virus	2.8	3.4	1.7	0.2	2
3. Membrane protein E_2	Semliki Forest virus	5.8	3.8	2.8	0.3	2
4. Membrane protein E_3	Semliki Forest virus	24.0	34.0	17.0	7.0	2
5. Membrane protein glycoprotein	Vesicular stomatitis virus	4.6	3.7	2.3	0.3	3, 4

Compiled by Morris Soodak.

REFERENCES

1. **Strauss, Burge, and Darnell,** *J. Mol. Biol.,* 47, 437 (1970).
2. **Garoff, Simons, and Renkonen,** *Virology,* 61, 493 (1974).
3. **Ethchison and Holland,** *Virology,* 60, 217 (1974).
4. **Ethchison and Holland,** *Proc. Natl. Acad. Sci. U.S.A.,* 71, 4011 (1974).

Table 5
GLYCOPROTEINS, BODY FLUIDS

Glycoprotein	Source	Carbohydrate content, g/100 g				Reference
		Neutral sugar	Acetyl hexosamine	Sialic acid	Fucose	
1. Tamm and Horsfall glycoprotein	Human urine	8.1	8.8	9.1	1.1	1–4
2. Tamm and Horsfall glycoprotein	Sheep urine	30.6	6.2	4.0	–	5
3. L-type Bence Jones protein	Human urine	6.1	4.4	3.1	1.6	6
4. Glycoprotein	Human plasmacytoma urine	16.6	12.5	6.8	4.5	7
5. B_{12}-binding proteins	Leukemic human urine	14–28	15–25	3.7–8.0	–	8
6. Pleural fluid glycoprotein	Human	14.2	13.9	11.9	–	9, 10
7. Synovial fluid glycoprotein	Human	11.5	11.6	4.9	1.0	11, 12
8. α-Globulin	Human amniotic fluid	47.0	24.5	9.0	–	13
9. Glycoprotein	Human cancer ascites fluid	19.8	12.3	30.5	5.0	14
10. Ascites fluid	Yoshida rat tumor	3.25	2.2	12.4	11.0	15, 16
11. Soluble glycoprotein	Boar seminal plasma	10.8	5.7	0.5	–	17
12. Insoluble glycoprotein	Boar seminal plasma	10.0	7.6	0.4	–	18
13. Tamm and Horsfall glycoprotein	Human urine	11.7	11.2	4.4	0.8	19
14. N_0 glycoprotein antigen	Normal human gastric juice	28.5	32.6	11.4	16.1	20
15. I Glycoprotein antigen	Normal human gastric juice	28.8	23.9	14.8	17.3	20
16. Sulfated glycoprotein, allantoic antigen	Chick allantoic fluid	26.0	32.1	1.0	4.3	21

Compiled by Morris Soodak.

Table 5 (continued)
GLYCOPROTEINS, BODY FLUIDS

REFERENCES

1. **Eylar,** *J. Theor. Biol.,* 10, 89 (1965).
2. **Gottschalk,** *Nature,* 170, 662 (1952).
3. **Maxfield,** in *Glycoproteins,* Gottschalk, Ed., Elsevier, Amsterdam, 1966, 446.
4. **Rosenfeld and Yusipova,** *Biokhimiya,* 32, 111 (1967).
5. **Cornelius, Bishop, Berger, and Pangborn,** *Am. J. Vet. Res.,* 22, 1000 (1961).
6. **Edmundson, Sheber, Ely, Simonds, Hutson, and Rossiter,** *Arch. Biochem. Biophys.,* 127, 725 (1968).
7. **Weicker, Huhnstock, and Grässlin,** *Clin. Chim. Acta,* 9, 19 (1964).
8. **Kallee, Debiasi, Karypidis, Heide, and Schwick,** *Acta Isot.,* 4, 103 (1964).
9. **Bourrillon, Michon, and Got,** *Biochim. Biophys. Acta,* 47, 243 (1961).
10. **Bourrillon, Got, and Meyer,** *Biochim. Biophys. Acta,* 74, 255 (1963).
11. **Miki and Noma,** in *Biochemistry and Medicine of Mucopolysaccharides,* Egami and Oshima, Eds., Maruzen, Tokyo, 1962, 191.
12. **Buddecke,** in *Glycoproteins,* Gottschalk, Ed., Elsevier, Amsterdam, 1966, 558.
13. **Lambotte and Gosselin-Rey,** *Arch. Int. Physiol. Biochem.,* 75, 109 (1967).
14. **Turumi, Takahashi, and Saito,** *Fukushima J. Med. Sci.,* 3, 31 (1956).
15. **Marcante,** *Clin. Chim. Acta,* 8, 799 (1963).
16. **Caputo, Marcante, and Zito,** *Br. J. Exp. Pathol.,* 47, 599 (1966).
17. **McIntosh and Boursnell,** *Biochim. Biophys. Acta,* 130, 252 (1966).
18. **Rottenberg and Boursnell,** *Biochim. Biophys. Acta,* 117, 157 (1966).
19. **Fletcher, Neuberger, and Ratcliffe,** *Biochem. J.,* 120, 417 (1970).
20. **Häkkinen,** *Transplant. Rev.,* 20, 61 (1974).
21. **How and Higginbothan,** *Carbohydr. Res.,* 12, 355 (1970).

Table 6
GLYCOPROTEINS, EGGS, MILK

Eggs	Source	Carbohydrate content, g/100 g			Reference
		Neutral sugar	Acetyl hexosamine	Sialic acid	
1. Avidin	Hen egg	5.8	14.7	0	1, 2
2. Ovalbumin	Hen egg	2.0	1.2	0	1, 3, 4
3. Ovoglycoprotein	Hen egg	13.6	17.0	3.0	5
4. Ovomucin	Hen egg	15.4	14.7	5.8	1, 6
5. Ovomucoid	Hen egg	5.7–10.5	11.3–21.0	0.4–2.0	1, 7–9
6. Conalbumin	Hen egg	0.8	1.7	0	10
7. Transferrin	Hen egg	0.8–1.4	1.5–1.7	0.4	11
8. Vitello mucoid	Hen egg	23.7	26.4	–	12
9. Ovoinhibitor of chymotrypsin	Hen egg	3.9	6.9	0.1	13
10. κ Casein	Cow milk	1.4	1.5	2.4	1, 14
11. Casein	Cow milk	0.3	0.22	0.3	15, 16
12. Casein	Human milk	2.5	2.7	0.8	16
13. Casein	Polar bear milk	2.8	1.3	1.9	15
14. Casein	Sheep milk	0.23	0.15	0.11	16
15. Casein	Goat milk	0.22	0.16	0.30	16
16. Casein	Whale milk	0.59	0.42	0.37	16
17. Casein	Horse milk	0.55	0.44	0.56	16

Table 6 (continued)
GLYCOPROTEINS, EGGS, MILK

Eggs	Source	Carbohydrate content, g/100 g			Reference
		Neutral sugar	Acetyl hexosamine	Sialic acid	
18. Casein	Reindeer milk	0.44	0.23	0.46	16
19. M_1-Glycoprotein	Cow colostrum	10.8	10.2	11.1	17
20. M_1-Glycoprotein	Cow milk	8.4	8.2	5.9	18
21. M_2-Glycoprotein	Cow colostrum	7.2	6.5	6.4	17
22. M_2-Glycoprotein	Cow milk	4.9	4.8	5.3	17
23. Phosphoglycoprotein	Cow milk	3.1	2.4	4.0	17
24. Lactotransferrin	Human milk	3.9	3.0	0.87	19
25. Glycoprotein a	Cow milk	3.1	–	–	20
26. Interfacial glycoprotein	Cow milk	5.6	4.8	4.7	21

Compiled by Richard J. Winzler.

REFERENCES

1. Eylar, *J. Theor. Biol.*, 10, 89 (1965).
2. Melamed and Green, *Biochem. J.*, 89, 591 (1963).
3. Fletcher, Marshall, and Neuberger, *Biochim. Biophys. Acta*, 74, 311 (1963).
4. Neuberger and Marshall, in *Glycoproteins*, Gottschalk, Ed., Elsevier, Amsterdam, 1966, 299.
5. Ketterer, *Biochem. J.*, 96, 372 (1965).
6. Gottschalk and Lind, *Br. J. Exp. Pathol.*, 30, 85 (1949).
7. Chatterjee and Montgomery, *Arch. Biochem. Biophys.*, 99, 426 (1962).
8. Bragg and Hough, *Biochem. J.*, 78, 11 (1961).
9. Montreuil, Castiglioni, Adam-Chosson, Caner, and Queval, *J. Biochem.*, 57, 514 (1965).
10. Williams, *Biochem. J.*, 83, 355 (1962).
11. Williams, *Biochem. J.*, 108, 57 (1968).
12. Osaki and Yosizawa, *Tohoku J. Exp. Med.*, 51, 62 (1949).
13. Davis, Zahnley, and Donovan, *Biochemistry*, 8, 2044 (1969).
14. Jolles, Alais, and Jolles, *Arch. Biochem. Biophys.*, 98, 56 (1962).
15. Baker, Huang, and Harington, *Biochem. Biophys. Res. Commun.*, 13, 227 (1963).
16. Johansson and Svennerholm, *Acta Physiol. Scand.*, 37, 324 (1956).
17. Bezkorovainy, *Arch. Biochem. Biophys.*, 110, 558 (1965).
18. Bezkorovainy, *J. Dairy Sci.*, 50, 1368 (1967).
19. Montreuil, Tonnelat, and Mullet, *Biochim. Biophys. Acta*, 45, 413 (1960).
20. Groves and Gordon, *Biochemistry*, 6, 2388 (1967).
21. Jackson, Coulson, and Clark, *Arch. Biochem. Biophys.*, 97, 373 (1962).

Table 7
GLYCOPROTEINS, PLANTS

	Protein	Source	Carbohydrate content, g/100 g		Reference
			Neutral sugar	Acetyl hexosamine	
1.	Hemagglutinin	Soybean	1.2–9.8	0	1–4
2.	Hemagglutinin	Phaseolus vulgaris	6.0–11.8	3.4	5, 6
3.	Taste modifying glycoprotein	Miracle fruit	6.7	–	7
4.	Globulin	Soybeans	3.8	1.5	8
5.	Stellacyanin	Japanese lac tree	20.0	25.0	9
6.	Glycopeptides from extensin	Various plant cell walls	70.5	–	10
7.	Agglutinin	Soybean (Glycine max)	4.5	1.2	11–13
8.	Agglutinin	Horse gram (Dolichos biflorus)	2.5	1.25	14–16
9.	Lectin A	Lotus tetragonolobulus	8	1.4	17, 18
10.	Lectin B	Lotus tetragonolobulus	4	0.8	17, 18
11.	Lectin C	Lotus tetragonolobulus	8	1.2	17, 18
12.	Agglutinin A	Red kidney bean (Phaseolus vulgaris)	10.0	2.9	19
13.	Agglutinin B	Red kidney bean (P. vulgaris)	10.5	2.4	19
14.	Agglutinin	Robina pseudoacacia	6.7	4.0	20
15.	Agglutinin	Lima bean (Phaseolus lunatus)	4.3	1.4	21, 22
16.	Agglutinin	Castor bean (Ricinus communis)	3.7	–	23
17.	Agglutinin	Wax bean (Phaseolus vulgaris)	9.5	1.1	24–26
18.	Agglutinin I	Goose (Ulex europeus)	3.8	1.7	27–29
19.	Agglutinin II	Goose (U. europeus)	19.9	2.2	27–29
20.	Lectin	Bandeiraea simplicifolia	7.8	2.8	30
21.	Agglutinin	Sophora japonica	5.9	2.2	31
22.	Mitogen	Wistaria floribunda	7.9	4.3	32, 33
23.	Agglutinin	Bauhinia purpurea alba	7.7	4.1	34
24.	Mitogen	Pokeweed	2.7	1.7	35, 36

Compiled by Morris Soodak.

REFERENCES

1. **Lis, Sharon, and Katchalski,** *J. Biol. Chem.,* 241, 684 (1966).
2. **Lis, Fridman, Sharon, and Katchalski,** *Arch. Biochem. Biophys.,* 117, 301 (1966).
3. **Liener and Pallansch,** *J. Biol. Chem.,* 197, 29 (1952).
4. **Wada, Pallansch, and Liener,** *J. Biol. Chem.,* 233, 395 (1958).
5. **Borjeson, Bouveng, Gardell, and Thunell,** *Biochim. Biophys. Acta,* 82, 158 (1964).
6. **Pusztai,** *Biochem. J.,* 101, 379 (1966).
7. **Kurihara and Beidler,** *Science,* 161, 1241 (1968).
8. **Koshiyama,** *Agric. Biol. Chem.,* 30, 646 (1966).
9. **Peisach, Levine, and Blumberg,** *J. Biol. Chem.,* 242, 2847 (1967).
10. **Lamport,** *Biochemistry,* 8, 1155 (1969).
11. **Lis, Sharon, and Katchalski,** *J. Biol. Chem.,* 241, 684 (1966).

Table 7 (continued)
GLYCOPROTEINS, PLANTS

12. Sharon and Lis, *Science,* 177, 949 (1972).
13. Lis and Sharon, *Annu. Rev. Biochem.,* 42, 541 (1973).
14. Font, Leseney, and Bourrillon, *Biochim. Biophys. Acta,* 243, 434 (1971).
15. Etzler, in *Methods in Enzymology,* Vol. 28B, Ginsburg, Ed., 1972, 340.
16. Etzler and Kabat, *Biochemistry,* 9, 899 (1970).
17. Kalb, *Biochim. Biophys. Acta,* 168, 532 (1968).
18. Yariv, Kalb, and Blumberg, in *Methods in Enzymology,* Vol. 28B, Ginsburg, Ed., 1972, 356.
19. Dahlgren, Porath, and Lindahl-Kiessling, *Arch. Biochem. Biophys.,* 137, 306 (1970).
20. Font and Bourrin, *Biochim. Biophys. Acta,* 243, 111 (1971).
21. Galbraith and Goldstein, in *Methods in Enzymology,* Vol. 28B, Ginsburg, Ed., 1972, 318.
22. Gould and Scheinberg, *Arch. Biochem. Biophys.,* 141, 607 (1970).
23. Waldschmidt-Leitz and Keller, *Z. Physiol. Chem.,* 351, 990 (1970).
24. Takahashi, Ramachandramurthy, and Liener, *Biochim. Biophys. Acta,* 133, 123 (1967).
25. Takahashi and Liener, *Biochim. Biophys. Acta,* 154, 560 (1968).
26. Sela, Lis, and Sharon, *Biochim. Biophys. Acta,* 310, 273 (1973).
27. Matsumoto and Osawa, *Biochim. Biophys. Acta,* 194, 180 (1969).
28. Matsumoto and Osawa, *Arch. Biochem. Biophys.,* 140, 484 (1970).
29. Osawa and Matsumoto, in *Methods in Enzymology,* Vol. 28B, Ginsburg, Ed., 1972, 323.
30. Hayes and Goldstein, *J. Biol. Chem.,* 249, 1904 (1974).
31. Poretz, Riss, Timberlake, and Chien, *Biochemistry,* 13, 250 (1974).
32. Osawa and Toyoshima, in *Methods in Enzymology,* Vol. 28B, Ginsburg, Ed., 1972, 328.
33. Toyoshima, Akiyama, Nakano, Tonomura, and Osawa, *Biochemistry,* 10, 4457 (1971).
34. Irimura and Osawa, *Arch. Biochem. Biophys.,* 151, 475 (1972).
35. Reisfield, Börjeson, Chessin, and Small, *Proc. Natl. Acad. Sci. U.S.A.,* 58, 2020 (1967).
36. Börjeson, Reisfeld, Chessin, Welsh, and Douglas, *J. Exp. Med.,* 124, 859 (1966).

Table 8
GLYCOPROTEINS, HORMONES

Hormone	Source	Carbohydrate content, g/100 g				Reference
		Neutral sugar	Acetyl hexosamine	Sialic acid	Fucose	
1. Chorionic gonadotropin	Human urine	11.0–11.2	10.8–11.1	8.5–9.0	1.2	1–4
2. Serum gonadotropin	Pregnant mare serum	18.6	17.5	10.4	1.4	1, 3, 6
3. Interstitial cell stimulating hormone	Ovine pituitary	5.0	9.6	0.5	1.1	3, 7, 8
4. Interstitial cell stimulating hormone	Human pituitary	1.1	2.5	1.0	0.4	3, 9
5. Follicle stimulating hormone	Ovine pituitary	4.0–5.7	8.6–5.8	12.9	–	3, 10
6. Follicle stimulating hormone	Porcine pituitary	2.6	4.6	–	1.1	3, 9
7. Follicle stimulating hormone	Human pituitary	3.9–12.2	2.9–9.0	1.4–5.1	–	11, 12
8. Lutinizing hormone	Human pituitary	11.3	4.9	2.0	–	11
9. Lutinizing hormone	Ovine pituitary	6.1–11.8	7.4–9.5	0.3–0.5	–	11, 13
10. Lutinizing hormone	Bovine pituitary	11.9	5.9	0.3	–	11
11. Thyrotropin	Bovine pituitary	9.4	13.2	–	1.3	3
12. Thyrotropin	Human pituitary	5.9	4.1	–	0.5	14
13. Erythropoietin	Ovine anemic plasma	29.2	21.5	13.0	–	3, 15
14. Gonadotropin transportation protein	Equine urine	14.0	13.3	12.0	–	1, 16
15. Thyroglobulin	Human thyroid	4.8	4.2	1.1	0.5	17, 18

Table 8 (continued)
GLYCOPROTEINS, HORMONES

Hormone	Source	Carbohydrate content, g/100 g				Reference
		Neutral sugar	Acetyl hexosamine	Sialic acid	Fucose	
16. Thyroglobulin	Porcine thyroid	4.0	3.4	1.2	0.5	17, 18
17. Thyroglobulin	Ovine thyroid	4.0	3.2	1.5	0.4	17, 18
18. Thyroglobulin	Bovine thyroid	3.7	3.2	1.4	0.4	17, 18
19. Chorionic gonadotropin (native)	Human urine	12.5	11.8	8.7	0.9	19
20. Chorionic gonadotropin α subunit	Human urine	13.8	11.6	8.8	0.06	19
21. Chorionic gonadotropin β subunit	Human urine	12.5	14.3	10.2	1.3	19
22. Follicle stimulating hormone (native)	Ovine pituitary	7.4	10.2	5.6	0.7	20
23. Follicle stimulating hormone α subunit	Ovine pituitary	7.4	10.3	3.6	0.4	20
24. Follicle stimulating hormone β subunit	Ovine pituitary	7.3	10.7	7.6	1.3	20
25. Follicle stimulating hormone (native)	Equine pituitary	8.7	9.2	6.8	1.0	21
26. Follicle stimulating hormone α subunit	Equine pituitary	7.0	8.1	5.6	0.5	21
27. Follicle stimulating hormone β subunit	Equine pituitary	8.4	9.4	7.4	1.4	21
28. Luteinizing hormone native	Human pituitary	7.3	7.2	2.8	0.9	22
29. Luteinizing hormone α subunit	Human pituitary	8.2	10.8	4.3	0.7	22
30. Luteinizing hormone β subunit	Human pituitary	4.0	2.7	1.8	1.6	22
31. Luteinizing hormone native	Ovine pituitary	6.5	9.4	–	1.4	23
32. Luteinizing hormone α subunit	Ovine pituitary	8.3	10.3	–	1.3	23
33. Luteinizing hormone β subunit	Ovine pituitary	4.3	7.0	–	1.0	23
34. Thyrotropin native	Human pituitary	4.4	6.6	1.8	0.5	24
35. Thyrotropin α subunit	Human pituitary	6.2	9.7	2.9	0.2	24
36. Thyrotropin β subunit	Human pituitary	2.8	4.1	0.4	0.5	24
37. Thyrotropin native	Bovine pituitary	5.6	11.0	–	0.76	25
38. Thyrotropin α subunit	Bovine pituitary	7.5	14.0	–	0.35	25
39. Thyrotropin β subunit	Bovine pituitary	3.2	7.3	–	1.1	25
40. Erythropoieten	Ovine anemic plasma	9.0	9.2	10.8	–	26, 27

Compiled by Morris Soodak.

Table 8 (continued)
GLYCOPROTEINS, HORMONES

REFERENCES

1. Eylar, *J. Theor. Biol.*, 10, 89 (1965).
2. Got, Bourrillon, and Michon, *Bull. Soc. Chim. Biol.*, 42, 41 (1960).
3. Papkoff, in *Glycoproteins,* Gottschalk, Ed., Elsevier, Amsterdam, 1966, 532.
4. Bahl, *J. Biol. Chem.*, 244, 567 (1969).
5. Got and Bourrillon, *Biochim. Biophys. Acta*, 42, 505 (1960).
6. Bourrillon, Michon, and Got, *Bull. Soc. Chim. Biol.*, 41, 493 (1959).
7. Li and Starman, *Nature*, 202, 291 (1964).
8. Li, *J. Natl. Cancer Inst. Monogr.*, 12, 181 (1963).
9. Steelman and Segeloff, *Recent Prog. Horm. Res.*, 15, 115 (1959).
10. Papkoff, Gospodarowicz, and Li, *Arch. Biochem. Biophys.*, 120, 434 (1967).
11. Kathan, Reichert, and Ryan, *Endocrinology*, 81, 45 (1967).
12. Papkoff, Mahlmann, and Li, *Biochemistry*, 6, 3976 (1967).
13. Walborg and Ward, *Biochim. Biophys. Acta,* 78, 304 (1963).
14. Kim, Shome, Liao, and Pierce, *Anal. Biochem.*, 20, 258 (1967).
15. Goldwasser, White, and Taylor, *Biochim. Biophys. Acta*, 64, 487 (1962).
16. Bourrillon, Got, and Marcy, *Bull. Soc. Chim. Biol.*, 40, 87 (1958).
17. McQuillan and Trikojus, in *Glycoproteins,* Gottschalk, Ed., Elsevier, Amsterdam, 1966, 516.
18. Spiro, *Fed. Proc.*, 22, 538 (1963).
19. Bahl, in *Hormonal Proteins and Peptides,* Vol. 1, Li, Ed., Academic Press, New York, 1973, 171.
20. Grimek and McShan, *J. Biol. Chem.*, 249, 5725 (1974).
21. Landefield and McShan, *J. Biol. Chem.*, 249, 3527 (1974).
22. Closset, Vandalem, Hennen, and Lequin, *Eur. J. Biochem.*, 57, 325 (1975).
23. Papkoff, in *Hormonal Proteins and Peptides,* Vol. 1, Li, Ed., Academic Press, New York, 1973, 68.
24. Cornell and Pierce, *J. Biol. Chem.*, 248, 4327 (1973).
25. Pierce, Liao, and Carlsen, in *Hormonal Proteins and Peptides,* Vol. 1, Li, Ed., Academic Press, New York, 1973, 28.
26. Goldwasser and Kung, *Fed. Proc.*, 30, 1128 (1971).
27. Winzler, in *Hormonal Proteins and Peptides,* Vol. 1, Li, Ed., Academic Press, New York, 1973, 2.

Table 9
GLYCOPROTEINS, ENZYMES

Enzyme	Source	Carbohydrate content, g/100 g				Reference
		Neutral sugar	Acetyl hexosamine	Sialic acid	Fucose	
1. Cholinesterase	Horse serum	–	–	3.2		1
2. Cholinesterase	Human serum	3.6–9.3	2.9–8.4	1.8–6.0		2–4
3. Non-specific esterase	Rat tissues	6.2	4.6	6.5		5
4. Glucose oxidase	*Aspergillus niger*	14.2	2.0	–		6
5. Peroxidase	Horse radish	16.2	1.9	–		7
6. Chloroperoxidase	Caldariomyces fumago	25–30	–	–		8
7. Monamine oxidase	Bovine plasma	4.6	–	–		9, 10
8. Ribonuclease B	Bovine pancreas	6.3	5.0	–		9, 11
9. Ribonuclease B	Bovine pancreatic juice	7.4	3.1	0.2		12
10. Ribonuclease C	Bovine pancreatic juice	6.4	5.5	2.5		12
11. Ribonuclease D	Bovine pancreatic juice	6.4	4.6	4.8		12
12. Ribonuclease	Porcine pancreatic juice	11.1–24.6	7.4–16.8	0–2.8		13
13. Deoxyribonuclease	Bovine pancreas	1.1–2.9	1.3–1.4	0		14, 15
14. Deoxyribonuclease	Hog spleen	–	4.1	–		16
15. DPN ase	Neurospora	82.2	2.0	–		17

Table 9 (continued)
GLYCOPROTEINS, ENZYMES

Enzyme	Source	Carbohydrate content, g/100 g				Reference
		Neutral sugar	Acetyl hexosamine	Sialic acid	Fucose	
16. DPN ase	*B. subtilis*	52.8	17.9	–		17
17. DPN ase inhibitor	*B. subtilis*	72.7	0	–		17
18. Alkaline phosphatase	Human placenta	9–11	10–11	4–6		18
19. Taka-amylase A	*Aspergillus oryzae*	2.74	0.91	–		19
20. α-Amylase	*Aspergillus oryzae*	3.0–7.5	0.9–1.4	0		20, 21
21. α-Galactosidase I	Vicia faba	25.0	–	–		22
22. α-Galactosidase II	Vicia faba	2.8	–	–		22
23. Invertase	Yeast	47.0	3.1	–		23
24. β-Glucuronidase	Bovine liver	2.8–3.8	–1%	–		24
25. Bromelain	Pineapple stem	1.5–2.1	2.7–3.6	0		25
26. Pepsinogen	Hog stomach	1.5	–	–		27
27. γ-Glutamyl transferase	–	21.0	9.5–12.0	3.5–5.3		9, 28, 29
28. Carboxylesterase	Porcine liver	1.8	0.2	–	0.2	30
29. Carboxylesterase	Porcine kidney	1.4	0.6	–	0.2	30
30. Carboxylesterase	Bovine liver	0.8	0.2	–	0.2	30
31. Carboxylesterase	Human liver microsomes	2.5	0.5	–	–	31
32. Desoxyribonuclease A	Bovine pancreas	3.1	1.2	–	–	32
33. Desoxyribonuclease B	Bovine pancreas	2.9	2.2	0.9	–	32
34. Desoxyribonuclease C	Bovine pancreas	2.5	1.2	–	–	32
35. Desoxyribonuclease D	Bovine pancreas	2.3	2.1	0.8	–	32
36. α-L-Fucosidase	Rat epididymus	0.3	0.1	–	0.02	33
37. Lactoperoxidase	Bovine milk	1.5	6.8	–	–	34
38. Lipase	Porcine pancreas	1.3	1.2	–	–	35
39. Monoamine oxidase	Bovine plasma	2.4	1.3	0.8	–	36
40. N-Acetyl-β-D-glucosaminidase	*Aspergillus oryzae*	3.8	1.1	–	–	37
41. β-Lactamase II	*Bacillus cereus*	10.4	8.0	–	1.4	38
42. Ribonuclease R$_1$	*Rhizopus oligosporus*	7.6	3.4	–	0.5	39
43. Ribonuclease R$_2$	*Rhizopus oligosporus*	7.0	2.7	–	0.5	39

Compiled by Morris Soodak.

REFERENCES

1. Heilbronn, *Biochim. Biophys. Acta*, 58, 222 (1962).
2. Schultze and Heremans, in *Molecular Biology of Human Proteins*, Vol. 1, Elsevier, 1965, 204.
3. Yamashina, *Arch. Kemi.*, 9, 225 (1956).
4. Haupt, Heide, Zwisler, and Schwick, *Blut*, 14, 65 (1966).
5. Dugan, Radhakrishnamurthy, and Berenson, *Enzymologia*, 33, 215 (1967).
6. Pazur, Kleppe, and Cepure, *Arch. Biochem. Biophys.*, 111, 351 (1965).
7. Shannon, Kay, and Lew, *J. Biol. Chem.*, 241, 2166 (1966).
8. Morris and Hager, *J. Biol. Chem.*, 241, 1763 (1966).
9. Eylar, *J. Theor. Biol.*, 10, 89 (1965).
10. Yamada, Gee, Ebata, and Yasunobu, *Biochim. Biophys. Acta*, 81, 165 (1964).
11. Plummer and Hirs, *J. Biol. Chem.*, 239, 2530 (1964).
12. Plummer, *J. Biol. Chem*, 243, 5961 (1968).
13. Reinhold, Dunne, Wriston, Schwarz, Sarda, and Hirs, *J. Biol. Chem.*, 243, 6482 (1968).

Table 9 (continued)
GLYCOPROTEINS, ENZYMES

14. Catley, Moore, and Stein, *J. Biol. Chem.*, 244, 933 (1969).
15. Price, Liu, Stein, and Moore, *J. Biol. Chem.*, 244, 917 (1969).
16. Bernardi, Appella, and Zito, *Biochemistry*, 4, 1725 (1965).
17. Everse and Kaplan, *J. Biol. Chem.*, 243, 6072 (1968).
18. Usategui-Gomez, *Proc. Soc. Exp. Biol. Med.*, 120, 385 (1965).
19. Hanafusa, Ikenaka, and Akabori, *J. Biochem.* (Tokyo), 42, 55 (1955).
20. McKelvy and Lee, *Arch. Biochem. Biophys.*, 132, 99 (1969).
21. Arai, Minoda, and Yamada, *Agric. Biol. Chem.*, 33, 922 (1969).
22. Dey and Pridham, *Biochem. J.*, 113, 49 (1969).
23. Neumann and Lampen, *Biochemistry*, 6, 468 (1967).
24. Plapp and Cole, *Biochemistry*, 6, 3676 (1967).
25. Ota, Moore, and Stein, *Biochemistry*, 3, 180 (1964).
26. Murachi, Suzuki, and Takahashi, *Biochemistry*, 6, 3730 (1967).
27. Neumann, Zehavi, and Tanksley, *Biochem. Biophys. Res. Commun.*, 36, 151 (1969).
28. Szewczuk and Connell, *Biochim. Biophys. Acta*, 83, 218 (1964).
29. Szewczuk and Baranowski, *Biochem. Z.*, 338, 317 (1963).
30. Klapp, Kirsch, and Borner, *Z. Physiol. Chem.*, 351, 81 (1970).
31. Junge, Heymann, Krisch, and Hollandt, *Arch. Biochem. Biophys.*, 165, 749 (1974).
32. Liao, *J. Biol. Chem.*, 249, 2354 (1974).
33. Carlsen and Pierce, *J. Biol. Chem.*, 247, 23 (1972).
34. Rombauts, Schroeder, and Morrison, *Biochemistry*, 6, 2965 (1967).
35. Garner and Smith, *J. Biol. Chem.*, 247, 561 (1972).
36. Watanabe and Yasunobu, *J. Biol. Chem.*, 245, 4612 (1970).
37. Mega, Ikenaka, and Matsushima, *J. Biochem.* (Tokyo), 68, 109 (1970).
38. Kuwahara, Adams, and Abraham, *Biochem. J.*, 118, 475 (1970).
39. Woodroof and Glitz, *Biochemistry*, 10, 1532 (1971).

METALLOPROTEINS AND METALLOENZYMES

Table 1
NONENZYMATIC METALLOPROTEINS

Protein	Metal	Molecular weight	Stoichiometry of metal	Source	Function	Reference
Vanadium						
Hemovanadin	V	Unknown	Unknown	Vanadocytes of Ascidians	Unknown	1–6
Manganese						
Manganin	Mn	57,000	1 Mn	Peanut seeds	Unknown	7
Iron						
Transferrin	Fe(Mn,Cu)	80,000	2 Fe	Human serum	Iron transport	9–14
	Fe	67,000	2 Fe	Rat serum	Iron transport	15
	Fe	70,000	2 Fe	Rabbit serum	Iron transport	16
	Fe	70,000–75,000	2 Fe	Snake serum (*Eunectes marinus*)	Iron transport	17
	Fe	67,000–72,400		Bovine serum	Iron transport	18
	Fe	75,000–80,000	2 Fe	Hagfish serum	Iron transport	19
Conalbumin	Fe(Mn,Cu)	77,000	2 Fe	Hen eggs	Iron transport	12, 20–22
Lactotransferrin	Fe	80,000–95,000	2 Fe	Human milk	Iron transport	23–26
	Fe	80,000	2 Fe	Cow milk	Iron transport	27
	Fe	70,000	2 Fe	Rabbit milk	Iron transport	16
Ferritin	Fe	Apoferritin 460,000	4,500 Fe^{3+}	Mammalian tissue	Iron storage	28–30
Hemerythrin	Fe	107,000	16 Fe	Blood cells of Polychaetes, Sipunculoids, Priapuloids, Brachiopodes, Coelomic fluid of marine worms	Respiratory protein	31–33
Gastroferrin	Fe	278,000	6% Fe	Human gastric juice	Iron absorption	34

Table 1 (continued)
NONENZYMATIC METALLOPROTEINS

Protein	Metal	Molecular weight	Stoichiometry of metal	Source	Function	Reference
Copper						
Hemocyanin	Cu	25,000–75,000	2 Cu	Blood of Molluscs arthropods	Respiratory protein	35–39
Pink protein	Cu	32,000		Human erythrocytes	Unknown	40
Neonatal hepatic mitochondrocuprein	Cu	Insoluble		Mammalian liver	Unknown	41, 42
Copper protein of *A. fecaelis*	Cu	90,000	4 Cu	*A. fecaelis*	Unknown	43
Zinc and cadmium						
Metallothionein	Zn Cd	6,600	8 Zn and Cd	Horse kidney, human kidney, human liver, rabbit liver, rat liver	Unknown	44–49
α_2-Macroglobulin	Zn		320–770 μg/g	Human serum	Unknown	50
Procarboxypeptidase A	Zn	90,000	1 Zn	Bovine pancreas	Precursor of Carboxypeptidase A	51–56
	Zn	52,500	1 Zn	Pacific spiny dog-fish pancreas	Precursor of Carboxypeptidase A	57
Procarboxypeptidase B	Zn	67,400	1 Zn	Bovine pancreas	Precursor of Carboxypeptidase B	58
Nickel						
Nickelplasmin	Ni	70,000	1 Ni	Rabbit serum	Unknown	59

Compiled by Bert L. Vallee and Warren E. C. Wacker.

Note: References follow Table 2.

Table 2
METALLOENZYMES

Enzyme	Metal	Cofactor	Molecular weight	Stoichiometry metal and cofactor	Source	Reference
Calcium proteins						
Amylase	Ca	None	50,000	1 Ca	Human saliva	60–62
	Ca	None	50,000	2–3 Ca	*Aspergillus oryzae*	60, 61
	Ca	None	50,000	1–2 Ca	Hog pancreas	60, 61
	Ca	None	50,000	4 Ca	*B. subtilis*	60–63
Pseudomonas protease	Ca	None	48,000	1–2 Ca	*Pseudomonas aeruginosa*	64
Manganese						
Pyruvate carboxylase	Mn	Biotin	655,000	3–4 Mn, 4 Biotin	Chicken liver Turkey liver	65–67
Superoxide dismutase	Mn	None	39,500	1.6–1.8 Mn	*E. coli*	68
	Mn	None	40,250	2 Mn	*Streptococcus mutans*	360
	Mn	None	80,000	2.3 Mn	Chicken liver mitochondria	8, 359
Iron sulfur proteins						
Rubredoxin	Fe	None	6,000	1 Fe	*Clostridium pasteurianum*	69
	Fe	None	6,000	1 Fe	*Desulfibvibrio desulfuricans*	70
	Fe	None	6,000	1 Fe	*Peptostreptococcus elsdenii*	71
	Fe	None	6,000	1 Fe	*Micrococcus aerogenes*	72
	Fe	None	19,000	1 or 2 Fe	*Pseudomonas oleovorans*	73
	Fe	None	—	—	*Clostridium sticklandii*	74
	Fe	None	—	—	*Desulfivibrio gigas*	75
	Fe	None	—	—	*Micrococcus lactolyticus*	76
	Fe	None	—	—	*Chlorobium thiosulfatophilum*	77
	Fe	None	—	—	*Chloropseudomonas ethylicum*	77

Table 2 (continued)
METALLOENZYMES

Enzyme	Metal	Cofactor	Molecular weight	Stoichiometry metal and cofactor	Source	Reference
Iron sulfur proteins (continued)						
Ferredoxin	Fe	None	11,600	2 Fe	Spinach	78–81
	Fe	None	11,500	2 Fe	Alfalfa	82
	Fe	None	11,000	2 Fe	*Anacystis nidulans*	83
	Fe	None	11,000	2 Fe	*Gossypium hirsutum*	84
	Fe	None	12,000	2 Fe	*Scenedesmus*	85
	Fe	None	12,000	2 Fe	Fern	86
	Fe	None	12,000	2 Fe	Amaranth	86
	Fe	None	12,000	2 Fe	*E. coli*	87
	Fe	None	10,000	8 Fe	*Chromatium*	88
	Fe	None	12,000	2 Fe	Taro	89
	Fe	None	12,000	2 Fe	*Euglena gracilis*	90
	Fe	None	10,660	2 Fe	Parsley	91
	Fe	None	11,000	2 Fe	*Anacystis nidulans*	83
	Fe	None	6,000	8 Fe	Various clostridia	93–98
	Fe	None	6,000	8 Fe	*Micrococcus aerogenes*	99
	Fe	None	6,000	2 Fe	*Methanobacillus omelianski*	100
	Fe	None	6,000	4–5 Fe	*Chlorobium thiosulfatophilum*	101
	Fe	None	13,000	6–7 Fe	*Azotobacter vinelandii*	102
	Fe	None	7,800	4 Fe	*B. polymyxa*	103
	Fe	None	7,800	4 Fe	*B. polymyxa*	104
	Fe	None	9,000	4 Fe	*Desulfovibrio gigas*	105, 106
	Fe	None	6,000–7,000	2 Fe	*Rhodospirillum rubrum*	107
	Fe	None	6,000–7,000	6 Fe	*Rhodospirillum rubrum*	107
Adrenodoxin	Fe	None	12,000	2 Fe	Hog adrenal medulla	108–111
Putida redoxin	Fe	None	12,000	2 Fe	*Pseudomonas putida*	112, 113
Iron sulfur protein I	Fe	None	21,000	2 Fe	*Azotobacter vinelandii*	114

Table 2 (continued)
METALLOENZYMES

Enzyme	Metal	Cofactor	Molecular weight	Stoichiometry metal and cofactor	Source	Reference
Iron sulfur proteins (continued)						
Iron sulfur protein II	Fe	None	24,000	2 Fe	*Azotobacter vinelandii*	114
Azoferredoxin	Fe	None	27,500	2 Fe	*Clostridium pasteurianium*	115
Molybdoferredoxin	Fe, Mo	None	110,000	12 Fe	*Clostridium pasteurianium*	116–118
High potential iron protein	Fe	None	12,000	4 Fe	*Chromatium*	119–121
Pyruvate ferredoxin oxidoreductase	Fe	None	240,000	6 Fe	*Clostridium acidurici*	122
Hydrogenase	Fe	None	60,000	4 Fe	*Clostridium pasteurianium*	123, 124
Glutamate synthetase	Fe	FAD FMN	800,000	38 Fe	*E. coli*	125
NADH-dehydrogenase	Fe	Flavin	8,000	4 Fe	Hog heart	126–129
Succinate dehydrogenase	Fe	Flavin	200,000	4 Fe	Bovine heart	130, 131
	Fe	Flavin	Unknown	8 Fe/Flavin	Hog heart	132
	Fe	Flavin	200,000	4 Fe 1 Flavin	Yeast	130, 133
Dihydrorotate dehydrogenase	Fe	2 FAD	115,000	4 Fe	*Zymobacterium ereticum*	134–136
Aldehyde oxidase	Fe Mo	2 FAD	300,000	8 Fe 2 Mo	Hog liver	137, 138
Xanthine oxidase	Fe Mo	2 FAD	300,000	8 Fe 2 Mo	Bovine milk	139, 140
	Fe Mo	2 FAD	300,000	8 Fe 2 Mo	Chicken liver	141
	Fe Mo	2 FAD	250,000	8 Fe 2 Mo	*Micrococcus lactilyticus*	142
Iron metalloenzymes						
Metapyrocatechase	Fe	None	140,000	3 Fe	*Pseudomonas arvilla*	143–145
Pyrocatechase	Fe	None	95,000	2 Fe	*Pseudomonas arvilla*	143–146

Table 2 (continued)
METALLOENZYMES

Enzyme	Metal	Cofactor	Molecular weight	Stoichiometry metal and cofactor	Source	Reference
Iron metalloenzymes (continued)						
Protocatechuate 3,4 dioxygenase	Fe	None	700,000	7 Fe	*Pseudomonas aerugenosa*	143–145, 147
Protocatechuate 4,5-dioxygenase	Fe	None	150,000	1 Fe	*Pseudomonas*	148
3,4-Dihydroxyphenylacetate 2,3 oxygenase	Fe	None	100,000	5 Fe	*Pseudomonas ovalis*	149
Agavain	Fe	None	52,000	1 Fe	Sisal extract	150
Sulfite reductase	Fe	FMN FAD	350,000	5 Fe 1 FMN 1 FAD	Yeast	151
Kojic acid oxidase	Fe	None	55,000	2 Fe	*Arthrobacter ureafaciens*	152, 153
Phosphoribosyl pyrophosphate amidotransferase	Fe	None	200,000	12 Fe	Pigeon liver	154
Ribonucleoside diphosphate reductase	Fe	None	82,000	2 Fe	*E. coli*	155, 156
Phenylalanine hydroxylase	Fe	None	100,000	1–2 Fe	Rat liver	157
Superoxide dismutase	Fe	None	38,700	1 Fe	*E. coli*	158
Acid phosphatase	Fe	None	23,000	1 Fe	Beef spleen	159
Lipoxygenase	Fe	None	100,000	1 Fe	Soy bean	160
Copper						
Ceruloplasmin	Cu	None	160,000	6 Cu	Human serum	161–166
	Cu	None	160,000	6 Cu	Porcine serum	167, 168
Superoxide dismutase	Cu, Zn	None	33–36,000	2 Cu 2 Zn	Mammalian erythrocytes, liver and brain, wheat germ, squash, garden peas	169–179
	Cu, Zn	None	33,000	1 Cu 2 Zn	*Photobacterium leiognathi*	180

Table 2 (continued)
METALLOENZYMES

Enzyme	Metal	Cofactor	Molecular weight	Stoichiometry metal and cofactor	Source	Reference
Copper (continued)						
Plastocyanin	Cu	None	21,000	2 Cu	Spinach, Chlorella Chemopodium	181, 182
	Cu	None	23,000	2 Cu	Corn leaves	183
	Cu	None	10,690	2 Cu	Phaseolus vulgaris	184
	Cu	None	10,000	1 Cu	Parsley	363
	Cu	None	14,000	1 Cu	Pseudomonas	185–190
Pseudomonas blue protein						
Azurin	Cu	None	14,000	1 Cu	Bordetella pertussis	191
Mung bean blue protein	Cu	None	22,000	1 Cu	Mung bean	192
Rhus blue protein (Stellacyanin)	Cu	None	16,000	1 Cu	Rhus vernicifera	193–195
Umecyanin	Cu	None	14,600	1 Cu	Horseradish	196
Monoamine oxidase	Cu	PLP	170,000	1 Cu, 1 PLP	Bovine plasma	197–202
Amine oxidase	Cu	?PLP	252,000	3 Cu	Aspergillus niger	203
	Cu	PLP	195,000	3 Cu, 3–4 PLP	Hog plasma	204, 205
Diamine oxidase	Cu	?PLP	96,000	1 Cu	Pea seedlings	206–208
	Cu	PLP	185,000	2 Cu, 2 PLP	Porcine kidney	209–212
D-Galactose oxidase	Cu	None	75,000	1 Cu	Dactylium dendroides	213
Uricase	Cu	None	120,000	1 Cu	Mammalian liver	214, 215
Tyrosinase	Cu	None	119,000	4 Cu	Mushroom	216
	Cu	None	Unknown	Unknown	Neurospora crassa	217
	Cu	None	Unknown	Unknown	Mammalian tissue	218–221
Dopamine-B-hydroxylase	Cu	Ascorbic acid	290,000	2 Cu	Bovine adrenal medulla	222–224
O-Diphenol oxidase	Cu	None	72,000	1 Cu	Potatoes	225
Laccase	Cu	None	120,000	6 Cu	Rhus vernicifera	226–232
	Cu	None	60,000	4 Cu	Polyporus versicolor	233
Ascorbic acid oxidase	Cu	None	140,000	8 Cu	Squash	234–239

Table 2 (continued)
METALLOENZYMES

Enzyme	Metal	Cofactor	Molecular weight	Stoichiometry metal and cofactor	Source	Reference
Copper (continued)						
Ribulose-diphosphate carboxylase	Cu	None	560,000	1 Cu	Spinach	240
Quercetinase	Cu	None	111,400	2 Cu	*Aspergillus flavus*	241
Tryptophan 2,3-dioxygenase	Cu	Heme	122,000	2 Cu	*Pseudomonas acidovarans*	242, 243
Cytochrome oxidase	Cu	Heme	167,000	2 Cu	Rat liver	242, 243
	Cu	Heme	Unknown	1 Cu/heme	Bovine heart	244–246
Zinc						
Carbonic anhydrase	Zn	None	29,000	1 Zn	Dog erythrocytes	247
	Zn	None	30,000	1 Zn	Equine erythrocytes	248
	Zn	None	30,000	1 Zn	Bovine erythrocytes	249–251
	Zn	None	28,000	1 Zn	Human erythrocytes	252–256
	Zn	None	30,375	1 Zn	Porcine erythrocytes	257
	Zn	None	29,000	1 Zn	*Macaca mulata* erythrocytes	258
	Zn	None	30,000	1 Zn	Sheep erythrocytes	259
	Zn	None	28,000	1 Zn	*Gallus domesticus* erythrocytes	260
	Zn	None	36,000–40,000	1 Zn	Tiger shark erythrocytes	261
	Zn	None	36,000–40,000	1 Zn	Bull shark erythrocytes	261
	Zn	None	≅500,000	1.7 µg Zn/mg/protein	Oyster	262
	Zn	None	180,000	6 Zn	Parsley	263
	Zn	None	180,000	6 Zn	Spinach	264
	Zn	None	188,000	6 Zn	*Pisium sativum*	265
	Zn	None	42,000	2 Zn	*Tradescantia albiflora*	265, 266
Carboxypeptidase A	Zn	None	28,000	1 Zn	*Neisseria sicca*	267, 268
	Zn	None	34,300	1 Zn	Bovine pancreas	269–275
	Zn	None	34,000	1 Zn	Human pancreas	276
	Zn	None	36,500	1 Zn	Pacific spiny dogfish	57, 277
	Zn	None	34,800	1 Zn	Porcine pancreas	278

Table 2 (continued)
METALLOENZYMES

Enzyme	Metal	Cofactor	Molecular weight	Stoichiometry metal and cofactor	Source	Reference
Zinc (continued)						
Carboxypeptidase B	Zn	None	34,300	1 Zn	Porcine pancreas	279, 280, 362
	Zn	None	34,300	1 Zn	Bovine pancreas	281
Carboxypeptidase G 1	Zn	None	92,000	4 Zn	Pseudomonas stutzeri	282
Neutral protease	Zn	None	44,700	1 Zn	B. subtilis	283–286
	Zn	None	—	1 Zn	B. polymyxa	287
	Zn	None	—	1 Zn	B. cereus	288
Megateriapeptidase	Zn	None	—	—	B. megateria	289
Thermolysin	Zn	None	37,500	1 Zn	B. thermoproteolyticus	290, 291
Protease	Zn	None	—	1 Zn	Serratia	292
Neutral protease	Zn	None	37,100	1 Zn	Streptomyces naraensis	293
Alcohol dehydrogenase	Zn	NAD	112,000	1.45 Zn	Peanuts	294
	Zn	NAD	150,000	4 Zn 4 NAD	Yeast	270, 272, 295
	Zn	NAD	80,000	4 Zn 2 NAD	Equine liver	270, 272, 296–299
	Zn	NAD	87,000	2 Zn 2 NAD	Human liver	302, 303
Glutamic dehydrogenase	Zn	NAD	65,000	2.9 Zn	Rat liver	304, 305
	Zn	NAD	1,000,000	2–6 Zn	Bovine liver	272, 306–310
α-Glyceraldehyde-PO₄ dehydrogenase	Zn	NAD	120,000	Unknown	Hog muscle, bovine muscle, yeast	311–313
D-Lactic cytochrome reductase	Zn	FAD	50,000	4–6 Zn	Yeast	314–318
Alkaline phosphatase	Zn	None	300,000	3–5 Zn	Guinea pig bone marrow	319
	Zn	None	89,000	4 Zn 2 Mg	E. coli	320–323 358
	Zn	None	140,000	4 Zn	Calf intestine	324
	Zn	None	116,000	2–3 Zn	Human placenta	325
Aspartate transcarbamylase	Zn	None	310,000	6 Zn	E. coli	326
Aldolase	Zn	None	80,000	1 Zn	Yeast	327, 328
	Zn	None	50,000	1 Zn	A. niger	329

Table 2 (continued)
METALLOENZYMES

Enzyme	Metal	Cofactor	Molecular weight	Stoichiometry metal and cofactor	Source	Reference
Zinc (continued)						
L-Rhammulose 1-Phosphate aldolase	Zn	None	135,000	2 Zn	E. coli	330
Phospholipase C	Zn	None	Unknown	Unknown	B. cereus	331
Dipeptidase	Zn	None	47,200	1 Zn	Hog kidney	332
Leucine amino peptidase	Zn	None	300,000	1 Zn	Porcine kidney	333, 334
Leucine amino peptidase	Zn	None	320,000	12 Zn	Beef lens	335–337
Particulate leucine amino peptidase	Zn	None	280,000	2 Zn	Hog kidney	338, 339
Amino peptidase	Zn	None	122,000	1 Zn	Human liver	365
Phosphomanase isomerase	Zn	None	45,000	1 Zn	Yeast	340
DNA polymerase	Zn	None	109,000	2 Zn	E. coli	342
	Zn	None	150,000	4 Zn	Sea urchin	342
RNA polymerase	Zn	None	370,000	2 Zn	E. coli	343
	Zn	None	—	Zn	E. gracilis	361
RNA-dependent DNA polymerase	Zn	None	160,000	1.8 Zn	Avian mycosis virus	344, 345
	Zn	None	160,000	~1.0–1.4 Zn	Murine leukemia virus, woolly monkey virus, feline leukemia virus	346
Pyruvate carboxylase	Zn	Biotin	600,000	3 Zn	Baker's yeast	347
Adenosine 5'-monophosphate amino hydrolase	Zn	None	278,000	2.6 Zn	Rabbit muscle	341
Mercaptopyruvate sulfur transferase	Zn	None	23,800	1 Zn	E. coli	348
Phosphoglucomutase	Zn	None	—	1 Zn	Yeast	349
δ-Amino levulinic dehydratase	Zn	None	280,000	0.5–2 Zn	Liver	350

Table 2 (continued)
METALLOENZYMES

Enzyme	Metal	Cofactor	Molecular weight	Stoichiometry metal and cofactor	Source	Reference
Molybdenum						
Nitrate reductase	Mo	FAD	228,000	1–2 Mo	*Neurospora crassa*	351
Sulfite oxidase	Mo	HEME	115,000	1 Mo 2 Heme	Beef liver	352–354
Zinc-cobalt						
Transcarboxylase	Zn Co	Biotin	670,000–790,000	6–8 Zn + Co	*Propionibacterium shermanii*	355
Selenium						
Glutathione peroxidase	Se	None	84,000	4 Se	Bovine blood	356, 357
Formate dehydrogenase	Se	—	—	—	Various clostridia	364
Glycine reductase protein A	Se	—	6,000	1 Se	*Clostridium sticklandii*	364

Compiled by Bert L. Vallee and Warren E. C. Wacker.

REFERENCES

1. Henze, *Hoppe Seyler's Z. Physiol. Chem.*, 72, 494 (1911).
2. Webb, *J. Exp. Biol.*, 16, 499 (1939).
3. Califano and Caselli, *Pubbl. Stn. Zool. Napoli*, 21, 235 (1948).
4. Califano and Caselli, *Pubbl. Stn. Zool. Napoli*, 23, 1 (1950).
5. Boeri, *Arch. Biochem. Biophys.*, 37, 449 (1952).
6. Bielig et al., in *Protides of the Biological Fluids*, Vol. 14, Peeters, Ed., Elsevier, Amsterdam, 1966, 197.
7. Dieckert and Rozacky, *Arch. Biochem. Biophys.*, 134, 273 (1969).
8. Scrutton, *Biochemistry*, 10, 3897 (1971).
9. Surgenor et al., *J. Am. Chem. Soc.*, 71, 1223 (1949).
10. Koechlin, *J. Am. Chem. Soc.*, 74, 2649 (1952).
11. Hazen, PhD thesis, Harvard, 1962.
12. Aasa et al., *Biochim. Biophys. Acta*, 75, 203 (1963).
13. Jeppson, *Acta Chem. Scand.*, 1, 1686 (1967).
14. Roberts et al., *J. Biol. Chem.*, 241, 4907 (1966).
15. Gordon and Louis, *Biochem. J.*, 83, 409 (1963).
16. Baker et al., *Biochemistry*, 7, 1371 (1968).
17. Makey and Seal, *Int. J. Biochem.*, 67, 76 (1970).
18. Efremov et al., *Biochem. Genet.*, 2, 159 (1971).
19. Aisen et al., *Biochemistry*, 11, 3461 (1972).
20. Warner and Weber, *J. Am. Chem. Soc.*, 75, 5095 (1953).
21. Wishnia et al., *J. Am. Chem. Soc.*, 83, 2071 (1961).
22. Feeney and Komatsu, *Struct. Bonding* (Berlin), 1, 149 (1966).
23. Blanc and Isliker, *Bull. Soc. Chim. Biol.*, 43, 7 (1961).
24. Masson and Heremans, *Eur. J. Biochem.*, 6, 579 (1968).
25. Montreuil et al., *Biochim. Biophys. Acta*, 45, 413 (1960).
26. Schade et al., in *Biological Fluids*, Vol. 16, Peeters, Ed., Elsevier, Amsterdam, 1968, 619.
27. Groves, *J. Am. Chem. Soc.*, 82, 3345 (1960).
28. Granick, *Chem. Rev.*, 38, 379 (1946).
29. Michaelis, *Adv. Protein Chem.*, 3, 53 (1947).
30. Farrant, *Biochim. Biophys. Acta*, 13, 569 (1954).
31. Klotz et al., *Arch. Biochem. Biophys.*, 68, 284 (1957).
32. Holleman and Biserte, *Bull. Soc. Chim. Biol.*, 40, 1417 (1958).
33. Ghiretti, in *Oxygenases*, Hayaishi, Ed., Academic Press, New York, 1962, 517.
34. Multani et al., *Biochemistry*, 9, 3970 (1970).
35. Rawlinson, *Aust. J. Exp. Biol. Med. Sci.*, 18, 131 (1940).
36. Polson and Wychoff, *Nature*, 160, 153 (1947).
37. Redfield, in *Copper Metabolism*, McElroy and Glass, Eds., Johns Hopkins, Baltimore, 1950, 174.
38. van Holde, in *Molecular Architecture in Cell Physiology*, Hayaishi, Ed., Prentice Hall, Englewood, 1964, 81.
39. Ghiretti-Magaldi et al., *Biochemistry*, 5, 1943 (1966).
40. Reed et al., *J. Biol. Chem.*, 245, 2954 (1970).
41. Porter, in *The Biochemistry of Copper*, Peisach et al., Eds., Academic Press, New York, 1966, 159.
42. Porter et al., *Biochim Biophys. Acta*, 65, 66 (1962).
43. Matsubara and Zwashi, *J. Biochem.* (Tokyo), 71, 747 (1972).
44. Margoshes and Vallee, *J. Am. Chem. Soc.*, 9, 4813 (1956).
45. Kagi and Vallee, *J. Biol. Chem.*, 235, 3460 (1960).
46. Kagi and Vallee, *J. Biol. Chem.*, 236, 2435 (1961).
47. Piscator, *Nord. Hyg. Tidskr.*, 44, 17 (1963).
48. Pulido et al., *Biochemistry*, 5, 1768 (1966).
49. Wrige and Rajagopolan, *Arch. Biochem. Biophys.*, 153, 755 (1972).
50. Parisi and Vallee, *Biochemistry*, 9, 2431 (1970).
51. Keller et al., *J. Biol. Chem.*, 223, 457 (1956).
52. Keller et al., *J. Biol. Chem.*, 230, 905 (1958).
53. Yamasaki et al., *Biochemistry*, 2, 859 (1963).
54. Brown et al., *Biochemistry*, 2, 877 (1963).
55. Brown et al., *Biochemistry*, 2, 867 (1963).
56. Piras and Vallee, *Biochemistry*, 6, 348 (1967).
57. Lacko and Neurath, *Fed. Proc.*, 26, 279 (1967).
58. Wintersberger et al., *Biochemistry*, 1, 1069 (1962).
59. Nomoto et al., *Biochemistry*, 10, 1647 (1971).
60. Fisher et al., *Proc. IVth Int. Congr. Biochem.* (Vienna), 8, 124 (1958).

61. Vallee et al., *J. Biol. Chem.*, 234, 2901 (1959).
62. Hsiu et al., *Biochemistry*, 3, 61 (1964).
63. Stein et al., *Biochemistry*, 3, 56 (1964).
64. Morihara and Tsuzuki, *Biochim. Biophys. Acta*, 92, 351 (1964).
65. Scrutton and Utter, *J. Biol. Chem.*, 240, 1 (1965).
66. Scrutton et al., *J. Biol. Chem.*, 241, 3480 (1966).
67. Mildvan et al., *J. Biol. Chem.*, 241, 3488 (1966).
68. Keele et al., *J. Biol. Chem.*, 245, 6176 (1970).
69. Lovenberg and Sobel, *Proc. Natl. Acad. Sci. U.S.A.*, 54, 193 (1965).
70. Newman and Postgate, *Eur. J. Biochem.*, 7, 48 (1968).
71. Bachmayer et al., *Proc. Natl. Acad. Sci. U.S.A.*, 57, 122 (1967).
72. Bachmayer et al., *Proc. Natl. Acad. Sci. U.S.A.*, 57, 122 (1967).
73. Lode and Coon, *J. Biol. Chem.*, 246, 791 (1971).
74. Stadtman and San Pietro, *A Symposium on Non-Heme Iron Proteins*, Antioch, Yellow Springs, 1965, 439.
75. LeGall and Dragoni, *Biochem. Biophys. Res. Commun.*, 25, 145 (1966).
76. Valentine et al., *J. Biol. Chem.*, 238, 1141, 1963.
77. Meyer et al., *Biochim. Biophys. Acta*, 234, 266 (1971).
78. San Pietro and Lang, *J. Biol. Chem.*, 231, 211 (1958).
79. Appella and San Pietro, *Biochem. Biophys. Res. Commun.*, 6, 379 (1961).
80. Fry and San Pietro, *Biochem. Biophys. Res. Commun.*, 9, 218 (1962).
81. Katoh and Takamiya, *Arch. Biochem. Biophys.*, 102, 189 (1963).
82. Keresztes-Nagy and Margoliash, *J. Biol. Chem.*, 241, 5955 (1966).
83. Yamanaka et al., *Biochim. Biophys. Acta*, 180, 196 (1969).
84. Newman et al., *Biochem. Biophys. Res. Commun.*, 36, 947 (1969).
85. Matsubara, *J. Biol. Chem.*, 243, 370 (1968).
86. Schurmann et al., *Biochim. Biophys. Acta*, 223, 450 (1970).
87. Velter and Knappe, *Hoppe-Seyler's Z. Physiol. Chem.*, 352, 433 (1971).
88. Matsubara et al., *J. Biol. Chem.*, 245, 2121 (1970).
89. Rao and Matsubara, *Biochem. Biophys. Res. Commun.*, 38, 500 (1970).
90. Yock et al., *Proc. Natl. Acad. Sci. U.S.A.*, 64, 1404 (1969).
91. Fee and Palmer, *Biochim. Biophys. Acta*, 245, 175 (1971).
92. Mitsui and San Pietro, *Plant Sci. Lett.*, 1, 157 (1973).
93. Mortensen et al., *Biochem. Biophys. Res. Commun.*, 7, 448 (1962).
94. Tagawa and Arnon, *Nature*, 195, 537 (1962).
95. Buchanan et al., *Proc. Natl. Acad. Sci. U.S.A.*, 49, 345 (1963).
96. Buchanan, *Struct. Bonding* (Berlin), 1, 109 (1966).
97. Benson et al., *Proc. Natl. Acad. Sci. U.S.A.*, 55, 1532 (1966).
98. Devanathan et al., *J. Biol. Chem.*, 244, 2846 (1970).
99. Tsunoda et al., *J. Biol. Chem.*, 243, 6262 (1969).
100. Buchanan and Rabinowitz, *J. Bacteriol.*, 88, 806 (1964).
101. Buchanan et al., *Biochim. Biophys. Acta*, 189, 46 (1969).
102. Shetna, *Biochim. Biophys. Acta*, 205, 58 (1960).
103. Yock and Valentine, *J. Bacteriol.*, 110, 1211 (1972).
104. Yock, *Arch. Biochem. Biophys.*, 158, 633 (1973).
105. Shetna et al., *Biochem. Biophys. Res. Commun.*, 42, 1108 (1971).
106. Laishley et al., *J. Bacteriol. Biol.*, 98, 302 (1964).
107. Shanmugum et al., *Biochim. Biophys. Acta*, 256, 447 (1971).
108. Suzuki and Kimura, *Biochem. Biophys. Res. Commun.*, 19, 340 (1965).
109. Kimura and Suzuki, *J. Biol. Chem.*, 242, 485 (1967).
110. Kimura et al., *Proc. 7th Int. Congr. Biochem.* (Tokyo), 111, 561 (1967).
111. Kimura et al., *Biochemistry*, 8, 4027 (1969).
112. Cushman and Gunsalus, *Bacteriol. Proc.*, p. 86 (1966).
113. Cushman et al., *Biochem. Biophys. Res. Commun.*, 26, 577 (1967).
114. Shetna et al., *Biochem. Biophys. Res. Commun.*, 31, 863 (1968).
115. Moustafa and Mortenson, *Biochim. Biophys. Acta*, 172, 106 (1969).
116. Nakos and Mortenson, *Biochem. Biophys. Acta*, 229, 411 (1971).
117. Mortensen et al., *Biochim. Biophys. Acta*, 141, 516 (1967).
118. Mortensen et al., *Biochemistry*, 10, 2066 (1971).
119. Bartsch et al., in *Bacterial Photosynthesis*, Antioch, Yellow Springs, 1963, 315.
120. DeKlerke and Kamen, *Biochim. Biophys. Acta*, 112, 175 (1966).
121. Dus et al., *Biochim. Biophys. Acta*, 140, 291 (1967).
122. Vyeda and Rabinowitz, *J. Biol. Chem.*, 245, 3111 (1971).

123. Nakos and Mortensen, *Biochim. Biophys. Acta,* 227, 576 (1971).
124. Nakos and Mortensen, *Biochemistry,* 10, 2442 (1971).
125. Miller and Stadtman, *J. Biol. Chem.,* 247, 7407 (1972).
126. Mahler et al., *J. Biol. Chem.,* 199, 385 (1952).
127. Mahler and Elowe, *J. Biol. Chem.,* 210, 165 (1954).
128. Singer, in *The Enzymes,* Vol. 7, Boyer, Lardy and Myrbäck, Eds., Academic Press, New York, 1963, 345.
129. Hatefi, in *The Enzymes,* Vol. 7, Boyer, Lardy, and Myrbäck, Eds., Academic Press, New York, 1963, 495.
130. Singer et al., *Adv. Enzymol.,* 18, 65 (1957).
131. Kearney, *J. Biol. Chem.,* 235, 865 (1960).
132. Zeylemaker et al., *Biochim. Biophys. Acta,* 99, 183 (1965).
133. Singer and Kearney, in *The Enzymes,* Vol. 7, Boyer, Lardy, and Myrbäck, Eds., Academic Press, New York, 1963, 383.
134. Friedman and Vennesland, *J. Biol. Chem.,* 235, 1526 (1960).
135. Handler et al., *Fed Proc.,* 23, 30 (1964).
136. Aleman-Aleman and Handler, *J. Biol. Chem.,* 242, 4087 (1967).
137. Fridovich and Handler, *J. Biol. Chem.,* 231, 899 (1958).
138. Rajagopalan et al., *J. Biol. Chem.,* 237, 922 (1962).
139. Avis et al., *J. Chem. Soc.,* p. 1100 (1955).
140. Massey et al., *J. Biol. Chem.,* 245, 2837 (1970).
141. Rajagopalan and Handler, *J. Biol. Chem.,* 242, 4097 (1967).
142. Smith et al., *J. Biol. Chem.,* 242, 4108 (1967).
143. Hayaishi, *Bacteriol. Rev.,* 30, 720 (1966).
144. Nozaki et al., in *Biological and Chemical Aspects of Oxygenases,* Block and Hayaishi, Eds., Maruzen, Tokyo, 1966, 347.
145. Nozaki et al., *Int. Oxidat. Symp.* (San Francisco), 2, 725 (1967).
146. Kojima et al., *J. Biol. Chem.,* 242, 3270 (1967).
147. Fujiwiwasa et al., *J. Biol. Chem.,* 247, 4422 (1972).
148. Ono et al., *Biochim. Biophys. Acta,* 220, 224 (1970).
149. Kita et al., *Biochem. Biophys. Res. Commun.,* 18, 66 (1965).
150. Tipton, *Biochim. Biophys. Acta,* 110, 414 (1965).
151. Yashimoto and Sato, *Biochim. Biophys. Acta,* 153, 555 (1968).
152. Nonamura and Tatsumi, *Agric. Biol. Chem.,* 33, 1223 (1969).
153. Imose et al., *Agric. Biol. Chem.,* 34, 1443 (1970).
154. Rowe and Wyngaarden, *J. Biol. Chem.,* 243, 6373 (1968).
155. Brown et al., *Eur. J. Biochem.,* 9, 512 (1969).
156. Brown et al., *Biochem. Biophys. Res. Commun.,* 30, 2522 (1968).
157. Fisher et al., *J. Biol. Chem.,* 247, 5161 (1972).
158. Yost and Fridovich, *J. Biol. Chem.,* 248, 4905 (1973).
159. Campbell and Zerner, *Biochem. Biophys. Res. Commun.,* 54, 1498 (1973).
160. Pistorius and Axelrod, *J. Biol. Chem.,* 249, 3183 (1974).
161. Holmberg and Laurell, *Acta Chem. Scand.,* 2, 550 (1958).
162. Blumberg et al., *J. Biol. Chem.,* 238, 1675 (1963).
163. Kasper and Deutsch, *J. Biol. Chem.,* 238, 2325 (1963).
164. Kasper et al., *J. Biol. Chem.,* 238, 2338 (1963).
165. Morrell et al., *J. Biol. Chem.,* 241, 3745 (1966).
166. Poillon and Bearn, *Biochim. Biophys. Acta,* 127, 407 (1966).
167. Shozo and Yoshiaki, *Arch. Biochem. Biophys.,* 130, 617 (1969).
168. Magdoff-Fairchild, *J. Biol. Chem.,* 244, 3497 (1969).
169. Mann and Keilin, *Proc. R. Soc. Lond. Ser. B,* 126, 303 (1968).
170. Stansell and Deutsch, *J. Biol. Chem.,* 240, 4306 (1965).
171. Porter, in *The Biochemistry of Copper,* Piesach et al., Eds., Academic Press, New York, 1966, 159.
172. Hartz and Deutsch, *J. Biol. Chem.,* 244, 4565 (1969).
173. Carrico and Deutsch, *J. Biol. Chem.,* 244, 6087 (1969).
174. McCord and Fridovich, *J. Biol. Chem.,* 244, 6049 (1969).
175. Carrico and Deutsch, *J. Biol. Chem.,* 245, 723 (1970).
176. Bannister et al., *Eur. J. Biochem.,* 18, 178 (1971).
177. Wood et al., *Eur. J. Biochem.,* 18, 187 (1971).
178. Weser et al., *Biochim. Biophys. Acta,* 243, 203 (1971).
179. Beauchamp and Fridovich, *Biochim. Biophys. Acta,* 193, 317 (1969).
180. Puget and Michelson, *Biochem. Biophys. Res. Commun.,* 58, 830 (1974).
181. Katoh, *Nature,* 186, 533 (1960).
182. Katoh et al., *J. Biochem.* (Tokyo), 51, 32 (1962).

183. Gamayunova, *Fiziol. Biochim.*, 4, 482 (1972).
184. Milne and Wells, *J. Biol. Chem.*, 245, 1566 (1970).
185. Horio, *J. Biochem.* (Tokyo), 45, 195 (1958).
186. Coval et al., *Biochim. Biophys. Acta*, 51, 246 (1961).
187. Suzuki and Iwasaki, *J. Biochem.* (Tokyo), 52, 931 (1962).
188. Mason, *Biochem. Biophys. Res. Commun.*, 10, 11 (1963).
189. Ambler, *Biochem. J.*, 89, 341 (1963).
190. Ambler and Brown, *Biochem. J.*, 104, 784 (1967).
191. Sutherland and Wilkinson, *J. Gen. Microbiol.*, 30, 105 (1963).
192. Shichi and Hackett, *Arch. Biochem. Biophys.*, 100, 185 (1963).
193. Omura, *J. Biochem.* (Tokyo), 50, 394 (1961).
194. Blumberg et al., *Biochem. Biophys. Res. Commun.*, 15, 277 (1964).
195. Peisach et al., *J. Biol. Chem.*, 242, 2847 (1967).
196. Paul and Stigbrand, *Biochim. Biophys. Acta*, 221, 255 (1970).
197. Yamada and Yasunobu, *J. Biol. Chem.*, 237, 1511 (1962).
198. Yamada and Yasunobu, *J. Biol. Chem.*, 237, 3077 (1962).
199. Yamada et al., *Nature*, 198, 1092 (1963).
200. Yamada and Nashamoto, *J. Biol. Chem.*, 238, 2669 (1963).
201. Achee et al., *Biochemistry*, 7, 4329 (1968).
202. Yasunobu et al., in *Symposium of Pyridoxal Enzymes*, Yamada et al., Eds., Maruzen, Tokyo, 1968, 139.
203. Yamada et al., in *Pyridoxal Catalysis*, Snell et al., Eds., Interscience, New York, 1968, 347.
204. Buffoni and Blaschko, *Biochem. J.*, 89, 111P (1963).
205. Buffoni et al., *Biochem. J.*, 106, 575 (1968).
206. Mann, *Biochem. J.*, 79, 623 (1961).
207. Zeller, in *The Enzymes*, Boyer, Lardy, and Myrbäck, Eds., Academic Press, New York, 1963, 313.
208. Hill and Mann, *Biochem. J.*, 91, 172 (1964).
209. Yamada et al., *Biochem. Biophys. Res. Commun.*, 29, 723 (1967).
210. Mondovi et al., *Arch. Biochem. Biophys.*, 119, 373 (1967).
211. Mondovi et al., in *Pyridoxal Catalysis*, Snell et al., Eds., Interscience, New York, 1968, 403.
212. Goryanchenkova et al., in *Pyridoxal Catalysis*, Snell et al., Eds., Interscience, New York, 1968, 391.
213. Amaral et al., *J. Biol. Chem.*, 238, 2281 (1963).
214. Mahler, in *Trace Elements*, Lamb et al., Eds., Academic Press, New York, 1958, 311.
215. Mahler, in *The Enzymes*, Vol. 8, Boyer, Lardy, and Myrbäck, Eds., Academic Press, New York, 1963, 285.
216. Bouchilloux et al., *J. Biol. Chem.*, 238, 1699 (1963).
217. Flung et al., *J. Biol. Chem.*, 238, 2045 (1963).
218. Lerner et al., *J. Biol. Chem.*, 178, 185 (1949).
219. Lerner et al., *J. Biol. Chem.*, 187, 793 (1950).
220. Lerner et al., *J. Biol. Chem.*, 191, 799 (1951).
221. Kertesz, *J. Natl. Cancer Inst.*, 14, 1081 (1954).
222. Goldstein et al., *J. Biol.Chem.*, 240, 2066 (1965).
223. Friedman and Kaufman, *J. Biol. Chem.*, 240, 4763 (1965).
224. Goldstein, in *The Biochemistry of Copper*, Peisach et al., Eds., Academic Press, New York, 1966, 443.
225. Balasingan and Ferdinand, *Biochem. J.*, 118, 15 (1970).
226. Keilin and Mann, *Nature*, 145, 304 (1940).
227. Tissieres, *Nature*, 162, 340 (1948).
228. Nakamura, *Biochim. Biophys. Acta*, 30, 44 (1958).
229. Omura, *J. Biochem.* (Tokyo), 50, 264 (1961).
230. Peisach and Levine, *J. Biol. Chem.*, 240, 2284 (1965).
231. Nakamura and Osura, in *The Biochemistry of Copper*, Peisach et al., Eds., Academic Press, New York, 1966, 389.
232. Levine, in *The Biochemistry of Copper*, Peisach et al., Eds., Academic Press, New York, 1966, 371.
233. Mosbach, *Biochim. Biophys. Acta*, 73, 204 (1963).
234. Lovett-Janison and Nelson, *J. Am. Chem. Soc.*, 62, 140 (1940).
235. Dunn and Dawson, *J. Biol. Chem.*, 189, 485 (1951).
236. Poillon and Dawson, *Biochim. Biophys. Acta*, 77, 27 (1963).
237. Stark and Dawson, in *The Enzymes*, Vol. 8, Boyer et al., Eds., Academic Press, New York, 1963, 294.
238. Tokayama et al., *Biochemistry*, 4, 1362 (1965).
239. Dawson, in *The Biochemistry of Copper*, Peisach et al., Eds., Academic Press, New York, 1966, 305.
240. Wishnick et al., *J. Biol. Chem.*, 244, 5761 (1969).
241. Oka and Simpson, *Biochem. Biophys. Res. Commun.*, 43, 1 (1971).
242. Brady et al., *J. Biol. Chem.*, 247, 7915 (1972).
243. Brady et al., *Arch. Biochem. Biophys.*, 157, 63 (1973).
244. Bienert et al., *J. Biol. Chem.*, 237, 2337 (1962).

245. Bienert et al., *Brookhaven Symp. Biol.*, 15, 229 (1962).
246. Morrison et al., *J. Biol. Chem.*, 283, 2220 (1963).
247. Byvoet and Gotti, *Mol. Pharmacol.*, 3, 142 (1967).
248. Furth, *J. Biol. Chem.*, 243, 4832 (1968).
249. Keilin and Mann, *Biochem. J.*, 34, 1163 (1940).
250. Davis, in *The Enzymes*, Vol. 5, Boyer et al., Eds., Academic Press, New York, 1961, 545.
251. Lindskog and Malmström, *J. Biol. Chem.*, 237, 1129 (1962).
252. Rickli and Edsall, *J. Biol. Chem.*, 237, PC 258 (1962).
253. Lindskog, *J. Biol. Chem.*, 238, 945 (1963).
254. Rickli et al., *J. Biol. Chem.*, 239, 1065 (1963).
255. Lindskog and Nyman, *Biochim. Biophys. Acta*, 85, 462 (1964).
256. Armstrong et al., *J. Biol. Chem.*, 241, 5137 (1966).
257. Ashworth et al., *Arch. Biochem. Biophys.*, 142, 122 (1971).
258. Duff and Coleman, *Biochemistry*, 5, 2009 (1966).
259. Tanis and Tashian, *Biochemistry*, 10, 26 (1971).
260. Bernstein and Schreer, *J. Biol. Chem.*, 247, 1306 (1972).
261. Traynaut and Coleman, *J. Biol. Chem.*, 240, 4455 (1971).
262. Nielsen and Frieden, *Comp. Biochem. Physiol.*, 41, 875 (1972).
263. Tobin, *J. Biol. Chem.*, 245, 2656 (1970).
264. Pocker and Ng, *Biochemistry*, 12, 5127 (1973).
265. Atkins et al., *Plant Physiol.*, 50, 218 (1972).
266. Kisell and Graf, *Phytochemistry*, 11, 113 (1972).
267. Brundell et al., *Biochim. Biophys. Acta*, 284, 311 (1972).
268. Brundell et al., *Biochim. Biophys. Acta*, 284, 298 (1972).
269. Vallee and Neurath, *J. Am. Chem. Soc.*, 76, 5006 (1954).
270. Vallee, *Adv. Protein Chem.*, 10, 317 (1955).
271. Neurath, in *The Enzymes*, Vol. 4, Boyer et al., Eds., Academic Press, New York, 1960, 11.
272. Vallee, in *The Enzymes*, Vol. 3, Boyer et al., Eds., Academic Press, New York, 1960, 225.
273. Vallee et al., *J. Biol. Chem.*, 234, 2621 (1959).
274. Coleman and Vallee, *J. Biol. Chem.*, 236, 2244 (1961).
275. Cox et al., *Biochemistry*, 1, 1067 (1962).
276. Peterson and Sokolovsky, *Fed. Proc.*, 33(5), 1529 (1974).
277. Neurath and Lacko, *Biochemistry*, 9, 4680 (1970).
278. Folk and Schirmer, *J. Biol. Chem.*, 238, 3884 (1963).
279. Folk and Gladner, *J. Biol. Chem.*, 235, 60 (1960).
280. Folk et al., *J. Biol. Chem.*, 237, 3100 (1962).
281. Wintersberger et al., *Biochemistry*, 1, 1069 (1962).
282. Chabner and Bertino, *Biochim. Biophys. Acta*, 276, 234 (1972).
283. McConn et al., *Arch. Biochem. Biophys.*, 120, 479 (1967).
284. Tsuru et al., *Biochem. Biophys. Res. Commun.*, 15, 367 (1964).
285. McConn et al., *J. Biol. Chem.*, 239, 3706 (1964).
286. Tsuru et al., *J. Biol. Chem.*, 240, 2415 (1965).
287. Fogarty and Griffin, *J. Biochem. Trans.*, 1, 400 (1973).
288. Feder et al., *Biochim. Biophys. Acta*, 251, 74 (1971).
289. Millet, *Eur. J. Biochem.*, 9, 456 (1969).
290. Latt et al., *Biochem. Biophys. Res. Commun.*, 37, 333 (1969).
291. Ohta et al., *J. Biol. Chem.*, 241, 5919 (1966).
292. Miyata et al., *Agric. Biol. Chem.*, 35, 460 (1971).
293. Hiramitsu and Ouchi, *J. Biochem.* (Tokyo), 71, 676 (1972).
294. Swaisgood and Pattee, *J. Food. Sci.*, 33, 400 (1968).
295. Vallee and Hoch, *Proc. Natl. Acad. Sci. U.S.A.*, 41, 327 (1955).
296. Theorell et al., *Acta Chem. Scand.*, 9, 1148 (1955).
297. Vallee and Hoch, *J. Biol. Chem.*, 225, 185 (1957).
298. Åkeson, *Biochem. Biophys. Res. Commun.*, 17, 211 (1964).
299. Drum et al., *Proc. Natl. Acad. Sci. U.S.A.*, 57, 1434 (1967).
300. Von Wartburg and Vallee, *Abstr. 145th Meet. Am. Chem. Soc.*, P100C, September 1963.
301. Bethune et al., *Abstr. 145th Meet. Am. Chem. Soc.*, P100C, September 1963.
302. Blair and Vallee, *Biochemistry*, 5, 2026 (1966).
303. Von Wartburg et al., *Biochemistry*, 3, 1775 (1964).
304. Arslanian et al., *Biochem. J.*, 125, 1039 (1971).
305. Markovic et al., *Acta Chem. Scand.*, 25, 195 (1971).

306. Vallee et al., *J. Am. Chem. Soc.*, 77, 5196 (1955).
307. Ito, *Chem. Abstr.*, 52, 4726B (1958).
308. Ito et al., *Chem. Abstr.*, 52, 15621C (1958).
309. Adelstein and Vallee, *J. Biol. Chem.*, 233, 589 (1958).
310. Kubo et al., *Bull. Soc. Chim. Biol.*, 40, 431 (1958).
311. Keleti et al., *Acta Physiol. Acad. Sci. Hung.*, 22, 11 (1964).
312. Keleti, *Biochim. Biophys. Acta*, 89, 422 (1964).
313. Keleti, *Biochem. Biophys. Res. Commun.*, 22, 640 (1966).
314. Curdel et al., *C.R. Soc. Biol.*, 249, 1959 (1959).
315. Itwatsubo and Curdel, *Biochem. Biophys. Res. Commun.*, 6, 385 (1961).
316. Gregolin and Singer, *Biochim. Biophys. Acta*, 67, 201 (1963).
317. Cremona and Singer, *J. Biol. Chem.*, 239, 1466 (1964).
318. Curdel, *Biochem. Biophys. Res. Commun.*, 22, 357 (1966).
319. Rosenblum et al., *Arch. Biochem. Biophys.*, 41, 303 (1970).
320. Plocke et al., *Biochemistry*, 1, 373 (1962).
321. Plocke and Vallee, *Biochemistry*, 1, 1039 (1962).
322. Simpson and Vallee, *Fed. Proc.*, 27, 291 (1968).
323. Simpson et al., *Biochemistry*, 7, 4336 (1968).
324. Fosset et al., *Biochemistry*, 13, 1783 (1974).
325. Gottlieb and Sussman, *Biochim. Biophys. Acta*, 161, 167 (1968).
326. Rosenbusch and Weber, *Proc. Natl. Acad. Sci. U.S.A.*, 68, 1019 (1971).
327. Rutter and Ling, *Biochim. Biophys. Acta*, 30, 71 (1958).
328. Kobes et al., *Biochemistry*, 8, 585 (1969).
329. Jagannathan et al., *Biochem. J.*, 63, 94 (1956).
330. Schwartz et al., *Biochemistry*, 13, 1726 (1974).
331. Ottolenghi, *Biochim. Biophys. Acta*, 106, 510 (1965).
332. Campbell et al., *Biochim. Biophys. Acta*, 118, 371 (1966).
333. Lisowski et al., *Acta Biochim. Poly.*, 17, 311 (1970).
334. Himmelhoch, *Arch. Biochem. Biophys.*, 134, 597 (1969).
335. Böttger et al., *Acta Biol. Med. Ger.*, 21, 143 (1968).
336. Hanson et al., *Z. Phys. Chem.*, 340, 107 (1961).
337. Carpenter and Vahl, *J. Biol. Chem.*, 248, 294 (1973).
338. Wacker et al., *Helv. Chim. Acta*, 54, 473 (1971).
339. Lehky et al., *Biochim. Biophys. Acta*, 321, 274 (1973).
340. Gracy and Noltman, *J. Biol. Chem.*, 243, 4109 (1968).
341. Zielke and Seulter, *J. Biol. Chem.*, 246, 2179 (1971).
342. Slater et al., *Biochem. Biophys. Res. Commun.*, 44, 37 (1971).
343. Scrutton et al., *Proc. Natl. Acad. Sci. U.S.A.*, 68, 2497 (1971).
344. Auld et al., *Biochem. Biophys. Res. Commun.*, 57, 967 (1974).
345. Auld et al., *Proc. Natl. Acad. Sci. U.S.A.*, 71, 2091 (1974).
346. Auld et al., *Biochem. Biophys. Res. Commun.*, 62, 296 (1975).
347. Scrutton et al., *J. Biol. Chem.*, 245, 6220 (1971).
348. Valchek and Wood, *Biochim. Biophys. Acta*, 258, 133 (1972).
349. Hirose et al., *Biochim. Biophys. Acta*, 289, 137 (1972).
350. Check and Neilands, *Biochem. Biophys. Res. Commun.*, 55, 1060 (1973).
351. Nicholas and Nason, *J. Biol. Chem.*, 207, 353 (1954).
352. Kessler and Rajogapolan, *J. Biol. Chem.*, 247, 6566 (1972).
353. Cohen and Fridovich, *J. Biol. Chem.*, 246, 367 (1971).
354. Cohen et al., *J. Biol. Chem.*, 246, 374 (1971).
355. Northrop and Wood, *J. Biol. Chem.*, 244, 5801 (1969).
356. Flohé et al., *FEBS Lett.*, 32, 132 (1973).
357. Sang-Hwan et al., *Biochemistry*, 13, 1825 (1974).
358. Bosran and Vallee, *Biochemistry*, in press.
359. Weisiger and Fridovich, *J. Biol. Chem.*, 248, 3582 (1974).
360. Vance and Keele, *J. Biol. Chem.*, 247, 4782 (1972).
361. Vallee, unpublished.
362. Folk et al., *J. Biol. Chem.*, 235, 2272 (1960).
363. Graziani et al., *Biochemistry*, 13, 804 (1974).
364. Stadtman, *Science*, 183, 915 (1974).
365. Garner and Behal, *Biochemistry*, 13, 3227 (1974).

CHARACTERIZATION OF HISTONES

Robert J. DeLange

The second edition of this handbook contained an extensive "Table of Histone Fractions" (C-56 to C-61) in which histone fractions prepared by various procedures were characterized. It is now generally accepted that almost all eukaryotic tissues contain only five major histone fractions (e.g., see the five calf thymus fractions below). In some tissues (e.g., nucleated erythrocytes) or organisms (e.g., trout) special histone fractions (5 and 6 below) are found. Each major histone fraction can often be separated into subfractions which originate through sidechain modifications (acetylation, methylation, phosphorylation, etc.) or through limited sequence heterogeneity.

Wherever possible the items in this table have been derived from sequence data (see Table of Histone Sequences). Other items have been selected and updated from the reviews by DeLange and Smith.[1-4]

CHARACTERIZATION OF HISTONES

Histone[a]	$\frac{Lys}{Arg}$ ratio	Total residues	Molecular weight	NH_2-Terminal	COOH-Terminal	Evolutionary conservation of sequence	Sequence heterogeneity (same tissue)	Sidechain modifications
			Calf Thymus				Eukaryotes in General	
1 (F1, I)	~21.0	~223	~22,130	Ac-Ser	Lys	Most variable	14% (Residue 1–73, rabbit thymus)	Phosphorylated, ADP-ribosylated in some species
2A (F2A2, IIb1)	1.17	129	14,004	Ac-Ser	Lys	Intermediate	Residue 16 in rat	Appear to be acetylated, methylated and phosphorylated in some species
2B (F2B, IIb2)	2.50	125	13,774	Pro	Lys	Intermediate	Not demonstrated	
3 (F3, III)	0.72	135	15,324	Ala	Ala	Highly conserved (3% difference, pea and calf)	Residue 96 in pea and calf	Acetylated and methylated in most tissues, phosphorylated in some species
4 (F2A1, IV)	0.79	102	11,282	Ac-Ser	Gly	Most conserved (2% difference, pea and calf)	Not demonstrated	
			Nucleated Erythrocytes					
5 (F2C, V)	~2.2	~197	~21,450	Thr	Lys	Insufficient data	Residue 15 in chicken	Phosphorylated
			Trout Tissues					
6 (T)	~3.0	~122	~14,500	Pro	?	Found only in trout thus far	Not demonstrated	?

Compiled by Robert J. DeLange.

[a] The histone nomenclature is that which was recently approved by a number of investigators in this field at the Ciba Foundation Symposium on "The Structure and Function of Chromatin," April 2 to 5, 1974, London. The two older systems of nomenclature, which were most commonly used, are shown in parentheses. Histone T has been given the designation 6 in this table, although this was not discussed at the symposium.

REFERENCES (see also Table of Histone Sequences)

1. **DeLange and Smith**, *Ann. Rev. Biochem.*, 40, 279 (1971).
2. **DeLange and Smith**, *Acc. Chem. Res.*, 5, 368 (1972).
3. **DeLange and Smith**, in *The Structure and Function of Chromatin*, Ciba Symp., April 1974, in press.
4. **DeLange and Smith**, in *The Proteins*, Vol. 4, 3rd ed., Neurath and Hill, Eds., Academic Press, New York, in press.

TABLE OF HISTONE SEQUENCES[a]

Histone 1[b] (I, F1)

		10			
Calf thymus-a[b](1)	Ac- Ser- Glu- Ala- Pro- Ala- Glu-Thr- Ala- Ala- Pro- Ala- Pro- Ala- Pro-
Rabbit thymus-d(1)	Ac- Ser- Glu- Ala- Pro- Ala- Glu-Thr- Ala- Ala- Pro- Ala- Pro- Ala-
Rabbit thymus-c(1)	Ac- Ser- Glu- Ala- Pro- Ala- Glu-Thr- Ala- Ala- Pro- Ala- Pro- Ala- Glu-
Trout testis(2)	————————————Unknown————————————

20			30		
Lys- Ser- Pro- Ala- Lys-Thr-Pro- Val-Lys- Ala- Ala- Lys-Lys- Lys-Lys- Pro- Ala-Gly- Ala- Arg-
Lys- Ser- Pro- Ala- Lys-Thr-Pro- Val-Lys- Ala- Arg- Lys-Lys- Lys- Ser- Ala- Gly- Ala- Ala- Lys-
Lys- Ser- Pro- Ala- Lys-Lys-Lys-   -Lys- Ala- Ala- Lys-Lys- Pro-Gly- Ala- Gly- Ala- Ala- Lys-
————————————————Unknown————————————————

40			50		
Arg-Lys- Ala- Ser- Gly-Pro-Pro- Val- Ser-Glu- Leu- Ile-Thr- Lys- Ala- Val- Ala- Ala- Ser- Lys-
Arg-Lys- Ala- Ser- Gly-Pro-Pro- Val- Ser-Glu- Leu- Ile-Thr- Lys- Ala- Val- Ala- Ala- Ser- Lys-
Arg-Lys- Ala- Ala- Gly-Pro-Pro- Val- Ser-Glu- Leu- Ile-Thr- Lys- Ala- Val- Ala- Ala- Ser- Lys-
————————————————Unknown————————————————

	60			70	
Glu-Arg- Ser- Gly- Val- Ser-Leu- Ala- Ala-Leu- Lys- Lys- Ala- Leu- Ala- Ala- Ala-Gly-Tyr
Glu-Arg- Ser- Gly- Val- Ser-Leu- Ala- Ala-Leu- Lys- Lys- Ala- Leu- Ala- Ala- Ala-Gly-Tyr
Glu-Arg- Asn- Gly- Leu- Ser-Leu- Ala- Ala-Leu- Lys- Lys- Ala- Leu- Ala- Ala- Gly-Gly-Tyr-Asp-
——Arg- Ser- Gly- Val- Ser-Leu- Ala- Ala-Leu- Lys- Lys- Ser- Leu- Ala- Ala- Gly-Gly-Tyr- Asp-

80			90		
Val-Glu- Lys- Asn- Asn- Ser-Arg- Ile-Lys-Leu- Gly- Leu-Lys- Ser-Leu- Val- Ser-Lys-Gly- Thr-
Val-Glu- Lys- Asn- Asn- Ser-Arg- Val-Lys- Ile- Ala- Val-Lys- Ser-Leu- Val- Thr-Lys-Gly- Thr-

	100			110	
Leu- Val- Glu- Thr- Lys-Gly-Thr- Gly- Ala- Ser- Gly- Ser-Phe- Lys-Leu-Asp- Lys-Lys- Ala- Ala-
Leu- Val- Glu- Thr- Lys-Gly-Thr- Gly- Ala- Ser- Gly- Ser-Phe- Lys-Leu-Asn- Lys-Lys- Ala-

	120			130	
Ser-Gly- Glu- Ala- Lys-Pro-Lys- Pro-   -Lys- Lys- Ala-Gly- Ala- Ala- Lys- Pro-Lys-Lys- Pro-
- Val- Glu- Ala- Lys-   -Lys- Pro- Ala-Lys- Lys- Ala-   - Ala- Ala-   - Pro- Lys- Ala- Lys-

	140			150	
Ala- Gly- Ala- Ala- Lys- Lys-Pro- Ala- Gly- Ala, Ala, Lys, Ala, Pro,Thr, Pro,Lys* Val- Ala-
Lys- Val- Ala- Ala- Lys- Lys-Pro- Ala-   - Ala- Ala-Lys- Ala-Pro-Lys-   - Lys- Val- Ala- Ala-

	160			170	
Lys*Lys- Ala- Val-Lys* Ala-Lys-Lys* Ser-Pro- Lys* Lys- Ala- Lys*Lys- Pro-Lys* Ala- Pro-Lys*
Lys-Lys- Ala- Val- Ala- Ala-Lys- Lys- Ser-Pro- Lys-   Lys- Ala Lys-Lys- Pro- Ala-Thr- Pro- Lys

	180			190	
Ser- Ala- Ala-Lys* Ser-Pro- Ala- Lys-Pro-   -Lys* Ala- Ala- Lys-Pro- Lys- Ala-Pro-Lys- Pro-
Lys- Ala- Ala- Lys- Ser-Pro-Lys- Lys- Ala-Thr- Lys- Ala- Ala- Lys-Pro- Lys- Ala- Ala-Lys- Pro-

[a] See the previous table for nomenclature and characterization of histones.
[b] Subfractions of histone 1 have been isolated and studied. These have usually been designated by numbers (1, 2, 3, etc.) but since these might be confused with the designations in the new system of nomenclature, letters (a for 1, b for 2, etc.) have been substituted here. The numbers above residues indicate relative positions only, but are not residue numbers (due to deletions, etc.). * indicates start and end of peptides.

TABLE OF HISTONE SEQUENCES (continued)

Histone 1^b(I, F1)

```
                         200                                              210
Lys*-Ala- Ala-Lys*-Lys*-Ala- Ala-Lys*-Ser-Pro- Ala-Lys*-Ala- Val-Lys- Pro-Lys*-Ala- Ala- Ala-
Lys- Ala- Ala- Lys- Lys- Ala- Ala- Lys- Ser-Pro- Lys- Lys-      - Val-Lys-      - Lys- Ala- Ala- Ala-

              220
Lys-Pro-Lys* Ala- Ala-Gly-Ala-Lys*-Lys-Lys-COOH
Lys-     - Lys- Ala- Pro-     - Ala- Lys-Lys-COOH
```

The positioning of peptides from calf thymus-a and rabbit thymus-d is by analogy with rabbit thymus-c.

The positioning of peptides in rabbit thymus-c beyond position 108 is based on analogy with trout testis histone and partially on overlapping thermolysin peptides. There is apparently serine-threonine heterogeneity at positions 163, 175, 203, and probably 179 in rabbit thymus-c.[3]

HISTONE 2A(IIb1, F2a2); CALF THYMUS SEQUENCE SHOWN[4,5]

```
                                    10
Ac- Ser-Gly-Arg-Gly-Lys-Gln-Gly- Gly- Lys- Ala-Arg- Ala- Lys- Ala-Lys-Thr-Arg- Ser- Ser-

 20                                 30
Arg- Ala-Gly-Leu-Gln-Phe- Pro- Val- Gly- Arg- Val- His- Arg-Leu-Leu-Arg-Lys-Gly-Asn-Tyr-

 40                                 50
Ala-Glu-Arg- Val-Gly- Ala-Gly- Ala- Pro- Val- Tyr-Leu- Ala- Ala- Val-Leu-Glu-Tyr-Leu-Thr-

 60                                 70
Ala-Glu- Ile-Leu-Glu-Leu- Ala-Gly-Asn- Ala- Ala-Arg-Asp-Asn- Lys-Lys-Thr-Arg- Ile- Ile-

 80                                 90
Pro-Arg- His-Leu-Gln-Leu- Ala- Ile- Arg-Asn-Asp-Glu- Glu-Leu-Asn-Lys-Leu-Leu-Gly- Lys-

100                                110
Val-Thr- Ile- Ala-Gln-Gly-Gly- Val- Leu- Pro-Asn- Ile- Gln- Ala- Val-Leu-Leu- Pro- Lys-Lys-

120                  129
Thr-Glu- Ser- His- His- Lys- Ala- Lys- Gly- Lys-COOH
```

In the rat histone, Residue 16 is Thr (some molecules) or Ser (other molecules) and Residue 99 is Arg.[6] Residue 6 is Thr in trout testis histone 2A;[7] only the first 11 residues are known.

TABLE OF HISTONE SEQUENCES[a] (continued)

HISTONE 2B(IIb2, F2B); CALF THYMUS SEQUENCE SHOWN[8]

10
HN- Pro- Glu- Pro- Ala- Lys- Ser- Ala- Pro- Ala- Pro- Lys- Lys- Gly- Ser- Lys- Lys- Ala- Val- Thr-

20   30
Lys- Ala- Gln- Lys- Lys- Asp- Gly- Lys- Lys- Arg- Lys- Arg- Ser- Arg- Lys- Glu- Ser- Tyr- Ser- Val-

40   50
Tyr- Val- Tyr- Lys- Val- Leu- Lys- Gln- Val- His- Pro- Asp- Thr- Gly- Ile- Ser- Ser- Lys- Ala- Met-

60   70
Gly- Ile- Met- Asn- Ser- Phe- Val- Asn- Asp- Ile- Phe- Glu- Arg- Ile- Ala- Gly- Glu- Ala- Ser- Arg-

80   90
Leu- Ala- His- Tyr- Asn- Lys- Arg- Ser- Thr- Ile- Thr- Ser- Arg- Glu- Ile- Gln- Thr- Ala- Val- Arg-

100   110
Leu- Leu- Leu- Pro- Gly- Glu- Leu- Ala- Lys- His- Ala- Val- Ser- Glu- Gly- Thr- Lys- Ala- Val- Thr-

120   125
Lys- Tyr- Thr- Ser- Ser- Lys-COOH

Residues 9 and 10 are not present and the sequence of Residues 21–23 is Ser (or Thr)-Ala-Gly in the trout testis histone;[9] only the first 22 residues are known.

TABLE OF HISTONE SEQUENCES[a] (continued)

HISTONE 3(III, F3); CALF THYMUS SEQUENCE SHOWN[10]

H$_2$N- Ala- Arg- Thr- Lys- Gln- Thr- Ala- Arg- Lys(CH$_3$)$_{0-3}$- Ser- Thr- Gly- Gly- Lys(Ac)$_{0,1}^-$ Ala- Pro- (10)

Arg- Lys- Gln- Leu- Ala- Thr- Lys(Ac)$_{0,1}^-$ Ala- Ala- Arg- Lys (CH$_3$)$_{0-3}$- Ser- Ala- Pro- Ala- Thr- Gly- (20, 30)

Gly- Val- Lys- Lys- Pro- His- Arg- Tyr- Arg- Pro- Gly- Thr- Val- Ala- Leu- Arg- Glu- Ile- Arg- Arg- (40, 50)

Tyr- Gln- Lys- Ser- Thr- Glu- Leu- Leu- Ile- Arg- Lys- Leu- Pro- Phe- Gln- Arg- Leu- Val- Arg- Glu- (60, 70)

Ile- Ala- Gln- Asp- Phe- Lys- Thr- Asp- Leu- Arg- Phe- Gln- Ser- Ser- Ala- Val- Met- Ala- Leu- Gln- (80, 90)

Glu- Ala- Cys- Glu- Ala- Tyr- Leu- Val- Gly- Leu- Phe- Glu- Asp- Thr- Asn- Leu- Cys- Ala- Ile- His- (100, 110)

Ala- Lys- Arg- Val- Thr- Ile- Met- Pro- Lys- Asp- Ile- Gln- Leu- Ala- Arg- Arg- Ile- Arg- Gly- Glu- (120, 130)

Arg- Ala- COOH (135)

In the pea seedling histone, Residue 41 is Phe, 53 is Lys, 90 is Ser, and 96 is Ala (60%) or Ser (40%). ε-N-Acetyllysine and ε-N-trimethyllysine are absent.[11]

In the fish testis histone, Residue 96 is Ser and Residues 14 and 23 are not acetylated.[12]

In the chicken erythrocyte histone, Residue 96 is Ser, and evidence was obtained for methylation of Residue 36 in addition to Residues 9 and 27.[13] The sites of ε-N-acetylation and several amides were not determined.

Some molecules of calf thymus histone contain Ser at Residue 96.[14]

TABLE OF HISTONE SEQUENCES[a] (continued)

HISTONE 4(IV, F2a1); CALF THYMUS SEQUENCE SHOWN[15,16]

                           10
Ac- Ser- Gly- Arg-Gly-Lys-Gly- Gly- Lys-Gly-Leu- Gly- Lys-Gly-Gly- Ala- Lys(Ac)$_{0,1}$ Arg-His-

  20   30
Arg-Lys(CH$_3$)$_{1,2}$ - Val-Leu- Arg- Asp- Asn- Ile- Gln- Gly- Ile- Thr- Lys- Pro- Ala- Ile-Arg-Arg-

               40   50
Leu- Ala- Arg- Arg-Gly-Gly- Val- Lys- Arg- Ile- Ser- Gly-Leu- Ile-Tyr- Glu- Glu- Thr-Arg-Gly-

               60   70
Val- Leu- Lys- Val-Phe-Leu-Glu-Asn- Val- Ile-Arg-Asp- Ala- Val-Thr-Tyr- Thr- Glu- His-Ala-

               80   90
Lys- Arg- Lys- Thr- Val-Thr- Ala-Met-Asp- Val- Val- Tyr- Ala-Leu-Lys- Arg- Gln- Gly-Arg-Thr-

              100
Leu- Tyr- Gly- Phe-Gly-Gly-COOH

In pea seedling histone 4,[15] Residue 60 is Ile, and Residue 77 is Arg; Residue 20 is not methylated and at least one other lysyl residue (Residue 8?) is ϵ-N-acetylated.
Other lysyl residues in calf thymus histone 4 can also be ϵ-N-acetylated to a small extent.[17]
The sequences of rat[18] and pig[19] histone 4 are identical to the calf thymus sequence. Sea urchin histone 4 apparently contains a cysteine residue.[20]

HISTONE 5(V, F2c) (INCOMPLETE); CHICKEN ERYTHROCYTE SEQUENCE SHOWN[21]

1   10   15   20
Thr- Glu- Ser- Leu- Val- Leu- Ser- Pro- Ala- Pro- Ala- Lys- Pro- Lys- Gln- Val- Lys- Ala- Ser- Arg- Arg-Ser- Ala- Ser- His-
                                                                   Arg

                           30   40   50
Pro- Thr- Tyr- Ser- Glu- Met- Ile- Ala- Ala- Ala- Ile- Arg- Ala- Glu- Lys- Ser- Arg- Gly- Gly- Ser- Ser- Arg-Gln- Ser- Ile-

                                        60   70
Gln- Lys- Tyr- Ile- Lys- Ser- His- Tyr- Lys- Val- Gly- His- Asn- Ala- Asp- Leu- Gln- Ile- Lys- Leu-

HISTONE 6(T) (INCOMPLETE); TROUT TESTIS[22]

                           10
HN- Pro- Lys- Arg- Lys-Ser- Ala-Thr- Lys-Gly- Asp- Glu-Pro- Ala- Arg- Arg-Ser- Ala- Arg-Leu-

  20
Ser- Gly- Arg- Pro- Val-Pro-Lys- Pro- Ala-Ala - - - - - -

Compiled by Robert J. Delange.

TABLE OF HISTONE SEQUENCES[a] (continued)

REFERENCES

1. **Rall and Cole,** *J. Biol. Chem.,* 246, 7175 (1971); Cole, personal communication.
2. Personal communicaiton from Professor G. H. Dixon; McLeod and Dixon, unpublished results.
3. **Jones, Rall, and Cole,** *J. Biol. Chem.,* 249, 2548 (1974).
4. **Yeoman, Olson, Sugano, Jordan, Taylor, Starbuck, and Busch,** *J. Biol. Chem.,* 247, 6018 (1972).
5. **Sautiere, Tyrou, Laine, Mizon, Ruffin, and Biserte,** *Eur. J. Biochem.,* 41, 563 (1974).
6. **Sautiere,** personal communication.
7. **Candido and Dixon,** *J. Biol. Chem.,* 247, 3868 (1972).
8. **Iwai, Hayashi, and Ishikawa,** *J. Biochem.,* 72, 357 (1972).
9. **Candido and Dixon,** *Proc. Natl. Acad. Sci. U.S.A.,* 69, 2015 (1972).
10. **DeLange, Hooper, and Smith,** *Proc. Natl. Acad. Sci. U.S.A.,* 69, 882 (1972); *J. Biol. Chem.,* 248, 3261 (1973).
11. **Patthy and Smith,** *J. Biol. Chem.,* 248, 6834 (1973).
12. **Hooper and Smith,** *J. Biol. Chem.,* 248, 3275 (1973).
13. **Brandt and von Holt,** *Eur. J. Biochem.,* 46, 419 (1974).
14. **Patthy and Smith,** *J. Biol. Chem.,* 250 (1975), in press.
15. **DeLange, Fambrough, Smith, and Bonner,** *Proc. Natl. Acad. Sci. U.S.A.,* 61, 1145 (1968); *J. Biol. Chem.,* 244, 319 (1969); *J. Biol. Chem.,* 244, 5669 (1969).
16. **Ogawa, Quagliarotti, Jordan, Taylor, Starbuck, and Busch,** *J. Biol. Chem.,* 244, 4387 (1969).
17. **Wangh, Ruiz-Carrillo, and Allfrey,** *Arch. Biochem. Biophys.,* 150, 44 (1972).
18. **Sautiere, Tyrou, Moschetto, and Biserte,** *Biochimie,* 53, 479 (1971).
19. **Sautiere, Lambelin-Breynaert, Moschetto, and Biserte,** *Biochimie,* 53, 711 (1971).
20. **Subirana,** *FEBS Lett.,* 16, 133 (1971); Sautiere, personal communication.
21. **Garel, Mazen, Champagne, Sautiere, Kmiecik, Loy, and Biserte,** *FEBS Lett.,* 50, 195 (1975).
22. **Huntley and Dixon,** *J. Biol. Chem.,* 247, 4916 (1972).

ENZYMES FOUND IN NORMAL HUMAN URINE

Enzyme	Mol wt	Reference
Lactic dehydrogenase, EC 1.1.1.27	144,000	2
Diamino-oxidase, EC 1.4.3.6	195–255,000	3
Dihydroxyphenylalanine oxidase	–	–
D-Glutamyltransferase, EC 2.3.2.1	–	–
Asparate aminotransferase (GOT), EC 2.6.1.1	90,000	3
Ribonuclease, EC 2.7.7.16	13,700 and 18,800	4
Triglyceride esterase (lipase), EC 3.1.1.3	39–200,000	3
Cholinesterase, EC 3.1.1.8	300,000	5
Alkaline phosphatase, EC 3.1.3.1	75,000	6
Acid phosphatase, EC 3.1.3.2	95,800	7
Acid deoxyribonuclease, EC 3.1.4.5	40–62,000	3
Neutral deoxyribonuclease, EC 3.1.4.6	38,000	3
Sulfatases, EC 3.1.6.1	107–411,000	3
Amylase, EC 3.2.1.1	55,000	8
Muramidase, EC 3.2.1.17	15,000	9
α-Glucosidase, EC 3.2.1.20	–	–
β-Glucosidase, EC 3.2.1.21	40–50,000	3
β-Galactosidase, EC 3.2.1.23	43–127,000	3
Trehalase, EC 3.2.1.28	–	–
β-Glucuronidase, EC 3.2.1.31	230,000	10
Leucine aminopeptidase, EC 3.4.1.2	300,000	3
Carboxypeptidase B, EC 3.4.2.2	34,000	3
Uropepsinogen, EC 3.4.4.1	40,000	11
Renin, EC 3.4.4.15	–	–
Kallikrein, EC 3.4.4.21	43,600	12
Urokinase, EC 3.4.4.a	31,500 and 54,700	13
Hyaluronidase, EC 4.2.99.1	–	–

Adapted from Raab[1] by Amadeo J. Pesce.

REFERENCES

1. Raab, *Clin. Chem.*, 18, 5 (1972).
2. Pesce, Fondy, Stolzenback, et al., *J. Biol. Chem.*, 242, 2151 (1967).
3. Barman, *Enzyme Handbook,* Springer-Verlag, New York, 1969.
4. Delaney, *Biochemistry,* 2, 438 (1963).
5. Surgenor and Ellis, *J. Am. Chem. Soc.,* 76, 6049 (1954).
6. Butterworth, *Biochem. J.,* 107, 467 (1968).
7. Ostrowski and Rybarsha, *Biochim. Biophys. Acta,* 105, 196 (1965).
8. Duane, Frerichs, and Levitt, *J. Clin. Invest.,* 51 156 (1971).
9. Osserman and Lawler, *J. Exp. Med.,* 124, 921 (1966).
10. Hygstedt and Jagenburg, *Scand. J. Clin. Lab. Invest.,* 17, 565 (1965).
11. **Keller, Agne, Mannebach, Leppla, and Dubach.** *Current Problems in Clinical Biochemistry,* Vol. 2, Williams and Wilkins, Baltimore, 1968.
12. Silva, Diniz, and Mares-Guia, *Biochemistry,* 13, 4304 (1974).
13. White, Barlow, and Mozen, *Biochemistry,* 5, 2160 (1966).

PROPERTIES OF UROKINASE

	S1	S2
Sedimentation coefficient $s_{20,w}$	2.66	3.27
Diffusion coefficient	7.41 × 10^{-7}	—
Partial specific volume	0.724	0.278
Molecular weight	31,300	54,700
Electrophoretic mobility ×10^5 cm^2/V at pH 4.8	+3.5	+2.2
$E_{1\,cm}^{1\%}$ 280 nm	13.2	13.6
Specific activity CTA units	218,000	93,500
Antibody to S1	+	+partial identity

Adapted from White, Barlow, and Mozen[1] by Amadeo J. Pesce.

REFERENCE

1. White, Barlow, and Mozen, *Biochemistry,* 5, 2160 (1966).

AMINO ACID COMPOSITION OF UROKINASE

	Amino acid residues per	
	31,500 g of protein (S1)	54,700 g of protein (S2)
Lysine	17.3	31.0
Histidine	10.2	19.7
Arginine	14.3	22.1
Aspartic acid	19.1	40.4
Threonine	19.0	31.6
Serine	21.8	34.8
Glutamic acid	28.3	44.1
Proline	16.4	28.7
Glycine	22.4	41.1
Alanine	10.1	20.2
Half-cystine	9.5	19.4
Valine	10.9	23.2
Methionine	4.5	7.7
Isoleucine	15.6	20.4
Leucine	20.6	33.6
Tyrosine	12.8	19.9
Phenylalanine	9.3	14.3
Tryptophan	4.9	9.6

Adapted from White, Barlow, and Mozen[1] by Amadeo J. Pesce.

REFERENCE

1. **White, Barlow, and Mozen,** *Biochemistry,* 5, 2160 (1966).

CARBOHYDRATE[a] AND PROTEIN COMPOSITION OF T-H GLYCOPROTEIN

	Composition (1% of dry wt)	
	T-H glycoprotein	Alkali-treated T-H glycoprotein
Hexose	11.7 ± 0.3	11.5 ± 0.2
Fucose	0.8 ± 0.2	N.D.
N-Acetylhexosamine	11.2 ± 0.2	11.0 ± 0.2
Sialic acid	4.4 ± 0.2	N.D.
Ash	2.3	N.D.
Protein	67.6	N.D.

[a] Expressed as free monosaccharides.
From Fletcher, Neuberger, and Ratcliffe, *Biochem. J.*, 120, 417 (1970). With permission.

AMINO ACID COMPOSITION OF T-H GLYCOPROTEIN

	Mol of residues/ 100,000 g of protein	Residues/100 residues in the protein
Asp	67.4 ± 0.9	10.90
Thr	47.2 ± 0.9	7.63
Ser	48.6 ± 1.2	7.86
Glu	52.2 ± 2.0	8.44
Pro	28.6 ± 1.8	4.26
Gly	52.0 ± 1.5	8.41
Ala	42.0 ± 1.8	6.79
Val	39.6 ± 1.3	6.40
Cys[a]	52.0 ± 0.5	8.41
Met	12.7 ± 0.9	2.05
Ile	15.2 ± 0.6	2.46
Leu	46.9 ± 1.4	7.58
Tyr	23.7 ± 0.8	3.83
Phe	19.4 ± 0.8	3.14
Lys	16.4 ± 0.7	2.65
His	16.5 ± 0.9	2.67
Arg	27.8 ± 0.9	4.49
Trp	10.4 ± 0.3[b]	1.68
Total	618.6	100.00

Adapted from Fletcher, Neuberger, and Ratcliffe[1] by Amadeo J. Pesce.

[a] Determined as cysteic acid.
[b] Average of two spectrophotometric methods.

REFERENCE

1. Fletcher, Neuberger, and Ratcliffe, *Biochem. J.*, 120, 417 (1970).

AMINO ACID COMPOSITION OF HUMAN RETINOL-BINDING PROTEIN

Amino acid	Residues/molecule
Lysine	10
Histidine	2
Ammonia	16
Arginine	13
Aspartic acid	22
Threonine	9
Serine	11
Glutamic acid	18
Proline	6
Glycine	11
Alanine	13
Half-cystine	5
Valine	12
Methionine	4
Isoleucine	4
Leucine	12
Tyrosine	8
Phenylalanine	10
Tryptophan	6
Total (excluding ammonia)	176

Adapted from Peterson and Berggard[1] by Amadeo J. Pesce.

REFERENCE

1. Peterson and Berggard, *J. Biol. Chem.*, 246, 25 (1971).

RETINOL-BINDING PROTEIN

Sedimentation coefficient $s^o_{20,w}$	2.3S
Molecular weight	21,400 d
Partial specific volume	0.720 ml/g
Isoelectric point	4.6–4.8
$E^{1\%}_{280}$	18.7
E_{330}/E_{280}	1.02
Moles of Vitamin A bound/mole protein	0.87
Serum	46 mg/liter
Urine excretion	0.11 mg/24 hr
Cerebral spinal fluid	0.35 mg/liter
Protein bound to prealbumin in serum	

Adapted from Peterson and Berggard[1] by Amadeo J. Pesce.

REFERENCE

1. Peterson and Berggard, *J. Biol. Chem.*, 246, 25 (1971).

TAMM-HORSFALL MUCOPROTEIN

		Reference
Sedimentation coefficient	29.5	1
	65	2
Diffusion coefficient	3.25×10^8 cm^2 sec^{-1}	3
Partial specific volume	0.685	1
	0.705 ml/g	4
Molecular weight	7,000,000	1
	28,000,000	5
	23,000,000	6
Isoelectric point	3.2	7
Subunit size	100,000	6
	76–82,000	8
Intrinsic viscosity	5 dl/g	2
	8 dl/g	2
	80 dl/g	9
f/f_0	5.32	1
Axial ratio	172	1
	95	1
Shape	rod 5,600 Å × 42 Å	1
	2,500 Å × 25Å, 110 Å periodicity	10
$E^{1\%}_{1\ cm}$	10.8	11

Compiled by Amadeo J. Pesce.

REFERENCES

1. **Tamm, et al.,** *J. Biol. Chem.*, 212, 125 (1955).
2. **Maxfield,** *Arch. Biochem. Biophys.*, 89, 281 (1960).
3. **Tamm and Horsfall,** *J. Exp. Med.*, 95, 71 (1952).
4. **Curtain,** *Aust. J. Exp. Biol. Med. Sci.*, 31, 615 (1953).
5. **Maxfield,** *Biochim. Biophys. Acta*, 49, 548 (1961).
6. **Stevenson and Kent,** *Biochem. J.*, 116, 791 (1970).
7. **Curtain,** *Aust. J. Exp. Biol. Med. Sci.*, 31, 255 and 623 (1953).
8. **Fletcher, et al.,** *Biochem. J.*, 120, 425 (1970).
9. **Stevenson,** *Biochim. Biophys. Acta*, 160, 296 (1968).
10. **Fletcher, et al.,** *Biochim. Biophys. Acta*, 214, 299 (1970).
11. **Maxfield,** in *Glycoproteins, Their Composition, Structure and Function*, Gottschalk, Ed., Elsiever, Amsterdam, 1966.

β_2 MICROGLOBULIN

Sedimentation coefficient s_{20w}	1.6 S
Partial specific volume V	0.727 ml/g
Molecular weight	11,815
E_{280}/mol	19,850

Adapted from Berggard and Bearn[1] by Amadeo J. Pesce.

REFERENCE

1. Berggard and Bearn, *J. Biol. Chem.*, 293, 4095 (1968).

AMINO ACID SEQUENCE OF β_2 MICROGLOBULIN

H- Ile- Gln- Arg- Thr- Pro- Lys- Ile- Gln- Val- Tyr- Ser- Arg- His- Pro- Ala- Glx- Asx- Gly- Lys-
(10)

Ser- Asx- Phe- Leu- Asn- Cys- Tyr- Val- Ser- Gly- Phe- His- Pro- Ser- Asp- Ile- Glu- Val- Asp-
(20 21) (30)

Leu- Leu- Lys- Asp- Gly- Glu- Arg- Ile- Glx- Lys- Val- (Asx, His, Ser, Glx)- Leu- Ser- Phe- Ser-
(40 41) (50)

Lys- Asn- Ser- Trp- Phe- Tyr- Leu- (Leu, Tyr, Ser)- Tyr- Thr- Glu- Phe- Thr- Pro- Thr- Glu- Lys-
(60 61) (70)

Asp- Glu- Tyr- Ala- Cys- Arg- Val- Asx- His- Val- Thr- Leu- Ser- Glx- Pro- Lys- Ile- Val- Lys-
(80 81) (90)

Trp- Asp- Arg- Asp- Met- OH
(100)

From Cunningham, et al., *Biochemistry*, 12, 4811 (1973). With permission. Copyright by the American Chemical Society.

MOLECULAR PARAMETERS OF THE CONTRACTILE PROTEINS[a]

Protein	Localization in myofibril	% Total protein	Intrinsic sedimentation coefficient (Svedberg)	Intrinsic viscosity (ml/g)	Molecular weight	Chain weight	% α-helix	References
Myosin	Thick filament	55	6.4	210	470,000	200,000 20,700[b] 18,000 16,500	57	1–8
C-protein	Thick filament	2	4.6	14	140,000	140,000	<10	9
Paramyosin	Thick filament	5–50	3.1	190	200,000	100,000	>90	10–12
M-proteins	M-line	<2	5.0	—	160,000	160,000	—	13, 14
			5.4	4.5	86,000	43,000	26	15, 16
G-actin	Thin filament	25	3.3	4	41,780	41,780[b]	26	17–20
Tropomyosin	Thin filament	5	2.6	34	65,000	32,760[b]	90	21–23
Troponin	Thin filament	5	4.0	4	80,000	37,000 24,000 17,850[b]	35	24–27
α-Actinin	Z-line	<2	6.2	9	180,000	90,000	60	28, 29

Compiled by S. Lowey.

[a]Parameters determined for proteins from vertebrate fast skeletal muscles, with the exception of paramyosin.
[b]Molecular weights based on sequence data.

REFERENCES

1. Lowey, Slayter, Weeds, and Baker, *J. Mol. Biol.*, 42, 1 (1969).
2. Gershman, Stracher, and Dreizen, *J. Biol. Chem.*, 244, 2726 (1969).
3. Gazith, Himmelfarb, and Harrington, *J. Biol. Chem.*, 245, 15 (1970).
4. Godfrey and Harrington, *Biochemistry*, 9, 894 (1970).
5. Frank and Weeds, *Eur. J. Biochem.*, 44, 317 (1974).
6. Lowey and Risby, *Nature*, 234, 81 (1971).
7. Sarkar, Sreter, and Gergely, *Proc. Natl. Acad. Sci. U.S.A.*, 68, 946 (1971).
8. Weeds and Lowey, *J. Mol. Biol.*, 61, 701 (1971).
9. Offer, Moos, and Starr, *J. Mol. Biol.*, 74, 653 (1973).
10. Bullard, Luke, and Winkelman, *J. Mol. Biol.*, 75, 359 (1973).
11. Stafford and Yphantis, *Biochem. Biophys. Res. Commun.*, 49, 848 (1972).
12. Lowey, Holtzer, and Kucera, *J. Mol. Biol.*, 7, 234 (1963).
13. Trinick, *Fed. Proc.*, in press.
14. Masaki and Takaiti, *J. Biochem.* (Tokyo), 75, 367 (1974).
15. Eaton and Pepe, *J. Cell Biol.*, 55, 681 (1972).
16. Morimoto and Harrington, *J. Biol. Chem.*, 247, 3052 (1972).
17. Elzinga, Collins, Kuehl, and Adelstein, *Proc. Natl. Acad. Sci. U.S.A.*, 70, 2687 (1973).
18. Cohen, *Arch. Biochem. Biophys.*, 117, 289 (1966).
19. Rees and Young, *J. Biol. Chem.*, 242, 4449 (1967).
20. Nagy, *Biochim. Biophys. Acta*, 115, 498 (1966).
21. Holtzer, Clark and Lowey, *Biochemistry*, 4, 2401 (1965).
22. Woods, *J. Biol. Chem.*, 242, 2859 (1967).
23. Stone, Sodek, Johnson, and Smillie, *Proc. IX FEBS Meet.*, Budapest, in press.
24. Collins, Potter, Horn, Wilshire, and Jackman, in *Calcium Binding Proteins*, Drabikowski, Strzelecka-Golaszewska, and Carafoli, Eds., Elsevier, Amsterdam, 1974.
25. Greaser and Gergely, *J. Biol. Chem.*, 246, 4226 (1971).
26. Perry, Cole, Head, and Wilson, *Cold Spring Harbor Sym. Quant. Biol.*, 37, 251 (1972).
27. Ebashi, Ohtsuki, and Mihaski, *Cold Spring Harbor Sym. Quant. Biol.*, 37, 215 (1972).
28. Goll, Suzuki, and Singh, *Biophys. J.*, 11, 107a (1971).
29. Suzuki, Goll, Stromer, Singh, and Temple, *Biochim. Biophys. Acta*, 295, 188 (1973).

PROTEINS IN NONMUSCLE CELLS

Table 1
IDENTIFICATION OF ACTIN, MYOSIN, AND ASSOCIATED PROTEINS IN NONMUSCLE CELLS*[†]

No.	Species	Cell type	Actomyosin	Vertebrate Actin	Myosin	Associated proteins
1	Cat	Brain	Isolation (131, 18)	Isolation (130, 132)	Isolation (19)	
2	Chicken	Brain (embryo and adult)		Gel electrophoresis and peptide map (43) Isolation (30, 83a) EM with HMM (72)	Isolation (83a)	Tropomyosin Isolation (42)
		Chondrogenic cell (embryo)		EM with HMM (72)		
		Epidermis (embryo)		EM with HMM (72)		
		Fibroblast (embryo)		EM with HMM (72)	Isolation (135)	Tropomyosin Isolation (169)
		Glial cell (embryo)		Isolation (168)		
		Intestinal epithelium (embryo and adult)		Gel electrophoresis and peptide map (30) Fluorescent HMM (136) EM with HMM (87) EM with HMM (72) Isolation (155)	Isolation (98)	α-Actinin Gel electrophoresis and antibody staining (98, 137)
		Kidney (embryo)		Gel electrophoresis and peptide map (30)		
		Lens (embryo)		Gel electrophoresis and peptide map (30)		
		Liver (embryo)	Antibody staining (41, 49, 52)	Gel electrophoresis and peptide map (30)		
		Lung (embryo)		Gel electrophoresis and peptide map (30) EM with HMM (72, 87)		
		Neuron (embryo)		Gel electrophoresis and peptide map (43)		

*This table lists reports of the identification of actin, myosin, and associated proteins in nonmuscle cells. In each case the method of identification is given using the following abbreviations: "EM with HMM" is electron microscopy of the complex between the presumed actin and bona fide muscle myosin heavy meromyosin; "fluorescent HMM" is the staining of cells with muscle myosin heavy meromyosin conjugated with fluorescein isothiocyanate. The criteria for the identification of "actomyosin" are less well defined than those for identification of actin and myosin, therefore the compiler of this table advises some skepticism regarding the identification of actomyosin, unless actin and/or myosin has been shown to be present. Some caution may also be necessary in accepting some of the identifications by immunological criteria alone. A review of the status of this field of investigation can be found in Reference 127.

[†] References follow Table 8.

Table 1 (continued)
IDENTIFICATION OF ACTIN, MYOSIN, AND ASSOCIATED PROTEINS IN NONMUSCLE CELLS

No.	Species	Cell type	Actomyosin	Actin	Myosin	Associated proteins
			Vertebrate (continued)			
	Chicken (continued)	Oviduct (embryo)		EM with HMM (142)		
		Pancreas (embryo)		Gel electrophoresis and peptide map (30)		
		Skin (embryo)		Gel electrophoresis and peptide map (30)		
		Stomach (embryo)		Gel electrophoresis and peptide map (30)		
		Trachea (embryo)		EM with HMM (72)		
3	Cow	Brain	Isolation (130, 132)	Isolation (130, 132)	Isolation (24)	Tropomyosin (24)
		Platelet	Isolation (112)	Isolation (143)	EM (173)	
		Thymus nuclei				
4	Dogfish	Brain		Gel electrophoresis and peptide map (30)	Isolation (126)	
		Liver		Gel electrophoresis and peptide map (30)		
5	Electric eel	Electric organ				Tropomyosin Isolation (78)
6	Guinea pig	Granulocyte		EM with myosin (147)	Isolation (147)	
		Lymphocyte		EM with HMM (10)	Isolation (126)	
		Macrophage				
7	Horse	Leukocyte	Isolation (139, 141)	Isolation (149)	Isolation (149)	
8	Human	Endothelium	Antibody staining (13–15)	Antibody staining (48)	Antibody staining (160)	
		Epithelial cell line (Bo-mat)				
		Erythrocyte	Isolation (74)	Isolation (140, 154)		Spectrin Isolation (29, 33, 91, 92, 154)
		Fibroblast cell line (HeLa)		EM with HMM (138) Gel electrophoresis and peptide map (53)		

Table 1 (continued)

IDENTIFICATION OF ACTIN, MYOSIN, AND ASSOCIATED PROTEINS IN NONMUSCLE CELLS

No.	Species	Cell type	Actomyosin	Actin	Myosin	Associated proteins
		Vertebrate (continued)				
	Human (continued)	Granulation tissue fibroblast	Antibody staining (15)			
		Granulocyte	Isolation (139, 141)	Isolation and EM with HMM (28)		
		Lung (fetus)	Antibody staining (41)			
		Megakaryocyte	Antibody staining (41)			
		Platelet	Isolation (20, 21, 39, 89, 90, 172)	Isolation (1, 21, 22, 25, 128)	Isolation (5, 6, 8, 21, 22, 26, 119, 126, 129)	Tropomyosin Isolation (35)
			Antibody staining (15, 27, 101)	EM with HMM (17, 23, 173)	EM (17, 173)	Calcium sensitive control proteins (36) ATPase inhibitor (151)
				Antibody staining (48)		
		Renal mesangial cell	Antibody staining (13)			
9	Mouse	Brain		Gel electrophoresis and peptide map (30) EM with HMM (83)		
		Ehrlich ascites			Isolation (7, 113)	
		Fibroblast cell line (L-929)		EM with HMM (93)		
		Fibroblast cell line (3T3)		Gel electrophoresis and peptide map (53)	Isolation (113)	
		Lung (embryo)		EM with HMM (142)	Antibody staining (160)	
		Neuroblastoma cell line		EM with HMM (31, 32)		
		Pancreas		Gel electrophoresis and peptide map (30) EM with HMM (142)		
		Salivary epithelium		Antibody staining (40)		
		Thymocyte				
10	Newt	Oocyte		EM with HMM (116)		

309

Table 1 (continued)
IDENTIFICATION OF ACTIN, MYOSIN, AND ASSOCIATED PROTEINS IN NONMUSCLE CELLS

No.	Species	Cell type	Actomyosin	Actin	Myosin	Associated proteins
		Vertebrate (continued)				
11	Pig	Platelet	Isolation (51)	Isolation (143)		
12	Rabbit	Capillary endothelium		Antibody staining (48)		
		Epidermis		Antibody staining (48)		
		Granulation tissue fibroblast		Antibody staining (48)		
		Hepatocyte		Isolation (29)	Isolation (29)	
		Macrophage	Isolation (54)	Isolation (54)	Isolation (54, 145)	Actin binding protein Isolation (54, 146) Cofactor Isolation (145, 146)
		Renal mesangial cell		Antibody staining (48)		
		Renal tubule epithelium		Antibody staining (48)		
13	Rat	Brain	Isolation (18, 19, 130–132)	Isolation (130)	Isolation (19, 130)	
		Epidermis		Antibody staining (48)		
		Fibroblast cell line (NRF)			Isolation (113)	
		Granulation tissue fibroblast		Antibody staining (48, 68)		
		Hepatocyte		Antibody staining (41, 48)		
		Intestinal epithelium		Antibody staining (48)		
		Lymphocyte		Antibody staining (48)		
		Megakaryocyte		Antibody staining (48)		
		Platelet		Antibody staining (48)		
		Polymorphonuclear leukocyte		Antibody staining (48)		
		Renal tubule epithelium		EM with HMM (133)		
		Sarcoma	Isolation (69)			
14	Torpedo	Electric organ	Isolation (79)			Tropomyosin Isolation (78)

Table 1 (continued)

IDENTIFICATION OF ACTIN, MYOSIN, AND ASSOCIATED PROTEINS IN NONMUSCLE CELLS

No.	Species	Cell type	Actomyosin	Actin	Myosin	Associated proteins
			Invertebrate			
1	Crane fly	Spermatocyte		EM with HMM (16, 45, 46)		
		Spermatozoa		EM with HMM (16, 44)		
2	Horseshoe crab	Spermatozoa		Isolation (152, 153a, 154)		α-Actinin Isolation and antibody staining (137, 154) Actin binding protein Isolation (152, 154)
3	Locust	Testis		EM with HMM (50)		
4	Scallop	Gill		Gel electrophoresis and peptide map (30)		
		Ovary		Gel electrophoresis and peptide map (30)		
		Testis		Gel electrophoresis and peptide map (30)		
5	Sea cucumber	Spermatozoa		Isolation and EM with HMM (156)		
6	Sea urchin	Oocyte		Isolation (56, 80, 94–96)	Isolation (88)	Actin binding protein Isolation (153, 154)
		Spermatozoa		EM with HMM (73)		
7	Starfish	Spermatozoa		Isolation and EM with HMM (156)		

Table 1 (continued)
IDENTIFICATION OF ACTIN, MYOSIN, AND ASSOCIATED PROTEINS IN NONMUSCLE CELLS

No.	Species	Cell type	Actomyosin	Actin	Myosin	Associated proteins
			Fungus			
1	*Dictyostelium discoideum*	Amoeba	Isolation (166)	Isolation (144, 167)	Isolation (34)	
2	*Physarum polycephalum*	Plasmodium	Isolation (3, 11, 12, 57, 61, 64, 81, 86, 109, 110, 159)	Isolation (2, 4, 47, 55, 58, 59, 62, 63, 157, 158) EM with HMM (9, 108) EM of paracrystals (65)	Isolation (2a, 4, 57, 60, 61, 66, 81, 103–106)	Calcium mediated control proteins Isolation (81, 107, 148)
		Nucleus		Isolation (75, 76, 85)	EM (77) Isolation (85)	
			Plant			
1	*Amaryllis belladona*	Pollen tube		EM with HMM (38)		
2	*Chara corallina*	Internode cell		EM with HMM (164)		
3	*Hydrilla*	Leaf	Isolation (170, 171)			
4	*Nitella*	Internode cell	Isolation (170, 171)	EM with HMM (82, 114, 115)		
5	Pumpkin	Leaf				
6	*Vicia faba*			EM with HMM (71)		
7	*Xylosma congestum*			EM with HMM (71)		

Table 1 (continued)

IDENTIFICATION OF ACTIN, MYOSIN, AND ASSOCIATED PROTEINS IN NONMUSCLE CELLS

No.	Species	Cell type	Actomyosin	Actin	Myosin	Associated proteins
		Protozoa				
1	*Acanthamoeba castellanii*		Isolation (118)	Isolation (118, 161–163) EM with HMM (125)	Isolation (119, 122–124)	Cofactor Isolation (122, 124)
2	*Amoeba proteus*			Isolation (99, 100) EM with HMM (121)	EM (67, 70, 117, 120) Isolation (67)	
3	*Chaos carolinensis*			EM with HMM (37)	EM (97, 102, 150)	
4	*Naegleria gruberii*		Isolation (84)			

Compiled by Thomas D. Pollard.

Table 2
PHYSICAL CHEMICAL PROPERTIES OF NONMUSCLE ACTINS

Cell type	Monomer molecular weight (daltons)	Reduced viscosity of polymers 0.1 M KCl (dl/g)	Reduced viscosity of polymers 0.1 M KCl + 1–2 mM Mg^{++} (dl/g)	Bound nucleotide	Bound divalent cation	N-Methylhistidine content	References*
				(Moles/42,000 g protein)			
Vertebrate							
Brain, cat						0.39	130
Brain, cow				0.91		0.97	132
Brain, rat							
Brain, chick embryo	42,000						30, 83a
Intestinal epithelium, chicken	42,000						155
Erythrocyte, human	42,000						140, 154
Fibroblast, chick embryo	42,000						168
Fibroblast, human (HeLa)	42,000					1	53
Fibroblast, mouse (3T3)	42,000					1	53
Hepatocyte, rabbit	42,000						29
Macrophage, rabbit	42,000	2.3		0.8			54
Neuron, chick embryo	42,000						43
Platelet, cow	42,000						143
human	42,000		3.1			0.88	25
pig	42,000		12	Present			128
	42,000						143
Invertebrate							
Oocyte, sea urchin		2.1					56
		0.8	0.27				94
		1.8	0.1 (Mg^{++} only)				94
	42,000						80

*References follow Table 8.

Table 2 (continued)
PHYSICAL CHEMICAL PROPERTIES OF NONMUSCLE ACTINS

Cell type	Monomer molecular weight (daltons)	Reduced viscosity of polymers (dl/g)		Bound nucleotide	Bound divalent cation	N-Methylhistidine content	References[*]
		0.1 M KCl	0.1 M KCl + 1–2 mM Mg^{++}		(Moles/42,000 g protein)		
Fungus							
Dictyostelium discoideum	48,000	3.5	2.7			0.92	167
	42,000						144
Physarum polycephalum	57,000	5.6	0.56	0.85	1		59, 62, 158
	42,000	3.6	3.4	0.76			2, 4
Protozoa							
Acanthamoeba castellanii	42,000	3.9				0.87	162, 163

Compiled by Thomas D. Pollard.

Table 3
INTERACTION OF NONMUSCLE ACTINS WITH MYOSIN

Myosin Mg^{++} ATPase (μmol/min·mg)

Cell type	Source of myosin	No. actin	Plus nonmuscle actin	Activation factor	Plus muscle actin	Activation factor	Actin concentration (mg/ml)	Ionic strength (M)	Temperature (°C)	References*
Vertebrate										
Brain	Rabbit myosin	0.02	0.10	5x			0.05	0.08	37	130
Fibroblast, chicken	Rabbit myosin	0.008	0.067	8x	0.072	9x	0.023	0.07	37	168
Hepatocyte, rabbit	Rabbit myosin	0.01	0.062	6x			0.53	0.06	37	29
Leucocyte, horse	Horse leukocyte	0.008	0.019	2.4x			0.21	0.09	25	149
Macrophage, rabbit	Rabbit myosin	0.01	0.41	40x			0.5	0.05	37	54, 146
Platelet, human	Rabbit myosin	0.09	0.43	5x			0.2–0.4	0.08	37	128
	Rabbit HMM	0.10	1.49	15x			0.2	0.03	37	5
	Rabbit myosin	0.02	0.028	1.4x	0.076	3.8x	0.11	0.07	22	25
	Rabbit myosin	0.02	0.034	1.7x	0.13	6.5x	0.22	0.07	22	25
Fungus										
Dictyostelium discoideum	Rabbit myosin	0.015	0.078	5x	0.35	23x	0.05	0.07	25	167
	Rabbit HMM		0.04		0.10		0.16	0.04	25	144
Physarum polycephalum	Rabbit myosin	0.019	0.28	14x			0.16	0.075	28	58
		0.013	0.11	9x			0.16	0.125	28	58
		0.006	0.013	2x			0.16	0.625	28	58
	Rabbit myosin	0.015	0.185	12x			0.14		24	4
		0.011	0.046	4x			0.14		24	4
Protozoa										
Acanthamoeba castellanii	Rabbit	0.03	0.20	7x			0.2	0.03	25	162
		0.03	0.14	5x	0.65	21.6x	0.2	0.03	25	162
		0.03	0.21	7x			0.4	0.03	25	162

Compiled by Thomas D. Pollard.

*References follow Table 8.

Table 4
PHYSICAL PROPERTIES OF NONMUSCLE MYOSINS

Cell type	Native molecular weight	Subunit composition (number × mol wt)	Stokes radius (nm)	Sedimentation coefficient (S)	References*
Vertebrate					
Brain, rat	—	? × 240,000	—	—	19
Chondrocyte	—	? × 200,000	—	—	135
Fibroblast, mouse L929	—	? × 200,000 ? × 20,000	19	—	7
Granulocyte, guinea pig	—	? × 200,000 ? × 20,000 ? × 16,000	19	—	147
Hepatocyte, rabbit	—	? × 215,000	19	—	29
Macrophage, rabbit	—	2 × 200,000 2 × 20,000 2 × 15,000	19	—	54
Platelet, human	460,000	2 × 200,000 2 × 19,000 2 × 16,000	19	6.2	5, 8, 126
	540,000	—	—	6.8	26
Invertebrate					
Oocyte, sea urchin	—	? × 200,000	—	6.3	88
Fungus					
Dictyostelium discoideum	—	2 × 210,000 ? × 18,000 ? × 16,000	—	—	34
Physarum polycephalum	458,000	2 × 240,000 4 × 21,000 2 × 17,000	17	6.4	4, 105, 106
	—	? × 220,000	—	—	66
Protozoa					
Acanthamoeba castellanii	180,000	1 × 140,000 1 × 16,000 1 × 14,000	5.5	8	118, 123
Amoeba proteus	—	? × 200,000	—	—	67

Compiled by Thomas D. Pollard.

*References follow Table 8.

Table 5
FILAMENTS OF NONMUSCLE MYOSINS

Cell type	KCl	pH	Me^{++}	Length (nm)	Width (nm)	References*
Vertebrate						
Fibroblast, mouse L-929	0.1 M	7.0	–	300	10	7
Granulocyte, guinea pig	0.1 M	7.0	1 mM Ca^{++} or Mg^{++}	300	10	147
Granulocyte, human	0.06 M	7.0	–	300	15	
Hepatocyte, rabbit	0.1 M	7.5	1 mM Mg^{++}	350	11	29
Macrophage, rabbit	0.1 M	7.0	1 mM EDTA	300	11	54
Platelet, bovine	0.1 M	6.5	–	300	10–12	173
Platelet, human	0.1 M	7.0	5 mM Mg^{++}	200–500	8–18	17
	0.1 M	7.0	–	325	10.7	118
Invertebrate						
Oocyte, sea urchin	0.2 M	7.7	–	370	17	88
Fungus						
Dictyostelium discoideum	0.03 M	7.5	10 mM Mg^{++}	600–800	–	34
Physarum polycephalum	0.03 M	6.8	6 mM Mg	400–500	13–25	64
	0.05 M	7	1 mM Ca^{++} or 10 mM Mg^{++}	450	25	103
	0.05 M	7.0	1 mM EGTA	No filaments		106
	0.05 M	7.0	5 mM Mg^{++} ± 0.1 mM Ca^{++}	1,000–2,500	17–28	106
	0.05 M	6.9	0.5 mM ATP 2 mM Mg^{++}	300–2,000	30	66
Protozoa						
Acanthamoeba castellanii	0.05 M	7.0	± 1 mM Ca^{++} or Mg^{++}	No filaments		123
Amoeba proteus	0.10 M	7	–	400–700	20–30	67

Compiled by Thomas D. Pollard.

*References follow Table 8.

Table 6
ENZYMATIC ACTIVITY OF NONMUSCLE MYOSINS

Cell type	ATPase activity (μmol Pi/min·mg) K$^+$EDTA	Ca^{++}	Mg^{++}	Temperature (°C)	Activation energy (Kcal/mol)	Substrate specificity	References[*]
Vertebrate							
Brain, rat	—	0.27	0.03	37	—	—	19
Chondrocyte, chick embryo	0.04	0.18	0.02	25	—	—	135
Fibroblast, mouse L-929	0.43	0.50	0.01	37	—	—	7
Fibroblast, mouse 3T3	0.16	—	—	37	—	—	113
Fibroblast, rat NRK (normal rat kidney)	0.14	—	—	37	—	—	113
Granulocyte, guinea pig	0.12	—	—	37	—	—	113
Hepatocyte, rabbit (60 mM KIc)	0.19	0.21	0.02	25	—	—	147
Macrophage, rabbit	0.14	0.09	0.04	37	—	—	29
Platelet, human	0.56	0.57	0.05	37	—	—	54
	0.55	0.44	0.02	37	—	—	8
Invertebrate							
Oocyte, sea urchin	—	0.018	0.008	25	—	—	88
Fungus							
Dictyostelium discoideum	0.02	0.10	0.01	25	—	—	34
Physarum polycephalum	0.03	0.87	0.03	20	12	ATP > ITP > GTP	4
	0.01	0.54	0.03	25	—	—	57
		1.2					106
		2.0		25			2a
Protozoa							
Acanthamoeba castellanii	3.5	0.4	0.05	29	9	ATP > GTP, CTP, ITP	123

Compiled by Thomas D. Pollard.

[*]References follow Table 8.

Table 7
INTERACTION OF NONMUSCLE MYOSINS WITH ACTIN

Cell type	Actin binding −ATP	Actin binding +ATP	Mg^{++} ATPase (μmol P$_i$/min·mg myosin) −Actin	+Actin	+Actin +cofactor	Actin* concentration (mg/ml)	Ionic strength (M)	Temperature (°C)	References[†]
Vertebrate									
Brain, rat	—	—	0.037	0.11	—	—	0.096	37	19
Fibroblast, mouse L-929	Yes	No	0.01	0.09	—	0.3	0.035	37	7
Granulocyte, guinea pig	Yes	No	0.03	0.08	—	0.7	0.100	25	147
Hepatocyte, rabbit	—	—	0.01	0.01	—	—	0.08	37	29
Macrophage, rabbit	Yes	No	0.01	0.01	0.40	1.5	0.05	37	146
Platelet, human (head fragment)	Yes	No	0.005	0.025	0.35	0.4	0.07	37	118, 126
	Yes	No	0.02	0.07	—	0.8	0.039	37	5
Invertebrate									
Oocyte, sea urchin	Yes	No	—	—	—	—	—	—	88
Fungus									
Dictyostelium discoideum	Yes	No	0.01	0.077	—	0.09	0.035	25	34
Physarum polycephalum	Yes	No	0.027	0.057	—	—	0.046	25	57
	—	—	0.03	0.16	—	2.5	0.045	23	106
				1.0	—	Extrapolated to infinite actin	0.025	24	2a
Protozoa									
Acanthamoeba castellanii	Yes	No	0.05	0.08	1.5	1.0	0.016	29	124

Compiled by Thomas D. Pollard.

*Rabbit muscle actin.
[†]References follow Table 8.

Table 8
PROPERTIES OF TROPOMYOSIN FROM NONMUSCLE CELLS

Cell type	Subunit molecular w (daltons)	α-Helix content (%)	Paracrystal period (nm)	Functional hybrid with muscle troponin	References
Vertebrate					
Brain, chicken embryo	30,000	~90	34	Yes	42
Brain, cow	30,000	–	–	–	24
Electric organ, electric eel	35,000	>90	40	–	78
Electric organ, Torpedo	35,000	>90	40	–	78
Fibroblast, chicken embryo	35,000	–	80	Yes	169
Platelet, human	30,000	~90	34	–	35

REFERENCES

1. Abramowitz, Stracher, and Detwiler, *Fed. Proc.*, 33, 1522 (1974).
2. Adelman, *Fed. Proc.*, 33, 1522 (1974).
2a. Adelman, *Biophys. J.*, 15, 161a (1975).
3. Adelman and Taylor, *Biochemistry*, 8, 4964 (1969a).
4. Adelman and Taylor, *Biochemistry*, 8, 4976 (1969b).
5. Adelstein and Conti, *Cold Spring Harbor Symp. Quant. Biol.*, 37, 599 (1973).
6. Adelstein, Conti, and Anderson, *Proc. Natl. Acad. Sci. U.S.A.*, 70, 3115 (1973).
7. Adelstein, Conti, Johnson, Pastan, and Pollard, *Proc. Natl. Acad. Sci. U.S.A.*, 69, 3693 (1972).
8. Adelstein, Pollard, and Kuehl, *Proc. Natl. Acad. Sci. U.S.A.*, 68, 2703 (1971).
9. Allera, Beck, and Wohlfarth-Botterman, *Cytobiologie*, 3, 437 (1971).
10. Allison, Davies, and dePetris, *Nat. New Biol.*, 232, 153 (1971).
11. Beck, Hinssen, Komnick, Stockem, and Wohlfarth-Botterman, *Cytobiologie*, 2, 259 (1970).
12. Beck, Komnick, Stockem, and Wohlfarth-Bottermann, *Cytobiologie*, 1, 99 (1969).
13. Becker, *Am. J. Pathol.*, 66, 97 (1972).
14. Becker and Murphy, *Am. J. Pathol.*, 55, 1 (1969).
15. Becker and Nachman, *Am. J. Pathol.*, 71, 1 (1973).
16. Behnke, Forer, and Emmerson, *Nature*, 234, 408 (1971).
17. Behnke, Kristensen, and Nielsen, *J. Ultrastruct. Res.*, 37, 361 (1971).
18. Berl and Puszkin, *Biochemistry*, 9, 2058 (1970).
19. Berl, Puszkin, and Nicklas, *Science*, 179, 441 (1973).
20. Bettex-Galland and Luscher, *Nature*, 184, 276 (1959).
21. Bettex-Galland and Luscher, *Adv. Protein Chem.*, 20, 1 (1965).
22. Bettex-Galland, Portzehl, and Luscher, *Nature*, 193, 777 (1962).
23. Bettex-Galland, Probst, and Behnke, *J. Mol. Biol.*, 68, 533 (1972).
24. Blitz and Fine, *Proc. Natl. Acad. Sci. U.S.A.*, 71, 4472 (1974).
25. Booyse, Hoveke, and Rafelson, *J. Biol. Chem.*, 248, 4083 (1973).
26. Booyse, Hoveke, Zschocke, and Rafelson, *J. Biol. Chem.*, 246, 4291 (1971).
27. Booyse, Sternberger, Zschocke, and Rafelson, *J. Histochem. Cytochem.*, 19, 540 (1971).
28. Boxer, Hedley-White and Stossel, *N. Engl. J. Med.*, 291, 1093 (1974).
29. Brandon, thesis, Harvard University, Cambridge, 1974.
30. Bray, *Cold Spring Harbor Symp. Quant. Biol.*, 37, 567 (1973).
31. Burton and Kirkland, *Nat. New Biol.*, 239, 244 (1972).
32. Chang and Goldman, *J. Cell Biol.*, 57, 867 (1973).
33. Clarke, *Biochem. Biophys. Res. Commun.*, 45, 1063 (1971).
34. Clarke and Spudich, *J. Mol. Biol.*, 86, 209 (1974).
35. Cohen and Cohen, *J. Mol. Biol.*, 68, 383 (1972).
36. Cohen, Kaminski, and de Vries, *FEBS Lett.*, 34, 315 (1973).
37. Comly, *J. Cell Biol.*, 58, 230 (1973).
38. Condeelis, *Exp. Cell Res.*, 88, 435 (1974).
39. Crawford, *Br. J. Haematol.*, 21, 53 (1971).
40. Fagraeus, The, and Biberfield, *Nat. New Biol.*, 246, 113 (1973).
41. Farrow, Holborow, and Brighton, *Nat. New Biol.*, 232, 186 (1973).

42. Fine, Blitz, Hitcock, and Kaminer, *Nat. New Biol.*, 245, 182 (1973).
43. Fine and Bray, *Nat. New Biol.*, 234, 115 (1971).
44. Forer and Behnke, *J. Cell Sci.*, 11, 491 (1972).
45. Forer and Behnke, *Chromosoma*, 39, 145 (1972).
46. Forer and Behnke, *Chromosoma*, 39, 175 (1972).
47. Fujime and Hatano, *J. Mechanochem. Cell Motility*, 1, 81 (1972).
48. Gabbiani, Ryan, Lamelin, Vassalli, Majno, Bouvier, Cruchaud, and Luscher, *Am. J. Pathol.*, 72, 473 (1974).
49. Garnett, Groschel-Stewart, Jones, and Kemp, *Cytobios*, 7, 163 (1973).
50. Gawadi, *Nature*, 234, 410 (1971).
51. Grette, *Acta Physiol. Scand. Suppl.*, 195, 46 (1962).
52. Groschel-Stewart, Jones, and Kemp, *Nature*, 227, 280 (1970).
53. Gruenstein, Rich, and Weihing, *J. Cell Biol.*, 64, 223 (1975).
54. Hartwig and Stossel, *J. Biol. Chem.*, 250, 5696 (1975).
55. Hatano, *J. Mechanochem. Cell Motility*, 1, 75 (1972).
56. Hatano, Kondo, and Miki-Noumura, *Exp. Cell Res.*, 55, 275 (1969).
57. Hatano and Ohnuma, *Biochim. Biophys. Acta*, 205, 110 (1970).
58. Hatano and Oosawa, *J. Cell Physiol.*, 68, 197 (1966).
59. Hatano and Oosawa, *Biochim. Biophys. Acta*, 127, 488 (1966).
60. Hatano and Takahashi, *J. Mechanochem. Cell Motility*, 1, 7 (1971).
61. Hatano and Tazawa, *Biochim. Biophys. Acta*, 154, 507 (1968).
62. Hatano and Totsuka, *J. Mechanochem. Cell Motility*, 1, 67 (1972).
63. Hatano, Totsuka, and Oosawa, *Biochim. Biophys. Acta*, 140, 109 (1967).
64. Hinssen, *Cytobiologie*, 2, 326 (1970).
65. Hinssen, *Cytobiologie*, 5, 146 (1972).
66. Hinssen and D'Haese, *J. Cell Sci.*, 15, 113 (1974).
67. Hinssen and D'Haese, *Cell Tissue Res.*, 151, 323 (1974).
68. Hirschel, Gabbiani, Ryan, and Majno, *Proc. Soc. Exp. Biol. Med.*, 138, 466 (1971).
69. Hoffmann-Berling, *Biochim. Biophys. Acta*, 19, 453 (1956).
70. Holberton and Preston, *Exp. Cell Res.*, 62, 473 (1970).
71. Ilker and Currier, *Planta*, 120, 311 (1974).
72. Ishikawa, Bischoff, and Holtzer, *J. Cell Biol.*, 43, 312 (1969).
73. Jessen, Behnke, Wingstrand, and Rostgaard, *Exp. Cell Res.*, 80, 47 (1973).
74. Jirgl, *Folia Biol.* (Prague), 17, 392 (1971).
75. Jockusch, Becker, Hindennach, and Jockusch, *Exp. Cell Res.*, 89, 241 (1974).
76. Jockusch, Brown, and Rusch, *J. Bacteriol.*, 108, 705 (1971).
77. Jockusch, Ryser, and Behnke, *Exp. Cell Res.*, 76, 464 (1973).
78. Kaminer and Szonyi, *J. Cell Biol.*, 55, 129a (1972).
79. Kaminer and Szonyi, *Biol. Bull.*, 145, 441 (1973).
80. Kane, *J. Cell Biol.*, 63, 161a (1974).
81. Kato and Tonomura, *J. Biochem.*, in press.
82. Kersey, *J. Cell Biol.*, 63, 165a (1974).
83. Kristensen, Simonsen, and Pape, *Virchows Arch. Abt. B Zellpathol.*, 13, 103 (1973).
83a. Kuczmarski and Rosenbaum, *J. Cell Biol.*, 63, 178a (1974).
84. Lastovica and Dingle, *Exp. Cell Res.*, 66, 337 (1971).
85. LeStourgeon, *J. Cell Biol.*, 63, 191a (1974).
86. Loewy, *J. Cell Comp. Physiol.*, 40, 127 (1952).
87. Luduena and Wessells, *Dev. Biol.*, 30, 427 (1973).
88. Mabuchi, *J. Cell Biol.*, 59, 542 (1973).
89. Malik, Abramowitz, Detwiler, and Stracher, *Arch. Biochem. Biophys.*, 161, 268 (1974).
90. Malik, Detwiler, and Stracher, *Biochem. Biophys. Res. Commun.*, 55, 912 (1973).
91. Marchesi and Steers, *Science*, 159, 203 (1968).
92. Marchesi, Steers, Marchesi, and Tillack, *Biochemistry*, 9, 50 (1970).
93. McNutt, Culp, and Black, *J. Cell Biol.*, 56, 412 (1973).
94. Miki-Noumura, *Dev. Growth Differ.*, 11, 219 (1969).
95. Miki-Noumura and Kondo, *Exp. Cell Res.*, 61, 31 (1970).
96. Miki-Noumura and Oosawa, *Exp. Cell Res.*, 56, 224 (1969).
97. Moore, Condeelis, Taylor, and Allen, *Exp. Cell Res.*, 80, 49 (1974).
98. Mooseker, *J. Cell Biol.*, 63, 231a (1974).
99. Morgan, *Exp. Cell Res.*, 65, 7 (1971).
100. Morgan, Fyfe, and Wolpert, *Exp. Cell Res.*, 48, 194 (1967).
101. Nachman, Marcus, and Safier, *J. Clin. Invest.*, 46, 1380 (1967).

102. Nachmias, *J. Cell Biol.*, 38, 40 (1968).
103. Nachmias, *Proc. Natl. Acad. Sci. U.S.A.*, 69, 2011 (1972).
104. Nachmias, *J. Cell Biol.*, 52, 648 (1972).
105. Nachmias, *Cold Spring Harbor Symp. Quant. Biol.*, 37, 607 (1973).
106. Nachmias, *J. Cell Biol.*, 62, 54 (1974).
107. Nachmias and Asch, *Biochem. Biophys. Res. Commun.*, 60, 656 (1974).
108. Nachmias, Huxley, and Kessler, *J. Mol. Biol.*, 50, 83 (1970).
109. Nachmias and Ingram, *Science,* 170, 743 (1970).
110. Nakajima, *Protoplasma,* 52, 412 (1960).
111. Ohnishi, Kawamura, Takeo, and Watanabe, *J. Biochem.*, 56, 273 (1964).
112. Ohnishi, Kawamura, and Tanaka, *J. Biochem.*, 56, 6 (1964).
113. Ostlund, Pastan, and Adelstein, *J. Biol. Chem.*, 249, 3903 (1974).
114. Palevitz, Ash, and Hepler, *Proc. Natl. Acad. Sci. U.S.A.*, 71, 363 (1973).
115. Palevitz and Hepler, *J. Cell Biol.*, 63, 257a (1974).
116. Perry, John, and Thomas, *Exp. Cell Res.*, 65, 249 (1971).
117. Pollard, *Acta Protozool.*, 11, 55 (1972).
118. Pollard, in *Molecules and Cell Movement,* Inoue and Stephens, Eds., Raven Press, New York, 1975, 259.
119. Pollard, Adelstein, and Korn, *Acta Protozool.*, 11, 59 (1972).
120. Pollard and Ito, *J. Cell Biol.*, 46, 267 (1970).
121. Pollard and Korn, *J. Cell Biol.*, 48, 216 (1971).
122. Pollard and Korn, *Cold Spring Harbor Symp. Quant. Biol.*, 37, 573 (1973).
123. Pollard and Korn, *J. Biol. Chem.*, 248, 4682 (1973).
124. Pollard and Korn, *J. Biol. Chem.*, 248, 4691 (1973).
125. Pollard, Shelton, Weihing, and Korn, *J. Mol. Biol.*, 50, 91 (1970a).
126. Pollard, Thomas, and Niederman, *Anal. Biochem.*, 60, 258 (1974).
127. Pollard and Weihing, *CRC Crit. Rev. Biochem.*, 2, 1 (1974).
128. Probst and Luscher, *Biochim. Biophys. Acta,* 278, 577 (1972).
129. Puszkin, Puszkin, Lo, and Tanenbaum, *J. Biol. Chem.*, 248, 7754 (1973).
130. Puszkin and Berl, *Biochim. Biophys. Acta,* 256, 695 (1972).
131. Puszkin, Berl, Puszkin, and Clarke, *Science,* 161, 120 (1968).
132. Puszkin, Nicklas, and Berl, *J. Neurochem.*, 19, 1319 (1972).
133. Rostgaard, Kristensen, and Nielsen, *Z. Zellforsch. Mikrosk. Anat.*, 132, 497 (1972).
134. Rostgaard, Kristensen, and Nielsen, *J. Ultrastruct. Res.*, 38, 207 (1972).
135. Rubinstein, Chi, and Holtzer, *Biochem. Biophys. Res. Commun.*, 57, 438 (1974).
136. Sanger, *J. Cell Biol.*, 63, 297a (1974).
137. Schollenmeyer, Goll, Tilney, Mooseker, Robson, and Stromer, *J. Cell Biol.*, 63a, 304 (1974).
138. Schroeder, *Proc. Natl. Acad. Sci. U.S.A.*, 70, 1688 (1973).
139. Senda, Shibata, Tatsumi, Kondo, and Hamada, *Biochim. Biophys. Acta,* 181, 191 (1969).
140. Sheetz, Painter, and Singer, unpublished results.
141. Shibata, Tatsumi, Tanaka, Okamura, and Senda, *Biochim. Biophys. Acta,* 256, 565 (1972).
142. Spooner, Ash, Wrenn, Frater, and Wessells, *Tissue Cell,* 5, 37 (1973).
143. Spudich, *Cold Spring Harbor Symp. Quant. Biol.*, 37, 585 (1973).
144. Spudich, *J. Biol. Chem.*, 249, 6013 (1974).
145. Stossel and Hartwig, *Fed. Proc.*, 33, 1581 (1974).
146. Stossel and Hartwig, *J. Biol. Chem.*, 250, 5706 (1975).
147. Stossel and Pollard, *J. Biol. Chem.*, 248, 8288 (1973).
148. Tanaka and Hatano, *Biochim. Biophys. Acta,* 257, 445 (1972).
149. Tatsumi, Shibata, Okamura, Takeuchi, and Senda, *Biochim. Biophys. Acta,* 305, 433 (1973).
150. Taylor, Condeelis, Moore, and Allen, *J. Cell Biol.*, 59, 378 (1973).
151. Thorens, Schaub, and Luscher, *Experientia,* 29, 349 (1973).
152. Tilney, *J. Cell Biol.*, 59, 346A (1973).
153. Tilney, *J. Cell Biol.*, 63, 349a (1974).
153a. Tilney, *J. Cell Biol.*, 64, 289 (1975).
154. Tilney, in *Molecules and Cell Movement,* Inoue and Stephens Eds., Raven Press, New York, 1975, 339.
155. Tilney and Mooseker, *Proc. Natl. Acad. Sci. U.S.A.*, 68, 2611 (1971).
156. Tilney, Hatano, Ishikawa, and Mooseker, *J. Cell Biol.*, 59, 109 (1973).
157. Totsuka, *Biochim. Biophys. Acta,* 234, 162 (1971).
158. Totsuka and Hatano, *Biochim. Biophys. Acta,* 223, 189 (1970).
159. Ts'o, Eggman, and Vinograd, *Biochim. Biophys. Acta,* 25, 532 (1957).
160. Weber and Groeschel-Stewart, *Proc. Natl. Acad. Sci. U.S.A.*, 71, 4561 (1974).
161. Weihing and Korn, *Biochem. Biophys. Res. Commun.*, 35, 906 (1969).

162. Weihing and Korn, *Biochemistry,* 10, 590 (1971).
163. Weihing and Korn, *Biochemistry,* 11, 1538 (1972).
164. Williamson, *Nature,* 248, 801 (1974).
165. Willingham, Ostlund, and Pastan, *Proc. Natl. Acad. Sci. U.S.A.,* 71, 4144 (1974).
166. Woolley, *J. Cell Physiol.,* 76, 185 (1970).
167. Woolley, *Arch. Biochem., Biophys.,* 150, 519 (1972).
168. Yang and Perdue, *J. Biol. Chem.,* 247, 4503 (1972).
169. Yang and Perdue, *J. Cell Biol.,* 63, 382a (1974).
170. Yen, Han, and Shih, *Kexue Tongboo,* 17, 138 (1966).
171. Yen and Shih, *Sci. Sin.,* 14, 490 (1965).
172. Zucker-Franklin, Nachman, and Marcus, *Science,* 157, 945 (1967).
173. Zucker-Franklin and Grusky, *J. Clin. Invest.,* 51, 419 (1972).

SUBUNIT CONSTITUTION OF PROTEINS

Dennis W. Darnall and Irving M. Klotz

The wide response from readers to our previously published protein subunit tables[1-4] indicates the usefulness of such compilations for teaching and research purposes. We have therefore prepared a new updated listing of proteins with subunits held together by noncovalent bonds (Table 1). Individual polypeptide chains held together by disulfide bridges have not been individually classified as subunits; insulin, for example, is listed as having a subunit molecular weight of 5733 even though this is the combined weight of the disulfide-linked A and B chains. Kleine[4a] has compiled a list of proteins whose subunits are associated through both noncovalent and covalent (disulfide) bonds.

In many instances, the subunit listed (Table 1) may not be the minimal subunit obtainable, but instead the minimal subunit that has been unequivocally obtained under conditions that eliminate cleavage of peptide or disulfide bonds. For some proteins, two or more stages of dissociation can be clearly recognized; in such instances two or more entries specifying the relations between the different aggregates are given. Parentheses around molecular weights or subunit numbers indicate uncertainty in the value.

The most accessible references are given for each entry; they do not necessarily indicate the source most deserving of credit for establishing the subunit stoichiometry. These sources are mentioned in the cited works.

Since Table 1 contains well over 500 entries, it seemed appropriate to make an alphabetical listing of the entries in Table 1. This listing is found in Table 2.

Table 1
SUBUNIT CONSTITUTION OF PROTEINS

Protein	Source	Organ	Molecular weight	Subunits No.	Subunits Molecular weight	References
Insulin	Bovine		11,466	2	5,733	5
S-100 protein	Escherichia coli	Brain	19,500	(4—3)	(4,100—7,000)	6
Mercaptopyruvate sulfur transferase (EC 2.8.1.2)	Escherichia coli		23,800	2	12,000	7
$\Delta_{5\to 4}$-3-Oxosteroid isomerase (EC 5.3.3.1)	Pseudomonas		26,300	2	13,000	8
Nerve growth factor	Mouse	Submaxillary gland	26,518	2	13,259	9
Leuteinizing hormone	Ovine		27,322	1	12,500	10
				1	14,830	
Cytochrome (CC')	Pseudomonas		28,000	2	14,000	11
Leuteinizing hormone	Human	Pituitary gland	28,260	1	13,853	12
				1	14,407	
Ribonuclease	Bull	Semen	29,000	2	14,000	13
Interstitial cell-stimulating hormone	Ovine	Pituitary gland	30,000	1	13,700	14
				1	16,300	
Phospholipase A_2	Crotalus	Venom	30,000	2	15,000	14a
Leuteinizing hormone	Rat	Pituitary gland	31,000	2	15,500	15
Superoxide dismutase	Neurospora		31,000	2	16,800	16
Superoxide dismutase	Human	Erythrocyte	32,000	2	16,000	17
Thyrotropin	Bovine		32,000	1	15,000	17a
				1	15,000	
Follicle stimulating hormone	Ovine		33,000	1	18,500	17b
				1	18,500	
Lactose specific factor III	E. coli and Staphylococcus		33,000	4	8,000	18
Follicle stimulating hormone	Equine	Pituitary gland	33,800	1	16,500	19
				1	16,000	
Adenine phosphoribosyltransferase (EC 2.4.2.7)	Human	Erythrocyte	34,000	3	11,000	20
Hemoglobin I	Blood clam	Erythrocyte	34,000	2	17,500	21
Follicle stimulating hormone	Human	Pituitary gland	35,000	2	17,500	22

*References follow Table 2.

Table 1 (continued)
SUBUNIT CONSTITUTION OF PROTEINS

Protein	Source	Organ	Molecular weight	Subunit No.	Subunit Molecular weight	References
β-Lactoglobulin	Bovine	Milk	35,000	2	17,500	23
Lactose-specific phosphocarrier protein	*Staphylococcus*		35,000	3	12,000	24
Agglutinin	Wheat	Germ	35,000	2	17,000	25
DNA dependent RNA polymerase	*Halobacterium*		36,000	1	18,000	26
				1	18,000	
β-Hydroxydecanoyl thioester dehydrase	*E. coli*		36,000	2	18,000	27
Rhodanese (EC 2.8.1.1)	Bovine	Liver	37,000	2	18,500	28
Chorionic gonadotropin	Human		37,900	2	14,900	29
				1	23,000	
Chymotrypsin inhibitor I	Potato		39,000	4	9,800	30
Superoxide dismutase	*E. coli*		39,500	2	21,600	31
Hemerythrin	*Phascolosoma*	Coelomic fluid	40,600	3	12,700	31a
Proteinase inhibitor I	Potato		42,000	2	19,300	32
			19,300	2	9,400	
Dethiobiotin synthetase	*E. coli*		42,000	2	24,500	32a
Dihydropteridine reductase	Sheep	Liver	42,000	2	21,000	33
Nucleoside phosphotransferase	Carrot		44,000	1	22,000	34
				1	22,000	
Biotin carboxyl carrier protein	*E. coli*		45,000	2	22,500	35
Catabolite gene-activator protein	*E. coli*		45,000	2	22,000	36
Cytoplasmic protein	*Neurospora*		45,000	3	15,000	37
Growth hormone	Bovine	Pituitary gland	48,000	2	25,000	38
Factor X	Bovine	Plasma	48,000	1	20,000	39
				1	30,000	
Hemagglutinin LcH	*Lens*		49,000	2	24,500	39a
Phycocyanin	*Chroomonas*		50,000	2	16,000	40
				2	10,000	
Phenylalanine hydroxylase-stimulating protein	Rat	Liver	51,500	4	12,500	41
Galactokinase (EC 2.7.1.6)	Human	Erythrocyte	53,000	2	27,000	42
Triosephosphate isomerase (EC 5.3.1.1)	Rabbit	Muscle	53,000	2	26,500	43

327

Table 1 (continued)
SUBUNIT CONSTITUTION OF PROTEINS

Protein	Source	Organ	Molecular weight	Subunit No.	Subunit Molecular weight	References
Malate dehydrogenase (EC 1.1.1.37)	*Neurospora*		54,000	4	13,500	44
Transglutaminase	Guinea pig	Hair follicle	54,000	2	27,000	45
Azoferredoxin	*Clostridium*		55,000	2	27,500	46
Hyaluronidase (EC 3.2.1.35)	Bovine	Testicle	55,000	4	14,000	47
5-10-Methylenetetrahydrofolate dehydrogenase (EC 1.5.1.5)	*Clostridium*		55,000	2	30,000	48
ω-Amidase	Rat	Liver	58,000	2	27,000	49
Alcohol dehydrogenase (EC 1.1.1.1)	*Drosophila*		60,000	8	7,400	49a
NADP-linked isocitrate dehydrogenase (EC 1.1.1.42)	Porcine	Heart	60,000	2	32,000	50
Deoxycytidylate deaminase	*Staphylococcus*		60,000	2	29,000	51
Hydrogenase (EC 1.12.1.1)	*Clostridium*		60,000	2	30,000	52
Lactose synthetase	Bovine	Milk	60,000	1	22,300	53
				1	36,600	
Transcobalamin II	Human	Plasma	60,000	1	38,000	54
				1	25,000	
Nuclear DNA polymerase (EC 2.7.7.7)	Rat	Liver	60,000	2	29,000	55
Aldose reductase (EC 1.1.1.21)	*Rhodotorula*		61,000	2	22,300	56
				1	36,600	
Prealbumin	Human	Plasma	62,000	4	15,500	57
Phosphoglucomutase (EC 2.7.5.a)	Rabbit	Muscle	62,000	1	31,000	58
				1	31,000	
Erythrocuprein	Bovine	Blood	64,000	4	16,000	59
Serine dehydratase (EC 4.2.1.13)	Rat	Liver	64,000	2	34,000	60
D-Galactose dehydrogenase (EC 1.1.1.48)	*Pseudomonas*		64,000	2	32,000	60a
T2 Phage induced thymidylate synthetase	*Lactobacillus*		64,400	2	31,500	61
Hemoglobin	Mammalian	Erythrocytes	64,500	4	16,000	62
T_u-T_s Complex	*E. coli*		65,000	1	41,500	63
				1	28,500	
L-3-Hydroxyacyl-CoA dehydrogenase (EC 1.1.1.35)	Pig	Heart	65,000	2	31,000	64
Inorganic pyrophosphatase (EC 3.6.1.1)	Yeast		65,000	2	32,000	65

Table 1 (continued)
SUBUNIT CONSTITUTION OF PROTEINS

Protein	Source	Organ	Molecular weight	Subunit No.	Subunit Molecular weight	References
Thiogalactoside transacetylase (EC 2.3.1.18)	E. coli		65,300	2	29,700	66
Phosphoglycerate mutase (EC 2.7.5.3)	Pig	Muscle	66,000	2	33,000	67
Thioredoxin reductase	E. coli		66,000	2	32,000	67a
Malate dehydrogenase (EC 1.1.1.37)	Rat	Liver	66,300	2	37,500	68
Malate dehydrogenase (EC 1.1.1.37)	Pig	Heart	67,000	2	35,000	69
Glucokinase (EC 2.7.1.12)	Bacillus		67,000	2	34,500	70
O-Acetylserine sulfhydrylase A	Salmonella		68,000	2	34,000	71
Tropomyosin B	Rabbit	Muscle	68,000	2	33,500	72
Transaldolase III (EC 2.2.1.2)	Candida		68,000	2	34,000	73
Adenylate kinase (EC 2.7.4.3)	Rat	Liver	68,000	3	23,000	74
Glycerol-3-phosphate dehydrogenase (EC 1.1.1.8)	Chicken, rabbit, honeybee	Muscle, thorax	68,000	2	34,000	75
17β-Estradiol dehydrogenase	Human	Placenta	68,000	2	33,000	76
Avidin	Chicken	Egg white	68,300	4	18,000	77
Hemoglobin III	Blood clam	Erythrocyte	69,000	4	17,500	78
D-Glycerate dehydrogenase	Beef	Liver	70,000	2	34,000	78a
D-Lactate dehydrogenase (EC 1.1.1.28)	Limulus	Muscle	70,000	2	35,000	79
Fructose diphosphate aldolase (EC 4.1.2.13)	E. coli		70,000	2	35,000	80
Malate-lactate transhydrogenase (EC 1.1.99.7)	Viellonella		70,000	2	35,000	81
Thymidylate synthetase (EC 2.1.1.6)	Lactobacillus		70,000	2	35,000	82
Hydroxypyruvate reductase (EC 1.1.1.29)	Pseudomonas		70,000	2	35,000	83
Nucleoside diphosphate kinase (EC 2.7.4.6)	Pea	Seed	70,000	4	17,000	84
NAD-Glycohydrolase (EC 3.2.2.5)	Mouse, rat, rabbit	Liver	70,000	2	38,000	85
Protein P11	Bacteriophage T4D		70,000	3	24,000	85a
2-Keto-3-deoxy-6-phosphogluconate aldolase (EC 4.1.2.12)	Pseudomonas		72,000	3	24,000	86

Table 1 (continued)
SUBUNIT CONSTITUTION OF PROTEINS

Protein	Source	Organ	Molecular weight	Subunit No.	Subunit Molecular weight	References
Malate dehydrogenase (EC 1.1.1.37)	Bovine	Heart	72,000	2	37,000	87
2,4-Diaminopentanoic acid C$_4$ dehydrogenase	Clostridium		72,000	2	40,000	88
Propylamine transferase	E. coli		73,000	2	37,000	89
L-Histidinol phosphate aminotransferase (EC 2.6.1.9)	Salmonella		74,000	2	37,000	90
Tryptophanyl tRNA synthetase (EC 6.1.1.2)	E. coli		74,000	2	37,000	91
Sedoheptulose 1,7-diphosphatase	Candida		75,000	2	35,000	92
Electron-transferring flavoprotein	Peptostreptococcus		75,000	1	41,000	93
				1	33,000	
Glutathione peroxidase (EC 1.11.1.9)	Rat	Liver	76,000	4	19,000	93a
Chorismate mutase-prephenate dehydrogenase	Aerobacter		76,000	2	40,000	94
Diacetyl reductase (EC 1.1.1.15)	Beef	Liver	76,000	3	26,000	95
Hypoxanthine-guanine phosphoribosyltransferase (EC 2.4.2.8)	Chinese hamster, human	Brain, erythrocyte	78,000	3	25,000	96
Histidyl tRNA synthetase	Salmonella		78,000	2	40,000	97
Cyclic AMP dependent protein kinase	Bovine	Sperm	78,000	1	35,000	98
				1	40,000	
Glycerol 1-phosphate dehydrogenase (EC 1.1.1.6)	Yeast, rabbit	Muscle	78,000	2	40,000	99
Hydroxyindole-O-methyl transferase (EC 2.1.1.4)	Bovine	Pineal gland	78,000	2	39,000	100
Uridine diphosphogalactose 4-epimerase (EC 5.1.3.2)	Yeast		79,000	2	39,000	101
Luciferase	Photobacterium		79,000	1	42,000	102
				1	37,000	
Phosphotransacetylase (EC 2.3.1.8)	Viellonella		80,000	2	(40,000)	103
Creatine kinase (EC 2.7.3.2)	Chicken	Muscle	80,000	2	40,000	104
Alcohol dehydrogenase (EC 1.1.1.1)	Horse	Liver	80,000	2	41,000	105

Table 1 (continued)
SUBUNIT CONSTITUTION OF PROTEINS

Protein	Source	Organ	Molecular weight	Subunit No.	Subunit Molecular weight	References
Aldolase (EC 4.1.2.13)	Yeast		80,000	2	40,000	106
Lombricine kinase (EC 2.7.3.5)	*Cancer* and *Homarus*		80,000	2	40,000	107
Taurocyamine kinase (EC 2.7.3.4)	*Lumbricus*		80,000	2	40,000	108
Prephenoloxidase	Silkworm	Hemolymph	80,000	2	40,000	109
Troponin	Rabbit	Muscle	80,000	1	37,000	110
				1	24,000	
L-Erythro-3,5-diaminohexanoate dehydrogenase	*Clostridium*		80,000	1	20,000	111
				2	39,300	
Histidinol dehydrogenase (EC 1.1.1.23)	*Salmonella*		80,000	2	40,000	112
Galactose 1-phosphate uridyltransferase (EC 2.7.7.12)	*E. coli*		80,000	2	40,000	113
Anthranilate synthase	*Pseudomonas*		81,000	1	63,000	114
				1	18,000	
Enolase (EC 4.2.1.11)	Rabbit	Muscle	82,000	2	42,000	115
ATP-Creatine transphosphorylase	Rabbit	Muscle	82,600	2	41,300	116
NADP-Specific isocitrate dehydrogenase	*E. coli*		83,000	2	43,000	116a
Glutathione peroxidase	Bovine	Blood	83,800	4	21,000	117
Purine nucleoside phosphorylase (EC 2.4.1.2)	Calf	Spleen	84,600	3	28,000	118
Histidyl tRNA synthetase	*E. coli*		85,000	2	42,500	119
Succinate dehydrogenase (EC 1.3.99.1)	*Rhodospirillum*		85,000	1	60,000	120
				1	25,000	
Glutamate decarboxylase (EC 4.1.1.15)	Mouse	Brain	85,000	2	44,000	121
Haptoglobin 1-1	Human	Serum	85,000	2	40,000	122
Acid phosphatase (EC 3.1.3.2)	*Neurospora*		85,000	2	42,000	123
Alkaline phosphatase (EC 3.1.3.1)	*E. coli*		86,000	2	43,000	124
Anthranilate synthetase	*Acinobacter*		86,000	1	70,000	125
				1	14,000	
Histidinol dehydrogenase (EC 1.1.1.23)	*E. coli*		87,000	2	40,000	126

Table 1 (continued)
SUBUNIT CONSTITUTION OF PROTEINS

Protein	Source	Organ	Molecular weight	Subunit No.	Subunit Molecular weight	References
Enolase (EC 4.2.1.11)	Yeast		88,000	2	44,000	127
Procarboxypeptidase A	Bovine	Pancreas	88,000	1	40,000	128
				2	23,000	
High density lipoprotein	Human	Serum	88,000	2	27,000	129
				2	17,000	
M-line Protein	Chicken	Muscle	88,000	2	43,000	130
Putrecine oxidase	*Micrococcus*		88,000	2	46,000	131
Enolase (EC 4.2.1.11)	*E. coli*		90,000	2	46,000	131a
Galactose 1-phosphate uridyl-transferase	Human	Liver	90,000	(3–4)	30,000	132
D-Erythulose reductase	Beef	Liver	90,000	4	22,000	133
2-Deoxycitrate synthase	*Penicillium*		90,000	2	45,000	134
Purine nucleoside phosphorylase	*Bacillus*		90,000	2	47,000	135
			47,000	2	24,000	
Pyruvate dehydrogenase (EC 1.2.4.1)	*E. coli*		90,000	2	45,000	136
Carboxypeptidase G₁	*Pseudomonas*		92,000	2	46,000	136a
Salycilate hydroxylase (EC 1.14.1.a)	*Pseudomonas*		92,000	2	52,000	137
Luciferase	Firefly	Lanterns	92,000	2	52,000	138
Isocitrate dehydrogenase	*Bacillus*		92,500	2	45,000	138a
Adenylate kinase (EC 2.7.4.3)	*Brevibacterium*		92,400	2	46,000	139
L-6-Hydroxynicotine oxidase	*Arthrobacter*		93,000	2	47,000	139a
6-Phosphogluconate dehydrogenase (EC 1.1.1.44)	Sheep	Liver	94,000	2	47,000	140
Dipeptidase M	*E. coli*		94,000	2	47,000	141
Prolyl tRNA synthetase	*E. coli*		94,000	2	47,000	141a
Tyrosyl tRNA synthetase	*Bacillus*		95,000	2	45,000	142
Ceramide trihexosidase	Human	Plasma	95,000	4	22,000	143
α-Amylase	*Bacillus*		96,000	2	48,000	144
			48,000	2	24,000	
Allophycocyanin	*Synechococcus*		96,000	(?)	17,250	145
				(?)	15,200	
Glyoxylic acid reductase	Spinach	Leaf	97,500	2	47,000	145a

Table 1 (continued)
SUBUNIT CONSTITUTION OF PROTEINS

Protein	Source	Organ	Molecular weight	Subunit No.	Subunit Molecular weight	References
L-Ribulokinase (EC 2.7.1.16)	*E. coli*		98,000	2	50,000	146
N-Formimino-L-glutamate imino-hydrolase	*Pseudomonas*		100,000	2	50,000	147
D-Amino acid oxidase (EC 1.4.3.3)	Pig	Kidney	100,000	2	50,000	148
Diacetyl (acetoin) reductase (EC 1.1.1.5)	*Aerobacter*		100,000	4	25,000	149
Cysteamine oxygenase (EC 1.13.1.22)	Horse	Kidney	100,000	2	50,000	150
Galactokinase (EC 2.7.1.6)	Yeast		100,000	4	23,000	151
Citrate synthase (EC 4.1.3.7)	Pig	Heart	100,000	2	50,000	152
Aspartate aminotransferase (EC 2.6.1.1)	Chicken	Heart	100,000	2	50,000	153
Seryl tRNA synthetase (EC 6.1.1.11)	*E. coli*		100,000	2	50,000	154
β-D-N-Acetylhexose amidase	Human	Placenta	100,000	6	17,000	155
3-α-Hydroxysteroid dehydrogenase (EC 1.1.1.50)	*Pseudomonas*		100,000	2	50,000	155a
6-Phosphogluconate dehydrogenase (EC 1.1.1.44)	*Bacillus*, rat	Liver	101,000	2	51,000	156
Glutamyl tRNA synthetase	*E. coli*		102,000	1	56,000	157
				1	46,000	
D-Galactose dehydrogenase	*Pseudomonas*		102,000	4	25,000	158
Glutathione reductase	Sea urchin	Egg	102,000	2	52,000	159
Acid phosphomonoesterase I (EC 3.1.3.2)	Human	Prostate	102,000	2	50,000	160
Hexokinase (EC 2.7.1.1)	Yeast		102,000	2	51,000	161
Nucleoside diphosphokinase (EC 2.7.4.6)	Yeast		102,000	6	17,000	162
Delta hemolysin	*Staphylococcus*		103,000	5	21,000	163
			21,000	4	5,000	
Uridine phosphorylase	Rat	Liver	103,000	4	26,000	164
6-Phosphogluconate dehydrogenase (EC 1.1.1.44)	Human	Erythrocyte	104,000	2	52,000	165
Aspartokinase III (EC 2.7.2.4)	*E. coli*		105,000	2	50,000	166
Ribosephosphate isomerase (EC 5.3.1.6)	*Candida*		105,000	4	26,000	167
Isocitrate dehydrogenase (TPN)	*Rhodopseudomonas*		105,000	2	50,000	168

Table 1 (continued)
SUBUNIT CONSTITUTION OF PROTEINS

Protein	Source	Organ	Molecular weight	Subunit No.	Subunit Molecular weight	References
Lipoamide dehydrogenase (EC 1.6.4.3)	E. coli		106,000	2	53,000	169
20-β-Hydroxysteroid dehydrogenase (EC 1.1.1.53)	Streptomyces		106,000	4	27,000	170
Glycyl-L-leucine hydrolase	Monkey	Small intestine	107,000	2	54,000	171
Sulphatase A (EC 3.1.6.1)	Ox	Liver	107,000	2	50,000	171a
Ornithine transcarbamylase (EC 2.1.3.3)	Streptococcus, bovine	Liver	108,000	3	36,000	172
Concanavalin A	Jack bean		108,000	2	54,000	173
			54,000	2	27,000	
Lipoxygenase (EC 1.99.2.1)	Soybean		108,000	2	54,000	174
Hemerythrin	Golfingia	Erythrocyte	108,000	8	13,500	175
Glutamine transaminase	Rat	Liver	110,000	2	54,000	176
Tubulin	Pig	Brain	110,000	1	56,000	177
				1	53,000	
Ribitol dehydrogenase (EC 1.1.1.56)	Klebsiella		110,000	4	27,000	177a
Tryptophanyl tRNA synthetase (EC 6.1.1.2)	Bovine	Pancreas	110,000	2	58,000	178
Phosphoglycerate mutase (EC 2.7.5.3)	Yeast		110,000	4	27,000	179
Urocanase	Pseudomonas		110,000	2	54,000	180
Histidine decarboxylase (EC 4.1.1.22)	Micrococcus		110,000	3	29,000	181
				3	7,000	
Phosphatase (EC 3.1.3.2)	Sweet potato	Tuber	110,000	2	55,000	182
Dihydrolipoyl dehydrogenase	E. coli		112,000	2	56,000	183
Canavalin	Canavalin		113,000	6	19,500	184
Adenosylmethionine decarboxylase	E. coli		113,000	(8)	(15,000)	184a
D-Ribulose 1,5-diphosphate carboxylase	Rhodospirillum		114,000	2	56,000	185
Lectin	Navy bean		114,000	4	30,000	186
DNA Modification methylase	E. coli		115,000	1	60,000	187
				1	55,000	

Table 1 (continued)
SUBUNIT CONSTITUTION OF PROTEINS

Protein	Source	Organ	Molecular weight	Subunit No.	Subunit Molecular weight	References
Monoamine oxidase	Pig	Liver	115,000	2	60,000	188
Sulfite oxidase (EC 1.8.3.1)	Bovine	Liver	115,000	2	55,000	188a
DNA Modification enzyme	Bacteriophage P1		115,000	1	70,000	189
				1	45,000	
Tyrosine aminotransferase (EC 2.6.1.5)	Rat	Liver	115,000	4	32,000	190
Tyrosine tRNA synthetase (EC 6.1.1.1)	Saccharomyces		116,000	4	31,500	191
Arginase (EC 3.5.3.1)	Human	Liver	118,000	4	30,000	192
Phosphoglucose isomerase	Yeast		119,400	4	30,000	193
α-Ketoglutaric semialdehyde dehydrogenase	Pseudomonas		120,000	2	60,000	194
Agglutinin	Soybean		120,000	4	30,000	195
Aspartokinase (EC 2.7.2.4)	Bacillus		120,000	2	43,000	196
				2	17,000	
L-Asparaginase (EC 3.5.1.1)	Proteus		120,000	4	30,000	197
Leucine tRNA synthetase (EC 6.1.1.4)	Yeast		120,000	2	60,000	198
Cyclic AMP-dependent protein kinase	Bovine	Sperm	120,000	1	78,000	199
				1	35,000	
Protein toxin B	Pasteurella		120,000	(5—6)	24,000	200
			24,000	2	12,000	
Pyrophosphatase (EC 3.6.1.1)	E. coli		120,000	6	20,000	201
Seryl tRNA synthetase (EC 6.1.1.11)	Yeast		120,000	2	60,000	201a
Anti-A1 lectin	Dolichos		120,000	4	30,000	201b
Aldolase (EC 4.1.2.13)	Spinach	Leaf	120,000	4	30,000	202
Oestradiol-receptor	Calf	Uterine	120,000	2	55,000	202a
DDT-Dehydrochlorinase (EC 4.5.1.1)	Housefly		120,000	4	30,000	203
Alkaline phosphatase (EC 3.1.3.1)	Bacillus		121,000	2	55,000	204
Tryptophan oxygenase (EC 1.13.1.12)	Pseudomonas		122,000	4	31,000	205
Fatty acylthiokinase I	E. coli		122,000	4	30,000	206
Cyclic AMP-dependent protein kinase	Rabbit	Muscle	123,000	1	82,000	207
				1	49,000	

Table 1 (continued)
SUBUNIT CONSTITUTION OF PROTEINS

Protein	Source	Organ	Molecular weight	Subunit No.	Subunit Molecular weight	References
Ceruloplasmin	Human	Serum	124,000	2	53,000	208
				2	16,000	
Methylmalonyl-CoA mutase (EC 5.4.99.2)	*Propionibacterium*		124,000	1	66,000	209
				1	61,000	
Glutathione reductase (EC 1.6.4.2)	*E. coli*		124,000	2	56,000	210
Deoxycytidylate deaminase (T-2 bacteriophage induced)	*E. coli*		124,000	6	20,200	211
Hybridase	Rat	Liver	125,000	1	85,000	212
				1	43,000	
Uricase (EC 1.7.3.3)	Pig	Liver	125,000	4	32,000	213
Xanthosine 5'-phosphate aminase (EC 6.3.4.1)	*E. coli*		126,000	2	63,000	214
Tyrosinase (EC 1.10.3.1)	Mushroom		128,000	4	32,000	215
Fructose diphosphatase (EC 3.1.3.11)	Swine	Kidney	130,000	4	34,000	216
Glucose 6-phosphate dehydrogenase (EC 1.1.1.49)	Rat	Mammary gland	130,000	2	63,000	217
D-Gluconate dehydratase (EC 4.2.1)	*Clostridium*		131,000	2	64,000	218
Ornithine aminotransferase (EC 2.6.1.13)	Rat	Liver	132,000	4	33,000	219
Methylmalonate semialdehyde dehydrogenase	*Pseudomonas*		132,000	2	59,000	219a
L-Asparaginase (EC 3.5.1.1)	*E. coli*		133,000	4	33,000	220
Aspartokinase (EC 2.7.2.4)	*Pseudomonas*		133,000	3	43,000	221
Phosphoglucose isomerase (EC 5.3.1.9)	Human, rabbit	Muscle	134,000	2	61,000	222
L-Amino acid oxidase (EC 1.4.3.2)	Rattlesnake	Venom	135,000	2	70,000	223
Myrokinase (EC 3.2.3.1)	Rape seed		135,000	2	65,000	224
L-Asparaginase (EC 3.5.1.1)	*Erwinia*		135,000	4	32,500	225
Aminotripeptidase (EC 3.4.1.3)	Swine	Kidney	137,200	2	71,100	226
Aspartate carbamoyltransferase (EC 2.1.3.2)	Yeast		138,000	(6)	21,000	227
Lysine tRNA synthetase (EC 6.1.1.6)	Yeast		138,000	2	72,000	228
Nucleoside diphosphokinase	Pig	Kidney	138,000	6	21,000	229
L-Asparaginase (EC 3.5.1.1)	*Achrombacteraceae*		138,000	4	35,000	230

Table 1 (continued)
SUBUNIT CONSTITUTION OF PROTEINS

Protein	Source	Organ	Molecular weight	Subunit No.	Subunit Molecular weight	References
Transketolase (EC 2.2.1.1)	Yeast		140,000	2	70,000	231
Protein phosphokinase (EC 2.7.1.37)	Bovine	Brain	140,000	1	80,000	232
				1	60,000	
Succinyl-CoA synthetase (EC 6.2.1.5)	E. coli		140,000	2	38,500	233
				2	29,500	
C-Reactive protein	Rabbit	Blood	140,000	6	23,000	234
Lactate dehydrogenase (EC 1.1.1.27)	Porcine	Heart	140,000	4	35,000	235
L-Rhamnulose 1-phosphate aldolase (EC 1.4.2.b)	E. coli		140,000	4	35,000	236
Ascorbate oxidase (EC 1.10.3.3)	Zucchini squash		140,000	2	65,000	237
Cyclic GMP protein kinase	Lobster	Muscle	140,000	1	100,000	238
				1	40,000	
Cyclic AMP-dependent protein kinase I	Beef	Brain	140,000	1	100,000	239
				1	40,000	
Exonuclease I	E. coli		140,000	2	70,000	239a
Phytohemagglutinin	Phaseolus		140,000	2	35,000	240
				2	36,000	
Fructose 1,6-diphosphatase (EC 3.1.3.11)	Rabbit	Liver, kidney, muscle	140,000	4	36,000	241
Esterase	Rat	Liver	140,000	2	70,000	242
3,4-Dihydroxyphenylacetate-2,3-dioxygenase (EC 1.13.1)	Pseudomonas		140,000	4	35,000	243
Aldolase (EC 4.1.2.13)	E. coli		140,000	4	35,000	244
L-Erythro-3,5-diaminohexanoate dehydrogenase	Clostridium		140,000	2	68,000	245
			68,000	2	37,000	
Alkaline phosphatase (EC 3.1.3.1)	Calf	Intestine	140,000	2	69,000	246
Alcohol dehydrogenase (EC 1.1.1.1)	Yeast		141,000	4	35,000	246a
Anthranilate synthetase	Serratia		141,000	2	60,000	247
				2	21,000	
Lectin	Navy bean		143,000	4	37,000	248
Tryptophan synthetase (EC 4.2.1.20)	Yeast		143,000	4	37,000	248a

Table 1 (continued)
SUBUNIT CONSTITUTION OF PROTEINS

Protein	Source	Organ	Molecular weight	Subunit No.	Subunit Molecular weight	References
Glyceraldehyde 3-phosphate dehydrogenase (EC 1.2.1.12)	Rabbit, lobster, pig	Muscle	144,000	2	72,000	249
			72,000	2	37,000	250
Tartaric acid dehydrase (EC 4.2.1.c)	Pseudomonas		145,000	4	39,000	251
Phosphofructokinase (EC 2.7.1.11)	Clostridium		145,000	4	35,000	252
Platelet factor XIII	Human	Platelet	146,000	2	75,000	252a
L-Asparaginase (EC 3.5.1.1)	Serratia		147,000	4	37,000	253
Malate dehydrogenase (EC 1.1.1.37)	Bacillus		148,000	4	37,000	254
Tryptophan synthetase (EC 4.2.1.20)	E. coli		148,000	2	45,000	255
				2	28,700	256
Aldolase C	Rabbit	Brain	148,000	4	37,000	257
5'-Nucleotidase (EC 3.1.3.5)	Mouse	Liver	150,000	2	75,000	258
Pyridoxamine pyruvate transaminase (EC 2.6.1.a)	Pseudomonas		150,000	4	38,000	259
Alcohol dehydrogenase (EC 1.1.1.1)	Yeast		150,000	4	37,000	260
Pyruvate dehydrogenase	Bovine	Heart, kidney	154,000	2	41,000	261
				2	36,000	261a
α-Acetylgalactosaminidase	Beef	Liver	155,000	4	42,000	262
Aspartic β-semialdehyde dehydrogenase (EC 1.2.1.11)	Yeast		156,000	4	41,000	263
Alkaline phosphate (EC 3.1.3.1)	Pig	Kidney	156,000	4	39,000	264
D-Xylose isomerase (EC 5.3.1.5)	Streptomyces		157,000	4	40,000	265
Crotonase (EC 4.2.1.17)	Clostridium		158,000	4	43,000	266
Aldolase	Rabbit	Muscle	160,000	4	40,000	266a
Lac repressor	E. coli		160,000	4	40,000	267
Cystathionine γ-synthetase (EC 4.2.1.21)	Salmonella		160,000	4	40,000	268
3-Deoxy-D-arabinoheptulosonate 7-Phosphate synthetase-chorismate mutase	Bacillus		160,000	4	38,500	269
Cystathionase (EC 4.2.1.15)	Rat	Liver	160,000	8	20,000	270
Trimethylamine dehydrogenase	Bacterium 4B 6		160,000	2	80,000	
Threonine deaminase (EC 4.2.1.16)	Clostridium		160,000	4	40,000	
			163,000	4	41,000	

Table 1 (continued)
SUBUNIT CONSTITUTION OF PROTEINS

Protein	Source	Organ	Molecular weight	Subunit No.	Subunit Molecular weight	References
Succinic semialdehyde dehydrogenase (EC 1.2.1.b)	*Pseudomonas*		164,000	3	54,500	271
Crotonase (EC 4.2.1.17)	Beef	Liver	164,000	6	28,000	272
Quinolate phosphoribosyltransferase	*Pseudomonas*		165,000	3	54,000	273
Carboxyl esterase (EC 3.1.1.1)	Beef	Liver	167,000	2	85,000	274
Tryptophan oxygenase (EC 1.13.1.12)	Rat	Liver	167,000	2	43,000	275
				2	44,000	
Palmitoyl-CoA synthetase (EC 6.2.1.3)	Rat	Liver	168,000	6	27,000	276
Molybdoferredoxin	*Clostridium*		168,000	2	59,000	277
				1	50,700	
Aspartokinase II-homoserine dehydrogenase II	*E. coli*		169,000	4	43,000	278
Acetoacetyl CoA thiolase (EC 2.3.1.9)	Avian	Liver	169,000	4	41,000	279
Thiolase (EC 2.3.1.9)	Pig	Heart	169,000	4	41,000	279a
Methionine tRNA synthetase (EC 6.1.1.10)	*E. coli*		170,000	2	85,000	280
β-Lysine mutase	*Clostridium*		170,000	2	52,000	281
				2	32,000	
Carbamylphosphate synthetase (EC 2.7.2.5)	*E. coli*		170,000	1	130,000	282
				1	42,000	
Aspartase (EC 4.3.1.1)	*E. coli*		170,000	4	45,000	283
Cyclic AMP-dependent protein kinase	Bovine	Heart	174,000	2	49,000	284
				2	38,000	
Carboxylesterase (EC 3.1.1.1)	Pig	Liver	180,000	3	60,000	285
D-α-Ornithine 5,4-aminomutase	*Clostridium*		180,000	2	95,000	286
Thetin homocysteine methylpherase (EC 2.1.1.10)	Horse	Liver	180,000	(3—4)	50,000	287
Carbonic anhydrase	Parsley	Leaf	180,000	6	29,000	287a
L-Leucine:2-oxoglutarate aminotransferase (EC 2.6.1.6)	*Salmonella*		183,000	6	81,500	288
L-Threonine dehydratase	*Clostridium*		184,000	4	46,000	289
Aminoacyl transferase I	Rabbit	Reticulocytes	186,000	3	62,000	290

Table 1 (continued)
SUBUNIT CONSTITUTION OF PROTEINS

Protein	Source	Organ	Molecular weight	Subunit No.	Subunit Molecular weight	References
α-Dialkyl amino acid transaminase	*Pseudomonas*		188,000	4	47,000	291
Histidine decarboxylase (EC 4.1.1.22)	*Lactobacillus*		190,000	5	9,000	292
				5	29,700	
Prohistidine decarboxylase	*Lactobacillus*		190,000	5	37,000	293
Fumarase (EC 4.2.1.2)	Swine	Heart	194,000	4	48,500	294
Pyruvate kinase (EC 2.7.1.40)	*Saccharomyces*		195,000	4	49,000	294a
Threonine deaminase (EC 4.2.1.16)	*Salmonella*		195,000	4	48,500	295
Dipeptidyl transferase (EC 3.4.4.9)	Beef	Spleen	197,000	2	100,000	296
			100,000	4	24,500	
Phosphoenolpyruvate carboxylase (EC 4.1.1.31)	*Salmonella*		198,000	4	49,200	297
Glutamine phosphoribosylpyrophosphate amidotransferase (EC 2.4.2.14)	Pigeon	Liver	200,000	2	100,000	298
			100,000	2	50,000	
Polynucleotide phosphorylase	*E. coli*		200,000	2	95,000	299
α-Isopropylmalate synthase	*Salmonella*		200,000	4	50,000	300
Aspartylkinase (lysine sensitive) (EC 2.7.2.4)	*E. coli*		200,000	2	100,000	301
			100,000	2	48,000	
Succinic dehydrogenase (EC 1.3.99.1)	Beef	Heart	200,000	2	100,000	302
				1	70,000	
			100,000	1	30,000	
Neuraminidase (EC 3.2.1.18)	Influenza		200,000	4	50,000	303
Tyrosinase (EC 1.10.3.1)	Frog	Epidermis	200,000	4	50,000	304
Mo-Fe Protein	Soybean	Nodule	200,000	4	50,000	305
Peroxidase	Pig	Thyroid	200,000	3	70,000	305a
Cytochrome oxidase	Bovine	Heart	200,000	2	100,000	305b
3-Hydroxy-3-methylglutaryl CoA reductase (EC 1.1.1.34)	Rat	Liver	200,000	3	65,000	305c
Aliphatic amidase (EC 3.5.1.4)	*Pseudomonas*		200,000	6	33,000	306
β-Amylase (EC 3.2.1.2)	Sweet potato	Tuber	201,000	4	50,000	307
Argininosuccinase (EC 4.3.2.1)	Steer	Liver	202,000	2	100,000	308
			100,000	2	50,000	

Table 1 (continued)
SUBUNIT CONSTITUTION OF PROTEINS

Protein	Source	Organ	Molecular weight	Subunit No.	Subunit Molecular weight	References
NAD-Linked malic enzyme (EC 1.1.1.38)	E. coli		200,000	4	52,500	309
Threonine deaminase (EC 4.2.1.16)	E. coli		204,000	4	51,000	310
Glucose 6-phosphate dehydrogenase	Human	Erythrocyte	204,800	2	101,400	311
(EC 1.1.1.49)			101,400	2	51,300	
Qβ Replicase	E. coli		205,000	1	70,000	312
				1	65,000	
				1	45,000	
				1	35,000	
Isocitrate lyase	Pseudomonas		206,000	4	48,200	313
Glucose 6-phosphate dehydrogenase	Neurospora		206,000	2	104,000	314
(EC 1.1.1.49)			104,000	2	57,000	
Tryptophanase (EC 4.2.1.e)	Bacillus		208,000	4	50,500	315
Pyruvate decarboxylase (EC 4.1.1.1)	Yeast		209,000	2	108,000	316
Invertase (EC 3.2.1.26)	Neurospora		210,000	4	51,500	317
Phenylalanine hydroxylase	Rat	Liver	210,000	4	51,000	317a
Cytidine triphosphate synthetase	E. coli		210,000	2	105,000	318
(EC 6.3.4.2)			105,000	2	50,000	
High density protein	Porcine	Plasma	210,000	4	28,000	319
Phosphoribosyl ATP pyrophosphate phosphoribosyl transferase	Salmonella		215,000	6	36,000	320
Serine transhydroxymethylase (EC 2.1.2.1)	Rabbit	Liver	215,000	4	47,000	321
Acetyl-CoA carboxylase (EC 6.4.1.2)	Rat	Liver	215,000	1	118,000	322
				1	125,000	
Pyruvate kinase (EC 2.7.1.40)	Bovine	Liver	215,000	4	54,000	323
α-Glucan phosphorylase (EC 2.4.1.1)	Potato		215,000	2	108,000	324
2-Oxoglutarate dehydrogenase	Pig	Heart	216,000	2	105,000	325
Tryptophanase (EC 4.2.1.e)	Aeromonas		216,000	4	54,000	326
α-L-Fucosidase	Rat	Epididymes	216,000	2	47,000	327
				2	59,300	
Glycerol kinase (EC 2.7.1.30)	E. coli		217,000	4	55,000	328

Table 1 (continued)
SUBUNIT CONSTITUTION OF PROTEINS

Protein	Source	Organ	Molecular weight	Subunit No.	Subunit Molecular weight	References
Adenosine deaminase (EC 3.5.4.4)	*Aspergillus*		217,000	2	105,000	328a
Acetol acetate-forming enzyme	*Aerobacter*		220,000	4	58,000	329
3-Aminopropanal dehydrogenase	*Pseudomonas*		220,000	3	74,000	329a
Tryptophanase (EC 4.2.1.e)	*Bacillus, E. coli*		220,000	2	110,000	330
Paramyosin	*Venus*	Muscle	110,000	2	55,000	331
			220,000	2	110,000	
Chorismate mutase prephenate dehydratase	*Salmonella*		220,000	2	109,000	332
Sucrase-isomaltase complex (EC 3.2.1)	Rabbit	Small intestine	220,000	1	110,000	333
				1	120,000	
ATPase (Na and K dependent)	Canine	Renal medulla	(220,000)	1	(135,000)	334
				2	(35,000)	
α-Acetohydroxyacid isomeroreductase	*Salmonella*		220,000	4	55,000	335
L(+) Hydroxybutyryl-CoA dehydrogenase (EC 1.1.1.35)	*Clostridium*		220,000	8	26,000	336
Pyruvate kinase (EC 2.7.1.40)	Frog	Muscle	220,000	4	55,000	337
Arylamidase	Human	Liver	223,500	6	38,100	338
Ornithine transcarbamylase (EC 2.1.3.3)	*Streptococcus*, bovine	Liver	223,000	3	74,000	339
			74,000	2	38,000	
Ribonucleoside diphosphate reductase	Bacteriophage T4		225,000	2	85,000	340
				2	35,000	
Glycyl tRNA synthetase (EC 6.1.1.e)			227,000	2	33,000	341
				2	80,000	
4-Aminobutanal dehydrogenase (EC 1.2.1.e)	*Pseudomonas*		228,000	3	75,000	342
Cysteine desulfhydrase	*Salmonella*		229,000	6	37,000	343
Pyruvate kinase (EC 2.7.1.40)	Bovine	Muscle	230,000	4	57,000	344
Glyoxylate carboligase (EC 4.1.1.b)	*Pseudomonas*		230,000	2	115,000	345
			115,000	2	61,000	
Collagen	Chicken	Leg tendon	231,000	12	18,500	345a
C-Phycocyanin	*Synechococcus*		232,000	6	19,000	346
				6	17,700	

Table 1 (continued)
SUBUNIT CONSTITUTION OF PROTEINS

Protein	Source	Organ	Molecular weight	Subunit No.	Subunit Molecular weight	References
Catalase (EC 1.11.1.6)	Bovine	Liver	232,000	4	57,500	347
Cytochrome b_2	Yeast		235,000	4	57,000	348
Pyruvate kinase (EC 2.7.1.40)	Rabbit	Muscle	237,000	4	57,200	349
Formyltetrahydrofolate synthetase (EC 6.3.4.3)	*Clostridium*		240,000	4	60,000	350
δ-Aminolevulinate dehydratase (EC 4.2.1.24)	*Rhodopseudomonas*		240,000	2	120,000	351
			120,000	3	40,000	
Anthranilate synthetase complex	*Neurospora*		240,000	6	40,000	352
Protein toxin A	*Pasteurella*		240,000	10–12	24,000	353
			24,000	2	12,000	
Protocollagen proline hydroxylase	Chick	Embryo	240,000	2	60,000	354
Glucocerebrosidase	Human	Placenta	240,000	2	65,000	355
			240,000	4	60,000	
Pyruvate kinase (EC 2.7.1.40)	*E. coli*		240,000	4	60,000	356
2,5-Dihydroxypyridine oxygenase	*Pseudomonas*		242,000	6	39,500	357
Citrate synthase (EC 4.1.3.7)	*Acinetobacter*		242,000	(4)	59,000	357a
Aldehyde dehydrogenase (EC 1.2.1.3)	Horse	Liver	245,000	4	57,000	358
Ribonucleotide diphosphate reductase	*E. coli*		245,000	1	78,000	359
				1	80,000	
				1	80,000	
Melilotate hydroxylase	Bacterial		250,000	4	64,000	360
Glycogen synthase (EC 2.1.1.11)	Rabbit	Muscle	(250,000)	(3)	90,000	361
DNA Restriction endonuclease	*E. coli*		250,000	1	135,000	362
				1	60,000	
				1	55,000	
Cystathionine synthetase (EC 4.2.1.21)	Rat	Liver	250,000	(2)	51,000	363
				(2)	73,000	
Uridine diphosphate galactose 4-epimerase (EC 5.1.3.2)	Yeast		250,000	2	125,000	364
			125,000	2	60,000	
Malic enzyme (EC 1.1.1.39)	*Ascaris*	Muscle	250,000	4	64,000	365
Citrate synthase (EC 4.1.3.7)	*Azotobacter*		250,000	(4)	59,000	365a

Table 1 (continued)
SUBUNIT CONSTITUTION OF PROTEINS

Protein	Source	Organ	Molecular weight	Subunit No.	Subunit Molecular weight	References
Phytochrome	Rye, oat	Shoots	252,000	6	42,000	366
Leucine aminopeptidase (EC 3.4.1.1)	Swine	Kidney	255,000	4	63,500	367
Malic enzyme (EC 1.1.1.40)	Pigeon	Liver	260,000	4	65,000	368
Butyrylcholinesterase (EC 3.1.1.8)	Horse	Serum	260,000	4–6	42,200–75,000	369
L-Phenylalanine tRNA synthetase (EC 6.1.1.b)	E. coli		267,000	2	94,000	370
				2	39,000	
Glutamine phosphoribuosylpyrophosphate amidotransferase (EC 2.4.2.14)	Human	Placenta	270,000	2	133,000	371
Glycollate oxidase (EC 1.1.3.1)	Spinach	Leaves	270,000	2	140,000	372
Phenylalanine tRNA synthetase (EC 6.1.1.b)	Yeast		276,000	2	75,000	373
				2	63,000	
AMP Deaminase (EC 3.5.4.6)	Rabbit, chicken	Muscle	278,000	4	69,000	374
Mandelate racemase (EC 5.1.2.2)	Pseudomonas		278,000	4	69,500	375
Acetylcholinesterase (EC 3.1.1.7)	Electrophorus	Tissue	280,000	4	70,000	376
Protein kinase	Bovine	Heart	280,000	(?)	42,000	377
				(?)	55,000	
β-Glucuronidase (EC 3.2.1.31)	Rat	Liver	280,000	4	75,000	378
Anthranilate synthetase complex	Salmonella		280,000	2	62,000	379
				2	62,000	
δ-Aminolevulinic acid dehydratase (EC 4.2.1.24)	Bovine	Liver	282,000	2	140,000	380
				4	35,000	
Glucose-6-phosphate dehydrogenase (EC 1.1.1.49)	Bovine	Adrenal gland	284,000	4	64,600	381
Lysine 2,3-amino mutase	Clostridium		285,000	6	48,000	382
Glutamate dehydrogenase (EC 1.4.1.3)	Neurospora		288,400	6	48,800	383
Dopamine-β-hydroxylase (EC 1.14.2.1)	Bovine	Adrenal gland	290,000	4	75,000	384
Nitrogenase	Klebsiella		295,000	1	229,000	385
				1	66,700	
			229,000	2	51,300	
				2	59,600	
			66,700	2	34,000	

Table 1 (continued)
SUBUNIT CONSTITUTION OF PROTEINS

Protein	Source	Organ	Molecular weight	Subunit No.	Subunit Molecular weight	References
Arginine decarboxylase (EC 4.1.1.19)	E. coli		296,000	4	75,000	386
Uridine diphosphate glucose dehydrogenase (EC 1.1.1.22)	Bovine	Liver	300,000	6	52,000	387
Glycogen synthetase (EC 2.4.1.11)	Yeast		300,000	4	77,000	388
Edestin	Hemp	Seed	300,000	6	50,000	389
Excelsin	Brazil nut		300,000	6	50,000	390
Isocitrate dehydrogenase (EC 1.1.1.41)	Yeast		300,000	8	39,000	391
Monoamine oxidase	Rat	Liver	300,000	4	75,000	391a
Cysteine synthetase (EC 4.2.1.22)	Salmonella		309,000	1	160,000	392
				2	68,000	
Aspartyl transcarbamylase (EC 2.1.3.2)	E. coli		310,000	2	100,000	393
				3	34,000	
			100,000	3	33,000	
			34,000	2	17,000	
Glutamate decarboxylase (EC 4.1.1.15)	E. coli		310,000	6	50,000	394
Cholinesterase (EC 3.1.1.8)	Horse	Serum	315,000	4	77,300	395
Carbamoylphosphate synthase (EC 2.7.2.5)	Rat	Liver	316,000	2	160,000	396
Glutamate dehydrogenase (EC 1.4.1.3)	Beef	Liver	320,000	6	57,000	397
Plasma factor VII	Human	Plasma	320,000	2	75,000	398
				2	88,000	
L-Phenylalanine ammonia lyase (EC 4.1.3.5)	Maize, potato, wheat		320,000	4	83,000	399
Chloroplast-coupling factor	Spinach	Chloroplasts	325,000	6	62,000	400
Leucine aminopeptidase (EC 3.4.1.1)	Bovine	Lens	327,000	6	54,000	401
Nitrogenase	Clostridium		330,000	1	220,000	402
			220,000	2	55,000	
				2	50,700	
				2	59,500	
			55,000	2	27,500	
Glutamate dehydrogenase (EC 1.4.1.3)	Neurospora		330,000	6	51,500	403
Glutamine synthetase (EC 6.3.1.2)	Hamster	Liver	335,000	8	42,000	404

Table 1 (continued)
SUBUNIT CONSTITUTION OF PROTEINS

Protein	Source	Organ	Molecular weight	Subunit No.	Subunit Molecular weight	References
Acetoacetate decarboxylase (EC 4.1.1.4)	Clostridium		340,000	6	62,000	405
Aspartokinase I-homoserine dehydrogenase I			62,000	2	29,000	406
			340,000	4	85,000	
Phosphoenolpyruvate carboxylase (EC 4.1.1.31)	Maize		340,000	2	160,000	407
Peptidase	Sheep	Erythrocyte	340,000	6	60,000	408
Glutamine synthetase (EC 6.3.1.2)	Chicken	Neural retina	340,000	8	42,000	409
Aminopeptidase	Clostridium		340,000	6	60,000	410
Arachin	Arachis		345,000	2	180,000	411
			180,000	6	30,000	
N-Methylglutamate synthetase	Pseudomonas		350,000	12	(30,000)	412
ATPase (EC 3.6.1.3)	Micrococcus		350,000	3	52,500	413
				3	47,000	
				1	41,000	
Glutamine synthetase (EC 6.3.1.2)	Rat	Liver	352,000	8	44,000	414
Enolase (EC 4.2.1.11)	Thermus		355,000	8	44,000	415
Phycocyanin	Anacystis, Nostic		360,000	12–24	15,000–30,000	416
Phosphofructokinase (EC 2.7.1.11)	Rabbit	Muscle	360,000	4	80,000	417
L-Arabinose isomerase (EC 5.3.1.4)	E. coli		360,000	6	60,000	418
Glutamine synthetase (EC 6.3.1.2)	Neurospora		360,000	4	90,000	419
Aspartate transcarbamylase (EC 2.1.3.2)	Pseudomonas		360,000	2	180,000	420
ATPase (EC 3.6.1.3)	Rat	Liver	360,000	6	53,000	421
				1	28,000	
				1	12,500	
				1	9,000	
Phosphoenolpyruvate carboxylase (EC 4.1.1.31)	E. coli W		361,000	4	88,200	422
D-Ribulose-1,5-biphosphate carboxylase (EC 4.1.1.39)	Chlorobium		361,000	6	53,000	423
Glycogen synthetase (EC 2.4.1.11)	Swine	Kidney	370,000	4	92,000	424

Table 1 (continued)
SUBUNIT CONSTITUTION OF PROTEINS

Protein	Source	Organ	Molecular weight	Subunit No.	Subunit Molecular weight	References
Phosphorylase A (EC 2.4.1.1)	Rabbit	Muscle	370,000	4	92,500	425
High density lipoprotein	Bovine	Serum	376,000	4	28,000	426
Polysaccharide depolymerase	Aerobacter		379,000	4	63,200	426a
				(3—4)	36,400	
ATPase (EC 3.6.1.3)	Streptococcus		385,000	12	33,000	427
Pyruvate phosphate dikinase	Maize		387,000	2	195,000	428
			195,000	2	94,000	
Phosphoenolpyruvate carboxylase (EC 4.1.1.31)	Salmonella		400,000	4	100,000	429
			100,000	2	50,000	
Aminopeptidase IA (EC 3.4.11.1)	Bacillus		400,000	10	36,000	430
				2	36,000	
Aminopeptidase IB (EC 3.4.11.1)	Bacillus		400,000	8	36,000	431
				4	36,000	
Aminopeptidase IC (EC 3.4.11.1)	Bacillus		400,000	6	36,000	432
				6	36,000	
RNA Polymerase (EC 2.7.7.6)	E. coli		400,000	2	39,000	433
				1	155,000	
				1	165,000	
RNA Polymerase (EC 2.7.7.6)	Rat	Liver	400,000	1	190,000	434
				1	150,000	
				1	35,000	
				1	25,000	
Cholesterol esterase (EC 3.1.1.13)	Rat	Pancreas	400,000	6	65,000	435
Lipovitellin	Chicken	Egg yolk	400,000	2	200,000	436
Glutamine synthetase	Ovine	Brain	400,000	8	49,000	437
Phosphoenolpyruvate carboxylase (EC 4.1.1.31)	E. coli B		402,000	4	99,600	438
Sucrose synthetase (EC 2.1.1.14)	Phaeolus		405,000	4	94,000	439
Phosphoenolpyruvate carboxytrans-phosphorylase (EC 4.1.1.38)	Entamoeba		408,000	2	200,000	440
			200,000	2	100,000	
Ribonucleotide reductase	Euglena		440,000	4	100,000	440a

347

Table 1 (continued)
SUBUNIT CONSTITUTION OF PROTEINS

Protein	Source	Organ	Molecular weight	Subunit No.	Subunit Molecular weight	References
Adenosine triphosphate sulfurylase (EC 2.7.7.4)	*Penicillium*		440,000	8	56,000	441
Apoferritin	Horse	Spleen	443,000	24	18,500	442
RNA Polymerase	*Physarum*		460,000	1	200,000	443
				1	135,000	
				1	45,000	
				2	24,000	
				1	17,000	
RNA Polymerase	*Bacillus*		466,000	1	160,000	444
				1	155,000	
				1	63,000	
				2	44,000	
Myosin	Rabbit	Muscle	468,000	2	212,000	445
				1	21,000	
				2	19,000	
				1	17,000	
Urease (EC 3.5.1.5)	Jack bean		480,000	2	240,000	446
			240,000	3	83,000	
Uridine diphosphate glucose pyrophosphorylase	Calf	Liver	480,000	8	60,000	447
Isocitrate lyase (EC 4.1.3.1)	*Turbatrix*		480,000	4	123,000	448
Fatty acid synthetase	Pigeon	Liver	480,000	1	240,000	449
				1	240,000	
Fatty acid synthetase	Chicken	Liver	500,000	2	240,000	450
Ribulose 1,5-diphosphate carboxylase (EC 4.1.1.39)	*Hydrogenomonas*		515,000	12–14	40,700	451
Pyruvate carboxylase	Pig	Liver	520,000	4	130,000	452
β-Galactosidase (EC 3.2.1.23)	*E. coli*		540,000	4	135,000	453
Ribulose diphosphate carboxylase (EC 4.1.1.39)	Spinach	Leaves	550,000	8	52,000	454
			24,500	6	24,500	
L-Malic enzyme (EC 1.1.1.40)	*E. coli*		550,000	2	12,000	455
				8	67,000	
Phosphoenolpyruvate carboxylase (EC 4.1.1.31)	Spinach	Leaf	560,000	4	130,000	455a

Table 1 (continued)
SUBUNIT CONSTITUTION OF PROTEINS

Protein	Source	Organ	Molecular weight	Subunit No.	Subunit Molecular weight	References
Citrate lyase (EC 4.1.3.6)	*Aerobacter*		575,000	2	290,000	456
			290,000	2	137,000	
			137,000	2	74,000	
Ribulose 1,5-diphosphate carboxylase (EC 4.1.1.39)	*Chlorella*		588,000	8	58,200	457
				8	15,300	
Phosphofructokinase (EC 2.7.1.11)	Yeast		590,000	6	100,000	458
Glutamine synthetase (EC 6.3.1.2)	*E. coli*		592,000	12	48,500	459
Glyceraldehyde 3-phosphate dehydrogenase	Spinach	Chloroplasts	600,000	4	145,000	460
Acetolactate synthase (EC 4.1.3.18)	*Pseudomonas*		600,000	(8)	60,000	461
				(8)	60,000	
Glutamine synthetase (EC 6.3.1.2)	*Bacillus*		600,000	12	50,000	462
Ovomacroglobulin	Chicken	Egg white	650,000	2	325,000	463
Globulin	French bean		654,000	4	163,000	464
			163,000	1	43,000	
				1	47,000	
				1	53,000	
Pyruvate carboxylase (EC 6.4.1.1)	Chicken	Liver	660,000	4	165,000	465
			165,000	4	45,000	
Thyroglobulin	Bovine	Thyroid	669,000	2	335,000	466
Isocitrate dehydrogenase (EC 1.1.1.41)	Bovine	Heart	670,000	2	330,000	467
			330,000	8	41,000	
L-Aspartate β-decarboxylase (EC 4.1.1.12)	*Alcaligines*		675,000	6	112,000	468
			112,000	2	57,000	
Sulfite reductase (EC 1.8.1.2)	*E. coli, Salmonella*		700,000	4	53,000	469
				8	58,000	
Nitrate reductase	*E. coli*		773,000	4	142,000	469a
				4	58,000	
Propionylcarboxylase (EC 6.4.1.3)	Pig	Heart	700,000	4	175,000	470
Lysine decarboxylase (EC 4.1.1.18)	*E. coli*		780,000	10	80,000	471
RNA Polymerase (EC 2.7.7.6)	*Azotobacter*		782,000	2	391,000	472

Table 1 (continued)
SUBUNIT CONSTITUTION OF PROTEINS

Protein	Source	Organ	Molecular weight	Subunit No.	Subunit Molecular weight	References
Transcarboxylase (EC 2.1.3.1)	*Propionibacterium*		792,000	1	360,000	473
				3	144,000	
			360,000	3	120,000	
			120,000	2	60,000	
			144,000	2	60,000	
				2	12,000	
Phosphofructokinase (EC 2.7.1.11)	Chicken	Liver	800,000	2	400,000	474
			400,000	2	210,000	
			210,000	2	100,000	
			100,000	2	60,000	
L-Aspartate β-decarboxylase (EC 4.1.1.12)			800,000	2	400,000	475
			400,000	4	100,000	
α-Crystallin			810,000	(30)	26,000	476
Arginine decarboxylase (EC 4.1.1.19)	*E. coli*		820,000	5	160,000	477
			160,000	2	82,000	
Cytochrome P-450	Bovine	Adrenocortical mitochondria	850,000	2	470,000	478
			470,000	8	53,000	
RNA Polymerase (EC 2.7.7.6)	*E. coli* B		880,000	2	440,000	479
Hemocyanin	*Loglio, Cancer*, ghost shrimp		300,000–9,000,000		760,000	480
					430,000	
					380,000	
					74,000	
					70,000	
Lipoate succinyltransferase	Pig	Heart	1,000,000	(24)	41,000	481
Monoamine oxidase	Pig	Liver	1,200,000	(8)	146,000	482
Phosphorylase kinase (EC 2.7.1.38)	Rabbit	Muscle	1,330,000	4	118,000	483
				4	108,000	
				8	41,000	
Fatty acid synthetase	*Mycobacterium*		1,390,000	(?)	250,000	484
Dihydrolipoyl transacetylase (EC 2.3.1.12)	*E. coli*		1,700,000	24	65,000	485
Chlorocruorin	*Spirographis*	Blood	2,750,000	12	250,000	486

Table 1 (continued)
SUBUNIT CONSTITUTION OF PROTEINS

Protein	Source	Organ	Molecular weight	Subunit No.	Subunit Molecular weight	Reference
α-Ketoglutarate dehydrogenase complex	E. coli		2,784,000	12	94,000	487
				12	54,000	
				24	42,000	
Hemoglobin	Arenicola	Blood	2,850,000	12	230,000	488
			230,000	4	54,000	
Erythrocruorin	Cirraformia	Plasma	3,000,000	162	18,500	489
Dihydrolipoyl transacetylase	Bovine	Kidney, heart	3,120,000	60	52,000	490
Phage fII			3,620,000	180	13,750	491
Pyruvate dehydrogenase complex	E. coli		4,600,000	24	96,000	492
				24	65,000	
				12	56,000	
Cowpea chlorotic mottle virus			4,608,000	180	19,600	492
Bromgrass mosaic virus			4,600,000	180	20,000	494
Broad bean mottle virus			4,800,000	180	20,900	495
Turnip yellow mosaic virus			5,000,000	180	20,133	496
Poliomyelitis virus			5,500,000	130	27,000	497
Cucumber mosaic virus			5,500,000	180	25,000	498
Alfalfa mosaic virus			7,400,000	(140)	51,600	499
			51,600	2	24,500	
Acetyl CoA carboxylase	Chicken	Liver	4–10,000,000	(?)	470,000	500
			470,000	2	117,000	
				1	129,000	
				1	139,000	
Rhinovirus 1A			8,400,000	60	96,000	501
Bushy stunt virus			9,000,000	180	42,000	502
Polyoma virus			24,000,000	420	50,200	503
Potato virus X			35,000,000	650	52,000	504
Tobacco mosaic virus			40,000,000	2130	17,500	505

Compiled by Dennis W. Darnall and Irving M. Klotz.

Table 2
ALPHABETICAL LISTING OF PROTEINS FROM TABLE 1

Protein	References
Acetoacetate decarboxylase	405
Acetacetyl-CoA thiolase	279
α-Acetohydroxy acid isomeroreductase	355
Acetol acetate-forming enzyme	329
Acetolactate synthase	461
β-D-N-Acetyl amidase	155
Acetylcholinesterase	376
Acetyl-CoA carboxylase	322, 500
α-Acetyl galactosaminidase	260
O-Acetylserine sulfhydrylase A	71
Acid phosphatase	123, 160, 182
Adenine phosphoribosyltransferase	20
Adenosine deaminase	328a
Adenosine triphosphate sulfurylase	441
Adenosylmethionine decarboxylase	184a
Adenylate kinase	74, 139
Agglutinin	25, 195
Alcohol dehydrogenase	49a, 105, 246a, 258
Aldehyde dehydrogenase	358
Aldolase	80, 106, 244, 255, 264
Aldose reductase	56
Alfalfa mosaic virus	499
Aliphatic amidase	306
Alkaline phosphatase	124, 204, 246, 261a
Allophycocyanin	145
ω-Amidase	49
D-Amino acid oxidase	148
L-Amino acid oxidase	223
Aminoacyl transferase	290
Aminobutanal dehydrogenase	342
δ-Aminolevulinate dehydratase	351, 380
Aminopeptidase	410, 430–432
3-Aminopropanal dehydrogenase	329a
Aminotripeptidase	226
AMP Deaminase	374
α-Amylase	144
β-Amylase	307
Anthranilate synthetase	114, 125, 247, 352, 379
Anti-Al lectin	201b
Apoferritin	442
L-Arabinose isomerase	418
Arachin	411
Arginase	192
Arginine decarboxylase	386, 477
Argininosuccinase	308

Table 2
ALPHABETICAL LISTING OF PROTEINS FROM TABLE 1 (continued)

Protein	References
Arylamidase	338
Ascorbate oxidase	237
L-Asparaginase	197, 220, 225, 230, 252a
Aspartase	283
Aspartate aminotransferase	153
L-Aspartate β-decarboxylase	468, 475
Aspartic β-semialdehyde dehydrogenase	261
Aspartokinase	166, 196, 221, 301
Aspartokinase-homoserine dehydrogenase	278, 406
Aspartyl transcarbamylase	227, 393, 420
ATPase	334, 413, 421, 427
Avidin	77
Azoferredoxin	46
Biotin carboxyl carrier protein	35
Broad bean mottle virus	495
Bromgrass mosaic virus	494
Bushy stunt virus	502
Butyryl cholinesterase	369
C-Reactive protein	234
Canavalin	184
Carbamoylphosphate synthase	282, 396
Carbonic anhydrase	287a
Carboxylesterase	274, 285
Carboxypeptidase G_1	136a
Catabolite gene activator protein	36
Catalase	347
Ceramide trihexosidase	143
Ceruloplasmin	208
Chlorocruorin	486
Chloroplast coupling factor	400
Cholesterol esterase	435
Cholinesterase	395
Chorionic gonadotropin	29
Chorismate mutase-prephenate dehydrogenase	94, 332
Chymotrypsin inhibitor	30
Citrate lyase	456
Citrate synthase	152, 357a, 365a
Collagen	345a
Concanavalin A	173
Creatine kinase	104
Creatine transphosphorylase	116
Crotonase	263, 272
α-Crystallin	476
Cowpea chlorotic mottle virus	493

Table 2
ALPHABETICAL LISTING OF PROTEINS FROM
TABLE 1 (continued)

Protein	References
Cucumber mosaic virus	498
Cyclic AMP dependent protein kinase	98, 199, 207, 232, 239, 284
Cyclic GMP protein kinase	238
Cystathionase	261
Cystathionine γ-synthetase	266, 363
Cysteamine oxygenase	150
Cysteine desulfhydrase	343
Cysteine synthetase	392
Cytidine triphosphate synthetase	318
Cytochrome b_2	348
Cytochrome (CC')	11
Cytochrome oxidase	305b
Cytochrome P-450	478
Cytoplasmic protein	37
DDT-Dehydrochlorinase	203
Delta hemolysin	163
2-Deoxycitrate synthase	134
Deoxycytidylate deaminase	51, 211
Dethiobiotin synthetase	32a
3-Deoxy-D-arabinoheptulosonate 7-phosphate synthetase-chorismate mutase	266a
Diacetyl reductase	95, 149
α-Dialkylamino acid transaminase	291
2,4-Diaminopentanoic acid C_4 dehydrogenase	88
Dihydrolipoyl dehydrogenase	183
Dihydrolipoyl transacetylase	485, 490
Dihydropteridine reductase	33
3,4-Dihydroxyphenylacetate-2,3-deoxygenase	243
Dihydroxypyridine oxygenase	357
Dipeptidase M	141
Dipeptidyltransferase	296
DNA Modification enzyme	189
DNA Modification methylase	187
DNA Polymerase	55
DNA Restriction endonuclease	362
Dopamine-β-hydroxylase	384
Edestin	389
Electron transferring flavoprotein	93
Enolase	115, 127, 131a, 415
Erythrocruorin	489
Erythrocuprein	59
Erythro-3,5-diaminohexanoate dehydrogenase	111, 243
Erythulose reductase	133
Esterase	242
Estradiol dehydrogenase	76

Table 2
ALPHABETICAL LISTING OF PROTEINS FROM TABLE 1 (continued)

Protein	References
Excelsin	390
Exonuclease I	239a
Factor X	39
Fatty acid synthetase	449, 450, 484
Fatty acylthiokinase	206
Follicle stimulating hormone	17b, 19, 22
N-Formimino-L-glutamate iminohydrolase	147
Formyltetrahydrofolate synthetase	350
Fructose diphosphatase	216, 241
α-L-Fucosidase	327
Fumarase	294
Galactokinase	42
D-Galactose dehydrogenase	60a, 158
Galactose-1-phosphate uridyltransferase	113, 132
β-Galactosidase	453
Globulin	464
α-Glucan phosphorylase	324
Glucocerebrosidase	355
Glucokinase	70, 151
D-Gluconate dehydratase	218
Glucose-6-phosphate dehydrogenase	217, 311, 314, 381
β-Glucuronidase	378
Glutamate decarboxylase	121, 394
Glutamate dehydrogenase	383, 397, 403
Glutamine phosphoribosylpyrophosphate amidotransferase	298, 371
Glutamine synthetase	404, 409, 414, 419, 437, 459, 462
Glutamine transaminase	176
Glutamyl tRNA synthetase	157
Glutathione peroxidase	93a, 117
Glutathione reductase	159, 210
Glyceraldehyde-3-phosphate dehydrogenase	249, 460
D-Glycerate dehydrogenase	78a
Glycerol kinase	328
Glycerol-1-phosphate dehydrogenase	99
Glycerol-3-phosphate dehydrogenase	75
Glycogen synthetase	361, 388, 424
Glycohydrolase	85
Glycollate oxidase	372
Glycyl-L-leucine hydrolase	171
Glycyl tRNA synthetase	341
Glyoxylate carboligase	345
Glyoxylic acid reductase	145a
Growth hormone	38
Haptoglobin	122

Table 2
ALPHABETICAL LISTING OF PROTEINS FROM
TABLE 1 (continued)

Protein	References
Hemagglutinin LcH	39a
Hemerythrin	31a, 175
Hemocyanin	480
Hemoglobin	21, 62, 78, 488
Hexokinase	161
High density lipoprotein	129, 319, 426
Histidine decarboxylase	181, 292
Histidinol dehydrogenase	112, 126
Histidinol phosphate aminotransferase	90
Histidyl tRNA synthetase	97, 119
Hyaluronidase	47
Hybridase	212
Hydrogenase	52
Hydroxyacyl-CoA dehydrogenase	64
Hydroxybutyryl-CoA dehydrogenase	336
β-Hydroxydecanoylthioester dehydrase	27
Hydroxyindole-O-methyl transferase	100
3-Hydroxy-3-methylglutaryl CoA reductase	305c
L-6-Hydroxynicotine oxidase	139a
Hydroxypyruvate reductase	83
3-Hydroxysteroid dehydrogenase	155a
20-β-hydroxysteroid dehydrogenase	170
Hypoxanthine-guanine phosphoribosyl transferase	96
Insulin	5
Interstitial cell-stimulating hormone	14
Invertase	317
Isocitrate dehydrogenase	50, 116a, 138a, 168, 391, 467
Isocitrate lyase	313, 488
α-Isopropylmalate synthase	300
2-Keto-3-deoxy-6-phosphogluconate aldolase	86
α-Ketoglutarate dehydrogenase	487
α-Ketoglutaric semialdehyde dehydrogenase	194
Lac repressor	265
Lactate dehydrogenase	79, 235
β-Lactoglobulin	23
Lactose specific factor	18
Lactose specific phosphocarrier protein	24
Lactose synthetase	53
Lectin	186, 248
Leucine aminopeptidase	367, 401
L-Leucine: 2-oxoglutarate aminotransferase	288
Leucine tRNA synthetase	198
Leuteinizing hormone	10, 12, 15
Lipoamide dehydrogenase	169

Table 2
ALPHABETICAL LISTING OF PROTEINS FROM
TABLE 1 (continued)

Protein	References
Lipoate succinyltransferase	481
Lipoxygenase	174
Lipovitellin	436
Lombricine kinase	107
Luciferase	102, 138
Lysine 2,3-aminomutase	382
Lysine decarboxylase	471
β-Lysine mutase	281
Lysine tRNA synthetase	228
M-line protein	130
Malate dehydrogenase	44, 68, 69, 89, 253
Malate-lactate transhydrogenase	81
Malic enzyme	30, 365, 367, 455
Mandelate racemase	375
Melilotate hydroxylase	360
Mercaptopyruvate sulfur transferase	7
Methionine tRNA synthetase	280
5,10-Methylenetetrahydrofolate dehydrogenase	48
Methylmalonate semialdehyde dehydrogenase	219a
Methylmalonyl-CoA mutase	209
N-Methylglutamate synthetase	412
Mo-Fe protein	305
Molybdoferredoxin	277
Monoamine oxidase	188, 391a, 482
Myosin	445
Myrokinase	224
Nerve growth factor	9
Neuraminidase	303
Nitrate reductase	469a
Nitrogenase	385, 402
5' Nucleosidase	256
Nucleoside diphosphate kinase	84, 162, 229
Nucleoside phosphotransferase	34
Oestradiol-receptor	202a
D-α-Ornithine 5,4-aminomutase	286
Ornithine aminotransferase	219
Ornithine transcarbamylase	172, 339
2-Oxoglutarate dehydrogenase	325
$\Delta_{5 \rightarrow 4}$-3-Oxosteroid isomerase	8
Ovomacroglobulin	463
Palmitoyl-CoA synthetase	276
Paramyosin	331
Peptidase	408
Peroxidase	305a

Table 2
ALPHABETICAL LISTING OF PROTEINS FROM
TABLE 1 (continued)

Protein	References
Phage F II	491
Phenylalanine ammonia lyase	399
Phenylalanine hydroxylase	317a
Phenylalanine hydroxylase stimulating protein	41
Phenylalanine tRNA synthetase	370, 373
Phosphofructokinase	251, 417, 458, 474
Phosphoenolpyruvate carboxylase	297, 407, 422, 429, 438, 452, 455a, 465
Phosphoenolpyruvate carboxytransphosphorylase	440
Phosphoglucomutase	58
6-Phosphogluconate dehydrogenase	140, 156, 165
Phosphoglucose isomerase	193, 222
Phosphoglycerate dehydrogenase	270
Phosphoglycerate mutase	67, 179
Phospholipase A_2	14a
Phosphoribosyl ATP pyrophosphate phosphoribosyl transferase	320
Phosphorylase A	425
Phosphorylase kinase	483
Phosphotransacetylase	103
Phycocyanin	40, 346, 416
Phytochrome	366
Phytohemagglutinin	240
Plasma factor XIII	398
Platelet factor XIII	252
Polio myelitus virus	497
Polynucleotide phosphorylase	299
Polyoma virus	503
Polysaccharide depolymerase	426a
Potato virus X	504
Prealbumin	57
Prephenoloxidase	109
Procarboxypeptidase A	128
Prohistidine decarboxylase	293
Prolyl tRNA synthetase	141a
Propionyl carboxylase	470
Propylamine transferase	89
Proteinase inhibitor I	32
Protein kinase	377
Protein P11	85a
Protein phosphokinase	232
Protein toxin A	353
Protein toxin B	200
Protocollagen proline hydroxylase	354

Table 2
ALPHABETICAL LISTING OF PROTEINS FROM
TABLE 1 (continued)

Protein	References
Purine nucleoside phosphorylase	118, 135
Putrecine oxidase	131
Pyridoxamine pyruvate transaminase	257
Pyrophosphatase	65, 201
Pyruvate carboxylase	452
Pyruvate decarboxylase	316
Pyruvate dehydrogenase	136, 492
Pyruvate kinase	294a, 323, 337, 344, 349, 356
Pyruvate phosphate dikinase	428
Quinolate phosphoribosyltransferase	273
Rhamnulose 1-phosphate aldolase	236
Rhinovirus 1A	501
Rhodanese	28
Ribitol dehydrogenase	177a
Ribonuclease	13
Ribonucleotide diphosphate reductase	340, 359
Ribonucleotide reductase	440a
Ribosephosphate isomerase	167
Ribulokinase	146
D-Ribose 1,5-diphosphate carboxylase	185, 423, 451, 454
RNA Polymerase	26, 433, 434, 443, 444, 472, 479
S-100 protein	6
Salycilate hydroxylase	137
Sedoheptulose 1,7-diphosphatase	92
Serine dehydratase	60
Serine transhydroxymethylase	321
Seryl tRNA synthetase	154, 201a
Succinic dehydrogenase	120, 302
Succinic semialdehyde dehydrogenase	271
Succinyl-CoA synthetase	233
Sucrase-isomaltase complex	333
Sucrose synthetase	439
Sulfite oxidase	188a
Sulfite reductase	469
Sulphatase A	171a
Superoxide dismutase	16, 17, 31
Tartaric acid dehydrase	250
Taurocyamine kinase	108
Thetin homocysteine methylpherase	287
Thiogalactoside transacetylase	66
Thiolase	279a
Threonine deaminase	269, 295, 310
Threonine dehydratase	289

Table 2
ALPHABETICAL LISTING OF PROTEINS FROM
TABLE 1 (continued)

Protein	References
Thioredoxin reductase	67a
Thymidylate synthetase	61, 82
Thyrotropin	17a
Thyroglobulin	466
Tobacco mosaic virus	505
Transaldolase III	73
Transcarboxylase	473
Transcobalamin	54
Transglutaminase	45
Transketolase	231
Trimethylamine dehydrogenase	268
Triosephosphate isomerase	43
Tropomyosin B	72
Troponin	110
Tryptophanase	315, 326, 330
Tryptophan oxygenase	205, 275
Tryptophan synthetase	248a, 254
Tryptophanyl tRNA synthetase	91, 178
Tubulin	177
Turnip yellow mosaic virus	496
T_u-T_s Complex	63
Tyrosinase	215, 304
Tyrosine aminotransferase	190
Tyrosyl tRNA synthetase	142, 191
Urease	446
Uricase	213
Uridine diphosphogalactose 4-epimerase	101, 364
Uridine diphosphate glucose dehydrogenase	387
Uridine diphosphate glucose pyrophosphorylase	447
Uridine phosphorylase	164
Urocanase	180
Xanthosine 5'-phosphate aminase	214
Xylose isomerase	262

Compiled by Dennis W. Darnall and Irving M. Klotz.

REFERENCES

1. Darnall and Klotz, *Arch. Biochem. Biophys.*, 149, 1 (1972).
2. Klotz, Langerman, and Darnall, *Annu. Rev. Biochem.*, 39, 25 (1970).
3. Klotz and Darnall, *Science,* 166, 126 (1969).
4. Klotz, *Science,* 155, 697 (1967).
4a. Kleine, *Fortschr. Arzneimittelforsch.*, 16, 365 (1972).
5. Crowfoot, *Proc. R. Soc. London, Ser. A,* 164, 580 (1938); Moody, Dissertation, University of Wisconsin, 1944; Waugh, *Adv. Protein Chem.*, 9, 325 (1954).
6. Stewart, *Biochem. Biophys. Res. Commun.*, 46, 1405 (1972); Dannies and Levine, *Biochem. Biophys. Res. Commun.*, 37, 587 (1969).
7. Vachek and Wood, *Biochim. Biophys. Acta,* 258, 133 (1972).
8. Weintraub, Vincent, Baulieu, and Alfsen, *FEBS Lett.*, 37, 82 (1973).
9. Angeletti and Bradshaw, *Proc. Natl. Acad. Sci. U.S.A.*, 68, 2417 (1971).
10. Liu, Nahm, Sweeney, Holcomb, and Ward, *J. Biol. Chem.*, 247, 4365 (1972); Liu, Nahm, Sweeney, Lamkin, Baker, and Ward, *J. Biol. Chem.*, 247, 4351 (1972).
11. Cusanovich, Tedro, and Kamen, *Arch. Biochem. Biophys.*, 141, 557 (1970).
12. Bishop and Ryan, *Biochemistry,* 12, 3077 (1973).
13. D'Alessio, Parente, Guida, and Leone, *FEBS Lett.*, 27, 285 (1972).
14. Sairam, Papkoff, and Li, *Arch. Biochem. Biophys.*, 153, 554 (1972); Sairam, Samy, Papkoff, and Li, *Arch. Biochem. Biophys.*, 153, 572 (1972).
14a. Wells, *Biochemistry,* 10, 4074 (1971).
15. Ward, Reichart, Fitak, Nahm, Sweeney, and Neill, *Biochemistry,* 10, 1796 (1971).
16. Misra and Fridovich, *J. Biol. Chem.*, 247, 3410 (1972).
17. Hartz and Deutsch, *J. Biol. Chem.*, 247, 7043 (1972).
17a. Hennen, Maghuin-Rogister, and Mamoir, *FEBS Lett.*, 9, 20 (1970).
17b. Grimek and McShan, *J. Biol. Chem.*, 249, 5725 (1974).
18. Schrecker and Hengstenberg, *FEBS Lett.*, 13, 209 (1971).
19. Landefeld and McShan, *J. Biol. Chem.*, 249, 3527 (1974).
20. Thomas, Arnold, and Kelley, *J. Biol. Chem.*, 248, 2529 (1973).
21. Ohnoki, Mitomi, Hata, and Satake, *J. Biochem.*, 73, 717 (1973).
22. Ryan, Jiang, and Hanlon, *Biochemistry,* 10, 1321 (1971); Saxena and Rathnam, *J. Biol. Chem.*, 246, 3549 (1971).
23. Bull, *J. Am. Chem. Soc.*, 68, 745 (1946); Townend and Timasheff, *J. Am. Chem. Soc.*, 79, 3613 (1957).
24. Hays, Simoni, and Roseman, *J. Biol. Chem.*, 248, 941 (1973).
25. Nagata and Burger, *J. Biol. Chem.*, 249, 3116 (1974).
26. Louis and Fitt, *Biochem. J.*, 127, 69 (1972).
27. Helmkamp and Bloch, *J. Biol. Chem.*, 244, 6014 (1969).
28. Volini, DeToma, and Westley, *J. Biol. Chem.*, 242, 5220 (1967).
29. Bellisario, Carlsen, and Bahl, *J. Biol. Chem.*, 248, 6796 (1973); Carlsen, Bahl, and Swaminathan, *J. Biol. Chem.*, 248, 6810 (1973).
30. Melville and Ryan, *Arch. Biochem. Biophys.*, 138, 700 (1970).
31. Keele, McCord, and Fridovich, *J. Biol. Chem.*, 245, 6176 (1970).
31a. Liberatore, Truby, and Klippenstein, *Arch. Biochem. Biophys.*, 160, 223 (1974).
32. Kiyohara, Iwasaki, and Yoshikawa, *J. Biochem.*, 73, 89 (1973).
32a. Krell and Eisenberg, *J. Biol. Chem.*, 245, 6558 (1970).
33. Craine, Hall, and Kaufman, *J. Biol. Chem.*, 247, 6082 (1972).
34. Rodgers and Chargaff, *J. Biol. Chem.*, 247, 5448 (1972).
35. Fall and Vagelos, *J. Biol. Chem.*, 247, 8005 (1972).
36. Riggs, Reiness, and Zubay, *Proc. Natl. Acad. Sci. U.S.A.*, 68, 1222 (1971).
37. Shannon and Hill, *Biochemistry,* 10, 3021 (1971).
38. Edelhoch, Condliffe, Lippoldt, and Burger, *J. Biol. Chem.*, 241, 5205 (1966).
39. Radcliffe and Barton, *J. Biol. Chem.*, 247, 7735 (1972).
39a. Howard, Sage, Stein, Young, Leon, and Dykes, *J. Biol. Chem.*, 246, 1590 (1971).
40. MacColl, Habig, and Berns, *J. Biol. Chem.*, 248, 7080 (1973).
41. Huang, Max, and Kaufman, *J. Biol. Chem.*, 248, 4235 (1973).
42. Blume and Buetler, *J. Biol. Chem.*, 246, 6507 (1971).
43. Hartman, *Biochemistry,* 10, 146 (1971).
44. Munkres, *Biochemistry,* 4, 2180, 2186 (1965).
45. Chung and Folk, *Proc. Natl. Acad. Sci. U.S.A.*, 69, 303 (1972).
46. Nakos and Mortenson, *Biochemistry,* 10, 455 (1971).

47. Khorlin, Vikha, and Milishnikov, *FEBS Lett.*, 31, 107 (1973).
48. O'Brien, Brewer, and Ljungdahl, *J. Biol. Chem.*, 248, 403 (1973).
49. Hersh, *Biochemistry*, 10, 2881 (1971).
49a. Jacobson and Pfuderer, *J. Biol. Chem.*, 245, 3938 (1970).
50. Magar and Robbins, *Biochim. Biophys. Acta*, 191, 173 (1969).
51. Bessman, Diamond, Debeer, and Duncan, *Fed. Proc., Fed. Am. Soc. Exp. Biol.*, 30, 1121 (1971); Duncan, Diamond, and Bessman, *J. Biol. Chem.*, 247, 8136 (1972).
52. Nakos and Mortenson, *Biochemistry*, 10, 2442 (1971).
53. Trayer and Hill, *J. Biol. Chem.*, 246, 6666 (1971); Magee, Mawal, and Ebner, *Fed. Proc., Fed. Am. Soc. Exp. Biol.*, 31, 499 (1972).
54. Allen and Majerus, *J. Biol. Chem.*, 247, 7709 (1972).
55. Hains, Wickremasinghe, and Johnston, *Eur. J. Biochem.*, 31, 119 (1972).
56. Sheys and Doughty, *Biochim. Biophys. Acta*, 235, 414 (1971).
57. Rask, Peterson, and Nilson, *J. Biol. Chem.*, 246, 6087 (1971); Gonzalez and Offord, *Biochem. J.*, 125, 309 (1971).
58. Duckworth and Sanwal, *Biochemistry*, 11, 3182 (1972).
59. Weser, Bunnenberg, Cammack, Djerassi, Flohé, Thomas, and Voelter, *Biochim. Biophys. Acta*, 243, 203 (1971).
60. Inoue, Kasper, and Pitol, *J. Biol. Chem.*, 246, 2626 (1971).
60a. Blachnitzky, Wengenmayer, and Kurz, *Eur. J. Biochem.*, 47, 235 (1974).
61. Galivan, Maley, and Maley, *Biochemistry*, 13, 2282 (1974).
62. Braunitzer, Hilse, Rudloff, and Hilschmann, *Adv. Protein Chem.*, 19, 1 (1964).
63. Hachmann, Miller, and Weissbach, *Arch. Biochem. Biophys.*, 147, 457 (1971).
64. Noyes and Bradshaw, *J. Biol. Chem.*, 248, 3060 (1973); Noyes, Glatthaar, Garavelli, and Bradshaw, *Proc. Natl. Acad. Sci. U.S.A.*, 71, 1334 (1974).
65. Heinrikson, Sterner, Noyes, Cooperman, and Bruckmann, *J. Biol. Chem.*, 248, 2521 (1973); Avaeva, Libedeva, Biesembaeva, and Egorov, *FEBS Lett.*, 24, 169 (1972); Ridling, Yang, and Butler, *Arch. Biochem. Biophys.*, 153, 714 (1972).
66. Brown, Brown, and Zabin, *J. Biol. Chem.*, 242, 4254 (1967).
67. Scopes and Penny, *Biochim. Biophys. Acta*, 236, 409 (1971).
67a. Thelander, *Eur. J. Biochem.*, 4, 407 (1968).
68. Mann and Vestling, *Biochemistry*, 8, 1105 (1969).
69. Noyes, Glatthaar, Garavelli, and Bradshaw, *Proc. Natl. Acad. Sci. U.S.A.*, 71, 1334 (1974).
70. Hengartner and Zuber, *FEBS Lett.*, 37, 212 (1973).
71. Becker, Kredich, and Tomkins, *J. Biol. Chem.*, 244, 2418 (1969).
72. Holtzer, Clark, and Lowey, *Biochemistry*, 4, 2401 (1965); Olander, Emerson, and Holtzer, *J. Am. Chem. Soc.*, 89, 3058 (1967); Woods, *J. Biol. Chem.*, 242, 2859 (1967).
73. Tsolas and Horecker, *Arch. Biochem. Biophys.*, 136, 303 (1970).
74. Criss, Sapico, and Litwack, *J. Biol. Chem.*, 245, 6346 (1970).
75. White, *Arch. Biochem. Biophys.*, 147, 123 (1971).
76. Burns, Engel, and Bethune, *Biochemistry*, 11, 2699 (1972); Jarabak and Street, *Biochemistry*, 10, 3831 (1971); Burns, Engel, and Bethune, *Biochem. Biophys. Res. Commun.*, 44, 786 (1971).
77. Green, *Biochem. J.*, 92, 16c (1964); Green and Ross, *Biochem. J.*, 110, 59 (1968).
78. Ohnoki, Mitomi, Hata, and Satake, *J. Biochem.*, 73, 717 (1973).
78a. Rosenblum, Antkowiak, Sallach, Flanders, and Fahien, *Arch. Biochem. Biophys.*, 144, 375 (1971).
79. Long and Kaplan, *Arch. Biochem. Biophys.*, 154, 696 (1973).
80. Stribling and Perham, *Biochem. J.*, 131, 833 (1973).
81. Allen, *Eur. J. Biochem.*, 35, 338 (1973).
82. Dunlap, Harding, and Huennekens, *Biochemistry*, 10, 88 (1971); Loeble and Dunlap, *Biochem. Biophys. Res. Commun.*, 49, 1671 (1972).
83. Utting and Kohn, *Fed. Proc., Fed. Am. Soc. Exp. Biol.*, 30, 1057 (1971).
84. Edlund, *FEBS Lett.*, 13, 56 (1971).
85. Green and Dobrjansky, *Biochemistry*, 10, 4533 (1971).
85a. Terzaghi and Terzaghi, *J. Biol. Chem.*, 249, 5119 (1974).
86. Hammerstedt, Mohler, Decker, and Wood, *J. Biol. Chem.*, 246, 2069, 2075 (1971); Vandlen, Ersfeld, Tulinsky, and Wood, *J. Biol. Chem.*, 248, 2251 (1973).
87. Wolfenstein, Englard, and Listowsky, *J. Biol. Chem.*, 244, 6415 (1969).
88. Somack and Costilow, *J. Biol. Chem.*, 248, 385 (1973).
89. Bowman, Tabor, and Tabor, *J. Biol. Chem.*, 248, 2480 (1973).
90. Henderson and Snell, *J. Biol. Chem.*, 248, 1906 (1973).
91. Joseph and Muench, *J. Biol. Chem.*, 246, 7610 (1971).
92. Traniello, Calcagno, and Pontremoli, *Arch. Biochem. Biophys.*, 146, 603 (1971).
93. Whitefield and Mayhew, *J. Biol. Chem.*, 249, 2801 (1974).
93a. Nakamura, Hosoda, and Hayashi, *Biochim. Biophys. Acta*, 358, 251 (1974).

94. Koch, Shaw, and Gibson, *Biochim. Biophys. Acta*, 212, 387 (1970); *Biochim. Biophys. Acta*, 229, 805 (1971).
95. Burgos and Martin, *Biochim. Biophys. Acta*, 268, 261 (1972).
96. Olsen and Milman, *J. Biol. Chem.*, 249, 4030, 4038 (1974); Arnold and Kelley, *J. Biol. Chem.*, 246, 7398 (1971).
97. DeLorenzo, Di Natale, and Schecter, *J. Biol. Chem.*, 249, 908 (1974).
98. Garbers, First, and Lardy, *J. Biol. Chem.*, 248, 875 (1973).
99. Pfeiderer and Auricchio, *Biochem. Biophys. Res. Commun.*, 16, 53 (1964); Deal and Holleman, *Fed. Proc., Fed. Am. Soc. Exp. Biol.*, 23, 264 (1964).
100. Jackson and Lovenberg, *J. Biol. Chem.*, 246, 4280 (1971).
101. Wilson and Hogness, *J. Biol. Chem.*, 244, 2132 (1969).
102. Meighen, Nicoli, and Hastings, *Biochemistry*, 10, 4062 (1971); Gunsalus-Miguel, Meighen, Nicoli, Nealson, and Hastings, *J. Biol. Chem.*, 247, 398 (1972).
103. Whiteley and Pelroy, *J. Biol. Chem.*, 247, 1911 (1972).
104. Dawson, Eppenberger, and Kaplan, *J. Biol. Chem.*, 242, 211 (1967); Bayley and Thomson, *Biochem. J.*, 104, 33c (1967).
105. Theorell and Winer, *Arch. Biochem. Biophys.*, 83, 291 (1959); Li and Vallee, *Biochemistry*, 3, 869 (1964); Drum, Harrison, Li, Bethune, and Vallee, *Proc. Natl. Acad. Sci. U.S.A.*, 57, 1434 (1967); Castellino and Barker, *Biochemistry*, 7, 2207 (1968); Weber and Osborn, *J. Biol. Chem.*, 244, 4406 (1969); Green and McKay, *J. Biol. Chem.*, 244, 5034 (1969).
106. Harris, Kobes, Teller, and Rutter, *Biochemistry*, 8, 2442 (1969).
107. Oriol, Landon, and Thoai, *Biochim. Biophys. Acta*, 207, 514 (1970).
108. Oriol, Landon, and Thoai, *Biochim. Biophys. Acta*, 207, 514 (1970).
109. Ashida, *Arch. Biochem. Biophys.*, 144, 749 (1971).
110. Ebashi, Wakabayashi, and Ebashi, *J. Biochem.*, 69, 441 (1971); Greaser and Gergely, *J. Biol. Chem.*, 248, 2124 (1973); Dabrowska, Barlyko, Nowak, and Drabikowski, *FEBS Lett.*, 29, 239 (1973).
111. Baker and van der Drift, *Biochemistry*, 13, 292 (1974).
112. Yang, Lee, and Haslam, *J. Molec. Biol.*, 81, 517 (1973).
113. Saito, Ozutsumi, and Kurahashi, *J. Biol. Chem.*, 242, 2362 (1967).
114. Queener, Queener, Meeks, and Gunsalus, *J. Biol. Chem.*, 248, 151 (1973).
115. Winstead and Wold, *Biochemistry*, 3, 791 (1964); Winstead and Wold, *Biochemistry*, 4, 2145 (1965); Cardenas and Wold, *Biochemistry*, 7, 2736 (1968).
116. Yue, Palmieri, Olsen, and Kuby, *Biochemistry*, 6, 3205 (1967).
116a. Burke, Johanson, and Reeves, *Biochim. Biophys. Acta*, 351, 333 (1974).
117. Flohé, Eisele, and Wendel, *Hoppe-Seyler's Z. Physiol. Chem.*, 352, 151 (1971).
118. Edwards, Edwards, and Hopkinson, *FEBS Lett.*, 32, 235 (1973).
119. Kalousek and Konigsberg, *Biochemistry*, 13, 999 (1974).
120. Hatefi, Davis, Baltscheffsky, Baltscheffsky, and Johansson, *Arch. Biochem. Biophys.*, 152, 613 (1972).
121. Wu, Matsuda, and Roberts, *J. Biol. Chem.*, 248, 3029 (1973).
122. Waks and Alfsen, *Arch. Biochem. Biophys.*, 123, 133 (1968).
123. Jacobs, Nye, and Brown, *J. Biol. Chem.*, 246, 1419 (1971).
124. Reynolds and Schlesinger, *Biochemistry*, 8, 588 (1969); Schlesinger and Barrett, *J. Biol. Chem.*, 240, 4284 (1965).
125. Sawula and Crawford, *J. Biol. Chem.*, 248, 3573 (1973).
126. Loper, *J. Biol. Chem.*, 243, 3264 (1968); Yourno, *J. Biol. Chem.*, 243, 3277 (1968); Lew and Roth, *Biochemistry*, 10, 204 (1971).
127. Mann, Castellino, and Hargrave, *Biochemistry*, 9, 4002 (1970).
128. Brown, Greenshields, Yamasaki, and Neurath, *Biochemistry*, 2, 867 (1963); Teller, *Biochemistry*, 9, 4201 (1970).
129. Scanu, Edelstein, and Lim, *Fed. Proc., Fed. Am. Soc. Exp. Biol.*, 31, 829 (1972).
130. Morimoto and Harrington, *J. Biol. Chem.*, 247, 3052 (1972).
131. DeSa, *J. Biol. Chem.*, 247, 5527 (1972).
131a. Spring and Wold, *J. Biol. Chem.*, 246, 6797 (1971).
132. Tedesco, *J. Biol. Chem.*, 247, 6631 (1972).
133. Uehara, Tanimoto, and Soto, *J. Biochem.*, 75, 333 (1974).
134. Måhlén, *Eur. J. Biochem.*, 22, 104 (1971).
135. Gilpin and Sadoff, *J. Biol. Chem.*, 246, 1475 (1971).
136. Reed, *Curr. Top. Cell. Regul.*, 1, 233 (1969).
136a. McCullough, Chabner, and Bertino, *J. Biol. Chem.*, 246, 7207 (1971).
137. White-Stevens and Kamin, *J. Biol. Chem.*, 247, 2358 (1972).
138. Travis and McElroy, *Biochemistry*, 5, 2170 (1966).
138a. Howard and Becker, *J. Biol. Chem.*, 245, 3186 (1971).
139. Takai, Kurashina, Suzuki-Hori, Okamoto, and Hayaishi, *J. Biol. Chem.*, 249, 1965 (1974).
139a. Dai, Decker, and Sund, *Eur. J. Biochem.*, 4, 95 (1968).
140. Silverberg and Dalziel, *Eur. J. Biochem.*, 38, 229 (1973).
141. Brown, *J. Biol. Chem.*, 248, 409 (1973).
141a. Lee and Muench, *J. Biol. Chem.*, 244, 223 (1969).

142. Koch, *Biochemistry*. 13, 2307 (1974).
143. Mapes, Suelter, and Sweeley, *J. Biol. Chem.*, 248, 2471 (1973).
144. Robyt and Ackerman, *Arch. Biochem. Biophys.*, 154, 445 (1973); Mitchell, Riquetti, Loring, and Carraway, *Biochim. Biophys. Acta*, 295, 314 (1973); Connellan and Shaw, *J. Biol. Chem.*, 245, 2845 (1970); Robyt, Chittenden, and Lee, *Arch. Biochem. Biophys.*, 144, 160 (1971).
145. Glazer and Cohen-Bazire, *Proc. Natl. Acad. Sci. U.S.A.*, 68, 1398 (1971).
145a. Kohn, Warren, and Carroll, *J. Biol. Chem.*, 245, 3821 (1970).
146. Lee, Patrick, and Barnes, *J. Biol. Chem.*, 245, 1357 (1970).
147. Wickner and Tabor, *J. Biol. Chem.*, 247, 1605 (1972).
148. Fonda and Anderson, *J. Biol. Chem.*, 243, 5635 (1968).
149. Hetland, Olsen, Christensen, and Stϕrmer, *Eur. J. Biochem.*, 20, 200 (1971).
150. Federici, Barra, Fiori, and Costa, *Physiol. Chem. Phys.*, 3, 448 (1971); Gavallini, Cannella, Federici, Dupre, Fiori, and DelGrosso, *Eur. J. Biochem.*, 16, 537 (1970).
151. Rustum and Barnard, *Fed. Proc., Fed. Am. Soc. Exp. Biol.*, 30, 1122 (1971).
152. Wu and Yang, *J. Biol. Chem.*, 245, 212 (1970); Singh, Books, and Srere, *J. Biol. Chem.*, 245, 4636 (1970); Moriyama and Srere, *J. Biol. Chem.*, 246, 3217 (1971).
153. Bertland and Kaplan, *Biochemistry*, 7, 134 (1968).
154. Katze and Konigsberg, *J. Biol. Chem.*, 245, 923 (1970); Boeker, Hays, and Cantorie, *Biochemistry*, 12, 2379 (1973).
155. Srivastava, Yoshida, Awasthi, and Beutler, *J. Biol. Chem.*, 249, 2049 (1974).
155a. Skålhegg, *Eur. J. Biochem.*, 46, 117 (1974).
156. Veronese, Boccu, Fontana, Benassi, and Scoffone, *Biochim. Biophys. Acta*, 334, 31 (1974); Proscal and Holten, *Biochemistry*, 11, 1310 (1972).
157. Lapointe and Söll, *J. Biol. Chem.*, 247, 4966 (1972).
158. Wengenmayer, Ueberschär, and Kurz, *Eur. J. Biochem.*, 43, 49 (1974).
159. Ii and Sakai, *Biochim. Biophys. Acta*, 350, 141 (1974).
160. Derechin, Ostrowski, Galka, and Barnard, *Biochim. Biophys. Acta*, 250, 143 (1971).
161. Pringle, *Biochem. Biophys. Res. Commun.*, 39, 46 (1970); Schmidt and Colowick, *Fed. Proc., Fed. Am. Soc. Exp. Biol.*, 29, 334 (1970).
162. Palmieri, Yue, Jacobs, Maland, Wu, and Kuby, *Fed. Proc., Fed. Am. Soc. Exp. Biol.*, 29, 914 (1970); Palmieri, Yue, Jacobs, Maland, Wu, and Kuby, *J. Biol. Chem.*, 248, 4486 (1973).
163. Kantor, Temples, and Shaw, *Arch. Biochem. Biophys.*, 151, 142 (1972).
164. Bose and Yamada, *Biochemistry*, 13, 2051 (1974).
165. Pearse and Rosemeyer, *Eur. J. Biochem.*, 42, 225 (1974).
166. Richard, Mazat, Gros, and Patte, *Eur. J. Biochem.*, 40, 619 (1973).
167. Domagk, Doering, and Chilla, *Eur. J. Biochem.*, 38, 259 (1973).
168. Chung and Braginski, *Arch. Biochem. Biophys.*, 153, 357 (1973).
169. Burleigh and Williams, *J. Biol. Chem.*, 247, 2077 (1972).
170. Blomquist, *Arch. Biochem. Biophys.*, 159, 590 (1973).
171. Das and Radhakrishnan, *Biochem. J.*, 135, 609 (1973).
171a. Roy and Jerfy, *Biochim. Biophys. Acta*, 207, 156 (1970).
172. Marshall and Cohen, *J. Biol. Chem.*, 247, 1641 (1972).
173. Wang, Cunningham, and Edleman, *Proc. Natl. Acad. Sci. U.S.A.*, 68, 1130 (1971); Hardman, Wood, Schiffey, Edmundson, and Ainsworth, *Proc. Natl. Acad. Sci. U.S.A.*, 68, 1393 (1971).
174. Stevens, Brown, and Smith, *Arch. Biochem. Biophys*, 136, 413 (1970).
175. Klotz and Keresztes-Nagy, *Biochemistry*, 2, 445, 923 (1963).
176. Cooper and Meister, *Biochemistry*, 11, 661 (1972).
177. Feit, Slusarek, and Shelanski, *Proc. Natl. Acad. Sci. U.S.A.*, 68, 2028 (1971).
177a. Taylor, Rigby, and Hartley, *Biochem. J.*, 141, 693 (1974).
178. Preddie, *J. Biol. Chem.*, 244, 3958 (1969); Gros, Lemaire, Rapenbusch, and Labouesse, *J. Biol. Chem.*, 247, 2931 (1972); Penneys and Muench, *Biochemistry*, 13, 560 (1974).
179. Sasaki, Sugimoto, and Chiba, *Agric. Biol. Chem.*, 34, 135 (1970); Campbell, Hodgson, Watson, and Scopes, *J. Mol. Biol.*, 61, 257 (1971).
180. Lynch and Phillips, *J. Biol. Chem.*, 247, 7799 (1972).
181. Prozorovski and Jörnvall, *Eur. J. Biochem.*, 42, 405 (1974).
182. Uehara, Fujimoto, and Taniguchi, *J. Biochem.*, 75, 627 (1974).
183. Reed, *Curr. Top. Cell. Regul.*, 1, 233 (1969).
184. McPherson and Rich, *J. Biochem.*, 74, 155 (1973).
184a. Wickner, Tabor, and Tabor, *J. Biol. Chem.*, 245, 2132 (1970).
185. Tabita and McFadden, *J. Biol. Chem.*, 249, 3459 (1974).
186. Andrews, *Biochem. J.*, 139, 421 (1974).
187. Lautenberger and Linn, *J. Biol. Chem.*, 247, 6176 (1972).
188. Oreland, Kinemuchi, and Stigbrand, *Arch. Biochem. Biophys.*, 159, 854 (1973).
188a. Cohen and Fridovich, *J. Biol. Chem.*, 246, 367 (1971).

189. Evans and Gurd, *Biochem. J.*, 133, 189 (1973).
190. Auricchio, Valeriote, Tomkins, and Riley, *Biochim. Biophys. Acta*, 221, 307 (1970).
191. Kucan and Chambers, *J. Biochem.*, 73, 811 (1973).
192. Carvajal, Venegas, Oestreicher, and Plaza, *Biochim. Biophys. Acta*, 250, 437 (1971).
193. Low and Reithel, *Fed. Proc., Fed. Am. Soc. Exp. Biol.*, 33, 1478 (1974).
194. Koo and Adams, *J. Biol. Chem.*, 249, 1704 (1974).
195. Lotan, Siegelman, Lis, and Sharon, *J. Biol. Chem.*, 249, 1219 (1974).
196. Biswas and Paulus, *J. Biol. Chem.*, 248, 2894 (1973).
197. Tosa, Sano, Yamamoto, Nakamura, and Chibata, *Biochemistry*, 12, 1075 (1973).
198. Chirikjian, Wright, and Fresco, *Proc. Natl. Acad. Sci. U.S.A.*, 69, 1638 (1972).
199. Garbers, First, and Lardy, *J. Biol. Chem.*, 248, 875 (1973).
200. Montie and Montie, *Biochemistry*, 10, 2094 (1971).
201. Wong, Hall, and Josse, *J. Biol. Chem.*, 245, 4335 (1970).
201a. Heider, Gottschalk, and Cramer, *Eur. J. Biochem.*, 20, 144 (1971).
201b. Pere, Font, and Bourrillon, *Biochim. Biophys. Acta*, 365, 40 (1974).
202. Rapoport, Davis, and Horecker, *Arch. Biochem. Biophys.*, 132, 286, (1969).
202a. Erdose and Fries, *Biochem. Biophys. Res. Commun.*, 58, 932 (1974).
203. Dinamarca, Levenbook, and Valdes, *Arch. Biochem. Biophys.*, 147, 374 (1971).
204. Hulett-Cowling and Campbell, *Biochemistry*, 10, 1371 (1971).
205. Poillon, Maeno, Koike, and Feigelson, *J. Biol. Chem.*, 244, 3447 (1969).
206. Bonner and Bloch, *J. Biol. Chem.*, 247, 3123 (1972).
207. Corbin, Bronstrom, King, and Krebs, *J. Biol. Chem.*, 247, 7791 (1972).
208. Freeman and Daniel, *Biochemistry*, 12, 4806 (1973).
209. Zagalak and Rétey, *Eur. J. Biochem.*, 44, 529 (1974).
210. Mavis and Stellwagen, *J. Biol. Chem.*, 243, 809 (1968).
211. Maley and Maley, *Fed. Proc., Fed. Am. Soc. Exp. Biol.*, 30, 1113 (1971).
212. Roewekamp and Sekeris, *Eur. J. Biochem.*, 43, 405 (1974).
213. Pitts, Priest, and Fish, *Biochemistry*, 13, 889 (1974).
214. Sakamoto, Hatfield, and Moyed, *J. Biol. Chem.*, 247, 5880 (1972).
215. Zito and Kertesz, in *Biological and Chemical Aspects of Oxygenases*, **Bloch and Hayaishi, Eds.**, Maruzen, Tokyo, 1966, 290; Bouchilloux, McMahill, and Mason, *J. Biol. Chem.*, 238, 1699 (1963).
216. Mendicino, Kratowich, and Oliver, *J. Biol. Chem.*, 247, 6643 (1972).
217. Levy, Raineri, and Nevaldine, *J. Biol. Chem.*, 241, 2181 (1966).
218. Bender and Gottschalk, *Eur. J. Biochem.*, 40, 309 (1973).
219. Peraino, Bunville, and Tahmisian, *J. Biol. Chem.*, 244, 2241 (1969).
219a. Bannerjee, Sanders, and Sokatch, *J. Biol. Chem.*, 245, 1828 (1970).
220. Frank, Pekar, Veros, and Ho, *J. Biol. Chem.*, 245, 3716 (1970).
221. **Dungan and Datta**, *J. Biol. Chem.*, 248, 8534 (1973).
222. **Carter and Yoshida**, *Biochim. Biophys. Acta*, 181, 12 (1969); **Yoshida and Carter**, *Biochim. Biophys. Acta*, 194, 151 (1969); **Blackburn and Noltman**, *J. Biol. Chem.*, 247, 5668 (1972).
223. deKok and Rawitch, *Biochemistry*, 8, 1405 (1969).
224. Lönnerdal and Janson, *Biochim. Biophys. Acta*, 315, 421 (1973).
225. Cammack, Marlborough, and Miller, *Biochem. J.*, 126, 361 (1972).
226. Chenoweth, Brown, Valenzuela, and Smith, *J. Biol. Chem.*, 248, 1684 (1973).
227. Aitken, Bhatti, and Kaplan, *Biochim. Biophys. Acta*, 309, 50 (1973).
228. Rymo, Lundvik, and Lagerkvist, *J. Biol. Chem.*, 247, 3888 (1972); **Lagerkvist, Rymo, Lindqvist, and Anderson**, *J. Biol. Chem.*, 247, 3897 (1972).
229. Hossler and Rendi, *Biochem. Biophys. Res. Commun.*, 43, 530 (1971).
230. Roberts, Holcenberg, and Dolowy, *J. Biol. Chem.*, 247, 84 (1972).
231. Heinrich and Wiss, *FEBS Lett.*, 14, 251 (1971).
232. Tao, Salas, and Lipmann, *Proc. Natl. Acad. Sci. U.S.A.*, 67, 408 (1970); **Miyamoto, Petzold, Harris, and Greengard**, *Biochem. Biophys. Res. Commun.*, 44, 305 (1971).
233. Bridger, *Biochem. Biophys. Res. Commun.*, 42, 948 (1971); **Leitzmann, Wu, and Boyer**, *Biochemistry*, 9, 2338 (1970).
234. Kushner and Somerville, *Biochim. Biophys. Acta*, 207, 105 (1970); **Gotschlich and Edelman**, *Proc. Natl. Acad. Sci. U.S.A.*, 54, 558 (1965).
235. Castellino and Barker, *Biochemistry*, 7, 2207 (1968); **Heck**, *J. Biol. Chem.*, 244, 4375 (1969); **Schwert, Miller, and Peanasky**, *J. Biol. Chem.*, 242, 3245 (1967); **Adams, Ford, Koekoek, Lentz, McPherson, Rossman, Smiley, Schevitz, and Wonacott**, *Nature*, 227, 1098 (1970).
236. Vance and Feingold, *Fed. Proc., Fed. Am. Soc. Exp. Biol.*, 30, 1057 (1971).
237. Strothkamp and Dawson, *Biochemistry*, 13, 434 (1974).

238. Miyamoto, Petzold, Kuo, and Greengard, *J. Biol. Chem.*, 248, 179 (1973).
239. Miyamoto, Petzold, Kuo, and Greengard, *J. Biol. Chem.*, 248, 179 (1973).
239a. Ray, Reuben, Molineux, and Gefter, *J. Biol. Chem.*, 249, 5379 (1974).
240. Oh and Conrad, *Arch. Biochem. Biophys.*, 152, 631 (1972).
241. Traniello, Melloni, Pontremoli, Sia, and Horecker, *Arch. Biochem. Biophys.*, 149, 222 (1972); Tashima, Tholey, Drummond, Bertrand, Rosenberg, and Horecker, *Arch. Biochem. Biophys.*, 149, 118 (1972); Black, Tol, Fernando, and Horecker, *Arch. Biochem. Biophys.*, 151, 576 (1972).
242. Haugen and Suttie, *J. Biol. Chem.*, 249, 2717 (1974).
243. Ono-Kamimoto and Senoh, *J. Biochem.*, 75, 321 (1974).
244. Stribling and Perham, *Biochem. J.*, 131, 833 (1973).
245. Baker, Jeng, and Barker, *J. Biol. Chem.*, 247, 7724 (1972).
246. Fosset, Chappelet-Tordo, and Lazdunski, *Biochemistry*, 13, 1783, (1974).
246a. Buhner and Sund, *Eur. J. Biochem.*, 11, 73 (1969).
247. Zalkin and Hwang, *J. Biol. Chem.*, 246, 6899 (1971).
248. Andrews, *Biochem. J.*, 139, 421 (1974).
248a. Wolf and Hoffmann, *Eur. J. Biochem.*, 45, 269 (1974).
249. Deal and Holleman, *Fed. Proc., Fed. Am. Soc. Exp. Biol.*, 23, 264 (1964); Harris and Perham, *J. Mol. Biol.*, 13, 876 (1965); Harrington and Karr, *J. Mol. Biol.*, 13, 885; Jaenicke, Schmid, and Knof, *Biochemistry*, 7, 919 (1968); Hoagland and Teller, *Biochemistry*, 8, 594 (1969).
250. Hurlbert and Jakoby, *J. Biol. Chem.*, 240, 2772 (1965).
251. Uyeda and Kurooka, *Fed. Proc., Fed. Am. Soc. Exp. Biol.*, 29, 399 (1970); Uyeda and Kurooka, *J. Biol. Chem.*, 245, 3315 (1970).
252. Schwartz, Pizzo, Hill, and McKee, *J. Biol. Chem.*, 248, 1395 (1973).
252a. Whelan and Wriston, *Biochim. Biophys. Acta*, 365, 212 (1974).
253. Yoshida, *J. Biol. Chem.*, 240, 1113 (1965).
254. Henning, Helinski, Chao, and Yanofsky, *J. Biol. Chem.*, 237, 1523 (1962); Carlton and Yanofsky, *J. Biol. Chem.*, 237, 1531 (1962); Wilson and Crawford, *Bacteriol. Proc.*, 1964, 92 (1964); Goldberg, Creighton, Baldwin, and Yanofsky, *J. Mol. Biol.*, 21, 71 (1966); Yanofsky, Drapeau, Guest, and Carlton, *Proc. Natl. Acad. Sci. U.S.A.*, 57, 296 (1967); Hathaway and Crawford, *Biochemistry*, 9, 1801 (1970).
255. Lee and Horecker, *Arch. Biochem. Biophys.*, 162, 401 (1974).
256. Evans and Gurd, *Biochem. J.*, 133, 189 (1973).
257. Kolb, Cole, and Snell, *Biochemistry*, 7, 2946 (1968).
258. Pfleiderer and Auricchio, *Biochem. Biophys. Res. Commun.*, 16, 53 (1964); Harris, *Nature*, 203, 30 (1964).
259. Barrera, Namihira, Hamilton, Munk, Eley, Linn, and Reed, *Arch. Biochem. Biophys.*, 148, 343 (1972).
260. Wang and Weissman, *Fed. Proc., Fed. Am. Soc. Exp. Biol.*, 29, 333 (1970); Wang and Weissman, *Biochemistry*, 10, 1067 (1971).
261. Holland and Westhead, *Biochemistry*, 12, 2264 (1974).
261a. Wachsmuth and Hiwada, *Biochem. J.*, 141, 273 (1974).
262. Berman, Rubin, Carrell, and Glusker, *J. Biol. Chem.*, 249, 3983 (1974).
263. Waterson, Castellino, Hass, and Hill, *J. Biol. Chem.*, 247, 5266 (1972).
264. Stellwagen and Schachman, *Biochemistry*, 1, 1056 (1962); Deal, Rutter, and Van Holde, *Biochemistry*, 2, 246 (1963); Schachman and Edelstein, *Biochemistry*, 5, 2681 (1966); Penhoet, Kochman, Valentine, and Rutter, *Biochemistry*, 6, 2940 (1967); Sia and Horecker, *Arch. Biochem. Biophys.*, 123, 186 (1968); Kawahara and Tanford, *Biochemistry*, 5, 1578 (1966).
265. Adler, Beyreuther, Fanning, Geisler, Cronenborn, Klemp, Muller-Hill, Pfahl, and Schmitz, *Nature*, 237, 322 (1972).
266. Kaplan and Flavin, *J. Biol. Chem.*, 241, 5781 (1966).
266a. Huang, Nakatsukasa, and Nester, *J. Biol. Chem.*, 249, 4467 (1974).
267. Churchich and Dupourque, *Biochem. Biophys. Res. Commun.*, 46, 524 (1972).
268. Colby and Zatman, *Biochem. J.*, 121, 9P (1971).
269. Whiteley, *J. Biol. Chem.*, 241, 4890 (1966).
270. Winicov and Pizer, *J. Biol. Chem.*, 249, 1348 (1974).
271. Rosemblatt, Callewaert, and Chen, *J. Biol. Chem.*, 248, 6014 (1973).
272. Hass and Hill, *J. Biol. Chem.*, 244, 6080 (1969).
273. Packman and Jakoby, *J. Biol. Chem.*, 242, 2075 (1967).
274. Benöhr, and Krisch, *Z. Physiol. Chem.*, 348, 1115 (1967).
275. Schutz and Feigelson, *J. Biol. Chem.*, 247, 5327 (1972).
276. Bar-Tana and Rose, *Biochem. J.*, 131, 443 (1973).
277. Dalton, Morris, Ward, and Mortenson, *Biochemistry*, 10, 2066 (1971).
278. Cohen, *Curr. Top. Cell. Regul.*, 1, 183 (1969).
279. Clinkenbeard, Sugiyama, Moss, Reed, and Lane, *J. Biol. Chem.*, 248, 2275 (1973).
279a. Gehring and Riepertinger, *Eur. J. Biochem.*, 6, 281 (1968).
280. Koch and Bruton, *FEBS Lett.*, 40, 180 (1974).

281. Baker, van der Drift, and Stadtman, *Biochemistry,* 12, 1054 (1973).
282. Trotta, Burt, Haschmeyer, and Meister, *Proc. Natl. Acad. Sci. U.S.A.,* 68, 2599 (1971); Matthews and Anderson, *Biochemistry,* 11, 1176 (1972).
283. Suzuki, Yamaguchi, and Tokushige, *Biochem. Biophys. Acta,* 321, 369 (1973); Rudolph and Fromm, *Arch. Biochem. Biophys.,* 147, 92 (1971).
284. Erlichman, Rubin, and Rosen, *J. Biol. Chem.,* 248, 7607 (1973).
285. Junge, Krisch, and Hollandt. *Eur. J. Biochem.,* 43, 379 (1974).
286. Somack and Costilow, *Biochemistry,* 12, 2597 (1973).
287. Durell and Cantoni, *Biochim. Biophys. Acta,* 35, 515 (1959); Klee, *Biochim. Biophys. Acta,* 59, 562 (1962).
287a. Tobin, *J. Biol. Chem.,* 245, 2656 (1970).
288. Lipscomb, Horton, and Armstrong, *Biochemistry,* 13, 2070 (1974).
289. Simon, Schorr, and Phillips, *J. Biol. Chem.,* 249, 1993 (1974).
290. McKeehan and Hardesty, *J. Biol. Chem.,* 244, 4330 (1969).
291. Lamartiniere and Dempey, *Fed. Proc., Fed. Am. Soc. Exp. Biol.,* 30, 1121 (1971).
292. Riley and Snell, *Biochemistry,* 9, 1485 (1970).
293. Recsei and Snell, *Biochemistry,* 12, 365 (1973).
294. Kanarek, Marler, Bradshaw, Fellows, and Hill, *J. Biol. Chem.,* 239, 4207 (1964); Penner and Cohen, *J. Biol. Chem.,* 246, 4261 (1971).
294a. Bornmann and Hess, *Eur. J. Biochem.,* 47, 1 (1974); Fell, Liddle, Peacocke, and Dwek, *Biochem. J.,* 139, 665 (1974).
295. Zarlengo, Robinson, and Burns., *J. Biol. Chem.,* 243, 186 (1968).
296. Metrione, Okuda, and Fairclough, *Biochemistry,* 9, 2427 (1970).
297. Maeba and Sanwal, *J. Biol. Chem.,* 244, 2549 (1969).
298. Rowe and Wyngaarden, *J. Biol. Chem.,* 243, 6373 (1968).
299. Lehrach, Schafer, and Scheit, *FEBS Lett.,* 14, 343 (1971).
300. Leary and Kohlhaw, *J. Biol. Chem.,* 247, 1089 (1972); Bartholomew and Calvo, *Biochim. Biophys. Acta,* 250, 568, 577 (1971).
301. Niles and Westhead, *Biochemistry,* 12, 1715 (1973).
302. Coles, Tisdale, Kenney, and Singer, *Biochem. Biophys. Res. Commun.,* 46, 1843 (1972).
303. Kendal and Eckert, *Biochim. Biophys. Acta,* 258, 484 (1972).
304. Barisas and McGuire, *J. Biol. Chem.,* 249, 3151 (1974).
305. Israel, Howard, Evans, and Russell, *J. Biol. Chem.,* 249, 500 (1974).
305a. Danner and Morrison, *Biochim. Biophys. Acta,* 235, 44 (1971).
305b. Love, Chan, and Stotz, *J. Biol. Chem.,* 245, 6664 (1970).
305c. Higgens, Brady, and Rudney, *Arch. Biochem. Biophys.,* 163, 271 (1974).
306. Brown, Symth, Clarke, and Rosemyer, *Eur. J. Biochem.,* 34, 177 (1973).
307. Colman and Matthews, *J. Mol. Biol.,* 60, 163 (1971); Spradlin and Thoma, *J. Biol. Chem.,* 245, 117 (1970).
308. Schulze, Lusty, and Ratner, *J. Biol. Chem.,* 245, 4534 (1970).
309. Yamaguchi, Tokushige, and Katsuki, *J. Biochem.,* 73, 169 (1973).
310. Calhoun, Rimerman, and Hatfield, *J. Biol. Chem.,* 248, 3511 (1973).
311. Bonsignore, Cancedda, Lorenzoni, Cosulich, and DeFlora, *Biochem. Biophys. Res. Commun.,* 43, 94 (1971).
312. Kamen, *Nature,* 228, 527 (1970); Kondo, Gallerani, and Weissman, *Nature,* 228, 525 (1970).
313. McFadden, Rao, Cohen, and Roche, *Biochemistry,* 7, 3574 (1968).
314. Scott, *J. Biol. Chem.,* 246, 6353 (1971).
315. Hoch and DeMoss, *J. Biol. Chem.,* 247, 1750 (1972).
316. Gounaris, Turkenkopf, Buckwald, and Young, *J. Biol. Chem.,* 246, 1302 (1971).
317. Meachum, Calvin, and Braymer, *Biochemistry,* 10, 326 (1971).
317a. Kaufman and Fisher, *J. Biol. Chem.,* 245, 4745 (1970).
318. Long, Levitzki, and Koshland, *J. Biol. Chem.,* 245, 80, (1970).
319. Cox and Tanford, *J. Biol. Chem.,* 243, 3083 (1968); Scanu, Reader, and Edelstein, *Fed. Proc., Fed. Am. Soc. Exp. Biol.,* 26, 435 (1967).
320. Voll, Appella, and Martin, *J. Biol. Chem.,* 242, 1760 (1967).
321. Martinez-Carrion, Critz, and Quashnock, *Biochemistry,* 11, 1613 (1972); Schirch, Edmiston, Chen, Barra, Bossa, Hinds, and Fassella, *J. Biol. Chem.,* 248, 6456 (1973).
322. Inoue and Lowenstein, *J. Biol. Chem.,* 247, 4825 (1972).
323. Cardenas and Dyson, *J. Biol. Chem.,* 248, 6938 (1973).
324. Iwata and Fukui, *FEBS Lett.,* 36, 322 (1973).
325. Koike, Hamada, Tanaka, Otsuka, Ogashara, and Koike, *J. Biol. Chem.,* 249, 3836 (1974).
326. Cowell and DeMoss, *J. Biol. Chem.,* 248, 6262 (1973).
327. Carlson and Pierce, *J. Biol. Chem.,* 247, 23 (1972).
328. Thorner and Paulus, *J. Biol. Chem.,* 246, 3885 (1971).
328a. Wolfenden, Tomozawa, and Bamman, *Biochemistry,* 7, 3965 (1968).

329. Huseby, Christensen, Olsen, and Størmer, *Eur. J. Biochem.*, 20, 209 (1971).
329a. Callewaert, Rosemblatt, and Chen, *Biochemistry*, 13, 4181 (1974).
330. Hoch and DeMoss, *Biochemistry*, 5, 3137 (1966); Morino and Snell, *J. Biol. Chem.*, 242, 5591 (1967).
331. Lowey, Kucera, and Holtzer, *J. Mol. Biol.*, 7, 234 (1963); Olander, Emerson, and Holtzer, *J. Am. Chem. Soc.*, 89, 3058 (1967); McCubbin and Kay, *Biochim. Biophys. Acta*, 154, 239 (1968).
332. Schmidt and Zalkin, *J. Biol. Chem.*, 246, 6002 (1971).
333. Cogoli, Eberle, Sigrist, Joss, Robinson, Mosimann, and Semenza, *Eur. J. Biochem.*, 33, 40 (1973); Mosimann, Semenza, and Sund, *Eur. J. Biochem.*, 36, 489 (1973).
334. Kyte, *J. Biol. Chem.*, 247, 7642 (1972).
335. Shematek, Divin, and Arfin, *Arch. Biochem. Biophys.*, 158, 126 (1973).
336. Madon, Hillmer, and Gottschalk, *Eur. J. Biochem.*, 32, 51 (1973).
337. Flanders, Bamburg, and Sallach, *Biochim. Biophys. Acta*, 242, 566 (1971).
338. Little, Riley, and Behal, *Fed. Proc., Fed. Am. Soc. Exp. Biol.*, 30, 1121 (1971); Little and Behal, *Biochim. Biophys. Acta*, 243, 312 (1971).
339. Marshall and Cohen, *J. Biol. Chem.*, 247, 1641 (1972).
340. Berglund, *J. Biol. Chem.*, 247, 7270 (1972).
341. Ostrem and Berg, *Proc. Natl. Acad. Soc. U.S.A.*, 67, 1967 (1970).
342. Callewaert, Rosenblatt, and Tchen, *J. Biol. Chem.*, 249, 1737 (1974).
343. Kredich, Keenan, and Foote, *J. Biol. Chem.*, 247, 7157 (1972).
344. Cardenas, Dyson, and Standholm, *J. Biol. Chem.*, 248, 6931 (1973).
345. Chung, Tan, and Suzuki, *Biochemistry*, 10, 1205 (1971).
345a. Kakiuchi and Kobayashi, *J. Biochem.*, 69, 43 (1971).
346. Glazer, Fang, and Brown, *J. Biol. Chem.*, 248, 5679 (1973).
347. Tanford and Lovrien, *J. Am. Chem. Soc.*, 84, 1892 (1962); Schroeder, Shelton, Shelton, and Olson, *Biochim. Biophys. Acta*, 89, 47 (1964); Weber and Sund, *Angew. Chem.*, 77, 621 (1965); Schroeder, Shelton, Shelton, Roberson, and Apell, *Arch. Biochem. Biophys.*, 131, 653 (1969).
348. Monteilhet and Risler, *Eur. J. Biochem.*, 12, 165 (1970); Lederer and Simon, *Eur. J. Biochem.*, 20, 469 (1971).
349. Morawiecki, *Biochim. Biophys. Acta*, 44, 604 (1960); Steinmetz and Deal, *Biochemistry*, 5, 1399 (1966).
350. Curthoys, Straus, and Rabinowitz, *Biochemistry*, 11, 345 (1972).
351. Heyningen and Shemin, *Biochemistry*, 10, 4676 (1971).
352. Gaertner and DeMoss, *J. Biol. Chem.*, 244, 2716 (1969).
353. Montie and Montie, *Biochemistry*, 10, 2094 (1971).
354. Berg, Olsen, and Kivirikko, *Fed. Proc., Fed. Am. Soc. Exp. Biol.*, 31, 479 (1972).
355. Pentchev, Brady, Hibbert, Gal, and Shapiro, *J. Biol. Chem.*, 248, 5256 (1973).
356. Waygood and Sanwal, *J. Biol. Chem.*, 249, 265 (1974).
357. Gauthier and Rittenberg, *J. Biol. Chem.*, 246, 3737 (1971).
357a. Johnson and Hanson, *Biochim. Biophys. Acta*, 350, 336 (1974).
358. Feldman and Weiner, *J. Biol. Chem.*, 247, 260 (1972).
359. Thelander, *J. Biol. Chem.*, 248, 4591 (1973).
360. Strickland and Massey, *J. Biol. Chem.*, 248, 2944 (1973).
361. Smith, Brown, and Larner, *Biochim. Biophys. Acta*, 242, 81 (1971).
362. Eskin and Linn, *J. Biol. Chem.*, 247, 6183 (1972).
363. Kashiwamata, Kotake, and Greenburg, *Biochim. Biophys. Acta*, 212, 501 (1970).
364. Darrow and Rodstrom, *J. Biol. Chem.*, 245, 2036 (1970).
365. Fodge, Gracy, and Harris, *Biochim. Biophys. Acta*, 268, 271 (1972).
365a. Johnson and Hanson, *Biochim. Biophys. Acta*, 350, 336 (1974).
366. Correll, Steers, Towe, and Shropshire, *Biochim. Biophys. Acta*, 168, 46 (1968).
367. Melius, Moseley, and Brown, *Biochim. Biophys. Acta*, 221, 62 (1970).
368. Nevaldine, Bassel, and Hsu, *Biochim. Biophys. Acta*, 336, 283 (1974).
369. Berman, *Biochemistry*, 12, 1710 (1973).
370. Hanke, Bartmann, Hennecke, Kosakowski, Jaenicke, Holler, and Böck, *Eur. J. Biochem.*, 43, 601 (1974).
371. Holmes, Wyngaarden, and Kelley, *J. Biol. Chem.*, 248, 6035 (1973).
372. Frigerio and Harbury, *J. Biol. Chem.*, 231, 135 (1958).
373. Schmidt, Wang, Stanfield, and Reid, *Biochemistry*, 10, 3264 (1971).
374. Boosman, Sammons, and Chilson, *Biochem. Biophys. Res. Commun.*, 45, 1025 (1971).
375. Fee, Hegeman, and Kenyon, *Biochemistry*, 13, 2529 (1974).
376. Rosenberry, Chen, and Bock, *Biochemistry*, 13, 3068 (1974).
377. Rubin, Erlichman, and Rosen, *J. Biol. Chem.*, 247, 36 (1972).
378. Stahl and Touster, *Fed. Proc., Fed. Am. Soc. Exp. Biol.*, 30, 1121 (1971).
379. Henderson and Zalkin, *J. Biol. Chem.*, 246, 6891 (1971).
380. Wu, Shemin, Richards, and Williams, *Proc. Natl. Acad. Sci. U.S.A.*, 71, 1767 (1974).
381. Singh and Squire, *Biochemistry*, 13, 1819 (1974).

382. Zappia and Barker, *Biochim. Biophys. Acta*, 207, 505 (1970).
383. Blumenthal and Smith, *J. Biol. Chem.*, 248, 6002 (1973).
384. Craine, Daniels, and Kaufman, *J. Biol. Chem.*, 248, 7838 (1973); Wallace, Krantz, and Lovenberg, *Proc. Natl. Acad. Sci. U.S.A.*, 70, 2253 (1973).
385. Eady and Postgate, *Nature*, 249, 805 (1974).
386. Wu and Morris, *J. Biol. Chem.*, 248, 1687 (1973).
387. Gainey, Pestell, and Phelps, *Biochem. J.*, 129, 821 (1972).
388. Huang and Cabib, *J. Biol. Chem.*, 249, 3851 (1974).
389. Schepman, Wichertjes, and Van Bruggen, *Biochim. Biophys. Acta*, 271, 279 (1972).
390. Schepman, Wichertjes, and Van Bruggen, *Biochim. Biophys. Acta*, 271, 279 (1972).
391. Barnes, Kuehn, and Atkinson, *Biochemistry*, 10, 3939 (1971).
391a. Youdim and Collins, *Eur. J. Biochem.*, 18, 73 (1971).
392. Kredich, Becker, and Tomkins, *J. Biol. Chem.*, 244, 2428 (1969).
393. Gerhart and Schachman, *Biochemistry*, 4, 1054 (1965); Schachman and Edelstein, *Biochemistry*, 5, 2681 (1966); Changeux, Gerhart, and Schachman, *Biochemistry*, 7, 531 (1968); Weber, *Nature*, 218, 116 (1968); Wiley and Lipscomb, *Nature*, 218, 1119 (1968).
394. Strausbauch and Fischer, *Biochemistry*, 9, 226 (1970).
395. Main, Tarkan, Aull, and Soucie, *J. Biol. Chem.*, 247, 566 (1972).
396. Virden, *Biochem. J.*, 127, 503 (1972).
397. Eisenberg and Tomkins, *J. Mol. Biol.*, 31, 37 (1968); Cassman and Schachman, *Biochemistry*, 10, 1015 (1971); Reisler and Eisenberg, *Biochemistry*, 10, 2659 (1971); Josephs, *J. Mol. Biol.*, 55, 147 (1971).
398. Schwartz, Pizzo, Hill, and McKee, *J. Biol. Chem.*, 248, 1395 (1973).
399. Nari, Mouttet, Fouchier, and Ricard, *Eur. J. Biochem.*, 41, 499 (1974); Havir and Hanson, *Biochemistry*, 12, 1583 (1973).
400. Farron, *Biochemistry*, 9, 3823 (1970).
401. Melbye and Carpenter, *J. Biol. Chem.*, 246, 2459 (1971).
402. Eady and Postgate, *Nature*, 249, 805 (1974).
403. Strickland, Jacobson, and Strickland, *Biochim. Biophys. Acta*, 251, 21 (1971).
404. Tiemeier and Milman, *J. Biol. Chem.*, 247, 2272 (1972).
405. Tagaki and Westheimer, *Biochemistry*, 7, 891 (1968); *Biochemistry*, 7, 895 (1968).
406. Starnes, Munk, Maul, Cunningham, Cox, and Shive, *Biochemistry*, 11, 677 (1972); Clark and Ogilvie, *Biochemistry*, 11, 1278 (1972).
407. Kerr and Robertson, *Biochem. J.*, 125, 34P (1971).
408. Witheiler and Wilson, *J. Biol. Chem.*, 247, 2217 (1972).
409. Sarkar, Fischman, Goldwasser, and Moscona, *J. Biol. Chem.*, 247, 7743 (1972).
410. Kessler and Yaron, *Biochem. Biophys. Res. Commun.*, 50, 405 (1973).
411. Tombs and Lowe, *Biochem. J.*, 105, 181 (1967).
412. Pollock and Hersch, *J. Biol. Chem.*, 246, 4737 (1971).
413. Andreu, Albendea, and Muñoz, *Eur. J. Biochem.*, 37, 505 (1973).
414. Tate and Meister, *Proc. Natl. Acad. Sci. U.S.A.*, 68, 781 (1971).
415. Stellwagen, Cronlund, and Barnes, *Biochemistry*, 12, 1552 (1973).
416. O'Carra and Killilea, *Biochem. Biophys. Res. Commun.*, 45, 1192 (1971); Hattori, Crespi, and Katz, *Biochemistry*, 4, 1225 (1965); Bloomfield, Van Holde, and Dalton, *Biopolymers*, 5, 149 (1967); Jennings, *Biopolymers*, 6, 1177 (1968); Glazer and Fang, *J. Biol. Chem.*, 248, 663 (1973); Glazer and Cohen-Bazire, *Proc. Natl. Acad. Sci. U.S.A.*, 68, 1398 (1973); Bennett and Bogorad, *Biochemistry*, 10, 3625 (1971); MacColl, Lee, and Berns, *Biochem. J.*, 122, 421 (1971); Berns, *Biochem. Biophys. Res. Commun.*, 38, 65 (1970); Neufeld and Riggs, *Biochim. Biophys. Acta*, 181, 234 (1969).
417. Aaronson and Frieden, *J. Biol. Chem.*, 247, 7502 (1972); Coffee, Aaronson, and Frieden, *J. Biol. Chem.*, 248, 1381 (1973); Uyeda, *Biochemistry*, 8, 2366 (1969); Paetkau, Younathan, and Lardy, *J. Mol. Biol.*, 33, 721 (1968).
418. Patrick and Lee, *J. Biol. Chem.*, 244, 4277 (1969).
419. Kapoor, Bray, and Ward, *Arch. Biochem. Biophys.*, 134, 423 (1969).
420. Adair and Jones, *J. Biol. Chem.*, 247, 2308 (1972).
421. Lambeth and Lardy, *Eur. J. Biochem.*, 22, 355 (1971); Catterall and Pedersen, *J. Biol. Chem.*, 246, 4987 (1971); Senior and Brooks, *Arch. Biochem. Biophys.*, 140, 257 (1970); Senior and Brooks, *FEBS Lett.*, 17, 327 (1971); Brooks and Senior, *Biochemistry*, 11, 4675 (1972).
422. Yoshinaga, Teraoka, Izui, and Katsuki, *J. Biochem.*, 75, 913 (1974).
423. Tabita, McFadden, and Pfennig, *Biochim. Biophys. Acta*, 341, 187 (1974).
424. Issa and Medicino, *J. Biol. Chem.*, 248, 685 (1973).
425. Seery, Fischer, and Teller, *Biochemistry*, 9, 3591 (1970); Madsen and Cori, *J. Biol. Chem.*, 223, 1055 (1956); Seery, Fischer, and Teller, *Biochemistry*, 6, 3315 (1967); DeVincenzi and Hedrick, *Biochemistry*, 6, 3489 (1967).
426. Jonas, *J. Biol. Chem.*, 247, 7767 (1972).
426a. Yurwicz, Ghalambor, Duckworth, and Heath, *J. Biol. Chem.*, 246, 5607 (1971).

427. Schnebli, Vatter, and Abrams, *J. Biol. Chem.*, 245, 1122 (1970).
428. Sugiyama, *Biochemistry*, 12, 2862 (1973).
429. Smando, Waygood, and Sanwal, *J. Biol. Chem.*, 249, 182 (1974).
430. Stoll, Ericsson, and Zuber, *Proc. Natl. Acad. Sci. U.S.A.*, 70, 3781 (1973).
431. Stoll, Ericsson, and Zuber, *Proc. Natl. Acad. Sci. U.S.A.*, 70, 3781 (1973).
432. Stoll, Ericsson, and Zuber, *Proc. Natl. Acad. Sci. U.S.A.*, 70, 3781 (1973).
433. Burgess, *J. Biol. Chem.*, 244, 6168 (1969); Johnson, DeBacker, and Boezi, *J. Biol. Chem.*, 246, 1222 (1971).
434. Weaver, Blath, and Rutler, *Proc. Natl. Acad. Sci. U.S.A.*, 68, 2994 (1971).
435. Hyun, Steinberg, Treadwell, and Vahouny, *Biochem. Biophys. Res. Commun.*, 44, 819 (1971).
436. Bernardi and Cook, *Biochim. Biophys. Acta*, 44, 96, 105 (1960); Burley and Cook, *Can. J. Biochem. Physiol.*, 40, 363 (1962).
437. Tate and Meister, *Proc. Natl. Acad. Sci. U.S.A.*, 68, 781 (1971).
438. Smith, *J. Biol. Chem.*, 246, 4234 (1971).
439. Delmer, *J. Biol. Chem.*, 247, 3822 (1972).
440. Haberland, Willard, and Wood, *Biochemistry*, 11, 712 (1972).
440a. Hamilton, *J. Biol. Chem.*, 249, 4428 (1974).
441. Tweedie and Segel, *J. Biol. Chem.*, 246, 2438 (1971).
442. Crichton, Eason, Barclay, and Bryce, *Biochem. J.*, 131, 855 (1973); Hoy, Harrison, and Hoare, *J. Mol. Biol.*, 84, 515 (1974); Bryce and Crichton, *J. Biol. Chem.*, 246, 4198 (1971); Bjork and Fish, *Biochemistry*, 10, 2844 (1971).
443. Gornick, Vuturo, West, and Weaver, *J. Biol. Chem.*, 249, 1792 (1973).
444. Spiegelman and Whiteley, *J. Biol. Chem.*, 249, 1476 (1974).
445. Holtzer and Lowey, *J. Am. Chem. Soc.*, 81, 1370 (1959); Mueller, *J. Biol. Chem.*, 239, 797 (1964); Tonomura, Appel, and Morales, *Biochemistry*, 5, 515 (1966); Richards, Chung, Menzel, and Olcott, *Biochemistry*, 6, 528 (1967); Gershman, Stracher, and Dreizen, *J. Biol. Chem.*, 244, 2726 (1969); Kominz, Carroll, Smith, and Mitchell, *Arch. Biochem. Biophys.*, 79, 191 (1959); Frederiksen and Holtzer, *Biochemistry*, 7, 3935 (1968); Weeds and Lowey, *J. Mol. Biol.*, 61, 701 (1971).
446. Contaxis and Reithel, *J. Biol. Chem.*, 246, 677 (1971); Bailey and Boulter, *Biochem. J.*, 113, 669 (1969); Creeth and Nichol, *Biochem. J.*, 77, 230 (1960); Reithel, Robbins, and Gorin, *Arch. Biochem. Biophys.*, 108, 409 (1964).
447. Levine, Gillett, Turnquist, and Hansen, *Fed. Proc., Fed. Am. Soc. Exp. Biol.*, 30, 1121 (1971); Levine, Gillett, Hageman, and Hansen, *J. Biol. Chem.*, 244, 5729 (1969).
448. Reiss and Rothstein, *Biochemistry*, 13, 1796 (1974).
449. Lornitzo, Qureshi, and Porter, *J. Biol. Chem.*, 249, 1654 (1974).
450. Yun and Hsu, *J. Biol. Chem.*, 247, 2689 (1973).
451. Kuehn and McFadden, *Biochemistry*, 8, 2403 (1969).
452. Warren and Tipton, *Biochem. J.*, 139, 297 (1974).
453. Craven, Steers, and Anfinsen, *J. Biol. Chem.*, 240, 2468 (1965); Fowler and Zabin, *J. Biol. Chem.*, 245, 5032 (1970).
454. Rutner, *Biochem. Biophys. Res. Commun.*, 39, 923 (1970); Kawashima and Wildman, *Biochem. Biophys. Res. Commun.*, 41, 1463 (1970); Trown, *Biochemistry*, 4, 908 (1965); Haselkorn, Fernández-Morán, Kieras, and van Bruggen, *Science*, 150, 1598 (1965).
455. Spina, Bright, and Rosenbloom, *Biochemistry*, 9, 3794 (1970).
455a. Miziorko, Nowak, and Mildvan, *Arch. Biochem. Biophys.*, 163, 378 (1974).
456. Bowen and Mortimer, *Biochem. J.*, 117, 71P (1970); Mahadik and SivaRaman, *Biochem. Biophys. Res. Commun.*, 32, 167 (1968).
457. Sugiyama, Ito, and Akazawa, *Biochemistry*, 10, 3406 (1971).
458. Wilgus, Pringle, and Stellwagen, *Biochem. Biophys. Res. Commun.*, 44, 89 (1971).
459. Woolfolk and Stadtman, *Arch. Biochem. Biophys.*, 122, 174 (1967); Valentine, Shapiro, and Stadtman, *Biochemistry*, 7, 2143 (1968).
460. Pupillo and Piccari, *Arch. Biochem. Biophys.* 154, 324 (1973).
461. Arfin and Koziell, *Biochim. Biophys. Acta*, 321, 356 (1973).
462. Tate and Meister, *Proc. Natl. Acad. Sci. U.S.A.*, 68, 781 (1971).
463. Donovan, Mapes, Davis, and Hamburg, *Biochemistry*, 8, 4190 (1969).
464. Sun, McLeester, Bliss, and Hall, *J. Biol. Chem.*, 249, 2118 (1974).
465. Valentine, Wrigley, Scrutton, Irias, and Utter, *Biochemistry*, 5, 3111 (1966).
466. Steiner and Edelhoch, *J. Am. Chem. Soc.*, 83, 1435 (1961); Edelhoch and de Crombrugghe, *J. Biol. Chem.*, 241, 4357 (1966).
467. Giorgio, Yip, Fleming, and Plaut, *J. Biol. Chem.*, 245, 5469 (1970); Harvey, Giorgio, and Plaut, *Fed. Proc., Fed. Am. Soc. Exp. Biol.*, 29, 532 (1970).
468. Bowers, Czubaroff, and Haschemyer, *Biochemistry*, 9, 2620 (1970); Tate and Meister, *Biochemistry*, 9, 2626 (1970).
469. Siegal and Kamin, *Fed. Proc., Fed. Am. Soc. Exp. Biol.*, 30, 1261 (1971); Siegel and Davis, *J. Biol. Chem.*, 249, 1587 (1974).
469a. MacGregor, Schnaitman, Normansell, and Hodgins, *J. Biol. Chem.*, 249, 5321 (1974).

470. Kaziro, Ochoa, Warner, and Chen, *J. Biol. Chem.*, 236, 1917 (1961).
471. Sabo, Boeker, Byers, Waron, and Fischer, *Biochemistry*, 13, 662 (1974).
472. Lee-Huang and Warner, *J. Biol. Chem.*, 244, 3793 (1969).
473. Green, Valentine, Wrigley, Ahmad, Jacobson, and Wood, *J. Biol. Chem.*, 247, 6284 (1972).
474. Kono, Uyeda, and Oliver, *J. Biol. Chem.*, 248, 8592 (1973).
475. Kakimoto, Kato, Shibatani, Nishimura, and Chibata, *J. Biol. Chem.*, 245, 3369 (1970).
476. Bloemendal, Bont, Jongkind, and Wisse, *Exp. Eye Res.*, 1, 300 (1962); Bloemendal, Bont, Benedett, and Wisse, *Exp. Eye Res.*, 4, 319 (1965).
477. Boeker, Fischer, and Snell, *J. Biol. Chem.*, 244, 5239 (1969); Boeker and Snell, *J. Biol. Chem.*, 243, 1678 (1968).
478. Shikita and Hall, *J. Biol. Chem.*, 248, 5605 (1973).
479. Stevens, Emery, and Sternberger, *Biochem. Biophys. Res. Commun.*, 24, 929 (1966); Richardson, *Proc. Natl. Acad. Sci. U.S.A.*, 55, 1616 (1966).
480. Carpenter and Van Holde, *Biochemistry*, 12, 2231 (1973); Loehr and Mason, *Biochem. Biophys. Res. Commun.*, 51, 741 (1973); Roxby, Miller, Blair, and Van Holde, *Biochemistry*, 13, 1662 (1974); DePhillips, Nickerson, Johnson, and Van Holde, *Biochemistry*, 8, 3665 (1969); Pickett, Riggs, and Larimer, *Science*, 151, 1005 (1966); Van Holde and Cohen, *Biochemistry*, 3, 1803 (1964); Fernández-Morán, van Bruggen, and Ohtsuki *J. Mol. Biol.*, 16, 191 (1966); Lontie and Witters, in *The Biochemistry of Copper*, Peisach, Aisen, and Blumberg, Eds., Academic Press, New York, 1966, 455.
481. Tanaka, Koike, Otsuka, Hamada, Ogasahara, and Koike, *J. Biol. Chem.*, 249, 191 (1974).
482. Carper, Stoddard, and Martin, *Biochim. Biophys. Acta*, 334, 287 (1974); Carper, Stoddard, and Martin, *Biochem. Biophys. Res. Commun.*, 54, 721 (1973).
483. Hayakawa, Perkins, Walsh, and Krebs, *Biochemistry*, 12, 567, 574 (1973); Cohen, *Eur. J. Biochem.*, 34, 1 (1973).
484. Vance, Mitsuhashi, and Block, *J. Biol. Chem.*, 248, 2303 (1973).
485. Eley, Namihira, Hamilton, Munk, and Reed, *Arch. Biochem. Biophys.*, 152, 655 (1972).
486. Guerritore, Bonacci, Brunori, Antonini, Wyman, and Rossi-Fanelli, *J. Mol. Biol.*, 13, 234 (1965).
487. Pettit, Hamilton, Munk, Namihira, Eley, Williams, and Reed, *J. Biol. Chem.*, 248, 5282 (1973).
488. Waxman, *J. Biol. Chem.*, 246, 7318 (1971).
489. Swaney and Klotz, *Arch. Biochem. Biophys.*, 147, 475 (1971).
490. Barrera, Namihira, Hamilton, Munk, Eley, Linn, and Reed, *Arch. Biochem. Biophys.*, 148, 343 (1972).
491. Hohn and Hohn, *Adv. Virus Res.*, 16, 43 (1970).
492. Eley, Namihira, Hamilton, Munk, and Reed, *Arch. Biochem. Biophys.*, 152, 655 (1972); Vogel, Hoehn, and Henning, *Proc. Natl. Acad. Sci. U.S.A.*, 69, 1615 (1972).
493. Bancroft, Hiebert, Rees, and Markham, *Virology*, 34, 224 (1968).
494. Bockstahler and Kaesberg, *Biophys. J.*, 2, 1 (1962).
495. Miki and Knight, *Virology*, 25, 478 (1965).
496. Markham, *Discuss. Faraday Soc.*, 11, 221 (1951); Harris and Hindley, *J. Mol. Biol.*, 3, 117 (1961); Finch and Klug, *J. Mol. Biol.*, 15, 344 (1966); Peter, Stehdin, Reinbolt, Collet, and Duranton, *Virology*, 49, 615 (1972).
497. Anderer and Restle, *Z. Naturforsch.*, 19b, 1026 (1964).
498. Yamazaki and Kaesberg, *Biochim. Biophys. Acta*, 53, 173 (1961); Van Regenmortel, Hendry, and Baltz, *Virology*, 49, 647 (1972).
499. Kelley and Kaesberg, *Biochim. Biophys. Acta*, 55, 236 (1962); Kelley and Kaesberg, *Biochim. Biophys. Acta*, 61, 865 (1962); Kruseman, Kraal, Jaspers, Bol, Brederode, and Veldstra, *Biochemistry*, 10, 447 (1971).
500. Guchhart, Zwergel, and Lane, *J. Biol. Chem.*, 249, 4776 (1974).
501. Medappa, McLean, and Rueckert, *Virology*, 44, 259 (1971).
502. Hersh and Schachman, *Virology*, 6, 234 (1958); Michelin-Lausarot, Ambrosino, Steere, and Reichmann, *Virology*, 41, 160 (1970); Weber and Rosenbusch, *Virology*, 41, 763 (1970).
503. Fine, Mass, and Murakami, *J. Mol. Biol.*, 36, 167 (1968).
504. Reichmann, *J. Biol. Chem.*, 235, 2959 (1960); Reichmann and Hatt, *Biochim. Biophys. Acta*, 49, 153 (1961).
505. Anderer, *Adv. Protein Chem.*, 18, 1 (1963); Caspar, *Adv. Protein Chem.*, 18, 37 (1963).

REFRACTIVE INDEX INCREMENTS OF PROTEINS

Protein	$\frac{dn}{dc'}$ ml/g	Wavelength nm	Conditions[a]	References
Actin	0.180	436	pH 8.0 veronal-HCl, 6 mM	1
	0.170	436	pH 7.8–8.0	2
	0.227		0.5 M KI	3
Actomyosin	0.200	436	Water, 23°	4
	0.195	436	Water, pH 4.26	5
	0.193	546	Water, pH 4.26	5
	0.191	578	Water, pH 4.26	5
Albumin, bovine serum (varies with source)	0.1954	436	Water, pH 5.05, 25°	6
	0.1883	546	Water, pH 5.05, 25°	6
	0.1869	578	Water, pH 5.05, 25°	6
	0.187	589	Water, pH 5.05, 25°	6
	0.193	578	Water, pH 5.05, 0.5°	6
	0.199	436	0.1 M NaCl, pH 5.35, 25°	6
	0.191	546	0.1 M NaCl, pH 5.35, 25°	6
	0.190	578	0.1 M NaCl, pH 5.35, 25°	6
	0.190	589	0.1 M NaCl, pH 5.35, 25°	6
	0.200	436	0.5 M NaCl, pH 5.31, 25°	6
	0.193	546	0.5 M NaCl, pH 5.31, 25°	6
	0.192	578	0.5 M NaCl, pH 5.31, 25°	6
	0.191	589	0.5 M NaCl, pH 5.31, 25°	6
	0.197	436	pH 7.67, phosphate, 0.1I, 25°	6
	0.190	546	pH 7.67, phosphate, 0.1I, 25°	6
	0.188	578	pH 7.67, phosphate, 0.1I, 25°	6
	0.188	589	pH 7.67, phosphate, 0.1I, 25°	6
	0.192	578	pH 7.67, phosphate, 0.1I, 0.5°	6
	0.198	578	pH 8.6 diethylbarbiturate, 0.5°	6
	0.204	436	NaOH, pH 10.72, 25°	6
	0.196	546	NaOH, pH 10.72, 25°	6
	0.195	578	NaOH, pH 10.72, 25°	6
	0.194	589	NaOH, pH 10.72, 25°	6
	0.1924	436	0.1 M NaCl, pH 5.2, 25°	7
	0.1854	546	0.1 M NaCl, pH 5.2, 25°	7
	0.188	546	pH 7.8, phosphate, 0.1I, 25°	8
	0.188	436	pH 7.8, phosphate, 0.18–0.38I, 25°	9
	0.1929	436	0.2 M acetate, pH 4.50, 25°	10
	0.185	436	0.1 M phosphate, pH 7, 25°	11
	0.185	578	Water, 25°	12
	0.185	578	0.1 M NaCl or KCl, 25°	12
	0.186	546	0.15 M NaCl, pH 4.0, 4.5, 5.3, 20°	13
	0.184	546	0.15 M NaCl, pH 3.52, 20°	13
	0.190	546	0.6 M NaCl, pH 5.40, 20°	13
	0.188	546	0.6 M NaCl, pH 3.71, 20°	13
	0.180	546	0.15 M NaCl, pH 1.9, 20°	13
	0.186	546	0.15 M NaCl, pH 5.3, 20°	13
	0.182	578	Water, 20°	14
	0.193	436	1 M NaCl, 20°	15
	0.187	546	1 M NaCl, 20°	15
	0.185	589	1 M NaCl, 20°	15
	0.194	546	Water	16

[a]I is ionic strength.

REFRACTIVE INDEX INCREMENTS OF PROTEINS (continued)

Protein	$\frac{dn}{dc'}$ ml/g	Wavelength nm	Conditions[a]	References
Albumin, bovine serum (continued)	0.185	546	0.05 M phosphate, pH 7	17
	0.185	546	0.2 M borate-HCl, pH 9	17
	0.186	546	Water, pH 7	18
	0.176	546	0.5 M NaCl, 25°	19
	0.174	578	0.5 M NaCl, 25°	19
	0.181	436	Water, 25°	19
	0.178	546	Water, 25°	19
	0.174	578	Water, 25°	19
	0.190	436	0.05 M NaCl, 25°	19
	0.183	546	0.05 M NaCl, 25°	19
	0.181	578	0.05 M NaCl, 25°	19
	0.189	436	Phosphate, pH 7, 20°	20
	0.185	546	Phosphate, pH 7, 20°	20
	0.1950	436	Water, 23°	21
	0.193	436	Water, 25°	22
	0.188	546	Water, 25°	22
	0.183	589	Water, 25°	22
	0.207(M)	436	2-Chloroethanol-water (20/80 vol), 25°	23
	0.197(M)	436	2-Chloroethanol-water (40/60 vol), 25°	23
	0.1485(M)	436	2-Chloroethanol-water (60/40 vol), 25°	23
	0.1295(M)	436	2-Chloroethanol-water (80/20 vol), 25°	23
	0.172(m)	436	2-Chloroethanol-water (20/40 vol), 25°	23
	0.1625(m)	436	2-Chloroethanol-water (40/60 vol), 25°	23
	0.1525(m)	436	2-Chloroethanol-water (60/40 vol), 25°	23
	0.1425(m)	436	2-Chloroethanol-water (80/20 vol), 25°	23
Albumin, horse serum	0.1912	436	0.1 M acetate buffer, pH 4.8, 25°	7
	0.1844	546	0.1 M acetate buffer, pH 4.8, 25°	7
	0.191	436	Water, 25°	22
Albumin, human serum	0.186	578	0.3 M NaCl, 20–25°	24
	0.189	578	Salt free, pH 4.85, 0.5°	6
	0.185	578	Salt free, pH 4.85, 25°	6
	0.190	578	pH 7.71, phosphate, 0.5°	6
	0.1863	546	Water	6
	0.1938	436	Water	6
	0.1854	578	Water	6
	0.189	589	Water, 17.5°	25
Amandin	0.1697	436	1.7 M NaCl, pH 5.6, 25°	26
	0.1686	546	1.7 M NaCl, pH 5.6, 25°	26
	0.1674	589	1.7 M NaCl, pH 5.6, 25°	26
	0.1678	653	1.7 M NaCl, pH 5.6, 25°	26

[b]M is at constant chemical potential; m is at constant molality.

REFRACTIVE INDEX INCREMENTS OF PROTEINS (continued)

Protein	$\frac{dn}{dc'}$ ml/g	Wavelength nm	Conditions[a]	References
α-Amylase (Bacillus subtilis)	0.197	436	0.1 M NaCl–0.005 M Ca acetate pH 6.9	27
Arachin	0.192	546	Phosphate-NaCl, 0.5I, pH 7.5	28
				8
S-Carboxymethyl-boxymethyl-kerateine 2	0.190	546	pH 6.7, phosphate-chloride,	29
	0.190	546	pH 6.7, phosphate-chloride, 0.2I	29
	0.166	546	pH 6.7, phosphate-chloride, 0.2I +0.5 g sodium dodecyl sulfate/g protein	29
	0.163	546	10 M acetic acid	29
	0.143	546	pH 6.7 phosphate-chloride	29
	0.126	546	6 M guanidine HCl	29
	0.164	546	Formic acid	29
	0.163	546	Formic acid + 0.1 M KCl	29
	0.160	546	Formic acid + 0.5 M KCl	29
	0.096	546	Dichloroacetic acid	29
	0.096	546	Dichloroacetic acid + 0.1 M KCl	29
$\alpha_{s1,2}$-Caseins	0.181	546	pH 12, phosphate	18
	0.189	436	Aqueous, various buffers, 25°	30
	0.120	436	2-Chloroethanol	30
	0.160	436	Formic acid	30
β-Casein	0.181		Water	31
Chlorocruorin	0.196	546	0.02 M phosphate, pH 7.0	32
α-Chymotrypsin	0.187	546	pH 6.1 Na phosphate, 0.2I-0.004 M β-phenylpropionate, 25°	33
	0.193	436	pH 7.1 phosphate, 0.1I, 25°	34
Chymotrypsinogen	0.194	436	Isoionic	35
	0.185	546	Isoionic	35
	0.185	436	pH 2.5 phosphate, 0.15I, 25°	9
Collagen	0.187	436	pH 3.7, 0.1 M citrate, 20°	36
	0.176	436	1.0 M KCl, pH 5.75, 40°	37
	0.192	436	Salt free, pH 5.75, 40°	38
Conalbumin	0.192	436	Isoionic, 25°	39
Cryoglobulin	0.190	436	pH 7.3, phosphate, 0.3I	40
Deoxyribo-nuclease, pancreatic	0.196	546	pH 7.6, phosphate, 20°	41
Edestin	0.198	546	Phosphate-NaCl, pH 7.5, 0.5I, 25°	8
				28
				42

REFRACTIVE INDEX INCREMENTS OF PROTEINS (continued)

Protein	$\frac{dn}{dc'}$ ml/g	Wavelength nm	Conditions[a]	References
Fibrinogen, bovine	0.1933	436	pH 6.3, NaCl, 0.1I	43
	0.1935	436	pH 6.3, NaCl, 0.2I	43
	0.1987	436	pH 6.3, NaCl, 0.3I	43
	0.1800	436	pH 6.3, NaCl, 0.5I	43
	0.1949	436	pH 9.2, 0.1 M glycine, 0.4 M NaCl, 0.05 M NaOH, 25°	44
	0.1898	436	pH 9.2, 0.1 M glycine, 0.4 M NaCl, 0.05 M NaOH, 0.5 M HMG (hexamethylene glycol), 25°	44
	0.1970	436	0.3 M NaCl, 25°	44
	0.1953	436	pH 6.2, phosphate (0.05I), 0.4 M NaCl, 25°	44
	0.1903	436	pH 6.2, phosphate (0.05I), 0.4 M NaCl, 0.5 M HMG, 25°	44
Fibrinogen, human	0.188	578	0.3 M NaCl, 20–25°	24
	0.180–0.214	436	NaCl 0.1–0.5I	45
	0.197	436	pH 5.57–6.66 phosphate, 7°	46
	0.201	436	pH 5.57–6.66 phosphate, 27°	46
	0.199	436	pH 6.60, NaCl	46
	0.198	436	pH 6.60, NaCl	46
	0.191	546	pH 5.57–6.66, phosphate, 7°	46
	0.195	546	pH 5.57, phosphate, 27°	46
	0.191	546	pH 6.6, phosphate, 27°	46
	0.1958	436	0.1I NaCl, 20°	43
	0.1924	436	0.2I NaCl, 20°	43
	0.2000	436	0.3I NaCl, 20°	43
	0.1986	436	0.5I NaCl, 20°	43
Fibrinogen, iodinated	0.195	437	pH 7.35 NH_4OH	47
Gelatin	0.187	436	pH 3.7, 0.1 M citrate, 20°	36
	0.173	436	2 M KCNS	36, 48
	0.194	436	H_2O, pH 5.1	48
	0.193	436	pH 5.1, 0.15 M NaCl	48
	0.186	436	pH 5.1, 1.0 M NaCl	48
	0.185	436	1.0 M KCNS	48
	0.1765	436	Formic acid	49
	0.192	436	pH 2.5, 0.02 M citrate, 0.2% NaCl, 25°	50
	0.172	436	1 M KCNS, 30°	51
	0.188	436	0.05 M phosphate, 30°	51
	0.191	436	Water, 25°	22
	0.186	546	Water, 25°	22
	0.184	589	Water, 25°	22
	0.1765	436	Formic acid	48
	0.1515	436	80% formic acid, 20% DMF (dimethylformamide)	48
	0.1423	436	60% formic acid, 40% DMF	48
	0.1267	436	40% formic acid, 60% DMF	48
	0.1223	436	10% formic acid, 90% DMF	48
Globin, fetal, human	0.190	546	Water, salt, 15°	52

REFRACTIVE INDEX INCREMENTS OF PROTEINS (continued)

Protein	$\frac{dn}{dc'}$ ml/g	Wavelength nm	Conditions[a]	References
Globin, horse	0.200	436	pH 2.5, 0.2 M NaCl	53
α_2-Globulin, human	0.183	589	0.3 M NaCl, 20–25°	24
β_1-Globulin, human	0.185	589	0.3 M NaCl, 20–25°	24
γ-Globulin human	0.188	578	0.3 M NaCl, 20–25°	24
	0.188	578	0.5°	6
	0.189	546		6
	0.196	436		6
Glutamic dehydrogenase	0.193	436	0.05 M, phosphate, pH 6.5, 11–3°	54
D-Glyceraldehyde-3-phosphate dehydrogenase (rabbit)	0.190	436	NaCl, pH 7.3, 0.15I K phosphate, pH 6.55, 0.1I	55
	0.185	546	K phosphate, pH 7.6, 0.135I	
	0.1885	436	Phosphate, pH 8, 20°	20
	0.1825	546	Phosphate, pH 8, 20°	20
D-Glyceraldehyde-3-phosphate dehydrogenase (yeast)	0.189	436	Phosphate, pH 8, 20°	20
	0.184	546	Phosphate, pH 8, 20°	20
Hemerythrin	0.190	546	Borate 0.5I, pH 8.0	56
Hemoglobin, bovine	0.1966	656	pH 5.6, acetate, 0°	57
	0.1949	656	pH 7.7, phosphate, 0°	57
	0.209	656	Water, 20°	58
Hemoglobin, canine	0.194	656	Water, 20°	58
	0.186	656	0.01 N NH$_4$OH, 20°	59
Hemoglobin, human (and oxyhemoglobin)	0.198	644	Distilled water	60
	0.197	546	Water, salt	61
Insulin	0.202	436	0.1 M KCl, various pH, 25°	62
	0.192	436	pH 2.6, various salts, 25°	9
	0.197(M)	436	2-Chloroethanol-water (20/80 vol), 25°	23
	0.186(M)	436	2-Chloroethanol-water (40/60 vol), 25°	23
	0.145(M)	436	2-Chloroethanol-water (60/40 vol), 25°	23
	0.117(M)	436	2-Chloroethanol-water (80/20 vol), 25°	23
	0.168(m)	436	2-Chloroethanol-water (20/80 vol), 25°	23
	0.149(m)	436	2-Chloroethanol-water (20/80 vol), 25°	23

REFRACTIVE INDEX INCREMENTS OF PROTEINS (continued)

Protein	$\frac{dn}{dc'}$ ml/g	Wavelength nm	Conditions[a]	References
Insulin (continued)	0.136(m)	436	2-Chloroethanol-water (20/80 vol), 25°	23
	0.124(m)	436	2-Chloroethanol-water (20/80 vol), 25°	23
α-Keratose	0.182	546	pH 6.7, phosphate chloride, 0.21	29
	0.156	546	Formic acid	29
	0.155	546	Formic acid – 0.1 M KCl	29
	0.152	546	Formic acid – 0.5 M KCl	29
	0.088	546	Dichloroacetic acid	29
	0.088	546	Dichloroacetic acid – 0.1 M KCl	29
α-Lactalbumin	0.195	436	Water, 25°	22
	0.198	546	Water, 25°	22
	0.188	589	Water, 25°	22
Lactic dehydrogenase, rabbit	0.183	547	pH 10.4, 20°	63
β-Lactoglobulin, bovine	0.187	578	0.5°	6
	0.184	578	25°	6
	0.202	578	Diethylbarbiturate, 0.5°	6
	0.189	578	Phosphate, 0.5°	6
	0.1890	436	0.1 M NaCl, pH 5.2, 25°	7
	0.1822	546	0.1 M NaCl, pH 5.2	7
	0.1856	546	Water, 25°	6
	0.1926	436	Water, 25°	6
	0.1842	578	Water, 25°	6
	0.196	366	0.5 M NaCl, 20°	64
	0.1892	436	pH 5.2, 0.5 M NaCl, 20°	64
	0.1818	546	pH 5.2, 0.5 M NaCl, 20°	64
	0.181	579	0.5 M NaCl, 20°	64
	0.187	436	Aqueous, 25°	22
	0.181	546	Aqueous, 25°	22
	0.180	589	Aqueous, 25°	22
	0.192(M)	436	2-Chloroethanol-water (5–95 vol), 25°	65
	0.195(M)	436	2-Chloroethanol-water (10–90 vol), 25°	65
	0.198(M)	436	2-Chloroethanol-water (20–80 vol), 25°	65
	0.202(M)	436	2-Chloroethanol-water (30–70 vol), 25°	65
	0.192(M)	436	2-Chloroethanol-water (40–60 vol), 25°	65
	0.175(M)	436	2-Chloroethanol-water (50–50 vol), 25°	65
	0.150(M)	436	2-Chloroethanol-water (60–40 vol), 25°	65
	0.117(M)	436	2-Chloroethanol-water (80–20 vol), 25°	65
	0.184(m)	436	2-Chloroethanol-water (5–95 vol), 25°	65
	0.179(m)	436	2-Chloroethanol-water (10–90 vol), 25°	65

REFRACTIVE INDEX INCREMENTS OF PROTEINS (continued)

Protein	$\frac{dn}{dc'}$ ml/g	Wavelength nm	Conditions[a]	References
β-Lactoglobulin, bovine (continued)	0.169(m)	436	2-Chloroethanol-water (20–80 vol), 25°	65
	0.159(m)	436	2-Chloroethanol-water (30–70 vol), 25°	65
	0.152(m)	436	2-Chloroethanol-water (40–60 vol), 25°	65
	0.143(m)	436	2-Chloroethanol-water (50–50 vol), 25°	65
	0.137(m)	436	2-Chloroethanol-water (60–40 vol), 25°	65
	0.131(m)	436	2-Chloroethanol-water (80–20 vol), 25°	65
	0.180(M)	436	Ethyleneglycol-water (20–80 vol), 25°	23
	0.168(M)	436	Ethyleneglycol-water (40–60 vol), 25°	23
	0.150(M)	436	Ethyleneglycol-water (60–40 vol), 25°	23
	0.136(M)	436	Ethyleneglycol-water (80–20 vol), 25°	23
	0.179(m)	436	Ethyleneglycol-water (20–80 vol), 25°	23
	0.166(m)	436	Ethyleneglycol-water (40–60 vol), 25°	23
	0.145(m)	436	Ethyleneglycol-water (60–40 vol), 25°	23
	0.129(m)	436	Ethyleneglycol-water (80–20 vol), 25°	23
	0.181(M)	436	Methyoxyethanol-water (20–80 vol), 25°	23
	0.167(M)	436	Methoxyethanol-water (40–60 vol), 25°	23
	0.157(M)	436	Methoxyethanol-water (60–40 vol), 25°	23
	0.148(M)	436	Methoxyethanol-water (80–20 vol), 25°	23
	0.180(m)	436	Methoxyethanol-water (20–80 vol), 25°	23
	0.167(m)	436	Methoxyethanol-water (40–60 vol), 25°	23
	0.149(m)	436	Methoxyethanol-water (60–40 vol), 25°	23
	0.143(m)	436	Methoxyethanol-water (80–20 vol), 25°	23
Legumin	0.197	546	Phosphate, NaCl, 0.5I, 25°	8, 28
	0.192	546	Phosphate, NaCl, 0.1–1.1I, 25°	66
α_1-Lipoprotein, human	0.178	589	0.3 M NaCl, 20–25°	24
β-Lipoprotein, human	0.177	578	0.3 M NaCl, 20–25°	24 24
β_1-Lipoprotein, human	0.171	589	0.3 M NaCl, 20–25°	24

REFRACTIVE INDEX INCREMENTS OF PROTEINS (continued)

Protein	$\frac{dn}{dc}$, ml/g	Wavelength nm	Conditions[a]	References
Lipovitellin	0.181	578	1 M formic acid	67
	0.181	578	Neutral salt	68
	0.198	578	4.0 M urea	68
	0.187	579		69
	0.197	436		69
α-Livetin	0.181	589		70
	0.189	436		70
β-Livetin	0.181	589		70
	0.189	436		70
γ-Livetin	0.188	589	25°	71
	0.194	436	25°	71
Lysozyme	0.1888	546	0.1 M NaCl, pH 6.2, 25°	7
	0.1955	436	0.1 M NaCl, pH 6.2, 25°	7
	0.185	589	Aqueous, 20–50°	72
	0.1888	546	Water, 25°	73
	0.184	586	Water, 25°	22
	0.196(M)	436	2-Chloroethanol-water (20–80 vol), 25°	23
	0.178(M)	436	2-Chloroethanol-water (40–60 vol), 25°	23
	0.144(M)	436	2-Chloroethanol-water (60–40 vol), 25°	23
	0.123(M)	436	2-Chloroethanol-water (80–20 vol), 25°	23
Myosin	0.195	436	0.5 M KCl, 4°	74
	0.195	436	0.1 M KSCN, 0.5 M KSCN	74
	0.209	436	0.01 M ADP/ATP, 4°	74
	0.206	436	0.6 M KCl, pH 7.2	75
	0.195	436	0.5 M KCl, 0.02 M Tris, pH 7.2, 20°	76
	0.209	436	Various conditions	77
	0.1910	436	0.5 M KCl,	78
	0.1887	546	0.5 M KCl, pH 6.8	78
	0.192	436	0.5 M KCl, pH 7, 3°	79
	0.198	436	0.6 M KCl, pH 7.3	80
	0.197	436	ATP-phosphate, pH 7, 20°	81
Ovalbumin	0.188	578	Salt free, 0.5°	6
	0.197	578	pH 8.60, diethylbarbiturate, 0.5°	6
	0.188	578	pH 7.74, phosphate, 0.5°	6
	0.1883	436	0.1 M NaCl, pH 4.8, 25°	7
	0.1820	546	0.1 M NaCl, pH 4.8, 25°	7
	0.1864	546	Water	6
	0.1935	436	Water	6
	0.1859	578	Water	6
	0.185	436	Water, 25°	22
	0.181	546	Water, 25°	22
	0.178	589	Water, 25°	22
	0.1854	589	Water, 23°	82
	0.179	589	Water, 22–23°	83

REFRACTIVE INDEX INCREMENTS OF PROTEINS (continued)

Protein	$\frac{dn}{dc'}$ ml/g	Wavelength nm	Conditions[a]	References
Pepsin	0.1928	436	pH 4.50, 0.2 M acetate, 25°	10
	0.1825	43	pH 5.00, 0.2 M acetate, 25°	10
	0.1863	436	pH 5.00, 0.2 M acetate, 25°	10
	0.1905	43	pH 4.50, 0.2 M acetate, 25°	10
	0.188	436	Water, 25°	22
	0.182	546	Water, 25°	22
	0.177	589	Water, 25°	22
Plasma proteins, human, pooled	0.183	589	0.3 M NaCl, 20–25°	24
Ribonuclease	0.192	436	Water, 25°	22
	0.186	546	Water, 25°	22
	0.185	589	Water, 25°	22
	0.187	546	Phosphate, pH 7.0, 20°	20
Soybean trypsin inhibitor	0.195	436	0.3 M KCl	84
T-2 Tail sheath	0.191	436	pH 7.0, ammonium acetate, 0.13I	85
Thyroglobulin	0.1949	436	Salt free or 0.10 M KNO$_3$, 23°	21
Tropomyosin	0.180	436	Water, 25°	86
	0.188	546	pH 2–12, 25°	87
	0.188	436	pH 7, phosphate, KCl, 1.1I	88
Tropomyosin B	0.1917	436	0.39 KCl, 0.01 M HCl, pH 2.0	89
	0.1870	436	1.0 M KCl, 0.1 M phosphate, pH 7.4	89
	0.147	436	5.0 M guanidine hydrochloride, 0.64 M phosphate, pH 6.1	89
Trypsin	0.193	436	0.3 M KCl	84
Trypsinogen	0.190	436	pH 7–12.5 buffers	90

Compiled by Serge N. Timasheff.

REFERENCES

1. **Ooi,** *J. Phys. Chem.,* 64, 984 (1960).
2. **Mommaerts,** *J. Biol. Chem.,* 198, 445 (1952).
3. **Steiner, Laki, and Spicer,** *J. Polymer Sci.,* 8, 23 (1952).
4. **Gergely,** *J. Biol. Chem.,* 220, 917 (1956).
5. **Wasserman and Harkness,** *J. Chem. Soc.,* 1954, 1344.
6. **Perlmann and Longsworth,** *J. Am. Chem. Soc.,* 70, 2719 (1948).
7. **Halwer, Nutting, and Brice,** *J. Am. Chem. Soc.,* 73, 2786 (1951).
8. **Goring and Johnson,** *Trans. Faraday Soc.,* 48, 367 (1952).
9. **Tietze and Neurath,** *J. Biol. Chem.,* 194, 1 (1952).
10. **Kronman and Stern,** *J. Phys. Chem.,* 59, 969 (1955).
11. **Ühlein and Stauff,** *Kolloid Z.,* 142, 150 (1955).
12. **Charlwood,** *J. Am. Chem. Soc.,* 79, 776 (1957).

REFRACTIVE INDEX INCREMENTS OF PROTEINS (continued)

13. Champagne, *J. Chim. Phys.,* 54, 378, 393 (1957).
14. Barer and Tkaczyk, *Nature,* 173, 84 (1954).
15. Kent, Record, and Wallis, *Philos. Trans. R. Soc. Lond.,* Ser. A, 250, No. 972 (1957).
16. Rhees and Foster, *Iowa State J. Sci.,* 27, 1 (1952).
17. Jaenicke and Stauff, *Kolloid Z.,* 178, 143 (1961).
18. Dreizen, Noble, and Waugh, *J. Am. Chem. Soc.,* 84, 4939 (1962).
19. Kratohvil, Deželić, and Deželić, *Arch. Biochem. Biophys.,* 106, 381 (1964).
20. Jaenicke, Schmid, and Knof, *Biochemistry,* 7, 919 (1968).
21. Edelhoch, *J. Biol. Chem.,* 235, 132, 1335 (1960).
22. McMeekin, Groves, and Hipp, *Adv. Chem. Ser.,* 44, 54 (1964).
23. Timasheff and Inoue, *Biochemistry,* 7, 2501 (1968).
24. Armstrong, Budka, Morrison, and Hasson, *J. Am. Chem. Soc.,* 69, 1747 (1947).
25. Schretter, *Biochem. Z,* 177, 335 (1926).
26. Putzeys and Brosteaux, *Bull. Soc. Chim. Biol.,* 18, 1681 (1936).
27. Kakiuchi, Hamaguchi, and Isemura, *J. Biochem.* (Tokyo) 57, 167 (1965).
28. Brand, Goring, and Johnson, *Trans. Faraday Soc.,* 51, 872 (1955).
29. Harrap and Woods., *Aust. J. Chem.,* 11, 581 (1958).
30. Swaisgood and Timasheff, *Arch. Biochem. Biophys.,* 125, 344 (1968).
31. Payens and Heremans, *Biopolymers,* 8, 335 (1969).
32. Antonini, Rossi-Fanelli, and Caputo, *Arch. Biochem. Biophys.,* 97, 343 (1962).
33. Sarfare, Kegeles, and Kwan-Rhee, *Biochemistry,* 5, 1389 (1966).
34. Tinoco, *Arch. Biochem. Biophys.,* 68, 367 (1957).
35. Wilcox, Kraut, Wade, and Neurath, *Biochim. Biophys. Acta,* 24, 72 (1957).
36. Boedtker and Doty, *J. Am. Chem. Soc.,* 78, 4267 (1956).
37. Veis and Cohen, *J. Am. Chem. Soc.,* 78, 6238 (1956).
38. Veis, Eggenberger, and Cohen, *J. Am. Chem. Soc.,* 77, 2368 (1955).
39. Timasheff and Tinoco, *Arch. Biochem. Biophys.,* 66, 427 (1957).
40. Guinand, *J. Polymer Sci.,* 29, 497 (1958).
41. Lindberg, *Biochemistry,* 6, 335 (1967).
42. Goring and Johnson, *Arch. Biochem. Biophys.,* 56, 448 (1955).
43. Sowinski, Oharenko, and Koenig, *J. Am. Chem. Soc.,* 81, 6193 (1959).
44. Casassa, *J. Phys. Chem.,* 60, 926 (1956).
45. Sowinski, Oharenko, and Koenig, *J. Am. Chem. Soc.,* 81, 6193 (1959).
46. Schultz and Ende, *Z. Physik. Chem. N.F.,* 36, 82 (1963).
47. Laki and Steiner, *J. Polymer Sci.,* 8, 457 (1952).
48. Boedtker and Doty, *J. Phys. Chem.,* 58, 968 (1954).
49. Veis and Anesey, *J. Phys. Chem.,* 63, 1720 (1959).
50. Gallop, *Arch. Biochem. Biophys.,* 54, 486 (1955).
51. Gouinlock, Flory, and Scheraga, *J. Polymer Sci.,* 16, 383 (1955).
52. Rossi-Fanelli, Antonini, and Caputo, *J. Biol. Chem.,* 234, 2906 (1959).
53. Richmann and Colvin, *Can. J. Chem.,* 34, 411 (1956).
54. Frieden, *J. Biol. Chem.,* 237, 2396 (1962).
55. Dandliker and Fox, *J. Biol. Chem.,* 214, 275 (1955).
56. Klotz and Keresztes-Nagy, *Biochemistry,* 2, 445 (1963).
57. Adair and Robinson, *Biochem. J.,* 24, 933 (1930).
58. Schönberger, *Biochem. Z.,* 267, 57 (1933).
59. Howard, *J. Biol. Chem.,* 41, 537 (1920).
60. Benhamou and Weill, *Biochim. Biophys. Acta,* 24, 548 (1957).
61. Rossi-Fanelli, Antonini, and Caputo, *J. Biol. Chem.,* 236, 391 (1961).
62. Doty, Gellert, and Rabinovitch, *J. Am. Chem. Soc.,* 74, 2065 (1952).
63. Agatova and Kurganov, *Dokl. Akad. Nauk SSSR,* 169, 1452 (1966).
64. Pedersen, *Biochem. J.,* 30, 961 (1936).
65. Inoue and Timasheff, *J. Am. Chem. Soc.,* 90, 1890 (1968).
66. Brand, Goring, and Johnson, *Trans. Faraday Soc.,* 54, 1911 (1958).
67. Kratohvil, Martin, and Cook, *Can. J. Biochem Physiol.,* 40, 855 (1962).
68. Jourbet and Cook, *Can. J. Biochem. Physiol.,* 36, 389 (1958).
69. Vandegaer, Reichmann, and Cook, *Arch. Biochem. Biophys.,* 62, 328 (1956).
70. Martin, Vandegaer, and Cook, *Can. J. Biochem. Physiol.,* 35, 241 (1957).
71. Martin, Vandegaer, and Cook, *Can. J. Biochem. Physiol.,* 36, 153 (1951).

REFRACTIVE INDEX INCREMENTS OF PROTEINS (continued)

72. Bourgoin and Jollès, *J. Chem. Phys.*, 63, 760 (1966).
73. Bruzzesi, Chiancone, and Antonini, *Biochemistry*, 4, 1796 (1965).
74. Brahms and Brezner, *Arch. Biochem. Biophys.*, 95, 219 (1961).
75. Ellenbogan, Iyengar, Stern, and Olson, *J. Biol. Chem.*, 235, 2642 (1960).
76. Kay and Pabst, *J. Biol. Chem.*, 237, 727 (1962).
77. Gellert, von Hippel, Morales, and Schachman, *J. Am. Chem. Soc.*, 81, 1384 (1959).
78. Rupp and Mommaerts, *J. Biol. Chem.*, 224, 277 (1957).
79. Gellert and Englander, *Biochemistry*, 2, 39 (1963).
80. Conway and Roberts, *Am. J. Physiol.*, 208, 243 (1965).
81. Mueller, Franzen, Rice, and Olson, *J. Biol. Chem.*, 239, 1447 (1964).
82. Barker, *J. Biol. Chem.*, 104, 667 (1934).
83. Haas, *J. Biol. Chem.*, 35, 119 (1918).
84. Steiner, *Arch. Biochem. Biophys.*, 49, 71 (1954).
85. Sarkar, Sarkar, and Kozloff, *Biochemistry*, 3, 511 (1964).
86. Ooi, Mihashi, and Kobayashi, *Arch. Biochem. Biophys.*, 98, 1 (1962).
87. Kay and Bailey, *Biochim. Biophys. Acta*, 40, 149 (1960).
88. Kay, *Biochim. Biophys. Acta*, 27, 469 (1958).
89. Holtzer, Clark, and Lowey, *Biochemistry*, 4, 2401 (1965).
90. Smillie, Kay, and Hilderman, *J. Biol. Chem.*, 236, 112, 118 (1961).

MOLAR ABSORPTIVITY AND $A_{1cm}^{1\%}$ VALUES FOR PROTEINS AT SELECTED WAVELENGTHS OF THE ULTRAVIOLET AND VISIBLE REGION

Protein	$\epsilon^a (\times 10^{-4})$	$A_{1cm}^{1\%}$ [b]	nm[c]	Ref.	Comments[d]
Acetoacetate decarboxylase (EC 4.1.1.4) *C. acetobutylicum*	3.05	10.5	280	567	MW = 29,000 (567) subunit
Acetolactate synthase (EC 4.1.3.18) *Aerobacter aerogenes*	–	8.3	280	568	pH 6, 50 mM P_i; data from Figure 1 (568)
Acetyl coenzyme A carboxylase (EC 6.4.1.2) Chicken liver	–	11.6	280	569	Calc. using OD$_{280}$ × 0.86 = mg/ml (569)
Acetylcholinesterase (EC 3.1.1.7) *Electrophorus electricus*	–	21.4	280	1322	Kjeldahl or Dumas[e]
	–	21.8	280	1322	Microninhydrin[e]
	–	18.8	280	1322	Nitrogen from amino acid anaylsis[e]
	–	17.6	280	1322	DR
	–	18.2	280	1322	Dry wt
	52.7	22.9	280	1	pH 7.0, 0.1 M NaCl, 0.03 M NaP$_i$, MW = 230,000 (1)
	–	16.1	280	2	0.02 M AcONH$_4$
	–	19.0	280	877	–
β-N-Acetyl-D-glucosaminidase (EC 3.2.1.30) *Aspergillus oryzae*	29.3	20.9	280	777	MW = 140,000 (777)
Beef spleen Enzyme A	–	12.8	278	1323	–
Enzyme B	–	12.7	278	1323	–
O-Acetylserine sulfhydrase A (cysteine synthase) (EC 4.2.99.8) *Salmonella typhimurium*	8.2	12.1	280	634	MW = 68,000 (634)
	0.76	1.12	412	634	MW = 68,000 (634)
Acetylserotonin methyltransferase (EC 2.1.1.4) (see hydroxyindole-O-methyltransferase)					

[a] ϵ is the molar absorption coefficient with units of M^{-1} cm^{-1} and is either the value reported in the reference cited or calculated from the $A_{1cm}^{1\%}$ value and the molecular weight.
[b] $A_{1cm}^{1\%}$ is the absorbance for a 1% solution in a 1-cm cuvette and is either the value reported in the reference cited or calculated from the ϵ and the molecular weight. The relationship between ϵ, $A_{1cm}^{1\%}$ and molecular weight, MW, is $10\epsilon = (A_{1cm}^{1\%})$ (MW).
[c] Refers to the wavelength cited and may not be the peak of the absorption band.
[d] Abbreviations used: SC, corrected for light scattering; P_i, phosphate; GdmCl, guanidinium chloride; PP$_i$, pyrophosphate; Gro-P, glycerophosphate; S$_2$ threitol, dithiothreitol; NaDodSO$_4$, sodium dodecyl sulfate; HSEtOH, 2-mercaptoethanol; Gly$_2$, glycyl-glycine; ImzAc, imidazoleacetate; Tes, N-Tris(hydroxymethyl)methyl-2-aminoethanesulfonic acid; SucNBr, N-bromosuccinimide; albumin, bovine serum albumin.
Methods of protein determination: Dry wt, dry weight; AA, amino acid analysis; Refr., R, refractometry; Biuret, colorimetric method; Folin, colorimetric method; N, nitrogen determination; UC, ultracentrifuge; FC, fringe counting (interferometry); DR, differential refractometry; Kjedahl: Lowry.
[e] Methods for determining nitrogen concentration in order to determine protein concentration.

MOLAR ABSORPTIVITY AND $A_{1cm}^{1\%}$ VALUES FOR PROTEINS AT SELECTED WAVELENGTHS OF THE ULTRAVIOLET AND VISIBLE REGION (continued)

Protein	$\epsilon^a (\times 10^{-4})$	$A_{1cm}^{1\%}$ [b]	nm[c]	Ref.	Comments[d]
Acid deoxyribonuclease (deoxyribonuclease II) (EC 3.1.4.6)					
Pig spleen	4.6	12.1	280	635	MW = 38,000 (635)
Aconitase (aconitate hydrase) (EC 4.2.1.3)					
Pig heart	–	13.7	280	778	–
Actin					
Muscle	–	11.08	280	1324	–
F-Actin					
Rabbit	–	11.08	280	1156	Kjeldahl
Rabbit muscle	–	9.65	280	3	0.1 M KCl
	–	11.49	280	4	–
	–	11.5	280	636	–
G-Actin					
Rabbit muscle	5.05	10.97	280	5	MW = 46,000 (5)
β-Actinin					
Rabbit muscle	–	9.8	278	878	
Acyl phosphatase (EC 3.6.1.7)					
Horse muscle	1.09	11.58	280	639	pH 5.3, 0.05 M; AcO⁻, MW = 9,400 (639)
Acyl-carrier-protein					
Escherichia coli	0.27	3.0[g]	275	637	MW = 9,100 (637); pH 7.0, 0.01 M KP$_i$, data from Figure 2 (638)
	0.18	–	275	638	
Acyl-CoA dehydrogenase (see fatty acyl-CoA dehydrogenase)					
Adenosine deaminase (EC 3.5.4.4)					
Calf spleen	2.7	8.15	278	6	Est. from Figure 1 (6) MW = 33,120 (6)
Aspergillus oryzae	27.8	13.0	280	879	MW = 214,000 (879)
Adenosine 5'-phosphate deaminase (EC 3.5.4.17)					
Rabbit muscle	–	9.13	280	880	Dry wt
	–	9.3	280	881	–
Adenosine 5'-phosphate nucleosidase (EC 3.2.2.4)					
Azotobacter vinelandii	5.58	9.73	280	1325	pH 8, 0.05 M triethanolamine · HCl containing 0.1 mM S$_2$ threitol and 1 mM EDTA; MW = 57,300 (1325)

[g]Optical density used for calculation corrected for light scattering by extrapolation from 350 nm.

MOLAR ABSORPTIVITY AND $A_{1cm}^{1\%}$ VALUES FOR PROTEINS AT SELECTED WAVELENGTHS OF THE ULTRAVIOLET AND VISIBLE REGION (continued)

Protein	$\epsilon^a (\times 10^{-4})$	$A_{1cm}^{1\%}$ [b]	nm[c]	Ref.	Comments[d]
Adenosine triphosphate sulfurylase (sulfate adenylyltransferase) (EC 2.7.7.4) see also ATP-sulfurylase.					
Penicillium chrysogenum	—	8.71	278	882	—
Adenovirus					
Hexon	—	14.6	279	1326	pH 7, 0.01 M NaP$_i$
Adenylic acid deaminase (EC 3.5.4.6)					
Rat muscle	28.5	9.84	280	1327	Dry wt, MW = 290,000 (1327)
Adrenodoxin					
Beef	1.14	—	276	1157	—
Beef adrenals	1.3	—	276	779	—
	1.26	—	325	779	—
	1.26	—	340	779	—
	0.98	—	414	779	—
	0.84	—	455	779	—
Beef adrenal cortex	0.579	—	276	1328	Values cited are per mole of Fe
	0.641	—	320	1328	
	0.496	—	414	1328	
	0.421	—	455	1328	
	—	6.75	276	1328	
	—	5.78	414	1328	
Apo-	0.76	—	276	779	
	0.35	—	276	1157	
Aequorin					
Aequorea	8.65	27.0[h]	280	780	MW = 32,000 (780); protein det. by dry wt
Apoaequorin-SH	—	18.2	280	780	
Apoaequorin-SO	—	18.2	280	780	
Agglutinin					
Wheat germ	10.9		280	1329	pH 7.0, 0.01 M NaP$_i$
	12		272	1329	
Alanine dehydrogenase (EC 1.4.1.1)					
Bacillus subtilis	51.2	22.3	280	883	pH 8, TrisCl, 0.05 M, MW = 230,000 (883)
Alanine racemase (EC 5.1.1.1)					
Pseudomonas putida	—	10.8	275	781	pH 7.4, 0.005 M KP$_i$, data from Figure 2 of Reference 781
Albocuprein					
Human, brain					
I	8.4	11.65	280	1158	pH 6, 0.05 M NaCl/0.05 M AcONa, dry wt, MW = 72,000 (1158)

[h] 1 mg of aequorin in 1 ml added to 10^{-4} EDTA gives an OD of 2.52 at 280 nm. After freeze-drying and redissolving, the OD is now 2.25 at 280 nm. (780)

MOLAR ABSORPTIVITY AND $A_{1cm}^{1\%}$ VALUES FOR PROTEINS AT SELECTED WAVELENGTHS OF THE ULTRAVIOLET AND VISIBLE REGION (continued)

Protein	$\epsilon^a(\times 10^{-4})$	$A_{1cm}^{1\%}$ b	nm^c	Ref.	Comments^d
Albocuprein (continued)					
II	1.2	8.63	280	1158	pH 6, 0.05 M NaCl/0.05 M AcONa, dry wt, MW = 14,000 (1158)
Albumin	—	10.6	278	1159	
Beef serum	—	6.49	280	1160	6 M GdmCl
	—	6.62	278	1161	pH 2, 0.01 N HCl
	—	3.58	255	1162	
	—	6.14	280	1162	pH 7.0, 0.01 M P$_i$
	—	0.50	310	1162	
	—	6.8	280	12	—
	—	6.67	279	13	—
	—	6.6	280	14	—
	—	6.6	279.5	15	—
	—	6.6	279	16	—
	3.96	—	280	17	pH 7
	4.36	6.61	280	11	MW = 66,000 (11)
	—	6.3	280	640	Water
	—	270	210	640	Water
	—	840	191	640	Water
	4.69	6.9	279	641	MW = 68,000 (641)
	—	6.7	278	642	—
	2.77	—	288	643	—
	4.24	—	279	643	—
	19.43	—	234	643	—
	—	6.75	278	644	—
	—	6.2	280	645	—
	—	3.7	253.7	645	—
Beef, mercapto- (see also mercaptoalbumin, beef)	—	6.82	279	894	—
	—	3.03	253	1330	pH 6.2
	—	6.54	278	1330	
	—	6.67	277.5	884	—
	4.37	—	280	885	2°C, alcohol-water mixtures
	4.6	—	280	886	—
	—	8.2	280	887	—
	—	650	191.4	887	—
	4.2	6.2	280	888	pH 7.96, TrisCl, MW = 68,000 (888), Water (889)
Fragment F$_2$	—	5.51	278	890	Dry wt
	1.71	5.51	278	647	MW = 31,000 (647)
Fragment F$_3$	2.74	7.55	278	647	MW = 36,300 (647)
	—	7.55	278	890	Dry wt
S-Carboxymethyl-	—	5.96	278	891	pH 8, 6 M GdmCl– 0.02 M EDTA
S-Cysteinyl-	—	5.96	278	891	pH 8, 6 M GdmCl– 0.02 M EDTA
	—	6.14	278	892	6 M GdmCl

MOLAR ABSORPTIVITY AND $A_{1cm}^{1\%}$ VALUES FOR PROTEINS AT SELECTED WAVELENGTHS OF THE ULTRAVIOLET AND VISIBLE REGION (continued)

Protein	$\epsilon^a (\times 10^{-4})$	$A_{1cm}^{1\%}$ [b]	nm[c]	Ref.	Comments[d]
Albumin (continued)					
Polypeptidyl derivatives					
Gly-261[i]	4.9	5.8	278	893	—
L-Phe-31	4.8	6.4	278	893	—
L-Phe-36	4.7	6.3	278	893	—
DL-Phe-48	4.9	6.3	278	893	—
L-Glu-13	4.8	6.8	278	893	—
L-Glu-41	4.8	6.4	278	893	—
L-Glu-73	4.7	5.9	278	893	—
L-Glu-218	4.6	4.7	278	893	—
L-Glu-275	4.9	4.7	278	893	—
L-Lys-2	4.9	7.0	278	893	—
L-Lys-14	5.0	7.1	278	893	—
Methylated	—	6.5	280	645	
	—	3.7	253.7	645	
Acetylated	—	6.9	280	645	80% acetylated
	—	3.7	253.7	645	on amino groups
Diazotized	—	7.5	280	645	—
	—	7.3	253.7	645	—
Guanidinated	—	9.2	280	645	—
	—	11.0	253.7	645	45 groups
Iodinated	—	12.0	312	645	
	—	7.5	280	645	32 mol I/mol
	—	18.6	253.7	645	
Glutaraldehyde modified	—	26.2	280	646	pH 8, borate
S-β-Pyridylethyl-	—	12.0	274	1161	pH 2, 0.01 N HCl
Cow's milk	4.55	6.6	280	901	MW = 69,000 (901)
Human serum	—	193	210	648	—
	—	143	215	648	—
	—	4.3	254	648	—
	—	7.15	280	648	—
	—	3.92	255	1162	
	—	5.94	280	1162	pH 7.0, 0.01 M P_i
	—	0.90	310	1162	
	—	6	280	7	
	4.0	5.8	280	8	MW = 69,000 (8)
	3.6	5.31	280	9	MW = 68,000 (10)
	3.5	5.3	280	11	MW = 66,000 (11)
	3.52	5.03	277.5	895	pH 2
					MW = 70,000 (895)
	—	7.15	280	896	—
	—	193	210	896	—
	—	143	215	896	—
	—	4.3	254	896	—
N₂ph-[q]	—	30.8	290	897	0.1 M NaOH
	—	11.7	360	897	0.1 M NaOH
Human mercapto-					
Fraction I	—	5.7	280	898	—
Fraction II	—	5.61	280	898	—
Fraction III	—	5.31	280	898	—
Fraction IV	—	5.8	280	898	—
Pig serum	—	6.72	280	899	pH 8.6, 0.2 M TrisCl, 22°C, dry wt
Rabbit serum	—	6.6	280	900	—

[i]Gly-261 means that 261 moles of glycine have been attached to albumin. Other derivatives have been prepared similarly using other amino acids and are so indicated.

[q]N₂ph- = 2,4-dinitrophenyl.

MOLAR ABSORPTIVITY AND $A_{1cm}^{1\%}$ VALUES FOR PROTEINS AT SELECTED WAVELENGTHS OF THE ULTRAVIOLET AND VISIBLE REGION (continued)

Protein	$\varepsilon^a (\times 10^{-4})$	$A_{1cm}^{1\%}$ [b]	nm[c]	Ref.	Comments[d]
Albuminoid (insoluble protein)					
Young rat, lens	–	18.4	280	782	
Young rat X-rayed, lens	–	18.6	280	782	
Old rat, lens	–	17.7	280	782	pH 9.8, 0.05 M borate/8 M urea
Medium-aged dogfish, lens	–	22.4	280	782	
Old dogfish, lens	–	21.2	280	782	
Albuminoid sulfonated					
Rat lens	–	17.3	280	902	–
	–	18.0	280	903	–
Dogfish lens	–	25.1	280	903	–
Beef lens	–	9.8	280	903	–
Human lens					
0–10 years old	–	11.1	280	903	–
11–20 years old	–	15.0	280	903	–
40–49 years old	–	15.9	280	903	–
50–59 years old	–	15.0	280	903	–
60–69 years old	–	16.5	280	903	–
70–79 years old	–	17.1	280	903	–
80– years old	–	18.0	280	903	–
Alcohol dehydrogenase (EC 1.1.1.1)					
Horse liver	–	4.55	280	904	–
	–	4.2	280	905	Dilute neutral buffer
	–	4.26	280	906	pH 7.2, 3 M GdmCl
	3.59	–	280	907	MW = 79,000 (907)
	3.82	4.55	280	18	–
	–	4.5	280	19	–
	3.83	4.6	280	20	MW = 83,300 (20)
	3.54	4.2	280	22	MW = 84,000 (21)
Zn-	3.44	4.3	280	908	
Co-	3.92	4.9	280	910	MW 80,000 (909)
Cd-	4.56	5.7	280	910	
	1.02	1.25	245	910	
Human liver	–	4.6	280	911	pH 7.0, 0.03 M P_i, 0.07 M NaCl
	5.3	6.1	280	23	pH 7.0, 0.1 μM NaP_i, MW = 87,000 (23)
Yeast	–	12.1	280	912	MW = 140,000 (913)
	20.78	14.8	278	913	MW = 141,000 (914)
	–	14.6	280	914	
	18.9	–	280	24	pH 8.1, 0.08 M glycine
	–	12.6	280	25	–
Arachis hypogea (peanuts)	7.2	6.4	278	915	MW = 112,000 (915)
Drosophila melanogaster	3.96	9.0	280	916	MW = 44,000 (916)
Aldehyde dehydrogenase (EC 1.2.1.3-.5)					
Horse liver	–	20.8	280	1163	pH 7, AcONH$_4$, dry wt
Pseudomonas aeruginosa	19.5	10.4	280	783	pH 7.0, 1 mM KP_i, protein con. det. by dry wt, MW = 187,000 (783)

MOLAR ABSORPTIVITY AND $A_{1cm}^{1\%}$ VALUES FOR PROTEINS AT SELECTED WAVELENGTHS OF THE ULTRAVIOLET AND VISIBLE REGION (continued)

Protein	$\epsilon^a (\times 10^{-4})$	$A_{1cm}^{1\%}$ [b]	nm[c]	Ref.	Comments[d]
Aldehyde dehydrogenase (continued)					
Yeast	13.4	6.7	280	26	Reference states: "1 mg enzyme ... OD equals 0.67," no volume given, assumed 1 ml, MW = 200,000 (26)
Aldehyde oxidase (EC 1.2.3.1)					
Rabbit liver	6.3	—	450	27	—
	2.2	—	550	27	—
Aldolase (EC 4.1.2.13)					
Rabbit muscle	—	9.1	280	28	—
	—	12.1	280	29	
	11.8	8.32	277	30	pH 2
	13.3	9.38	280	30	pH 5.7, MW = 142,000 (31)
	—	7.8	276	32	pH 2
	11.8	7.4	280	917	pH 7.5–13, MW = 160,000 (918)
	—	8.4	280	919 }	pH 12.5, 0.1 M borate
	—	9.6	289.5	919	
	—	8.16[j]	276	920	3 M GdmCl
	—	8.20[j]	276	920	5 M GdmCl
	—	8.21[j]	276	920	6 M GdmCl
	—	8.23[j]	276	920	7 M GdmCl
Succinyl-	—	8.2	276.5	921	Dry wt/KN[k]
Rabbit liver	13.09	8.5	280	33	MW = 154,000 (33)
	13.3	8.40	280	922	MW = 158,000 (922)
Gallus domesticus, muscle (chicken)	16	10.3	280	923	pH 6.5, MW = 158,000 (923)
Liver	13.7	8.6	280	924	pH 7.5, 2 mM Tris, 0.2 mM EDTA MW = 160,000 (925)
Rat muscle	15.0	9.39	280	926	MW = 160,000 (926)
Gradus morhua (codfish) muscle	15.2	9.5	280	927	pH 7.5, MW = 160,000 (927)
Spinach	20.76	17.3	280	34	pH 7.4, 0.05 M P_i, MW = 120,000 (34)
	—	13.3	280	930	pH 7.5, 0.1 M P_i, Lowry[l]
	—	11.0	280	930	Dry wt
Yeast	7.95	10.6	280	35	MW = 75,000 (35)
	8	10	280	928	MW = 80,000 (928), dry wt
	8.15	10.2	280	929	MW = 80,000 (929)
	8.0	10.1	280	929	Dry wt
	8.5	10.6	280	929	DR[m]
	7.9	9.9	280	929	FD[n]

[j] Calculated from an equation in Reference 920.
[k] KN, protein concentration determined by Kjeldahl nitrogen.
[l] Lowry, protein concentration determined by Lowry method using bovine serum albumin as standard.
[m] DR, protein concentration determined by differential refractometry.
[n] FD, protein concentration determined by fringe displacement method.

MOLAR ABSORPTIVITY AND $A_{1cm}^{1\%}$ VALUES FOR PROTEINS AT SELECTED WAVELENGTHS OF THE ULTRAVIOLET AND VISIBLE REGION (continued)

Protein	$\epsilon^a (\times 10^{-4})$	$A_{1cm}^{1\%}$ [b]	nm[c]	Ref.	Comments[d]
Aldolase (continued)					
Lobster muscle					
(*Homarus americanus*)	17.9	11.2	280	1164	MW = 160,000 (1164)
Shark muscle					
(*Mustelus canis*)	–	8.64	278.4	1165	Dry wt, Kjeldahl
	–	8.60	280	1165	
Rabbit liver	–	8.9	280	1166	Dry wt
Rabbit brain	–	8.8	280	1166	Refract
Aldolase, L-Rhamnulose 1-phosphate (EC 4.1.2.19)					
Escherichia coli	23.4	17.3	280	1167	MW = 135,000 (1167)
Aldolase, 3-Deoxy-2-keto-6-phosphogluconate (EC 4.1.2.14)					
Pseudomonas putida	–	8.63	280	1168	0.1 N NaOH
Aldose 1-epimerase (EC 5.1.3.3)					
Escherichia coli K12	–	10.8	280	1331	–
Allantoicase (EC 3.5.3.4), 0.9S					
Pseudomonas aeruginosa	–	27.3	280	1169	Calcd from data in Figure 4 (1169), pH 7.7
	–	26.0	280	1169	Calcd from data in Figure 4 (1169), pH 4.6
	–	24.3	280	1169	Calcd from data in Figure 4 (1169), pH 4.6, in the presence of 0.1 M glycolate
	–	31.7	280	1169	Lowry
Allergen					
Short ragweed pollen					
Antigen E	–	11.3	280	1170	pH 7.15
Antigen Ra.3	16.4	10.9	280	1171	pH 7.3, 0.005 M NH$_4$ HCO$_3$, MW = 15,000 (1171)
Atopic					
Rye grass pollen					
I-B	–	15.0	280	1332	pH 7, MW =
	–	2.18	305	1332	34,000 (1332)
II-B	–	10.3	280	1332	pH 7, MW =
	–	0.88	305	1332	11,000 (1332)
B	–	14.1	280	1332	
	–	3.10	305	1332	pH 7
D[IEP]	–	14.7	280	1332	
	–	4.75	305	1332	
K	–	14.8	280	1332	pH 7, MW = 38,200 (1332)
Pool Cc	–	7.63	280	1332	
	–	1.05	305	1332	
Trifidin A	–	4.1	280	1332	
	–	1.20	305	1332	
Ipecac IPC-D	–	10.5	280	1332	pH 7
	–	4.10	305	1332	
Liquorice SL-F	–	11.0	305	1332	
Pyrethrum					
Whole dialysate	–	76.8	280	1332	
	–	69.8	305	1332	

MOLAR ABSORPTIVITY AND $A_{1cm}^{1\%}$ VALUES FOR PROTEINS AT SELECTED WAVELENGTHS OF THE ULTRAVIOLET AND VISIBLE REGION (continued)

Protein	$\epsilon^a (\times 10^{-4})$	$A_{1cm}^{1\%}$ [b]	nm[c]	Ref.	Comments[d]
Allergen (continued)					
Kapok KP-E	—	76.2	280	1332	
	—	64.4	305	1332	pH 7
Cotton CL-E	—	20.6	280	1332	
	—	13.0	305	1332	
Cotton seed CS 60C	—	6.58	255	1162	—
	—	6.85	280	1162	—
	—	4.89	310	1162	—
Castor bean [CB-1A] SRI	—	3.38	280	1332	pH 7
Human dandruff HD-E	—	5.21	305	1332	
	—	9.32	255	1162	—
	—	10.02	280	1162	—
	—	4.36	310	1162	—
Horse dandruff	—	8.28	255	1162	—
	—	9.62	280	1162	—
	—	3.44	310	1162	—
Whole dialysate	—	6.40	280	1332	
	—	1.20	305	1332	
Feathers FE-B	—	58.0	280	1332	
	—	43.3	305	1332	
Caddis fly Pool 2	—	35.2	280	1332	pH 7
	—	15.5	305	1332	
Alternaria	—	8.00	280	1332	
	—	4.14	305	1332	
Trichophytin	—	5.10	305	1332	
	—	7.96	255	1162	—
	—	7.94	280	1162	—
	—	4.80	310	1162	—
House dust HE-E	—	8.1	305	1332	pH 7
	—	13.80	255	1162	—
	—	13.64	280	1162	—
	—	8.36	310	1162	—
Tomato TO-G	—	21.0	280	1332	pH 7
	—	12.5	305	1332	
	—	14.56	255	1162	—
	—	13.80	280	1162	—
	—	9.40	310	1162	—
Cow's milk VM-5	—	9.20	280	1332	pH 7, MW = 36,000 (1332)
	—	1.95	305	1332	
Egg white VE$_9$	—	4.44	280	1332	pH 7, MW = 31,500 (1332)
	—	0.40	305	1332	
Hay HH-C	—	84.2	280	1332	pH 7
	—	67.0	305	1332	
Succus liquiritiae	—	15.80	255	1162	—
	—	14.02	280	1162	—
	—	10.58	310	1162	—
Radix ipecacuanhae	—	9.66	255	1162	—
	—	11.02	280	1162	—
	—	3.84	310	1162	—
Alliin lyase (EC 4.4.1.4) Garlic (*Allium sativum*)	—	16.6	280	1172	Calcd from data in Figure 4 (1172), pH 7.5, 10% glycerol-0.02 M P$_i$

MOLAR ABSORPTIVITY AND $A_{1cm}^{1\%}$ VALUES FOR PROTEINS AT SELECTED WAVELENGTHS OF THE ULTRAVIOLET AND VISIBLE REGION (continued)

Protein	ϵ^a (× 10^{-4})	$A_{1cm}^{1\%}$ [b]	nm[c]	Ref.	Comments[d]
Amandin					
Almonds	–	7	280	1173	From Figure 1 (1173), pH 5.7
Amidophosphoribosyltransferase (EC 2.4.2.14) (see glutamine phosphoribosylpyrophosphate-amidotransferase and phosphoribosyldiphosphate amidotransferase)					
Amine dehydrogenase (amine oxidase)					
Pseudomonas AM 1	11.3	8.46	280	1174	pH 7.5, 0.05 M P$_i$, Lowry, MW = 133,000 (1174)
Amine oxidase (EC 1.4.3.4)					
Aspergillus niger	–	11.8	280	1179	–
Beef plasma	–	9.8	280	1180	–
Amine oxidase (EC 1.4.3.6) (see diamine oxidase and monoamine oxidase)					
D-Amino-acid oxidase (EC 1.4.3.3)					
Pig kidney	–	15.6	277	1175	–
	–	126	220	1175	–
	7.31	16.0	280	36	MW = 182,000 (36)
Batch I enzyme	–	23.0	274	37	–
Apo-	–	15.4	278	37	pH 8.5, 0.1 M PP$_i$
Batch II enzyme	–	19.8	274	37	pH 8.5, 0.1 M PP$_i$
Apo-	15.1	15.1	280	38	pH 8.3, M/60 PP$_i$, MW = 100,000 (39)
Apo-	17.5	14.0	280	40	pH 8.3, 0.1 M PP$_i$, MW = 125,000 (40)
Apo-	–	14	280	1176	pH 8.3, 0.1 M PP$_i$
L-Amino-acid oxidase (EC 1.4.3.2)					
Crotalus adamanteus	23.6	17.9	275	41	0.1 M KCl, MW = 132,000 (41)
	2.35	1.78	390	41	0.1 M KCl, MW = 132,000 (41)
	2.26	1.71	462	41	0.1 M KCl, MW = 132,000 (41)
Rat kidney	8.5	9.55	275	42	MW = 89,000 (42)
	1.07	1.20	358	42	MW = 89,000 (42)
	1.27	1.43	455	42	MW = 89,000 (42)
Amino-acid racemase (EC 5.1.1.10)					
Pseudomonas striata	–	8.3	280	1177	Data from Figure 3 (1177), pH 7.0, 0.01 M KP$_i$
Aminoacyl-tRNA: ribosome binding enzyme					
Rabbit reticulocytes	17.9	9.6	280	1178	Calc. from data in Figure 3 (1178), pH 7.5, 0.01 M KP$_i$, MW = 186,000 (1178)

MOLAR ABSORPTIVITY AND $A_{1cm}^{1\%}$ VALUES FOR PROTEINS AT SELECTED WAVELENGTHS OF THE ULTRAVIOLET AND VISIBLE REGION (continued)

Protein	$\epsilon^a (\times 10^{-4})$	$A_{1cm}^{1\%}$ [b]	nm[c]	Ref.	Comments[d]
Aminopeptidase (EC 3.4.11.1-.2) (see also leucine aminopeptidase)					
Pig kidney	—	16.3	280	43	pH 7.2, 0.06 M P$_i$
	—	12.28	266	43	pH 7.2, 0.06 M P$_i$
	—	125	225	43	pH 7.2, 0.06 M P$_i$
	—	168	215	43	pH 7.2, 0.06 M P$_i$
Pig kidney, particulate (EC 3.4.11.2)	—	16	280	1182	Estimated from figure in Reference 1182
	47.3	16.9	280	1183	MW = 280,000 (1183), refr. and Lowry
Rat kidney	—	16.1	280	44	—
Aeromonas proteolytica	4.18	14.4	278.5	1181	MW = 29,000 (1181)
B. stearothermophilus	41	10.2	280	1184	pH 7.2, 0.05 M TrisCl, 0.001 M Co$^+$, MW = 400,000 (1184)
Aminopeptidase (microsomal) (EC 3.4.11.2)					
Pig kidney	45.5	16.2	280	1186	MW = 280,000 (1186)
Aminopeptidase P (aminoacyl-proline aminopeptidase, EC 3.4.11.9)					
Escherichia coli B.	—	10.3	280	1187	Kjeldahl
Aminotransferase alanine[o] (EC 2.6.1.2)					
Rat liver	—	6.85	278.7	1333	DR, pH 7.0, 50 mM KP$_i$ containing 0.5 mM S$_2$ threitol
Amylase (EC 3.2.1.1-.3)					
Human plasma	—	9	280	1188	Water, calcd from data in Figure 4 (1188)
Rat pancreas	—	20	280	1189	—
B. subtulis Takamine	—	25.2	280	1190	—
B. subtilis Kalle	—	25.2	280	1190	—
α-Amylase (EC 3.2.1.1)	—	24.4	290	1191	0.1 N NaOH
Human saliva	—	26	280	45	—
Pig pancreas	—	26	280	45	—
	12.8	25	280	46	Water, MW = 51,300 (47)
B. subtilis	—	25.3	280	45	—
A. oryzae	—	19.7	280	45	—
B. macerans	9.9	7.11	280	48	pH 6.2, 0.01 M P$_i$, MW = 139,000 (48)
Pirkka barley	8.55	15.0	280	1334	Dry wt, pH 7.0, 0.05 M NaP$_i$, MW = 57,200 (1334)
B. subtilis	9.35	19.8	280	1192	MW = 47,300 (1193)
	12.5	25.6	280	1195	MW = 49,000 (1195)

[o]Pyridoxal or pyridoxamine form.

MOLAR ABSORPTIVITY AND $A_{1cm}^{1\%}$ VALUES FOR PROTEINS AT SELECTED WAVELENGTHS OF THE ULTRAVIOLET AND VISIBLE REGION (continued)

Protein	$\epsilon^a (\times 10^{-4})$	$A_{1cm}^{1\%}$ [b]	nm[c]	Ref.	Comments[d]
α-Amylase (continued)					
B. stearothermophilus	13.8	28.7	280	1194	MW = 48,000 (1194)
Rat pancreas	–	16.4	–[p]	1196	
Pig pancreas	12.0	24	280	1197	MW = 50,000 (1197), pH 7.4, Tris
B. subtilis	8.3	19.8	280	1198	MW = 41,900 (1198)
B. subtilis var. saccharitikus					
Fukumoto	8.2	20.0	280	1199	MW = 41,000 (1199) pH 6.8, 0.01 M AcO⁻
	7.9	19.3	280	1199	0.1 N NaOH
β-Amylase (EC 3.2.1.2)					
Sweet potato	–	17.7	280	1201	–
	26	17.1	280	49	pH 4.8, 0.016 M AcO⁻
Iodosobenzoate oxidized		17.05	–	1202	
α-Amylase inhibitor					
Wheat					
I	2.7	15.0	280	1200	MW = 18,215 (1200)
II	2.6	10.0	280	1200	MW = 26,200 (1200)
Anaphylatoxin					
Rat serum		4.1	280	1335	pH 7.2
Anthranilate synthase (EC 4.1.3.27)					
Salmonella typhimurium					
Component I	3.3	5.2	278	1202	pH 7.4, 0.05 M TrisCl, MW = 64,000 (1202)
Escherichia coli	2.18	3.64	280	1204	pH 7.0, 0.05 M KP$_i$, MW = 60,000 (1204), calcd from data in Figure 5 (1204)
	2.94	4.91	295	1204	0.1 N NaOH, MW = 60,000 (1204), calcd from data in Figure 5 (1204)
Anthranilate synthase: anthranilate phosphoribosyl-transferase complex (EC 4.1.3.27: 2.4.2.18)					
Salmonella typhimurium	16.1	5.75	280	1205	pH 7.4, 0.05 M KP$_i$, MW = 280,000 (1205)
Antibody					
Rabbit					
Anti-N$_2$ ph[q]	–	15.7	279	1206	–
Anguilla rostrata (eel)					
Anti-human blood group H[O]	–	12.696	278	1207	Water

[p] Wavelength not cited.
[q] N$_2$ ph, 2,4-dinitrophenyl-.

MOLAR ABSORPTIVITY AND $A_{1cm}^{1\%}$ VALUES FOR PROTEINS AT SELECTED WAVELENGTHS OF THE ULTRAVIOLET AND VISIBLE REGION (continued)

Protein	$\epsilon^a (\times 10^{-4})$	$A_{1cm}^{1\%}$ [b]	nm[c]	Ref.	Comments[d]
Antigen					
Human					
Hepatitis associated					
(Australia)	—	9.42	260	1208	—
Blood group N active					
Erythrocyte, NN	65	10.90	274	1209	MW = 595,000 (1209)
Meconium	23.2	4.49	274	1209	MW = 520,000 (1209)
Paramecium aurelia					
Immobilization	—	11.9	277	1210	From Figure 1 (1210)
Apocytochrome *c*					
Horse heart	—	9.2	277	1336	—
Apolipoprotein Glu-II					
Human plasma	1.91	10.97	276	1337	pH 8.0, 0.01% EDTA,
	1.80	10.35	280	1337	MW = 17,380 (1337)
α-L-Arabinofuranosidase					
(EC 3.2.1.55)					
Aspergillus niger	12.7	23.1	280	1211	pH 7.0, 0.02 M NaP$_i$, MW = 53,000 (1211), data from Figure 8 (1211)
Arachin	—	7.98	278	50	8 M urea, 0.1 M sulfite
	—	7.85	278	50	6 M GdmCl, 0.1 M sulfite
Arachis hypogaea					
(peanut)	—	8.8	281.5	1338	pH 10.5, 0.1 M P$_i$
Arginase (EC 3.5.3.1)					
Rat liver	—	10.9	280	51	
Beef liver	—	9.6	278	1212	Dry wt
Pig liver	—	13.0	280	1213	
Chicken liver	—	260	210	1339	pH 7.5, 0.05 M
	—	22	340	1339	TrisCl
Arginine decarboxylase					
(EC 4.1.1.19)					
Escherichia coli	133	15.7	280	1214	MW = 850,000 dry wt and refr.
Arginine kinase (EC 2.7.3.3)					
Sipunculus nudus					
(Marine worm)	8.4	9.8	280	1215	MW = 86,000 (1215)
Cancer pagurus					
(Crab)	2.9	7.35	278	1216	MW = 39,500 (1216)
Homarus vulgaris					
(Lobster)	2.9	7.35	278	1216	MW = 39,500 (1216)
	—	7.8	275	1217	
	—	8.1	280	1218	pH 8.0 0.01 M P$_i$
	—	6.1	271	1219	Alkaline sol.
Arginine racemase					
(EC 5.1.1.9)					
Pseudomonas graveolens	15.5	9.3	280	1220	Dry wt, MW = 167,000 (1220)

MOLAR ABSORPTIVITY AND $A_{1cm}^{1\%}$ VALUES FOR PROTEINS AT SELECTED WAVELENGTHS OF THE ULTRAVIOLET AND VISIBLE REGION (continued)

Protein	$\epsilon^a (\times 10^{-4})$	$A_{1cm}^{1\%}$ [b]	nm[c]	Ref.	Comments[d]
Argininosuccinase (EC 4.3.2.1)					
Beef liver	25.8	13.0[r]	280	1221	pH 7.5, 0.05 M KP$_i$,
	–	7.1[r]	260	1221	MW = 202,000 (1221)
Beef kidney	25.0	12.5[r]	280	1221	
	–	6.8[r]	260	1221	
Aromatic α-ketoacid reductase (diiodophenyl pyruvate reductase, EC 1.1.1.96)					
Rat kidney	–	10	280	1222	pH 6.5, 0.005 M NaP$_i$
Ascorbate oxidase (EC 1.10.3.3)					
Cucumis sativas	1480[s]	1120[s]	280	1340	pH 7.0, 0.1 M P$_i$,
	61.6[s]	46.8[s]	607	1340	MW = 132,000 (1340)
Cucumber	0.53	–	330	1223	–
	0.97	–	607	1223	–
	0.36	–	760	1223	–
Corcubita pepo condensa (yellow crookneck squash)	28.5	–	280	1224	Dry wt
Corcubita pepo medullosa (green zucchini)	28.5	–	280	1224	Dry wt
Asparaginase (EC 3.5.1.1)					
Proteus vulgaris	–	6.6	280	1225	Dry wt, pH 7.0, 0.05 M NaP$_i$
Escherichia coli	9.9	7.46	280	1226	MW = 133,000 (1226)
	–	7.2	278	1342	pH 7
	–	9.9	292	1342	pH 13
E. coli HAP	8.83	6.26	278	1227	MW = 141,000 (1227)
E. coli B	9.2	7.1	278	649	MW = 130,000 (649) pH 7.3, P$_i$
	–	7.1	278	650	pH 5, 0.05 M AcONa
	–	7.1	278	650	pH 8.5, 0.05 M Tris
	–	6.5	276	650	7 M urea
	–	6.5	276	650	5 M guanidine
Succinylated monomer	–	6.7	276	1228	–
Erwinia carotovora	8.2	6.1	280	1341	pH 7.4
Asparaginase A	–	7.5	278	652	pH 7, P$_i$, calcd from opt. factor of 1.325
	–	9.5	290.5	652	0.1 N NaOH, calcd from opt. factor of 1.059
Escherichia coli ATCC 9637	–	7.9	277	651	pH 7, M/15 P$_i$, data from Figure 2 (651)
Deaminated	–	7.9	277	651	pH 7, M/15 P$_i$, data from Figure 2 (651) 651

[r]Amino acid analysis used for determining protein concentration.
[s]Values calculated from data and are for 8 Cu atoms per molecule.

MOLAR ABSORPTIVITY AND $A_{1cm}^{1\%}$ VALUES FOR PROTEINS AT SELECTED WAVELENGTHS OF THE ULTRAVIOLET AND VISIBLE REGION (continued)

Protein	$\epsilon^a (\times 10^{-4})$	$A_{1cm}^{1\%}$ [b]	nm[c]	Ref.	Comments[d]
Aspartate aminotransferase (EC 2.6.1.1)					
Pig heart muscle	11.1	14.1	280	1230	MW = 78,600 (1230)
Apo-	10.6	13.5	280	1231	pH 7.4, 0.1 M P_i, MW = 78,600 (1230)
Pig heart	13.5	15	280	1232 }	MW = 90,000 (1232)
Apo-	12.8	14.2	280	1232 }	
Pig heart	—	12.6	280	1233	pH 5.3, 0.1 M AcO⁻ data from Figure 1 (1233)
α form	—	14.8	278	1234	—
Chicken heart	—	14.0	280	1235	pH 7.4, 0.1 M TrisCl, pH 7.5, 0.05 M KP$_i$
Soluble mixture	—	14.2	280	1235	—
		15	280	53	pH 7.5
α soluble	—	13.7	280	1235	—
β soluble	—	14.5	280	1235	—
γ soluble	—	14.0	280	1235	—
Mitochondrial	—	13.2	280	1235	—
Apo-	—	14.5	280	53	pH 7.5
Rat brain					
Cytoplasmic	10.8	13.5	280	1236 }	pH 7.5, MW = 80,000 (1236)
Mitochondrial	8.65	10.8	280	1236 }	
Ox heart	—	14.40	278	1237	—
Apo-	—	14.14	278	1238	—
Beef kidney	12.5	13.4	280	1343	pH 8, 10 mM P_i
L-Aspartate β-decarboxylase (EC 4.1.1.12)					
A. faecalis	88	11	278	52	pH 6.8, 0.1 M P_i, MW = 800,000 (52)
Pseudomonas dacunhae	—	10.0	280	1229	pH 6.8, 0.1 M KP$_i$
D-Aspartate oxidase (EC 1.4.3.1)					
Octopus vulgaris	—	18.25	275	1344	Dry wt
Aspartate kinase (see aspartokinase)					
Aspartate transcarbamoylase (EC 2.1.3.2)					
Escherichia coli	18.2	5.9	280	54	MW = 310,000 (54)
	18.2	5.9	279	55	MW = 310,000 (54)
	18.4	5.9	284	56	MW = 310,000 (54)
Mercury derivative	—	7.6	280	1239	—
Regulatory subunit	—	8.0	280	1240	—
	—	7.2	280	1239	—
	—	12	284	57	In presence of HSEtOH MW = 30,000 (56)
Mercury derivative	—	8.3	280	1239	—
Zinc derivative	—	3.2	280	1239	—
Apo-	—	3.2	280	1239	—
Catalytic subunit	—	7.0	280	1240	—
	7.2	7.2	284	56	MW = 100,000 (56)
Permanganate modified	—	7.0	280	1241	—
5-Thio-2-nitrobenzoate derivative	—	9.2	280	1242	—

MOLAR ABSORPTIVITY AND $A_{1cm}^{1\%}$ VALUES FOR PROTEINS AT SELECTED WAVELENGTHS OF THE ULTRAVIOLET AND VISIBLE REGION (continued)

Protein	$\epsilon^a (\times 10^{-4})$	$A_{1cm}^{1\%}$ [b]	nm[c]	Ref.	Comments[d]
Aspartokinase (EC 2.7.2.4)					
Bacillus polymyxa	7.74	6.7	280	1243	pH 6.5, 6 M GdmCl/0.02 M KP_i MW = 116,000 (1243)
Escherichia coli					
Lysine sensitive	4.7	3.6	276	1244	MW = 130,000 (1244)
Aspartokinase I: homoserine dehydrogenase I (EC 2.7.2.4:1.1.1.3)					
Escherichia coli	5.4	6.3	278	1345	AA, pH 7.2, 20 mM KP_i containing 0.15 M KCl, 2 mM Mg titriplex, 1 mM L-threonine, and 1 mM S_2 threitol, MW = 86,000 (1345)
Threonine-sensitive	–	4.6	280	1245	–
	–	4.7	280	1246	6 M GdmCl
Aspartokinase II: homoserine dehydrogenase II (EC 2.7.2.4:1.1.1.3)					–
Escherichia coli K12	–	8.7	280	1247	–
Aspergillopeptidase A (EC 3.4.23.6)					
A. saitoi	4.5	13.15	280	58	MW = 34,500 (58)
Aspergillopeptidase B (EC 3.4.21.15)					
A. oryzae	1.73	9.08	278	59	pH 5, 0.1 M AcO$^-$ MW = 18,000 (59)
	1.62	9.00	280	59	pH 5, 0.1 M AcO$^-$ MW = 18,000 (59)
ATP:AMP Phosphotransferase (EC 2.7.4.3)					
Human	1.43	6.67	279	1346	Dry wt, MW = 22,000 (1346)
ATP citrate-lyase (EC 4.1.3.8)					
Rat liver	–	11.8	280	1248	Calcd from data in Figure 5 (1248)
ATP Phosphoribosyltransferase (see phosphoribosyltransferase and phosphoribosy-ATP)	–	7.1	280	1248	Calcd from data in Figure 5 (1248), corrected for light scattering
ATP-Sulfurylase (sulfate adenylyltransferase) (EC 2.7.7.4)					
Penicillium chrysogenum	–	8.71	278	1347	Kjeldahl[e]
Avidin					
Egg white	8.3	15.7	280	1249	MW = 53,000 (1249)
		53	233	1250	–

MOLAR ABSORPTIVITY AND $A_{1cm}^{1\%}$ VALUES FOR PROTEINS AT SELECTED WAVELENGTHS OF THE ULTRAVIOLET AND VISIBLE REGION (continued)

Protein	$\epsilon^a (\times 10^{-4})$	$A_{1cm}^{1\%}$ [b]	nm[c]	Ref.	Comments[d]
Avimanganin					
Chicken liver					
Mitochondria	0.0508	0.057	480	1251	MW = 89,000 (1251)
	0.0250	0.028	600	1251	
Azobacterflavoprotein					
Azobacter vinelandii,					
oxidized	1.06	—	452	60	pH 7.0, 25 mM KP$_i$
	—	16.5	274	60	pH 7.0, 25 mM KP$_i$
Azurin					
Pseudomonas fluorescens	0.0285	—	459	1252	—
	0.350	—	625	1252	—
	0.032	—	781	1252	—
Pseudomonas aeruginosa	0.027	—	467	1252	—
	0.350	—	625	1252	—
	0.039	—	820	1252	—
Bacteriochlorophyll-protein complex					
Chloropseudomonas ethylicum	—	90	371	61	MW = 37,940/subunit (61)
Bacteriocin DF 13					
Enterobacter cloacae DF 13	5.5	9.87	280	931	pH 7.0, 0.06 M P$_i$, MW = 56,000 (931)
Biotin carboxylase (EC 6.3.4.14)					
Escherichia coli	—	6.25	280	932	—[t]
	—	6.25[u]	280	785	—
Brain					
Pig, basic	—	4	280	439	pH 6.8, 0.05 N P$_i$, est. from Figure 7 (439) no value cited for wavelength
Bromelain (EC 3.4.22.4-.5)	6.33	19.0	280	62	MW = 33,315 (62)
	6.68	20.1	280	63	MW = 33,000 (64)
C1 Inactivator					
Human plasma	—	4.5	280	1253	—
C1q component of complement					
Human serum	26.4	6.8	278	1254	1% NaDodSO$_4$ MW = 388,000 (1254)
C2 component of complement					
Guinea pig	—	13.9	280	786	—
Caerulein					
Hyla caerulae	0.725	—	280	787	80% ethanol
Calcitonin					
Salmon	0.15	4.5	280	788	—

[t] OD$_{280}$ = (mg protein/ml)/1.6.
[u] Calculated from equation: $\frac{\text{mg protein}}{\text{ml}} = 1.6 \times \text{OD}_{280}$. (785)

MOLAR ABSORPTIVITY AND $A_{1cm}^{1\%}$ VALUES FOR PROTEINS AT SELECTED WAVELENGTHS OF THE ULTRAVIOLET AND VISIBLE REGION (continued)

Protein	$\epsilon^a (\times 10^{-4})$	$A_{1cm}^{1\%}$ [b]	nm[c]	Ref.	Comments[d]
Calcium-binding phosphoprotein					
Pig brain	0.36	3.1	Not cited	1348	Kjeldahl[e], MW = 11,500 (1348)
Chicken intestinal mucosa	–	21.9	280	451	pH 8.1, 0.2% TrisCl, 0.77% glycine, 0.1 mM glutathione, est. from Figure 8 (451)
Carbamate kinase (EC 2.7.2.2)					
S. faecalis	1.798	5.8	280	789	pH 7.5, 50 mM Na P$_i$, MW = 31,000 (789)
Carbamoylphosphate synthase (EC 2.7.2.5-.9)					
Rat liver	–	8.4	280	1349	DR
Carbonic anhydrase (carbonate dehydratase, EC 4.2.1.1)					
Beef					
B	5.7	19.0	280	1255	MW = 30,000 (1255)
	5.6	18.0	280	66	MW = 31,000 (66)
Beef erythrocyte					
B	5.2	16.8	280	656	MW = 31,000 (656), pH 7.4
Ox, rumen					
Isozyme a	–	17	280	1256	–
Isozyme b	–	17	280	1256	–
Gallus domesticus (Chicken)	5.6	20	280	1257	MW = 28,000 (1258)
Carcharhinus leucas (Bull shark)	7.5	20.9	280	1259	MW = 36,000 (1259)
Galeocerdo cuvieri (Tiger shark)	6.3	16.0	280	1259	MW = 39,500 (1259)
Guinea pig Blood/mucosa G.I. tract					
High activity	–	17.1	280	1260	Dry wt
	–	16.9[r]	280	1260	–
	–	16.7[v]	280	1260	–
	–	17.0	280	1260	Value used
Low activity	–	16.4	280	1260	Dry wt
	–	16.5[r]	280	1260	–
	–	16.1[v]	280	1260	–
	–	16.5	280	1260	Value used
Human erythrocytes					
A	4.67	16.3	280	653	MW = 28,600 (653)
A, generated from B in vitro	4.29	16.0	280	653	MW = 26,800 (653)
B	4.56	16.3	280	653	MW = 28,000 (653)
	4.9	16.3	280	65	pH 7.0, 0.1 ionic strength Na P$_i$, MW = 30,000 (65)
	4.89	16.3[w]	280	1262	MW = 30,000 (1262)

[v] Calculated according to Wetlaufer, *Adv. Protein Chem.*, 17, 362 (1962).

[w] The value of $A_{1cm}^{1\%}$ of 16.3 and $\epsilon = 4.89 \times 10^4$ taken from Rickli, Ghazanfar, Gibbons, and Edsall, *J. Biol. Chem.*, 239, 1065 (1964), and a MW of 30,000 assumed (Reference 1262).

MOLAR ABSORPTIVITY AND $A_{1cm}^{1\%}$ VALUES FOR PROTEINS AT SELECTED WAVELENGTHS OF THE ULTRAVIOLET AND VISIBLE REGION (continued)

Protein	$\epsilon^a (\times 10^{-4})$	$A_{1cm}^{1\%}$ [b]	nm[c]	Ref.	Comments[d]
Carbonic anhydrase (continued)	2.05	–	250	1262	
B	5.74	–	240	1262	
	12.8	–	235	1262	
	24.5	–	230	1262	MW = 30,000 (1262)
	36.0	–	225	1262	
	45.5	–	220	1262	
	54.5	–	215	1262	
B, nitrated	–	17.5	280	654	pH 9
C	5.34	17.8	280	65	pH 7.0, 0.1 ionic strength Na P_i, MW = 30,000 (68)
	5.05	18.7	280	653	MW = 27,000 (653)
	–	17.8	280	655	–
	5.34	17.8[x]	280	1262	
	2.01	–	250	1262	
	5.47	–	240	1262	
	12.9	–	235	1262	MW = 30,000 (1262)
	25.0	–	230	1262	
	37.0	–	225	1262	
	48.0	–	220	1262	
	60.0	–	215	1262	
D	4.52	16.0	280	653	MW = 28,200 (653)
D, generated from B in vitro	4.14	16.1	280	653	MW = 25,700 (653)
G	4.9	18.5	280	653	MW = 26,500 (653)
H	5.14	18.5	280	653	MW = 27,800 (653)
O	3.98	15.4	280	653	MW = 25,800 (653)
P	4.21	15.4	280	653	MW = 27,300 (653)
Pmut	4.47	16	280	1261	MW = 28,000 (1261)
Horse erythrocyte	4.38	15.9	280	653	MW = 27,500 (653)
B	3.95	13.6	280	657	MW = 29,007 (657)
C	3.73	13.4	280	657	MW = 27,918 (657)
Rat erythrocyte					
1a	5.22	18	280	658	MW = 29,000 (658)
2	5.22	18	280	658	MW = 29,000 (658)
3	4.93	17	280	658	MW = 29,000 (658)
Prostate					
1b	5.22	18	280	658	MW = 29,000 (658)
Parsley (*Petroselinum crispum* var. *latifolium*)	3.18	11.3	280	659	MW = 28,150 (659)
Apocarbonic anhydrase	5.7	–	280	660	–
Precursor of above	5.7	–	280	660	
Pig erythrocytes					
B	5.3	17.4	280	1263	MW = 30,375 (1263)
	5.6	18.5	280	1264	MW = 30,000 (1264)
C	4.6	15.2	280	1264	MW = 30,000 (1264)
Spinach	–	8.6	280	1265	pH 7.2, 0.03 $M P_i$
Monkey					
B	4.88	–	280	67	–
C, *M. mulata*	5.35	17.8	280	67	MW = 30,000 (67)
Neisseria sicca	3.575	12.5	280	1350	Kjeldahl,[e] MW = 28,600 (1350)

[x]The value of $A_{1cm}^{1\%}$ of 17.8 and an ϵ of 5.34 × 10⁴ taken from Nyman and Lindskog, *Biochim. Biophys. Acta,* 85, 141 (1964).

MOLAR ABSORPTIVITY AND $A_{1cm}^{1\%}$ VALUES FOR PROTEINS AT SELECTED WAVELENGTHS OF THE ULTRAVIOLET AND VISIBLE REGION (continued)

Protein	$\epsilon^a(\times 10^{-4})$	$A_{1cm}^{1\%}$ [b]	nm[c]	Ref.	Comments[d]
Carboxylesterase (EC 3.1.1.1)					
Ox liver	20.1	13.40	280	1266	pH 7.92, 0.15 M Tris, MW = 150,000 (1267)
Pig kidney					
Microsomes	15.3	9.4	280	1268	MW = 163,000 (1268)
Liver	20.27	13.05	280	69	pH 8.16, 0.15 M Tris, MW = 172,000 (69)
S-Carboxymethylalbumin, beef (see Albumin, beef)					
Carboxypeptidase (EC 3.4.12.1-.3)					
Beef pancreas	–	19.4	278	70	10% LiCl
	6.32	18.1	278	71	–
	6.42	–	278	72	–
	6.45	–	278	73	–
	6.49	–	278	74	–
	6.67	–	278	75	–
	–	23	280	76	–
	7.9	–	280	77	–
	8.6	25	278	78	MW = 34,300 (78)
Carboxypeptidase A					
Ox					
Squalis acanthias	6.48	18.5	280	790	MW = 35,000 (790)
Co II	–	0.0205	530	1351	–
	–	0.0195	572	1351	–
Co III	–	0.0500	503	1351	–
A_1, Pig	6.72	19.6	278	79	MW = 34,800 (79)
A_2, Pig	6.72	19.6	278	79	–
A_α, Beef	6.49	18.8	278	80	pH 7.0, 0.5 M NaCl, 0.01 M Tris, MW = 34,600 (80)
A, Acetyl-, beef	6.17	–	280	72	In the absence of β-phenylpropionate
	6.01	–	278	73	In the absence of β-phenylpropionate
	5.9	–	280	72	In the presence of β-phenylpropionate
	5.78	–	278	73	In the presence of β-phenylpropionate
A, Arsonilazo-, beef	7.32	–	278	1353	–
A, Succinyl-, beef	–	18.3	280	81	–
	6.47	–	278	71	–
Carboxypeptidase B					
Pig	7.34	21.4	278	82	pH 8.0, 0.005 M Tris
Beef	7.35	21	280	83	MW = 34,600 (83)
Barley	–	16.5	280	791	Water
P. omnivorum	5.53	17.6	278	792	pH 6.5, MW = 31,400 (792)
Gossypium hirsutum	17.3	20.5	280	1352	MW = 84,500 (1352)
Carnitine acetyltransferase (EC 2.3.1.7)					
Pigeon breast muscle	4.8	8.25	280	1269	–

MOLAR ABSORPTIVITY AND $A_{1cm}^{1\%}$ VALUES FOR PROTEINS AT SELECTED WAVELENGTHS OF THE ULTRAVIOLET AND VISIBLE REGION (continued)

Protein	$\epsilon^a(\times 10^{-4})$	$A_{1cm}^{1\%}$ [b]	nm[c]	Ref.	Comments[d]
Carotenoid-protein complex					
Pecten maximus					
ovary	—	9.7	280	1270	pH 7, 0.2 M P$_i$
Plesionika edwardsi	—	13	280	1270	pH 7, 0.2 M P$_i$, data from Figure 3 (1270)
Carotenoproteins[y]					
Aristeus antennatus					
Carapace					
α (+ salt)	12.3	—	593	1271	—
Stomach					
α (+ salt)	12.4	—	588	1271	—
Scyllarus arctus					
Carapace					
α (+ salt)	12.0	—	616	1271	—
Clibanarius erythropus					
Exoskeleton					
α (+ salt)	12.6	—	620	1271	—
Labidocera acutifrons	—	152	640	1354	pH 7, 0.02 M P$_i$
Casein	—	10.0	280	84	—
French Friesian cows	—	6.7	278	1272	—
Bovine					
βA	—	4.6	280	1273	—
βB	—	4.7	280	1273	—
βC	—	4.6	280	1273	—
30% acid prep.	—	8.4	278	85	—
2% acid prep.	—	8.6	278	85	—
3% Am. sulfate prep.	—	8.7	278	85	—
20% Am. sulfate prep.	—	8.7	278	85	—
Calcium gel prep.	—	8.1	278	85	—
α$_s$	2.73	10.1	280	86	MW = 27,000 (85)
α$_{s1}$	2.73	10.1	280	87	MW = 27,000 (85)
β	1.15	4.6	280	88	MW = 25,000 (85)
k	2.44	12.2	280	89	MW = 29,000 (85)
Catalase (EC 1.11.16)					
Human erythrocyte	28	12.5	280	90	MW = 225,000 (90), $A_{1cm}^{1\%}$ = 16.9–18.7 at 405 nm (90)
	—	14.4	280	91	—
	—	17.8	405	91	—
	30.8	13.2	280	1355	MW = 232,400 (1355)
	39.7	17.1	405	1355	
Beef erythrocyte	43.1	16.8	405	661	MW = 257,000 (661)
Liver	82.2	36.5	276	662	MW = 225,000 (662), pH 7.3, 0.05 M P$_i$
	63.6	28.2	404	662	MW = 225,000 (662), pH 7.3, 0.05 M P$_i$
	31	—	405	92	pH 7.4
	38	—	280	92	pH 7.4
	32.4	13.5	405	93	MW = 240,000 (93)
	31.4	12.9	276	1274	MW = 240,000 (1274)
	16.6	6.9	405	1274	

[y] All species are decapods.

MOLAR ABSORPTIVITY AND $A_{1cm}^{1\%}$ VALUES FOR PROTEINS AT SELECTED WAVELENGTHS OF THE ULTRAVIOLET AND VISIBLE REGION (continued)

Protein	$\epsilon^a (\times 10^{-4})$	$A_{1cm}^{1\%}$ [b]	nm[c]	Ref.	Comments[d]
Catalase (continued)					
Horse liver	67.5	27	275	663	pH 8, 0.004 M P_i, data from Figure 1 (663)
	8.3	–	623	664	
	9.4	–	536	664	
	78.5	–	400	664	–
	85.0	–	280	664	–
	65	–	280	665	pH 7.15, P_i, MW = 225,000, data from Figure 4 (665)
Horse blood	65	–	280		pH 7.15, P_i, MW = 225,000 data from Figure 4 (665)
Rat liver	–	15.8	276	1356	From Figure 1 (1356)
	–	16	407	666	–
	39.7	15.5	276	667	MW = 256,000 (667)
	43.0	16.8	407	667	MW = 256,000 (667)
Spinach	–	1.97	502	668	pH 7.4, 0.1 M Tricene-NaOH
	–	1.38	620	668	
	–	14.8	278	94	pH 7.4, 0.1 M Tricene-NaOH
	–	14.9	404	94	pH 7.4, 0.1 M Tricene-NaOH
Commercial					
Crystalline	41	16.4	276	669	
	30	12.0	406	669	
Lyophilized-A	32	12.8	276	669	MW = 250,000 (669)
	20	8.0	406	669	
Lyophilized-B	34	13.6	276	669	
	7.3	2.9	406	669	
Cathepsin B (EC 3.4.22.1)					
Beef spleen	–	–[z]	280	1357	pH 7.6, 50 mM P_i
Cellulase (EC 3.2.1.4)					
Trichoderma koningi					
I	5.73	22	280	95	MW = 26,000 (95)
II	12.5	25	280	95	MW = 50,000 (95)
Stereum sanguinolentum					
P 1 Fraction	8	38	280	670	pH 5.4, 0.1 M, AmAc, MW = 21,000 (670)
Penicillium notatum	–	26	280	671	–
	9.1	–	280	672	pH 5, 0.05 M AcO$^-$ MW = 35,500 (672)
Cerebrocuprein					
Human brain	–	7.35	265	1275	–
	–	0.075	675	1275	–
Beef brain	–	0.18	660	1276	–
Ceruloplasmin					
Human	–	0.684	610	1277	–
	1.13	–	610	1278	–
	–	15.03	280	1277	–
	0.40	–	332	1278	–

[z]14–15.

MOLAR ABSORPTIVITY AND $A_{1cm}^{1\%}$ VALUES FOR PROTEINS AT SELECTED WAVELENGTHS OF THE ULTRAVIOLET AND VISIBLE REGION (continued)

Protein	$\epsilon^a (\times 10^{-4})$	$A_{1cm}^{1\%}$ [b]	nm[c]	Ref.	Comments[d]
Ceruloplasmin (continued)					
Human (continued)					
	0.12	—	459	1278	—
	0.22	—	794	1278	—
	—	16.3	280	8	—
	23.7	14.9	280	97	MW = 159,000 (97)
	8.15	—	280	98	—
	10	—	605	98	—
	—	14.6	280	99	MW = 160,000 (99)
	—	14.4	279	100	—
	—	0.68	610	100	—
Form I	—	15.5	280	1279	—
IIa/IIb	—	16.2	280	1279	—
IIIb	—	16.0	280	1279	—
Asialoceruloplasmin	—	15.0	280	1277	—
	—	0.64	610	1277	—
Pig	1.01	0.63	610	1280	MW = 160,000 (1281)
2 days old	—	0.52	610	1282	—
	—	11.5	280	1282	—
10 weeks old	—	0.61	610	1282	—
	—	12.8	280	1282	—
Rat, copper deficient	—	0.64	610	1283	—
Rabbit	—	13.1	280	101	—
	—	0.618	610	101	—
Chitinase (EC 3.2.1.14)					
Streptomyces	—	15.0	280	1358	—
Chloride peroxidase (EC 1.11.1.10)					
Caldariomyces fumago	7.53	—	403	102	—
	1.15	—	515	102	—
	1.08	—	542	102	—
	0.42	—	650	102	—
Choleragen (Cholera enterotoxin)	9.6	11.41	280	1284	pH 7.5, 0.2 M Tris, MW = 84,000 (1284)
	8.75	10.39	280	1284	pH 8.0, 5 M GdmCl, MW = 84,000 (1284), dry wt
Choleragenoid (Toxoid)	1.43	9.56	280	1284	pH 7.5, 0.2 M Tris MW = 15,000 (1284), dry wt
	1.36	9.09	280	1284	pH 8.0, 5 M GdmCl, MW = 15,000 (1284), dry wt
Cholinesterase (EC 3.1.1.8)					
Amiarus nebulosus (a fish)	—	3.5	275	1285	pH 7, 0.02 M KP_i/0.13 M KCl, data from Figure 4 (1285)

MOLAR ABSORPTIVITY AND $A_{1cm}^{1\%}$ VALUES FOR PROTEINS AT SELECTED WAVELENGTHS OF THE ULTRAVIOLET AND VISIBLE REGION (continued)

Protein	$\epsilon^a (\times 10^{-4})$	$A_{1cm}^{1\%}$ [b]	nm[c]	Ref.	Comments[d]
Chorionic gonadotropin					
Human	1.4	5.2	280	1286	MW = 27,000 (1286)
	–	3.88	278	1287	–
Chorismate mutase (EC 5.4.99.5): Prephenate dehydrogenase (EC 1.3.1.12)					
E. coli	7.8	9.5	280	1288	MW = 82,000 (1289)
Aerobacter aerogenes	7.2[aa]	9.5[aa]	278	1290	pH 8.0, 0.1 M TrisCl, MW = 76,000 (1290)
Chymopapain (EC 3.4.22.6)					
I	6.16	18.40	280	104	pH 5.0, AcO$^-$, MW = 33,500 (104)
II	6.18	18.45	280	104	pH 5.0, AcO$^-$, MW = 33,500 (104)
III	6.23	18.60	280	104	pH 5.0, AcO$^-$, MW = 33,500 (104)
IV	6.16	18.40	280	104	pH 5.0, AcO$^-$, MW = 33,500 (104)
B	–	19.6	280	103	pH 7.2, 0.1 M cacodylate, est. from Figure 3 (103)
Chymosin (see rennin)					
Chymotrypsin (EC 3.4.21.1)					
Purified, commercial	5.205	–	281	1292	–
Commercial					
Diazoacetyl derivative	2.77	–	250	1293	–
Above photolytically modified	1.75	–	250	1293	–
Chymotrypsin II					
Human	4.75	19.0	280	1299	MW=25,000(1299)
Chymotrypsin A (EC 3.4.21.1)					
Beef pancreas	–	20.2	280	110	0.001 M HCl
Chymotrypsin C (EC 3.4.21.2)					
Pig pancreas	5.95	25	278	111	MW=23,800 (111)
Chymotrypsin, iPr$_2$P.[bb]					
Beef pancreas	5.0	–	280	112	–
α-Chymotrypsin					
Commercial	–	20.8	280	1294	–
	–	300	210	1294	–
	–	920	191	1294	–
	–	19.95	280	1295	pH 6.2, P$_i$
Polyvalyl-	–	18.51	280	1295	
Trans-cinnamoyl-	1.78	–	292	1296	–
Denatured derivative	1.78	–	281	1296	–
Furylacryloyl-	1.98	–	320	1296	–
Denatured derivative	2.00	–	310	1296	–
Indolacryloyl-	1.78	–	360	1296	–
Denatured derivative	1.90	–	335	1296	–
Beef	–	18.7	280	1297	–
Pancreas	–	21.5	280	1298	–

[aa] The same values used at 280 nm in Reference 1291.
[bb] iPr$_2$P, -diisopropylphospho.

MOLAR ABSORPTIVITY AND $A_{1cm}^{1\%}$ VALUES FOR PROTEINS AT SELECTED WAVELENGTHS OF THE ULTRAVIOLET AND VISIBLE REGION (continued)

Protein	$\epsilon^a(\times 10^{-4})$	$A_{1cm}^{1\%}$ [b]	nm[c]	Ref.	Comments[d]
α-Chymotrypsin (continued)	4.46	20.75	282	105	MW=21,500 (105)
	–	20.7	282	106	–
	5.0	20	280	107	MW=25,000 (107)
	–	20.4	282	108	–
	–	18.9	280	109	–
Chymotrypsinogen					
Commercial	–	19.7	282	1300	Dry wt
Carbon disulfide derivative	6.0	–	285	1301	Data from Figure 4 (301)
Alkaline denatured	3.66	–	292.8	1302	–
	4.56	–	285.5	1302	–
	23.07	–	230	1302	–
Acid denatured	3.66	–	293	1302	–
	4.56	–	285.5	1302	–
	4.75	–	276	1302	–
	23.06	–	230.5	1302	–
Rat pancreas	4.0	16	280	1303	MW=25,000 (1303)
Chymotrypsinogen A					
Commercial	–	20.3	282	1304	pH 2.0
Beef	–	20.3	282	1305	pH 9.3, glycine buffer, 0.1 μM, dry wt
	5.15	20	280	115	MW=25,761 (115)
Pig	4.43	18	280	116	MW=24,600 (116)
Spiny Pacific dogfish	–	21.4[r]	280	1306	–
(*Squalus acanthias*)	–	21.7	280	1306	Refractometry
Chymotrypsinogen B					
Beef	–	18.7	280	117	0.001 *M* HCl
	–	18.4	280	115	–
Pig, pancreas	6.0	23.8	278	1307	MW=26,000 (1307)
Chymotrypsinogen C					
Pig	7.6	23.8	278	118	MW=31,800 (118)
Chymotrypsinogen, *S*-sulfo					
Beef pancreas	–	17.9	280	119	–
α-Chymotrypsinogen					
Beef pancreas	–	20.6	282	113	–
	5.02	20.0	282	114	MW=25,100 (114)
Citramalate hydrolase (citramalate lyase) (EC 4.1.3.22)					
Clostridium tetanomorphum					
Component I	–	9.0	280	673	–
Component II	–	9.5	280	673	–
Citrate condensing enzyme [citrate (*re*)-synthase] (EC 4.1.3.28)					
Pig heart	13.15	15.5	280	120	MW=85,000 (120)
Citrate synthase (EC 4.1.3.7)					
Pig heart	17.1	17.8	280	1308	MW=96,000 (1308)
	15	15	280	674	MW=100,000 (674)

MOLAR ABSORPTIVITY AND $A_{1cm}^{1\%}$ VALUES FOR PROTEINS AT SELECTED WAVELENGTHS OF THE ULTRAVIOLET AND VISIBLE REGION (continued)

Protein	$\epsilon^a (\times 10^{-4})$	$A_{1cm}^{1\%}$ [b]	nm[c]	Ref.	Comments[d]
Cobratoxin					
Naja naja atra	1.45	13.2	280	675	MW=11,000 (675)
	–	12.1	280	676	Data from Figure 1 (676)
Cocoonase (EC 3.4.21.4)					
Silkmoth					
Antherea polyphemus	–	11	280	1309	1 N NaOH
	–	13	280	1310	0.1 N NaOH
Bombyx mori	–	9.8	280	1310	0.1 N NaOH
Cocosin					
Coconuts	–	7.0	280	677	–
	–	2.7	255	677	–
Cocytotaxin					
Rat serum	–	2.90	280	1363	pH 7.2, 0.05 M NaKP$_i$
Colicin					
D, Escherichia coli K12	–	7.26	280	1359	Neutral sol
E_1, E. coli	–	7.36	280	1311	pH 7.0, 0.01 M KP$_i$
E_2, E. coli W 3110	5.84	9.73	280	121	MW=60,000
E_3, E. coli W 3110	7.45	12.42	280	121	MW=60,000
1_a, E. coli W 3110-r	–	10.3	280	678	–
1_b, E. coli W 3110-r	–	10.9	280	678	–
Colipase					
Pig pancreas	–	4.0	280	1360	–
Coliphage N4					
Phenolic subunits	1.29	7.0	280	679	pH 7.2, 0.1 M Tris, MW=18,500 (680)
Collagen proline hydroxylase					
Rat skin, newborn	–	12.3	280	1313	pH 7.0
Collagenase (EC 3.4.24.3)					
Clostridium histolyticum					
A	–	14.7	280	1312	–
B	–	13.8	280	1312	–
C	–	16.8	280	1312	–
Complement system (see also C1, C1q, and C2)					
Guinea pig serum	–	13.2	280	1314	–
Conalbumin					
Hen	8.5	11.1	280	122	0.02 M HCl, MW=76,600
Egg	–	12.0	280	681	–
Egg white	8.8	–	280	1315	–
Iron	12.2	–	280	1315	–
	–	0.62	470	1316	pH 5–10
Copper	12.2	–	280	1315	–

MOLAR ABSORPTIVITY AND $A_{1cm}^{1\%}$ VALUES FOR PROTEINS AT SELECTED WAVELENGTHS OF THE ULTRAVIOLET AND VISIBLE REGION (continued)

Protein	$\epsilon^a (\times 10^{-4})$	$A_{1cm}^{1\%}$ [b]	nm[c]	Ref.	Comments[d]
Concanavalin A					
Canavalia ensiformis	7.98	11.4	280	933	pH 7.0, 1 N NaCl, MW=70,000 (933)
	7.75	11.4	280	934	1 M NaCl, MW = 68,000 (935)
	—	13.7	280	936	pH 6.8, 0.05 M P$_i$, 0.2 M NaCl
	—	12.4	280	936	pH 5.2, 0.05 M AcONa, 0.2 M NaCl
Copper, blue					
Pseudomonas	—	6.7	280	444	—
C-Reactive protein					
Human	—	20	280	123	—
	—	18	280	124	—
Creatine kinase (EC 2.7.3.2)					
Rabbit muscle	3.28	—	250	1317	0.1 M Tris, 0.02 M EDTA
	3.7	8.88	280	793	pH 7.0, 0.05 M P$_i$, MW=41,300 (793)
	7.3	8.9	280	125	pH 7, MW=81,000 (125)
	7.1	8.7	280	125	pH 9.8, MW=81,000 (125)
Cyrrinus carpio L.	8.0	9.38	280	794	MW=85,100 (794)
Calf brain	—	8.24	280	795	pH 7.0, 0.05 M NaP$_i$
Human muscle	7.1	8.8	280	796	pH 8.0, MW=81,000 (796)
Crotonase (Enoyl-CoA hydratase, EC 4.2.1.17)					
Beef liver	—	5.76	280	797	—
	—	5.56	280	797	pH 7.4, 6 M GdmCl or 8 M urea
Clostridium acetobutylicum	—	8.90	280	1362	pH 8, 0.05 M TrisCl, 0.1 M KCl
Crotoxin					
Crotalus terrificus terrificus	—	19	280	798	Neutral sol.
Crustacyanin					
Homarus grammarus	—	11.5	278	937	pH 7
	—	37.2	633	937	pH 7
	—	35.6[cc]	600	937	Water
	—	5.4[cc]	320	937	pH 7
	—	5.3[cc]	360	937	Water
	—	5.7[cc]	370	937	pH 7
Cryoglobulin, 6.6S					
Human blood	—	13.3	280	938	Dry wt

[cc] Data from Figure 1 of Reference 937.

MOLAR ABSORPTIVITY AND $A_{1cm}^{1\%}$ VALUES FOR PROTEINS AT SELECTED WAVELENGTHS OF THE ULTRAVIOLET AND VISIBLE REGION (continued)

Protein	$\epsilon^a (\times 10^{-4})$	$A_{1cm}^{1\%}$ [b]	nm[c]	Ref.	Comments[d]
Cryptocytochrome c					
Pseudomonas denitrificans					
Aerobic cells					
Ferri form	16.0	–	–[dd]	1371	
	1.97	–	500	1371	
	0.65	–	642	1371	
Ferro form	18.3	–	426	1371	
	17.3	–	438	1371	
	1.97	–	–[ee]	1371	
CO-ferro form	44.1	–	419	1371	
	2.2	–	540	1371	
	1.86	–	570	1371	
NO-ferro form	17.2	–	396	1371	pH 7, dry wt
	2.1	–	490	1371	
	2.0	–	–[ff]	1371	
Anaerobic cells	1.9	–	570	1371	
Ferri form	16.0	–	–[dd]	1371	
	1.8	–	500	1371	
	0.6	–	642	1371	
Ferro form	17.8	–	426	1371	
	17.1	–	438	1371	
	1.8	–	–[ee]	1371	
Crystallin					
Beef					
α	–	9.6	280	1318	–
	–	8.85	280	126	Value drops to 8.0 after aging 7 months
	73.1	8.7	280	127	MW=840,000 (127)
α'	–	8.0	280	129	–
α"	–	8.3	280	129	–
γ	–	21.0	280	1318	–
	–	21.0	280	130	–
B_s	5.28	18.6	278	131	pH 8.2, 0.0005 M P_i, MW=28,402 (131)
	5.25	18.5	280	131	pH 8.2, 0.0005 M P_i, MW=28,402 (131)
Calf					
α	–	9.9	280	1319	7 M urea
Dogfish					
α	–	8.5	280	1318	–
Medium age	–	8.80	280	939	pH 9.8, 0.05 M borate–8 M urea
Old	–	7.88	280	939	
β					
Medium age	–	15.9	280	939	pH 9.8, 0.05 M borate–8 M urea
Old	–	15.9	280	939	
γ		22.4	280	1318	
Medium age	–	23.1	280	939	pH 9.8, 0.05 M borate–8 M urea
Old	–	22.2	280	939	
Fox					
α	55.2	8.9	280	127	MW=620,000 (127)

[dd] 400–402 nm.
[ee] 550–560 nm.
[ff] 530–540 nm.

MOLAR ABSORPTIVITY AND $A_{1cm}^{1\%}$ VALUES FOR PROTEINS AT SELECTED WAVELENGTHS OF THE ULTRAVIOLET AND VISIBLE REGION (continued)

Protein	$\epsilon^a (\times 10^{-4})$	$A_{1cm}^{1\%}$ [b]	nm[c]	Ref.	Comments[d]
Crystallin (continued)					
Horse					
α	103.8	9.6	280	127	MW = 1,050,000 (127)
Human					
α, 0–10 years in age	—	10.2	280	1318	—
β, 0–10 years in age	—	16.7	280	1318	—
γ, 0–10 years in age	—	15.7	280	1318	—
α, 11–20 years in age	—	11.6	280	1318	—
β, 11–20 years in age	—	16.5	280	1318	—
γ, 11–20 years in age	—	15.6	280	1318	—
α, 40–49 years in age	—	16.2	280	1318	—
β, 40–49 years in age	—	16.0	280	1318	—
γ, 40–49 years in age	—	15.4	280	1318	—
α, 50–59 years in age	—	16.0	280	1318	—
β, 50–59 years in age	—	16.8	280	1318	—
γ, 50–59 years in age	—	15.2	280	1318	—
α, 60–69 years in age	—	17.3	280	1318	—
β, 60–69 years in age	—	15.7	280	1318	—
γ, 60–69 years in age	—	15.9	280	1318	—
α, 70–79 years in age	—	15.7	280	1318	—
β, 70–79 years in age	—	16.3	280	1318	—
γ, 70–79 years in age	—	15.9	280	1318	—
α, 80–89 years in age	—	15.6	280	1318	—
β, 80–89 years in age	—	14.8	280	1318	—
γ, 80–89 years in age	—	17.3	280	1318	—
Mink					
α	55.4	8.8	280	127	MW = 630,000 (127)
Pig					
α	70.6	8.5	280	127	MW = 830,000 (127)
Rabbit					
α	43.8	8.4	280	127	MW - 575,000 (127)
	—	8.3	280	128	—
β	—	21.5	280	128	—
γ	—	17.6	280	128	—
Rat					
α	—	8.0	280	1318	—
Young	—	7.73	280	939	
X-Rayed	—	7.49	280	939	pH 9.8, 0.05 M borate–8 M urea
Old	—	7.88	280	939	
β					
Young	—	18.6	280	939	
X-Rayed	—	16.8	280	939	pH 9.8, 0.05 M borate–8 M urea
Old	—	16.5	280	939	
γ	—	18.0	280	1318	—
Young	—	19.9	280	939	
Young, X-rayed	—	19.8	280	939	pH 9.8, 0.05 M borate–8 M urea
Old	—	16.6	280	939	
Sulfonated	—	17.1	280	132	—
Sheep					
α	100.1	8.4	280	127	MW = 840,000 (127)
Cyanocobalamin-protein					
Pig pyloric mucosa	—	9	278	1320	—
Cyanocobalamin-binding factor					
Pig pyloris	—	8	278	1321	Water, dry wt

MOLAR ABSORPTIVITY AND $A_{1cm}^{1\%}$ VALUES FOR PROTEINS AT SELECTED WAVELENGTHS OF THE ULTRAVIOLET AND VISIBLE REGION (continued)

Protein	$\epsilon^a (\times 10^{-4})$	$A_{1cm}^{1\%}$ [b]	nm[c]	Ref.	Comments[d]
Cystathione γ-lyase (EC 4.4.1.1) (see homoserine deaminase)					
Cysteamine dioxygenase (EC 1.13.11.19)					
Horse kidney	11.2	13.5	280	940	—
Cysteine desulfyhydrase (cystathionine γ-lyase, EC 4.4.1.1)					
S. typhimurium	7.85	21.2	280	1503a	MW = 37,000 (1503a)
Cysteine synthase (EC 4.2.99.8)					
S. typhimurium	28.4	9.2	280	575	MW = 309,000 (575)
S-Cysteinylalbumin (see albumin, beef)					
Cytochrome b					
B. anitratum	—	9.5	280	570	pH 7, 0.06 M P_i, data from Figure 5 (570)
Beef heart	11.4[gg]	—	429	1364	pH 7.4, 0.01 M P_i, 0.001 M
	2.07[gg]	—	—[hh]	1364	NaDodSO$_4$, values based on
	1.32[jj]	—	562.5	1364	pyridine hemechromogen value
	—	36.6	418	1364	From Figure 2 (1364)
	—	7.2	562	1364	
	—	3.7	532	1364	
Cytochrome b_2					
Yeast, oxidized	0.92	—	560	571	
	1.13	—	530	571	
	12.95	—	413	571	
	3.44	—	—[kk]	571	
	8.35	—	280	571	pH 7.0, 0.2 M P_i, 0.2 mM
Reduced with 19.5 mM lactate	3.09	—	557	571	EDTA
	1.56	—	528	571	
	18.3	—	424	571	
	3.9	—	328	571	
	8.8	—	269	571	
Yeast					
Fraction B	—	11.0	275	1365	10% AcOH
Fraction C	—	6.7	275	1365	
Core, ferri-form	12.0	—	413	1366	—
Apo-	4.81	2.05	278	1367	MW = 235,000 (1368)
Polypeptide, oxidized	2.3[ll]	—	260	572	
	2.2[ll]	—	280	572	
	11.2[ll]	—	413	572	MW = 11,000 (572)
Reduced	15.8[ll]	—	423	572	
	13.4[ll]	—	528	572	
	2.68[ll]	—	557	572	

[gg] Absolute reduced.
[hh] 562.5–600 nm.
[jj] Reduced-oxidized.
[kk] 360–365 nm.
[ll] Per heme.

MOLAR ABSORPTIVITY AND $A_{1cm}^{1\%}$ VALUES FOR PROTEINS AT SELECTED WAVELENGTHS OF THE ULTRAVIOLET AND VISIBLE REGION (continued)

Protein	$\epsilon^a (\times 10^{-4})$	$A_{1cm}^{1\%}$ [b]	nm[c]	Ref.	Comments[d]
Cytochrome b_5					
Rat liver					
Oxidized	11.7	—	413	1369	—
Reduced	17.1	—	423	1369	—
	1.34	—	526	1369	—
	2.56	—	556	1369	—
Cytochrome b_{562}					
E. coli B					
Reduced	3.16	—	562	1370	—
	1.74	—	531.5	1370	—
	18.01	—	427	1370	—
Oxidized	0.97	—	564	1370	—
	1.06	—	530	1370	—
	11.74	—	418	1370	—
	2.1	—	280	1370	From OD_{562}/OD_{280} = 1.5 (1370)
Cytochrome c	—	19.5	280	1372	Dry wt
	—	290	210	1372	
	—	870	191	1372	
Beef heart	2.77	—	550	137	—
	2.42	—	280	138	Calcd from OD_{550}/OD_{280} = 1.26
	2.9	—	550	139	Reduced with dithionite
	3.053	23.94	550	138	MW = 12,750 (138)
Dithionite reduced	—	23.94	550	1373	pH 6.8, 40 mM $NH_4\ P_i$
Horse heart	—	17.1	280	1374	—
	—	650	—[mm]	1374	—
	1.12	—	528	1375	—
	—	9.05	528	1376	—
	2.77	—	550	140	pH 6.8, 0.1 M P_i, reduced
	3.18	—	—	140	pH 6.8, 0.1 M P_i, reduced, wavelength 268–272 nm
	—	1.12	528	140	pH 6.8, 0.1 M P_i, oxidized
	—	2.32	280	140	pH 6.8, 0.1 M P_i, oxidized
NO-ferro form	0.54	—	570	1377	
	0.65	—	540	1377	
NO-ferri form	0.56	—	571	1377	Wavelength determined from frequency
	0.69	—	561	1377	
	0.67	—	540	1377	
	0.67	—	527	1377	
Human heart	—	23.1	550	135	—
	2.77	21.9	550	136	Reduced, MW = 12,600 (136)
	2.33	18.5	280	136	Oxidized, MW = 12,600 (136)
Micrococcus dentrificans					
$NaBH_4$ reduced	2.68	—	550	1378	—
	2.23	—	280	1378	From OD_{550}/OD_{280} = 1.2 (1378)
Camelus dromedarius (camel)					
Reduced	2.4	—	550	1379	pH 7.5
	0.646	—	535	1379	
	1.382	—	520	1379	
	10.95	—	417	1379	

[mm] 191–194 nm.

MOLAR ABSORPTIVITY AND $A_{1cm}^{1\%}$ VALUES FOR PROTEINS AT SELECTED WAVELENGTHS OF THE ULTRAVIOLET AND VISIBLE REGION (continued)

Protein	$\epsilon^a(\times 10^{-4})$	$A_{1cm}^{1\%}$ [b]	nm[c]	Ref.	Comments[d]
Cytochrome c (continued)					
Oxidized	0.056	–	695	1379	
	0.946	–	530	1379	
	9.197	–	409	1379	pH 7.5
	2.440	–	360	1379	
	2.033	–	280	1379	
Cytochrome c_1					
Beef heart	–	14	276	1380	–
Cytochrome c_2					
Rhodospirillum rubrum					
Reduced	2.81[nn]	–	550	1381	
	1.7[nn]	–	521	1381	
	14.3[nn]	–	415	1381	
	3.71[nn]	–	316	1381	
	3.38[nn]	–	272	1381	pH 6.90, 0.1 M NaP$_i$
Oxidized	1.05[nn]	–	525	1381	
	11.5[nn]	–	410	1381	
	2.95[nn]	–	357	1381	
	2.47[nn]	–	275	1381	
Cytochrome c_{550}					
Bacillus subtilis					
Reduced	2.55	–	550	1384	–
	1.31	–	520	1384	–
	12.75	–	414	1384	–
	3.14	–	316	1384	–
	3.94	–	279	1384	–
Oxidized	1.18	–	528	1384	–
	9.25	–	407	1384	–
	3.06	–	279	1384	–
Spirillum itersonii					
Reduced[oo]	2.76	26.4	550	1385	
	1.61	15.5	522	1385	MW = 10,411 (1385)
	14.6	14.0	416	1385	
Oxidized[oo]	11.9	11.4	412	1385	
Thiobacillus novellus					
Reduced	2.58	–	550	1386	
	13.4	–	414.5	1386	At 77°K
Oxidized	2.92	–	280	1386	
Cytochrome c_{551}					
Thiobacillus novellus					
Reduced	1.96	–	551	1386	–
	13.9	–	416	1386	–
Oxidized	15.1	–	280	1386	–
Cytochrome $c_{551.5}$					
Chloropseudomonas ethylica					
Reduced[pp]	19.91	–	418	1387	pH 7.0, 0.05 M TrisCl
	1.58	–	523	1387	
	3.08	–	551.5	1387	

[nn] The value given here is 1/1000 of the value cited in Reference 1364. "I believe that a concentration of moles/1000 ml was used rather than mole/liter, which resulted in a very large value." [Kirschenbaum, *Anal. Biochem.*, 55, 166 (1973)].
[oo] Based on one atom of Fe per molecule.
[pp] All values per heme.

MOLAR ABSORPTIVITY AND $A_{1cm}^{1\%}$ VALUES FOR PROTEINS AT SELECTED WAVELENGTHS OF THE ULTRAVIOLET AND VISIBLE REGION (continued)

Protein	$\epsilon^a (\times 10^{-4})$	$A_{1cm}^{1\%}$ [b]	nm[c]	Ref.	Comments[d]
Cytochrome $c_{551.5}$ (continued)					
Oxidized[pp]	2.75	—	351	1387	
	12.44	—	408	1387	pH 7.0, 0.05 M
	1.03	—	528	1387	TrisCl
Cytochrome c_{552}					
Chromatium strain D					
Reduced α peak	3.12	4.35	550	1388	MW = 72,000 (1388)
Cytochrome c_{552} [I]					
P. stutzeri, reduced	3.1[qq]	—	552	573	—
	1.63[qq]	—	523	573	—
	15.08[qq]	—	418	573	—
Oxidized	1.05[qq]	—	530	573	—
	10.04[qq]	—	410	573	—
	2.51[qq]	—	284	573	—
Cytochrome c_{552} [II]					
P. stutzeri, reduced	1.95[qq]	—	552	573	—
	1.66[qq]	—	523	573	—
	15.35[qq]	—	417	573	—
Oxidized	0.99[qq]	—	525	573	—
	12.21[qq]	—	410	573	—
	1.69[qq]	—	277	573	—
Cytochrome c_{553}					
Petalonia fascia (an alga)					
Ferro form	2.68	25.5	273	1389	
	2.19	20.9	293	1389	
	4.40	41.9	317.5	1389	
	19.71	187.5	415.5	1389	
	0.38	3.6	471	1389	
	1.86	17.8	521.5	1389	MW = 10,500 (1389)
	2.85	27.1	553	1389	
Ferri form	2.68	25.5	269	1389	
	2.21	21.0	292	1389	
	3.62	34.5	360	1389	
	13.53	129.0	409	1389	
	1.29	12.3	528	1389	
Monochrysis lutheri (an alga)					
Reduced	2.59		553	1390	pH 7, P_i value
	15.65		416	1390	based on iron determination
Cytochrome c_{554}					
Bacillus subtilis					
Reduced	3.39	—	279	1384	—
	2.0	—	554	1384	—
	1.45	—	550	1384	—
	1.48	—	521	1384	—
	14.36	—	417	1384	—
	4.0	—	316	1384	—
Oxidized	2.4	—	280	1384	—
	1.02	—	523	1384	—
	10.67	—	409	1384	—

[qq] All ϵ values are based on heme content.

MOLAR ABSORPTIVITY AND $A_{1cm}^{1\%}$ VALUES FOR PROTEINS AT SELECTED WAVELENGTHS OF THE ULTRAVIOLET AND VISIBLE REGION (continued)

Protein	$\epsilon^a (\times 10^{-4})$	$A_{1cm}^{1\%}$ [b]	nm[c]	Ref.	Comments[d]
Cytochrome c_{555}					
Crithidia fasciculata					
Reduced[pp]	2.97	24.8	555.5	1391	
	1.68	14.0	525	1391	
	15.4	128	420	1391	
Oxidized[pp]	11.2	93	413	1391	MW = 12,051 (1391)
	1.22	10.2	533	1391	
	0.9	7.5	555.5	1391	
	0.84	7.0	565	1391	
Cytochrome $c_{555(550)}$					
Chloropseudomonas ethylica					
Reduced[pp]	15.34	—	417.5	1387	
	1.71	—	523	1387	
	2.04	—	555	1387	
Oxidized[pp]	4.02	—	275	1387	pH 7.0, 0.05 M TrisCl
	3.01	—	358	1387	
	13.20	—	412	1387	
	1.13	—	525	1387	
Cytochrome $c_{557(551)}$					
Alcaligenes faecalis					
Reduced[rr]	4.46	—	557	1383	Dry wt
	3.72	5.7	557	1383	
	2.89	4.44	525	1383	
	28.3	43	420	1383	
Reduced-CO[rr]	3.72	5.7	557	1383	pH 7, 0.05 M P_i,
	2.89	4.44	525	1383	MW = 65,000 (1383)
	40.0	61.4	416	1383	
Oxidized[rr]	2.16	3.32	530	1383	
	26.2	40.2	408	1383	
Cytochrome cc'					
Pseudomonas denitrificans					
Ferri form[pp]	0.37	—	635	1382	
	1.02	—	495	1382	
	8.0	—	400	1382	
	3.08	—	280	1382	
Ferro form[pp]	0.71	—	550	1382	pH 7.3, 0.02 M
	8.75	—	434	1382	Tris, 0.5 M NaCl
	9.70	—	426	1382	
CO-ferro form[pp]	1.05	—	564	1382	
	1.18	—	534	1382	
	21.0	—	418	1382	
Ferri form[pp]	0.63	—	575	1382	
	0.94	—	538	1382	
	9.90	—	413	1382	
	2.51	—	348	1382	3 N NaOH
Ferro form[pp]	2.49	—	550	1382	
	1.32	—	522	1382	
	15.62	—	416	1382	
Ferri form[pp]	0.245	—	635	1382	pH 5
	0.41	—	635	1382	pH 10

[rr] As diheme derivative.

MOLAR ABSORPTIVITY AND $A_{1cm}^{1\%}$ VALUES FOR PROTEINS AT SELECTED WAVELENGTHS OF THE ULTRAVIOLET AND VISIBLE REGION (continued)

Protein	$\epsilon^a (\times 10^{-4})$	$A_{1cm}^{1\%}$ [b]	nm[c]	Ref.	Comments[d]
Cytochrome *cd*					
Alcaligenes faecalis					
Reduced[pp]	18.9	21.0	418	1383	
	4.45	4.95	460	1383	
	2.87	3.18	556	1383	pH 7, 0.05 M P_i,
	2.65	2.94	525	1383	MW = 90,000 (1383)
	1.85	2.06	625	1383	
Oxidized[pp]	15.1	16.8	412	1383	
	2.12	2.35	640	1383	
	13.8	15.3	280	1383	pH 7, 0.05 M P_i, MW = 90,000 (1383), value calcd from $A_{412}/A_{280} = 1.2$ (1383)
Cytochrome P-450					
Rat liver microsomes					
Males	7.96	—	450	1392	Biuret
Females	9.41	—	450	1392	Biuret
Cytochrome oxidase (EC 1.9.3.1-.2)					
Pseudomonas (EC 1.9.3.2)	17,600	—	280	574	Data from Figure 2 (574)
Reduced	18,200	—	418	574	
	3,600	—	625	574	pH 6.0, 0.1 M NaP$_i$
Oxidized	14,900	—	408	574	
	3,020	—	630	574	
Pseudomonas aeruginosa	22	18.5	280	1393	Dry wt
Beef heart	—	17.4	280	133	—
Cytochrome *c* peroxidase (EC 1.11.1.5)					
Pseudomonas	6.48	12.1	280	1394	Dry wt, MW = 53,500 (1394)
Yeast	8.1	—	408	941	Dimethyl protoheme
	9.3	23.2	408	134	pH 6, MW = 40,000 (134)
	7.4	18.5	282	134	pH 6, MW = 40,000 (134)
Apo-	5.5	13.75	282	134	pH 6, MW = 40,000 (134)
Cytocuprein					
Human	—	210	210	576	
	—	5.8	268	576	
	—	5.5	280	576	
	—	0.08	675	576	Data from Figure 5 (576)
Apocytocuprein					
Human	—	210	210	576	
	—	4.3	268	576	
	—	4.2	280	576	
3-Deoxy-2-keto-6-phosphogluconate aldolase (see aldolase)					
Deoxyribonuclease I (EC 3.1.4.5) (DNase)					
Beef pancreas	—	11.5	280	141	—
	—	12.8	280	942	—

MOLAR ABSORPTIVITY AND $A_{1cm}^{1\%}$ VALUES FOR PROTEINS AT SELECTED WAVELENGTHS OF THE ULTRAVIOLET AND VISIBLE REGION (continued)

Protein	$\epsilon^a (\times 10^{-4})$	$A_{1cm}^{1\%}$ [b]	nm[c]	Ref.	Comments[d]
Deoxyribonuclease I (EC 3.1.4.5) (DNase)					
Beef pancreas (continued)	–	11.1	280	942	Nbs inactivated
	–	11.8	280	799	2.5 mM HCl
	3.72	12.0	–	577	pH 4.7, 0.2 M AcONa, MW = 31,000 (577)
	–	12.3	280	578	pH 7.6, 0.1 M KP$_i$
	–	13.9	280	578	pH 13, 0.1 N NaOH
	–	12.3	280	142	pH 7.6, 0.1 M KP$_i$
	–	13.9	280	142	pH 13, 0.1 N NaOH
Deoxyribonuclease inhibitor					
Calf spleen	6.05	10.2	280	579	pH 7.6, 0.1 M KP$_i$, MW = 59,400 (579)
	6.2	10.4	280	579	0.1 N NaOH, MW = 59,400 (579)
Deoxyribonucleic acid polymerase (DNA polymerase, EC 2.7.7.7)					
E. coli	9.26	8.5	280	580	10 mM NH$_4$HCO$_3$, MW = 109,000 (580)
	11.5	10.5	290	580	0.1 M NaOH–5 mM NH$_4$HCO$_3$, data from Figure 3 (580)
Large fragment	–	9.3	278	1401	
Ehrlich ascites tumor	–	23.9	280	581	pH 7.0, 0.2 M KP$_i$–0.001 M EDTA, 0.01 M 2–HSEtOH, data from Figure 4 (581)
		19.2	290	581	
	–	20.4	280	581	pH 10, data from Figure 4 (581)
	–	17.5	290	581	
Dextranase (EC 3.2.1.11)					
Aspergillus carneus	–	17.8	280	1395	Dry wt
Diamine oxidase (amine oxidase, EC 1.4.3.6)					
Pig kidney	10.6	12.8	280	143	MW = 87,000 (143)
Dihydrofolate reductase (tetrahydrofolate dehydrogenase) (EC 1.5.1.3)					
Streptomyces faecium	4.47	22.0	280	1396	DR, MW = 20,300 (1396)
Lactobacillus casei	2.15	–	278	1397	pH 7
	2.76	–	268	1397	+ NADPH[ss]
	0.72	–	340	1397	
E. coli, Methotrexate resistant	4	23.8	280	1398	Microbiuret and dry wt, MW = 16,810 (1398)
E. coli	–	19.1	280	1399	pH 7.0, 0.04 M
T4 phage	–	12.0	280	1399	KP$_i$
Chicken liver	3.4	15.5	278	682	MW = 22,000 682
Streptococcus faecium	–	20	280	683	

[ss]NADPH, nicotinamide adenine dinucleotide phosphate, reduced.

MOLAR ABSORPTIVITY AND $A_{1cm}^{1\%}$ VALUES FOR PROTEINS AT SELECTED WAVELENGTHS OF THE ULTRAVIOLET AND VISIBLE REGION (continued)

Protein	$\epsilon^a (\times 10^{-4})$	$A_{1cm}^{1\%}$ [b]	nm[c]	Ref.	Comments[d]
Dihydrolipoyl transacetylase (lipoate acetyltransferase, EC 2.3.1.12)					
E. coli	—	4.5	280	684	pH 7.0, P_i
Dihydroorotic dehydrogenase (orotate reductase, EC 1.3.1.14)					
Zymobacterium oroticum	3.5	—	450	27	—
	0.59	—	550	27	—
3,4-Dihydroxy-9,10-secoandrosta-1,3,5[10]-triene-9,17-dione 4,5-dioxygenase (steroid 4,5-dioxygenase) (EC 1.13.11.25)					
Nocardia restrictus	26.8	9.3	280	800	MW = 286,000 (800)
Dimethylglycine dehydrogenase (EC 1.5.99.2)					
Rat liver, mitochondria	—	17	280	685	pH 7.5, 0.0075 M KP_i, data from Figure 4 (685)
Dipeptidase (EC 3.4.13.11)					
Pig kidney	4.2	8.96	280	686	pH 8.0, 0.002 M Tris, MW = 47,200 (686)
p-Diphenol oxidase (monophenol monooxygenase) (EC 1.14.18.1)					
Polyporus versicolor	—	11.5	280	1400	—
Diphtheria toxin	—	12.7	278	145	pH 6.8, $M/15$ P_i, 0.175 M NaCl
	—	12.9	278	145	0.1 N HCl
	—	14.1	293	145	0.1 N NaOH
DPNase (NAD nucleosidase, EC 3.2.2.5)					
Pig brain	8.0	3.1	280	144	MW = 26,000 (144)
Edeine A					
B. brevis Vm 4	0.131	—	270	582	Water
Edeine B					
B. brevis Vm 4	0.131	—	270	582	Water
Elastase (EC 3.4.21.11)	—	11.0	280	943	—
Pig	5.74	22.2	280	801	0.1 M NaOH
	5.23	20.2	280	801	pH 5.0, 0.05 M AcONa
	—	18.5	280	583	—
	4.85	—	280	944	$\epsilon = [M/2.06] \times 10^5$
	—	22.0	280	945	—
Elastase-like enzyme					
Streptomyces griseus					
I	2.94	10.5	280	946	MW = 28,000 (946)
II	0.85	12.1	280	946	MW = 7,000 (946)
III	1.58	11.3	280	946	MW = 14,000 (946)

MOLAR ABSORPTIVITY AND $A_{1cm}^{1\%}$ VALUES FOR PROTEINS AT SELECTED WAVELENGTHS OF THE ULTRAVIOLET AND VISIBLE REGION (continued)

Protein	$\epsilon^a (\times 10^{-4})$	$A_{1cm}^{1\%}$ [b]	nm[c]	Ref.	Comments[d]
Elastoidin, soluble					
Prionace glauca, pectoral fins					
(Great blue shark)	–	1.86	277	947	0.5 M AcOH, data from Figure 2 (947)
Elinin					
Human erythrocyte	4420	22.1	274	948	MW = 20 × 10⁶ (948)[tt]
Elongation factor 2					
ADP ribosylated					
Rat liver					
Aminoethylated	9.7	–	276	959	pH 7.8, 5 M GdmCl,
	9.4	–	280	959	50 mM Tris
	9.9	–	273	959	
	8.9	–	280	959	pH 3.3, 50 mM
Endopolygalacturonase					
Verticillium albo-atrum	–	12.7[uu]	280	802	–
	–	11.2[uu]	280	802	–
Aspergillus niger	–	1.29	280	1404	–
Enolase (2-phosphoglycerate hydro-lyase) (EC 4.2.1.11)					
Oncorhynchus kisutch					
(Coho salmon)	7.8	7.4	280	949	MW = 105,560 (949)
Oncorhynchus keta					
(Chun salmon)	8.6	8.7	280	949	MW = 99,260 (949)
Yeast	7.8	8.9	280	949	MW = 88,000 (949)
	–	9.0	280	584	–
Rabbit	–	9.0	280	949	–
Rabbit muscle	–	8.95	280	141	–
	7.65	9	280	146	MW = 85,000 (146)
	6.1	9	280	147	MW = 67,200 (148)
	–	8.85	280	1407	Kjeldahl[e]
E. coli	–	5.7	280	1405	DR
	–	6.1	277	1405	
Rhesus monkey	7.2	8.8	280	1406	MW = 82,000 (1406)
Trout	–	7.9	280	949	–
Enterokinase (Enteropeptidase) (EC 3.4.21.9)					
Pig duodena	–	17.8	280	1408	–
Enterotoxin A					
Staphylococcus aureus	4.09	14.6	277	1409	MW = 28,000 (1409)
Enterotoxin B					
Staphylococci	–	12.1	277	149	–
	–	15	277	149	N = 16.1% (149)
	–	14	277	150	–

[tt]In Figure 3 of Reference 948: OD = 0.375 for a 0.4% solution. This is equivalent to an $A_{1cm}^{1\%}$ of 9.35. The value cited in the paper was "$E_{0.1cm}^{0.1\%} = 0.221$" from which an $A_{1cm}^{1\%}$ was obtained.
[uu]Two different preparations.

MOLAR ABSORPTIVITY AND $A_{1cm}^{1\%}$ VALUES FOR PROTEINS AT SELECTED WAVELENGTHS OF THE ULTRAVIOLET AND VISIBLE REGION (continued)

Protein	$\epsilon^a (\times 10^{-4})$	$A_{1cm}^{1\%}$ [b]	nm[c]	Ref.	Comments[d]
Enterotoxin B, nitrated					
Staphylococcal	6.29	20.6	277	687	pH 8, MW = 30,500 (687)
	2.42	7.9	428	687	
	2.16	7.1	360	687	pH 6.2, MW = 30,500 (687)
Enterotoxin C					
Staphylococcus aureus	–	12.1	277	688	–
Enterotoxin E					
Staphylococcus aureus	–	11.9	280	1410	–
	–	12.5	277	1410	–
Enzyme, thrombin-like					
Crotalus adamanteus	4.84	14.8	280	1411	Dry wt, MW = 32,700 (1411)
trans-Epoxysuccinate hydratase (see tartrate epoxidase)					
Erabutoxin a					
Laticauda semifasciata	0.7	–	280	803	Water
Erythrocruorin					
Lumbricus terrestris	1.18	5.13	504	804	
	11.36	49.4	430	804	
Oxygenated	1.37	5.95	542	804	
	11.27	49.0	417	804	pH 7.0, 0.1 M P$_i$,
	5.22	22.7	283	804	MW = 23,230 (804)
Plus CO	1.374	5.97	538	804	
	18.83	81.9	420	804	
Ferric derivative	1.165	5.07	500	804	
	10.0	43.7	395	804	
Erythrocruorin, oxy					
Cirraformia grandis (annelid worm)	4.09	22.1	280	1412	
	2.58	13.9	345	1412	Dry wt, MW = 18,500 (1412)
	9.68	52.2	415	1412	
	1.21	6.59	539	1412	
	1.18	6.38	574	1412	
Erythrocuprein (superoxide dismutase, EC 1.15.1.1)					
Beef	0.0313	–	680	1413	–
	0.984	–	259	1414	–
Apo-	0.241	–	252	1415	
	0.367	–	259	1415	
	0.330	–	262	1415	
	0.419	–	264	1415	pH 7.2, 10 mM P$_i$
	0.413	–	268	1415	
	0.383	–	275	1415	
	0.330	–	252	1415	
	0.420	–	259	1415	
	0.420	–	261	1415	GdmCl, pH 5.9
	0.540	–	269	1415	
	0.540	–	275	1415	
	0.460	–	281	1415	
	0.790	–	295	1415	pH 11.7, GdmCl
	3.81	–	246	1415	

MOLAR ABSORPTIVITY AND $A_{1cm}^{1\%}$ VALUES FOR PROTEINS AT SELECTED WAVELENGTHS OF THE ULTRAVIOLET AND VISIBLE REGION (continued)

Protein	$\epsilon^a (\times 10^{-4})$	$A_{1cm}^{1\%}$ [b]	nm[c]	Ref.	Comments[d]
Erythrocuprein (continued)					
Human	–	5.06	265	805	
	–	0.075	655	805	pH 6.5, 0.15 M NaCl
	–	0.077	675	805	
Human blood	0.0284	0.085	655	98	MW = 33,200 (432)
	1.84	5.5	265	98	MW = 33,200 (432)
	0.0350	0.104	655	151	MW = 33,600 (151)
	1.87	5.58	265	151	MW = 33,600 (151)
Erythropoietin					
Human urine	–	9.26	279	1416	–
Esterase (carboxylesterase, EC 3.1.1.1)					
Pig liver	–	13.8	280	585	pH 8.0
Goat intestine	–	14.73	275	586	–
	–	13.75	280	586	–
Excelsin					
Brazil nuts	–	9	279	806	pH 5.5
	–	14	279	806	pH 12.2
Exo-1,3-3-glucosidase (EC 3.2.1.58) (see glucanase)					
Factor					
Antihemorrhagic *Trimeresurus flavoviridis* (snake) serum	–	8.8	280	1417	–
Direct lytic (acetate) *Haemachatus haemachates* (snake) venom	0.29	4.2	278	1418	MW = 7,000 (1418)
Epidermal growth Mice (adult male, albino), submaxillary gland	1.81	30.9	280	1419	pH 5.6, 0.1 M AcONa or water DR, MW = 6,045 (1419)
X					
Beef plasma	–	12.4	280	1420	–
X, activated					
Beef plasma	4.1	8.6	280	1421	Dry wt, MW = 48,000 (1421)
X, thrombokinase, activated Stuart factor					
Human	–	5.8	280	587	–
XIII					
Human plasma	–	13.8	280	1503b	Dry wt
Fatty acid synthetase (EC 2.3.1.41?) (see also 3-oxoacyl-[ACP] reductase)					
Chicken liver	48.3	9.65	279	588	MW = 500,000 (588)
Pigeon liver	46.4	8.35	279	589	MW = 545,000 (589)
	38.7	8.6	280	153	MW = 450,000 (153)

MOLAR ABSORPTIVITY AND $A_{1cm}^{1\%}$ VALUES FOR PROTEINS AT SELECTED WAVELENGTHS OF THE ULTRAVIOLET AND VISIBLE REGION (continued)

Protein	$\epsilon^a (\times 10^{-4})$	$A_{1cm}^{1\%}$ [b]	nm[c]	Ref.	Comments[d]
Fatty acyl-CoA dehydrogenase (acyl-CoA dehydrogenase) (EC 1.3.1.8 or .9 or EC 1.3.99.3)					
Pig liver	–	13.8	275	152	pH 7.5, 0.036 M P$_i$, est. from Figure 3 (152)
Ferredoxin					
Alfalfa	0.95	–	277	154	Per mole iron
	1.83	16.6	277	154	MW = 11,000 (154)
	–	7.2	465	154	–
	–	7.9	422	154	–
	–	10.6	331	154	–
Scenedesmus	1.5	13.1	276	155	MW = 11,500 (155)
	1.33	–	330	155	–
	0.98	–	421	155	–
Apo-	0.855	15.8	280	156	pH 5.4, est. from Figure 1 (156), apoprotein made with α,α'-dipyridyl
	0.765	14.2	280	156	pH 7.4, est. from Figure 1 (156), apoprotein made with mersalyl
Azotobacter vinelandii					
Oxidized, FdI	2.7	19.2	400	1425	MW = 14,140 (1425)
Bacillus polymyxa					
Oxidized	–	21	279	950	
	–	12.3	325	950	
	–	11.1	400	950	pH 7.3, 25 mM TrisCl
	–	5.1	500	950	Data from Figure 2 (950)
Reduced with Na$_2$S$_2$O$_4$	–	6.9	400	950	
	–	3.1	500	950	
Chlorobium thiosulfatophilum	2.04	34	280	951	pH 7.3, 0.3 M Tris, 0.54 M NaCl, Data from Figure 1 (951), MW = 6,000 (951)
E. coli	0.96	7.6	416	952	MW = 12,600 (952)
	1.78	13.3	277	952	
Clostridium pasteurianum	2.16	41.5	285	952	
	2.16	41.5	300	952	MW = 5,200
	1.73	33.2	390	952	
	3.0	–	390	952	
	–	34.0	390	811	–
	2.1	35.0	390	812	MW = 6,000 (812)
Monomer	2.600	–	390	1428	Dry wt
	3.126	–	390	1428	Kjeldahl[e]
Dimer	1.600	–	390	1428	Dry wt
	1.540	–	390	1428	Kjeldahl[e]
lyophilized	1.617	–	390	1428	Dry wt
	1.260	–	390	1428	Kjeldahl[e]
C. tartarivorum	3.106	–	280	810	–
	2.422	–	290	810	–
C. thermosaccharolyticum	3.09	–	280	810	–
	2.413	–	390	810	–

MOLAR ABSORPTIVITY AND $A_{1cm}^{1\%}$ VALUES FOR PROTEINS AT SELECTED WAVELENGTHS OF THE ULTRAVIOLET AND VISIBLE REGION (continued)

Protein	ϵ^a (× 10^{-4})	$A_{1cm}^{1\%}$ [b]	nm[c]	Ref.	Comments[d]
Ferredoxin (continued)					
C. acidi urici	–	37.0	390	811	–
	3.06	–	390	813	–
	3.15	58.4	280	156	pH 7.4, 0.1 M Tris, est. from Figure 1 of Reference 156, MW = 5,400 (156)
	3.06[vv]	–	390	955	–
	2.98	–	390	955	Dry wt[ww]
	3.36[xx]	–	390	955	–
	3.02[yy]	–	390	955	–
C. tetanomorphum	–	35.0	390	811	–
C. butyricun	–	31.0	390	811	–
C. cylindrosporum	–	29.6	390	811	–
Cyperus rotundus L.	0.88	–	465	807	–
	0.98	–	420	807	–
	1.45	–	330	807	–
	2.42	–	275	807	–
Cotton, type I	0.655	–	460	808	–
	0.758	–	419	808	–
	1.082	–	325	808	–
Bacterial	2.45	–	390	809	–
Horsetail leaves	0.88	–	421	956	–
	1.17	–	276	956	From $A_{421}/A_{276} = 0.75$
Methanobacterium omelianskii	–	39	280	957	pH 7.3, 0.07 M TrisCl, data from Figure 1 (957)
Pseudomonas oleovorans	–	3.36	497	958	pH 7.3, 0.1 M Tris, one atom Fe
	–	20.2	280	958	From $A_{280}/A_{497} = 6.3$
	–	5.74	495	958	pH 7.3, 0.1 M Tris, two atoms Fe
	–	21.2	280	958	
Apo-	–	18	277	958	From $A_{280}/A_{495} = 3.7$, 50% AcOH
Rhodospirillum rubrum					
Type I	2.43	27.9	385	1426	MW = 8,700 (1426)
Type II	0.88	11.7	385	1426	MW = 7,500 (1426)
Parsley	1.24	11.47	255	1427	
	1.27	11.75	260	1427	
	1.5	13.9	277	1427	
	0.98	9.09	294	1427	
	1.22	11.34	330	1427	MW = 10,800 (1427)
	0.74	6.92	390	1427	
	0.92[zz]	8.65	422	1427	
	0.82	7.59	448	1427	
	0.84	7.79	463	1427	
	0.97	0.90	690	1427	
	1.01	–	420	814	–
Spinach	0.88	–	420	812	–
	0.94	–	420	814	–
Urea-oxygen denat.	2.0	–	275	815	Contains Fe
	1.3	–	275	815	Lacks Fe

[vv] Protein concentration determined by release of C-terminal amino acid by carboxypeptidase A.
[ww] MW = 6232, 8 eq. Fe and 8 eq. S.
[xx] Protein concentration determined by aspartic acid analysis.
[yy] Protein concentration determined by glutamic acid analysis.
[zz] This value cited in Reference 1427. "I calculated 9350 as ϵ." [Kirschenbaum, Anal. Biochem., 55, 166 (1973)].

MOLAR ABSORPTIVITY AND $A_{1cm}^{1\%}$ VALUES FOR PROTEINS AT SELECTED WAVELENGTHS OF THE ULTRAVIOLET AND VISIBLE REGION (continued)

Protein	$\epsilon^a (\times 10^{-4})$	$A_{1cm}^{1\%}$ b	nm^c	Ref.	Comments^d
Ferredoxin: NADP reductase (EC 1.6.7.1)					
Spinach	1.074	–	456	1429	–
Ferritin					
Apo-, horse spleen	–	–	280	157	$A_{1cm}^{1\%}$ = 8.6–9.7
Rat liver	–	2	320	1430	–
	–	302	260	1430	–
Fetuin					
Commercial	–	4.5	278	1431	This value good for native, reduced and carboxymethylated, oxidized, and neuraminidase-treated fetuin
Calf serum	–	4.1	278	1432	Water
	–	5.3	278	1432	–
Calf spleen	–	4.5	278	158	Australian sample
	–	4.8	278	158	Colorado sample
Fibrin					
Beef	–	16.16	283	816	6 M alkaline urea
	–	16.19	290	816	
	–	16.84	282	1433	40% urea, 0.2 N NaOH
Human	–	15	280	1434	–
Fibrin stabilizing factor					
Beef plasma	–	14.15	283	1422	8 M urea, 0.2 M NaOH
Fibrinogen					
Beef					
α chain	–	11.8	280	1435	0.1 M NaOH
β chain	–	17.4	280	1435	
γ chain	–	20.4	280	1435	
Beef	–	16.51	282	1433	8 M urea, 0.2 N NaOH
	–	15.4	279	1433	pH 2, 0.3 M NaCl
	–	14.0	279	1433	pH 6, 0.3 M NaCl
	–	15.6	282	1433	pH 11, 0.3 M NaCl
	–	15.04^{a*}	280	1436	0.3 M NaCl
	–	15.49^{a*}	280	1436	pH 6.9, 0.05 M P$_i$
	–	15.16^{a*}	280	1436	2% AcOH
	–	15.29^{a*}	280	1436	pH 5.3, 1 M NaBr
	–	15.45^{a*}	280	1436	30% urea
	–	15.60^{a*}	280	1436	pH 5.45, 6 N GdmCl
	–	15.94^{a*}	280	1436	pH 5.8, 2 M KCNS

^{a*}Read against value at 320 nm.

MOLAR ABSORPTIVITY AND $A_{1cm}^{1\%}$ VALUES FOR PROTEINS AT SELECTED WAVELENGTHS OF THE ULTRAVIOLET AND VISIBLE REGION (continued)

Protein	$\epsilon^a (\times 10^{-4})$	$A_{1cm}^{1\%\,b}$	nm^c	Ref.	Comments^d
Fibrinogen (continued)					
Beef (continued)	–	15.06	280	161	pH 7.1, ionic strength = 0.3
	–	16.01	282.5	161	pH 12.8, 0.1 M KOH
	–	15.92	289.5	161	pH 12.8, 0.1 M KOH
	–	15.17	278.5	161	pH 5.8, 5 M GdmCl
	–	15.06	279	161	pH 7.4, 5 M urea
	–	15.00	278	161	pH 7.6
	–	15.1	280	162	–
	–	15	280	689	–
	–	15.87	283	816	6 M alkaline urea
	–	15.88	290	816	
	–	15.04	280	816	3 M NaCl
	–	15.5	280	816	–
Fragment D	–	20.7	280	816	–
Fragment E	–	13.0	280	816	–
Calf	–	15.9	280	1439	
Dog	–	15.8	280	1439	0.2 M KCl, corrected for scattering
Elephant	–	15.7	280	1439	
Goat	–	15.6	280	1439	
Human	–	13.9	280	1437	pH 7.1, 0.055 M Na$_3$Cit·2H$_2$O
	–	17.65	282	1437	Alkaline urea
	46.4	13.6	280	96	MW = 341,000 (159)
	45.3	15.1	280	690	MW = 300,000 (690)
Fraction I-4	–	14.5	280	819	
Fraction I-8	–	15.6	280	819	
Fragment X	34.1	14.2	280	690	MW = 240,000 (690)
Fragment Y	27.3	17.6	280	690	MW = 155,000 (690)
Fragment D	16.7	20.8	280	690	MW = 83,000 (690)
	–	20.8	280	1438	
Fragment E	5.1	10.2	280	690	MW = 50,000 (690)
	–	10.2	280	1438	
Knot	–	11.8	280	1438	
Human, high solubility	–	16.8	282	160	0.1 N NaOH–5 M urea
Low solubility	–	16.7	280	160	0.1 N NaOH–5 M urea
Panulirus interruptus	52.5	12.5	280	820	MW = 420,000 (820)
Sheep	–	15.5	280	1439	0.2 M KCl, corrected for scattering
Fibroin					
Bombyx mori L	–	11.3	276	691	pH 4.3–7.3, 0.2 M NaCl
Ficin (EC 3.4.22.3)	5.8	22.4	280	163	MW = 26,000 (163)
	5.9	22.6	280	164	MW = 26,000 (163)
	4.6	–	280	821	
Fraction III	5.1	21	280	165	0.05 M NH$_4$HCO$_3$, MW = 24,500 (165)
Ficus glabrata					
Component G	5.4	–	280	1440	–
F. carica var. *kodata*	–	20.2	280	822	–
Flavodoxin					
Desulfovibrio gigas					
Oxidized	4.7	–	273	959	–
	0.82	–	374	959	–
	1.02	–	456.5	959	–

MOLAR ABSORPTIVITY AND $A_{1cm}^{1\%}$ VALUES FOR PROTEINS AT SELECTED WAVELENGTHS OF THE ULTRAVIOLET AND VISIBLE REGION (continued)

Protein	$\epsilon^a (\times 10^{-4})$	$A_{1cm}^{1\%}$ [b]	nm[c]	Ref.	Comments[d]
Flavodoxin (continued)					
Semiquinone	0.87	—	349	959	—
	0.41	—	580	959	—
Desulfovibrio vulgaris					
Oxidized	4.8	—	273	959	—
	0.87	—	375.5	959	—
	1.07	—	456–7	959	—
Semiquinone	0.9	—	349	959	—
	0.47	—	580	959	—
Apo-	2.00	—	278	1442	—
E. coli	0.825	—	467	960	—
	0.382	—	580	960	—
	5.0	—	274	960	From $\epsilon_{274}/\epsilon_{467} = 6.67$ (960), MW = 14,500 (960)
Clostridium MP					
Oxidized	4.68	—	272	961	—
	0.91	—	376	961	—
	1.04	—	445	961	—
Semiquinone	0.84	—	350	961	—
	0.485	—	376	961	—
	0.24	—	445	961	—
	0.462	—	575	961	—
Reduced	0.175	—	445	961	—
Clostridium pasteurianum	4	27.4	272	166	MW = 14,600 (166)
	0.79	5.3	372	166	MW = 14,600 (166)
	0.91	6.2	443	166	MW = 14,600 (166)
Oxidized	4.58	—	272	961	—
	0.847	—	374	961	—
	1.04	—	443	961	—
Semiquinone	0.766	—	350	961	—
	0.465	—	374	961	—
	0.208	—	443	961	—
	0.455	—	575	961	—
Reduced	0.16	—	443	961	—
	—	36.8	274	167	pH 7.3, 0.02 M TrisCl
Apo-	2.5	—	278	962	—
	2.52	—	282	962	—
Peptostreptococcus elsdenii					
Oxidized	4.76	—	272	963	pH 7.8, 0.002 M NaP$_i$
	0.63	—	350	963	—
	0.875	—	377	963	—
	1.02	—	445	963	—
Semiquinone	0.765	—	350	963	—
	0.5	—	377	963	—
	0.21	—	445	963	—
	0.45	—	580	963	—
Reduced	0.16–0.18	—	445	963	—
Apo-	2.67	—	278	962	—
Rhodospirillum rubrum					
Oxidized	1.12	—	460	1441	—
	1.13	—	376	1441	—
	5.42	—	272	1441	—
Semiquinone	0.5	—	627	1441	—
	0.45	—	588	1441	—
	1.09	—	353	1441	—
	6.07	—	273	1441	—

MOLAR ABSORPTIVITY AND $A_{1cm}^{1\%}$ VALUES FOR PROTEINS AT SELECTED WAVELENGTHS OF THE ULTRAVIOLET AND VISIBLE REGION (continued)

Protein	ϵ^a (× 10^{-4})	$A_{1cm}^{1\%}$ [b]	nm[c]	Ref.	Comments[d]
Flavodoxin (continued)					
Apo-	3.50	—	276	1442	—
Chlorella fuscea	1.0	—	464	1443	—
	5.46	—	275	1443	—
	0.905	—	379	1443	—
Flavoprotein	7.65	—	276	964	
Egg yolk	1.0	—	375	964	pH 7.2, 0.1 M NaP$_i$
	1.32	—	458	964	
Azobacter vinelandii	—	13.4	280	445	Biuret
Apo-	—	13.4	280	445	Lowry
Follicle stimulating hormone (FSH)					
Sheep	1.23	4.9	275	168	Neutral and acidic solutions, MW = 25,000 (168)
Human	2.3	9.2	276	169	MW = 25,000 (169)
Formiminoglutamase (EC 3.5.3.8)					
Pseudomonas ATCC 11,299b	—	14.7	280	1444	Dry wt, pH 7.4, 1 mM KP$_i$, 20 mM NaCl, 1 mM HSEtOH
Formylglycinamide ribonucleotide amidotransferase (phosphoribosylformyl-glycinamidine synthetase, EC 6.3.5.3)					
Chicken liver	18.6	—	280	170	pH 6.5, 0.1 M P$_i$
Formyltetrahydrofolate synthetase (EC 6.3.4.3)					
Clostridium acidi-urici	—	3.70	280	1445	Dry wt and DR
C. cylindrosporum	12.7	5.3	280	1446	—
	12.2	5.3	280	171	MW = 230,000 (171)
C. thermoaceticum	—	7.37	280	823	pH 8.1, TrisCl
β-Fructofuranosidase (see invertase)					
Fructokinase					
Rat liver	—	20	280	1447	pH 7, 0.12 M P$_i$, data from Figure 9 (1447)
Fructose-1,6-bisphosphatase (hexose bisphosphatase) (EC 3.1.3.11); (see also aldolase)					
Chicken					
Muscle	—	7.4	280	1448	—
Liver	—	7.4	280	1448	—
Pig kidney	—	8.9	280	1448	pH 8.0, 0.05 M TrisCl
	—	7.55	280	172	—
Rabbit					
Liver	—	8.9	280	824	—
	—	5.3	260	824	—
	—	8.3	280	174	—

MOLAR ABSORPTIVITY AND $A_{1cm}^{1\%}$ VALUES FOR PROTEINS AT SELECTED WAVELENGTHS OF THE ULTRAVIOLET AND VISIBLE REGION (continued)

Protein	$\epsilon^a (\times 10^{-4})$	$A_{1cm}^{1\%}$ [b]	nm[c]	Ref.	Comments[d]
Fructose-1,6-bisphosphatase (continued)					
Liver (neutral)	5.18	3.70	280	1450	Dry wt, MW = 140,000 (1450)
Liver (alkaline)	11.6	8.9	280	1451	MW = 130,000 (1451)
Muscle	–	6.1	260	1451	–
	–	9.4	280	173	–
Kidney	9.7	6.9	280	1452	Dry wt, MW = 140,000 (1452)
L-Fucose binding					
Lotus tetragonolobus					
A	21.4	17.8	280	452	MW = 120,000 (452)
B	12.2	20.9	280	452	MW = 58,000 (452)
C	20.4	17.4	280	452	MW = 117,000 (452)
Fumarase (fumarate hydratase) (EC 4.2.1.2)					
Pig heart	9.9	5.1	280	175	pH 7.3, MW = 194,000 (175)
Pig heart muscle	11.65	5.3	280	1334	pH 7.2, 0.005 M P_i, MW = 220,000 (1453)
α-Galactosidase (EC 3.2.1.22)					
Sweet almonds	8.32	25.5	276	176	MW = 33,000 (176)
	7.99	24.3	280	176	MW = 33,000 (176)
Vicia sativa	2.7	9	280	591	pH 7.2, TrisCl 10 mM MW = 30,000 (591), data from Figure 4 (591)
α-Galactosidase I					
Vicia faba	37.6	18	280	590	MW = 209,000 (590)
α-Galactosidase II					
Vicia faba	7.6	20	280	590	MW = 38,000 (590)
	7.2	19	278	590	MW = 38,000 (590)
β-Galactosidase (EC 3.2.1.23)					
E. coli ML 309	143.3	19.1	280	177	MW = 750,000 (178)
E. coli K12		19.1	280	1454	–
Galactosyl transferase A protein (lactose synthase, EC 2.4.1.22)					
Milk	5.3	12	279	1455	MW = 44,000 (1455)
Galactothermin		7.4	280	1456	pH 6.8, 0.01 M P_i
Human milk	1.04	7.4	277	965	0.1 M HCl, MW = 14,000 (965)
	1.20	8.55	290	965	0.1 M NaOH,
	1.13	8.15	283	965	MW = 14,000 (965)
Gastricsin					
Human gastric juice	–	4.56	247	1457	–
(pepsin C, EC 3.4.23.3)	–	12.83	278	1457	–
Gastrin					
Human	4.83	15.32	278	179	MW=31,500 (179)
Pig	5.02	15.3	280	180	MW=32,500 (180)

MOLAR ABSORPTIVITY AND $A_{1cm}^{1\%}$ VALUES FOR PROTEINS AT SELECTED WAVELENGTHS OF THE ULTRAVIOLET AND VISIBLE REGION (continued)

Protein	ϵ^a(× 10^{-4})	$A_{1cm}^{1\%}$ [b]	nm[c]	Ref.	Comments[d]
Gliadin					
α-, Wheat	2.9	5.8	276	181	pH 0.7, 6 M HCl, or pH 3.0, 0.001 M HCl, MW=50,000 (181)
	4.45	8.9	290	181	0.1 N NaOH, MW=50,000 (181)
Hard, red winter wheat	–	5.7	276	1458	Dry wt, corrected for light scattering
α$_{1b}$-	–	5.6	276	1458	–
α$_{1c}$-	–	5.6	276	1458	–
α$_2$-	–	5.6	276	1458	–
Globin					
Human	–	8.75	280	182	–
	–	8.5	280	183	
	–	8.74	280	184	pH 4.8, ionic strength 0.05
	3.36	8.0	280	185	MW=42,000 (185)
	–	10.62	280	1459	Dry wt
Horse	–	8.5	280	1460	–
Beef	–	7.9	280	1460	–
Dog	–	8.9	280	1460	–
Rabbit	–	8.5	280	1460	–
Rat	–	8.7	280	1460	–
Chironomus thummi thummi	–	7.23	282	1461	Dry wt
Globulin					
α-, Human hepatoma	–	5.26	278	1462	Dry wt
α-, Human	1.0		280	1463	–
β-, Human	1.54		280	1463	–
β$_{1c}$-, Human	–	10.0	280	186	–
β$_2$-, micro-, Human	1.99	16.8	280	187	pH 7.0, 0.1 M P$_i$, MW=11,815 (187)
γ-, low MW Human	1.72	10.1	280	188	MW=17,000 (188)
Corticosteroid binding					
Human	3.78	7.4	279	189	MW=51,700 (189)
Rabbit	3.43	8.4	279	190	MW-40,700 (190)
Rat	3.84	6.2	279	191	MW=61,000 (191)
Thyroxine binding					
Human	5.1	8.9	280	192	MW=58,000 (192)
Rabbit lens					
1	–	8.5	–[k†]	440	–
2	–	4.5	–[k†]	440	–
3	–	4.5	–[k†]	440	–
4	–	4.4	–[k†]	440	–
5	–	4.2	–[k†]	440	–
Globulin, 7S, soybean	–	6.9	280	1464	Dry wt
Globulin, 11 S (Glycinin)	9.85	5.47	280	446	MW=180,000 (447)
Soybean seed (*Glycine max*)	–	8.04	280	1465	–
Globulin, cold-insoluble					
Human plasma	–	12.8	280	825	pH 7.0, 0.05 M P$_i$, 0.15 M NaCl
	–	14.8	282	825	0.1 N NaOH–5 M urea

MOLAR ABSORPTIVITY AND $A_{1cm}^{1\%}$ VALUES FOR PROTEINS AT SELECTED WAVELENGTHS OF THE ULTRAVIOLET AND VISIBLE REGION (continued)

Protein	$\epsilon^a (\times 10^{-4})$	$A_{1cm}^{1\%}$ [b]	nm[c]	Ref.	Comments[d]
Glucagon					
Commercial	—	23.0	280	1466	
	0.72	—	260	1466	pH 10.0, 0.2 M P_i
	1.1	—	250	1466	
Pig	—	23.0	279	1467	0.1 M glycine/0.1 M NaCl/0.1 N NaOH, pH 10.4
Beef	0.83	23.8	278	193	pH 2, MW=3647 (193)
Glucanase, exo-β-D-[1 →, 3] (EXO-1-β- glucosidase, EC 3.2.1.58)					
Basidomycete QM 806	9.2	18.1	280	194	MW=51,000 (194)
Gluconolactonase (see lactonase)					
Glucose dehydrogenase (EC 1.1.1.47 or .118 or .119)					
Bacterium anitratum	—	9.1	280	966	pH 7.0, 0.06 M P_i, data from Figure 1(966)
Soluble	1.56	—	350	195	Oxidized state
	3.89	—	339	195	Reduced state
Glucose isomerase (EC 5.3.1.18)					
Streptomyces	15.7	10	280	692	MW=157,000 (692)
Bacillus coagulans	17.0	10.6	280	826	pH 6.0, 0.1 M AcONa, MW=160,000 (826)
	17.0	10.6	280	826	0.1 M NaOH, data from Figure 7 (826), MW=160,000.
	23.8	14.9	290	826	
Glucose oxidase (EC. 1.1.3.4)					
A. niger	31.1	16.7	280	196	MW=186,000 (196)
Penicillium amagasukiense	18.8	11.9	278	967	MW=158,000 (967)
Apo-	19.2	—	278	1468	—
Glucose-6-phosphate dehydrogenase (EC 1.1.1.49)					
Human erythrocyte	—	6.15	280	693	—
Leuconostoc mesenteroides	11.9	11.5	280.5	968	MW=103,700 (968)
D-Glucose-6-phosphate ketol isomerase (EC 5.2.1.9)					
Pea	9.6	8.75	280	197	MW=110,000 (197)
α-Glucosidase (EC 3.2.1.20)					
Beef liver	14.35	13.4	280	694	MW=107,000 (694)
	11.2	10.5	288	694	
β-Glucosidase (EC 3.2.1.21)					
Aspergillus wentii	33.8	19.1	278	1469	MW=170,000 (1470)
Sweet almonds	—	21.8	278	696	—
A	—	18.8	278	695	—
B	—	18.2	278	695	—

MOLAR ABSORPTIVITY AND $A_{1cm}^{1\%}$ VALUES FOR PROTEINS AT SELECTED WAVELENGTHS OF THE ULTRAVIOLET AND VISIBLE REGION (continued)

Protein	$e^a(\times 10^{-4})$	$A_{1cm}^{1\%}$ [b]	nm[c]	Ref.	Comments[d]
β-Glucuronidase (EC 3.2.1.31)					
Beef liver	47.6	17	280	198	pH 5, MW=280,000 (198)
Glutamate decarboxylase (EC 4.1.1.15)					
E. coli	51.0	17	280	827	pH 7.0, MW=300,000 (827)
Glutamate dehydrogenase (EC 1.4.1.2-4)					
Commercial	46.5	—[b*]	279	1471	MW=56,100 (1472)
Clostridium SB4	—	10.7	280	1473	DR
Beef liver	—	9.3	280	1474	Dry wt, pH 7.0, 0.11 M P_i
	—	8.9	280		Dry wt, pH 7.0, 0.11 M P_i,
	—	9.5	280	1474 } 1475 }	corrected for light scattering, pH 7.6, M/15 KNaP$_i$
Frog liver	—	9.5	280	828	—
Pig liver	—	9.7	279	829	—
Glutamate dehydrogenase, (NAD dependent) (EC 1.4.1.2)					
Cl. SB$_4$	29.4	10.7	280	830	pH 7, 50 mM KP$_i$ or pH 7.4, 50 mM TrisCl, MW=275,000 (830)
Glutamate mutase (methylaspartate mutase, EC 5.4.99.1)					
C. tetanomorphum H 1					
Component E	7.94	6.2	280	831	MW=128,000 (831)
Component S	1.1	6.44	280	199	MW=17,000 (199)
Glutamic dehydrogenase (EC 1.4.1.4)					
E. coli	—	12.9	278	1476	Dry wt, Kjeldahl, and ash[e]
Beef liver	—	10.0	280	200	—
	—	9.73	279	201	pH 7, 0.2 M P_i
	—	8.55	276	202	5.1 M GdmCl
	—	9.71	279	203	—
	—	8.20	279	203	6 M GdmCl
Frog, liver	23.8	9.5	280	204	pH 8.0, 0.1 M Tris- AcO$^-$, 0.0001 M EDTA, MW=250,000 (204)
Glutamin-(asparagin-)ase (EC 3.5.1.38)					
Achromobacteracae	—	10.2	280	1477	pH 7.2, 0.01 M NaP$_i$
Glutaminyl-peptide glutaminase (see peptidoglutaminase II)					

[b*] Duplicate analyses: 8.31 and 8.36.

MOLAR ABSORPTIVITY AND $A_{1cm}^{1\%}$ VALUES FOR PROTEINS AT SELECTED WAVELENGTHS OF THE ULTRAVIOLET AND VISIBLE REGION (continued)

Protein	$\epsilon^a (\times 10^{-4})$	$A_{1cm}^{1\%}$ [b]	nm[c]	Ref.	Comments[d]
Glutamine cyclotransferase (glutaminyl-tRNA cyclotransferase, EC 2.3.2.5)	3.6	14.3	280	205	MW=25,000 (205)
Glutamine phosphoribosylpyrophosphateamidotransferase (amidophosphoribosyl transferase, EC 2.4.2.14)					
Pigeon liver	1.02	–	415	206	–
	8.18	8.18	279	207	MW=100,000 (207)
Glutamine synthetase (EC 6.3.1.2)					
Pig brain	51	13.8	279	1478	pH 7.2, MW= 370,000 (1478)
Sheep brain	–	10.0	280	208	
	–	13.5	280	832	–
E. coli	52.4	7.7	280	209	pH 7.0, 0.01 M imidazole HCl, MW=680,000 (210)
Glutamine transaminase (EC 2.6.1.15)					
Rat liver	–	6.5	280	1479	Lowry[e], pH 7.2, 0.005 M KP$_i$
	–	3.26	260	1479	
	–	0.78	415	1479	
γ-Glutamylcysteine synthetase (EC 6.3.2.2)					
Rat kidney	10.6	11.5	280	1480	Lowry[e], MW= 92,000 (1480)
Glutathione peroxidase (EC 1.11.1.9)					
Beef blood	6.3	7.5	280	969	MW=84,000 (969)
Glutathione reductase (EC 1.6.4.2)					
Yeast	12.0	10.5	280	211	MW = 118,000 (211), protein det. by dry wt
	–	18.6	280	212	Used biuret method for protein det.
	–	14.5	280	212	Protein det. by dry wt
	18.6	15.4	280	213	MW=121,000 (213)
Human red blood cell	19.5	16.3	275	214	Estimated from Figure 8 (214), Reference 214, MW=120,000 (214)
Penicillium chrysogenum	–	18.6	280	833	
Rice embryos	9.51	9.1	275	1481	
	18.3	17.6	280	1481	
	1.09	10.5	370	1481	MW=104,000 (1481)
	1.10	10.6	379	1481	
	1.16	11.1	463	1481	
Gluten					
Wheat	–	6.20	276	1482	
S-β-(1-Pyridylethyl)-	–	7.07	275	1482	pH 2, 0.01 N HCl
Acrylonitrile derivative	–	6.58	276	1482	

MOLAR ABSORPTIVITY AND $A_{1cm}^{1\%}$ VALUES FOR PROTEINS AT SELECTED WAVELENGTHS OF THE ULTRAVIOLET AND VISIBLE REGION (continued)

Protein	ϵ^a ($\times 10^{-4}$)	$A_{1cm}^{1\%}$ [b]	nm[c]	Ref.	Comments[d]
Glyceraldehyde-3-phosphate dehydrogenase (EC 1.2.1.12)					
Pig	—	9.6	280	970	5 M GdmCl
	14	10	280	971	0.1 N NaOH, MW=140,000 (971)
	—	9.6	280	972	—
Apo-	12.7	9.1	280	971	0.1 N NaOH, MW=140,000 (971)
Pig, skeletal	—	9	280	221	0.1 NaOH, coenzyme free
Rabbit	17.5	12.7	280	973	MW=145,000 (973)
	—	10.6	276	974	—
	—	8.15	280	974	Charcoal treated enzyme
Rabbit muscle	—	10.3	280	1483	—
Denatured	3.0	—	337	1484	—
Furylacryloyl-	3.0	—	344	1484	—
Chicken heart	14	10.2	280	1485	MW=137,600 (1485)
Lobster	—	5.55	290	975	—
	14	10	280	976	MW=140,000 (977)
	—	9.6	280	978	With 4 NAD$^+$
	—	8.0	280	979	Charcoal treated
Lobster, muscle	14.8	10.1	276	218	MW=146,650 (219)
	—	9.6	280	220	pH 8.5, 0.05 M, Na PP$_i$
Yeast	—	8.94	280	215	—
	10.9	9.08	280	216	MW=120,000 (216)
	12.4	8.6	280	217	MW=144,700 (217)
	—	9.4	280	980	—
	—	8.6	280	981	—
Apo-	—	8.94	280	982	—
	—	8.2	280	1485	—
Human erythrocyte	13.6	9.9	280	983	pH 8.0, 0.05 M TrisCl, MW=137,000 (983)
Beef liver	13.0	9.13	280	984	MW=142,000 (984)
E. coli	14.4	10	280	1486	pH 8, 20 mM TrisCl, 2 mM EDTA
Rabbit muscle	11.5	9.8	280	222	MW=118,000 (222)
	11.4	8.29	280	223	MW=138,000 (223) Charcoal treated
	14.9	10.3	280	217	pH 8, PP$_i$, MW=144,000 (217)
Glycerol dehydrogenase (EC 1.1.1.6)					
Aspergillus niger	—	17	280	985	pH 7.2, P$_i$
Glycerol kinase (EC 2.7.1.30)					
E. coli	—	15.6	280	986	pH 6.5, 6 M GdmCl, data from Figure 7 (986)
	—	14.1	280	986 }	pH 12.5, data from Figure 7 (986)
	—	14.7	290	986 }	
Glycerol-3-phosphate dehydrogenase (EC 1.1.1.8, EC 1.1.99.5)					
Apis mellifera, thorax	—	3.3	280	226	pH 7.8
	2.5	—	280	227	pH 7.5

MOLAR ABSORPTIVITY AND $A_{1cm}^{1\%}$ VALUES FOR PROTEINS AT SELECTED WAVELENGTHS OF THE ULTRAVIOLET AND VISIBLE REGION (continued)

Protein	$\epsilon^a (\times 10^{-4})$	$A_{1cm}^{1\%}$ [b]	nm[c]	Ref.	Comments[d]
Glycerol-3-phosphate dehydrogenase (continued)					
Chicken					
Liver	3.59[d*]	–	280	1488	–
Muscle	2.83[d*]	–	280	1488	–
Rabbit muscle	4.2	7.0	280	987	
	4.4	7.3	280	988	MW = 60,000 (987)
	3.8	6.3	280	989	
	–	6.3	280	224	–
	–	5.3	280	224	Charcoal treated
	–	7.5	280	141	–
	3.8	6.5	280	225	pH 6.6, 0.02 M P_i, MW = 58,300 (225)
	3.52	6.0	280	225	pH 7.2, 0.1 M P_i, MW = 58,300 (225)
Rat liver					
Fraction 1 enzyme	11.0	18.3	280	1487	pH 7.2, 0.1 M P_i,
Fraction 2 enzyme	4.0	6.7	280	1487	MW = 60,000 (1487)
Glycinin					
Soybean flakes	32.2	9.2	280	990	MW = 350,000 (991)
Glycocyaminekinase (guanidinoacetate kinase, EC 2.7.3.1)					
Nephthys coeca	7.5	8.6	278	228	pH 8.1, Tris-AcO$^-$ 0.1 M 10^{-4} M EDTA, MW = 87,500 (228)
Glyco-α-lactalbumin					
Cow's milk	–	17.7	280	1489	–
Glycollate oxidase (EC 1.1.3.1)					
Spinach	–	7.7	280	1490	Calcd from Figure 3 (1490), pH 8.3
Glycoprotein					
Envelope-specific, E. coli	–	11	278	229	–
Tamm-Horsfall					
Rabbit urine	–	6.7	277	1503e	Water and 6 M GdmCl
Human urine	–	13	277	1491	Calcd from Figure 1 (1491)
	–	10.8	277	1492	–
	–	9.5	277	1493	6 M GdmCl
	–	9.4	277	1493	Water
Acceptor of glycosyl transferase					
Rat intestinal mucosa	–	4.8	278	1494	–
Glycoprotein (α-globulin)					
Mouse plasma	–	8.8	278	1495	–
Tumor	–	11.6	278	1495	–
α_1-Glycoprotein					
Human					
Easily precipitable	4.5	9.0	280	230	MW = 50,000 (230)
Tryptophan poor	–	6.0	280	231	–

[d*] Average of values obtained for two different preparations.

MOLAR ABSORPTIVITY AND $A_{1cm}^{1\%}$ VALUES FOR PROTEINS AT SELECTED WAVELENGTHS OF THE ULTRAVIOLET AND VISIBLE REGION (continued)

Protein	$\epsilon^a (\times 10^{-4})$	$A_{1cm}^{1\%}$ [b]	nm[c]	Ref.	Comments[d]
α_1-Glycoprotein (continued)					
Acid					
(orosomucoid)	3.9	8.9	280	232	MW = 44,100 (233)
Liver	4.0	9.16	279	1497	MW = 43,600 (1497)
Blood					
Variant pI 3.0	–	9.38	278	1498	–
Variant pI 3.2	–	9.31	278	1498	–
Variant pI 3.4	–	9.32	278	1498	–
Pool	–	9.33	278	1498	–
Serum	–	18.2	280	1503c	pH 7.0, 1/15 M P$_i$
Anti-trypsin	2.3	5.3	280		MW = 45,000 (234)
Chimpanzee plasma	–	8.52	278	1496	–
α_2-Glycoprotein					
Neuramino	–	5.0	280	235	–
Zinc	–	18	280	236	–
Histidine-rich 3.8S					
Human serum	–	5.85	280	1499	–
Macroglobulin					
Heat labile	68.9	8.4	280	8	MW = 820,000 (238)
α_{2HS}-Glycoprotein	2.7	5.6	280	230	MW = 49,000 (237)
α_2-β_1-Glycoprotein					
Human plasma	–	11.5	278	1500	pH 6, 0.1 M NaCl
α_3-Glycoprotein, 8S					
Human serum	–	10.0	280	1501	–
α_{IX}-Glycoprotein	–	6.0	280	230	–
β-Glycoprotein					
Glycine-rich					
Human serum	–	6.2	280	1502	–
β_1-Glycoprotein					
Sialic acid free					
Human plasma	–	11.2	278	1503	pH 7
γ-Glycoprotein					
Glycine-rich					
Human serum	–	10.0	280	1503d	–
Glyoxylate reductase					
(EC 1.1.1.26)					
Spinach leaves	–	9.76	280	834	Protein det. by Lowry method
	–	10.7	280	834	Protein det. by dry wt and N
Isozyme R$_f$ 0.22	–	10.8	280	834	–
Isozyme R$_f$ 0.19	–	14.5	280	834	–
Isozyme R$_f$ 0.17	–	13.8	280	834	–
Gramicidin A, HO-NBzl[e*]	7.25[f*]	–	410	1503f	}
	7.27[g*]	–	410	1503f	} 0.1 N Na$_2$CO$_3$
	3.58[f*]	–	270	1503f	}

[e*] HO-NBzl, 2-hydroxy-5-nitrobenzyl group.
[f*] Same fraction of two fractions.
[g*] Same fraction of two fractions.

MOLAR ABSORPTIVITY AND $A_{1cm}^{1\%}$ VALUES FOR PROTEINS AT SELECTED WAVELENGTHS OF THE ULTRAVIOLET AND VISIBLE REGION (continued)

Protein	$\epsilon^a (\times 10^{-4})$	$A_{1cm}^{1\%}$ [b]	nm[c]	Ref.	Comments[d]
Gramicidin A, HO-NBzl[e]* (continued)	3.73[g]*	—	270	1503f	0.1 N Na$_2$CO$_3$
Gramicidin D, HO-NBzl[e]*	7.18	—	410	1503f	
	4.17	—	270	1503f	
Growth hormone					
Beef	—	7.1	280	239	—
	—	6.5	280	835	—
Haptoglobin					
Human, type I					
Plasma	—	11.0	278	697	0.1 N HCl, pH 1
	—	11.9	278	697	0.02 M HCO$_3^-$, pH 8.6
	—	11.1	278	697	0.1 N NaOH, pH 12
Urine	—	12.7	278	697	0.02 M HCO$_3^-$, pH 8.6
Type II	—	15.6	278	698	—
	—	11.2	280	697	0.1 N HCl, pH 1
	—	12.1	280	697	0.02 M HCO$_3^-$, pH 8.6
	—	11.5	280	697	0.1 N NaOH, pH 12
Type II complex with hemoglobin	—	14.5	280	698	
	—	29	408	698	—
Canine	9.4	11.6	—	699	MW = 81,000 (699)
Haptoglobin 1-1, human	10.5	—	280	1503g	—
Haptoglobulin					
Human	12.1	12.1	280	240	MW = 100,000 (240)
	—	11.6	278	241	—
Hemagglutinin					
P. lunatus	—	12.3	280	836	—
Agaricus campestris	10.5	16.4	280	837	MW = 64,000 (837)
Pisum sp (peas)		10	280	1504	From Figure 1 (1504)
Phascolus vulgaris	9.6	10.5[h]*	280	1505	pH 2, 0.01 M HCl
Lens culinaris	5.6	12.5	280	1506	MW = 44,050 (1506)
L. culinaris A	6.2	12.6	280	1507	MW = 49,000 (1507)
L. culinaris B	6.2	12.6	280	1507	
Hemerythrin					
Phascolosoma gouidii					
Coelomic fluid					
Methemerythrin					
Azide	0.019	—	680	1508	
	0.370	—	446	1508	
	0.675	—	326	1508	
Bromide	0.0165	—	677	1508	ϵ/dimeric iron unit
	0.540	—	387	1508	
	0.650	—	331	1508	
Chloride	0.018	—	656	1508	
	0.600	—	380	1508	
	0.660	—	329	1508	

[h]*Calculated from the amino acid analysis.

MOLAR ABSORPTIVITY AND $A_{1cm}^{1\%}$ VALUES FOR PROTEINS AT SELECTED WAVELENGTHS OF THE ULTRAVIOLET AND VISIBLE REGION (continued)

Protein	$\epsilon^a (\times 10^{-4})$	$A_{1cm}^{1\%}$ [b]	nm[c]	Ref.	Comments[d]
Hemerythrin (continued)					
Cyanate	0.0166	—	650	1508	
	0.650	—	377	1508	
	0.655	—	334	1508	
Cyanide	0.014	—	695	1508	
	0.077	—	493	1508	
	0.530	—	374	1508	
	0.640	—	330	1508	
Fluoride	0.500	—	362	1508	ϵ/dimeric iron unit
	0.560	—	317	1508	
Hydroxide	0.016	—	597	1508	
	0.590	—	362	1508	
	0.680	—	320	1508	
Thiocyanate	0.020	—	674	1508	
	0.510	—	452	1508	
	0.720	—	327	1508	
Water	0.640	—	355	1508	
	0.630	—	340	1508	ϵ/dimeric iron unit, plus iodide
Oxyhemerythrin	0.220	—	500	1508	ϵ/dimeric iron unit
	0.680	—	330	1508	
Dendrostomum pyroides	—	30.3	280	838	Protein det. by Biuret method
	—	31.0	280	838	Protein det. by Lowry method
Golfingia gouldii	—	25.8	280	839	—
Hemocyanin					
Busycon canaliculatum	—	15.1	280	1512	
	—	3.28	345	1512	FC, pH 8.2 Tris cont
Cancer borcalis	—	14.0	280	1512	0.01 M MgCl$_2$,;[9] SC
	—	2.29	336	1512	
Cancer magister	—	15.0	280	1512	
25S particle	—	14.7	279	843	pH 7.0, Tris
5S particle	—	14.1	—	843	pH 10, Bic
Carcinus meanas	—	14.2	280	1512	FC, pH 8.2 Tris cont
	—	2.33	335	1512	0.01 M MgCl$_2$, SC
C. sapidus	—	12.4	278	243	—
Dolabella auricularia (gastropod)	9.84	17.6	278	1509	MW = 55,800 (1509)
Eriphia spinifrons (arthropod)	—	12.07	278	1510	pH 7.2, 0.01 M P$_i$
	—	12.7	278	1510	pH 9.7, 0.05 M CO$_3^{2-}$
	—	16.1	278	243	—
E. moschata	—	14.9	278	243	—
Homarus americanus	—	14.3	280	1512	FC, pH 8.2 Tris cont
	—	2.69	335	1512	0.01 M MgCl$_2$, SC
	—	13.4	280	840	pH 9.6, 0.014 M Ca^{++}, glycine buffer, protein det. by dry wt
H. vulgaris	—	14.4	278	243	—
Levantina hierosolima	—	4.0	345	846	pH 6.6, oxygenated
	—	3.4	348	846	pH 12
Limulus polyphemus	—	13.9	280	1512	FC, pH 8.2 Tris cont
	—	2.23	340	1512	0.01 M MgCl$_2$, SC
	—	11.2	278	243	—
Loligo pealii	—	2.79	345	1512	FC, pH 8.2 Tris cont 0.01 M MgCl$_2$, SC
	—	15.8	278	243	—

MOLAR ABSORPTIVITY AND $A_{1cm}^{1\%}$ VALUES FOR PROTEINS AT SELECTED WAVELENGTHS OF THE ULTRAVIOLET AND VISIBLE REGION (continued)

Protein	$\epsilon^a (\times 10^{-4})$	$A_{1cm}^{1\%}$ [b]	nm[c]	Ref.	Comments[d]
Hemocyanin (continued)					
Helix pomatia	—	16.1	278	841	pH 9.2, borate
α-	—	13.8	278	242	—
β-	—	14.1	278	242	—
Succinylated	—	14.16	278	842	pH 9.2, 0.1 M Na$_2$B$_4$O$_7$
Megathura crenulata	—	17.7	280	844	Protein det. by dry wt
	—	15.5	280	845	
Murex trunculus (whelk)	—	13.9	280	1511	pH 9.2
	—	18.9	278	243	—
Apo-	—	13.9	280	1511	pH 9.2
M. brandacis	—	18.1	278	243	—
O. vulgaris	—	13.5	278	243	—
O. macropus	—	16.6	278	243	—
Pagarus pollicarus	—	15.6	280	1512	FC, pH 8.2 Tris cont
	—	2.58	335	1512	0.01 M MgCl$_2$, SC
P. vulgaris	—	13.8	278	243	—
Hemoglobin					
Aphrodite aculeata (polychaet annelid)[i*]					
nerves/ganglia	12.6	—	425	1513	—
	1.41	—	549	1513	—
	1.48	—	566	1513	—
HbO$_2$	14.7	—	414	1513	—
	1.70	—	541	1513	—
	1.84	—	577	1513	—
HbCO	22.0	—	419	1513	—
	1.82	—	537	1513	—
	1.84	—	567	1513	—
HbCN	14.6	—	434	1513	—
	1.5	—	536	1513	—
	1.9	—	564	1513	—
Aplysia californica (mollusc)[i*] nerves	12.0	—	435	1513	—
	1.3	—	560	1513	—
HbO$_2$	13.0	—	416	1513	—
	1.5	—	543	1513	—
	1.4	—	578	1513	—
HbCO	17.0	—	423	1513	—
	1.5	—	541	1513	—
	1.4	—	571	1513	—
HbCN	15.0	—	437	1513	—
	1.6	—	539	1513	—
	1.8	—	568	1513	—
Ascaris, bodywall					
Oxygenated	0.22		621	700	
	1.11		578	700	
	1.25		543	700	
	11.7		412	700	
Deoxygenated	0.3		615	700	0.1 M P$_i$, pH 7.0, ϵ_M values expressed per mole heme
	1.32		556	700	
	11.1		429	700	
Carbon monoxide	0.31		614	700	
	1.32		566	700	
	1.33		538	700	
	18.2		419	700	

[i*] ϵ/mole heme.

MOLAR ABSORPTIVITY AND $A_{1cm}^{1\%}$ VALUES FOR PROTEINS AT SELECTED WAVELENGTHS OF THE ULTRAVIOLET AND VISIBLE REGION (continued)

Protein	$\epsilon^a (\times 10^{-4})$	$A_{1cm}^{1\%}$ [b]	nm[c]	Ref.	Comments[d]
Hemoglobin (continued)					
Cyanide	0.31		614	700	
	2.1		562	700	
	1.47		532	700	
	14.7		429	700	
Methemoglobin, acid	0.23		635	700	0.1 M P_i, pH 7.0, ϵ_M values expressed per mole heme
	1.02		499	700	
	14.1		405	700	
Methemoglobin, cyanide	1.2		542	700	
	10.9		417	700	
Ascaris lumbricoides, perienteric fluid[i]*	1.22	—	553.5	1515	
	10.9	—	429.5	1515	
HbO$_2$	1.04	—	576.5	1515	pH 7.0, 0.1 M P_i
	1.23	—	542	1515	
	10.95	—	412	1515	
HbO$_2$[j]*	11.0	—	412	1514	
	1.23	—	541	1514	
	1.04	—	575	1514	At 273°K
Hb IV[j]*	9.75	—	411	1514	
	1.06	—	542	1514	
	0.87	—	576	1514	
HbCO	1.28	—	568	1515	
	1.27	—	538	1515	
	16.8	—	417.5	1515	
HbCN	2.02	—	564	1515	
	1.48	—	534	1515	
	14.9	—	431	1515	
MetHb					
Acid	0.36	—	632	1515	pH 7.0, 0.1 M P_i
	1.06	—	501	1515	
	15.5	—	404	1515	
CN$^-$	1.22	—	540	1515	
	11.4	—	417	1515	
N$_3^-$	0.31	—	630	1515	
	1.08	—	540	1515	
	11.7	—	414	1515	
Biomphalaria glabrata (mollusc)	—	23.8	280	1516	Lowry, dry wt
Glycera dibranchiata (common bloodworm)					
Fe^{2+}-O$_2$ (P)[k]*	1.51	—	576	1517	
	1.47	—	540	1517	
	14.2	—	414	1517	
Fe^{2+}-O$_2$ (M)[l]*	1.71	—	575	1517	
	1.65	—	540	1517	
	14.2	—	420	1517	pH 7, 0.1 M P_i, ϵ/mole hematin
Fe^{2+}-CO (P)	1.46	—	569	1517	
	1.51	—	538	1517	
	19.8	—	419	1517	
Fe^{2+}-CO (M)	1.62	—	569	1517	
	1.63	—	538	1517	
	23.2	—	422	1517	

[j]*Prepared from deoxyhemoglobin.
[k]*(P) = polymer.
[l]*(M) = monomer.

MOLAR ABSORPTIVITY AND $A^{1\%}_{1cm}$ VALUES FOR PROTEINS AT SELECTED WAVELENGTHS OF THE ULTRAVIOLET AND VISIBLE REGION (continued)

Protein	$\epsilon^a (\times 10^{-4})$	$A^{1\%\ b}_{1cm}$	nm^c	Ref.	Comments^d
Hemoglobin (continued)					
Fe^{2+}-H_2O (P)	1.34	—	555	1517	
	13.5	—	428	1517	
Fe^{2+}-H_2O (M)	1.47	—	564	1517	
	11.7	—	429	1517	
Fe^{3+}-H_2O (P)	0.32	—	632	1517	
	0.95	—	503	1517	
	15.0	—	408	1517	
Fe^{3+}-H_2O (M)	0.32	—	637	1517	
	1.48	—	505	1517	
	12.9	—	391	1517	pH 7, 0.1 M P_i,
Fe^{3+}-CN^- (P)	1.11	—	536	1517	ϵ/mole hematin
	13.9	—	418	1517	
Fe^{3+}-CN^- (M)	1.20	—	545	1517	
	14.3	—	420	1517	
Fe^{3+}-N_3^- (P)	0.89	—	573	1517	
	1.13	—	539	1517	
	12.5	—	416	1517	
Fe^{3+}-N_3^- (M)	1.14	—	575	1517	
	1.20	—	542	1517	
	13.1	—	419	1517	
Fe^{3+}-OH^- (P)	0.87	—	574	1517	
	0.97	—	537	1517	pH 10.5 glycine buffer,
	11.4	—	411	1517	ϵ/hematin
Fe^{3+}-OH^- (M)	0.76	—	576	1517	
	0.99	—	534	1517	pH 9.9, methylamine
	9.5	—	398	1517	buffer, ϵ/hematin
Fe^{3+}-F^- (M)	1.07	—	593	1517	pH 7, 0.1 M P_i,
	12.3	—	395	1517	ϵ/hematin
Cucumaria miniata Brandt (echinoderm)					
HbCO	1.33	—	570	1518	
	1.31	—	539	1518	
	12.33	—	416	1518	
Cucumaria piperata Stimpson (echinoderm)					
HbCO	1.66	—	570	1518	
	1.65	—	539	1518	AA
	14.89	—	417	1518	
Molpadia intermedia Ludwig (echinoderm)					
HbCO^{m*}	1.31	—	570	1518	
	1.46	—	540	1518	
	14.69	—	418	1518	
Thunnus orientalis (tuna)^{n*}					
	12.4	—	428–9	1519	—
	1.37	—	555	1519	—
HbO$_2$	13.3	—	411	1519	—
	1.49	—	540	1519	—
	1.53	—	575	1519	—
HbCO	20.3	—	418	1519	—
	1.51	—	538	1519	—
	1.44	—	568	1519	—

^{m*}Half-molecule of Hb-2.
^{n*}Hb concentration determined using the value of $\epsilon = 1.15 \times 10^4$ at 540 nm for MetHbCN.

MOLAR ABSORPTIVITY AND $A_{1cm}^{1\%}$ VALUES FOR PROTEINS AT SELECTED WAVELENGTHS OF THE ULTRAVIOLET AND VISIBLE REGION (continued)

Protein	$\epsilon^a (\times 10^{-4})$	$A_{1cm}^{1\%}$ [b]	nm[c]	Ref.	Comments[d]
Hemoglobin (continued)					
MetHb	14.4	–	405	1519	–
	0.98	–	500	1519	–
	0.42	–	630	1519	–
MetHbCN	11.5	–	418	1519	–
Anguilla japonica (eel)[n]*					
Component F					
	12.6	–	430–1	1519	–
	1.33	–	555	1519	–
HbO$_2$	11.5	–	415	1519	–
	1.39	–	540	1519	–
	1.42	–	575	1519	–
HbCO	17.8	–	420–1	1519	–
	1.46	–	537–9	1519	–
	1.34	–	568–70	1519	–
MetHb	9.3	–	408–9	1519	–
	0.9	–	520–2	1519	–
	0.3	–	630–2	1519	–
MetHbCN	8.5	–	421	1519	–
Component S					
	13.5	–	430–1	1519	–
	1.35	–	555	1519	–
HbO$_2$	13.2	–	410–2	1519	–
	1.39	–	540	1519	–
	1.43	–	575	1519	–
HbCO	17.5	–	418–9	1519	–
	1.43	–	538–9	1519	–
	1.38	–	567–9	1519	–
MetHb	15.3	–	404–5	1519	–
	0.99	–	500	1519	–
	0.44	–	630	1519	–
MetHbCN	10.1	–	419	1519	–
Misgurnus anguillicaudalus (loach)[n]*					
Component F					
	12.4	–	430	1520	–
	1.20	–	555	1520	–
HbO$_2$	13.2	–	413	1520	–
	1.48	–	540–1	1520	–
	1.53	–	576	1520	–
HbCO	20.6	–	418	1520	–
	1.53	–	538	1520	–
	1.46	–	568	1520	–
MetHb	19.4	–	404–5	1520	–
	0.43	–	630	1520	–
	0.96	–	–[o]*	1520	–
MetHbCN	12.8	–	419	1520	–
Component S					
	12.2	–	430	1520	–
	1.18	–	555	1520	–
HbO$_2$	12.4	–	413	1520	–
	1.47	–	541	1520	–
	1.53	–	576–7	1520	–
HbCO	20.4	–	419	1520	–
	1.5	–	538	1520	–
	1.43	–	568	1520	–

[o]*499–502 nm.

MOLAR ABSORPTIVITY AND $A_{1cm}^{1\%}$ VALUES FOR PROTEINS AT SELECTED WAVELENGTHS OF THE ULTRAVIOLET AND VISIBLE REGION (continued)

Protein	$\epsilon^a (\times 10^{-4})$	$A_{1cm}^{1\%}$ [b]	nm[c]	Ref.	Comments[d]
Hemaglobin (continued)					
MetHb	18.2	—	405	1520	—
	0.96	—	—p*	1520	—
	0.43	—	630	1520	—
MetHbCN	11.7	—	419	1520	—
Oncorhynchus tshawytscha (chinook salmon)					
HbCS-1					
MetHb	0.44	—	628	1521	
	0.88	—	498	1521	
	13.2	—	405	1521	
	4.6	—	276	1521	
CarboxyHb	1.34	—	566	1521	
	1.38	—	537	1521	
	16.4	—	419	1521	
HbCS-2					Dry wt
MetHb	0.41	—	625	1521	
	0.89	—	499	1521	
	12.3	—	404	1521	
	3.8	—	276	1521	
CarboxyHb	1.29	—	564	1521	
	1.36	—	534	1521	
	16.9	—	417	1521	
Salmo gairdneri (rainbow trout)					
Hemoglobin RT-1					
MetHb	0.39	—	628	1521	
	0.71	—	497	1521	
	13.7	—	406	1521	
	4.5	—	276	1521	
CarboxyHb	1.32	—	567	1521	
	1.39	—	536	1521	
	17.5	—	419	1521	
Hemoglobin RT-3					Dry wt
MetHb	0.35	—	628	1521	
	0.79	—	500	1521	
	13.6	—	405	1521	
	4.2	—	277	1521	
CarboxyHb	1.28	—	566	1521	
	1.34	—	536	1521	
	16.2	—	419	1521	
Chironomus plumosus, larvae haemolymph					
MetHbCN	1.28	4.0	540	1522	MW = 32,000 (1522)
Sheep					
Hb A, β chain	1.53	9.5	280	1523	MW = 16,134 (1523)
Hb B, β chain	1.46	9.0	280	1523	MW = 16,245 (1523)
Hb C, β chain	1.56	9.9	280	1523	MW = 15,788 (1523)
Beef					
MetHb	—	5.32	550	1524	FC
Horse					
Hb	13.5[n*]	—	430–1	1519	—
	1.37[n*]	—	555	1519	—
	59.6	—	406	1526	—
	—	16.8	280	1527	—
	—	630	191–4	1527	—

p*499–503 nm.

MOLAR ABSORPTIVITY AND $A_{1cm}^{1\%}$ VALUES FOR PROTEINS AT SELECTED WAVELENGTHS OF THE ULTRAVIOLET AND VISIBLE REGION (continued)

Protein	$\epsilon^a (\times 10^{-4})$	$A_{1cm}^{1\%}$ [b]	nm[c]	Ref.	Comments[d]
Hemoglobin (continued)					
Apo-	–	8.5	280	1528	–
	4.95	–	280	1529	Corrected for scattering
HbO$_2$	12.4[n*]	–	412–3	1519	–
	1.52[n*]	–	541	1519	–
	1.59[n*]	–	576	1519	–
HbCO	20.6[n*]	–	419–20	1519	See footnote q*
	1.49[n*]	–	538	1519	–
	1.50[n*]	–	568	1519	–
MetHb	66	–	406	1525	–
	13.9[n*]	–	405–6	1519	–
	0.97[n*]	–	500	1519	–
	0.43[n*]	–	630	1519	–
MetHbCN	9.5[n*]	–	420	1519	–
Mouse					
Hb	–	17.5	280	1530	–
Human					
Hb A	12.7	–	430	709	pH 6.4, 0.1 M P$_i$
Oxy	0.43	–	523	710	–
	1.57	–	542	710	–
	1.64	–	578	710	–
Deoxy	1.36	–	558	710	–
	0.27	–	583	710	–
Hb	3.14	–	275	1531	–
	11.8	–	430	1531	–
	1.29	–	552.5	1531	–
	0.0396	–	755	1531	–
Apo-	–	8.5	280	702	0.1 N NaOH
HbO$_2$	3.60	–	275	1531	–
	2.88	–	350	1531	–
	12.85	–	415	1531	–
	1.42	–	541.5	1531	–
	1.54	–	576	1531	–
	–	8.5	541	703	–
	5.6	–	540	706	–
	5.9	–	576	706	–
Hb FII	–	22.0	280	1532	
Hb AII	–	21.8	280	1532	From Figure 5 (1532)
Hb β_4	–	22.2	280	1532	
Hb γ_4	–	23.9	280	1532	
HbCO	6.63	–	429	1533	–
	20.8	–	419	1533	–
	17.4	–	419	1534	–
	1.43	–	540	702	–
	–	8.4	540	703	–
Hb-BuNC[r*]	18.75	–	429	1533	–
	9.65	–	419	1533	–
MesoHb	11.5	–	421	1535	–
	1.2	–	550	1535	–
MesoHbO$_2$	1.28	–	543	1535	–
	1.06	–	568	1535	–
MesoHbCO	21.0	–	410	1535	–
	1.39	–	532	1535	–
	1.26	–	560	1535	–

q*ϵ = 18.5–19.0 × 10^4 at 419 nm; 1.25–1.43 × 10^4 at 538 nm; 1.22–1.37 × 10^4 at 569 nm. See Reference 1534.
r*BuNC, butylisocyanide.

MOLAR ABSORPTIVITY AND $A_{1cm}^{1\%}$ VALUES FOR PROTEINS AT SELECTED WAVELENGTHS OF THE ULTRAVIOLET AND VISIBLE REGION (continued)

Protein	$\epsilon^a (\times 10^{-4})$	$A_{1cm}^{1\%\,b}$	nm^c	Ref.	Comments^d
Hemoglobin (continued)					
MesoHb$^+$	14.4	—	396	1535	
	0.89	—	495	1535	pH 7
	0.38	—	620	1535	
DeuteroHb	11.5	—	421	1535	—
	1.13	—	544	1535	—
DeuteroHbO$_2$	11.6	—	403	1535	—
	1.21	—	532	1535	—
	0.91	—	565	1535	—
DeuteroHbCO	20.0	—	408	1535	—
	1.22	—	528	1535	—
	0.93	—	556	1535	—
DeuteroHb$^+$	11.6	—	394	1535	
	0.71	—	500	1535	pH 7
	0.28	—	620	1535	
ChloroHb	13.0	—	443	1535	—
	1.85	—	567	1535	—
ChloroHbCo	14.0	—	432	1535	—
	1.70	—	550	1535	—
	2.00	—	590	1535	—
ChloroHb$^+$	13.0	—	421	1535	
	1.45	—	550	1535	pH 7
	1.30	—	598	1535	
HematoHb	11.3	—	423	1535	—
	1.33	—	552	1535	—
HematoHbCO	18.3	—	412	1535	—
	1.33	—	536	1535	—
	1.10	—	564	1535	—
HematoHb$^+$	15.0	—	398	1535	
	1.00	—	501	1535	pH 7
	0.41	—	628	1535	
Fetal (7% adult)	—	9.87	576	1536	
	—	10.0	541	1536	
	—	18.8	290	1536	pH 7, 0.02 M P$_i$
	—	24.0	280	1536	
	—	24.8	270	1536	
Adult	—	10.2	576	1536	
	—	9.6	541	1536	
	—	18.8	290	1536	pH 7, 0.02 M P$_i$
	—	23.3	280	1536	
	—	25.8	270	1536	
MetHbCn	4.350	—	540	1537	pH 7.4
	10.800	—	281	1537	
	4.388	—	540	1538	Fe det
	4.284	—	540	1539	Fe det
	4.360	—	540	1540	Nitrogen det
Cyanmethemoglobin	1.1	—	540	702	—
	—	20	280	704	—
	—	80	416	704	—
	4.59	—	541	705	—
	12.02	—	280	705	—
MetHb	—	5.97	540	708	—
No source cited					
60% MetHb	—	15.6	280	701	—
	—	330	210	701	—
	—	910	191	701	—

MOLAR ABSORPTIVITY AND $A_{1cm}^{1\%}$ VALUES FOR PROTEINS AT SELECTED WAVELENGTHS OF THE ULTRAVIOLET AND VISIBLE REGION (continued)

Protein	$\epsilon^a(\times 10^{-4})$	$A_{1cm}^{1\%}$ [b]	nm[c]	Ref.	Comments[d]
Hemoglobin chains					
α-HgBzO^{-S*}	11.1	—	428	709	
β-HgBzO$^-$	11.2	—	428	709	
α-HgBzO$^-$ + β-HgBzO$^-$	12.6	—	430	709	pH 6.4, 0.1 M P_i
α-SH^{t*}	11.1	—	429	709	
β-SH^{t*}	11.0	—	428	709	
α-SH + β-SH	12.4	—	430	709	
Hemoglobins, synthetic					
ProtoHb	14.0	—	430	1541	
	1.34	—	555	1541	
ProtoHbO$_2$	13.5	—	414	1541	pH 7.0, 0.1 M P_i
	1.48	—	541	1541	
	1.57	—	577	1541	
ProtometHb	15.4	—	406	1541	
	0.95	—	500	1541	pH 6.5, 0.1 M P_i
	0.38	—	630	1541	
DimethylprotoHb	14.0	—	430	1541	
	1.36	—	555	1541	
DimethylprotoHbO$_2$	13.3	—	414	1541	pH 7.0, 0.1 M P_i
	1.46	—	540	1541	
	1.53	—	577	1541	
DimethylprotoMetHb	15.1	—	405	1541	
	0.94	—	500	1541	pH 6.5, 0.1 M P_i
	0.39	—	630	1541	
MesoHb	13.5	—	421	1541	
	1.33	—	546	1541	
MesoHbO$_2$	13.9	—	404	1541	pH 7.0, 0.1 M P_i
	1.38	—	534	1541	
	1.24	—	567	1541	
MesometHb	18.5	—	395	1541	pH 6.5, 0.1 M P_i
	0.73	—	620	1541	
Deoxy Hb	—	4.25	524.5	707	—
DimethylmesoHb	13.5	—	421	1541	
	1.37	—	545	1541	
DimethylmesoHbO$_2$	13.8	—	404	1541	
	1.40	—	534	1541	
	1.30	—	568	1541	pH 7.0, 0.1 M P_i
EtioHb	13.4	—	420	1541	
	1.20	—	540	1541	
EtioHbO$_2$	13.8	—	401	1541	
	1.33	—	532	1541	
	1.15	—	566	1541	
EtiometHb	18.1	—	395	1541	
	1.32	—	490	1541	pH 6.5, 0.1 M P_i
	0.65	—	620	1541	
HematoHb	13.7	—	426	1541	
	1.42	—	551	1541	
HematoHbO$_2$	13.4	—	409	1541	
	1.47	—	537	1541	
	1.28	—	572	1541	pH 7.0, 0.1 M P_i
DeuteroHb	11.9	—	420	1541	
	1.29	—	542	1541	
DeuteroHbO$_2$	11.5	—	402	1541	
	1.35	—	531	1541	
	0.94	—	564	1541	

[s*]-HgBzO$^-$, paramercuribenzoate.

[t*]SH, free sulfhydryl group.

MOLAR ABSORPTIVITY AND $A_{1cm}^{1\%}$ VALUES FOR PROTEINS AT SELECTED WAVELENGTHS OF THE ULTRAVIOLET AND VISIBLE REGION (continued)

Protein	$\epsilon^a (\times 10^{-4})$	$A_{1cm}^{1\%}$ [b]	nm[c]	Ref.	Comments[d]
Human Hb, subunits					
αCO	—	8.4	540	1542	—
α-HgBzO$^-$-CO[u*]	1.4	—	540	1543	—
Ferric α-HgBzO^{-}[u*]	11.65	—	410	1544	ϵ/mole heme
	1.04	—	533	1544	
α(Mn^{2+})	1.19	—	585	1545	—
	1.23	—	570	1545	—
	2.03	—	555	1545	—
	1.09	—	535	1545	—
	15.7	—	430	1545	—
	10.0	—	420	1545	—
Apo-α	1.62	—	280	1545	—
βCO	—	8.4	540	1542	—
β-Fe^{2+}-CO	0.55	—	585	1545	—
	1.52	—	570	1545	—
	1.30	—	555	1545	—
	1.54	—	535	1545	—
	6.78	—	430	1545	—
	20.9	—	420	1545	—
Apoβ	1.62	—	280	1545	—
Hemoglobin-reductase complex, Candida-Mycoderma (yeast)	12	—	278	1546	pH 6.0, 0.1 M K P$_i$
	14	—	415	1546	
	17	—	420	1546	+ Na$_2$S$_2$O$_4$ + CO
	11	—	423	1546	+ Na$_2$S$_2$O$_4$
Hemopexin	13.5	16.9	280	246	MW = 80,000 (246)
Rabbit	—	21.8	280	1547	pH 7.1, 0.05 M KP$_i$
	—	19.2	414	1547	0.05 M KCl, dry wt
Rabbit blood	—	23.9	280	992	—
	—	26.4	280	992	Contains 1 heme
	—	23.2	413.5	992	
Apo-	—	19.7	280	1547	pH 7.1, 0.05 M KP$_i$, 0.05 M KCl, dry wt
Human blood	—	23.8	280	992	—
	—	26.4	280	992	Contains 1 heme
	—	23.0	414	992	
Hemoprotein					
Chironomus thummi	—	71.5	415	244	Ferric form, Calcd from equation in Reference 244, mg = absorbance × f where f = 0.14
559, Beef heart	6.95	—	400	245	pH 9.5, oxidized
	8.84	—	266	245	pH 9.5, oxidized
	6.94	—	394	245	pH 12, oxidized
	1.37	—	559	245	pH 9.5, reduced
	0.958	—	530	245	pH 9.5, reduced
	8.95	—	423	245	pH 9.5, reduced
	2.5	—	557	245	pH 12, reduced
	1.48	—	528	245	pH 12, reduced
	1.33	—	422	245	pH 12, reduced, also data for oxidized and KCN and reduced and CO in Reference 245

[u*]-HgBzO$^-$, *p*-mercuribenzoate.

MOLAR ABSORPTIVITY AND $A_{1cm}^{1\%}$ VALUES FOR PROTEINS AT SELECTED WAVELENGTHS OF THE ULTRAVIOLET AND VISIBLE REGION (continued)

Protein	$\epsilon^a (\times 10^{-4})$	$A_{1cm}^{1\%}$ [b]	nm[c]	Ref.	Comments[d]
Hepatocuprein					
Human liver	–	0.075	675	592	–
	–	5.87	265	592	–
	–	7.00	265	592	–
	–	5.6	278	593	pH 8.6
Hexokinase					
B, Yeast	–	9.16	280	247	pH 5.0, 5 mM, sodium succinate
	–	9.20	278	247	pH 5.0, 5 mM sodium succinate
Yeast	12.5	13	278	248	MW = 96,000 (248)
	10	9	280	1549	MW = 106,000 (1549) refractometry
Rat brain	5.1	–	280	1548	–
Hexosebisphosphatase (see fructose-1,6-bisphosphatase)					
High potential iron protein					
Chromatium, strain D	1.61	16.1	388	993	MW = 10,074 (993), reduced with HSEtOH
	1.86	18.6	450	993	MW = 10,074 (993),
	2.0	20.0	375	993	oxidized with ferricyanide,
	2.18	21.8	325	993	pH 7.0, 0.05 M P$_i$
	41.3	4.1	283	443	Reduced, MW = 10,074 (443)
	39.3	3.9	283	443	Oxidized, MW = 10,074 (443)
Rhodopseudomonas gelatinosa	1.53	16	388	993	MW = 9,579, pH 7.0, 0.05 M P$_i$ reduced with HSEtOH
	1.69	17.6	450	993	MW = 9,579
	1.88	19.6	375	993	pH 7.0, 0.05 M P$_i$,
	2.11	22.2	325	993	oxidized with ferricyanide
	35.4	3.7	283	443	Reduced, MW = 9,579 (443)
	33.8	3.5	283	443	Oxidized, MW = 9,579 (443)
Histaminase					
Pig plasma	–	9.2	278	994	Data from Figure 4 (994)
	–	0.169	470	994	
Histidine ammonia lyase					
Pseudomonas ATCC 11299b	–	5.6	280	995	Data from Figure 7 (995)
	–	4.0	280	995	Data from Figure 7 (995), NaBH$_4$ added
	–	3.9	280	995	Data from Figure 7 (995), cysteine added
Pseudomonas	–	5.0	280	996	pH 7.2, 0.1 M KP$_i$
	10.3	4.80	279	1550	MW = 215,000 (1550), dry wt
Histidine decarboxylase (EC 4.1.1.22)					
Lactobacillus 30a	–	16.1	280	249	pH 4.8, 0.2 M AcONH$_4$
	–	16.2	280	249	pH 8.0, 0.05 M NH$_4$HCO$_3$

MOLAR ABSORPTIVITY AND $A_{1cm}^{1\%}$ VALUES FOR PROTEINS AT SELECTED WAVELENGTHS OF THE ULTRAVIOLET AND VISIBLE REGION (continued)

Protein	$\epsilon^a (\times 10^{-4})$	$A_{1cm}^{1\%}$ b	nm^c	Ref.	Comments^d
Histidine decarboxylase (EC 4.1.1.22)					
Lactobacillus 30a (continued)	—	17.3	280	250	pH 4.8, 0.2 M AcONH$_4$
	6.3	16.1	280	997	MW = 38,800 (997)
Chain I	1.1	12.1	280	997	MW = 9,000 (997)
Chain II	5.1	17.2	280	997	MW = 29,700 (997)
Histidinol dehydrogenase (EC 1.1.1.23)					
S. typhimurium	3.6	4.8	280	251	pH 6, 0.05 M P$_i$, MW = 75,000 (251), data from Figure 3 (251)
S. typhimurium LT-2	3.8	4.78	280	252	pH 7.5, MW = 80,000 (253)
LT-7	—	4.63	280	254	pH 8.0, 0.01 M NH$_4$HCO$_3$
N. crassa	—	12.11	280	255	Water
Histidinol dehydrogenase — Histidinolphosphate amino transferase (EC 1.1.1.23, EC 2.6.1.9)					
Salmonella typhimurium LT 2 Strain TM 220	—	15.3	279	998	pH 7.5, 0.1 M triethanolamine ·HCl Data from Figure 9 (998)
Histone					
Chicken erythocyte	2.9	18.6	275	999	MW = 15,714 HCl der. used
Calf thymus F-8$_{2a}$	—	1.5	276	1000	—
F-8$_{b[z]}$	—	4.0	276	1000	—
F-6$_{3bb}$	—	3.8	276	1000	—
P-4$_4$C$_b$	—	4.5	276	1000	—
P-8$_{a[z]}$	—	4.6	276	1000	—
Histone f2					
Calf thymus	—	12	276	1552	pH 6.5
Histone IV					
Calf thymus	0.0047	—	230	1551	ϵ/mole of residue/l
Homoserine deaminase (cystathione γ-lyase, EC 4.4.1.1)					
Rat liver	—	6.64	280	1001	pH 7.5, 0.2 M KP$_i$
Homoserine dehydrogenase (EC 1.1.1.3)					
Rhodospirillum rubrum	—	3.85^l	280	1002	pH 7.5, 0.05 M KP$_i$ 0.05 M KCl, 0.001 M EDTA

MOLAR ABSORPTIVITY AND $A_{1cm}^{1\%}$ VALUES FOR PROTEINS AT SELECTED WAVELENGTHS OF THE ULTRAVIOLET AND VISIBLE REGION (continued)

Protein	$\epsilon^a (\times 10^{-4})$	$A_{1cm}^{1\%}$ [b]	nm[c]	Ref.	Comments[d]
Homoserine dehydrogenase − Aspartokinase (EC 1.1.1.3, EC 2.7.2.4)					
E. coli K-12	15.8	4.4	278	256	pH 7.2, 20 mM KP$_i$, 0.15 M KCl, 2 mM Mg titriplex and 4 mM DL-threonine
	16.5	4.6	278	1003	MW = 360,000 (1003)
Hormone					—
Chorionic gonadotropin, human	—	5.47	276	1553	—
	—	5.72	276	1553	Asialo der.
Chorionic somatomammotropin, human	—	8.22	277	1554	pH 8.2, 0.1 M Tris
Follicle-stimulating, human	—	4.40	250	1555	
	—	5.09	260	1555	—
	—	6.54	270	1555	—
	—	7.17	277	1555	—
	—	7.10	280	1555	—
	—	4.96	290	1555	—
Growth					
Beef pituitary	—	7.30	277	1556	0.1 N AcOH
Human pituitary	2	9.31	277	1557	pH 2.0−8.5, MW = 21,500 (1557)
Sheep pituitary	—	7.30	277	1556	0.1 N AcOH
Lactogen (MPL-2), monkey placenta	1.92	9.12	277	1558	MW = 21,000 (1558)
Lactogenic, sheep pituitary	—	9.09	277	1559	pH 8.0−8.5, dil NH$_4$OH
	—	8.94	278	1560	—
Interstitial cell stimulating, sheep	—	4.39	276	1561	—
α subunit	—	5.86	276	1560	—
β subunit	—	3.01	276	1561	—
Parathyroid					
Pig	0.53	5.6	280	1562	MW = 9423 (1562)
Beef	0.63	6.6	280	1562	MW = 9563 data from Figure 5 (562) 0.1 N AcOH
Prolactin					
Sheep	—	9.71	278	1563	—
Hyaluronidase inhibitor					
Human blood	—	8.5	280	1005	—
Hyaluronoglucuronidase (EC 3.2.1.36)					
Beef, testicular	—	9.6	280	1004	—
L-α-Hydroxyacid oxidase (EC 1.1.3.15)					
Rat liver	—	2.3	330	257	pH 7.9, 0.005 M NaP$_i$
	—	14.7	280	257	pH 7.9, 0.005 M NaP$_i$

MOLAR ABSORPTIVITY AND $A_{1cm}^{1\%}$ VALUES FOR PROTEINS AT SELECTED WAVELENGTHS OF THE ULTRAVIOLET AND VISIBLE REGION (continued)

Protein	$\epsilon^a(\times 10^{-4})$	$A_{1cm}^{1\%}$ [b]	nm[c]	Ref.	Comments[d]
p-Hydroxybenzoate hydroxylase (EC 1.14.99.13)					
Pseudomonas desmolytica	–	10.8	280	1008	Data from Figure on p. 331, Reference 1008
Pseudomonas putida	10.4	–	278	1009	pH 7.5, 0.05 M KP$_i$, data from Figure (1009), 2°C
Hydrogenase (EC 1.12.2.1)					
Desulfovibrio vulgaris	4.1	9.1	277	1006	–
Clostridium pasteurianum	0.82	–	400	1007	MW = 45,000 (1006)
(EC 1.12.7.1)	2.45	–	280	1007	–
Hydroxyindole-O-methyl-transferase (acetylserotonin methyltransferase, EC 2.1.1.4)					
Beef pineal gland					
Fraction A	6.2	7.68	280	1564	pH 7.7, 0.05 M TrisCl, data from Figure 6 (1564), MW = 81,000 (1564), SC
Fraction B	5.8	7.68	280	1564	pH 7.7, 0.05 M TrisCl, data from Figure 6 (1564), MW = 76,000 (1564), SC
L-6-Hydroxynicotine oxidase (EC 1.5.3.5)					
Arthrobacter oxidans	19.3	20.7	274	258	pH 7.5, 0.1 M P$_i$, wavelength est. from Figure 3 (258), MW = 93,000 (258)
Hydroxynitrilase (oxynitrilase, mandelonitrite lyase, EC 4.1.2.10)					
Prunaceae	11.2	15	275	259	pH 7.5, 0.1 M PP$_i$
Prunus communis Stokes	11.4	14.3	275	260	MW = 80,000 (260)
Prunoideae amygdalus	–	15.8	275	1565	pH 7.5, 0.1 M P$_i$, data from Figure 1 (1565)
Maloideae communis	–	18.3	275	1565	
Almonds	–	1.500	460	1010	–
	–	1.654	390	1010	–
Isoenzyme I	–	14.14	275	1010	–
Isoenzyme II	–	13.93	275	1010	–
Isoenzyme III	–	14.22	275	1010	–
Hydroxyproline 2-epimerase (EC 5.1.1.8)					
Pseudomonas putida	–	11.89	280	1011	Dry wt
Hydroxypyruvate reductase (EC 5.1.1.1.81)					
Pseudomonas acidovorans	6.6	8.8	280	261	pH 7.5, MW = 75,000 (261)

MOLAR ABSORPTIVITY AND $A_{1cm}^{1\%}$ VALUES FOR PROTEINS AT SELECTED WAVELENGTHS OF THE ULTRAVIOLET AND VISIBLE REGION (continued)

Protein	$\epsilon^a (\times 10^{-4})$	$A_{1cm}^{1\%}$ [b]	nm[c]	Ref.	Comments[d]
Imidazole acetate monooxygenase (EC 1.14.13.5)					
Pseudomonas ATCC 11299B	9.61	10.7	270	847	MW = 90,000 (847)
	1.08	1.2	383	847	
	1.07	1.19	442	847	
Imidazolylacetolphosphate: L-glutamate aminotransferase					
Salmonella typhimurium	–	9.54	279	1012	pH 6, 0.1 M TEA, pyridoxamine enzyme, data from Figure 3 (1012)
	6.49	11.8	295	262	0.1 N NaOH, est. from Figure 9 (262), MW = 59,000 (262)
	5.78	9.86	280	262	pH 7.5, 0.01 M Tris, MW = 59,000 (262)
Immunoglobulins					
Human					
Bence-Jones Protein	6.4	14.2	280	1013	MW = 45,000 (1013)
	2.7	12.2	280	1014	MW = 22,000 (1014)
Hac	–	10.5	280	1567	Neutral pH
Sch	–	10.5	280	1567	Neutral pH
Nu	–	13.0	280	1567	Neutral pH
MIg	–	14.5	280	1567	Neutral pH
λ chain	–	14.6	280	1568	–
Normal light chain	–	14.3	280	1568	–
½ normal light chain	–	14.5	280	1568	–
Constant half of λ chain	–	14.6	280	1568	–
Variable half of λ chain	–	14.6	280	1568	–
Variable fragment	1.3	11.6	280	1014	MW = 11,000 (1014)
L chain	–	9.8	280	1569	pH 6.8, 0.01 M P_i
J chain	–	6.5	280	1569	pH 6.8, 0.01 M P_i
H chain	–	10.7	280	1569	pH 6.8, 0.01 M P_i
IgG	–	13.3	278	1570	pH 7.4, 0.01 M NaP_i – 0.15 M NaCl
γ chain	–	13.8	282	1570	0.1 N NaOH
Light chain	–	11.0	282	1570	0.1 N NaOH
Heavy chain	7	14	280	1014	MW = 50,000 (1014)
Doty, type γ_1, K	–	14.10	280	1019	Dry wt
Sackfield	–	10.69	280	1019	Dry wt
Atypical	–	10.52	280	1018	–
Serum	21.6	13.3	277.5	1017	pH 2, MW = 162,000 (1017)
	–	13.42	280	595	–
IgG	21.2	13.8	280	1571	MW = 153,000 (1571)
IgA	21.7	13.4	280	1571	MW = 162,000 (1571)
Serum	–	13.4	280	1015	–
Secretory	–	13.9	280	1016	–
Colostrum	48.3	12.37	280	595	0.01 N HCl, MW = 390,000 (595)
IgE(PS)	–	12.5	280	1572	–
IgE(ND)	–	15.33	280	1572	–

MOLAR ABSORPTIVITY AND $A_{1cm}^{1\%}$ VALUES FOR PROTEINS AT SELECTED WAVELENGTHS OF THE ULTRAVIOLET AND VISIBLE REGION (continued)

Protein	$\epsilon^a (\times 10^{-4})$	$A_{1cm}^{1\%}$ [b]	nm[c]	Ref.	Comments[d]
Immunoglobulins (continued)					
IgM	–	14.5	280	1573	–
Serum	–	13.3	280	1020	–
Waldenstrom	–	13.5	280	600	0.25 M AcOH
IgG	–	13.5	280	1574	–
γ chain	–	14.0	280	1574	–
Light chain	–	12.0	280	1574	–
IgND, Myeloma	–	14	280	1021	–
Kappa chain					
urine	1.8	10.7	280	594	MW = 17,000 (594)
Chicken					
Anti-N_2 ph[v*]	–	17.7	280	1022	Neutral buffer
	–	24.6	290	1022	0.1 N NaOH
	–	13.0	290	1022	0.1 N NaOH, another sample
Pig					
Anti-N_2 ph-BGG[w*] λ chain	–	11.3	280	1023	–
π chain	–	9.8	280	1023	–
ρ chain	–	12.6	280	1023	From antibody
ρ chain	–	12.2	280	1023	Non-specific
γ chain	–	14.2	280	1023	–
Light chain fraction	–	11.3	280	1023	From antibody
	–	11.0	280	1023	Non-specific
Epinephelus itaiva					
(Giant grouper)					
IgG 6.4S	–	16.57	280	1024	0.3 M KCl
	–	17.82	280	1024	0.1 M NaOH
	–	16.50	280	1024	5.0 M GdmCl
IgG 16S	–	13.78	280	1024	0.3 M KCl
	–	15.03	280	1024	0.1 N NaOH
	–	13.53	280	1024	5.0 M GdmCl
Lepisosteus osseus [Gar]					
IgG 14S	–	15	280	1025	0.85% NaCl
Rabbit	21.9	14.6	280	1026	pH 8, Tris,
	8.98	6.0	250	1026	MW = 150,000 (1026)
IgG	–	14.0	280	1027	–
IgM	–	13.0	280	1029	–
Anti-N_2 ph[v*]	23.4	14.6	279	1030	MW = 160,000 (1030)
Anti azobenzenearsonate					
IgG	–	14.6	280	1031	–
IgM	–	13.4	280	1031	–
Polyalanylated Ab	–	14.6	280	1032	–
Goat					
IgG	18.7	13.0	280	1033	pH 7.0, MW = 144,000 (1033)
	–	14	280	1034	
	20.2	13.1	280	1579	pH 7.2, 5 mM NaP_i, 0.2 M NaCl, MW = 146,000 (1579)
Heavy chain	6.4	12.0	280	1033	pH 3.0 0.1 M AcOH MW = 53,600 (1033)
Light chain	2.9	12.7	280	1033	MW = 23,000 (1033)
IgM	–	13	280	1034	–

[v*] N_2 ph, dinitrophenyl.
[w*] N_2 ph-BGG, dinitrophenyl bovine γ-globulin.

MOLAR ABSORPTIVITY AND $A_{1cm}^{1\%}$ VALUES FOR PROTEINS AT SELECTED WAVELENGTHS OF THE ULTRAVIOLET AND VISIBLE REGION (continued)

Protein	$\epsilon^a (\times 10^{-4})$	$A_{1cm}^{1\%\ b}$	nm[c]	Ref.	Comments[d]
Immunoglobulins (continued)					
Horse	–	14.4	277	1580	–
IgG [Anti-tick borne encephalitis]	–	15.06	278	1034	–
	–	1.37	330	1035	–
	–	1.27	333	1035	–
	–	1.22	335	1035	–
	–	1.20	338	1035	–
	–	1.18	340	1035	–
Heavy chain, reduced and alkylated	–	14.3	280	1036	–
Light chain, reduced and alkylated	–	12.7	280	1036	–
Anti-polysaccharide	–	15.0	287	1566	0.1 N NaOH
Type II	–	14.6	280	1566	pH 8.0, borate-NaCl
Mouse, myeloma protein with antibody activity					
IgA, MOPC-315 tumor	–	13.5	275	1578	pH 7.4, 0.01 M KP$_i$ – 0.15 – NaCl
Fraction III	–	14	278	1037	–
7S Monomer	15	12.5	278	1037	MW = 120,000 (1037)
Fab fragment	7.7	14.0	278	1037	MW = 55,000 (1037)
Rabbit					
IgG	–	13.5	280	1574	–
γ chain	–	14.0	280	1574	–
Light chain	–	12.0	280	1574	–
IgA	–	13.5	280	1575	0.1 N NaOH
Anti-benzylpenicilloyl-	–	13.8	280	1576	Saline + 0.25 N AcOH
	–	15.4	294	1576	0.1 N NaOH
Anti-N$_2$ ph^{V*}	–	15.5	280	1577	Dry wt
Heavy chain	–	15.4	280	1577	Refractometry
Light chain	–	13.2	280	1577	
Zebra	–	15.2	277	1580	–
Donkey	–	13.2	277	1580	–
Mule	–	14.7	277	1580	–
Hinny	–	14.1	277	1580	–
Dog					
Colostral IgA	–	11.80	280	1581	pH 7.4, phosphate buffered saline, FC
α chain	–	10.07	280	1581	6 M GdmCl, FC
L chain	–	7.32	280	1581	
Serum IgA	–	14.08	280	1581	pH 7.4, phosphate buffered saline, FC
α chain	–	10.63	280	1581	6 M GdmCl, FC
L chain	–	8.37	280	1581	
Beef					
IgG	–	13.7	280	1582	–
Sheep					
IgA	–	12	280	596	
Rat					
IgG	–	14.6	280	599	
IgM	–	12.5	280	599	
Guinea pig					
γM	–	10.5	280	848	
γ$_2$G	–	12.3	280	848	0.5% (NH$_4$)$_2$CO$_3$
γ$_1$G	–	13.0	280	848	
γ$_2$ chain	–	13.2	280	848	0.1 M AcOH
γ$_1$ chain	–	13.5	280	848	

MOLAR ABSORPTIVITY AND $A_{1cm}^{1\%}$ VALUES FOR PROTEINS AT SELECTED WAVELENGTHS OF THE ULTRAVIOLET AND VISIBLE REGION (continued)

Protein	$\epsilon^a (\times 10^{-4})$	$A_{1cm}^{1\%}$ [b]	nm[c]	Ref.	Comments[d]
Immunoglobulins (continued)					
μ chain	—	9.2	280	848	0.1 M AcOH
Light chain	—	12.0	280	848	
F ab [trypsin]	—	15	280	597	
F c [trypsin]	—	15	280	597	
IgGs, colostrum	—	13.68	278	598	P_i buffered saline
	—	13.70	278	598	0.1 N HCl
	—	14.77	290	598	0.1 N NaOH, after 10–20
	—	14.73	283	598	min
IgG$_1$, serum	—	13.57	278	598	P_i buffered saline
	—	13.44	278	598	0.1 N HCl
	—	14.77	290	598	0.1 N NaOH, after 10–20
	—	14.65	283	598	min
IgG$_2$, serum	—	13.52	278	598	P_i buffered saline
	—	13.32	278	598	0.1 N HCl
	—	14.87	290	598	0.1 N NaOH, after 10–20
	—	14.73	283	598	min
Immunoglobulins, specific					
Rabbit anti-HGG[x*]	—	14.4	279	263	0.25 M acetic acid
Anti-BSA[y*]	—	15.0	279	263	0.25 M acetic acid
Anti-N$_2$ph[v*]	—	15.8	278	263	pH 7.4, 0.15 M NaCl, 0.02 M P$_i$
Anti-N$_2$ph[v*] pepsin frag.	—	16.9	278	263	pH 7.4, 0.15 M NaCl, 0.02 M KP$_i$
Anti-N$_2$ph[v*]	—	16.8	278	263	pH 7.4, 0.15 M NaCl, 0.02 M KP$_i$
Anti-N$_2$ph[v*] Fab	—	16.5	278	263	pH 7.4, 0.15 M NaCl, 0.02 M KP$_i$
Anti-N$_2$ph,[v*] Fc	—	13.5	278	263	pH 7.4, 0.15 M NaCl, 0.02 M KP$_i$
Anti-N$_2$ph,[v*] pepsin frag.	—	18.1	278	263	pH 7.4, 0.15 M NaCl, 0.02 M KP$_i$
Anti-N$_2$ph,[v*]	—	15.4	278	263	pH 7.4, 0.04 M KP$_i$
	—	15.7	278	263	pH 7.4, 0.04 M KP$_i$
	—	15.5	278	263	pH 7.4, 0.04 M KP$_i$
	—	15.6	278	263	pH 7.4, 0.04 M KP$_i$
	—	16.4	278	263	pH 7.4, 0.04 M KP$_i$
	—	13.6	278	263	pH 7.4, 0.15 M NaCl, 0.01 M P$_i$
	—	5.3	251	263	pH 7.4, 0.15 M NaCl, 0.01 M P$_i$
γG-Anti-N$_2$ph[v*]	—	16.2	278	263	pH 7.4, 0.04 M KP$_i$
	—	16.0	278	263	pH 7.4, 0.04 M KP$_i$
	—	15.9	278	263	pH 7.4, 0.02 M KP$_i$
γG-Anti-N$_3$ph[z*]	—	15.8	278	263	pH 7.4, 0.15 M NaCl 0.02 M KP$_i$
	—	14.9	278	263	pH 7.4, 0.15 M NaCl, 0.02 M KP$_i$
	—	15.3	278	263	pH 7.4, 0.15 M NaCl, 0.02 M KP$_i$
Anti-p-azobenzenearsonate	—	14.8	278	263	pH 7.4, 0.04 M KP$_i$
Light chains	—	12.8	278	263	pH 8.0, 0.05 M, Na dodecyl sulfate

[x*]HGG, human gamma globulin.
[y*]Albumin, bovine serum albumin.
[z*]N$_3$ph, trinitrophenyl.

MOLAR ABSORPTIVITY AND $A_{1cm}^{1\%}$ VALUES FOR PROTEINS AT SELECTED WAVELENGTHS OF THE ULTRAVIOLET AND VISIBLE REGION (continued)

Protein	$\epsilon^a (\times 10^{-4})$	$A_{1cm}^{1\%}$ [b]	nm[c]	Ref.	Comments[d]
Immunoglobulins, specific (continued)					
γG-Anti-phenyl[p-aminobenzoylamino]acetate	–	13.9	279	263	pH 7.4, 0.15 M NaCl, 0.02 M P_i
Horse, γG-Anti-lac	–	14.7	280	263	Neutral solvent
γA-Anti-lac	–	14.7	280	263	Neutral solvent
Pepsin fragment	–	14.6	280	263	Neutral solvent
γG, Human	–	14.3	280	263	pH 7.5, 0.2 M NaCl
γ_1-globulin, Human	–	14.7	280	263	pH 6.0, 0.1 M NaCl
γM, Human	–	11.85	280	263	pH 7.5, 0.2 M NaCl
γM, Human, subunit	–	12.0	280	263	pH 7.5, 0.2 M NaCl
γG, Horse	–	13.8	280	263	pH 6.5, 0.0175 M NaP_i
	–	13.8	280	263	8 M urea, neutral solution
Heavy chains	–	15.4	280	263	0.04 M $NaDodSO_4$
	–	15.2	280	263	1 N propionic acid
Light chains	–	14.0	280	263	0.04 M $NaDodSO_4$
	–	13.6	280	263	1 N propionic acid
γG, Rabbit	–	14.6	280	263	–
	–	15.0	278	263	pH 7.4, 0.04 M KP_i
	–	15.4	278	263	pH 7.4, 0.04 M KP_i
	–	14.5	278	263	pH 7.4, 0.04 M KP_i
	–	15.1	278	263	pH 7.4, 0.04 M KP_i
	–	14.7	278	263	pH 7.4, 0.04 M KP_i
	–	14.9	278	263	pH 7.4, 0.04 M KP_i
	–	14.6	278	263	pH 7.4, 0.15 M NaCl, 0.02 M P_i
Heavy chains	–	13.5	280	263	0.01 N HCl
Light chains	–	13.7	280	263	0.01 N HCl
Fd fragment	–	11.8	280	263	0.01 N HCl
	–	14.4	280	263	0.01 N HCl
Heavy chains, mildly reduced and alkylated	–	14.5	280	263	pH 7.2, 0.04 M $DodSO_4^-$ 0.01 M P_i
Light chains mildly reduced and alkylated	–	13.2	280	263	pH 7.2, 0.04 M $DodSO_4^-$ 0.01 M P_i
γG, Rabbit	–	13.6	280	263	pH 7, 5 M GdmCl
Heavy chains	–	13.7	280	263	pH 7, 5 M GdmCl
Light chains	–	11.4	280	263	pH 7, 5 M GdmCl
Fab fragment	7.5	15.0	278	263	MW=50,000 (263)
γG, Rabbit	–	13.8	278	263	–
Fab fragment	–	15.3	278	263	–
Fc fragment	–	12.2	278	263	
Fab fragment	–	15.0	280	263	pH 7.5, 0.1 M P_i
5S pepsin fragment	–	14.8	280	263	–
γA, Rabbit, colostrum	–	13.5	280	263	0.1 N NaOH
	–	12.8	280	263	5 M GdmCl
α-chains	–	10.6	280	263	5 M GdmCl
γ_1, Guinea pig	–	15	278	263	
γ_2, Guinea pig	–	13.2	278	263	–
γ-globulin fraction, chicken	–	13.5	280	263	–
γG, Lemon shark	–	13.85	280	263	0.3 M KCl
	–	14.04	280	263	0.1 N NaOH
	–	12.82	280	263	5 M GdmCl

MOLAR ABSORPTIVITY AND $A_{1cm}^{1\%}$ VALUES FOR PROTEINS AT SELECTED WAVELENGTHS OF THE ULTRAVIOLET AND VISIBLE REGION (continued)

Protein	$\epsilon^a (\times 10^{-4})$	$A_{1cm}^{1\%}$ [b]	nm[c]	Ref.	Comments[d]
Immunoglobulins, specific (continued)					
γM, Lemon shark	–	13.39	280	263	0.3 M KCl
	–	13.75	280	263	0.1 N NaOH
	–	12.79	280	263	5 M GdmCl
γG Heavy chains	–	11.74	280	263	5 M GdmCl
γG Light chains	–	13.1	280	263	5 M GdmCl
Indole-3-glycerophosphate synthase (EC 4.1.1.48)	3.6	8	280	264	pH 7.0, 5 mM P_i
	4.3	9.5	280	264	0.1 N NaOH, est. from Figure 8 (264)
Inhibitor					
Amylase, *Colocasia esculenta*	–	10.7	280	1583	pH7
Phospholipase A, *Bothrops neuwiedii* (snake) venom	–	9.09	280	1584	–
	–	16.36	260	1584	–
	–	201.60	230	1584	–
Proteinase, potato	–	9.18	280	1585	–
Proteinase IIa, potato	–	10.03	278	1586	pH 6, 0.1 M NaCl, dry wt
Proteinase IIb, potato	–	10.06	278	1586	
Trypsin, *Phaseolus aureus* Roxb. (mungbean)					
Type A	–	3.7	280	1587	–
Type B	–	3.7	280	1587	–
Trypsin-chymotrypsin *Arachis hypogaea* (groundnut)	0.1958	2.5	280	1038	MW=7832 (1038)
Protease					
Barley	–	8.82	280	1039	–
Insulin	–	10.4	276	711	–
	0.553	–	277.5	712	–
	–	10.3	275	713	0.01 N HCl
	–	10.52	276	714	pH 7.0, 0.03 M P_i
	0.608	–	277	849	
Beef	0.57	10	280	265	MW=5733 (265)
	–	10.4	277	11	–
	0.61	10.6	278	266	pH 7.0, 0.025 M P_i, MW=5734 (266)
	0.52	–	280	267	–
	–	9.91	276	1588	pH 7.2, 0.01 M NaP_i–0.1 N NaCl
Crystalline	0.6740	–	276	1589	–a†
$N^{\alpha A1}$-Acetyl-	0.6110	–	276	1589	–a†
	0.5790	–	276	1589	–b†
I_{A-a}	0.6480	–	276	1589	–a†
Amorphous	0.6870	–	276	1589	–a†
	0.6230	–	276	1589	–b†
$N^{\epsilon B29}$-Acetyl-	0.6630	–	276	1589	–a†
	0.6180	–	276	1589	–b†
$N^{\alpha A1}, N^{\epsilon B29}$-Diacetyl-	0.6440	–	275	1589	–a†
	0.6020	–	275	1589	–b†

a†Concentration calculated from the optical density at 210 nm.
b†Weighed sample used.

MOLAR ABSORPTIVITY AND $A_{1cm}^{1\%}$ VALUES FOR PROTEINS AT SELECTED WAVELENGTHS OF THE ULTRAVIOLET AND VISIBLE REGION (continued)

Protein	$\epsilon^a (\times 10^{-4})$	$A_{1cm}^{1\%}$ b	nm^c	Ref.	Comments^d
Insulin (continued)					
$\qquad$ N$^{\alpha A1}$, N$^{\alpha B1}$, N$^{\epsilon B29}$-	0.6390	–	275	1589	_a†
$\qquad\qquad$ Triacetyl-					
	0.6270	–	275	1589	_b†
$\qquad$ Diacetyl-	0.3420	–	264	1589	_a†
	0.3020	–	264	1589	_b†
$\qquad$ Triacetyl-	0.3460	–	264	1589	_a†
	0.3480	–	264	1589	_b†
$\qquad$ β-chain	0.31	–	276	268	–
Interferon					
$\qquad$ Chick embryo	–	8.6	280	269	–
Intrinsic factor, pig pylorus	–	9.2	278	1590	Dry wt
Invertase (β-fructofuranosidase, EC 3.2.1.26)					
$\qquad$ Neurospora crassa	–	18.6	280	850	–
	–	18.6	280	270	–
$\qquad$ Yeast	62.1	23	280	271	MW=270,000 (271)
Isoamylase (EC 3.2.1.68)					
$\qquad$ Pseudomonas sp. str Sb-15	–	22.6	280	851	pH 4.5, 0.01 M AcO$^-$
Isocitrate dehydrogenase (EC 1.1.1.41-.42)					
$\qquad$ Saccharomyces cerevisiae	–	6.9	280	1591	–
$\qquad$ (baker's yeast)	–	3.5	260	1591	–
$\qquad$ Pig heart	5.3	9.1	280	272	MW=58,000 (272)
$\qquad$ Pig liver					
$\qquad\qquad$ Cytoplasm	4.73	12.6	280	716	MW=37,500 (716)
$\qquad$ Acobacter vinelandii	7.12	8.9	280	715	MW=80,000 (715)
Isocitrate lyase (EC 4.1.3.1)					
$\qquad$ P. indigofera	38	17.1	280	852	MW=222,000 (852)
Isomerase [N-(5-phospho-D-ribosylformimino)-5-amino-(5″-phosphoribosyl)-4-imidazolecarboxamide isomerase, EC 5.3.1.16]					
$\qquad$ S. typhimurium	3.3	11.4	280	273	pH 8, 0.05 M TrisCl
C$_{55}$-Isoprenoid alcohol phosphokinase	3.188	18.7	280	1592	MW=17,000 (1592)
$\qquad$ Staphylococcus aureus	2.200	12.9	288	1592	
β-Isopropylmalate dehydrogenase (EC 1.1.1.85)					
$\qquad$ Salmonella typhimurium	5.34	7.63	278	853	MW=70,000 (853)
Kallikrein (kininogenin, EC 3.4.21.8)	–	16.6	280	1593	pH 7 P$_i$ buffer, dry wt
$\qquad$ Pig pancreas	4.92	20.5	280	1594	MW=24,000 (98)
Kallikrein A,					
$\qquad$ Cinnamoyl-	1.80	–	298	1594	pH 8.8
$\qquad$ Indoleacrylyl-	2.24	–	353	1594	

MOLAR ABSORPTIVITY AND $A_{1cm}^{1\%}$ VALUES FOR PROTEINS AT SELECTED WAVELENGTHS OF THE ULTRAVIOLET AND VISIBLE REGION (continued)

Protein	$\epsilon^a (\times 10^{-4})$	$A_{1cm}^{1\%}$ [b]	nm[c]	Ref.	Comments[d]
Kallikrein B					
Cinnamoyl-	1.80	—	298	1594	pH 8.8
Indoleacrylyl-	2.17	—	353	1594	
Kallikrein inactivator					
Lung	5.5	8.4	276	274	MW=6511 (274)
Kerateine fractions					
Wool, S-carboxymethyl-					
SCMKA2	—	8.6	276	1595	—
SCMKB1	—	5.5	276	1595	—
SCMKB2	—	5.9	276	1595	—
Feather rachis, S-carboxymethyl-					
SCMK	—	7.0	276	1595	—
Keratinase	—	8.4	280	275	pH 8, 0.1 M Tris
S. fradiae (EC 3.4.99.8)	—	10.42	280	276	pH 5.0, AcO⁻
Trichophyton mentagrophytes (EC 3.4.99.12)	—	13.2	280	854	—
Keratinase (EC 3.4.99.11) conjugate					
S. fradiae	—	58.1	280	277	—
Keratins, zinc precipitatable fractions					
Hair					
Lincoln sheep	—	6.6	277	1596	—
Pig	—	6.5	277	1596	—
Cattle	—	8.3	277	1596	—
Macaca irus (monkey)	—	8.7	277	1596	—
Lama glauca (llama)	—	9.8	277	1596	—
Guinea pig	—	9.9	277	1596	—
Merino sheep	—	10.2	277	1596	—
Rat	—	11.7	277	1596	—
Mouse	—	14.2	277	1596	—
Horny keratin					
Diceros bicornis (rhinocerous) horn	—	6.4	277	1596	—
Fingernail	—	7.0	277	1596	—
Sheep horn	—	7.6	277	1596	—
Cattle horn	—	8.0	277	1596	—
Sheep hoof	—	8.8	277	1596	—
Erethizon dorsatum (porcupine) quill	—	9.3	277	1596	—
Balaenoptera muscelus (whale) baleen	—	12.1	277	1596	—
Erinaceus europaeus (hedgehog) quill	—	12.3	277	1596	—
Histrix cristata (porcupine) quill	—	15.3	277	1596	—
Tachyglossus aculeatus aculeatus (echidna) quill	—	20.4	277	1596	—

MOLAR ABSORPTIVITY AND $A_{1cm}^{1\%}$ VALUES FOR PROTEINS AT SELECTED WAVELENGTHS OF THE ULTRAVIOLET AND VISIBLE REGION (continued)

Protein	$\epsilon^a (\times 10^{-4})$	$A_{1cm}^{1\%}$ [b]	nm[c]	Ref.	Comments[d]
α-Keto acid dehydrogenase complex					
Pig heart	–	5.1	280	278	pH 7, 0.05 M KP$_i$, est. from Figure 6 in Reference 278,
3-Keto-Δ5-steroid isomerase					
P. testosteroni	–	4.13	280	279	–
Kininogenin (see kallikrein)					
Lac repressor, *E. coli*	–	6.9	280	1569	pH 6.8, 0.01 M P$_i$
Laccase (monophenol monooxygenase, EC 1.14.18.1)					
Rhus vernicfera	0.26	–	330	856	–
	0.52	–	614	856	–
	0.09	–	788	856	–
Oxidized	9.35	–	280	717	–
Reduced	0.57	–	614	717	–
	0.28	–	333	717	–
	0.55	–	614	717	–
Polyporus versicolor	8.4	13.7	280	718	MW=61,000 (718)
	0.41	–	610	718	–
	0.33	–	330	858	–
	0.08	–	440	858	–
	0.46	–	610	858	–
	0.2	–	720	858	–
Rhus succedanea	0.32	–	325	857	–
	0.45	–	610	857	–
	0.12	–	770	857	–
Fungal	7.4	11.6	280	859	MW=64,000 (859)
	7.44	–	280	1040	
α-Lactalbumin	–	20.9	280	280	–
	–	20.1	280	281	–
Human	2.04	–	280	1041	–
	–	16.2	280	1597	–
	–	19.0	–[c†]	1598	–
American Indian	–	14.1	280	1042	–
Caucasian	–	15.3	280	1042	–
Negro	–	15.2	280	1042	–
Japanese	–	15.0	280	1042	–
Beef	–	20.1	280	1042	–
	–	20.5	280	601	–
A, Droughtmaster	–	20.2	281.5	1043	–
B, Droughtmaster	–	20.9	281.5	1043	–
Cow's milk	2.9	–	280	719	–
	–	20.9	280	720	20 mM Tris, pH 7.4
Carboxyl modified with glucineamide	3.0	–	280	719	–
Goat	–	17.3	280	1042	–
Sheep	–	16.7	280	1042	–
Pig	–	18.1	280	1042	–
Guinea pig	–	16.7	–[c†]	1598	–
Camel	–	19.0	–[c†]	1598	–

[c†] No wavelength cited.

MOLAR ABSORPTIVITY AND $A_{1cm}^{1\%}$ VALUES FOR PROTEINS AT SELECTED WAVELENGTHS OF THE ULTRAVIOLET AND VISIBLE REGION (continued)

Protein	$\epsilon^a (\times 10^{-4})$	$A_{1cm}^{1\%}$ [b]	nm[c]	Ref.	Comments[d]
β-Lactamase II (cephalosporinase, EC 3.5.2.8)					
Bacillus cereus 569H	—	8.7	277	1044	MW = 22,500
Hp	2.41	10.7	—[c†]	1599	MW = 22,500 refractometry
β-Lactamase I B. cereus					
569/H	2.44	8.76	—[c†]	1599	MW = 27,800 (1599), refractometry
Lactate dehydrogenase (EC 1.1.1.27 and/or .28)					
Beef					
Heart	—	14.55	280	1600	5 M GdmCl, Kj
	21	15	280	284	—
	20	—	280	285	—
	—	15	280	1050	—[e†]
	—	13.8	280	1051	—
	—	14.2	280	1052	—
	—	14.5	280	1053	—
	—	14.9	280	1054	—
H$_1$ [92%]	—	14.6	280	1048	—
H$_2$M$_2$ [80%]	—	14.4	280	1048	—
Skeletal muscle	20	—	280	285	—
Muscle	18.1	12.9	280	284	—
Isozyme A	21	15.6	280	286	MW = 134,000 (286)
Isozyme B	19.6	14.5	280	286	MW = 134,000 (286)
Duck, M$_4$	18.9	13.5	280	287	MW = 140,000 (287)
Turkey, M$_4$	20.4	14.6	280	287	MW = 140,000 (287)
Pheasant, M$_4$	19.1	13.7	280	287	MW = 140,000 (287)
Ostrich, M$_4$	18.6	13.3	280	287	MW = 140,000 (287)
Rhea, M$_4$	18.1	13	280	287	MW = 140,000 (287)
Halibut, M$_4$	18.9	13.5	280	287	MW = 140,000 (287)
Skeletal muscle	20	—	280	285	—
Bullfrog, M$_4$	19.3	13.8	280	287	MW = 140,000 (287)
Tuna, M$_4$	17.5	12.5	280	287	MW = 140,000 (287)
Dogfish, M$_4$	20.8	14.8	280	287	MW = 140,000 (287)
	19.7	—	280	287	Using Kjeldahl N
Chicken heart muscle	18	—	280	285	—
	19	13.6	280	284	—
Skeletal muscle	22	—	280	285	—
	21.8	15.6	280	284	—
Rabbit	20	—	280	285	—
M$_4$	—	14.4	280	286	—
	20.1	—	280	287	—
Muscle	16.2	12.3	280	288	pH 7.6, 0.1 M NaP$_i$, MW = 132,000 (288)
	16.7	12.6	280	289	0.2 N NaOH, MW = 132,000 (289)
	—	14.0	280	141	—
Skeletal muscle					
V[96%]	20.68	14.6	279	1049	IV[2%], II/III [0.5%], I [1.5%], MW = 142,000 (1049)
	—	8.9	280	1050	—[d†]

[d†] mg protein/ml = (1.13) (OD$_{280}$).
[e†] mg/ml = (0.67) (OD$_{280}$).

MOLAR ABSORPTIVITY AND $A_{1cm}^{1\%}$ VALUES FOR PROTEINS AT SELECTED WAVELENGTHS OF THE ULTRAVIOLET AND VISIBLE REGION (continued)

Protein	$\epsilon^a (\times 10^{-4})$	$A_{1cm}^{1\%}$ [b]	nm[c]	Ref.	Comments[d]
Lactate dehydrogenase (continued)					
Pig heart	–	12.9	280	1045	0.1 NaOH, coenzyme free
	–	14.5	280	1046	–
	–	151	215	1047	0.9% NaCl
M₄ [96%]	–	14.0	280	1048	–
I [97%]	19.59	13.8	279	1049	II [2.5%], III [0.5%], MW = 142,000 (1049)
Heart [H₄]	–	13.7	280	286	–
Human	–	14.6	280	290	pH 7, P_i
Heart	–	16.4	280	1055	–
Uterus	18.7	12.3	280	295	MW = 152,000 (295)
Uterine myoma	17.7	12.4	280	295	MW = 143,000 (295)
Rat liver	–	12.6	280	291	MW = 132,000 (292)
	–	12.8	280	1058	–
Jensen sarcoma	–	11.7	280	1059	pH 7.4
Yeast	23.2	–	424	293	–
	–	29	423	1056	–
E. coli B	–	0.727[f†]	280	294	–
Homarus americanus	21.8	15.6	280	1057	MW = 140,000 (1057)
Lactate-malate dehydrogenase (see malate-lactate transhydrogenase)					
L-Lactate oxidase (lactate 2-monooxygenase, EC 1.13.12.4)					
M. smegmatis	–	21.9	280	282	pH 7.0, NaP_i
	–	2.62	452	282	–
M. phlei	81.5	20.4	280	283	pH 7.0, 0.1 $M P_i$, MW = 399,000 (283)
	–	2.26	454	283	pH 7.0, 0.1 $M P_i$
Lactoferrin					
Beef	–	14.5	280	1060	–
	–	11.3	280	1060	6 M GdmCl
Cow's milk	–	15.1	280	306	–
	–	0.547	450	307	–
	–	0.460	470	1603	pH 7
Apo-	–	12.7	280	1603	–
Human milk	–	0.540	465	602	pH 8.2
	11.096	–	280	1601	–
	–	14.6	280	1602	–
	–	0.510	470	1603	pH 7
Apo-	8.512	–	280	1601	–
	–	11.2	280	1602	–
	–	10.9	280	1603	–
β-Lactoglobulin					
Beef	–	9.5	280	296	–
	–	9.66	278.5	297	–
	–	9.7	280	298	–
	3.66	–	280	299	–
	–	9.6	278	300	pH 5.3, ionic strength 0.1, AcO⁻

[f†]This may be in error. Value should be 7.27.

MOLAR ABSORPTIVITY AND $A_{1cm}^{1\%}$ VALUES FOR PROTEINS AT SELECTED WAVELENGTHS OF THE ULTRAVIOLET AND VISIBLE REGION (continued)

Protein	$\epsilon^a (\times 10^{-4})$	$A_{1cm}^{1\%}$ b	nm^c	Ref.	Comments^d
β-Lactoglobulin (continued)					
β-A, β-B	–	9.6	278	301	–
β-C	–	9.5	278	301	–
β-A, β-B	–	9.4	278	302	–
β-1	–	9.5	290	303	0.1 N NaOH
	–	9.5	282	303	0.1 N NaOH
β-2	–	9.9	288	303	0.1 N NaOH
Sheep, β-A	–	9.2	278	304	0.1 M NaCl
β-B	–	8.35	278	304	0.1 M NaCl
Goat	–	9.4	278	305	–
Pig's milk	1.05	5.65	280	603	0.1 M NaOH
Buffalo	–	9.4	279	1061	–
Cow's milk	–	5.12	255	1604	⎫
	–	10.80	280	1604	⎬ pH 7.0, 0.01 M P$_i$
β-Lactoglobulin-B	–	1.00	310	1604	⎭
Cow's milk	3.5	10.0	280	1062	MW = 35,000 (1062)
MalNEt^{g†}	3.5	10.0	280	1062	–
ClHgBzO$^{-h\dagger}$	3.7	10.56	280	1062	–
Lactollin					
Cow's milk	7.1	16.5	280	1063	MW = 43,000 (1063)
Lactonase (gluconolactonase) (EC 3.1.1.17)					
Actinoplanes missouriensis	–	4	280	1064	pH 7.0, 0.07 M KP$_i$, data from Figure 8 (1064)
Lactoperoxidase (peroxidase, EC 1.11.1.7)					
Cow's milk					
B-1	–	14.9	280	1065	–
B-2$_I$	–	15.0	280	1065	–
B-2$_{II}$	–	14.9	280	1065	–
B-3	–	14.9	280	1065	–
A	–	15.5	280	1065	–
Cow's milk	–	15.2	280	1066	–
	–	15.41	280	1067	–
	–	1.37	497	1067	–
Cow's milk	11.6	–	412	604	⎫
+ hydrogen peroxide	8.89	–	425	604	⎬ Data from Figure 2 (604)
+ Na$_2$S$_2$O$_4$	7.26	–	437	604	⎭
Cow's milk	16.1	–	280	604	⎫
+ hydrogen peroxide	12.3	–	280	604	⎬ Calculated from
+ Na$_2$S$_2$O$_4$	10.1	–	280	604	⎭ $\epsilon_{412}/\epsilon_{280} = 0.72$
Lactose synthetase, A protein, (see also galactosyl transferase)					
Cow's milk	–	16.1	280	1605	Refractometry
Lactosiderophilin lactotransferrin					
Human milk	10.9	11.7	280	308	pH 7, sat. with iron, MW = 93,000 (308)
	–	0.500	452.5	308	–

^{g†}MalNEt, *N*-ethylmaleimide.
^{h†}ClHgBzO$^-$, *p*-chloromercuribenzoate.

MOLAR ABSORPTIVITY AND $A_{1cm}^{1\%}$ VALUES FOR PROTEINS AT SELECTED WAVELENGTHS OF THE ULTRAVIOLET AND VISIBLE REGION (continued)

Protein	$\epsilon^a (\times 10^{-4})$	$A_{1cm}^{1\%}$ [b]	nm[c]	Ref.	Comments[d]
Leghemoglobin					
Soybean	1.51	—	574	1068	
	1.5	—	541	1068	pH 6.4, 0.01 M P_i
	13.9	—	411	1068	
Lupinus luteus (yellow lupin),					
Root nodules	3.8	—	275	1606	—
$Lb^{3+}H_2O/OH^-$	16.0	—	403.5	1607	
	1.03	—	498	1607	pH 6.0
	0.44	—	624	1607	
	14.3	—	404	1607	
	1.05	—	495	1607	
	0.98	—	535	1607	pH 8.5
	0.7	—	574	1607	
	0.35	—	622	1607	
	12.3	—	411	1607	
	1.35	—	544	1607	pH 10.5
	1.12	—	574	1607	
$LL^{3+}F^-$	15.1	—	402.5	1607	
	0.99	—	495	1607	pH 6.0
	0.47	—	615	1607	
	14.5	—	403	1607	
	1.4	—	490	1607	
	0.84	—	538	1607	pH 8.5
	0.75	—	575	1607	
$Lb^{2+}NCS^-$	12.4	—	410	1607	pH 6.0
	1.33	—	534	1607	
	12.3	—	410	1607	pH 8.5
	1.14	—	534	1607	
	12.3	—	410	1607	pH 10.5
	1.63	—	534	1607	
$Lb^{2+}N_2^-$	13.0	—	414	1607	pH 6.0
	1.28	—	545	1607	
	13.3	—	414	1607	pH 8.5
	1.37	—	545	1607	
$Lb^{2+}CN^-$	11.9	—	416	1607	pH 6.0
	1.38	—	545	1607	
	12.0	—	416	1607	pH 8.5
	1.56	—	544	1607	
	12.0	—	416	1607	pH 10.5
	1.65	—	543	1607	
Lb^{2+} imidazole	13.2	—	407.5	1607	pH 6.0
	1.32	—	533	1607	
	13.0	—	408	1607	pH 8.5
	1.38	—	535	1607	
	12.6	—	410	1607	pH 10.5
	1.4	—	533	1607	
Lb^{2+} pyridine	13.5	—	406	1607	
	1.27	—	530	1607	pH 6.0
	1.1	—	562	1607	
	13.3	—	406.5	1607	
	1.26	—	537	1607	pH 8.5
	1.08	—	574	1607	
	12.9	—	408	1607	
	1.52	—	530	1607	pH 10.5
	1.22	—	558	1607	

MOLAR ABSORPTIVITY AND $A_{1cm}^{1\%}$ VALUES FOR PROTEINS AT SELECTED WAVELENGTHS OF THE ULTRAVIOLET AND VISIBLE REGION (continued)

Protein	$\epsilon^a (\times 10^{-4})$	$A_{1cm}^{1\%}$ [b]	nm[c]	Ref.	Comments[d]
Leghemoglobin (continued)					
Lb^{2+}(desoxy)	10.7	—	421	1607	
	10.5	—	428	1607	pH 6.0
	1.38	—	555	1607	
	10.5	—	421	1607	
	10.5	—	428	1607	pH 8.5
	1.4	—	555	1607	
	10.3	—	421	1607	
	10.3	—	428	1607	pH 10.5
	1.3	—	555	1607	
$Lb^{2+}NCS^-$	11.2	—	421	1607	
	10.7	—	427	1607	pH 6.0
	1.43	—	555	1607	
	10.8	—	421.5	1607	
	10.0	—	427	1607	pH 8.5
	1.23	—	555	1607	
	11.3	—	421	1607	
	11.3	—	428	1607	pH 10.5
	1.43	—	555	1607	
$Lb^{2+}N_3^-$	13.2	—	416.5	1607	
	1.52	—	525	1607	pH 6.0
	1.47	—	561.5	1607	
	11.1	—	417.5	1607	pH 8.5
	1.39	—	550	1607	
	13.4	—	416.5	1607	
	1.35	—	525	1607	pH 10.5
	1.74	—	553	1607	
$Lb^{2+}CN^-$	11.3	—	421	1607	
	9.4	—	425	1607	
	1.18	—	535	1607	pH 6.0
	1.29	—	560	1607	
	1.13	—	561	1607	
	13.1	—	430	1607	
	1.47	—	535	1607	pH 8.5
	2.0	—	562.5	1607	
	16.6	—	431.5	1607	pH 10.5
	2.0	—	534	1607	pH 10.5
	2.43	—	562.5	1607	
Lb^{2+} imidazole	12.4	—	422	1607	
	1.09	—	527	1607	pH 6.0
	1.6	—	556.5	1607	
	16.0	—	422	1607	
	1.43	—	527	1607	pH 8.5
	2.62	—	556.5	1607	
	15.1	—	422	1607	
	1.46	—	527	1607	pH 10.5
	2.90	—	556.5	1607	
Lb^{2+} pyridine	19.6	—	419	1607	
	1.26	—	467	1607	pH 6.0
	2.2	—	523.5	1607	
	3.72	—	555	1607	
	18.5	—	420	1607	
	2.26	—	523.5	1607	pH 8.5
	3.68	—	555	1607	
	17.7	—	419.5	1607	
	1.98	—	523.5	1607	pH 10.5
	3.46	—	555	1607	

MOLAR ABSORPTIVITY AND $A_{1cm}^{1\%}$ VALUES FOR PROTEINS AT SELECTED WAVELENGTHS OF THE ULTRAVIOLET AND VISIBLE REGION (continued)

Protein	$\epsilon^a (\times 10^{-4})$	$A_{1cm}^{1\%}$ [b]	nm[c]	Ref.	Comments[d]
Leghemoglobin (continued)					
Lb^{2+} nicotinic acid	11.6	—	418	1607	
	1.46	—	524	1607	pH 6.0
	2.53	—	554.5	1607	
	9.4	—	420	1607	
	1.36	—	524	1607	pH 8.5
	1.7	—	554.5	1607	
	10.0	—	420	1607	
	1.1	—	524	1607	pH 10.5
	1.5	—	554.5	1607	
Lb^{2+} ethyl isocyanide	15.7	—	426	1607	
	1.78	—	526	1607	pH 6.0
	1.98	—	556	1607	
	14.0	—	426	1607	
	1.68	—	526	1607	pH 8.5
	1.98	—	556	1607	
	14.8	—	426	1607	
	1.5	—	526	1607	pH 10.5
	1.98	—	556	1607	
Lb^{2+}CO	17.1	—	417.5	1607	
	1.34	—	540	1607	pH 6.0
	1.33	—	562	1607	
	19.0	—	417	1607	
	1.29	—	536	1607	pH 8.5
	1.24	—	562	1607	
	20.6	—	417	1607	
	1.7	—	537.5	1607	pH 10.5
	1.7	—	562	1607	
Legumin					
Vicia sativa	—	7.5	280	1069	pH 6.5 and pH 12.4
Leucine aminopeptidase (amino peptidase, EC 3.4.11.1-.2)					
Pig kidney	—	8.4	280	309	pH 8.0, 0.005 M Tris, 0.005 M MgCl$_2$, est. from Figure 5 (309)
Beef lens	—	9.2	280	605	—
	—	10	280	606	pH 8.0, 0.1 M Tris
Leucine binding					
E. coli	—	6.5	280	453	pH 6.9
Leucine dehydrogenase (EC 1.4.1.9)					
B. subtilis SJ-2	103.5	45	280	1070	MW = 230,000 (1070)
Bacillus sphacricus		6.44	280	1608	—
Lipase (triacylglycerol lipase, EC 3.1.1.3)					
Pig pancreas	6.65	13.3	280	1071	MW = 50,000 (1071)
N$_3$ ph-[i†]	—	14.2	280	1071	—
		11	280	1609	—
Rat pancreas	—	12	280	1072	—
Lipoamide dehydrogenase					
C. krusei	6.2	11.76	273	310	pH 7.0, MW = 53,000 (310)

[i†] N$_3$ ph, trinitrophenyl.

MOLAR ABSORPTIVITY AND $A_{1cm}^{1\%}$ VALUES FOR PROTEINS AT SELECTED WAVELENGTHS OF THE ULTRAVIOLET AND VISIBLE REGION (continued)

Protein	$\epsilon^a (\times 10^{-4})$	$A_{1cm}^{1\%}$ [b]	nm[c]	Ref.	Comments[d]
Lipoate acetyltransferase (EC 2.3.1.12) (see dihydrolipoyl transacetylase)					
Lipoprotein, HDL$_2$, human					
Apo-	—	18.2	280	1610	—
Fraction III	—	12.2	280	1610	—
Fraction IV	—	9.2	280	1611	—
Carboxymethyl-low density	—	9.2	280	1611	—
Apo-	—	8.0	—[c†]	1612	Dry wt, 7.5 M GdmCl
	—	7.7	—[c†]	1612	
Lipoprotein, very low density					
Human, ApoLP-Val	—	9.1	280	1073	pH 7.5, 0.02 M KP$_i$
ApoLP-Ala	—	16.1	280	1073	
Low density					
Rat serum	—	11.7	280	1074	pH 8.6, data from
	—	8.0	290	1074	Figure 6 (1074)
	—	11.3	280	1074	pH 11.6, data from
	—	9.7	290	1074	Figure 6 (1074)
Apo-	—	10.0	280	1074	pH 8.6, data from
	—	7.3	290	1074	Figure 6 (1074)
	—	10.4	280	1074	pH 11.6, data from
	—	10.0	290	1074	Figure 6 (1074)
High density					
Rat serum	—	12.5	280	1074	pH 8.6, data from
	—	9.6	290	1074	Figure 5 (1074)
	—	12.3	280	1074	pH 11.6, data from
	—	11.0	290	1074	Figure 5 (1074)
Apo-	—	11.3	280	1074	pH 8.6, data from
	—	8.8	290	1074	Figure 6 (1074)
	—	11.7	280	1074	pH 11.6, data from
	—	11.2	290	1074	Figure 6 (1074)
Human skin, high density	—	10.8	278	1074	
Lipovitellin					
Leucophaea maderae (cockroach)	—	8.3	280	1613	—
Lipoxygenase (EC 1.13.11.12)					
Soybean	18.8	17.4	280	721	MW = 108,000 (721)
Isoenzyme		14	280	1614	—
Peas (*Pisum sativum L*)	9.5	13.2	278	722	0.05 M TrisCl, pH 7.2, MW = 72,000 (722)
Lombricine kinase (EC 2.7.3.5)					
Lumbricus terrestris	—	11.4	280	607	—
Photinus pyralis	7.5	7.5	278	608	MW = 100,000 (608)
Renilla reniformis	—	10.4	280	860	pH 7.5, 100 mM KP$_i$ plus 1.4 mM HSEtOH, plus 2.2 mM EDTA,
	—	0.38	500	860	data from Figure 4 (860)
Diplocardia longa (earthworm)	54	18	278	1615	Biuret, MW = 300,000 (1615)
Bacterial	6.3	8.3	280	311	MW = 76,000 (311)

MOLAR ABSORPTIVITY AND $A_{1cm}^{1\%}$ VALUES FOR PROTEINS AT SELECTED WAVELENGTHS OF THE ULTRAVIOLET AND VISIBLE REGION (continued)

Protein	ϵ^a (× 10^{-4})	$A_{1cm}^{1\%}$ ^{b}	nm^c	Ref.	Comments^d
Lutenizing hormone- releasing factor					
Synthetic					
[Gly^2]LRF	1.22	—	244	1423	
	0.574	—	280	1423	
	0.561	—	288	1423	
	0.410	—	294	1423	0.1 N NaOH
des-His^2-LRF	1.0465	—	244	1423	
	0.5310	—	280	1423	
	0.5177	—	288	1423	
	0.3706	—	294	1423	
Lysin					
Tegula pfeifferi					
Egg, membrane	2.1	23.8	280	861	pH 6.0, 0.03 M P$_i$
ϵ-Lysine acylase (lysine acetyltransferase, EC 2.3.1.32)					
A. pestifer	—	12.0	280	862	—
Lysine decarboxylase (EC 4.1.1.18)					
E. coli B	—	11.3	280	724	Data from Figure 2 (724), pH 6
	—	15.2	290	724	Data from Figure 2 (724), 2 N NaOH
Bacterium cadaveris	—	10.1	280	725	pH 6.2, 0.01 M KP$_i$
L-Lysine 6-aminotransferase (EC 2.6.1.36)					
Achromobacter liquidum	8.5	7.35	280	723	pH 7.4, MW = 116,000 (723)
L-Lysine 2-monooxygenase (EC 1.13.12.2)					
Psuedomonas fluorescens	34.2	17.9	280	726	MW = 191,000 (726)
Lysozyme (EC 3.2.1.17)	—	22.8	280	1616	—
	—	320	210	1616	—
	—	910	191	1616	—
Egg white	3.79		—^{c†}	1617	—
	—	26.04	280	1618	—^{j†}
	—	12.60	255	1604	—
	—	24.80	280	1604	—
	—	0.72	310	1604	—
	3.60	24.7	280	1619	MW = 14,600 (1620)
	—	25.5	277.5	1621	Neutral pH
	—	20.2	290	1621	
	—	27.4	281	1622	pH 5–6, 0.1 M KCl, dry wt
	—	27.2	280	1623	0.02 N HCl
	3.9		280	1624	0.2 M AcO$^-$, pH 4.75
	3.8		280	1624	Carboxyl modified with aminomethylsulfonic acid
	—	26	280	11	—
	—	25.32	280	312	—
	—	26.9	280	313	—
	3.88	—	281	314	0.1 N HCl, MW = 14,700 (314)
	—	23.05	281	315	—
	—	27.3	282	316	—
	—	26.35	280	317	pH 5.4
	3.65	25.5	280	318	MW = 14,307 (318)
	—	26.5	280	319	pH 7.0, 0.2 M NaP$_i$
	390	—	281	320	pH 3.9
	—	23.3	290	321	4.8 M guanidine-0.01 M HCl

^{j†}Values given for photo-oxidation products in Reference 1618.

MOLAR ABSORPTIVITY AND $A_{1cm}^{1\%}$ VALUES FOR PROTEINS AT SELECTED WAVELENGTHS OF THE ULTRAVIOLET AND VISIBLE REGION (continued)

Protein	$\epsilon^a (\times 10^{-4})$	$A_{1cm}^{1\%}$ [b]	nm[c]	Ref.	Comments[d]
Lysozyme (continued)					
Oxidized, tryptophan-108 oxidized to oxindole	—	22.7	280	1625	—
N-Bromosuccinimide Modified	3.37	—	280	1626	—
Reduced, carboxy-methylated and N-bromosuccinimide modified	3.37	—	280	1626	See Figure 1 of Reference 1626
Azophenyl-p-sulfonic acid modified					
S-1	1.85	12.6	500	1627	0.1 N NaOH
S-2	3.60	24.5	330	1627	
Azophenyl-p-carboxylic acid modified					
B-1	8.58	58.4	340	1627	pH 7
B-1	2.28	15.5	500	1627	0.1 N NaOH
B-2	8.54	58.1	330	1627	pH 6
	2.84	19.3	478	1627	0.1 N NaOH
Azotetrazole modified					
T-1-2	5.58	38.0	330	1627	pH 7
	2.87	19.6	510	1627	0.1 N NaOH
T-2	1.96	13.3	478	1627	
Azophenyldiethyl-methylammonium chloride modified					
A-1	4.09	27.8	330	1627	pH 7
A-2	1.36	9.25	330	1627	pH 6
Mouse	—	21.7	280	1628	pH 6.5, 0.2 M P_i, dry wt
Human					
Milk	—	25.1	280	1629	—
	—	25.65	280	1633	pH 6.0, 0.1 M AcONa – 0.1 M NaCl
	3.1	20.7	280	322	Est. from Figure 3 (322), MW = 15,000
Tear	3.6	24.2	280	323	pH 5.5, est. from Figure (323), MW = 14,900,
Urine	—	25.1	280	1629	—
	3.51	24.7	280	1630	MW = 14,200 (1631)
	—	24.6	281	1632	pH 5.8
Duck	—	26.6	—[c†]	1634	—
Goose	—	14.8	—[c†]	1634	—
Chalaropis sp.	—	24.8	280	1634	Dry wt
Bacteriophage T4	2.4	12.8	280	324	MW = 19,000 (324)
Papaya	5.95	23.8	280	325	MW = 25,000 (325)
Bacteriophage λ	1.9	10.7	280	326	MW = 17,900 (326)
Lysyl-tRNA synthetase (EC 6.1.1.6)					
Yeast	—	6.4	280	1075	pH 7, 0.1 M P_i, 0.1 M EDTA, data from Figure 4 (1075)
α_2-Macroglobulin					
Beef plasma	80	10	280	863	pH 8.0 MW = 800,000 (863)

MOLAR ABSORPTIVITY AND $A_{1cm}^{1\%}$ VALUES FOR PROTEINS AT SELECTED WAVELENGTHS OF THE ULTRAVIOLET AND VISIBLE REGION (continued)

Protein	$\epsilon^a (\times 10^{-4})$	$A_{1cm}^{1\%}$ [b]	nm[c]	Ref.	Comments[d]
α_2-Macroglobulin (continued)					
Human	66.5	8.1	280	864	MW = 820,000 (864)
Pig	98	10.2	280	865	MW = 960,000 (865)
Mouse	–	7.42	280	1635	Dry wt
Rabbit	–	9	277	1028	–
Malate dehydrogenase (EC 1.1.1.37)					
Pig heart	1.78	–	280	867	–
	–	4.6	280	327	–
	–	2.8	280	328	–
	–	3.8	280	329	–
Mitochondria	1.98	3.05	280	337	pH 8.0, 0.05 μM Tris-AcO$^-$, MW = 65,000 (337)
	–	2.5	280	1636	–
5S protein	1.97	2.9	280	868	MW = 68,000 (868)
9S protein	7.74	5.6	280	868	MW = 138,000 (868)
Supernatant	6.9	9.3	280	869	MW = 74,000 (869)
Chicken					
Intramitochondrial	–	2.9	280	870	–
Extramitochondrial	–	13.1	280	870	–
Rat liver					
Mitochondrial	3.4	5.08	280	871	MW = 66,300 (871)
Horse heart	–	2.8	280	327	–
Ox heart	–	8.5	280	330	–
Beef heart	3.25	5.0	280	331	MW = 65,000 (331)
B. subtilis	2.44	6.6	280	332	pH 7.7, 0.05 M P$_i$, MW = 37,000 (332)
	7.8	6.67	280	333	MW = 117,000 (333)
B. stearothermophilus	–	5.82	280	333	–
E. coli	2.03	3.39	280	333	MW = 60,000 (333)
Ostrich heart	8.96	12.8	280	334	pH 7.5, 0.1 M P$_i$, MW = 70,000 (334)
Tuna heart	20.8	31	280	335	pH 7.5, 0.1 M P$_i$, MW = 67,000 (335)
P. acidovorans	3.4	8.0	–[k†]	336	MW = 43,000 (336)
Malate-lactate transhydrogenase (EC 1.1.99.7) (lactate-malate dehydrogenase)					
Veillonelia alcalescens	–	12.7	280	1637	–
Malic enzyme [malate dehydrogenase (decarboxylating)] (EC 1.1.1.38-.40)					
Pigeon liver	25.8	9.2	278	338	pH 7.0, 0.042 M Tris, MW = 280,000 (338), protein contains NADP[l†]
	24.1	8.6	278	338	pH 7.0, 0.042 M Tris, MW = 280,000 (338)

[k†]No value cited for wavelength.
[l†]NADP, nicotinamide adenine dinucleotide phosphate.

MOLAR ABSORPTIVITY AND $A_{1cm}^{1\%}$ VALUES FOR PROTEINS AT SELECTED WAVELENGTHS OF THE ULTRAVIOLET AND VISIBLE REGION (continued)

Protein	$\epsilon^a (\times 10^{-4})$	$A_{1cm}^{1\%}$ [b]	nm[c]	Ref.	Comments[d]
Malic enzyme (continued)					
E. coli K 10 [HfrC]	26.5	4.8	279	872	MW = 550,000 (872), protein con. det. from tyr/trp content
	28.1	5.1	279	872	MW = 550,000 (872), protein con. det. by Lowry
Malic enzyme, NAD-linked [malate dehydrogenase (decarboxylating)] (EC 1.1.1.38-.39)					
E. coli	—	10.2	278	1638	Dry wt
Mandelonitrite lyase (see hydroxynitrilase)					
α-Mannosidase (EC 3.2.1.24)					
Vicia sativa	—	13.3	280	727	Data from Figure 3 (727), pH 7.2, Tris-0.01 N HCl
Soybeans	—	20	280	1639	
α-Melanotropin; 5-glutamine, nitrophenyl sulfenyl					
Synthetic	1.65	—	282	873	0.001 N HCl
	0.4	—	365	873	
Melilotate hydroxylase (melilotate 3-monooxygenase, EC 1.14.13.4)					
Arthrobacter, apoenzyme	—	11.1	277	339	pH 7.3, 0.15 M P_i, 0.1 M KCl, 1 mM cysteine, est. from Figure 6 (339)
Holoenzyme	—	18.9	280	339	pH 7.3, 0.15 M P_i, 0.1 M KCl, 1 mM cysteine, est. from Figure 6 (339)
Mercaptoalbumin, human see albumin, human					
Mercaptopyruvate sulfurtransferase (EC 2.8.1.2)					
E. coli	2.23	9.3	280	1640	pH 7.5, 0.05 M Tris, 0.8 M KCl, MW = 24,000 (1640)
Meromyosin, heavy	—	6	280	1641	Kjeldahl, absorption corrected for light scattering
Rabbit	—	—	—	1642	See footnote[m†]
Metallothionein					
Chicken liver	8.06	—	205	1643	pH 6.6
Rat liver	8.06	—	205	1643	pH 6.6

[m†] $\epsilon_{211} - \epsilon_{350} = 776$ cm²/g in 0.5 N NaOH.

MOLAR ABSORPTIVITY AND $A_{1cm}^{1\%}$ VALUES FOR PROTEINS AT SELECTED WAVELENGTHS OF THE ULTRAVIOLET AND VISIBLE REGION (continued)

Protein	$\epsilon^a (\times 10^{-4})$	$A_{1cm}^{1\%}$ [b]	nm[c]	Ref.	Comments[d]
Metapyrocatechase (catechol 2,3-dioxygenase, (EC 1.1.3.11.2))					
Pseudomonas arvilla	18.9	13.5	280	1076	MW = 140,000 (1076)
	–	13.2	280	1644	See footnote[n†]
Methemoglobin (see hemoglobin)					
Methemoglobin, N_3^-	0.34	–	630	700	0.1 M P_i, pH 7.0,
	1.07	–	542	700	ϵ_M values expressed per mole
	11.7	–	414	700	heme
Methemoglobin reductase (EC 1.6.2.1.-2?)					
Human, erythrocytes	1.13	–	462	1645	Oxidized enzyme
Form I	–	7.45	278	728	–
Form II	–	17.9	268	728	–
Methionyl-tRNA synthetase (EC 6.1.1.10)					
E. coli K 12	–	13.9	283	1077	Native enzyme
	–	16.2	283	1077	Trypsin modified
	–	20	280	874	pH 7.4, 0.02 M KP_i
β-Methylaspartase (methylaspartate ammonia-lyase, EC 4.3.1.2)					
C. tetranomorphum	5.63	5.63	279	340	pH 7.0, 0.5 M Me_4NCl, MW = 100,000 (340)
	–	6.60	280	341	pH 6.5, 0.005 M P_i
Methylaspartate mutase (EC 5.4.99.1) (see glutamate mutase)					
Metmyoglobins (see myoglobin)					
β_2-Microglobulin					
Human urine	1.985	17	280	866	pH 7.0, MW = 11,600 (866)
Mitogenic components Phaseolus vulgaris					
A	–	6	280	1646	pH 7.0, 5 mM NaP_i,
B	–	3.6	280	1646	0.1 M NaCl
Molybdoferredoxin					
Clostridium pasteurianum	22.5	13.4	280	1078	pH 7.0, 0.1 M Tes buffer, data from Figure 7 (1078), MW = 168,000 (1078)
Monellin					
Dioscoreophyllum cumminsil	1.47	13.7	277	1647	pH 7.2
	1.83	17	290	1647	pH 12.8

[n†] Absorption at 280 nm taking the value of 1.32 ml/cm-mg.

MOLAR ABSORPTIVITY AND $A_{1cm}^{1\%}$ VALUES FOR PROTEINS AT SELECTED WAVELENGTHS OF THE ULTRAVIOLET AND VISIBLE REGION (continued)

Protein	$\epsilon^a (\times 10^{-4})$	$A_{1cm}^{1\%}$ [b]	nm[c]	Ref.	Comments[d]
Monoamine oxidase (amine oxidase, EC 1.4.3.4)					
Beef kidney	53.9	18.6	280	342	pH 7.6, 0.05 M P_i, MW = 290,000 (342)
	4.7	1.62	455	342	pH 7.6, 0.05 M P_i, MW = 290,000 (342)
Plasma	25.0	9.8	280	343	MW = 255,000 (343)
Myeloperoxidase (peroxidase, EC 1.11.1.7?)					
Canine uterine pus	22	—	280	875	Data from Figure 3 (875)
Human leukocyte	—	24	280	344	pH 7.0, 0.2 M P_i
Myoglobin	17.1	—	409	609	
	0.97	—	500	609	
Synthetic					
Proto-Mb$^+$	18.8	—	409	1648	
	1.16	—	502	1648	
	0.47	—	630	1648	
+CN$^-$	12.6	—	422	1648	
	1.30	—	540	1648	
+N$_3^-$	12.5	—	419	1648	pH 6, 0.1 M P_i
	1.22	—	540	1648	
	0.97	—	574	1648	
+F$^-$	16.0	—	406	1648	
	1.09	—	490	1648	
	0.98	—	605	1648	
+H$_2$O$_2$	12.3	—	421	1648	—
	1.25	—	546	1648	—
+OH$^-$	11.7	—	413	1648	—
	1.16	—	542	1648	—
	1.10	—	583	1648	—
+Na$_2$S$_2$O$_4$	13.5	—	434	1648	
	1.54	—	556	1648	
+Na$_2$S$_2$O$_4$ + CO	20.1	—	422	1648	pH 6, 0.1 M P_i
	1.78	—	541	1648	
	1.56	—	578	1648	
Meso-Mb$^+$	17.2	—	395	1648	
	0.88	—	495	1648	
	0.41	—	622	1648	
+CN$^-$	12.4	—	411	1648	
	0.992	—	531	1648	
+N$_3^-$	12.0	—	409	1648	pH 6, 0.1 M P_i
	0.9	—	530	1648	
	0.68	—	564	1648	
+F$^-$	15.4	—	394	1648	
	0.83	—	486	1648	
	0.69	—	598	1648	
+H$_2$O$_2$	11.6	—	408	1648	—
	1.02	—	536	1648	—
+OH$^-$	11.4	—	398	1648	—
	0.91	—	531	1648	—
	0.82	—	569	1648	—
	0.85	—	587	1648	—

MOLAR ABSORPTIVITY AND $A_{1cm}^{1\%}$ VALUES FOR PROTEINS AT SELECTED WAVELENGTHS OF THE ULTRAVIOLET AND VISIBLE REGION (continued)

Protein	$\epsilon^a (\times 10^{-4})$	$A_{1cm}^{1\%}$ [b]	nm[c]	Ref.	Comments[d]
Myoglobin (continued)					
$+Na_2S_2O_4$	11.1	—	421	1648	
	1.27	—	544	1648	
$+Na_2S_2O_4 + CO$	20.6	—	409	1648	
	1.46	—	530	1648	
	1.16	—	558	1648	
Deutero-Mb⁺	12.8	—	392	1648	
	0.67	—	495	1648	
	0.29	—	620	1648	
$+CN^-$	10.8	—	409	1648	
	0.83	—	532	1648	
$+N_3^-$	10.5	—	408	1648	pH 6, 0.1 $M P_i$
	0.84	—	530	1648	
	0.61	—	560	1648	
$+F^-$	10.5	—	392	1648	
	0.69	—	483	1648	
	0.54	—	595	1648	
$+Na_2S_2O_2$	9.76	—	419	1648	
	1.0	—	542	1648	
$+Na_2S_2O_4 + CO$	17.2	—	408	1648	
	1.21	—	528	1648	
	0.81	—	556	1648	
$+H_2O_2$	9.64	—	409	1648	—
	0.72	—	534	1648	—
$+OH^-$	8.87	—	400	1648	—
	0.67	—	532	1648	—
	0.53	—	568	1648	—
	0.53	—	587	1648	—
Hemato-Mb⁺	14.5	—	400	1648	
	0.94	—	499	1648	
	0.39	—	627	1648	
$+CN^-$	10.8	—	415	1648	
	1.05	—	537	1648	
$+N_3^-$	10.3	—	412	1648	
	0.96	—	535	1648	
	0.74	—	566	1648	pH 6, 0.1 $M P_i$
$+F^-$	13.2	—	400	1648	
	0.88	—	489	1648	
	0.66	—	600	1648	
$+Na_2S_2O_4$	11.0	—	426	1648	
	1.30	—	550	1648	
$+Na_2S_2O_4 + CO$	16.3	—	413	1648	
	1.50	—	534	1648	
$+H_2O_2$	10.2	—	413	1648	—
	1.07	—	540	1648	—
$+OH^-$	10.6	—	402	1648	—
	0.92	—	536	1648	—
	0.83	—	589	1648	—
Proto monomethyl Mb⁺	16.6	—	406	1648	
	1.05	—	501	1648	
	0.45	—	630	1648	
$+CN^-$	11.7	—	418	1648	pH 6, 0.1 $M P_i$
	1.18	—	540	1648	
$+N_3^-$	11.5	—	418	1648	
	1.11	—	540	1648	
	0.90	—	572	1648	

MOLAR ABSORPTIVITY AND $A_{1cm}^{1\%}$ VALUES FOR PROTEINS AT SELECTED WAVELENGTHS OF THE ULTRAVIOLET AND VISIBLE REGION (continued)

Protein	$\epsilon^a (\times 10^{-4})$	$A_{1cm}^{1\%}$ [b]	nm[c]	Ref.	Comments[d]
Myoglobin (continued)					
+F$^-$	13.9	--	404	1648	
	1.00	--	488	1648	
	0.95	--	604	1648	
+Na$_2$S$_2$O$_4$	12.7	--	431	1648	
	1.49	--	556	1648	pH 6, 0.1 M P$_i$
+Na$_2$S$_2$O$_4$ + CO	17.5	--	419	1648	
	1.64	--	539	1648	
	1.37	--	573	1648	
+H$_2$O$_2$	11.1	--	418	1648	
	1.08	--	544	1648	
OH$^-$	11.1	--	410	1648	
	1.03	--	540	1648	
	0.96	--	580	1648	
Proto dimethyl					
Mb$^+$	14.5	--	407	1648	
	0.88	--	502	1648	
	0.40	--	630	1648	
+CN$^-$	10.5	--	420	1648	
	1.00	--	541	1648	
+N$_3^-$	10.5	--	418	1648	pH 6, 0.1 M P$_i$
	1.01	--	541	1648	
	0.83	--	572	1648	
+F$^-$	11.7	--	405	1648	
	0.86	--	486	1648	
	0.83	--	605	1648	
+Na$_2$S$_2$O$_4$	10.5	--	432	1648	
	1.30	--	556	1648	
+Na$_2$S$_2$O$_4$ + CO	15.9	--	421	1648	pH 6, 0.1 M P$_i$
	1.41	--	540	1648	
	1.23	--	570	1648	
+H$_2$O$_2$	9.15	--	418	1648	
	0.94	--	545	1648	
+OH$^-$	7.88	--	410	1648	
	0.96	--	538	1648	
	0.86	--	578	1648	
Oxyforms					
Proto	14.5		417	1649	
	1.73		543	1649	
	1.79		581	1649	
Meso	13.9		404	1649	
	1.50		533	1649	
	1.32		568	1649	
Deutero	11.4		402	1649	
	1.27		532	1649	
	0.89		565	1649	
Hemato	11.9		410	1649	
	1.40		538	1649	
	1.16		574	1649	
Protomonoester	11.4		415	1649	
	1.40		541	1649	
	1.35		579	1649	
Protodiester	9.65		416	1649	
	1.64		542	1649	
	1.68		579	1649	

MOLAR ABSORPTIVITY AND $A_{1cm}^{1\%}$ VALUES FOR PROTEINS AT SELECTED WAVELENGTHS OF THE ULTRAVIOLET AND VISIBLE REGION (continued)

Protein	ϵ^a (× 10^{-4})	$A_{1cm}^{1\%}$ [b]	nm[c]	Ref.	Comments[d]
Myoglobin (continued)					
Albacore tuna					
(*Thunnus germo*)	—	6.4	555	1081	—
	—	61	430	1081	—
MetMb	—	2.1	630	1081	—
	—	1.7	580	1081	—
	—	4.8	502	1081	—
	—	81	408	1081	—
	—	15.5	280	1081	—
Myoglobin, carboxy-	—	6.7	570	1081	—
	—	7.5	538	1081	—
	—	95	421	1081	—
Beef, MetMbCN	2.96	—	280	1086	pH 7.2, 0.01 M P$_i$ containing
	0.945	—	340	1086	0.01% KCN
Bluefin tuna					
(*Thunnus thynnus*)	—	7.1	558	1081	—
	—	56	431	1081	—
MetMb	—	2.1	634	1081	—
	—	1.7	580	1081	—
	—	4.8	504	1081	—
	—	86	407	1081	—
	—	13.3	275	1081	—
Myoglobin, carboxy-	—	7.6	570	1081	—
	—	8.4	540	1081	—
	—	104	420	1081	—
Camel	3.13	—	280	615	
	17.2	—	409	615	
	0.796	—	470	615	Ferri form, acidic
	0.987	—	503	615	
	0.360	—	580	615	
	0.366	—	630	615	
	3.43	—	280	615	
	10.4	—	414	615	Basic
	0.948	—	542	615	
	0.93	—	587	615	
	3.33	—	280	615	
	2.87	—	360	615	Cyanide form
	11.5	—	423	615	
	1.132	—	542	615	
MetMbCN	2.90	—	280	1086	pH 7.2, 0.01 M P$_i$
	0.915	—	340	1086	containing 0.01% KCN
Chicken					
(*Gallus gallus*)	3.1	—	280	1651	
Muscle					
MetMbCN	0.96	—	534	1650	—
	10.2	—	423	1650	—
MetMb	0.32	—	632	1650	—
	0.84	—	504	1650	—
Muscle, distrophic					
MetMbCN	0.95	—	543	1650	—
	10.0	—	423	1650	—
	0.32	—	632	1650	—
	0.84	—	504	1650	—
Chinook salmon					
(*Oncorhynchus tschawtscha*)	—	7.0	558	1081	—
	—	50	428	1081	—

MOLAR ABSORPTIVITY AND $A^{1\%}_{1cm}$ VALUES FOR PROTEINS AT SELECTED WAVELENGTHS OF THE ULTRAVIOLET AND VISIBLE REGION (continued)

Protein	$\epsilon^a (\times 10^{-4})$	$A^{1\%}_{1cm}{}^b$	nm^c	Ref.	Comments^d
Myoglobin (continued)					
MetMb	—	2.2	632	1081	—
	—	1.9	580	1081	—
	—	5.1	502	1081	—
	—	85	404	1081	—
	—	20.4	280	1081	—
Myoglobin, carboxy-	—	7.4	569	1081	—
	—	7.7	539	1081	—
	—	95	420	1081	—
Cormorant (*Phalacrocorax*)	—	6.4	558	1081	—
	—	64	435	1081	—
MetMb	—	1.8	634	1081	—
	—	1.4	580	1081	—
	—	4.8	504	1081	—
	—	90	409	1081	—
	—	17.1	280	1081	—
Myoglobin, carboxy-	—	6.5	579	1081	—
	—	7.7	542	1081	—
	—	99	432	1081	—
Fin whale, component VII	3.2	18.28	280	1083	MW = 17,504 (1083)
Goat					
MetMbCN	2.95	—	280	1086	pH 7.2, 0.01 M P$_i$ containing 0.01% KCN
	1.01	—	340	1086	
Hamster					
MetMbCN	—	53.9	420	1087	
Habor seal	2.99	—	280	614	Ferri form, pH 6.2
	16.2	—	409	614	
Horse	1.33	—	555	1082	—
	11.3	—	434	1082	
MetMbCN	2.95	—	280	1086	pH 7.2, 0.01 M P$_i$ containing 0.01% KCN
	0.95	—	340	1086	
MetMb	—	17.9	280	1088	—
Myoglobin, carboxy-	1.18	—	578	1082	—
	1.4	—	540	1082	—
	17.8	—	423	1082	—
Mb$^+$	16.0	—	408	1082	—
	1.02	—	505	1082	—
	0.42	—	630	1082	—
Mb-OH	9.0	—	414	1082	—
	0.91	—	540	1082	—
	0.86	—	580	1082	—
Human	3.05	17.4	280	610	MW = 17,510 (610)
Apo-	1.55	9.2	280	610	MW = 16,900 (610)
MetMbCN	3.07	—	280	1086	Ph 7.2, 0.01 M P$_i$ containing 0.01% KCN
	1.04	—	340	1086	
Humpback whale (*Megaptera nodosa*)	—	6.5	558	1081	—
	—	61	434	1081	—
MetMb	—	1.9	634	1081	—
	—	1.5	580	1081	—
	—	5.0	504	1081	—
	—	85	409	1081	—
	—	15.9	281	1081	—
Myoglobin, carboxy-	—	6.1	577	1081	—
	—	7.2	543	1081	—
	—	96	423	1081	—

MOLAR ABSORPTIVITY AND $A_{1cm}^{1\%}$ VALUES FOR PROTEINS AT SELECTED WAVELENGTHS OF THE ULTRAVIOLET AND VISIBLE REGION (continued)

Protein	$\epsilon^a (\times 10^{-4})$	$A_{1cm}^{1\%}$ [b]	nm[c]	Ref.	Comments[d]
Myoglobin (continued)					
Lamb					
MetMbCN	3.09	—	280	1086	pH 7.2, 0.01 $M P_i$
	1.08	—	340	1086	containing 0.01% KCN
Molluscs					
(*Aplysia depilans* and *Aplysia limacina*)	1.3	—	555	1082	—
	11.3	—	438	1082	—
(*Acanthopleura granulata*)					
Myoglobin, carboxy-					
Type 1'	1.22	—	570	1094	—
	1.38	—	538	1094	—
	14.94	—	419	1094	—
Type 2'	1.24	—	572	1094	—
	1.32	—	538	1094	—
	17.90	—	419	1094	—
Type 3'	1.30	—	572	1094	—
	1.33	—	538	1094	—
	17.93	—	419	1094	—
(*Buccinum undatum L.*)					
Myoglobin, carboxy-	1.35	—	570	1095	—
	1.42	—	538	1095	—
	18.44	—	418	1095	—
	3.4	—	280	1095	—
(*Aplysia depilans* and *Aplysia limacina*)					
Myoglobin, carboxy-	1.37	—	571	1082	—
	1.42	—	541	1082	—
	17.6	—	424	1082	—
Mb$^+$	9.9	—	400	1082	—
	1.31	—	505	1082	—
	0.38	—	640	1082	—
Mb-OH	9.1	—	412	1082	—
	0.9	—	543	1082	—
	0.87	—	580	1082	—
	0.87	—	600	1082	—
Mb-O$_2$	10.8	—	416	1082	—
	1.32	—	542	1082	—
	1.33	—	578	1082	—
Monkey					
MetMbCn	3.01	—	280	1086	pH 7.2, 0.01 $M P_i$
	1.02	—	340	1086	containing 0.01% KCN
Pelican					
(*Pelecanus occidentalis*)	—	6.2	556	1081	—
	—	57	433	1081	—
MetMb	—	1.6	632	1081	—
	—	1.3	581	1081	—
	—	4.8	504	1081	—
	—	78	409	1081	—
	—	14.6	280	1081	—
Myoglobin, carboxy-	—	6.2	577	1081	—
	—	7.3	540	1081	—
	—	87	423	1081	—
Penguin					
(*Aptenodytes forsteri*)	3.2	—	280	1651	—
Porpoise	2.98	—	280	614	
	16.2	—	409	614	Ferri form, pH 6.2

MOLAR ABSORPTIVITY AND $A_{1cm}^{1\%}$ VALUES FOR PROTEINS AT SELECTED WAVELENGTHS OF THE ULTRAVIOLET AND VISIBLE REGION (continued)

Protein	$\epsilon^a (\times 10^{-4})$	$A_{1cm}^{1\%}$ [b]	nm[c]	Ref.	Comments[d]
Myoglobin (continued)					
Skip jack tuna					
(*Katsuwonus pelamis*)	—	6.4	556	1081	—
	—	62	431	1081	—
MetMb	—	2.0	632	1081	—
	—	1.5	578	1081	—
	—	4.7	502	1081	—
	—	89	406	1081	—
	—	12.2	275	1081	—
Myoglobin, carboxy-	—	6.7	569	1081	—
	—	7.3	539	1081	—
	—	102	421	1081	—
Sperm whale	10.5	—	434	1079	pH 9.1, 0.1 M borax, data from
	1.08	—	558	1079	
	17.9	100	408	1080	Figure 14 (1079), MW = 17,800
	3.45	19	280	1080	(1080)
	3.79	—	280	611	Fe 0.31% (611)
	3.34	—	280	612	
	16.8	—	409	612	Ferri form
	0.367	—	634	612	
	3.06	—	280	614	Ferri form, pH 6.2
	16.4	—	409	614	
	30.6	—	589	1652	—
	1.15	—	557	1652	—
	0.64	—	521	1652	—
Apo-	1.59	8.9	280	1080	MW = 17,800 (1080)
	1.54	—	280	1084	Dry wt, corrected for light scattering
	—	9.3	280	1085	Neutral
	—	9.2	280	1085	0.1 N NaOH
	1.58	—	280	613	pH 6.8, 0.1 M P_i
MbNO	0.7	—	583	1652	—
	1.0	—	546	1652	—
MetMbCN	0.2	—	583	1652	—
	0.95	—	539	1652	—
MbCO	0.65	—	582	1652	—
	0.78	—	572	1652	—
	1.36	—	544	1652	—
	0.4	—	522	1652	—
MbO_2	1.08	—	582	1652	—
	0.41	—	571	1652	—
	1.06	—	550	1652	—
	0.79	—	533	1652	—
MetMbNO	0.77	—	575	1652	—
	0.215	—	562	1652	—
$MetMbN_3$	0.36	—	587	1652	—
	0.38	—	570	1652	—
	0.62	—	546	1652	—
	0.38	—	514	1652	—
MetMb	—	93.4	410	1088	—
	14.4	—	407	1079	pH 6.5, P_i buffer containing 25
	0.82	—	502	1079	mM $Na_2S_2O_4$, data from
	0.34	—	633	1079	Figure 14 (1079) Water/0.1 N NaOH
	—	18.0	280	1089	Water/0.1 N NaOH
	0.35	—	630	1090	—
	0.93	—	505	1090	—
	16.0	—	409	1090	—

MOLAR ABSORPTIVITY AND $A_{1cm}^{1\%}$ VALUES FOR PROTEINS AT SELECTED WAVELENGTHS OF THE ULTRAVIOLET AND VISIBLE REGION (continued)

Protein	$\epsilon^a (\times 10^{-4})$	$A_{1cm}^{1\%}$ b	nm^c	Ref.	Comments^d
Myoglobin (continued)					
MetMb (continued)	3.1	–	280	1090	–
	0.35	–	630	1090	
	0.91	–	505	1090	
	15.9	–	409	1090	Regenerated
	3.2	–	280	1090	
	–	17.9	280	1085	0.1 N NaOH
Myoglobin, carboxy-	3.24	18.2	280	1091	pH 8.8, dry wt, MW = 17,800 (1092), data from Figure 2 (1091)
	1.21	6.8	578	1093	
	1.38	7.2	540	1093	MW = 17,816 (1093)
	18.6	104	423	1093	
Mb-Fe^{3+}	0.9	5.05	503	1093	–
	16.8	94	409	1093	–
	3.42	19.3	280	1093	–
MbO$_2$	1.45	8.13	581	1093	–
	1.36	7.6	543	1093	–
	12.5	70	418	1093	–
Mb-Fe^{2+}	1.17	6.6	555	1093	–
	11.6	65	434	1093	–
Mb-Fe^{3+}-N$_3$	11.1	62.1	421	1093	–
Yellowfin tuna					
(*Neothunnis macropterus*)	–	6.6	556	1081	–
	–	60	431	1081	–
MetMb	–	2.1	631	1081	–
	–	1.6	578	1081	–
	–	4.8	501	1081	–
	–	85	406	1081	–
	–	13.9	275	1081	–
Myoglobin, carboxy-	–	7.0	568	1081	–
	–	7.7	538	1081	–
	–	107	420	1081	–
Myokinase	1.15	5.38	277	876	pH 7.0, 0.15 M KCl–0.01 M KP$_i$, MW = 21,400 (876) Protein con. det. by dry wt
Rabbit muscle	–	5.2	280	141	–
	1.1	5.3	279	345	pH 6.9, MW = 21,000 (345)
	–	11.8	279	346	pH 7.0, 0.01 M P$_i$
Myosin					
Rabbit muscle	–	5.2	280	347	pH 7.0, 0.5 M KCl, 20 mM Tris
	–	2.5	250	347	pH 7.0, 0.5 M KCl, 20 mM Tris
	–	6.47	280	348	–
	–	5.07	276	348	pH 7.5, 5 M GdmCl
	–	5.88	280	348	0.5 M KCl^{o†}
	–	4.83	280	348	5 M GdmCl^{o†}
	–	5.43	280	349	–
	–	5.9	280	350	pH 7, 0.3 M KCl, 0.01 M Tris

^{o†}These values have been corrected for light scattering.

MOLAR ABSORPTIVITY AND $A_{1cm}^{1\%}$ VALUES FOR PROTEINS AT SELECTED WAVELENGTHS OF THE ULTRAVIOLET AND VISIBLE REGION (continued)

Protein	$\epsilon^a (\times 10^{-4})$	$A_{1cm}^{1\%}$ [b]	nm[c]	Ref.	Comments[d]
Myosin (continued)					
Rabbit muscle (continued)	—	5.87	277	351	Neutral KCl
	—	5.50	280	616	pH 7.3, 0.5 M KCl–0.01 M
	—	5.52	280	616	EDTA + 0.05 M P$_i$
	—	5.55	280	616	+ 0.2 M P$_i$
	—	5.58	280	616	+ 0.5 M P$_i$
	—	5.35	280	1097	Free of low molecular weight protein
Light chains	0.7	3.5	280	1098	MW = 20,200 (1098)
Meromyosin, heavy	—	6.35	280	1099	—
Meromyosin, light	—	3.0	280	1100	—
Rabbit filament	—	2.00	278	1653	—
Skeletal muscle	—	5.6	280	1654	—
Heavy meromyosin	—	6.25	280	352	—
	—	6.47	279	353	—c†
	—	6.47	280	1654	—
Beef heart	—	5.7	280	618	—
Light meromyosin, fraction I	—	3.69	279	353	Trypsin digestion for 25 sec
	—	3.29	279	353	Trypsin digestion for 25 min
Light component	—	3.5	280	1654	—
Light subfragment I					
Beef heart	—	6.4	280	618	—
Meromyosin, light					
Rabbit	—	2.97	278	619	pH 3, 0.1 M NaCl–HCl
Light chain	—	4.03	280	1655	N
Slime mold	—	5.2	278	617	pH 7.4, 0.5 M KCl–0.01 M Tris, data from Figure 3 (617)
Human heart	—	5.1	280	1096	pH 6.8, 0.4 M KCl-borate, corrected for scattering
Paramyosin	—	3.04	277	354	—
Heavy alkali subunit	—	5.77	277	351	Neutral KCl
	—	5.19	277	351	6 M GdmCl
	—	5.29	277	351	5 M GdmCl
Light alkali subunit	—	4.34	277	351	Neutral KCl
	—	4.79	277	351	5 M GdmCl
S-1 fragment	—	7.9	280	355	—
S-n fragment	—	8.0	280	355	—
Chicken					
Red/slow muscle	—	5.33	280	356	—
White/fast muscle	—	5.05	280	356	—
Pectoral muscle	—	5.25	280	1655	—
Pig muscle					
Chesterwhite	—	4.9	280	347	pH 7.0, 0.5 M KCl, 20 mM Tris
	—	2.4	250	347	pH 7.0, 0.5 M KCl, 20 mM Tris
Poland China	—	4.9	280	347	pH 7.0, 0.5 M KCl, 20 mM Tris
	—	2.5	250	347	pH 7.0, 0.5 M KCl, 20 mM Tris
Poland China [PSE]	—	5.2	280	347	pH 7.0, 0.5 M KCl, 20 mM Tris
Myrosinase (thioglucosidase) (EC 3.2.3.1)					
Sinapsis alba (white mustard seed)	—	15	278	1656	—

MOLAR ABSORPTIVITY AND $A_{1cm}^{1\%}$ VALUES FOR PROTEINS AT SELECTED WAVELENGTHS OF THE ULTRAVIOLET AND VISIBLE REGION (continued)

Protein	$\epsilon^a (\times 10^{-4})$	$A_{1cm}^{1\%}$ [b]	nm[c]	Ref.	Comments[d]
NAD nucleosidase (EC 3.2.2.5) (see DPNase)					
NADH-cytochrome b_5 reductase					
Rat liver, microsomes	–	23	280	729	pH 7.5, 0.01 M KP$_i$
NADH:FAD oxidoreductase					
E. coli B	16.4	13	272	1101	MW = 126,000 (1101) pH 7.0, 0.05 M KP$_i$, data from Figure 1 (1101)
	2.33	1.85	380	1101	–p†
	2.24	1.78	448	1101	–p†
NADPH-adrenodoxin reductase					
Beef adrenocortical mitochondria	–	18	272	1657	From Figure 2 (1657)
	–	2.0	378	1657	
NADPH oxidase					
Rabbit liver, microsomes	–	19.1	280	730	pH 7.4, 0.34 M P$_i$
NADPH-sulfite reductase (EC 3.2.2.5)					
Bull semen	–	10.9	278	1658	pH 7.4, 0.01 M NaP$_i$, dry wt
NADPH-sulfite reductase (EC 1.8.1.2)					
E. coli B	–	4.60	386	1659	Dry wt
Nagarse [BPN']	–	8.8	280	357	–
Nerve growth factor					
Mouse, 2.5S	2.2	16.4	280	1660	AA, MW = 13,259
Salivary gland	–	14.7	280	1661	pH 4, 0.1 M AcO$^-$ or pH 4, 0.1 M AcO$^-$ 8 M urea, from Figure 1 (1661)
Submaxillary gland	4.15	13.86	280	1102	pH 5.0, 0.05 M AcO$^-$, MW = 30,000 (1102)
Naja naja	–	11.8	280	1661	pH 4, 0.1 M AcO$^-$ from Figure 1 (1661)
	–	12.9	280	1661	pH 4, 0.1 M AcO$^-$ 8 M urea, from Figure 1 (1661)
Vipera russelli venom	–	9.9	282	1662	
Cobra venom	3.22	12.7	280	620	Water MW = 25,300 (620)
Neurophysin-II					
Beef pituitary	0.395	–	260	621	–
Neurotoxin					
Naja naja siamensis (Thialand cobra)	0.83	10.6	279	1103	Neutral sol MW = 7820 (1130)

p†Calculated from OD$_{272}$/OD$_{448}$ = 7.33, and OD$_{380}$/OD$_{448}$ = 1.04.

MOLAR ABSORPTIVITY AND $A_{1cm}^{1\%}$ VALUES FOR PROTEINS AT SELECTED WAVELENGTHS OF THE ULTRAVIOLET AND VISIBLE REGION (continued)

Protein	$\epsilon^a (\times 10^{-4})$	$A_{1cm}^{1\%}$ [b]	nm[c]	Ref.	Comments[d]
Neurotoxin (continued)					
Reduced and carboxy-methylated	0.68	–	279	1103	10% AcOH
Toxin 3	–	10.6	279	1664	–
Toxin 3C	0.89	12.9	279	1103	MW = 6793 (1103)
Toxin 5	1.3	19.1	280	1103	MW = 6875 (1103)
Toxin 7C	0.9	12.9	279	1103	MW = 6985 (1103)
Naja naja naja (Indian cobra)					
Toxin 3	0.85	10.9	279	1103	MW = 7834 (1103)
	–	10.9	279	1664	
Reduced and carboxy-methylated	0.66	–	279	1103	–
Toxin 4	0.85	10.9	279	1103	MW = 7807 (1103)
Reduced and carboxy-methylated	0.65	–	279	1103	–
Naja haje (cobra)					
Toxin I[q†]	0.917	13.28	280	1104	MW = 6843 (1104)
Toxin I[r†]	0.923	13.49	279	1104	MW = 6843 (1104)
Toxin II[q†]	0.828	12.07	280	1104	MW = 6857 (1104)
Toxin II[r†]	1.123	16.31	278	1104	MW = 6887 (1104)
Toxin III[r†]	0.898	11.48	280	1104	MW = 7806 (1104)
Naja nigricollis	0.87	12.8	279	731	pH 7.3, 0.03 M NaP$_i$, MW = 6787 (731)
	0.84	12.4	279	731	pH 2.3, MW = 6787 (731)
Carboxymethyl-	0.66	–	279	731	–
I	0.901	13.26	280	1663	MW = 6794
II	0.886	13.03	280	1663	MW = 6796
Notechis scutatus scutatus (Australian tiger snake)					
Venom[s†]	2.8	20.6	278	1665	Acid pH
Androctonus australis Hector (scorpion)					
Toxin I	1.071	15.75	275	1105	MW = 6808 (1105)
	1.03	15.12	280	1105	
Toxin II	1.801	24.80	276	1105	MW = 7249 (1105)
	1.67	23.08	280	1105	
Toxin III	1.191	17.45	277.5	1105	MW = 6826 (1105)
	1.169	17.13	280	1105	
Buthus occitanus tunetanus (scorpion)					
Toxin I	1.888	–	278	1105	MW[t†]
	–	26.75	280	1105	–
Toxin II	2.126	28.20	278	1105	MW = 7539 (1105)
	2.079	27.58	280	1105	
Toxin III	1.915	26.34	278	1105	MW = 7270 (1105)
	1.857	25.55	280	1105	
Leirus quinquestriatus quinquestriatus (scorpion)					
Toxin I	1.511	21.81	275	1105	MW = 6928 (1105)
	1.467	21.17	280	1105	
Toxin II	1.340	–	277	1105	MW[u†]
	–	19.86	280	1105	–

[q†] From Miami Serpentarium.
[r†] From Institute Pasteur.
[s†] Protein called "Notexin," Reference 1665.
[t†] MW = 6919–6933.
[u†] MW = 6511–6545.

MOLAR ABSORPTIVITY AND $A_{1cm}^{1\%}$ VALUES FOR PROTEINS AT SELECTED WAVELENGTHS OF THE ULTRAVIOLET AND VISIBLE REGION (continued)

Protein	$\epsilon^a (\times 10^{-4})$	$A_{1cm}^{1\%}$ [b]	nm[c]	Ref.	Comments[d]
Neurotoxin (continued)					
Toxin III	2.153	–	276.5	1105	MW[v†]
	–	30.53	280	1105	–
Toxin IV	2.002	27.38	278	1105 } MW = 7313 (1105)
	1.966	26.89	280	1105	
Toxin V	2.135	28.61	275	1105 } MW = 7462 (1105)
	2.082	27.90	280	1105	
Nicotin oxidase (nicotine dehydrogenase, EC 1.5.99.4)					
A. oxydans	–	23.6	280	358	pH 7.9, 0.1 M PyP$_i$, mM EDTA
Nicotinic acid hydroxylase (nicotinate dehydrogenase, EC 1.5.1.13)					
Clostridium	33	11	275	1106	MW = 300,000 (1106) data from Figure 6 (1106)
Nitrate reductase					
Micrococcus dentrificans	41.9	26	280	1107	pH 7.4, 0.2 M P$_i$, MW = 161,129 (1107) data from Figure 3 (1107)
Nitrite reductase (EC 1.7.99.3)					
Achromobacter cycloclastes	0.16	–	400	1666	
	0.40	–	464	1666 } Dry wt
	0.20	–	590	1666	
	0.17	–	700	1666	
Achromobacter fisheri	1.49	–	525	1667	
	16.6	–	409	1667 } pH 7.0, oxidized
	11.8	–	278–80	1667	
	4.26	–	551	1667	
	2.6	–	523	1667 } pH 7.0, reduced
	21.9	–	420	1667	
Chlorella fusca	2.2	–	384	1668	
Nitrogenase					
Azotobacter vinelandii	47.0	17.4	280	1108 } MW = 270,000 (1108)
	8.5	3.15	412	1108	
Mo-Fe protein	47	17.4	280	1669	pH 7.4, 0.25 M NaCl 0.01 M TrisCl under N$_2$ MW = 270,000 (1669)
	8.5	3.2	418	1669	
Klebsiella pneumonia					
Kp 1	26	–	258.5	1670	
	30	–	269	1670	
	33	–	277.5	1670 } Reduced and oxidized
	32	–	282	1670	
	25	–	289	1670	
	35	–	430	1670	
	50	–	430	1670	Oxidized
Kp 2	11	–	258	1670 } Reduced and oxidized
	11.2	–	268	1670	
	0.4–0.5	–	460	1670	Reduced
	1.0	–	460	1670	Oxygen inactivated

[v†]MW = 6764–6792.

MOLAR ABSORPTIVITY AND $A_{1cm}^{1\%}$ VALUES FOR PROTEINS AT SELECTED WAVELENGTHS OF THE ULTRAVIOLET AND VISIBLE REGION (continued)

Protein	$\epsilon^a (\times 10^{-4})$	$A_{1cm}^{1\%}{}^b$	nm^c	Ref.	Comments^d
Non-histone					
Rat liver	1.8	12.3	275	442	pH 8, 0.01 M Tris, 0.01% NaDodSO$_4$, est. from Figure 1 (442)
Nuclease					
Bacteriophage T4	–	14.8	280	1109	pH 7.5^{w†}
Staphylococcal	–	9.7	277	1110	–
Performic acid oxidized	–	7.0	274	1110	–
Lightly acetylated	–	9.7	277	1110	–
Heavily acetylated	–	9.7	277	1110	–
Trifluoroacetylated	–	9.0	277	1110	–
S. aureus	–	11.6	280	359	pH 7.2, 0.15 N NaP$_i$
3-Nucleotidase (see phosphodiesterase, 2′:3′-cyclic)					
Nucleoside triphosphate–adenylate kinase (EC 2.7.4.10)					
Beef heart mitochondria	4.55	8.75	280	1111	pH 8.5, 0.05 M N(EtOH)$_3$HCl, MW = 52,000 (1111), data from Figure 3 (1111)
Octopine dehydrogenase (EC 1.5.1.11)	4.33	–	280	1671	–
Pecten maximus	–	11.4	280	732	–
Ornithine-oxoacid aminotransferase (EC 2.6.1.13)					
Rat liver	–	10.6	280	736	Data from Figure 3 (736), ph 8, 0.2 M TrisCl
With added ornithine	–	11.0	280	736	
Apo-	–	10.8	280	736	
Ornithine transcarbamylase (EC 2.1.3.3)					
S. faecalis	–	8.32	280	1672	pH 7.0, 50 mM NaP$_i$, dry wt and nitrogen
Beef liver		12.3	280	1672	
Orosomucoid					
Rabbit serum	1.93	6.05	276	1673	MW = 32,000
	1.94	6.06	277	1673	
	1.93	6.04	278	1673	
	1.92	6.00	280	1673	
Orotate reductase (EC 1.3.1.14) (see dihydroorotic dehydrogenase)					
Ovalbumin	2.85	–	280	17	pH 4.9
	–	7.5	280	16	–
	–	7.35	280	361	–
	–	7.5	280	733	
	–	260	210	733	0.1 M NaOH
	–	750	191	733	

^{w†}14 mM Tris-HCl, 140 mM KCl, 1.4 mM β-mercaptoethanol (HSEtOH), and 7.1% glycerol. Protein determined by Lowry method and absorption corrected for scattering.

MOLAR ABSORPTIVITY AND $A_{1cm}^{1\%}$ VALUES FOR PROTEINS AT SELECTED WAVELENGTHS OF THE ULTRAVIOLET AND VISIBLE REGION (continued)

Protein	$\epsilon^a (\times 10^{-4})$	$A_{1cm}^{1\%}$ [b]	nm[c]	Ref.	Comments[d]
Ovalbumin (continued)	–	7.15	280	734	
	2.94	7.35	280	1674	MW = 45,000
	–	3.40	255	1675	
	–	7.60	280	1675	pH 7.0, 0.01 M P$_i$
	–	0.48	310	1675	
	–	7.37	280	1676	Native and SDS denatured. Refr.
	–	7.54	280	1676	6 M GdmCl Refr.
	1.579	–	293	1112	–
	2.792	–	287	1112	–
	21.42	–	232	1112	–
Chicken	3.15	7.01	280	1113	MW = 45,000 (1114)
Chicken egg	3.218	–	280	1677	–
	–	6.9	280	1678	–
	–	7.14	280	1679	–
Turkey	3.74	8.32	280	1113	MW = 45,000 (1114)
Duck	3.72	8.28	280	1113	MW = 45,000 (1114)
Ovoglycoprotein					
Chicken egg	0.93	3.8	280	1680	MW = 24,440 (1680)
Ovoinhibitor					
Chicken	–	7.1	280	1681	pH 6.8, 0.01 M P$_i$
Chicken egg white	–	6.5–6.9	278	1682	–
	–	7.4	278	1115	
Quail egg white	–	7.1	278	1116	pH 8, 0.1 M Tris
	–	7.1	278	1116	
Ovomacroglobulin					
Chicken egg white	–	8.6	278	1117	Dry wt
Ovomucin					
Chicken egg white	–	10.3	290	1118	pH 13
	–	9.3	277.5	1118	Neutral sol.
	–	5.6	290	1118	
Ovomucoid	–	4.55	280	362	–
	1.19	4.13	280	363	MW = 28,800 (363)
Chicken		4.1	280	1681	pH 6.8, 0.01 M P$_i$
	1.17	4.3	277	1683	pH 7.5, water: glycerol, 1:1
Chicken egg white	1.1	4.10	278	1119	MW = 27,300 (1119)
Turkey	–	4.15	280	1684	pH 7.8, Tris
Ovorubin					
Pomacea canaliculata australis	–	10.5	280	735	Data from Figure 3 (735)
Ovotransferrin (conalbumin)					
Chicken	0.475	0.620	470	1685	pH 6–9 MW = 76,600
	8.5	11.1	280	1685	0.02 M HCl, MW = 76,600 (1685)
	–	11.6	280	1686	–
	7.96	–	280	1687	pH 6.5, 6 M Gdn 0.02 M P$_i$

MOLAR ABSORPTIVITY AND $A_{1cm}^{1\%}$ VALUES FOR PROTEINS AT SELECTED WAVELENGTHS OF THE ULTRAVIOLET AND VISIBLE REGION (continued)

Protein	$\epsilon^a (\times 10^{-4})$	$A_{1cm}^{1\%}$ [b]	nm[c]	Ref.	Comments[d]
Ovotransferrin (continued)					
Chicken (continued)	9.03	—	280	1687	pH 10.5, 0.1 M glycine-NaOH
	—	11.2	280	1688	—
Iodate oxidized, pH 8.5	9.2	—	280	1687	pH 6.5, 6 M Gdn 0.02 M P_i
Iodate oxidized at pH 5.0	9.5	—	280	1687	
Iron free, oxidized	7.96	—	280	1687	
Iodate oxidized at pH 8.5	9.33	—	280	1687	pH 10.5, 0.1 M glycine-NaOH
Iodate oxidized at pH 5.0	9.80	—	280	1687	
Iron free, oxidized	9.33	—	280	1687	
Oxaloglycolate reductase (decarboxylating) (EC 1.1.1.92)					
P. putida	7.2	11.8	280	364	MW = 61,000 (364)
3-Oxoacyl-[acyl-carrier-protein] reductase (EC 1.1.1.100) (see also fatty acid synthetase)					
Pig liver	46	—	279	1424	—
3-Oxosteroid $\Delta^{4,5}$-isomerase (steroid Δ-isomerase) (EC 5.3.3.1) (see also 3-keto-Δ^5-steroid isomerase)					
		3.28	280	1689	Refr.
Beef adrenals	11.3	10.1	—	855	MW = 112,000 (855)
Oxynitrilase (see hydroxynitrilase)	—	3.72	277	1689	Refr.
Oxytyrosinase (monophenol monoxygenase or tyrosinase, EC 1.14.18.1, or tyrosine 3-monooxygenase, EC 1.14.16.2)					
Mushroom	0.9	—	345	1690	Value per mole Cu
	0.6	—	600	1690	
Papain (EC 3.4.22.2)	—	25.0	278	365	—
	4.9	—	280	367	MW = 20,700 (366)
	5.1	—	280	368	—
	1.2	—	295	365	pH 5–8
Papaya	—	21.5	280	1691	0.01 N HCl, from Figure 1 (1691)
	—	27	278	1692	—
Papaya latex					
Trans-cinnamoyl-	2.6	—	326	1120	—
Furylacryloyl-	3.0	—	360	1121	—
	3.0	—	337	1121	Denatured
Indolacryloyl-	4.3	—	398	1121	—
	4.3	—	373	1121	Denatured
Mercury derivative	5.1	—	280	1122	—

MOLAR ABSORPTIVITY AND $A_{1cm}^{1\%}$ VALUES FOR PROTEINS AT SELECTED WAVELENGTHS OF THE ULTRAVIOLET AND VISIBLE REGION (continued)

Protein	$\epsilon^a (\times 10^{-4})$	$A_{1cm}^{1\%}$ [b]	nm[c]	Ref.	Comments[d]
Paramyosin					
Venus mercenaria	–	3.24[x†]	277	1123	pH 7.4, 1 M KCl, 0.1 M KP$_i$
Reduced and carboxy-methylated	–	3.39[y†]	277	1123	–
Crassostrea commercialis (oyster)	5.95	2.86	276	1124	pH 7.0, 1.1 M NaCl – 25 mM NaP$_i$ MW = 208,000 (1124)
Aulacomya magellanica (mollusc)	8.76	3.4	280	1125	pH 7.5, 1 M KCl
	9.55	3.7	276	1125	MW = 258,000 (1125)
Parathyroid hormone					
Beef	0.72	7.6	280	369	pH 4.7, AcONH$_4$ MW = 9,500 (369)
Parvalbumin					
Chondrostoma nasus	–	1.84	260	1693	Water, from Figure 7 (1693)
	–	1.57	260	1693	0.1 N NaOH, from Figure 7 (1693)
Merluccius merluccius (hake)	0.202	1.8	259	1694	MW = 11,500 (1694) AA
	0.215	–	259	1695	–
Raja clavata (thornback ray)	0.142	–	275	1695	–
Penicillinase (EC 3.5.2.6)					
Bacillus cereus 569	–	6.0[z†]	280	1126	–
	3.2	10.5	280	1127	Biuret
Bacillus cereus 569/H	–	6.35[z†]	280	1126	–
Bacillus cereus 5/B	–	7.35[z†]	280	1126	–
B. licheniformis 6346/C	–	5.65[z†]	280	1126	–
B. licheniformis 749/C[c]	–	5.45[z†]	280	1126	–
B. licheniformis 749/C	–	4.75[z†]	280	1126	–
Staphylococcus aureus PC 1	–	7.38[z†]	280	1126	–
E. coli K 12	6.1	21.0	280	1128	pH 6.8, 0.01 M KP$_i$ MW = 29,000 (1128)
Penicillocarboxypeptidase-S (peptidase B)					
Penicillium janthinelium	–	26	280	1696	–
Penicillopepsin (EC 3.4.23.7)					
Penicillium janthinellum	4.32	13.5	280	1129	MW = 32,000 (1129)
Pepsin	–	5.38	255	1675	
	–	12.20	280	1675	pH 7.0, 0.01 M P$_i$
	–	0.28	310	1675	
	1.94	–	292.8	1697	
	3.68	–	286	1697	Denatured at pH
	5.02	–	279	1697	1.45, 3 hr, 39°C
	23.87	–	230	1697	

[x†] In GdmCl: $\epsilon_{277} = 0.035$ (M GdmCl) + 3.24.
[y†] In GdmCl: $\epsilon_{277} = 0.34$ (M GdmCl) + 3.39.
[z†] $A_{1cm}^{1\%}$/mg N.

MOLAR ABSORPTIVITY AND $A_{1cm}^{1\%}$ VALUES FOR PROTEINS AT SELECTED WAVELENGTHS OF THE ULTRAVIOLET AND VISIBLE REGION (continued)

Protein	$\epsilon^a (\times 10^{-4})$	$A_{1cm}^{1\%}$ [b]	nm[c]	Ref.	Comments[d]
Pepsin (continued)	—	13.1	280	1698	
	—	290	210	1698	Water, dry wt
	—	920	191	1698	
	5.247	—	278	1699	—
Beef	4.94	14.81	280	1700	MW = 33,367 (1700)
	5.17	14.3	280	370	—
	5.09	—	278	371	—
Pig	—	14.1	280	1701	—
	—	780	191–4	1701	—
Chicken	5.34	15.2	276	1702	pH 7.5, 0.1 M P$_i$,
	—	150	220	1702	MW = 35,000 (1702)
Human	6.2	17.3	278	179	pH 5, MW = 34,000 (179)
Pepsinogen					
Beef	5.6	12.5	278	373	MW = 41,000 (373)
	5.1	13.05	280	374	MW = 39,000 (374)
Succinyl-	5.5	—	278	375	pH 7.7, 0.032 M P$_i$
Bovine 1	—	13.45	280	1703	—
2	—	13.05	280	1703	—
4	—	13.25	280	1703	—
Dog	—	12.79	280	1704	—
Chicken	5.59	13.0	278	1702	pH 1.83, 0.02 N HCl, MW = 43,000 (1702)
	—	12.66	280	1702	—
	—	139.9	220	1702	
Peptidoglutaminase I peptidyl-glutaminase, EC 3.5.1.43)	—	7.27	280	1705	—
Peptidoglutaminase II (glutaminyl-peptide glutaminase) (EC 3.5.1.44)	—	11.62	280	1705	—
Peroxidase (EC 1.11.1.7) (see also lactoperoxidase, myeloperoxidase)					
Raphanus sativus (Japanese radish)	1.175	—	500	739	—
Isoenzyme 3	2.72	—	280	1130	—
	11.14	—	404	1130	—
	1.17	—	502	1130	—
	0.323	—	644	1130	—
Isoenzyme 5	3.82	—	280	1130	—
	11.66	—	405	1130	—
	1.156	—	502	1130	—
	0.338	—	644	1130	—
Isoenzyme 16	2.97	—	280	1130	—
	10.42	—	403	1130	—
	1.186	—	504	1130	—
	0.339	—	645	1130	—
Apo-	1.72	—	—	739	—
Brassica napus L					
P$_1$	3.8	—	276	740	—
P$_2$	3.4	—	276	740	—

MOLAR ABSORPTIVITY AND $A_{1cm}^{1\%}$ VALUES FOR PROTEINS AT SELECTED WAVELENGTHS OF THE ULTRAVIOLET AND VISIBLE REGION (continued)

Protein	$\epsilon^a(\times 10^{-4})$	$A_{1cm}^{1\%}$ [b]	nm[c]	Ref.	Comments[d]
Peroxidase (continued)					
P_3	3.0	—	277	740	—
P_6	3.0	—	278	740	—
P_7	2.9	—	278	740	—
Japanese radish, a					
(see also peroxidase a, apo-)					
Oxidized	3.38	—	276	379	—
	11.10	—	405	379	—
	1.18	—	500	379	—
	0.33	—	645	379	—
Japanese radish, c					
Oxidized	3.0	—	280	379	—
	10.64	—	420	379	—
	1.17	—	540	379	—
Reduced	10.38	—	425	379	—
	1.26	—	560	379	—
Horseradish	—	22	403	1706	—
	9.1	—	403	1707	—
	10.2	—	403	737	—
	9.1	—	403	376	—
	—	13.4	275	377	pH 7.0
	10.4	—	498	378	Ferric derivative
Nitroso-	11.0	—	420	378	—
Carbon monoxide	15.3	—	422	378	—
Apoenzyme-A1	0.92	—	276	738	pH 6.8, 50 mM NaP_i
Apoenzyme-C	1.3	—	277	738	
A_1	10.2	—	401	1708	
A_2	10.2	—	401	1708	
A_3	9.7	—	401	1708	MW = 40,000 (1708)
B	9.5	—	401	1708	
C	9.5	—	401	1708	
Component I	1.31[a‡]	—	440	1709	—
	1.74[a‡]	—	432	1709	—
	1.86[a‡]	—	430	1709	—
	2.83[a‡]	—	420	1709	—
	4.12[a‡]	—	410	1709	—
	4.80[a‡]	—	400	1709	—
	4.62[a‡]	—	390	1709	—
	4.30[a‡]	—	380	1709	—
	4.00[a‡]	—	370	1709	—
	3.60[a‡]	—	360	1709	—
	3.28[a‡]	—	350	1709	—
Fraction Ib	2.78[b‡]	—	280	1710	—
	11.5[b‡]	—	403	1710	—
Fraction IIIb	2.87[b‡]	—	280	1710	—
	9.98[b‡]	—	403	1710	—
Fraction Vb	3.68[b‡]	—	280	1710	—
	9.50[b‡]	—	280	1710	—
Fraction VI	4.08[b‡]	—	403	1710	—
	12.29[b‡]	—	403	1710	—
Native, oxidized	0.323	—	641	1711	
	1.095	—	498	1711	pH 7.0, 10 mM NaP_i
	10.00	—	402	1711	

[a‡] All values based on $\epsilon_m = 9.14 \times 10^4$ at 430 nm.
[b‡] Based on hemin content.

MOLAR ABSORPTIVITY AND $A_{1cm}^{1\%}$ VALUES FOR PROTEINS AT SELECTED WAVELENGTHS OF THE ULTRAVIOLET AND VISIBLE REGION (continued)

Protein	$\epsilon^a (\times 10^{-4})$	$A_{1cm}^{1\%}$ [b]	nm[c]	Ref.	Comments[d]
Peroxidase (continued)					
Protoperoxidase	0.289	—	641	1711	pH 7.0, 10 mM NaP$_i$, oxidized
	1.046	—	499	1711	
	9.43	—	402	1711	
Dipropenyldeutero-hematinperoxidase	0.287	—	635	1711	
	1.122	—	500	1711	
	10.22	—	402	1711	
Dibutenyldeutero-hematinperoxidase	0.298	—	636	1711	
	1.123	—	502	1711	
	11.18	—	404	1711	
Deuterohematinperoxidase	0.272	—	627	1711	
	0.708	—	497	1711	pH 7.0, 10 mM NaP$_i$
	11.26	—	393	1711	
Mesoperoxidase	0.275	—	634	1711	
	0.942	—	492	1711	
	8.80	—	395	1711	
Diacetyldeutero-hematinperoxidase	0.226	—	642	1711	
	0.732	—	508	1711	
	8.01	—	412	1711	
Hematoperoxidase	0.299	—	633	1711	
	0.850	—	497	1711	
	10.27	—	397	1711	
Native, reduced	1.260	—	556	1711	
	8.86	—	436	1711	
Protoperoxidase	1.255	—	555	1711	
	7.53	—	433	1711	
Dipropenyldeutero-hematinperoxidase	1.213	—	555	1711	
	8.45	—	433	1711	
Dibutenyldeutero-hematinperoxidase	1.251	—	556	1711	pH 7.0, 10 mM NaP$_i$, reduction by dithionite
	8.49	—	433	1711	
Deuterohematinperoxidase	1.000	—	546	1711	
	7.90	—	427	1711	
Mesoperoxidase	1.156	—	549	1711	
	8.80	—	428	1711	
Diacetyldeutero-hematinperoxidase	1.037	—	562	1711	
	7.82	—	447	1711	
Hematoperoxidase	1.093	—	549	1711	
	8.91	—	431	1711	
Fig latex (*Ficus carica*)	10.1	—	403	1713	—
	1.16	—	500	1713	—
	0.33	—	640	1713	—
Reduced	9.21	—	438	1713	—
	1.26	—	556	1713	—
Reduced + CO	15.52	—	423	1713	—
	1.31	—	543	1713	—
	1.38	—	573	1713	—
+OH$^-$	10.6	—	417	1713	—
	1.02	—	544	1713	—
	0.79	—	574	1713	—
+NaN$_3$	11.9	—	415	1713	—
	0.93	—	535	1713	—
	0.23	—	637	1713	—

MOLAR ABSORPTIVITY AND $A_{1cm}^{1\%}$ VALUES FOR PROTEINS AT SELECTED WAVELENGTHS OF THE ULTRAVIOLET AND VISIBLE REGION (continued)

Protein	$\epsilon^a (\times 10^{-4})$	$A_{1cm}^{1\%}$ [b]	nm[c]	Ref.	Comments[d]
Peroxidase (continued)					
+NaF	15.55	—	404	1713	—
	0.93	—	490	1713	—
	0.57	—	561	1713	—
	0.78	—	613	1713	—
+NaCN	10.25	—	421	1713	—
	1.17	—	540	1713	—
Complex I	5.8	—	402	1713	—
	0.76	—	562	1713	—
	0.62	—	657	1713	—
Complex II	9.6	—	419	1713	—
	0.9	—	527	1713	—
	0.95	—	558	1713	—
Complex III	10.45	—	418	1713	—
	1.17	—	546	1713	—
	1.01	—	583	1713	—
FPO-A	3.78	—	280	1713	
FPO-B	3.31	—	280	1713	Calculated
FPO-C	3.45	—	280	1713	
Pig intestinal mucosa	—	15.5	280	1714	—
	—	14.07	417	1714	—
	—	1.54	490	1714	—
	—	1.27	543	1714	—
	—	1.59	596	1714	—
	—	0.89	642	1714	—
Peroxidase II					
Horseradish, oxidized	3.27	—	270	279	—
	10.77	—	403	279	—
	1.19	—	497	379	—
	0.34	—	641.5	379	—
Reduced	9.17	—	437	379	—
	1.33	—	556	379	—
Peroxidase a, apo- Japanese radish roots (*Raphanus salivus*)	1.22	2.7	280		MW = 45,300 (77)
Phenolase, mushroom (monophenol monooxygenase, EC 1.14.18.1)	—	26.92	282	380	—
Phenolhydroxylase (phenol 2-monoxygenase) (EC 1.14.13.7)					
Trichosporon cutaneum	—	9.87	276	1715	pH 7.6, 0.1 M KP$_i$, from Figure 5 (1715)
Phenoloxidase, pre- (ct. pre-phenoloxidase)					
L-Phenylalanine ammonia lyase (EC 4.3.1.5)					
Maize (*Zea mays*)	—	8.9	280	1716	AA

MOLAR ABSORPTIVITY AND $A_{1cm}^{1\%}$ VALUES FOR PROTEINS AT SELECTED WAVELENGTHS OF THE ULTRAVIOLET AND VISIBLE REGION (continued)

Protein	$\epsilon^a (\times 10^{-4})$	$A_{1cm}^{1\%}$ [b]	nm[c]	Ref.	Comments[d]
Phenylalanine 4-monooxygenase (EC 1.14.16.1)					
Rat liver	–	5.6	280	1717	pH 8.6, 0.01 M TrisCl, from Figure 3 (1717)
Phenylalanine monooxygenase-stimulating protein					
Rat liver	4.67	9.06	280	1718	Biuret MW = 51,500 (1718)
Phenylpyruvate tautomerase (EC 5.3.2.1)					
Pig thyroid	5.95	13.5	280	741	pH 6.2, 0.1 M P_i, MW = 44,000 (741)
Phosphatase, acid (EC 3.1.3.2)					
Human prostate	–	24	280	1746	–
Phaseolus mungo	8.75	16	278	1747	pH 5.6, MW = 55,000 (1747)
Rat, liver	4.7	4.7	278	387	MW = 100,000 (387)
	6.18	6.18	278	1724	MW = 100,000 (1725)
Sweet potato	–	9.1	280	1726	pH 6.0, 0.01 M KP_i,
	–	0.21	555	1726	from Figure 2 (1726)
N. crassa	9.18	10.8	280	1727	Water, MW = 85,000 (1727)
Phosphatase, alkaline (EC 3.1.3.1)					
Escherichia coli	6.4	7.2	278	742	MW = 89,000 (742)
	0.0260	–	640	1722	–
	0.0220	–	605	1722	–
	0.0378	–	555	1722	–
	0.0335	–	510	1722	–
	–	7.2	278	1721	–
E. coli C90	6.2	7.7	280	381	MW = 80,000 (381)
E. coli K 12	5.6	7.0	278	382	MW = 80,000 (382)
	6.6	7.7	280	743	
Azatryptophan substituted for tryptophan	11	12.8	280	743	MW = 86,000 (744)
Tryptazan substituted for tryptophan	9.6	11.1	280	743	
B. licheniformis	7.5[c‡]	6.2	278	1719	MW = 121,000 (1719)
B. subtilis	0.0305	–	620	1720	–
	0.0335	–	596	1720	–
	0.0500	–	567	1720	–
	0.0382	–	517	1720	–
Micrococcus sodonensis	13.7	17.3	280	1723	–
N. crassa	17.4	11.3	280	383	pH 8.3, 0.01 M Tris, HCl, MW = 154,000 (383)
A. nidulans	–	10.2	278	384	pH 7.0
Intestinal	7.0	–	280	385	–
Placental, human	9.75	7.8	278	386	pH 7.0, 0.05 M P_i, MW = 125,000 (386)

[c‡] On p. 1369 of Reference 1719, value is given as 7.25×10^4.

MOLAR ABSORPTIVITY AND $A_{1cm}^{1\%}$ VALUES FOR PROTEINS AT SELECTED WAVELENGTHS OF THE ULTRAVIOLET AND VISIBLE REGION (continued)

Protein	$\epsilon^a (\times 10^{-4})$	$A_{1cm}^{1\%}$ [b]	nm[c]	Ref.	Comments[d]
Phosphatase, alkaline (continued)					
Calf intestine	—	10	280	745	pH 7.7
Milk	—	11.5	280	745	—
Pig kidney	—	13.9	280	746	—
	—	12.0	260	746	—
Phosphodiesterase 2′:3′-cyclic (2′:3′-cyclic-nucleoside monophosphate phosphodiesterase, EC 3.1.4.16)					
B. subtilis	—	11	280	1728	pH 7.5, 0.05 M TrisCl
Phospho*enol*pyruvate carboxykinase (EC 4.1.1.32, .38, .49)					
Baker's yeast	32	12.7	280	1729	pH 7.01 M EDTA – 0.025 M Na-Borate MW = 252,000 (1724)
Phospho*enol*pyruvate carboxylase (EC 4.1.1.31)					
E. coli	63.4	15.5	280	1730	MW = 402,000 (1730) dry wt
E. coli, str. B	—	10.9	280	1731	pH 8.5, 0.1 M TrisCl, 10 mM $MgCl_2$, 10 mM $KHCO_3$ 0.01 N NaOH
S. typhimurium	—	15.5	280	1732	—
Phosphofructokinase					
Rabbit muscle	—	10.2	279	388	—
	—	9.4	279	389	pH 7
	—	8.7	283	389	0.1 N NaOH
	—	8.7	290	389	0.1 N NaOH
	—	10.9	290	390	0.1 N NaOH
Sheep heart	—	10.0	280	391	
Yeast	72.4	12.4	279	392	MW = 584,000 (392)
	54.2	9.5	279	1733	pH 7.0, 0.1 M P_i, MW = 570,000 (1733)
Phosphoglucoisomerase					
Rabbit muscle	17.2	13.2	280	393	pH 7.0, 0.01 M P_i, MW = 130,000 (393)
Phosphoglucomutase (EC 2.7.5.1)					
Rabbit muscle	—	7.7	278	394	—
	5.98	7.8	278	395	MW = 77,000 (395)
	4.77	7.7	278	1734	MW = 62,000 (1735)
Yeast	8.2	11.8	280	1736	MW = 69,500 (1736), turbidimetric
6-Phosphogluconate dehydrogenase (EC 1.1.1.44)					
C. utilis, Type I	12.8	12.7	280	396	MW = 101,000 (396)
Type II	14.1	12.7	280	396	MW = 111,000 (396)
Sheep liver	—	10.3	280	622	—
	10.7	11.4	280	1737	MW = 94,000 (1737), dry wt

MOLAR ABSORPTIVITY AND $A_{1cm}^{1\%}$ VALUES FOR PROTEINS AT SELECTED WAVELENGTHS OF THE ULTRAVIOLET AND VISIBLE REGION (continued)

Protein	$\epsilon^a (\times 10^{-4})$	$A_{1cm}^{1\%}$ [b]	nm[c]	Ref.	Comments[d]
Phosphoglucose isomerase (glucosephosphate isomerase, EC 5.3.19)					
Rabbit muscle	17.4	13.2	280	623	pH 7.0, 0.01 M P$_i$, MW = 132,000 (623)
Human muscle	—	12	280	624	pH 7.2
Human erythrocytes	—	13.1	280	1738	Neutral pH, refr.
Phosphoglycerate dehydrogenase					
E. coli	11.2	6.7	280	397	MW = 165,000 (397)
3-Phosphoglycerate kinase (EC 2.7.2.3)					
Rabbit muscle	3.12	6.9	280	1739	MW = 45,200 (1739)
Yeast	2.24	4.9	280	1739	MW = 45,800 (1739)
	—	5.0	280	141	—
Rabbit muscle	2.16	5.7	280	1740	MW = 38,000 (1740)
Yeast	—	5.35[d‡]	278	1741	Dry wt
Phosphoglycerate phosphomutase (EC 5.4.2.1)					
Yeast, Component I	—	14.2	280	1742	—
Component II	—	14.9	280	1742	—
Phosphoglyceromutase (EC 2.7.5.3)					
Rabbit muscle	—	12.5	280	398	—
Yeast	—	14.2	280	399	—
Sheep muscle	—	7.1	280	1743	pH 7.0, 0.1 μm P$_i$
Phospholipase A$_2$ (EC 3.1.1.4)					
Agkistrodon halys blomhoffi (snake)					
A-I	2.06[e‡]	14.9	278.5	1744	pH 7.2, 0.1 M NaP$_i$, MW = 13,800 (1744)
A-II	2.06[f‡]	15.1	278.5	1744	pH 7.2, 0.1 M NaP$_i$, MW = 13,700 (1744)
Pig pancreas	—	14.2	280	1745	pH 8. AA
Crotalus adamanteus	6.76	22.7	280	401	MW = 29,864 (401)
Phosphomonoesterase, acid (see phosphatase, acid)					

[d‡] Value may be in error by 10%. (1741).

[e‡] From Figure 12 of Reference 1744 the following values were estimated: ϵ_M = 7500 (277 nm, pH 1.2, 0.1 M HCl); ϵ_M = 9500 (278.5 nm, pH 7.2, 0.1 M NaP$_i$); ϵ_M = 9800 (290 nm, pH 12.5, 0.1 M NaOH). The value at pH 7.2 does not agree with the value calculated from the molecular weight, 13,800, and the $A_{1cm}^{1\%}$ value of 14.9.

[f‡] From Figure 13 of Reference 1744, the following values were estimated: ϵ_M = 8800 (277 nm, pH 1.2, 0.1 M HCl); ϵ_M = 10,000 (278.5 nm, pH 7.2, 0.1 M NaP$_i$); ϵ_M = 11,200 (290 nm, pH 12.5, 0.1 N NaOH). The value at pH 7.2 does not agree with the value calculated from the molecular weight, 13,700, and the $A_{1cm}^{1\%}$ of 15.

MOLAR ABSORPTIVITY AND $A_{1cm}^{1\%}$ VALUES FOR PROTEINS AT SELECTED WAVELENGTHS OF THE ULTRAVIOLET AND VISIBLE REGION (continued)

Protein	$\epsilon^a (\times 10^{-4})$	$A_{1cm}^{1\%}$ [b]	nm[c]	Ref.	Comments[d]
5-Phosphoriboisomerase (ribosephosphate isomerase, EC 5.3.1.6)					
Spinach	–	4.35	280	1748	Ref./Biuret
Phosphoribosyl-ATP pyrophosphorylase (ATP phosphoribosyl transferase, EC 2.4.2.17)					
S. typhimurium	16.0	7.45	280	400	MW = 215,000 (400)
Phosphoribosyldiphosphate amidotransferase					
Chicken liver	–	12.5	280	1749	pH 8.0, 0.05 *M* TrisCl from Figure 1 (1749)
N-(5'-Phospho-D-ribosyl formimino)-5-amino-(5''-phosphoribosyl)-4-imidazolecarboxamide isomerase (see isomerase)					
Phosphoribosyl formylglycinamidine synthetase (see formylglycinamide ribonucleotide amidotransferase)					
Phosphoribosyltransferase (ATP phosphoribosyltransferase, EC 2.4.2.17)					
S. typhimurium	–	10.7	280	1750	pH 7.5, 6 *M* GdmCl– 0.025 *M* TrisCl
Phosphorylase					
Rabbit muscle	–	11.8	277	405	pH 8.0, 0.1 *M* HCO$_3^-$
Pig liver	–	11.9	277	406	–
Frog muscle	–	12.8	288	407	–
Rabbit muscle	–	13.2	280	407	–
	–	12.3[g‡]	280	408	pH 6.8, 5 m*M* P$_i$ dry wt
	–	13.1	279	409	pH 7, 0.01 *M* NaP$_i$
	–	11.5	278	410	–
	–	11.9	278	411	–
	–	13.2[h‡]	280	404	–
Apo-	–	12.0	278	411	–
Human	28.8	11.9	278	410	MW = 242,000 (410)
Potato	–	11.7	278	1757	pH 7.5, 0.005 *M* TrisCl
Rat muscle	–	12.5	280	1751	–
Beef spleen	21.8	11.5	278	1752	pH 7.4, 0.01 *M* TrisCl, 2 m*M* HSEtOH, MW = 190,000 (1752), dry wt
	1.1	0.565	333	1752	

[g‡] $A_{1cm}^{1\%}$ by amino acid anaylsis, 13.2.

[h‡] $A_{1cm}^{1\%}$ by Biuret and amino acid analysis, 13.5; by refractive index increment, 13.2.

MOLAR ABSORPTIVITY AND $A_{1cm}^{1\%}$ VALUES FOR PROTEINS AT SELECTED WAVELENGTHS OF THE ULTRAVIOLET AND VISIBLE REGION (continued)

Protein	$\epsilon^a (\times 10^{-4})$	$A_{1cm}^{1\%}$ [b]	nm[c]	Ref.	Comments[d]
Phosphorylase (continued)					
Rabbit muscle	–	13.1	280	1753	pH 7, 50 mM P_i, 1.0 mM EDTA, 1.5 mM HSEtOH MW = 200,000 (1753)
Pacific dogfish (*Squalus sucklii*)	25.8	12.9	280	1753	
Baker's yeast (*Saccharomyces cerevisiae*)					
a	15.3	14.9	280	1754	Refr. MW = 103,000 (1754)
b	15.3	14.9	280	1754	Refr. MW = 103,000 (1754)
Silky shark (*Carcharhinus falciformis*)					
b	–	13.0	280	1755	pH 7.5, 0.15 M Tris, 0.01 M EDTA, N
Lobster (*Homarus americanus*)					
b	–	13.5	280	1756	pH 6.9, 0.04 M Tris, 0.01 M EDTA
Phosphorylase kinase (EC 2.7.1.38)					
Rabbit muscle	–	12.4	280	1758	Refr.
	–	11.8	280	1759	pH 7.0, 5 mM P_i – 0.2 mM EDTA, Biuret
Phosphorylase, purine-nucleoside					
Beef liver	–	16	280	402	–
Human erythrocyte	8.91	11	280	403	pH 7.5, MW = 81,000 (403)
Phosphotransacetylase (phosphate acetyltransferase, EC 2.3.1.8)					
C. kluyveri	–	4.2	280	415	Calculated from data
Phosphovitin					
Salmon	1.8	9.4[i‡]	280	1760	pH 2, from Figure 2 (1760), MW = 19,000 (1760)
Trout	1.8	9.3[i‡]	280	1760	pH 2, from Figure 2 (1760), MW = 19,350 (1760)
Phosphovitin kinase					
Calf brain	–	15.7	280	1761	Lowry
Phycocyanin					
Chroomonas sp.	–	114	645	1763	pH 6.0, dry wt
Synechococcus sp.	10.40	–	352	1764	MW = 36,700 (1764)
	10.65	–	662.5	1764	
α-Subunit	3.26	–	352	1764	MW = 17,700 (1764)
	3.32	–	662.5	1764	
β-Subunit	6.63	–	352	1764	MW = 19,000 (1764)
	6.95	–	662.5	1764	
Allophycocyanin	3.17	–	352	1764	MW = 16,500 (1764)
	3.22	–	662.5	1764	

[i‡] Same values at pH 7.

MOLAR ABSORPTIVITY AND $A_{1cm}^{1\%}$ VALUES FOR PROTEINS AT SELECTED WAVELENGTHS OF THE ULTRAVIOLET AND VISIBLE REGION (continued)

Protein	$\epsilon^a (\times 10^{-4})$	$A_{1cm}^{1\%}$ [b]	nm[c]	Ref.	Comments[d]
Phycocycenin (continued)		72	610	1765	
Synechcococcus lividus	–	72	610	1765	pH 7.0, 0.01 M P_i,
Plectonema calothricoides	–	72	620	1765	from Figure 1 (1765)
Phormidium luridum	–	19	620	1765	pH 6.2 M urea, from Figure 2 (1765)
	–	9.2	620	1765	pH 6, 4 M urea
	–	10	620	1765	pH 6, 6 M and 8 M urea
Coccochloris elabens	–	60	620	1766	pH 6.0
Phycocyanin, anacystis	9.9	7.9	615	412	MW = 12,500 (412)
C-Phycocyanin					
Alga (*Anacystis nidulans*)					
Monomer	23	–	615	1762	pH 7.0, 0.05 M P_i
Hexamer	33	–	621	1762	pH 5.5, 0.2 M AcOH
α-Subunit	9.8	–	620	1762	pH 7.0, 0.05 M P_i
β-Subunit	14.3	–	608	1762	
Phycoerythrin					
Porphyridium cruentin	2.4	2.73	565	412	MW = 87,000 (412)
Phytochrome					
Oat seedlings	–	12.5	280	1767	–
	–	12.5	660	1767	–
Winter rye					
(*Secale cereale*)	9	–	280	1768	Red-absorbing
	9	–	280	1768	Far-red-absorbing
	7	–	665	1768	Red-absorbing
	4	–	730	1768	Far-red-absorbing
Phytohemagglutinin					
Robinia pseudoaccacia	–	9.65	278	1769	–
Pigeon droppings					
Old	–	27.5	280	1770	pH 7, dialyzed and
Fresh	–	30.5	280	1770	lyophilized extracts
Pigment					
Serum, eel					
(*Anguilla japonica*)	4.41	–	279	1771	
	0.948	–	383–4	1771	MW = 89,100 (1771)
	0.222	–	704–5	1771	
Pinguinain (EC 3.4.99.18)	–	22.0	280	413	–
	4.72	24.6	280	414	pH 7.3 and pH 4.6, MW = 19,200 (414)
Plasma					
Human, basic B_2	0.299	3.32	278	438	pH 6.0, 0.1 M NaCl, MW = 9,000 (438)
Plasmin (EC 3.4.21.7)					
Human, urokinase activated	–	16.7	280	418	0.1 N NaOH
Streptokinase activated	–	17.3	280	418	0.1 N NaOH
Human	–	19.4	280	419	0.01 N HCl
	–	20.0	280	419	0.01 N HCl
Plasmin, iPr$_2$$P$-human	–	16.8	280	1772	AA

MOLAR ABSORPTIVITY AND $A_{1cm}^{1\%}$ VALUES FOR PROTEINS AT SELECTED WAVELENGTHS OF THE ULTRAVIOLET AND VISIBLE REGION (continued)

Protein	$\epsilon^a (\times 10^{-4})$	$A_{1cm}^{1\%}$ [b]	nm[c]	Ref.	Comments[d]
Plasminogen					
Human	—	17.1	280	418	0.1 N NaOH
	—	18.4	280	419	0.01 N HCl
	—	21.7	280	419	0.01 N HCl, ascending portion of peak on DEAE-cellulose
	—	19.8	280	419	0.01 N HCl, descending portion of peak on DEAE-cellulose
	—	16.1	280	1773	—
Human, A	—	16.8	280	1772	AA
B	—	16.8	280	1772	AA
Plastocyanin					
Comfrey (*Symphytum officinale*)	0.45	—	597	1774	Extinction coefficient of copper chromophore
Elder (*Sambucus nigra*)	0.45	—	597	1774	
Nettle (*Urtica dioica*)	0.45	—	597	1774	
Dog's mercury (*Mercurialis perennis*)	0.45	—	597	1774	
Goose grass (*Galium aparine*)	0.45	—	597	1774	
Lettuce (*Lactuca sativa*)	4.5[j‡]	—	597	1774	
Spinach (*Spinacea oleracea*)	0.118	—	460	1775	—
	0.980	—	597	1775	—
	0.330	—	770	1775	—
French beau (*Phaseolus vulgaris*)	0.45	4.2	597	1776	MW = 10,690 (1776)
	0.45	4.2	278	1776	
Pokeweed mitogen *Phytolacca americana*	5.9	18.5	280	416	pH 6 and water, MW = 32,000 (416)
Polymerase, DNA (EC 2.7.7.7)	92.7	8.5	280	417	10 mM NH$_4$HCO$_3$, MW = 109,000 (417)
Polynucleotide phosphorylase (EC 2.7.7.8)					
M. luteus	—	4.30	280	1777	N: assumed 16.5%
Form I	—	5.30	280	1777	Lowry
Form T	—	4.40	280	1777	N: assumed 16.5%
	—	4.40	280	1777	Lowry
Polyphenol oxidase (monophenol monooxygenase, EC 1.14.18.1)					
Mushroom	—	26.92	280	420	—
Camellia sinensis L.	—	13.5	279	421	pH 6.8, 0.3 M NaP$_i$
	—	0.84	611	421	—

[j‡] This is given as 45 × 10³ M (1774). It may be a typographical error.

MOLAR ABSORPTIVITY AND $A_{1cm}^{1\%}$ VALUES FOR PROTEINS AT SELECTED WAVELENGTHS OF THE ULTRAVIOLET AND VISIBLE REGION (continued)

Protein	$\epsilon^a (\times 10^{-4})$	$A_{1cm}^{1\%}$ [b]	nm[c]	Ref.	Comments[d]
Polysaccharide depolymerase					
Aerobacter aerogenes	41	10.8	280	1778	MW = 379,000 (1778)
Porphyrenglobin					
Human, fast moving	13.3	—	403	1779	—
	1.01	—	506	1779	—
	0.91	—	542	1779	—
	0.57	—	568	1779	—
	0.37	—	621.5	1779	—
Slow moving	13.3	—	403	1779	—
	0.99	—	506	1779	—
	0.92	—	542	1779	—
	0.58	—	568	1779	—
	0.38	—	621.5	1779	—
Postalbumin, 4.6S					
Human serum	—	8.0	280	1781	—
Post-γ-globulin					
Human urine	—	9.1	280	1780	pH 7
Prealbumin					
Human serum	8.5	13.3	280	1782	MW = 64,000 (1782)
	—	14.4	280	1783	pH 7.0
	—	14.1	280	1784	Dry wt
	—	12.2	280	1785	—
Tryptophan-rich	—	13.2	280	1786	—
Thyroxine-binding	—	12.3	280	1787	—
	9.93	13.6	280	1788	MW = 73,000 (1788)
Cynomolgus monkey					
(*Macaca irus*)	8.71	15	280	1789	MW = 58,000 (1789)
Rhesus monkey					
Pt-1-1	—	14.4	280	1790	—
Pt-2-2	—	14.4	280	1790	—
Chicken egg yolk	—	18.5	280	1791	pH 7.0, 0.05 M NaP$_i$
Mouse					
Urinary	—	6.0	280	422	—
Serum	—	8.0	280	422	—
Pre-phenoloxidase					
Bombyx mori	—	13.0	280	1792	—
	0.029	—	650	1792	Per atom copper
Principle, sweet					
Dioscoreophyllum cumminsii	—	16.2	278	1793	pH 5.6
	—	17.6	288	1793	pH 13
Procarboxypeptidase					
A					
Cobalt derivative	0.0110	—	500	1794	—
	0.0140	—	555	1794	—
	0.0140	—	570	1794	—
Shrimp					
(*Penaeus setiferus*)	—	25.8	280	1795	AA
Spiny Pacific dogfish					
(*Squalus acanthias*)	—	16.5	280	1796	pH 8.0, 0.1 M Tris, 0.01 M CoCl$_2$, Refr.

MOLAR ABSORPTIVITY AND $A_{1cm}^{1\%}$ VALUES FOR PROTEINS AT SELECTED WAVELENGTHS OF THE ULTRAVIOLET AND VISIBLE REGION (continued)

Protein	$\epsilon^a (\times 10^{-4})$	$A_{1cm}^{1\%}$ [b]	nm[c]	Ref.	Comments[d]
Procarboxypeptidase (continued)					
Beef pancreas	18.2	19	280	77	MW = 96,000 (77)
S_5	11.5	17.7	280	1797	Refr. MW = 63,000 (1791)
S_6	16.5	19	280	77	MW = 87,000 (77)
B					
Shrimp					
(*Penaeus setiferus*)	—	27.8	280	1795	UC
African lungfish					
(*Protopterus aethiopicus*)	7.3	16.2	280	1798	Refr. MW = 45,000 (1798)
Beef	9.2	16	280	83	MW = 57,400 (83)
Proelastase					
Pig pancreas	4.4	17.0	280	1799	MW = 25,840 (1799)
	—	15.8	280	1800	—
Progesterone-binding globulin					
Pregnant guinea pig serum	—	7.3	280	1801	—
Proinsulin					
Beef	—	7.0	276	1802	—
	—	5.9	276	1803	pH 7.2, Ph, 0.1 N NaCl, 10^{-5} M EDTA
Pig	—	6.67	276	423	pH 7.0, 0.03 M P_i
Prolactin					
Sheep	2.05	9.09[k‡]	280	1804	—
Proline hydroxylase (EC 1.14.11.2), collagen (see collagen)					
Protease (proteinase)					
Chinese gooseberry					
(*Actinidia chinensis*)		21.2	280	1809	—
Acremonium Kiliense	2.73	10.1	280	1826	1mM HCl, dry wt, MW = 27,000
Agkistrodonhalys blomhoffi					
a	4.54	9.08	280	424	Water, MW = 50,000 (424)
b	7.0	7.4	280	425	MW = 95,000 (425)
c	7.7	10.98	280	424	Water, MW = 70,000 (424)
Alternaria tenuissima	3.2	13	277	426	0.1 M AcONa, MW = 24,750 (426)
Aspergillus candidus		7.1	280	1827	Dry wt
Aspergillus flavus	1.61	9.04	280	1828	pH 5, MW = 17,800 (1828)
Aspergillus sojae	2.03	8.98	280	1829	MW = 22,600 (1829)
I	6.96	16.7	280	1810	pH 7.3, 50 mM TrisCl, dry wt, MW = 41,700 (1810)
II	1.78	9.0	280	1810	pH 7.3, 50 mM TrisCl, dry wt, MW = 19,800 (1810)

[k‡] Personal communication from Dr. Li. The value cited in Reference 1804 is in error. The correct value is 9.09.

MOLAR ABSORPTIVITY AND $A_{1cm}^{1\%}$ VALUES FOR PROTEINS AT SELECTED WAVELENGTHS OF THE ULTRAVIOLET AND VISIBLE REGION (continued)

Protein	$\epsilon^a (\times 10^{-4})$	$A_{1cm}^{1\%}$ [b]	nm[c]	Ref.	Comments[d]
Protease (continued)					
Bacillus natto	2.38	8.8	280	1811	–
B. subtilis	2.7	10	278	427	MW = 27,000 (427)
Neutral	–	13.6	280	428	–
B. subtilis NRRL B3411					
A	–	14.8	280	1825	–
B	–	14.7	280	1825	–
B. subtilis var. *amylosaccchariticus*	–	13.6–8	280	1812	–
B. thermoproteolyticus	6.63	17.65	280	429	pH 7.0, 0.05 M Tris, MW = 37,500 (429)
	7.34	19.6	280	429	0.1 N NaOH
Lotus seed (*Nelumbo nucifera* Gaertn)	–	10	278	1813	pH 4.0, 50 mM AcO⁻
Mouse, submaxillary					
A	7.5	24.9	280	1814	MW = 30,000 (1814)
B	7.25	25.9	280	1814	MW = 28,000 (1814)
Myxobacter, α-lytic	1.94		280	1815	–
Myxobacter, AL-1	2.2	15.8	280	430	0.1 M P_i, MW = 14,000 (430)
Rhizopus chinensis	–	12.6	280	431	–
S. fradiae	–	8.4	280	275	pH 8, 0.1 M Tris
Streptomyces griseus str. K		8.1	280	1820	Folin
Streptomyces naraensis	4.16	11.22	280	1824	pH 7.5, 0.005 M TrisCl, MW = 37,000 (1824)
Streptomyces rectus var. *proteolyticus*	3.92	18.2	280	1830	MW = 21,500 (1830)
S. griseus K1					
I	–	16.2	278	433	–
III	–	11.5	278	433	–
IV	–	10.2	278	433	–
S. maraensis, neutral	–	9.15	280	434	pH 7.5, 0.005 M Tris, HCl, 0.005 M Ca(OH)$_2$
Tricophyton granulosum extracellular	7.6	22.15	274	435	pH 6.0, 0.1 M AcO⁻, MW = 34,300 (435)
Sorangium, α-lytic	1.94	–	280	1816	–
	–	9.7	280	1817	–
Worm (*Schistosoma mansoni*)	2.42[1‡]	–	280	1818	pH 3.95
	3.75[1‡]	–	280	1818	pH 13
Vibrio B-30	–	10	280	1819	–
Streptococcal	–	16.4	280	1821	–
Zymogen	–	13.7	280	1821	–
Yeast, A	–	11.9	280	1822	–
	–	11.9	280	436	pH 6.2, 0.01 M NaP_i
C	–	16.6	280	1822	–
Yeast	–	16.6	280	436	pH 6.2, 0.01 M NaP_i
(*Saccharomyces cerevisiae*)	–	14.82	280	1823	pH 7.0, 0.01 M NaP_i, 0.1 M KCl, dry wt
Protease, acid					
Cladosporium sp. 45-2	–	10.7	280	1805	Dry wt
Rhizopus chimensis	–	12.1	280	1806	From Figure 1 (1806)
Rhodotorula glutinis K-24	–	12.9	280	1807	–
	–	14.0	280	1808	pH 4, 0.01 M citrate

[1‡] MW = 28,000.

MOLAR ABSORPTIVITY AND $A_{1cm}^{1\%}$ VALUES FOR PROTEINS AT SELECTED WAVELENGTHS OF THE ULTRAVIOLET AND VISIBLE REGION (continued)

Protein	$\epsilon^a (\times 10^{-4})$	$A_{1cm}^{1\%}$ [b]	nm[c]	Ref.	Comments[d]
Protease inhibitor, soybean	0.35	4.4	280	537	MW = 7975 (437)
Proteinase (see protease)					
Proteolipid					
Beef brain white matter					
Crude	—	7.6	278	1831	—
Purified	—	14	278	1831	—
Prothrombin					
Beef	—	13.4	280	454	pH 6. Est. from Figure 1 (454)
	—	10.8	280	455	0.1 N NaOH
	—	15.3	280	456	pH 6, P_iNaCl
	—	14.8	280	456	0.1 N NaOH
Prep. I	—	13.2	280	1832	—
Prep. II	—	14.4	280	1832	—
Human	—	15.2	280	457	
	—	13.6	280	458	pH 6 and 7, 0.1 M P_i
A-DEAE purified	—	11.7	280	459	pH 7.4, 0.02 M Tris, 0.1 M NaCl
B-DEAE purified	—	12.6	280	459	pH 7.4, 0.02 M Tris, 0.1 M NaCl
C-Disc electrophoresis prep.	—	13.8	280	459	pH 7.4, 0.02 M Tris, 0.1 M NaCl
D,E-NIH-DEAE purified	—	13.3	280	459	pH 7.4, 0.02 M Tris, 0.1 M NaCl
F-NIH-DEAE purified	—	11.6	280	459	pH 7.4, 0.02 M Tris, 0.1 M NaCl
G-NIH=Disc electrophoresis prep.	—	12.9	280	459	pH 7.4, 0.02 M Tris, 0.1 M NaCl
Protocatechuate 3,4-dioxygenase (EC 1.13.11.3)					
Pseudomonas aeruginosa	—	13.2	280	1833	pH 8.5
	92.4	13.2	280	460	pH 8.5, 50 mM TrisCl, MW = 700,000 (460)
Protocatechuate 4,5-dioxygenase (EC 1.13.11.8)					
Pseudomonad	11.15	7.45	280	1834	pH 75., 0.05 M KP_i + 10% EtOH, MW = 150,000 (1834), from Figure 3 (1834)
Protocollagen hydroxylase (proline, 2-oxoglutarate dioxygenase, EC 1.14.11.2)					
Chicken embryo	—	49.5	230	1835	—
Protoheme, P-450 particle					
Rabbit liver	9.6	—	414	441	Reduced
	2.15	—	543	441	Reduced
	13.8	—	415–8	441	Oxidized
	1.9	—	532	441	Oxidized
	1.85	—	567	441	Oxidized

MOLAR ABSORPTIVITY AND $A_{1cm}^{1\%}$ VALUES FOR PROTEINS AT SELECTED WAVELENGTHS OF THE ULTRAVIOLET AND VISIBLE REGION (continued)

Protein	$\epsilon^a (\times 10^{-4})$	$A_{1cm}^{1\%}$ [b]	nm[c]	Ref.	Comments[d]
Putrescine oxidase (EC 1.4.3.10)					
Micrococcus rubens	1.08	–	458	1836	–
	11.45	–	275	1836	–
Apo-		10.0	280	1836	–
Pyocin R					
P. aeruginosa R	–	16.9	280	461	–
Sheath	–	18.1	280	1837	0.1 N NaOH – 0.1 M NaCl
S-β-Pyridylethylalbumin (see albumin, beef)					
Pyridoxamine-pyruvate transaminase (EC 2.6.1.30)					
Pseudomonas MA-1	–	9.75	280	462	–
Pyrocatechase (catechol 1,2-dioxygenase, EC 1.13.11.1)					
P. arvilla	0.47	0.52	440	463	pH 8.0, 0.05 M Tris, MW = 90,000 (463)
	8.04	8.93	280	463	pH 8.0, 0.05 M Tris, MW = 90,000 (463)
Pyrophosphatase, inorganic					
Yeast	–	14.5	280	464	0.1 M HCl or water
Pyruvate dehydrogenase (EC 1.2.4.1)					
E. coli K-12	30	10	276	747	pH 7, 0.05 M KP$_i$ MW 3 × 10⁶ (747), data from Figure 3A (747)
	1.07	0.40	355	465	MW = 265,000 (465)
	1.13	0.43	370	465	MW = 265,000 (465)
	1.05	0.40	415	465	MW = 265,000 (465)
	1.46	0.55	438	465	MW = 265,000 (465)
	1.27	0.47	460	465	MW = 265,000 (465)
Pyruvate kinase					
Rabbit muscle	–	5.4	280	466	–
Human	–	5.4	280	466	–
Rat	–	5.4	280	466	–
Yeast	10.8	6.53	280	748	MW = 166,000 (748)
Quinolinate phosphoribosyltransferase (nicotinatemononucleotide pyrophosphorylase, EC 2.4.2.19)					
Pseudomonad	6.04	3.4	278	467	pH 7, 0.05 M KP$_i$, MW = 178,000 (467)
RNA polymerase (EC 2.7.7.6)					
E. coli	24.0	5.41	280	1135	MW = 440,000 (1135)
E. coli B	–	5.9	280	749	Con. by Biuret and Lowry (749)
	–	6.7	280	749	Con. det by refract. incr. with BSA as standard (749)
	–	6.5	280	749	–
	–	11.8	278	750	Calc. from data in Figure 1 (750)

MOLAR ABSORPTIVITY AND $A_{1cm}^{1\%}$ VALUES FOR PROTEINS AT SELECTED WAVELENGTHS OF THE ULTRAVIOLET AND VISIBLE REGION (continued)

Protein	$\epsilon^a (\times 10^{-4})$	$A_{1cm}^{1\%}$ [b]	nm[c]	Ref.	Comments[d]
Relaxing protein					
Rabbit muscle	—	3.32	278	1131	—
Rennin (chymosin, EC 3.4.23.4)	—	14.3	278	468	—
Retinol-binding protein					
Human serum	3.91	18.5	280	1132	MW = 21,000 (1132)
Retinol-transporting protein					
Human urine	—	18.7	280	1133	—
Rhodanese, (thiosulfate sulfur transferase, EC 2.8.1.1)					
Beef liver	—	17.5	280	469	—
Rhodopsin	4.2	—	498	470	pH 6.5 $M/15$ P_i, 1% Emulphogene BC720
	1.1	—	350	470	pH 6.5 $M/15$ P_i, 1% Emulphogene BC720
	7.4	—	280	470	pH 6.5 $M/15$ P_i, 1% Emulphogene BC720
	4.06	—	498	471	—
	—	9.8	278	471	—
Cattle	4.06	10.1	500	625	MW = 40,000 (626)
	8.12	20.2	278	625	Calc. from $\epsilon_{278}/\epsilon_{500} = 2.0$
	7.17	—	279	627	In cetyltrimethylammonium bromide
	1.06	—	345	627	
	3.97	—	498	627	
Ribitol dehydrogenase					
Aerobacter aerogenes	—	11.1	280	628	pH 7.4, data from Figure (628)
Ribonuclease	1.19	—	278	472	0.2 M NaCl, 99.85% D_2O
Beef pancreatic	0.88	6.95	280	473	—
	—	6.9	280	474	—
	0.98	7.2	227.5	475	—
	—	7.2	280	476	—
	0.98	7.2	278	477	pH 6.5, MW = 13,700 (477)
	1.13	8.3	278	478	MW = 13,683 (478)
	1.06	—	279.5	479	Ethylene glycol
	1.14	—	278	480	Ethylene glycol
Corn	4.32	18.8	280	485	MW = 23,000 (485)
Ribonuclease I (EC 3.1.4.22)					
Beef pancreatic	0.91	7.14	277.5	481	—
	—	6.95	280	482	—
41-N_2ph	—	11.2	280	488	—
	—	6.7	280	484	—
Sheep pancreas	—	7.1	280	483	—

MOLAR ABSORPTIVITY AND $A_{1cm}^{1\%}$ VALUES FOR PROTEINS AT SELECTED WAVELENGTHS OF THE ULTRAVIOLET AND VISIBLE REGION (continued)

Protein	$\epsilon^a (\times 10^{-4})$	$A_{1cm}^{1\%}$ [b]	nm[c]	Ref.	Comments[d]
Ribonuclease II (T_2) (EC 3.1.4.23)					
A. oryzae	7.16	19.9	281	487	MW = 36,000 (487)
Ribonuclease P					
Pig pancreas	0.41	3	280	486	MW = 13,500 (486)
Ribonuclease S	–	6.95	280	482	–
S Protein	–	7.84	280	482	–
Ribonucleotide-diphosphate reductase (EC 1.17.4.1) E. coli B					
Protein B-1	8.2	10.5	280	751	MW = 78,000 (751)
Protein B-2	11.5	14.8	280	751	
	0.4	–	410	752	G-200 prep.
	0.6	–	360	752	
	0.33	–	410	752	Gel electrophoresis prep.
	0.56	–	360	752	
Ribosephosphate isomerase (see 5-phosphoriboisomerase)					
L-Ribulokinase, E. coli	–	15.1	280	489	–
E. coli B/r	15.2	15.5	280	753	pH 7.6, MW = 98,000 (753)
Ribulosebisphosphate carboxylase (EC 4.1.1.39)					
Spinacea oleracea	–	16	280	490	–
Hydrogenomonas eutropha	80.0	15.51	280	1134	pH 8.0, 0.02 TrisSO$_4$ 0.1 M MgCl$_2$, MW = 515,000 (1134)
Hydrogenomonas facilis	62.6	12.28	280	1134	pH 8.0, 0.02 TrisSO$_4$, 0.01 M MgCl$_2$, MW = 551,000 (1134)
L-Ribulose phosphate 4-epimerase (EC 5.1.3.4)					
E. coli	16.2	15.7	280	491	pH 7.75, 10 mM NH$_4$HCO$_3$ MW = 103,000 (491)
L-Rhamnulose-1-phosphate (see aldolase)					
Rubredoxin					
Desulfovibrio desulfurican	1.342	–	280	492	Oxidized
	0.596	–	380	492	Oxidized
	0.512	–	490	492	Oxidized
	1.413	–	277	492	Reduced
	0.543	–	312	492	Reduced
	0.219	–	335	492	Reduced
Peptostreptococcus elsdenii	1.83	–	280	493	–
	0.819	–	350	493	–
	0.94	–	378	493	–
	0.763	–	390	493	–
	0.38	–	566	493	–

MOLAR ABSORPTIVITY AND $A_{1cm}^{1\%}$ VALUES FOR PROTEINS AT SELECTED WAVELENGTHS OF THE ULTRAVIOLET AND VISIBLE REGION (continued)

Protein	$\epsilon^a (\times 10^{-4})$	$A_{1cm}^{1\%}$ [b]	nm[c]	Ref.	Comments[d]
Rubredoxin (continued)					
M. lactilyticus	2.2	36.7	280	494	MW = 6,000 (494)
	0.91	15.3	490	494	—
M. aerogenes	1.83	30.5	280	495	MW = 6,000 (495)
	0.84	14	350	495	MW = 6,000 (495)
	0.92	15.3	378	495	MW = 6,000 (495)
	0.765	12.8	490	495	MW = 6,000 (495)
	0.35	5.8	570	495	MW = 6,000 (495)
P. oleovarans	1.11	8.7	495	496	MW = 12,800 (496)
	1.08	8.4	280	496	MW = 12,800 (496)
Rubredoxin, S-aminoethyl					
M. aerogenes	0.85	14.1	280	495	MW = 6,000 (495)
Rubredoxin, Apo					
C. pasteurianum	1.85	30.8	280	497	MW = 6,000 (497)
Rubredoxin reductase (EC 1.6.7.2)					
Pseudomonas oleovorans	7.3	—	272	1402	—
	1.0	—	378	1402	—
	1.1	—	450	1402	—
Serine dehydratase (EC 4.2.1.13, .14, .16)					
Rat liver	6.66	10.4	280	629	pH 7.2, 0.025 M KP$_i$, 0.001 M EDTA, 0.001 S$_2$ threitol,
Aposerine dehydratase	6.02	9.4	280	629	data from Figure 14 (629)
D-Serine dehydratase (EC 4.2.1.14)					
E. coli K 12	5.33	14.2	280	1136	MW = 37,300 (1136)
	0.533	1.42	415	1136	
E. coli	4.78	10.5	280	1137	pH 6.5–7.8, 0.1 M KP$_i$, dry wt/R, MW = 45,500 (1137)
	4.11	9.02	280	1137	pH 6.5, 0.4 M imidazole-citrate, MW = 45,500 (1137)
	4.35	9.56	280	1137	0.1 N NaOH, MW = 45,500 (1137)
Apo-	—	9.78	280	1137	pH 7.8, 0.1 M KP$_i$
	—	9.56	280	1137	0.1 N NaOH
Serine hydroxymethylase					
Rabbit liver, soluble	—	9	278	630	pH 7.1, 0.05 M KP$_i$,
Mitochondrial	—	7	278	630	data from Figure 4 (630)
Siderophilin					
Human	—	14	280	7	See transferrin
	—	10.9	280	8	—
	—	11.2	280	308	—
Pig	—	11.0	278	498	0.1 N HCl
	—	13.8	280	498	pH 7
	—	13.3	290	498	0.1 N NaOH

MOLAR ABSORPTIVITY AND $A_{1cm}^{1\%}$ VALUES FOR PROTEINS AT SELECTED WAVELENGTHS OF THE ULTRAVIOLET AND VISIBLE REGION (continued)

Protein	$\epsilon^a (\times 10^{-4})$	$A_{1cm}^{1\%}$ [b]	nm[c]	Ref.	Comments[d]
Spectrin					
Human erythrocytes	–	8.8	280	631	–
Stellacyanin					
Rhus vernicifera	2.6	–	280	499	0.1 N NaOH
	2.8	–	290	499	0.1 N NaOH
Oxidized	2.32	–	280	717	–
	0.096	–	450	717	–
	0.48	–	604	717	–
	0.079	–	850	717	–
Reduced	0.088	–	450	717	–
	0.408	–	604	717	–
	0.079	–	850	717	–
Steroid 4,5-dioxygenase (EC 1.13.11.25) (see 3,4-dihydroxy-9,10-secoandrosta-1,3,5[10]-triene-9,17-dione 4,5-dioxygenase)					
Streptavidin					
S. avidinii	–	34	280	500	–
	–	31	282	501	–
Streptokinase, Streptococcus	–	9.49	280	502	–
Subtilisin (EC 3.4.21.14)	–	8.6	280	357	–
Subtilisin, Thiol-	3.31	–	278	506	–
Subtilisin BPN'	–	8.8	280	754	–
Subtilisin BPN', $iPr_2 P^{bb}$	3.23	11.7	278	504	pH 7.0, 0.05 M AcONa, MW = 27,600 (504)
Subtilisin Carlsberg, $iPr_2 P^{bb}$	–	8.6	278.1	505	–
Subtilisin, Nova	3.11	–	278	506	–
Subtilopeptidase (subtilisin, EC 3.4.21.14)	–	10.7	280	507	–
Succinyl-Co A synthetase (EC 6.2.1.4, .5)					
E. coli	7.2	5.11	280	508	pH 7.2, 1 mM KP_i, MW = 141,000 (508)
Sulfate adenylyltransferase (see ATP-sulfurylase)					
Sulphatase, (aryl sulfatase, EC 3.1.6.1)					
Beef liver					
A	7.5	7.0	280	509	pH 7.5, MW = 107,000 (509)
B	3.5	14.0	280	509	pH 7.5, MW = 25,000 (509)

MOLAR ABSORPTIVITY AND $A_{1cm}^{1\%}$ VALUES FOR PROTEINS AT SELECTED WAVELENGTHS OF THE ULTRAVIOLET AND VISIBLE REGION (continued)

Protein	ϵ^a (× 10^{-4})	$A_{1cm}^{1\%}$ [b]	nm[c]	Ref.	Comments[d]
Sulphatase (continued)					
$B_{\alpha 2}$	—	20	280	509	Protein det. by Folin Method
	—	13.3	280	509	Refractometric method used to det. protein
B_β	—	19.9	280	509	Protein det. by Folin method
	—	13.8	280	509	Refractometric method used to det. protein
Tartrate dehydrogenase (EC 1.1.1.93)					
P. putida	20.9	14.4	280	510	MW = 145,000 (510)
Tartrate epoxidase (trans-epoxysuccinate hydratase, EC 4.2.1.37)					
Pseudomonas putida	—	14.0	280	632	—
Tartronic semialdehyde reductase (EC 1.1.1.60)					
P. putida	7.1	6.83	280	511	MW = 104,000 (511)
Taurocyamine kinase					
Arenicola marinae	—	9.7	280	607	—
Tetanus toxin					
C. tetani	—	7.8	280	512	—
Thermolysin (EC 3.4.24.4)					
B. thermoproteolyticus Rokko	6.63	17.6	280	1138	MW = 37,500 (1138)
Thioredoxin					
Bacteriophage T4	6.3	6.06[m‡]	280	1139	MW = 104,000 (1139)
E. coli	1.37	11.4	280	513	MW = 12,000 (513)
Thioredoxin reductase (EC 1.6.4.5)					
E. coli	0.33	14.0	280	514	MW = 66,000 (515)
Threonine aldolase (EC 4.1.2.5)					
Candida humicola	11.6	4.17	280	755	pH 6.4, 0.03 M KP$_i$, 0.005 M HSEtOH, 0.001 M EDTA, MW = 277,000 (755), calc. from Figure 2 (755)
Threonine deaminase (threonine dehydratase, EC 4.2.1.1.6)					
S. typhimurium	18.0	9.3	278	516	pH 7.4, 0.05 M KPh, 0.8 mM L-isoleucine, 0.5 mM EDTA 0.5 mM S$_2$ threitol
E. coli	2.6	1.75	415	517	MW = 147,000 (517)
	8.1	5.5	277	517	MW = 147,000 (517)
Rhodospirillum rubrum	—	3.82	278	1140	pH 6.8, 25 mM KP$_i$,
	—	1.31[n‡]	412	1140	data from Figure 7 (1140)

[m‡] OD = 1 when con. = 1.65 mg/ml.
[n‡] OD$_{412}$ = OD$_{278}$/2.9.

MOLAR ABSORPTIVITY AND $A_{1cm}^{1\%}$ VALUES FOR PROTEINS AT SELECTED WAVELENGTHS OF THE ULTRAVIOLET AND VISIBLE REGION (continued)

Protein	$\epsilon^a (\times 10^{-4})$	$A_{1cm}^{1\%}$ [b]	nm[c]	Ref.	Comments[d]
Thrombin (EC 3.4.21.5)					
Human	–	16.2	280	518	–
Beef	–	19.5	280	519	–
Thyrocalcitonin	0.76	21	280	520	MW = 3,604 (520)
Pig	0.757	21	280	756	0.1 M AcOH
Thyroid stimulating hormone					
Beef	2.6	10.5	292	521	0.1 N NaOH, est. from Figure 1 (521), MW = 25,000 (521)
Human	2.5	9.9	292	521	0.1 N NaOH, est. from Figure 1 (521), MW = 25,000 (521)
Thyroglobulin, beef					
19S	–	10.0	280	503	pH 7.4, KCl-P_i
19S	–	10.5	280	96	–
19S	65	10	280	757	MW = 650,000
	1310	201	210	757	
27S	132	10.8	280	757	MW = 1,220,000
	2700	221	210	757	
27S Iodoprotein	–	10.8	280	503	pH 7.4, KCl-P_i
Human	–	10.5	280	360	–
Lamprey, 12S	–	8.8	280	372	–
Transaldolase (EC 2.2.1.2), *C. utilis*	–	11	280	522	0.1 N NaOH
Transcortin, Human plasma	–	7.4	280	523	–
Transferrin, Human plasma	–	14.1	280	524	Iron saturated
	–	11.2	280	525	–
Human apoenzyme	–	11.4	280	524	–
Pig	–	14.1	280	526	Iron saturated
	–	0.6	470	526	–
Apoenzyme	–	11.4	280	526	–
Transglutaminase					
Pig heart	–	15.8	280	527	–
Guinea pig liver	14.2	15.8	280	528	MW = 90,000 (528)
Tripeptide synthetase (glutathione synthetase, EC 6.3.2.3)					
Yeast	18.5	15	280	529	MW = 123,000 (529)
Tropomyosin B					
Rabbit muscle	–	3.3	277	530	pH 7.0, ionic strength 1.1
	–	3.1	276	530	8 M urea
Troponin					
Rabbit muscle	–	42	260	758	pH 7.5, 2 mM TrisCl, data from Figure 1 (758)
Trypsin (see also cocoonase) (EC 3.4.21.4)					
Beef pancreas	0.154	–	280	531	–
	–	16.6	280	532	–

MOLAR ABSORPTIVITY AND $A_{1cm}^{1\%}$ VALUES FOR PROTEINS AT SELECTED WAVELENGTHS OF THE ULTRAVIOLET AND VISIBLE REGION (continued)

Protein	$\epsilon^a (\times 10^{-4})$	$A_{1cm}^{1\%}$ [b]	nm[c]	Ref.	Comments[d]
Trypsin (continued)					
Beef pancreas (continued)	—	15.6	280	533	—
	—	15.0	280	534	—
	—	17.1	280	535	Acid sol.
	—	14.4	280	536	—
	—	16	—[k†]	537	1 mM HCl
	—	12.9	280	538	—
	—	17.24	280	539	—
	—	15.5	280	540	—
Acetyltrypsin B	—	14.4	280	541	pH 7.5
N₂ ph-Trypsin	—	14.9	280	535	Acid sol.
Sheep pancreas	—	17.4	280	540	—
Pig pancreas	—	15.0	280	540	—
Trypsin inhibitor					
Inter-α, human	—	7.1	280	542	—
Beef pancreas	—	7.9	280	535	Acid sol.
	—	8.2	280	535	Neutral sol.
	—	8.25	280	543	Acid sol.
	—	8.35	280	543	Neutral sol.
	0.38	6.2	276.1	544	MW = 6,155 (544)
	0.36	5.9	280	544	MW = 6,155 (544)
Kazals	—	6.5	280	535	Neutral sol.
Pig pancreas					
I	—	5.18	280	545	pH 7.8
II	—	6.06	280	545	pH 7.8
Soybean	—	9.1	280	535	Acid sol.
	—	9.54	280	535	Neutral sol.
	—	10.5	280	546	—
	—	4.8	280	547	—
A₂	—	9.94	280	548	—
I	—	9.44	280	549	—
F₁	—	7.16	280	550	—
F₂	—	10.4	280	550	—
F₃	—	6.34	280	550	—
Colostrum, beef	—	5.0	280	535	Acid sol.
Barley	1.78	12.7	280	759	MW = 14,000 (759)
Trypsin-Trypsin inhibitor complex					
Pancreatic inhibitor, beef	—	12.3	280	535	Acid sol.
Soybean inhibitor	—	13.1	280	535	—
Colostrum inhibitor, beef	—	11.9	280	535	Acid sol.
Trypsinogen					
Beef pancreas	—	13.9	280	536	—
Pig pancreas	—	13.9	280	536	—
Sheep pancreas	—	14.1	280	551	—
Beef pancreas, S-Sulfo-	—	14.2	280	552	—
Tryptophan oxygenase (EC 1.13.11.11)					
Pseudomonas acidovorans	14.6	12.0	280	760	MW = 121,000 (760)
	22.9	18.8	405	760	

MOLAR ABSORPTIVITY AND $A_{1cm}^{1\%}$ VALUES FOR PROTEINS AT SELECTED WAVELENGTHS OF THE ULTRAVIOLET AND VISIBLE REGION (continued)

Protein	$\epsilon^a (\times 10^{-4})$	$A_{1cm}^{1\%}$ [b]	nm[c]	Ref.	Comments[d]
Tryptophan synthase (EC 4.2.1.20) *E. coli*					
A protein	1.37	4.05	278	761	pH 7.2, 0.01 M KP$_i$ MW = 29,500 (761)
	1.4	4.75	293	761	0.1 M NaOH, MW = 29,500 (761)
B protein	–	6.5	280	762	pH 7.3, 0.05 M KPh conc. 1γ/ml, pyridoxal-P and 0.001 M HSEtOH
	–	1.11	300	762	
	–	0.71	335	762	Calc. from ratios of ODs
	–	1.14	414	762	
	–	3.2	290	762	
	6.2	–	278	763	pH 7.5, 0.01 M KP$_i$, 0.01 M HSEtOH
Apo B protein	–	5.8	280	762	
B component	–	5.7	278	554	–
Tryptophanase (EC 4.1.99.1) *E. coli*					
Apo-	1.74	7.95	278	553	pH 7.5, 0.1 M KP$_i$ 2 mM, EDTA, 2 mM mercapto-ethanol, MW = 22,000 (553)
Reduced holoenzyme	0.34	1.5	336	553	MW = 22,000 (553)
	1.79	8.14	277	553	MW = 22,000 (553)
Tryptophanyl-tRNA synthetase (EC 6.1.1.2)					
Beef pancreas	9	8.4	280	764	pH 7.5, 0.2 M KCl, MW = 108,000 (764)
	–	8.0	280	764	8 M urea, pH 8
Tyramine oxidase (amine oxidase, EC 1.4.3.4)					
Sarcina lutea	34.8	27	280	555	MW = 129,000 (555)
β-Tyrosinase (tyrosine phenol-lyase) (EC 4.1.99.2)					
E. intermedia	–	83.7	280	556	–
Tyrosine aminotransferase (EC 2.6.1.5)					
Rat liver	9.1	10	277	557	pH 7.0, 0.1 M P$_i$, MW = 91,000 (557) est. from Figure 6 (557)
UDPG dehydrogenase (EC 1.1.1.22)					
Beef liver	–	9.8	277	633	–
Umecyanin *Armoacia lapathiofolia* (Horseradish root)	0.34	2.3	610	765	
	0.012	0.083	400	765	pH 5.75, 30 mM AcONa
	0.021	0.144	330	765	MW = 14,600 (765)
	1.27	8.7	280	765	

MOLAR ABSORPTIVITY AND $A_{1cm}^{1\%}$ VALUES FOR PROTEINS AT SELECTED WAVELENGTHS OF THE ULTRAVIOLET AND VISIBLE REGION (continued)

Protein	ϵ^a ($\times 10^{-4}$)	$A_{1cm}^{1\%}$ [b]	nm[c]	Ref.	Comments[d]
Urease (EC 3.5.1.5)					
Beans	—	5.5	280	766	
Cajanus indicus	—	20.1	280	767	pH 7.0, 0.05 M Tris-AcO⁻
	—	20.6	277	767	
Jack bean	28.4	5.89	280	768	MW = 483,000 (768)
Carboxymethyl-	—	7.2	—[o‡]	769	—
Jack bean meal	37	7.7	272	558	MW = 480,000 (558)
	—	7.71	272	559	—
	—	7.54	278	560	pH 7, 0.02 M P$_i$
	—	6.4	278	560	pH 7, 0.02 M P$_i$, HSEtOH added
Uricase, pig liver	—	11.3	276	561	1% sodium carbonate
Urocanase (EC 4.2.1.99)					
Pseudomonas putida	—	8.3	280	1141	pH 7.5, 0.2 M KP$_i$ data from Figure 6 (1141) and corrected for scattering
	—	8.03	280	770	Calc. from data in Figure 4 (770), corrected for scattering of light
Urokinase (EC 3.4.99.26)					
Human urine, S$_1$	4.23	13.2	280	1142	pH 6.5, MW=32,000 (1142)
S$_2$	7.48	13.6	280	1142	pH 6.5, MW=55,000 (1142)
Human placenta	—	10.7	280	771	—
Virus					
Barley stripe mosaic	—	34	240	562	
	—	26	260	562	
	—	25	280	562	
	—	17	280	449	
	—	28	260	449	
	—	17	280	1143	Dry wt, Lowry, and Biuret
Broad bean mottle	—	3.2	276.5	1144	Water
	—	5.5	292	1144	0.1 N NaOH, data from Figure 1 (1144)
Brome mosaic	—	36	240	562	—
	—	48	260	562	—
	—	31	280	562	—
Bromegrass mosaic	—	4.6[p‡]	260	1145	—
	—	7.6[p‡]	280	1145	—
	2.0	5.0	260	1145	pH 6.0, 1 M CaCl$_2$ –0.05 M Nacacodylate MW=40,000 (1145), data from Figure 1 (1145)
	3.52	8.8	280	1145	
Foot-and-mouth disease	—	11.1	276	1146	—
Mouse-Elberfeld (ME)	—	14.9	280	1147	0.002 M AcOH
Southern bean mosaic	—	48	240	562	
	—	58	260	562	
	—	37	280	562	
	—	5.85	260	1148	0.1 N NaOH
	—	12[p‡]	260	1148	

[o‡] Wavelength not cited; may be 277 nm.
[p‡] Calculated from the content of tyrosine, tryptophan, and phenylalanine.

MOLAR ABSORPTIVITY AND $A^{1\%}_{1cm}$ VALUES FOR PROTEINS AT SELECTED WAVELENGTHS OF THE ULTRAVIOLET AND VISIBLE REGION (continued)

Protein	$\epsilon^a (\times 10^{-4})$	$A^{1\%}_{1cm}$ [b]	nm[c]	Ref.	Comments[d]
Virus (continued)					
Tobacco etch	–	9.5	280	1149	pH 6.5, 0.02 M P_i, 6 M GdmCl
Tobacco mosaic	–	57	240	562	–
	–	32.4	260	562	–
	–	27	280	562	–
	–	27	260	563	pH 7.5, 0.033 M P_i
	–	26	263	774	
	–	27	260	1150	
	–	13	281	1150	
	–	13	281	774	pH 7.1, 0.033 M NaP_i
	–	13	281	563	pH 7.5, 0.033 M P_i
PM 2					
Non-functioning	–	13.7	280	450	–
Tomato ringspot	–	10.3	260	1151	–
Turnip yellow mosaic					
Artificial top component	–	11.8	275	1152	pH 7.0, 0.01 M NaP_i
White clover mosaic, str.					
WCD-17	1.88	13.2	280	1153	Water, MW=14,300 (1153)
Mengo					
L-	–	17.1	280	772	–
M-	–	16.8	280	772	–
S-	–	17.0	280	772	–
Alfalfa mosaic	–	52	260	773	–
Potato X	–	12.3	280	775	pH 7.5, 0.05 M P_i
Visual pigment, beef	3.7	13.7	280	564	MW=27,000 (564)
	2.3	8.5	500	564	MW=27,000 (564)
Vitellogenin					
Xenopus laevis	–	7.5	280	1154	Dry wt
Wool, helix-rich fraction	–	5.9–6.5	277	776	–
Xanthine oxidase (EC 1.2.3.2)					
milk	–	2.3	450	565	pH 7.8, 0.05 M P_i
	–	11.5	280	565	–
	20.4	11.26	280	566	pH 8.0, 0.02 M P_i, MW=181,000 (566)
	2.2	–	550	27	–
	–	2.41	450	1155	–
High mol. wt. fract.	–	0.87	450	1155	–

Compiled with the assistance of Waldo E. Cohn, and material supplied by Donald M. Kirschenbaum.

REFERENCES

1. Kremzner and Wilson, *Biochemistry*, 3, 1902 (1964).
2. Leuzinger, Baker, and Cauvin, *Proc. Natl. Acad. Sci. U.S.A.*, 59, 620 (1968).
3. Nanninga, *Biochim. Biophys. Acta*, 82, 507 (1964).
4. Eisenberg and Moos, *J. Biol. Chem.*, 242, 2945 (1967).
5. Rees and Young, *J. Biol. Chem.*, 242, 4449 (1967).
6. Pfrogner, *Arch. Biochem. Biophys.*, 119, 147 (1967).
7. Tombs, Souter, and Maclagan, *Biochem. J.*, 73, 167 (1959).
8. Schonenberger, *Z. Naturforsch.*, 10b, 474 (1955).
9. Hunter and McDuffie, *J. Am. Chem. Soc.*, 81, 1400 (1959).
10. Phelps and Putnam, in *The Plasma Proteins*, Putnam, Ed., Academic Press, New York, 1960, 143.
11. Wetlaufer, *Adv. Protein Chem.*, 17, 378 (1962).
12. Van Kley and Stahmann, *J. Am. Chem. Soc.*, 81, 4374 (1959).
13. Everett, *J. Biol. Chem.*, 238, 2676 (1963).
14. Tanford and Roberts, *J. Am. Chem. Soc.*, 74, 2509 (1952).
15. Kolthoff, Shore, Tan, and Matsuoka, *Anal. Biochem.*, 12, 497 (1965).
16. Foster and Yang, *J. Am. Chem. Soc.*, 76, 1015 (1954).
17. Weber, in *The Biochemists Handbook*, Long, Ed., E. and F. N. Spon, Ltd., 1961, 82.
18. Bonnichsen, *Acta Chem. Scand.*, 4, 715 (1950).
19. Bonnichsen and Brink, *Methods Enzymol.*, 1, 495 (1955).
20. Theorell, Taniguchi, Akeson, and Skursky, *Biochem. Biophys. Res. Commun.*, 24, 603 (1966).
21. Sund and Theorell, *Enzymology*, 7, 25 (1963).
22. Rosenberg, Theorell, and Yonetani, *Arch. Biochem. Biophys.*, 110, 413 (1965).
23. Mourad and Woronick, *Arch. Biochem. Biophys.*, 121, 431 (1967).
24. Ohta and Ogura, *J. Biochem.* (Tokyo), 58, 73 (1965).
25. Hayes and Velick, *J. Biol. Chem.*, 207, 225 (1954).
26. Steinman and Jakoby, *J. Biol. Chem.*, 242, 5019 (1967).
27. Rajagopalan and Handler, *J. Biol. Chem.*, 239, 1509 (1964).
28. Baranowski and Niederland, *J. Biol. Chem.*, 180, 543 (1949).
29. Taylor, Green, and Cori, *J. Biol. Chem.*, 173, 591 (1948).
30. Donovan, *Biochemistry*, 3, 67 (1964).
31. Stellwagen and Schachman, *Biochemistry*, 1, 1056 (1962).
32. Sia and Horecker, *Arch. Biochem. Biophys.*, 123, 186 (1968).
33. Rajkumar, Woodfin, and Rutter, *Methods Enzymol.*, 9, 491 (1966).
34. Fluri, Ramasarma, and Horecker, *Eur. J. Biochem.*, 1, 117 (1967).
35. Rutter, Hunsley, Groves, Calder, Rajkumar, and Woodfin, *Methods Enzymol.*, 9, 479 (1966).
36. Massey, Palmer, and Bennett, *Biochim. Biophys. Acta*, 48, 1 (1961).
37. Antonini, Brunori, Bruzzesi, Chiancone, and Massey, *J. Biol. Chem.*, 241, 2358 (1960).
38. Yagi, Naoi, Harada, Okamura, Hidaka, Ozawa, and Kotaki, *J. Biochem.* (Tokyo), 61, 580 (1967).
39. Kotaki, Harada, and Yagi, *J. Biochem.* (Tokyo), 61, 598 (1967).
40. Miyake, Aki, Hashimoto, and Yamano, *Biochim. Biophys. Acta*, 105, 86 (1965).
41. Wellner and Meister, *J. Biol. Chem.*, 235, 2013 (1960).
42. Nakano and Danowski, *J. Biol. Chem.*, 241, 2075 (1966).
43. Wachsmuth, Fritze, and Pfleiderer, *Biochemistry*, 5, 169 (1966).
44. Hanson, Hutter, Mannsfeldt, Kretschmer, and Sohr, *Hoppe-Seyler's Z. Physiol. Chem.*, 348, 680 (1967).
45. Fischer and Stein, in *The Enzymes*, Vol. 4, 2nd ed., Bover, Lardy, and Myrbäck, Eds., Academic Press, New York, 1960, 319.
46. Caldwell, Adams, Kung, and Toralballa, *J. Am. Chem. Soc.*, 74, 4033 (1952).
47. Caldwell, Dickey, Hanrahan, Kung, Kung, and Misko, *J. Am. Chem. Soc.*, 76, 143 (1954).
48. DePinto and Campbell, *Biochemistry*, 7, 114 (1968).
49. Englard and Singer, *J. Biol. Chem.*, 187, 213 (1950).
50. Tombs and Lowe, *Biochem. J.*, 105, 181 (1967).
51. Schmike, *J. Biol. Chem.*, 239, 3808 (1964).
52. Wilson and Meister, *Biochemistry*, 5, 1166 (1966).
53. Bertland and Kaplan, *Biochemistry*, 7, 134 (1968).
54. Gerhart and Schachman, *Biochemistry*, 4, 1054 (1965).
55. Dratz and Calvin, *Nature*, 211, 497 (1966).
56. Gerhart and Holoubek, *J. Biol. Chem.*, 242, 2886 (1967).
57. Gerhart and Schachman, *Biochemistry*, 7, 538 (1968).
58. Ichishima and Yoshida, *Biochim. Biophys. Acta*, 110, 155 (1965).
59. Subramanian and Kalnitsky, *Biochemistry*, 3, 1868 (1964).

60. Hinkson and Bulen, *J. Biol. Chem.*, 242, 3345 (1967).
61. Thornber and Molson, *Biochemistry*, 7, 2242 (1968).
62. Murachi and Yasui, *Biochemistry*, 4, 2275 (1965).
63. Murachi, Inagami, and Yasui, *Biochemistry*, 4, 2815 (1965).
64. Murachi, Yasui, and Yasuda, *Biochemistry*, 3, 48 (1964).
65. Rickli, Ghazanfar, Gibbons, and Edsall, *J. Biol. Chem.*, 239, 1065 (1964).
66. Lindskog, *Biochim. Biophys. Acta*, 39, 218 (1960).
67. Duff and Coleman, *Biochemistry*, 5, 2009 (1966).
68. Edsall, Mehta, Meyers, and Armstrong, *Biochem. Z.*, 345, 9 (1966).
69. Horgan, Webb, and Zerner, *Biochem. Biophys. Res. Commun.*, 23, 23, (1966).
70. Vallee, Rupley, Coombs, and Neurath, *J. Biol. Chem.*, 235, 64 (1960).
71. Bethune, Ulmer, and Vallee, *Biochemistry*, 6, 1955 (1967).
72. Simpson, Riordan, and Vallee, *Biochemistry*, 2, 616 (1963).
73. Riordan and Vallee, *Biochemistry*, 2, 1460 (1963).
74. McClure, Neurath, and Walsh, *Biochemistry*, 3, 1897 (1964).
75. Smith and Stockell, *J. Biol. Chem.*, 207, 501 (1954).
76. Blostein and Rutter, *J. Biol. Chem.*, 238, 3280 (1963).
77. Keller, Cohen, and Neurath, *J. Biol. Chem.*, 223, 457 (1956).
78. Neurath, *Methods Enzymol.*, 1, 77 (1955).
79. Folk and Schirmer, *J. Biol. Chem.*, 238, 3884 (1963).
80. Bargetzi, Sampathkumar, Cox, Walsh, and Neurath, *Biochemistry*, 2, 1468 (1963).
81. Freisheim, Walsh, and Neurath, *Biochemistry*, 6, 3010 (1967).
82. Folk, Piez, Carroll, and Gladner, *J. Biol. Chem.*, 235, 2272 (1960).
83. Cox, Wintersberger, and Neurath, *Biochemistry*, 1, 1078 (1962).
84. Herskovits, *Biochemistry*, 5, 1018 (1966).
85. McKenzie, *Adv. Protein Chem.*, 22, 55 (1967).
86. Herskovits, *J. Biol. Chem.*, 240, 628 (1965).
87. Thompson and Kiddy, *J. Dairy Sci.*, 47, 626 (1964).
88. Thompson and Pepper, *J. Dairy Sci.*, 47, 633 (1964).
89. Zittle and Custer, *J. Dairy Sci.*, 46, 1183 (1963).
90. Bonnichsen, *Methods Enzymol.*, 2, 781 (1955).
91. Stansell and Deutsch, *J. Biol. Chem.*, 240, 4299 (1965).
92. Hiraga, Anan, and Abe, *J. Biochem.* (Tokyo), 56, 416 (1964).
93. Samejima and Yang, *J. Biol. Chem.*, 238, 3256 (1963).
94. Gregory, *Biochim. Biophys. Acta*, 159, 429 (1968).
95. Iwasaki, Hayashi, and Funatsu, *J. Biochem.* (Tokyo), 55, 209 (1964).
96. Edelhoch, *J. Biol. Chem.*, 235, 1326 (1960).
97. Kasper and Deutsch, *J. Biol. Chem.*, 238, 2325 (1963).
98. Markowitz, Cartwright, and Wintrobe, *J. Biol. Chem.*, 234, 40 (1959).
99. Schwick and Heide, in *Protides of the Biological Fluids*, Vol. 14, Peeters, Ed., Elsevier, Amsterdam, 1967, 55.
100. Sgouris, Coryell, Gallick, Storey, McCall, and Anderson, *Vox Sang.*, 7, 394 (1962).
101. Morell, Irvine, Sternlieb, Scheinberg, and Ashwell, *J. Biol. Chem.*, 243, 155 (1968).
102. Morris and Hager, *J. Biol. Chem.*, 241, 1763 (1966).
103. Tsunoda and Yasunobu, *J. Biol. Chem.*, 241, 4610 (1966).
104. Kunimitsu and Yasunobu, *Biochim. Biophys. Acta*, 139, 405 (1967).
105. Schwert and Kaufman, *J. Biol. Chem.*, 190, 807 (1951).
106. Narasinga, Rao, and Kegeles, *J. Am. Chem. Soc.*, 80, 5724 (1958).
107. Dixon and Neurath, *J. Biol. Chem.*, 225, 1049 (1957).
108. Morimoto and Kegeles, *Biochemistry*, 6, 3007 (1967).
109. Moon, Sturtevant, and Hess, *J. Biol. Chem.*, 240, 4204 (1965).
110. Laskowski, *Methods Enzymol.*, 2, 8 (1955).
111. Folk and Cole, *J. Biol. Chem.*, 240, 193 (1965).
112. Wootton and Hess, *J. Am. Chem. Soc.*, 84, 440 (1962).
113. Schwert, *J. Biol. Chem.*, 190, 799 (1951).
114. Wilcox, Cohen, and Tan, *J. Biol. Chem.*, 228, 999 (1957).
115. Guy, Gratecos, Rovery, and Desnuelle, *Biochim. Biophys. Acta*, 115, 404 (1966).
116. Charles, Gratecos, Rovery, and Desnuelle, *Biochim. Biophys. Acta*, 140, 395 (1967).
117. Smillie, Enenkel, and Kay, *J. Biol. Chem.*, 241, 2097 (1966).
118. Folk and Schirmer, *J. Biol. Chem.*, 240, 181 (1965).
119. Pechere, Dixon, Maybury, and Neurath, *J. Biol. Chem.*, 233, 1364 (1958).
120. Srere, *J. Biol. Chem.*, 241, 2157 (1966).
121. Herschman and Helinski, *J. Biol. Chem.*, 242, 5360 (1967).
122. Warner and Weber, *J. Am. Chem. Soc.*, 75, 5094 (1953).

123. Gotschlich and Edelman, *Proc. Natl. Acad. Sci. U.S.A.*, 54, 558 (1965).
124. Wood and McCarty, *J. Clin. Invest.*, 30, 616 (1951).
125. Noda, Kuby, and Lardy, *J. Biol. Chem.*, 209, 203 (1954).
126. Wisse, Zweers, Jongkind, Bont, and Bloemendal, *Biochem. J.*, 99, 179 (1966).
127. Bjork, *Exp. Eye Res.*, 7, 129 (1968).
128. Mason and Hines, *Invest. Ophthalmol.*, 5, 601 (1966).
129. Bjork, *Exp. Eye Res.*, 2, 339 (1963).
130. Bjork, *Exp. Eye Res.*, 3, 254 (1964).
131. Van Dam, *Exp. Eye Res.*, 5, 255 (1966).
132. Zigman and Lerman, *Biochim. Biophys. Acta*, 154, 423 (1968).
133. Yonetani, *J. Biol. Chem.*, 236, 1680 (1961).
134. Yonetani, *J. Biol. Chem.*, 242, 5008 (1967).
135. Paleus, *Arch. Biochem. Biophys.*, 96, 60 (1962).
136. Matsubara, Chu, and Yasunobu, *Arch. Biochem. Biophys.*, 101, 209 (1962).
137. Paul, *Acta Chem. Scand.*, 5, 389 (1951).
138. Flatmark, *Acta Chem. Scand.*, 18, 1517 (1964).
139. Flatmark, *Acta Chem. Scand.*, 20, 1476 (1966).
140. Margoliash and Frohwirt, *Biochem. J.*, 71, 570 (1959).
141. Jirgensons, *J. Biol. Chem.*, 240, 1064 (1965).
142. Linberg, *Biochemistry*, 6, 335 (1967).
143. Mondovi, Rotilio, Costa, Finazzi-Agro, Chiancone, Hansen, and Beinert, *J. Biol. Chem.*, 242, 1160 (1967).
144. Swislocki and Kaplan, *J. Biol. Chem.*, 242, 1083 (1967).
145. Raynaud, *Proc. 2nd Meet. Fed. Eur. Biochem. Soc.*, Vienna 1965, 1, 199 (1967).
146. Holt and Wold, *J. Biol. Chem.*, 236, 3227 (1961).
147. Bucher, *Methods Enzymol.*, 1, 427 (1955).
148. Malmstrom, Kimmel, and Smith, *J. Biol. Chem.*, 234, 1108 (1959).
149. Bergdoll, Chu, Huang, Rowe, and Shih, *Arch. Biochem. Biophys.*, 112, 104 (1965).
150. Schantz, Roessler, Wagman, Spero, Dunnery, and Bergdoll, *Biochemistry*, 4, 1011 (1965).
151. Stansell and Deutsch, *J. Biol. Chem.*, 240, 4306 (1965).
152. Crane, Mii, Hauge, Green, and Beinert, *J. Biol. Chem.*, 218, 701 (1956).
153. Yang, Butterworth, Bock, and Potter, *J. Biol. Chem.*, 242, 3501 (1967).
154. Keresztes-Nagy and Margoliash, *J. Biol. Chem.*, 241, 5955 (1966).
155. Matsubara, *J. Biol. Chem.*, 243, 370 (1968).
156. Malkin and Rabinowitz, *Biochemistry*, 5, 1262 (1966).
157. Hofmann and Harrison, *J. Mol. Biol.*, 6, 256 (1963).
158. Verpoorte, Green, and Kay, *J. Biol. Chem.*, 240, 1156 (1965).
159. Caspary and Kekwick, *Biochem. J.*, 56, 35 (1954).
160. Mosesson, Alkjaersig, Sweet, and Sherry, *Biochemistry*, 6, 3279 (1967).
161. Mihalyi, *Biochemistry*, 7, 208 (1968).
162. Mihalyi and Godfrey, *Biochim. Biophys. Acta*, 67, 73 (1963).
163. Gould and Liener, *Biochemistry*, 4, 90 (1965).
164. Hornby, Lilly, and Crook, *Biochem. J.*, 98, 420 (1966).
165. Englund, King, Craig, and Walti, *Biochemistry*, 7, 163 (1968).
166. Knight and Hardy, *J. Biol. Chem.*, 241, 2752 (1966).
167. Knight, D'Eustachio, and Hardy, *Biochim. Biophys. Acta*, 113, 626 (1966).
168. Papkoff, Gospodarowicz, and Li, *Arch. Biochem. Biophys.*, 120, 434 (1967).
169. Papkoff, Mahlmann, and Li, *Biochemistry*, 6, 3976 (1967).
170. Mizobuchi and Buchanan, *J. Biol. Chem.*, 243, 4842 (1968).
171. Himes and Cohn, *J. Biol. Chem.*, 242, 3628 (1967).
172. Marcus and Hubert, *J. Biol. Chem.*, 243, 4923 (1968).
173. Fernando, Enser, Pontremoli, and Horecker, *Arch. Biochem. Biophys.*, 126, 599 (1968).
174. Pontremoli, Grazi, and Accorsi, *Biochemistry*, 7, 3628 (1968).
175. Kanarek and Hill, *J. Biol. Chem.*, 239, 4202 (1964).
176. Malhotra and Dey, *Biochem. J.*, 103, 508 (1967).
177. Wallenfells and Golker, *Biochem. Z.*, 346, 1 (1966).
178. Wallenfells and Malhotra, in *The Enzymes*, Vol. 4, 2nd ed., Boyer, Lardy, and Myrback, Eds., Academic Press, New York, 1960, 413.
179. Mills and Tang, *J. Biol. Chem.*, 242, 3093 (1967).
180. Chiang, Sanchez-Chiang, Mills, and Tang, *J. Biol. Chem.*, 242, 3098 (1967).
181. Bernardin, Kasarda, and Mecham, *J. Biol. Chem.*, 242, 445 (1967).
182. Vodrazka, Hrkal, Cejka, and Sipalova, *Collect. Czech. Chem. Commun.*, 32, 3250 (1967).
183. Gibson and Antonini, *J. Biol. Chem.*, 238, 1384 (1963).

184. Hrkal and Vodrazka, *Biochim. Biophys. Acta*, 133, 527 (1967).
185. Rossi-Fanelli, Antonini, and Caputo, *J. Biol. Chem.*, 234, 2906 (1959).
186. Schultze, Heide, and Haupt, *Klin. Wochenschr.*, 40, 729 (1962).
187. Berggard and Bearn, *J. Biol. Chem.*, 243, 4095 (1968).
188. Deutsch, *Science*, 141, 435 (1963).
189. Muldoon and Westphal, *J. Biol. Chem.*, 242, 5636 (1967).
190. Chader and Westphal, *J. Biol. Chem.*, 243, 928 (1968).
191. Chader and Westphal, *Biochemistry*, 7, 4272 (1968).
192. Giorgio and Tabachnick, *J. Biol. Chem.*, 243, 2247 (1968).
193. Gratzer, Bailey, and Beaven, *Biochem. Biophys. Res. Commun.*, 28, 914 (1967).
194. Huotari, Nelson, Smith, and Kirkwood, *J. Biol. Chem.*, 243, 952 (1968).
195. Hauge, *Methods Enzymol.*, 9, 107 (1966).
196. Swoboda and Massey, *J. Biol. Chem.*, 240, 2209 (1965).
197. Takeda, Hizukuri, and Nikuni, *Biochim. Biophys. Acta*, 146, 568 (1967).
198. Plapp and Cole, *Arch. Biochem. Biophys.*, 116, 193 (1966).
199. Switzer and Barker, *J. Biol. Chem.*, 242, 2658 (1967).
200. Tomkins, Yielding, Curran, Summers, and Bitensky, *J. Biol. Chem.*, 240, 3793 (1965).
201. Eisenberg and Tomkins, *J. Mol. Biol.*, 31, 37 (1968).
202. Olson and Anfinsen, *J. Biol. Chem.*, 197, 67 (1952).
203. Reithel and Sakura, *J. Phys. Chem.*, 67, 2497 (1963).
204. Fahien, Wiggert, and Cohen, *J. Biol. Chem.*, 240, 1083 (1965).
205. Messer and Ottesen, *C. R. Trav. Lab. Carlsberg Ser. Chim.*, 35, 1 (1965).
206. Rowe and Wyngaarden, *Fed. Proc.*, 27, No. 2, 340 (1968).
207. Rowe and Wyngaarden, *J. Biol. Chem.*, 243, 6373 (1968).
208. Pamiljans, Krishnaswamy, Dumville, and Meister, *Biochemistry*, 1, 153 (1962).
209. Shapiro and Stadtman, *J. Biol. Chem.*, 242, 5069 (1967).
210. Woolfolk, Shapiro, and Stadtman, *Arch. Biochem. Biophys.*, 116, 177 (1966).
211. Colman and Black, *J. Biol. Chem.*, 240, 1796 (1965).
212. Massey and Williams, *J. Biol. Chem.*, 240, 4470 (1965).
213. Mavis and Stellwagen, *J. Biol. Chem.*, 243, 809 (1968).
214. Icen, *Scand. J. Clin. Lab. Invest., Suppl.*, 20, 96 (1967).
215. Kirschner and Voigt, *Hoppe-Seyler's Z. Physiol. Chem.*, 349, 632 (1968).
216. Krebs, *Methods Enzymol.*, 1, 407 (1955).
217. Jaenicke, Schmid, and Knof, *Biochemistry*, 7, 919 (1968).
218. Trentham, *Biochem. J.*, 109, 603 (1968).
219. Davidson, Sajgo, Noller, and Harris, *Nature*, 216, 1181 (1967).
220. Allison, *Methods Enzymol.*, 9, 212 (1966).
221. Mora and Elodi, *Eur. J. Biochem.*, 5, 574 (1968).
222. Velick, Hayes, and Harting, *J. Biol. Chem.*, 203, 527 (1953).
223. Fox and Dandliker, *J. Biol. Chem.*, 221, 1005 (1958).
224. Ankel, Bucher, and Czok, *Biochem. Z.*, 332, 315 (1960).
225. Fondy, Levin, Sollohub, and Ross, *J. Biol. Chem.*, 243, 3148 (1968).
226. Marquardt and Brosemer, *Biochim. Biophys. Acta*, 128, 454 (1966).
227. Brosemer and Marquardt, *Biochim. Biophys. Acta*, 128, 464 (1966).
228. Pradel, Kassab, Conlay, and Thoai, *Biochim. Biophys. Acta*, 154, 305 (1968).
229. Okuda and Weinbaum, *Biochemistry*, 7, 2819 (1968).
230. Schultze and Heremans, in *Molecular Biology of Human Proteins*, Vol. 1, Elsevier, New York, 1966, 176.
231. Haupt and Heide, *Clin. Chim. Acta*, 10, 555 (1964).
232. Schmidt, *J. Am. Chem. Soc.*, 72, 2816 (1950).
233. Smith, Brown, Weimer, and Winzler, *J. Biol. Chem.*, 185, 569 (1950).
234. Bundy and Mehl, *J. Biol. Chem.*, 234, 1124 (1959).
235. Schultze, Heide, and Haupt, *Naturwissenschaften*, 49, 133 (1962).
236. Burgi and Schmid, *J. Biol. Chem.*, 236, 1067 (1961).
237. Schmid and Burgi, *Biochim. Biophys. Acta*, 47, 440 (1961).
238. Schonenberger, Schmidtberger, and Schultze, *Z. Naturforsch.*, 13b, 761 (1958).
239. Edelhoch, Condliffe, Lippoldt, and Burger, *J. Biol. Chem.*, 241, 5205 (1966).
240. Polonovski and Sayle, *Bull. Soc. Chim. Biol.*, 21, 661 (1939).
241. Lisowska and Dobryszycka, *Biochim. Biophys. Acta*, 133, 338 (1967).
242. Heirwegh, Borginon, and Lontie, *Biochim. Biophys. Acta*, 48, 517 (1961).
243. Ghiretti-Magaldi, Nuzzolo, and Ghiretti, *Biochemistry*, 5, 1943 (1966).
244. Formanek and Engel, *Biochim. Biophys. Acta*, 160, 151 (1968).
245. Schichi and Kuroda, *Arch. Biochem. Biophys.*, 118, 682 (1967).

246. Schultze, Heide, and Haupt, *Naturwissenschaften.*, 48, 696 (1961).
247. Derechin, Ramel, Lazarus, and Barnard, *Biochemistry*, 5, 4017 (1966).
248. McDonald, *Methods Enzymol.*, 1, 269 (1955).
249. Riley and Snell, *Biochemistry*, 7, 3520 (1968).
250. Chang and Snell, *Biochemistry*, 7, 2005 (1968).
251. Loper and Adams, *J. Biol. Chem.*, 240, 788 (1965).
252. Yourno and Ino, *J. Biol. Chem.*, 243, 3273 (1968).
253. Yourno, *J. Biol. Chem.*, 243, 3277 (1968).
254. Loper, *J. Biol. Chem.*, 243, 3264 (1968).
255. Bennett, Creaser, and McDonald, *Biochem. J.*, 109, 307 (1968).
256. Truffa-Bachi, Van Rapenbusch, Jannin, Gros, and Cohen, *Eur. J. Biochem.*, 5, 73 (1968).
257. Nakano, Ushijima, Saga, Tsutsumi, and Asami, *Biochim. Biophys. Acta*, 167, 9 (1968).
258. Dai, Decker, and Sund, *Eur. J. Biochem.*, 4, 95 (1968).
259. Becker and Pfeil, *Biochem. Z.*, 346, 301 (1966).
260. Becker, Benthin, Eschenhof, and Pfeil, *Biochem. Z.*, 337, 156 (1963).
261. Kohn and Jakoby, *J. Biol. Chem.*, 243, 2494 (1968).
262. Martin and Goldberger, *J. Biol. Chem.*, 242, 1168 (1967).
263. Little and Donahue, *Meth. Immunol. Immunochem.*, 2, 343 (1968).
264. Creighton and Yanofsky, *J. Biol. Chem.*, 241, 4616 (1966).
265. Porter, *Biochem. J.*, 53, 320 (1953).
266. Weil, Seibles, and Herskovits, *Arch. Biochem. Biophys.*, 111, 308 (1965).
267. Praissman and Rupley, *Biochemistry*, 7, 2431 (1968).
268. Nakaya, Horinishi, and Shibata, *J. Biochem. (Tokyo)*, 61, 345 (1967).
269. Lampson, Tytell, Nemes, and Hilleman, *Proc. Soc. Exp. Biol. Med.*, 112, 468 (1963).
270. Metzenberg, *Arch. Biochem. Biophys.*, 100, 503 (1963).
271. Neumann and Lampen, *Biochemistry*, 6, 468 (1967).
272. Colman, *J. Biol. Chem.*, 243, 2454 (1968).
273. Margolies and Goldberger, *J. Biol. Chem.*, 242, 256 (1967).
274. Anderer and Hornle, *J. Biol. Chem.*, 241, 1568 (1966).
275. Morihara, Oka, and Tsuzuki, *Biochim. Biophys. Acta*, 139, 382 (1967).
276. Nickerson and Durand, *Biochim. Biophys. Acta*, 77, 87 (1963).
277. Nickerson, Noval, and Robison, *Biochim. Biophys. Acta*, 77, 73 (1963).
278. Hirashima, Hayakawa, and Koike, *J. Biol. Chem.*, 242, 902 (1967).
279. Kawahara, Wang, and Talalay, *J. Biol. Chem.*, 237, 1500 (1962).
280. Wetlaufer, *C. R. Trav. Lab. Carlsberg Ser. Chim.*, 32, 125 (1960).
281. Kronman and Andreotti, *Biochemistry*, 3, 1145 (1964).
282. Sullivan, *Biochem. J.*, 110, 363 (1968).
283. Takemori, Nakazawa, Nakai, Suzuki, and Katagiri, *J. Biol. Chem.*, 243, 313 (1968).
284. Pesce, McKay, Stolzenbach, Cahn, and Kaplan, *J. Biol. Chem.*, 239, 1753 (1964).
285. DiSabato, Pesce, and Kaplan, *Biochim. Biophys. Acta*, 77, 135 (1963).
286. Markert and Appella, *Ann. N. Y. Acad. Sci.*, 94, 678 (1961).
287. Pesce, Fondy, Stolzenbach, Castillo, and Kaplan, *J. Biol. Chem.*, 242, 2151 (1967).
288. Fromm, *J. Biol. Chem.*, 238, 2938 (1963).
289. Schellenberg, *J. Biol. Chem.*, 242, 1815 (1967).
290. Jaenicka, *Biochem. Z.*, 338, 614 (1963).
291. Gibson, Davisson, Bachhawat, Ray, and Vestling, *J. Biol. Chem.*, 203, 397 (1953).
292. Wieland and Pfleiderer, *Ann. N.Y. Acad. Sci.*, 94, 691 (1961).
293. Appleby and Morton, *Biochem. J.*, 73, 539 (1969).
294. Tarmy and Kaplan, *J. Biol. Chem.*, 243, 2579 (1968).
295. Okabe, Hayakawa, Hamada, and Koike, *Biochemistry*, 7, 79 (1968).
296. Polis, Schmuckler, Custer, and McMeekin, *J. Am. Chem. Soc.*, 72, 4965 (1950).
297. Baker and Saroff, *Biochemistry*, 4, 1670 (1965).
298. Wetlaufer and Lovrien, *J. Biol. Chem.*, 239, 596 (1964).
299. Gordon, Basch, and Kalan, *J. Biol. Chem.*, 236, 2908 (1961).
300. Townend, Winterbottom, and Timasheff, *J. Am. Chem. Soc.*, 82, 3161 (1960).
301. Townend, Herskovits, Swaisgood, and Timasheff, *J. Biol. Chem.*, 239, 4196 (1964).
302. Tanford and Nozaki, *J. Biol. Chem.*, 234, 2874 (1959).
303. Ogston and Tombs, *Biochem. J.*, 66, 399 (1957).
304. Bell and McKenzie, *Biochim. Biophys. Acta*, 147, 123 (1967).
305. Ghose, Chaudhuri, and Sen, *Arch. Biochem. Biophys.*, 126, 232 (1968).
306. Groves, *J. Am. Chem. Soc.*, 82, 3345 (1960).
307. Masson and Heremans, in *Protides of the Biological Fluids*, Peeters, Ed., Vol. 14, Elsevier, Amsterdam, 1967, 115.

308. Montreuil, Tonnelat, and Mullet, *Biochim. Biophys. Acta*, 45, 413 (1960).
309. Spackman, Smith, and Brown, *J. Biol. Chem.*, 212, 255 (1955).
310. Kawahara, Misaka, and Nakanishi, *J. Biochem.* (Tokyo), 63, 77 (1968).
311. Hastings, Riley, and Massa, *J. Biol. Chem.*, 240, 1473 (1965).
312. Bruzzesi, Chiancone, and Antonini, *Biochemistry*, 4, 1796 (1965).
313. Hamaguchi and Kurono, *J. Biochem.* (Tokyo), 54, 111 (1963).
314. Fromageot and Schnek, *Biochim. Biophys. Acta*, 6, 113 (1950).
315. Chandan, Parry, and Shahani, *Biochim. Biophys. Acta*, 110, 389 (1965).
316. Glazer, *Aust. J. Chem.*, 12, 304 (1959).
317. Sophianopoulos, Rhodes, Holcomb, and Van Holde, *J. Biol. Chem.*, 237, 1107 (1962).
318. Praissman and Rupley, *Biochemistry*, 7, 2446 (1968).
319. Canfield, *J. Biol. Chem.*, 238, 2691 (1963).
320. Wetlaufer and Stahmann, *J. Am. Chem. Soc.*, 80, 1493 (1958).
321. Yutani, Yutani, Imanishi, and Isemura, *J. Biochem.* (Tokyo), 64, 449 (1968).
322. Jolles and Jolles, *Biochemistry*, 6, 411 (1967).
323. Bonavida, Sapse, and Sercarz, *J. Lab. Clin. Med.*, 70, 951 (1967).
324. Tsugita, Inouye, Terzaghi, and Streisinger, *J. Biol. Chem.*, 243, 391 (1968).
325. Howard and Glazer, *J. Biol. Chem.*, 242, 5715 (1967).
326. Black, Ph.D. thesis, Stanford University, 1967.
327. Wolfe and Neilands, *J. Biol. Chem.*, 221, 61 (1956).
328. Thorne, *Biochim. Biophys. Acta*, 59, 624 (1962).
329. Pfliederer and Hohnholz, *Biochem. Z.*, 331, 245 (1959).
330. Davies and Kun, *Biochem. J.*, 66, 307 (1957).
331. Grimm and Doherty, *J. Biol. Chem.*, 236, 1980 (1961).
332. Yoshida, *J. Biol. Chem.*, 240, 1113 (1965).
333. Murphey, Barnaby, Lin, and Kaplan, *J. Biol. Chem.*, 242, 1548 (1967).
334. Kitto, *Biochim. Biophys. Acta*, 139, 16 (1967).
335. Kitto and Lewis, *Biochim. Biophys. Acta*, 139, 1 (1967).
336. Kohn and Jakoby, *J. Biol. Chem.*, 243, 2472 (1968).
337. Harada and Wolfe, *J. Biol. Chem.*, 243, 4123 (1968).
338. Hsu and Lardy, *J. Biol. Chem.*, 242, 520 (1967).
339. Levy, *J. Biol. Chem.*, 242, 747 (1967).
340. Hsiang and Bright, *J. Biol. Chem.*, 242, 3079 (1967).
341. Barker, Smyth, Wilson, and Weissbach, *J. Biol. Chem.*, 234, 320 (1959).
342. Erwin and Hellerman, *J. Biol. Chem.*, 242, 4230 (1967).
343. Yamada and Yasunobu, *J. Biol. Chem.*, 237, 1511 (1962).
344. Rohrer, van Wartburg, and Aebi, *Biochem. Z.*, 344, 478 (1966).
345. Noda and Kuby, *J. Biol. Chem.*, 226, 551 (1957).
346. Callaghan and Weber, *Biochem. J.*, 73, 473 (1959).
347. Quass and Briskey, *J. Food Sci.*, 33, 180 (1968).
348. Kielley and Harrington, *Biochim. Biophys. Acta*, 41, 401 (1960).
349. Gellert and Englander, *Biochemistry*, 2, 39 (1963).
350. Nanninga, *Biochim. Biophys. Acta*, 82, 507 (1964).
351. Frederiksen and Holtzer, *Biochemistry*, 7, 3935 (1968).
352. Morita and Yagi, *Biochem. Biophys. Res. Commun.*, 22, 297 (1966).
353. Young, Himmelfarb, and Harrington, *J. Biol. Chem.*, 239, 2822 (1964).
354. Riddiford, *J. Biol. Chem.*, 241, 2792 (1966).
355. Yagi, Yazawa, and Yasui, *Biochem. Biophys. Res. Commun.*, 29, 331 (1967).
356. Wu, *Biochemistry*, 8, 39 (1969).
357. Hagihara, in *The Enzymes*, Vol. 4, 2nd ed., Boyer, Lardy, and Myrbäck, Eds., Academic Press, New York, 1960, 193.
358. Hochstein and Dalton, *Biochim. Biophys. Acta*, 139, 56 (1967).
359. Taniuchi and Anfinsen, *J. Biol. Chem.*, 241, 4366 (1966).
360. Edelhoch and Lippoldt, *J. Biol. Chem.*, 237, 2788 (1962).
361. Cunningham and Nuenke, *J. Biol. Chem.*, 234, 1447 (1959).
362. Edelhoch and Steiner, *J. Biol. Chem.*, 240, 2877 (1965).
363. Chatterjee and Montgomery, *Arch. Biochem. Biophys.*, 99, 426 (1962).
364. Kohn and Jakoby, *J. Biol. Chem.*, 243, 2486 (1968).
365. Glazer and Smith, *J. Biol. Chem.*, 236, 2948 (1961).
366. Smith, Light, and Kimmel, *Biochem. Soc. Symp.*, 21, 88 (1962).
367. Finkle and Smith, *J. Biol. Chem.*, 230, 669 (1958).
368. Whitaker and Bender, *J. Am. Chem. Soc.*, 87, 2728 (1965).
369. Potts, Aurbach, and Sherwood, *Recent Prog. Horm. Res.*, 22, 114 (1966).

370. Edelhoch, *J. Am. Chem. Soc.*, 79, 6100 (1957).
371. Perlman, *J. Biol. Chem.*, 241, 153 (1966).
372. Aloj, Salvatore, and Roche, *J. Biol. Chem.*, 242, 3810 (1967).
373. Perlman, Oplatka, and Katchalsky, *J. Biol. Chem.*, 242, 5163 (1967).
374. Chow and Kassel, *J. Biol. Chem.*, 243, 1718 (1968).
375. Gounaris and Perlman, *J. Biol. Chem.*, 242, 2739 (1967).
376. Keilin and Hartree, *Biochem. J.*, 49, 88 (1951).
377. Maehly, *Methods Enzymol.*, 2, 801 (1955).
378. Wittenberg, Antonini, Brunori, Noble, Wittenberg, and Wyman, *Biochemistry*, 6, 1970 (1967).
379. Paul, in *The Enzymes*, Vol. 8, 2nd ed., Boyer, Lardy, and Myrbäck, Eds., Academic Press, New York, 1963, 233.
380. Kertesz and Zito, in *Oxygenases*, Hayashi, Ed., Academic Press, New York, 1962, 307.
381. Rothman and Byrne, *J. Mol. Biol.*, 6, 330 (1963).
382. Plocke, Levinthal, and Vallee, *Biochemistry*, 1, 373 (1962).
383. Kadner, Nye, and Brown, *J. Biol. Chem.*, 243, 3076 (1968).
384. Dorn, *J. Biol. Chem.*, 243, 3500 (1968).
385. Neumann, *J. Biol. Chem.*, 243, 4671 (1968).
386. Harkness, *Arch. Biochem. Biophys.*, 126, 503 (1968).
387. Igarashi and Hollander, *J. Biol. Chem.*, 243, 6084 (1968).
388. Parmeggiani, Luft, Love, and Krebs, *J. Biol. Chem.*, 241, 4625 (1966).
389. Paetkau and Lardy, *J. Biol. Chem.*, 242, 2035 (1967).
390. Younathan, Paetkau, and Lardy, *J. Biol. Chem.*, 243, 1603 (1968).
391. Froede, Geraci, and Mansour, *J. Biol. Chem.*, 243, 6021 (1968).
392. Lindell and Stellwagen, *J. Biol. Chem.*, 243, 907 (1968).
393. Chatterjee and Noltman, *Eur. J. Biochem.*, 2, 9 (1967).
394. Najjar, *J. Biol. Chem.*, 175, 281 (1948).
395. Najjar, *Methods Enzymol.*, 1, 294 (1955).
396. Rippa, Signorini, and Pontremoli, *Eur. J. Biochem.*, 1, 170 (1967).
397. Sugimoto and Pizer, *J. Biol. Chem.*, 243, 2090 (1968).
398. Zwaig and Milstein, *Biochem. J.*, 98, 360 (1966).
399. Sugimoto, Sasaki, and Chiba, *Arch. Biochem. Biophys.*, 113, 444 (1966).
400. Voll, Appella, and Martin, *J. Biol. Chem.*, 242, 1760 (1967).
401. Wells and Hanahan, *Biochemistry*, 8, 414 (1969).
402. Korn and Buchanan, *J. Biol. Chem.*, 217, 183 (1955).
403. Agarwal and Parks, *Fed. Proc.*, 27, 585, Abstr. No. 2072 (1968).
404. Buc and Buc, in *Symposium on Regulation of Enzyme Activity and Allosteric Interactions*, 4th Federation of European Biochemical Societies, Oslo, 1967. Academic Press, 1967, 109.
405. Velick and Wicks, *J. Biol. Chem.*, 190, 741 (1951).
406. Appleman, Krebs, and Fischer, *Biochemistry*, 5, 2101 (1966).
407. Metzger, Glaser, and Helmreich, *Biochemistry*, 7, 2021 (1968).
408. Kastenschmidt, Kastenschmidt, and Helmreich, *Biochemistry*, 7, 3590 (1968).
409. Gold, *Biochemistry*, 7, 2106 (1968).
410. Appleman, Yunis, Krebs, and Fischer, *J. Biol. Chem.*, 238, 1358 (1963).
411. Shaltiel, Hedrick, and Fischer, *Methods Enzymol.*, 11, 675 (1967).
412. Brody and Brody, *Biochim. Biophys. Acta*, 50, 348 (1961).
413. Toro-Goyco and Matos, *Nature*, 210, 527 (1966).
414. Toro-Goyco, Maretzki, and Matos, *Arch. Biochem. Biophys.*, 126, 91 (1968).
415. Bergmeyer, Holz, Klotzsch, and Lang, *Biochem. Z.*, 338, 114 (1963).
416. Reisfeld, Borjeson, Chessin, and Small, *Proc. Natl. Acad. Sci. U.S.A.*, 58, 2020 (1967).
417. Jovim, Englund, and Bertsch, *J. Biol. Chem.* (1969).
418. Robbins, Summaria, Elwyn, and Barlow, *J. Biol. Chem.*, 240, 541 (1965).
419. Robbins and Summaria, *J. Biol. Chem.*, 238, 952 (1963).
420. Kertesz and Zito, *Biochim. Biophys. Acta*, 96, 447 (1965).
421. Gregory and Bendall, *Biochem. J.*, 101, 569 (1966).
422. Reuter, Hamoir, Marchand, and Kennes, *Eur. J. Biochem.*, 5, 233 (1968).
423. Frank and Veros, *Biochem. Biophys. Res. Commun.*, 32, 155 (1968).
424. Oshima, Matsuo, Iwanaga, and Suzuki, *J. Biochem.* (Tokyo), 64, 227 (1968).
425. Oshima, Iwanaga, and Suzuki, *J. Biochem.* (Tokyo), 54, 215 (1968).
426. Jonsson, *Arch. Biochem. Biophys.*, 129, 62 (1969).
427. Ganno, *J. Biochem.* (Tokyo), 58, 556 (1965).
428. McConn, Tsuru, and Yasunobu, *J. Biol. Chem.*, 239, 3706 (1964).
429. Ohta, Ogura, and Wada, *J. Biol. Chem.*, 241, 5919 (1966).
430. Jackson and Wolfe, *J. Biol. Chem.*, 243, 879 (1968).
431. Fukumoto, Tsuru, and Yamamoto, *Agric. Biol. Chem.*, 31, 710 (1967).

432. Kimmel, Markowitz, and Brown, *J. Biol. Chem.*, 234, 46 (1959).
433. Narahashi, Shibuya, and Yanagita, *J. Biochem.* (Tokyo), 64, 427 (1968).
434. Hiramatsu, *J. Biochem.* (Tokyo), 62, 353 (1967).
435. Day, Toncic, Stratman, Leeman, and Harmon, *Biochim. Biophys. Acta*, 167, 597 (1968).
436. Hata, Hayashi, and Doi, *Agric. Biol. Chem.*, 31, 357 (1967).
437. Frattali, *J. Biol. Chem.*, 244, 274 (1969).
438. Iwasaki and Schmid, *J. Biol. Chem.*, 242, 5247 (1967).
439. Tomasi and Kornguth, *J. Biol. Chem.*, 242, 4933 (1967).
440. Wood, Massi, and Solomon, *J. Biol. Chem.*, 234, 329 (1959).
441. Miyake, Gaylor, and Mason, *J. Biol. Chem.*, 243, 5788 (1968).
442. Marushige, Britlag, and Bonner, *Biochemistry*, 7, 3149 (1968).
443. Dus, De Klerk, Sletten, and Bartsch, *Biochim. Biophys. Acta*, 140, 291 (1967).
444. Tang and Coleman, *J. Biol. Chem.*, 243, 4286 (1968).
445. Hinkson, *Biochemistry*, 7, 2666 (1968).
446. Koshiyama, *Cereal Chem.*, 45, 405 (1968).
447. Koshiyama, *Cereal Chem.*, 45, 394 (1968).
448. Shore and Shore, *Biochemistry*, 6, 1962 (1967).
449. Gumpf and Hamilton, *Virology*, 35, 87 (1968).
450. Zaitlin and McCaughey, *Virology*, 26, 500 (1965).
451. Wasserman, Corradino, and Taylor, *J. Biol. Chem.*, 243, 3978 (1968).
452. Kalb, *Biochim. Biophys. Acta*, 168, 532 (1968).
453. Penrose, Nichoalds, Piperno, and Oxender, *J. Biol. Chem.*, 243, 5921 (1968).
454. Lamy and Waugh, *J. Biol. Chem.*, 203, 489 (1953).
455. Shulman and Hearon, *J. Biol. Chem.*, 238, 155 (1963).
456. Tishkoff, Williams, and Brown, *J. Biol. Chem.*, 243, 4151 (1968).
457. Aronson, *Thromb. Diath. Haemorrh.*, 16, 491 (1966).
458. Shapiro and Waugh, *Thromb. Diath. Haemorrh.*, 16, 469 (1966).
459. Lanchantin, Hart, Friedman, Saavedra, and Mehl, *J. Biol. Chem.*, 243, 5479 (1968).
460. Fujisawa and Hayashi, *J. Biol. Chem.*, 243, 2673 (1968).
461. Yui, Ishii, and Egami, *J. Biochem.* (Tokyo), 65, 37 (1969).
462. Ayling and Snell, *Biochemistry*, 7, 1616 (1968).
463. Kojima, Fujisawa, Nakazawa, Nakazawa, Kanetsuna, Taniuchi, Nozaki, and Hayaishi, *J. Biol. Chem.*, 242, 3270 (1967).
464. Kunitz, *J. Gen. Physiol.*, 35, 423 (1952).
465. Williams and Hager, *Methods Enzymol.*, 9, 265 (1966).
466. Bucher and Pfleiderer, *Methods Enzymol.*, 1, 435 (1955).
467. Packman and Jakoby, *J. Biol. Chem.*, 240, PC4107 (1965).
468. Foltman, *C. R. Trav. Lab. Carlsberg Ser. Chim.*, 34, 319 (1964).
469. Wang and Volini, *J. Biol. Chem.*, 243, 5465 (1968).
470. Shichi, Lewis, Irreverre, and Stone, *J. Biol. Chem.*, 244, 529 (1969).
471. Shields, Dinovo, Henriksen, Kimbel, and Millar, *Biochim. Biophys. Acta*, 147, 238 (1967).
472. Meadows and Jardetzky, *Proc. Natl. Acad. Sci. U.S.A.*, 61, 406 (1968).
473. Hill and Schmidt, *J. Biol. Chem.*, 237, 389 (1962).
474. Taborsky, *J. Biol. Chem.*, 234, 2915 (1959).
475. Sela, Anfinsen, and Harrington, *Biochim. Biophys. Acta*, 26, 502 (1957).
476. Jaenicke, Schmid, and Knof, *Biochemistry*, 7, 919 (1968).
477. Tanford, Hauenstein, and Rands, *J. Am. Chem. Soc.*, 77, 6409 (1955).
478. Blumenfeld and Levy, *Arch. Biochem. Biophys.*, 76, 97 (1958).
479. Sage and Singer, *Biochim. Biophys. Acta*, 29, 663 (1958).
480. Sage and Singer, *Biochemistry*, 1, 305 (1962).
481. Bigelow, *J. Biol. Chem.*, 236, 1706 (1961).
482. Sherwood and Potts, *J. Biol. Chem.*, 240, 3799 (1965).
483. Keller, Cohen, and Neurath, *J. Biol. Chem.*, 233, 344 (1958).
484. Aqvist and Anfinsen, *J. Biol. Chem.*, 234, 1112 (1959).
485. Wilson, *J. Biol. Chem.*, 242, 2260 (1967).
486. Yamasaki, Murakami, Irie, and Ukita, *J. Biochem.* (Tokyo), 63, 25 (1968).
487. Uchida, *J. Biochem.* (Tokyo), 60, 115 (1966).
488. Ettinger and Hirs, *Biochemistry*, 7, 3374 (1968).
489. Lee and Bendet, *J. Biol. Chem.*, 242, 2043 (1967).
490. Akoyunoglou, Argyroudi-Akoyunoglou, and Methenitou, *Biochim. Biophys. Acta*, 132, 481 (1967).
491. Lee, Patrick, and Masson, *J. Biol. Chem.*, 243, 4700 (1968).
492. Newman and Postgate, *Eur. J. Biochem.*, 7, 45 (1968).

493. Bachmayer, Yasunobu, Peel, and Mayhew, *J. Biol. Chem.*, 243, 1022 (1968).
494. Lovenberg, in *Protides of the Biological Fluids*, Vol. 14, Peeters, Ed., Elsevier, Amsterdam 1967, 165.
495. Bachmayer, Benson, Yasunobu, Garrard, and Whiteley, *Biochemistry*, 7, 986 (1968).
496. Peterson and Coon, *J. Biol. Chem.*, 243, 329 (1968).
497. Lovenberg and Williams, *Biochemistry*, 8, 141 (1969).
498. Laurell, *Acta Chem. Scand.*, 7, 1407 (1953).
499. Peisach, Levine, and Blumberg, *J. Biol. Chem.*, 242, 2841 (1967).
500. Green and Malamed, *Biochem. J.*, 100, 614 (1966).
501. Chaiet and Wolf, *Arch. Biochem. Biophys.*, 106, 1 (1964).
502. Taylor and Botts, *Biochemistry*, 7, 232 (1968).
503. Salvatore, Vecchio, Salvatore, Cahnman, and Robbins, *J. Biol. Chem.*, 240, 2935 (1965).
504. Matsubara, Kaspar, Brown, and Smith, *J. Biol. Chem.*, 240, 1125 (1965).
505. Landon, Evans, and Smith, *J. Biol. Chem.*, 243, 2165 (1968).
506. Neet, Nanci, and Koshland, *J. Biol. Chem.*, 243, 6392 (1968).
507. Gounaris and Ottesen, *C. R. Trav. Lab. Carlsberg Ser. Chim.*, 35, 37 (1965).
508. Ramaley, Bridger, Moyer, and Boyer, *J. Biol. Chem.*, 242, 4287 (1967).
509. Allen and Roy, *Biochim. Biophys. Acta*, 168, 243 (1968).
510. Kohn, Packman, Allen, and Jakoby, *J. Biol. Chem.*, 243, 2479 (1968).
511. Kohn, *J. Biol. Chem.*, 243, 4426 (1968).
512. Murphy, Plummer, and Miller, *Fed. Proc.*, 27, 268 Abstr. No. 298 (1968).
513. Holmgren and Reichard, *Eur. J. Biochem.*, 2, 187 (1967).
514. Thelander, *J. Biol. Chem.*, 242, 852 (1967).
515. Thelander and Baldesten, *Eur. J. Biochem.*, 4, 420 (1968).
516. Burns and Zarlengo, *J. Biol. Chem.*, 243, 178 (1968).
517. Shizuta, Nakazawa, Tokushige, and Hayaishi, *J. Biol. Chem.*, 244, 1883 (1969).
518. Kezdy, Lorand, and Miller, *Biochemistry*, 2302 (1965).
519. Winzor and Scheraga, *Arch. Biochem. Biophys.*, 104, 202 (1964).
520. Brewer, Keutmann, Potts, Reisfeld, Schlueter, and Munson, *J. Biol. Chem.*, 243, 5739 (1968).
521. Shome, Brown, Howard, and Pierce, *Arch. Biochem. Biophys.*, 126, 456 (1968).
522. Lai, Chen, and Tsolas, *Arch. Biochem. Biophys.*, 121, 790 (1967).
523. Seal and Doe, *J. Biol. Chem.*, 237, 3136 (1962).
524. Aisen, Aasa, Malstrom, and Vanngard, *J. Biol. Chem.*, 242, 2484 (1967).
525. Koechlin, *J. Am. Chem. Soc.*, 74, 2649 (1952).
526. Leibman and Aisen, *Arch. Biochem. Biophys.*, 121, 717 (1967).
527. Folk, Mullooly, and Cole, *J. Biol. Chem.*, 242, 1838 (1967).
528. Folk and Cole, *J. Biol. Chem.*, 241, 5518 (1966).
529. Mooz and Meister, *Biochemistry*, 6, 1722 (1967).
530. Woods, *J. Biol. Chem.*, 242, 2859 (1967).
531. Mee, Navon, and Stein, *Biochim. Biophys. Acta*, 104, 151 (1965).
532. Mihalyi and Harrington, *Biochim. Biophys. Acta*, 36, 447 (1959).
533. Green, *Biochem. J.*, 66, 407 (1957).
534. Kassell, Radicevic, Berlow, Peanasky, and Laskowski, *J. Biol. Chem.*, 238, 3274 (1963).
535. Laskowski and Laskowski, *Adv. Protein Chem.*, 9, 203 (1954).
536. Davie and Neurath, *J. Biol. Chem.*, 212, 515 (1955).
537. Labeyrie, Groudinsky, Jacquot-Armand, and Naslin, *Biochim. Biophys. Acta*, 128, 492 (1966).
538. Shaw, Mares-Guia, and Cohen, *Biochemistry*, 4, 2219 (1965).
539. Meloun, Fric, and Sorm, *Eur. J. Biochem.*, 4, 112 (1968).
540. Buck, Vithayathil, Bier, and Nord, *Arch. Biochem. Biophys.*, 97, 417 (1962).
541. Labouesse and Gervais, *Eur. J. Biochem.*, 2, 215 (1967).
542. Heide, Heilburger, and Haupt, *Clin. Chim. Acta*, 11, 82 (1956).
543. Pharo, Sordahl, Edelhoch, and Sanadi, *Arch. Biochem. Biophys.*, 125, 416 (1968).
544. Greene, Rigbi, and Fackre, *J. Biol. Chem.*, 241, 5610 (1966).
545. Greene, DiCarlo, Sussman, Bartelt, and Roark, *J. Biol. Chem.*, 243, 1804 (1968).
546. Kunitz, *J. Gen. Physiol.*, 30, 291 (1947).
547. Birk, Gertler, and Khalef, *Biochem. J.*, 87, 281 (1963).
548. Rackis, Sasame, Mann, Anderson, and Smith, *Arch. Biochem. Biophys.*, 98, 471 (1962).
549. Wu and Scheraga, *Biochemistry*, 1, 698 (1962).
550. Frattali and Steiner, *Biochemistry*, 7, 521 (1968).
551. Schyns, Bricteux-Gregoire, and Florkin, *Biochim. Biophys. Acta*, 175, 97 (1969).
552. Pechere, Dixon, Maybury, and Neurath, *J. Biol. Chem.*, 233, 1364 (1958).
553. Morino and Snell, *J. Biol. Chem.*, 242, 2800 (1967).
554. Hathaway, Kida, and Crawford, *Biochemistry*, 8, 989 (1969).

555. Kumagai, Matsui, Ogata, and Yamada, *Biochim. Biophys. Acta,* 171, 1 (1968).
556. Yamada, Kumagi, Matsui, Ohgishi, and Ogata, *Biochem. Biophys. Res. Commun.,* 33, 10 (1968).
557. Hayashi, Granner, and Tomkins, *J. Biol. Chem.,* 242, 3998 (1967).
558. Gorin and Chin, *Biochim. Biophys. Acta,* 99, 418 (1965).
559. Gorin, Fuchs, Butler, Chopra, and Hersh, *Biochemistry,* 1, 911 (1962).
560. Gorin and Chin, *Anal. Biochem.,* 17, 49 (1966).
561. Mahler, Hubscher, and Baum, *J. Biol. Chem.,* 216, 625 (1955).
562. *Iscotables,* Instrumentation Specialities Co., Inc., Lincoln, Nebraska, 1967, 9.
563. Shalaby, Banerjee, and Lauffer, *Biochemistry,* 7, 955 (1968).
564. Heller, *Biochemistry,* 7, 2906 (1968).
565. Avis, Bergel, and Bray, *J. Chem. Soc.* (Lond.), 1219 (1956).
566. Massey, Brumby, Komai, and Palmer, *J. Biol. Chem.,* 224, 1682 (1969).
567. O'Leary and Westheimer, *Biochemistry,* 7, 913 (1968).
568. Stormer, *J. Biol. Chem.,* 243, 3740 (1968).
569. Gregolin, Ryder, and Lane, *J. Biol. Chem.,* 243, 4227 (1968).
570. Hauge, *Arch. Biochem. Biophys.,* 94, 308 (1961).
571. Pajot and Groudinsky, *Eur. J. Biochem.,* 12, 158 (1970).
572. Labeyrie, Groudinsky, Jacquot-Armand, and Naslin, *Biochim. Biophys. Acta,* 128, 492 (1966).
573. Kodama and Shidara, *J. Biochem.* (Tokyo), 65, 356 (1969).
574. Horio, Higashi, Yamanaka, Matabara, and Okunuki, *J. Biol. Chem.,* 23, 944 (1961).
575. Kredich, Becker, and Tomkin, *J. Biol. Chem.,* 244, 2428 (1969).
576. Carrico and Deutsch, *J. Biol. Chem.,* 245, 723 (1970).
577. Price, Liu, Stein, and Moore, *J. Biol. Chem.,* 244, 917 (1969).
578. Lindberg, *Biochemistry,* 6, 335 (1967).
579. Lindberg, *Biochemistry,* 6, 323 (1967).
580. Jovin, Englund, and Bertsch, *J. Biol. Chem.,* 244, 2996 (1969).
581. Roychoudhury and Bloch, *J. Biol. Chem.,* 244, 3359 (1969).
582. Roncari, Kurylo-Borowska, and Craig, *Biochemistry,* 5, 2153 (1966).
583. Bender, Begue-Canton, Blakeley, Brubacher, Feder, Gunter, Kesdy, Killhefer, Jr., Marshall, Miller, Roeske, and Stoops, *J. Am. Chem. Soc.,* 88, 5890 (1966).
584. Warburg and Christian, *Biochem. Z.,* 310, 384 (1941).
585. Barker and Jencks, *Biochemistry,* 8, 3879 (1969).
586. Malhotra and Philip, *Indian J. Biochem.,* 3, 7 (1966).
587. Lanchantin, Friedman, and Hart, *J. Biol. Chem.,* 244, 865 (1969).
588. Hsu and Yun, *Biochemistry,* 9, 239 (1970).
589. Hsu, Wasson, and Porter, *J. Biol. Chem.,* 240, 3736 (1965).
590. Dey and Pridham, *Biochem. J.,* 113, 49 (1969).
591. Petek, Villarroya, and Courtois, *Eur. J. Biochem.,* 8, 395 (1969).
592. Carrico and Deutsch, *J. Biol. Chem.,* 244, 6087 (1969).
593. Porter, Sweeney, and Porter, *Arch. Biochem. Biophys.,* 105, 319 (1964).
594. Deutsch, *Immunochemistry,* 2, 207 (1965).
595. Newcomb, Normansell, and Stanworth, *J. Immunol.,* 101, 905 (1968).
596. Heimer, Jones, and Maurer, *Biochemistry,* 8, 3937 (1969).
597. Doi and Jirgensons, *Biochemistry,* 9, 1066 (1970).
598. Kickhofen, Hammer, and Scheel, *Hoppe-Seylers Z. Physiol. Chem.,* 349, 1755 (1968).
599. Binaghi and Oriol, *Bull. Soc. Chim. Biol.,* 50, 1035 (1968).
600. Mihaesco and Mihaesco, *Biochem. Biophys. Res. Commun.,* 33, 869 (1968).
601. Krigbaum and Kugler, *Biochemistry,* 9, 1216 (1970).
602. Johansson, *Acta Chem. Scand.,* 23, 683 (1969).
603. Kessler and Brew, *Biochim. Biophys. Acta,* 200, 449 (1970).
604. Morrison, Hamilton, and Stotz, *J. Biol. Chem.,* 228, 767 (1957).
605. Frohne and Hanson, *Hoppe-Seylers Z. Physiol. Chem.,* 350, 207 (1969).
606. Kretschmer, *Hoppe-Seylers Z. Physiol. Chem.,* 349, 846 (1968).
607. Oriol-Audit, Landon, Robin, and van Thoai, *Biochim. Biophys. Acta,* 188 132 (1969).
608. Lee and McElroy, *Biochemistry,* 8, 130 (1969).
609. Antonini, *Physiol. Rev.,* 45, 123 (1965).
610. Harris and Hill, *J. Biol. Chem.,* 244, 2195 (1969).
611. Straus, Gordon, and Wallach, *Eur. J. Biochem.,* 11, 201 (1969).
612. Willick, Schonbaum, and Kay, *Biochemistry,* 8, 3729 (1969).
613. Stryer, *J. Mol. Biol.,* 13, 482 (1965).
614. Hapner, Bradshaw, Hartzell, and Gurd, *J. Biol. Chem.,* 243, 683 (1968).
615. Awad and Kotite, *Biochem. J.,* 98, 909 (1966).

616. Godfrey and Harrington, *Biochemistry,* 9, 886 (1970).
617. Adelman and Taylor, *Biochemistry,* 8, 4976 (1969).
618. Tada, Bailin, Barany, and Barany, *Biochemistry,* 8, 4842 (1969).
619. Woods, *Int. J. Protein Res.,* 1, 29 (1969).
620. Angeletti, *Proc. Natl. Acad. Sci., U.S.A.,* 65, 668 (1970).
621. Furth and Hope, *Biochem. J.,* 116, 545 (1970).
622. Villet and Dalziel, *Biochem. J.,* 115, 639 (1969).
623. Pon, Schnackerz, Blackburn, Chatterjee, and Noltmann, *Biochemistry,* 9, 1506 (1970).
624. Carter and Yoshida, *Biochim. Biophys. Acta,* 181, 12 (1969).
625. Wald and Brown, *J. Gen. Physiol.,* 37, 189 (1953–54).
626. Hubbard, *J. Gen. Physiol.,* 37, 381 (1953–54).
627. Schichi, *Biochemistry,* 9, 1973 (1970).
628. Nordlie and Fromm, *J. Biol. Chem.,* 234, 2522 (1959).
629. Nakagawa and Kinnura, *J. Biochem.* (Tokyo), 66, 669 (1969).
630. Fojioka, *Biochim. Biophys. Acta,* 185, 338 (1969).
631. Marchesi, Steers, Marchesi, and Tillack, *Biochemistry,* 9, 50 (1970).
632. Allen and Jakoby, *J. Biol. Chem.,* 244, 2078 (1968).
633. Zalitis and Feingold, *Arch. Biochem. Biophys.,* 132, 457 (1969).
634. Becker, Kredich, and Tomkins, *J. Biol. Chem.,* 244, 2418 (1969).
635. Bernardi, *Adv. Enzymol.,* 31, 1 (1968).
636. Eisenberg, Zobel, and Moos, *Biochemistry,* 7, 3186 (1968).
637. Pugh and Wakil, *J. Biol. Chem.,* 240, 4727 (1965).
638. Sauer, Pugh, Wakil, Delany, and Hill, *Proc. Natl. Acad. Sci., U.S.A.,* 52, 1360 (1964).
639. Ramponi, Guerritore, Treves, Nassi, and Baccari, *Arch. Biochem. Biophys.,* 130, 362 (1969).
640. Webster, *Biochim. Biophys. Acta,* 207, 371 (1970).
641. Koberstein, Weber, and Jaenicke, *Z. Naturforsch. B,* 23, 474 (1968).
642. Andersson, *Int. J. Protein Res.,* 1, 151 (1969).
643. Glazer and Smith, *J. Biol. Chem.,* 235, PC43 (1960).
644. Bonewell and Rossini, *Ital. J. Biochem.,* 18, 457 (1969).
645. Kaldor, Saifer, and Westley, *Arch. Biochem. Biophys.,* 99, 275 (1962).
646. Habeeb and Hiramoto, *Arch. Biochem. Biophys.,* 126, 16 (1968).
647. Pederson and Foster, *Biochemistry,* 8, 2357 (1969).
648. Groulade, Chicault, and Waltzinger, *Bull. Soc. Chim. Biol.,* 48, 1609 (1967).
649. Frank and Veros, *Fed. Proc.,* 28, 728 (1969).
650. Frank, Peker, Veros, and Ho, *J. Biol. Chem.,* 245, 3716 (1970).
651. Wagner, Irion, Arens, and Bauer, *Biochem. Biophys. Res. Commun.,* 37, 383 (1969).
652. Arens, Rauenbusch, Irion, Wagner, Bauer, and Kaufmann, *Hoppe-Seylers Z. Physiol Chem.,* 351, 199 (1970).
653. Funakoshi and Deutsch, *J. Biol. Chem.,* 244, 3438 (1969).
654. Verpoorte and Linnblow, *J. Biol. Chem.,* 243, 5993 (1968).
655. Armstrong, Myers, Verpoorte, and Edsall, *J. Biol. Chem.,* 241, 5137 (1966).
656. Carpy, *Biochem. Biophys. Acta,* 151, 245 (1968).
657. Furth, *J. Biol. Chem.,* 243, 4832 (1968).
658. McIntosh, *Biochem. J.,* 114, 463 (1969).
659. Tobin, *J. Biol. Chem.,* 245, 2656 (1970).
660. Emery, *Biochemistry,* 8, 877 (1969).
661. Deisseroth and Dounce, *Arch. Biochem. Biophys.,* 131, 18 (1969).
662. Petit and Tauber, *J. Biol. Chem.,* 195, 703 (1952).
663. Stern and Lavin, *Science,* 88, 263 (1938).
664. Agner, *Biochem. J.,* 32, 1702 (1938).
665. Bonnichsen, *Arch. Biochem.,* 12, 83 (1947).
666. Ushijima and Nakano, *Biochim. Biophys. Acta,* 178, 429 (1969).
667. Greenfield and Price, *J. Biol. Chem.,* 220, 607 (1956).
668. Gregory, *Biochim. Biophys. Acta,* 159, 429 (1968).
669. Tanford and Lovrien, *J. Am. Chem. Soc.,* 84, 1892 (1962).
670. Bjorndal and Eriksson, *Arch. Biochem. Biophys.,* 124, 149 (1968).
671. Eriksson and Pettersson, *Arch. Biochem. Biophys.,* 124, 160 (1968).
672. Pettersson and Eaker, *Arch. Biochem. Biophys.,* 124, 154 (1968).
673. Wang and Barker, *Methods Enzymol.,* 13, 331 (1969).
674. Wu and Yang, *J. Biol. Chem.,* 245, 212 (1970).
675. Yang, *J. Biol. Chem.,* 240, 1616 (1965).
676. Chang and Hayashi, *Biochem. Biophys. Res. Commun.,* 37, 841 (1969).
677. Sjogren and Spychalski, *J. Am. Chem. Soc.,* 52, 4400 (1930).

678. Konisky and Richards, *J. Biol. Chem.*, 245, 2973 (1970).
679. Schito, *G. Microbiol.*, 14, 77 (1966).
680. Schito, Molina, and Pesce, *Biochem. Biophys. Res. Commun.*, 28, 611 (1967).
681. Tan and Woodworth, *Biochemistry*, 8, 3711 (1969).
682. Freisheim and Huennekens, *Biochemistry*, 8, 2271 (1969).
683. Nixon and Blakley, *J. Biol. Chem.*, 243, 4722 (1968).
684. Schwartz and Reed, *J. Biol. Chem.*, 243, 639 (1968).
685. Frisell and Mackenzie, *J. Biol. Chem.*, 237, 94 (1962).
686. Rene and Campbell, *J. Biol. Chem.*, 244, 1445 (1969).
687. Chu, *J. Biol. Chem.*, 243, 4342 (1968).
688. Borja, *Biochemistry*, 8, 71 (1969).
689. Huseby and Murray, *Biochem. Biophys. Res. Commun.*, 35, 169 (1969).
690. Marder, Shulman, and Carroll, *J. Biol. Chem.*, 244, 2111 (1969).
691. Iizuka and Yang, *Biochemistry*, 7, 2218 (1968).
692. Takasaki, Kosugi, and Kambayashi, *Agric. Biol. Chem.*, 33, 1527 (1969).
693. Rattazzi, *Biochim. Biophys. Acta*, 181, 1 (1969).
694. Bruni, Auricchio, and Covelli, *J. Biol. Chem.*, 244, 4735 (1969).
695. Legler, *Hoppe-Seylers Z. Physiol. Chem.*, 349, 1755 (1970).
696. Legler, *Hoppe-Seylers Z. Physiol. Chem.*, 348, 1359 (1967).
697. Herman-Boussier, Moretti, and Jayle, *Bull. Soc. Chim. Biol.*, 42, 837 (1960).
698. Guinand, Tonnelat, Boussier, and Jayle, *Bull. Soc. Chim. Biol.*, 38, 329 (1956).
699. Dobryszycka, Elwyn, and Kukral, *Biochim. Biophys. Acta*, 175, 220 (1969).
700. Okazaki, Wittenberg, Briehl, and Wittenberg, *Biochim. Biophys. Acta*, 140, 258 (1967).
701. Webster, *Biochim. Biophys. Acta*, 207, 371 (1970).
702. Bucci, Fronticelli, and Ragatz, *J. Biol. Chem.*, 243, 241 (1968).
703. Chiancone, Curell, Vecchini, Antonini, and Wyman, *J. Biol. Chem.*, 245, 4105 (1970).
704. Peacock, Pastewka, Reed, and Ness, *Biochemistry*, 9, 2275 (1970).
705. Atassi, Brown, and McEwan, *Immunochemistry*, 2, 379 (1965).
706. Benesch and Benesch, *J. Biol. Chem.*, 236, 405 (1961).
707. Bucci, Fronticelli, Bellelli, Antonini, Wyman, and Rossi-Fanelli, *Arch. Biochem. Biophys.*, 100, 364 (1963).
708. Li and Johnson, *Biochemistry*, 3, 2083 (1969).
709. Antonini, Bucci, Fronticelli, Chiancone, Wyman, and Rossi-Fanelli, *J. Mol. Biol.*, 17, 29 (1966).
710. Ueda, Shiga, and Tyuma, *Biochem. Biophys. Acta*, 207, 18 (1970).
711. Ozawa, *Biochemistry*, 9, 2158 (1970).
712. Harrison and Garratt, *Biochem. J.*, 113, 733 (1969).
713. Moller, Castleman, and Terhorst, *FEBS Lett.*, 8, 192 (1970).
714. Frank and Veros, *Biochem. Biophys. Res. Commun.*, 32, 155 (1968).
715. Chung and Franzen, *Biochemistry*, 8, 3175 (1969).
716. Illingworth and Tipton, *Biochem. J.*, 118, 253 (1970).
717. Malmstrom, Reinhammar, and Vanngard, *Biochim. Biophys. Acta*, 205, 48 (1970).
718. Tang, Coleman, and Myer, *J. Biol. Chem.*, 243, 4286 (1968).
719. Lin, *Biochemistry*, 9, 984 (1970).
720. Ebner, Denton, and Brodbeck, *Biochem. Biophys. Res. Comm.*, 22, 232 (1966).
721. Stevens, Brown, and Smith, *Arch. Biochem. Biophys.*, 136, 413 (1970).
722. Eriksson and Svensson, *Biochim. Biophys. Acta*, 198, 449 (1970).
723. Soda and Misone, *Biochemistry*, 7, 4110 (1968).
724. Sher and Mallette, *Arch. Biochem. Biophys.*, 53, 354 (1954).
725. Soda and Moriguchi, *Biochem. Biophys. Res. Comm.*, 34, 34 (1969).
726. Takeda, Yamamoto, Kojima, and Hayaishi, *J. Biol. Chem.*, 244, 2935 (1969).
727. Petek and Villarroya, *Bull. Soc. Chim. Biol.*, 50, 725 (1968).
728. Kajita, Kerwar, and Huennekens, *Arch. Biochem. Biophys.*, 130, 662 (1969).
729. Takesue and Omura, *J. Biochem.* (Tokyo), 67, 267 (1970).
730. Nishibayashi-Yamashita and Sato, *J. Biochem.* (Tokyo), 67, 199 (1970).
731. Karlsson, Eaker, and Porath, *Biochim. Biophys. Acta*, 127, 505 (1966).
732. Pho, Olomucki, Huc, and Thoai, *Biochim. Biophys. Acta*, 206, 46 (1970).
733. Webster, *Biochim. Biophys. Acta*, 207, 371 (1970).
734. Willumsen, *C. R. Trav. Lab. Carlsberg*, 37, 21 (1969).
735. Cheesman, *Proc. R. Soc. Lond. (Biol.)*, 149, 571 (1958).
736. Peraino, Bunville, and Tahmisian, *J. Biol. Chem.*, 244, 2241 (1969).
737. Willick, Schonbaum, and Kay, *Biochemistry*, 8, 3729 (1969).
738. Hardin Strickland, Kay, and Shannon, *J. Biol. Chem.*, 243, 3560 (1968).
739. Hamaguchi, Ikeda, Yoshida, and Morita, *J. Biochem.* (Tokyo), 66, 191 (1969).

740. Mazza, Charles, Bouchet, Ricard, and Raynaud, *Biochim. Biophys. Acta,* 167, 89 (1968).
741. Blasi, Fragomele, and Corelli, *J. Biol. Chem.,* 244, 4866 (1969).
742. Simpson, Vallee, and Taft, *Biochemistry,* 7, 4336 (1968).
743. Schlesinger, *J. Biol. Chem.,* 243, 3877 (1968).
744. Rothman and Byrne, *J. Mol. Biol.,* 6, 330 (1963).
745. Morton, *Biochem. J.,* 60, 573 (1955).
746. Alvarez and Lora-Tamayo, *Biochem. J.,* 69, 312 (1958).
747. Dennert and Hoglund, *Eur. J. Biochem.,* 12, 502 (1970).
748. Hunsley and Suelter, *J. Biol. Chem.,* 244, 4185 (1969).
749. Richardson, *Proc. Natl. Acad. Sci., U.S.A.,* 55, 1616 (1966).
750. Neuhoff, Schill, and Sternbach, *Hoppe-Seylers Z. Physiol. Chem.,* 350, 767 (1969).
751. Brown, Canellakis, Lundin, Reichard, and Thelander, *Eur. J. Biochem.,* 9, 561 (1969).
752. Brown, Eliasson, Reichard, and Thelander, *Eur. J. Biochem.,* 9, 512 (1969).
753. Lee, Patrick, and Barnes, *J. Biol. Chem.,* 245, 1357 (1970).
754. Matsubara and Nishimura, *J. Biochem.* (Tokyo), 45, 503 (1958).
755. Yamada, Kumagi, Nagate, and Yoshida, *Biochem. Biophys. Res. Commun.,* 39, 53 (1970).
756. Brewer and Edelhoch, *J. Biol. Chem.,* 245, 2402 (1970).
757. Salvatore, Vecchio, Salvatore, Cahnmann, and Robbins, *J. Biol. Chem.,* 240, 2935 (1965).
758. Han and Benson, *Biochem. Biophys. Res. Commun.,* 38, 378 (1970).
759. Mikola and Suolinna, *Eur. J. Biochem.,* 9, 555 (1969).
760. Poillon, Maeno, Koike, and Feigelson, *J. Biol. Chem.,* 244, 3447 (1968).
761. Henning, Helinski, Chao, and Yanofsky, *J. Biol. Chem.,* 237, 1523 (1962).
762. Hathaway and Crawford, *Biochemistry,* 9, 1801 (1970).
763. Wilson and Crawford, *J. Biol. Chem.,* 240, 4801 (1965).
764. Lemaire, van Rapenbusch, Gros, and Labouesse, *Eur. J. Biochem.,* 10, 334 (1969).
765. Paul and Stigbrand, *Biochim. Biophys. Acta,* 221, 255 (1970).
766. Haas, Lumfrom, and Goldblatt, *Arch. Biochem. Biophys.,* 44, 79 (1953).
767. Malhorta and Roni, *Indian J. Biochem.,* 6, 15 (1969).
768. Blakeley, Webb, and Zerner, *Biochemistry,* 8, 1984 (1969).
769. Bailey and Boulter, *Biochem. J.,* 112, 669 (1969).
770. George and Phillip, *J. Biol. Chem.,* 245, 529 (1970).
771. Kawano, Morimoto, and Uemura, *J. Biochem.* (Tokyo), 67, 333 (1970).
772. Scraba, Hostvedt, and Colter, *Can. J. Biochem.,* 47, 165 (1969).
773. Hull, Hills, and Markham, *Virology,* 37, 416 (1969).
774. Stevens and Lauffer, *Biochemistry,* 4, 31 (1965).
775. Reichman, *J. Biol. Chem.,* 235, 2959 (1966).
776. Crewther and Harrap, *J. Biol. Chem.,* 242, 4310 (1967).
777. Mega, Ikenaka, and Matsushima, *J. Biochem.* (Tokyo), 68, 109 (1970).
778. Villafranca and Mildvan, *J. Biol. Chem.,* 246, 772 (1971).
779. Kimura and Huang, *Arch. Biochem. Biophys.,* 137, 357 (1963).
780. Shimomura and Johnson, *Biochemistry,* 8, 3991 (1969).
781. Rosso, Takashima, and Adams, *Biochem. Biophys. Res. Commun.,* 34, 134 (1969).
782. Hamlin, *Exp. Gerontol.,* 4, 189 (1969).
783. Von Tigerstron and Razzell, *J. Biol. Chem.,* 243, 2691 (1968).
784. Von Tigerstron and Razzell, *J. Biol. Chem.,* 243, 6495 (1968).
785. Dimroth, Guchhait, Stoll, and Lane, *Proc. Natl. Acad. Sci., U.S.A.,* 67, 1353 (1970).
786. Mayer and Miller, *Anal. Biochem.,* 36, 91 (1970).
787. Anastasi, Erspamer, and Endean, *Arch. Biochem. Biophys.,* 125, 57 (1968).
788. Keutmann, Parsons, Potts, Jr., and Schlueter, *J. Biol. Chem.,* 245, 1491 (1970).
789. Marshall and Cohen, *J. Biol. Chem.,* 241, 4197 (1970).
790. Lacko and Neurath, *Biochemistry,* 9, 4680 (1970).
791. Visuri, Mikola, and Enari, *Eur. J. Biochem.,* 7, 193 (1969).
792. Boston and Prescott, *Arch. Biochem. Biophys.,* 128, 88 (1968).
793. Yue, Palmieri, Olson, and Kuby, *Biochemistry,* 6, 3204 (1967).
794. Gosselin-Rey and Gerday, *Biochim. Biophys. Acta,* 221, 241 (1970).
795. Yue, Jacobs, Okabe, Keutel, and Kuby, *Biochemistry,* 7, 4291 (1968).
796. Kumudavalli, Moreland, and Watts, *Biochem. J.,* 117, 513 (1970).
797. Hass and Hill, *J. Biol. Chem.,* 244, 6080 (1969).
798. Fraenkel-Conrat and Singer, *Arch. Biochem. Biophys.,* 60, 64 (1956).
799. Zimmerman and Coleman, *J. Biol. Chem.,* 246, 309 (1971).
800. Tai and Sih, *J. Biol. Chem.,* 245, 5062 (1970).
801. Shotton, *Methods Enzymol.,* 19, 113 (1970).

802. Wang and Keen, *Arch. Biochem. Biophys.*, 141, 749 (1970).
803. Seto, Sato, and Tamiya, *Biochim. Biophys. Acta*, 214, 483 (1970).
804. Fanelli, Chiancone, Vecchini, and Antonini, *Arch. Biochem. Biophys.*, 141, 278, (1970).
805. Hartz and Deutsch, *J. Biol. Chem.*, 244 4565 (1969).
806. Svedberg and Sjogren, *J. Am. Chem. Soc.*, 52, 279 (1930).
807. Lee, Travis, and Black, Jr., *Arch. Biochem. Biophys.*, 141, 676 (1970).
808. Newman, Ihle, and Dure, III, *Biochem. Biophys. Res. Commun.*, 36, 947 (1969).
809. Mayhew, Petering, Palmer, and Foust, *J. Biol. Chem.*, 244, 2830 (1969).
810. Devanathan, Akagi, Hersh, and Himes, *J. Biol. Chem.*, 244, 2846 (1969).
811. Lovenberg, Buchanan, and Rabinowitz, *J. Biol. Chem.*, 238, 3899 (1963).
812. Bayer, Eckstein, Hagenmaier, Josef, Koch, Krauss, Roder, and Schretzmann, *Eur. J. Biochem.*, 8, 33 (1969).
813. Hong and Rabinowitz, *J. Biol. Chem.*, 245, 4982 (1970).
814. Moss, Petering, and Palmer, *J. Biol. Chem.*, 244, 2275 (1969).
815. Petering, Fee, and Palmer, *J. Biol. Chem.*, 246, 643 (1971).
816. Hormann and Gollwitzer, *Z. Physiol. Chem.*, 346, 21 (1966).
817. Mihalyi, *Biochemistry*, 7, 208 (1968).
818. Budzynski, *Biochim. Biophys. Acta*, 229, 663 (1971).
819. Huseby, Mosesson, and Murray, *Physiol. Chem. Phys.*, 2, 374 (1970).
820. Fuller and Doolittle, *Biochemistry*, 10, 1305 (1971).
821. Holloway, Antonini, and Brunori, *FEBS Lett.*, 4, 299 (1969).
822. Kramer and Whitaker, *J. Biol. Chem.*, 239, 2178 (1964).
823. Brewer, Ljungdahl, Spencer, and Neece, *J. Biol. Chem.*, 245, 4798 (1970).
824. Pontremoli, *Methods Enzymol.*, 9, 625 (1966).
825. Mosesson and Umfleet, *J. Biol. Chem.*, 245, 5728 (1970).
826. Danno, *Agric. Biol. Chem.*, 34, 1795 (1970).
827. Strausbauch and Fischer, *Biochemistry*, 9, 226 (1970).
828. Fahien and Cohen, *Methods Enzymol.*, 17A, 839 (1970).
829. Dessen and Pantaloni, *Eur. J. Biochem.*, 8, 292 (1969).
830. Winnacker and Barker, *Biochim. Biophys. Acta*, 212, 225 (1970).
831. Barker, *Methods Enzymol.*, 13, 319 (1969).
832. Rowe, Ronzio, Wellner, and Meister, *Methods Enzymol.*, 17A, 900 (1970).
833. Woodin and Segel, *Biochim. Biophys. Acta*, 167, 64 (1968).
834. Kohn, Warner, and Carroll, *J. Biol. Chem.*, 245, 3820 (1970).
835. Edelhoch and Lippoldt, *J. Biol. Chem.*, 245, 4199 (1970).
836. Gould and Scheinberg, *Arch. Biochem. Biophys.*, 137, 1 (1970).
837. Sage and Connett, *J. Biol. Chem.*, 244, 4713 (1969).
838. Ferrell and Kitto, *Biochemistry*, 9, 3053 (1970).
839. Keresztes-Nagy, Ph.D. thesis, Northwestern University, 1962; cited in Subramanian, Holleman, and Klotz, *Biochemistry*, 7, 2859 (1968).
840. Morimoto and Kegeles, *Arch. Biochem. Biophys.*, 142, 247 (1971).
841. Gruber, in *Physiology and Biochemistry of Hemocyanins*, Ghiretti, Ed., Academic Press, New York, 1968, 49.
842. Konings, Dijk, Wichertjes, Beuvery, and Gruber, *Biochim. Biophys. Acta*, 188, 43 (1969).
843. Ellerton, Carpenter, and Van Holde, *Biochemistry*, 9, 2225 (1970).
844. Joniau, Grossberg, and Pressman, *Immunochemistry*, 7, 755 (1970).
845. Amkraut, personal communication.
846. Shaklai and Daniel, *Biochemistry*, 9, 564 (1970).
847. Maki, Yamamoto, Nozaki, and Hayaishi, *J. Biol. Chem.*, 244, 2942 (1969).
848. Leslie and Cohen, *Biochem. J.*, 120, 787 (1970).
849. Herskovits, *Arch. Biochem. Biophys.*, 130, 19 (1969).
850. Meachum, Jr., Colvin, Jr., and Braymer, *Biochemistry*, 10, 326 (1971).
851. Yokobayashi, Misaki, and Harada, *Biochim. Biophys. Acta*, 212, 458 (1970).
852. McFadden, *Methods Enzymol.*, 13, 163 (1969).
853. Parsons and Burns, *J. Biol. Chem.*, 244, 996 (1969).
854. Yu, Harmon, Wachter, and Blank, *Arch. Biochem. Biophys.*, 135, 363 (1969).
855. Alfsen, Baulieu, Claquin, and Falcoz-Kelly, *Proc. 2nd Int. Congr. Hormonal Steroids*, 1967, 508.
856. Malkin and Malmstrom, *Adv. Enzymol.*, 33, 178 (1970).
857. Nakamura and Ogura, *J. Biochem. (Tokyo)*, 59, 449 (1966).
858. Malkin, Malmstrom, and Vanngard, *Eur. J. Biochem.*, 10, 324 (1969).
859. Malkin, Malmstrom, and Vanngard, *Eur. J. Biochem.*, 7, 253 (1969).
860. Karkhanis and Cormier, *Biochemistry*, 10, 317 (1971).
861. Haino, *Biochim. Biophys. Acta*, 229, 459 (1971).
862. Chibata, Ishikawa, and Tosa, *Methods Enzymol.*, 19, 675 (1970).

863. Nagasawa, Sugihara, Han, and Suzuki, *J. Biochem.* (Tokyo), 67, 809 (1970).
864. Schonenberger, Schmidtberger, and Schultze, *Z. Naturforsch.*, 136, 761 (1958).
865. Jacquot-Armand and Guinand, *Biochim. Biophys. Acta*, 133, 289 (1967).
866. Berggard and Bearn, *J. Biol. Chem.*, 243, 4095 (1968).
867. Gregory and Harrison, *Biochem. Biophys. Res. Commun.*, 40, 995 (1970).
868. Covelli, Consiglio, and Varrone, *Biochim. Biophys. Acta*, 184, 678 (1969).
869. Gerding and Wolfe, *J. Biol. Chem.*, 244, 1164 (1969).
870. Kitto, *Methods Enzymol.*, 13, 106 (1969).
871. Mann and Vestling, *Biochemistry*, 8, 1105 (1969).
872. Spina, Jr., Bright, and Rosenbloom, *Biochemistry*, 9, 3794 (1970).
873. Ramachandran, *Biochem. Biophys. Res. Commun.*, 41, 353 (1970).
874. Lawrence, *Eur. J. Biochem.*, 15, 436 (1970).
875. Agner, *Acta Chem. Scand.*, 12, 89 (1958).
876. Schirmer, Schirmer, Schulz, and Thuma, *FEBS Lett.*, 10, 333 (1970).
877. Leuzinger, *Biochem. J.*, 123, 139 (1971).
878. Muruyama, *J. Biochem.* (Tokyo), 69, 369 (1971).
879. Minato, *J. Biochem.* (Tokyo), 64, 813 (1969).
880. Zielke and Suelter, *J. Biol. Chem.*, 246, 2179 (1971).
881. Zielke and Suelter, *Fed. Proc.*, 28, 2624 (1969).
882. Tweedie and Segel, *J. Biol. Chem.*, 246, 2438 (1971).
883. Lebeault, Zevaco, and Hermier, *Bull. Soc. Chim. Biol.*, 52, 1073 (1970).
884. Sterman and Foster, *J. Am. Chem. Soc.*, 78, 3656 (1956).
885. Frigerio and Hettinger, *Biochim. Biophys. Acta*, 59, 228 (1962).
886. Emery, *Biochemistry*, 8, 877 (1969).
887. Mayer and Miller, *Anal. Biochem.*, 36, 91 (1970).
888. King and Spencer, *J. Biol. Chem.*, 245, 6134 (1970).
889. Webster, *Biochim. Biophys. Acta*, 207, 371 (1970).
890. Pederson and Foster, *Biochemistry*, 8, 2357 (1969).
891. Noelken, *Biochemistry*, 9, 4117 (1970).
892. Noelken, *Biochemistry*, 9, 4122 (1970).
893. Van Kley and Stahmann, *J. Am. Chem. Soc.*, 81, 4374 (1959).
894. Janatova, Fuller, and Hunter, *J. Biol. Chem.*, 243, 3612 (1968).
895. Lerner and Barnum, *Arch. Biochem. Biophys.*, 10, 417 (1946).
896. Groulade, Chicault, and Waltzinger, *Bull. Soc. Chim. Biol.*, 49, 1609 (1967).
897. Warner and Schumaker, *Biochemistry*, 9, 451 (1970).
898. Petersen and Foster, *J. Biol. Chem.*, 240, 3861 (1965).
899. Laggner, Kratky, Palm, and Holasek, *FEBS Lett.*, 15, 220 (1971).
900. Joniau, Grossberg, and Pressman, *Immunochemistry*, 7, 755 (1970).
901. Polis, Shmukler, and Custer, *J. Biol. Chem.*, 187, 349 (1950).
902. Zigman and Lerman, *Biochim. Biophys. Acta*, 154, 423 (1968).
903. Lerman, *Can. J. Biochem.*, 47, 1115 (1969).
904. Bonnichsen, *Acta Chem. Scand.*, 4, 715 (1950).
905. Ehrenberg and Dalziel, *Acta Chem. Scand.*, 12, 465 (1958).
906. Green and McKay, *J. Biol. Chem.*, 244, 5034 (1969).
907. Cannon and McKay, *Biochem. Biophys. Res. Commun.*, 35, 403 (1969).
908. Drum, Li, and Vallee, *Biochemistry*, 8, 3783 (1969).
909. Drum, Harrison, Li, Bethune, and Vallee, *Proc. Natl. Acad. Sci., U.S.A.*, 57, 1434 (1967).
910. Drum and Vallee, *Biochem. Biophys. Res. Commun.*, 41, 33 (1970).
911. Von Wartburg, Bethune, and Vallee, *Biochemistry*, 3, 1775 (1964).
912. Negelein and Wulff, *Biochem. Z.*, 293, 351 (1937).
913. Koberstein, Weber, and Jaenicke, *Z. Naturforsch. B*, 23, 474 (1968).
914. Buhner and Sund, *Eur. J. Biochem.*, 11, 73 (1969).
915. Swaisgood and Pattee, *J. Food Sci.*, 33, 400 (1968).
916. Sofer and Ursprung, *J. Biol. Chem.*, 243, 3110 (1968).
917. Biszku, Boross, and Szabolcsi, *Acta Physiol. Acad. Sci. Hung.*, 25, 161 (1964).
918. Kawahara and Tanford, *Biochemistry*, 5, 1578 (1966).
919. Sine and Hass, *J. Biol. Chem.*, 244, 430 (1969).
920. Reisler and Eisenberg, *Biochemistry*, 8, 4572 (1969).
921. Hass, *Biochemistry*, 3, 535 (1964).
922. Gracy, Lacko, and Horecker, *J. Biol. Chem.*, 244, 3913 (1969).
923. Marquardt, *Can. J. Biochem.*, 47, 517 (1969).
924. Marquardt, *Can. J. Biochem.*, 49, 647 (1971).

925. Marquardt, *Can. J. Biochem.,* 49, 658 (1971).
926. Suh, *Fed. Proc.,* 30, 1157 (1971).
927. Lai and Chen, *Arch. Biochem. Biophys.,* 144, 467 (1971).
928. Kobes, Simpson, Vallee, and Rutter, *Biochemistry,* 8, 585 (1969).
929. Harris, Kobes, Teller, and Rutter, *Biochemistry,* 8, 2442 (1969).
930. Rapoport, Davis, and Horecker, *Arch. Biochem. Biophys.,* 132, 286 (1969).
931. DeGraaf, Goedvolk-DeGroot, and Stouthamer, *Biochem. Biophys. Acta,* 221, 566 (1971).
932. Dimroth, Guchhait, and Lane, *Hoppe-Seylers Z. Physiol. Chem.,* 352, 351 (1971).
933. Doyle, Pittz, and Woodside, *Carbohydr. Res.,* 8, 89 (1968).
934. Agrawal and Goldstein, *Biochim. Biophys. Acta,* 133, 376 (1967).
935. Olson and Liener, *Biochemistry,* 6, 105 (1967).
936. Yariv, Kalb, and Levitzki, *Biochim. Biophys. Acta,* 165, 303 (1968).
937. Cheesman, Zagalsky, and Ceccaldi, *Proc. R. Soc. Lond. (Biol.),* 164, 130 (1966).
938. Cummings, *Biochem. Biophys. Res. Commun.,* 33, 165 (1968).
939. Hamlin, *Exp. Gerentol.,* 4, 189 (1969).
940. Cavallini, Scandurra, and Dupre, in *Biological and Chemical Aspects of Oxygenases,* Bloch and Hayaishi, Eds., Maruzen, Tokyo, 1966, 73.
941. Mochan, *Biochim. Biophys. Acta,* 216, 80 (1970).
942. Poulos and Price, *J. Biol. Chem.,* 246, 4041 (1971).
943. Jargenson, *J. Biol. Chem.,* 240, 1064 (1965).
944. Kaplan and Dugas, *Biochem. Biophys. Res. Commun.,* 34, 681 (1969).
945. Wasi and Hofmann, *Biochem. J.,* 106, 926 (1968).
946. Gertler and Trop, *Eur. J. Biochem.,* 19, 90 (1971).
947. Kimura and Kubata, *Bull. Jap. Soc. Sci. Fish.,* 34, 535 (1968).
948. Vulpis, Vulpis, and Santoro, *Ital. J. Biochem.,* 15, 189 (1966).
949. Ruth, Soja, and Wold, *Arch. Biochem. Biophys.,* 140, 1 (1970).
950. Shethna, Stombaugh, and Burris, *Biochem. Biophys. Res. Commun.,* 42, 1108 (1971).
951. Buchanan, Matsubara, and Evans, *Biochim. Biophys. Acta,* 189, 46 (1969).
952. Vetter and Knappe, *Hoppe-Seylers Z. Physiol. Chem.,* 352, 433 (1970).
953. Mortenson, *Biochim. Biophys. Acta,* 81, 71 (1964).
954. Sobel and Lovenberg, *Biochemistry,* 5, 6 (1966).
955. Hong and Rabinowitz, *J. Biol. Chem.,* 245, 4982 (1970).
956. Aggarwal, Rao, and Matsubara, *J. Biochem. (Tokyo),* 69, 601 (1971).
957. Buchanan and Rabinowitz, *J. Bacteriol.,* 88, 806 (1964).
958. Lode and Coon, *J. Biol. Chem.,* 246, 791 (1971).
959. Dubourdieu and Le Gall, *Biochem. Biophys. Res. Commun.,* 38, 965 (1970).
960. Vetter, Jr. and Knappe, *Hoppe-Seylers Z. Physiol. Chem.,* 352, 433 (1970).
961. Mayhew, *Biochim. Biophys. Acta,* 235, 276 (1971).
962. Mayhew, *Biochim. Biophys. Acta,* 235, 289 (1971).
963. Mayhew and Massey, *J. Biol. Chem.,* 244, 794 (1969).
964. Zak, Steczko, and Ostrowski, *Bull. Soc. Chim. Biol.,* 51, 1065 (1969).
965. Schade and Reinhart, *Biochem. J.,* 118, 181 (1970).
966. Hauge, *Arch. Biochem. Biophys.,* 94, 308 (1961).
967. Yoshimura and Isemura, *J. Biochem. (Tokyo),* 69, 839 (1971).
968. Olive and Levy, *J. Biol. Chem.,* 246, 2043 (1971).
969. Flohe, Eisele, and Wendel, *Hoppe-Seylers Z. Physiol. Chem.,* 352, 151 (1971).
970. Harrington and Karr, *J. Mol. Biol.,* 13, 885 (1965).
971. Cseke and Boross, *Acta Biochim. Biophys. Acad. Sci. Hung.,* 2, 39 (1967).
972. Parker and Allison, *J. Biol. Chem.,* 244, 180 (1969).
973. Koberstein, Weber, and Jaenicke, *Z. Naturforsch. (B),* 23, 474 (1968).
974. Murdock and Koeppe, *J. Biol. Chem.,* 239, 1983 (1964).
975. McMurray and Trentham, *Biochem. J.,* 115, 913 (1959).
976. Wassarman, Watson, and Major, *Biochim. Biophys. Acta,* 191, 1 (1969).
977. Davidson, Sajgo, Noller, and Harris, *Nature,* 216, 1181 (1962).
978. Allison, *Methods Enzymol.,* 9, 210 (1966).
979. Devijlder, Boers, and Slater, *Biochim. Biophys. Acta,* 191, 214 (1969).
980. Warburg and Christian, *Biochem. Z.,* 303, 40 (1939).
981. Jaenicke, in *Pyridine Nucleotide Dependent Dehydrogenases,* Sund, Ed., Springer-Verlag, Berlin, 1970, 70.
982. Durchschlag, Puchwein, Kratky, Schuster, and Kirschner, *Eur. J. Biochem.,* 19, 9 (1971).
983. Oguchi, *J. Biochem. (Tokyo),* 68, 427 (1970).
984. Heinz and Kulbe, *Hoppe-Seylers Z. Physiol. Chem.,* 351, 249 (1970).
985. Baliga, Bhatnagar, and Jagannathan, *Indian J. Biochem.,* 1, 86 (1964).

986. Thorner and Paulus, *J. Biol. Chem.*, 246, 3385 (1971).
987. Fondy, Ross, and Sollohub, *J. Biol. Chem.*, 244, 1631 (1969).
988. Beisenherz, Bucher, and Gorbade, *Methods Enzymol.*, 1, 397 (1955).
989. Ankel, Bucher, and Czok, *Biochem. Z.*, 332, 315 (1960).
990. Catsimpoolas, Berg, and Meyer, *Int. J. Protein Res.*, 3, 63 (1971).
991. Wolf and Briggs, *Arch. Biochem. Biophys.*, 63, 40 (1959).
992. Hrkal and Muller-Eberhard, *Biochemistry*, 10, 1746 (1971).
993. Dus, DeKlerk, Sletten, and Bartsch, *Biochem. Biophys. Acta*, 140, 291 (1967).
994. Buffoni and Blaschko, *Proc. R. Soc. Lond. (Biol.)*, 161, 153 (1965).
995. Klee, *J. Biol. Chem.*, 245, 3143 (1970).
996. Rechler, *J. Biol. Chem.*, 244, 551 (1969).
997. Riley and Snell, *Biochemistry*, 9, 1485 (1970).
998. Rechler and Bruni, *J. Biol. Chem.*, 246, 1806 (1971).
999. Champagne, Pouyet, Ouellet, and Garel, *Bull. Soc. Chim. Biol.*, 52, 377 (1970).
1000. Oh, *J. Biol. Chem.*, 245, 6404 (1970).
1001. Matsuo and Greenberg, *J. Biol. Chem.*, 230, 545 (1958).
1002. Datta, *J. Biol. Chem.*, 245, 5779 (1970).
1003. Janin, van Rapenbusch, Truffa-Bachi, and Cohen, *Eur. J. Biochem.*, 8, 128 (1969).
1004. Rhodes, Dodgson, Olavesen, and Hogberg, *Biochem. J.*, 122, 575 (1971).
1005. Newman, Berenson, Mathews, Goldwasser, and Dorfman, *J. Biol. Chem.*, 217, 31 (1955).
1006. Haschke and Campbell, *J. Bacteriol.*, 105, 249 (1971).
1007. Nakos and Mortenson, *Biochemistry*, 10, 2442 (1971).
1008. Yano, Morimoto, Higashi, and Arima, in *Biological and Chemical Aspects of Oxygenases*, Bloch and Hayaishi, Eds., Maruzen, Tokyo, 1966, 331.
1009. Hesp, Calvin, and Hosokawa, *J. Biol. Chem.*, 244, 5644 (1968).
1010. Aschhoff and Pfeil, *Hoppe-Seylers Z. Physiol. Chem.*, 351, 818 (1970).
1011. Finlay and Adams, *J. Biol. Chem.*, 245, 5248 (1970).
1012. Martin, *Arch. Biochem. Biophys.*, 138, 239 (1970).
1013. Hamaguichi and Migita, *J. Biochem. (Tokyo)*, 56, 512 (1964).
1014. Ruffilli and Givol, *Eur. J. Biochem.*, 2, 429 (1967).
1015. Heimburger, Heide, and Haupt, *Clin. Chim. Acta*, 10, 293 (1964).
1016. Tomasi and Bienenstock, *Adv. Immunol.*, 9, 1 (1968).
1017. Lerner and Barnum, *Arch. Biochem.*, 10, 417 (1946).
1018. Lewis, Bergsagel, Bruce-Robertson, Schachter, and Connell, *Blood*, 32, 189 (1968).
1019. Connell, Dorington, Lewis, and Parr, *Can. J. Biochem.*, 48, 784 (1970).
1020. Schultze and Heremans, *Molecular Biology of Human Protein*, Vol. I, Elsevier, Amsterdam, 1966, 234.
1021. Bennich and Johansson, in *Gamma Globulins, Nobel Symposium No. 3*, Killander, Ed., Interscience, New York, 1967, 200.
1022. Orlans, *Immunology*, 14, 61 (1968).
1023. Yamashita, Franek, Skvaril, and Simek, *Eur. J. Biochem.*, 6, 34 (1968).
1024. Clem, *J. Biol. Chem.*, 246, 9 (1971).
1025. Acton, Weinheimer, Dupree, Evans, and Bennett, *Biochemistry*, 10, 2028 (1971).
1026. Freedman, Grossberg, and Pressman, *Biochemistry*, 7, 1941 (1968).
1027. Porter, *Biochem. J.*, 66, 677 (1957).
1028. Knight and Dray, *Biochemistry*, 7, 3830 (1968).
1029. Van Dalen, Seijen, and Gruber, *Biochim. Biophys. Acta*, 147, 421 (1967).
1030. Day, Sturtevant, and Singer, *Ann. N.Y. Acad. Sci.*, 103, 611 (1963).
1031. Onoue, Yagi, Grossberg, and Pressman, *Immunochemistry*, 2, 401 (1965).
1032. Freedman, Grossberg, and Pressman, *Immunochemistry*, 5, 367 (1968).
1033. Givol and Hurwitz, *Biochem. J.*, 115, 371 (1969).
1034. Haimovich, Schechter, and Sela, *Eur. J. Biochem.*, 4, 537 (1969).
1035. Sokol, Hana, and Albrecht, *Folia Microbiol. (Praha)*, 6, 145 (1961).
1036. Montgomery, Dorrington, and Rockey, *Biochemistry*, 8, 1247 (1969).
1037. Eisen, Simms, and Potter, *Biochemistry*, 7, 4126 (1968).
1038. Tur-Sinai, Birk, Gertler, and Rigbi, *Isr. J. Chem.*, 8, 176 (1970).
1039. Mikola and Suolinna, *Arch. Biochem. Biophys.*, 144, 566 (1971).
1040. Malmström, Agro, and Antonini, *Eur. J. Biochem.*, 9, 383 (1969).
1041. Phillips and Jenness, *Biochim. Biophys. Acta*, 229, 407 (1971).
1042. Schmidt and Ebner, *Biochim. Biophys. Acta*, 243, 273 (1971).
1043. Bell, Hopper, McKenzie, Murphy, and Shaw, *Biochim. Biophys. Acta*, 214, 437 (1970).
1044. Kuwabara and Lloyd, *Biochem. J.*, 124, 215 (1971).
1045. Möra and Elödi, *Eur. J. Biochem.*, 5, 574 (1968).

1046. Gutfreund, Cantwell, McMurray, Criddle, and Hathaway, *Biochem. J.*, 106, 683 (1968).
1047. Reeves and Fimognari, *Methods Enzymol.*, 9, 289 (1966).
1048. Jaenicke, in *Pyridine Dependent Dehydrogenases*, Sund, Ed., Springer-Verlag, New York, 1970, 70.
1049. Koberstein, Weber, and Jaenicke, *Z. Naturforsch. (B)*, 23, 474 (1968).
1050. Foye and Solis, *J. Pharm. Sci.*, 58, 352 (1969).
1051. Pfleiderer and Jeckel, *Biochem. Z.*, 329, 370 (1957).
1052. Velick, *J. Biol. Chem.*, 233, 1455, (1958).
1053. Hakala, Glaid, and Schwert, *J. Biol. Chem.*, 221, 191 (1956).
1054. Neilands, *J. Biol. Chem.*, 199, 373 (1952).
1055. Nisselbaum and Bodansky, *J. Biol. Chem.*, 236, 323 (1961).
1056. Symons and Burgoyne, *Methods Enzymol.*, 9, 314 (1966).
1057. Kaloustian, Stolzenbach, Everse, and Kaplan, *J. Biol. Chem.*, 244, 2891 (1969).
1058. Vestling and Kunsch, *Arch. Biochem. Biophsy.*, 127, 568 (1968).
1059. Kubowitz and Ott, *Biochem. Z.*, 314, 94 (1943).
1060. Castellino, Fish, and Mann, *J. Biol. Chem.*, 245, 4269 (1970).
1061. Ghosh, Chaudhuri, Roy, Sinha, and Sen, *Arch. Biochem. Biophys.*, 144, 6 (1971).
1062. Joniau, Bloemmen, and Lontie, *Biochim. Biophys. Acta*, 214, 468 (1970).
1063. Groves, in *Milk Proteins, Chemistry and Molecular Biology*, McKenzie, Ed., Vol. 2, Academic Press, New York, 1971, 367.
1064. Hou and Perlman, *J. Biol. Chem.*, 245, 1289 (1970).
1065. Carlstrom, *Acta Chem. Scand.*, 23, 185 (1969).
1066. Theorell and Pedersen, in *The Svedberg*, Tiselius and Pedersen, Eds., Almqvist and Wiksells Boktryckeri, Uppsala and Stockholm, 1944, 523.
1067. Polis and Shmukler, *J. Biol. Chem.*, 201, 475 (1953).
1068. Appleby, *Biochim. Biophys. Acta*, 188, 222 (1969).
1069. Sjögren and Svedberg, *J. Am. Chem. Soc.*, 52, 3279 (1930).
1070. Hermier, Lebeault, and Zevaco, *Bull. Soc. Chim. Biol.*, 52, 1089 (1970).
1071. Verger, Sarda, and Desnuelle, *Biochim. Biophys. Acta*, 242, 580 (1971).
1072. Vandermeers and Christophe, *Biochim. Biophys. Acta*, 154, 110 (1968).
1073. Brown, Levy, and Fredrickson, *J. Biol. Chem.*, 245, 6588 (1970).
1074. Koga, Horwitz, and Scanu, *J. Lipid Res.*, 10, 577 (1969).
1075. Chlumecká, Tigerstrom, D'Obrenan, and Smith, *J. Biol. Chem.*, 245, 5481 (1969).
1076. Hayaishi, in *Oxidases and Related Systems*, King, Mason, and Morrison, Eds., John Wiley & Sons, New York, 1965, 286.
1077. Cassio and Waller, *Eur. J. Biochem.*, 20, 283 (1971).
1078. Dalton, Morris, Ward, and Mortensen, *Biochemistry*, 10, 2066 (1971).
1079. Keilin and Hartree, *Biochem. J.*, 61, 153 (1953).
1080. Harrison and Blout, *J. Biol. Chem.*, 240, 299 (1965).
1081. Brown, Martinez, Johnstone, and Olcott, *J. Biol. Chem.*, 237, 81 (1962).
1082. Rossi-Fanelli, Antonini, and Povoledo, in *Symposium on Protein Structure*, Neuberger, Ed., Methuen and Co., London, 1958, 144.
1083. Atassi and Saplin, *Biochem. J.*, 98, 82 (1966).
1084. Herskovits, *Arch. Biochem. Biophys.*, 130, 19 (1969).
1085. Crumpton and Wilkinson, *Biochem. J.*, 94, 545 (1965).
1086. Atassi, *Biochim. Biophys. Acta*, 221, 612 (1970).
1087. Cameron, Azzam, Kotite, and Awad, *J. Lab. Clin. Med.*, 65, 883 (1965).
1088. Crumpton and Polson, *J. Mol. Biol.*, 11, 722 (1965).
1089. Boegman and Crumpton, *Biochem. J.*, 120, 373 (1970).
1090. Breslow, *J. Biol. Chem.*, 239, 486 (1964).
1091. Hermans, Jr., *Biochemistry*, 1, 193 (1962).
1092. Edmundson and Hirs, *Nature*, 190, 663 (1961).
1093. Ray and Gurd, *J. Biol. Chem.*, 242, 2062 (1967).
1094. Terwilliger and Read, *Comp. Biochem. Physiol.*, 29, 551 (1969).
1095. Terwilliger and Read, *Comp. Biochem. Physiol.*, 31, 55 (1969).
1096. Kritcher, Thyrum, and Luchi, *Biochim. Biophys. Acta*, 221, 264 (1970).
1097. Gazith, Himmelfarb, and Harrington, *J. Biol. Chem.*, 245, 15 (1970).
1098. Gershman, Stracher, and Dreizen, *J. Biol. Chem.*, 244, 2726 (1969).
1099. Shimizu, Morita, and Yagi, *J. Biochem.* (Tokyo), 69, 447 (1971).
1100. Young, Blanchard, and Brown, *Proc. Natl. Acad. Sci. U.S.A.*, 61, 1087 (1968).
1101. Otaiza and Jaenicke, *Hoppe-Seylers Z. Physiol. Chem.*, 352, 385 (1971).
1102. Bocchini, *Eur. J. Biochem.*, 15, 127 (1970).
1103. Karlsson, Arnberg, and Eaker, *Eur. J. Biochem.*, 21, 1 (1971).

1104. Miranda, Kupeyan, Rochat, Rochat, and Lissitzky, *Eur. J. Biochem.*, 17, 477 (1970).
1105. Miranda, Kupeyan, Rochat, Rochat, and Lissitzky, *Eur. J. Biochem.*, 16, 514 (1970).
1106. Holcenberg and Stadtman, *J. Biol. Chem.*, 244, 1194 (1969).
1107. Forget, *Eur. J. Biochem.*, 18, 442 (1971).
1108. Burns, Holsten, and Hardy, *Biochem. Biophys. Res. Commun.*, 39, 90 (1970).
1109. Nossal and Hershfield, *J. Biol. Chem.*, 246, 541 (1971).
1110. Omenn, Onjes, and Anfinsen, *Biochemistry*, 9, 304 (1970).
1111. Albrecht, *Biochemistry*, 9, 2462 (1970).
1112. Glazer and Smith, *J. Biol. Chem.*, 235, PC43 (1960).
1113. Weintraub and Schlamowitz, *Comp. Biochem. Physiol.*, 38B, 513 (1971).
1114. Weintraub and Schlamowitz, *Comp. Biochem. Physiol.*, 37, 49 (1970).
1115. Tomimatsu, Clary, and Bartulovich, *Arch. Biochem. Biophys.*, 115, 536 (1966).
1116. Liu, Means, and Feeny, *Biochim. Biophys. Acta*, 229, 176 (1971).
1117. Donovan, Mapes, Davis, and Hamburg, *Biochemistry*, 8, 4190 (1969).
1118. Donovan, Davis, and White, *Biochim. Biophys. Acta*, 207, 190 (1970).
1119. Davis, Mapes, and Donovan, *Biochemistry*, 10, 39 (1971).
1120. Bender and Brubacher, *J. Am. Chem. Soc.*, 86, 5333 (1964).
1121. Hinkle and Kirsch, *Biochemistry*, 9, 4633 (1970).
1122. Arnon and Shapira, *J. Biol. Chem.*, 244, 1033 (1969).
1123. Olander, *Biochemistry*, 10, 601 (1971).
1124. Woods, *Biochem. J.*, 113, 39 (1969).
1125. Milstein, *Biochem. J.* 103, 634 (1967).
1126. Citri and Pollock, *Adv. Enzymol.*, 28, 237 (1966).
1127. Imsande, Gillin, Tanis, and Atherly, *J. Biol. Chem.*, 245, 2205 (1970).
1128. Lindstrom, Boman, and Steele, *J. Biol. Chem.*, 101, 218 (1970).
1129. Sodek and Hofmann, *Methods Enzymol.*, 19, 372 (1970).
1130. Morita, Toshida, and Maeda, *Agric. Biol. Chem.*, 35, 1074 (1971).
1131. Staprans and Watanabe, *J. Biol. Chem.*, 245, 5962 (1970).
1132. Peterson, *J. Biol. Chem.*, 246, 34 (1971).
1133. Peterson and Berggard, *J. Biol. Chem.*, 246, 25 (1971).
1134. Kuehn and McFadden, *Biochemistry*, 8, 2394 (1968).
1135. Nicholson, *Biochem. J.*, 123, 117 (1971).
1136. Labow and Robinson, *J. Biol. Chem.*, 241, 1239 (1966).
1137. Dowhan, Jr. and Snell, *J. Biol. Chem.* 245, 4618 (1970).
1138. Matsubara, *Methods Enzymol.*, 19, 642 (1970).
1139. Berglund and Sjöberg, *J. Biol. Chem.*, 245, 6030 (1970).
1140. Fedberg and Datta, *Eur. J. Biochem.*, 21, 438 (1971).
1141. Hug and Roth, *Biochemistry*, 10, 1397 (1971).
1142. White and Barlow, *Methods Enzymol.*, 19, 665 (1970).
1143. Gumpf and Hamilton, *Virology*, 35, 87 (1968).
1144. Yamazaki and Kaesberg, *J. Mol. Biol.*, 6, 465 (1963).
1145. Stubbs and Kaesberg, *J. Mol. Biol.*, 8, 314 (1964).
1146. Bachrach and Van den Woude, *Virology*, 34, 282 (1968).
1147. Rueckert, *Virology*, 26, 345 (1965).
1148. Ghabrial, Shepherd, and Grogan, *Virology*, 33, 17 (1967).
1149. Damirdagh and Shepherd, *Virology*, 40, 84 (1970).
1150. Budzynski and Fraenkel-Conrat, *Biochemistry*, 9, 3300 (1970).
1151. Tremaine and Stace-Smith, *Virology*, 35, 102 (1968).
1152. Dorne, Jonard, Witz, and Hirth, *Virology*, 43, 279 (1971).
1153. Miki and Knight, *Virology*, 31, 55 (1967).
1154. Wallace, *Biochim. Biophys. Acta*, 215, 176 (1970).
1155. Bray, Chisholm, Hart, Meriwether, and Watts, in *Flavins and Flavoproteins*, Slater, Ed., Elsevier, Amsterdam, 1966, 117.
1156. West, Nagy, and Gergely, in *Symposium on Fibrous Proteins*, Crewther, Ed., Plenum Press, New York, 1968, 164.
1157. Kimura and Ting, *Biochem. Biophys. Res. Commun.*, 45, 1227 (1971).
1158. Fushimi, Hamison, and Ravin, *J. Biochem.* (Tokyo), 69, 1041 (1971).
1159. Shinowara, in *Blood Platelets, Henry Ford International Symposium No. 10,* Johnson, Monto, Rebuck, and Horn, Jr., Eds., Little Brown, Boston, 1961, 347.
1160. Reynolds and Johnson, *Biochemistry*, 10, 2821 (1971).
1161. Wu, Cluskey, Krull, and Friedman, *Can. J. Biochem.*, 49, 1042 (1971).
1162. Berrens and Bleumink, *Int. Arch. Allergy*, 28, 150 (1965).

1163. Feldman and Weiner, *J. Biol. Chem.*, 247, 260 (1972).
1164. Guha, Lai, and Horecker, *Arch. Biochem. Biophys.*, 147, 692 (1971).
1165. Caban and Hass, *J. Biol. Chem.*, 246, 6807 (1971).
1166. Penhoet, Kochman and Rutter, *Biochemistry*, 8, 4396 (1969).
1167. Chiu and Feingold, *Biochemistry*, 8, 98 (1969).
1168. Robertson, Hammerstedt, and Wood, *J. Biol. Chem.*, 246, 2075 (1971).
1169. 'S-Gravenmade, Drift, Van Der, and Vogels, *Biochim. Biophys. Acta*, 251, 393 (1971).
1170. King and Norman, *Biochemistry*, 1, 709 (1962).
1171. Underdown and Goodfriend, *Biochemistry*, 8, 980 (1969).
1172. Mazelis and Crews, *Biochem. J.*, 108, 725 (1968).
1173. Svedberg and Sjögren, *J. Am. Chem. Soc.*, 52, 279 (1930).
1174. Eady and Large, *Biochem. J.*, 123, 757 (1971).
1175. Henn and Ackers, *J. Biol. Chem.*, 244, 465 (1969).
1176. Miyake and Yamano, *Biochim. Biophys. Acta*, 198, 438 (1970).
1177. Soda and Osumi, *Biochem. Biophys. Res. Commun.*, 35, 363 (1969).
1178. McKeehan and Hardesty, *J. Biol. Chem.*, 244, 4330 (1969).
1179. Yamada, Adachi, and Ogata, in *Pyridoxyl Catalysis: Enzymes and Model Systems*, Snell, Braunstein, Severin, and Torchinsky, Eds., Interscience, New York, 1968, 347.
1180. Wang, Achee, and Yasunobu, *Arch. Biochem. Biophys.*, 128, 106 (1968).
1181. Prescott, Wilkes, Wagner, and Wilson, *J. Biol. Chem.*, 246, 1756 (1971).
1182. Hanson, Hutter, Mansfeldt, Kretschmer, and Sohr, *Hoppe-Seylers Z. Physiol. Chem.*, 348, 680 (1967).
1183. Wacker, Lehky, Fischer, and Stein, *Helv. Chim. Acta*, 54, 473 (1971).
1184. Roncari and Zuber, *Int. J. Protein Res.*, 1, 45 (1969).
1185. Pfleiderer and Femfert, *FEBS Lett.*, 4, 265 (1969).
1186. Auricchio and Bruni, *Biochem. Z.*, 340, 321 (1964).
1187. Yaron and Mlynar, *Biochem. Biophys. Res. Commun.*, 32, 658 (1968).
1188. Grszkiewicz, *Acta Biochem. Pol.*, 9, 301 (1962).
1189. Vandermeers and Christophe, *Biochim. Biophys. Acta*, 154, 110 (1968).
1190. Menzi, Stein, and Fischer, *Helv. Chim. Acta*, 40, 534 (1957).
1191. Yutani, Yutani, and Isemura, *J. Biochem.* (Tokyo), 66, 823 (1969).
1192. Nishida, Fukumoto, and Yamamoto, *Agric. Biol. Chem.*, 31, 682 (1967).
1193. Nishida, Ph. D. thesis; cited in Nishida, Fukumoto, and Yamamoto, *Agric. Biol. Chem.*, 31, 682 (1967).
1194. Ogasahara, Imanishi, and Isemura, *J. Biochem.* (Tokyo), 67, 65 (1970).
1195. Junge, Stein, Neurath, and Fischer, *J. Biol. Chem.*, 234, 556 (1959).
1196. Sanders and Rutter, *Biochemistry*, 11, 130 (1972).
1197. Krysteva and Erodi, *Acta Biochim. Biophys. Acad. Sci. Hung.*, 3, 275 (1968).
1198. Yoshida, Hiroshi, and Ono, *J. Biochem.* (Tokyo), 65, 741 (1969).
1199. Yutani, Yutani, and Isemura, *J. Biochem.* (Tokyo), 65, 201 (1969).
1200. Shainkin and Berk, *Biochim. Biophys. Acta*, 221, 502 (1970).
1201. Takeda and Hizukuri, *Biochim. Biophys. Acta*, 185, 469 (1969).
1202. England, Sorof, and Singer, *J. Biol. Chem.*, 189, 217 (1951).
1203. Nagano and Zalkin, *J. Biol. Chem.*, 245, 3097 (1970).
1204. Ito, Cox, and Yanofsky, *J. Bacteriol.*, 97, 725 (1969).
1205. Henderson and Zalkin, *J. Biol. Chem.*, 246, 6891 (1971).
1206. Warner and Schumaker, *Biochemistry*, 9, 451 (1970).
1207. Bezkorovainy, Springer, and Dese, *Biochemistry*, 10, 3761 (1971).
1208. Kim, *Vox Sang*, 20, 461 (1971).
1209. Bezkorovainy, Springer, and Hotti, *Biochim. Biophys. Acta*, 115, 501 (1966).
1210. Preer, *J. Immunol.*, 33, 385 (1959).
1211. Kaji and Tagawa, *Biochim. Biophys. Acta*, 207, 456 (1970).
1212. Harell and Sokolovsky, *Eur. J. Biochem.*, 25, 102 (1972).
1213. Sakai and Murachi, *Physiol. Chem. Phys.*, 1, 31 (1969).
1214. Blethen, Boeker, and Snell, *J. Biol. Chem.*, 243, 1671 (1968).
1215. Regnouf, Pradel, Kassab, and Thoai, *Biochim. Biophys. Acta*, 194, 540 (1969).
1216. Oriol-Audit, Landon, Robin, and Thoai, *Biochim. Biophys. Acta*, 188, 132 (1969).
1217. Kassab, Fattoum, and Pradel, *Eur. J. Biochem.*, 12, 264 (1970).
1218. Kassab, Roustan, and Pradel, *Biochim. Biophys. Acta*, 167, 308 (1968).
1219. Landon, Oriol, and Thoai, *Biochim. Biophys. Acta*, 214, 168 (1970).
1220. Yorifuji, Ogata, and Soda, *J. Biol. Chem.*, 246, 5085 (1971).
1221. Bray and Ratner, *Arch. Biochem. Biophys.*, 146, 531 (1971).
1222. Nakano, Tsutsumi, and Danowski, *J. Biol. Chem.*, 245, 4443 (1974).
1223. Nakamura, Makino, and Ogura, *J. Biochem.* (Tokyo), 64, 189 (1969).

1224. Penton and Dawson, in *Oxidases and Related Redox Systems,* King, Mason, and Morrison, Eds., John Wiley and Sons, New York, 1965, 221.
1225. Tosa, Sano, Yamamoto, Nakamura, and Chibata, *Biochemistry,* 11, 21 (1972).
1226. Lu and Handschumacher, *J. Biol. Chem.,* 247, 66 (1972).
1227. Nishumara, Makino, Takenaka, and Inada, *Biochim. Biophys. Acta,* 227, 171 (1971).
1228. Shifrin and Grochowski, *J. Biol. Chem.,* 247, 1048 (1972).
1229. Kakimoto, Kato, Shibatani, Nishimura, and Chibata, *J. Biol. Chem.,* 244, 353 (1969).
1230. Banks, Doonan, Lawrence, and Vernon, *Eur. J. Biochem.,* 5, 528 (1968).
1231. Banks and Vernon, *J. Chem. Soc.,* p. 1968 (1961).
1232. Arrio-Dupont, Cournil, and Duie, *FEBS Lett.,* 11, 144 (1970).
1233. Martinez-Carrion, Kuczenski, Tiemeier, and Peterson, *J. Biol. Chem.,* 245, 799 (1970).
1234. Bergami, Marino, and Scardi, *Biochem. J.,* 110, 471 (1968).
1235. Bertlund and Kaplan, *Biochemistry,* 9, 2653 (1970).
1236. Magee and Phillips, *Biochemistry,* 10, 3397 (1971).
1237. Marino, Greco, Scardi, and Zito, *Biochem. J.,* 99, 589 (1966).
1238. Scardi, in *Pyridoxal Catalysis: Enzymes and Model Systems,* Snell, Braunstein, Severin, and Torchinsky, Eds., Interscience, New York, 1968, 179.
1239. Nelbach, Pigiet, Gerhart, and Schachman, *Biochemistry,* 11, 315 (1972).
1240. Meighen, Pigiet, and Schachman, *Proc. Natl. Acad. Sci., U.S.A.,* 65, 234 (1970).
1241. Benisek, *J. Biol. Chem.,* 246, 3151 (1971).
1242. Vanaman and Stark, *J. Biol. Chem.,* 245, 3565 (1970).
1243. Biswas, Gray, and Paulus, *J. Biol. Chem.,* 245, 4900 (1970).
1244. Lafuma, Gros, and Patte, *Eur. J. Biochem.,* 15, 111 (1970).
1245. Janin, van Rapenbusch, Truffa-Bachi, Cohen, and Gros, *Eur. J. Biochem.,* 8, 128 (1969).
1246. Truffa-Bachi, van Rapenbusch, Janin, Gros, and Cohen, *Eur. J. Biochem.,* 7, 401 (1969).
1247. Falcoz-Kelly, van Rapenbusch, and Cohen, *Eur. J. Biochem.,* 8, 146 (1969).
1248. Inoue, Suzuki, Fukunishi, Adachi, and Takeda, *J. Biol. Chem.,* 60, 543 (1966).
1249. Melamed and Green, *Biochem. J.,* 89, 591 (1963).
1250. Green, *Biochem. J.,* 89, 599 (1963).
1251. Scrutton, *Biochemistry,* 10, 3897 (1971).
1252. Brill, Bryce, and Maria, *Biochim. Biophys. Acta,* 154, 342 (1968).
1253. Haupt, Heimburger, Krantz, and Schwick, *Eur. J. Biochem.,* 17, 254 (1970).
1254. Vonemasu, Stroud, Niedermeier, and Butler, *Biochem. Biophys. Res. Commun.,* 43, 1388 (1971).
1255. Nilsson and Lindskog, *Eur. J. Biochem.,* 2, 309 (1967).
1256. Carter, *Biochim. Biophys. Acta,* 235, 222 (1971).
1257. Bernstein and Schraer, *J. Biol. Chem.,* 247, 1306 (1972).
1258. Bernstein and Schraer, *Fed. Proc.,* Abstr. 1387, 30 (1291).
1259. Maynard and Coleman, *J. Biol. Chem.,* 246, 4455 (1971).
1260. Carter and Parsons, *Biochem. J.,* 120, 797 (1970).
1261. Funakoshi and Deutsch, *J. Biol. Chem.,* 245, 4913 (1970).
1262. Edsall, Mehta, Myers, and Armstrong, *Biochem. Z.,* 345, 9 (1966).
1263. Ashworth, Spencer, and Brewer, *Arch. Biochem. Biophys.,* 142, 122 (1971).
1264. Tanis, Tashian, and Yu, *J. Biol. Chem.,* 245, 6003 (1970).
1265. Rossi, Chersi, and Cortivo, in CO_2: *Chemical, Biochemical and Physiological Aspects,* Forster, Edsall, Otis, and Roughton, Eds., NASA SP-188, 1969, 131.
1266. Runnegar, Scott, Webb, and Zerner, *Biochemistry,* 8, 2013 (1969).
1267. Runnegar, Webb, and Zerner, *Biochemistry,* 8, 2018 (1969).
1268. Franz and Krisch, *Hoppe-Seylers Z. Physiol. Chem.,* 149, 575 (1968).
1269. Chase and Tubbs, *Biochem. J.,* 111, 225 (1969).
1270. Zagalsky, Cheesman, and Ceccaldi, *Comp. Biochem. Physiol.,* 22, 851 (1967).
1271. Zagalsky, Ceccaldi, and Daumas, *Comp. Biochem. Physiol.,* 34, 579 (1970).
1272. Dumas and Garnier, *J. Dairy Res.,* 37, 269 (1970).
1273. Thompson and Pepper, *J. Dairy Sci.,* 47, 633 (1964).
1274. Herskovits, *Arch. Biochem. Biophys.,* 130, 19 (1969).
1275. Carrico and Deutsch, *J. Biol. Chem.,* 244, 6087 (1969).
1276. Porter and Folch, *J. Neurochem.,* 1, 260 (1957).
1277. Ashwell and Morell, in *Red Cross Scientific Symposium on Glycoproteins of Blood Cells and Plasma,* Jamieson and Greenwalt, Eds., J. B. Lippincott, Philadelphia, 1971, 173.
1278. Blumberg, Eisinger, Aisen, Morell, and Scheinberg, *J. Biol. Chem.,* 238, 1675 (1963).
1279. Ryden, *Int. J. Protein Res.,* 3, 131 (1971).
1280. Matsunaga and Nosoh, *Biochim. Biophys. Acta,* 215, 280 (1970).
1281. Osaki, *J. Biochem.* (Tokyo), 48, 190 (1960).

1282. Milne and Matrone, *Biochim. Biophys. Acta,* 212, 43 (1970).
1283. Holtzman and Gaumnitz, *J. Biol. Chem.,* 245, 2350 (1970).
1284. Lospalluto and Finkelstein, *Biochim. Biophys. Acta,* 257, 158 (1972).
1285. Kover, Szaboic, and Csabal, *Arch. Biochem. Biophys.,* 106, 333 (1964).
1286. Schumberger, *Z. Naturforsch. (B),* 23, 1412 (1968).
1287. Bahl, *J. Biol. Chem.,* 244, 567 (1969).
1288. Koch, Shaw, and Gibson, *Biochim. Biophys. Acta,* 229, 805 (1971).
1289. Koch, Shaw, and Gibson, *Biochim. Biophys. Acta,* 229, 795 (1971).
1290. Koch, Shaw, and Gibson, *Biochim. Biophys. Acta,* 212, 375 (1970).
1291. Koch, Shaw, and Gibson, *Biochim. Biophys. Acta,* 212, 387 (1970).
1292. Nakagawa and Bender, *Biochemistry,* 9, 259 (1970).
1293. Singh, Thornton, and Westheimer, *J. Biol. Chem.,* 237, PC3006 (1962).
1294. Webster, *Biochim. Biophys. Acta,* 207, 371 (1970).
1295. Krausz and Becker, *J. Biol. Chem.,* 243, 4606 (1968).
1296. Oliver, Viswanatha, and Whish, *Biochem. Biophys. Res. Commun.,* 27, 107 (1967).
1297. Babul and Stellwagen, *Anal. Biochem.,* 28, 216 (1969).
1298. Rovery, *Methods Enzymol.,* 11, 231 (1967).
1299. Coan, Roberts, and Travis, *Biochemistry,* 10, 2711 (1971).
1300. Jackson and Brandts, *Biochemistry,* 9, 2294 (1970).
1301. Chervenka and Wilcox, *J. Biol. Chem.,* 222, 621 (1956).
1302. Glazer and Smith, *J. Biol. Chem.,* 235, PC43 (1960).
1303. Vandermeers and Christophe *Biochim. Biophys. Acta,* 188, 101 (1969).
1304. Brandts and Lumry, *J. Phys. Chem.,* 67, 1484 (1963).
1305. Nichol, *J. Biol. Chem.,* 243, 4065 (1968).
1306. Prahl and Neurath, *Biochemistry,* 5, 2131 (1966).
1307. Gratecos, Guy, Rovery, and Desnuelle, *Biochim. Biophys. Acta,* 175, 82 (1969).
1308. Singh, Brooks, and Srere, *J. Biol. Chem.,* 245, 4636 (1970).
1309. Berger, Kafatos, Felsted, and Law, *J. Biol. Chem.,* 246, 4131 (1971).
1310. Hruska and Law, *Methods Enzymol.,* 19, 221 (1970).
1311. Schwartz and Helinski, *J. Biol. Chem.,* 246, 6318 (1971).
1312. Grant and Alburn, *Arch. Biochem. Biophys.,* 82, 245 (1959).
1313. Rhoads and Udenfriend, *Arch. Biochem. Biophys.,* 139, 329 (1970).
1314. Shin and Mayer, *Biochemistry,* 7, 2991 (1968).
1315. Emery, *Biochemistry,* 8, 877 (1969).
1316. Ehrenpreis and Warner, *Arch. Biochem. Biophys.,* 61, 38 (1956).
1317. Grant-Greene and Friedberg, *Int. J. Protein Res.,* 2, 235 (1970).
1318. Lerman, *Can. J. Biochem.,* 47, 1115 (1969).
1319. Augusteyn and Spector, *Biochem. J.,* 124, 345 (1971).
1320. Gregory, Holdsworth, and Ottesen *C. R. Trav. Lab. Carlsberg Ser. Chim.,* 30, 147 (1957).
1321. Holdsworth, *Biochim. Biophys. Acta,* 51, 295 (1961).
1322. Rosenberry, Chang, and Chen, *J. Biol. Chem.,* 247, 1555 (1972).
1323. Verpoorte, *J. Biol. Chem.,* 247, 4787 (1972).
1324. West, Nagy, and Gergely, in *Symposium on Fibrous Proteins,* Crewther, Ed., Plenum Press, New York, 1968, 164.
1325. Schramm and Hochstein, *Biochemistry,* 11, 2777 (1972).
1326. Day, Franklin, Pettersson, and Philipson, *Eur. J. Biochem.,* 29, 537 (1972).
1327. Ronca-Testoni, Ranieri, Raggi, and Ronca, *Ital. J. Biochem.,* 19, 262 (1970).
1328. Suhara, Takemori, and Katagiri, *Biochim. Biophys. Acta,* 263, 272 (1972).
1329. Levine, Kaplan, and Greenaway, *Biochem. J.,* 129, 847 (1972).
1330. Claesson, *Ark. Kem.,* 10, 4 (1956).
1331. Wallenfels and Herrmann, *Methods Enzymol.,* 9, 608 (1966).
1332. Berrens, in *The Chemistry of Atopic Allergens,* Karger, Basel, 1971, 205.
1333. Marsuzawa and Segal, *J. Biol. Chem.,* 243, 5929 (1968).
1334. Visuri and Nummi, *Eur. J. Biochem.,* 28, 555 (1972).
1335. Wissler, *Eur. J. Immunol.,* 2, 73 (1972).
1336. Stellwagen, Rysavy, and Babul, *J. Biol. Chem.,* 247, 8074 (1972).
1337. Lux, John, and Brewer, *J. Biol. Chem.,* 247, 7510 (1972).
1338. Tombs, *Biochem. J.,* 96, 119 (1965).
1339. Grazi and Magri, *Biochem. J.,* 126, 667 (1972).
1340. Nakamura, Makino and Ogura, *J. Biochem.* (Tokyo), 64, 189 (1968).
1341. Cammack, Marlborough, and Miller, *Biochem. J.,* 126, 316 (1972).
1342. Laboureur, Langlois, Labrousse, Boudon, Emeraud, Samain, Ageron, and Dumesnil, *Biochimie,* 53, 1147 (1971).

1343. Scandurra and Cannella, *Eur. J. Biochem.*, 27, 196 (1972).
1344. D'Aniello and Rocca, *Comp. Biochem. Physiol.*, B41, 625 (1972).
1345. Falcoz-Kelly, Janin, Saari, Veron, Truffa-Bachi, and Cohen, *Eur. J. Biochem.*, 28, 507 (1972).
1346. Thuma, Schirmer, and Schirmer, *Biochim. Biophys. Acta*, 268, 81 (1972).
1347. Tweedie and Segel, *Prep. Biochem.*, 1, 91 (1971).
1348. Wolff and Siegel, *J. Biol. Chem.*, 247, 4180 (1972).
1349. Virden, *Biochem. J.*, 127, 503 (1972).
1350. Brundell, Falkbring, and Nyman, *Biochim. Biophys. Acta*, 284, 311 (1972).
1351. Kang, Storm, and Carson, *Biochim. Biophys. Res. Commun.*, 49, 621 (1972).
1352. Ihle and Dure, III, *J. Biol. Chem.*, 247, 5034 (1972).
1353. Johansen, Livingston, and Vallee, *Biochemistry*, 11, 2584 (1972).
1354. Zagalsky and Herring, *Comp. Biochem. Physiol.*, B41, 397 (1972).
1355. Bonaventura, Schroeder, and Fang, *Arch. Biochem. Biophys.*, 150, 606 (1972).
1356. Price, Sterling, Tarantola, Hartley, and Rechcigl, *J. Biol. Chem.*, 237, 3468 (1962).
1357. Otto and Bhakdi, *Hoppe-Seyler's Z. Physiol. Chem.*, 350, 1577 (1969).
1358. Skujins, Pukite, and McLaren, *Enzymologia*, 39, 353 (1970).
1359. Timmis, *J. Bacteriol.*, 109, 12 (1972).
1360. Maylie, Charles, Gache, and Desnuelle, *Biochim. Biophys. Acta*, 229, 286 (1971).
1361. Barth, Bunnenberg, and Djerassi, *Anal. Biochem.*, 48, 471 (1972).
1362. Waterson, Castellino, Hass, and Hill, *J. Biol. Chem.*, 247, 5266 (1972).
1363. Wissler, *Eur. J. Immunol.*, 2, 84 (1972).
1364. Goldberger, Smith, Tisdale, and Bomstein, *J. Biol. Chem.*, 236, 2788 (1961).
1365. Lederer and Simon, *Eur. J. Biochem.*, 20, 469 (1971).
1366. Groudinsky, *Eur. J. Biochem.*, 18, 480 (1971).
1367. Mevel-Ninio, Pajot, and Labeyrie, *Biochimie*, 53, 35 (1971).
1368. Monteilhet and Risler, *Eur. J. Biochem.*, 12, 165 (1970).
1369. Strittmatter and Velick, *J. Biol. Chem.*, 221, 253 (1956).
1370. Itagaki and Hager, *J. Biol. Chem.*, 241, 3687 (1966).
1371. Iwasaki and Shidara, *Plant Cell Physiol.*, 10, 291 (1969).
1372. Webster, *Biochim. Biophys. Acta*, 207, 371 (1970).
1373. Flatmark and Sletten, *J. Biol. Chem.*, 243, 1623 (1968).
1374. Mayer and Miller, *Anal. Biochem.*, 36, 91 (1970).
1375. Herskovits, *Arch. Biochem. Biophys.*, 130, 19 (1969).
1376. Herskovits, Jaillet, and Gadegbeku, *J. Biol. Chem.*, 245, 4544 (1970).
1377. Bolard and Garnier, *Biochim. Biophys. Acta*, 263, 535 (1972).
1378. Scholes, McLain, and Smith, *Biochemistry*, 10, 2072 (1971).
1379. Schejter, Grosman, and Sokolovsky, *Isr. J. Chem.*, 10, 37 (1972).
1380. Yu, Yu, and King, *J. Biol. Chem.*, 247, 1012 (1972).
1381. Horio and Kamen, *Biochim. Biophys. Acta*, 48, 266 (1961).
1382. Cusanovich, Tedro, and Kamen, *Arch. Biochem. Biophys.*, 141, 557 (1970).
1383. Iwasaki and Matsubara, *J. Biochem.* (Tokyo), 69, 847 (1971).
1384. Miki and Okunuki, *J. Biochem.* (Tokyo), 66, 831 (1969).
1385. Clark-Walker and Lascelles, *Arch. Biochem. Biophys.*, 136, 153 (1970).
1386. Yamanaka, Takenami, Akijama, and Okunuki, *J. Biochem.* (Tokyo), 70, 349 (1971).
1387. Shioi, Takamiya, and Nishimura, *J. Biochem.* (Tokyo), 71, 285 (1972).
1388. Yong and King, *J. Biol. Chem.*, 245, 1331 (1970).
1389. Sugimura and Yakushiji, *J. Biochem.* (Tokyo), 63, 281 (1968).
1390. Laycock and Craigie, *Can. J. Biochem.*, 49, 641 (1971).
1391. Kusel, Suriano, and Weber, *Arch. Biochem. Biophys.*, 133, 293 (1969).
1392. Stripp, Greene, and Gillette, *Pharmacology*, 6, 56 (1971).
1393. Kuronen and Ellfolk, *Biochim. Biophys. Acta*, 275, 308 (1972).
1394. Ellfolk and Soininen, *Acta Chem. Scand.*, 25, 1535 (1971).
1395. Hiraoka, Fukumoto, and Tsuru, *J. Biochem.* (Tokyo), 71, 57 (1972).
1396. D'Souza, Warwick, and Freisheim, *Biochemistry*, 11, 1528 (1972).
1397. Gunderson, Dunlap, Harding, Freisheim, Otting, and Huennekens, *Biochemistry*, 11, 1018 (1972).
1398. Greenfield, Williams, Poe, and Hoogsteen, *Biochemistry*, 11, 4706 (1972).
1399. Erickson and Mathews, *Biochemistry*, 12, 372 (1973).
1400. Butzow, *Biochim. Biophys. Acta*, 168, 490 (1968).
1401. Setlow, Brutlag, and Kornberg, *J. Biol. Chem.*, 247, 224 (1972).
1402. Ueda, Lode, and Coon, *J. Biol. Chem.*, 247, 2109 (1972).
1403. Robinson and Maxwell, *J. Biol. Chem.*, 247, 7023 (1972).
1404. Rexova-Benkova and Slezarik, *Collect. Czech. Chem. Commun.*, 35, 1255 (1970).

1405. Spring and Wold, *J. Biol. Chem.*, 246, 6797 (1971).
1406. Winstead, *Biochemistry*, 11, 1046 (1972).
1407. Malmstrom, *Arch. Biochem. Biophys. Suppl.*, 1, 247 (1962).
1408. Maroux, Baratti, and Desnuelle, *J. Biol. Chem.*, 246, 5031 (1971).
1409. Schantz, Roessler, Woodburn, Lyach, Jacoby, Silverman, Gorman, and Spero, *Biochemistry*, 11, 360 (1972).
1410. Borja, Fanning, Huang, and Bergdoll, *J. Biol. Chem.*, 247, 2456 (1972).
1411. Markland, and Damus, *J. Biol. Chem.*, 246, 6460 (1971).
1412. Swaney, and Klotz, *Arch. Biochem. Biophys.*, 147, 475 (1971).
1413. Weser, Bunnenberg, Cammack, Djerassi, Flohe, Thomas, and Voelter, *Biochim. Biophys. Acta*, 243, 203 (1971).
1414. Weser, and Hartmann, *Fed. Eur. Biochem. Soc. Lett.*, 17, 78 (1971).
1415. Weser, Barth, Djerassi, Hartmann, Krauss, Voelker, Voelter, and Voetsch, *Biochem. Biophys. Acta*, 278, 28 (1972).
1416. Espada, Langton, and Dorado, *Biochim. Biophys. Acta*, 285, 427 (1972).
1417. Omori-Satoh, Sadahiro, Ohsaka, and Murata, *Biochim. Biophys. Acta*, 285, 414 (1972).
1418. Aloof-Hirsch, DeVries, and Berger, *Biochim. Biophys. Acta*, 154, 53 (1968).
1419. Taylor, Mitchell, and Cohen, *J. Biol. Chem.*, 247, 5928 (1972).
1420. Fujikawa, Legaz, and Davie, *Biochemistry*, 11, 4882 (1972).
1421. Radcliffe, and Barton, *J. Biol. Chem.*, 247, 7735 (1972).
1422. Takagi, and Konishi, *Biochim. Biophys. Acta*, 271, 363 (1972).
1423. Monahan, Rivier, Vale, Guillemin, and Burgus, *Biochim. Biophys. Res. Commun.*, 47, 551 (1972).
1424. Dutler, Coon, Kull, Vogel, Waldvogel, and Prelog, *Eur. J. Biochem.*, 22, 203 (1971).
1425. Yoch, and Arnon, *J. Biol. Chem.*, 247, 4514 (1972).
1426. Shanmugan, Buchanan, and Arnon, *Biochim. Biophys. Acta*, 256, 477 (1972).
1427. Fee, and Palmer, *Biochim. Biophys. Acta*, 245, 175 (1971).
1428. Gersonde, Trittelvitz, Schlaak, and Stabel, *Eur. J. Biochem.*, 22, 57 (1971).
1429. Nakamura, and Kimura, *J. Biol. Chem.*, 245, 6235 (1971).
1430. Jackson, Munro, and Korner, *Biochim. Biophys. Acta*, 91, 666 (1964).
1431. Murray, Oikawa, and Kay, *Biochim. Biophys. Acta*, 175, 331 (1969).
1432. Graham, in *Glycoproteins*, Gottschalk, Ed., Elsevier, Amsterdam, 1966, 361.
1433. Blomback, *Ark, Kem.*, 12, 99 (1958).
1434. Pisano, Finlayson, Peyton, and Nagai, *Proc. Natl. Acad. Sci. U.S.A.*, 68, 770 (1971).
1435. Gollwitzer, Timpl, Becker, and Furthmayr, *Eur. J. Biochem.*, 28, 497 (1972).
1436. Gollwitzer, Karges, Hormann, and Kuhn, *Biochim. Biophys. Acta*, 207, 445 (1970).
1437. Kazal, Amsel, Miller, and Tocantins, *Proc. Soc. Exp. Biol. Med.*, 113, 989 (1963).
1438. Marker, Budzynski, and James, *J. Biol. Chem.*, 247, 4775 (1972).
1439. Bion, Marguérie, Hudry, and Chagniel, *C.R. Acad. Sci. (D)* (Paris), 273, 901 (1971).
1440. Whitaker, *Biochemistry*, 8, 1896 (1969).
1441. Cusanovich, and Edmondson, *Biochem. Biophys. Res. Commun.*, 45, 327 (1971).
1442. D'Anna, and Tollin, *Biochemistry*, 11, 1073 (1972).
1443. Zumft, and Spiller, *Biochem. Biophys. Res. Commun.*, 45, 112 (1971).
1444. Wickner, and Tabor, *J. Biol. Chem.*, 247, 1605 (1972).
1445. Curthoys, and Rabinowitz, *J. Biol. Chem.*, 246, 6942 (1971).
1446. Welch, Buttlaire, Hersh, and Himes, *Biochim. Biophys. Acta*, 236, 599 (1971).
1447. Sanchez, Gonzalez, and Pontis, *Biochim. Biophys. Acta*, 227, 67 (1971).
1448. Olson, and Marquardt, *Biochim. Biophys. Acta*, 268, 453 (1972).
1449. Mendicino, Kratowich, and Oliver, *J. Biol. Chem.*, 247, 6643 (1972).
1450. Traniello, Melloni, Pontremoli, Sia, and Horecker, *Arch. Biochem. Biophys.*, 149, 222 (1972).
1451. Fernando, Pontremoli, and Horecker, *Arch. Biochem. Biophys.*, 129, 370 (1969).
1452. Tashima, Tholey, Drummond, Bertrand, Rosenberg, and Horecker, *Arch. Biochem. Biophys.*, 149, 118 (1972).
1453. Frieden, Bock, and Alberty, *J. Am. Chem. Soc.*, 76, 2482 (1953).
1454. Loontiens, Wallenfels, and Weil, *Eur. J. Biochem.*, 14, 138 (1970).
1455. Klee, and Klee, *J. Biol. Chem.*, 247, 2336 (1972).
1456. Barth, Bunnenberg, and Djerassi, *Anal. Biochem.*, 48, 471 (1972).
1457. Tang, Wolf, Caputto, and Trucco, *J. Biol. Chem.*, 234, 1174 (1959).
1458. Platt, and Kasarda, *Biochim. Biophys. Acta*, 243, 407 (1971).
1459. Konieczny and Domanski, *Acta Biochim. Pol.*, 10, 325 (1963).
1460. Vodrazka, Hrkal, Kodicek, and Jandova, *Eur. J. Biochem.*, 31, 296 (1972).
1461. Amiconi, Antonini, Brunori, Formaneck, and Huber, *Eur. J. Biochem.*, 31, 52 (1972).
1462. Nashi, *Cancer Res.*, 30, 2507 (1970).
1463. Yip, Waks, and Beychok, *J. Biol. Chem.*, 247, 7237 (1972).
1464. Marshall and Pensky, *Arch. Biochem. Biophys.*, 146, 76 (1971).
1465. Kohsiyama, *Int. J. Pept. Protein Res.*, 4, 167 (1972).

1466. Swann, and Hammes, *Biochemistry*, 8, 1 (1969).
1467. Kay, and Marsh, *Biochim. Biophys. Acta*, 33, 251 (1959).
1468. D'Anna, Jr. and Tollin, *Biochemistry*, 11, 1073 (1972).
1469. Legler, von Radloff, and Kempfle, *Biochim. Biophys. Acta*, 257, 40 (1971).
1470. Legler, *Hoppe-Seyler's Z. Physiol. Chem.*, 348, 1359 (1967).
1471. Malcolm, *Hoppe-Seyler's Z. Physiol. Chem.*, 352, 883 (1971).
1472. Smith, Langdon, Piszkiewicz, Brattin, Langley, and Melamed, *Proc. Natl. Acad. Sci. U.S.A.*, 67, 724 (1970).
1473. Winnacker, and Barker, *Biochim. Biophys. Acta*, 212, 225 (1970).
1474. Egan, and Dalziel, *Biochim. Biophys. Acta*, 250, 47 (1971).
1475. Sund, and Akeson, *Biochem. Z.*, 340, 421 (1964).
1476. Miller, and Stadtman, *J. Biol. Chem.*, 247, 7407 (1972).
1477. Roberts, Holcenberg, and Dolowy., *J. Biol. Chem.*, 247, 84 (1972).
1478. Stahl, and Jaenicke, *Eur. J. Biochem.*, 29, 401 (1972).
1479. Cooper, and Meister, *Biochemistry*, 11, 661 (1972).
1480. Orlowski, and Meister, *J. Biol. Chem.*, 246, 7095 (1971).
1481. Ida, and Morita, *Agric. Biol. Chem.*, 35, 1542 (1971).
1482. Wu, Cluskey, Krull, and Friedman, *Can. J. Biochem.*, 49, 1042 (1971).
1483. Jaenicke, in *Pyridine Nucleotide-dependent Dehydrogenases*, Sund, Ed., Springer-Verlag, Berlin, 1970, 70.
1484. Malhotra, and Bernhard, *J. Biol. Chem.*, 243, 1243 (1968).
1485. Aune and Timasheff, *Biochemistry*, 9, 1481 (1970).
1486. D'Alessio and Josse, *J. Biol. Chem.*, 246, 4326 (1971).
1487. Ross, Curry, Schwartz, and Fondy, *Arch. Biochem. Biophys.*, 145, 591 (1971).
1488. White, III, and Kaplan, *J. Biol. Chem.*, 244, 6031 (1969).
1489. Barel, Turneer, and Dolmans, *Eur. J. Biochem.*, 30, 26 (1972).
4190. Frigerio and Harbury, *J. Biol. Chem.*, 231, 135 (1958).
1491. Tamm and Horsfall, *J. Exp. Med.*, 95, 71 (1952).
1492. Maxfield, *Arch. Biochem. Biophys.*, 85, 382 (1959).
1493. Fletcher, Neuberger, and Ratcliffe, *Biochem. J.*, 120, 417 (1970).
1494. Frot-Coutaz, Louisot, and Got, *Biochim. Biophys. Acta*, 264, 362 (1972).
1495. Nisselbaum and Bernfeld, *J. Am. Chem. Soc.*, 78, 687 (1965).
1496. Li and Li, *J. Biol. Chem., Soc.*, 245, 825 (1970).
1497. Patrito and Martin, *Hoppe-Seyler's Z. Physiol. Chem.*, 352, 89 (1971).
1498. Ryan and Westphal, *J. Biol. Chem.*, 247, 4050 (1972).
1499. Heimburger, Haupt, Kranz, and Baudner, *Hoppe-Seyler's Z. Physiol. Chem.*, 353, 1133 (1972).
1500. Iwasaki and Schmid, *J. Biol. Chem.*, 245, 1814 (1970).
1501. Haupt, Baudner, Kranz, and Heimburger, *Eur. J. Biochem.*, 23, 242 (1971).
1502. Boenisch and Alper, *Biochim. Biophys. Acta*, 221, 529 (1970).
1503. Labat, Ishiguro, Fujisaki, and Schmid, *J. Biol. Chem.*, 244, 4975 (1969).
1503a. Kredich, Keenan, and Foote, *J. Biol. Chem.*, 247, 7157 (1972).
1503b. Schwartz, Pizzo, Hill, and McKee, *J. Biol. Chem.*, 248, 1395 (1973).
1503c. Haupt, Heimburger, Kranz, and Boudner, *Hoppe-Seyler's Z. Physiol. Chem.*, 353, 1841 (1972).
1503d. Boenisch and Alper, *Biochim. Biophys. Acta*, 214, 135 (1970).
1503e. Marr, Neuberger, and Ratcliffe, *Biochem. J.*, 122, 623 (1971).
1503f. Rambhar and Ramachandran, *Indian J. Biochem. Biophys.*, 9, 21 (1972).
1503g. Waks, Kahn, and Beychok, *Biochem. Biophys. Res. Commun.*, 45, 1232 (1971).
1504. Huprikar and Sohonie, *Enzymologia*, 28, 333 (1965).
1505. Dahlgren, Porath, and Lindahl-Kiessling, *Arch. Biochem. Biophys.*, 37, 306 (1970).
1506. Howard and Sage, *Biochemistry*, 8, 2436 (1969).
1507. Howard, Sage, Stein, Yound, Leon, and Dyckes, *J. Biol. Chem.*, 246, 1590 (1971).
1508. Garbett, Darnall, Klotz, and Williams, *Arch. Biochem. Biophys.*, 135, 419 (1969).
1509. Makino, *J. Biochem.* (Tokyo), 70, 149 (1971).
1510. Giamberardino, *Arch. Biochem. Biophys.*, 118, 273 (1967).
1511. Bannister and Wood, *Comp. Biochem. Physiol.*, B40, 7 (1971).
1512. Nickerson and Van Holde, *Comp. Biochem. Physiol.*, B39, 855 (1971).
1513. Wittenberg, Briehl, and Wittenberg, *Biochem. J.*, 96, 363 (1965).
1514. Wittenberg, Wittenberg, and Noble, *J. Biol. Chem.*, 247, 4008 (1971).
1515. Wittenberg, Ozazaki, and Wittenberg, *Biochim. Biophys. Acta*, 111, 485 (1965).
1516. Figueiredo, Gomez, Heneine, Santos, and Hargreaves, *Comp. Biochem. Physiol.*, B44, 481 (1973).
1517. Seamonds, Forster, and George, *J. Biol. Chem.*, 246, 5391 (1971).
1518. Terwilliger and Read, *Comp. Biochem. Physiol.*, 36, 339 (1970).
1519. Yamaguchi, Kochiyama, Hashimoto, and Matsuura, *Bull. Jap. Soc. Sci. Fish.*, 28, 184 (1962).
1520. Yamaguchi, Kochiyama, Hashimoto, and Matsuura, *Bull. Jap. Soc. Sci. Fish.*, 29, 174 (1963).

1521. Buhler, *J. Biol. Chem.*, 238, 1665 (1963).
1522. Mohr, Scheler, Schumann, and Muller, *Eur. J. Biochem.*, 3, 158 (1967).
1523. Boyer, Hathaway, Pascasio, Bordley, Orton, and Naughton, *J. Biol. Chem.*, 242, 2211 (1967).
1524. Babul and Stellwagen, *Anal. Biochem.*, 28, 216 (1969).
1525. Inada, Kurozumi, and Shibata, *Arch. Biochem. Biophys.*, 93, 30 (1961).
1526. Herskovits, Gabegbeku, and Jzillet, *J. Biol. Chem.*, 245, 2588 (1970).
1527. Mayer and Miller, *Anal. Biochem.*, 36, 91 (1970).
1528. Javahezian and Beychok, *J. Mol. Biol.*, 37, 1 (1968).
1529. Herskovits, *Arch. Biochem. Biophys.*, 130, 19 (1969).
1530. Malchy and Dixon, *Can. J. Biochem.*, 48, 192 (1970).
1531. Sidwell, Munch, Guzman Barron, and Hogness, *J. Biol. Chem.*, 123, 335 (1938).
1532. Jones and Schroeder, *Biochemistry*, 2, 1357 (1963).
1533. Olson and Gibson, *J. Biol. Chem.*, 246, 5241 (1971).
1534. Allis and Steinhardt, *Biochemistry*, 9, 2286 (1970).
1535. Antonini, Brunori, Caputo, Chiancone, Rossi-Fanelli, and Wyman, *Biochim. Biophys. Acta*, 79, 284 (1964).
1536. Beaven, Hoch, and Holiday, *Biochem. J.*, 49, 374 (1951).
1537. Itano, Fogarty, Jr., and Alford, *Am. J. Clin. Pathol.*, 55, 135 (1971).
1538. Morningstar, Williams, and Suutarinen, *Am. J. Clin. Pathol.*, 46, 603 (1966).
1539. Zettner and Mensch, *Am. J. Clin. Pathol.*, 49, 196 (1968).
1540. Tentori, Vivaldi, and Salvati, *Clin. Chim. Acta*, 14, 276 (1966).
1541. Sugita and Yoneyama, *J. Biol. Chem.*, 246, 389 (1971).
1542. Bucci and Fronticelli, *J. Biol. Chem.*, 240, PC 551 (1965).
1543. DeBruin and Bucci, *J. Biol. Chem.*, 246, 5228 (1971).
1544. Bucci and Fronticelli, *Biochim. Biophys. Acta*, 243, 170 (1971).
1545. Waterman and Yonetani, *J. Biol. Chem.*, 245, 5842 (1970).
1546. Oshino, Asakura, Tamura, Oshino, and Chance, *Biochem. Biophys. Res. Commun.*, 46, 1055 (1972).
1547. Seery, Hathaway, and Eberhard, *Arch. Biochem. Biophys.*, 150, 269 (1972).
1548. Chou and Wilson, *Arch. Biochem. Biophys.*, 151, 48 (1972).
1549. Easterby and Rosemeyer, *Eur. J. Biochem.*, 28, 241 (1972).
1550. Klee, *J. Biol. Chem.*, 247, 1398 (1972).
1551. Wickett, Li, and Isenberg, *Biochemistry*, 11, 2952 (1972).
1552. Pieri and Kergueris, *C. R. Acad. Sci., Paris*, 2740, 2366 (1973).
1553. Mori and Hollands, *J. Biol. Chem.*, 246, 7223 (1971).
1554. Bewley and Li, *Arch. Biochem. Biophys.*, 144, 589 (1971).
1555. Donini, Puzzuoli, D'Alessio, and Donini, in *Pharmacology of Hormonal Polypeptides and Proteins*, Beck, Martini, and Paoletti, Eds., Plenum Press, New York, 1968, 229.
1556. Bewley and Li, *Biochemistry*, 11, 927 (1972).
1557. Bewley, Brovetto-Cruz, and Li, *Biochemistry*, 8, 4701 (1969).
1558. Shome and Friesen, *Endocrinology*, 89, 631 (1971).
1559. Bewley and Li, *Biochemistry*, 11, 884 (1972).
1560. Ma, Brovetto-Cruz, and Li, *Biochemistry*, 9, 2302 (1970).
1561. Bewley, Sairam, and Li, *Biochemistry*, 11, 932 (1971).
1562. Woodhead, O'Riordan, Keutmann, Stolz, Dawson, Niall, Robinson, and Potts, Jr., *Biochemistry*, 10, 2787 (1971).
1563. Bewley and Li, *Int. J. Protein Res.*, 1, 117 (1969).
1564. Jackson and Lovenberg, *J. Biol. Chem.*, 246, 4280 (1971).
1565. Gerstner and Pfeil, *Hoppe-Seyler's Z. Physiol. Chem.*, 353, 271 (1972).
1566. Corneil and Wofsy, *Immunochemistry*, 4, 183 (1967).
1567. Pollet, Rossi, and Edelhoch, *J. Biol. Chem.*, 247, 5921 (1972).
1568. Anders Karlsson, Peterson, and Berggard, *J. Biol. Chem.*, 247, 1065 (1972).
1569. Barth, Bunnenberg, and Djerassi, *Anal. Biochem.*, 48, 471 (1972).
1570. Evans, Herron, and Goldstein, *J. Immunol.*, 101, 915 (1968).
1571. Grey, Abel, and Zimmerman, *Ann. N.Y. Acad. Sci.*, 190, 37 (1972).
1572. Kochwa, Terry, Capra, and Yang, *Ann. N.Y. Acad. Sci.*, 190, 49 (1971).
1573. Kaygorodova and Kaversneva, *Mol. Biol. USSR*, 1, 224 (1967); cited in Egaroy, Chernyak, Dunaevsky, Gavrilova, and Moiseev, *Immunochemistry*, 8, 157 (1971).
1574. Stevenson and Dorrington, *Biochem. J.*, 118, 703 (1970).
1575. O'Daly and Cebra, in *Protides of the Biological Fluids*, Peeters, Ed., Pergamon Press, New York, 1969, 205.
1576. Levine and Levytska, *J. Immunol.*, 102, 647 (1969).
1577. Painter, Sage, and Tanford, *Biochemistry*, 11, 1327 (1972).
1578. Underdown, Simms, and Eisen, *Biochemistry*, 10, 4359 (1971).
1579. Weintraub and Schlamowitz, *Comp. Biochem. Physiol.*, B38, 513 (1971).

1580. Helms and Allen, *Comp. Biochem. Physiol.,* B38, 439 (1971).
1581. Reynolds and Johnson, *Biochemistry,* 10, 2821 (1971).
1582. Butler, *Biochim Biophys. Acta.,* 251, 435 (1971).
1583. Narayana, Shurpalekab, and Sundarvalli, *Indian J. Biochem.,* 7, 241 (1970).
1584. Vidal and Stoppani, *Arch. Biochim. Biophys.,* 147, 66 (1971).
1585. Kiyohara, Iwasaki, and Yoshikawa, *J. Biochem.* (Tokyo), 73, 89 (1972).
1586. Iwasaki, Kiyohara, and Yoshikawa, *J. Biochem.* (Tokyo), 70, 817 (1971).
1587. Chu and Chi, *Sci. Sin.,* 14, 1441 (1965).
1588. Markussen, *Int. J. Protein Res.,* 3, 201 (1971).
1589. Brandenberg, Gattner, and Wollmer, *Hoppe-Seyler's Z. Physiol. Chem.,* 353, 599 (1972).
1590. Holdsworth, *Biochim. Biophys. Acta,* 51, 295 (1961).
1591. Illingworth, *Biochem. J.,* 129, 1119 (1972).
1592. Sanderman, Jr., and Strominger, *J. Biol. Chem.,* 247, 5123 (1972).
1593. Kutzbach and Schmidt-Kastner, *Hoppe-Seyler's Z. Physiol. Chem.,* 353, 1099 (1972).
1594. Fielder, Muller, and Werle, *Fed. Eur. Biochem. Soc. Lett.,* 22, 1 (1972).
1595. Crewther, Fraser, Lennox, and Lindley, *Adv. Protein Chem.,* 20, 191 (1965).
1596. Gillespie, *Comp. Biochem. Physiol.,* B41, 723 (1972).
1597. Barel, Prieels, Maes, Looze, and Leonis, *Biochim. Biophys. Acta,* 257, 288 (1972).
1598. Cowburn, Brew, and Gratzer, *Biochemistry,* 11, 1228 (1972).
1599. Dagleish and Peacocke, *Biochem. J.,* 125, 155 (1971).
1600. Apella and Markert, *Biochem. Biophys. Res. Commun.,* 6, 171 (1961).
1601. Thuwissen, Masson, Osinski, and Heremans, *Eur. J. Biochem.,* 31, 239 (1972).
1602. Masson, *La Lactoferrine,* Arsica, Brussels and Maloine, Paris, 1970.
1603. Aisen and Leibman, *Biochim. Biophys. Acta,* 257, 314 (1972).
1604. Berrens and Bleumink, *Int. Arch. Allergy,* 28, 150 (1965).
1605. Trayer and Hill, *J. Biol. Chem.,* 246, 6666 (1971).
1606. Peive, Atanasov, Zhiznevskaya, and Krasnobaeva, *Dokl. Akad. Nauk SSR Biochem. Sect. (Transl.),* 202, 39 (1972).
1607. Atanasov, Bulgarian Academy of Sciences, Bulgaria, submitted.
1608. Soda, Misono, Mori, and Sakato, *Biochem. Biophys. Res. Commun.,* 44, 931 (1971).
1609. Garner, Jr. and Smith, *J. Biol. Chem.,* 247, 561 (1972).
1610. Edelstein, Lim, and Scanu, *J. Biol. Chem.,* 247, 5842 (1972).
1611. Scanu, Lim, and Edelstein, *J. Biol. Chem.,* 247, 5850 (1972).
1612. Smith, Dawson, and Tanford, *J. Biol. Chem.,* 247, 3376 (1972).
1613. Dejmal and Brookes, *J. Biol. Chem.,* 247, 869 (1972).
1614. Christopher, Pistorius, and Axelrod, *Biochim. Biophys. Acta,* 198, 12 (1970).
1615. Bellisario, Spencer, and Cormier, *Biochemistry,* 11, 2256 (1972).
1616. Webster, *Biochim. Biophys. Acta,* 207, 371 (1970).
1617. Bradshaw and Deranleau, *Biochemistry,* 9, 3310 (1970).
1618. Kravchenko and Lapuk, *Biokhimia,* 34, 832 (1969).
1619. Davies, Neuberger, and Wilson, *Biochim. Biophys. Acta,* 178, 294 (1969).
1620. Blake, Johnson, Mair, North, Phillips, and Sarma, *Proc R. Soc. Lond. (Biol.),* 167, 378 (1967).
1621. Donovan, Davis, and White, *Biochim. Biophys. Acta,* 270, 190 (1970).
1622. Roxby and Tanford, *Biochemistry,* 10, 3348 (1971).
1623. Ehrenpreis and Warner, *Arch Biochem. Biophys.,* 61, 38 (1956).
1624. Lin and Koshland, *J. Biol. Chem.,* 244, 505 (1969).
1625. Teichberg, Kay, and Sharon, *Eur. J. Biochem.,* 16, 55 (1970).
1626. Hayashi, Imoto, Funatsu, and Funatsu, *J. Biochem.,* (Tokyo), 58, 227 (1965).
1627. Franek and Pechan, *Scr. Fac. Sci. Nat. Ujep. Brunensis Chem.,* 1, 67 (1971).
1628. Riblet and Herzenberg, *Science,* 168, 45 (1970).
1629. Barel, Prieels, Maes, Looze, and Leonis, *Biochim. Biophys. Acta,* 257, 288 (1972).
1630. Fawcett, Limbird, Oliver, and Borders, *Can. J. Biochem.,* 49, 816 (1971).
1631. Latovitzki, Halper, and Beychok, *J. Biol. Chem.,* 246, 1457 (1971).
1632. Parry, Jr., Chandan, and Shahani, *Arch. Biochem. Biophys.,* 130, 59 (1969).
1633. Cowburn, Brew, and Gratzer, *Biochemistry,* 11, 1228 (1972).
1634. Mitchell and Hash, *J. Biol. Chem.,* 244, 17 (1969).
1635. Greene, Damian, and Hubbard, *Biochim. Biophys. Acta,* 236, 659 (1971).
1636. Humphries, Rohrbach, and Harrison, *Biochem. Biophys. Res. Commun.,* 50, 493 (1973).
1637. Allen, *Eur. J. Biochem.,* 35, 338 (1973).
1638. Yamaguchi, Tokushige, and Katsuki, *J. Biochem.* (Tokyo), 73, 169 (1973).
1639. Saita, Ikenaka, and Mutsushima, *J. Biochem.* (Tokyo), 70, 827 (1971).
1640. Vachek and Wood, *Biochim. Biophys. Acta,* 258, 133 (1972).

1641. Heazlitt, Conway, and Montag, *Biochim. Biophys. Acta,* 317, 316 (1973).
1642. Lymn and Taylor, *Biochemistry,* 10, 4617 (1971).
1643. Weser, Donay, and Rupp, *FEBS Lett.,* 32, 171 (1973).
1644. Hirata, Nakazawa, Nozaki, and Hayaishi, *J. Biol. Chem.,* 246, 5882 (1971).
1645. Kuma and Inomata, *J. Biol. Chem.,* 247, 556 (1972).
1646. Oh and Conrad, *Arch. Biochem. Biophys.,* 146, 525 (1971).
1647. Morris, Martenson, Deibler, and Cagan, *J. Biol. Chem.,* 248, 534 (1973).
1648. Tamura, Asakura, and Yonetani, *Biochim. Biophys. Acta,* 295, 467 (1973).
1649. Tamura, Woodrow, and Yonetani, *Biochim. Biophys. Acta,* 317, 34 (1973).
1650. Goldbloom and Brown, *Arch. Biochem. Biophys.,* 147, 367 (1971).
1651. Deconinck, Peiffer, Schnek, and Leonis, *Biochimie,* 54, 969 (1972).
1652. Bolard and Garnier, *Biochim. Biophys. Acta,* 263, 535 (1972).
1653. Harrington and Himmelfarb, *Biochemistry,* 11, 2945 (1972).
1654. Kakol, *Biochem. J.,* 125, 261 (1972).
1655. Katoh, Kubo, and Takahashi, *J. Biochem.* (Tokyo), 74, 771 (1973).
1656. Bjorkman and Janson, *Biochim. Biophys. Acta,* 276, 508 (1972).
1657. Suhara, Ikeda, Takemori, and Katagiri, *FEBS Lett.,* 28, 45 (1972).
1658. Yuan, Barnett, and Anderson, *J. Biol. Chem.,* 247, 511 (1972).
1659. Siegel, Murphy, and Kamin, *J. Biol. Chem.,* 248, 251 (1973).
1660. Frazier, Hogue-Angeletti, Sherman, and Bradshaw, *Biochemistry,* 12, 328 (1973).
1661. Angeletti, *Biochim. Biophys. Acta,* 214, 478 (1970).
1662. Pearce, Banks, Banthorpe, Berry, Davies, and Vernon, *Eur. J. Biochem.,* 29, 417 (1972).
1663. Kopeyan, vanRietschoten, Martinez, Rochat, and Miranda, *Eur. J. Biochem.,* 35, 244 (1973).
1664. Karlsson, Eaker, and Ponterius, *Biochim. Biophys. Acta,* 257, 235 (1972).
1665. Karlsson, Eaker, and Ryden, *Toxicon,* 10, 405 (1972).
1666. Iwasaki and Matsubara, *J. Biochem.* (Tokyo), 71, 645 (1972).
1667. Prakash and Sadana, *Arch. Biochem. Biophys.,* 148, 614 (1972).
1668. Zumft, *Biochim. Biophys. Acta,* 276, 363 (1972).
1669. Burns and Hardy, *Methods Enzymol.,* 24B, 480 (1972).
1670. Eady, Smith, Cook, and Postgate, *Biochem. J.,* 128, 655 (1972).
1671. Luisi, Olomucki, Baici, and Karlovic, *Biochemistry,* 12, 4100 (1973).
1672. Marshall and Cohen, *J. Biol. Chem.,* 247, 1641 (1972).
1673. Marcais, Nicot, and Moretti, *Bull. Soc. Chim. Biol.,* 52, 741 (1970).
1674. Joniau, Bloemmen, and Lontie, *Biochim. Biophys. Acta,* 214, 468 (1970).
1675. Berrens and Bleumink, *Int. Arch. Allergy,* 28, 150 (1965).
1676. Holt and Creeth, *Biochem. J.,* 129, 665 (1972).
1677. Willumsen, *C.R. Trav. Lab. Carlsberg,* 36, 247 (1967).
1678. Ifft, *C.R. Trav. Lab. Carlsberg,* 38, 315 (1971).
1679. Babul and Stellwagen, *Anal. Biochem.,* 28, 216 (1969).
1680. Ketterber, *Biochem. J.,* 96, 372 (1965).
1681. Barth, Bunnenberg, and Djerassi, *Anal. Biochem.,* 48, 471 (1972).
1682. Davis, Zahnley, and Donovan, *Biochemistry,* 8, 2044 (1966).
1683. Kay, Strickland, and Billups, *J. Biol. Chem.,* 249, 797 (1974).
1684. Sjoberg, and Feeney, *Biochim. Biophys. Acta,* 168, 79 (1968).
1685. Warner and Weber, *J. Biol. Chem.,* 191, 173 (1951).
1686. Rhodes, Azari, and Feeney, *J. Biol. Chem.,* 230, 399 (1958).
1687. Azari, and Phillips, *Arch. Biochem. Biophys.,* 138, 32 (1970).
1688. Azari, and Baugh, *Arch. Biochem. Biophys.,* 118, 138 (1967).
1689. Weintraub, Vincent, Baulieu, and Alfsen, *FEBS Lett.,* 37, 82 (1973).
1690. Jolley, Evans, Makino, and Mason, *J. Biol. Chem.,* 249, 335 (1974).
1691. Darby, *J. Biol. Chem.,* 139, 721 (1941).
1692. Lauwers, in *West European Symp. on Clin. Chem. – Symp. on Enzymes in Clin. Chem.,* Ruyssen and Vandendriessche, Eds., Elsevier, New York, 1965, 19.
1693. Piront and Gerday, *Comp. Biochem. Physiol.,* 46B, 349 (1973).
1694. Pechere, Capony, and Ryden, *Eur. J. Biochem.,* 23, 421 (1971).
1695. Parello and Pechere, *Biochimie,* 53, 1079 (1971).
1696. Jones, and Hofmann, *Can. J. Biochem.,* 50, 1297 (1972).
1697. Glazer and Smith, *J. Biol. Chem.,* 235, PC43 (1960).
1698. Webster, *Biochim. Biophys. Acta,* 207, 371 (1970).
1699. Blumenfeld and Perlmann, *J. Gen. Physiol.,* 42, 563 (1959).
1700. Lang and Kassell, *Biochemistry,* 10, 2296 (1971).
1701. Mayer and Miller, *Anal. Biochem.,* 36, 91 (1970).

1702. **Bohak**, *J. Biol. Chem.*, 244, 4638 (1969).
1703. **Meitner and Kassell**, *Biochem. J.*, 121, 249 (1971).
1704. **Marcinszyn, and Kassell**, *J. Biol. Chem.*, 246, 6560 (1971).
1705. **Kikuchi and Sakaguchi**, *Agric. Biol. Chem.*, 37, 827 (1973).
1706. **Ternynck and Avrameas**, *FEBS Lett.*, 23, 24 (1972).
1707. **Herskovits**, *Arch. Biochem. Biophys.*, 130, 19 (1969).
1708. **Shih, Shannon, Kay, and Lew**, *J. Biol. Chem.*, 246, 4546 (1971).
1709. **Roman and Dunford**, *Biochemistry*, 11, 2076 (1972).
1710. **Paul and Stigbrand**, *Acta Chem. Scand.*, 24, 3607 (1970).
1711. **Ohlsson and Paul**, *Biochim. Biophys. Acta*, 315, 293 (1973).
1712. **Morita and Yoshida**, *Agric. Biol. Chem.*, 34, 590 (1970).
1713. **Ex-Fekih and Kertesz**, *Bull. Soc. Chim. Biol.*, 50, 547 (1968).
1714. **Stelmaszynska and Zgliczynski**, *Eur. J. Biochem.*, 19, 56 (1971).
1715. **Neujahr and Gaal**, *Eur. J. Biochem.*, 35, 386 (1973).
1716. **Havir and Hanson**, *Biochemistry*, 12, 1583 (1973).
1717. **Fisher, Kirkwood, and Kaufman**, *J. Biol. Chem.*, 247, 5161 (1972).
1718. **Huang, Max, and Kaufman**, *J. Biol. Chem.*, 248, 4235 (1973).
1719. **Hulett-Cowling and Campbell**, *Biochemistry*, 10, 1364 (1971).
1720. **Yoshizumi and Coleman**, *Arch. Biochem. Biophys.*, 160, 255 (1974).
1721. **Halford, Benneti, Trentham, and Gutfreund**, *Biochem. J.*, 114, 243 (1969).
1722. **Taylor, Lau, Applebury, and Coleman**, *J. Biol. Chem.*, 248, 6216 (1973).
1723. **Glew and Heath**, *J. Biol. Chem.*, 246, 1556 (1971).
1724. **Igarashi, Takahashi, and Tsuyama**, *Biochim. Biophys. Acta*, 220, 85 (1970).
1725. **Igarashi and Hollander**, *J. Biol. Chem.*, 243, 6084 (1968).
1726. **Uehara, Fujimoto, and Taniguchi**, *J. Biochem.* (Tokyo), 70, 183 (1971).
1727. **Jacobs, Nyc, and Brown**, *J. Biol. Chem.*, 246, 1419 (1971).
1728. **Shimada and Sugino**, *Biochim. Biophys. Acta*, 185, 367 (1969).
1729. **Cannata**, *J. Biol. Chem.*, 245, 792 (1970).
1730. **Smith**, *J. Biol. Chem.*, 246, 4234 (1971).
1731. **Wohl and Markus**, *J. Biol. Chem.*, 237, 5785 (1972).
1732. **Maeba and Sanwal**, *J. Biol. Chem.*, 244, 2549 (1969).
1733. **Kopperschlager, Lorenz, Diezel, Marquardt, and Hofmann**, *Acta Biol. Med. Ger.*, 29, 561 (1972).
1734. **Najjar**, in *The Enzymes* Vol. 6, Boyer, Lardy, and Myrbach, Eds., Academic Press, New York, 1962, 161.
1735. **Filmer and Koshland**, *Biochim. Biophys. Acta*, 77, 334 (1963).
1736. **Hirose, Sugimoto, and Chiba**, *Biochim. Biophys. Acta*, 250, 514 (1971).
1737. **Silverberg and Dalziel**, *Eur. J. Biochem.*, 38, 229 (1973).
1738. **Tsuboi, Fukunaga, and Chervenka**, *J. Biol. Chem.*, 246, 7586 (1971).
1739. **Krietsch and Bucher**, *Eur. J. Biochem.*, 17, 568 (1970).
1740. **Scopes**, *Biochem. J.*, 113, 551 (1969).
1741. **Scopes**, *Biochem. J.*, 122, 89 (1971).
1742. **Sasaki, Sugimoto, and Chiba**, *Biochim. Biophys. Acta*, 227, 584 (1971).
1743. **James, Hurst, and Flynn**, *Can. J. Biochem.*, 49, 1183 (1971).
1744. **Kawauchi, Iwanaga, Samejima, and Suzuki**, *Biochim. Biophys. Acta*, 236, 142 (1971).
1745. **Janssen, deBruin, and Haas**, *Eur. J. Biochem.*, 28, 156 (1972).
1746. **Boman**, *Ark. Kem.* 12, 453 (1958).
1747. **Felenbok**, *Eur. J. Biochem.*, 17, 165 (1970).
1748. **Rutner**, *Biochemistry*, 9, 178 (1970).
1749. **Hartman**, *J. Biol. Chem.*, 238, 3024 (1963).
1750. **Blasi, Aloj, and Goldberger**, *Biochemistry*, 10, 1409 (1971).
1751. **Sevilla and Fischer**, *Biochemistry*, 8, 2161 (1969).
1752. **Kamogawa and Fukui**, *Biochim. Biophys. Acta*, 242, 55 (1971).
1753. **Cohen, Duewer, and Fischer**, *Biochemistry*, 10, 2683 (1971).
1754. **Fosset, Muir, Nielsen, and Fischer**, *Biochemistry*, 10, 4105 (1971).
1755. **Assaf and Yunis**, *Biochemistry*, 12, 1423 (1973).
1756. **Assaf and Graves**, *J. Biol. Chem.*, 244, 5544 (1969).
1757. **Kamogawa, Fukui, and Nikuni** *J. Biochem.* (Tokyo), 63, 361 (1968).
1758. **Cohen**, *Eur. J. Biochem.*, 34, 1 (1972).
1759. **Hayakawa, Perkins, Walsh, and Krebs**, *Biochemistry*, 12, 567 (1973).
1760. **Mano and Yoshida**, *J. Biochem.* (Tokyo), 66, 105 (1969).
1761. **Walinder**, *Biochim. Biophys. Acta*, 258, 411 (1972).
1762. **Glazer, Fang, and Brown**, *J. Biol. Chem.*, 248, 5679 (1973).
1763. **Maccoll, Habig, and Berns**, *J. Biol. Chem.*, 248, 7080 (1973).

1764. Glazer and Fang, *J. Biol. Chem.*, 248, 659 (1973).
1765. Boucher, Crespi, and Katz, *Biochemistry*, 5, 3796 (1968).
1766. Kao, Berns, and Town, *Biochem. J.*, 131, 39 (1973).
1767. Anderson, Jenner, and Mumford, *Biochim. Biophys. Acta*, 221, 69 (1970).
1768. Tobin and Briggs, *Photochem. Photobiol.*, 18, 487 (1973).
1769. Bourrillon and Font, *Biochim. Biophys. Acta*, 154, 28 (1968).
1770. Berrens and Maesen, *Clin. Exp. Immunol.*, 10, 383 (1972).
1771. Kochiyama, Yamaguchi, Hashimoto, and Matsuura, *Bull. Jap. Soc. Sci. Fish.*, 32, 867 (1966).
1772. Sjoholm, Wiman, and Wallen, *Eur. J. Biochem.*, 39, 471 (1973).
1773. Wallen and Wiman, *Biochim. Biophys. Acta*, 221, 20 (1970).
1774. Ramshaw, Brown, Scawen, and Boulter, *Biochim. Biophys. Acta*, 303, 269 (1973).
1775. Katoh, Shiratori, and Takamiya, *J. Biochem.* (Tokyo), 51, 32 (1962).
1776. Milne and Wells, *J. Biol. Chem.*, 245, 1566 (1970).
1777. Klee, *J. Biol. Chem.*, 244, 2558 (1969).
1778. Yurewicz, Ghalambor, Duckworth, and Heath, *J. Biol. Chem.*, 246, 5607 (1971).
1779. Treffry and Ainsworth, *Biochem. J.*, 137, 319 (1974).
1780. Cejka and Fleischmann, *Arch. Biochem. Biophys.*, 157, 168 (1973).
1781. Heide, Haupt, and Schultze, *Nature*, 201, 1218 (1964).
1782. Peterson, *J. Biol. Chem.*, 246, 34 (1971).
1783. Schultze, Schonenberger, and Schwick, *Biochem. Z.*, 328, 267 (1956).
1784. Raz and Goodman, *J. Biol. Chem.*, 244, 3230 (1969).
1785. Seal and Doe, in *Proc. 2nd Int. Congr. Endocrinology*, Vol. 19, Sect. 3, Part 1, Excerpta Medica, Amsterdam, 1964, 229.
1786. Schultze and Heremans, *Molecular Biology of Human Proteins*, Vol. 1, Elsevier, Amsterdam, 1966, 234.
1787. Tritsch, *J. Med.* (Basel), 3, 129 (1972).
1788. Oppenheimer, Surks, Smith, and Squef, *J. Biol. Chem.*, 240, 173 (1965).
1789. Vahlquist and Peterson, *Biochemistry*, 11, 4526 (1972).
1790. Jaarsveld, Branch, Robbins, Morgan, Kanda, and Canfield, *J. Biol. Chem.*, 248, 7898 (1973).
1791. Stratil, *Animal Blood Groups Biochem. Genet.*, 3, 63 (1972).
1792. Ashida, *Arch. Biochem. Biophys.*, 144, 749 (1973).
1793. Van der Wel, *FEBS Lett.*, 21, 88 (1972).
1794. Behnke and Vallee, *Fed. Proc.*, 31, 435 (1972).
1795. Gates and Travis, *Biochemistry*, 12, 1867 (1973).
1796. Lacko and Neurath, *Biochemistry*, 9, 4680 (1970).
1797. Uren and Neurath, *Biochemistry*, 11, 4483 (1972).
1798. Reeck and Neurath, *Biochemistry*, 11, 3947 (1972).
1799. Gertler and Birk, *Eur. J. Biochem.*, 12, 170 (1970).
1800. Uram and Lamy, *Biochim. Biophys. Acta*, 194, 102 (1969).
1801. Lea, *Biochim. Biophys. Acta*, 317, 351 (1973).
1802. Frank, Veros, and Pekar, *Biochemistry*, 11, 4926 (1972).
1803. Markussen, *Int. J. Protein Res.*, 3, 201 (1971).
1804. Dixon, Schmidt, and Pankov, *Arch. Biochem. Biophys.*, 141, 705 (1970).
1805. Murao, Funakoshi, and Oda, *Agric. Biol. Chem.*, 36, 1327 (1972).
1806. Tsuru, Hattori, Tsuji, and Fukumoto, *J. Biochem.* (Tokyo), 67, 15 (1970).
1807. Sodek and Hofmann, *Methods Enzymol.*, 19, 372 (1970).
1808. Oda, Kamada, and Murao, *Agric. Biol. Chem.*, 36, 1103 (1972).
1809. McDowell, *Eur. J. Biochem.*, 14, 214 (1970).
1810. Sekine, *Agric. Biol. Chem.*, 36, 198 (1972).
1811. Yoshimoto, Fukumoto, and Tsuru, *Int. J. Protein Res.*, 3, 285 (1971).
1812. Tsuru, Yoshimoto, Yoshida, Kira, and Fukumoto, *Int. J. Protein Res.*, 2, 75 (1970).
1813. Shinano and Fukushima, *Agric. Biol. Chem.*, 33, 1236 (1969).
1814. Levy, Fishman, and Schenkein, *Methods Enzymol.*, 19, 672 (1970).
1815. Jurasek and Whitaker, *Can. J. Biochem.*, 43, 1955 (1965).
1816. Kaplan and Whitaker, *Can. J. Biochem.*, 47, 305 (1969).
1817. Paterson and Whitaker, *Can. J. Biochem.*, 47, 317 (1969).
1818. Sauer and Senft, *Comp. Biochem. Physiol.*, 42B, 205 (1972).
1819. Merkel and Sipos, *Arch. Biochem. Biophys.*, 145, 126 (1971).
1820. Siegel, Brady, and Awad, *J. Biol. Chem.*, 247, 4155 (1972).
1821. Liu, Neumann, Elliott, Moore, and Stein, *J. Biol. Chem.*, 238, 251 (1963).
1822. Hata, Hayashi, and Doi, *Agric. Biol. Chem.*, 31, 357 (1967).
1823. Aibard, Hayashi, and Hata, *Agric. Biol. Chem.*, 35, 658 (1971).
1824. Hiramatsu and Ouchi, *J. Biochem.* (Tokyo), 71, 676 (1972).

1825. Pangburn, Burstein, Morgan, Walsh, and Neurath, *Biochem. Biophys. Res. Commun.*, 54, 371 (1973).
1826. Van Heyningen, *Eur. J. Biochem.*, 27, 436 (1972).
1827. Nasuno and Ohara, *Agric. Biol. Chem.*, 36, 1791 (1972).
1828. Turkova, Mikes, Gancev, and Boublik, *Biochim. Biophys. Acta,* 178, 100 (1969).
1829. Gertler and Hayashi, *Biochim. Biophys. Acta,* 235, 378 (1971).
1830. Mizusawa and Yoshida, *J. Biol. Chem.*, 247, 6978 (1972).
1831. Lees, Leston, and Marfey, *J. Neurochem.*, 16, 1025 (1969).
1832. Cox and Hanahan, *Biochim. Biophys. Acta,* 207, 49 (1970).
1833. Fujisawa and Hayaishi, *J. Biol. Chem.*, 243, 2673 (1968).
1834. Ono, Nozaki, and Hayaishi, *Biochim. Biophys. Acta,* 220, 224 (1970).
1835. Berg and Prockop, *J. Biol. Chem.*, 248, 1175 (1973).
1836. DeSa, *J. Biol. Chem.*, 247, 5527 (1972).
1837. Yui, *J. Biochem.* (Tokyo), 69, 101 (1971).

PROPERTIES OF PURIFIED LECTINS*

No.	Source	Molecular weight	No. of sub-units	Carbo-hydrate content %	Method of puri-fication[a]	Specificity Human blood type	Specificity Mono- or disaccha-ride	Specificity Association constant K_a	Refer-ences[b]
	Plants								
1	*Agaricus bisporus* golden white mushroom	58,000			C				2, 3
2	*Agaricus campestris* meadow mushroom	64,000	4	4	C				1
3	*Abrus precatorius* crab's eye vine	134,000	4[c]		F		D-Gal		4
4	*Bandeiraea simplicifolia*	114,000	4	6.7	F	B	α-D-Gal		5
5	*Bauhinia purpurea alba* orchid tree	195,000		11.1	F	None	GalNAc		6
6	*Canavalia ensiformis*[d] jack bean	53,000	2	0	F	None	α-D-Man α-D-Glc	$3.8 \times 10^3\ M^{-1}$ [e]	1, 7–10
7	*Crotolaria juncea* Sunn hemp	120,000		5	F	None	D-Gal		13
8	*Cytisus sessilifolius* broom	110,000			F	H(O)	(GlcNAc)$_2$		14
9	*Dolichos biflorus* horse gram	113,000–120,000	4	3.8–4.3	C,F	A	α-D-GalNAc		1, 15, 16
10	*Glycine max* soybean	120,000	4	6.0	F	None	D-GalNAc	$3.0 \times 10^4\ M^{-1}$	1, 17–19
11	*Lens culinaris*[f] lentil	49,000–63,000	2 or 4	2–3.3	F	None	α-D-Man α-D-Glc	$2.5 \times 10^2\ M^{-1}$	1, 20–22
12	*Lotus tetragonolobus* winged pea I II III	120,000 58,000 117,000		9.4 4.8 9.2	F	H(O) H(O) H(O)	L-Fuc L-Fuc L-Fuc	$1.2 \times 10^4\ M^{-1}$ $0.6 \times 10^4\ M^{-1}$ $3.2 \times 10^4\ M^{-1}$	1, 23
13	*Maackia amurensis* I II	130,000 130,000	4 2[g]	9.5 8.7	F	None None			24
14	*Phaseolus coccineus L* scarlet runner	120,000	4	7.8	C	None	D-GalNAc		25

*Abbreviations used: D-Gal = D-galactose; D-GalNAc = *N*-acetyl-D- galactosamine; D-Man = D-mannose; D-Glc = D-glucose; L-Fuc = L-fucose; (GlcNAc)$_2$ = di-*N*-acetylchitobiose; NeuNAc = *N*-acetylneuraminic acid.

[a] C = conventional methods; F = affinity chromatography.

[b] Only references not included in Table 4 of Reference 1 are given. Procedures for the isolation of a number of lectins are described in Volumes 28 and 34 of *Methods in Enzymology*.

[c] Number of subunits obtained after reduction of interchain S–S bonds.

[d] A lectin with very similar physico-chemical properties has been purified from *Canavalia gladiata* (nata bean) (References 11 and 12).

[e] For methyl α-D-glucoside.

[f] Also known as *Lens esculenta*.

[g] Each subunit composed of two polypeptide chains bound by an S–S bond(s).

PROPERTIES OF PURIFIED LECTINS (continued)

No.	Source	No. of binding sites	Metal requirement	Mitogenic activity on lymphocytes from				No. of isolectins	Crystalline	References[b]
				Mouse	Human	Other	Not specified			
	Plants (continued)									
1	*Agaricus bisporus* golden white mushroom				−					2, 3
2	*Agaricus campestris* meadow mushroom				+					1
3	*Abrus precatorius* crab's eye vine	2					+			4
4	*Bandeiraea simplicifolia*		Ca^{2+}							5
5	*Bauhinia purpurea alba* orchid tree									6
6	*Canavalia ensiformis*[d] jack bean	2	Mn^{2+}, Ca^{2+}	+	+	+			+	1, 7–10
7	*Crotolaria juncea* Sunn hemp									13
8	*Cytisus sessilifolius* broom									14
9	*Dolichos biflorus* horse gram						−	2		1, 15, 16
10	*Gylcine max* soybean	2	Mn^{2+}			+		4		1, 17–19
11	*Lens culinaris*[f] lentil							2		1, 20–22
12	*Lotus tetragonolobus* winged pea I II III	4 2 4								1, 23
13	*Maackia amurensis*									24
14	*Phaseolus coccineus L* scarlet runner		Mn^{2+}, Ca^{2+}	+						25

PROPERTIES OF PURIFIED LECTINS (continued)

No.	Source	Molecular weight	No. of sub-units	Carbo-hydrate content %	Method of puri-fication[a]	Specificity Human blood type	Specificity Mono- or disaccha-ride	Specificity Association constant K_a	Refer-ences[b]
	Plants (continued)								
15	*Phaseolus limensis*[h] lima bean I	110,000–130,000	2[g]	5	F	A	D-GalNAc		1, 26–28
	II	195,000–265,000	4[g]			A	D-GalNAc		
16	*Phaseolus vulgaris* black kidney bean	128,000		5.7					1
17	*Phaseolus vulgaris* navy (haricot) bean	114,000	4	8.13	C	None			29
18	*Phaseolus vulgaris*[i] red kidney bean I	128,000–150,000	4	6.6–8.9	C	None	D-GalNAc[j]	$4.7 \times 10^3\ M^{-1}$	1, 30–32
	II	126,000	4	10.2	C	None	D-GalNAc[j]		1, 30, 32, 33
19	*Phaseolus vulgaris* wax bean	120,000	4	8–10.4	C	None			1, 34, 35
20	*Phytolacca americana*[k] pokeweed	32,000		4.6	C				36
21	*Pisum sativum* garden pea	49,000–53,000	4	0.3	F	None	D-Man D-Glc	$1.4 \times 10^3\ M^{-1}$	1, 37–39
22	*Ricinus communis*[l] castor bean	120,000	4[c]	8–9	F	None	D-Gal	$1.2 \times 10^3\ M^{-1}$	1, 4, 40–44
23	*Robinia pseudoacacia* black locust	90,000		10.7	C	None			1, 45, 46
24	*Solanum tuberosum* potato	120,000	2	50	C	None	(GlcNAc)$_2$		47
25	*Sophora japonica* Japan pagoda tree	135,000	4	18.4	F	B > A	β-D-Gal β-D-GalNAc		48, 49
26	*Triticum vulgaris* wheat	36,000	2	0	C, F	None	(D-GlcNAc)$_2$	$0.45 \times 10^4\ M^{-1} -$ $2 \times 10^4\ M^{-1}$	1, 46, 50–55

[h] Also known as *Phaseolus lunatus*
[i] Two lectins with distinct activities appear to be present in the red kidney bean, an erythroagglutinin (I) and a leucoagglutinin (II).
[j] Acts only at high concentrations (~10 mg/ml).
[k] This lectin is a mitogen with relatively weak hemagglutinating and leucoagglutinating activities.
[l] Castor beans also contain ricin, a powerful toxin, mol wt 60,000. Although it does not agglutinate cells, it binds to sugars on cell surfaces and it has been suggested to classify it as a lectin (Reference 4). The castor bean agglutinin seems to be a dimer of ricin.

PROPERTIES OF PURIFIED LECTINS (continued)

No.	Source Plants (continued)		No. of binding sites	Metal requirement	Mitogenic activity on lymphocytes from				No. of isolectins	Crystalline	References[b]
					Mouse	Human	Other	Not specified			
15	*Phaseolus limensis*[h] lima bean	I II	2 4	Mn^{2+} Mn^{2+}	– –	± +					1, 26–28
16	*Phaseolus vulgaris* black kidney bean										1
17	*Phaseolus vulgaris* navy (haricot) bean										29
18	*Phaseolus vulgaris*[i] red kidney bean	I II		Mn^{2+}, Ca^{2+}	+ +	+ +			3	+	1, 30–32 1, 30, 32, 33
19	*Phaseolus vulgaris* wax bean			Ca^{2+}, Cu^{2+} Zn^{2+}	+				2		1, 34, 35
20	*Phytolacca americana*[k] pokeweed				+	+					36
21	*Pisum sativum* garden pea		2	Mn^{2+}, Ca^{2+}	+				2		1, 37–39
22	*Ricinus communis*[l] castor bean		2					+			1, 4, 40–44
23	*Robinia pseudoacacia* black locust				+						1, 45, 46
24	*Solanum tuberosum* potato										47
25	*Sophora japonica* Japan pagoda tree					+			3		48, 49
26	*Triticum vulgaris* wheat		4		–				3	+	1, 46, 50–55

PROPERTIES OF PURIFIED LECTINS (continued)

No.	Source	Molecular weight	No. of sub-units	Carbo-hydrate content %	Method of puri-fication[a]	Specificity Human blood type	Specificity Mono- or disaccha-ride	Specificity Association constant K_a	Refer-ences[b]
	Plants (continued)								
27	*Ulex europeus* I	46,000		7.2	F	H(O)	L-Fuc		56
	gorse (furze)	170,000		5.2	C	H(O)	L-Fuc		1, 46
	II			21.7	C	H(O)	(GlcNAc)$_2$		1, 46
28	*Vicia faba*	53,000	3–4	+	F	None	D-Man		57
	fava bean						D-Glc		
29	*Wistaria floribunda*[k]	70,000		11.4	C		D-GalNAc		58, 59
	Japanese wisteria								
	Microorganisms								
30	*Pseudomonas aeruginosa*	65,000			C, F	None	D-Gal		60, 61
	Invertebrates								
31	*Anguilla anguilla* eel	123,000–140,000	3–4	0.39	F	H(O)	L-Fuc		14, 62
32	*Crassostrea virginica* oyster	(20,000)$_n$	n	8.8					63, 64
33	*Electrophorus electricus* electric eel	33,000			F	None	Lactose	$1.0 \times 10^3\ M^{-1} - 5.5 \times 10^3\ M$	65
34	*Helix pomatia* garden snail	79,000	6[c]	7.3	F	A	α-GalNAc α-GlcNAc	$5.0 \times 10^3\ M^{-1}$ [m]	1, 66, 67
35	*Homarus americanus* American lobster	(55,000)$_n$	n	+	C	None			68
36	*Limulus polyphemus* horseshoe crab	420,000	18	24	C, F	None	NeuNAc		1, 69, 70
37	*Panulirus argus* spiny lobster	(68,000)$_n$	n	4.6					64
	Mammals								
38	Rabbit	$4 \times 10^5 - 4 \times 10^6$		10	F	None	D-GalNAc D-Gal		71, 72

[m] For methyl α-N-acetyl-D-galactosaminide.

PROPERTIES OF PURIFIED LECTINS (continued)

No.	Source	No. of binding sites	Metal requirement	Mitogenic activity on lymphocytes from				No. of iso-lectins	Crystaline	References[b]
				Mouse	Human	Other	Not specified			
	Plants (continued)									
27	*Ulex europeus* I		Mn^{2+}, Zn^{2+}	−						56
	gorse (furze)			+						1, 46
	II			−						1, 46
28	*Vicia faba*			+	+				+	57
	fava bean									
29	*Wistaria floribunda*[k]				+					58, 59
	Japanese wisteria									
	Microorganisms									
30	*Pseudomonas aeruginosa*		Mn^{2+}, Mg^{2+}, Ca^{2+}	−						60, 61
	Invertebrates									
31	*Anguilla anguilla* eel									14, 62
32	*Crassostrea virginica* oyster									63, 64
33	*Electrophorus electricus* electric eel									65
34	*Helix pomatia* garden snail	6								1, 66, 67
35	*Homarus americanus* American lobster		Ca^{2+}					2–3		68
36	*Limulus polyphemus* horseshoe crab		Ca^{2+}							1, 69, 70
37	*Panulirus argus* spiny lobster		Ca^{2+}							64
	Mammals									
38	Rabbit		Ca^{2+}							71, 72

Compiled by Halina Lis and Nathan Sharon.

PROPERTIES OF PURIFIED LECTINS (continued)

REFERENCES

1. Sharon and Lis, *Science,* 177, 949 (1972).
2. Presant and Kornfeld, *J. Biol. Chem.,* 247, 6937 (1972).
3. Ahmann and Sage, *Immunol. Commun.,* 1, 553 (1972).
4. Olsnes, Saltvedt, and Pihl, *J. Biol. Chem.,* 249, 803 (1974).
5. Hayes and Goldstein, *J. Biol. Chem.,* 249, 1904 (1974).
6. Irimura and Osawa, *Arch. Biochem. Biophys.,* 151, 475 (1972).
7. Edelman, Cunningham, Reeke, Becker, Waxdal, and Wang, *Proc. Natl. Acad. Sci. U.S.A.,* 69, 2580 (1972).
8. Hardman and Ainsworth, *Biochemistry,* 11, 4910 (1972).
9. Hardman and Ainsworth, *Biochemistry,* 12, 4442 (1973).
10. Goldstein, Reichert, Misaki, and Gorin, *Biochim. Biophys. Acta,* 317, 500 (1973).
11. Akedo, Mori, Tanigaki, Shinkai, and Morita, *Biochim. Biophys. Acta,* 271, 378 (1972).
12. Surolia, Prakash, Bishayee, and Bachhawat, *Indian J. Biochem. Biophys.,* 10, 145 (1973).
13. Ersson, Aspberg, and Porath, *Biochim. Biophys. Acta,* 310, 446 (1973).
14. Matsumoto and Osawa, *Biochemistry,* 13, 582 (1974).
15. Pére, Font, and Bourrillon, *Biochim. Biophys. Acta,* 365, 40 (1974).
16. Carter and Etzler, *J. Biol. Chem.,* 250, 2756 (1975).
17. Gordon, Blumberg, Lis, and Sharon, *FEBS Lett.,* 24, 193 (1972).
18. Novogrodsky and Katchalski, *Proc. Natl. Acad. Sci. U.S.A.,* 70, 2515 (1973).
19. Lotan, Siegelman, Lis, and Sharon, *J. Biol. Chem.,* 249, 1219 (1974).
20. Kornfeld, Rogers, and Gregory, *J. Biol. Chem.,* 246, 6581 (1971).
21. Stein, Sage, and Leon, *Exp. Cell Res.,* 75, 475 (1972).
22. Fliegerova, Salvetova, Ticha, and Kocourek, *Biochim. Biophys. Acta,* 351, 416 (1974).
23. Blumberg, Hildesheim, Yariv, and Wilson, *Biochim. Biophys. Acta,* 264, 171 (1972).
24. Kawaguchi, Matsumoto, and Osawa, *J. Biol. Chem.,* 249, 2786 (1974).
25. Novakova and Kocourek, *Biochim. Biophys. Acta,* 359, 320 (1974).
26. Galbraith and Goldstein, *Biochemistry,* 11, 3976 (1972).
27. Reichert, Pan, Mathews, and Goldstein, *Nat. New Biol.,* 242, 146 (1973).
28. Ruddon, Weisenthal, Lundeen, Bessler, and Goldstein, *Proc. Natl. Acad. Sci. U.S.A.,* 71, 1848 (1974).
29. Andrews, *Biochemistry,* 139, 421 (1974).
30. Weber, Aro, and Nordman, *Biochim. Biophys. Acta,* 263, 94 (1972).
31. Kornfeld, Gregory, and Kornfeld, *Methods Enzymol.,* 28, 344 (1972).
32. Miller, Noyes, Heinrikson, Kingdon, and Yachnin, *J. Exp. Med.,* 138, 939 (1973).
33. Räsänen, Weber, and Gräsbeck, *Eur. J. Biochem.,* 38, 193 (1973).
34. Sela, Lis, Sharon, and Sachs, *Biochim. Biophys. Acta,* 310, 273 (1973).
35. Takahashi, Shimabayashi, Iwamoto, Izutsu, and Liener, *Agric. Biol. Chem.,* 35, 1274 (1971).
36. Reisfeld, Börjeson, Chessin, and Small, *Proc. Natl. Acad. Sci. U.S.A.,* 58, 2020 (1967).
37. Trowbridge, *Proc. Natl. Acad. Sci. U.S.A.,* 70, 3650 (1973).
38. Trowbridge, *J. Biol. Chem.,* 249, 6004 (1974).
39. Marik, Entlicher, and Kocourek, *Biochim. Biophys. Acta,* 336, 53 (1974).
40. Nicolson and Blaustein, *Biochim. Biophys. Acta,* 266, 543 (1972).
41. Lugnier and Dirheimer, *FEBS Lett.,* 35, 117 (1973).
42. Gürtler and Horstmann, *Biochim. Biophys. Acta,* 295, 582 (1973).
43. Nicolson, Blaustein, and Etzler, *Biochemistry,* 13, 196 (1974).
44. Podder, Surolia, and Bachhawat, *Eur. J. Biochem.,* 44, 151 (1974).
45. Leseney, Bourrillon, and Kornfeld, *Arch. Biochem. Biophys.,* 153, 831 (1972).
46. Schumann, Schnebli, and Dukor, *Int. Arch. Allergy Appl. Immunol.,* 45, 331 (1973).
47. Allen and Neuberger, *Biochem. J.,* 135, 307 (1973).
48. Terao and Osawa, *J. Biochem.,* 74, 199 (1973).
49. Poretz, Riss, Timberlake, and Chien, *Biochemistry,* 13, 250 (1974).
50. Allen, Neuberger, and Sharon, *Biochem. J.,* 131, 155 (1973).
51. Lotan, Gussin, Lis, and Sharon, *Biochem. Biophys. Res. Commun.,* 52, 656 (1973).
52. Lotan and Sharon, *Biochem. Biophys. Res. Commun.,* 55, 1340 (1973).
53. Nagata and Burger, *J. Biol. Chem.,* 249, 3116 (1974).
54. Rice and Etzler, *Biochem. Biophys. Res. Commun.,* 59, 414 (1974).
55. Privat, Delmonte, Mialonier, Bouchard, and Monsigny, *Eur. J. Biochem.,* 47, 5 (1974).
56. Horejsi and Kocourek, *Biochim. Biophys. Acta,* 336, 329 (1974).
57. Wang, Becker, Reeke, and Edelman, *J. Mol. Biol.,* 88, 259 (1974).

PROPERTIES OF PURIFIED LECTINS (continued)

58. Toyoshima, Akiyama, Nakano, Tonomura, and Osawa, *Biochemistry,* 10, 4457 (1971).
59. Toyoshima, Fukuda, and Osawa, *Biochemistry,* 11, 4000 (1972).
60. Gilboa-Garber, *Biochim. Biophys. Acta,* 273, 165 (1972).
61. Gilboa-Garber, Mizrahi, and Garber, *FEBS Lett.,* 28, 93 (1972).
62. Desai and Springer, *Methods Enzymol.,* 28, 383 (1972).
63. Acton, Bennett, Evans, and Schrohenloher, *J. Biol. Chem.,* 244, 4128 (1969).
64. Acton, Weinheimer, and Niedermeier, *Comp. Biochem. Physiol.,* 44B, 185 (1973).
65. Teichberg, Silman, Beitsch, and Resheff, *Prod. Natl. Acad. Sci. U.S.A.,* 72, 1383 (1975).
66. Hammarström, Westöö, and Björk, *Scand. J. Immunol.,* 1, 295 (1972).
67. Hammarström, *Methods Enzymol.,* 28, 368 (1972).
68. Hall and Rowlands, *Biochemistry,* 13, 821 (1974).
69. Finstad, Good, and Litman, *Ann. N.Y. Acad. Sci.,* 234, 170 (1974).
70. Oppenheimer, Nachbar, Salton, and Aull, *Biochem. Biophys. Res. Commun.,* 58, 1127 (1974).
71. Hudgin, Priecer, Ashwell, Stockert, and Morell, *J. Biol. Chem.,* 249, 5536 (1974).
72. Stockert, Morell, and Scheinberg, *Science,* 186, 365 (1974).

LIGAND BINDING TO PLASMA ALBUMIN

Colin F. Chignell

The data in this table have been taken from reviews by Goldstein[1] and Meyer and Guttman[2] and from a monograph by Steinhardt and Reynolds.[3] Supplemental data were obtained by searching MEDLINE (1970 to 1974) and CBAC (1968 to 1974) for references pertaining to ligand binding to plasma albumin. Only papers that describe attempts to make quantitative measurements on purified plasma albumin have been included. Many articles contained binding measurements made with a large number of structurally related ligands. In order to keep the length of this list as short as possible, only one representative or parent ligand has been listed with the number of related ligands appearing in parentheses.

Nomenclature proved to be a difficult problem to resolve, since many ligands, particularly those which are drug molecules, are usually known by their generic names rather than their chemical names. An attempt to standardize nomenclature has been made by employing the cross index of names which appears in *The Merck Index* (8th ed., Merck and Co. Inc., Rahway, N.J., 1968, pp. 1401–1710). Wherever possible, the preferred name appearing in this cross index has been used.

While every attempt has been made to compile a comprehensive list of all the compounds that bind to plasma albumin, omissions will have occurred. It is hoped that the users of this handbook will bring such omissions to the attention of the compiler so that they can be included in future editions.

In order to make it possible to locate a given ligand as quickly as possible, the list has been subdivided as follows:

A) Endogenous ligands
B) Exogenous ligands
    1) Miscellaneous (including fatty acids)
    2) Metal ions
    3) Inorganic anions
    4) Steroids (natural and synthetic)
    5) Visible dyes
    6) Drugs
        a) Antibiotics
            (i) Sulfonamides
            (ii) Penicillins
            (iii) Miscellaneous
        b) Barbiturates
        c) Analgesic (anti-inflammatory agents)
        d) Tranquilizers
        e) Diuretics
        f) Anticoagulants
        g) Hypoglycemic agents
        h) Radio-opaque agents
        i) Cardiac glycosides
        j) Miscellaneous

The abbreviations used to describe the techniques employed to measure binding are as follows

A	Autoradiography
Ad	Adsorption (including differential adsorption)
B	Biological assay
C	Column chromatography
Cal	Calorimetry
CD	Circular dichroism
Con	Conductivity
CP	Paper chromatography
D	Equilibrium dialysis
DC	Continuous flow dialysis
DF	Diffusion
E	Electrophoresis
EA	Enzyme assay
EG	Gel electrophoresis
EMF	EMF measurements
EP	Paper electrophoresis
F	Fluorescence
Fi	Filtration
Fp	Freezing point
G	Gel filtration
I	Immunological assay
L	Light scattering
NMR	Nuclear magnetic resonance
OP	Osmotic pressure
OR	Optical rotation or optical rotatory dispersion
P	Precipitation
PG	Polarography
PR	Pulse radiolysis
Pt	Partition
R	Refractive index
Ram	Raman spectroscopy
Res	Resin adsorption
S	Solubility
Sp	Spectrophotometric
SR	Sedimentation rate
St	Stabilization of ligand or protein
ST	Surface tension
T	Tubidometric analysis
Ti	Titration
UF	Ultrafiltration
UC	Ultracentrifugation
V	Viscosity
Vp	Vapor press
X	X-Ray crystallography

The abbreviations used for the sources of the different plasma albumins are as follows

A	Alligator		Go	Goose
B	Bovine		Gp	Guinea pig
Bb	Baboon		H	Human
C	Chicken		M	Mouse
Ca	Cat		P	Pig
D	Dog		R	Rabbit
E	Horse		Rt	Rat
F	Frog		S	Sheep
G	Goat			

Finally, it is a pleasure to acknowledge the help of Mr. R. H. Sik in compiling this list and to thank Mrs. D. M. Sherwood for typing the final manuscript.

LIGAND BINDING TO PLASMA ALBUMIN

Ligand	Source	Technique	Reference
Endogenous			
Acetate	B	D	5
	H	E	4
N-Acetyl-L-tryptophan	A, B, C, Ca, D, E, F, Go, H, P, R, S	D	11
	B	D	7, 12
	B, H	St, V	6
	H	D	8, 10
	H	D, Sp	9
L-Arabinose	D	R	13
Bilirubin	B	Sp, UF, D	17
	B	F	30
	B	Sp	35
	B	Sp, OR	39
	B, H	Sp	16
	B, H	UC	31
	B, H	F, UC, CD	32
	H	B	14
	H	Sp, EP	15
	H	UF, G	18
	H	D, Sp	19
	H	EA	20, 44
	H	–	22
	H	D	23
	H	G	24, 25, 26, 27, 33, 36
	H	UC, Sp, CD	37
	H	Sp	28, 46
	H	D, OR, V	29
	H	CD	40–43
	H	L, DC, UC, Sp	34
	H	D, F	478
	H, B	G	38
	H, R, G	Res	21
	H, B, R, P, S	F	45
Biliverdine	B	CD	47
β-Carotene	H	C	48
Cholic acid	H	G	49
Epinephrine	B	NMR	51
	H	D, G, UF	50
D-Fructose	B	R	13
D-Galactose			
L-Galactose	B	R	13
D-Glucose			
L-Glucose	B	R	13
Hemin	H	CD, F, Sp	52
Heparin	B	E	54, 55
	H	E, T, B	53
L-Histidine	H	D, Sp, Ti	420
Hyaluronic acid	B	E, UC	56
Indole	A, B, C, Ca, D, E, F, Go, H, P, R, S	D	11
Indole-3-propionate	A, B, C, Ca, D, E, F, Go, H, P, R, S	D	11
	H	D	10
meso-Inositol	B	R	13
L-Kynurenine	B	F	57
Lactose	B	R	13

LIGAND BINDING TO PLASMA ALBUMIN (continued)

Ligand	Source	Technique	Reference
Endogenous (continued)			
Lysolecithin	B	Cal	58
Maltose	B	R	13
Mannitol	B	R	13
Nicotinic acid	H, B, R, E, P, M, D, S, Gp	D, Sp	59
Norepinephrine	H	D, G, UF	50
Phosphatidylserine	H	T	60
Prostaglandin α_2	H	C, E	174
Prostaglandin E_2	H	C, E	174
	H	D, EG	475
Prostaglandin $F_{2\alpha}$	H	C, E	174
Protoporphyrin	H	G	61
Pyridoxal-5-phosphate	B	Sp	62
Retinoic acid	H	C	48
D-Ribose	B	R	13
Serotonin	H	—	63
Skatole	H	D	8, 10
	A, B, C, Ca, D, E, F, Go, H, P, R, S	D	11
	B	D	7
	H	D, Sp	9
L-Sorbose	B	R	13
L-Thyroxine	B, Rt	Sp	77
	C	EA	68
	H	D	23, 64–66, 76
	H	D, Sp	70
	H	E	72, 73
	H	F	69, 78
	H	G	71
	H	Sp	74, 75
L-Thyroxine (16)	H	D	67
Tocopherol	B	Pt	79
3,3′,5-Triiodothyronine	C	EA	68
	H, R, Gp, Rt	Res	80
D-Tryptophan	A, B, C, Ca, D, E, F, Go, H, P, R, S	D	11
	B	D	81
L-Tryptophan	A, B, C, Ca, D, E, F, Go, H, P, R, S	D	11
	B	D	84, 86
	H	CD	85
	H	D	8, 10, 81, 82
	H	D, Sp	9
	H	UF	87
	H, B	D	83, 88
	H, Rt	EG	212
Tryptamine	H	D	8, 10
	H	D, Sp	9
Tyramine	R	D	89
Uric acid	H	D	93
	H	D, F	391
	H	UF	90–92
Vitamin A	H	C	48
D-Xylose			
L-Xylose	B	R	13

LIGAND BINDING TO PLASMA ALBUMIN (continued)

Ligand	Source	Technique	Reference
Exogenous – Miscellaneous			
Acetophenone	B	D	94
Acetylene	B	OR	95
Acridyl benzoic acid	B	F	96
m-Aminobenzene sulfonic acid	B	D	97
o-Aminobenzene sulfonic acid	B	D	97
o-Aminobenzoic acid	B	D	97
m-Aminobenzoic acid	B	D	97
Aminonaphthalene disulfonic acid	B	D	96
Aminonaphthalene trisulfonic acid	B	D	96
1-Anilinonaphthalene-8-sulfonic acid	B	CD	102, 103
	B	F	100, 101, 105–111
	H	F	98
	H, B	F	99
	H, B, R, Rt, E	F	104
Anisole	B	D	94
2-Anthraquinone sulfonic acid	B	D	97
Anthraquinone	B	Pt	112
Benzenesulfonic acid	B	D	97
	B	Ti	113
Benzoic acid	B	Sp	114
	B, H	St, V	6
Benzonitrile	B	D	94
3,4-Benzpyrene	B	G	27
4-Bromoacetanilide	B	D	94
4-Bromoaniline	B	D, OR	94
Bromochlorotrifluoroethane	B	OR	95
2-(*p*-Bromophenyl)-1,3-indanedione	H	D	116
Butane	B	OR	95
	B	S	119
	H	S	117, 118
N-n-Butyl[5(-dimethylamino)-1-naphthalene]sulfonamide	B	F	120
n-Butyl sulfonic acid	H	NMR	121
Camphorquinone	B	D	94
Caprylic acid	B	D	81, 127
	B	V, Sp	124, 125
	B, H	St, V	6
	H	D	123
	H	UF, G	126
	H, B, R	D	122
9-(4'-Carboxyanilino)-6-chloro-2-methoxyacridine	B	F	128
	H, B, R, Rt, E	F	104
4-Chloroaniline	B	D, OR	94
o-Chlorobenzene sulfonic acid	B	D	97
Chloroform	B	OR	95
	H	S	129

LIGAND BINDING TO PLASMA ALBUMIN (continued)

Ligand	Source	Technique	Reference
Exogenous — Miscellaneous (continued)			
4-Chloronitrobenzene	B	D	94
p-Chlorophenol	H, B, R, E, P, M, D, S, Gp	D, Sp	59
4-Chlorophenoxy-acetic acid	H, B, R, E, P, M, D, S, Gp	Sp, D	59
Chlorophenoxy*iso*butyric acid	H	D	222
L-α-4-Chlorophenoxy propionic acid	H, B, R, E, P, M, D, S, Gp	Sp, D	59
α-(4-Chlorophenoxy)-α-methylpropionic acid	B	D, PG, UF	130
2-(*p*-Chlorophenyl)-1,3-indanedione	H	D	116
Chromone-2-carboxylic acid	H	D	472
Cinnamic acid	B	D	131
Coumarin	H	D	132
Cyclopropane	H	S	129
Dansyl-L-glutamic acid	B	F	477
Dansylglycine	B	F	477
	H	F	133–136
Dansyl-L-proline	B	F	477
Dansyl-DL-tryptophan	B	F	477
Decanoic acid	B	D, Sp, V	124
n-Decyl alcohol	B	D, S	137
	B	D, V, Sp	124
Decylsulfonic acid	B	D	140
	B	Sp	139
	B	Sp, D, V, OR	138
	H, B	D, F, V, OR	141
	H, B	Sp, F	142
Dextran	H	E	143
2,5-Dichlorobenzene-sulfonic acid	B	D	97
2,4-Dichloronitrobenzene	B	D	94
2,4-Dichlorophenol	B	D, Sp	145
	H	UC, G	144
2,4-Dichlorophenoxy-acetic acid	B	G	146
5-Dimethylaminonaphthalene-1-sulfonamide	H	F	147–149
1-Dimethylaminonaphthalene-5-sulfonic acid	B	F	109, 110
2,4-Dinitrobenzene-sulfonic acid	B	D	97
	B	Ti	113
3,5-Dinitrobenzoic acid	B, H	St, V	6
2,4-Dinitronaphthol	B	D	97
	B	Ti	113
2,4-Dinitrophenol	B	D	131
Dioxan	B	R	13
1-Dodecanol	B	D, S	137
p-Dodecylbenzene-sulfonic acid	B	Sp	150, 151
	B	D, Sp	137

LIGAND BINDING TO PLASMA ALBUMIN (continued)

Ligand	Source	Technique	Reference
Exogenous — Miscellaneous (continued)			
Dodecyl sulfate	B	CD, OR, Sp, UC, D, Ti, V	163
	B	D	140
	B	D, Sp	137, 138
	B	E	155
	B	EMF	156
	B	OR	157–159
	B	Sp	139
	B	Sp, D	138
	B	Ti	113, 161
	E	E, V	152–154
	H	F	162
	H	NMR	121
	H	OR	160
	H, B	D, F, V, OR	141
	H, B	Sp, F	142
	H, B	St	6
Dodecylsulfonic acid	B	Sp	139
	B	Sp, V, D, OR	138
	H, B	D, OR, V, F	141
Ethane	B	OR	95
Ethyl chloride	B	OR	95
Ethyl ether	B	OR	95
	H	S	129
Ethylene	H	S	129
Fatty acids (C_1–C_{12})	H, B	UF, B	164
N-(2-Fluorenyl) acetamide	B, Rt	CP, E	165
3-Fluoroaniline	B	D	94
Gentisic acid	B	Sp	114
Halothane	B	OR	95
Heptanoic acid	H	NMR	121
n-Heptylsulfonic acid	H	NMR	121
n-Hexanoic acid	H	D	123
1-Hexanol	B	D	94
	B	Sp	166, 167
n-Hexyl sulfate	B	Ti	113, 161
	H, B	Sp, F	142
Hexylsulfonic acid	B	Sp	139
Hippuric acid	B	D	131
1-Hydroxyadamantane	B	D, OR	94
o-Hydroxybenzoic acid	B	D	97
p-Hydroxybenzoic acid	B	Sp	114
p-Hydroxybenzoic acid methyl ester (4)	B	F	120
4-Hydroxycoumarin	H	CD	168
	H	D	132
7-Hydroxycoumarin	H	D	132
Indole	B	D	94
Lauric acid	B	D, Sp, V	124
	H	Pt	169
	H, B	D, OR, V, F	141
	H, B	Sp, F	142
Linoleic acid	H	Pt	169
	H, B	D	170
Methane	B	OR	95
Methyl chloride	B	OR	95
4-Methoxyaniline	B	D	94

LIGAND BINDING TO PLASMA ALBUMIN (continued)

Ligand	Source	Technique	Reference
Exogenous — Miscellaneous (continued)			
4-Methoxybenzyl alcohol	B	D	94
Methoxyflurane	B	OR	95
2-(p-Methoxyphenyl)-1,3-indandione	H	D	116
4-Methylaniline	B	D	94
4-Methyl-7-diethyl aminocoumarin	H	F	162
Methyl ethyl ketone	B	R	13
$unsymm$-Methylphenyl-urea	B	D	94
Myristamidopropyldimethylbenzyl-ammonium chloride	H	P	171
Myristic acid	H	Pt, Sp	169
Myristyl sulfate	B	D, V, Sp, OR	138
	B	Sp	139
	B	Ti	113
	H, B	D, F, V, OR	141
	H, B	Sp, F	142
Naphthalene	B	D	94
Naphthalene-1-sulfonic acid	B	D	97
Naphthalene-2-sulfonic acid	B	D	97
1-Naphthylamine	B	D, OR	94
Neopentyl alcohol	B	D	94
4-Nitroanisole	B	D	94
Nitrobenzene	B	D	94
p-Nitrobenzene sulfinic acid	B	D	172
p-Nitrobenzene sulfonic acid	B	D	172
m-Nitrobenzoic acid	B, H	St, V	6
Nitrobenzoic acid	B	D	97
3-Nitrobenzonitrile	B	D	94
m-Nitrophenol	B	D	131
	B	D, Sp	145
o-Nitrophenol	B	D	131
	B	D, Sp	145
p-Nitrophenol	B	D	131
	B	D, Sp	145
p-Nitrophenylmethyl-sulfide	B	D	172
p-Nitrophenylmethyl-sulfone	B	D	172
Nitrous oxide	B	OR	95
	H	S	129
2-Nonanone	B	D	94
Ochratoxin A	B	D	173
iso-Octane	B	S	174
1-Octanol	B	D, S	137
	B	E	175
p-Octylbenzenesulfonic acid	B	Ti	151

LIGAND BINDING TO PLASMA ALBUMIN (continued)

Ligand	Source	Technique	Reference
Exogenous — Miscellaneous (continued)			
Octylsulfate	B	D	140
	B	D, Sp	137
	B	Sp	139
	B	Ti	113, 161
	B	V, Sp, D, OR	138
	H	NMR	121
	H, B	Sp, F	142
Octylsulfonic acid	B	D, Sp	137
	B	Sp	139
	B	V, D, Sp, OR	138
	H	NMR	121
Oleic acid	H	Pt	169
	H, B	D	170
	H, B, R	D	122
Palmitic acid	B	E	176
	B	UF, G	126
	H	G	33
	H	Pt	169, 177
	H, B	D	170
	H, B, R	D	122
n-Pentane	B	S	178
	H	S	117, 118
p-$tert$-Pentylphenol	H	UC, G	144
Phenol (19)	D	D, OR	94
Phenol	B	R	13
	H	UC, G	144
Phenoxyacetic acid	B	D	131
α-Phenoxy-α-methyl-propionic acid	H, B, R, E, P, M, D, S, Gp	D, Sp	59
Phenylacetic acid	B	D	131
Phenylbutyric acid	B	D	131
Phenylethyl carbamate	B	D	94
2-Phenyl-1,3-indandione	H	D	116
Phthalic acid	B	D	97
Phytic acid	H	P, Sp	179, 180
Picric acid	B	D	97, 131, 181
	B	D, Sp	145
	B	Ti	113
n-Propane	B	OR	95
	H	S	117, 118
γ-Resorcylic acid	B	Sp	114
Stearic acid	H	Pt	169
	H, B	D	170
Toluene	B	Sp	166
2-p-Toluidinylnaph-thalene-6-sulfonic acid	B	F	182
Thymol	B	D, OR	94
Trichloroacetic acid	B	L	185
	H	EMF	183, 184
	H, B	St	6
2,4,5-Trichlorophe-noxyacetic acid	B	G	146
2,4,6-Trinitro-m-cresolate	B	D	97
Triton X-100	B	D	186
Vinyl ether	B	OR	95
Xenon	B	OR	95
	H	S	129

LIGAND BINDING TO PLASMA ALBUMIN (continued)

Ligand	Source	Technique	Reference

Exogenous — Metal Ions

Ligand	Source	Technique	Reference
Ca^{++}	H	Ti, D	187
Cd^{++}	B	NMR	191
	B	P	188
	B	PG	189, 190
Cu^{++}	B	Cal	197
	B	D	193
	B	D, Sp	194, 195
	B	Ti, Sp	192
	H	D, Sp, Ti	420
	H, D	Sp	196
Hg^{++}	H	NMR	191
	H	P, L, UC	198
Mn^{++}	B	NMR	199, 200
Ni^{++}	H, D, P, R, Rt	D	362
Pb^{++}	H	D	201
Tc^{++++}	H	D, G	387
Zn^{++}	B	NMR	191
	E	St, E	203
	H	D, Sp	202

Exogenous — Inorganic Anions

Ligand	Source	Technique	Reference
Bromide	H	NMR	204
Chloride	B	EMF	209
	B	NMR	191
	B, H	OP	205, 206
	E	Con	203
	H	D, Con	207, 208
	H	D, EMF	183
Ferrocyanide	H	E	210
Iodide	B	D	5
	H	EMF	183, 208
Phosphate	H	D	211
	H	E	4
Pyrophosphate	H	D	211
Thiocyanate	B	D	213
	H	D	206
	H	EMF	183, 184, 208

Exogenous — Steroids

Ligand	Source	Technique	Reference
Aldosterone-18-glucuronide	H	D	213
Androsterone	B	Pt	112
5α-Androstane-3α,17β-diol	H	D	214
Androstanedione	B	Pt	112
Androstane-17-one	B	Pt	112
Δ⁴-Androstene-3-one	B	Pt	112
Androstenedione	B	Pt	112
Androsterone sulfate	H	D, UF	215
Corticosterone	H	D	216
Cortisone	B	Pt	112
	H	D	216
Dehydroepiandrosterone	B	Pt	112
	H	E	217

LIGAND BINDING TO PLASMA ALBUMIN (continued)

Ligand	Source	Technique	Reference
Exogenous – Steroids (continued)			
Dehydroepiandrosterone sulfate	H	D, UF	215
Deoxycorticosterone	B	Pt	112
	H	Sp	218
Estradiol	B	Pt	112
	S	D	257
Estriol	S	D	257
Estrone sulfate	H	UF	219
Etiocholanolone	B	Pt	112
Etiocholanone sulfate	H	D, UF	215
Hydrocortisone	B, H	D	220
	H	D	216, 222, 223
	H	D, G	221
	H	Sp	218, 225
	H	Sp, D	224
Hydrocortisone-21-succinate (18)	B	D	225
Methyltestosterone	B	Pt	112
Prednisolone	H	D	216
Prednisone	H	D	216
Progesterone	H	D	226
	H	Sp	218
Testosterone (14)	B	F	228
Testosterone	B	D, Pt	230
	B	S	112, 231
	B, H	G	227
	H	D, G	229
	H	G	232
	H	Sp	218, 225
	H	Sp, D	224
Tetrahydroaldosterone glucuronide	H	D	213
Triamcinolone	H	D	223
Exogenous – Visible Dyes			
Alizarin red	B	D	97
Amaranth	B	D	97
	B	Sp	233
Amidoschwartz-10B	H	CP	235
4-Aminoazobenzene	H	Sp	235
	Rt	Sp	236
p-Aminoazobenzene-p'-sulfonic acid	B	D	97
1-Amino-2-naphthol-4-sulfonic acid	B	D	97
Azobenzene	B	D	94
	H	Sp	235
Azocarmine-B	H	Sp	235
Azorubin	H	C, Sp	237
Bromchlorophenyl blue	B, H	Sp, F	238
Bromcresol green	B	D, UC, Sp	244
	B	E	239
	B	Sp	243
	–	D	242
	H	D, Sp	19
	H	D, Sp, E	241
	H	Sp	235, 240
	H, B	Sp, F	238

LIGAND BINDING TO PLASMA ALBUMIN (continued)

Ligand	Source	Technique	Reference
Exogenous – Visible Dyes (continued)			
Bromcresol purple	B	D	97
	B, H	Sp, F	238
Bromphenol blue	B	D	97
	B	D, Sp	246
	B	E	239, 245
	B	Sp	250
	B, H	Sp, F	238
	H	CP	249
	H	E	247
	H	E, Sp	248
	H	Sp	235
Bromthymol blue	–	Sp	251
Chlorophenyl red	H, B	Sp, F	238
Congo red	–	Ad	254
	–	Ad, Fp	252
	B	D	255
	B	Sp	253
	–	DF	253
	D	Sp	256
Cresol red	H, B	Sp, F	238
4,4-Dibenzoylamido-stilbene-2,2'-disulfonic acid	H	F	479
p-Diazobenzene sulfonic acid	B	Sp	258
Dihydroxyazobenzene-p-sulfonate	–	D	259
4-Diethylaminoazobenzene	Rt	Sp	236
4-Dimethylaminoazobenzene	H	Sp	235
	Rt	Sp	236
2,4-Dihydroxyazobenzene-4'-sulfonate	B	D	97
p-Dimethylaminobenzeneazobenzoylacetate	B	D, Sp	260
m-Dimethylaminoazobenzene-p-carboxylate	B, H	D, Sp	261
o-Dimethylaminoazobenzene-p-carboxylate	B, H	D, Sp	261
p-Dimethylaminoazobenzene-p-carboxylate	B, H	D, Sp	261
Dinitro-1-naphthol-7-sulfonate	B	D	97
Eosine yellowish	B	D, UC, Sp	244
	B	F	96
Evans blue	R	PG	262
Flavinate	–	D	259
	B	D	181
Fluorescein	B	D	265
	–	D	263
	B	F	96
Gurr's lissamine green	B	Sp	243
Gurr's light green	B	Sp	243
4-Hydroxyazobenzene	H	Sp	235
Hydroxyazobenzene-p-sulfonic acid	–	D	259
p-Hydroxyazobenzene p'-sulfonate	B	D	97

LIGAND BINDING TO PLASMA ALBUMIN (continued)

Ligand	Source	Technique	Reference
Exogenous – Visible Dyes (continued)			
2-(4'-Hydroxybenzene azo)-benzoic acid	B	Sp	271, 272
	B	UF, D, PG	130
	B, Rt	Sp	77, 115
	H	D	267
	H	D, Sp	266
	H	Sp	268, 269
	H, B, R, E, P, M, D, S, Gp	Sp	59
	P, E, S, C, B, D	Sp	273
	Rt, R, E, P, H, B, D	D, Sp	270
p(2-Hydroxy-5-methyl-phenylazo)benzoic acid	B	D, Sp	274, 275
Indigo	B	D	181
Indigo disulfonate	B	D	97
Indigo trisulfonate	B	D	97
	–	D	259
Malachite green	H	–	276
Metamethyl red	B, H	D, Sp	276
Metanil yellow	–	D	259
(2-Methoxy-6-chloro-9-acridinyl)-N-p-aminobenzenearsonic acid	B	F, D	277
Methylene blue	–	D, P, SR, V	278
	–	–	279
	B	Sp	250
	H, E	Sp	280
2-Methyl-4-dimethyl-aminoazobenzene	Rt	Sp	236
3'-Methyl-4-dimethyl-aminoazobenzene	R	E, Sp, EP	281
	Rt	Sp	236
Methyl orange	B	D	283, 286, 287, 294, 298–300
	B	D, Sp	145
	B	G	296
	B	P	282
	B	Ram	295
	B	Sp	292
	B	Sp, Ad	233
	B	Sp, D	291
	B	UC	284, 293
	B, H	D	288
	–	D, E, Sp	289
	B, H	D, Sp, OR	290
	B, H	E	276
	–	D, P, SR, V	278
	H	D, Sp	19
	H	G	297
	H	Sp	235
	H	UF	90
	H, E	Sp	280
	H, E, B, C, R	Sp, D, E	285
Methyl red	B	Sp, E	301
	B, H	D	288
	B, H	D, Sp	276

LIGAND BINDING TO PLASMA ALBUMIN (continued)

Ligand	Source	Technique	Reference
Exogenous — Visible Dyes (continued)			
Naphthalene-1-5-disulfonate	B	D	97
Naphthalein black	H	Sp	235
1-Naphthol-4-[4-(4'-azobenzene azo) phenyl] arsonic acid	B	Sp	302
2-Naphthol-7-sulfonic acid	B	D	97
1-Naphthol-2-sulfonic acid-4-[4-(4'-azobenzene azo) phenyl] arsonic acid	B	Sp	302
Neutral red	H, E	Sp	280
Niagara sky blue	H, D	SR, Sp, E, D, UC	303
Niagara sky blue 6B	H, D	SR, Sp, E, D, UC	303
Orange II	B	D	97, 181
	—	D	259
	B	Sp	292
Para methyl red	B, H	D	246
Phenolphthalein	H, B, R	D	122
Phenolsulfonphthalein	B	D	97, 299, 300
	B	UC	312
	B	UF	310
	B	UF, Sp	313
	B, H	Sp, F	238
	H	CD	304
	H	D	268, 305, 316
	—	D	259
	H	D, Sp	19
	H	G, Sp	311
	H	Ti, I	315
	H	UF	317
	—	UF	318
	H	UF, Sp	314
	H	UF, Sp, Con	306–309
	H, E	Sp	280
Phenyl-*p*-dimethyl-aminobenzeneazobenzoylacetate	B	D, Sp	260
Ponceau red	B	D	97
Rosaniline	B	Sp	250
Rose bengal	D	UF, ST	264
Sulfobromophthalein	B	D	322
	H	D	323
	H	D, Sp	19
	H	EP	319
	H	UF	321, 324
	H, B	D, Sp	320
Tartrazine	B	D	97
Thymol blue	B	D	97
Toluidine blue	B	Sp	250
	H, E	Sp	280
Tropeoline	B	D	97, 181

LIGAND BINDING TO PLASMA ALBUMIN (continued)

Ligand	Source	Technique	Reference
Exogenous — Visible Dyes (continued)			
Trypan blue	H, D	SR, Sp, E, D, UC	303
	R, Gp, Bb	Sp	325
	B	Sp	480
Vital red	B	Sp	326
	H	Sp	327
Exogenous — Drugs			
Antibiotics			
Sulfonamides			
Azosulfathiazole	B	D	291
	B	Sp	292
5-Ethylsulfadiazine	H	UF	329
Sulfacetamide	B	NMR	328, 331
Sulfacetamide (3)	H	D	330
Sulfadiazine	B	D	322
	B, H, R, D, Rt, M	D	334
Sulfadiazine (4)	H	D	336
Sulfadiazine (6)	H	D	333
Sulfadiazine (21)	H	D	332
Sulfadiazine (25)	H	D	335
Sulfadiazine (6)	H	UF, G	337
Sulfadiazine	H, B, R	D	122
Sulfadimethoxine (10)	B	F	105
Sulfadimethoxine	B, H, R, D, Rt, M	D	334
	H	F	149
Sulfaethidole	B	CD	340
	B	CD, D	342
	B	D	338
	B, R	D	339
	H	UC, G, D	341
Sulfaguanidine	H	F	147
Sulfamerazine	B	NMR	328
Sulfamerazine (12)	H	D	343
Sulfameter	B, H, R, D, Rt, M	D	334
	H	D	343
	H	UC, G, D	341
Sulfamethizole	H	—	344
Sulfamethoxazole	B	NMR	331
Sulfamethoxypyridiazine	H	F	147
Sulfamethoxypyridiazine	B	D, G	346
	B	UF	310
Sulfamethoxypyridiazine (17)	B	St, D	345
Sulfanilamide	B	D, G	346
5-Sulfanilamido-1-ethyl-pyrazole	H	D	347
Sulfanilothiocarbamide	H	UC, G, D	341
Sulfanilycytosine (9)	H	D	348
Sulfaphenazole	H	D	349
	H	NMR	331
Sulfapyridine	B	D, G	346

LIGAND BINDING TO PLASMA ALBUMIN (continued)

Ligand	Source	Technique	Reference
Exogenous – Drugs (continued)			
Sulfapyrimidine (14)	H	OR	350
Sulfasomizole (2)	H	CD, D	351
Sulfathiazole (4)	B	D	356
Sulfathiazole (5)	B	D	353
Sulfathiazole	B	St	354
	H	D	352, 355
	H	F	147
	H	NMR	331
Sulfathiourea	H	UC, G, D	341
Penicillins			
α-Aminophenylpenicillin	H	D	357
Benzylpenicillin (5)	B	D, S	360
Benzylpenicillin (6)	B	F	105
Benzylpenicillin	B	NMR	328, 364
	Gp, H, E	B	363
	H	B	366
	H	D	365
Benzylpenicillin (8)	H	D, B	359
Benzylpenicillin (10)	H	PR	367, 368
Benzylpenicillin	H	UF	361
Benzylpenicillin (3)	H, B	B	358
Methicillin	H	D	357
Phenbenicillin	B	D	310
	H	D	357
Phenethicillin	H	D	357
Phenylthiodiphenylpenicillin	H	D	357
Miscellaneous			
Cephalosporin (3)	B	PR	367, 368
Chloramphenicol	B	D	369
	H	D, B	359
Chlortetracycline	Rt	E	370
Dapsone	H, M	D	371
Monoacetyldapsone	H	D	371
Novobiocin	E, R, D, S, P, Rt, H, B	CD	373
	H	CD, UF	372
	Rt	D	221
Oxytetracycline	Rt	E	370
Streptomycin	B	D	374
Streptomycin (3)	H	D, B	359
Tetracycline	B	F	376
	H	D, G	375
	Rt	E	370
Tetracycline (6)	H	D, B	359
Barbiturates			
Amobarbital	H	D	377
Ethylamylbarbituric acid	B	D	94
Mephobarbital	H	UF	378
Pentobarbital	H	D	377

LIGAND BINDING TO PLASMA ALBUMIN (continued)

Ligand	Source	Technique	Reference
Exogenous – Drugs (continued)			
Phenobarbital (4)	B	UF	379
Phenobarbital	B, H	D	380
Phenobarbital (6)	H	CD	381
Phenobarbital	H	D	377
Secobarbital	H	UF	361
Thiopental	B	D	322, 382
	H, B	UF	383
Analgesic			
Acetylsalicylic acid	B, H	St, V	6
	H	CD	381
	H	G	384, 385
	H	NMR	386
Flufenamic acid	B	Sp	114
	H	CD, D	135
	H	CD, D, F	134
Indomethacin	B	Sp	114
	H	CD	168, 381
	H	D	388
Indomethacin amide	B	Sp	114
Meclofenamic acid	H	CD	304
	H	CD, D, F	134
Mefenamic acid	B	Sp	114
	H	CD	304
	H	CD, D, F	134
N-Methylanthranilic acid	B	Sp	114
Oxyphenbutazone	B	Sp	114
	H	CD	304, 389
	H	D, CD	133
Phenylbutazone	B	D	322
	B	F	390
	B	Sp	114
	B, H	CD	304
Phenylbutazone (2)	H	CD	381
Phenylbutazone	H	CD	389
	H	CD, D	135
	H	D	222, 392, 393
	H	D, F	391
	H	F	149
	H, B, P, E, R	D, CD, Sp, F	133
	H, B, R	D	122
Salicylic acid	B	D	299, 300, 322, 397–399
	B	Sp	114
	B, H	D, UF, G	395
	B, H	St, V	6
	H	CD	168, 381
	H	D	396
	H	D, C	394
	H	D, F	391
	H	G	384, 385
	H	UF	90, 361, 400
	H, B	D	88, 170
	H, B, R	D	122

LIGAND BINDING TO PLASMA ALBUMIN (continued)

Ligand	Source	Technique	Reference
Exogenous – Drugs (continued)			

Tranquilizers

Ligand	Source	Technique	Reference
Amitriptyline	B	G, D	401
Chlorimipramine	B	G, D	401
3-Chlorodesmethyl-imipramine	H, B	D	402
Chlorpromazine (12)	B	D	407
Chlorpromazine	B	G	403, 404, 406
Chlorpromazine (12)	B, H	Sp	405
Chlorpromazine (3)	B, H, D, R, Rt, P, E, S, G, C	F, Sp	454
Chlorpromazine (12)	H	CD	381
Chlorpromazine	H	D	234
	H	G, Sp	408
Clonazepam	H	CD	409
Desipramine	B	D, UC, Sp	244
	B	G, D	401
	H, B	D	402
Desmethylchlorpromazine	B	D, G	410
Desmethylpromazine	B	D, G	410
Diazepam (3)	H	CD	381
Diazepam	H	CD	409
	H	G	411
Diethazine	H	G, Sp	408
Flurazepam	H	CD	409
Imipramine	B	D, UC, Sp	244
	B	G, D	401
	H, B	D	402
Imipramine (8)	H	CD	381
Nitrazepam	H	G	411
Nortriptyline	H	D	344
Oxazepam	H	G	411
Perazine	B	D, G	410
Phenothiazone	B	Sp, D	412
Promazine	B	G	403, 404, 413
Promethazine	B	G	410
	H	G, Sp	408
Prothipendyl	B	G, D	401
Thional	B	Sp, D	412
Thiopropazate	H	G, Sp	408
Trimeprazine	B	D, G	410

Diuretics

Ligand	Source	Technique	Reference
Bendroflumethiazide	H	D	414
Chlorthiazide	B	D	416
	H	D, UF, CD	415
Hydroflumethiazide	H	D	414

Anticoagulants

Ligand	Source	Technique	Reference
Acenocoumarin	H	D	132, 222, 393
Acenocoumarol	H	D	216
Anisindione	H	D	216
Diphenadione	H	D	216

LIGAND BINDING TO PLASMA ALBUMIN (continued)

Ligand	Source	Technique	Reference
Exogenous − Drugs (continued)			
Bishydroxycoumarin	B	CD	418
	B	D	322
	H	D	132
	H	D, F, CD	136
	H	D, S, Sp	417
	H, B, R	D	122
Bishydroxycoumarin (6)	H, D	D	419
Ethyl biscoumacetate	H	D	132, 216, 393, 421
Phenindione	H	D	393
Phenprocoumarol	H	D	421
Phenprocoumon	H	D	216
Warfarin	B	D	299, 300
	B	F	390
	H	B	423
	H	Cal	425
	H	CD	168
	H	D	132, 222, 424−426
	H	F	149, 136
Warfarin (3)	H	D	422
Warfarin (6)	H, D	D	419
Hypoglycemic agents			
Acetohexamide	H	D	216
	H, B, R, Rt, E	F	104
Carbutamide	B, H, D, E, P, R	D, UF	427
Chlorpropamide	H	D	216, 428
	H, B, R, Rt, E	F	104
Chlorpromide	B, H, D, E, P, R	D, UF	427
Glipizide	H, B, R, Rt, E	F	104
Glyburide	H, B, R, Rt, E	F	104
Metahexamide	B, H, D, E, P, R	D, UF	427
Tolbutamide	B, H, D, E, P, R	D, UF	427
	H	D	216, 428
	H	F	147
	H, B	UC	429
	H, B, R, Rt, E	F	104
	Rt	D	221
Radio-opaque agents			
Acetrizoate	H	D	440
Iodipamide	−	Fi	441
o-Iodohippuric acid	H	UF	317
Iodopyracet	H	UF	317
Iopanoic acid	H	F	98
Iophenoxic acid	H	F	98
Methiodal	H	UF	317
Cardiac glycosides			
Convallatoxin	H	UC	459
Digitoxigenin	H	G	430

LIGAND BINDING TO PLASMA ALBUMIN (continued)

Ligand	Source	Technique	Reference
\multicolumn{4}{c}{Exogenous – Drugs (continued)}			
Digitoxin	B	D	431
	H	D, P	432, 433
	H	G	430, 434
	H	UC	459
	H	UF	435
Digoxin	H	D, G, E	436
	H	G	430
	H	UC	459
Gitaloxin	H	UC	459
Ouabagenin	B	D	431
Ouabain	H	G	430
	H	UC	459
Quinidine	H	D	437–439
Strophanthin	B	D	431
	H	D, P	432, 433

Miscellaneous

Ligand	Source	Technique	Reference
Acetanilide (14)	B	D	442
Acetophenetidin	B	D	443
Aflatoxin B1	H	D	444
	H	G	456
Aminopyrine	B	Sp	114
Aminopyrine (2)	H	CD	381
p-Aminosalicylic acid	B	D	445
	B	UF	446
Antipyrine	B	Sp	114
Antipyrine (7)	B	D	443
Aptin	H	D	344
Atropine	B	B, UF, Sp	447
	B	U, D	448
Caffeine	B	D	449
	B, H	Sp	450
Caffeine (2)	B	D	299, 300
Camptothecin	H	F	451
Carbenoxolone	H	UF, F	452
Chloranil	E	Sp	453
Chloroquine	B	Sp	114, 434
Cinchophen	B	Sp	114
Clofibrate	B, H, Rt	Sp, D	77
Cycloserine	H	D, B	359
Diazoxide	H	D	455
Diethylstilbestrol	S	D	257
Diphenylhydantoin	B	D	322
	H	CD	381
	H	D	457
	H	D, UF	460
	H	E	72
	H	UF	458
	H, B, R	D	122
Ethacridine	B	E, D, P	461
Ethacrynic acid	B	D, EP	462
Fenoprofen	H	CD	463
Hexamethonium	H	Pt	464
7-Hydroxy-Δ-1-tetra-cannabinol	H	D, UF	465

LIGAND BINDING TO PLASMA ALBUMIN (continued)

Ligand	Source	Technique	Reference
Exogenous – Drugs (continued)			
Hyoscyamine	H	D	466
Methadone	H	D	467, 468
Methotrexate	B	UF	469
N-Methylsaccharin	H	D	414
Metronidazole	H, B	D	470
Niflumic acid	H	D	471
Practolol	H	D	473
Probenecid	H	F	147, 148
Procaine (3)	H	D	474
Propranolol	H	D	473
Saccharin	H	D	414
Sotalol	H	D	473
Succinyldicholine	H	D	476
Succinylmonocholine	H	D	476
Sulfinpyrazone	B	Sp	114
	H	CD	133, 304
Theophylline (12)	B	D	449
	B, H	Sp	450
Tripelennamine	H	D	344

Compiled by Colin F. Chignell.

REFERENCES

1. Goldstein, *Pharmacol. Rev.*, 1, 102 (1949).
2. Meyer and Guttman, *J. Pharm. Sci.*, 57, 895 (1968).
3. Steinhardt and Reynolds, *Multiple Equilibria in Proteins*, Academic Press, New York, 1969.
4. Ballou, Boyer, and Luck, *J. Biol. Chem.*, 159, 111 (1945).
5. Saifer, Westley, and Stergman, *Biochemistry*, 3, 1624 (1964).
6. Duggan and Luck, *J. Biol. Chem.*, 172, 205 (1948).
7. McMenamy and Krasner, *J. Biol. Chem.*, 241, 4186 (1966).
8. McMenamy, *J. Biol. Chem.*, 239, 2835 (1964).
9. McMenamy, *J. Biol. Chem.*, 240, 4235 (1965).
10. McMenamy, *Arch. Biochem. Biophys.*, 103, 409 (1963).
11. McMenamy and Watson, *Comp. Biochem. Physiol.*, 26, 329 (1968).
12. Krasner and McMenamy, *J. Biol. Chem.*, 241, 4186 (1966).
13. Giles and McKay, *J. Biol. Chem.*, 237, 3388 (1962).
14. Manken, Waggoner, and Berlin, *J. Neurochem.*, 13, 1241 (1966).
15. Otsuki, *Okayama Igakkai Zasshi*, 71, 7155 (1959).
16. Barac, *Arch. Int. Physiol. Biochim.*, 61, 129 (1953).
17. Odell, *J. Clin. Invest.*, 38, 823 (1959).
18. Josephson and Furst, *Scand. J. Clin. Lab. Invest.*, 18, 51 (1966).
19. Watson, *Clin. Chim. Acta*, 15, 121 (1967).
20. Jacobsen, *Eur. J. Biochem.*, 27, 513 (1972).
21. Schmid, Diamond, Hammaker, and Gundersen, *Nature*, 206, 1041 (1965).
22. Tilstone and Khan, *Clin. Biochem.*, 6, 5 (1973).
23. Tabachnick, Downs, and Giorgio, *Proc. Soc. Exp. Biol. Med.*, 118, 1180 (1965).
24. Liem and Muller-Eberhard, *Biochem. Biophys. Res. Commun.*, 42, 634 (1971).
25. Bradlid and Fog, *Scand. J. Clin. Lab Invest.*, 25, 257 (1970).
26. Stephan, Welin, Mainard, and Amory, *C.R. Acad. Sci. Ser. D*, 273, 2009 (1971).
27. Anghileri, *Biochim. Biophys. Acta*, 136, 386 (1967).
28. Starinsky and Shafrir, *Clin. Chim. Acta*, 29, 311 (1970).
29. Scholtan and Gloxhuber, *Arzneim. Forsch.*, 16, 520 (1966).
30. Krasner, *Biochem. Med.*, 7, 135 (1973).

31. Lee and Cowger, *Res. Commun. Chem. Pathol. Pharmacol.*, 8, 327 (1974).
32. Lee and Cowger, *Res. Commun. Chem. Pathol. Pharmacol.*, 6, 621 (1973).
33. Svenson, Holmer, and Andersson, *Biochim. Biophys. Acta,* 342, 54 (1974).
34. Brodersen, Funding, Pederson, and Roigaard-Petersen, *Scand. J. Clin. Lab. Invest.,* 29, 433 (1972).
35. Wennberg and Cowger, *Clin. Chim. Acta,* 43, 55 (1973).
36. Tuilie and Larinois, *Biol. Neonatorum,* 21, 447 (1972).
37. Woolley and Hunter, *Arch. Biochem. Biophys.,* 140, 197 (1970).
38. Kalisker and Dixon, *Proc. West. Pharmacol. Soc.,* 12, 65 (1969).
39. Blauer and King, *J. Biol. Chem.,* 245, 372 (1970).
40. Blauer, Harmatz, and Snir, *Biochim. Biophys. Acta,* 278, 68 (1972).
41. Blauer, Harmatz, and Snir, *Biochim. Biophys. Acta,* 278, 89 (1972).
42. Blauer, Blondheim, Harmatz, Kapitulnik, Kaufmann, and Zulichevsky, *FEBS Lett.,* 33, 320 (1973).
43. Blauer, Harmatz, and Noparstek, *FEBS Lett.,* 9, 53 (1970).
44. Jacobsen, *FEBS Lett.,* 5, 112 (1969).
45. Chen, *Arch. Biochem. Biophys.,* 160, 106 (1972).
46. Coutinho, Lucek, Cheripko, and Kuntzman, *Ann. N.Y. Acad. Sci.,* 226, 238 (1973).
47. Lee and Cowger, *Res. Commun. Chem. Pathol. Pharmacol.,* 5, 505 (1973).
48. Moffa and Krause, *Proc. Soc. Exp. Biol. Med.,* 134, 406 (1970).
49. Burke, Lewis, Panveliwalla, and Tabaqchali, *Clin. Chim. Acta,* 32, 207 (1971).
50. Danon and Sapira, *J. Pharmacol. Exp. Ther.,* 182, 295 (1972).
51. Zia, Cox, and Luzzi, *J. Pharm. Sci.,* 60, 89 (1971).
52. Beaven, Chen, D'Albis, and Gratzer, *Eur. J. Biochem.,* 41, 539 (1974).
53. Gorter and Nanninga, *Discuss. Faraday Soc.,* 13, 205 (1953).
54. Gorter and Nanninga, *K. Ned. Akad. Wer. Proc. 55C Biol. Med. Sci.,* p.341 (1952).
55. Clarke and Monkhouse, *Can. J. Med. Sci.,* 31, 394 (1953).
56. Pigman, *Biochim. Biophys. Acta,* 46, 100 (1961).
57. Churchich, *Biochim. Biophys. Acta,* 285, 91 (1972).
58. Klopfenstein, *Biochim. Biophys. Acta,* 181, 323 (1969).
59. Witiak and Whitehouse, *Biochem. Pharmacol.,* 18, 971 (1969).
60. Therriault and Taylor, *J. Am. Oil Chem. Soc.,* 41, 490 (1964).
61. Marecek, Jirsa, and Korinek, *Clin. Chim. Acta,* 45, 409 (1973).
62. Dempsey and Christensen, *J. Biol. Chem.,* 237, 1113 (1962).
63. Kerp and Kasemir, *Naunyn-Schmiedebergs Arch. Exp. Pathol. Pharmakol.,* 243, 147 (1962).
64. Tabachnik, *J. Biol. Chem.,* 234, 1242 (1964).
65. Lein, *Fed. Proc.,* 11, 91 (1952).
66. Bezkorovainy and Doherty, *Biochim. Biophys. Acta,* 58, 124 (1962).
67. Tabachnik and Giorgio, *Arch. Biochem. Biophys.,* 105, 563 (1964).
68. Tritsch and Tritsch, *J. Biol. Chem.,* 240, 3789 (1965).
69. Hocman and Hegedus, *Physiol. Bohemoslov.,* 18, 435 (1970).
70. Okubo, *J. Biochem.* (Tokyo), 69, 803 (1971).
71. Elewaut, *Clin. Chim. Acta,* 45, 37 (1973).
72. Lightfoot and Christian, *J. Clin. Endocrinol. Metab.,* 26, 305 (1966).
73. Britton, Webster, Ezrin, and Volpe, *Can. J. Biochem.,* 44, 1234 (1965).
74. Tritsch and Tritsch, *J. Biol. Chem.,* 238, 138 (1963).
75. Tritsch, *Arch. Biochem. Biophys.,* 127, 384 (1968).
76. Tabachnik, Downs, and Giorgio, *Arch. Biochem. Biophys.,* 136, 467 (1970).
77. Nazareth, Sokoloski, Witiak, and Hopper, *J. Pharm. Sci.,* 63, 199 (1974).
78. Steiner, Roth, and Robbins, *J. Biol. Chem.,* 241, 560 (1966).
79. Voth and Miller, *Arch. Biochem. Biophys.,* 77, 191 (1958).
80. Nathanielsz, *J. Endocrinol.,* 45, 489 (1969).
81. King and Spencer, *J. Biol. Chem.,* 235, 6134 (1970).
82. McMenamy and Seder, *J. Biol. Chem.,* 238, 3241 (1963).
83. McMenamy and Oncley, *J. Biol. Chem.,* 233, 1436 (1958).
84. Fuller and Roush, *Res. Commun. Chem. Pathol. Pharmacol.,* 8, 563 (1974).
85. Sjoholm and Grahnen, *FEBS Lett.,* 22, 109 (1972).
86. McMenamy, Madeja, and Watson, *J. Biol. Chem.,* 243, 2625 (1968).
87. McArthur, Dawkins, and Smith, *J. Pharm. Pharmacol.,* 23, 393 (1971).
88. McArthur and Dawkins, *J. Pharm. Pharmacol.,* 21, 744 (1969).
89. Cohen and Jullien, *C. R. Soc. Biol.,* 165, 1040 (1971).

90. Campion and Olsen, *J. Pharm. Sci.*, 63, 249 (1974).
91. Campion, Bluestone, and Klinenberg, *J. Clin. Invest.*, 52, 2383 (1973).
92. Campion, Bluestone, and Klinenberg, *Biochem. Pharmacol.*, 23, 1653 (1974).
93. Schlosstein, Kippen, Whitehouse, Bluestone, Paulus, and Klinenberg, *J. Lab. Clin. Med.*, 82, 412 (1973).
94. Helmer, Kiehs, and Hansch, *Biochemistry*, 7, 2858 (1968).
95. Balasubramanian and Wetlaufer, *Proc. Natl. Acad. Sci. U.S.A.*, 55, 762 (1966).
96. Laurence, *Biochem. J.*, 51, 168 (1952).
97. Fredericq, *Bull. Soc. Chim. Belg.*, 63, 158 (1954).
98. Sudlow, Birkett, and Wade, *Mol. Pharmacol.*, 9, 649 (1973).
99. Ma, Jun, and Luzzi, *J. Pharm. Sci.*, 62, 2038 (1973).
100. Layton and Symmons, *FEBS Lett.*, 30, 325 (1973).
101. Santos and Spector, *Biochemistry*, 11, 2299 (1972).
102. Daniel and Yang, *Biochemistry*, 12, 508 (1973).
103. Anderson, *Biochemistry*, 8, 4838 (1969).
104. Hsu, Ma, and Luzzi, *J. Pharm. Sci.*, 63, 570 (1974).
105. Hsu, Ma, and Luzzi, *J. Pharm. Sci.*, 63, 27 (1974).
106. Anderson and Weber, *Biochemistry*, 8, 371 (1969).
107. Daniel and Weber, *Biochemistry*, 5, 1893 (1966).
108. Daniel and Weber, *Biochemistry*, 5, 1900 (1966).
109. Ainsworth and Flanagan, *Biochim. Biophys. Acta*, 194, 213 (1969).
110. Flanagan and Ainsworth, *Biochim. Biophys. Acta*, 168, 16 (1968).
111. Nakatani, Haga, and Hiromi, *FEBS Lett.*, 43, 293 (1974).
112. Schellman, Lumry, and Samuels, *J. Am. Chem. Soc.*, 76, 2808 (1954).
113. Cassel and Steinhardt, *Biochemistry*, 8, 2603 (1969).
114. Skidmore and Whitehouse, *J. Pharm. Pharmacol.*, 17, 671 (1965).
115. Nazareth, Sokoloski, Witiak, and Hopper, *J. Pharm. Sci.*, 63, 203 (1974).
116. Brandys and Danek, *Diss. Pharm. Pharmacol.*, 24, 521 (1972).
117. Wishnia, *Proc. Natl. Acad. Sci. U.S.A.*, 48, 2200 (1962).
118. Wishnia and Pinder, *Biochemistry*, 3, 1377 (1964).
119. Wetlaufer and Lovrien, *J. Biol. Chem.*, 239, 596 (1966).
120. Jun, Mayer, Himel, and Luzzi, *J. Pharm. Sci.*, 60, 1821 (1971).
121. Gillberg, *Biochem. Biophys. Res. Commun.*, 38, 137 (1970).
122. Rudman, Bixler, and Del Rio, *J. Pharmacol. Exp. Ther.*, 176, 261 (1971).
123. Ashbrook, Spector, and Fletcher, *J. Biol. Chem.*, 247, 7038 (1972).
124. Reynolds, Herbert, and Steinhardt, *Biochemistry*, 7, 1357 (1968).
125. Teresi and Luck, *J. Biol. Chem.*, 194, 823 (1952).
126. Andersson, Brandt, and Johansson, *Arch. Biochem. Biophys.*, 146, 428 (1971).
127. King, *Arch. Biochem. Biophys.*, 153, 627 (1972).
128. Ma, Hsu, and Luzzi, *J. Pharm. Sci.*, 63, 32 (1974).
129. Featherstone, Muehlbaecher, and DeBon, *Anesthesiology*, 22, 977 (1961).
130. Chien, Sokoloski, Olson, Witiak, and Nazareth, *J. Pharm. Sci.*, 62, 440 (1974).
131. Teresi and Luck, *J. Biol. Chem.*, 174, 653 (1948).
132. Garten and Wosilait, *Biochem. Pharmacol.*, 20, 1661 (1971).
133. Chignell, *Mol. Pharmacol.*, 5, 244 (1969).
134. Chignell, *Mol. Pharmacol.*, 5, 455 (1969).
135. Chignell and Starkweather, *Mol. Pharmacol.*, 7, 229 (1971).
136. Chignell, *Mol. Pharmacol.*, 6, 1 (1970).
137. Ray, Reynolds, Polet, and Steinhardt, *Biochemistry*, 5, 2606 (1966).
138. Reynolds, Herbert, Polet, and Steinhardt, *Biochemistry*, 6, 937 (1967).
139. Polet and Steinhardt, *Biochemistry*, 7, 1348 (1968).
140. Karush and Sonenberg, *J. Am. Chem. Soc.*, 71, 1369 (1949).
141. Steinhardt, Krijn, and Leidy, *Biochemistry*, 10, 4005 (1971).
142. Steinhardt, Leidy, and Mooney, *Biochemistry*, 11, 1809 (1972).
143. Ponder and Ponder, *Nature*, 190, 277 (1961).
144. Starr and Judis, *J. Pharm. Sci.*, 57, 768 (1968).
145. Teresi, *J. Am. Chem. Soc.*, 72, 3972 (1950).
146. Kolberg, Helgeland, and Jonsen, *Acta. Pharmacol. Toxicol.*, 33, 470 (1973).
147. Dunn, *J. Med. Chem.*, 62, 1575 (1973).
148. Dunn, *J. Med. Chem.*, 16, 484 (1973).
149. Costanzo, Gaut, Benedict, and Solomon, *Pharmacology*, 6, 164 (1971).
150. Yang and Foster, *J. Am. Chem. Soc.*, 75, 5560 (1953).

151. Decker and Foster, *Biochemistry*, 5, 1242 (1966).
152. Putnam and Neurath, *J. Biol. Chem.*, 150, 263 (1943).
153. Putnam and Neurath, *J. Biol. Chem.*, 150, 195 (1945).
154. Putnam, *Adv. Protein Chem.*, 4, 79 (1948).
155. Pallansch and Briggs, *J. Am. Chem. Soc.*, 76, 1396 (1954).
156. Strauss and Strauss, *J. Phys. Chem.*, 62, 1321 (1958).
157. Markus and Karush, *J. Am. Chem. Soc.*, 79, 3624 (1957).
158. Bigelow and Sonenberg, *Biochemistry*, 1, 197 (1962).
159. Mullen, *University Microfilms*, Ann Arbor, Michigan, Order No. 4356, 1963.
160. Markus, Love, and Wissler, *J. Biol. Chem.*, 239, 3687 (1964).
161. Halfman and Steinhardt, *Biochemistry*, 10, 3564 (1971).
162. Takenaka, Aizawa, Tamaura, Hirano, and Inada, *Biochim. Biophys. Acta*, 263, 696 (1972).
163. Avruch, Reynolds, and Reynolds, *Biochemistry*, 8, 1855 (1969).
164. Boyer, Lum, Ballou, Luck, and Rice, *J. Biol. Chem.*, 162, 181 (1946).
165. Bahl and Gutmann, *Biochim. Biophys. Acta*, 90, 391 (1964).
166. Karush, *J. Am. Chem. Soc.*, 72, 2705 (1950).
167. Karush, *J. Am. Chem.. Soc.*, 73, 1246 (1951).
168. Perrin and Nelson, *Life Sci.*, 11, 277 (1972).
169. Goodman, *J. Am. Chem. Soc.*, 80, 3892 (1958).
170. Dawkins, McArthur, and Smith, *J. Pharm. Pharmacol.*, 22, 405 (1970).
171. Chinard, *J. Biol. Chem.*, 176, 1439 (1948).
172. Burkhard, Lothers, and Smith, *Arch. Biochem. Biophys.*, 81, 1 (1959).
173. Chu, *Arch. Biochem. Biophys.*, 147, 359 (1971).
174. Raz, *Biochem. J.*, 130, 631 (1972).
175. Schmid, *J. Biol. Chem.*, 234, 3163 (1959).
176. Kessler, Demeny, and Sobotka, *J. Lipid Res.*, 8, 185 (1967).
177. Arvidsson, Green, and Laurell, *J. Biol. Chem.*, 246, 5373 (1971).
178. Mohammadzadeh, Feeney, Samuels, and Smith, *Biochim. Biophys. Acta*, 147, 583 (1967).
179. Barré and Nguyen van Huot, *Bull. Soc. Chim. Biol.*, 47, 1399 (1965).
180. Barré and Nguyen van Huot, *Bull. Soc. Chim. Biol.*, 47, 1419 (1965).
181. Fredericq, *Bull. Soc. Chim. Belg.*, 64, 639 (1955).
182. Brand and Gohlke, *J. Biol. Chem.*, 246, 2317 (1971).
183. Scatchard, Coleman, and Shen, *J. Am. Chem.Soc.*, 79, 12 (1957).
184. Scatchard, Wu, and Shen, *J. Am. Chem. Soc.*, 81, 6104 (1959).
185. Scatchard and Zaromb, *J. Am. Chem. Soc.*, 81, 610 (1959).
186. Sukow and Sandberg, *FEBS Lett.*, 42, 36 (1974).
187. Saroff and Lewis, *J. Phys. Chem.*, 67, 1211 (1963).
188. Tanford, *J. Am. Chem. Soc.*, 73, 2066 (1951).
189. Tanford, *J. Am. Chem. Soc.*, 74, 211 (1952).
190. Rao and Lal, *J. Am. Chem. Soc.*, 80, 3222, 3226 (1958).
191. Bryant, *J. Am. Chem. Soc.*, 91, 4096 (1969).
192. Peters and Blumenstock, *J. Biol. Chem.*, 242, 1574 (1967).
193. Peters, *Biochim. Biophys. Acta*, 39, 546 (1960).
194. Klotz and Curme, *J. Am. Chem. Soc.*, 70, 939 (1948).
195. Klotz and Fiess, *J. Phys. Colloid Chem.*, 55, 101 (1951).
196. Appleton and Sarkar, *J. Biol. Chem.*, 246, 5040 (1971).
197. Reynolds, Burkhard, and Mueller, *Biochemistry*, 12, 359 (1973).
198. Hughes, *J. Am. Chem. Soc.*, 69, 1836 (1947).
199. Mildvan and Cohn, *Biochemistry*, 2, 910 (1963).
200. Cohn and Hughes, *J. Biol. Chem.*, 237, 176 (1962).
201. Gurd and Murray, *J. Am. Chem. Soc.*, 76, 187 (1954).
202. Gurd and Goodman, *J. Am. Chem. Soc.*, 74, 670 (1952).
203. Pauli and Schön, *Biochem. Z.*, 153, 253 (1924).
204. Zeppezauer, Lindman, Forsen, and Lindqvist, *Biochem. Biophys. Res. Commun.*, 37, 137 (1969).
205. Scatchard, Batchelder, and Brown, *J. Am. Chem. Soc.*, 68, 2320 (1946).
206. Scatchard and Black, *J. Phys. Colloid Chem.*, 53, 88 (1949).
207. Scatchard, *Am. Sci.*, 40, 61 (1952).
208. Scatchard and Yap, *J. Am. Chem. Soc.*, 86, 3434 (1964).
209. Carr, *Arch. Biochem. Biophys.*, 46, 417, 424 (1953).

210. Ott, *Z. Gesamte Exp. Med.*, 122, 346 (1953).
211. Barré and Nguyen van Huot, *Bull. Soc. Chim. Biol.*, 45, 661 (1964).
212. Fadda, Biggio, and Liguori, *Experientia*, 30, 635 (1974).
213. McMenamy, Madeja, and Watson, *J. Biol. Chem.*, 243, 2328 (1968).
214. Clark and Bird, *J. Endocrinol.*, 57, 289 (1973).
215. Plager, *J. Clin. Invest.*, 44, 1234 (1965).
216. Judis, *J. Pharm. Sci.*, 62, 232 (1973).
217. Schirazi, Schwarz, and Oertel, *Steroids Lipids Res.*, 4, 98 (1973).
218. Ryan, *Arch. Biochem. Biophys.*, 126, 407 (1968).
219. Rosenthal, Pietrzak, Slaunwhite, and Sandberg, *J. Clin. Endocrinol. Metab.*, 34, 805 (1972).
220. Brinkhorst and Hess, *Arch. Biochem. Biophys.*, 111, 54 (1965).
221. Wosilait and Eisenbrandt, *Res. Commun. Chem. Pathol. Pharmacol.*, 5, 109 (1973).
222. Tillement, Zini, Athis, and Vassent, *Eur. J. Clin. Pharmacol.*, 7, 307 (1974).
223. Florini and Buyske, *J. Biol. Chem.*, 236, 247 (1961).
224. Starinsky and Shafrir, *Clin. Chim. Acta*, 29, 311 (1970).
225. Ryan and Gibbs, *Biochim. Biophys. Acta*, 222, 206 (1970).
226. Westphal and Harding, *Biochim. Biophys. Acta*, 310, 518 (1973).
227. Pearlman and Crepy, *J. Biol. Chem.*, 242, 182 (1967).
228. Attallah and Lata, *Biochim. Biophys. Acta*, 168, 321 (1968).
229. Vermeulen and Verdonck, *Steroids*, 11, 609 (1968).
230. Alfsen, *Proc. 2nd Int. Congr. Hormonal Steroids*, Milan, 508 (1966).
231. Bischoff and Pilhorn, *J. Biol. Chem.*, 174, 663 (1948).
232. Pearlman, *Acta Endocrinol. Suppl.*, 147, 225 (1970).
233. Klotz and Walker, *J. Am. Chem. Soc.*, 69, 1609 (1947).
234. Noval and Champion, *Res. Commun. Chem. Pathol. Pharmacol.*, 6, 123 (1973).
235. Franglen and Martin, *Biochem. J.*, 57, 626 (1954).
236. Dijkstra and Louw, *Br. J. Cancer*, 16, 757 (1962).
237. Westphal, Stets, and Priest, *Arch. Biochem. Biophys.*, 43, 463 (1953).
238. Nishikimi and Yoshino, *J. Biochem.* (Tokyo), 72, 1237 (1972).
239. Sarkar, *J. Sci. Ind. Res.* (India), 20, 239 (1961).
240. Rodkey, *Arch. Biochem. Biophys.*, 108, 510 (1964).
241. Franglen, *Clin. Chim. Acta*, 3, 63 (1958).
242. Lee and Hong, *Yonsei Med. J.*, 1, 22 (1960).
243. Osborn, *Clin. Chim. Acta*, 5, 777 (1960).
244. Weder and Bickel, *J. Pharm. Sci.*, 59, 1563 (1970).
245. Butler, Jackson, Polya, and Tetlow, *Enzymologia*, 20, 119 (1958).
246. Waldmann, Meyer, and Schilling, *Arch. Biochem. Biophys.*, 64, 291 (1956).
247. Izuoka, Oshima, Kawamura, Iida, and Nishimura, *Kyoto Furitsu Ika Daigaku Zasshi*, 63, 665 (1958).
248. Sunderman and Sunderman, *Clin. Chem.*, 5, 171 (1959).
249. Scardi and Bonavita, *Clin. Chim. Acta*, 4, 322 (1959).
250. Kusonoki, *J. Biochem.* (Tokyo), 40, 277 (1953).
251. Varetskaya and Ryabohon, *Ukr. Biokhim. Zh.*, 32, 507 (1960).
252. Zemplenyi, *Rev. Czech. Med.*, 4, 189 (1958).
253. Torii, *Seikagaku*, 29, 94 (1957).
254. Zemplenyi, *Cas. Lek. Cesk.*, 97, 1230 (1958).
255. Torii, *Nara Igaku Zasshi*, 8, 423 (1957).
256. Makarchenko, Roitrub, and Zlatin, *Zh. Vyssh. Nerv. Deyat. im. I. P. Pavlova*, 15, 838 (1965).
257. Challis, *J. Endocrinol.*, 56, 319 (1973).
258. Higgins and Harrington, *Arch. Biochem. Biophys.*, 85, 409 (1959).
259. Fredericq, *Bull. Soc. Chim. Belg.*, 65, 631 (1956).
260. Karush, *J. Phys. Chem.*, 56, 70 (1952).
261. Klotz, Burkhard, and Urquhart, *J. Phys. Chem.*, 56, 77 (1952).
262. Markus and Baumberger, *J. Biol. Chem.*, 206, 59 (1954).
263. Genau, *Abh. Dtsch. Akad. Wiss. Berlin, Kl. Chem. Geol. Biol.*, p.248 (1964).
264. Rosenthal, *J. Pharmacol. Exp. Ther.*, 29, 521 (1926).
265. Andersson, Rehnström, and Eaker, *Eur. J. Biochem.*, 20, 371 (1971).
266. Porter and Waters, *J. Lab. Clin. Med.*, 67, 660 (1966).
267. Spector and Imig, *Mol. Pharmacol.*, 7, 511 (1971).
268. Koslowski, Braun, and Weidemann, *Z. Gesamte. Inn. Med. Ihre Grenzgeb.*, 26, 779 (1971).

269. Bruce, *Anesthesiology*, 36, 588 (1972).
270. Baxter, *Arch. Biochem. Biophys.*, 108, 375 (1964).
271. Terada, *Life Sci.*, 11, 417 (1972).
272. Terada, Koichiro, and Kamada, *Biochim. Biophys, Acta*, 342, 41 (1974).
273. Pemberton and de Jong, *Anal. Biochem.*, 43, 575 (1971).
274. Karush, *J. Am. Chem. Soc.*, 72, 2705 (1950).
275. Karush, *J. Am. Chem. Soc.*, 72, 2714 (1950).
276. Pavkovskaya and Pasynskii, *Dokl. Adad. Nauk Arm. SSR*, 149, 976 (1963).
277. Berns and Singer, *Immunochemistry*, 1, 209 (1964).
278. Prokopova and Munk, *Collect. Czech. Chem. Commun.*, 28, 957 (1963).
279. Sponar and Vodrazka, *Chem. Listy*, 50, 853 (1956).
280. Kusunoki, *J. Biochem.* (Tokyo), 39, 245 (1952).
281. Dijkstra and Joubert, *Br. J. Cancer.*, 15, 168 (1961).
282. Karush, *J. Am. Chem. Soc.*, 73, 1246 (1951).
283. Stein, *Anal. Biochem.*, 13, 305 (1965).
284. Ugman, *Nature*, 171, 653 (1953).
285. Kusunoki, *J. Biochem.* (Tokyo), 39, 349 (1952).
286. Klotz and Luborsky, *J. Am. Chem. Soc.*, 81, 5119 (1959).
287. Burkhard, *J. Am. Chem. Soc.*, 75, 229 (1953).
288. Klotz, Burkhard, and Urquhart, *J. Am. Chem. Soc.*, 74, 202 (1952).
289. Colvin and Briggs, *J. Phys. Chem.*, 56, 717 (1952).
290. Colvin, *Can. J. Chem.*, 31, 734 (1953).
291. Klotz, Walker, and Pivan, *J. Am. Chem. Soc.*, 68, 1486 (1946).
292. Klotz, *J. Am. Chem. Soc.*, 68, 2299 (1946).
293. Kerp and Steinhilber, *Klin. Wochenschr.*, 40, 540 (1962).
294. Klotz and Urquhart, *J. Am. Chem. Soc.*, 71, 847 (1949).
295. Carey, Schneider, and Bernstein, *Biochem. Biophys. Res. Commun.*, 47, 588 (1972).
296. Nichol, Jackson, and Smith, *Arch. Biochem. Biophys.*, 144, 438 (1971).
297. Hirose and Kano, *Biochim. Biophys. Acta*, 251, 376 (1971).
298. Klotz and Shikama, *Arch. Biochem. Biophys.*, 128, 551 (1968).
299. Meyer and Guttman, *J. Pharm. Sci.*, 59, 33 (1970).
300. Meyer and Guttman, *J. Pharm. Sci.*, 59, 39 (1970).
301. Burkhard, Moore, and Louloudes, *Arch. Biochem. Biophys.*, 94, 291 (1961).
302. Froese, Sehon, and Eigen, *Can. J. Chem.*, 40, 1786 (1962).
303. Rawson, *Am. J. Physiol.*, 138, 708 (1943).
304. Chignell, *Life Sci.*, 7, 1181 (1968).
305. Beeck, Braun, Damm, Erttmann, and Gerhardt, *Naunyn-Schmiedebergs Arch. Pharmakol. Exp. Pathol.*, 275, 277 (1972).
306. Kragh-Hansen and Moller, *Biochim. Biophys. Acta*, 295, 438 (1973).
307. Kragh-Hansen and Moller, *Biochim. Biophys. Acta*, 295, 447 (1973).
308. Kragh-Hansen and Moller, *Biochim. Biophys, Acta*, 336, 30 (1974).
309. Kragh-Hansen, Moller, and Sheikh, *Pfluegers Arch. Gesamte Physiol. Menschen Tiere*, 337, 163 (1972).
310. Keen, *Br. J. Pharmacol. Chemother.*, 26, 704 (1966).
311. Lee and Debro, *J. Chromatogr.*, 10, 68 (1963).
312. Nishida, *Igaku To Seibutsugaku*, 59, 33 (1961).
313. Rodkey, *Arch. Biochem. Biophys.*, 94, 526 (1961).
314. Rodkey, *Arch. Biochem. Biophys.*, 94, 38 (1961).
315. Blondheim, *J. Lab. Clin. Med.*, 45, 740 (1955).
316. Huggins, Jensen, Player, and Hospelhorn, *Cancer Res.*, 9, 753 (1949).
317. Smith and Smith, *J. Biol. Chem.*, 124, 107 (1938).
318. Grollman, *J. Biol. Chem.*, 64, 141 (1925).
319. Pezold, *Z. Gesamte Exp. Med.*, 121, 600 (1953).
320. Baker and Bradley, *J. Clin. Invest.*, 45, 281 (1966).
321. Crawford, Jones, Thompson, and Wells, *Br. J. Pharmacol. Chemother.*, 44, 80 (1972).
322. Rudman, Bixler, and Del Rio, *J. Pharmacol. Exp. Ther.*, 176, 261 (1971).
323. Thompson, *Br. J. Pharmacol. Chemother.*, 47, 133 (1973).
324. Thompson, *Br. J. Pharmacol. Chemother.*, 47, 133 (1973).
325. Brenner, *S. Afr. J. Med. Sci.*, 17, 61 (1952).
326. Dow, *Fed. Proc.*, 4, 16 (1945).
327. Gregersen and Gibson, *Am. J. Physiol.*, 120, 494 (1937).
328. Zia, Cox, and Luzzi, *J. Pharm. Sci.*, 60, 45 (1971).

329. Buttner and Portwich, *Antimicrob. Agents Chemother.*, p.177 (1965).
330. Scholtan, *Arzneim. Forsch.*, 14, 1139 (1964).
331. Jardetsky and Wade-Jardetsky, *Mol. Pharmacol.*, 1, 214 (1965).
332. van Dyke, Tupikova, Chow, and Walker, *J. Pharmacol. Exp. Ther.*, 83, 203 (1945).
333. Scholtan, *Makromol. Chem.*, 54, 24 (1962).
334. Scholtan, *Arzneim. Forsch.*, 14, 469 (1964).
335. Elofsson, Nilsson, and Kluczykowska, *Acta Pharm. Suec.*, 8, 465 (1971).
336. Garn and Kimbel, *Arzneim. Forsch.*, 11, 701 (1961).
337. Josephson and Furst, *Scand. J. Clin. Lab. Invest.*, 18, 51 (1966).
338. Genazzini, Bononi, Pagnine, and Di Carlo, *Antimicrob. Agents Chemother.*, p. 192 (1965).
339. Anton, *J. Pharmacol. Exp. Ther.*, 134, 291 (1961).
340. Perrin and Nelson, *J. Pharm. Pharmacol.*, 25, 125 (1973).
341. Scholtan, *Arzneim. Forsch.*, 15, 1433 (1965).
342. Kostenbauder, Jawad, Perrin, and Averhart, *J. Pharm. Sci.*, 60, 1658 (1971).
343. Agren, Elofsson, and Nilsson, *Acta Pharmacol. Toxicol. Suppl.*, 29, 48 (1971).
344. Agren, Elofsson, Meresaar, and Nilsson, *Acta Pharm. Suec.*, 7, 105 (1970).
345. Anton and Boyle, *Can. J. Physiol. Pharmacol.*, 42, 809 (1964).
346. Cooper and Wood, *J. Pharm. Pharmacol.*, 20, 150S (1968).
347. Krueger-Thiemer, Roesch, Rohmer, and Wempe, *Chemotherapia*, 12, 321 (1967).
348. Doub, Krolls, Vandenbelt, and Fisher, *J. Med. Chem.*, 13, 242 (1970).
349. Chignell, Vesell, Starkweather, and Berlin, *Clin. Pharamacol. Ther.*, 12, 897 (1971).
350. Scholtan, *Arzneim. Forsch.*, 14, 1234 (1964).
351. Wood and Stewart, *J. Pharm. Pharmacol.*, 23, 248S (1971).
352. Davis, *J. Clin. Invest.*, 22, 753 (1943).
353. Nakagaki, Koga, and Terada, *Yakugaku Zasshi*, 84, 516 (1964).
354. Genazzani, Di Rosa, and Graziani, *Chemotherapia*, 6, 117 (1963).
355. Anton and Corey, *Acta Pharmacol. Toxicol. Suppl.*, 3, 134 (1971).
356. Nakagaki, Koga, and Terada, *Yakugaku Zasshi*, 83, 586 (1963).
357. Rollo, *Can. J. Physiol. Pharmacol.*, 50, 976 (1972).
358. Joos and Hall, *J. Pharmacol. Exp. Ther.*, 166, 113 (1969).
359. Scholtan, *Arzneim. Forsch.*, 13, 347 (1963).
360. Klotz, Urquhart, and Weber, *Arch Biochem. Biophys.*, 26, 420 (1950).
361. Moskowitz, Somani, and McDonald, *Clin. Toxicol.*, 6, 247 (1973).
362. Callan and Sunderman, *Res. Commun. Chem. Pathol. Pharmacol.*, 5, 459 (1973).
363. Pindell, Tisch, Hoekstra, and Reiffenstein, *Antibiot. Annu.* 119 (1959).
364. Fischer and Jardetsky, *J. Am. Chem. Soc.*, 87, 3237 (1965).
365. Halpern, Dolkart, Lesh, Kutz, Dey, and Wolnak, *J. Pharmacol. Exp. Ther.*, 103, 202 (1951).
366. Kerp, *Protides Biol. Fluids Proc. Colloq. Bruges*, 10, 262 (1962).
367. Phillips, Power, Robinson, and Davies, *Biochim. Biophys. Acta*, 215, 491 (1970).
368. Phillips, Power, Robinson, and Davies, *Biochim. Biophys. Acta*, 295, 8 (1973).
369. Smith, Joslyn, Gruhzit, McLean, Penner, and Ehrlich, *J. Bacteriol.*, 55, 425 (1948).
370. Dessi and Gianni, *Boll. Soc. Ital. Biol. Sper.*, 31, 753 (1955).
371. Riley and Levy, *Proc. Soc. Exp. Biol. Med.*, 142, 1168 (1973).
372. Brand and Toribara, *Mol. Pharmacol.*, 8, 751 (1972).
373. Brand and Toribara, *Biochem. Biophys. Red. Commun.*, 52, 511 (1973).
374. Klotz, Gelewitz, and Urquhart, *J. Am. Chem. Soc.*, 74, 209 (1952).
375. Powis, *J. Pharm. Pharmacol.*, 26, 113 (1974).
376. Popov, Vaptazarova, Kossekov, and Nikolov, *Biochem. Pharmacol.*, 21, 2363 (1972).
377. Branstad, Meresaar, and Agren, *Acta Pharm. Suec.*, 9, 129 (1972).
378. Buech, Knabe, Buzello, and Rummel, *J. Pharmacol. Exp. Ther.*, 175, 709 (1970).
379. Goldbaum and Smith, *J. Pharmacol. Exp. Ther.*, 111, 197 (1954).
380. Waddel and Butler, *J. Clin. Invest.*, 36, 1217 (1957).
381. Sjöholm and Sjödin, *Biochem. Pharmacol.*, 21, 3041 (1972).
382. Yoshikawa and Loehning, *Experientia*, 21, 376 (1965).
383. Taylor, Richards, Davin, and Asher, *J. Pharmacol. Exp. Ther.*, 112, 40 (1954).
384. Ma and Routh, *Clin. Chem.*, 15, 1027 (1969).
385. Kramer and Routh, *Clin. Biochem.*, 6, 98 (1973).
386. Sykes, *Biochem. Biophys. Res. Commun.*, 39, 508 (1970).
387. Perrin, Nelson, and Tyson, *J. Pharm. Sci.*, 61, 1667 (1972).
388. Airaksinen and Airaksinen, *Ann. Clin. Res.*, 4, 361 (1972).

389. Rosen, *Biochem. Pharmacol.,* 19, 2075 (1970).
390. Jun, Luzzi, and Hsu, *J. Pharm. Sci.,* 61, 1835 (1972).
391. Whitehouse, Kippen, and Klinenberg, *Biochem. Pharmacol.,* 20, 3309 (1971).
392. Solomon, Schrogie, and Williams, *Biochem. Pharmacol.,* 17, 143 (1968).
393. Tillement, Zini, Mattei, and Singlas, *Eur. J. Clin. Pharmacol.,* 6, 15 (1973).
394. Keresztes-Nagy, Mais, Oester, and Zaroslinski, *Anal. Biochem.,* 48, 80 (1972).
395. McArthur and Smith, *J. Pharm. Pharmacol.,* 21, 589 (1969).
396. Stafford, *Lancet,* I, 243 (1962).
397. Fredericq, *Bull. Soc. Chim. Belg.,* 63, 158 (1954).
398. Lindenbaum and Schubert, *J. Phys. Chem.,* 60, 1663 (1956).
399. Davison and Smith, *J. Pharmacol. Exp. Ther.,* 133, 161 (1961).
400. Spector, Korkin, and Lorenzo, *J. Pharm. Pharmacol.,* 24, 786 (1972).
401. Glasser and Krieglstein, *Naunyn-Schmiedebergs Arch. Pharmakol. Exp. Pathol.,* 265, 321 (1970).
402. Weder and Bickel, *J. Pharm. Sci.,* 59, 1505 (1970).
403. Jähnchen, Krieglstein, and Kuschinsky, *Naunyn-Schmiedebergs Arch. Pharmakol. Exp. Pathol.,* 263, 375 (1969).
404. Franz, Jähnchen, and Krieglstein, *Naunyn-Schmiedebergs Arch. Pharmakol. Exp. Pathol.,* 264, 462 (1969).
405. Huang and Gabay, *Biochem. Pharmacol.,* 23, 957 (1974).
406. Krieglstein and Kuschinsky, *Naunyn-Schmiedebergs Arch. Pharmakol. Exp. Pathol.,* 262, 1 (1969).
407. Nambu and Nagai, *Chem Pharm. Bull. (Japan),* 20, 2463 (1972).
408. Jindrova, Sipal, and Jindra, *Cesk. Farm.,* 13, 393 (1964).
409. Mueller and Wollert, *Naunyn-Schmiedebergs Arch. Pharmakol. Exp. Pathol.,* 278, 301 (1973).
410. Krieglstein, Meiler, and Staab, *Biochem. Pharmacol,* 21, 985 (1972).
411. Mueller and Wollert, *Naunyn-Schmiedebergs Arch. Pharmakol. Exp. Pathol.,* 280, 229 (1973).
412. Allenby and Collier, *Arch. Biochem. Biophys.,* 38, 147 (1957).
413. Jähnchen, Krieglstein, and Kuschinsky, *Naunyn-Schmiedebergs Arch. Pharmakol. Exp. Pathol.,* 260, 147 (1968).
414. Agren and Back, *Acta Pharm. Suec.,* 10, 223 (1973).
415. Breckenridge and Rosen, *Biochim. Biophys. Acta,* 229, 610 (1971).
416. Dollery, Emslie-Smith, and Muggleton, *Br. J. Pharmacol. Chemother.,* 17, 488 (1961).
417. Cho, Mitchell, and Penarowski, *J. Pharm. Sci.,,* 196 (1971).
418. Perrin and Idsvoog, *J. Pharm. Sci.,* 60, 602 (1971).
419. O'Reilly and Motley, *Mol. Pharmacol.,* 7, 209 (1971).
420. Lau and Sarkar, *J. Biol. Chem.,* 246, 5938 (1971).
421. Brodie, Weiner, Burns, Simson, and Yale, *J. Pharmacol. Exp. Ther.,* 106, 453 (1952).
422. O'Reilly, *J. Clin. Invest.,* 48, 193 (1969).
423. Aggeler, O'Reilly, Leong, and Kowitz, *N. Engl. J. Med.,* 276, 496 (1967).
424. O'Reilly, *J. Clin. Invest.,* 46, 829 (1967).
425. O'Reilly, *J. Clin. Invest.,* 48, 193 (1969).
426. Sellers and Koch-Weser, *Clin. Pharmacol. Ther.,* 11, 524 (1970).
427. Wishinsky, Glasser, and Perkal, *Diabetes,* 11, 18s (1962).
428. Crooks and Brown, *J. Pharm. Sci.,* 62, 1904 (1973).
429. Buttner and Portwich, *Klin. Wochenschr.,* 45, 225 (1967).
430. Kuschinsky, *Naunyn-Schmiedebergs Arch. Pharmakol. Exp. Pathol.,* 262, 388 (1969).
431. Gennazzani and Santamaria, *Pharmacol. Res. Commun.,* 1, 249 (1969).
432. **Haarmann, Hagemeier, and Lendle,** *Naunyn-Schmiedebergs Arch. Exp. Pathol. Pharmakol.,* 194, 205 (1940).
433. **Haarman, Korfmacher, and Lendle,** *Naunyn-Schmiedebergs Arch. Exp. Pathol. Pharmakol.,* 194, 229 (1940).
434. Kuschinsky, *Arznem. Forsch.,* 20, 842 (1970).
435. Solomon, Reich, Spirt, and Abrams, *Ann. N.Y. Acad. Sci.,* 179, 362 (1971).
436. Evered, *Eur. J. Pharmacol.,* 18, 236 (1972).
437. Conn and Luchi, *J. Clin. Invest.,* 39, 978 (1960).
438. Conn and Luchi, *J. Clin. Invest.,* 40, 509 (1961).
439. Conn and Luchi, *J. Pharmacol. Exp. Ther.,* 133, 76 (1961).

440. Pinckard, Hawkins, and Farr, *Ann. N.Y. Acad. Sci.*, 226, 341 (1973).
441. Lajos, *Fortschr. Geb. Roentgenstr. Nuklearmed.*, 85, 292 (1956).
442. Dearden and Tomlinson, *J. Pharm. Pharmacol.*, 22, 53S (1970).
443. Nery, *Biochem. J.*, 122, 311 (1971).
444. Bassir and Bababunmi, *Biochem. Pharmacol.*, 22, 132 (1973).
445. Parker and Weed, *J. Biol. Chem.*, 204, 289 (1953).
446. Way, Smith, Howie, Weiss, and Swanson, *J. Pharmacol. Exp. Ther.*, 93, 368 (1948).
447. Oroszlan and Maengwyn-Davies, *Biochem. Pharmacol.*, 11, 1203 (1962).
448. Oroszlan and Maengwyn-Davies, *Biochem. Pharmacol.*, 11, 1213 (1962).
449. Eichman, Guttman, Van Winkle, and Guth, *J. Pharm. Sci.*, 51, 66 (1962).
450. Guttman and Gadzala, *J. Pharm. Sci.*, 54, 742 (1965).
451. Guarino, Anderson, Starkweather, and Chignell, *Cancer Chemother. Rep.*, 57, 125 (1973).
452. Parke and Lindup, *Ann. N.Y. Acad. Sci.*, 226, 200 (1973).
453. Birks and Slifkin, *Nature*, 197, 4862 (1963).
454. Gabay and Huang, *Adv. Biochem. Psychopharmacol.*, 9, 175 (1974).
455. Sellers and Koch-Weser, *N. Engl. J. Med.*, 281, 1141 (1969).
456. Lu and Ling, *J. Formosan Med. Assoc.*, 72, 434 (1973).
457. Shoeman, Benjamin, and Azarnoff, *Ann. N.Y. Acad. Sci.*, 226, 127 (1973).
458. Rane, Lunde, Jalling, Yaffe, and Sjoqvist, *J. Pediat.*, 78, 877 (1971).
459. Scholtan, Schlossman, and Rosenkranz, *Arzneim. Forsch.*, 16, 109 (1966).
460. Hooper, Sutherland, Bochner, Tyrer, and Eadie, *Aust. N. Z. J. Med.*, 3, 377 (1973).
461. Kaldar, Saifer, and Vecsler, *Arch. Biochem. Biophys.*, 94, 207 (1961).
462. Ronwin and Zacchei, *Can. J. Biochem.*, 45, 1433 (1967).
463. Perrin, *J. Pharm. Pharmacol.*, 25, 208 (1973).
464. Wassermann, *Arzneim. Forsch.*, 22, 1993 (1972).
465. Widman, Nilsson, Nilsson, Agurell, Borg, and Granstrand, *J. Pharm. Pharmacol.*, 25, 453 (1973).
466. Akopyan, Kramarenko, and Lisevich, *Farm. Zh.*, 28, 73 (1973).
467. Olsen, *Science*, 176, 525 (1972).
468. Olsen, *Clin. Pharmacol. Ther.*, 14, 338 (1973).
469. Dixon, Henderson, and Rall, *Fed. Proc.*, 24, 454 (1965).
470. Bamgbose and Bababunmi, *Biochem. Pharmacol.*, 22, 2926 (1973).
471. Plessas, *Chem. Chron.*, 1, 175 (1972).
472. Paubel and Niviere, *Chim. Ther.*, 8, 469 (1973).
473. Guidicelli, Tillement, Boissier, and Garnier, *J. Pharmacol.*, 4, 129 (1973).
474. Sawinski and Rapp, *J. Dent. Res.*, 42, 1429 (1963).
475. Raz, *Biochim. Biophys. Acta*, 280, 602 (1972).
476. Dal Santo, *Anesthesiology*, 29, 435 (1968).
477. Chen, *Arch. Biochem. Biophys.*, 120, 609 (1967).
478. Chen, *Arch. Biochem. Biophys.*, 160, 106 (1974).
479. Kotaki, Naoi, and Yagi, *Biochim. Biophys. Acta*, 229, 547 (1971).
480. Lang and Lasser, *Biochemistry*, 6, 2403 (1967).

INTRODUCTION TO PROTEINASE INHIBITORS*

Beatrice Kassell and Monica June Williams

The tabulations include properties, specificities, amino acid and carbohydrate compositions, and sequences. Most of the inhibitors selected are those that have been highly purified and for which evidence of near homogeneity has been presented. However, a great deal of available data on specificity has been collected using inhibitors that are or may be mixtures of closely related inhibitors; these have been included. As far as possible, only inhibition tests against purified enzymes are reported. Information has been combined and several references listed when it was clear that different laboratories were dealing with the same inhibitor and when the data were in reasonable agreement. Where inhibitors from similar sources were possibly different, they are listed separately.

Only protein inhibitors are included; for peptide inhibitors Umezawa[15] may be consulted. For further information, the following partial list of recent review articles is included.

We are grateful to Professor Yehudith Birk for reading part of the manuscript. Beatrice Kassell thanks Professor Michel Lazdunski and the Centre de Biochimie, Université de Nice, Nice, France, for hospitality and the use of facilities during the preparation of these tables.

*Supported by grant BMS 74-22102 from the National Science Foundation.

REVIEW ARTICLES

General Reviews
1. **Fritz, Tschesche, Greene, and Truscheit, Eds.**, *Bayer Symposium V, Proteinase Inhibitors,* Springer-Verlag, Berlin, 1974.
2. **Uhlig and Kleine,** *Die Pharmazie,* 29, 81 (1974).
3. **Means, Ryan, and Feeney,** *Acc. Chem. Res.,* 7, 315 (1974).
4. **Tschesche,** *Angew. Chem. Int. Ed. Engl.,* 13, 11 (1974).
5. **Dayhoff,** *Atlas of Protein Sequence and Structure,* Vol. 5, National Biomedical Research Foundation, Washington, D.C., 1972, D-165.
6. **Werle and Zickgraf-Rüdel,** *Z. Klin. Chem. Klin. Biochem.,* 10, 139 (1972).
7. **Fritz and Tschesche, Eds.,** *Proc. 1st Int. Res. Conf. Proteinase Inhibitors,* Walter de Gruyter, Berlin, 1971.
8. **Laskowski, Jr. and Sealock,** in *The Enzymes,* Vol. 3, 3rd ed. Boyer, Ed., Academic Press, New York, 1971, 375.
9. **Vogel and Werle,** in *Handbook of Experimental Pathology,* Vol. 25, Erdös, Ed., Springer-Verlag, Berlin, 1970, 213.
10. **Ikenaka and Koide,** *Kagaku To Seibutsu,* 8, 1 (1970).
11. **Kassell,** in *Methods in Enzymology,* Vol. 19, Perlmann and Lorand, Eds., Academic Press, New York, XIX, 1970, 839.
12. **Sawaryn,** *Postepy Hig. Med. Dosw.,* 23, 787 (1969).
13. **Vogel, Trautschold, and Werle,** *Natural Proteinase Inhibitors,* Academic Press, New York, 1968.
14. **Métais, Bieth, and Warter,** *Probl. Actuels Biochim. Appl.,* 2, 219 (1968).
15. **Umezawa,** *Enzyme Inhibitors of Microbial Origin,* University Park, Baltimore, 1972.

Plant Inhibitors
16. **Ryan,** *Ann. Rev. Plant. Physiol.,* 24, 173 (1973).
17. **Whitaker and Feeney,** in *Toxicants Occurring Naturally in Foods,* National Academy of Sciences, Washington, D.C., 1973, 276.
18. **Pressey,** *J. Food Sci.,* 37, 521 (1972).
19. **Weder and Hory,** *Lebensm. Wiss. Technol.,* 5, 54 (1972).
20. **Nilova, Yalovich, and Khotyanovich,** *Prikl. Biokhim. Mikrobiol.,* 7, 373 (1971).
21. **Dechary,** *Econ. Bot.,* 24, 113 (1970).
22. **Liener and Kakade,** in *Toxic Constituents of Plant Foodstuffs,* Liener, Ed., Academic Press, New York, 1969, 7.
23. **Lorenc-Kubis,** *Wiad. Bot.,* 13, 133 (1969).
24. **Birk,** *Ann. N.Y. Acad. Sci.,* 146, 388 (1968).
25. **Pusztai,** *Nutr. Abstr. Rev.,* 37, 1 (1967).

Egg White Inhibitors
26. **Lin and Feeney,** in *Glycoproteins. Part B,* Vol. 5, 2nd ed., Gottschalk, Ed., Elsevier, New York, 1972, 762.
27. **Feeney and Allison,** *Evolutionary Biochemistry of Proteins,* John Wiley & Sons, New York, 1969.

Inhibitors of Animal Tissues and Blood

28. **Lebreton de Vonne and Mouray,** *Ann. Biol. Clin.,* 32, 185 (1974).
29. **Barton and Yin,** in *Metabolic Inhibitors,* Vol. 4, Hochster, Kates, and Quastel, Eds., Academic Press, New York, 1973, 215.
30. **Steinbuch,** *Rev. Fr. Transfus.,* 14, 61 (1971).
31. **Burck,** in *Methods in Enzymology,* Vol. 19, Perlmann and Lorand, Eds., Academic Press, New York, 1970, 906.

SPECIFICITIES AND SOME PROPERTIES OF PLANT PROTEASE INHIBITORS

Table 1*
SOLANUM TUBEROSUM, POTATO TUBERS

Source	Name	Molecular weight[a]	I.P.	Trypsin	α-Chymotrypsin	Plasmin	Plasma or serum kallikrein	Porcine pancreatic kallikrein	Urinary kallikrein	Salivary kallikrein	Thrombin
var. Russet Burbank	Chymotrypsin inhibitor I[c]	9,300 sub, el 9,900 sub, gel 39,000 ulc		+ 0[d,e]	I_{sub}-E +[f]	0[d]	0[d]	0			0[d]
Commercial	Polyvalent protease inhibitor										
	K1a	23,500 gel[g]		+[g]	+	+	+[h]	0	0[d]		
	K2	24,000 gel		+	+	+	+[h]	0	0[d]		
	K3	23,000 gel		+	+	+	+[h]	0	0[d]		
var. Danshaku-Imo	Kallikrein inhibitor										
	1	24,200 ulc 22,689 aa	5.6	±	±	±[d]	+	I-E	±[d]	±[d]	±[d]
	2	28,700 ulc 22,148 aa	6.4					I-E			
var. Maritta	Proteinase inhibitor										
	K4		>9.2	+	+						
	A1a		8.9	+	+						
	A1b		8.7	+	0						
	A2		8.2	+	0						
	A4		7.2	+	+						
	A5	6,000 sub, ulc 6,400 aa 22,000–26,000 gel	6.3	+	I_{sub}-E $I_{-2\,sub}$-E						
	A6		6.9	+	0						
	A7a		6.4	+	+						
	A7b		5.1	+	+						
	A8		5.8	+	+						
var. Danshaku-Imo	Proteinase inhibitor										
	I	9,400 aa 19,300 gel[i] 42,000 gel[j]		±	I_{sub}-E[k]						
	IIa	10,350 aa 10,600 gel	8.5	I-E[l]	I-E[m]						
	IIb	10,350 aa 10,600 gel	9.1	±	I-E[m]						

Source	Name	Molecular weight[a]	I.P.	Trypsin	α-Chymotrypsin	Plasmin	Bovine carboxypeptidase A	Porcine carboxypeptidase B	Shrimp carboxypeptidase A and B
var. Russett Burbank	Carboxypeptidase inhibitor	4,100–4,300 aa 3,100 gel 3,600 l 3,000 eq		0	0		I-E[o]	I-E[p]	+

*Footnotes and references follow Table 4.

Table 1 (continued)
SOLANUM TUBEROSUM, POTATO TUBERS

Source	Name	Papain	Subtilopeptidases	Streptomyces griseus proteases	Aspergillus spp. proteases	Penicillium spp. proteases	Nagarse	Other	Amino acid table	Sequence table	Reference
					Specificity[b]				Information in		
var. Russet Burbank	Chymotrypsin inhibitor I[c]	0	+	+				Ficin, 0 Bromelin, 0 Pepsin, 0	*	*	1–9
Commercial	Polyvalent protease inhibitor K1a K2 K3								*		10
var. Danshaku-Imo	Kallikrein inhibitor										
	1	±		0				Pepsin, 0			8, 11, 12
	2										
var. Maritta	Proteinase inhibitor										
	K4		+		+				*		13–16
	A1a										
	A1b										
	A2		0		0				*		
	A4		+		+				*		
	A5		+	+	+			Fungal proteinase K, +			
	A6		0		0				*		
	A7a		0		0						
	A7b		+		+				*		
	A8		0		0				*		
var. Danshaku-Imo	Proteinase inhibitor										
	I	0	0		+	+	I-E[n]	Thermolysin, 0 Pepsin, 0	*		17
	IIa	0	0		±	±	±[n]	Thermolysin, 0 Pepsin, 0	*		17–19
	IIb	0	0		±	±	+[n]	Thermolysin, 0 Pepsin, 0	*		17–19

Source	Name	Pseudomonas stutzeri carboxypeptidase G1	Human serum carboxypeptidase N	Bovine spleen cathepsin A	Yeast protease C	Yeast protease α-S6	Nagarse	Other	Amino acid table	Sequence table	Ref.
				Specificity[b]					Information in		
var. Russett Burbank	Carboxypeptidase inhibitor	+	0	0	0	0			*	*	4, 6, 21

Compiled by Beatrice Kassell and Monica June Williams. Supported by grant BMS74-221-2 from the National Science Foundation.

Table 2*
LEGUMES

Source	Name	Molecular weight[a]	I.P.	K_I (M) Trypsin	K_I (M) Chymotrypsin	Specificity[b] Trypsin	α-Chymotrypsin	Plasmin	Plasma or serum kallikrein	Organ kallikrein	Thrombin
Glycine soya Soybean	Kunitz inhibitor	20,100 aa 21,500 ulc	4.0	10^{-10}		I-E[e,g,q,r]	+[f,s]	+	+[d,q]	±	I-E[g] +[d]
	Bowman-Birk inhibitor	7,900 aa 8,000–16,400 ulc 8,250 gel	4.0–4.3	1.1×10^{-11} [u] $10^{-8}-10^{-9}$	$10^{-7}-10^{-9}$	I-E ±[d]	I-E	+[d]	±[d,q]	±	0[d]
	Elastase inhibitors 1, 2, 3, 4					+					
	V					+	+				
	Tribolium protease inhibitor					0	0				

*Footnotes and References follow Table 4.

Table 2 (continued)
LEGUMES

Source	Name	Specificity[b]							Information in		References
		Papain	Acrosin	B. subtilis proteases	Streptomyces griseus proteases	Elastase	Pepsin	Other	Amino acid table	Sequence table	
Glycine soya Soybean	Kunitz inhibitor	0	+[d,q,t]	0		0	0	Angiotensin converting enzyme, + Prekallikrein activator, + Cocoonase, + Cathepsins A, B, C, D, 0 Aminopeptidase, 0 Rabbit liver lysosomal acid proteinase, 0 Human sperm "chymotrypsin," 0 Human anionic trypsin I, + Human anionic trypsin II, I-E Human cationic trypsin, + Human erythrocyte insulin-degrading proteinase, 0 Urinary kallikrein, 0 Human pancreatic protease E, + Carboxypeptidase A, 0 *Flavobacterium elastolyticum* elastase, 0 Human activated factor X, + Human activated factor IX, 0	*	*	7, 9, 20, 22–64
	Bowman-Birk inhibitor	0	+[d,q]		+	0	0	Urinary kallikrein, + *Tribolium* larval proteinase, 0 Human sperm "chymotrypsin," 0 Human erythrocyte insulin-degrading proteinase, + Human pancreatic protease E, ±	*	*	7, 20, 22, 26, 30–33, 36, 45, 47, 55, 57, 64–73
	Elastase inhibitors 1, 2, 3, 4					+					74
	V					+[v]					75
	Tribolium protease inhibitor							*Tribolium castaneum* larval proteinase, + *Tribolium confusum* larval proteinase, +			76

Table 2 (continued)
LEGUMES

Source	Name	Molecular weight[a]	I.P.	K_I (M) Trypsin	K_I (M) Chymotrypsin	Specificity[b] Trypsin	α-Chymotrypsin	Plasmin	Plasma or serum kallikrein	Organ kallikrein	Thrombin
Phaseolus vulgaris L Navy bean	Trypsin inhibitor	7,900 eq				+[d]	+				
	Trypsin inhibitor I	23,000 ulc				I-Eg		±[d]			0[d]
	II			2×10^{-9}		I-Eg					
Kidney Bean var. Haricot	Protease inhibitor Isoinhibitors	10,000 ulc	5			I-Eg ±[d]	I-E	I-E ±[d]			±[d] 0[d]
Garden bean var. Great Northern	I	8,100 aa				I-E	±				
	II	8,400 aa				I-E	±				
	III[b]	8,900 aa				I-E	I-E				
Bush bean var. Borlotto	PVI 2					I-E	±				
	PVI 3					I-E	+				
	PVI 4					I-E	+				
Phaseolus aureus Mung bean	Trypsin inhibitor	18,000 os				I-E	0				
Green gram	Trypsin inhibitor B	9,200 os	5	10^{-9}–10^{-10}		I-E$_2$	±				
		9,000 ulc									
		8,000 aa									
Phaseolus coccineus	Trypsin-chymotrypsin inhibitor										
Fire bean	2	8,800 aa				I-E	+				
Scarlet runner var. Prize-winner	3	8,700 aa				I-E$_2$	+				
		9,000 gel									
	4	8,700 aa				I-E$_2$	+				
		9,600 gel									
	5	9,300 aa				I-E	+				
		10,700 gel									

Table 2 (continued)
LEGUMES

Source	Name	Papain	Acrosin	B. subtilis proteases	Streptomyces griseus proteases	Elastase	Pepsin	Other	Amino acid table	Sequence table	References
Phaseolus vulgaris L Navy bean	Trypsin inhibitor I										7, 77, 78
	Trypsin inhibitor II										79
Kidney Bean var. Haricot	Protease inhibitor	0		0		I-E	0	Carboxypeptidase A, 0; Carboxypeptidase B, ±	*		7, 80, 81
Garden Bean var. Great Northern	Isoinhibitors I					±			*		82, 83
	II					+			*	*	
	III[b]					±			*		
Bush bean var. Borlotto	PVI 2								*		84
	PVI 3								*		
	PVI 4								*		
Phaseolus aureus Mung bean	Trypsin inhibitor	±					0				85
Green gram	Trypsin inhibitor B								*		86—88
Phaseolus coccineus Fire bean	Trypsin-chymotrypsin inhibitor 2								*		89
	3								*		
	4								*		
Scarlet runner var. Prize winner	5								*		

Table 2 (continued)
LEGUMES

Source	Name	Molecular weight[a]	I.P.	K_I (M) Trypsin	K_I (M) Chymotrypsin	Specificity[b] Trypsin	Specificity[b] α-Chymotrypsin	Plasmin	Plasma or serum kallikrein	Organ kallikrein	Thrombin
Phaseolus lunatus											
Lima bean	Trypsin inhibitor										
	1	8,400 aa				I-E	+				
	2	8,300 aa				I-E	+				
	3	9,900 aa				I-E	+				
	4	9,400 aa				I-E	I-E				
	Trypsin-chymotrypsin inhibitor										
	4	9,700 aa				I-E	+				
	6	16,200 ulc				I-E	+				
	Lima bean inhibitor (mixture of components)		3.6			I-E[d,g] +e	+[d,f,g]	±[d]	0[d,q]	±	0
Vigna sinensis											
Black-eyed pea	Trypsin and chymotrypsin inhibitor	13,300 gel 10,700 os 12,300 ulc 9,500 aa		1.4×10^{-8}	7.6×10^{-9}	I-E[g] +d	I-E	±[d]			0[d]
Vigna unguiculata											
Yard-long bean (cotyledons)	Trypsin inhibitor	12,000				+					

Table 2 (continued)
LEGUMES

Source	Name	Specificity[b]								Information in		References
		Papain	Acrosin	B. subtilis proteases	Streptomyces griseus proteases	Elastase	Pepsin	Other		Amino acid table	Sequence table	
Phaseolus lunatus	Trypsin inhibitor											
Lima bean	1									*		90–92
	2									*		
	3									*	*	
	4										*	
	Trypsin-chymotrypsin inhibitor											
	4				0	0			*Aspergillus oryzae* proteinase, 0	*		93
	6			0	0				*Aspergillus oryzae* proteinase, 0	*		
	Lima bean inhibitor (mixture of components)	0	+t		+	0	0	Human chymotrypsins I and II, + *Tribolium* larval proteinase, 0 Cathepsins A, B, C, D, 0 Human plasma prekallikrein activator, + Urinary kallikrein, + Human sperm "chymotrypsin," 0 Human cationic and anionic trypsin, + Human pancreatic protease E, ±		*		7, 9, 20, 22, 28, 31, 33, 47, 51, 54, 67, 72, 94–101
Vigna sinesis Black-eyed pea	Trypsin and chymotrypsin inhibitor									*		7, 102
Vigna unguiculata Yard-long bean (cotyledons)	Trypsin inhibitor							Proteolytic enzymes of cotyledon of same plant, +				103

Table 2 (continued)
LEGUMES

Source	Name	Molecular weight[a]	I.P.	K_I (M)		Specificity[b]					
				Trypsin	Chymo-trypsin	Trypsin	α-Chymo-trypsin	Plas-min	Plasma or serum kalli-krein	Organ kalli-krein	Thrombin
Pisum sativum	Trypsin and chymo-trypsin inhibitors										
Pea	1	9,800 el				I-E	+				
		11,300 aa									
var. Schnee-bergeri	3	11,450 el				+					
		10,300 aa^w									
	4	8,700 el	5.2			I-E	I-E				
		11,200 aa^w									
	6	10,100 el	5.9			I-E	I-E				
		8,000 aa									
	8	10,050 el	7.7			I-E	I-E				
		12,800 aa									
Lathyrus odoratus Sweet pea	Trypsin and chymo-trypsin inhibitor	11,800 aa				+	+				
Lathyrus sativus Field pea	Trypsin inhibitors 1 to 5					+					
Vicia faba L. Double bean	Trypsin inhibitor	23,000 os	8.6			+	+				
Broad bean	Trypsin inhibitors 1					+					
	2					+					

Table 2* (continued)
LEGUMES

Source	Name	Specificity[b]							Information in		References
		Pa-pain	Acro-sin	B. sub-tilis pro-teases	Strepto-myces griseus pro-teases	Elas-tase	Pep-sin	Other	Amino acid table	Sequence table	
Pisum sativum	Trypsin and chymo-trypsin inhibitors										
Pea var. Schnee-bergeri	1								*		104
	3								*		
	4								*		
	6								*		
	8								*		
Lathyrus odoratus											
Sweet pea	Trypsin and chymo-trypsin inhibitor								*		105
Lathyrus sativus											
Field pea	Trypsin inhibitors 1 to 5										106
Vicia faba L											
Double bean	Trypsin inhibitor	±					0				107
Broad bean	Trypsin inhibitors										
	1								*		108
	2								*		

Table 2 (continued)
LEGUMES

Source	Name	Molecular weight[a]	I.P.	K_I (M) Trypsin	K_I (M) Chymotrypsin	Specificity[b] Trypsin	α-Chymotrypsin	Plasmin	Plasma or serum kallikrein	Organ kallikrein	Thrombin
Dolichos lab lab Field bean	Trypsin inhibitor	23,500 eq	4.7	3.3×10^{-10}		I-E	I-E				
Hakuhenzu bean	Trypsin inhibitor	6,400 aa 9,500 gel[x]				+	+				
Arachis hypogaea Peanut	Proteinase inhibitor	7,500 ulc 7,700 gel 8,100 aa	8-9			I-E	I-E	+	+[d,q]		
Ground nut	Proteinase inhibitor PI and PII	17,000 gel 5,300 aa				+	+	+	+	0	
	Proteinase inhibitor	17,000 aa	6.75			+	+	+			
Cicer arietinum Chick pea	Inhibitor CI	12,000 aa				+	+				
Bauhenia purpura seed (shrub)	Trypsin and chymotrypsin inhibitor			$10^{-8} - 10^{-10}$		+	+				
Bauhenia seed	Multi-functional protease inhibitor	24,300– 25,400 ulc				+	+				
Medicago sativa Alfalfa meal	Trypsin inhibitor A'					+					

Table 2. (continued)
LEGUMES

Source	Name	Specificity[b]							Information in		References
		Pa-pain	Acro-sin	B. sub-tilis pro-teases	Strepto-myces griseus pro-teases	Elas-tase	Pep-sin	Other	Amino acid table	Sequence table	
Dolichos lab lab Field bean	Trypsin inhibitor	0					0				109
Hakuhenzu bean	Trypsin inhibitor										110
Arachis hypogaea Peanut	Proteinase inhibitor		+q			0q		*	*		20, 32, 33, 111–113
Ground nut	Proteinase inhibitor PI and PII								*		55, 114, 115
	Proteinase inhibitor							Human erythrocyte insulin-degrading proteinase, +	*		116
Cicer arietinum Chick pea	Inhibitor CI								*		117
Bauhenia purpura seed (shrub)	Trypsin and chymo-trypsin inhibitor	0			+y,z	0	0	Carboxypeptidase A, 0			118
Bauhenia seed	Multi-functional protease inhibitor	+			+y	0	+	Ficin, 0 Bromelain, 0	*		119
Medicago sativa Alfalfa meal	Trypsin inhibitor A'										120

Compiled by Beatrice Kassell and Monica June Williams. Supported by grant BMS74-22102 from the National Science Foundation.

Table 3*
GRAMINEAE CEREAL GRAINS

					Specificity[b]								Information in		
Source	Name	Molecular weight[a]	I.P.	Carbohydrate content	Trypsin	α Chymotrypsin	Organ kallikrein	Papain	B. subtilis proteases	Aspergillus sp. proteases	Streptomyces griseus proteases	Pepsin	Other	Amino acid table	Sequence table	References
Barley var. Pirkka endosperm	Trypsin inhibitor	14,055 aa 14,400 ulc			+	0			0	0	0	0	Barley endopeptidases, 0 Barley carboxypeptidase, 0 Green malt endopeptidase, 0 *Pseudomonas fluorescens* protease, 0	*		121–123
Embryo	Trypsin inhibitor	18,500 gel			+											122
Grains	Microbial alkaline protease iso-inhibitors	25,000 gel	4.6–5.4		0			0	+	+	+	0	*Alternaria tenuissima* protease, + Barley endopeptidases, 0			124
Rye																
Flour	Trypsin inhibitor	18,700 aa 17,000 gel			+									*		125
Embryo	Trypsin inhibitor	18,500 gel			+											122
Endosperm	Trypsin inhibitor	14,000 gel			+											122
Endosperm var. Smolickie	Trypsin inhibitor	10,000 gel			+	+	0[q]	0				0				126

*Footnotes and references follow Table 4.

Table 3 (continued)
GRAMINEAE CEREAL GRAINS

Source	Name	Molecular weight[a]	I.P.	Carbohydrate content	Trypsin	α-Chymotrypsin	Organ kallikrein	Papain	B. subtilis proteases	Aspergillus ssp. proteases	Streptomyces griseus proteases	Pepsin	Other	Amino acid table	Sequence table	References
Wheat Whole wheat Flour	Trypsin inhibitor			+	+	0						0				127
Wheatgerm	Trypsin inhibitor 1	18,300 aa; 17,000 gel			+									*		125
	2	12,000 gel			+											
Embryo	Trypsin inhibitor	18,500 gel			+											122
Whole wheat Flour	Tribolium larval protease inhibitor				0	0		0					Tribolium castaneum protease, +; Carboxypeptidase B, 0			128
Corn seeds Zea mays	Trypsin inhibitor	7,000 aa; 18,500 gel; 18,500 ulc		0	I_{sub}-E									*		129, 130
var. saccharata	Trypsin inhibitor			+	+	+		0								131
Rice embryo	Protease inhibitor				+									Rice seed protease, +			132
Grain	Subtilisin inhibitor				0	0			+							133

Compiled by Beatrice Kassell and Monica June Williams. Supported by grant BMS74-22102 from the National Science Foundation.

Table 4
OTHER PLANTS

Source	Name	Molecular weight[a]	I.P.	$K_I(M)$ Trypsin	Specificity[b] Trypsin	α-Chymo-trypsin	Plasmin	Papain	Bromelain	Ficin	B. subtilis proteases	Aspergillus proteases	Streptomyces griseus proteases	Pepsin	Amino acid table	Sequence table	References
Raphanus sativus Radish seed	Proteinase inhibitor																
	R-I	8,000 gel 7,200 aa		10^{-10}	I-E	0		0	0				+	0	*		134, 135
	R-III	11,200 ulc 11,000 aa		10^{-10}	I-E	+		0	0		+		+	0	*		
Ipomoea batas Sweet potato Tubers	Trypsin inhibitor																
	II	23,000 gel	4.24		I-E	0	+, I-E[a]	0	0		0	0		0			136–138
	III	24,000 gel	4.39	1.3×10^{-7}	I-E	0		0	0		0	0		0			
Brassica juncea Rape seed	Proteinase inhibitor																
	I				+	0											139
	II	20,000 gel			+	+											
	III	10,000 gel			+	+											
	IV				+	0											
Ananas comosus Pineapple stem	Proteinase inhibitor																
	III	5,500 aa			+			+	E-I[b']	+					*		140
	VII	5,630 aa			+			+	E-I[b']	+					*	*	

*Footnotes and references follow Table 4.

Compiled by Beatrice Kassell and Monica June Williams. Supported by grant BMS74-22102 from the National Science Foundation.

Table 4 (continued)

[a]Symbols used to designate the determination of molecular weight are: gel, gel filtration; sub, subunit molecular weight; aa, minimum molecular weight by amino acid composition; l, light scattering; eq, equivalence to an enzyme; el, SDS-gel electrophoresis; ulc, various ultracentrifugal methods; os, osmotic pressure. The molecular weights calculated from amino acid compositions are approximations when the asparagine and glutamine contents are not known. Unless otherwise indicated, molecular weight calculated from amino acid composition does not include carbohydrate.

[b]Inhibition is indicated as follows: stoichiometric inhibition, complex I-E (inhibitor-enzyme); sub, subunit; +, strong inhibition; ±, weak inhibition; 0, no inhibition. Bovine trypsin and bovine α-chymotrypsin are understood when species is not specified.

[c]Fraction II inhibits porcine and human serum kallikrein.[20]

[d]Human.

[e]Turkey.

[f]Chicken.

[g]Bovine.

[h]Canine.

[i]0.1 M acetic acid.

[j]0.1 M NaCl.

[k]$K_I = 5 \times 10^{-10}$ M.

[l]$K_I = 4.5 \times 10^{-9}$ M [soybean inhibitor (kunitz) used for reference gave a $K_I = 3.1 \times 10^{-8}$ M, not in agreement with other data (see Table 3, Legumes)].

[m]$K_I = 3.2 \times 10^{-8}$ M for inhibitor IIa and 2.7×10^{-9} M for inhibitor IIb.

[n]$K_I = 1 \times 10^{-10}$ M, 1.1×10^{-7} M, 2.2×10^{-8} M, for inhibitors I, IIa, and IIb, respectively.

[o]$K_I = 5 \times 10^{-9}$ M.

[p]$K_I = 5 \times 10^{-8}$ M.

[q]Porcine.

[r]Starfish.

[s]Bovine α and B.

[t]Rabbit.

[u]Bovine β-trypsin.

[v]Apparent $K_I = 9.6 \times 10^{-8}$ M.

[w]Minimum molecular weight × 2.

[x]This inhibitor contains more than 20% carbohydrate.

[y]Pronase trypsin.

[z]Pronase elastase.

[a']$K_I = 2.7 \times 10^{-7}$ M.

[b']$K_I = 10^{-7}$ M at pH 3 to 5.

REFERENCES

1. **Balls and Ryan,** *Science,* 138, 983 (1962).
2. **Balls and Ryan,** *J. Biol. Chem.,* 238, 2976 (1963).
3. **Ryan,** *Biochemistry,* 5, 1592 (1966).
4. **Ryan,** *Biochem. Biophys. Res. Commun.,* 44, 1265 (1971).
5. **Melville and Ryan,** *J. Biol. Chem.,* 247, 3445 (1972).
6. **Ryan, Hass, and Kuhn,** *J. Biol. Chem.,* 249, 5495 (1974).
7. **Feeney, Means, and Bigler,** *J. Biol. Chem.,* 244, 1957 (1969).
8. **Hojima, Moriya, and Moriwaki,** *J. Biochem.* (Tokyo), 69, 1027 (1971).
9. **Ryan, Clary, and Tomimatsu,** *Arch. Biochem. Biophys.,* 110, 175 (1965).
10. **Hochstrasser, Werle, Siegelmann, and Schwarz,** *Hoppe Seyler's Z. Physiol. Chem.,* 350, 897 (1969).
11. **Moriya, Hojima, Moriwaki, and Tajima,** *Experientia,* 26, 720 (1970).
12. **Hojima, Moriwaki, and Moriya,** *J. Biochem.* (Tokyo), 73, 923 (1973).
13. **Santarius and Belitz,** *Chem. Mikrobiol. Technol. Lebensm.,* 2, 56 (1972).
14. **Belitz, Kaiser, and Santarius,** *Biochim. Biophys. Res. Commun.,* 42, 420 (1971).
15. **Kaiser and Belitz,** *Z. Lebensm. Unters. Forsch.,* 151, 18 (1973).
16. **Santarius and Belitz,** *Chem. Mikrobiol. Technol. Lebensm.,* 3, 180 (1975).

17. Kiyohara, Iwasaki, and Yoshikawa, *J. Biochem.* (Tokyo), 73, 89 (1973).
18. Iwasaki, Kiyohara, and Yoshikawa, *J. Biochem.* (Tokyo), 70, 817 (1971).
19. Iwasaki, Kiyohara, and Yoshikawa, *J. Biochem.* (Tokyo), 72, 1029 (1972).
20. Wunderer, Kummer, and Fritz, *Hoppe Seyler's Z. Physiol. Chem.*, 353, 1646 (1972).
21. Ryan, Hass, Kuhn, and Neurath, in *Bayer Symposium V, "Proteinase Inhibitors,"* Fritz, Tschesche, Greene, and Truscheit, Eds., Springer-Verlag, Berlin, 1974, 565.
22. Back and Steger, *Fed. Proc.*, 27, 96 (1968).
23. Bagdasarian, Talamo, and Colman, *J. Biol. Chem.*, 248, 3456 (1973).
24. Barrett, *Biochem. J.*, 104, 601 (1967).
25. Boaz, Wyatt, and Fitz, *Fed. Proc.*, 33, 1234 (1974).
26. Catsimpoolas, *Sep. Sci.*, 4, 483 (1969).
27. Engleman and Greenbaum, *Biochem. Pharmacol.*, 20, 922 (1971).
28. Feinstein, Hoffstein, Koifmann, and Sokolovsky, *Eur. J. Biochem.*, 43, 569 (1974).
29. Figarella, Negri, and Guy, in *Bayer Symposium V, "Proteinase Inhibitors,"* Fritz, Tschesche, Greene, and Truscheit, Eds., Springer-Verlag, Berlin, 1974, 213.
30. Fink, Fritz, Jaumann, Schiessler, Förg-Brey, and Werle, in *Protides of the Biological Fluids, 20th Colloq.*, Peeters, Ed., Pergamon Press, Oxford, 1973, 425.
31. Fritz, Arnhold, Förg-Brey, Zaneveld, and Schumacher, *Hoppe Seyler's Z. Physiol. Chem.*, 353, 1651 (1972).
32. Fritz, Förg-Brey, Schiessler, Arnhold, and Fink, *Hoppe Seyler's Z. Physiol. Chem.*, 353, 1010 (1972).
33. Gertler and Feinstein, *Eur. J. Biochem.*, 20, 547 (1971).
34. Green, *J. Biol. Chem.*, 205, 535 (1953).
35. Greenbaum and Fruton, *J. Biol. Chem.*, 226, 173 (1957).
36. Harry and Steiner, *Eur. J. Biochem.*, 16, 174 (1970).
37. Kafatos, Law, and Tarkakoff, *J. Biol. Chem.*, 242, 1488 (1967).
38. Koide, Tsunasawa, and Ikenaka, *J. Biochem.* (Tokyo), 71, 165 (1972).
39. Kozlovskaya and Elyakova, *Biochim. Biophys. Acta*, 371, 63 (1974).
40. Kunitz, *J. Gen. Physiol.*, 29, 149 (1946); 30, 291 (1947).
41. Lanchantin, Friedmann, and Hart, *J. Biol. Chem.*, 244, 865 (1969).
42. Laskowski, Jr. and Sealock, in *The Enzymes*, Vol. 3, 3rd ed., Boyer, Ed., Academic Press, New York, 1971, 375.
43. Lebowitz and Laskowski, Jr., *Biochemistry*, 1, 1044 (1962).
44. Levilliers, Péron-Renner, and Pudles, in *Bayer Symposium V, "Proteinase Inhibitors,"* Fritz, Tschesche, Greene, and Truscheit, Eds., Springer-Verlag, Berlin, 1974, 432.
45. Liener and Kakade, in *Toxic Constituents of Plant Foodstuffs*, Liener, Ed., Academic Press, New York, 1969, 7.
46. Lundblad and Davie, *Biochemistry*, 3, 1720 (1964); 4, 113 (1965).
47. Mallory and Travis, in *Bayer Symposium V, "Proteinase Inhibitors,"* Fritz, Tschesche, Greene, and Truscheit, Eds., Springer-Verlag, Berlin, 1974, 250.
48. Mandl and Cohen, *Arch. Biochem. Biophys.*, 91, 47 (1960).
49. Mansfeld, Rybák, Horáková, and Hladovec, *Hoppe Seyler's Z. Physiol. Chem.*, 318, 6 (1960).
50. Matsubara and Nishimura, *J. Biochem.* (Tokyo), 45, 413, 503 (1958).
51. Misaka and Tappel, *Comp. Biochem. Physiol. B*, 38, 651 (1971).
52. Rackis, Sasame, Mann, Anderson, and Smith, *Arch. Biochem. Biophys.*, 98, 471 (1962).
53. Sach and Thély, *C.R. Acad. Sci. Ser. D.*, 266, 1200 (1968).
54. Stambaugh, Brackett, and Mastroianni, *Biol. Reprod.*, 1, 223 (1969).
55. Tschesche, Dietl, Kolb, and Standl, in *Bayer Symposium V, "Proteinase Inhibitors,"* Fritz, Tschesche, Greene, and Truscheit, Eds., Springer-Verlag, Berlin, 1974, 586.
56. Walford and Kickhöfen, *Arch. Biochem. Biophys.*, 98, 191 (1962).
57. Wang, *Cereal Chem.*, 48, 303 (1971).
58. Werle and Kaufmann-Boetsch, *Naturwissenschaften*, 19, 559 (1959).
59. Werle and Mayer, *Biochem. Z.*, 323, 279 (1952).
60. Wu and Laskowski, Sr., *J. Biol. Chem.*, 213, 609 (1955).
61. Wu and Scheraga, *Biochemistry*, 1, 698 (1962).
62. Zaneveld, Dragoje, and Schumacher, *Science*, 177, 702 (1972).
63. Zaneveld, Polakoski, Robertson, and Williams, in *Proc. 1st Int. Res. Conf. Proteinase Inhibitors*, Fritz and Tschesche, Eds., Walter de Gruyter, Berlin, 1970, 236.
64. Zaneveld, Schumacher, Tauber, and Propping, in *Bayer Symposium V, "Proteinase Inhibitors,"* Fritz, Tschesche, Greene, and Truscheit, Eds., Springer-Verlag, Berlin, 1974, 136.
65. Birk, *Ann. N.Y. Acad. Sci.*, 146, 388 (1968).
66. Birk, Gertler, and Khalef, *Biochem. J.*, 87, 281 (1963).
67. Birk, Harpaz, Ishaaya, and Bondi, *J. Insect Physiol.*, 8, 417 (1962).
68. Frattali, *J. Biol. Chem.*, 244, 274 (1969).
69. Kakade, Simons, and Liener, *Biochim. Biophys. Acta*, 200, 168 (1970).
70. Millar, Willick, Steiner, and Frattali, *J. Biol. Chem.*, 244, 281 (1969).

71. Odani and Ikenaka, *J. Biochem.* (Tokyo), 71, 839 (1972).
72. Trop and Birk, *Biochem. J.*, 109, 475 (1968).
73. Yamamoto and Ikenaka, *J. Biochem.* (Tokyo), 62, 141 (1967).
74. Bieth and Frechin, *Biochim. Biophys. Acta*, 364, 97 (1974).
75. Bieth and Frechin, in *Bayer Symposium V, "Proteinase Inhibitors,"* Fritz, Tschesche, Greene, and Truscheit, Eds., Springer-Verlag, Berlin, 1974, 291.
76. Birk, Gertler, and Khalef, *Biochim. Biophys. Acta*, 67, 326 (1963).
77. Bowman, *Arch. Biochem. Biophys.*, 144, 541 (1971).
78. Wagner and Riehm, *Arch. Biochem. Biophys.*, 121, 672 (1967).
79. Whitley, Jr. and Bowman, *Fed. Proc.*, 33, 1530 (1974) and personal communication, *Arch. Biochem. Biophys.*, 169, 42 (1975).
80. Pusztai, *Biochem. J.*, 101, 379 (1966).
81. Pusztai, *Eur. J. Biochem.*, 5, 252 (1968).
82. Wilson and Laskowski, Sr., *J. Biol. Chem.*, 248, 756 (1973).
83. Wilson and Laskowski, Sr., in *Bayer Symposium V "Proteinase Inhibitors,"* Fritz, Tschesche, Greene, and Truscheit, Eds., Springer-Verlag, Berlin, 1974, 286.
84. Belitz, Fuchs, Nitsche, and Al-Sultan, *Z. Lebensm. Unters. Forsch.*, 150, 215 (1972).
85. Honavar and Sohonie, *J. Sci. Ind. Res. Sec. C*, 18, 202 (1959).
86. Wang, Chi, and Tsao, *Sheng Wu Hua Hsueh Yu Sheng Wu Wu Li Hsueh Pao*, 5, 510 (1965).
87. Chü and Chi, *Sci. Sin.*, 14, 1441 (1965).
88. Chü, Lo, Jen, Chi, and Tsao, *Sci. Sin.*, 14, 1444 (1965).
89. Belitz, Fuchs, and Grimm, *Lebensm. Wiss. Technol.*, 4, 89 (1971).
90. Jones, Moore, and Stein, *Biochemistry*, 2, 66 (1963).
91. Krahn and Stevens, *FEBS Lett.*, 13, 339 (1971).
92. Krahn and Stevens, *FEBS Lett.*, 28, 313 (1972).
93. Haynes and Feeney, *J. Biol. Chem.*, 242, 5378 (1967).
94. Coan and Travis, in *Proc. 1st Int. Res. Conf. Proteinase Inhibitors,* Fritz and Tschesche, Eds., Walter de Gruyter, Berlin, 1971, 294.
95. Feinstein, Hoffstein, and Sokolovsky, in *Bayer Symposium V, "Proteinase Inhibitors,"* Fritz, Tschesche, Greene, and Truscheit, Eds., Springer-Verlag, Berlin, 1974, 199.
96. Fraenkel-Conrat, Bean, Ducay, and Olcott, *Arch. Biochem. Biophys.*, 37, 393 (1952).
97. Jirgensons, Ikenaka, and Gorguraki, *Makromol. Chem.*, 39, 149 (1960).
98. Lewis and Ferguson, *J. Biol. Chem.*, 204, 503 (1953).
99. Robbins and Summaria, *Immunochemistry*, 3, 29 (1966).
100. Ryan and Clary, *Arch. Biochem. Biophys.*, 108, 169 (1964).
101. Skoza, Tse, Semar, and Johnson, *Ann. N.Y. Acad. Sci.*, 146, 659 (1968).
102. Ventura, Xavier Filho, Azevedo Moreira, Menezes Aquino, and Augusto Pinheiro, *An. Acad. Bras. Cienc.*, 43, 233 (1971).
103. Royer, Miège, Grange, Miège, and Mascherpa, *Planta*, 119, 1 (1974).
104. Weder and Hory, *Lebensm. Wiss. Technol.*, 5, 86 (1972).
105. Weder and Belitz, *Dtsch. Lebensm. Rundsch.*, 78, 1969.
106. Roy, *J. Agric. Food Chem.*, 20, 778 (1972).
107. Sohonie, Huprikar, and Joshi, *J. Sci. Ind. Res. Ser. C*, 18, 95 (1959).
108. Warsy and Stein, *Qual. Plant. Plant Foods Hum. Nutr.*, 23, 157 (1973).
109. Banerji and Sohonie, *Enzymologia*, 36, 137 (1969).
110. Furusawa, Kurosawa, and Chuman, *Agric. Biol. Chem.*, 38, 1157 (1974).
111. Birk and Gertler, in *Proc. 1st Int. Res. Conf. Proteinase Inhibitors,* Fritz and Tschesche, Eds., Walter de Gruyter, Berlin, 1971, 142.
112. Egeblad, *Thromb. Diath. Haemorrh.*, 27, 31 (1967).
113. Tur-Sinai, Birk, Gertler, and Rigbi, *Biochim. Biophys. Acta*, 263, 666 (1972).
114. Hochstrasser, Illchmann, and Werle, *Hoppe Seyler's Z. Physiol. Chem.*, 350, 929 (1969).
115. Hochstrasser, Illchmann, Werle, Hössl, and Schwarz, *Hoppe Seyler's Z. Physiol. Chem.*, 351, 1503 (1970).
116. Tixier, *C. R. Acad. Sci.*, 266, 2498 (1968).
117. Birk, in *Bayer Symposium V, "Proteinase Inhibitors,"* Fritz, Tschesche, Greene, and Truscheit, Eds., Springer-Verlag, Berlin, 1974, 355.
118. Pinsky and Schwimmer, *Phytochemistry*, 13, 779 (1974).
119. Goldstein, Trop, and Birk, *Nat. New Biol.*, 246, 29 (1973).
120. Chien and Mitchell, *Phytochemistry*, 9, 717 (1970).
121. Mikola and Suolinna, *Eur. J. Biochem.*, 9, 555 (1969).
122. Mikola and Kirsi, *Acta Chem. Scand.*, 26, 787 (1972).
123. Kirsi and Mikola, *Planta*, 96, 281 (1971).
124. Mikola and Suolinna, *Arch. Biochem. Biophys.*, 144, 566 (1971).
125. Hochstrasser, Werle, Schwarz, and Siegelmann, *Hoppe Seyler's Z. Physiol. Chem.*, 350, 249 (1969).

126. Polanowski, *Acta Soc. Bot. Pol.*, 43, 27 (1974).
127. Shyamala and Lyman, *Can. J. Biochem.*, 42, 1825 (1964).
128. Applebaum and Konijn, *J. Insect Physiol.*, 12, 665 (1966).
129. Hochstrasser, Muss, and Werle, *Hoppe Seyler's Z. Physiol. Chem.*, 348, 1337 (1967).
130. Hochstrasser, Illchmann, and Werle, *Hoppe Seyler's Z. Physiol. Chem.*, 351, 721 (1970).
131. Chen and Mitchell, *Phytochemistry*, 12, 327 (1973).
132. Horiguchi and Kitagishi, *Plant Cell Physiol.*, 12, 907 (1971).
133. Kato, Tominaga, and Kihara, in Symposium on Protein Structure, Maebashi, Japan, 1972, taken from **Laskowski, Jr., Kato, Leary, Schrode, and Sealock**, in *Bayer Symposium V, "Proteinase Inhibitors,"* Fritz, Tschesche, Greene, and Truscheit, Eds., Springer-Verlag, Berlin, 1974, 597.
134. Hata, Ogawa, and Hayashi, *Bull. Res. Inst. Kyoto Univ.*, 30, 51 (1967).
135. Ogawa, Higasa, and Hata, *Agric. Biol. Chem.*, 32, 484 (1968); 35, 712, 717 (1971).
136. Sohonie and Honavar, *Sci. Cult.*, 21, 538 (1956).
137. Sugiura, Ogiso, Takeuti, Tamura, and Ito, *Biochem. Biophys. Acta*, 328, 407 (1973).
138. Ogiso, Tamura, Kato, and Sugiura, *J. Biochem.* (Tokyo), 76, 147 (1974).
139. Ogawa, Higasa, and Hata, *Memoirs Res. Inst. Food Sci. Kyoto Univ.*, 32, 1 (1971).
140. Perlstein and Kézdy, *J. Supramol. Struct.*, 1, 249 (1973).

AMINO ACID COMPOSITION OF PLANT PROTEINASE INHIBITORS*

Amino acid, residues/mole

Source	Designation	Trp	Lys	His	Arg	Asp[a]	Thr	Ser	Glu[a]	Pro	Gly	Ala	1/2 Cys	Val	Met	Ile	Leu	Tyr	Phe	Total No.	References
Solanum tuberosum								Potato Tubers													
Var. Russet Burbank	Chymotrypsin inhibitor I subunit[b] A	—[c]	5.2	0.4	3.9	11.6	3.6	5.1	11.1	6.6	6.9	6.5	2.5	9.6	1.1	7.7	8.0	2.3	3.1		1
	B	—	5.0	0.4	3.6	10.4	2.8	4.0	9.3	7.1	5.5	2.3	2.1	8.0	0.6	6.3	9.8	0.6	3.2		
	C	—	5.9	0.7	4.6	10.5	3.1	4.2	8.6	5.4	7.1	2.4	2.6	8.7	0.6	7.7	10.4	1.0	2.3		
	D	—	5.8	1.0	3.9	8.4	3.3	3.4	8.0	4.3	8.1	2.5	2.1	6.1	1.4	5.6	6.9	1.3	1.9		
Var. Ulster Prince	Chymotrypsin inhibitor I subunit[d,e] A	1	4	1	4	10	2	4	9	5	6	2	2	11	1	8	11	0	3	84	2, 2a
	B	1	5	1	4	10	2	3	7	5	7	2	2	11	1	9	10	2	2	84	
	C	1	6	0	4	10	2	3	8	5	7	2	2	11	1	9	10	2	2	85	
	D	1	6	0	4	11	2	4	8	5	7	2	2	11	1	9	9	1	2	85	
Var. Danshaku-Imo	Proteinase inhibitor I	1	5	1	4	11	3	4	8	5	7	2	2	9	0	9	10	1	3	85	3
	Proteinase inhibitor IIa	0	8	0	3	11	6	7	8	6	13	5	12	2	0	4	3	6	3	97	4
	Proteinase inhibitor IIb	0	10	1	2	9	5	6	8	9	12	6	12	1	0	4	3	6	3	97	4
Potato tuber	Polyvalent proteinase inhibitor K1a	—[c]	14	0	8	25	11	16	17	12	26	10	12	5	0	7	13	10	8	194	5
	K1c	—	19	0	10	18	10	14	16	14	26	12	12	1	0	6	8	12	6	184	
	K2	—	13	1	14	28	10	23	13	11	24	10	12	1	0	9	24	8	12	213	
	K3	—	11	3	7	23	10	21	16	11	23	10	12	2	0	13	22	5	9	198	

*Supported by grant BMS74-22102 from the National Science Foundation.

[a]Includes amides.
[b]Based on a molecular weight of 9,500 for all subunits (1).
[c]— = not determined.
[d]Composition based on the sequence. Other major variants of the chymotrypsin inhibitor are shown in the sequence table.

605

AMINO ACID COMPOSITION OF PLANT PROTEINASE INHIBITORS (continued)

Source	Designation	Trp	Lys	His	Arg	Asp[a]	Thr	Ser	Glu[a]	Pro	Gly	Ala	1/2 Cys	Val	Met	Ile	Leu	Tyr	Phe	Total No.	References
									Amino acid, residues/mole												
								Potato Tubers (continued)													
Var. Maritta	Proteinase inhibitor K4	—[c]	4	1	2	5	2	3	4	3	4	3	3	2	0	2	2	1	1	42	6
	A2	—	6	1	3	11	4	6	6	6	8	4	2	6	0	5	8	3	4	83	
	A4	—	6	1	3	7	5	5	6	4	8	4	8	2	0	3	3	4	2	71	
	A5	0	4	0	2	6	3	4	5	3	6	3	7	1	0	2	2	4	2	54	
	A6	—	4	1	3	9	4	7	5	4	7	3	1	5	1	4	6	3	4	71	
	Proteinase inhibitor A7b	—	4		2	6	3	4	5	3	6	3	5	1	0	2	2	3	2	51	6
	A8	—	5	1	3	10	5	9	6	4	8	4	0	6	0	4	7	3	4	79	
Var. Russet Burbank	Carboxypeptidase inhibitor[d]	2	2	2	1	5	2	2	3	3	3	4	6	1	0	1	0	1	1	39	7, 8
Var. Danshaku-Imo	Kallikrein inhibitor 1	5	10	2	7	23	12	26	14	10	21	9	4	16	2	11	20	7	10	209	9
	2	4	11	5	6	23	12	23	13	9	21	10	3	15	2	10	20	7	10	204	
								Legumes													
Glycine max Soybean	Kunitz inhibitor	2	10	2	9	26	7	11	18	10	16	8	4	14	2	14	15	4	9	181	10, 11
	Bowman-Birk inhibitor	0	5	1	2	11	2	9	7	6	0	4	14	1	1	2	2	2	2	71	12
Phaseolus vulgaris Navy bean	Inhibitor 1	—[c,e]	5	3	3	10	3	9	6	6	1	3	9	1	0	3	3	1	1	67	13, 14
	2	—	8	3	5	8	4	10	5	4	2	2	8	2	1	3	2	1	0	68	
Kidney bean, var. Haricot	Protease[f] inhibitor	0.8	4.0	3.3	3.1	12.5	6.0	12.3	6.4	5.7	2.4	3.7	12.0	1.5	1.0	3.7	2.6	1.3	1.2		15, 16
Garden bean, var. Great Northern	Isoinhibitor I	0	4	3	3	11	3	11	6	6	1	3	14	1	0	3	3	1	1	74	17
	II	0	4	3	3	10	5	13	5	6	2	2	14	2	0	2	2	2	0	77	
	IIIb	0	4	5	3	11	5	12	7	6	2	3	14	0	0	4	2	1	2	81	
Bush bean,	PVI 2	—[c]	4	4	3	12	5	11	10	6	4	4	10	2	0	4	3	1	2	85	18
Borlotto	PVI 3	—	4	4	3	12	5	12	9	5	4	4	12	2	0	4	3	1	2	86	
var. nanus	PVI 4	—	5	4	3	11	5	10	10	5	5	5	8	3	0	3	3	1	2	83	

[e] A previously isolated navy bean inhibitor had no tryptophan (14).
[f] Calculated on the basis of a molecular weight of 10,000.

AMINO ACID COMPOSITION OF PLANT PROTEINASE INHIBITORS (continued)

Source	Designation	Trp	Lys	His	Arg	Asp[a]	Thr	Ser	Glu[a]	Pro	Gly	Ala	1/2 Cys	Val	Met	Ile	Leu	Tyr	Phe	Total No.	References
Legumes (continued)																					
Phaseolus aureus Mung bean	Trypsin inhibitor	0	6	4	4	10	3	10	7	6	2	3	8	1	2	2	2	1	1	72	19
Phaseolus coccineus Fire bean, scarlet runner	Trypsin and chymotrypsin inhibitor 2	—[c]	6	3	4	11	5	9	11	4	7	6	6	3	1	3	3	1	1	84	20
	3	—	5	4	4	11	5	9	10	5	6	4	8	2	1	3	3	1	1	82	
	4	—	6	4	4	10	5	9	11	4	6	5	8	3	1	2	3	1	1	83	
	5	—	7	3	4	11	6	10	11	4	7	7	4	4	1	3	4	1	2	89	
Phaseolus lunatus Lima bean, var. Fordhook	Trypsin inhibitor 1	0	4	5	2	12	4	12	6	6	1	3	12	1	0	4	3	1	1	77	21
	2	0	4	3	2	14	3	12	5	6	0	3	14	1	0	4	3	1	1	76	
	3	0	4	6	2	13	5	15	7	7	1	4	16	1	0	5	3	2	2	93	
	4	0	4	6	2	13	5	13	7	7	1	3	14	1	0	4	3	1	2	86	
var. Wilbur	Trypsin and chymotrypsin inhibitor 4	0	4	6	2	14	5	12	7	7	1	3	12	1	0	4	4	1	1	84	22
	6	0	4	6	2	14	5	13	8	7	1	3	14	1	0	4	4	1	2	89	
Commercial lima bean inhibitor	Protease inhibitor IV[g]	0	4	6	2	13	4	15	6	6	1	2	14	1	0	4	3	1	2	84	23
	I	0	4	3	2	12	3	11	5	6	0	3	14	1	0	4	3	1	1	73	
Vigna sinensis Black-eyed pea var. Seridó	Trypsin and chymotrypsin inhibitor	1	5	4	4	12	3	10	7	8	2	5	10	1	0	4	2	3	3	84	24

[g]Some other variants are shown in the sequence tables. Many others have been detected (23).

AMINO ACID COMPOSITION OF PLANT PROTEINASE INHIBITORS (continued)

Amino acid, residues/mole

Legumes (continued)

Source	Designation	Trp	Lys	His	Arg	Asp[a]	Thr	Ser	Glu[a]	Pro	Gly	Ala	1/2 Cys	Val	Met	Ile	Leu	Tyr	Phe	Total No.	References
Pisum sativum Pea	Trypsin and chymotrypsin inhibitor 1	—[c]	7.7	3.4	2.8	11.3	7.7	11.9	11.1	5.9	26.2	9.7	0	5.6	0	2.6	4.2	1.0	3.0		25
	Trypsin and chymotrypsin inhibitor 3[h]	—	7.4	2.6	2.4	10.2	7.6	15.0	8.8	4.4	24.2	9.0	0	2.2	0	2.2	3.2	2.2	3.2		25
	4[h]	—	7.8	5.2	1.8	14.0	8.6	11.8	8.6	7.6	16.4	9.6	0	6.2	0	2.4	4.2	2.0	3.8		
	6	—	6.8	3.9	2.0	9.3	5.2	7.6	6.4	2.9	20.6	5.7	0	4.9	0	1.7	1.9	1.0	2.3		
	8	—	14.4	6.0	3.5	15.8	8.5	10.3	10.0	7.4	22.4	10.8	2.5	5.8	0	2.1	2.6	1.0	3.3		
Lathyrus odoratus Sweet pea	Trypsin and chymotrypsin inhibitor	—[c]	10	5	3	16	8	10	7	6	3	6	18	4	0	3	3	2	3	107	26
Vicia faba L. Broad bean	Trypsin inhibitor[f] 1	—[c]	4.4	3.1	3.2	8.0	4.3	3.1	3.1	+	2.9	2.9	—	—	—	1.3	0.4	1.5	1.5		27
	2	—	5.1	2.4	3.0	6.8	4.0	5.2	3.4	+	3.0	3.9	—	4.4	—	1.0	2.4	2.0	1.7		
Arachis hypogaea Peanuts Groundnuts	Trypsin and chymotrypsin inhibitor	0	2	2	7	8	7	5	6	7	4	3	14	5	0	0	1	1	2	74	28, 29
		0	7	3	14	14	14	10	16	13	12	7	22	9	1	1	4	3	4	154	30
		0	1	1	4	5	5	3	4	6	1	2	10	3	0	0	0	1	2	48	31
Cicer arietinum Chick pea	Inhibitor CI	0	9	0	4	12	8	8	11	6	8	7	10	5	1	4	3	2	1	99	32
Bauhinia seeds of the tree	Multifunctional proteinase inhibitor	1	6	4	4	16	9	18	23	14	17	10	4	14	1	11	16	4	8	180	33

[h]Values corresponding to molecular weight times two.

AMINO ACID COMPOSITION OF PLANT PROTEINASE INHIBITORS (continued)

Amino acid, residues/mole

Source	Designation	Trp	Lys	His	Arg	Asp[a]	Thr	Ser	Glu[a]	Pro	Gly	Ala	1/2 Cys	Val	Met	Ile	Leu	Tyr	Phe	Total No.	References
								Cereal Grains													
Gramineae																					
Barley var. Pirkka	Trypsin inhibitor	3	2	3	9	10	7	8	14	11	10	10	10	6	2	5	9	5	3	127	34
Rye	Trypsin inhibitor	—[c]	10	1	15	15	11	17	12	25	7	12	24	7	2	5	2	3	5	168	35
Wheat	Trypsin inhibitor	—	10	0	15	15	11	15	10	23	9	9	24	9	2	5	4	3	5	163	35
Corn	Trypsin inhibitor	0	1	1	8	3	3	3	5	10	7	4	6	2	1	4	6	1	0	65	36
								Other Plants													
Raphanus sativus	Proteinase inhibitor RI	0	4	1	4	7	3	3	6	3	10	5	6	3	0	3	3	3	3	67	37
Radish seeds	RIII	1	7	1	9	9	4	6	8	6	10	5	12	2	1	2	4	3	5	95	
Ananas sativus	Protease inhibitor III	0	7	0	1	5	2	3	4	3	2	2	8	2	0	2	3	3	2	49	38, 39
Pineapple stem	VII	0	6	0	0	6	2	3	4	3	2	3	10	2	0	2	3	4	2	52	

Compiled by Beatrice Kassell and Monica June Williams.

AMINO ACID COMPOSITION OF PLANT PROTEINASE INHIBITORS (continued)

REFERENCES

1. Melville and Ryan, *J. Biol. Chem.*, 247, 3445 (1972).
2. Richardson, *Biochem. J.*, 137, 101 (1974).
2a. Richardson and Cossins, *FEBS Lett.*, 45, 11 (1974).
3. Kiyohara, Iwasaki, and Yoshikawa, *J. Biochem.* (Tokyo), 73, 89 (1973).
4. Iwasaki, Kiyohara, and Yoshikawa, *J. Biochem.* (Tokyo), 72, 1029 (1972).
5. Hochstrasser, Werle, Siegelmann, and Schwarz, *Hoppe-Seyler's Z. Physiol. Chem.*, 350, 897 (1969).
6. Kaiser and Belitz, *Chem. Mikrobiol. Technol. Lebensm.*, 1, 191 (1972).
7. Ryan, Hass, and Kuhn, *J. Biol. Chem.*, 249, 5495 (1974).
8. Ryan, Hass, Kuhn, and Neurath, in *Bayer Symp. V Proteinase Inhibitors,* Fritz, Tschesche, Greene, and Truscheit, Eds., Springer-Verlag, Berlin, 1974, 565.
9. Hojima, Moriwaki, and Moriya, *J. Biochem.* (Tokyo), 73, 923 (1973).
10. Koide, Tsunasawa, and Ikenaka, *J. Biochem.* (Tokyo), 71, 165 (1972).
11. Brown, Lerman, and Bohak, *Biochem. Biophys. Res. Commun.*, 23, 561 (1966).
12. Odani and Ikenaka, *J. Biochem.* (Tokyo), 71, 839 (1972).
13. Whitley, Jr. and Bowman, personal communication, 1974; *Arch. Biochem. Biophys.* 169, 42 (1975).
14. Wagner and Riehm, *Arch. Biochem. Biophys.*, 121, 672 (1967).
15. Pusztai, *Biochem. J.*, 101, 379 (1966).
16. Pusztai, *Eur. J. Biochem.*, 5, 252 (1968).
17. Wilson and Laskowski, *J. Biol. Chem.*, 248, 757 (1973).
18. Belitz, Fuchs, Nitsche, and Al-Sultan, *Z. Lebensm. Unters. Forsch.*, 150, 215 (1972).
19. Chü, Lo, Jen, Chi, and Tsao, *Sci. Sin.*, 14, 1454 (1965).
20. Belitz, Fuchs, and Grimm, *Lebensm. Wiss. Technol.*, 4, 89 (1971).
21. Jones, Moore, and Stein, *Biochemistry*, 2, 66 (1963).
22. Haynes and Feeney, *J. Biol. Chem.*, 242, 5378 (1967).
23. Stevens, Wuerz, and Krahn, in *Bayer Symp. V Proteinase Inhibitors,* Fritz, Tschesche, Greene, and Truscheit, Eds., Springer-Verlag, Berlin, 1974, 344.
24. Ventura, Filho, Moreira, Aquino, and Pinheiro, *Ann. Acad. Bras. Cienc.*, 43, 234 (1971).
25. Weder and Hory, *Lebensm. Wiss. Technol.*, 5, 86 (1972).
26. Weder and Belitz, *Dtsch. Lebensm. Rundsch.*, 65, 78 (1969).
27. Warsy and Stein, *Qual. Plant. Plant Foods Hum. Nutr.*, 23, 157 (1973).
28. Birk and Gertler, in *Proc. Int. Res. Conf. Proteinase Inhibitors, 1st,* Fritz and Tschesche, Eds., Walter de Gruyter, Berlin, 1970, 142.
29. Tur-Sinai, Birk, Gertler, and Rigbi, *Biochim. Biophys. Acta*, 63, 666 (1972).
30. Tixier, *C. R. Acad. Sci.*, 266, 2498 (1968).
31. Hochstrasser, Illchmann, Werle, Hössl, and Schwarz, *Hoppe-Seyler's Z. Physiol. Chem.*, 351, 1503 (1970).
32. Birk, in *Bayer Symp. V Proteinase Inhibitors,* Fritz, Tschesche, Greene, and Truscheit, Eds., Springer-Verlag, Berlin, 1974, 355.
33. Goldstein, Trop, and Birk, *Nat. New Biol.*, 246, 29 (1973).
34. Mikola and Suolinna, *Eur. J. Biochem.*, 9, 555 (1969).
35. Hochstrasser, Werle, Schwarz, and Siegelmann, *Hoppe-Seyler's Z. Physiol. Chem.*, 350, 249 (1969).
36. Hochstrasser, Illchmann, and Werle, *Hoppe-Seyler's Z. Physiol. Chem.*, 351, 721 (1970).
37. Ogawa, Higasa, and Hata, *Agric., Biol. Chem.*, 35, 712 (1971).
38. Perlstein and Kézdy, *J. Supramol. Struct.*, 1, 249 (1973).
39. Reddy, Keim, Heinrikson, and Kézdy, personal communication; *J. Biol. Chem.* 250, 1741 (1975).

AMINO ACID SEQUENCES OF PLANT[a] PROTEINASE INHIBITORS[**]

Inhibitor from

Soybean, Kunitz (1–4)

                                                                 10

Asp- Phe- Val- Leu- Asp- Asn- Glu- Gly- Asn- Pro- Leu- Glu- Asn- Gly- Gly- Thr- Tyr- Tyr-

            20   30

Ile- Leu- Ser- Asp- Ile- Thr- Ala- Phe- Gly- Gly- Ile- Arg- Ala- Ala- Pro- Thr- Gly- Asn-

            40   50

Glu- Arg- Cys- Pro- Leu- Thr- Val- Val- Gln- Ser- Arg- Asn- Glu- Leu- Asp- Lys- Gly- Ile-

                60   *†   70

Gly- Thr- Ile- Ile- Ser- Pro- Ser- Tyr- Arg- Ile- Arg- Phe- Ile- Ala- Glu- Gly- His- Pro-

                        80   90

Leu- Ser- Leu- Lys- Phe- Asp- Ser- Phe- Ala- Val- Ile- Met- Leu- Cys- Val- Gly- Ile- Pro-

                                    100

Thr- Glu- Trp- Ser- Val- Val- Glu- Asp- Leu- Pro- Glu- Gly- Pro- Ala- Val- Lys- Ile- Gly-

        110   120

Glu- Asn- Lys- Asp- Ala- Met- Asp- Gly- Trp- Phe- Arg- Leu- Glu- Arg- Val- Ser- Asp- Asp-

            130   140

Glu- Phe- Asn- Asn- Tyr- Lys- Leu- Val- Phe- Cys- Pro- Gln- Gln- Ala- Glu- Asp- Asp- Lys-

                150   160

Cys- Gly- Asp- Ile- Gly- Ile- Ser- Ile- Asp- Asp- Asp- Gly- His- Thr- Arg- Arg- Leu- Val-

                170   180

Val- Ser- Lys- Asn- Lys- Pro- Leu- Val- Val- Gln- Phe- Gln- Lys- Leu- Asp- Lys- Glu- Ser-

181
Leu

Disulfide bridges: 39–86, 136–145

[a]See table of properties for Latin names of the plants.

*Reactive site for trypsin
†Reactive site for chymotrypsin
**Supported by grant BMS 74-22102 from the National Science Foundation.

AMINO ACID SEQUENCES OF PLANT[a] PROTEINASE INHIBITORS (continued)

										10					
Soybean, Bowman-Birk (5–9)	BB								Asp-	Asp-	Glu-	Ser-	Ser-	Lys-	Pro- Cys-
Lima bean, components	LB IV	Ser-	Gly-	His-	His-	Glu-	His-	Ser-	Thr-	Asp-	Glx-	Pro-	Ser-	Glx	— — — —
IV[b] and I (10–13)	LB I									—	—	—	—	—	— — — —
Garden bean, isoinhibitors	GB II	(Asx,—	Asx,	Glx,	—,	—,	Ser,	Asx,	—,	—,	—,	—,	—,	—,	Pro, —, —,
II and II′ (14–17)	GB II′								—	—	—	—	—	—	— —

		10				*			20					
BB	Cys-	Asp-	Gln-	Cys-	Ala-	Cys-	Thr-	Lys-	Ser-	Asn-	Pro-	Pro-	Gln-	Cys-	Arg- Cys- Ser- Asp-
LB IV	(—,	Asx)	His(—,	Leu, —)	—	—	—	Ile	—	—	—	—	—	—	— Thr — / Ser
			Ala												
LB I	(— —)	—	(— Ala —)	—	—	—	—	—	—	—	—	—	—	—	— Thr —
GB II	—,	Asx,	Ile, —)	Val	—	—	Ala	—	—	—	—	—	—	—	(Ile, —, —, Asx,
GB II′	—	—	—	—	—	—	—	—	—	—	—	—	—	—	(Ile, —, —, Asx,

		30								40			†	
BB	Met-	Arg-	Leu-	Asn-	Ser-	Cys-	His-	Ser-	Ala-	Cys-	Lys-	Ser- Cys-	Ile- Cys-	Ala- Leu- Ser-
LB IV	Leu — — / Phe	—	Asp	—	—	—	—	—	—	—	—	50 — —	— — Thr	† — —
LB I	Leu	—	—	—	—	—	—	—	—	—	—	—	—	—
GB II	Val)	—	—	Asx	—	—	—	—	—	—	Met	50 — — Arg	* —
GB II′	Val)	—	—	Asx	—	—	—	—	—	—	Met	50 — — Arg	* —

(Sequence continued on next page)

Note: Dashes indicate that the 3-letter symbol is the same as the one above it.
†Reactive site for elastase.

AMINO ACID SEQUENCES OF PLANT[a] PROTEINASE INHIBITORS (continued)

						50							60		
Garden bean, isoinhibitors (continued)	BB	Tyr-	Pro-	Ala-	Gln-	Cys-	Phe-	Cys-	Val-	Asp-	Ile-	Thr-	Asp-	Phe-	Cys-	Tyr-	Glu-	Pro-	Cys-

						60						70	
	LB IV	Ile	—	—	—	—	Val	(—, Thr, Asx) Asx	—	Asx Thr	—	—	—	—	—	—	—
	LB I	—	—	—	—	—	—	(—, Asx, —)	—	Asx	—	—	—	—	—	—	—

						60						70	
	GB II	Met	—	Gly-	Lys	—	Arg	—	Leu	—	Thr-	Thr-	Asx-	Tyr	—	—	Lys-	Ser	—
	GB II′	Met	—	Gly-	Lys	—	Arg	—	Leu	—	Thr-	Thr-	Asx-	Tyr	—	—	Lys-	Ser	—

						70	71
	BB	Lys-	Pro-	Ser-	Glu-	Asp-	Asp-	Lys-	Glu-	Asn

						80		83
	LB IV	—	Ser	—	His-	Ser	—	Asp-	Asp	—	Asn-	Asn
	LB I	—	(—, —, —, —, Asx, Asx, Asx, Asx)

					79
	GB II	—	—	Asx-	Ser-	Gly-	Glx	—	—
	GB II′	—	—	Asx-	Ser-	Gly-	Glx	—	—

Disulfide bridges: BB 8–62, 9–24, 12–58, 14–22, 32–39, 36–51, 41–49

					10				
Peanut (ground nut) (18)	Cys-	Thr-	Asx-	Lys-	Thr-	Glx-	Gly-	Arg-	Cys-	Pro-	Val-	Thr-	Glx-	Cys-	Arg-	Ser-	Asx-	Pro-
		20					30				
	Pro-	Glx-	Cys-	Arg-	Ala-	Pro-	Pro-	Tyr-	Phe-	Glx-	Cys-	Val-	Cys-	Val-	Asx-	Thr-	Phe-	Asx-
			40				48			
	His-	Cys-	Pro-	Ala-	Ser-	Cys-	Asx-	Ser-	Cys-	Cys-	Thr-	Arg

					10				
Maize seeds (19)	Ser-	Ala-	Gly-	Thr-	Ser-	Cys-	Val-	Pro-	Pro	(Pro, Ser, Asx, Gly, Cys, His)	Ala-	Ile- Leu-
		20			*		30			
	Arg-	Thr-	Gly-	Ile-	Pro-	Gly-	Arg-	Leu-	Pro-	Leu-	Glx-	Lys-	Thr-	Cys-	Gly-	Ile- Gly-
			40				50			
	Pro-	Arg-	Gln-	Val-	Glx-	Arg-	Leu-	Gln-	Asx-	Leu-	Pro-	Cys-	Pro-	Gly-	Arg-	Arg-	Gln- Leu-
			60			65			
	Ala-	Asx-	Met-	Ile-	Ala-	Tyr-	Cys-	Pro-	Arg-	Cys-	Arg

Note: Dashes indicate that the 3-letter symbol is the same as the one above it.

AMINO ACID SEQUENCES OF PLANT[a] PROTEINASE INHIBITORS (continued)

Potato

Chymotrypsin inhibitor I subunits[b] (20, 21)

								10							
A		Glu-	Phe-	Glu-	Cys-	Asp-	Gly-	Lys-	Leu-	Gln-	Trp-	Pro-	Glu-	Leu-	Ile-	Gly-	Val-	Pro-
B		—	—	—	—	Lys	—	—	—	—	—	—	—	—	—	—	—	—
C	Lys	—	—	—	—	—	—	—	—	—	—	—	—	—	—	—	—	—
D	Lys	—	—	—	—	Asn	—	—	—	—	—	—	—	—	—	—	—	—

		20								30				
A	Thr-	Lys-	Leu-	Ala-	Lys-	Glu-	Ile-	Ile-	Glu-	Lys-	Gln-	Asn-	Ser	-Leu-	Ile-	Ser-	Asn-
B	—	—	—	—	—	Gly	—	—	—	—	—	—	—	—	Thr	—	Ser
C	—	—	—	—	—	—	—	—	—	—	—	—	—	—	Thr	—	Ser
D	—	—	—	—	—	—	—	—	—	—	—	—	—	—	Thr	—	Ser

				40							50			
A	Val-	His-	Ile-	Leu-	Leu-	Asn-	Gly-	Ser-	Pro-	Val-	Thr-	Leu-	Asp-	Ile-	Leu-	Gly-	Asp-	Val-
B	—	—	—	—	—	—	—	—	—	—	—	—	—	—	—	—	—	—
C	—	Gln	—	—	—	—	—	—	—	—	—	—	—	—	—	—	—	—
D	—	—	—	—	Lys	—	—	—	—	—	—	—	—	—	—	—	Asn	—

					60†							
A	Val-	Gln-	Leu-	Pro-	Val-	Val-	Gly-	Met	-Asp-	Phe-	Arg-	Cys-	Asp-	Arg-	Val-	Arg-
B	—	Asp-	Ile	—	—	—	—	†	—	Tyr	—	—	Asn	—	—	—
C	—	—	—	—	—	—	60 †	—	—	—	—	—	—	—	—	—
D	—	—	—	—	—	—	†	—	—	—	—	—	—	—	—	—

(Sequence continued on next page)

Note: Dashes indicate that the 3-letter symbol is the same as the one above it.

AMINO ACID SEQUENCES OF PLANT[a] PROTEINASE INHIBITORS (continued)

Potato, Chymotrypsin inhibitor I (continued)

							70							80				84
A		Leu-	Phe-	Asp-	Asp-	Ile-	Leu-	Gly-	Ser-	Val-	Val-	Gln-	Ile-	Pro-	Arg-	Val-	Ala
B		—	—	—	—	—	—	—	Tyr	—	—	—	—	—	—	—	84 —
C		70 —	—	—	Asn	—	—	—	Asn / Tyr	—	—	—	80 —	—	—	—	85 —
D		—	—	—	—	—	—	—	Asn / Ser	—	—	—	—	—	—	—	85 —

Potato

Carboxypeptidase inhibitor (22)

Pyr (Glx, His, Ala) Asp- Pro- Ile- Cys- Asn- Lys- Pro- Cys- Lys- Thr- His- Asp- Asp- Cys-
(10)

Ser- Gly- Ala- Trp- Phe- Cys- Gln- Ala- Cys- Trp- Asn- Ser- Ala- Arg- Thr- Cys- Gly- Pro-
(20) (30)

Tyr- Val- Gly
(39)

Pineapple stem

Bromelain inhibitors[c]

Fraction VII isoinhibitors (23–24)

A Chain

Major I: Asp- Glu- Tyr- Lys- Cys- Tyr- Cys- Ala- Asp- Thr- Tyr- Ser- Asp- Cys- Pro- Gly- Phe- Cys-
(10)

Major II: Pyr — — — — — — Thr — — — — — — — — — —
(10)

Minor: Glu — — — — — — — — — — — — — — — — —
(10)

Major I: Lys- Lys- Cys- Lys- Ala- Glu- Phe- Gly- Lys- Tyr- Ile- Cys- Leu- Asp- Leu- Ile- Ser- Pro-
(20) (30)

Major II: — — — — — — — — — — — — — — — — — —
(20) (30)

Minor: — — — — — — — — — — — — — — — — — —
(20) (30)

Major I: Asn- Asp- Cys- Val- Lys
(40) (41)

Major II: — — — — —
(40)

Minor: — — — — —
(40)

(Sequence continued on next page)

Note: Dashes indicate that the 3-letter symbol is the same as the one above it.

[c] Six of the half-cystine residues can be aligned with the basic pancreatic and related inhibitors.

AMINO ACID SEQUENCES OF PLANT[a] PROTEINASE INHIBITORS (continued)

Pineapple stem (continued)

B Chain

									10	11
Major I	Thr-	Ala-	Cys-	Ser-	Glu-	Cys-	Val-	Cys-	Pro-	Leu-	Gln
Major II	—	—	—	—	—	—	—	—	—	—	Arg
Minor	—	—	—	—	—	—	—	—	—	—	
Minor	—	—	—	—	—	—	—	—	—	—	Gln

Note: Dashes indicate that the 3-letter symbol is the same as the one above it.

Compiled by Beatrice Kassell and Monica June Williams.

AMINO ACID SEQUENCES OF PLANT[a] PROTEINASE INHIBITORS (continued)

REFERENCES

1. Koide, Tsunasawa, and Ikenaka, *J. Biochem.*, 71, 165 (1972).
2. Koide and Ikenaka, *Eur. J. Biochem.*, 32, 417 (1973).
3. Ozawa and Laskowski, Jr., *J. Biol. Chem.*, 241, 3955 (1966).
4. Bidlingmeyer, Leary, and Laskowski, Jr., *Biochemistry*, 11, 3303 (1972).
5. Odani, Koide, and Ikenaka, *Proc. Jap. Acad.*, 47, 621 (1971).
6. Odani and Ikenaka, *J. Biochem.*, 71, 839 (1972).
7. Odani and Ikenaka, *J. Biochem.*, 74, 697 (1973).
8. Seidl, Liener, and Jarvis, *Biochim. Biophys. Acta*, 251, 83 (1971).
9. Seidl, Liener, and Jarvis, *Biochim. Biophys. Acta*, 258, 303 (1972).
10. Stevens, in *Proc. Int. Res. Conf. Proteinase Inhibitors, 1st,* Fritz and Tschesche, Eds., Walter de Gruyter, Berlin, 1970, 149.
11. Tan and Stevens, *Eur. J. Biochem.*, 18, 515 (1971).
12. Krahn and Stevens, *Biochemistry*, 11, 1804 (1972).
13. Stevens, Wuerz, and Krahn, in *Bayer Symp. V Proteinase Inhibitors,* Fritz, Tschesche, Greene, and Truscheit, Eds., Springer-Verlag, Berlin, 1974, 344.
14. Wilson and Laskowski, Sr., *J. Biol. Chem.*, 248, 756 (1973).
15. Wilson and Laskowski, Sr., *Fed. Proc.*, 33, 1530 (1974).
16. Wilson and Laskowski, Sr., in *Bayer Symp. V Proteinase Inhibitors,* Fritz, Tschesche, Greene, and Truscheit, Eds., Springer-Verlag, Berlin, 1974, 286.
17. Laskowski, Sr., personal communication; Wilson and Laskowski, Sr., *J. Biol. Chem.*, 250, 4261 (1975).
18. Hochstrasser, Illchmann, Werle, Hössl, and Schwarz, *Hoppe-Seyler's Z. Physiol. Chem.*, 351, 1503 (1970).
19. Hochstrasser, Illchmann, and Werle, *Hoppe-Seyler's Z. Physiol. Chem.*, 351, 721 (1970).
20. Richardson, *Biochem. J.*, 137, 101 (1974).
21. Richardson and Cossins, *FEBS Lett.*, 45, 11 (1974).
22. Ryan, Hass, Kuhn, and Neurath, in *Bayer Symp. V Proteinase Inhibitors,* Fritz, Tschesche, Greene, and Truscheit, Eds., Springer-Verlag, Berlin, 1974, 565.
23. Reddy, Keim, Heinrickson, and Kézdy, *Fed. Proc.*, 33, 1530 (1974).
24. Reddy, Keim, Heinrickson, and Kézdy, personal communication; *J. Biol. Chem.*, 250, 1741 (1975).

SPECIFICITIES AND SOME PROPERTIES OF ANIMAL PROTEINASE INHIBITORS

Table 1*
LOWER ANIMALS

Source	Name	Molecular weight[a]	I.P.	Carbohydrate content	$K_I(M)$ Trypsin	$K_I(M)$ α-Chymotrypsin	Specificity[b] Trypsin	α-Chymotrypsin	Plasmin	Plasma or serum kallikrein	Organ kallikrein	Thrombin
Anemonia sulcata Sea anemone	Polyvalent isoinhibitors											
	B	6,480					I-E	+	I-E		±[c]	
	C2a	6,620					I-E	+	I-E		+	
	C4f	6,580					+	±	+	+[c,d]	+	
	5-II[e]				3×10^{-10}		+	±	+		+	
Ascaris[f] *lumbricoides* Whole ascarids	Trypsin inhibitor											
	CM-1	7,190 aa / 7,000 gel		0			+[c,g]					
	CM-2	6,500 aa		0			+[c,g]					
Body walls[h]	Trypsin inhibitor											
	I	5,520 aa			$9 \times 10^{-8\ c}$		I-E	0				
	II				$1.3 \times 10^{-8\ c}$		I-E	0				
Body walls[h]	Chymotrypsin inhibitor	7,955 aa / 8,000 ulc	>10.75	0		7×10^{-9}	0	I-E				
Whole ascarids	Chymotrypsin inhibitor	8,600 gel				5×10^{-8}	0	I-E				
Body walls	Pepsin inhibitor											
	I	17,500 aa		0			0	0				
	II	15,500 el / 15,580 aa		0								
	III	15,500 el / 16,120 aa		0								
	IV	15,500 el / 31,720 aa		0								
Body walls	Carboxypeptidase A inhibitors I, II, III											
Loligo vulgaris Cuttle fish	Proteinase isoinhibitors											
	A	6,630 aa					I-E	+	0[c]	+[c]		
Squid	B	6,690 aa					I-E	+	0	+		
	E	6,930 aa					I-E	I-E	+	+		
	L	7,440 aa					I-E	I-E	+	+		

*Footnotes and References follow Table 4.

Table 1 (continued)
LOWER ANIMALS

Source	Name	Specificity[b]						Information in		
		Papain	Ficin and bromelain	B. subtilis proteases	Streptomyces griseus proteases	Pepsin	Gastricsin	Other	Amino acid table	Sequence table	References
Anemonia sulcata											
Sea anemone	Polyvalent isoinhibitors										
	B			0				Boar acrosin, +	*		1–5
	C2a			0				Human erythrocyte insulin-degrading proteinase, 0	*		
	C4f			0					*		
	5-II[e]										
Ascaris lumbricoides											
Whole ascarids	Trypsin inhibitor										
	CM-1	0			0	0[c]			*	*	6–8
	CM-2	0			0	0[c]			*		
Body walls[h]	Trypsin inhibitor										
	I					0[c]		Human anionic trypsin I: inhibitor I, 0; inhibitor II, +	*		9–15
	II							Human anionic trypsin II: inhibitor I, ±; inhibitor II, +			
								Human cationic trypsin, 0			
								Elastase, 0			
Body walls[h]	Chymotrypsin inhibitor	0	0	+		0[c]		Leucine aminopeptidase, 0	*		13, 14, 16–18
								Carboxypeptidases A and B, 0			
								Elastase,[i] +			
Whole ascarids	Chymotrypsin inhibitor							Chymotrypsin B: I-E($K_I = 3 \times 10^{-8}$ M)			19
Body walls	Pepsin inhibitor										
	I					I-E[c,d,g,j]	I-E[c,k] ±[d]	Rennin,[l] 0	*		14, 20, 21
	II					I-E[c,d,g,j]	I-E[c,k] 0[d]	Cathepsin,[l] E of bovine spleen, +	*		
	III					I-E[c,d,g,j]	I-E[c,k] 0[d]	Cathepsin[l] B1, C, D, of bovine spleen, 0	*		
	IV					I-E[c,d,g,j]	I-E[c,k] 0[d]		*		
Body walls	Carboxypeptidase A inhibitors I, II, III					0[c]		Carboxypeptidase A,[m] +	*		14
Loligo vulgaris											
Cuttle fish	Proteinase isoinhibitors										
Squid	A								*	*	22
	B								*		
	E								*		
	L								*		

619

Table 1 (continued)
LOWER ANIMALS

Source	Name	Molecular weight[a]	I.P.	Carbohydrate content	$K_I(M)$ Trypsin	$K_I(M)$ α-Chymotrypsin	Specificity[b] Trypsin	α-Chymotrypsin	Plasmin	Plasma or serum kallikrein	Organ kallikrein	Thrombin
Helix pomatia Snail	Proteinase isoinhibitors											
	B	6,463 aa			0.8×10^{-9}		I-E	+	+[c]	+[c]	+[c]	
	E	6,431 aa			2×10^{-9}		I-E	+	+	+	+	
	G	6,591 aa										
	H	6,575 aa					I-E		+[c]		+[c]	
	K	6,463 aa					I-E		+		+	
Achatina fulica Giant African snail	Hemocyanin subunit	70,000 gel					+					
Pomacea canaliculata Eggs (mollusk)	Ovorubin	335,000 minimum					I-E	+				
Hirudo medicinalis Leech[o]	Hirudin	10,800 aa 9,060 ulc	3.9									I-E[p]
	Bdellins											
	A-1a	5,970 aa		0			I-E	0	I-E[c]	0[c,d]	0[c]	
	A-2,3	6,340 aa		0			I-E	0	I-E	0[c,d]	0	
	A-4	6,220 aa		0			I-E	0	I-E	0	0[c]	
	B-1			0			I-E	0	I-E[c]	0[c,d]	0[c]	
	B-2			0			I-E	0	I-E	0	0	
	B-3	4,970 aa		0			I-E	0	I-E	0	0	
	B4, 5, 6			0			I-E	0	I-E	0	0	
Bombyx mori L Silkworm Hemolymph	Trypsin inhibitor	($s_{20,w} = 4.5S$)					+	0				
Aedes aegypti Adult mosquito Thorax	Chymotrypsin inhibitor							0				
Drosophila melanogaster Fly	Larval trypsin inhibitor						+	0				
	Larval chymotrypsin inhibitor						±	+				

Table 1 (continued)
LOWER ANIMALS

Source	Name	Specificity[b]							Information in		References
		Papain	Ficin and bromelain	*B. subtilis* proteases	*Streptomyces griseus* proteases	Pepsin	Gastricsin	Other	Amino acid table	Sequence table	
Helix pomatia Snail	Proteinase isoinhibitors										4, 23–25
	B							Rat liver insulin-specific proteinase,[n] +	*		
	E								*		
	G								*		
	H							Human erythrocyte insulin-degrading proteinase,[l] +	*		
	K									*	
Achatina fulica Giant African snail	Hemocyanin subunit								*		26
Pomacea canaliculata Eggs (mollusk)	Ovorubin			+	±						27, 28
Hirudo medicinalis Leech[o]	Hirudin							Factor Xa, 0	*		2, 29–33
	Bdellins										2, 3, 34, 35
	A-1a			0				Acrosin,[c,l] +	*		
	A-2, 3			0					*		
	A-4			0					*		
	B-1			0					*		
	B-2			0					*		
	B-3			0				Acrosin,[c,q] +	*		
	B4, 5, 6			0					*		
Bombyx mori L Silkworm Hemolymph	Trypsin inhibitor										36
Aedes aegypti Adult mosquito Thorax	Chymotrypsin inhibitor							Mosquito and black fly chymotrypsins, +			37
Drosophila melanogaster Fly	Larval trypsin inhibitor										38
	Larval chymotrypsin inhibitor										

Table 1 (continued)
LOWER ANIMALS

Source	Name	Molecular weight[a]	I.P.	Carbohydrate content	$K_I(M)$ Trypsin	$K_I(M)$ α-Chymotrypsin	Specificity[b] Trypsin	α-Chymotrypsin	Plasmin	Plasma or serum kallikrein	Organ kallikrein	Thrombin
Apis mellifera Bee venom	Protease inhibitor	9,000 gel					+	+				+
Vipera russelli Russell's viper venom	Proteinase inhibitors I	6,300 aa 7,200 el		0			I-E					
	II	6,841 aa 7,200 el 6,900 eq		0	7.6×10^{-10}	1.4×10^{-10}	I-E		+g,r	+g,s	+c	0d,g
Dendroapsis polylepis polylepis Black mamba venom	Toxin I	7,147 aa					0	±				
	K	6,557 aa					+	0				
Hemachatus hemachatus Ringhals cobra venom	Proteinase inhibitor II	6,300 aa		0			+		+g	+g		0d,g
Red sea turtle	Inhibitor						+					

Table 1 (continued)
LOWER ANIMALS

Source	Name	Specificity[b]						Information in		References
		Papain	Ficin and bromelain	B. subtilis proteases	Streptomyces griseus proteases	Pepsin	Gastricsin	Other	Amino acid table	Sequence table	
Apis mellifera Bee venom	Protease inhibitor	+	+			+[c]		Streptokinase, + Collagenase, + Leucine aminopeptidase, +			39, 40
Vipera russelli Russell's viper venom	Proteinase inhibitor I	0	0	0					*		41–44
	II								*	*	
Dendroapsis polylepis polylepis Black mamba venom	Toxin I							Carboxypeptidase A and B, 0 Thermolysin, 0	*	*	45
	K								*	*	
Hemachatus hemachatus Ringhals cobra venom	Proteinase inhibitor II	0	0	0	0			Carboxypeptidase A and B, 0 Thermolysin, 0	*		44
Red sea turtle	Inhibitor			+						*	46

Compiled by Beatrice Kassell and Monica June Williams. Supported by grant BMS74-22102 from the National Science Foundation.

Table 2*
THE EGG WHITES OF BIRDS

Source	Name	Molecular weight[a]	I.P.	Carbohydrate content	$K_I(M)$ Trypsin	$K_I(M)$ Chymotrypsin	Specificity[b] Trypsin	Specificity[b] α-Chymotrypsin	Specificity[b] Human trypsin	Specificity[b] Turkey trypsin	Specificity[b] Chicken chymotrypsin
Gallus gallus Chicken	Ovomucoid	28,800 os 27,000–31,000 ulc 27,800 el	3.8–4.8	see table	$5-6 \times 10^{-9}$ 1.3×10^{-10} [w]		I-E[c,g,x]	0	0[t,u,v] +[y]	I-E	0
Meleagris gallopavo Turkey	Ovomucoid	28,500 ulc	4.28	+	$<10^{-9}$	$<10^{-9}$	I-E	I-E	0[t,v] +[v]	I-E	I-E
Casuarius casuarius Cassowary	Ovomucoid		3.90	+	10^{-7}		I-E	0			
Dromicaeus novaehollandiae Emu	Ovomucoid		3.78	+	10^{-7}		I-E_2	+			
Struthio camelus Ostrich	Ovomucoid		3.97	+			I-E	±			
Rhea americana Rhea	Ovomucoid		3.82	+	10^{-7}		I-E	±			

*Footnotes and References follow Table 4.

Table 2 (continued)
THE EGG WHITES OF BIRDS

Source	Name	Specificity[b]							Information in		References
		Plasmin	Thrombin	Papain	Subtilisin	*Aspergillus* spp. proteases	Elastase	Other	Amino acid table	Sequence table	
Gallus gallus Chicken	Ovomucoid	0[d]	0[d]			0	0	*Clostridium histolyticum* collagenase, 0 Amylase, 0 Phosphatase, 0 Chymotrypsin B, 0 Urinary kallikrein, ± Human erythrocyte insulin-specific protease, 0 Lipase, 0 Peroxidase, 0 Rabbit acrosin, ± Human acrosin, ± Human pancreatic protease E, ± *Flavobacterium elastolyticum* elastase, 0 Organ kallikrein, 0[c] Serum kallikrein, 0[c,d] Rat liver cathepsins, A, B, C, D, 0 Human chymotrypsin I and II, 0	*		2, 4, 15, 47–72
Meleagris gallopavo Turkey	Ovomucoid	0	0		+[z]	0	I-E	Rat liver cathepsins A, B, C, D, 0 Human chymotrypsin I, ± Human chymotrypsin II, +	*		50–54, 57, 63 65, 67, 73–77
Casuarius casuarius Cassowary	Ovomucoid				0	0			*		
Dromicaeus novaehollandiae Emu	Ovomucoid				±	0			*		
Struthio camelus Ostrich	Ovomucoid				0	0			*		
Rhea americana Rhea	Ovomucoid				±	0			*		

625

Table 2 (continued)
THE EGG WHITES OF BIRDS

Source	Name	Molecular weight[a]	I.P.	Carbohydrate content	$K_I(M)$ Trypsin	$K_I(M)$ Chymotrypsin	Specificity[b] Trypsin	α-Chymotrypsin	Human trypsin	Turkey trypsin	Chicken chymotrypsin
Eudromia elegans Tinamou	Ovomucoid		4.73	+		$<10^{-9}$	±	I-E		0^v	
Anas platyrhynckas Duck	Ovomucoid	28,000 ulc		+	$<10^{-9}$	$<10^{-9}$	I-E$_2$	I-E			
Numida meleagris Guinea fowl	Ovomucoid						+	+			
Anser anser Goose	Ovomucoid						I-E$_2$	0			
Chrysolophus amherstiae Lady Amherst's pheasant	Ovomucoid						0	+			
Phasianus colchicus Ring-necked pheasant	Ovomucoid						+	+			
Chrysolophus pictus Golden pheasant	Ovomucoid					$<10^{-9}$	0	I-E			
Lophortyx californica California valley quail	Ovomucoid						I-E or I-E$_2$	I-E			
Oreortyx picti Painted quail	Ovomucoid						I-E	0			
Coturnix coturnix japonica Japanese quail	Ovomucoid				10^{-8}		I-E	0	$+^v$		

Table 2 (continued)
THE EGG WHITES OF BIRDS

Source	Name	Specificity[b]							Information in		References
		Plasmin	Thrombin	Papain	Subtilisin	*Aspergillus* spp. proteases	Elastase	Other	Amino acid table	Sequence table	
Eudromia elegans Tinamou	Ovomucoid	0	0		+[z]	0			*		50—54, 57, 63 65, 67, 73—77
Anas platyrhynckas Duck	Ovomucoid				+[a']	0			*		
Numida meleagris Guinea fowl	Ovomucoid										
Anser anser Goose	Ovomucoid										
Chrysolophus amherstiae Lady Amherst's pheasant	Ovomucoid										
Phasianus colchicus Ring-necked pheasant	Ovomucoid										
Chrysolophus pictus Golden pheasant	Ovomucoid				+[a']	0					
Lophortyx californica California valley quail	Ovomucoid				0	0					
Oreortyx picti Painted quail	Ovomucoid										
Coturnix coturnix japonica Japanese quail	Ovomucoid	0	0		0	0		Rat liver cathepsins A, B, C, D, 0			

Table 2 (continued)
THE EGG WHITES OF BIRDS

Source	Name	Molecular weight[a]	I.P.	Carbohydrate content	$K_I(M)$ Trypsin	$K_I(M)$ Chymotrypsin	Specificity[b] Trypsin	α-Chymotrypsin	Human trypsin	Turkey trypsin	Chicken chymotrypsin
Pygoscelis adelia Penguin	Ovomucoid										
Adelia		27,600 ulc		+	8×10^{-8}	10^{-6}					
Emperor		28,100 ulc		+			+	±			
Crested		28,200 ulc		+			+	±			
Royal		27,800 ulc		+			+	±			
White-flippered		28,900 ulc		+			+	±			
Yellow-eyed		28,200 ulc		+			+	±			
Chicken	Ovoinhibitor	44,000 ulc 49,000 ulc 49,000 l 52,400 ulc	5.1	see table	$10^{-7}-10^{-9}$	$10^{-8}-10^{-9}$	I-E or I-E$_2$	I-E$_2$	$0^{t,v}$	I-E	I-E
Turkey	Ovoinhibitor	48,000					I-E$_2$	I-E$_2$			
Duck	Ovoinhibitor						I-E$_2$	I-E$_2$			
Penguin (Adelie)	Ovoinhibitor						I-E$_2$	0			
Japanese quail	Ovoinhibitor	48,000		+			I-E$_2$	I-E$_2$			
Ostrich	Ovoinhibitor						0	0			
Chicken	Ficin-papain inhibitor	12,700 gel		+			0	0			

Table 2 (continued)
THE EGG WHITES OF BIRDS

Source	Name	Specificity[b]							Information in		References
		Plasmin	Thrombin	Papain	Subtilisin	*Aspergillus* spp. proteases	Elastase	Other	Amino acid table	Sequence table	
Pygoscelis adelia	Ovomucoid							Cathepsins A, B, C, D of rat liver lysosomes, 0*			50, 74, 78
Penguin											
Adelia					+r				*		
Emperor					+r				*		
Crested					+r				*		
Royal					+r				*		
White-flippered					+r				*		
Yellow-eyed					+r				*		
Chicken	Ovoinhibitor	0	0	0	+	+	I-E	Human chymotrypsins I and II, +; *Streptomyces griseus* protease, +	*		50, 52–54, 65, 73, 77, 79–85; 24, 40
Turkey	Ovoinhibitor				+	+b'					
Duck	Ovoinhibitor				+	+					
Penguin (Adelie)	Ovoinhibitor				+	+					
Japanese quail	Ovoinhibitor				+	+			*		
Ostrich	Ovoinhibitor				+	+					
Chicken	Ficin-papain inhibitor			I-E	0		0 c'	Proteases from *Pseudomonas aeruginosa*, 0; *Proteus vulgaris*, 0; *Bacillus cereus*, 0; Bovine spleen cathepsins B1 and C, +; Human liver cathepsins B1 and C, +; Cathepsins D and E, 0; Collagenase,[c] 0; Ficin, I-E[d]; Bromelain, ±	*		86–89

Compiled by Beatrice Kassell and Monica June Williams. Supported by grant BMS74-22102 from the National Science Foundation.

Table 3*
MAMMALIAN ORGANS AND SECRETIONS

Source	Name	Molecular weight[a]	I.P.	$K_I(M)$ Trypsin	$K_I(M)$ α-Chymotrypsin	Specificity[b] Trypsin	Specificity[b] α-Chymotrypsin	Specificity[b] Human trypsin	Specificity[b] Plasmin	Specificity[b] Thrombin	Specificity[b] Organ kallikrein	Specificity[b] Plasma or serum kallikrein
Bovine[e']	Basic trypsin kallikrein inhibitor — Kunitz	6,512 aa	10–10.5	$3 \times 10^{-11} -$ 6×10^{-14}	$1-8 \times 10^{-8}$	I-E[c,g,f']	I-E	I-E[t,y]	±[c,d,g']	0[d]	+[d,g]	+[e,d,g]
pancreas, lung, parotid gland, liver, ovary, heart, thyroid gland, uterus		6,500–11,800 ulc					±[d]	+[u]			+[c,r]	
Bovine pancreas	Secretory inhibitor–Kazal	6,153 aa 6,100 ulc		3.3×10^{-11}		I-E	0	0[t,u,y]	±[c,d]	0[g,i'] +[g,j]	0[c] 0[d]	
Ovine pancreas	Secretory inhibitor	6,137 aa				I-E	0		0[c]	0[i']	0	
Porcine pancreas	Secretory inhibitor I	6,023 aa 6,040 ulc	8.27	$<10^{-11}$		I-E	0	0[t] +[u,y]	0[c,d]	0[d,g]	0[c]	
	II	5,609 aa 5,400 ulc	8.35			I-E	0	0[t] +[u,y]	0[c,d]	0[d,g]	0[c]	

*Footnotes and References follow Table 4.

Table 3 (continued)
MAMMALIAN ORGANS AND SECRETIONS

Source	Name	Urinary kallikrein	Urokinase	Acrosin	Papain	Elastase	*Aspergillus* spp. protease	*Streptomyces griseus* proteases	Other	Amino acid table	Sequence table	References
Bovine[e] pancreas, lung, parotid gland, liver, ovary, heart, thyroid gland, uterus	Basic trypsin kallikrein inhibitor — Kunitz	+[d]	+	+[c,d]	0	0	0	+	Chymotrypsin B, ± Chicken chymotrypsin, + Rabbit plasmin, + Organ kallikreins of dog, cat, rat, mouse, and guinea pig, 0 Thermolysin, 0 Ficin, 0 Chymosin (rennin), 0 Bromelain, 0 Collagenase,[h] 0 Cathepsins A, B, C, D, 0 Pepsin, 0 Carboxypeptidase A and B, 0 Leucine aminopeptidase, 0 Renin, 0 Angiotensin-converting enzyme, 0 Angiotensinase, 0 Human pancreatic protease E, + Human erythrocyte insulin-degrading proteinase, 0 Human neutral leucocyte proteinase, 0 *Bacillus subtilis* proteases, 0 Rabbit liver lysosomal acid protease, 0 Human sperm "chymotrypsin," 0 Human serum lipase, 0 Microbial acid and alkaline phosphatases, 0 Liver esterase, 0 Enterokinase, 0 Ribonuclease, 0 Lysozyme, 0 Clostripain,[h] 0	*	*	2, 4, 15, 30, 47, 52–54, 60, 65, 69, 71, 72, 90–126
Bovine pancreas	Secretory inhibitor — Kazal	0[c]	0	+[c,d]					Chymotrypsin B, 0	*	*	29, 46, 52, 69, 111, 127–136
Ovine pancreas	Secretory inhibitor	0[c]	0	+[c]						*		131, 135, 137
Porcine pancreas	Secretory inhibitor I	0[c]	0	+[c]					Chymotrypsin B, 0 Human erythrocyte insulin-degrading proteinase, +	*		52, 60, 127, 130, 131, 135, 138–140
	II	0[c]	0	+[c]							*	

631

Table 3 (continued)
MAMMALIAN ORGANS AND SECRETIONS

Source	Name	Molecular weight[a]	I.P.	$K_I(M)$		Specificity[b]						
				Trypsin	α-Chymotrypsin	Trypsin	α-Chymotrypsin	Human trypsin	Plasmin	Thrombin	Organ kallikrein	Plasma or serum kallikrein
Canine pancreas	Secretory inhibitor	6,100 aa	8			+			±[c]	0[g]	0[c]	
Feline pancreas	Secretory inhibitor	6,097 aa				+						
Human pancreas	Secretory inhibitor	6,242 aa 6,300 ulc 6,000 el				I-E[c,g]	±	I-E[t,u,y]	±[c]	0[g]		
Human pancreas	Inhibitor 1	5,240 aa	8.7			I-E	±	I-E[v] +,t +,t,v				
Canine submandibular gland	Proteinase inhibitor A₂	4,330 aa 12,750 aa 12,000 gel 11,900 ulc	6.5			+ I-E	I-E		+[c]		0[c]	0[c,d]
Elastase inhibitor						+	+					
Human placenta	Trypsin inhibitor					+						
Urokinase inhibitor Major		100,000 gel 105,000 ulc	5.2						0			
Minor			5.8						0			
Urokinase inhibitor I		43,000 gel	4.8–4.9			0						
II		43,000 gel	4.8–4.9			+						
Human skin	Trypsin inhibitor					+						
	Trypsin-elastase inhibitor					+						
Rabbit skin	Protease inhibitor	12,500 ulc	6.6			0	0					
Bovine colostrum	Trypsin inhibitor	7,505 aa (contains carbohydrate in addition)	4.2			I-E	+	+,v	+[d]	0[d]		0[c,d]
Porcine colostrum	Isoinhibitors of trypsin					I-E[c,g]	+					
Human nasal secretion	Proteinase inhibitor	13,000 gel				+	+		0		0[c]	
Human bronchial secretion	Proteinase inhibitor I	14,800 aa 13,000–14,000 gel				I-E	+		0		0[c]	
	Proteinase inhibitor II	20,000 gel				+	+					

Table 3 (continued)
MAMMALIAN ORGANS AND SECRETIONS

Source	Name	Specificity[b]								Information in		References
		Urinary kallikrein	Urokinase	Acrosin	Papain	Elastase	*Aspergillus* spp. protease	*Streptomyces griseus* proteases	Other	Amino acid table	Sequence table	
Canine pancreas	Secretory inhibitor	0[c]		+[c]	+				Human pancreatic protease E, +	*	*	130, 131, 135, 141, 142
Feline pancreas	Secretory inhibitor									*		135, 143
Human pancreas	Secretory inhibitor	0[c]							Human chymotrypsin, 0	*	*	15, 130, 135, 144, 145
Human pancreas	Inhibitor 1 2									*		53, 54
Canine submandibular gland	Proteinase inhibitor A$_2$	0[c]				I-E[c]	+[b']	+	Subtilisin, novo, I-E Collagenase, 0 Human sperm "chymotrypsin," 0	*		2, 101, 146—148
	Elastase inhibitor					+						149
Human placenta	Trypsin inhibitor											150
	Urokinase inhibitor											
	Major		+									
	Minor		+									
	Urokinase inhibitor I		+						Streptokinase, 0			151
	II								Streptokinase, 0			152
Human skin	Trypsin inhibitor					0						153
	Trypsin-elastase inhibitor					+						153
Rabbit skin	Protease inhibitor				+				Arthus protease, +			154
Bovine colostrum	Trypsin inhibitor			+[c]					Cathepsins A, B, C, D, 0 Human erythrocyte insulin-degrading proteinase, 0 Chymotrypsin B, 0 Subtilisin, ±	*	*	2—4, 35, 52, 69, 74, 114, 155, 156
Porcine colostrum	Isoinhibitors of trypsin								Carboxypeptidase B, 0 Pepsin, 0 Chymotrypsin B, + *B. subtilis* proteases, 0	Carbohydrate table		157, 158
Human nasal secretion	Proteinase inhibitor					0		+	*B. subtilis* proteases, 0 Leucocyte protease, +			159
Human bronchial secretion	Proteinase inhibitor I					0		+	Leucocyte protease, +	*		160
	Proteinase inhibitor II											161

633

Table 3 (continued)
MAMMALIAN ORGANS AND SECRETIONS

Source	Name	Molecular weight[a]	I.P.	$K_I(M)$ Trypsin	$K_I(M)$ α-Chymotrypsin	Trypsin	α-Chymotrypsin	Human trypsin	Plasmin	Thrombin	Organ kallikrein	Plasma or serum kallikrein
Human pregnancy urine	Trypsin inhibitor (mingin)	30,000 eq 28,000 ulc	2.1	7.7×10^{-10}		+	+		±[d,g]		±[d,g]	
	Trypsin inhibitor	70,000 gel				+						
Human urine	Proteolytic inhibitor	16,400 ulc	2.8			I-E	+					
Human cervical mucus	Acid-stable proteinase inhibitor	11,500 gel				+	+					
Human seminal plasma	Inhibitor I	12,700 gel 10,130 aa				+	+	+[u,y]	0[c]		0[c]	
	Inhibitor II	5,400 gel 6,200 aa				+	0	+[u,y]	0[c]		0[c]	
	Inhibitor	11,500 gel				+	0		0	0	±[c]	
	Inhibitor II	4,000 gel				+	+		0[d]	0		
	III	6,635 aa				+			0[d]	0		
	Inhibitor I	6,600 gel		2.3×10^{-9}		I-E	0		0	0	0[c]	
Guinea pig seminal vesicles	II	6,687 aa		$<10^{-10}$		I-E	0		I-E			
Boar seminal plasma (sperm plasma)	Proteinase inhibitor III	6,700 gel				I-E[c]						
	Trypsin-plasmin inhibitor	6,781 aa 13,400 gel				+			+			0[c,d]
	Isoinhibitor A	11,000 aa[m]				+			+[c]			

Table 3 (continued)
MAMMALIAN ORGANS AND SECRETIONS

Source	Name	Specificity[b]								Information in		References
		Urinary kallikrein	Urokinase	Acrosin	Papain	Elastase	Aspergillus spp. protease	Streptomyces griseus proteases	Other	Amino acid table	Sequence table	
Human pregnancy urine	Trypsin inhibitor (mingin)	+k'							Plasminogen activator, ±			4, 35, 162–165
Human urine	Trypsin inhibitor											166
Human urine	Proteolytic inhibitor											167
Human cervical mucus	Acid-stable proteinase inhibitor			0					Neutral leucocyte proteinases,[d] +			168
Human seminal plasma	Inhibitor I			0[c] +[d]					Neutral leucocyte proteinases[d]: inhibitor I, +; inhibitor II, 0	*		3, 15, 101, 118, 169, 170
	Inhibitor II			+[c] I-E[d,r]					Human sperm "chymotrypsin," 0	*		169, 170
	Inhibitor III		0									171
	Inhibitor II III		0 0									172
Guinea pig seminal vesicles	Inhibitor I			+c,I'						*		3, 72, 173–175
	Inhibitor II			+c,I'								
Boar seminal plasma	Proteinase inhibitor III			+[c]								176
Boar seminal plasma (sperm plasma)	Trypsin-plasmin inhibitor								Neutral leucocyte proteinases,[d] 0			2, 3, 175
	Isoinhibitor A			I-E[c,r]					Erythrocyte insulin-degrading enzyme, +[d]	*		177

Compiled by Beatrice Kassell and Monica June Williams. Supported by grant BMS74-22102 from the National Science Foundation.

Table 4*
MAMMALIAN AND CHICKEN BLOOD

Source	Name	Molecular weight[a]	I.P.	Concentration in serum mg/100 ml	$K_I(M)$ Trypsin	$K_I(M)$ α-Chymotrypsin	Specificity[b] Trypsin	Specificity[b] α-Chymotrypsin	Specificity[b] Plasmin	Specificity[b] Organ kallikrein	Specificity[b] Plasma or serum kallikrein
Human plasma or serum	α₁-Antitrypsin	54,200 aa 49,500– 59,900 ulc 50,300 el	4 5.1	200–400			I-E g,u,y I-E₂ c,d,g +t,v,o′	I-E or I-E₂ +o′	+d	+c	+c,d
	α₁-Antichymo- trypsin	69,000 ulc		30–60			0	+	0d	0c	
	Inter-α-trypsin inhibitor	160,000 ulc		20–70	1.6×10^{-11}	8×10^{-10}	I-E g,t,y +u	I-E	0d	0c	0
	Antithrombin III′ activated factor X inhibitor, heparin cofactor, autoprothrombin c	63,700 ulc		22–39			+g,u,y	0	±d	0c	
	C̄1-inactivator	104,000 ulc	2.7–2.8	15–35			±g 0u,y	±	+d	0c	+c,d
	α₂-Macroglobulin[s]	725,000– 800,000 ulc 845,000 eq 363,000 sub 185,000 sub	5.0–5.5	150–350♂ 175–400♀	10^{-8}		I-E₂ or I-E, +t,o′	I-E′ +o′	I-E d	0c	+

*Footnotes and References follow Table 4.

Table 4 (continued)
MAMMALIAN AND CHICKEN BLOOD

Source	Name	Specificity[b]						Information in		References[n']
		Thrombin	Elastase	Acrosin	Leucocyte protease	Cl-esterase	Other	Amino acid table	Carbohydrate table	
Human plasma or serum	α_1-Antitrypsin	±	+c,d	+c,d	+d,o',p'	0	Synovial gland[l'] collagenase, 0; Human granulocyte elastase, +; Cathepsin D and E, 0; Human pancreatic protease E, I-E; Chymotrypsin B, +; Human chymotrypsin II, +; Human sperm "chymotrypsin," 0; Leucine aminopeptidase, 0; *B. subtilis* proteases, +; *Streptomyces griseus* proteases, +; Carboxypeptidases A and B, 0; Papain, +; Bromelain, +; Human granulocyte collagenase, +; Human sperm "chymotrypsin," 0	*	*	15, 60, 71, 86, 101, 178–202
	α_1-Antichymotrypsin	0	0[c]	0[d]		0		*	*	71, 101, 187, 188
	Inter-α-trypsin inhibitor	0	± 0[c]	+c,d		0	Human cationic trypsin; I-E[q]	*	*	15, 71, 184, 185, 187, 188, 199, 203
	Antithrombin III[r'] activated factor X inhibitor, heparin cofactor, autoprothrombin c	+	0[c]	+c,d		0	Human chymotrypsin II; I-E[z]; Activated Stuart factor (Xa), +	*	*	15, 71, 184, 185, 187, 188, 204–208
	Cl-inactivator	0	0[c]	±[d]		+	Complement factors Cl̄r and Cl̄s, +; Activated Hageman factor (XIIa), +; *Clostridium histolyticum* proteinase, +	*		15, 186–188, 199, 200, 209–213
	α_2-Macroglobulin[s']	+[d]	+[c]	+[c]	+d,l	0	*Pseudomonas aeruginosa* proteinase, +; Tadpole collagenase, +; Synovial gland collagenase, l'+; Urokinase,[d] +; Aminopeptidase M, 0; Human granulocyte elastase, +; Human granulocyte collagenase, +; *Staphylococcus aureus* neutral protease, +; *Proteus vulgaris* neutral protease, +; *Trichophyton mentagrophytes* keratinase, +	*	*	187, 188, 190, 193–195, 199, 202, 203, 212, 214–237

Table 4 (continued)
MAMMALIAN AND CHICKEN BLOOD

Source	Name	Molecular weight[a]	I.P.	Concentration in serum mg/100 ml	$K_I(M)$ Trypsin	$K_I(M)$ α-Chymotrypsin	Specificity[b] Trypsin	Specificity[b] α-Chymotrypsin	Specificity[b] Plasmin	Specificity[b] Organ kallikrein	Specificity[b] Plasma or serum kallikrein
	$α_2$-Macroglobulin[s] (continued)										
Human serum	Kallikrein inhibitor	53,000 gel					0	0	+		+
	Group-specific component						0	0			
	Inhibitors I to IV	4,000–7,000 gel					0				
Bovine plasma	$α_1$-Antitrypsin	72,000 ulc	<3.85		$2 × 10^{-10}$		I-E	+	+	0[c]	+[c]
	Acid-labile trypsin inhibitor	71,000 ulc 39,000 eq					I-E		+		
	$α_1$-Antitrypsin Antithrombin	45,000					+		+		
Calf serum	$α_2$-Macroglobulin[s] Fetuin	800,000 43,000–48,300 ulc	5.07 3.3		$7 × 10^{-7}$		I-E$_3$ I-E		I-E[g]		0[g]
Sheep serum	Inhibitor	40,600 ucl	4.3		$4 × 10^{-9}$		I-E	+	+[d,o']		
Pig serum	$α_2$-Macroglobulin[s]	960,000 l	5.4		$4.5 × 10^{-9}$		+		+		
Platelets	Soluble trypsin inhibitor	70,000 gel	4.62				+		+		
Rabbit serum	$α_2$-Macroglobulin[s] $α_1$-Macroglobulin[s]	850,000 ulc	4.6–5.1 4.35–4.5				+ I-E$_2$ +	+ I-E$_2$	+ + +		
Plasma	Acid stable inhibitor Activated factor X inhibitor, heparin cofactor, antithrombin III										+

Table 4 (continued)
MAMMALIAN AND CHICKEN BLOOD

Source	Name	Thrombin	Elastase	Acrosin	Leucocyte protease	Cl-esterase	Other	Amino acid table	Carbohydrate table	References[n']
	α_2-Macroglobulin[s'] (continued)						Clostridiopeptidase A,[d,i] +			
							Fusiformis nodosus proteinase, +			
							Cathepsin[d] B_1, I-E			
							Papain, I-E			
							Human chymotrypsin II, +			
							Leucine aminopeptidase, +			
							B. subtilis proteases, 0			
							Streptomyces griseus proteases, 0			
Human serum	Kallikrein inhibitor	+								238
	Group-specific component		0[c]			0				239
	Inhibitors I to IV						Rat liver insulin-specific protease, +			240
Bovine plasma	α_1-Antitrypsin Acid-labile trypsin inhibitor		+							241
										242—244
	α_1-Antitrypsin	+								245
	Antithrombin	+			+					246
	α_2-Macroglobulin[s']	±					Prethrombin E, +	*		244, 247—249
Calf serum	Fetuin						Thrombin E, +			250—254
Sheep serum	Inhibitor	0					Mosquito trypsin, I-E			255
Pig serum	α_2-Macroglobulin[s']		I-E[c]				Rat skin proteinase A, +	*	*	178, 218, 256
Platelets	Soluble trypsin inhibitor									257
										258
Rabbit serum	α_2-Macroglobulin[s']	+						*		259—261
	α_1-Macroglobulin[s']	+						*		261, 262
	Acid-stable Inhibitor									263
Plasma	Activated factor X inhibitor, heparin cofactor, antithrombin III	+					Activated factor X, +			208, 264
							Activated factors XI and XII, 0			

Table 4 (continued)
MAMMALIAN AND CHICKEN BLOOD

Source	Name	Molecular weight[a]	I.P.	Concentration in serum mg/100 ml	$K_I(M)$ Trypsin	$K_I(M)$ α-Chymotrypsin	Specificity[b] Trypsin	Specificity[b] α-Chymotrypsin	Plasmin	Organ kallikrein	Plasma or serum kallikrein
Guinea pig serum	Protease inhibitors										
	Major	73,000 el					I-E	0			
	Minor	78,000 el					I-E	+	+		
	C1̄-inactivator	170,000 gel									
	Elastase inhibitor	120,000 ulc					+	+			
Rat plasma	$α_1$-Antitrypsins f and s	58,200 ulc	4.6—4.7				+				
Mouse serum	$α_2$-Proteinase inhibitor						+	+			
Chicken serum											

Table 4 (continued)
MAMMALIAN AND CHICKEN BLOOD

Source	Name	Specificity[b]							Information in		References[n]
		Thrombin	Elastase	Acrosin	Leucocyte protease	Cl-esterase		Other	Amino acid table	Carbohydrate table	
Guinea pig serum	Protease inhibitors										
	Major										265, 266
	Minor										
	Cl̄-inactivator					+					267, 268
Rat plasma	Elastase inhibitor		I-E								269
Mouse serum	α_1-Antitrypsins f and s									*	270
Chicken serum	α_2-Proteinase inhibitor							*Aspergillus* proteinase, +			271

Compiled by Beatrice Kassell and Monica June Williams. Supported by grant BMS74-22102 from the National Science Foundation.

Table 4 (continued)

[a] Symbols used to designate the determination of molecular weight are: gel, gel filtration; sub, subunit molecular weight; aa, minimum molecular weight by amino acid composition; l, light scattering; eq, equivalence to an enzyme; el, SDS-gel electrophoresis; ulc, various ultracentrifugal methods; os, osmotic pressure. The molecular weights calculated from amino acid compositions are approximations when the asparagine and glutamine contents are not known. Unless otherwise indicated, molecular weight calculated from amino acid composition does not include carbohydrate.
[b] Inhibition is indicated as follows: stoichiometric inhibition, complex I-E, (inhibitor-enzyme), sub, subunit; +, strong inhibition; ± weak inhibition; 0, no inhibition.
[c] Porcine.
[d] Human.
[e] Isoinhibitors II to IX have similar inhibition spectra; isoinhibitors X and XI have little effect on pancreatic kallikrein.
[f] Trypsin and chymotrypsin inhibitors have also been purified from *Ascaris galli*.[6,7]
[g] Bovine.
[h] For earlier studies, see References 9 to 11.
[i] K_I approximately 10^{-8} M for the four inhibitors.
[j] $K_I = 10^{-10}$ M at pH 5.3 for all the inhibitors vs. porcine and bovine pepsins and 2×10^{-9} M vs. human pepsin.
[k] $K_I = 0.4-4 \times 10^{-9}$ M for the four inhibitors.
[l] The mixture of inhibitors.
[m] For fraction II, $K_I = 1.3 \times 10^{-9}$ M for peptidase action and 2.3×10^{-7} M for esterase action.
[n] Isoinhibitors F, G, and K.
[o] Inhibitors have been isolated from other blood-sucking insects, but these have not been studied in purified form (29).
[p] $K_I = 8 \times 10^{-11}$ M.
[q] $K_I = <10^{-10}$ M.
[r] $K_I = 10^{-9}$ M.
[s] $K_I = 2.9 \times 10^{-10}$ M.
[t] Human cationic trypsin.
[u] Human anionic trypsin I (15).
[v] Human anionic trypsin (52 or 53).
[w] Bovine β-trypsin (58).
[x] Ovine.
[y] Human anionic trypsin II (15).
[z] $K_I = 10^{-7}$ M.
[a'] $K_I = 10^{-8}$ M.
[b'] *Aspergillus oryzae* alkaline protease.
[c'] From human polymorphonuclear leucocytes.
[d'] $K_I = 1.5 \times 10^{-8}$ M.
[e'] The inhibitor from ovine pancreas is identical. See Reference 124 for other ruminants. A low molecular weight protein from pituitary gland (112) is not included here.
[f'] Turkey.
[g'] $K_I = 4 \times 10^{-9}$ M for porcine plasmin.
[h'] *Clostridium histolyticum*.
[i'] Esterase activity.
[j'] Clotting activity.
[k'] $K_I = 8 \times 10^{-7}$ M.
[l'] Rabbit.
[m'] Includes carbohydrate.
[n'] Many of the earlier references are in References 187 and 199.
[o'] Dog.
[p'] Horse.
[q'] $K_I = 9 \times 10^{-9}$ M.
[r'] There is some suspicion that antithrombin III is an enzyme.
[s'] Action on proteolytic activities are recorded. There is generally little or no inhibition with synthetic substrates.

REFERENCES

1. Fritz, Brey, and Béress, *Hoppe Seyler's Z. Physiol. Chem.*, 353, 19 (1972).
2. Wunderer, Kummer, and Fritz, *Hoppe Seyler's Z. Physiol. Chem.*, 353, 1646 (1972).
3. Fink, Fritz, Jaumann, Schiessler, Förg-Brey, and Werle, in *Protides of the Biological Fluids, 20th Colloquium*, Peeters, Ed., Pergamon Press, Oxford, 1973, 425.
4. Tschesche, Dietl, Kolb, and Standl, in *Bayer Symposium V, "Proteinase Inhibitors,"* Fritz, Tschesche, Greene, and Truscheit, Eds., Springer-Verlag, Berlin, 1974, 586.
5. Wunderer, Kummer, Fritz, Béress, and Machleidt, in *Bayer Symposium V, "Proteinase Inhibitors,"* Fritz, Tschesche, Greene, and Truscheit, Eds., Springer-Verlag, Berlin, 1974, 277.
6. Kenynya, *Latv. PSR Zinat. Akad. Vestis*, p. 88 (1971).
7. Kenynya and Polyakova, *Tr. Vses. Inst. Gel'mintol.*, p. 121 (1971).
8. Portmann and Fraefel, *Helv. Chim. Acta*, 50, 2078 (1967).
9. Collier, *Can. J. Res. Sect. B*, 19, 90 (1941).
10. Green, *Biochem. J.*, 66, 416 (1957).
11. Rhodes, Marsh, and Kelley, Jr., *Exp. Parasitol.*, 13, 266 (1963).
12. Kucich and Peanasky, *Biochim. Biophys. Acta*, 200, 47 (1970).
13. Peanasky and Abu-Erreish, in *Proc. 1st Int. Res. Conf. Proteinase Inhibitors*, Fritz and Tschesche, Eds., Walter de Gruyter, Berlin, 1971, 281.
14. Peanasky, Abu-Erreish, Gaush, Homandberg, O'Heeron, Linkenheil, and Kucich, in *Bayer Symposium V, "Proteinase Inhibitors,"* Fritz, Tschesche, Greene, and Truscheit, Eds., Springer-Verlag, Berlin, 1974, 649.
15. Figarella, Negri, and Guy, in *Bayer Symposium V, "Proteinase Inhibitors,"* Fritz, Tschesche, Greene, and Truscheit, Eds., Springer-Verlag, Berlin, 1974, 213.
16. Peanasky and Laskowski, Sr., *Biochim. Biophys. Acta*, 37, 167 (1960).
17. Peanasky and Szucs, *J. Biol. Chem.*, 239, 2525 (1964).
18. Kassell, *Methods Enzymol.*, 19, 872 (1970).
19. Rola and Pudles, *Arch. Biochem. Biophys.*, 113, 134 (1966).
20. Abu-Erreish and Peanasky, *J. Biol. Chem.*, 249, 1558, 1566 (1974).
21. Keilová and Tomášek, *Biochim. Biophys. Acta*, 284, 461 (1972).
22. Tschesche and von Rücker, *Hoppe Seyler's Z. Physiol. Chem.*, 354, 1447 (1973).
23. Tschesche and Dietl, *Eur. J. Biochem.*, 30, 560 (1972).
24. Brush and Tschesche, in *Bayer Symposium V, "Proteinase Inhibitors,"* Fritz, Tschesche, Greene, and Truscheit, Eds., Springer-Verlag, Berlin, 1974, 581.
25. Dietl and Tschesche, in *Bayer Symposium V, "Proteinase Inhibitors,"* Fritz, Tschesche, Greene, and Truscheit, Eds., Springer-Verlag, Berlin, 1974, 254.
26. Kareem, Gilles, Nguyen-Thanh, and Keil, *Comp. Biochem. Physiol. B*, 44, 963 (1973).
27. Cheesman, *Proc. R. Soc. London Ser. B*, 149, 571 (1958).
28. Norden, *Comp. Biochem. Physiol. B*, 42, 569 (1972).
29. Barton and Yin, in *Metabolic Inhibitors*, Vol. IV, Hochster, Kates, and Quastel, Eds., Academic Press, New York, 1973, 215.
30. Jutisz, Charbonnel-Bérault, and Martinoli, *Bull. Soc. Chim. Biol.*, 45, 55 (1963).
31. Markwardt, Hoffmann, and Landmann, *Thromb. Diath. Haemorrh.*, 11, 230 (1964).
32. Markwardt, *Methods Enzymol.*, 19, 924 (1970).
33. Landmann, *Folia Haematol.* (Leipzig), 98, 437 (1972).
34. Fritz, Gebhardt, Meister, and Fink, in *Proc. 1st Int. Res. Conf. Proteinase Inhibitors*, Fritz and Tschesche, Eds., Walter de Gruyter, Berlin, 1971, 271.
35. Fritz, Förg-Brey, Schiessler, Arnhold, and Fink, *Hoppe Seyler's Z. Physiol. Chem.*, 353, 1010 (1972).
36. Umetsu and Shimura, *Nippon Nogei Kagaku Kaishi*, 46, 385 (1972).
37. Yang and Davies, *Comp. Biochem. Physiol. B*, 43, 137 (1972).
38. Kang and Fuchs, *Comp. Biochem. Physiol. B*, 46, 367 (1973).
39. Shkenderov, *FEBS Lett.*, 33, 343 (1973).
40. Shkenderov, *Abstr. 9th FEBS Meet.*, Budapest, August 1974, 442.
41. Takahashi, Iwanaga, and Suzuki, *FEBS Lett.*, 27, 207 (1972).
42. Takahashi, Iwanaga, Hokama, Suzuki, and Kitagawa, *FEBS Lett.*, 38, 217 (1974).
43. Takahashi, Iwanaga, and Suzuki, *J. Biochem.* (Tokyo), 76, 709 (1974).
44. Takahashi, Iwanaga, Kitagawa, Hokama, and Suzuki, in *Bayer Symposium V, "Proteinase Inhibitors,"* Fritz, Tschesche, Greene, and Truscheit, Eds., Springer-Verlag, Berlin, 1974, 265.
45. Strydom, *Nat. New Biol.*, 243, 88 (1973).
46. Laskowski, Jr., Kato, Leary, Schrode, and Sealock, in *Bayer Symposium V, "Proteinase Inhibitors,"* Fritz, Tschesche, Greene, and Truscheit, Eds., Springer-Verlag, Berlin, 1974, 597.
47. Back and Steger, *Fed. Proc.*, 27, 96 (1968).
48. Buck, Bier, and Nord, *Arch. Biochem. Biophys.*, 98, 528 (1962).

49. Davis, Mapes, and Donovan, *Biochemistry*, 10, 39 (1971).
50. Feeney, in *Proc. 1st Int. Res. Conf. Proteinase Inhibitors*, Fritz and Tschesche, Eds., Walter de Gruyter, Berlin, 1971, 189.
51. Feeney and Allison, *Evolutionary Biochemistry of Proteins*, John Wiley & Sons, New York, 1969, 85.
52. Feeney, Means, and Bigler, *J. Biol. Chem.*, 244, 1957 (1969).
53. Feinstein, Hoffstein, and Sokolovsky, in *Bayer Symposium V, "Proteinase Inhibitors,"* Fritz, Tschesche, Greene, and Truscheit, Eds., Springer-Verlag, Berlin, 1974, 199.
54. Feinstein, Hoffstein, Koifmann, and Sokolovsky, *Eur. J. Biochem.*, 43, 569 (1974).
55. Fredericq and Deutsch, *J. Biol. Chem.*, 181, 499 (1949).
56. Green, *J. Biol. Chem.*, 205, 535 (1963).
57. Kassell, *Methods Enzymol.*, 19, 890 (1970).
58. Levilliers, Péron-Renner, and Pudles, in *Bayer Symposium V, "Proteinase Inhibitors,"* Fritz, Tschesche, Greene, and Truscheit, Eds., Springer-Verlag, Berlin, 1974, 432.
59. Lineweaver and Murray, *J. Biol. Chem.*, 171, 565 (1947).
60. Mallory and Travis, in *Bayer Symposium V, "Proteinase Inhibitors,"* Fritz, Tschesche, Greene, and Truscheit, Eds., Springer-Verlag, Berlin, 1974, 250.
61. Mandl and Cohen, *Arch. Biochem. Biophys.*, 91, 47 (1960).
62. Mandl, Zipper, and Ferguson, *Arch. Biochem. Biophys.*, 74, 465 (1958).
63. Misaka and Tappel, *Comp. Biochem. Physiol. B.* 38, 651 (1971).
64. Oegema and Jourdian, *Arch. Biochem. Biophys.*, 160, 26 (1974).
65. Ryan, Clary, and Tomimatsu, *Arch. Biochem. Biophys.*, 110, 175 (1965).
66. Stambaugh, Brackett, and Mastroianni, *Biol. Reprod.*, 1, 223 (1969).
67. Vithayathil, Buck, Bier, and Nord, *Arch. Biochem. Biophys.*, 92, 532 (1961).
68. Werle and Zickgraf-Rüdel, *Z. Klin. Chem. Klin. Biochem.*, 10, 139 (1972).
69. Wu and Laskowski, Sr., *J. Biol. Chem.*, 213, 609 (1955).
70. Yenson, *Bull. Fac. Med. Istanbul*, 17, 481 (1954).
71. Zaneveld, Dragoje, and Schumacher, *Science*, 177, 702 (1972).
72. Zaneveld, Polakoski, Robertson, and Williams, in *Proc. 1st Int. Conf. Proteinase Inhibitors*, Fritz and Tschesche, Eds., Walter de Gruyter, Berlin, 1971, 236.
73. Gertler and Feinstein, *Eur. J. Biochem.*, 20, 547 (1971).
74. Osuga, Bigler, Uy, Sjöberg, and Feeney, *Comp. Biochem. Physiol. B*, 48, 519 (1974).
75. Osuga and Feeney, *Arch. Biochem. Biophys.*, 118, 340 (1967).
76. Rhodes, Bennett, and Feeney, *J. Biol. Chem.*, 235, 1686 (1960).
77. Tomimatsu, Clary, and Bartulovich, *Arch. Biochem. Biophys.*, 115, 536 (1966).
78. Lin and Feeney, in *Glycoproteins*, Vol. 5b, 2nd ed., Gottschalk, Ed., Elsevier, New York, 1972, 762.
79. Chu and Hsu, *Sheng Wu Hua Hsueh Yu Sheng Wu Wu Li Hsueh Pao*, 5, 434 (1965).
80. Davis, Zahnley, and Donovan, *Biochemistry*, 8, 2044 (1969).
81. Feinstein, *Biochim. Biophys. Acta*, 236, 74 (1971).
82. Haynes and Feeney, *J. Biol. Chem.*, 242, 5378 (1967).
83. Liu, Means, and Feeney, *Biochim. Biophys. Acta*, 229, 176 (1971).
84. Matsushima, *Science*, 127, 1178 (1958); *J. Agric. Chem. Soc. Jap.*, 32, 211 (1958).
85. Zahnley and Davis, *Biochemistry*, 9, 1428 (1970).
86. Barrett, in *Bayer Symposium V, "Proteinase Inhibitors,"* Fritz, Tschesche, Greene, and Truscheit, Eds., Springer-Verlag, Berlin, 1974, 574.
87. Fossum and Whitaker, *Arch. Biochem. Biophys.*, 125, 367 (1968).
88. Sen and Whitaker, *Arch. Biochem. Biophys.*, 158, 623 (1973).
89. Keilová and Tomášek, *Biochim. Biophys. Acta*, 334, 179 (1974).
90. Amris, *Scand. J. Haematol.*, 3, 19 (1965).
91. Anderer and Hörnle, *Z. Naturforsch. Teil B*, 20, 457 (1965).
92. Astrup, *Ann. N.Y. Acad. Sci.*, 146, 601 (1968).
93. Avineri-Goldman, Snir, Blauer, and Rigbi, *Arch. Biochem. Biophys.*, 121 107 (1967).
94. Bakhle, *Nature*, 220, 219 (1968).
95. Barrett, *Biochem. J.*, 104, 601 (1967).
96. Chauvet, Nouvel, and Acher, *Biochim. Biophys. Acta*, 92, 200 (1964).
97. Coan and Travis, in *Proc. 1st Int. Res. Conf. Proteinase Inhibitors*, Fritz and Tschesche, Eds., Walter de Gruyter, Berlin, 1971, 294.
98. Egeblad and Astrup, *Scand. J. Clin. Lab. Invest.*, 18, 567 (1966).
99. Finkenstadt, Hamid, Mattis, Schrode, Sealock, Wang, and Laskowski, Jr., in *Bayer Symposium V, "Proteinase Inhibitors,"* Fritz, Tschesche, Greene, and Truscheit, Eds., Springer-Verlag, Berlin, 1974, 389.
99a. Fritz, Hutzel, and Werle, *Hoppe Seyler's Z. Physiol. Chem.*, 348, 950 (1967).
100. Fritz, Schult, Meister, and Werle, *Hoppe Seyler's Z. Physiol. Chem.*, 350, 1531 (1969).
101. Fritz, Arnhold, Förg-Brey, Zaneveld, and Schumacher, *Hoppe Seyler's Z. Physiol. Chem.*, 353, 1651, (1972).

102. Green and Work, *Biochem. J.*, 54, 257, 347 (1953).
103. Kassell, *Methods Enzymol.*, 19, 844 (1970).
104. Kassell and Laskowski, Sr., *J. Biol. Chem.*, 219, 203 (1956).
105. Kassell, Radicevic, Berlow, Peanasky, and Laskowski, Sr., *J. Biol. Chem.*, 238, 3274 (1963).
106. Kraut and Bhargava, *Hoppe Seyler's Z. Physiol. Chem.*, 334, 236 (1963); 348, 1498, 1500 (1967).
107. Kraut, Frey, and Bauer, *Hoppe Seyler's Z. Physiol. Chem.*, 175, 97 (1928).
108. Kraut, Körbel, Scholtan, and Schultz, *Hoppe Seyler's Z. Physiol. Chem.*, 321, 90 (1960).
109. Kunitz and Northrop, *J. Gen. Physiol.*, 19, 991 (1936).
110. Laskowski, Jr. and Sealock, in *The Enzymes*, Vol. 3, 3rd ed. Boyer, Ed., Academic Press, New York, 1971, 375.
111. Lazdunski, Vincent, Schweitz, Péron-Renner, and Pudles, in *Bayer Symposium V, "Proteinase Inhibitors,"* Fritz, Tschesche, Greene, and Truscheit, Eds., Springer-Verlag, Berlin, 1974, 420.
112. Lisowski and Sawaryn, *Acta Biochim. Pol.*, 19, 125 (1972).
113. Lomako, Wilusz, and Mejbaum-Katzenellenbogen *Arch. Immunol. Ther. Exp.*, 20, 277 (1972).
114. Misaka and Tappel, *Comp. Biochem. Physiol. B*, 38, 651 (1971).
115. Pütter, *Hoppe Seyler's Z. Physiol. Chem.*, 348, 1197 (1967).
116. Ripa and Gilli, *Boll. Soc. Ital. Biol. Sper.*, 44, 1297 (1968).
117. Sach and Thély, *C.R. Acad. Sci.*, 266, 1200 (1968).
118. Schiessler, Arnhold and Fritz, in *Bayer Symposium V, "Proteinase Inhibitors,"* Fritz, Tschesche, Greene, and Truscheit, Eds., Springer-Verlag, Berlin, 1974, 147.
119. Trautschold, Werle, and Zickgraf-Rüdel, *Arzneimittelforschung*, 16, 1507 (1966).
120. Trop and Birk, *Biochem. J.*, 109, 475 (1968).
121. Vairel and Thély, *Ann. Biol. Clin.* (Paris), 18, 363 (1960).
122. Vincent and Lazdunski, *Biochemistry*, 11, 2967 (1972); *Eur. J. Biochem.*, 38, 365 (1973).
123. Vogel, Trautschold, and Werle, *Natural Proteinase Inhibitors*, Academic Press, New York, 1968, 76.
124. Vogel and Werle, in *Handbook of Experimental Pharmacology*, Eichler, Farah, Herken, and Welch, Eds., Springer-Verlag, Berlin, 1970, 236.
125. Wang and Kassell, *Biochem. Biophys. Res. Commun.*, 40, 1039 (1970).
126. Wilusz, *Arch. Immunol. Ther. Exp.*, 19, 735 (1971).
127. Burck, *Methods Enzymol.*, 19, 906 (1970).
128. Burck, Hamill, Cerwinsky, and Grinnan, *Biochemistry*, 6, 3180 (1967).
129. Cerwinsky, Burck, and Grinnan, *Biochemistry*, 6, 3175 (1967).
130. Fritz, Hüller, Wiedemann, and Werle, *Hoppe Seyler's Z. Physiol. Chem.*, 348, 405 (1967).
131. Fritz, Schiessler, Förg-Brey, Tschesche, and Fink, *Hoppe Seyler's Z. Physiol. Chem.*, 353, 1013 (1972).
132. Greene, Rigbi, and Fackre, *J. Biol. Chem.*, 241, 5610 (1966).
133. Kazal, Spicer, and Brahinsky, *J. Am. Chem. Soc.*, 70, 3034 (1948).
134. Schweitz, Vincent, and Lazdunski, *Biochemistry*, 12, 2841 (1973).
135. Tschesche, Wachter, Kupfer, Obermaier, Reidel, Haenisch, and Schneider, in *Proc. 1st Int. Res. Conf. Proteinase Inhibitors*, Fritz and Tschesche, Eds., Walter de Gruyter, Berlin, 1971, 207.
136. Zaneveld, Schumacher, Tauber, and Propping, in *Bayer Symposium V, "Proteinase Inhibitors,"* Fritz, Tschesche, Greene, and Truscheit, Eds., Springer-Verlag, Berlin, 1974, 136.
137. Fritz, Schramm, Greif, Hochstrasser, Fink, and Werle, *Hoppe Seyler's Z. Physiol. Chem.*, 351, 145 (1970).
138. Greene, DiCarlo, Sussman, and Bartelt, *J. Biol. Chem.*, 243, 1804 (1968).
139. Tschesche, *Angew. Chem.*, 81, 122 (1969).
140. Tschesche, Wachter, Kupfer, and Niedermeier, *Hoppe Seyler's Z. Physiol. Chem.*, 350, 1247 (1969).
141. Fritz, Hartwich, and Werle, *Hoppe Seyler's Z. Physiol. Chem.*, 345, 150 (1966).
142. Fritz, Hutzel, Müller, Wiedmann, Stahlheber, Lehnert, and Forell, *Hoppe Seyler's Z. Physiol. Chem.*, 348, 1561 (1967).
143. Tschesche and Kupfer, *Hoppe Seyler's Z. Physiol. Chem.*, 352, 764 (1971).
144. Greene, Roark, and Bartelt, in *Bayer Symposium V, "Proteinase Inhibitors,"* Fritz, Tschesche, Greene, and Truscheit, Eds., Springer-Verlag, Berlin, 1974, 188.
145. Pubols, Bartelt, and Greene, *J. Biol. Chem.*, 249, 2235 (1974).
146. Fritz, Trautschold, Haendle, and Werle, *Ann. N.Y. Acad. Sci.*, 146, 400 (1968).
147. Fritz, Jaumann, Meister, Pasquay, Hochstrasser, and Fink, in *Proc. 1st Int. Res. Conf. Proteinase Inhibitors*, Fritz and Tschesche, Eds., Walter de Gruyter, Berlin, 1971, 257.
148. Trautschold, Werle, Haendle, and Sebening, *Hoppe Seyler's Z. Physiol. Chem.*, 332, 328 (1963).
149. Geokas, Silverman, and Rinderknecht, *Experientia*, 26, 942 (1970).
150. Scevola, *G. Biochim.*, 5, 44 (1956).
151. Uszynski and Abildgaard, *Thromb. Diath. Haemorrh.*, 25, 580 (1971).
152. Kawano, Morimoro, and Uemura, *J. Biochem.* (Tokyo), 67, 333 (1970).
153. Junnila, Jansén, and Hopsu-Havu, *Acta Derm. Venereol.*, 51, 251 (1971).
154. Udaka and Hayashi, *Biochim. Biophys. Acta*, 97, 251 (1965); 104, 600 (1965).
155. Laskowski, Jr. and Laskowski, Sr., *J. Biol. Chem.*, 190, 563 (1951).

156. Laskowski, Jr., Mars, and Laskowski, Sr., *J. Biol. Chem.*, 198, 745 (1952).
157. Laskowski, Sr., Kassell, and Hagerty, *Biochim. Biophys. Acta*, 24, 300 (1957).
158. Kress, Martin, and Laskowski, Sr., *Biochim. Biophys. Acta*, 229, 836 (1971).
159. Hochstrasser, Haendle, Reichert, Werle, and Schwarz, *Hoppe Seyler's Z. Physiol. Chem.*, 352, 954 (1971).
160. Hochstrasser, Reichert, Schwarz, and Werle, *Hoppe Seyler's Z. Physiol. Chem.*, 353, 221 (1972).
161. Hochstrasser, Reichert, Schwarz, and Werle, *Hoppe Seyler's Z. Physiol. Chem.*, 354, 923 (1973).
162. Astrup and Sterndorff, *Scand. J. Clin. Lab. Invest.*, 7, 239 (1955).
163. Astrup, Alkjaer, and Soardi, *Scand. J. Clin. Lab. Invest.*, 11, 181 (1959).
164. Egeblad and Astrup, *Scand. J. Clin. Lab. Invest.*, 18, 181 (1966).
165. Faarvang, *Scand. J. Clin. Lab. Invest. Suppl.*, 83 (1965).
166. Proksch and Routh, *J. Lab. Clin. Med.*, 79, 491 (1972).
167. Schulman, *J. Biol. Chem.*, 213, 656 (1955).
168. **Wallner and Fritz,** *Hoppe Seyler's Z. Physiol. Chem.*, 355, 709 (1974).
169. Fink, Jaumann, Fritz, Ingrisch, and Werle, *Hoppe Seyler's Z. Physiol. Chem.*, 352, 1591 (1971).
170. Zaneveld, Schumacher, Fritz, Fink, and Jaumann, *J. Reprod. Fertil.*, 32, 525 (1973).
171. Hirschhäuser and Kionke, *Fertil. Steril.*, 22, 360 (1971).
172. Suominen and Niemi, *J. Reprod. Fertil.*, 29, 163 (1972).
173. Fink, *7th Int. Congr. Clin. Chem., Geneva/Evian 1969; Clinical Enzymology*, Vol. 2, Karger, Basel, 1970, 74.
174. Fritz, Fink, Meister, and Klein, *Hoppe Seyler's Z. Physiol. Chem.*, 351, 1344 (1970).
175. Fritz, Förg-Brey, Fink, Schiessler, Jaumann, and Arnhold, *Hoppe Seyler's Z. Physiol. Chem.*, 353, 1007 (1972).
176. Polakoski and Williams, in *Bayer Symposium V, "Proteinase Inhibitors,"* Fritz, Tschesche, Greene, and Truscheit, Eds., Springer-Verlag, Berlin, 1974, 156.
177. Tschesche, Kupfer, Lengel, Klauser, Meier, and Fritz, in *Bayer Symposium V, "Proteinase Inhibitors,"* Fritz, Tschesche, Greene, and Truscheit, Eds., Springer-Verlag, Berlin, 1974, 164.
178. Bieth, Pichoir, and Métais, *FEBS Lett.*, 8, 319 (1970).
179. Blatrix, Steinbuch, Vignal, and Lopes, *Bibl. Haematol.* (Basel) 38, 504 (1971).
180. Bundy and Mehl, *J. Biol. Chem.*, 234, 1124 (1959).
181. Cohen, *Fed. Proc.*, 33, 1311 (1974).
182. Crawford, *Arch. Biochem. Biophys.*, 156, 215 (1973).
183. Fritz, Brey, Schmal, and Werle, *Hoppe Seyler's Z. Physiol. Chem.*, 350, 1551 (1969).
184. Fritz, Heimburger, Meier, Arnhold, Zaneveld, and Schumacher, *Hoppe Seyler's Z. Physiol. Chem.*, 353, 1953 (1972).
185. Fritz, Schleuning, and Schill, in *Bayer Symposium V, "Proteinase Inhibitors,"* Fritz, Tschesche, Greene, and Truscheit, Eds., Springer-Verlag, Berlin, 1974, 118.
186. Fritz, Wunderer, Kummer, Heimburger, and Werle, *Hoppe Seyler's Z. Physiol. Chem.*, 353, 906 (1972).
187. Heimburger, Haupt, and Schwick, in *Proc. 1st Int. Res. Conf. Proteinase Inhibitors,* Fritz and Tschesche, Eds., Walter de Gruyter, Berlin, 1971, 1.
188. Heimburger, in *Bayer Symposium V, "Proteinase Inhibitors,"* Fritz, Tschesche, Greene, and Truscheit, Eds., Springer-Verlag, Berlin, 1974, 14.
189. Johnson, Pannell, and Travis, *Biochem. Biophys. Res. Commun.*, 57, 584 (1974).
190. Koj, Chudzik, Pajdak, and Dubin, *Biochim. Biophys. Acta*, 268, 199 (1972).
191. Kress and Laskowski, Sr., *Prep. Biochem.*, 3, 541 (1973).
192. Kress and Laskowski, Sr., in *Bayer Symposium V, "Proteinase Inhibitors,"* Fritz, Tschesche, Greene, and Truscheit, Eds., Springer-Verlag, Berlin, 1974, 23.
193. Ohlsson, *Scand. J. Clin. Lab. Invest.*, 28, 225 (1971).
194. Ohlsson, in *Fund. Probl. Cystic Fibrosis Relat. Dis.*, Selec. Pap. 6th Int. Cystic Fibrosis Congr., Mangos, Ed., Intercontinental, New York, 1973, 173.
195. Ohlsson, in *Bayer Symposium V "Proteinase Inhibitors,"* Fritz, Tschesche, Green, and Truscheit, Eds., Springer-Verlag, Berlin, 1974, 96.
196. Pannell, Johnson, and Travis, *Biochemistry*, 13, 5439 (1974).
197. Rimon, Shamash, and Shapiro, *J. Biol. Chem.*, 241, 5102 (1966).
198. Schultze, Heide, and Haupt, *Klin. Wochenschr.*, 40, 427 (1962).
199. Steinbuch, *Rev. Fr. Transfus.*, 14, 61 (1971).
200. Steinbuch and Audran, in *Bayer Symposium V., "Proteinase Inhibitors,"* Fritz, Tschesche, Greene, and Truscheit, Eds., Springer-Verlag, Berlin, 1974, 78.
201. Travis, Johnson, and Pannell, in *Bayer Symposium V, "Proteinase Inhibitors,"* Fritz, Tschesche, Greene, and Truscheit, Eds., Springer-Verlag, Berlin, 1974, 31.
202. Werb, *Biochem. Soc. Trans.*, 1, 382 (1973).
203. Bieth, Aubry, and Travis, in *Bayer Symposium V, "Proteinase Inhibitors,"* Fritz, Tschesche, Greene, and Truscheit, Ed., Springer-Verlag, Berlin, 1974, 53.
204. Abildgaard, *Scand. J. Clin. Lab. Invest.*, 19, 190 (1967).
205. Monkhouse, *Methods Enzymol.*, 19, 915 (1970).

206. Seegers, Cole, Harmison, and Monkhouse, *Can. J. Biochem.*, 42, 359 (1964).
207. Yin and Wessler, *Thromb. Diath. Haemorrh.*, 21, 398 (1969).
208. Yin, Wessler, and Stoll, *J. Biol. Chem.*, 246, 3712 (1971).
209. Ratnoff, Pensky, Ogston, and Naff, *J. Exp. Med.*, 129, 315 (1969).
210. Forbes, Pensky, and Ratnoff, *J. Lab. Clin. Med.*, 76, 809 (1970).
211. Haupt, Heimburger, Kranz, and Schwick, *Eur. J. Biochem.*, 17, 254 (1970).
212. McConnell, *J. Clin. Invest.*, 51, 1611 (1972).
213. Schumacher and Zaneveld, in *Bayer Symposium V, "Proteinase Inhibitors,"* Fritz, Tschesche, Greene, and Truscheit, Eds., Springer-Verlag, Berlin, 1974, 178.
214. Abe and Nagai, *Biochim. Biophys. Acta*, 278, 125 (1972).
215. Barrett and Starkey, *Biochem. J.*, 133, 709 (1973).
216. Barrett, Starkey, and Munn, in *Bayer Symposium V, "Proteinase Inhibitors,"* Fritz, Tschesche, Greene, and Truscheit, Eds., Springer-Verlag, Berlin, 1974, 72.
217. Baumstark, *Biochim. Biophys. Acta*, 207, 318 (1970).
218. Baumstark, *Biochim. Biophys. Acta*, 309, 181 (1973).
219. Brown, Hook, and Tragakis, *Invest. Opthalmol.*, 11, 149 (1972).
220. Frénoy and Bourrillon, *Biochim. Biophys. Acta*, 371, 168 (1974).
221. Gentou, Yon, and Filitti-Wurmser, *Bull. Soc. Chim. Biol.*, 50, 2003 (1968).
222. Hamberg, Stelwagen, and Ervast, *Eur. J. Biochem.*, 40, 439 (1973).
223. Harpel, *J. Exp. Med.*, 132, 329 (1970).
224. Harpel, *J. Exp. Med.*, 138, 508 (1973).
225. Hochstrasser, Theopold, and Brandl, *Hoppe Seyler's Z. Physiol. Chem.*, 354, 1013 (1973).
226. Iwamoto and Abiko, *Biochim. Biophys. Acta*, 214, 402 (1970).
227. Jones, Creeth, and Kekwick, *Biochem. J.*, 127, 187 (1972).
228. Kueppers and Bearn, *Proc. Soc. Exp. Biol. Med.*, 121, 1207 (1966).
229. Lanchantin, Plesset, Friedmann, and Hart, *Proc. Soc. Exp. Biol. Med.*, 121, 444 (1966).
230. Loeven, *Acta Physiol. Pharmacol. Neerl.*, 11, 350 (1962).
231. Luzzati, Goldlust, and Levine, *J. Bacteriol.*, 96, 1969 (1968).
232. Ogston, Bennett, Herbert, and Douglas, *Clin. Sci.*, 44, 73 (1973).
233. Roberts, Riesen, and Hall, in *Bayer Symposium V, "Proteinase Inhibitors,"* Fritz, Tschesche, Greene, and Truscheit, Eds., Springer-Verlag, Berlin, 1974, 63.
234. Schwick, Heimburger, and Haupt, *Z. Gesamte Inn. Med. Ihre Grenzgeb.*, 21, 193 (1966).
235. Starkey, *Biochem. Soc. Trans.*, 1, 381 (1973).
236. Starkey and Barrett, *Biochem. J.*, 131, 823 (1973).
237. Yu, Grappel, and Blank, *Experientia*, 28, 886 (1972).
238. Sumi and Fujii, *J. Biochem. (Tokyo)*, 75, 541 (1974).
239. Jungfer, *Naturwissenschaften*, 60, 54 (1973).
240. Brush and Shah, *Biochem. Biophys. Res. Commun.*, 53, 894 (1973).
241. Gray, Priest, Blatt, Westphal, and Jensen, *J. Biol. Chem.*, 235, 56 (1960).
242. Wu and Laskowski, Sr., *J. Biol. Chem.*, 235, 1680 (1960).
243. Habermann, *Ann. N.Y. Acad. Sci.*, 146, 479 (1968).
244. Huang, *Insect Biochem.*, 1, 207 (1971).
245. Gans, *Pharm. Weekbl.*, 104, 717 (1969).
246. Seegers and Andary, *Thromb. Res.*, 4, 869 (1974).
247. Nagasawa, Sugihara, Han, and Suzuki, *J. Biochem. (Tokyo)*, 67, 809 (1970).
248. Nagasawa, Han, Sugihara, and Suzuki, *J. Biochem. (Tokyo)*, 67, 821 (1970).
249. Sugihara, Nagasawa, and Suzuki, *J. Biochem. (Tokyo)*, 70, 649 (1971).
250. Deutsch, *J. Biol. Chem.*, 208, 669 (1954).
251. Galembeck and Cann, *Arch. Biochem. Biophys.*, 164, 326 (1974).
252. Green and Kay, *Arch. Biochem. Biophys.*, 102, 359 (1963).
253. Verpoorte, Green, and Kay, *J. Biol. Chem.*, 240, 1158 (1965).
254. Spiro, *J. Biol. Chem.*, 235, 2860 (1960).
255. Martin, *J. Biol. Chem.*, 236, 2672 (1961); 237, 2099 (1962).
256. Jacquot-Armand and Guinand, *Biochim. Biophys. Acta*, 133, 289 (1967).
257. Demaille, Dautrevaux, Havez, and Biserte, *Bull. Soc. Chim. Fr.*, p. 3506, 1965.
258. Woelk, Nagel, and Weber, *Acta Univ. Carol. Med. Monogr.*, 53/54, 157 (1972).
259. Lebreton, de Vonne, and Mouray, *C. R. Acad. Sci., Ser. D*, 266, 1076 (1968).
260. Picard, Roels, Carbonara, and Heremans, in *Protides of the Biological Fluids, 12th Colloq.*, Peeters, Ed., Pergamon Press, Oxford, 1964, 353.
261. Scharfman, Hayem, Moschetto, and Havez, *C. R. Acad. Sci. Ser. D*, 271, 2412 (1970).
262. Berthillier, Got, and Bertagnolio, *Biochim. Biophys. Acta*, 170, 140 (1968).

263. Paskhina and Nartikova, *Vopr. Med. Khim.*, 12, 325 (1966).
264. Yin, Wessler, and Stoll, *J. Biol. Chem.*, 246, 3694, 3703 (1971).
265. Norman and Hill, *J. Exp. Med.*, 108, 639 (1958).
266. Kobayashi and Nagasawa, *Biochim. Biophys. Acta,* 342, 372 (1974).
267. Loos, Wagner, and Opferkuch, in *Protides of the Biological Fluids, 17th Colloq.*, Peeters, Ed., Pergamon Press, Oxford, 1969, 311.
268. Loos, Opferkuch, and Ringelman, *Z. Med. Mikrobiol. Immunol.*, 156, 194 (1971).
269. Miyake and Ito, *Biochim. Biophys. Acta,* 328, 173 (1973).
270. Myerowitz, Chrambach, Rodbard, and Robbins, *Anal. Biochem.*, 48, 394 (1972).
271. Barrett, *Biochim. Biophys. Acta,* 371, 52 (1974).

AMINO ACID COMPOSITION OF ANIMAL PROTEINASE INHIBITORS

Amino acid, residues per mole

Lower Animals

Source	Designation	Trp	Lys	His	Arg	Asp[a]	Thr	Ser	Glu[a]	Pro	Gly	Ala	½ Cys	Val	Met	Ile	Leu	Tyr	Phe	Total no.	Reference
Anemonia sulcata Sea anemone	Polyvalent isoinhibitors																				
	A	0	4	2	5	6	1	5	5	3	6	2	6	4	0	2	2	3	2	58	1
	B	0	4	2	5	6	1	5	5	3	6	2	6	4	0	2	2	3	2	58	
	C2a	0	4	1	6	6	1	4	5	3	7	2	6	4	0	2	2	3	3	59	
	C4b	0	3	1	7	6	1	3	5	2	7	2	6	4	0	2	3	3	3	58	
	D	0	3	1	6	6	7	3–4	4	2	6	2	6	4	0	1	2	3	2–3	52–54	
	S-II	0	3.8	1.0	6.3	6.0	1.0	3.9	5.0	2.2	7.0	1.8	6.0	3.9	0	1.8	2.7	3.9	3.0		2
Ascaris lumbricoides Whole ascarids	Trypsin inhibitor																				
	CM-1	1	7	0	3	5	4	1	11	6	6	5	10	2	0	3	0	0	2	66	3
	CM-2	1	6	0	3	4	4	1	9–10	5	6	5	8	2	0	3	0	0	2	59–60	
Body walls	Trypsin inhibitor																				
	I	1	5	0	3	3	3	1	8	4	4	4	8	2	0	2	0	0	2	50	4
Body walls	Chymotrypsin inhibitors																				
	I	1.0	5.2	1.0	5.8	3.9	3.5	2.7	7.2	8.1	5.8	1.1	8.1	2.9	2.5	0	1.9	0	0		5
	II	1.0	4.0	0.9	4.0	3.0	2.7	1.9	5.3	5.7	4.2	1.0	5.7	2.2	1.8	0.1	1.4	0	0.2		
	III	1.0	7.2	0.9	6.1	4.0	4.9	2.0	8.7	9.1	7.1	2.3	10.8	3.6	2.4	0.5	1.8	0	0.9		
	IV	1.0	5.1	0.9	5.0	2.8	4.1	1.9	6.3	6.8	5.2	1.0	7.7	2.6	2.3	0	1.8	0	0		
Body walls	Pepsin inhibitors																				
	I	1	10	3	3	15	9	7	22	18	11	12	6	13	5	4	8	1	12	160	6
	II	0	9	3	3	14	8	7	17	17	9	10	6	11	4	4	7	1	11	142	
	III	1	9	3	3	15	8	6	19	18	10	10	6	12	4	4	7	1	11	147	
	IV	2	22	5	5	30	16	12	42	27	17	31	8	19	6	10	16	4	18	290	
Loligo vulgaris Cuttlefish Squid	Proteinase isoinhibitors																				
	A	0	3	0	1	11	5	3	6	2	7	3	8	2	2	2	3	1	3	62	7, 8
	B	0	4	0	2	11	5	3	5	2	7	3	8	1	2	2	3	1	3	62	
	E	0	8	3	3	11	2	2	4	0	6	3	8	3	2	2	2	1	3	62	
	L	0	7	3	5	11	4	1	5	1	7	3	8	2	3	0	2	1	3	66	
Helix pomatia Snail	Proteinase isoinhibitors																				
	B	0	1	0	4	7	2	5	6	5	6	3	6	0	1	2	2	4	4	58	9
	E	0	1	0	4	6	2	5	7	5	7	3	6	0	1	2	1	4	4	58	
	G	0	2	0	4	7	2	2	8	3	6	2	6	3	0	2	2	4	4	58	
	H	0	2	2	4	6	2	4	9	3	6	2	6	3	0	2	1	4	4	58	
	K	0	2	0	4	5	3	4	9	3	8	2	6	2	0	1	1	4	4	58	10
Achatina fulica African giant snail	Haemocyanin subunit[b] (trypsin inhibitor)	—[c]	28.6	32.0	27.7	74.7	33.9	37.6	67.5	35.5	39.2	47.6	—[c]	34.5	3.4	29.6	68.4	29.9	37.6		11
Hirudo medicinalis Leech	Hirudin	0	4	1	0	10	4	4	13	3	9	1	6	3	0	2	4	2	2	68	12, 13
Salivary gland	Bdellins A-1a	0	3	3	3	6	6	3	6	2	5	4	10	4	0	2	2	1	1	56	14, 15
	A2, 3, 4	0	5	3	0	8	3	3	5	3	4	4	10	5	1	1	1	1	2	59	
	B1	0	2	5	0.6	5	2	2	5	3	4	3.5	6	4	0	0	1.6	1	0.4		
	B2	0	1.5	5	1	5	3.7	2.7	6	1.4	4	4	6	4	0	0.3	2	1	1	45	
	B3	0	1	5	1	5	4	2	6	0	4	4	6	4	0	0	2	1	0		
	B4, 5, 6	0	2	5	1	7	4	4	8	1	4	4	6	4	0	0.3	2	1	0.5		

[a] Includes amides.
[b] Based on a molecular weight of 70,000.
[c] —, not determined.

AMINO ACID COMPOSITION OF ANIMAL PROTEINASE INHIBITORS (continued)

Amino acid, residues per mole

Source	Designation	Trp	Lys	His	Arg	Asp[a]	Thr	Ser	Glu[a]	Pro	Gly	Ala	½ Cys	Val	Met	Ile	Leu	Tyr	Phe	Total no.	Reference
Proteinase inhibitors																					
										Lower Animals (continued)											
Vipera russelli Russell's viper venom	I	0	2	2	5	7	2	2	5	5	7	3	4	1	0	2	1	1	3	52	16–18
	II	0	3	2	7	8	3	2	5	2	8	2	6	1	0	1	3	3	4	60	
Dendroaspis polylepis polylepis Black mamba venom	Toxin I (chymotrypsin inhibitor)	1	7	1	7	4	3	2	7	3	5	1	6	0	0	4	3	3	3	60	19
	Toxin K (trypsin inhibitor)	1	8	0	5	3	2	2	3	4	5	4	6	1	0	3	3	4	3	57	
Hemachatus haemachatus Ringhals cobra venom	Inhibitor II	0	2	1	4	6	3	2	6	2	6	5	6	1	0	3	4	2	4	57	17
										The Egg Whites of Birds											
Gallus gallus Chicken	Ovomucoid[d]	0	12.8	4.4	6.0	30.1	13.8	11.5	13.8	7.2	14.7	10.8	16.3	14.9	1.8	2.9	10.8	5.7	5.1		20
	Ovomucoid[e]	0	13.1	4.4	5.5	28.7	13.0	12.0	14.0	8.0	14.0	11.2	16.0	13.9	1.5	2.6	11.9	5.6	4.9		21
	Ovomucoid[f]	0	13.6	4.3	6.3	31.9	14.6	12.5	14.9	7.7	16.1	11.7	17.5	16.0	1.9	3.2	12.2	6.7	5.3		
Meleagris gallopavo Turkey	Ovomucoid[f]	0	11.2	5.2	5.8	27.2	14.2	10.0	19.0	8.8	17.3	8.6	16.7	15.7	1.8	4.4	13.5	6.8	3.2		22, 23
Casuarius casuarius Cassowary	Ovomucoid[f]	0	16.5	3.1	1.0	25.8	11.8	14.2	17.7	10.4	14.3	7.3	19.9	15.4	1.0	6.0	12.5	10.6	3.1		
Dromicaeus novaehollandiae Emu	Ovomucoid[f]	0	16.2	3.0	0	27.1	12.2	13.4	16.6	8.9	14.9	7.0	16.7	15.1	0.9	6.6	13.4	6.4	3.1		
Struthio camelus Ostrich	Ovomucoid[f]	0	14.8	2.2	3.4	27.3	17.0	18.3	18.0	11.4	17.0	5.6	19.9	16.9	1.0	4.5	15.3	11.1	3.3		
Rhea americana Rhea	Ovomucoid[f]	0	15.8	4.1	0	28.7	16.2	16.7	20.1	9.7	14.6	5.2	17.4	17.9	0.9	3.7	12.6	8.2	5.0		
Eudromia elegans Tinamou	Ovomucoid[f]	0	17.1	3.4	3.5	30.8	14.0	12.5	18.1	14.0	17.6	8.5	19.6	17.6	0	4.7	9.4	10.4	4.7		
Anas platyrhynckos Duck	Ovomucoid[f]	0	17.2	3.5	1.2	33.0	19.9	13.3	20.1	10.4	19.2	8.4	20.2	17.4	7.9	2.6	13.5	11.5	4.7		

[d] Based on a molecular weight of 27,300.
[e] Based on a molecular weight of 27,800.
[f] Based on a molecular weight of 28,000.

AMINO ACID COMPOSITION OF ANIMAL PROTEINASE INHIBITORS (continued)

Amino acid, residues per mole

Source	Designation	Trp	Lys	His	Arg	Asp[a]	Thr	Ser	Glu[a]	Pro	Gly	Ala	½ Cys	Val	Met	Ile	Leu	Tyr	Phe	Total no.	Reference
Pygoscelis adeliae Penguin	Ovomucoids[f]																				
Sp. Adelia		0	13.8	2.4	3.2	29.5	14.8	13.7	19.1	10.4	15.0	5.8	15.2	19.6	2.0	4.3	13.4	10.1	2.8		24
Sp. Emperor		0	14.8	2.0	3.1	29.8	14.0	12.6	18.8	10.1	15.4	5.3	16.5	19.6	2.0	6.4	12.9	9.8	2.2		
Sp. Crested		0	14.6	3.1	2.0	29.4	14.3	12.6	18.2	10.1	14.8	5.3	16.2	17.6	2.0	7.0	12.3	9.5	2.0		
Sp. Royal		0	14.8	3.1	2.2	30.2	14.3	12.6	19.3	9.5	15.4	5.6	16.8	18.2	2.0	7.3	13.2	9.8	2.2		
Sp. White-flippered		0	15.1	2.2	3.1	28.8	13.7	12.3	17.1	10.6	16.5	5.6	15.7	17.9	2.0	7.3	12.6	9.8	2.2		
Sp. Yellow-eyed		0	15.1	2.0	3.1	28.6	13.4	12.0	18.2	10.1	15.1	5.3	16.0	17.7	1.4	8.4	12.6	9.8	2.2		
Gallus gallus Chicken	Ovoinhibitor	1	23	14	21	47	33	27	39	18	32	20	34	27	4	17	22	17	6	401	25
	Ovoinhibitor[g]	1.0	23.7	12.7	20.1	46.3	27.4	25.5	39.9	16.9	31.6	19.7	34.0	25.4	3.5	16.9	22.0	14.8	6.2		26
Coturnix coturnix japonica Japanese quail	Ovoinhibitor[g]	1.0	26.3	17.7	18.7	46.4	25.0	27.9	38.5	15.3	28.5	14.9	29.0	22.7	5.0	19.6	18.5	16.3	4.7		27
Gallus gallus Chicken	Papain and ficin inhibitor[h]	—[c]	5.68	0.98	5.65	8.88	2.78	8.60	14.2	1.95	5.10	6.69	2.64	7.49	1.61	4.40	7.94	3.77	2.42		28

Mammalian Organs and Secretions

Source	Designation	Trp	Lys	His	Arg	Asp[a]	Thr	Ser	Glu[a]	Pro	Gly	Ala	½ Cys	Val	Met	Ile	Leu	Tyr	Phe	Total no.	Reference
Bovine organs	Basic trypsin-kallikrein inhibitor (Kunitz)	0	4	0	6	5	3	1	3	4	6	6	6	1	1	2	2	4	4	58	29–34
Sheep lung	Polyvalent protease inhibitor	0	4	0	6	5	3	1	3	4	6	6	6	1	1	2	2	4	4	58	35
Bovine pancreas	Secretory inhibitor isoinhibitors A and C	0	3	0	3	7	4	2	7	4	5	1	6	4	1	3	4	2	0	56	36–38
		0	3	0	3	7	4	2	7	4	5	1	6	4	1	3	4	2	0	56	39
Sheep pancreas	Secretory inhibitor	0	3	0	3	7	4	1	7	4	5	2	6	4	1	3	4	2	0	56	40
Porcine pancreas	Secretory inhibitor I	0	4	0	2	4	6	6	7	5	4	1	6	4	0	3	2	2	0	56	41–44
	II	0	4	0	2	4	5	5	6	4	4	1	6	4	0	3	2	2	0	52	
Canine pancreas	Secretory inhibitor	0	5	0	2	7	3	4	6	3	4	2	6	3	0	4	5	2	0	56	45
Feline pancreas	Secretory inhibitor	0	6	0	2	6	3	4	6	4	4	2	6	3	0	3	5	2	0	56	45, 46
Human pancreas	Secretory inhibitor Inhibitor 1[i]	0[c]	4	0	3	8	4	3	6	3	5	1	6	2	0	3	4	3	1	56	47, 48
	2	—[c]	3	0	3	6	3	3	6	3	5	2	4	2	0	2	3	2	1	48	49
		—[c]	3	1	2	6	2	2	4	2	3	1	4	2	0	2	3	1	1	39	

[g]Based on a molecular weight of 48,000.
[h]Calculated as residues per 10,000 g.
[i]Prepared by affinity chromatography on a trypsin resin.

AMINO ACID COMPOSITION OF ANIMAL PROTEINASE INHIBITORS (continued)

Amino acid, residues per mole

Mammalian Organs and Secretions (continued)

Source	Designation	Trp	Lys	His	Arg	Asp[a]	Thr	Ser	Glu[a]	Pro	Gly	Ala	½ Cys	Val	Met	Ile	Leu	Tyr	Phe	Total no.	Reference
Canine submandibular glands	Proteinase inhibitors																				50
	I[j]	0	10	3	5	13	7	8	9	6	9	6	12	4	3	5	6	5	4	115	
	II	0	11	3	5	13	7	8	8	6	9	6	12	4	3	5	6	5	4	115	
	C	0	11	3	5	13	7	8	8	5	8	6	12	4	3	5	6	5	4	113	
Porcine colostrum	Isoinhibitor																				51
	II	0	1.2	0	3.6	5.8	4.3	2.2	5.8	6.5	4.7	5.1	6.2	2.0	0.9	1.0	4.1	2.4	3.8		
	III	0	2.1	0	3.7	6.1	4.3	2.2	6.1	6.8	4.9	5.2	6	2.2	0.7	1.0	4.3	2.5	3.7		
	IV	0	2.2	0	3.7	6.1	4.3	2.1	5.9	6.7	4.8	5.2	6	2.4	0.8	1.0	4.2	2.5	3.6		
	V	0	2.0	0	3.8	6.0	4.2	2.0	5.7	6.9	4.9	5.3	6	2.1	0.8	1.0	4.0	2.6	3.3		
Bovine colostrum	Trypsin inhibitor Component B2 and isoinhibitors 4, 5, 6	0	2	0	3	8	6	3	10	7	4	4	6	0	1	1	5	3	4	67	52–54
	Isoinhibitor 3	0	3.0	0	3.0	8.5	5.1	3.1	10.2	6.9	4.0	4.1	6.0	0	1.0	0.8	5.2	2.6	3.5		
Human bronchial secretion	Proteinase inhibitor	0	8	4	4	16	9	15	12	12	6	5	12	2	2	6	6	3	3	125	55
Human seminal plasma	Inhibitor																				56
	I C	0	12	1	4	8	4	6	8	12	9	3	12	5	—[c]	1	5	2	2	94	
	I D	0	12	0	4	8	4	6	7	12	9	3	12	5	3	1	4	2	2	54	
	II G1	0	3	2	6	6	3	3	3	5	5	1	6	1	1	3	2	3	1	54	
	II G2	0	3	2	5	6	3	4	3	5	5	1	6	1	1	3	2	3	1	54	
Boar seminal plasma	Acrosin isoinhibitor																				57
	A	1	5	3	5	7	4	5	5	3	6	2	6	1	1	2	2	3	5	66	
	B	1	6	3	6	7	4	5	5	3	6	2	6	1	1	2	2	3	5	68	
	A1	1	4	3	6	7	4	5	5	3	6	2	6	1	1	2	2	3	5	66	
Boar spermatozoa	Inhibitor	—[c]	4	3	5	7	4	5	5	3	6	2	6	1	1	2	2	3	5		57
Boar sperm plasma	Inhibitor	2	8	4	8	11	6	7	8	4	8	4	8	2	2	4	4	4	7	101	15
Guinea pig seminal vesicles	Trypsin inhibitor																				
	a[1]	0	1	2[k]	6	6	1	2	10	5	6	1	6	3	0	4	5	2	0	60	15
	α	0	5	5	9	6	2	7	17	5–6	8	2	6	7	0	4	5	1	1	90–91	58, 59
	β	0	2	2–3	6	5	1	3	9	4	5	1	6	3	0	3	4	1	0	55–56	
	Trypsin-plasmin inhibitor																				
	d[i]	0	4	3	4	6	4	5	4	2	5	0	6	3	1	1	3	4	3	58	15, 58
	e₁[i]	0	5	3	4	6	4	6	4	3	5	1	6	3	1	1	3	4	3	62	
	e₂[i]	0	5	3	4	6	4	5	4	3	5	1	6	3	1	1	3	4	4	63	
	f₁[i]	0	5	3	4	6	4	6	4	2	5	0	6	3	1	1	3	4	3	59	
	f₂[i]	0	5	3	4	6	4	6	4	2	5	0	6	3	1	1	3	4	3	60	

[j] Inhibitor A₂ has the same composition.
[k] In some preparations, one residue of histidine was found.

AMINO ACID COMPOSITION OF ANIMAL PROTEINASE INHIBITORS (continued)

Amino acid, residues per mole

Mammalian Blood

Source	Designation	Trp	Lys	His	Arg	Asp[a]	Thr	Ser	Glu[a]	Pro	Gly	Ala	½ Cys	Val	Met	Ile	Leu	Tyr	Phe	Total no.	Reference
Human plasma	α_1-Antitrypsin[l]	3	36	12	8	44	28	21	52	19	23	24	2	27	8	21	51	6	29	414	60, 61
	α_1-Antitrypsin	2	31	12	6	38	28	20	43	18	20	22	2	21	8	16	41	6	26	360	62
	α_1-Antitrypsin[m]	2	32	13	7	44	31	22	50	16	22	24	1	23	8	18	45	7	29	394	63
	α_1-Antitrypsin	1	41	13	7	47	26	23	55	23	23	25	2	25	6	20	48	6	27	418	64, 65
	α_1-Antitrypsin[n]	2	37	12	8	43	33	27	54	19	22	27	1	24	9	18	44	6	28	414	66
	α_1-Antichymotrypsin[o]	3	29	9	15	48	32	32	55	16	16	32	0	25	12	21	55	10	26	436	67, 68
	α_2-Macroglobulin	75	414	181	224	516	444	504	800	356	398	403	126	578	107	236	586	239	276	6,463	69
	α_2-Macroglobulin	49	408	172	219	499	431	542	769	336	415	404	144	579	100	207	640	252	270	6,436	70
	Inter-α-trypsin inhibitor[o]	15	70	26	53	132	73	90	148	61	81	76	27	95	24	55	105	39	54	1,224	67, 68
	Antithrombin III[o]	7	34	5	20	49	24	33	59	22	20	32	6	28	11	21	42	12	25	450	67, 68
	C̄1-Inactivator	7	36	12	18	54	60	52	62	38	17	37	6	38	14	21	65	17	29	583	71
Bovine plasma	α_2-Macroglobulin	56	334	169	272	506	422	434	891	410	358	447	85	609	105	264	632	256	269	6,519	72
Porcine serum	α_1-Macroglobulin[o]	62	432	149	207	522	415	461	870	361	437	468	97	593	117	281	675	224	284	6,655	73
	α_2-Macroglobulin[o]	54	341	120	130	525	559	541	998	322	448	450	114	572	36	241	573	202	268	6,494	74
Rabbit plasma	α_1-Macroglobulin	—[c]	343	106	222	490	642	627	626	220	401	375	162	448	48	196	454	198	291		75
	α_2-Macroglobulin	—[c]	318	149	200	402	394	314	582	309	286	317	204	458	82	286	513	184	237		75
Mouse serum	α_1-Antitrypsin	—[c]	21	9	7	25	19	17	35	10	13	19	2	14	4	10	30	1	13		76
Calf serum (Fetal)	Fetuin[g]	2	16	10	12	33	25	26	34	34	24	33	12	40	0	15	27	7	11	361	77
		2	16	11	16	27	16	21	31	35	18	28	<18	33	0	12	28	8	9		78

[l]Molecular weight from analysis, 54,200, including carbohydrate.
[m]Based on a molecular weight of 50,000.
[n]From MM-type plasma.
[o]Calculated to the nearest whole residue from percentage data in the reference given. Molecular weight used: α_1-antichymotrypsin, 69,000; inter-α-trypsin inhibitor, 160,000; antithrombin III, 65,000, α_1-macroglobulins, 820,000.
[p]Based on a molecular weight of 820,000.

Compiled by Beatrice Kassell and Monica June Williams. Supported by grant BMS 74-22102 from the National Science Foundation.

AMINO ACID COMPOSITION OF ANIMAL PROTEINASE INHIBITORS (continued)

REFERENCES

1. Fritz, Brey, and Béress, *Hoppe Seyler's Z. Physiol. Chem.*, 353, 19 (1972).
2. Wunderer, Kummer, Fritz, Béress, and Machleidt, in *Bayer Symposium V, Proteinase Inhibitors,* Fritz, Tschesche, Greene, and Truscheit, Eds., Springer-Verlag, Berlin, 1974, 277.
3. Portmann and Fraefel, *Helv. Chim. Acta,* 50, 2078 (1967).
4. Kucich and Peanasky, *Biochim. Biophys. Acta,* 200, 47 (1970).
5. Peanasky and Abu-Erreish, in *Proc. 1st Int. Res. Conf. Proteinase Inhibitors,* Fritz and Tschesche, Eds., Walter de Gruyter, Berlin, 1971, 281.
6. Abu-Erreish and Peanasky, *J. Biol. Chem.,* 249, 1558 (1974).
7. Tschesche and von Rücker, *Hoppe Seyler's Z. Physiol. Chem.,* 354, 1447 (1973).
8. Tschesche and von Rücker, in *Bayer Symposium V, Proteinase Inhibitors,* Fritz, Tschesche, Greene, and Truscheit, Eds., Springer-Verlag, Berlin, 1974, 284.
9. Tschesche and Dietl, *Eur. J. Biochem.* 30, 560 (1972).
10. Dietl and Tschesche, in *Bayer Symposium V, Proteinase Inhibitors,* Fritz, Tschesche, Greene, and Truscheit, Eds., Springer-Verlag, Berlin, 1974, 254.
11. Kareem, Gilles, Nguyen-Thanh, and Keil, *Comp. Biochem. Physiol.,* 44B, 963 (1973).
12. Markwardt, *Methods Enzymol.,* 19, 924 (1970).
13. Landmann, *Folia Haematol.,* 98, 437 (1972).
14. Fritz, Gebhardt, Meister, and Fink, in *Proc. 1st Int. Res. Conf. Proteinase Inhibitors,* Fritz and Tschesche, Eds., Walter de Gruyter, Berlin, 1971, 271.
15. Fink, Fritz, Jaumann, Schiessler, Förg-Brey, and Werle, in *Protides of the Biological Fluids, Proc. 20th Colloq.* Peters, Ed., Pergamon Press, Oxford, 1973, 425.
16. Takahashi, Iwanaga, Hokama, Suzuki, and Kitagawa, *FEBS Lett.,* 38, 217 (1974).
17. Takahashi, Iwanaga, Kitagawa, Hokama, and Suzuki, in *Bayer Symposium V, Proteinase Inhibitors,* Fritz, Tschesche, Greene, and Truscheit, Eds., Springer-Verlag, Berlin, 1974, 265.
18. Takahashi, Iwanaga, and Suzuki, *J. Biochem.* (Tokyo), 76, 709 (1974).
19. Strydom, *Nat. New Biol.,* 243, 88, (1973).
20. Davis, Mapes, and Donovan, *Biochemistry,* 10, 39 (1971).
21. Oegema, Jr. and Jourdian, *Arch. Biochem. Biophys.,* 160, 26 (1974).
22. Osuga and Feeney, *Arch Biochem. Biophys.,* 124, 560 (1968).
23. Feeney and Allison, *Evolutionary Biochemistry of Proteins,* Interscience, New York, 1969, 86.
24. Osuga, Bigler, Uy, Sjöberg, and Feeney, *Comp. Biochem. Physiol.,* 48B, 519 (1974).
25. Davis, Zahnley, and Donovan, *Biochemistry,* 8, 2044 (1969).
26. Zahnley and Davis, *Biochem. J.* 135, 59 (1973).
27. Liu, Means, and Feeney, *Biochim. Biophys. Acta,* 229, 176 (1971).
28. Sen and Whitaker, *Arch. Biochem. Biophys.,* 158, 623 (1973).
29. Kassell, Radicevic, Berlow, Peanasky, and Laskowski, Sr., *J. Biol. Chem.,* 238, 3274 (1963).
30. Kassell, Radicevic, Ansfield, and Laskowski, Sr., *Biochem. Biophys. Res. Commum.,* 18, 255 (1965).
31. Anderer and Hörnle, *Z. Naturforsch.,* 20b, 457 (1965).
32. Anderer, *Z. Naturforsch.,* 20b, 499 (1965).
33. Chauvet and Acher, *Int. J. Protein Res.,* 2, 165 (1970).
34. Chauvet and Acher, *FEBS Lett.,* 23, 317 (1972).
35. Fritz, Greif, Schramm, Hochstrasser, and Werle, *Hoppe Seyler's Z. Physiol. Chem.,* 351, 139 (1970).
36. Greene, Rigbi, Fackre, and Broich, *J. Biol. Chem.,* 241, 5610 (1966).
37. Cerwinsky, Burck, and Grinnan, *Biochemistry,* 6, 3175 (1967).
38. Greene and Bartelt, *J. Biol. Chem.,* 244, 2646 (1969).
39. Schneider, Stasiuk, and Laskowski, Sr., in *Bayer Symposium V, Proteinase Inhibitors,* Fritz, Tschesche, Greene, and Truscheit, Eds., Springer-Verlag, Berlin, 1974, 223.
40. Hochstrasser, Schramm, Fritz, Schwarz, and Werle, *Hoppe Seyler's Z. Physiol. Chem.,* 350, 893 (1969).
41. Greene, DiCarlo, Sussman, Bartelt, and Roark, *J. Biol. Chem.,* 243, 1804 (1968).
42. Tschesche, Wachter, Kupfer, and Niedermeier, *Hoppe Seyler's Z. Physiol. Chem.,* 350, 1247 (1969).
43. Tschesche and Wachter, *Eur. J. Biochem.,* 16, 187 (1970).
44. Bartelt and Greene, *J. Biol. Chem.,* 246, 2218 (1971).
45. Tschesche, Wachter, Kupfer, Obermeier, Reidel, Haenisch, and Schneider, in *Proc. 1st Int. Res. Conf. Proteinase Inhibitors,* Fritz and Tschesche, Eds., Walter de Gruyter, Berlin, 1971, 207.
46. Tschesche and Kupfer, *Hoppe Seyler's Z. Physiol. Chem.,* 352, 764 (1971).
47. Pubols, Bartelt, and Greene, *J. Biol. Chem.,* 249, 2235 (1974).

AMINO ACID COMPOSITION OF ANIMAL PROTEINASE INHIBITORS (continued)

48. Greene, Roark, and Bartelt, in *Bayer Symposium V, Proteinase Inhibitors,* Fritz, Tschesche, Greene, and Truscheit, Eds., Springer-Verlag, Berlin, 1974, 188.
49. Feinstein, Hoffstein, and Sokolovsky, in *Bayer Symposium V, Proteinase Inhibitors,* Fritz, Tschesche, Greene, and Truscheit, Eds., Springer-Verlag, Berlin, 1974, 199.
50. Fritz, Jaumann, Meister, Pasquay, Hochstrasser, and Fink, in *Proc. 1st Int. Res. Conf. Proteinase Inhibitors,* Fritz and Tschesche, Eds., Walter de Gruyter, Berlin, 1971, 257.
51. Kress, Martin, and Laskowski, Sr., *Biochim. Biophys. Acta,* 229, 836 (1971).
52. Laskowski, Jr., Mars, and Laskowski, Sr., *J. Biol. Chem.,* 198, 745 (1952).
53. Čechová, Jonáková, and Šorm, *Coll. Czech. Chem. Commun.,* 36, 3342 (1971).
54. Čechová, *Coll. Czech. Chem. Commun.,* 39, 647 (1974).
55. Hochstrasser, Reichert, Schwarz, and Werle, *Hoppe Seyler's Z. Physiol. Chem.,* 353, 221 (1972).
56. Schiessler, Arnhold, and Fritz, in *Bayer Symposium V, Proteinase Inhibitors,* Fritz, Tschesche, Greene, and Truscheit, Eds., Springer-Verlag, Berlin, 1974, 147.
57. Tschesche, Kupfer, Lengel, Klauser, Meier, and Fritz, in *Bayer Symposium. V, Proteinase Inhibitors,* Fritz, Tschesche, Greene, and Truscheit, Eds., Springer-Verlag, Berlin, 1974, 164.
58. Fritz, Fink, Meister, and Klein, *Hoppe Seyler's Z. Physiol Chem.,* 351, 1344 (1970).
59. Fink, Klein, Hammer, Müller-Bardorff, and Fritz, in *Proc. 1st Int. Res. Conf. Proteinase Inhibitors,* Fritz and Tschesche, Eds., Walter de Gruyter, Berlin, 1971, 225.
60. Kress and Laskowski, *Prep. Biochem.,* 3, 541 (1973).
61. Kress and Laskowski, in *Bayer Symposium V, Proteinase Inhibitors,* Fritz, Tschesche, Greene, and Truscheit, Eds., Springer-Verlag, Berlin, 1974, 23.
62. Jeppsson and Laurell, in *Bayer Symposium V, Proteinase Inhibitors,* Fritz, Tschesche, Greene, and Truscheit, Eds., Springer-Verlag, Berlin, 1974, 47.
63. Crawford, *Arch. Biochem. Biophys.,* 156, 215 (1973).
64. Travis, Johnson, and Pannell, in *Bayer Symposium V, Proteinase Inhibitors,* Fritz, Tschesche, Greene, and Truscheit, Eds., Springer-Verlag, Berlin, 1974, 31.
65. Pannell, Johnson, and Travis, *Biochemistry,* 13, 5439 (1974).
66. Chan, Luby, and Wu, *FEBS Lett.,* 35, 79 (1973).
67. Schwick, Heimburger, and Haupt, *Z. Gesamte Inn. Med. Ihre Grenzgeb.,* 21, 193 (1966).
68. Heimburger, Haupt, and Schwick, in *Proc. 1st Int. Res. Conf. Proteinase Inhibitors,* Fritz and Tschesche, Eds., Walter de Gruyter, Berlin, 1971, 1.
69. Dunn and Spiro, *J. Biol. Chem.,* 242, 5549 (1967).
70. Demaille, Broussal, Colette, Guilleux, Magnan and de Bornier, *C. R. Soc. Biol.,* 164, 626 (1970).
71. Haupt, Heimburger, Kranz, and Schwick, *Eur. J. Biochem.,* 17, 254 (1970).
72. Nagasawa, Sugihara, Han, and Suzuki, *J. Biochem.* (Tokyo), 67, 809 (1970).
73. Armand and Guinand, *Biochim. Biophys. Acta,* 133, 289 (1967).
74. Demaille, Dautrevaux, Havez, and Biserte, *Bull. Soc. Chim.,* p. 3506 (1965).
75. Lebreton de Vonne and Mouray, *Ann. Biol. Clin.,* 32, 185 (1974).
76. Myerowitz, Chrambach, Rodbard, and Robbins, *Anal. Biochem.,* 48, 394 (1972).
77. Spiro and Spiro, *J. Biol. Chem.,* 237, 1507 (1962).
78. Fisher, O'Brien, and Puck, *Arch. Biochem. Biophys.,* 99, 241 (1962).

AMINO ACID SEQUENCE OF ANIMAL PROTEINASE INHIBITORS**

Inhibitor from

Ascaris (1–3)

Ascaris lumbricoides

                                                                            10

Glu- Ala- Glu- Lys- Cys- Asx- Glx- Glx- Pro- Gly- Trp- Thr- Lys- Gly- Gly- Cys- Glu- Thr-

          20   30

Cys- Gly- Cys- Ala- Gln- Lys- Ile- Val- Pro- Cys- Thr- Arg- Glu- Thr- Lys- Pro- Asn- Pro-

             40   50

Glu- Cys- Pro- Arg- Lys- Gln- Cys- Cys- Ile- Ala- Ser- Ala- Gly- Phe- Val- Arg- Asp- Ala-

             60   66

Gln- Gly- Asn- Cys- Ile- Lys- Phe- Glu- Asp- Cys- Pro- Lys

Disulfide bridges: 5–28, 16–44, 19–58, 21–38, 43–64

Pancreatic secretory inhibitors

                                                           10   *

Bovine: Kazal's (4–7)   Asn- Ile- Leu- Gly- Arg- Glu- Ala- Lys- Cys- Thr- Asn- Glu- Val- Asn- Gly- Cys- Pro- Arg-

Ovine (8)   — — — — — — — — — — — — — — — — — —

                                                                                                                                                                  *

Porcine I (9–11, 13, 14)   Thr- Ser- Pro- Gln — — — Thr — — Ser — — Ser — — — Lys

Porcine II (9, 12)   — — — — — — — — — — — — — — — — — —

Human (15)   Asp- Ser- Leu- Gly — — — Lys (—, —, Asn, —, Leu, Asn, —, —, Thr) —

                                                  20   30

Bovine   Ile- Tyr- Asn- Pro- Val- Cys- Gly- Thr- Asp- Gly- Val- Thr- Tyr- Ser- Asn- Glu- Cys- Leu-

Ovine   — — — — — — — — — — — — — Ala — — — —

Porcine I   — — — — — — — — — Ile — — Ser — — — Val

Porcine II   — — — — — — — — — — — — — — — — — —

Human   — — (—, —, —, —, —, —, —, —, Asp, —, —, Pro, —, —, —, —)

(Sequence continued on next page)

Note: Dashes indicate that the 3-letter symbol is the same as the one above it.

*Reactive site for trypsin.
**Supported by grant BMS 74-22102 from the National Science Foundation.

AMINO ACID SEQUENCE OF ANIMAL PROTEINASE INHIBITORS (continued)

					40									50				
Bovine	Leu-	Cys-	Met-	Glu-	Asn-	Lys-	Glu-	Arg-	Gln-	Thr-	Pro-	Val-	Leu-	Ile-	Gln-	Lys-	Ser-	Gly-
Ovine	—	—	—	—	—	—	—	—	—	—	—	—	—	—	—	—	—	—
Porcine I	—	—	Ser	—	—	—	Lys	—	—	—	—	—	—	—	—	—	—	—
Porcine II	—	—	—	—	—	—	—	—	—	—	—	—	—	—	—	—	—	—
Human	(—,	—,	Phe,	—,	—)	Arg	—	—	—	—	(Ser,	Ile,	—,	—,	—)	—	—	—

		56
Bovine	Pro-	Cys
Ovine	—	—
Porcine I	—	—
Porcine II	—	—
Human	—	—

Disulfide bridges: 9–38, 16–35, 24–56

Note: Dashes indicate that the 3-letter symbol is the same as the one above it.

AMINO ACID SEQUENCE OF ANIMAL PROTEINASE INHIBITORS (continued)

Inhibitors of bovine organs (A)
Pancreas (16–19), lung (20), parotid gland (21), liver (22), ovary (23) [reactive site (24–28)]
Sheep lung (A) (29)
Bovine colostrum (B) (30, 31)
Snail isoinhibitor K (C) (32, 33)
Inhibitor II from venom of Russell's viper (D), *Vipera russelli* (34, 35)
Toxins from Black Mamba venom (36), *Dendroaspis polylepis polylepis*
Toxin I (E), weak chymotrypsin inhibitor
Toxin K (F), weak trypsin inhibitor
Red sea turtle egg white (G) (37)

```
                    1                                     10                          *†
A        Arg- Pro- Asp- Phe- Cys- Leu- Glu- Pro- Pro- Tyr- Thr- Gly- Pro- Cys- Lys-
                                                                                  *
B        Phe- Gln- Thr- Pro __ __ Leu __ Gln- Leu __ Gln- Ala- Arg __ __ __ __

C        Pyr- Gly- Arg __ Ser- Phe __ Asn __ __ Ala- Glu- Thr __ __ __ __

D        His- Asp __ __ Thr __ __ __ __ Ala- Pro __ Ser __ Arg __ Arg-

E        (Gln,Pro)- Leu- Arg- Lys- Leu __ Ile __ His- Arg- Asn- Pro __ __ __ Tyr-

F        Ala- Ala __ Tyr __ Lys __ Pro- Leu- Arg- Ile __ Pro __ Lys-

Gᵃ       X(Lys, Glx, Asx, Gly, Arg) Asp- Ile __ Arg __ __ Pro- Glu- Gln __ __ __


                          20                              30
A        Ala- Arg- Ile- Ile- Arg- Tyr- Phe- Tyr- Asn- Ala- Lys- Ala- Gly- Leu- Cys- Gln- Thr- Phe-

B        __ Ala- Leu- Leu __ __ __ __ Asx- Ser- Thr- Ser- Asn- Ala __ Glu- Pro __

C        __ Ser- Phe- Arg- Gln __ Tyr __ Asn __ Lys __ Gly- Gly __ Gln- Gln __

D        Gly- His- Leu __ Arg- Ile __ __ __ Leu- Glu __ Asn- Lys __ Lys- Val __

E        Gln- Lys- Ile- Pro- Ala- Phe __ __ Gln- Lys- Lys- Lys- Gln __ (Glx,Gly) __

F        Arg __ __ __ Ser __ __ __ Lys- Trp- __ Ala __ __ __ Leu- Pro __

G        Gly- Arg __ __ Arg- Tyr- Phe __ Asn- Pro- Ala- Ser- Arg- Met __ Glu- Ser __
```

(Sequence continued on next page)

Note: Dashes indicate that the 3-letter symbol is the same as the one above it.

ᵃX indicates a blocked amino terminus.

†Reactive site for chymotrypsin.

AMINO ACID SEQUENCE OF ANIMAL PROTEINASE INHIBITORS (continued)

(Sequence continued from previous page)

							40										50	
A	Val-	Tyr-	Gly-	Gly-	Cys-	Arg-	Ala-	Lys-	Arg-	Asn-	Asn-	Phe-	Lys-	Ser-	Ala-	Glu-	Asp-	Cys-
B	Thr	___	___	___	___	Gln-	Gly-	Asn-	Asn-	Asx	___	___	Glu-	Thr-	Thr	___	Met-	___
C	Ile	___	___	___	___	Arg	___	___	Gln-	Asn-	Arg	___	Asp	___	___	Gln-	Gln-	___
D	Phe	___	___	___	___	Gly	___	___	Ala	___	Asn	___	Glu	___	Arg-	Asp-	Glu-	___
E	Thr-	Trp-	Ser	___	___	___	___	___	Ser	___	Arg	___	Lys	___	Ile-	Glu	___	___
F	Asp-	Tyr	___	___	___	___	___	___	Ala	___	___	___	___	___	___	___	___	___
G	Ile	___	Gly	___	___	Lys	___	___	Lys	___	Asn	___	___	___	Lys-	Ala	___	___

							58						
A	Met-	Arg-	Thr-	Cys-	Gly-	Gly-	Ala						
													67
B	Leu	___	Ile	___	Glu-	Pro-	Pro-	Gln-	Gln-	Thr-	Asp-	Lys-	Ser
					58								
C	Gln-	Gly-	Val	___	Val								
							60						
D	Arg-	Glu-	Thr	___	Gly-	Gly-	Lys						
							60						
E	___	Arg	___	___	Ile-	Arg	___						
						57							
F	___	___	___	___	Val-	Gly							
										66			
G	Val	___	___	___	Gly-	Pro-	Gly-	Ile-	Cys-	Leu(+60 more residues)			

Disulfide bridges: A 5–55, 14–38, 30–51; B and D are homologous

Note: Dashes indicate that the 3-letter symbol is the same as the one above it.

Compiled by Beatrice Kassell and Monica June Williams.

AMINO ACID SEQUENCE OF ANIMAL PROTEINASE INHIBITORS (continued)

REFERENCES

1. Fraefel and Acher, *Biochim. Biophys. Acta,* 154, 615 (1968).
2. Induni, Dissertation, University of Freiburg, Switzerland, 1969.
3. Kassell, *Methods Enzymol.,* 19, 872 (1970).
4. Greene and Giordano, *J. Biol. Chem.,* 244, 285 (1969).
5. Greene and Bartelt, *J. Biol. Chem.,* 244, 2646 (1969).
6. Guy, Shapanka, and Greene, *J. Biol. Chem.,* 246, 7740 (1971).
7. Rigbi and Greene, *J. Biol. Chem.,* 243, 5457 (1968).
8. Hochstrasser, Schramm, Fritz, Schwarz, and Werle, *Hoppe-Seyler's Z. Physiol. Chem.,* 350, 893 (1969).
9. Tschesche and Wachter, *Eur. J. Biochem.,* 16, 187 (1970).
10. Bartelt and Greene, *J. Biol. Chem.,* 246, 2218 (1971).
11. Tschesche, *Angew. Chem.,* 86, 21 (1974).
12. Greene, DiCarlo, Sussman, Bartelt, and Roark, *J. Biol. Chem.,* 243, 1804 (1968).
13. Tschesche, *Hoppe-Seyler's Z. Physiol. Chem.,* 348, 1216 (1967).
14. Tschesche and Klein, *Hoppe-Seyler's Z. Physiol Chem.,* 349, 1645 (1968).
15. Greene, Roark, and Bartelt, in *Bayer Symp. V Proteinase Inhibitors,* Fritz, Tschesche, Greene, and Truscheit, Eds., Springer-Verlag, Berlin, 1974, 188.
16. Kassell, Radicevic, Ansfield, and Laskowski, Sr., *Biochem. Biophys. Res. Commun.,* 18, 255 (1965).
17. Kassell, and Laskowski, Sr., *Biochem. Biophys. Res. Commun.,* 20, 463 (1965).
18. Acher and Chauvet, *Bull. Soc. Chim. Fr.,* p. 3954 (1967).
19. Dlouhá, Pospíšilová, Meloun, and Šorm, *Collect. Czech. Chem. Commun.,* 33, 1363 (1968).
20. Anderer, *Z. Naturforsch.,* 20b, 462 (1965).
21. Anderer, *Z. Naturforsch.,* 20b, 499 (1965).
22. Chauvet and Acher, *Int. J. Protein Res.,* 2, 165 (1970).
23. Chauvet and Acher, *FEBS Lett.,* 23, 317 (1972).
24. Chauvet and Acher, *J. Biol. Chem.,* 242, 4274 (1967).
25. Kress and Laskowski, Sr., *J. Biol. Chem.,* 243, 3548 (1968).
26. Fritz, Fink, Gebhardt, Hochstrasser, and Werle, *Hoppe-Seyler's Z. Physiol. Chem.,* 350, 933 (1969).
27. Rigbi, in *Proc. Int. Res. Conf. Proteinase Inhibitors, 1st,* Fritz and Tschesche, Eds., Walter de Gruyter, Berlin, 1971, 74.
28. Tschesche, Jering, Schorp, and Dietl, in *Bayer Symp. V Proteinase Inhibitors,* Fritz, Tschesche, Greene, and Truscheit, Eds., Springer-Verlag, Berlin, 1974, 362.
29. Fritz, Greif, Schramm, Hochstrasser, and Werle, *Hoppe-Seyler's Z. Physiol. Chem.,* 351, 139 (1970).
30. Čechová, Jonáková, and Šorm, *Collect. Czech. Chem. Commun.,* 36, 3342 (1971).
31. Čechová and Ber, *Collect. Czech. Chem. Commun.,* 39, 680 (1974).
32. Dietl and Tschesche, in *Bayer Symp. V Proteinase Inhibitors,* Fritz, Tschesche, Greene, and Truscheit, Eds., Springer-Verlag, Berlin, 1974, 254.
33. Tschesche, *Angew. Chem.,* 86, 21 (1974).
34. Takahashi, Iwanaga, Hokama, Suzuki, and Kitagawa, *FEBS Lett.,* 38, 217 (1974).
35. Takahashi, Iwanaga, Kitagawa, Hokama, and Suzuki, in *Bayer Symp. V Proteinase Inhibitors,* Fritz, Tschesche, Greene, and Truscheit, Eds., Springer-Verlag, Berlin, 1974, 265.
36. Strydom, *Nat. New Biol.,* 243, 88 (1973).
37. Laskowski, Jr., Kato, Leary, Schrode, and Sealock, in *Bayer Symp. V Proteinase Inhibitors,* Fritz, Tschesche, Greene, and Truscheit, Eds., Springer-Verlag, Berlin, 1974, 597.

SPECIFICITIES AND SOME PROPERTIES OF MICROBIAL PROTEINASE INHIBITORS

Source	Name	Molecular weight[a]	I.P.	Trypsin	α-Chymotrypsin	Plasmin	Thrombin	Papain	Pepsin	Subtilisin	Yeast proteinases A	B	C	Other	Amino acid table	Sequence table	Ref.
Saccharomyces cerevisiae yeast	Proteinase A inhibitor I_A^2 I_A^3	6,100 el, sub 6,100 el, sub 23,000 gel							0		I_{sub}-E	0	0				1
	Proteinase B inhibitor I II	10,000 gel, el 10,000 gel, el	8.0 7.0	0	0				0	0	0 0	I:E +	0 0				2, 3
	Carboxypeptidase Y (proteinase C) inhibitor	25,000 gel									0	0	+	Carboxypeptidases A and B, 0			4
	Inhibitor of chitin synthetase activating factor[c]	8,500 gel 8,500 gel 7,600–8,000 ulc	7.1											Chitin synthetase activating factor, +	*		5

[a] Symbols used to designate the determination of molecular weight are: gel, gel filtration; sub, subunit molecular weight; aa, minimum molecular weight by amino acid composition; el, SDS-gel electrophoresis; ulc, various ultracentrifugal methods. The molecular weights calculated from amino acid compositions are approximations when the asparagine and glutamine contents are not known.
[b] Inhibition is indicated as follows: stoichiometric inhibition, complex I-E (inhibitor-enzyme), sub, subunit; +, strong inhibition; ±, weak inhibition; 0, no inhibition.
[c] Believed to be similar to proteinase B.

SPECIFICITIES AND SOME PROPERTIES OF MICROBIAL PROTEINASE INHIBITORS (continued)

| Source | Name | Molecular weight[a] | I.P. | Specificity[b] ||||||| Yeast proteinases ||| Other | Information in |||
				Tryp-sin	α-Chymo-trypsin	Plas-min	Throm-bin	Papain	Pepsin	Subtil-isin	A	B	C		Amino acid table	Sequence table	Ref.
Streptomyces albogriseolus S-3253	Microbial alkaline protease inhibitor S-SI	11,445 aa 23,000 ulc 12,000 el	4.3	0	0	0	0	0	0	I_{sub}-E				Ficin, 0 *Pseudomonas aeruginosa* neutral proteinase, 0 *Streptomyces griseus* alkaline protease, + *Cephalosporium* alkaline protease, + *Rhodotorula glutinis* acid protease, 0 *Cladosporium* acid protease, 0	*	*	6–10

Compiled by Beatrice Kassell and Monica June Williams. Supported by grant BMS74-22102 from the National Science Foundation.

SPECIFICITIES AND SOME PROPERTIES OF MICROBIAL PROTEINASE INHIBITORS (continued)

REFERENCES

1. Saheki, Matsuda, and Holzer, *Eur. J. Biochem.*, 47, 325 (1974).
2. Lenney and Dalbec, *Arch. Biochem. Biophys.*, 129, 407 (1969).
3. Betz, Hinze, and Holzer, *J. Biol. Chem.*, 249, 4515 (1974).
4. Matern, Hoffmann, and Holzer, *Proc. Natl. Acad. Sci. U.S.A.*, 71, 4874 (1974).
5. Ulane and Cabib, *J. Biol. Chem.*, 249, 3418 (1974).
6. Murao and Sato, *Agric. Biol. Chem.*, 36, 160 (1972).
7. Sato and Murao, *Agric. Biol. Chem.*, 37, 1067 (1973).
8. Satow, Mitsui, Iitaka, Murao, and Sato, *J. Mol. Biol.*, 75, 745 (1973).
9. Sato and Murao, *Agric. Biol. Chem.*, 38, 587 (1974).
10. Ikenaka, in *Bayer Symposium V, "Proteinase Inhibitors,"* Fritz, Tschesche, Greene, and Truscheit, Eds., Springer-Verlag, Berlin, 1974, 305.

AMINO ACID COMPOSITION OF MICROBIAL PROTEINASE INHIBITORS*

Amino acid, residues/mole

Source	Designation	Trp	Lys	His	Arg	Asp[a]	Thr	Ser	Glu[a]	Pro	Gly	Ala	1/2 Cys	Val	Met	Ile	Leu	Tyr	Phe	Total No.	References
Saccharomyces cerevisiae	Inhibitor of activating factor of chitin synthetase[b]	0	11	7	0	13	5	3	7	2	8	2	0	9	0	5	5	1	3	81	1
Streptomyces albogriseolus S-3253	Alkaline protease inhibitor S-SI[c]	1	2	2	4	9	8	8	6	8	11	19	4	13	3	0	9	3	3	113	2

Compiled by Beatrice Kassell and Monica June Williams.

*Supported by grant BMS 74-22102 from the National Science Foundation.

[a] Includes amides.
[b] Similar to the inhibitor of yeast proteinase B.
[c] Per 12,000 mol wt.

REFERENCES

1. Ulane and Cabib, *J. Biol. Chem.*, 249, 3418 (1974).
2. Sato and Murao, *Agric. Biol. Chem.*, 38, 587 (1974).

AMINO ACID SEQUENCE OF A MICROBIAL PROTEINASE INHIBITOR*†

Streptomyces albogriseolus Alkaline Proteinase Inhibitor S-SI (1,2)

```
                                    10                                      20
Asp- Ala- Pro- Ser- Ala- Leu- Tyr- Ala- Pro- Ser- Ala- Leu- Val- Leu- Thr- Val- Gly- Lys- Gly- Val-

                                    30                                      40
Ser- Ala- Thr- Thr- Ala- Ala- Pro- Glu- Arg- Ala- Val- Thr- Leu- Thr- Cys- Ala- Pro- Gly- Pro- Ser-

                                    50                                      60
Gly- Thr- His- Pro- Ala- Ala- Gly- Ser- Ala- Cys- Ala- Asp- Leu- Ala- Ala- Val- Gly- Gly- Asp- Leu-

                                    70                                      80
Asn- Ala- Leu- Thr- Arg- Gly- Glu- Asp- Val- Met- Cys- Pro- Met- Val- Tyr- Asp- Pro- Val- Leu- Leu-

                                    90                                      100
Thr- Val- Asp- Gly- Val- Trp- Gln- Gly- Lys- Arg- Val- Ser- Tyr- Glu- Arg- Val- Phe- Ser- Asn- Glu-

                                    110       113
Cys- Glu- Met- Asn- Ala- His- Gly- Ser- Ser- Val- Ala- Phe- Phe
```

Disulfide bridges: 35–50, 71–101

Compiled by Beatrice Kassell and Monica June Williams.

*Supported by grant BMS 74-22102 from the National Science Foundation.
†Small peptide inhibitors are not included.

REFERENCES

1. **Ikenaka**, in *Bayer Symp. V Proteinase Inhibitors,* Fritz, Tschesche, Greene, and Truscheit, Eds., Springer-Verlag, Berlin, 1974, 305.
2. **Ikenaka, Odani, Sakai, Nabeshima, Sato and Murao,** *J. Biochem.* (Tokyo), 76, 1191 (1974).

CARBOHYDRATE COMPOSITION OF SELECTED PROTEINASE INHIBITORS

Residues per mole[b]

Source[a]	Designation	Hexose	Glucose	Galactose	Mannose	Fucose	Sialic acid	N-Acetyl hexosamine	Glucosamine	N-acetyl glucosamine	Galactosamine	N-Acetyl neuraminic acid
Chicken egg white	Ovomucoid	16.7		3.0	12.0		0.3		21.0			
	Ovomucoid[c]								29.4			<0.2
	Ovomucoid[d]	11.6–15.9		0.8–6.1	9.6–12.9				14.3–26.7			0.02–1.95
	Ovoinhibitor[d]	5.6–9.7							7.6–15.2			
Human plasma	α_1-Antitrypsin			6.8	7.8	0	0.2–0.5		13.6		0	8.0
	α_1-Antichymotrypsin											
	Inter-α-trypsin inhibitor											
	Antithrombin III											
	C1-Inactivator			62	94	13				146		48
	α_2-Macroglobulin											
	α_2-Macroglobulin[d]											
Boar seminal plasma	Acrosin isoinhibitor A		4.14	1.98	1.84		0.99		3.40		2.27	
porcine colostrum	Trypsin inhibitors[d]		0–0.95	1.06–1.42	1.79–2.22	0.24–0.42						

[a]Many inhibitors are known to be free of carbohydrate. These include some inhibitors from legumes, e.g., soybean, peanut, kidney bean, and garden bean, from corn and potatoes, trypsin and chymotrypsin inhibitors from *Ascaris*, Bdellins from leeches, inhibitors from Russell's viper venom, and the basic and secretory pancreatic inhibitors. References are in the table of properties and amino acid composition.
[b]See the table of properties for molecular weights.
[c]An ovomucoid isolated under nondenaturing conditions.
[d]The range of carbohydrate content is given for fractions of similar amino acid compositions.

CARBOHYDRATE COMPOSITION OF SELECTED PROTEINASE INHIBITORS (continued)

Source[a]	Designation	Percent									Total carbohydrate		Reference
		Hexose	Galactose	Mannose	Fucose	Sialic acid	N-Acetyl hexosamine	Glucosamine	N-Acetyl glucosamine	N-Acetyl neuraminic acid	%	Approximate number of residues	
Chicken egg white	Ovomucoid											38	1
	Ovomucoid[c]											44	2
	Ovomucoid[d]	7.8–10.6						9.5–17.7		0.03–2.23			3
	Ovoinhibitor[d]												4
Human plasma	α_1-Antitrypsin	5.0			0.2	3.4	3.6				12.2		5
												37	6
	α_1-Antichymotrypsin	9.9			0.7	6.6	7.4				24.6		5
	Inter-α-trypsin inhibitor	3.1			0.1	2.0	3.2				8.4		5
	Antithrombin III	6.2				3.1	4.1				13.4		5
	C1-Inactivator	10.8			0.4	14.3	9.2				34.7		5
	α_2-Macroglobulin		1.2	1.9	0.2				3.6	1.7	8.6		7
	α_2-Macroglobulin[d]		1.3–2.8	3.0–3.1					3.2–3.6	1.6–2.7		363	8
Boar seminal plasma	Acrosin isoinhibitor A												9
Porcine colostrum	Trypsin inhibitors[d]												10

Compiled by Beatrice Kassell and Monica June Williams. Supported by grant BMS 74-22102 from the National Science Foundation.

CARBOHYDRATE COMPOSITION OF SELECTED PROTEINASE INHIBITORS (continued)

REFERENCES

1. Osuga and Feeney, *Arch. Biochem. Biophys.,* 124, 560 (1968).
2. Oegema, Jr. and Jourdian, *Arch. Biochem. Biophys.,* 160, 26 (1974).
3. Beeley, *Biochem. J.,* 123, 399 (1971).
4. Davis, Zahnley, and Donovan, *Biochemistry,* 8, 2044 (1969).
5. Heimburger, Haupt, and Schwick, in *Proc. 1st Int. Res. Conf. Proteinase Inhibitors,* Fritz and Tschesche, Eds., Walter de Gruyter, Berlin, 1971, 1.
6. Kress and Laskowski, Sr., *Prep. Biochem.,* 3, 541 (1973).
7. Dunn and Spiro, *J. Biol. Chem.,* 242, 5549 (1967).
8. Frénoy and Bourrillon, *Biochim. Biophys. Acta,* 371, 168 (1974).
9. Tschesche, Kupfer, Lengel, Klauser, Meier, and Fritz, in *Bayer Symposium V, Proteinase Inhibitors,* Fritz, Tschesche, Greene, and Truscheit, Eds., Springer-Verlag, Berlin, 1974, 164.
10. Kress, Martin, and Laskowski, Sr., *Biochim. Biophys. Acta,* 229, 836 (1971).

ENDO-γ-GLUTAMINE:ε-LYSINE TRANSFERASES
ENZYMES WHICH CROSS-LINK PROTEINS

Laszlo Lorand and Pál Stenberg

Significance of Biological Transamidation

There is a group of transamidating enzymes with the specialized function of bringing about the post-translational modification of some native proteins enabling them to form covalently linked polymeric assemblies. Characteristically, these enzymes contain cysteine-thiol active centers and require calcium ions for catalyzing the nucleophilic displacement reaction depicted in Figure 1. The scheme illustrates the dimeric cross-linking of a protein molecule through a γ-glutamyl-ε-lysine peptide bridge. Inasmuch as the protein substrate is at least bifunctional in terms of carrying both the acceptor (i.e., γ-glutamine carbonyl) and the electron donor (i.e., ε-lysine amino) groups required for the reaction, there is a possibility for extended polymerization. In various physiological situations this may either result in the direct formation of a specific precipitate [lobster fibrinogen → clot (1, 2); seminal vesicle secretion protein → copulation plug (3)] or in the covalent fusion ("ligation") of individual molecules within an already existing ordered protein aggregate [vertebrate fibrin gel → cross-linked or ligated gel (4)]. Even in the latter case, the introduction of a few covalent cross-bridges into the gel network [ca. 1 γ-glutamyl-ε-lysine peptide per 500 amino acid residues on the average (5, 6)] contributes greatly to the physical stiffening of the gel network, as expressed by an increase in elastic storage modulus (7, 8) and to its resistance to lytic enzymes (9). In some systems (e.g., guinea pig copulation plug) the average cross-link density may reach values as high as 1/23 amino acid residues (3). Historically, cross-linking of the fibrin gel was the first physiological example (4).

Recognition of Enzymatic Cross-linking

In addition to being inhibited, for example, by iodoacetamide and calcium chelators (EDTA), enzymatic reactions of the type discussed can be recognized by the fact that addition of a suitable pseudo or substitute donor compound (H_2NR) inhibits cross-linking and results in the incorporation of the added amine into the protein substrate (Figure 2). As best exemplified by fibrin, incorporation of a labeled amine tracer permits the titration of the acceptor cross-linking sites of the protein (ca. 6 amines become specifically incorporated per 330,000 g or mol of fibrin), evaluation of the ordered reactivity and localization of sites in terms of its constituent chains (5) (γ-chains of fibrin react first, then α-chains; the β-chains do not seem to be involved), and finally, makes it possible to work out the primary amino acid sequences around these sites (10, 11). A Lossen-type of rearrangement performed following the incorporation of hydroxylamine (12) or enzymatic degradation of the protein labeled with a suitable amine (13) permits the identification of γ-glutamyl residues as acceptors.

Using proteolytic enzymes (e.g., pronase, leucine amino peptidase, and carboxypeptidase in succession), the cross-linked proteins may be digested to a point where the protease-resistant γ-glutamyl-ε-lysine dipeptide can be directly isolated and quantitated by means of standard amino acid analytical procedures. Significant amounts of this isopeptide have been isolated from clots of bovine and human fibrin (14–17), guinea pig copulation plug (3) (see Figure 3), lobster fibrin (2), and sheep wool (18, 19). Minute quantities (ca $1/10^6$ g of protein) have been reported in a variety of organisms (20). It should be mentioned that γ-glutamyl-ε-lysine may be produced artificially during handling of proteins (21).

Cross-linked proteins may also be conveniently analyzed for the accessibility of ε-lysine amino groups to reagents such as nitrous acid (22, 23) or acrylonitrile (6).

However, results with these methods can be interpreted only if, as with fibrin and the copulation plug, independent proof is available that all inaccessible ϵ-lysine is in fact γ-glutamyl bonded. Ammonia liberation could also be studied either in conjunction with cross-linking (Figure 1) or amine incorporation (Figure 2); in general, however, literature data pertaining to the biological systems have been rather tenuous thus far.

In conclusion, there are a number of biological systems in which protein assemblies occur by the enzymatic formation of γ-glutamyl-ϵ-lysine bridges. The breaking of the γ-amide in glutamine, and probably also the precipitation of the polymerized protein product, would seem to provide the energy for the synthesis of the intermolecular peptide bonds without the participation of nucleoside triphosphates.

Detection of Transamidase Activity

The simplest methods for measuring transamidase activity in biological media rest on the incorporation of amines into nonspecific protein acceptors. Amines labeled with a fluorescent group [e.g., dansylcadaverine, i.e., N-(5-aminopentyl)-5-dimethylamino-naphthalene-1-sulfonamide (14, 24–28)] or ^3H and ^{14}C [e.g., putrescine or histamine (22, 26, 29, 30)] have been most widely used. In addition to native proteins [α, α_s, κ-casein (24); β-lactoglobulin (22)], their chemically modified forms [acetyl (24), succinyl (24), N, N-dimethyl (25)], in which lysine side chains are excluded from participating in cross-link formation, are particularly useful. Dephosphorylation of casein also aids in increasing its solubility in the presence of calcium ions (28).

Since incorporation of the fluorescent amine into casein and some other proteins is accompanied by an increase in quantum yield and a shift in the wavelength of the emitted light (24, 28), the reaction may be followed in a continuous manner (Figure 4). Alternately, protein-bound fluorescence or radioactivity may be quantitated after separation by chromatography (26) or after precipitation with trichloroacetic acid (27, 29). A rapid version of the latter technique was devised using a filter paper disc method (30).

With hydroxylamine as substrate, hydroxamate formation may be measured (22, 31). For some of the enzymes (e.g., liver transglutaminase), peptides [e.g., N-benzyloxycarbonyl-L-glutaminylglycine (32)] may be employed as amine acceptors in lieu of proteins.

Table 1 lists representative proteins modified by the enzymatic incorporation of amines, and it is assumed that replacements always occur on γ-glutamine side chains. In addition to the amines already mentioned, isotopic glycine ethylester (33), methylamine (32), and even isoniazid (34) have been used for measuring the extent of incorporation.

The Fibrin Stabilizing Factor System of Blood Plasma

Transamidase activity measured, for example, by the incorporation of putrescine into casein does not occur in normal plasma as such, but the precursor form (called fibrin stabilizing factor, Factor XIII) of the enzyme is present. Conversion of the zymogen occurs during coagulation, and the process is governed by thrombin and calcium ions (41, 42) in a consecutive manner

$$\text{zymogen (Factor XIII)} \xrightarrow{\text{thrombin}} \text{zymogen' (FXIII')} \xrightarrow{Ca^{2+}} \text{enzyme (FXIII}_a) \rightarrow \quad (1)$$

The Factor XIII zymogen has an (ab) protomeric structure, and modification by thrombin consists of the hydrolytic removal of an N-terminal peptide from the (a) subunit (43, 44). However, the proteolytically modified zymogen (a'b) is still devoid of transamidase activity which is generated only after reaction of the protein with calcium ions. In a specific manner, the latter cause both the dissociation of the subunits of the zymogen and the unmasking of the active center SH group in the (a′) subunit (25, 42)

$$(ab) \xrightarrow[\text{fragment}]{\text{thrombin}} (a'b) \xrightarrow{Ca^{2+}} a^* - SH + b \tag{2}$$

The calcium-dependent dissociation can be readily demonstrated by electrophoresis (Figure 5). Though there is some evidence that the catalytically inert carrier (b) subunit may also be modified by itself becoming a substrate for the (a*) enzyme activity (45), it is seen that addition of EDTA can still reverse dissociation (25).

Platelets contain about one third as much potential transamidase activity as the plasma compartment of blood and the isolated platelet zymogen contains only the (a) type of subunits (46). As seen in Figure 6, hybridization of this protein with carrier (b) subunit reproduces the behavior of the plasma zymogen. The results suggest that of the two steps involved in the calcium activation of the proteolytically modified zymogen [i.e., dissociation of subunits (I) and the unmasking of the active center cysteine, paralleling the generating of enzyme activity (II)],

$$a'b \xrightarrow[(I)]{Ca^{2+}} a' + b \atop \Big\downarrow Ca^{2+} \text{ (II)} \atop a^* - SH \tag{3}$$

step I is by far the slower, rate limiting one.

The a*-SH active enzyme is referred by the name "fibrinoligase." The term "plasma transglutaminase" has also been used.

Sources of Transamidase Activity

Though the biological significance and specific protein substrates of only very few endo-γ-glutamine:ϵ-lysine transamidases are known, the ability to incorporate primary amines into proteins in a nonspecific manner is characteristic for enzymes derived from many tissues. Guinea pig liver transglutaminase, first described by Waelsch and collaborators (22, 38), typifies this group of enzymes with as yet unknown biological function. Though liver transglutaminase is interchangeable with some of the specific enzymes in the sense, for example, that like fibrinoligase it is capable of cross-linking vertebrate fibrin, a detailed analysis (4) reveals that the action of transglutaminase even on this protein is quite nonspecific. While fibrinoligase incorporates ca. 6 mol of amine into fibrin, transglutaminase readily promotes the incorporation of more than 20; while the former enzyme reacts quite sluggishly with fibrinogen, transglutaminase reacts with this protein even faster than with fibrin. There are also significant differences in the order of the reactivity of glutamines in the γ- and α-chains of fibrin vis-a-vis the two enzymes (5).

With liver transglutaminase and some of the related intracellular enzymes, it is not even clear whether their biological role actually lies in catalyzing the formation of γ-glutamyl-ϵ-lysine bridges in protein systems or rather in promoting some acyl group transfer of other metabolic significance. In the latter context, it may be worth mentioning that guinea pig liver transglutaminase was shown to catalyze acyl transfer reactions, for example, with butyrylcoenzyme A as substrate rather well (47).

Table 2 gives a representative summary of tissues possessing calcium-dependent enzymes with amine incorporating activities. Of the guinea pig organs, liver, spleen, kidney, adrenal gland, heart and lung (22, 48), and prostate (49) seemingly contain the highest activities. However, the distribution of activity appears to be different in human (48), with lung and uterus being the highest. It has either been demonstrated or may be assumed that all, except sheep follicle enzyme (50), are inhibited by EDTA.

Purified Transamidases

Most intracellular transamidases (e.g., from guinea pig liver and red blood cell) appear to consist of a single-chain protein with a molecular weight in the range of about 70,000 to 90,000 (56, 60). The guinea pig hair follicle enzyme seems to be an exception, comprising two identical subunits of 27,000 mol wt each (61).

Human placenta, like platelets, yielded apparently a thrombin-sensitive zymogen with a postulated a_2 structure of $2 \times 75,000$ (51), while the plasma zymogen contains heterologous subunits (51, 54). Pertinent data are summarized in Table 3.

Active Site Titrations

Iodoacetamide is known to inhibit several of the enzymes tested so far. Human fibrinoligase (42, 63), as well as its counterpart from platelets (64), the liver (65), and coagulation gland enzymes (49) from the guinea pig, combine with stoichiometric amounts of the reagent as enzymatic activity is neutralized. This makes it possible to determine the concentration of active sites in solutions and to evaluate the purity of these enzymes. Titration of the first two is totally dependent on the presence of calcium ions which also enhance the specific active site reactivity of the two latter guinea pig enzymes. Cysteine was identified as the residue reacting with iodoacetamide for the enzyme from human plasma (42, 63), platelets (64), and guinea pig liver (65), with a sequence around this active center of TyrGlyGln*Cys*Trp (65, 66) in the liver and plasma proteins.

As seen in Table 4, 5, 5'-dithiobis(2-nitrobenzoic acid) and organomercurials were also employed as titrants for the human plasma and platelet enzymes, and various active-center-directed alkylating agents were used for guinea pig liver transglutaminase. The latter could also be titrated by means of the "kinetic burst" using *p*-nitrophenyl trimethylacetate (67).

Synthetic Substrates and the Pathway of Catalysis

In keeping with the selective nature of the reactions catalyzed by endo-γ-glutamine:ε-lysine transamidases, these enzymes display a great deal of kinetic specificity for synthetic substrates of both the amine donor and acceptor type. In the former category, only primary amines need to be considered, whereas three classes of acceptor compounds have been investigated in depth: amides, *p*-nitrophenylesters, and thiolesters. These studies have been performed mostly in conjunction with only two of the enzymes, so far: fibrinoligase and guinea pig liver transglutaminase. Substituted glutamine peptides [e.g., *N*-benzyloxycarbonyl-L-glutaminyl-L-valine methylester (70) or *N*-benzyloxycarbonyl-L-glutaminylglycine (32)], related amides, and also some *p*-nitrophenylesters could be used to good advantage only with the liver enzyme; these data have been summarized in a recent review by Folk and Chung (71). Though fibrinoligase could also be shown to hydrolyze *p*-nitrophenyl acetate in a Ca^{2+}-dependent manner (72), as far as the synthetic acceptor type of substrates is concerned, only the water-soluble thiolesters can be considered as being really versatile at present (73). β-Phenylpropionylthiocholine appears to be the best, with thiocholine esters of acetate, propionate, trimethylacetate, and butyrate being approximately 60- to 20-fold poorer substrates; the benzoate and transcinnamate esters are nearly as good, whereas the β-phenylbutyrate ester is about the same. The observations would seem to suggest that the β-phenylpropionyl residue fits optimally into perhaps a crevicelike domain of the enzyme surface.

Methods were developed for studying the steady-state kinetics of enzymatic hydrolysis and aminolysis of these esters. By selecting appropriate substrates (as illustrated in Figure 7) a variety of approaches could be used for monitoring the disappearance of substrates and formation of products. Most attractive among these are the ones which readily lend themselves to continuous recording, e.g., formation of thiocholine (74) or the water insoluble amide product (75). Analysis of the steady-state measurements reveals that the enzymatic reaction proceeds through a pathway of consecutive acylation and deacylation. The latter may occur by hydrolysis or, in the presence of an amine substrate, by aminolysis. Altogether, the following distinct steps should be formulated

$$R_2COSR_1 + HS-E \overset{K_a}{\rightleftharpoons} [R_2COSR_1; HS-E] \xrightarrow[\text{acylation}]{k_2} R_2COS-E + HS-R_1 \quad (4)$$

$$R_2COS-E + H_2O \xrightarrow[\text{hydrolysis}]{k_3} HS-E + R_2COOH \quad (5)$$

$$R_2COS-E + H_2NR_3 \overset{K_b}{\rightleftharpoons} [R_2COS-E; H_2NR_3] \xrightarrow[\text{aminolysis}]{k_5} HS-E + R_2CONHR_3 \quad (6)$$

Data pertaining to the fibrinoligase-catalyzed hydrolysis and aminolysis of thiolesters are given in Tables 5 and 6.

One of the most significant and unique results emerging from the kinetic studies was the recognition that, as indicated in Table 6, the acylenzyme intermediates displayed a very high degree of specificity for the amine substrate. The limiting maximal velocities (V_{lim}^{am}) obtainable in the presence of the different amines remain essentially identical, but the concentration required to bring about a half-maximal aminolytic enhancement varies significantly even after allowance is made for the varying degrees of protonation of the added amines. Interestingly, despite the fact that there have been numerous reports in the literature on systematically exploring the side-chain specificity of primary amines (see 14, 76, 77), the recent review by Folk and Chung (71) seems to have completely overlooked their significance. Observations obtained with compounds in the monotosylated or dansylated alkyldiamine series (76–78) indicated that, as expected from the ε-lysine requirement, there is an optimal alkyl side-chain length for pentamine substrates, perhaps reflecting a requirement to reach the acylated cysteine active center within a crevice on the surface of the enzyme intermediate.

With the β-phenylpropionylthiocholine and dansylcadaverine substrate pair (Figure 7), it has already been possible to draw kinetic comparisons between human fibrinoligase and guinea pig liver transglutaminase (75). On the basis of functional normality of active sites, k_{cat} values for amide formation gave 1.8 and 0.9 sec^{-1} for the plasma and liver enzymes, respectively. They showed identical affinities for the first substrate, β-phenylpropionyl-thiocholine, with a $K_a = 4 \times 10^{-4} M$.

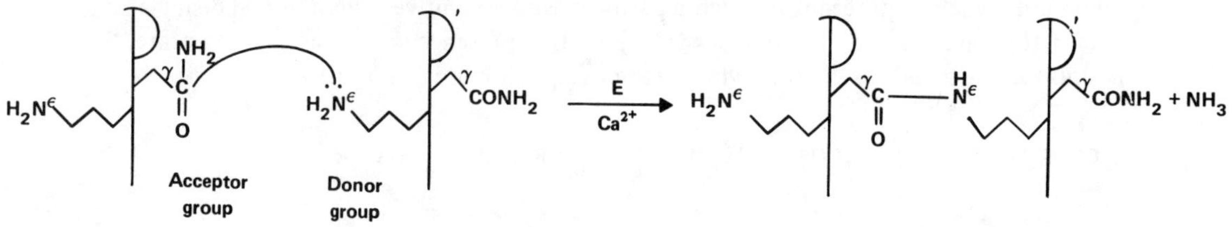

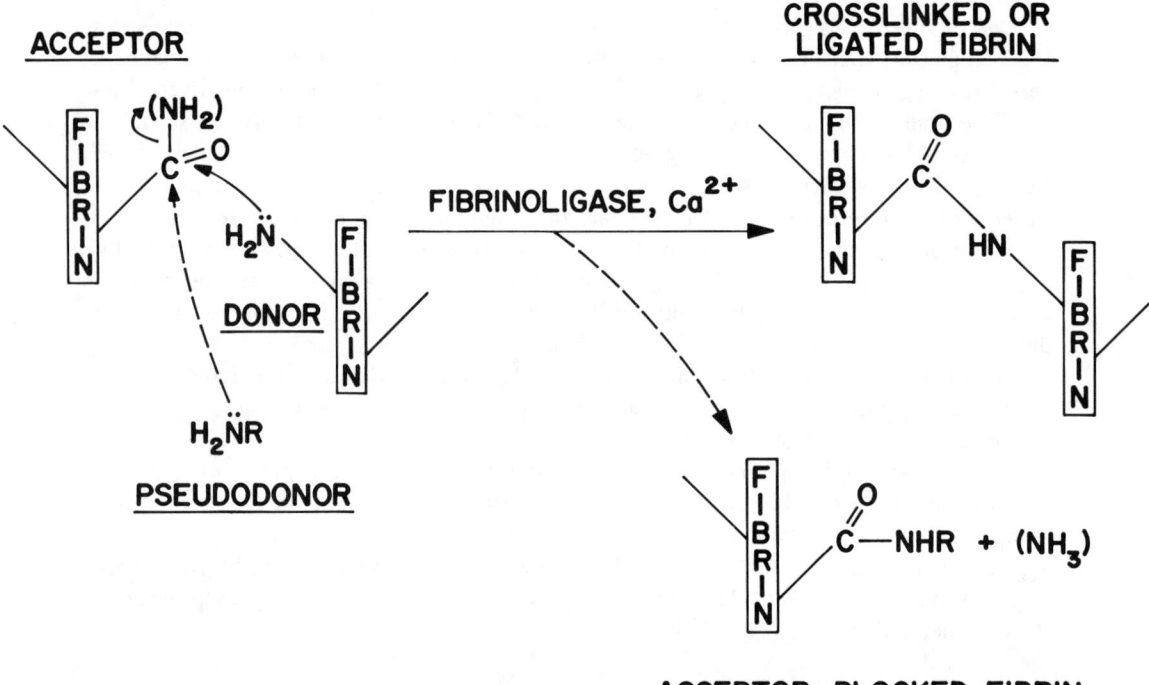

FIGURE 1. Scheme for a transamidase (E) catalyzed protein assembly. Dimerization by forming a single intermolecular γ-glutamyl-ε-lysine peptide bridge is shown. Typically, P and P' represent molecules of the same protein species in a given biological system (e.g., vertebrate fibrin) (4).

FIGURE 2. Inhibition of a transamidase-catalyzed protein assembly reaction by primary amines. Inhibition of cross-linking goes together with incorporation of the pseudo or substitute donor into the parent protein, as illustrated for fibrin (4).

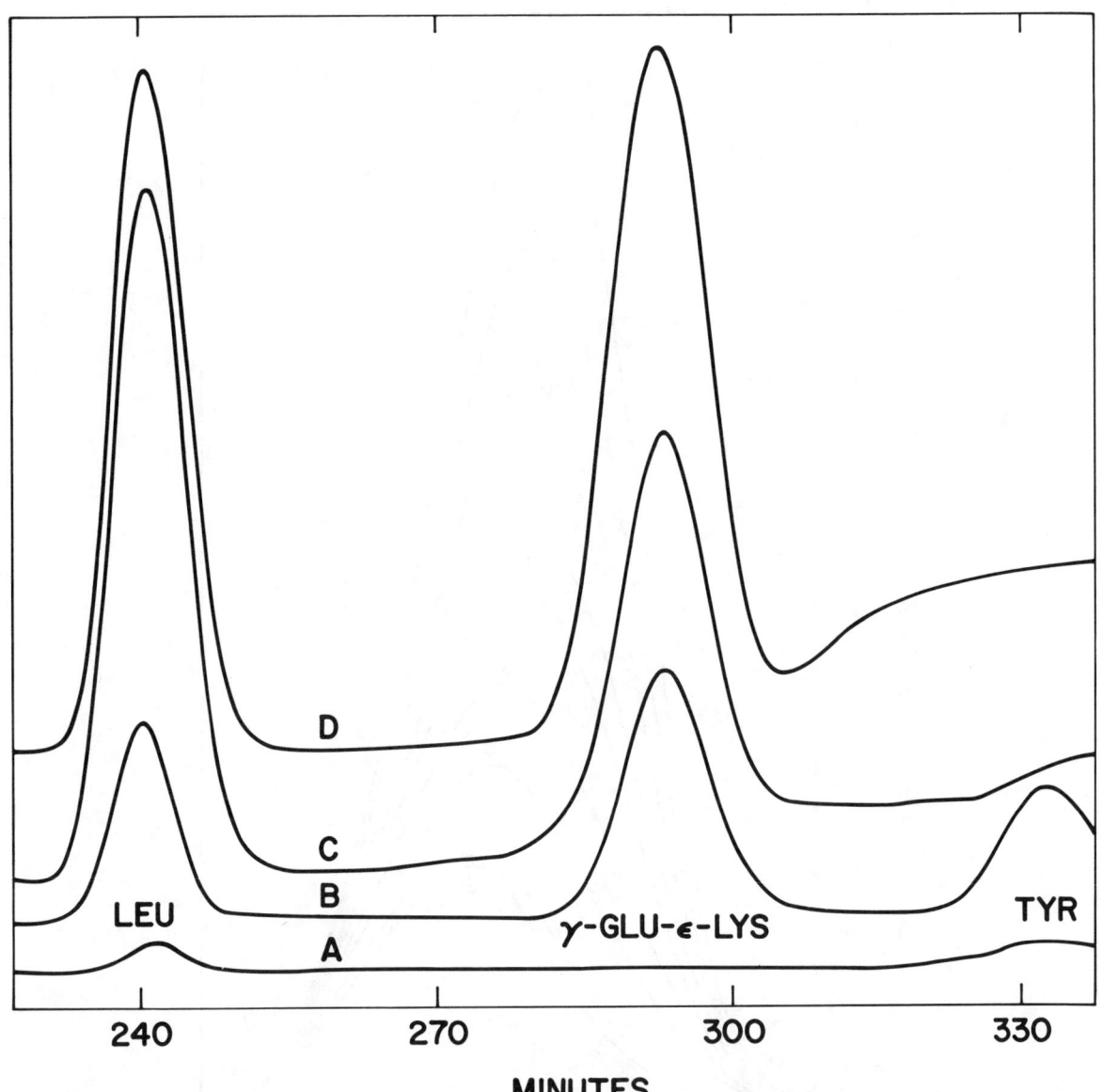

FIGURE 3. Identification of γ-glutamyl-ε-lysine in proteolytic digest of clotted seminal vesicle secretion protein. (A) Incubated proteolytic enzyme control exhibited no peak at 296 min. (B) Beckman standard amino-acid mixture, with 111 nmol of authentic γ-glutamyl-ε-lysine dipeptide added. (C) Proteolytic enzyme digest of 1.1 mg of clotted basic vesicular secretion protein gave rise to a γ-glutamyl-ε-lysine dipeptide peak at 296 min that contains 273 nmol. Note the absence of tyrosine in the sample. (D) Same as C, except that 222 nmol of authentic γ-glutamyl-ε-lysine was mixed with the digest as reference dipeptide. The peak at 296 min corresponded to 477 nmol of the total dipeptide recovered (calculated value, 495 nmol) (3).

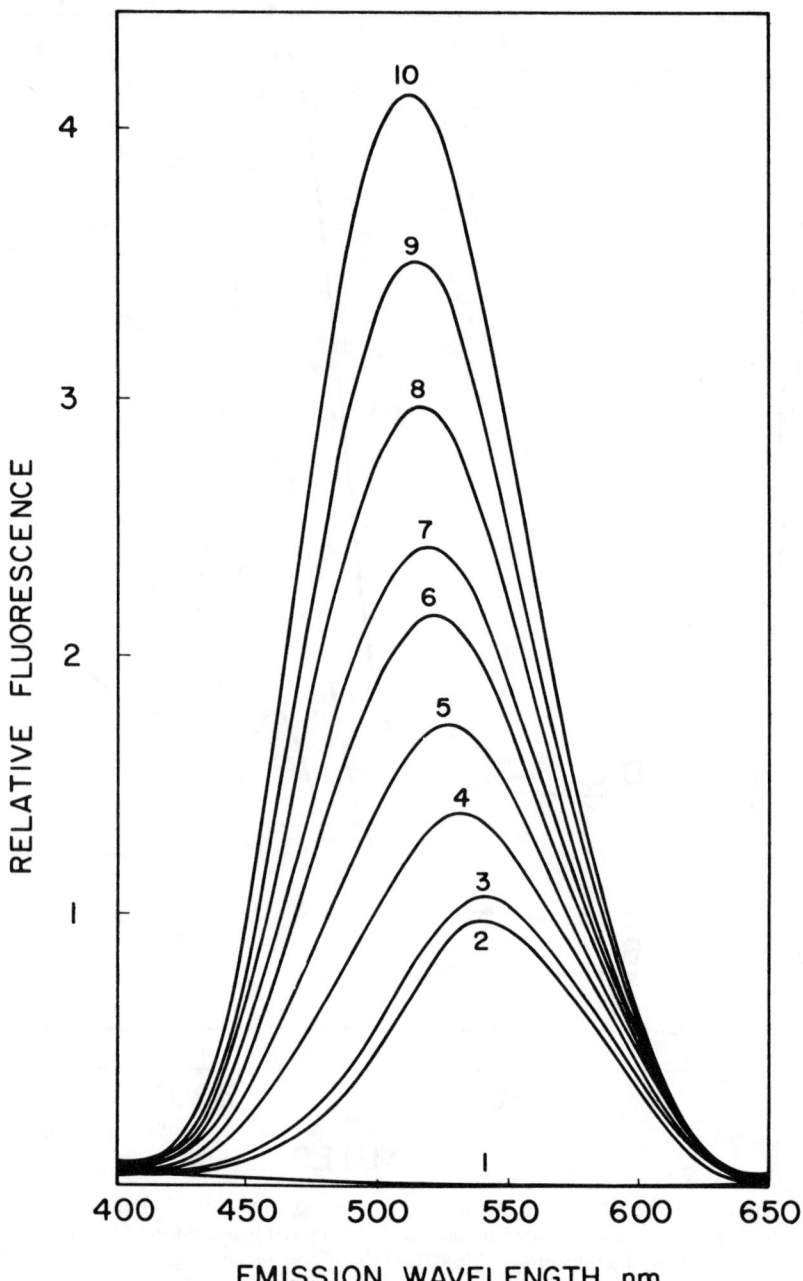

FIGURE 4. (A) Fluorescence changes during enzymic incorporation of dansylcadaverine into α-casein. Reaction mixture contained 2.6×10^{-5} M α-casein, 2×10^{-5} M dansylcadaverine, 3 mM calcium chloride, 1% glycerol, and muscle transamidase (55) at 1.2 units/ml, 0.05 M tris-HCl, pH 7.5. Scans of fluorescence emissions were obtained in an Amino-Bowman spectrophotofluorometer, with excitation set to 340 nm. Control mixtures: (Curve 1) Without dansylcadaverine. (Curve 2) Without enzyme (at 0 and at 120 min). Emission spectra on the complete reaction mixture were taken at 1, 10, 20, 30, 40, 60, 90, and 120 min (curves 2 to 10) (24). (B) Measurement of fibrin-stabilizing factor (Factor XIII) in normal human plasma by direct recording of fluorescence changes (ordinate) in the dansylcadaverine-acetylcasein reaction. Graph at top is from complete system; at bottom, without dansylcadaverine. Horizontal bar represents 3 min. Vertical marker corresponds to 0.5 mV (24).

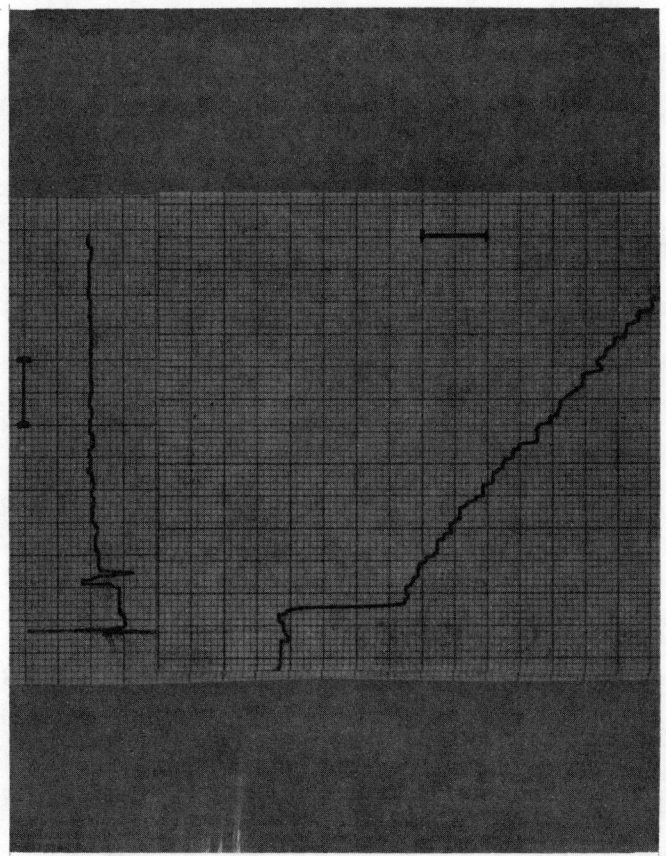

FIGURE 4B.

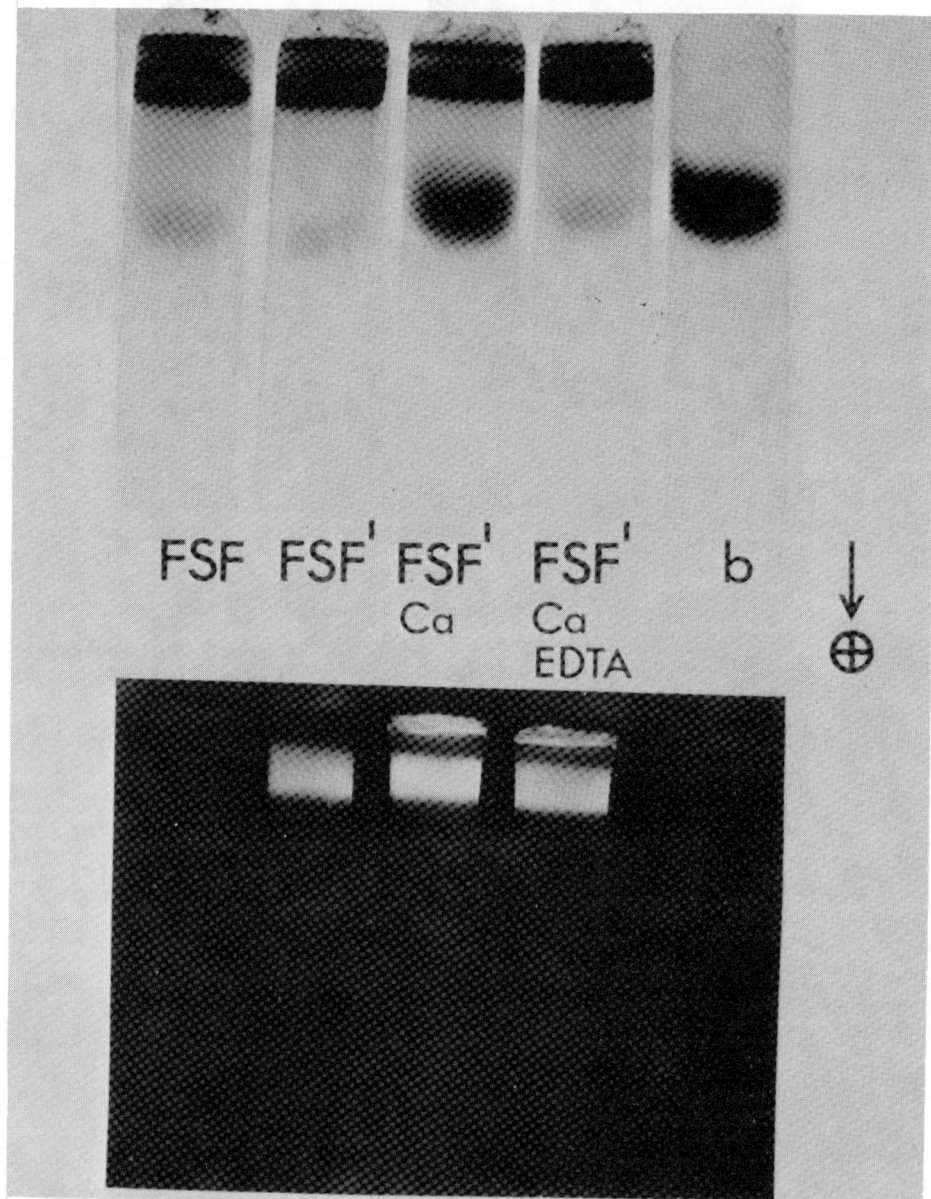

FIGURE 5. Disc gel electrophoresis, pH 7.8, of fibrin stabilizing factor (FSF) at various stages of activation. (Top) Coomassie blue protein stain. (Bottom) Transamidase specific fluorescent activity stain, comprising dansylcadaverine, dimethylcasein, and calcium chloride. Approximately 30 μg protein was applied to each gel. The thrombin-activated zymogen is denoted as FSF'; b stands for this subunit isolated from the zymogen (25).

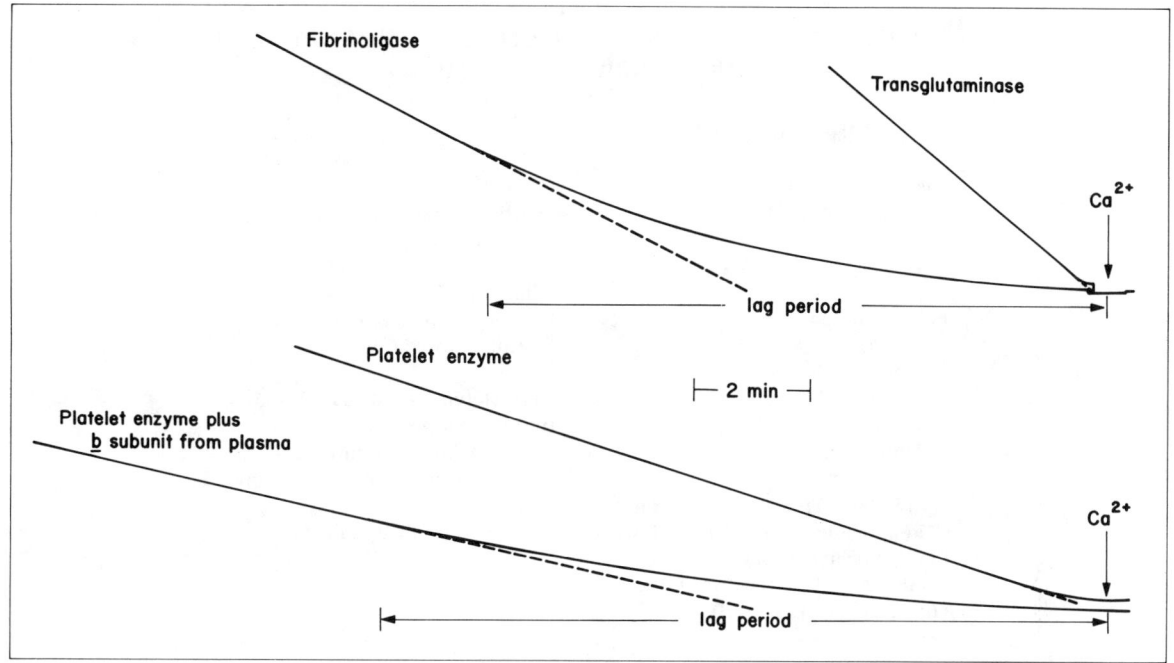

FIGURE 6. Progression curves for the formation of the amide product in enzyme-catalyzed reactions between β-phenylpropionylthiocholine iodide and dansylcadaverine. Approximately 0.11 mg/ml of platelet Factor XIII and 0.08 mg/ml of b were used. Ordinate: relative fluorescence (25).

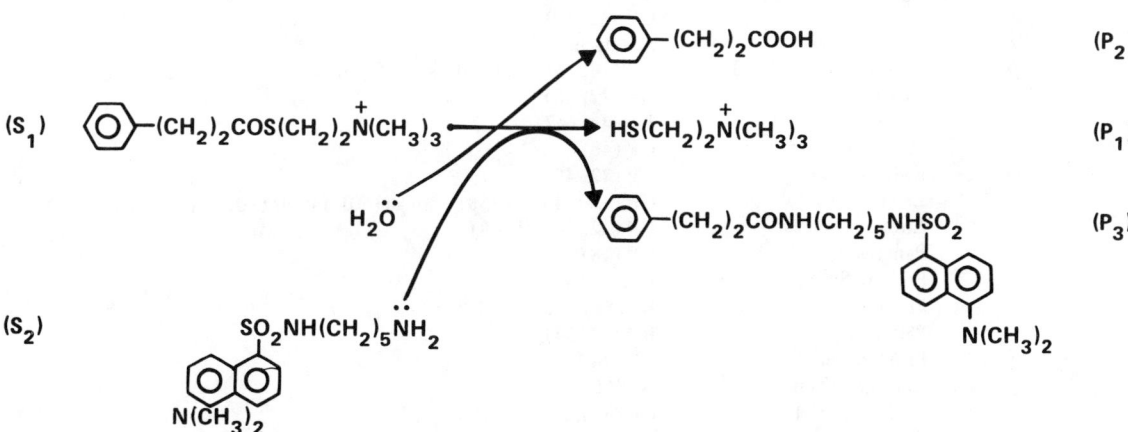

FIGURE 7. Reaction of β-phenylpropionylthiocholine (S_1) with dansylcadaverine (S_2). Formation of thiocholine (P_1) is measured by reaction with 5,5'-dithiobis(2-nitrobenzoic acid) (74); production of acid (P_2) can be followed in a pH' stat (79) and the water-insoluble coupling product (P_3) is quantitated by fluorescence after extraction into a heptane phase (75). In addition, chromatographic procedures are available for measuring S_2 and P_3 (73). Hydrolysis of *trans*-cinnamoylthiocholine is accompanied by large spectral changes (73).

Table 1
PROTEINS SERVING AS SUBSTRATES FOR THE ENZYMATIC INCORPORATION OF AMINES

Native proteins	Altered proteins
Casein (α, α_s, and Hammarsten) (24)	Casein (Hammarsten): partially digested (22), acetyl (24), dephospho and acetyl (28), dimethyl (25)
Ferredoxin (35)	Ferredoxin: oxidized (35)
Fibrin (5)	Fibrin: oxidized (39)
Fibrinogen (5)	
Glucagon (36)	Hemoglobin: α- and β-chains (35, 37)
Insulin (38)	Insulin: oxidized (35)
β-Lactoglobulin	β-Lactoglobulin: succinyl (24); lysozyme: oxidized (35), succinyl (24)
Myosin (80); Tropomyosin; Actin (81)	
Cold insoluble plasma globulin (40)	Ribonuclease: oxidized (35)
Embryonic hyaline layer (47)	
Erythrocyte ghost spectrin and band 3; human (82), mouse (83)	

Table 2
TRANSAMIDASE ACTIVITY IN TISSUES

Source	Species and reference
Adrenal gland	GP (48)
Amebocyte	Limulus (47)
Brain	GP (22, 48)
Egg	Sea urchin[a] (47)
Hair follicle	GP[b] (48, 50), rat (50), sheep[c] (50)
Heart	GP (22, 48)
Hemocyte[d]	Lobster (47)
Intestine	GP (48)
Kidney	GP (22, 48)
Liver	GP[b] (22, 48), H (48), mouse (50), rabbit (50), rat (50)
Lung	GP (22, 48), H (48)
Pancreas	GP (48)
Placenta[e]	B[b], H (51)
Plasma[e]	B[b] (52), GP (22), H[b] (52)
Platelets[e]	H[b] (53, 54)
Prostate gland	GP[b] (49)
Red blood cells	GP[b], H (56)
Skeletal muscle	GP (48), lobster[a] (55)
Skin	B[a], H, frog, rat, toad, turtle (57)
Spleen	B[a] (58), GP (22, 48)
Testis	GP (48)
Tumors	Mouse ascites (47), YPC-1 (59)
Uterus[e]	GP[b], H (48)

Note: GP, guinea pig; B, bovine; H, human.

[a] Partially purified.
[b] Purified and characterized.
[c] Not inhibited by EDTA (50).
[d] Lobster hemocyte enzyme released under clotting conditions (47).
[e] Occurs as a zymogen, requiring proteolytic activation by thrombin.

Table 3
PURIFIED TRANSAMIDASES

Source and reference to method of isolation	Approx. mol wt and subunit structure
Guinea pig	
Hair follicle (50, 61)	$54{,}000 = 2 \times 27{,}000$ (61), a_2
Liver (62)	85,000 (60)
Prostate gland (49)	$70{,}000 \pm 10{,}000$ (49)[a]
Red blood cells (56)	$80{,}000 \pm 5{,}000$ (56)
Human	
Placenta[b] (Factor XIII) (51)	165,000 (51), a_2
Plasma[b] (Factor XIII) (52)	$326{,}000 = 2 \times (75{,}000 + 88{,}000)$ (54), $a_2 b_2$
	156,000–195,000 (15), ab
Platelets[b] (Factor XIII) (54)	$150{,}000 = 2 \times 75{,}000$ (54), a_2
Red blood cells (56)	$80{,}000 \pm 5{,}000$ (56)
Uterus (48)	$76{,}000 \pm 4{,}000$ (56)
Sea urchin	
Egg (47)	95,000–100,000 (47)[a]

[a]Tentative values.
[b]Zymogen forms.

Table 4
ACTIVE SITE TITRATIONS

	Reagent	Reference
Human enzymes		
Plasma Factor XIII$_a$	$ICH_2{}^{14}CONH_2$	42, 63
(fibrinoligase)	$ICH_2{}^{14}COOH$	66
	HOOC–C$_6$H$_3$(NO$_2$)–S–S–C$_6$H$_3$(NO$_2$)–COOH	68
	HgCl–C$_6$H$_3$(O$_2$N)–OH	68
	HO–C$_6$H$_3$(O$_2$N)–COCH$_2$Br	64
Platelet Factor XIII$_a$	$ICH_2{}^{14}CONH_2$	64
Guinea pig enzymes	$ICH_2{}^{14}CONH_2$	65
Liver		
	O$_2$N–C$_6$H$_4$–OOCC(CH$_3$)$_3$	67
	HO–C$_6$H$_3$(O$_2$N)–COCH$_2$Br	69

Table 4 (continued)
ACTIVE SITE TITRATIONS

	Reagent	Reference
Guinea pig enzymes Liver (continued)	(3-hydroxy-5-nitrobenzyl structure with CH$_2$CCH$_2$Cl and COOCH$_3$ groups)	69
Coagulating gland	ICH$_2{}^{14}$CONH$_2$	49

Table 5
KINETIC CONSTANTS FOR THE NONENZYMATIC AND FOR THE FIBRINOLIGASE-CATALYZED STEADY-STATE HYDROLYSIS OF THIOL ESTERS AT pH 7.5 AND 25° (74)

	Nonenzymatic sec^{-1} × 10^6	Enzymatic		
		$K^{hyd}_{m,app}$ M × 10^3	k^{hyd}_{cat} sec^{-1}	$k^{hyd}_{cat}/K^{hyd}_{m,app}$ M^{-1}sec^{-1} × 10^{-3}
C$_6$H$_5$(CH$_2$)$_2$COS(CH$_2$)$_2$N(CH$_3$)$_3{}^+$I$^-$ (β-phenylpropionylthiocholine iodide)	1.3	0.29	0.21	0.74
C$_6$H$_5$(CH$_2$)$_2$COS(CH$_2$)$_2$N(CH$_3$)$_2$·HCl (2-dimethylaminoethanethiol-β-phenylpropionate hydrochloride)	58.5	0.18	0.13	0.74
C$_6$H$_5$CH=CHCOS(CH$_2$)$_2$N(CH$_3$)$_3{}^+$I$^-$ (*trans*-cinnamoylthiocholine iodide)	1.3	0.10	0.08	0.08
C$_6$H$_5$CH=CHCOS(CH$_2$)$_2$N(CH$_2$C$_6$H$_5$)(C$_2$H$_5$)$_2{}^+$Br$^-$ (2-benzyldiethylaminoethanethiol *trans*-cinnamate bromide)	8.0	0.25	0.08	0.32

Table 6
AMINE SPECIFICITY

Enhancement of the Steady-state Formation of Thiocholine in the Reactions of Five Different Amines with 1.5 mM of β-Phenylpropionylthiocholine Iodide at 50 mM Calcium Chloride, $E_0 = 0.12$ μM, pH 7.5, and 25° (74)

Amine substrates	v^{am}_{lim}/E_0 sec^{-1}	$K^{am}_{m, app}$ $M \times 10^4$
N-(5-Aminopenyl)-5-dimethyl-amino-1-naphthalenesulfonamide; dansylcadaverine [structure: naphthalene with SO$_2$NH(CH$_2$)$_5$NH$_2$ and N(CH$_3$)$_2$]	1.7	8
N-(5-Amino-3-thiapentyl)-5-dimethylamino-1-naphthalenesulfonamide; dansylthiacadaverine [structure: naphthalene with SO$_2$NH(CH$_2$)$_2$S(CH$_2$)$_2$NH$_2$ and N(CH$_3$)$_2$]	1.7	4
N-α-*p*-Tosyl-L-lysine methyl ester [structure: benzene with SO$_2$NHCH(CH$_2$)$_4$NH$_2$, COOCH$_3$, and CH$_3$]	1.4	11
NH$_2$(CH$_2$)$_4$NH$_2$ (putrescine)	2.1	77
CH$_3$(CH$_2$)$_3$NH$_2$ (*n*-butylamine)	1.2	430

REFERENCES

1. Lorand, Doolittle, Konishi, and Riggs, *Arch. Biochem. Biophys.*, 102, 171 (1963).
2. Fuller and Doolittle, *Biochemistry*, 10, 1311 (1971).
3. Williams-Ashman, Notides, Pabalan, and Lorand, *Proc. Natl. Acad. Sci. U.S.A.* 69, 2322 (1972).
4. Lorand, *Ann. N.Y. Acad. Sci.*, 202, 6 (1972).
5. Lorand, Chenoweth, and Gray, *Ann. N.Y. Acad. Sci.*, 202, 155 (1972).
6. Pisano, Bronzert, and Peyton, *Ann. N.Y. Acad. Sci.*, 202, 98 (1972).
7. Roberts, Lorand, and Mockros, *Biorheology*, 10, 29 (1973).
8. Mockros, Roberts, and Lorand, *Biophys. Chem.*, 2, 164 (1974).
9. Bruner-Lorand, Lorand, and Pilkington, *Nature*, 210, 1273 (1966).
10. Chen and Doolittle, *Biochemistry*, 10, 4486 (1971).
11. Chenoweth, Ph.D. dissertation, Northwestern University, Evanston, 1971.
12. Lorand and Ong, *Biochem. Biophys. Res. Commun.*, 23, 188 (1966).
13. Matacic and Loewy, *Biochem. Biophys. Res. Commun.*, 24, 858 (1966).
14. Lorand, Rule, Ong, Furlanetto, Jacobsen, Downey, Oner, and Bruner-Lorand, *Biochemistry*, 7, 1214 (1968).
15. Lorand, Downey, Gotoh, Jacobsen, and Tokura, *Biochem. Biophys. Res. Commun.*, 31, 222 (1968).
16. Matacic and Loewy, *Biochem. Biophys. Res. Commun.*, 30, 356 (1968).
17. Pisano, Finlayson, and Peyton, *Science*, 160, 892 (1968).
18. Asquith, Otterburn, Buchanan, Cole, Fletcher, and Gardner, *Biochim. Biophys. Acta*, 207, 342 (1970).
19. Harding and Rodgers, *Biochim. Biophys. Acta*, 257, 37 (1972).
20. Birckbichler, Dowben, Matacic, and Loewy, *Biochim. Biophys. Acta*, 291, 149 (1973).
21. Asquith and Otterburn, *Appl. Polym. Symp.*, 18, 277 (1971).
22. Clarke, Mycek, Neidle, and Waelsch, *Arch. Biochem. Biophys.*, 79, 338 (1959).
23. Lorand, Ong, Lipinski, Rule, Downey, and Jacobsen, *Biochem. Biophys. Res. Commun.*, 25, 629 (1966).
24. Lorand, Lockridge, Campbell, Myhrman, and Bruner-Lorand, *Anal. Biochem.*, 44, 221 (1971).
25. Lorand, Gray, Brown, Credo, Curtis, Domanik, and Stenberg, *Biochem. Biophys. Res. Commun.*, 56, 914 (1974).
26. Lorand and Campbell, *Anal. Biochem.*, 44, 207 (1971).
27. Lorand, Urayama, Atencio, deKiewiet, and Nossel, *J. Clin. Invest.*, 48, 1054 (1969).
28. Cooke and Holbrook, *Biochem. J.*, 141, 71 (1974).
29. Dvilansky, Britten, and Loewy, *Br. J. Haematol.*, 18, 399 (1970).
30. Lorand, Campbell-Wilkes, and Cooperstein, *Anal. Biochem.*, 50, 623 (1972).
31. Lorand and Ong, *Biochemistry*, 5, 1747 (1966).
32. Chung and Folk, *J. Biol. Chem.*, 247, 2798 (1972).
33. Lorand and Jacobsen, *Biochemistry*, 3, 1939 (1964).
34. Lorand, Campbell-Wilkes, and Robertson, *Biochemistry*, 11, 434 (1972).
35. Toda and Folk, *Biochim. Biophys. Acta*, 175, 427 (1969).
36. Folk and Cole, *J. Biol. Chem.*, 240, 2951 (1965).
37. Pincus and Waelsch, *Arch. Biochem. Biophys.*, 34, 126 (1968).
38. Mycek, Clarke, Neidle, and Waelsch, *Arch. Biochem. Biophys.*, 84, 528 (1959).
39. Lorand, *Fed. Proc.*, 24, 784 (1965).
40. Mosher, *Fed. Proc.*, Abstr. 1567, 34 (1975).
41. Lorand and Konishi, *Arch. Biochem. Biophys.*, 105, 58 (1964).
42. Curtis, Brown, Credo, Domanik, Gray, Stenberg, and Lorand, *Biochemistry*, 13, 3774 (1974).
43. Mikuni, Iwanaga, and Konishi, *Biochem. Biophys. Res. Commun.*, 54, 1393 (1973).
44. Takagi and Doolittle, *Biochemistry*, 13, 750 (1974).
45. Gray and Lorand, *Fed. Proc.*, Abstr. 1419, 33 (1974).
46. Schwartz, Pizzo, Hill, and McKee, *J. Biol. Chem.*, 248, 1395 (1973).
47. Campbell-Wilkes, Ph.D. dissertation, Northwestern University, Evanston, 1973.
48. Chung, *Ann. N.Y. Acad. Sci.*, 202, 240 (1972).
49. Wing, Curtis, Lorand, and Williams-Ashman, *Fed. Proc.*, Abstr. 486, 33 (1974).
50. Harding and Rogers, *Biochemistry*, 11, 2858 (1972).
51. Bohn and Schwick, *Arzneim. Forsch.*, 21, 1432 (1971).
52. Lorand and Gotoh, *Methods Enzymol.*, 19, 770 (1970).
53. Buluk, *Pol. Tyg. Lek.*, 10, 191 (1955).
54. Schwartz, Pizzo, Hill, and McKee, *J. Biol. Chem.*, 246, 5851 (1971).
55. Myhrman and Bruner-Lorand, *Methods Enzymol.*, 19, 765 (1970).

56. Chung and Folk, *Fed. Proc.*, Abstr. 234, 34 (1975).
57. Goldsmith, Baden, Roth, Colman, Lee, and Fleming, *Biochim. Biophys. Acta,* 351, 113 (1974).
58. Zuch, Kaminski, and Buluk, *Thromb. Res.,* 5, 571 (1974).
59. Yancey and Laki, *Ann. N.Y. Acad. Sci.,* 202, 344 (1972).
60. Folk, *Ann. N.Y. Acad. Sci.,* 202, 59 (1972).
61. Chung and Folk, *Proc. Natl. Acad. Sci. U.S.A.,* 69, 303 (1972).
62. Folk and Cole, *J. Biol. Chem.,* 241, 5518 (1966).
63. Curtis, Stenberg, Chou, Gray, Brown, and Lorand, *Biochem. Biophys. Res. Commun.,* 52, 51 (1973).
64. Chung, Lewis, and Folk, *J. Biol. Chem.,* 249, 940 (1974).
65. Folk and Cole, *J. Biol. Chem.,* 241, 3238 (1966).
66. Holbrook, Cooke, and Kingston, *Biochem. J.,* 135, 901 (1973).
67. Folk, Cole, and Mullooly, *J. Biol. Chem.,* 242, 4329 (1967).
68. Cooke, Pestell, and Holbrook, *Biochem. J.,* 141, 675 (1974).
69. Folk and Cole, *J. Biol. Chem.,* 246, 6683 (1971).
70. Neidle and Acs, *Fed. Proc.,* 20, 234 (1961).
71. Folk and Chung, *Adv. Enzymol.,* 38, 109 (1973).
72. Lockridge, Ph.D. dissertation, Northwestern University, Evanston, 1971.
73. Lorand, Chou, and Simpson, *Proc. Natl. Acad. Sci. U.S.A.,* 69, 2645 (1972).
74. Curtis, Stenberg, Brown, Baron, Chen, Gray, Simpson, and Lorand, *Biochemistry,* 13, 3257 (1974).
75. Stenberg, Curtis, Wing, Tong, Credo, Gray, and Lorand, *Biochem J.,* 147, 155 (1975).
76. Lorand and Nilsson, in *Drug Design,* Vol. 3, Ariens, Ed., Academic Press, New York, 1972, 415.
77. Nilsson, Stenberg, Ljunggren, Hoffmann, Lunden, Eriksson, and Lorand, *Ann. N.Y. Acad. Sci.,* 202, 286 (1972).
78. Nilsson, Stenberg, Ljunggren, Eriksson, and Lunden, *Acta Pharm. Suec.,* 8, 415 (1971).
79. Chen, Ph.D. dissertation, Northwestern University, Evanston, 1974.
80. Lorand and Campbell-Wilkes, unpublished data.
81. Derrick and Laki, *Biochem. Biophys. Res. Commun.,* 22, 82 (1966).
82. Lorand, Shishido, Parameswaran, and Steck, *Biochem. Biophys. Res. Commun.,* 67, 1158 (1975).
83. Dutton and Singer, *Proc. Natl. Acad. Sci. U.S.A.,* 72, 2568 (1975).

CONNECTIVE TISSUE POLYSACCHARIDES (GLYCOSAMINOGLYCANS, MUCOPOLYSACCHARIDES)[a]

Name	Repeating units[b]	Carbohydrate-protein linkage compound (11,12)	Other monosaccharide components (11)	Occurrence	$[\alpha]_D$, degrees[c]	Infrared, cm^{-1},[d]	Intrinsic viscosity (1)[c]	Molecular weight (1)[c]
Chondroitin 4-sulfate (Chondroitin sulfate A)	(1 → 4)-O-β-D-glucopyranosyluronic acid-(1 → 3)-(2-acetamido-2-deoxy-β-D-galactopyranosyl 4-sulfate) (4)	3-O-(β-D-Xylopyranosyl)-L-serine	D-Galactose	Cartilage, bone, cornea, notochord, skin	−25 to −30 (4)	724, 851, 930 (13,14)	0.2–1.0	5–50 × 10^3
Chondroitin-6-sulfate (Chondroitin sulfate C)	(1 → 4)-O-β-D-Glucopyranosyluronic acid-(1 → 3)-(2-acetamido-2-deoxy-β-D-galactopyranosyl 6-sulfate) (5)	3-O-(β-D-Xylopyranosyl)-L-serine	D-Galactose	Cartilage, aorta, umbilical cord	−12 to −22 (5)	775, 820, 1000 (13,14)	0.2–1.3	5–50 × 10^3
Dermatan sulfate (Chondroitin sulfate B, β-heparin)	1. (1 → 4)-O-α-L-Idopyranosyluronic acid-(1 → 3)-2-acetamido-2-deoxy-β-D-galactopyranosyl 4-sulfate), and 2. (1 → 4)-O-β-D-Glucopyranosyluronic acid-(1 → 3)-(2-acetamido-2-deoxy-β-D-galactopyranosyl 4-sulfate) (3,6)	3-O-(β-D-Xylopyranosyl)-L-serine	D-Galactose	Skin, aorta, umbilical cord, tendon	−55 to −63 (15)	724, 851, 930 (14)	0.5–1.0	1.5–4 × 10^4
Heparan sulfate	1. (1 → 4)-O-β-D-Glucopyranosyluronic acid-(1 → 4)-(2-acetamido- [or 2-sulfoamino- or 2-amino-] 2-deoxy-α-D-glucopyranosyl [or glucopyranosyl 6-sulfate]), and 2. (1 → 4)-O-α-L-Idopyranosyluronic acid- [or (α-L-idopyranosyluronic acid 2-sulfate)-] (1 → 4)-(2-acetamido- [or 2-sulfoamino- or 2-amino-] 2-deoxy-α-D-glucopyranosyl 6-sulfate) (3,7)	3-O-(β-D-Xylopyranosyl)-L-serine	D-Galactose	Aorta, lung, cell surfaces of many types of cells	+39 to +69 (16)	920, 1050 (19)	—	1.5 × 10^4
Heparin	1. (i → 4)-O-(α-L-Idopyranosyluronic acid 2-sulfate)- [or α-L-idopyranosyluronic acid-] (1 → 4)-(2-sulfoamino- [or 2-acetamido- or 2-amino-] 2-deoxy-α-D-glucopyranosyl 6-sulfate [or glucopyranosyl], and 2. (1 → 4)-O-β-D-Glucopyranosyluronic acid- (1 → 4)-(2-sulfoamino- [or 2-acetamido- or 2-amino-] 2-deoxy-α-D-glucopyranosyl 6-sulfate [or glucopyranosyl] (3,8)	3-O-(β-D-Xylopyranosyl)-L-serine	D-Galactose	Lung, intestine, liver, mast cells	+48 (17)	890, 940 (17)	0.1–0.2	7–16 × 10^3
Hyaluronic acid	(1 → 4)-O-β-D-Glucopyranosyluronic acid-(1 → 3)-2-acetamido-2-deoxy-β-D-glucopyranosyl (9)		Traces of various monosaccharides are often found, particularly glucose, galactose, and arabinose; none has been proven to be an integral part of the molecule.	Synovial fluid, vitreous humor, umbilical cord, skin, aorta	−68 (18)	900, 950 (18)	2.0–48	0.5–80 × 10^3

[a] In their native form, several polysaccharides exist as components of macromolecular proteoglycans which contain many polysaccharide chains linked covalently to a protein core (e.g., the proteoglycan population of bovine nasal cartilage consists of subunits with an average mol wt of 2.5 × 10^6; these subunits may form aggregates (mol wt 30–50 × 10^6) by association with hyaluronic acid and specific "link proteins"). Current knowledge of the properties of proteoglycans is reviewed elsewhere.[1-3]
[b] Certain polysaccharides (dermatan sulfate, heparin, and heparan sulfate) exhibit considerable variation in their repeating units which expresses itself both in the uronic acid composition (D-glucuronic acid vs. L-iduronic acid) as well as in the position and degree of sulfate substitution. In the chondroitin sulfates, there is some variation in the degree of sulfation and, in addition, 4- and 6-substituted residues are often present in the same chain.
[c] The values for specific rotation and intrinsic viscosity are given as ranges because these depend to some extent on the method of isolation and original tissue source. The references cited are review articles which refer to research papers.
[d] Fingerprint region.

CONNECTIVE TISSUE POLYSACCHARIDES (GLYCOSAMINOGLYCANS, MUCOPOLYSACCHARIDES) (continued)

Name	Repeating units[b]	Carbohydrate-protein linkage compound (11,12)	Other monosaccharide components (11)	Occurrence	$[\alpha]_D$, degrees[c]	Infrared, cm^{-1},[d]	Intrinsic viscosity (1)[c]	Molecular weight (1)[c]
Keratan sulfate I[e] (Keratosulfate I)	(1 → 3)-O-β-D-Galactopyranosyl-(1 → 4)-(2-acetamido-2-deoxy-β-D-glucopyranosyl 6-sulfate) (2,7,10)	2-Acetamido-1-N-(4'-L-aspartyl)-2-deoxy-β-D-glucopyranosylamine	D-Mannose, L-fucose, and N-acetylneuraminic acid	Cornea	+4.5 (7)	775, 820, 998 (20)	0.2–0.5	8–12 × 10^3
Keratan sulfate II[e] (Keratosulfate II)	(1 → 3)-O-β-D-Galactopyranosyl-(1 → 4)-(2-acetamido-2-deoxy-β-D-glucopyranosyl 6-sulfate) (2,7,10)	3-O-(2-Acetamido-2-deoxy-α-D-glucopyranosyl)-L-threonine 3-O-(2-Acetamido-2-deoxy-α-D-glucopyranosyl)-L-serine	D-Mannose, L-fucose, N-acetylneuraminic acid, and N-acetyl-D-galactosamine	Cartilage, nucleus pulposus, annulus fibrous				

[e]Some galactose residues of the repeating units are sulfated in position 6. Additional galactose units (sulfated or nonsulfated) are also present in this polysaccharide.

Compiled by L. Roden.

CONNECTIVE TISSUE POLYSACCHARIDES (GLYCOSAMINOGLYCANS, MUCOPOLYSACCHARIDES)
(continued)

REFERENCES

1. **Mathews,** *Connective Tissue,* Springer-Verlag, Berlin, Heidelberg, New York, 1975, 93.
2. **Muir and Hardingham,** *Biochemistry of Carbohydrates,* Whelan, Ed., Medical and Technical Publishing Co. and Butterworths, London, 1975.
3. **Lindahl,** *MTP International Review of Science, Organic Chemistry Series 2 – Carbohydrate Chemistry,* Aspinall, Ed., Medical and Technical Publishing Co. and Butterworths London, 1975.
4. **Jeanloz,** in *Methods Carbohydr. Chem.,* 5, 110 (1965).
5. **Jeanloz,** in *Methods Carbohydr. Chem.,* 5, 113 (1965).
6. **Jeanloz,** in *Methods Carbohydr. Chem.,* 5, 114 (1965).
7. **Brimacombe and Webber,** *Mucopolysaccharides,* Elsevier, Amsterdam, 1964, 136.
8. **Brimacombe and Webber,** *Mucopolysaccharides,* Elsevier, Amsterdam, 1964, 117.
9. **Brimacombe and Webber,** *Mucopolysaccharides,* Elsevier, Amsterdam, 1964, 43.
10. **Bhavanandan and Meyer,** *Science,* 151, 1404 (1966).
11. **Lindahl and Roden,** in *Glycoproteins,* Gottschalk, Ed., Elsevier, Amsterdam, 1972, 491.
12. **Neuberger, Gottschalk, Marshall, and Spiro,** in *Glycoproteins,* Gottschalk, Ed., Elsevier, Amsterdam, 1972, 450.
13. **Orr,** *Biochim. Biophys. Acta,* 14, 173 (1954).
14. **Mathews,** *Nature,* 181, 421 (1958).
15. **Brimacombe and Webber,** *Mucopolysaccharides,* Elsevier, Amsterdam, 1964, 82.
16. **Linker, Hoffman, Sampson, and Meyer,** *Biochim. Biophys. Acta,* 29, 443 (1958).
17. **Danishefsky, Eiber, and Carr,** *Arch. Biochem. Biophys,* 90, 114 (1960).
18. **Danishefsky and Bella,** *J. Biol. Chem.,* 241, 143 (1966).
19. **Schiller,** *Biochim. Biophys. Acta,* 32, 315 (1959).
20. **Brimacombe and Webber,** *Mucopolysaccharides,* Elsevier, Amsterdam, 1964, 138.

PROTEIN pK VALUES

Lynne H. Botelho and Frank R. N. Gurd

The general techniques for determining individual pK values in proteins usually depend on NMR,[1,2] absorption,[3] or kinetic[4] procedures. Effects of neighboring groups may be evident in chemical shift influences[5-7] or in electrostatic influences on the hydrogen ion equilibria proper.[8]

1. Markley, Finkenstadt, Dugas, Leduc, and Drapeau, *Biochemistry,* 14, 998 (1975).
2. Markley, *Acc. Chem. Res.,* 8, 70 (1975).
3. Tanford, Hanenstein, and Rands, *J. Am. Chem. Soc.,* 77, 6409 (1955).
4. Garner, Bogardt, and Gurd, *J. Biol. Chem.,* in press.
5. Sachs, Schechter, and Cohen, *J. Biol. Chem.,* 246, 6576 (1971).
6. Shrager, Cohen, Heller, Sachs, and Schechter, *Biochemistry,* 11, 541 (1972).
7. Deslauriers, McGregor, Sarantakis, and Smith, *Biochemistry,* 13, 3443 (1974).
8. Roxby and Tanford, *Biochemistry,* 10, 3348 (1971).

Table 1A
SPECIFIC His pK ASSIGNMENTS IN RIBONUCLEASE A

Protein	pK values	Reference
Bovine		
His 12	6.3	1
His 48	5.8	1
His 73	6.4	1
His 105	6.7	1
Rat		
His 12	6.6	1
His 48	6.2	1
His 73	7.6	1
His 105	6.3	1
His 119	6.1	1

Compiled by Lynne H. Botelho and Frank R. N. Gurd.

REFERENCES

1. Migchelsen and Beintema, *J. Mol. Biol.,* 79, 25 (1973).

Table 1B
NONSPECIFIC His pK VALUES FOR RIBONUCLEASE A

Protein	pK values				Reference
Coypu	5.8	6.3	6.3	8.0	1
Chinchilla	4.9	6.0	6.1	7.2	1
Bovine	6.01	6.17	6.72	6.9	2

Compiled by Lynne H. Botelho and Frank R. N. Gurd.

REFERENCES

1. Migchelsen and Beintema, *J. Mol. Biol.,* 79, 25 (1973).
2. Markley, *Acc. Chem. Res.,* 8, 70 (1975).

Table 1C
pK VALUES FOR HISTIDINE RESIDUES IN MYOGLOBIN

Species	pK observed		Reference
Sperm whale	5.37	5.33	1, 2
	5.53	5.39	
	6.34	6.21	
	6.44	6.31	
	6.65	6.55	
	6.83	6.72	
	8.05	7.97	
Horse	5.7	5.5	1, 2
	6.0	5.8	
	6.6	6.5	
	6.9	6.8	
	7.0	6.9	
	7.6	7.6	
California grey whale	5.7		2
	6.2		
	6.6		
	6.8		
	7.8		
Inia geoffrensis	5.53		2
	5.95		
	6.17		
	6.31		
	6.45		
	6.66		
	8.05		
Tursiops truncatus	5.50		2
	5.95		
	6.24		
	6.26		
	6.42		
	6.60		
	7.82		
Balaenoptera acutorostrata	5.46		2
	5.65		
	6.10		
	6.23		
	6.41		
	6.59		
	7.86		

Compiled by Lynne H. Botelho and Frank R. N. Gurd.

REFERENCES

1. Cohen, Hagenmaier, Pollard, and Schechter, *J. Mol. Biol.*, 71, 513 (1972).
2. **Botelho, Hanania, and Gurd,** unpublished observations.

Table 1D
HISTIDINE pK VALUES IN HUMAN HEMOGLOBIN

Hemoglobin	pK values	Reference
Human	6.8	1, 2
	7.0	
	7.0	
	7.1	
	7.2	
	7.13	
	7.7	
	8.1	
	8.1	

Compiled by Lynne H. Botelho and Frank R. N. Gurd.

REFERENCES

1. Donovan, *Methods Enzymol.*, 27, 497 (1973).
2. Mandel, *Proc. Natl. Acad. Sci. U.S.A.*, 52, 736 (1964).

Table 1E
pK VALUE FOR HUMAN Hb His 146 β

Protein	His 146 β pK value	Reference
Human hemoglobin Deoxy, His 146 β	8.0	1
	8.1	2
	7.4	3
Human hemoglobin Carboxy, His 146 β	7.1	1
	6.8	2

Compiled by Lynne H. Botelho and Frank R. N. Gurd.

REFERENCES

1. Kilmartin, Breen, Roberts, and Ho, *Proc. Natl. Acad. Sci. U.S.A.*, 70, 1246 (1973).
2. Greenfield and Williams, *Biochim. Biophys. Acta*, 257, 187 (1972).
3. Huestis and Raftery, *Biochemistry*, 11, 1648 (1972).

Table 1F
SPECIFIC HISTIDINE pK VALUE FOR CYTOCHROME c

Species	pK values	Reference
Horse		
His 33	6.41	1
Yeast		
His 33	6.74	2
His 39	6.56	2

Compiled by Lynne H. Botelho and Frank R. N. Gurd.

REFERENCES

1. Cohen, Fisher, and Schechter, *J. Biol. Chem.,* 249, 1113 (1974).
2. Cohen and Hayes, *J. Biol. Chem.,* 249, 5472 (1974).

Table 1G
pK VALUES FOR HISTIDINE RESIDUES IN CARBONIC ANHYDRASE

Species	pK values			Reference
Human, B	5.91	5.88	6	1–3
	6.04	6.09	6.98	
	7.00	6.93	7.23	
	7.23	7.23	8.2	
		8.2	8.24	
Human, C	5.87	5.74		2, 3
	5.96	6.43		
	6.10	6.49		
	6.20	6.5		
	6.63	6.57		
	7.20	6.63		
	7.28	7.25		

Compiled by Lynne H. Botelho and Frank R. N. Gurd.

REFERENCES

1. King and Roberts, *Biochemistry,* 10, 558 (1971).
2. Cohen, Yem, Kandel, Gornall, Kandell, and Friedman, *Biochemistry,* 11, 327 (1972).
3. Pesando, *Biochemistry,* 14, 675 (1975).

Table 1H
HISTIDINE pK VALUES FROM NUCLEASE

Protein	pK values		Reference
Staphylococcus aureus Foggi	5.46	5.37	1, 2
	5.76	5.71	
	5.66, 5.74, 6.54[a]	5.74	
	6.57	6.50	
Staphylococcus aureus V8	5.55		1
	5.80, 6.10[a]		
	6.50		

Compiled by Lynne H. Botelho and Frank R. N. Gurd.

[a]pK values of one histidine existing in multiple conformational forms of the enzyme which slowly interconvert.

REFERENCES

1. Markley, *Acc. Chem. Res.,* 8, 70 (1975).
2. Cohen, Shrager, McNeel, and Schechter, *Nature,* 228, 642 (1970).

Table 1I
HISTIDINE pK VALUES IN VARIOUS PROTEINS

Protein	Number of His resolved	pK	Specific assignment	Reference
Adenylate kinase (pig)	2 of 2	<5.5		1
		6.3		
Chymotrypsin A_δ (cow)	1 of 2	7.2	His 57	2, 3
Chymotrypsinogen A (cow)	1 of 2	7.2	His 57	2, 3
Lysozyme				
chicken		5.8		4, 5
human		7.1		4, 6
Neurophysin II (cow)	1 of 1	6.87		7
Ovomucoid (chicken)	4 of 4	5.94		8
		6.71		
		6.75		
		8.07		
Protease				
α-Lyter (Myxobacter 495)	1 of 1	<4		9, 10
(*Staphylococcus aureus*, V8)	3 of 3	6.69		11
		6.85		
		7.19		
Ribonuclease T_1 (*Aspergillus oryzae*)	2 of 3	7.9		12
		8.0		
Serine esterase		6.5–7.5		13
Trypsin (pig, β form)	4 of 4	5.0		14
		6.54		
		6.66		
		7.20		
Trypsin inhibitor (soybean, Kunitz)	2 of 2	5.27		15
		7.00		

Compiled by Lynne H. Botelho and Frank R. N. Gurd.

REFERENCES

1. Cohn, Leigh, and Reed, *Cold Spring Harbor Symp. Quant. Biol.,* 36, 533 (1972).
2. Robillard and Shulman, *J. Mol. Biol.,* 71, 507 (1972).
3. Robillard and Shulman, *Ann. N.Y. Acad. Sci.,* 69, 599 (1972).
4. Meadows, Markley, Cohen, and Jardetsky, *Proc. Natl. Acad. Sci. U.S.A.,* 58, 1307 (1967).
5. Cohen, Hagenmaier, Pollard, and Schechter, *J. Mol. Biol.,* 71, 513 (1972).
6. Cohen, *Nature* (Lond.), 223, 43 (1969).
7. Cohen, Griffen, Camier, Caizergues, Fromageot, and Cohen, *FEBS Lett.,* 25, 282 (1972).
8. Markley, *Ann. N.Y. Acad. Sci.,* 222, 347 (1973).
9. Hunkapiller, Smallcombe, Whitaker, and Richards, *J. Biol. Chem.,* 248, 8306 (1973).
10. Hunkapiller, Smallcombe, Whitaker, and Richards, *Biochemistry,* 12, 4732 (1973).
11. Markley, Finkenstadt, Dugas, Leduc, and Drapeau, *Biochemistry,* 14, 998 (1975).
12. Riterjans and Pongs, *Eur. J. Biochem.,* 18, 313 (1971).
13. Polgar and Bender, *Proc. Natl. Acad. Sci. U.S.A.,* 69, 599 (1972).
14. Markley, *Acc. Chem. Res.,* 8, 70 (1975).
15. Markley, *Biochemistry,* 12, 2245 (1973).

Table 2
pK VALUES FOR α-AMINO GROUPS IN PROTEINS

Protein	pK' value	Reference
Human hemoglobin		
Carboxy		
α chain	6.72	1
	6.95	2
β chain	7.05	2
Cyano		
α chain	6.74	2
β chain	6.93	2
Deoxy		
α chain	7.79	2
β chain	6.84	2
Myoglobin		
Sperm whale	7.77	2
	7.96	3
California grey whale	7.74	3
Pilot whale	7.43	3
Dall porpoise	7.22	3
Harbor seal	7.66	3
Bovine pancreatic Ribonuclease A	8.14	4
Horse hemoglobin		
Oxy		
α chain	7.3	5
Deoxy		
α chain	7.7	5

Compiled by Lynne H. Botelho and Frank R. N. Gurd.

REFERENCES

1. Hill and Davis, *J. Biol. Chem.*, 242, 2005 (1967).
2. Garner, Bogardt, Jr., and Gurd, *J. Biol. Chem.*, in press.
3. Garner, Garner, and Gurd, *J. Biol. Chem.*, 248, 5451 (1973).
4. Carty and Hirs, *J. Biol. Chem.*, 243, 5254 (1968).
5. Kilmartin and Rossi-Bernardi, *Biochem. J.*, 124, 31 (1971).

Table 3
pK VALUES FOR ε-AMINO GROUPS IN PROTEINS

Protein	pK' value	Method	Reference
Bovine pancreatic Ribonuclease A			
Lys-41	9.11	Kinetic	1
Other Lys	10.1	Kinetic	1
All Lys	10.2	Titration	2
Sperm Whale Myoglobin			
All Lys except one	10.6	Titration (intrinsic pK)	3
Hen egg white Lysozyme			
Lys 97	10.1	NMR	4
Lys 116	10.2		
Lys 13	10.3		
Lys 33	10.4		
Lys 1	10.6		
Lys 96	10.7		

Compiled by Lynne H. Botelho and Frank R. N. Gurd.

REFERENCES

1. Carty and Hirs, *J. Biol. Chem.*, 243, 5254 (1968).
2. Tanford and Hanenstein, *J. Am. Chem. Soc.*, 78, 5287 (1956).
3. Shire, Hanania, and Gurd, *Biochemistry*, 13, 2967 (1974).
4. Bradbury and Brown, *Eur. J. Biochem.*, 40, 565 (1973).

Table 4A
TYROSINE pK VALUES

Protein	No of groups	pK	Reference
Ribonuclease A	3 of 6	9.9	1
Insulin		9.7	2
Pepsin		9.5	3
Serum Albumin		10.35	3
Lysozyme		10.8	4
Trypsin inhibitor (BPTI)			
Bovine, tyrosines		10.6	5
		10.8	
		11.1	
		11.6	

Compiled by Lynne H. Botelho and Frank R. N. Gurd.

REFERENCES

1. Tanford, Hauenstein, and Rands, *J. Am. Chem. Soc.*, 77, 6409 (1955).
2. Tanford and Epstein, *J. Am. Chem. Soc.*, 76, 2163 (1954).
3. Tanford and Roberts, Jr., *J. Am. Chem. Soc.*, 74, 2509 (1952).
4. Fromageot and Schnek, *Biochem. Biophys. Acta*, 6, 113 (1950); Tanford and Wagner, *J. Am. Chem. Soc.*, 76, 2331 (1954).
5. Karplus, Snyder, and Sykes, *Biochemistry*, 12, 1323 (1973).

Table 4B
TYROSINE pK VALUES IN Hb

Species	No. of residues	pK	Specific residue pK	Reference
Horse	8 of 12	10.6		1
	4 of 12	>12		
Human A	8 of 12	10.6		1
	4 of 12	>12		
Human A carboxy	8 of 12	10.60	β145 10.6	2
			β130 10.6	
	4 of 12	>10.6	β35 >10.6	
Human a deoxy	6 of 12	10.77		2
Human F carboxy	6 of 10	10.45		2
Human F deoxy	4 of 10	10.65		2

Compiled by Lynne H. Botelho and Frank R. N. Gurd.

REFERENCES

1. Hermans, Jr., *Biochemistry*, 1, 193 (1962).
2. Nagel, Ranney, and Kucinskis, *Biochemistry*, 5, 1934 (1966).

Table 4C
TYROSINE pK VALUES IN Mb

Species	pK	Reference
Sperm whale	10.3	1
	11.5	
	>12.8	
Horse	10.3	1
	11.5	
	>12.8	

Compiled by Lynne H. Botelho and Frank R. N. Gurd.

REFERENCE

1. Hermans, Jr., *Biochemistry*, 1, 193 (1962).

Table 5
pK VALUES FOR HUMAN Hb Cys β 93 SH

Human hemoglobin	pK	Reference
Deoxy, cys β 93 SH	>11	1
	>10	2
	>9.5	3
Carboxy, cys β 93 SH	>11	1

Compiled by Lynne H. Botelho and Frank R. N. Gurd.

REFERENCES

1. Janssen, Willekens, De Bruin, and van Os, *Eur. J. Biochem.,* 45, 53 (1974).
2. Snow, *Biochem. J.,* 84, 360 (1962).
3. Guidotti, *J. Biol. Chem.,* 242, 3673 (1967).

Table 6
CARBOXYL SIDE CHAIN pK VALUES ESTIMATED IN LYSOZYMES

Residue	Range of pK values[1]
Gln 35	6–6.5
Asp 101	4.2–4.7
Asp 66	1.5–2
Asp 52	3–4.6

Compiled by Lynne H. Botelho and Frank R. N. Gurd.

REFERENCE

1. Imoto, Johnson, North, Phillips, and Rupley, in *The Enzymes,* Vol. VII, 3rd ed., Bayer, Ed., Academic Press, New York, 1972, 665.

ENZYMATIC ACTIVITIES OF SUBCELLULAR FRACTIONS

Walter C. Schneider

The following tables list enzymatic activities that have been found to be associated with subcellular components isolated from tissues by differential centrifugation. Wherever possible, the percentage of the total tissue activity recovered in a fraction and the relative specific activity (RSA), defined as the ratio of the specific activity of the isolated fraction to that of the unfractionated tissue, are given. RSA values greater than 1.0 indicate that the enzyme is concentrated in the isolated fraction. For rat or mouse liver, exclusive localization of an enzyme in a fraction would yield RSA values of approximately ten for nuclei, four for mitochondria or microsomes, and three for the soluble fraction.

In addition to tables on nuclei, mitochondria, microsomes, and the soluble fraction, data on nucleoli, lysosomes, microbodies, zymogen granules, Golgi material, and plasma membranes are also presented. The first four are the major fractions obtained when a tissue homogenate is fractionated. Nuclei can be isolated by two methods: Centrifugal separation from tissue homogenates prepared in aqueous solutions (A) or separation from frozen-dried tissue powders using flotation in organic solvents (F). Nucleoli are obtained by subfractionation of disrupted nuclei (A). Zymogen (secretion) granules are present only in specialized glandular tissues such as the pancreas. Lysosomes and microbodies (peroxisomes) are normally recovered in the mitochondrial fraction or in an intermediate fraction between mitochondria and microsomes and must be purified by special methods. Lysosomes and peroxisomes have been calculated to account for 2 and 2.5%, respectively, of total rat liver protein [Leighton, Poole, Beaufay, Baudhuin, Coffey, Fowler, and de Duve., *J. Cell Biol.*, 37, 482 (1968)]. Exclusive localization of an enzyme would therefore yield an RSA value of 50 for lysosomes and 40 for peroxisomes. Plasma (cell) membranes and Golgi material are isolated by special methods that usually involve flotation in sucrose density gradients. When tissue homogenates are fractionated by the usual procedure, plasma membranes would be recovered in the nuclear fraction with fragments also occurring in the microsomal fraction while the Golgi material would be present in the microsomal fraction. These subcellular components like lysosomes and microbodies also represent small fractions of the total cellular protein; e.g., 1 to 2% in rat liver.

No attempt has been made to provide references to all of the papers that have appeared with regard to a single enzyme or the variety of tissues or species studied. The tables are intended as a general view of the intracellular localization of enzymes. Some data have been excluded for lack of completeness, and other data on intracellular distribution may have been overlooked because of their inclusion in papers as ancillary parts of other projects. The interested reader is referred to reviews by de Duve, Wattiaux, and Baudhuin [*Adv. Enzymol. Relat. Areas Mol. Biol.*, 24, 291 (1962)] and by Schneider and Kuff [*Cytology and Cell Physiology*, 3rd ed., Bourne, Ed., Academic Press, New York, 1964, 19)] for further details and discussions.

Table 1
NUCLEI

Enzyme	Tissue	Isolation method[b]	RSA[a]	Reference	Remarks
Glycolytic Enzymes					
Aldolase	Liver, rat	F	2.4	4	—
Enolase	Liver, rat	F	1.0	4	—
Hexokinase	Liver, rat	F	0.3	4	—
Phosphofructokinase	Liver, rat	F	3.0	4	—
Phosphoglycerate kinase	Liver, rat	F	0.62	4	See Table 10 Soluble Fraction
Pyruvate kinase	Liver, rat	F	1.1	4	See Table 10 Soluble Fraction
Triose phosphate isomerase	Liver, rat	F	2.8	4	See Table 10 Soluble Fraction
Hydrolytic and Degradative Activities					
Acid phosphatase	Liver, rat	F	0.43	2	See Table 6 Lysosomes
Adenosine deaminase	Liver, heart, calf	F	2.1, 6.0	1	See Table 10 Soluble Fraction
Adenosine triphosphatase	Liver, calf	F	0.12	1	—
Alkaline phosphatase	Liver, calf	F	0.12	1	See Table 8 Zymogen Granules
Amylase	Pancreas, bovine, horse	F	0.03, 0.07	1	See Table 8 Zymogen Granules
Arginase	Liver, calf, horse	F	0.54, 0.57	1	Recovered in nuclear and microsomal fractions in rat liver (6,7), localization dependent upon aqueous medium used
Dipeptidase	Liver, rat	F	1.7–3.5	2	Range with 6 substrates
Esterase	Liver, calf	F	0.76	1	See Table 4 Microsomes and Table 7 Cell Membranes
Glucose-6-phosphatase	Liver, rat	A	0.25, 0.43	2, 5	See Table 4 Microsomes
β-Glucuronidase	Liver, calf	F	0.17	1	See Table 6 Lysosomes
Lipase	Pancreas, bovine, horse	F	0.02, 0.08	1	See Table 8 Zymogen Granules
Nucleoside phosphorylase	Liver, heart, calf	F	1.0, 4.4	1	See Table 10 Soluble Fraction
5'-Nucleotidase	Liver, calf	F	0.18	1	See Table 4 Microsomes and Table 7 Cell Membranes

[a]RSA = relative specific activity, the ratio of fraction to whole tissue.
[b]A = isolation in aqueous media, F = isolation by flotation in organic solvents.

Table 1 (continued)
NUCLEI

Enzyme	Tissue	Isolation method[b]	RSA[a]	Reference		Remarks
		Oxidoreductase Activities				
Catalase	Liver, calf, horse	F	0.32, 0.71		1	Enzyme appears to be localized in microbodies, see Table 5
Catalase	Liver, rat	F	0.05		2	Enzyme appears to be localized in microbodies, see Table 5
Cytochrome oxidase	Liver, rat	A	0.07		5	See Table 3 Mitochondria
Glucose-6-phosphate dehydrogenase	Liver, calf, rat	F	0.94, 0.56		3, 4	See Table 10 Soluble Fraction
Glutamic dehydrogenase	Liver, rat	F	0.05		2	See Table 3 Mitochondria
α-Glycerophosphate dehydrogenase	Liver, rat	F	1.4		4	See Table 10 Soluble Fraction
Isocitric dehydrogenase (NADP)	Liver, rat	F	2.0		4	See Table 10 Soluble Fraction
Lactic dehydrogenase	Liver, rat	F	1.2		4	See Table 10 Soluble Fraction
Malic dehydrogenase	Liver, rat	F	0.6		4	See Table 10 Soluble Fraction
NADH-cytochrome c reductase	Liver, rat	A	0.17		5	See Table 4 Microsomes
6-Phosphogluconic dehydrogenase	Liver, rat	F	0.73		4	See Table 10 Soluble Fraction
Testosterone-5α-Reductase	Prostate, rat	A	14.2		18	RSA of nuclear membranes, 89.6
Triose phosphate dehydrogenase	Liver, rat	F	1.5		4	See Table 10 Soluble Fraction
Uricase	Liver, calf, horse	F	0, 0.05		1	Enzyme appears to be localized in microbodies, see Table 5
		Synthetic Activities				
Adenylate polymerase	Thymus, calf Liver, guinea pig	A	—		11, 12	Activity stated to be localized in nuclei
Cytidyl-n acetyl neuraminic acid ligase	Brain, calf	A	14.7		14	
DNA polymerase	Liver, regenerating rat	F	1.6		9	Activity recovered in soluble fraction when tissue was fractionated in aqueous media (10)
DNA polymerase	Liver, regenerating; hepatoma, rat	F	1.3, 3.7		10	
DNA polymerase	Embryo, sea urchin	A	3.2		15	62% recovered in nuclei
DNA polymerase	Brain, liver, rat	A	—		16, 17	41–79% recovered in nuclei with remainder in soluble fraction

Table 1 (continued)
NUCLEI

Enzyme	Tissue	Isolation method[b]	RSA[a]	Reference	Remarks
NAD-pyrophosphorylase	Liver, rat	F,A	8.0, 5.5	2, 8	92% recovered in nuclei (8)
Orotidine-5'-phosphate pyrophosphorylase	Liver, rat	F,A	1.7, 3.0	13	—
RNA-polymerase	Liver, rat	F,A	—	13	Activity 4–5 times as great in aqueous nuclei activity could not be measured in whole tissue

Compiled by Walter C. Schneider.

REFERENCES

1. Stern, Mirsky, Allfrey, and Saetren, *J. Gen. Physiol.*, 35, 559 (1952).
2. Siebert, Humphrey, Theman, and Kersten, *Z. Physiol. Chem.*, 340, 51 (1965).
3. Stern and Mirsky, *J. Gen. Physiol.*, 37, 177 (1953).
4. Siebert, *Biochem. Z.*, 334, 369 (1961).
5. Widnell and Tata, *Biochem. J.*, 92, 313 (1964).
6. Ludewig and Chanutin, *Arch. Biochem.*, 29, 441 (1950).
7. Rosenthal, Gottlieb, Gorry, and Vars, *J. Biol. Chem.*, 223, 469 (1956).
8. Hogeboom and Schneider, *J. Biol. Chem.*, 197, 611 (1952).
9. Keir, Smellie, and Siebert, *Nature*, 196, 752 (1962).
10. Behki and Schneider, *Biochim. Biophys. Acta*, 68, 34 (1963).
11. Edmonds and Abrams, *J. Biol. Chem.*, 235, 1142 (1960).
12. Heppel, Ortiz, and Ochoa, *Science*, 123, 415 (1956).
13. Reid, El Aaser, Turner, and Siebert, *Z. Physiol. Chem.*, 339, 135 (1964).
14. Van den Eijnden, *J. Neurochem.*, 21, 949 (1973).
15. Loeb, *J. Biol. Chem.*, 224, 1672 (1969).
16. Chin and Sung, *J. Neurochem.*, 20, 617 (1973).
17. Murthy and Barucha, *Can. J. Biochem.*, 49, 1284 (1971).
18. Moore and Wilson, *J. Biol. Chem.*, 247, 958 (1972).

Table 2
NUCLEOLI

Enzyme	Tissue	RSA[a]	Reference	Remarks
NAD-pyrophosphorylase	Eggs, starfish	48.1	1	–
NAD-pyrophosphorylase	Liver, rat	3.0[b]	2	–
Nucleoside phosphorylase	Eggs, starfish	30.6	1	–
Pyruvate kinase	Liver, rat	2.0[b]	2	–
Ribonuclease	Liver, rat	6.1[b]	2	–
RNA polymerase	Liver, rat	23.6[b]	2	–

Compiled by Walter C. Schneider.

[a]RSA = relative specific activity, the ratio of fraction to whole tissue.
[b]Ratio of specific activity of nucleoli to specific activity of unfractionated nuclei.

REFERENCES

1. Baltus, *Biochim. Biophys. Acta,* 15, 263 (1954).
2. Siebert, Villalobos, Ro, Steels, Lindenmayer, Adams, and Busch, *J. Biol. Chem.,* 241, 71 (1966).

Table 3
ISOLATED MITOCHONDRIA

Enzyme	Tissue	RSA[a]	Reference	Remarks
Glycolytic Enzymes				
Hexokinase	Brain, calf	8.0	30	8 other glycolytic enzymes present mainly in soluble fraction
Phosphofructokinase	Brain, calf	6.3	30	—
Hydrolytic or Degradative Enzymes				
Acetyl CoA carboxylase	Liver, chick embryo	3.3	41	67% recovered in mitochondria
Adenosine triphosphatase	Liver, mouse	2.2	26	Ca^{2+} activated; 50% recovered in mitochondria
Deoxyribonuclease (alkaline)	Liver, rat	3.2	17	53% recovered in mitochondria
Glutaminase I	Liver, mouse	3.7	27	78% recovered in mitochondrial fraction
Glutamate decarboxylase	Brain, rat	—	42	55% recovered in mitochondria
Inorganic pyrophosphatase	Liver, mouse	3.0	28	Subsequently shown to catalyze transfer of phosphate to glucose (29), see Table 4 Microsomes
Pyruvate carboxylase	Brain, rat	1.9	43	80% recovered in mitochondria
5' Ribonuclease	Liver, rat	4.5	46	64% recovered in mitochondria, Poly A substrate
Oxidative and Related Enzymes				
Aconitase	Liver, rat	—	7	16% recovered in isolated mitochondria
Aconitase	Brain, rabbit	2.2	8	86% recovered in isolated mitochondria
Betaine aldehyde dehydrogenase	Liver, rat	—	14	50% recovered in isolated mitochondria
Choline oxidase	Liver, rat	3.7	13	78% recovered in isolated mitochondria
Coenzyme Q oxidase	Liver, rat	4.3	19	78% of liver activity recovered in isolated mitochondria
Coproporphyrinogen oxidase	Liver, beef	2.2	18	67% recovered in isolated mitochondria
Cytochrome c	Liver, rat	2.3	23	50% recovered in mitochondrial fraction
Cytochrome oxidase	Liver, rat	3.1	10	75% recovered in isolated mitochondria
Fumarase	Brain, rabbit	2.4	8	67% recovered in isolated mitochondria
Fumarase	Liver, mouse	2.0	9	55% recovered in isolated mitochondria
Glutamic dehydrogenase	Liver, mouse	3.3	12	73% recovered in isolated mitochondria

[a] RSA = relative specific activity, the ratio of fraction to whole tissue.

Table 3 (continued)
ISOLATED MITOCHONDRIA

Enzyme	Tissue	RSA[a]	Reference	Remarks
α-Glycerophosphate oxidase	Liver, rat	—	15	60% recovered in isolated mitochondria
D-β-Hydroxybutyric dehydrogenase	Liver, rat	3.4	17	55% recovered in mitochondria
Isocitric dehydrogenase (NAD)	Heart, guinea pig	10	6	—
Isocitric dehydrogenase (NAD)	Brain, rat	2.1	43	80% recovered in mitochondria
Isocitric dehydrogenase (NADP)	Brain, rat	1.2	43	51% recovered in mitochondria
Isocitric dehydrogenase (NADP)	Brain, rabbit	1.6	5	58% recovered in isolated mitochondria
Isocitric dehydrogenase (NADP)	Liver, mouse	0.5	4	Most of activity present in soluble fraction
α-Ketoglutaric oxidase	Liver, rat	1.5	3	—
Malate dehydrogenase	Liver, rat	3.7	17	60% recovered in mitochondria
NAD-NADP transhydrogenase	Liver, rat	—	25	Activity found only in mitochondrial fraction
NADH cytochrome c reductase	Liver, mouse	1.2	20	Inhibited by Antimycin A (22), see Table 4 Microsomes
NADPH cytochrome c reductase	Liver, mouse	2.3	21	Inhibited by Antimycin A (22), see Table 4 Microsomes
Octanoic oxidase	Liver, rat	2.3	11	80% recovered in isolated mitochondria
Oxalacetic oxidase	Liver, rat	1.8	2	45% recovered in isolated mitochondria
Peroxidase	Ehrlich ascites tumor	6.1	24	—
Succinoxidase	Liver, rat	2.7	1	67% recovered in isolated mitochondria
Tyramine oxidase	Liver, rat	2.2	16	57% recovered in isolated mitochondria
Synthetic Activities				
Acetyl-CoA ligase	Liver, rat	—	37	55–72% of activity recovered in isolated mitochondria
Acetyl-CoA ligase	Liver, rat	3.4	44	51% recovered in mitochondria
Butyryl-CoA ligase	Liver, rat	4.7	44	71% recovered in mitochondria
Adenylate kinase	Liver, rat	—	39	29% recovered in mitochondria, see Table 10 Soluble Fraction
γ-Amino butyrate transaminase	Brain, rat	—	42	72% recovered in mitochondria
p-Aminohippurate synthesis	Liver, mouse	2.8	32	Distribution of individual enzymes involved has not been tested
CDP-diglyceride synthetase	Liver, rat	5.3	47	92% recovered in mitochondria; results disagree with those on guinea pig liver microsomes
δ-Aminolevulinate synthetase	Liver, guinea pig	—	38	Enzyme induced in mitochondria by feeding 3,5-dicarbethoxy-1,4-dihydrocollidine

Table 3 (continued)
ISOLATED MITOCHONDRIA

Enzyme	Tissue	RSA[a]	Reference	Remarks
Glutamic-oxaloacetic transaminase	Liver, rat	2.3	31	—
p-Hydroxy cinnamate synthetase	Liver, rat	11.9	45	89% recovered in mitochondria; RSA suggests anomalous homogenate activity
β-Hydroxy-β-methyl glutaryl CoA condensing enzyme	Liver, rat	2.5	36	68% of condensing and cleavage activity recovered in isolated mitochondria
Palmityl CoA-carnitine-palmityl transferase	Liver, calf	3.1	35	—
Rhodanese	Liver, rat	2.5	40	62% recovered in nitrochondria
Tetrahydrofolate synthesis	Liver, rat	—	34	78% of activity recovered in mitochondria in 0.25 M sucrose; loss of one or more enzymes from mitochondria occurred in saline
Ureidosuccinate formation	Liver, rat	2.9	33	47% recovered in mitochondrial fraction

Compiled by Walter C. Schneider.

Table 3 (continued)
ISOLATED MITOCHONDRIA

REFERENCES

1. Schneider and Hogeboom, *J. Biol. Chem.*, 183, 123 (1950).
2. Schneider and Potter, *J. Biol. Chem.*, 177, 893 (1949).
3. Siekevitz, *J. Biol. Chem.*, 195, 549 (1952).
4. Hogeboom and Schneider, *J. Biol. Chem.*, 186, 417 (1950).
5. Shepherd, *J. Histochem. Cytochem.*, 4, 47 (1955).
6. Plaut and Sung, *J. Biol. Chem.*, 207, 305 (1954).
7. Dickman and Speyer, *J. Biol. Chem.*, 206, 67 (1954).
8. Shepherd and Kalnitsky, *J. Biol. Chem.*, 207, 605 (1954).
9. Kuff, *J. Biol. Chem.*, 207, 361 (1954).
10. Schneider and Hogeboom, *J. Natl. Cancer Inst.*, 10, 969 (1950).
11. Schneider, *J. Biol. Chem.*, 176, 259 (1948).
12. Hogeboom and Schneider, *J. Biol. Chem.*, 186, 417 (1950).
13. Kensler and Langeman, *J. Biol. Chem.*, 192, 551 (1951).
14. Williams, *J. Biol. Chem.*, 195, 37 (1952).
15. Dianzani, *Arch. Fisiol.*, 50, 187 (1951).
16. Cotzias and Dole, *Proc. Soc. Exp. Biol. Med.*, 78, 157 (1951).
17. Beaufay, Bendall, Baudhuin, and de Duve, *Biochem. J.*, 73, 623 (1959).
18. Sano and Granick, *J. Biol. Chem.*, 236, 1173 (1961).
19. Sastry, Jayaraman, and Ramasarma, *Nature*, 189, 577 (1961).
20. Hogeboom and Schneider, *J. Natl. Cancer Inst.*, 10, 983 (1950).
21. Hogeboom and Schneider, *J. Biol. Chem.*, 186, 417 (1950).
22. de Duve, Pressman, Gianetto, Wattiaux, and Appelmans, *Biochem. J.*, 60, 604 (1955).
23. Schneider and Hogeboom, *J. Biol. Chem.*, 183, 123 (1950).
24. Sauer, Martin, and Stotz, *Cancer Res.*, 20, 251 (1960).
25. Reynafarje and Potter, *Cancer Res.*, 17, 1112 (1957).
26. Schneider, Hogeboom, and Ross, *J. Natl. Cancer Inst.*, 10, 977 (1950).
27. Shepherd and Kalnitsky, *J. Biol. Chem.*, 192, 1 (1951).
28. Rafter, *J. Biol. Chem.*, 230, 643 (1958).
29. Rafter, *J. Biol. Chem.*, 235, 2475 (1960).
30. Beattie, Sloan, and Basford, *J. Cell Biol.*, 19, 309 (1963).
31. Müller and Leuthardt, *Helv. Chim. Acta*, 33, 628 (1950).
32. Kielley and Schneider, *J. Biol. Chem.*, 185, 869 (1950).
33. Reichard, *Acta Chem. Scand.*, 8, 795 (1954).
34. Noronha and Sreenivaasan, *Biochim. Biophys. Acta*, 44, 64 (1960).
35. Norum, *Biochim. Biophys. Acta*, 89, 95 (1964).
36. Bucher, Overath, and Lynen, *Biochim. Biophys. Acta*, 40, 491, (1960).
37. Aisenberg and Potter, *J. Biol. Chem.*, 215, 737 (1955).
38. Granick and Urata, *J. Biol. Chem.*, 238, 821 (1963).
39. Novikoff, Hecht, Podber, and Ryan, *J. Biol. Chem.*, 194, 153 (1952).
40. Ludewig and Chanutin, *Arch. Biochem.*, 29, 441 (1950).
41. Donaldson, Mueller, and Mason, *Biochim. Biophys. Acta*, 248, 34 (1971).
42. Van Kempen, Van den Berg, Van den Helm, and Veldstra, *J. Neurochem.*, 12, 581 (1965).
43. Salganicoff and Koeppe, *J. Biol. Chem.*, 243, 3416 (1968).
44. Aas and Bremer, *Biochim. Biophys. Acta*, 164, 157 (1968).
45. Ranganathan and Ramasarma, *Biochem. J.*, 122, 487 (1971).
46. de Lamirande, Morais, and Blackstein, *Arch. Biochem. Biophys.*, 118, 347 (1967).
47. Vorbeck and Martin, *Biochem. Biophys. Res. Commun.*, 40, 901 (1970).

Table 4
MICROSOMES

Hydrolytic or Degradative Activities

Enzyme	Tissue	RSA[a]	Reference	Remarks
Alkaline phosphatase	Liver, rat	—	22	No added Mg^{2+}, 42% recovered in microsomes
Aryl sulfatase C	Liver, rat	—, 4.7	21, 52	62% recovered in microsomes; p-acetylphenyl sulfate as substrate
Cholesterol esterase	Liver, rat	—	20	112% recovered in microsomes
Choline esterase	Liver, rat	3.1	19	48% recovered in microsomes
Cysteinyl glycinase	Kidney, rat	—	30	Activity recovered mainly in microsomal fraction, overall recovery of activity was greater than 100%
N-2-Difluorenylacetamide deacylase	Liver, rat	—	25	100% of activity recovered in microsomes
Esterase	Liver, mouse	4.2	16	47% recovered in microsomes; methyl butyrate substrate
Glucose-6-phosphatase	Liver, rat	3.0	17	74% recovered in microsomes
Glutathionase	Kidney, rat	—	30	Activity recovered mainly in microsomal fraction, overall recovery of activity was greater than 100%
Glyceryl-1-decanoate esterase	Liver, rat	2.5	40	—
Inorganic pyrophosphatase	Liver, rat	2.6	26	81% of activity recovered in microsomes; activity apparently identical with glucose-6-phosphatase and pyrophosphate glucose phosphotransferase (27–29)
Lactonase	Kidney, rat	3.7	18	71% recovered in microsomes
Lipoprotein lipase	Liver, rat	1.8	41	RSA of lysosomes higher
Lysolecithinase	Liver, rat	3.0	41	RSA of lysosomes higher
NAD-nucleosidase	Liver, mouse	2.3	24	Other tissues and species also tested with similar results
NAD-pyrophosphatase	Liver, mouse	2.2	24	
5'-Nucleotidase	Liver, rat	2.2	23	Microsomal activity may be due to presence of plasma membranes
Plasmalogen hydrolase	Liver, rat	—	31	Microsomes had highest specific activity; whole tissue not tested
Phosphodiesterase I	Mammary gland, rabbit	3.4	32	RSA calculated with reference to nuclei free homogenate
Phosphodiesterase I	Liver, rat	2.1, 1.5	42, 43	Measured at pH 9; 36% recovered in microsomes, 41% in nuclear fraction

[a]RSA = Relative specific activity, the ratio of fraction to whole tissue.

Table 4 (continued)
MICROSOMES

Enzyme	Tissue	RSA[a]	Reference	Remarks
Triphosphoinositide phosphomonoesterase	Kidney, rat	3.4	44	57% recovered in microsomes
Uronolactonase	Liver, rat	—	33	94% recovered in microsomes
Vitamin A ester esterase	Liver, rat	3.0	41	RSA of lysosomes higher

Oxidative and Related Activities

Enzyme	Tissue	RSA[a]	Reference	Remarks
L-Amino acid oxidase	Liver, chicken	—	9	53% of cytoplasmic activity recovered in microsomes
Cytochrome b_5	Liver, rat	—	15	Presence detected only in microsomes
Diamine oxidase	Liver, rabbit	3.0	45	81% recovered in microsomes
Dimethylamino-azobenzene reductase	Liver, rat	—	10	Activity mainly in microsomes and required NADPH
Glucuronolactone reductase	Liver, rat	—	7, 8	Activity recovered in microsomes greater than 100%
Gulonolactone dehydrogenase				
Hexobarbital hydroxylase	Liver, rabbit	—	12	Enzyme required NADPH and O_2 and was localized in microsomes; same or similar microsomal enzymes metabolize a variety of drugs by a similar mechanism
α-Hydroxystearic acid decarboxylase	Brain, rat	22	14	Microsomes required unidentified dialyzable supernatant factor; factor not tested on other cell fraction
17β-Hydroxysteroid dehydrogenase	Liver, guinea pig	3.0, 6.7	4.5	NAD specific, see Table 10 Soluble Fraction for NADP specific enzyme
11β-Hydroxysteroid dehydrogenase	Liver, rat	—	6	Enzyme only in microsomes and required NADH generating system
NAD-cytochrome c reductase	Liver, mouse	2.5	1	59% recovered in isolated microsomes and see Table 3 Mitochondria
NADP-cytochrome c reductase	Liver, mouse	2.0	2	39% recovered in isolated microsomes, see Table 3 Mitochondria
NADPH-diaphorase	Liver, rat	—	11	80% of cytoplasmic activity with neotetrazolium as acceptor recovered in microsomes
Steroid-β-ol-dehydrogenase	Adrenal medulla, bovine	4.6	13	39% recovered in microsomes

Table 4 (continued)
MICROSOMES

Enzyme	Tissue	RSA[a]	Reference	Remarks
Synthetic Activities				
Aspartate carbamoyl transferase	Liver, rat	—	46	68% recovered in microsomes
Cytidine diphosphate diglyceride synthetase	Liver, guinea pig	—	37	62% of activity present in separate fractions recovered in microsomes
Diglyceride acyl transferase	Liver, rat	4.7	39	—
Glutamine synthetase	Liver, rat	3.2	47	47% recovered in microsomes
γ-Glutamyl transpeptidase	Kidney, rat	4.8	35	RSA compared to that of cytoplasmic extract
Lactose synthetase	Mammary gland, rat	—	48	74% of A protein recovered in microsomes, see Table 10 Soluble Fraction
Nucleoside phosphotransferase	Liver, rat	1.7	34	49% recovered in microsomes same RSA in mitochondria
Palmityl: CoA ligase	Liver, rat	2.5	49	58% recovered in microsomes
Phosphatidyl inositol synthetase	Thyroid, pig	325	50	RSA indicates anomalous homogenate activity
Phosphoryl choline glyceride transferase	Liver, rat	6.3	38	Overall recovery of activity >100%
Phosphoryl choline glyceride transferase	Liver, rat	3.3	39	—
Pyrophosphate-glucose phosphotransferase	Liver, rat	2.7	26	85% of glucose-6-phosphate forming activity recovered in microsomes; see inorganic pyrophosphatase above
UDP-galactosyl transferase	Brain, guinea pig	—	36	Psychosine formation showed highest RSA in heavy microsomes
UDP-glucuronyl-bilirubin transferase	Liver, rat	5.2	51	—
Vitamin A ester synthetase	Liver, cat	2.2	3	—

Compiled by Walter C. Schneider.

Table 4 (continued)
MICROSOMES

REFERENCES

1. Hogeboom and Schneider, *J. Natl. Cancer Inst.*, 10, 983 (1950).
2. Hogeboom and Schneider, *J. Biol. Chem.*, 186, 417 (1950).
3. Futterman and Andrews, *J. Biol. Chem.*, 239, 4077 (1964).
4. Endahl, Kochakian, and Hamm, *J. Biol. Chem.*, 235, 2792 (1960).
5. Villee and Spencer, *J. Biol. Chem.*, 235, 3615 (1960).
6. Hurlock and Talalay, *Arch. Biochem.*, 80, 468 (1959).
7. Chatterjee, Chatterjee, Ghosh, Ghosh, and Guha, *Biochem. J.*, 76, 279 (1960).
8. Isherwood, Mapson, and Chen, *Biochem. J.*, 76, 157 (1960).
9. Struck and Sizer, *Arch. Biochem.*, 90, 22 (1960).
10. Mueller and Miller, *J. Biol. Chem.*, 180, 1125 (1949).
11. Williams, Gibbs, and Kamin, *Biochim. Biophys. Acta,* 32, 568 (1959).
12. Cooper and Brodie, *J. Pharmacol. Exp. Ther.*, 114, 409 (1955).
13. Beyer and Samuels, *J. Biol. Chem.*, 219, 69 (1956).
14. Levis and Mead, *J. Biol. Chem.*, 239, 77 (1964).
15. Strittmatter and Ball, *Proc. Natl. Acad. Sci. U.S.A.*, 38, 19 (1952).
16. Omachi, Barnum, and Glick, *Proc. Soc. Exp. Biol. Med.*, 67, 133 (1948).
17. de Duve, Pressman, Gianetto, Wattiaux, and Appelman, *Biochem. J.*, 60, 604 (1955).
18. Meister, *Science,* 115, 521 (1952).
19. Goutier and Goutier-Pirotte, *Biochim. Biophys. Acta,* 16, 361 (1955).
20. Schotz, Rice, and Alfin-Slater, *J. Biol. Chem.*, 207, 665 (1954).
21. Dodgson, Spencer, and Thomas, *Biochem. J.*, 59, 29 (1955).
22. Allard, de Lamirande, Faria, and Cantero, *Can. J. Biochem. Physiol.*, 32, 383 (1954).
23. Segal and Brenner, *J. Biol. Chem.*, 235, 471 (1960).
24. Jacobson and Kaplan, *J. Biophys. Biochem. Cytol.*, 3, 31 (1957).
25. Seal and Gutman, *J. Biol. Chem.*, 234, 648 (1959).
26. Stetten, *J. Biol. Chem.*, 239, 3576 (1964).
27. Stetten, *J. Biol. Chem.*, 240, 2248 (1965).
28. Nordlie and Arion, *J. Biol. Chem.*, 239, 1680 (1964).
29. Nordlie and Arion, *J. Biol. Chem.*, 240, 2155 (1965).
30. Binkley, *J. Biol. Chem.*, 236, 1075 (1961).
31. Warner and Lands, *J. Biol. Chem.*, 236, 2404 (1961).
32. Smith, Easter, and Dils, *Biochim. Biophys. Acta,* 125, 445 (1966).
33. Winkelman and Lehninger, *J. Biol. Chem.*, 233, 794 (1958).
34. Brawerman and Chargaff, *Biochim. Biophys. Acta,* 16, 524 (1955).
35. Avi Dor, *Biochem. J.*, 76, 370 (1960).
36. Cleland and Kennedy, *J. Biol. Chem.*, 235, 45 (1960).
37. Carter and Kennedy, *J. Lipid Res.*, 7, 678 (1966).
38. Schneider, *J. Biol. Chem.*, 238, 3572 (1963).
39. Wilgram and Kennedy, *J. Biol. Chem.*, 238, 2615 (1963).
40. Hayase and Tappel, *J. Biol. Chem.*, 244, 2269 (1969).
41. Shibko and Tappel, *Arch. Biochem. Biophys.*, 106, 259 (1964).
42. de Lamirande, Morais, and Blackstein, *Arch. Biochem. Biophys.*, 118, 347 (1967).
43. Brightwell and Tappel, *Arch. Biochem. Biophys.*, 124, 325 (1968).
44. Lee and Huggins, *Arch. Biochem. Biophys.*, 126, 206 (1968).
45. Argento-Ceru, Sartori, and Autori, *Eur. J. Biochem.*, 34, 369 (1973).
46. Bottomley and Lovig, *Biochim. Biophys. Acta,* 148, 588 (1967).
47. Banay-Schwartz and Strecker, *Int. J. Biochem.*, 1, 371 (1970).
48. Brodbeck and Ebner, *J. Biol. Chem.*, 241, 5526 (1966).
49. Farstad, Bremer, and Norum, *Biochim. Biophys. Acta,* 132, 492 (1967).
50. Jungalwala, Freinkel, and Dawson, *Biochem. J.*, 123, 19 (1971).
51. Wong, *Biochem. J.*, 125, 27 (1971).
52. Dolly, Dodgson, and Rose, *Biochem. J.*, 128, 337 (1972).

Table 5
MICROBODIES (PEROXISOMES)

Enzyme	Tissue	RSA[a]	Reference	Remarks
d-Amino acid oxidase	Liver, rat	28.8, 30.0	1, 2	–
Catalase	Liver, rat	32.6, 36.3	1, 2	–
1-α Hydroxy acid oxidase	Liver, rat	36.8, 35.8	1, 2	–
Urate oxidase	Liver, rat	57.4, 50.0	1, 2	–

Compiled by Walter C. Schneider.

[a]RSA = relative specific activity, the ratio of fraction to whole tissue.

REFERENCES

1. Leighton, Poole, Lazarow, and de Duve, *J. Cell Biol.*, 41, 521 (1969).
2. Leighton, Poole, Beaufay, Baudhuin, Coffey, Fowler, and de Duve, *J. Cell Biol.*, 37, 482 (1968).

Table 6
LYSOSOMES

Enzyme	Tissue	RSA[a]	Reference	Remarks
β-N-Acetyl galactosaminidase	Liver, guinea pig	23.2	1	—
β-N-Acetyl glucosaminidase	Liver, guinea pig	22.0	1	—
Acid deoxyribonuclease	Liver, rat	18.1	2	—
Acid phosphatase	Liver, guinea pig, rat	25.0, 48.5	1,3	—
Acid ribonuclease	Liver, guinea pig	23.0	1	—
Arylamidase	Liver, guinea pig	19.2	1	—
Aryl sulfatase A and B	Liver, rat	5.5	4	Activity of a light mitochondrial fraction
Aryl sulfatase C	Liver, rat	4.0	4	Activity of a light mitochondrial fraction, see Table 4 Microsomes
β-Aspartyl glucosylamine aminohydrolase	Liver, guinea pig	17.0	1	—
Cathepsin B	Liver, guinea pig	20.0	1	—
Cathepsin C	Liver, guinea pig	23.0	1	—
Cathepsin D	Liver, guinea pig	25.0	1	—
α-1 Fucosidase	Liver, guinea pig	30.0	1	—
α-Galactosidase	Liver, guinea pig	21.4	1	—
β-Galactosidase	Liver, guinea pig	17.0	1	—
α-Glucosidase	Liver, guinea pig	22.0	1	—
β-Glucosidase	Liver, guinea pig	14.8	1	—
β-Glucuronidase	Liver, guinea pig	28.0	1	—
Hyaluronidase	Liver, guinea pig	18.5	1	—
Lipase	Liver, guinea pig	15.0	1	—
α-Mannosidase	Liver, guinea pig	25.0	1	—
Neuraminidase	Liver, rat	2.0	5	Activity of a light mitochondrial fraction
Phosphodiesterase	Liver, guinea pig	18.3	1	See Table 7 Plasma Membranes and Table 4 Microsomes
Phosphodiesterase II	Liver, rat	6.2	7	Activity of a light mitochondrial fraction
Phospholipase	Liver, guinea pig	23.4	1	—
Proteinase	Liver, rat	47.3	1	Activity at pH 6.9 with rat liver soluble proteins as substrate

[a]RSA = relative specific activity, the ratio of fraction to whole tissue.

Table 6 (continued)
LYSOSOMES

Enzyme	Tissue	RSA[a]	Reference	Remarks
o-Seryl-*N*-acetyl galactosaminide glycosidase	Liver, guinea pig	18.0	1	
Sialidase	Liver, guinea pig	18.3	1	—
Sphingomyelinase	Liver, rat	8.0	6	Activity of a light mitochondrial fraction
β-Xylosidase	Liver, guinea pig	17.8	1	—

Compiled by Walter C. Schneider.

REFERENCES

1. **Patel and Tappel,** *Biochim. Biophys. Acta,* 208, 163 (1970).
2. **Dulaney and Touster,** *J. Biol. Chem.,* 247, 1424 (1972).
3. **Leighton, Poole, Beaufay, Baudhuin, Coffey, Fowler, and de Duve,** *J. Cell Biol.,* 37, 482 (1968).
4. **Dolly, Dodgson, and Rose,** *Biochem. J.,* 128, 337 (1972).
5. **Taha and Carubelli,** *Arch. Biochem. Biophys.,* 119, 55 (1967).
6. **Fowler,** *Biochim. Biophys. Acta,* 191, 481 (1969).
7. **Erecinska, Sierakowska, and Shugar,** *Eur. J. Biochem.,* 11, 465 (1969).

Table 7
PLASMA (CELL) MEMBRANES

Enzyme	Tissue	RSA[a]	Reference	Remarks
Adenyl cyclase	Kidney, rat	24.5	1	Activity hormone responsive
ADP-ase	Liver, rat	25.5	2	—
Alkaline phosphatase	Kidney, rabbit, rat	19.0, 9.9	3.8	—
Aminopeptidase	Kidney, rabbit, rat	15.0	3	—
5' Nucleotidase	Liver, rat	23.9	2	—
5' Nucleotidase	Liver, rat	30.5	4	—
5' Nucleotidase	Liver, rat	39.1	5	—
ATP-ase	Liver, rat	11.9	6	Na^+-K^+ activation
ATP-ase	Kidney, rabbit	5.0	7	Na^+-K^+ activation
ATP-ase	Intestine, rat	4.0	7	Na^+-K^+ activation
Dephospho CoA pyrophosphatase	Liver, rat	45.6	4	—
Glyceryl phosphoryl choline phosphodiesterase	Liver, rat	8.1	5	—
Leucine β-naphthyl amidase	Kidney, rat	8.9	8	—
Maltase	Kidney, rabbit	16.3	3	—
Phosphodiesterase I	Liver, rat	19.8	9	RSA of microsomes 2.2, see Table 4 Microsomes
Trehalase	Kidney, rabbit	13.7	3	—

Compiled by Walter C. Schneider.

[a]RSA = relative specific activity, the ratio of fraction to whole tissue.

REFERENCES

1. **Marx, Fedak, and Aurbach,** *J. Biol. Chem.,* 247, 6913 (1972).
2. **Wattiaux-De Coninck and Wattiaux,** *Biochim. Biophys. Acta,* 183, 118 (1969).
3. **Quirk and Robinson,** *Biochem. J.,* 128, 1319 (1972).
4. **Skrede,** *Eur. J. Biochem.,* 38, 401 (1973).
5. **Lloyd-Davies, Mitchell, and Coleman,** *Biochem. J.,* 127, 357 (1972).
6. **Victoria, van Golde, Hostetler, Scherpof, and van Deenen,** *Biochim. Biophys. Acta,* 239, 443 (1971).
7. **Douglas, Kerley, and Isselbacher,** *Biochem. J.,* 128, 1329 (1972).
8. **Price, Taylor, and Robinson,** *Biochem. J.,* 129, 919 (1972).
9. **Erecinska, Sierakowska, and Shugar,** *Eur. J. Biochem.,* 11, 465 (1969).

Table 8
ZYMOGEN GRANULES

Enzyme	Tissue	RSA[a]	Reference	Remarks
Alkaline protease	Pituitary, cow	15.0	4	Hormone storage granules
Amylase	Pancreas, dog	2.4	1	—
Amylase	Pancreas, rat	9.6	5	—
Amylase	Parotid, rat	1.6	2	—
Amylase	Parotid, rat	2.2	3	—
Lipase	Pancreas, dog	1.9	1	—
Proteinase	Pancreas, dog	95	1	—

Compiled by Walter C. Schneider.

[a]RSA = relative specific activity, the ratio of fraction to whole tissue.

REFERENCES

1. Hokin, *Biochim. Biophys. Acta,* 18, 379 (1955).
2. Schramm and Danon, *Biochim. Biophys. Acta,* 50, 102 (1961).
3. Grommet-Elhannan and Winnick, *Biochim. Biophys. Acta,* 69, 85 (1963).
4. Tesar, Koenig, and Hughes, *J. Cell Biol.,* 40, 225 (1969).
5. Van Lancker and Holtzer, *J. Biol. Chem.,* 234, 2359 (1959).

Table 9
GOLGI MATERIAL

Enzyme	Tissue	RSA[a]	Reference	Remarks
N-Acetyl glucosaminyl transferase	Liver, rat	8	1	42% recovered in Golgi fraction
Galactosyl transferase	Liver, rat	9	1	41% recovered in Golgi fraction
Galactosyl transferase	Liver, cow	42.1	2	2.1% recovered in Golgi fraction
Galactosyl transferase	Liver, rat	97.5	3	2.9% recovered in Golgi fraction
Galactosyl transferase	Liver, rat	54.6	4	32% recovered in Golgi fraction
Galactosyl transferase	Testis, rat	13.3	5	4.4% recovered in Golgi fraction
Sialyl transferase	Liver, rat	8.5	1	38% recovered in Golgi fraction

Compiled by Walter C. Schneider.

[a]RSA = relative specific activity, the ratio of fraction to whole tissue.

REFERENCES

1. Schacter, Jubbal, Hudgin, Pinteric, McGuire, and Roseman, *J. Biol. Chem.,* 245, 1090 (1970).
2. Fleischer, Fleischer, and Ozawa, *J. Cell Biol.,* 43, 59 (1969).
3. Fleischer and Fleischer, *Biochim. Biophys. Acta,* 219, 301 (1970).
4. Bergeron, Ehrenreich, Siekevitz, and Palada, *J. Cell Biol.,* 59, 73 (1973).
5. Cunningham, Mollenbauer, and Nyquist, *J. Cell Biol.,* 51, 273 (1971).

Table 10
SOLUBLE FRACTION

Enzyme	Tissue	RSA[a]	Reference	Remarks
Glycolytic Enzymes				
Aldolase	Brain, calf	2.0	9	—
Aldolase	Liver, rat	—	44	96% recovered in soluble fraction
Enolase	Brain, calf	1.6	9	—
Glucose-6-phosphate isomerase	Brain, calf	1.7	9	—
Hexokinase	Brain, calf	0.7	9	—
Phosphofructokinase	Brain, calf	2.6	9	—
Phosphoglucomutase	Liver, rat	—	29	97% recovered in soluble fraction
Phosphoglycerate kinase	Brain, calf	1.8	9	—
Phosphoglycerate mutase	Brain, calf	2.2	9	—
Phosphorylase	Liver, rat	—	29	>80% recovered in soluble fraction
Pyruvate kinase	Brain, calf	1.8	9	—
Triosephosphate isomerase	Brain, calf	1.8	9	—
Hydrolytic or Degradative Activities				
Acetyl CoA-deacylase	Liver, rat	—	30	74% recovered in soluble fraction
Adenine deaminase	Liver, rat	2.5	27	101% recovered in soluble fraction
Adenosine deaminase	Liver, mouse	2.7	18	90% recovered in soluble fraction: see Table 1 Nuclei
Aldonolactonase	Liver, rat	—	23	89% recovered in soluble fraction
Benzoylcholine esterase	Pancreas, dog	—	19	40% recovered in soluble fraction
Carbonic anhydrase	Liver, rat	2.0	28	86% recovered in soluble fraction
γ-Glutamyl lactamase	Liver, rabbit	1.6	22	85% recovered in soluble fraction
Guanine deaminase	Liver, rat	2.5	27	99% recovered in soluble fraction

[a] RSA = relative specific activity, the ratio of fraction to whole tissue.

Table 10 (continued)
SOLUBLE FRACTION

Enzyme	Tissue	RSA[a]	Reference	Remarks
Hexose diphosphatase	Liver, rat	—	29	80–96% recovered in soluble fraction
Histaminase	Kidney, pig	—	20	49–75% recovered in soluble fraction
Insulinase	Liver, rat	2.1	26	70% recovered in soluble fraction
Leucine amino peptidase	Liver, rat	2.1	25	—
Nucleoside phosphorylase	Liver, mouse	2.8	18	93% recovered in soluble fraction: see Table 1 Nuclei
Phosphatidyl inositolase	Thyroid, pig	2.0	45	—
Phosphomonoesterase I	Liver, rat	—	21	71% recovered in soluble fraction in presence of 0.01 M Mg^{2+}; see Table 4 Microsomes
Phosphoprotein phosphatase	Liver, rat	2.7	24	Soluble fraction contains all activity insensitive to molybdate
Oxidative and Related Enzymes				
Aconitase	Liver, rat	—	2	95% recovered in soluble fraction when assayed at pH 7.3; see Table 3 Mitochondria
Alcohol dehydrogenase	Liver, rat	2.4	7	95% recovered in soluble fraction
L-Amino acid oxidase	Kidney, rat	—	47	62% recovered in soluble fraction, remainder in mitochondrial fractions
4,5-Cortisone reductase	Liver, rat	—	14	77% recovered in soluble fraction
Cysteine desulfhydrase	Liver, rat	—	3	126% recovered in soluble fraction
Glucose-6-phosphate dehydrogenase	Liver, rat	—	6	Over 90% recovered in soluble fraction
Glutathione peroxidase	Liver, rat	—	48	60% recovered in soluble fraction
Glutathione reductase	Liver, rat	2.0	15	90% recovered in soluble fraction
α-Glycerophosphate dehydrogenase	Liver, rat	—	8	96% recovered in soluble fraction

Table 10 (continued)
SOLUBLE FRACTION

Enzyme	Tissue	RSA[a]	Reference	Remarks
Isocitric (NADP) dehydrogenase	Liver, mouse	2.2	1	82% recovered in soluble fraction; see Table 3 Mitochondria
17β-Ketosteroid dehydrogenase (NADP)	Liver, guinea pig	1.4, 1.7	11,12	80–96% recovered in soluble fraction
Lactic dehydrogenase	Brain, calf	3.4	9	96% recovered in soluble fraction
Malic dehydrogenase (decarboxylating)	Liver, pigeon	2.8	10	96% recovered in soluble fraction
NADPH diaphorase	Liver, rat	—	16	88% of cytoplasmic activity with dichlorophenolindophenol as acceptor recovered in soluble fraction
6-Phosphogluconate dehydrogenase	Liver, rat	—	6	Over 90% recovered in soluble fraction
Triose phosphate dehydrogenase	Brain, rat	—	13	75% recovered in soluble fraction
Tryptophan pyrrolase	Liver, rat	—	17	35% recovered in soluble fraction; activated by mitochondria and microsomes
Uracil oxidation	Liver, rat	6.8	4	Overall activity recovered was ca. 220%
Xanthine dehydrogenase	Liver, rat	—	5	104% recovered in soluble fraction
Synthetic Activities				
Acetate CoA ligase	Mammary gland, rat	1.8	32	—
Acetyl CoA carboxylase	Liver, rat	1.9	32	—
N-Acetyl glucosamine kinase	Liver, rat	—	46	100% recovered in soluble fraction
N-Acetyl glucosamine phosphomutase	Liver, rat	—	46	100% recovered in soluble fraction
Adenylate kinase	Liver, rat	—	36	31% recovered in soluble fraction; see Table 3 Mitochondria
Adenylate kinase	Heart, rat	2.9	37	95% recovered in soluble fraction
Alanine-α-keto-glutarate transaminase	Liver, rat	2.0	42	90% recovered in soluble fraction

Table 10 (continued)
SOLUBLE FRACTION

Enzyme	Tissue	RSA[a]	Reference	Remarks
β-Aminolevulinate dehydrase	Liver, rat	—	40	105% recovered in soluble fraction
DNA polymerase	Brain, liver, rat	—	49	48 and 51%, respectively, recovered in soluble fraction. Baril et. al. (50) report chromatographic differentiation of large fraction of this polymerase from the nuclear enzyme
Fatty acid synthetase	Mammary gland, rat	1.5–17	31	Synthesis from acetate
Fatty acid synthetase	Mammary gland, rat	2.6	32	Synthesis from mevalonyl CoA
Fatty acid synthetase	Liver, chick embryo	2.6	51	97% recovered in soluble fraction
Flavin adenine dinucleotide pyrophosphorylase	Liver, rat, mouse	—	38	Soluble fraction contained only activity; overall recovery of activity >400%
Glutamic-oxaloacetate transaminase	Liver, rat	1.1	34	46% recovered in soluble fraction; see Table 3 Mitochondria
Glycerokinase	Liver, rat	4.4	33	RSA of pH 5 supernatant of homogenate which contained over 90% of homogenate activity
Glucosamine acetyl transferase	Liver, rat	—	46	100% recovered in soluble fraction
Glucosamine kinase	Liver, rat	—	46	80% recovered in soluble fraction
Glucosamine phosphate isomerase	Liver, rat	—	46	82% recovered in soluble fraction
Glucosamine phosphomutase	Liver, rat	—	46	69% recovered in soluble fraction
Glucosamine phosphate acetyl transferase	Liver, rat	—	46	100% recovered in soluble fraction
Hexose phosphate glutamine amino transferase	Liver, rat	—	46	100% recovered in soluble fraction
Lactose synthetase (B protein)	Mammary gland, rat	—	52	54% recovered in soluble fraction; 31% in microsomes; see Table 4 Microsomes for A protein
Phosphorylcholinecytidyl transferase	Liver, rat	—	43	>90% recovered in soluble fraction; enzyme displayed latent behavior

Table 10 (continued)
SOLUBLE FRACTION

Enzyme	Tissue	RSA[a]	Reference	Remarks
Phosphoprotein kinase	Kidney, rat	—	53	74% recovered in soluble fraction
Pyridoxal kinase	Liver, rat	30	35	Inhibitors present in particulate fractions
Riboflavin kinase	Liver, rat	12	41	>400% recovery of activity
tRNA nucleotidyl transferase	Liver, rat	2.3	54	98% recovered in soluble fraction; mitochondria contain 35–40% as much latent activity
UDP-N-acetyl glucosamine pyrophosphorylase	Liver, rat	—	46	100% recovered in soluble fraction
UDP-glucosamine pyrophosphorylase	Liver, rat	—	46	100% recovered in soluble fraction
UDP-glucose pyrophosphorylase	Liver, rat	—	39	100% recovered in soluble fraction

Compiled by Walter C. Schneider.

Table 10 (continued)
SOLUBLE FRACTION

REFERENCES

1. Hogeboom and Schneider, *J. Biol. Chem.*, 186, 417 (1950).
2. Dickman and Speyer, *J. Biol. Chem.*, 206, 67 (1954).
3. Jackson, Albert, and Reeves, *Proc. Soc. Exp. Biol. Med.*, 88, 594 (1955).
4. Rutman, Cantarow, and Paschkis, *J. Biol. Chem.*, 210, 321 (1954).
5. Villela, Mitidieri, and Affonson, *Nature*, 175, 1087 (1955).
6. Glock and McLean, *Biochem. J.*, 55, 400 (1953).
7. Nyberg, Schuberth, and Angaard, *Acta Chem. Scand.*, 7, 1170 (1953).
8. Young and Pace, *Arch. Biochem. Biophys.*, 76, 112 (1958).
9. Beattie, Sloan, and Basford, *J. Cell Biol.*, 19, 309 (1963).
10. Rutter and Lardy, *J. Biol. Chem.*, 233, 374 (1958).
11. Endahl, Kochakian, and Hamm, *J. Biol. Chem.*, 235, 2792 (1960).
12. Villee and Spencer, *J. Biol. Chem.*, 235, 3615 (1960).
13. Johnson, *Biochem. J.*, 77, 610 (1960).
14. McGuire and Tompkins, *Nature*, 182, 261 (1958).
15. Rall and Lehninger, *J. Biol. Chem.*, 194, 119 (1952).
16. Williams, Gibbs, and Kamin, *Biochim. Biophys. Acta*, 32, 568 (1959).
17. Feigelson and Greengaard, *J. Biol. Chem.*, 236, 153 (1961).
18. Schneider and Hogeboom, *J. Biol. Chem.*, 195, 161 (1952).
19. Goutier and Goutier-Pirotte, *Biochim. Biophys. Acta*, 16, 558 (1955).
20. Valette, Cohen, and Burkard, *C. R. Soc. Biol.*, 147, 1762 (1954).
21. Allard, de Lamirande, Faria, and Conter, *Can. J. Biochem. Physiol.*, 32, 383 (1954).
22. Cliffe and Waley, *Biochem. J.*, 79, 118 (1961).
23. Winkelman and Lehninger, *J. Biol. Chem.*, 233, 794 (1958).
24. Paigen and Griffiths, *J. Biol. Chem.*, 234, 299 (1959).
25. Maver and Greco, *J. Natl. Cancer Inst.*, 12, 37 (1951).
26. Narahura, Tomizawa, Miller, and Williams, *J. Biol. Chem.*, 217, 675 (1955).
27. de Lamirande, Allard, and Cantero, *Cancer Res.*, 18, 952 (1958).
28. Datta and Shepard, *Arch. Biochem.*, 81, 124, (1959).
29. Hers, Berthet, Berthet, and de Duve, *Bull. Soc. Chim. Biol.*, 33, 21 (1951).
30. Szekeley, *Acta Physiol. Acad. Sci. Hung.*, 8, 291 (1955).
31. Popjak and Tietz, *Biochem. J.*, 60, 147 (1960).
32. Smith, Easter, and Dils, *Biochim. Biophys. Acta*, 125, 445 (1966).
33. Bublitz and Kennedy, *J. Biol. Chem.*, 211, 951 (1954).
34. Gaull and Villee, *Biochim. Biophys. Acta*, 39, 560 (1960).
35. McCormick, Gregory, and Snell, *J. Biol. Chem.*, 236, 2076 (1961).
36. Novikoff, Hecht, Podber, and Ryan, *J. Biol. Chem.*, 194, 153 (1952).
37. Cleland and Slater, *Biochem. J.*, 53, 547 (1953).
38. Deluca and Kaplan, *Biochim. Biophys. Acta*, 30, 7 (1958).
39. Reid, *Biochim. Biophys. Acta*, 32, 253 (1959).
40. Gibson, Neuberger, and Scott, *Biochem. J.*, 61, 618 (1955).
41. McCormick, *J. Biol. Chem.*, 237, 959 (1962).
42. Swick, Barnstein, and Stange, *J. Biol. Chem.*, 240, 3334 (1965).
43. Schneider, *J. Biol. Chem.*, 238, 3572 (1963).
44. Kennedy and Lehninger, *J. Biol. Chem.*, 179, 957 (1949).
45. Jungalawala, Freinkel, and Dawson, *Biochem. J.*, 123, 19 (1971).
46. Izumi, *J. Biochem.*, 57, 539 (1965).
47. Nakano, Saga, and Tsutsumi, *Biochim. Biophys. Acta*, 185, 19 (1969).
48. Green and O'Brien, *Biochim. Biophys. Acta*, 197, 31 (1970).
49. Murthy and Bhorucha, *Can. J. Biochem.*, 49, 1284 (1971).
50. Baril, Brown, Jenkins, and Laszlo, *Biochemistry*, 10, 1981 (1971).
51. Donaldson, Mueller, and Mason, *Biochim. Biophys. Acta*, 248, 34 (1971).
52. Brodbeck and Ebner, *J. Biol. Chem.*, 241, 5526 (1966).
53. Jackson, Jackson, and Freeman, *Biochim. Biophys Acta*, 105, 483 (1965).
54. Mukerji and Deutscher, *J. Biol. Chem.*, 247, 481 (1972).

INTRINSIC VISCOSITY OF PROTEINS IN NATIVE AND DENATURED STATES[a]

Protein	Native state ml/g	Denatured state in concentrated guanidine hydrochloride solution	
		S–S bond intact ml/g	S–S bond broken ml/g
Insulin	–	–	6.1
Ribonuclease	3.3	9.4	16.6 (16.3)
Lysozyme	2.7	6.5	17.1
Hemoglobin	3.6	–	18.9
Myoglobin	3.1	–	20.9
β-Lactoglobulin	3.4	19.1	22.8
Ovomucoid	5.5[2]	8.1[3]	16.0[3]
Ovalbumin A_1 [4]	3.5	27.0	31.0
Papain[5]	3.5	–	24.5
Chymotrypsinogen	2.5	11.0	26.8
Phosphoribosyl transferase	–	–	31.9
Glyceraldehyde-3-phosphate dehydrogenase	–	–	34.5
Tropomyosin	45	–	33.0
Pepsinogen	–	27.2	31.5
Aldolase	4.0	–	33.5
Serum albumin	3.7	22.9	52.2
Thyroglobulin	4.7	–	82.0
Myosin	217	–	92.6
Paramyosin	103	–	65.6
Ovomucin[6]	210	–	78.0

Compiled by V. S. Ananthanarayanan.

[a]Unless otherwise indicated, the data have been taken from Tanford.[1]

REFERENCES

1. Tanford, *Adv. Protein Chem.*, 23, 121 (1968).
2. Donavan, *Biochemistry*, 6, 3918 (1967).
3. Ahmad and Salahuddin, *Int. J. Pept. Protein Res.*, in press.
4. Ahmad and Salahuddin, to be published.
5. Ahmad and Salahuddin, *Biochemistry*, 13, 245 (1974).
6. Donavan, Davis, and White, *Biochim. Biophys. Acta*, 207, 190 (1970).

METHODS FOR THE IMMOBILIZATION OF ENZYMES

L. Goldstein

Enzymes immobilized on or within a solid matrix, by conjugation with synthetic water-insoluble polymeric supports, can serve as reusable reagents that often possess improved storage and operational stability.[1-16] Such preparations can be easily removed from the reaction mixture by filtration or centrifugation; continuous reactions can be carried out with immobilized-enzyme columns. Such processes can be controlled by removing the insoluble reagent at any desired stage of the reaction or by monitoring the rate of flow of substrate through an enzyme column.

Four basic approaches have been used for the immobilization of enzymes and other biologically active macromolecules:[1,4-10,13,14]

a. Adsorption onto inert supports or ion-exchange resins; adsorption methods are the easiest, they are however of limited reliability when absolute immobilization is desired.

b. Entrapment by occlusion within cross-linked gels or by encapsulation within microcapsules, hollow fibers, or liposomes; the generality of occlusion techniques is limited by the fact that they are suitable mainly for enzymes which utilize substrates of molecular weight low enough to diffuse through the constraining matrix.

c. Covalent binding to polymeric supports via functional groups on the protein which are nonessential for its biological activity; this approach is the most general — it offers the most permanent mode of fixation and allows a wide choice of coupling reactions and polymeric supports.

d. Cross-linking by bi- or multifunctional reagents, often following adsorption or entrapment onto a structure of defined geometry.

In most cases a trial and error approach has to be used to find the method best suited for immobilization of a given enzyme. Several methods have gained wider acceptance due to the ease of their application in the biochemical laboratory, or the availability of the parent polymer from easily accessible sources,[8,9,13-15] e.g., entrapment within polyacrylamide gels;[17-23] coupling to polysaccharide supports such as cellulose, dextrane, or agarose which have been activated with cyanogen bromide;[25-36] coupling to derivatized polyacrylamide;[37] coupling to derivatized porous glass beads.[38-55]

Immobilized enzymes can be used for the controlled degradation of proteins and other biological macromolecules;[8-10,14,56-62] to carry out specific transformations;[8-11,14] in analysis, for continuous assays with enzyme columns or tubes, enzyme electrotrodes, or enzyme thermistors;[8,9,12,14,16,300-304] for the study of subunit interactions and structure-function relationships in multisubunit enzymes;[63-65] as specific bio-adsorbents for the isolation and purification of enzyme inhibitors and related substances;[66] more generally immobilized proteins and other low- or high-molecular-weight ligands can be used as specific immuno-adsorbents, and in affinity chromatography.[66] Immobilized enzymes are also beginning to be used in large scale industrial processes.[11,12,67-69] From the theoretical viewpoint, immobilized enzymes can serve as relatively simple and well-characterized model systems for the investigation of the effects of the microenvironment on the kinetic behavior and stability of an enzyme, the effects of diffusional restrictions on enzymic reactions taking place in physically constrained systems (e.g., particulate and membranal enzymes), and for the study of multi-enzyme processes and compartmentalization phenomena.[5-9,11,17-19,70-78]

The preparation, properties, and applications of immobilized enzyme systems have been extensively reviewed.[1-11,14,15]

Tables 1 to 4 list the more common methods of enzyme immobilization. The methods are classified according to type of support material. The tables also include the type of treatment or chemical modification used to activate or derivatize the parent polymer, the procedure employed for immobilization of a protein, and where available, the protein-binding capacity of the support.

Comprehensive lists of the various enzymes for which immobilization procedures have been reported, are available in several publications.[8,13-15]

Table 1
SUPPORTS FOR IMMOBILIZATION OF ENZYMES BY ADSORPTION

Adsorbent	Capacity mg protein/gm adsorbent	References
Silica gel	—	79, 80
Carbon	—	81
Calcium phosphate gel	—	82
Kaolinite	—	83–86
Hydroxylapatitie	—	87
Glass, porous (pore diameter (900 Å)	0.1–0.33	88–91
Alumina	—	81
Clay	—	81, 92
Stainless steel particles (100–200 μm) activated by coating with titanium oxide (TiO_2)	17	93
Sephadex, diethylaminoethyl ether	~100	94–101
Sephadex, carboxymethyl ether	—	81, 102–106
Cellulose	—	105, 106
Cellulose (millipore or sartorius filters)	—	107
Cellulose, carboxymethyl ether	—	81
Cellulose diethylaminoethyl ether (DEAE cellulose)	50–150	81, 94–96, 101–103, 108–112
Collodion membrane	~200 μgs/cm²	113
Amberlite® CG-4B type II	—	81
Amberlite IR-45 (OH^-)	—	111
Dowex 1-X10 (Cl^-)	—	111
Collagen Collagen-enzyme conjugates prepared by direct impregnation of preswollen collagen membrane or by electrodeposition	6–50	114–116, 118–121
Specific adsorption		
Sepharose, n-butylamino derivative	—	122
Sepharose, n-octylamino derivative	—	122
Conconavalin A-sepharose	—	123

Compiled by L. Goldstein.

Note: References follow Table 4.

Table 2
SUPPORTS FOR THE IMMOBILIZATION OF ENZYMES BY ENTRAPMENT

Entrapment matrix	Capacity mg/gm conjugate	References
Polyacrylamide, cross-linked gel	6–100	17–24, 110, 124–134
Polyacrylamide, cross-linked beads	2–5	126, 135, 136
Polyvinyl alcohol gel (radiation cross-linked)	5–10	137
Polysiloxane (silicon rubber, silastic)	–	134, 138, 139
Starch gel	–	131, 140–142
Silica gel	–	143
Nylon microcapsules	–	144–155
Liposomes	–	156

Compiled by L. Goldstein.

Note: References follow Table 4.

Table 3
POLYMERIC SUPPORTS FOR THE IMMOBILIZATION OF ENZYMES BY COVALENT BONDS

Parent polymer	Modification of polymer	Method of coupling	Capacity mg protein/gm conjugate	References
		Organic Supports		
Cellulose		Cyanogen bromide activation	60—300	25, 157—161
Cellulose		Activation with 2,4,6-trichloro-s-triazine (cyanuric chloride); arylation of amino groups on protein	200—300	162—165
Cellulose		Activation with 2-amino-4,6-dichloro-s-triazine; arylation of amino groups on protein	4	166
Cellulose	Coated with diazotized m-diamino-benzene (Bismarck brown)	Azo-bond formation	50—60	167
Cellulose		Activated with dichloro-s-triazinyl dyestuffs (Procion® dyes); arylation of amino groups on protein	—	168, 169
Cellulose	Carboxymethyl ether (CM-cellulose)	Activation with 2-amino-4,6-dichloro-s-triazine; arylation of amino groups on protein	113	166
Cellulose	Carboxymethyl ether (CM-cellulose)	Activation of support carboxyl groups with N-ethyl-5-phenylisoxazolium-3'-sulfonate (Woodwards Reagent K); peptide bond formation with protein amino groups	300—500	170—172
Cellulose	Carboxymethyl ether (CM-cellulose)	Activation of support carboxyls with N,N'-disubstituted carbodiimides; peptide bond formation with protein amino groups	40	173, 174
Cellulose	Carboxymethyl ether (CM-cellulose)	Activation of support carboxyls with N-ethoxycarbonyl-2-ethoxy-1,2-dihydroquinoline	—	175

Note: References follow Table 4.

Table 3 (continued)
POLYMERIC SUPPORTS FOR THE IMMOBILIZATION OF ENZYMES BY COVALENT BONDS

Parent polymer	Modification of polymer	Method of coupling	Capacity mg protein/gm conjugate	References
Cellulose	Carboxymethyl ether hydrazide (CM-cellulose hydrazide)	Activation of support hydrazide groups by conversion to azide; peptide bond formation with protein amino groups	50–400	176–181
Cellulose	Aminoethyl ether	Activation of support amino groups with glutaraldehyde	130–220	182, 183
Cellulose	Diethylaminoethyl ether (DEAE-cellulose)	Activation with 2-amino-4,6-dichloro-s-triazine; arylation of amino groups on protein	166	166
Cellulose	Diethylaminoethyl ether (DEAE-cellulose)	Activation with dichloro-s-triazinyl dyestuffs (Procion dyes); arylation of amino groups of protein	~15	169, 184
Cellulose	p-Aminobenzyl ether (PAB-cellulose)	Activation by diazotization; azo-bond formation mainly with tyrosyl residues on protein	100	185–190
Cellulose	(p-Aminobenzylsulfonyl) ethyl ether	Activation by diazotization; azo-bond formation, mainly with tyrosyl residues and protein	—	191
Cellulose (microcrystalline)	3-(p-Aminophenoxy)-hydroxypropyl ether	Activation by diazotization; azo-bond formation mainly with tyrosyl residues on protein	10–40	187,188
Cellulose	3-(m-Aminobenzyloxy) methyl ether	Activation by diazotization; azo-bond formation, mainly with tyrosyl residue on protein	180–200	162
Cellulose (microcrystalline)	3-(p-Isothiocyanatophenoxy)-2-hydroxypropyl ether	Thiocarbamylation of amino groups on protein	10–18	187,188
Cellulose	Bromoacetyl ester	Alkylation, mainly of amino groups of protein	—	158, 159, 192–194
Cellulose	trans-2,3-Cyclic carbonate	Formation of urethan bonds with amino groups of protein	10–15	195–198
Cellulose	Periodate oxidized (dialdehyde cellulose)	Schiff's base formation with amino groups of protein	—	199–202
Cellulose		Activation with transition metal salts (e.g., $TiCl_4$)	—	203

Table 3 (continued)
POLYMERIC SUPPORTS FOR THE IMMOBILIZATION OF ENZYMES BY COVALENT BONDS

Parent polymer	Modification of polymer	Method of coupling	Capacity mg protein/gm conjugate	References
Starch	Periodate oxidized (dialdehyde-starch)	Schiff's base formation with amino groups of protein	—	204
Starch	Aminoaryl derivative (prepared from oxidized starch and p,p'-diaminodiphenylmethane)	Activation by diazotization; azo-bond formation, mainly with tyrosyl residues of protein	100	189
Agarose beads (Sepharose®)		Cyanagen bromide activation	70—330	25—34, 36
Agarose beads (Sepharose)		Activation with 2-amino-4,6-dichloro-s-triazine; arylation of amino groups of protein	390	166
Agarose beads (Sepharose)	Polylysine conjugate converted to the bromoacetyl derivative	Coupling via alkylation of amino groups of protein	—	34, 205
Agarose beads (Sepharose)	Polylysine conjugate converted to the p-aminobenzamido derivative	Activation by diazotization; azo-bonds mainly with tyrosyl residues of protein	—	34, 205
Agarose beads (Sepharose)	Bis-oxirane (1,4-bis 2,3 epoxy propoxy butane) derivative	Alkylation of amino groups of protein	100—120	206, 207
Agarose beads (Sepharose)	Cross-linked with epichlorohydrine	Cyanogen bromide activation	100—120	36, 176
Agarose beads (Sepharose)	Arylamino derivative	Coupling to protein carboxyl groups by four component condensation, in the presence of acetaldehyde and dimethylaminopropyl isocyanide	—	36, 117
Agarose beads (Sepharose)	Glutathione-2-pyridyl-disulfide derivative	Coupling to cystein residues of protein by a thiol-disulfide interchange reaction	100—180	209
Agarose beads (Sepharose)	Dithiobisnitrobenzoic acid (DTNB) derivative	Coupling to cystein residues of protein by a thiol-disulfide interchange	—	210
Agarose beads (Sepharose)		Activation with p-benzoquinone; coupling by arylation of amino groups of protein	70—80	211

Table 3 (continued)
POLYMERIC SUPPORTS FOR THE IMMOBILIZATION OF ENZYMES BY COVALENT BONDS

Parent polymer	Modification of polymer	Method of coupling	Capacity mg protein/gm conjugate	References
Dextrane, cross-linked beads (Sephadex®)		Cyanogen bromide activation	100–280	25, 26, 30, 32, 35, 212
Dextrane, cross-linked beads (Sephadex)	(p-Isothiocyanato-phenoxy)-hydroxy-propyl ether	Thiocarbomylation of amino groups of protein	—	213, 214
Dextrane, cross-linked beads (Sephadex)	Carboxymethyl ether	Activation of support carboxyls with N-ethoxycarbonyl-2-ethoxy-1,3-dihydroquinoline; peptide bond formation with protein amino groups	—	175
Dextrane cross-linked beads (Sephadex)		Activation with 2-amino-4,6-dichloro-s-triazine; arylation or amino groups of protein	212	166
Polygalacturonic acid		Activation of support carboxyls with N-ethyl-5-phenylisoxazolium-3'-sulfonate (Woodward's Reagent K); peptide bond formation with amino groups of protein	600–800	170–172
Alginic acid		Activation with transition metal salts (TiCl₄)	—	215
Chitin		Activation with transition metal salts (TiCl₄)	—	215
Chitin		Activation with glutaraldehyde	30	216
Polyacrylamide, cross-linked beads		Activation with glutaraldehyde	30–100	217, 218, 135, 136
Polyacrylamide, cross-linked beads (Bio-Gel®)	Acyl-hydrazide derivative	Activation of support hydrazide groups by conversion to azide; peptide bond formation with amino groups of protein	160	37
Polyacrylamide, cross-linked beads (Bio-Gel)	p-Aminobenzamidoethyl derivative	Activation by diazotization; azo-bond formation, mainly with tyrosine residues of protein	300	37, 190, 219
Copolymer of acrylamide and acrylic acid, cross-linked		Activation of support carboxyls with water-soluble carbodiimide	114	18, 220

Table 3 (continued)
POLYMERIC SUPPORTS FOR THE IMMOBILIZATION OF ENZYMES BY COVALENT BONDS

Parent polymer	Modification of polymer	Method of coupling	Capacity mg protein/gm conjugate	References
Copolymer of acrylamide and 2-hydroxyethylmethacrylate, cross-linked		Cyanogen bromide activation	30	18, 19, 135, 136
Copolymer of acrylamide and methacrylic acid anhydride, cross-linked		Peptide bond formation with amino groups of protein	200—500	221
Copolymer of acrylamide and p-amino acrylanilide; cross-linked (Enzacryl AA)		Activation by diazotization; azo bond formation, mainly with tyrosine residues of protein	10—30	222, 223
Copolymer of acrylamide and p-aminoacrylanilide, cross-linked (Enzacryl AA)	Conversion to the isothiocyanato derivative by treatment with thiophosgene	Thiocarbamylation of amino groups of protein	20—30	222, 223
Copolymer of acrylamide and acryloyl hydrazide, cross-linked (Enzacryl AH)		Activation of support hydrazide groups by conversion to azide; peptide bond formation with amino groups of protein	2	222, 223
Copolymer of acrylamide and N-acryloylcystein, cross-linked (Enzacryl polythiol)		Formation of —S—S— bonds by oxidative coupling in the presence of potassium ferricyanide	—	224
Poly (N-acryloylaminoacetaldehyde dimethyl acetal), cross-linked (Enzacryl polyacetal)		Coupling to amino groups of protein to give aminol and possibly azomethine (Schiff's base) linkages	—	225
Poly (N-acryloyl4-amino salicylic acid), cross-linked		Activation with transition metal salts (TiCl$_4$)	—	226
Polyacrylic acid, cross-linked	Acid hydrazide derivative	Activation of support hydrazide groups by conversion to azide; peptide bond formation with amino groups of protein	—	227

Table 3 (continued)
POLYMERIC SUPPORTS FOR THE IMMOBILIZATION OF ENZYMES BY COVALENT BONDS

Parent polymer	Modification of polymer	Method of coupling	Capacity mg protein/gm conjugate	Reference
Polyacrylic acid		Activation of support carboxyls with N-ethyl-5-phenylisoxazolium-3'-sulfonate (Woodward's Reagent K); peptide bond formation with amino groups of protein	450	170, 171
Copolymer of acrylic acid and 3-isothiocyanatostyrene, cross-linked		Thiocarbamylation of amino groups of protein	200–450	228–230
Polymethacrylic acid anhydride		Peptide bond formation with amino groups of protein	—	231
Polymethacrylic acid	Esters of ω-iodo-n-alcohols (polyiodals)	Coupling by alkylation of cystein -SH groups of protein	10–20	232–235
Poly(hydroxyalkyl-methacrylate), cross-linked gels (Spheron®)		Activation with cyanogen bromide	—	236
Copolymer of methacrylic acid and 3-fluoro-4,6-dinitrostyrene		Coupling by arylation of amino groups of protein	300–500	228–230, 237
Copolymer of methacrylic acid and methacrylic acid-3-fluoro-4,6-dinitroanilide, cross-linked		Coupling by arylation of amino groups of protein	150–400	228, 238–242
Copolymer of methacrylic acid and -3-isothiocyanato styrene		Coupling by thiocarbomylation of amino groups of protein	150–350	228–230
Poly(allyl carbonate)		Formation of urethane bonds with amino groups of protein	2–5	196
Copolymer of allyl alcohol and vanilin methacrylate		Coupling to amino groups of protein to give aminol and possible azomethine (Schiff's base) linkages	20–140	232, 243, 244
Copolymer of maleic anhydride and ethylene (EMA)		Peptide bond formation with amino groups of protein	100–800	7, 40, 75, 245–251

Table 3 (continued)
POLYMERIC SUPPORTS FOR THE IMMOBILIZATION OF ENZYMES BY COVALENT BONDS

Parent polymer	Modification of polymer	Method of coupling	Capacity mg protein/gm conjugate	References
Copolymer of maleic anhydride and butandiol divinyl ether		Peptide bond formation with amino groups of protein	200	181
Copolymer of maleic anhydride and methyl vinyl ether		Peptide bond formation with amino groups of protein	—	251
Copolymer of maleic anhydride and styrene		Peptide bond formation with amino groups of protein	—	252, 253
Copolymer of maleic anhydride and acrylamide, cross-linked		Peptide bond formation with amino groups of protein	0.5–15	254
Copolymer of maleic acid and ethylene, cross-linked	1-Amino-6-hexamido derivative	Activation with glutaraldehyde	100	255
Copolymer of maleic acid and ethylene, cross-linked	1-Amino-6-hexamido derivative	Coupling to carboxyl groups of protein by carbodiimide activation	100	255
Copolymer of maleic acid and ethylene, cross-linked	4-Amino-4'-amidodiphenylmethane derivative	Activation by diazotization; azo bond formation, mainly with tyrosine residues of protein	100–300	256
Copolymer of maleic acid and ethylene, cross-linked	Acid hydrazide derivative	Activation of support hydrazide groups by conversion to azide; peptide bond formation with amino groups of protein	100–400	256
Polyglutamic acid		Activation of support carboxyls with N-ethyl-5-phenylisoxazolium-3'-sulfonate (Woodward's Reagent K); peptide bond formation with amino groups of protein	600–800	170, 171
Copolymer of L-glutamic acid and L-alanine		Activation of support carboxyls with N-ethyl-5-phenylisoxazalium-3'-sulfonate (Woodward's Reagent K); peptide bond formation with amino groups of protein	330	170, 171
Copolymer of p-amino-DL-phenylalanine and L-leucine		Activation by diazotization; azo bond formation, mainly with tyrosine residues of protein	100–300	56, 257–259

Table 3 (continued)
POLYMERIC SUPPORTS FOR THE IMMOBILIZATION OF ENZYMES BY COVALENT BONDS

Parent polymer	Modification of polymer	Method of coupling	Capacity mg protein/gm conjugate	References
Collagen	Acid-hydrazide derivative	Activation of support hydrazide groups by conversion to azide; peptide bond formation with amino groups of protein	—	260, 261
Polyamides (nylon 6; nylon 6,6)		Activation with transition metal salts (TiCl$_4$)	—	203
Polyamides (nylon 6; nylon 6,6)	Acid-hydrazide derivative of partially hydrolyzed nylon	Activation of support hydrazide groups by conversion to azide; peptide bond formation with amino groups of protein	—	300
Polyamides (nylon 6; nylon 6,6)	4-Amino-4'-amido diphenyl derivative of partially hydrolyzed nylon	Activation by diazotization; azo-bond formation, mainly with tyrosine residues of protein	—	300
Polyamide (nylon 6; nylon 6,6)	Isocyanate derivative of partially hydrolyzed nylon	Coupling by carbamylation of amino groups of protein	—	262
Polyamides (nylon 6; nylon 6,6)	Partial hydrolysis	Activation with 2,4,6-trichloro-S-triazine (cyanuric chloride); arylation of amino groups of protein	—	262
Polyamides (nylon 6; nylon 6,6)	Partial hydrolysis	Activation with glutaraldehyde	—	263–266
Polyamides (nylon 6; nylon 6,6)	Partial depolymerization by transamidation with N,N-dimethyl-1,3-propane diamine	Activation with glutaraldehyde	—	301, 302
Polyamides (nylon 6; nylon 6,6)	O-Alkylation of peptide bonds with dimethyl sulfate to form polymeric imidate salts	Coupling via formation of amidines with amino groups of protein	—	303
Polyamides (nylon 6; nylon 6,6)	O-Alkylation of peptide bonds with triethyloxonium tetrafluoroborate to form imidate salts	Coupling via formation of amidines with amino groups of protein	—	304

Table 3 (continued)
POLYMERIC SUPPORTS FOR THE IMMOBILIZATION OF ENZYMES BY COVALENT BONDS

Parent polymer	Modification of polymer	Method of coupling	Capacity mg protein/gm conjugate	References
Polyamides (nylon 6; nylon 6,6)	Introduction of isocyanide side-chains by N-alkylation by peptide bonds	Coupling by four component condensation reaction in the presence of acetaldehyde and acetate (coupling to amino groups of protein) or Tris (coupling to carboxyl groups of protein)	80—100	267, 268
Polyamides (nylon 6; nylon 6,6)	Introduction of aminoaryl side chains by N-alkylation of peptide bonds	Activation by diazotization. Azo bond formation, mainly with tyrosine residues of protein	80	268
Inorganic Supports				
Porous glass	Arylamino derivative	Activation by diazotization; azo bond formation mainly with tyrosine residues of protein	10—20	38—51
Porous glass	Alkylamino derivative	Activation with glutaraldehyde	12—16	47, 52
Porous glass	Alkylamino derivative	Coupling to carboxyl groups of protein by carbodiimide activation	10	53, 55
Porous glass	N-Hydroxysuccinimide ester derivative	Coupling by formation of peptide bonds with amino groups of protein	25	269
Porous glass	Isothiocyanato derivative	Coupling by thiocarbamylation of amino groups of protein	—	40
Nickel/nickel oxide screens	Isothiocyanato derivative	Coupling by thiocarbamylation of amino groups of protein	0.4—0.5	270
Silica-alumina impregnated with nickel	Isothiocyanato derivative	Coupling by thiocarbamylation of amino groups of protein	—	271, 272
Bentonite		Activation of support with 2,4,6-trichloro-s-triazine (cyanuric chloride); coupling by arylation of amino groups of protein	3—15	273

Table 3 (continued)
POLYMERIC SUPPORTS FOR THE IMMOBILIZATION OF ENZYMES BY COVALENT BONDS

Parent polymer	Modification of polymer	Method of coupling	Capacity mg protein/gm conjugate	References
Iron oxide powder (magnetite, Fe_2O_3)	Silanization with γ-aminopropyltriethoxysilane and conversion to isocyanate	Coupling by carbamylation of amino groups of protein	7	274
Iron oxide powder (magnetite, Fe_2O_3)	Particles coated with cellulose	Cyanogen bromide activation	4	275
Iron oxide powder (magnetite, Fe_2O_3)	Silanization with γ-aminopropyltriethoxysilane	Activation with glutaraldehyde	4	275
Iron oxide powder (magnetite, Fe_2O_3)	Particles coated with polyacrylamide and then converted to the acid hydrazide derivative	Activation of support hydrazide groups by conversion to azide; peptide bond formation with amino groups of protein	—	276

Compiled by L. Goldstein.

Table 4
MULTIFUNCTIONAL REAGENTS FOR THE FIXATION OF ENZYMES ON SOLID SUPPORTS

Multifunctional reagent	Method of immobilization	Capacity mg protein/gm	References
Glutaraldehyde	Impregnation of cellophane membranes with enzyme followed by cross-linking	0.1 mg. protein/1 cm^2 membrane	277, 278
Glutaraldehyde	Enzyme cross-linked in solution, then included in agarose-polyacrylamide gel	7	278
Glutaraldehyde	Enzyme co-cross-linked in solution with inert protein, e.g., albumin then spread on glass plate to obtain membrane	7–8	278
Glutaraldehyde	Enzyme co-cross-linked in solution with inert protein, e.g., albumin then frozen at –30 and warmed slowly to obtain sponge-like conjugate	70–80	278
Glutaraldehyde	Enzyme co-cross-linked with inert protein, e.g., gelatin in the presence of fillers (bentonite, alumina, silica gel, or celite)	50–500	101
Glutaraldehyde	Enzyme co-cross-linked with chitin	30	216
Glutaraldehyde	Enzyme cross-linked with inert protein bovine serum albumin	—	279, 280
Glutaraldehyde	Enzyme adsorbed on magnetite (Fe$_2$O$_3$) followed by cross-linking	4–36	274
Glutaraldehyde	Adsorption on collodion membranes followed by cross-linking	—	281
Glutaraldehyde	Adsorption on colloidal silica particles (210–230 m^2/gm) followed by cross-linking	300	284, 285
Glutaraldehyde	Microencapsulation within collodion or nylon microcapsules followed by cross-linking	—	150, 151
Glutaraldehyde	Cross-linking of whole crystals	—	286–289
Glutaraldehyde	Cross-linking of protein aggregates	—	290
Glutaraldehyde	Enzyme cross-linked in the presence of sodium sulfate, ammonium sulfate, or acetone	—	182, 291–294
Bisdiazobenzidine 3,3'-disulfonic acid	Adsorption on collodion membranes followed by cross-linking	—	282, 283
Diphenyl-4,4'-dissothiocyanate 2,2'-disulfonic acid	Cross-linking	—	229, 230
1,5-Difluoro-2,4-dinitrobenzene	Cross-linking	—	295
dimethyl adipimidate	Cross-linking	—	296
4,4'-Difluoro-3,3'-dinitropheyl sulfone	Cross-linking	—	297, 298
Toluene-2-isocyanate-4-isothiocyanate	Cross-linking	—	299

Compiled by L. Goldstein.

REFERENCES

1. Silman and Katchalski, *Annu. Rev. Biochem.*, 35, 873 (1966).
2. Manecke, *Pure Appl. Chem.*, 4, 507 (1962).
3. Manecke, *Naturwissenschaften*, 51, 25 (1964).
4. Manecke, in *Proc. Int. Symp. Macromolecules*, Rio de Janeiro, July 26–31, 1974, Mano, Ed., Elsevier, Amsterdam, 1975, 397.
5. Goldstein and Katchalski, *Z. Anal. Chem.*, 243, 375 (1968).
6. Goldstein, in *Fermentation Advances*, Perlman, Ed., Academic Press, New York, 1969, 391.
7. Goldstein, *Methods Enzymol.*, 19, 935 (1970).
8. Goldman, Goldstein, and Katchalski, in *Biochemical Aspects of Reactions on Solid Supports*, Stark, Ed., Academic Press, New York, 1971, 1.
9. Katchalski, Silman, and Goldman, *Adv. Enzymol.*, 34, 445 (1971).
10. Smiley and Strandberg, *Adv. Appl. Microbiol.*, 14, 13 (1972).
11. Weetall, Ed., *Immobilized Enzymes, Antigens, Antibodies and Peptides*, Marcel Dekker, New York, 1975.
12. Pye and Wingard, Eds., *Enzyme Engineering*, Vol. 2, Plenum Press, New York, 1974.
13. Melrose, *Rev. Pure Appl. Chem.*, 21, 83 (1971).
14. Zaborsky, *Immobilized Enzymes*, CRC Press, Cleveland, Ohio, 1973.
15. Royer, Andrews, and Uy, *Enzyme Technol. Dig.*, 1(3), 99 (1973).
16. Salmona, Saronio, and Garattini, Eds., *Insolubilized Enzymes*, Raven Press, New York, 1974.
17. Mosbach and Mosbach, *Acta Chem. Scand.*, 20, 2807 (1966).
18. Mosbach, *Acta Chem. Scand.*, 24, 2084 (1970).
19. Mosbach and Mattiasson, *Acta Chem. Scand.*, 24, 2093 (1970).
20. Bernfeld and Wan, *Science*, 142, 678 (1963).
21. Bernfeld, Bieber, and McDonnell, *Arch. Biochem. Biophys.*, 127, 779 (1968).
22. Bernfeld, Bieber, and Watson, *Biochim. Biophys. Acta*, 131, 587 (1969).
23. Bernfeld and Bieber, *Arch. Biochem. Biophys.*, 131, 587 (1969).
24. Degani and Miron, *Biochim. Biophys. Acta*, 212, 362 (1970).
25. Axén and Ernback, *Eur. J. Biochem.*, 18, 351 (1971).
26. Porath, Axén, and Ernback, *Nature*, 215, 1491 (1967).
27. Axén, Heilbron, and Winter, *Biochim. Biophys. Acta*, 191, 478 (1969).
28. Cuatrecasas, *J. Biol. Chem.*, 245, 3059 (1970).
29. Gabel, Vretblad, Axén, and Porath, *Biochim. Biophys. Acta*, 214, 561 (1970).
30. Gabel, Steinberg, and Katchalski, *Biochemistry*, 10, 4661 (1971).
31. March, Parikh, and Cuatrecasas, *Anal. Biochem.*, 60, 149 (1974).
32. Wilchek, Oka, and Topper, *Proc. Natl. Acad. Sci. U.S.A.*, 72, 1055 (1975).
33. Jost, Miron, and Wilchek, *Biochim. Biophys. Acta*, 362, 75 (1974).
34. Wilchek, in *Immobilized Biochemicals and Affinity Chromatography*, Dunlop, Ed., Plenum Press, New York, 1974, 15.
35. Svensson, *FEBS Lett.*, 29, 167 (1973).
36. Axén, Carlsson, Janson, and Porath, *Enzymologia*, 41, 359 (1971).
37. Inman and Dintzis, *Biochemistry*, 8, 4074 (1969).
38. Weetall, *Nature*, 223, 959 (1969).
39. Weetall, *Science*, 106, 616 (1969).
40. Weetall, *Biochim. Biophys. Acta*, 212, 1 (1970).
41. Weetall and Hersh, *Biochim. Biophys. Acta*, 185, 464 (1969).
42. Weetall and Baum, *Biotechnol. Bioeng.*, 12, 399 (1970).
43. Royer and Green, *Biochem. Biophys. Res. Commun.*, 44, 426 (1971).
44. Grove, Strandberg, and Smiley, *Biotechnol. Bioeng.*, 13, 709 (1971).
45. Mason and Weetall, *Biotechnol. Bioeng.*, 14, 637 (1972).
46. Weetall and Messing, in *The Chemistry of Biosurfaces*, Hair, Ed., Marcel Dekker, New York, 1972, 563.
47. Dixon, Stolzenbach, Berenson, and Kaplan, *Biochem. Biophys. Res. Commun.*, 52, 905 (1973).
48. Weetall and Mason, *Biotechnol. Bioeng.*, 15, 455 (1973).
49. Royer and Uy, *J. Biol. Chem.*, 248, 2627 (1973).
50. Royer, Andrews, and Uy, *Enzyme Technol. Dig.*, 1(3), 99 (1973).
51. Weibel and Bright, *Biochem. J.*, 124, 801 (1971).
52. Robinson, Dunnill, and Lilly, *Biochim. Biophys. Acta*, 242, 659 (1971).
53. Baum, Ward, and Weetall, *Biochim. Biophys. Acta*, 268, 411 (1971).
54. Line, Kwong, and Weetall, *Biochim. Biophys. Acta*, 242, 194 (1971).
55. Cho and Swaisgood, *Biochim. Biophys. Acta*, 258, 675 (1972).

56. Cebra, Givol, Silman, and Katchalski, *J. Biol. Chem.*, 236, 1720 (1961).
57. Cebra, Givol, and Katchalski, *J. Biol. Chem.*, 237, 751 (1962).
58. Cebra, *J. Immunol.*, 92, 977 (1964).
59. Lowey, Goldstein, Cohen, and Luck, *J. Mol. Biol.*, 23, 287 (1967).
60. Lowey, Slayter, Weeds, and Baker, *J. Mol. Biol.*, 42, 1 (1968).
61. Slayter and Lowey, *Proc. Natl. Acad. Sci. U.S.A.*, 58, 1611 (1967).
62. Wolodko and Kay, *Can. J. Biochem.*, 53, 175 (1975).
63. Chan, *Biochem. Biophys. Res. Commun.*, 41, 1198 (1970).
64. Chan, Schutt, and Brand, *Eur. J. Biochem.*, 40, 533 (1973).
65. Fukui, Ikeda, Fujimura, Yamada, and Kumagai, *Eur. J. Biochem.*, 51, 155 (1975).
66. Jacoby and Wilchek, Eds., *Methods in Enzymology*, Vol. 34, Academic Press, New York, 1974.
67. Olson and Cooney, Eds., *Immobilized Enzymes in Food and Microbial Processes*, Plenum Press, New York, 1974.
68. Messing, Ed., *Immobilized Enzymes for Industrial Reactors*, Academic Press, New York, 1975.
69. Weetall and Suzuki, Eds., *Immobilized Enzyme Technology*, Plenum Press, New York, 1975.
70. Srere and Mosbach, *Annu. Rev. Microbiol.*, 28, 61 (1974).
71. Srere, Mattiasson, and Mosbach, *Proc. Natl. Acad. Sci. U.S.A.*, 70, 2534 (1973).
72. Mosbach, Mattiasson, Gestrelius, and Srere, in *Enzyme Engineering* Vol. 2, Pye and Wingard, Eds., Plenum Press, New York, 1974, 143.
73. Gestrelins, Mattiasson, and Mosbach, *Biochim. Biophys. Acta*, 276, 339 (1972).
74. Gestrelins, Mattiasson, and Mosbach, *Eur. J. Biochem.*, 36, 89 (1973).
75. Goldstein, Levin, and Katchalski, *Biochemistry*, 3, 1913 (1964).
76. Goldstein, *Biochemistry*, 11, 4072 (1972).
77. Goldstein, *Anal. Biochem.*, 50, 40 (1972).
78. Goldstein, *Israel J. Chem.*, 11, 379 (1973).
79. Tveritinova, Kirai, Chukhrai, and Poltorak, *Vestn. Mosk. Univ. Khim.*, 24, 16 (1969).
80. Zhirkov, Chukhrai, Poltorak, *Vestn. Mosk. Univ. Khim.*, 12, 405 (1971).
81. Miyamoto, Fuji, and Miura, *J. Ferment. Technol.*, 49, 565 (1971).
82. Koelsch, Lasch, and Hanson, *Acta Biol. Med. Ger.*, 24, 833 (1970).
83. McLaren and Estermann, *Arch. Biochem. Biophys.*, 61, 158 (1956).
84. McLaren and Estermann, *Arch. Biochem. Biophys.*, 68, 157 (1957).
85. McLaren, *Enzymologia*, 21, 356 (1960).
86. McLaren and Packer, *Adv. Enzymol.*, 33, 245 (1970).
87. Traub, Kaufmann, and Teitz, *Anal. Biochem.*, 28, 469 (1969).
88. Messing, *J. Am. Chem. Soc.*, 91, 2370 (1969).
89. Messing, *Enzymologia*, 39, 12 (1970).
90. Messing, *Enzymologia*, 38, 39 (1970).
91. Messing, *Enzymologia*, 38, 370 (1970).
92. Usami and Sherasaki, *J. Ferment. Technol.*, 48, 506 (1970).
93. Hasselberger, Allen, Paruchuri, Charles, and Coughlin, *Biochem. Biophys. Res. Commun.*, 57, 1054 (1974).
94. Tosa, Mori, Fuse, and Chibata, *Enzymologia*, 31, 214 (1966).
95. Tosa, Mori, Fuse, and Chibata, *Enzymologia*, 31, 225 (1966).
96. Tosa, Mori, Fuse, and Chibata, *Enzymologia*, 32, 153 (1967).
97. Tosa, Mori, Fuse, and Chibata, *Biotechnol. Bioeng.*, 9, 603 (1967).
98. Tosa, Mori, and Chibata, *Agric. Biol. Chem.*, 33, 1053 (1969).
99. Tosa, Mori, Fuse, and Chibata, *Agric. Biol. Chem.*, 33, 1047 (1969).
100. Tosa, Mori, and Chibata, *Enzymologia*, 40, 49 (1971).
101. Solomon and Levin, *Biotechnol. Bioeng.*, 16, 1161 (1974).
102. Mitz and Schlueter, *J. Am. Chem. Soc.*, 81, 4024 (1959).
103. Nikolaev, *Biokhimiya*, 27, 843 (1962); Engl. transl., 27, 713 (1962).
104. Nikolaev and Mardashev, *Biokhimiya*, 26, 641 (1961); Engl. transl., 26, 565 (1961).
105. Fletcher and Okada, *Nature*, 176, 882 (1955).
106. Fletcher and Okada, *Radiat. Res.*, 11, 291 (1959).
107. Wheeler, Edwards, and Whittam, *Biochim. Biophys. Acta*, 191, 187 (1969).
108. Suzuki, Ozawa, and Maeda, *Agric. Biol. Chem.*, 30, 807 (1966).
109. Smiley, *Biotechnol. Bioeng.*, 13, 309 (1971).
110. Maeda, Yamauchi, and Suzuki, *Biochim. Biophys. Acta*, 315, 18 (1973).
111. Bachler, Strandberg, and Smiley, *Biotechnol. Bioeng.*, 12, 85 (1970).
112. Becker and Pfeil, *J. Am. Chem. Soc.*, 88, 4299 (1966).
113. Goldman and Lenhoff, *Biochim. Biophys. Acta*, 242, 514 (1971).
114. Saini, Vieth, and Wang, *Trans. N.Y. Acad. Sci.*, 34, 664 (1972).

115. Suzuki, Karube, and Watanabe, in *Fermentation Technology Today,* Terui, Ed., Society for Fermentation Technology, Japan, 1972, 375.
116. Suzuki, Sonobe, Karube, and Aizawa, *Chem. Lett.,* 1, 9 (1974).
117. Vretblad and Axén, *Acta Chem. Scand.,* 27, 2769 (1973).
118. Karube and Suzuki, *Biochem. Biophys. Res. Commun.,* 47, 51 (1972).
119. Karube and Suzuki, *Biochem. Biophys. Res. Commun.,* 48, 320 (1972).
120. Wang and Vieth, *Biotechnol. Bioeng.,* 15, 93 (1973).
121. Constantinides, Vieth, and Fernandes, *Mol. Cell. Biochem.,* 1, 127 (1973).
122. Hofstee and Otillio, *Biochem. Biophys. Res. Commun.,* 53, 1137 (1973).
123. Sulkowski and Laskowski, *Biochem. Biophys. Res. Commun.,* 57, 463 (1974).
124. Walton and Eastman, *Biotechnol. Bioeng.,* 15, 951 (1973).
125. Mori, Sato, Tosa, and Chibata, *Enzymologia,* 43, 213 (1972).
126. Nilsson, Mosbach, and Mosbach, *Biochim. Biophys. Acta,* 268, 253 (1972).
127. Hicks and Updike, *Anal. Chem.,* 38, 726 (1966).
128. Strandberg and Smiley, *Appl. Microbiol.,* 21, 588 (1971).
129. Wieland, Determann, and Buenning, *Z. Naturforsch.,* 21, 1003 (1966).
130. Nadler and Updike, *Enzyme,* 18, 150 (1974).
131. Guilbault and Das, *Anal. Biochem.,* 33, 341 (1970).
132. Guilbault and Montalvo, *J. Am. Chem. Soc.,* 92, 2533 (1970).
133. Brown, Patel, and Chattopadhyay, *J. Chromatogr.,* 35, 103 (1968).
134. Brown, Patel, and Chattopadhyay, *J. Biomed. Mater. Res.,* 2, 231 (1968).
135. Johansson and Mosbach, *Biochim. Biophys. Acta,* 370, 339 (1974).
136. Johansson and Mosbach, *Biochim. Biophys. Acta,* 370, 348 (1974).
137. Maeda, Suzuki, and Yamauchi, *Biotechnol. Bioeng.,* 15, 607 (1973).
138. Pennington, Brown, Patel, and Chattopadhyay, *J. Biomed. Mater. Res.,* 2, 443 (1968).
139. Pennington, Brown, Patel, and Knowles, *Biochim. Biophys. Acta,* 167, 479 (1968).
140. Bauman, Goodson, Guilbault, and Kramer, *Anal. Chem.,* 37, 1378 (1965).
141. Bauman, Goodson, and Thomson, *Anal. Biochem.,* 19, 587 (1967).
142. Guilbault and Kramer, *Anal. Chem.,* 37, 1675 (1965).
143. Johnson and Whateley, *J. Colloid Interface Sci.,* 37, 557 (1971).
144. Chang, *Science,* 146, 524 (1964).
145. Chang, McIntosh, and Mason, *Can. J. Physiol. Pharmacol.,* 44, 115 (1966).
146. Chang, *Trans. Am. Soc. Artif. Intern. Organs,* 12, 13 (1966).
147. Chang and Poznansky, *J. Biomed. Mater. Res.,* 2, 187 (1968).
148. Chang, *Sci. Tools LKB Instrum. J.,* 16, 33 (1969).
149. Chang, *Sci. J.* (Lond.), 3, 62 (1967).
150. Chang, *Biochim. Biophys. Res. Commun.,* 44, 1531 (1971).
151. Chang, *Nature,* 229, 117 (1971).
152. Chang, *Artificial Cells,* Charles C. Thomas, Springfield, Ill., 1972.
153. Chang, *Proc. Eur. Dialysis Transpl. Assoc.,* 9, 568 (1972).
154. Chang, *J. Dent. Res.,* 51, 319 (1972).
155. Chang, in *Immobilized Enzymes,* Salmona, Saronio, and Garattini, Eds., Raven Press, New York, 1974, 15.
156. Gregoriadis, in *Insolubilized Enzymes,* Salmona, Saronio, and Garattini, Eds., Raven Press, New York, 1974, 165.
157. Patel, Stasiw, Brown, and Ghiron, *Biotechnol. Bioeng.,* 14, 1031 (1972).
158. Maeda and Suzuki, *Agric. Biol. Chem.,* 36, 1581 (1972).
159. Maeda and Suzuki, *Agric. Biol. Chem.,* 36, 1839 (1972).
160. Bartling, Brown, Forrester, Koes, Mather, and Stasiw, *Biotechnol. Bioeng.,* 14, 1039 (1972).
161. Coughlan and Johnson, *Biochim. Biophys. Acta,* 302, 200 (1973).
162. Surinov and Manoylov, *Biokhimiya,* 31, 387 (1966); Engl. transl., 31, 387 (1966).
163. Kay and Crook, *Nature,* 216, 514 (1967).
164. Kay, Lilly, Sharp, and Wilson, *Nature,* 217, 641 (1968).
165. Self, Kay, and Lilly, *Biotechnol. Bioeng.,* 11, 337 (1969).
166. Kay and Lilly, *Biochim. Biophys. Acta,* 198, 276 (1970).
167. Gray, Livingstone, Jones, and Baker, *Biochim. Biophys. Acta,* 341, 457 (1974).
168. Wilson, Kay, and Lilly, *Biochem. J.,* 108, 845 (1968).
169. Stasiw, Patel, and Brown, *Biotechnol. Bioeng.,* 14, 629 (1972).
170. Patel, Lopiekes, Brown, and Price, *Biopolymers,* 5, 577 (1967).
171. Patel and Price, *Biopolymers,* 5, 883 (1967).
172. Patel, Pennington, and Brown, *Biochim. Biophys. Acta,* 178, 626 (1969).
173. Weliky and Weetall, *Immunochemistry,* 2, 293 (1965).

174. Weliky, Brown, and Dale, *Arch. Biochem. Biophys.*, 131, 1 (1969).
175. Sundaram, *Biochem. Biophys. Res. Commun.*, 61, 667 (1974).
176. Porath, Janson, and Laas, *J. Chromatogr.*, 60, 167 (1971).
177. Epstein and Anfinsen, *J. Biol. Chem.*, 237, 2175 (1962).
178. Hornby, Lilly, and Crook, *Biochem. J.*, 98, 420 (1966).
179. Wharton, Crook, and Brocklehurst, *Eur. J. Biochem.*, 6, 565 (1968).
180. Wharton, Crook, and Brocklehurst, *Eur. J. Biochem.*, 6, 572 (1968).
181. Bruemmer, Hennrich, Klockow, Long, and Orth, *Eur. J. Biochem.*, 25, 129 (1972).
182. Habeeb, *Arch. Biochem. Biophys.*, 119, 264 (1967).
183. Glassmeyer and Ogle, *Biochemistry*, 10, 786 (1971).
184. Stasiw, Brown, and Hasselberger, *Can. J. Biochem.*, 48, 1314 (1970).
185. Mitz and Summaria, *Nature*, 189, 576 (1961).
186. Lilly, Money, Hornby, and Crook, *Biochem. J.*, 95, 451 (1965).
187. Barker, Somers, and Epton, *Carbohydr. Res.*, 8, 491 (1968).
188. Barker, Somers, and Epton, *Carbohydr. Res.*, 9, 257 (1969).
189. Goldstein, Pecht, Blumberg, Atlas, and Levin, *Biochemistry*, 29, 2322 (1970).
190. Datta, Armiger, and Ollis, *Biotechnol. Bioeng.*, 15, 993 (1973).
191. Li, Chang, Sun, Ku, Yang, Li, and Yang, *Wei Sheng Wu Hsueh Pao*, 13, 31 (1973).
192. Jagendorf, Patchornik, and Sela, *Biochim. Biophys. Acta*, 78, 516 (1963).
193. Shaltiel, Mizrahi, Stupp, and Sela, *Eur. J. Biochem.*, 14, 509 (1970).
194. Sato, Mori, Tosa, and Chibata, *Arch. Biochem. Biophys.*, 147, 788 (1971).
195. Barker, Doss, Gray, Kennedy, Stacey, and Yeo, *Carbohydr. Res.*, 20, 1 (1971).
196. Kennedy, Barker, and Rosevear, *J. Chem. Soc. Perkin Trans. I*, p. 2568 (1972).
197. Kennedy, Barker, and Rosevear, *J. Chem. Soc. Perkin Trans. I*, p. 2293 (1973).
198. Kennedy and Rosevear, *J. Chem. Soc. Perkin Trans. I*, p. 755 (1974).
199. Flemming, Gabert, and Roth, *Acta Biol. Med. Ger.*, 30, 177 (1973).
200. Flemming, Gabert, and Roth, *Acta Biol. Med. Ger.*, 31, 365 (1973).
201. Flemming, Gabert, and Roth, *Acta Biol. Med. Ger.*, 31, 449 (1973).
202. Van Leemputten and Horisberger, *Biotechnol. Bioeng.*, 16, 385 (1974).
203. Barker, Emery, and Novais, *Process Biochem.*, 5, 11 (1971).
204. Weakley and Mehltretter, *Biotechnol. Bioeng.*, 15, 1189 (1973).
205. Wilchek and Miron, *Mol. Cell Biochem.*, 4, 181 (1974).
206. Sundberg and Porath, *J. Chromatogr.*, 90, 87 (1974).
207. Vretblad, *FEBS Lett.*, 47, 86 (1974).
208. Vretblad and Axén, *FEBS Lett.*, 18, 254 (1971).
209. Carlsson, Axén, Brocklehurst, and Crook, *Eur. J. Biochem.*, 44, 189 (1974).
210. Lin and Foster, *Anal. Biochem.*, 63, 485 (1975).
211. Brandt, Andersson, and Porath, *Biochim. Biophys. Acta*, 386, 196 (1975).
212. Axén, Porath, and Ernback, *Nature*, 214, 1302 (1967).
213. Axén and Porath, *Nature*, 210, 367 (1966).
214. Sundberg and Kristiansen, *FEBS Lett.*, 22, 175 (1972).
215. Kennedy and Doyle, *Carbohydr. Res.*, 28, 89 (1973).
216. Stanley, Waters, and Chan, *Biotechnol. Bioeng.*, 17, 315 (1975).
217. Weston and Avrameas, *Biochem. Biophys. Res. Comm.*, 45, 1574 (1971).
218. Ternynck and Avrameas, *FEBS Lett.*, 23, 24 (1972).
219. Zabriskie, Ollis, and Burger, *Biotechnol. Bioeng.*, 15, 981 (1973).
220. Martensson and Mosbach, *Biotechnol. Bioeng.*, 14, 715 (1972).
221. Krämer, Lehmann, Plainer, Reisner, and Sprössler, *J. Polym. Sci. Symp.*, No. 47, 77 (1974).
222. Barker, Somers, and Epton, *Carbohydr. Res.*, 14, 323 (1970).
223. Barker, Somers, Epton, and McLaren, *Carbohydr. Res.*, 14, 287 (1970).
224. Barker and Epton, *Process Biochem.*, 5, 14 (1970).
225. Epton, McLaren, and Thomas, *Carbohydr. Res.*, 22, 301 (1972).
226. Kennedy and Epton, *Carbohydr. Res.*, 27, 11 (1973).
227. Erlanger, Isamber, and Michelson, *Biochem. Biophys. Res. Commun.*, 40, 70 (1970).
228. Manecke, Günzel, and Forster, *J. Polym. Sci. Part C*, 30, 607 (1970).
229. Manecke and Günzel, *Naturwissenschaften*, 54, 531 (1967).
230. Manecke and Günzel, *Naturwissenschaften*, 54, 647 (1967).
231. Conte and Lehman, *Hoppe-Seyler's Z. Physiol. Chem.*, 352, 533 (1971).
232. Brown and Racois, *Bull. Soc. Chem. Fr.*, p. 4357 (1971).
233. Brown and Racois, *Bull. Soc. Chem. Fr.*, No. 12, 4351 (1971).
234. Brown, Racois, and Gueniffey, *Tetrahedron Lett.*, No. 25, 2139 (1970).
235. Brown, Racois, and Gueniffey, *Bull. Soc. Chim. Fr.*, No. 12, 4341 (1971).

236. Coupek, Krivakova, and Pokorny, *J. Polym. Sci. Symp.*, No. 42, 182 (1973).
237. Manecke and Forster, *Makromol., Chem.*, 91, 136 (1966).
238. Manecke, *Pure Appl. Chem.*, 4, 507 (1962).
239. Manecke, *Naturwissenschaften*, 51, 25 (1964).
240. Manecke and Singer, *Makromol. Chem.*, 39, 13 (1960).
241. Manecke and Günzel, *Makromol. Chem.*, 51, 199 (1962).
242. Manecke, Singer, and Gilbert, *Naturwissenschaften*, 47, 63 (1960).
243. Brown and Racois, *Tetrahedron*, 30, 683 (1974).
244. Brown and Racois, *Tetrahedron*, 30, 675 (1974).
245. Levin, Pecht, Goldstein, and Katchalski, *Biochemistry*, 3, 1905 (1964).
246. Ong, Tsang, and Perlmann, *J. Biol. Chem.*, 241, 5661 (1966).
247. Westman, *Biochem. Biophys. Res. Commun.*, 35, 313 (1969).
248. Fritz, Hochstrasser, Werle, Brey, and Gebhardt, *Z. Anal. Chem.*, 243, 452 (1968).
249. Fritz, Gebhardt, Fink, Schramm, and Werle, *Hoppe-Seyler's Z. Physiol. Chem.*, 350, 129 (1969).
250. Fritz, Brey, Schmal, and Werle, *Hoppe-Seyler's Z. Physiol. Chem.*, 350, 617 (1969).
251. Zingaro and Uziel, *Biochim. Biophys. Acta*, 213, 371 (1970).
252. Goldstein, Lifshitz, and Sokolovsky, *Int. J. Biochem.*, 2, 448 (1971).
253. Solomon and Levin, *Biotechnol. Bioeng.*, 16, 1393 (1974).
254. Jaworek, in *Insolubilized Enzymes,* Salmona, Saronio, and Garattini, Eds., Raven Press, New York, 1974, 65.
255. Goldstein, *Biochim. Biophys. Acta,* 327, 132 (1973).
256. Goldstein, *Biochim. Biophys. Acta,* 315, 1 (1973).
257. Bar Eli and Katchalski, *Nature,* 188, 856 (1960).
258. Bar Eli and Katchalski, *J. Biol. Chem.,* 238, 1690 (1963).
259. Silman, Albu-Weissenberg, and Katchalski, *Biopolymers,* 4, 441 (1966).
260. Julliard, Godinot, and Gautheron, *FEBS Lett.,* 14, 185 (1971).
261. Coulet, Julliard, and Gautheron, *Biotechnol. Bioeng.,* 16, 1055 (1974).
262. Horvath and Solomon, *Biotechnol. Bioeng.,* 14, 885 (1972).
263. Sundaram and Hornby, *FEBS Lett.,* 10, 325 (1970).
264. Allison, Davidson, Gutierrez-Hartman, and Kitto, *Biochem. Biophys. Res. Commun.,* 47, 66 (1972).
265. Filippusson, Hornby, and McDonald, *FEBS Lett.,* 20, 291 (1972).
266. Bunting and Laidler, *Biotechnol. Bioeng.,* 16, 119 (1974).
267. Goldstein, Freeman, and Sokolovsky, in *Enzyme Engineering,* Vol. 2, Pye and Wingard, Eds., Plenum Press, New York, 1974, 97.
268. Goldstein, Freeman, and Sokolovsky, *Biochem. J.,* 143, 497 (1974).
269. Cuatrecasas and Parikh, *Biochemistry,* 11, 2291 (1972).
270. Weetall and Hersh, *Biochim. Biophys. Acta,* 206, 54 (1970).
271. Herring, Laurence, and Kittrell, *Biotechnol. Bioeng.,* 14, 975 (1972).
272. Traher and Kittrell, *Biotechnol. Bioeng.,* 16, 413 (1974).
273. Monsan and Durand, *FEBS Lett.,* 16, 39 (1971).
274. Van Leemputten and Horisberger, *Biotechnol. Bioeng.,* 16, 997 (1974).
275. Robinson, Dunnill, and Lilly, *Biotechnol. Bioeng.,* 15, 603 (1973).
276. Dunnill and Lilly, *Biotechnol. Bioeng.,* 16, 987 (1974).
277. Broun, Selegny, Avrameas, and Thomas, *Biochim. Biophys. Acta,* 185, 260 (1969).
278. Broun, Thomas, Gellf, Domurado, Berjonneau, and Guillon, *Biotechnol. Bioeng.,* 15, 359 (1973).
279. Avrameas and Ternynck, *Immunochemistry,* 6, 53 (1969).
280. Avrameas and Gilbert, *Biochimie,* 53, 603 (1971).
281. Goldman, Kedem, and Katchalski, *Biochemistry,* 10, 165 (1971).
282. Goldman, Kedem, Silman, Caplan, and Katchalski, *Biochemistry,* 7, 486 (1968).
283. Goldman, Silman, Caplan, Kedem, and Katchalski, *Science,* 150, 758 (1965).
284. Haynes and Walsh, *Biochem. Biophys. Res. Commun.,* 36, 235 (1969).
285. Walsh, Hauston, and Kenner, in *Structure-Function Relationships of Proteolytic Enzymes,* Desnuelle, Neurath, and Ottesen, Eds., Academic Press, New York, 1970, 56.
286. Quiocho and Richards, *Proc. Natl. Acad. Sci. U.S.A.,* 52, 833 (1964).
287. Quiocho and Richards, *Biochemistry,* 5, 4062 (1966).
288. Bishop, Quiocho, and Richards, *Biochemistry,* 6, 4077 (1966).
289. Sluyterman and DeGraaf, *Biochim. Biophys. Acta,* 171, 277 (1969).
290. Josephs, Eisenberg, and Reisler, *Biochemistry,* 12, 4060 (1973).
291. Ogata, Ottesen, and Svendsen, *Biochim. Biophys. Acta,* 159, 403 (1968).

292. Jansen and Olson, *Arch. Biochem. Biophys.,* 129, 221 (1969).
293. Jansen, Tomimatsu, and Olson, *Arch. Biochem. Biophys.,* 144, 394 (1971).
294. Ottesen and Svensson, *C.R. Trav. Lab. Carlsberg,* 38, 171 (1971).
295. Marfrey and King, *Biochim. Biophys. Acta,* 105, 178 (1965).
296. Hartman and Wold, *Biochemistry,* 6, 2439 (1967).
297. Wold, *J. Biol. Chem.,* 236, 106 (1961).
298. Zahn, *Angew. Chem.,* 67, 56 (1955).
299. Schick and Singer, *J. Biol. Chem.,* 236, 2447 (1961).
300. Hornby and Filippusson, *Biochim. Biophys. Acta,* 220, 343 (1970).
301. Hornby, Inman, and McDonald, *FEBS Lett.,* 23, 114 (1972).
302. Inman and Hornby, *Biochem. J.,* 137, 25 (1974).
303. Campbell, Hornby, and Morris, *Biochim. Biophys. Acta,* 384, 307 (1975).
304. Morris, Campbell, and Hornby, *Biochem. J.,* 147, 593 (1975).

DIMENSIONS OF THE AMINO ACID GROUP, AMINO ACID SIDE CHAINS, AND THE PEPTIDE LINKAGE

M. Vijayan

The dimensions of the amino acid group and amino acid side chains have been compiled from single crystal X-ray and neutron diffraction studies on free amino acids and their salts. Salts containing atoms "heavier" than chlorine have been excluded. Only neutron results have been considered for calculating average bond lengths and angles involving hydrogen atoms. When more than one structure determination has been reported on the same crystal, only the most accurate one has been used for the compilation.

The bond lengths (in Å) and angles (in degrees) given in the figures represent the weighted average values. The weighted average values of some selected bond lengths and angles are given in tabular form also. The bond lengths used for averaging are those uncorrected for thermal motion. The weighted average standard deviations are not given as, in many cases, they are of doubtful significance in view of the differences in the thermal vibration amplitudes in various structures and the different types of systematic effects present in different structure determinations.

The distributions of different dihedral angles are presented individually; no effort has been made to bring out the correlations between different dihedral angles in a molecule. Such correlations have been discussed extensively by Sasisekharan and Ponnuswamy.[53,54] Again, only neutron results have been used in calculating the dihedral angles involving hydrogen atoms. The dimensions of the peptide linkage presented here have been taken from Ramachandran et al.[55] and Kolaskar et al.[56]

I thank Professors G.N. Ramachandran and V. Sasisekharan for many useful discussions. I am indebted to Dr. Marjore Harding and Mr. H. Howieson for making available to me their unpublished results on L- and DL-leucine and also the first chapter of Mr. Howieson's Master of Philosophy thesis. My thanks are also due to Mr. T. N. Bhat for computational assistance.

AVERAGE VALUES OF SOME SELECTED BOND LENGTHS AND ANGLES IN AMINO ACID SIDE CHAINS

$C(Sp^3)-C(Sp^3)$	1.524Å
$C(Sp^3)-H$	1.090
C–S	1.810
S–S	2.036
$C(Sp^3)-C(Sp^3)-C(Sp^3)$	113.0°
$C(Sp^3)-C(Sp^3)-H$	109.3
$H-C(Sp^3)-H$	107.2
C–S–C	100.4
C–S–S	104.5

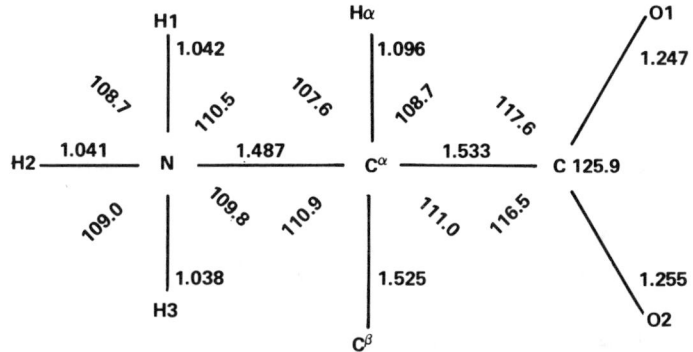

FIGURE 1. Bond lengths and angles in the zwitterionic amino acid grouping.[1-5,7-9,11-15,18-22,24-26,30-46,48,49]

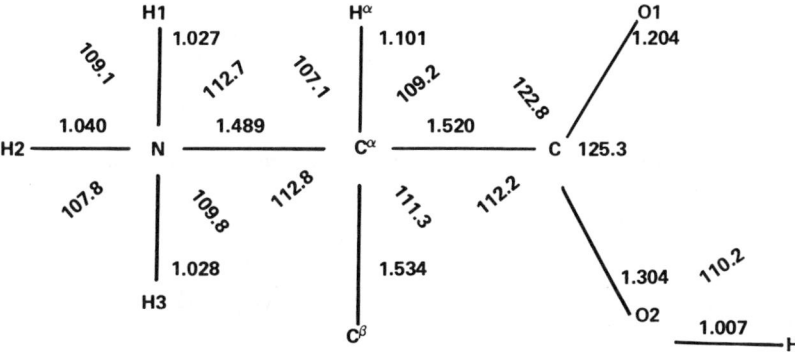

FIGURE 2. Bond lengths and angles in the cationic amino acid grouping.[6,10,16,17,23,24,27-29,47]

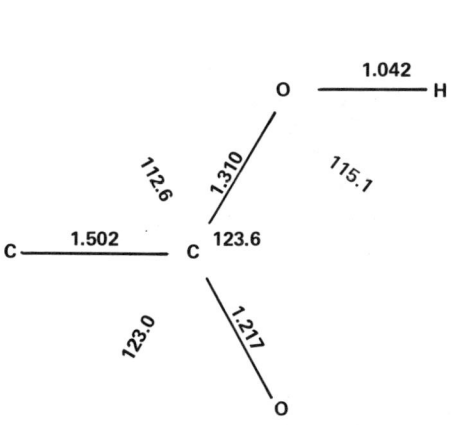

FIGURE 3. Bond lengths and angles in the carboxyl group of the side chains in aspartic and glutamic acids.[44-47]

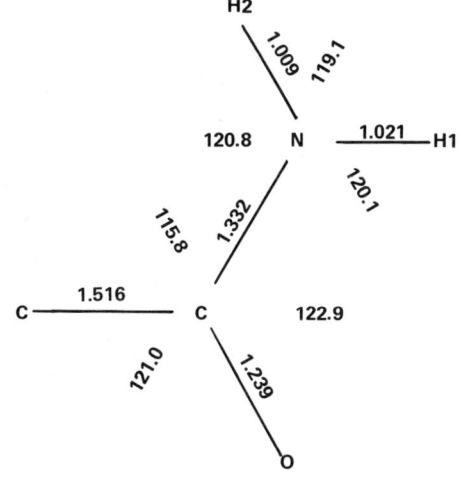

FIGURE 4. Bond lengths and angles in the amide group of the side chains in asparagine and glutamine.[48,49]

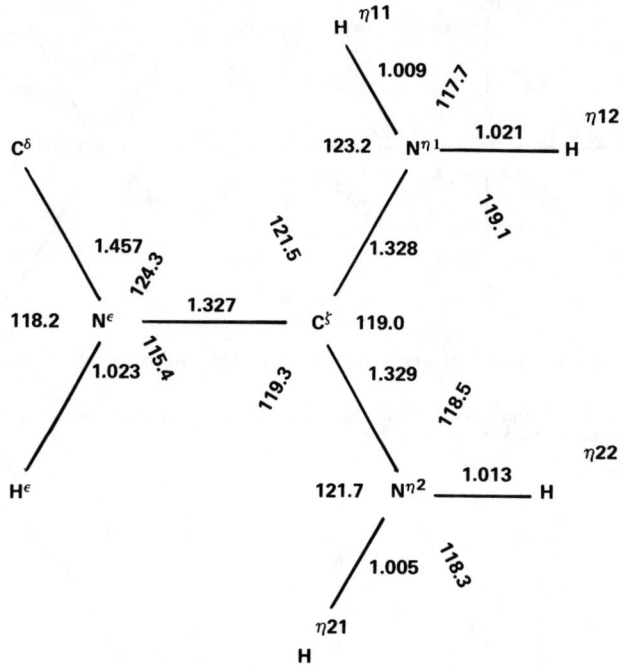

FIGURE 5. Bond lengths and angles in the guanidinium group in arginine.[33-37]

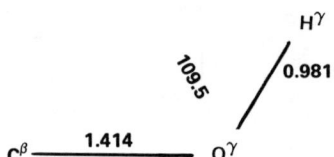

FIGURE 6. Bond lengths and angles in the hydroxyl group of serine and threonine.[20-22]

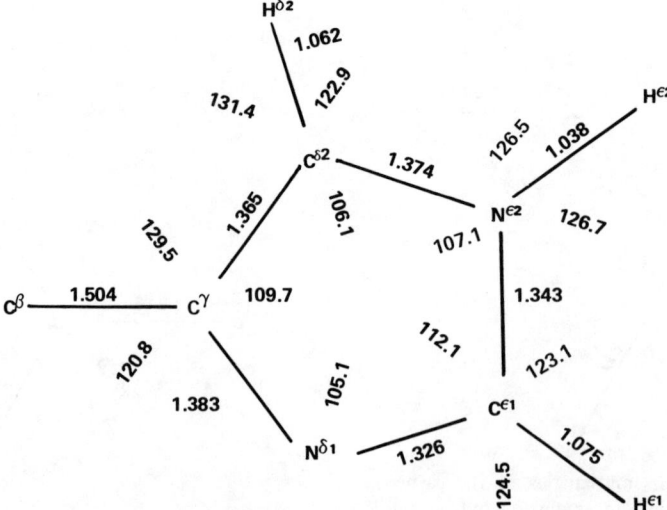

FIGURE 7. Bond lengths and angles in the unprotonated imidazole group in histidine.[38-41]

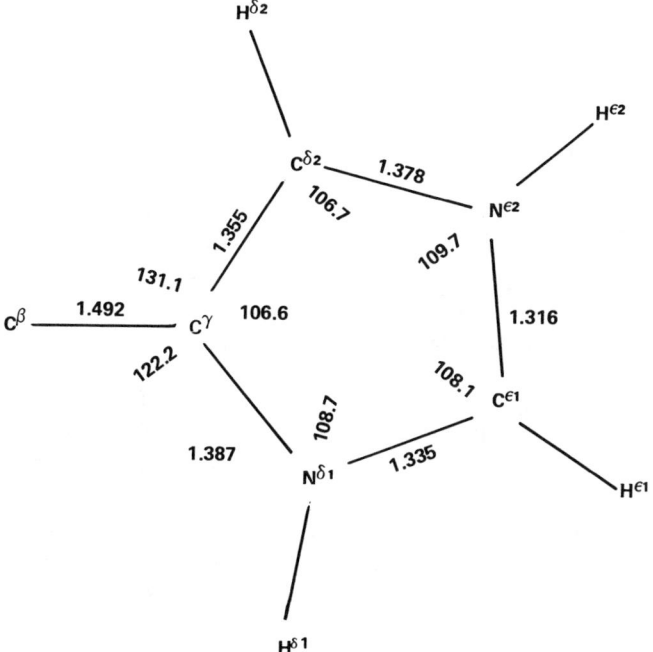

FIGURE 8. Bond lengths and angles in the protonated imidazole group in histidine.[42-43]

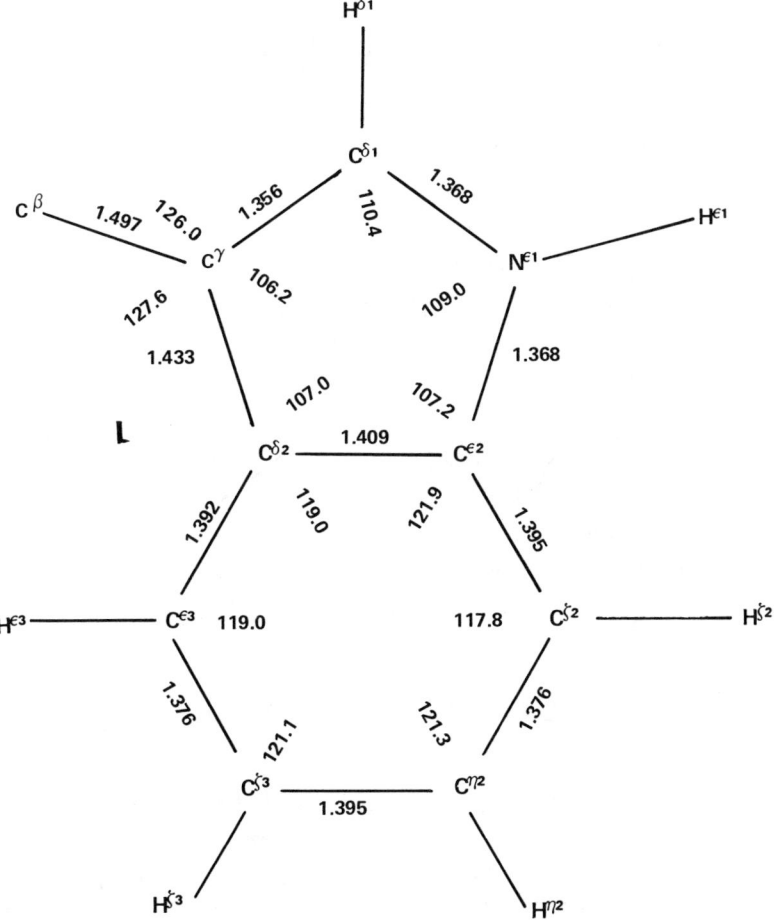

FIGURE 9. Bond lengths and angles in the indole group in tryptophan.[27-29]

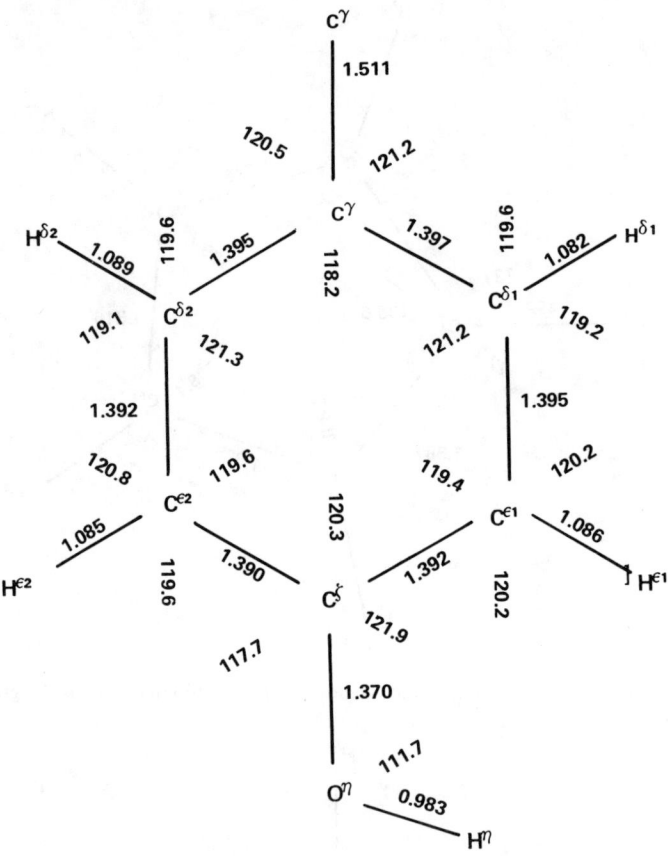

FIGURE 10. Bond lengths and angles in the phenyl group in phenylalanine and the phenol group in tyrosine.[23-26]

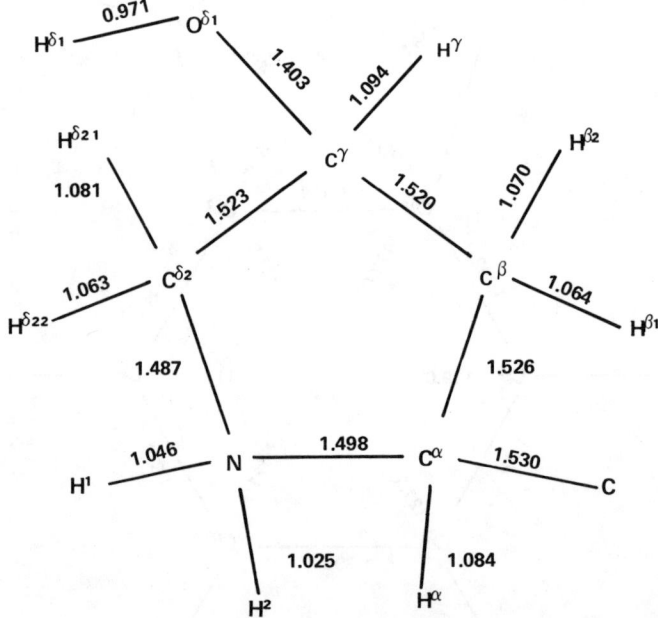

FIGURE 11. Bond lengths in pyrrolidine in proline and γ-hydroxypyrrolidine in hydroxyproline. The average bond angles are not given as they, especially the internal angles, are dependent on the puckering of the ring.[50-52]

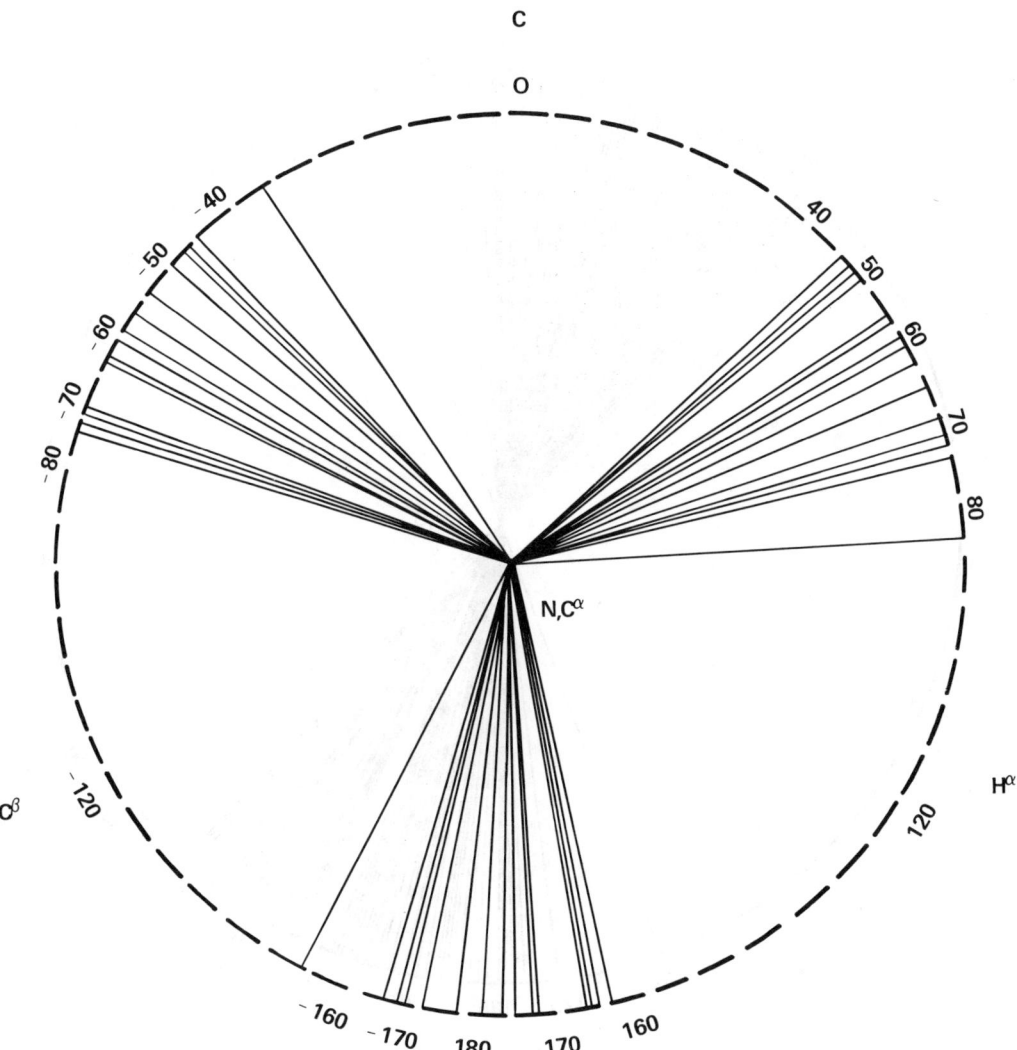

FIGURE 12. Distribution of dihedral angles ϕ_1, ϕ_2, and ϕ_3.[1,4,6,17,20,22,24,30,31,46-49]

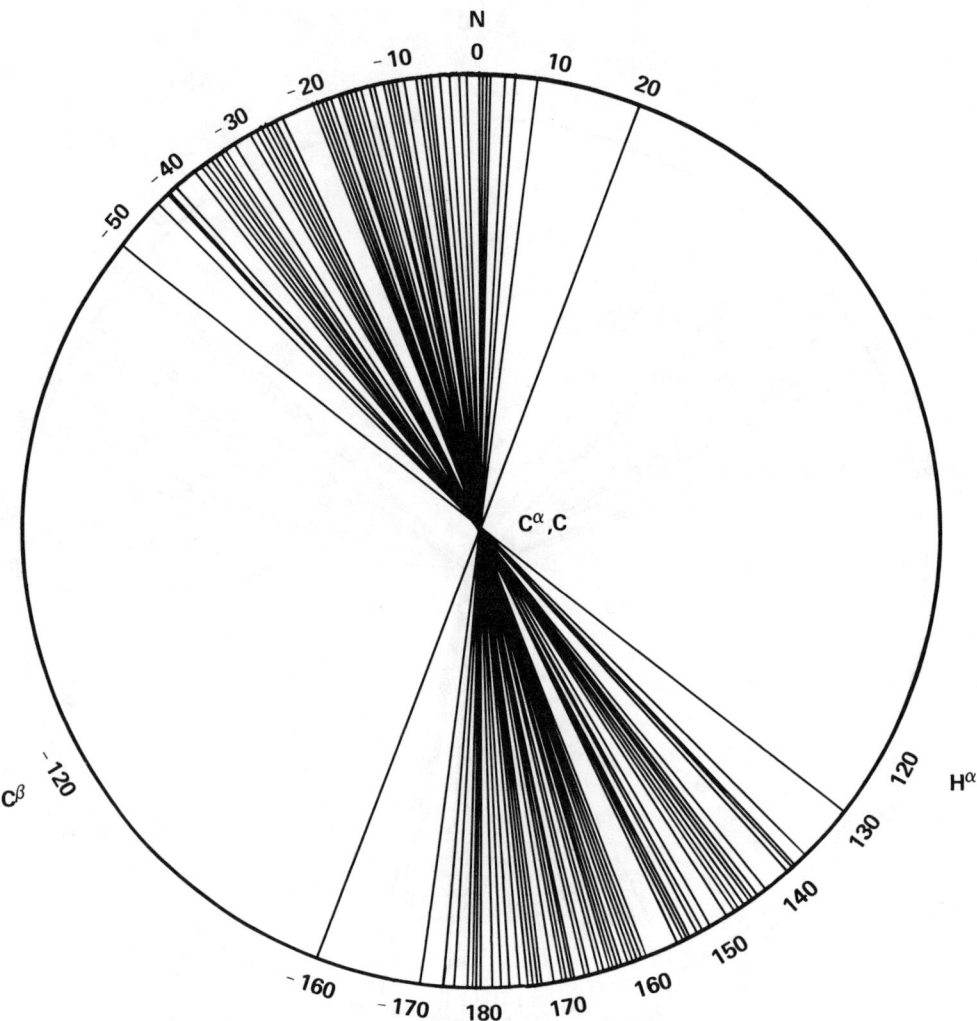

FIGURE 13. Distribution of dihedral angles ψ_1 and ψ_2 in amino acids other than glycine.[4-49]

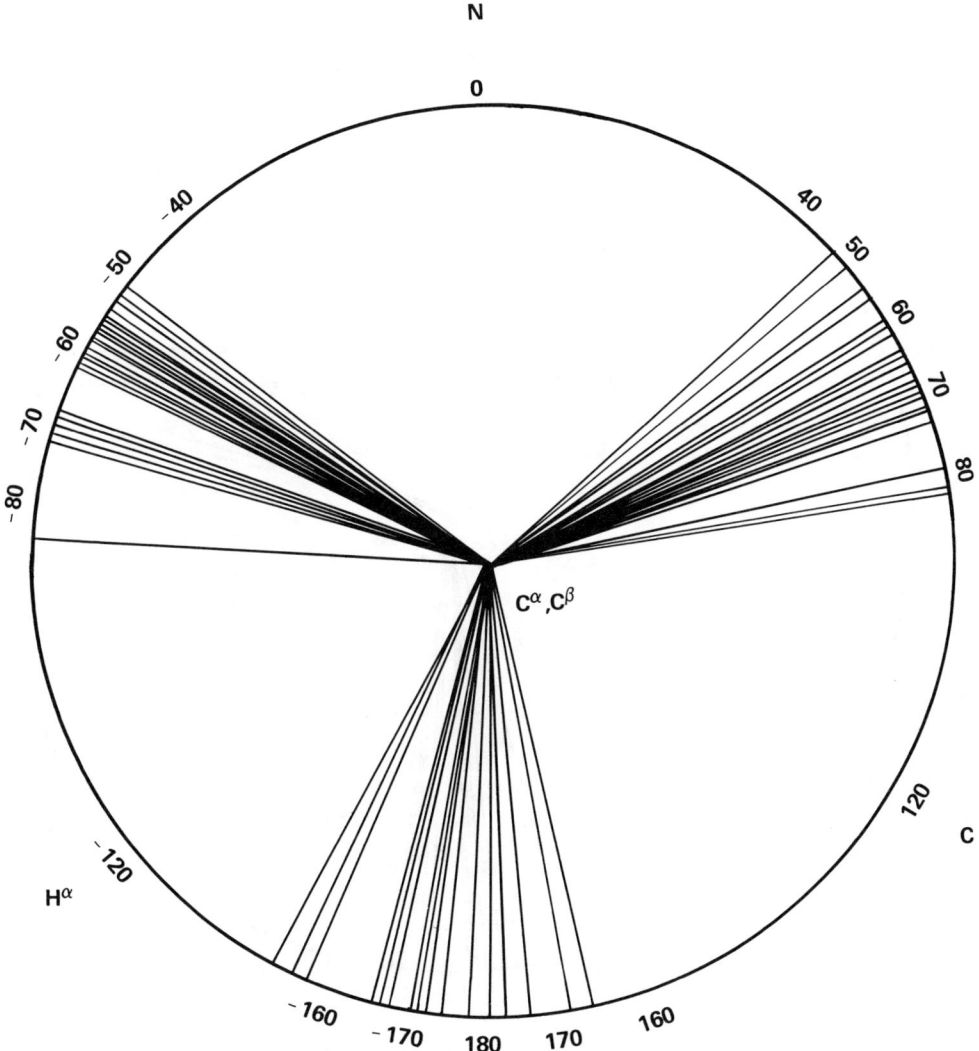

FIGURE 14. Distribution of the dihedral angle χ_1.[6-49]

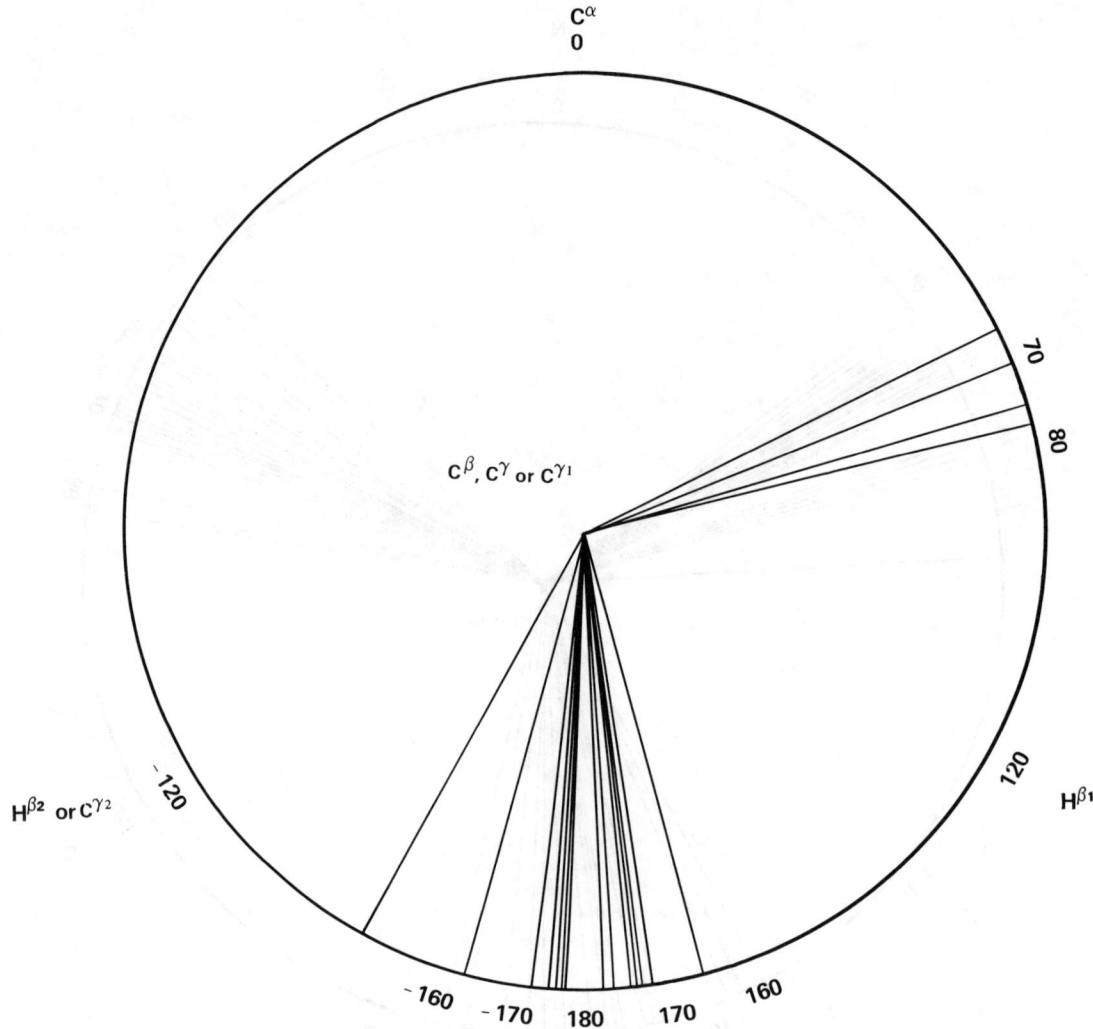

FIGURE 15. Distribution of the dihedral angle χ_2 in leucine, isoleucine, methionine, lysine, and argine.[9-13,30-37]

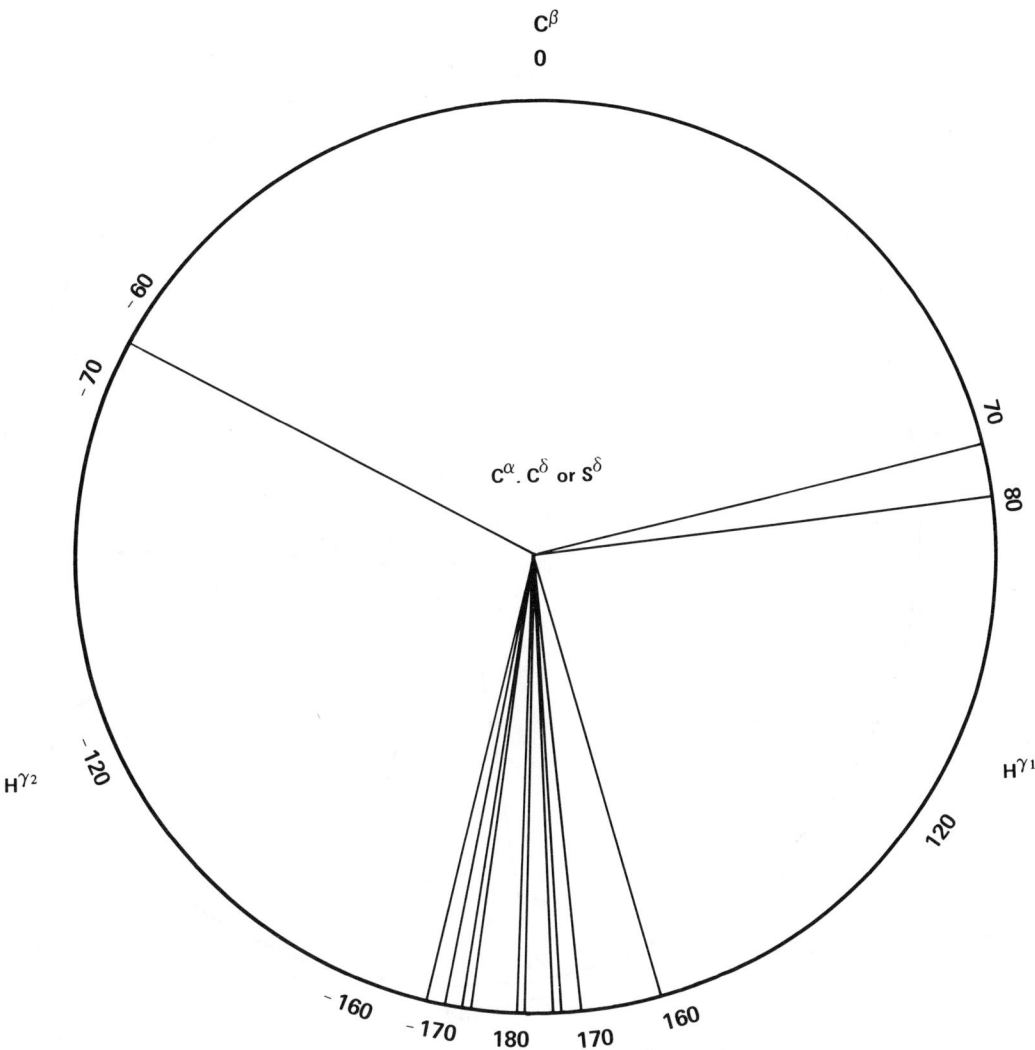

FIGURE 16. Distribution of the dihedral angle χ_3 in methionine, lysine, and arginine.[12,13,30-37]

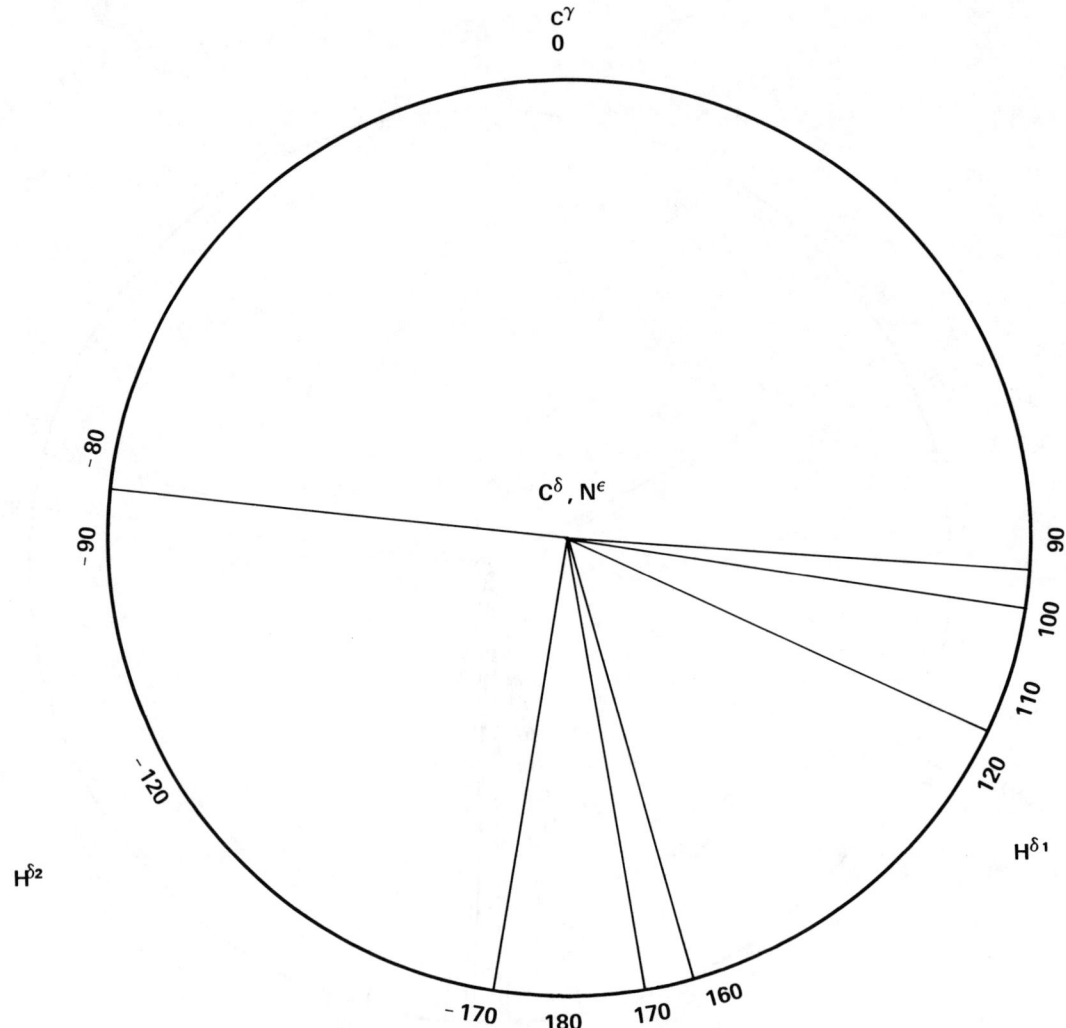

FIGURE 17. Distribution of the dihedral angle χ_4 in arginine.[33-37]

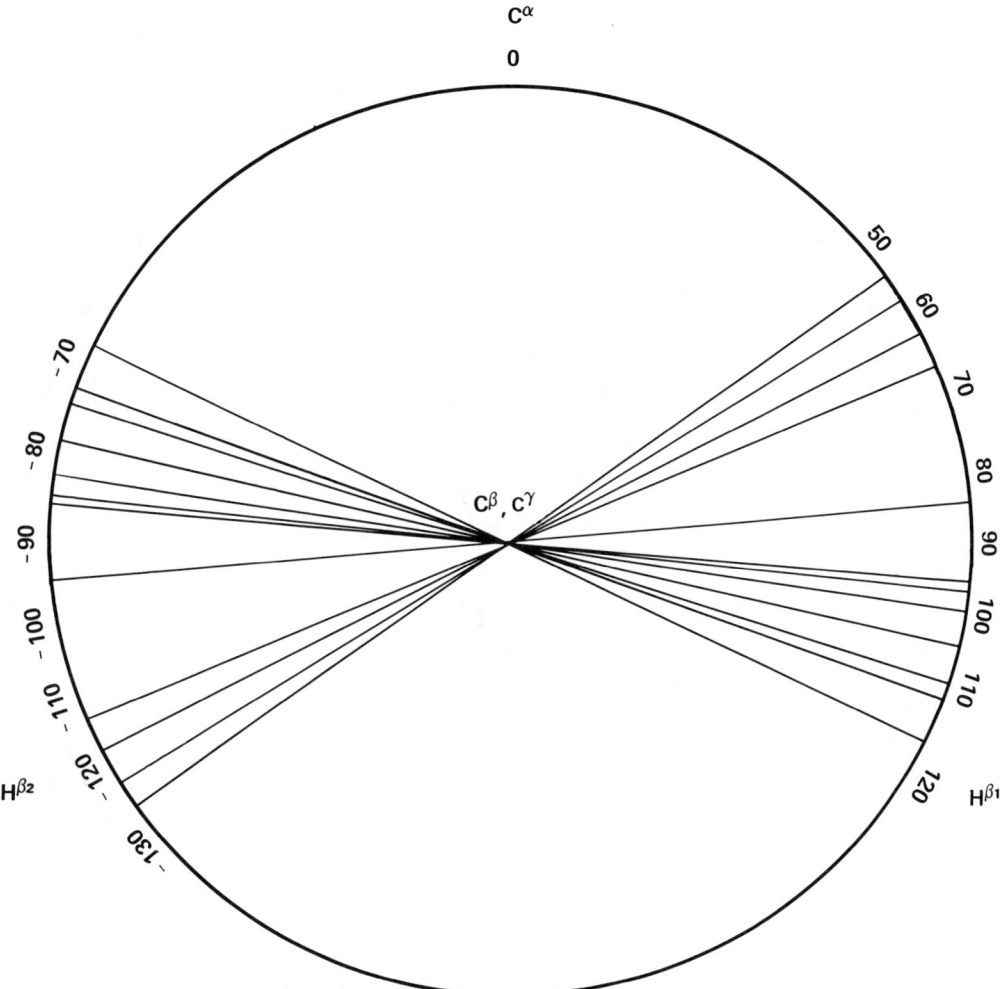

FIGURE 18. Distribution of dihedral angles χ_{21} and χ_{22} in phenylalanine, tyrosine, tryptophan, and histidine.[23-29,38-43]

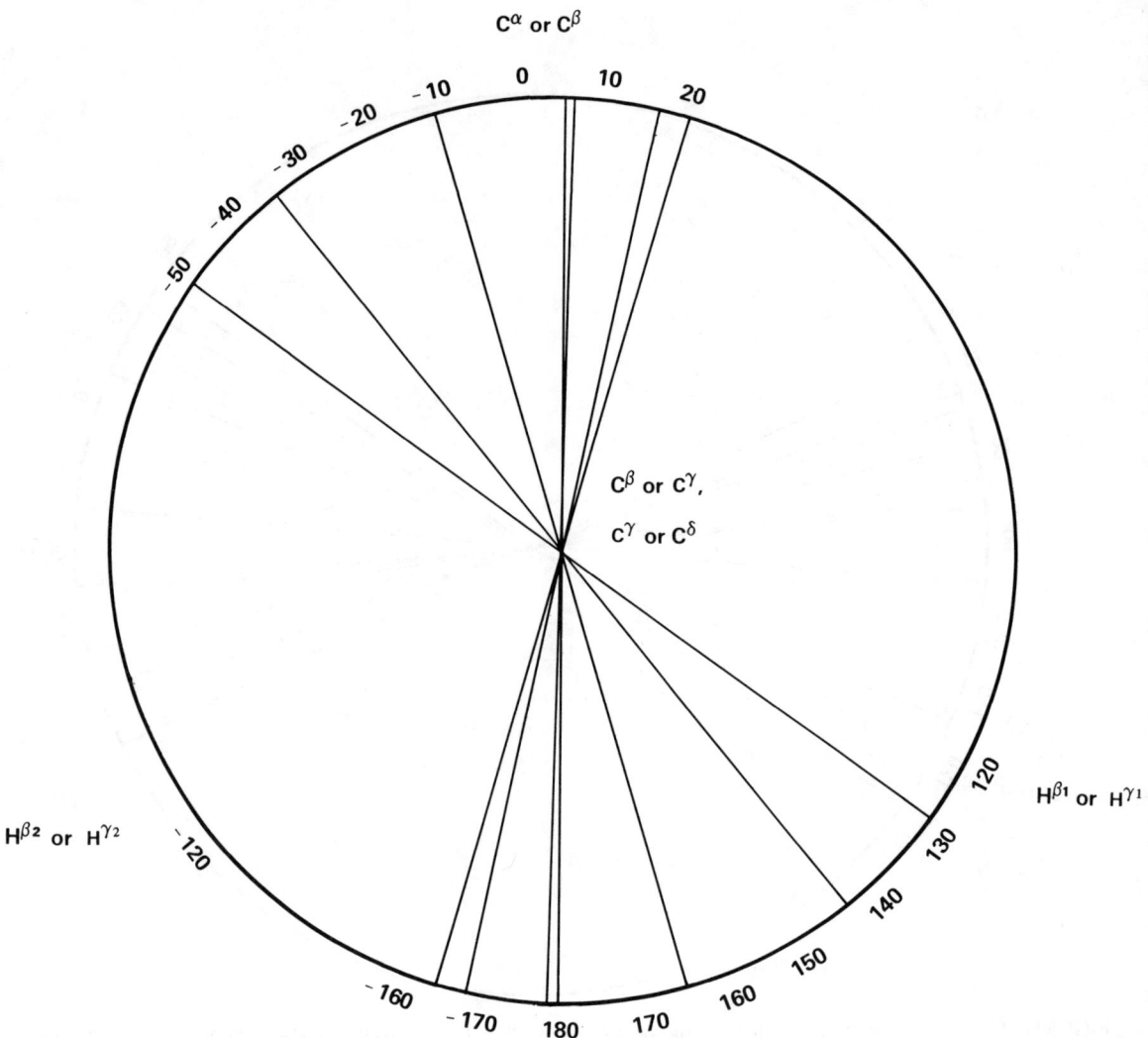

FIGURE 19. Distribution of dihedral angles χ_{21} and χ_{22} in aspartic acid and asparagine, and χ_{31} and χ_{32} in glutamic acid and glutamine.[44-49]

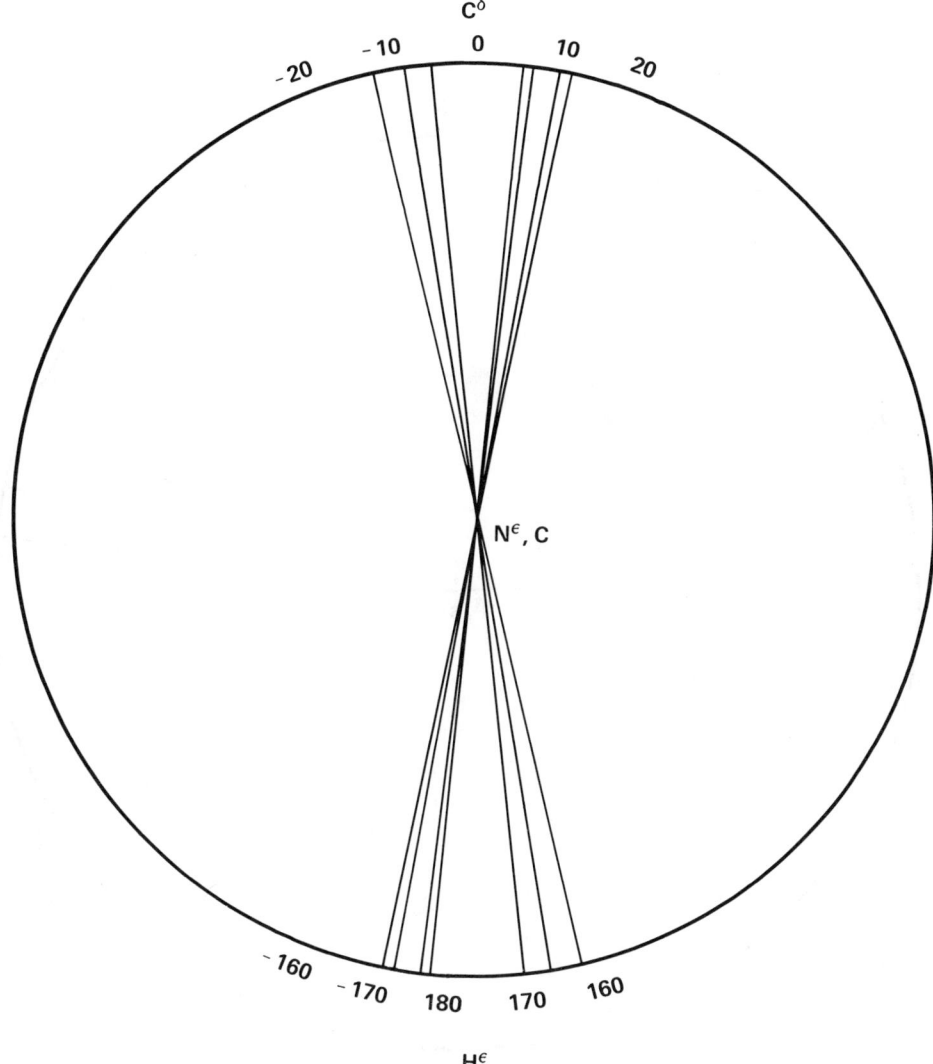

FIGURE 20. Distribution of dihedral angles $\chi_{5\,1}$ and $\chi_{5\,2}$ in arginine.[33-37]

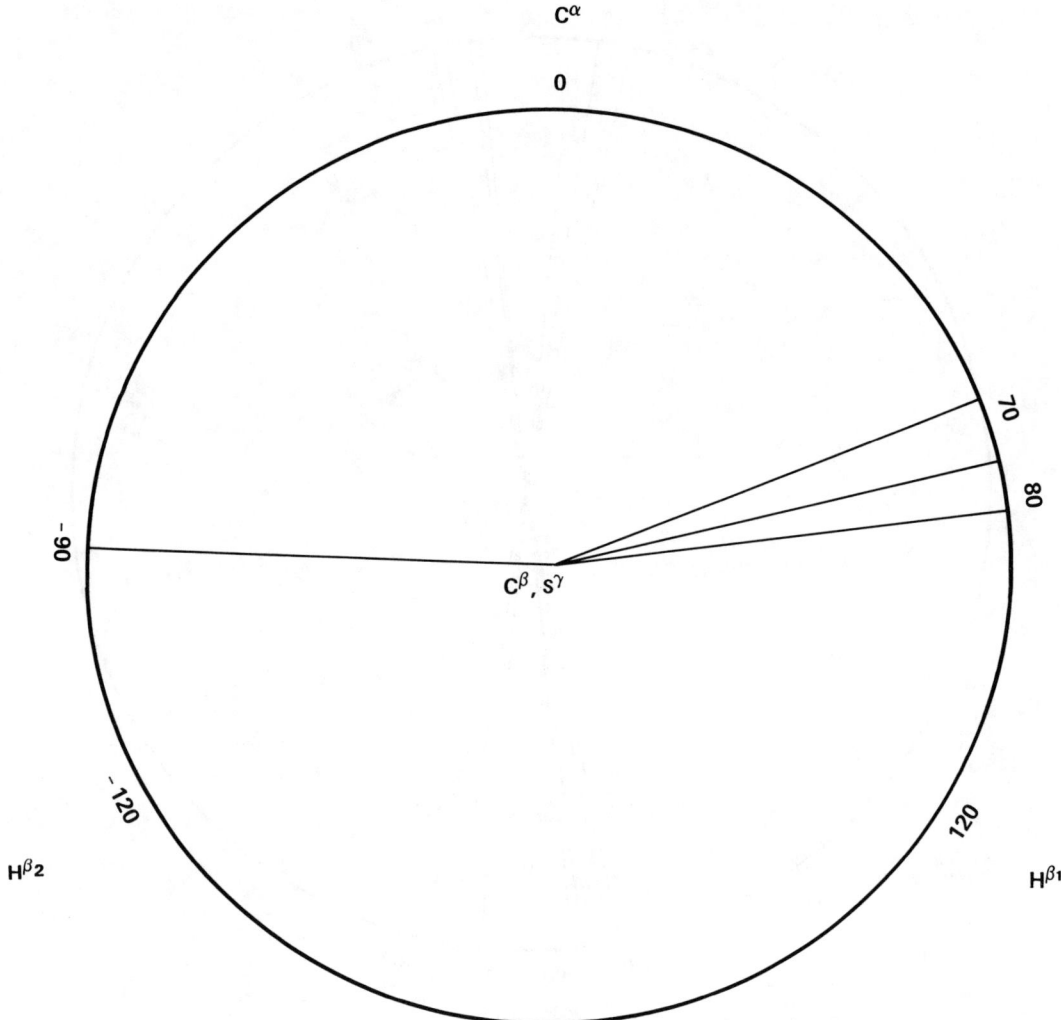

FIGURE 21. Distribution of the dihedral angle χ_2 in cystine.[17-19]

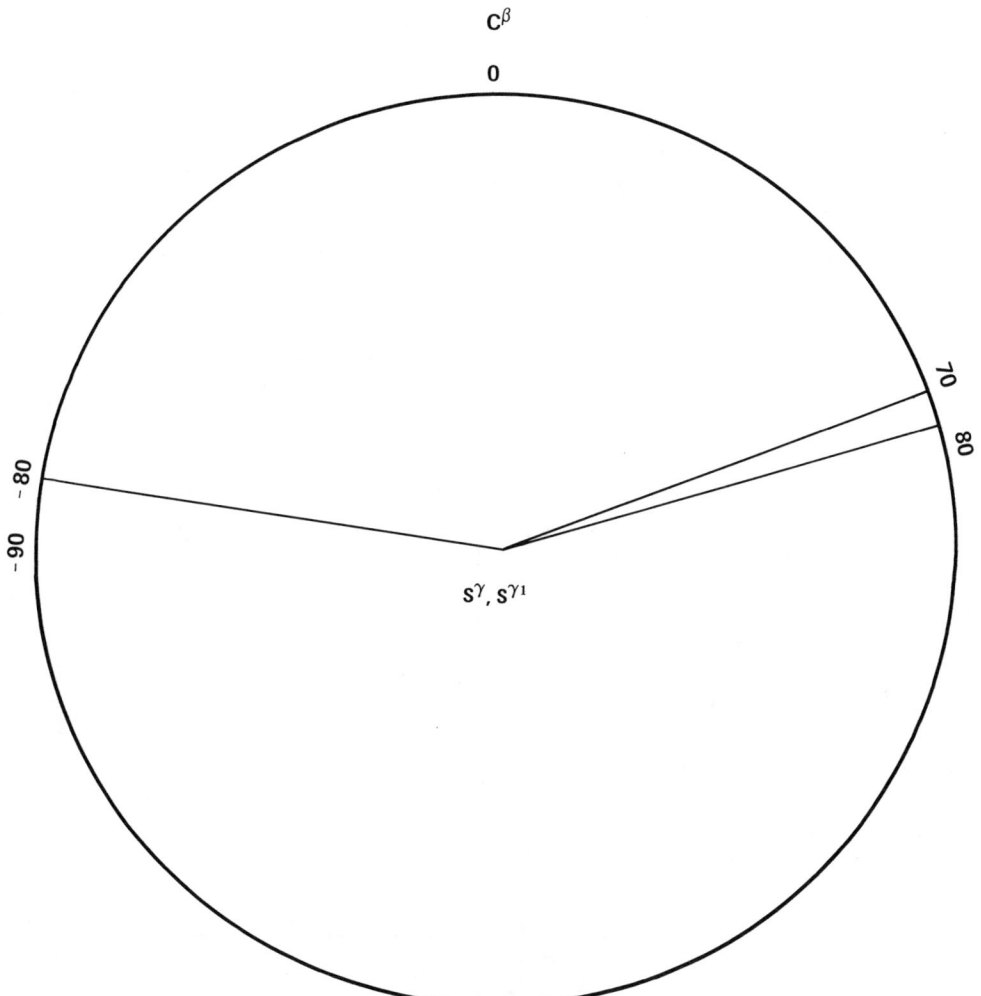

FIGURE 22. Distribution of the dihedral angle χ_3 in cystine.[17-19]

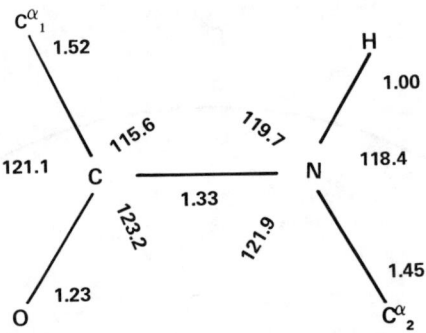

FIGURE 23. Bond lengths and angles in the peptide unit. (Adapted from Figure 1 in Ramachandran et al.[55])

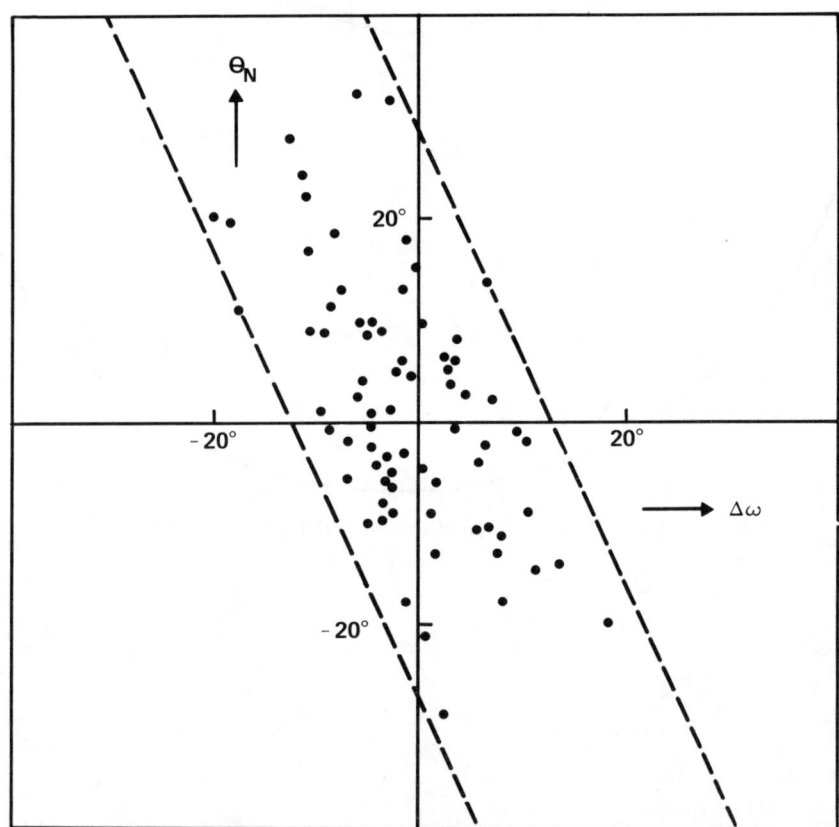

FIGURE 24. Distribution of dihedral angles θ_N and $\Delta\omega$, which define the nonplanarity of the peptide unit, found in small peptide structures. (Adapted from Figure 4a in Kolaskar et al.[56]) Note that most of the points roughly follow the correlated variation given by $\theta_N = -2\Delta\omega$.

Compiled by M. Vijayan.

REFERENCES

1. Jönsson and Kvick, *Acta Crystallogr. (B) (Kbh.)*, 28, 1827 (1972).
2. Iitaka, *Acta Crystallogr.*, 13, 35 (1960).
3. Iitaka, *Acta Crystallogr.*, 14, 1 (1961).
4. Lehmann, Koetzle, and Hamilton, *J. Am. Chem. Soc.*, 94, 2657 (1972).
5. Donohue, *J. Am. Chem. Soc.*, 72, 949 (1950).
6. Koetzle, Golic, Lehmann, Verbist, and Hamilton, *J. Chem. Phys.*, 60, 4690 (1974).
7. Torii and Iitaka, *Acta Crystallogr. (B) (Kbh.)*, 26, 1317 (1970).
8. Mallikarjunan and Thyagaraja Rao, *Acta Crystallogr. (B) (Kbh.)*, 25, 296 (1969).
9. Torii and Iitaka, *Acta Crystallogr. (B) (Kbh.)*, 27, 2237 (1971).
10. Trommel and Bijvoet, *Acta Crystallogr.*, 7, 703 (1954).
11. Howieson, M. Phil. thesis, University of Edinburgh, 1974.
12. Torii and Iitaka, *Acta Crystallogr. (B) (Kbh.)*, 29, 2799 (1973).
13. Mathieson, *Acta Crystallogr.*, 5, 332 (1952).
14. Kerr and Ashmore, *Acta Crystallogr. (B) (Kbh.)*, 29, 2124 (1973).
15. Harding and Lang, *Acta Crystallogr. (B) (Kbh.)*, 24, 1096 (1968).
16. Iyer, *Z. Kristallogr.*, 126, 227 (1968).
17. Gupta, Sequeira, and Chidambaram, *Acta Crystallogr. (B) (Kbh.)*, 30, 562 (1974).
18. Chaney and Steinrauf, *Acta Crystallogr. (B) (Kbh.)*, 30, 711 (1974).
19. Oughton and Harrison, *Acta Crystallogr.*, 12, 396 (1959).
20. Frey, Lehmann, Koetzle, and Hamilton, *Acta Crystallogr. (B) (Kbh.)*, 29, 876 (1973).
21. Benedetti, Pedone, and Sirigo, *Cryst. Struct. Commun.*, 1, 35 (1972).
22. Ramanadham, Sikka, and Chidambaram, *Pramana*, 1, 247 (1973).
23. Gurskaya, *Sov. Phys. Crystallogr.*, 9, 709 (1965).
24. Frey, Koetzle, Lehmann, and Hamilton, *J. Chem. Phys.*, 58, 2547 (1973).
25. Mostad, Nissen, and Rømming, *Acta Chem. Scand.*, 26, 3819 (1972).
26. Mostad and Rømming, *Acta Chem. Scand.*, 27, 401 (1973).
27. Bye, Mostad, and Rømming, *Acta Chem. Scand.*, 27, 471 (1973).
28. Takigawa, Ashida, Sasada, and Kakudo, *Bull. Chem. Soc. Jap.*, 39, 2369 (1966).
29. Gortland, Freeman, and Bugg, *Acta Crystallogr. (B) (Kbh.)*, 30, 1841 (1974).
30. Koetzle, Lehmann, Verbist, and Hamilton, *Acta Crystallogr. (B) (Kbh.)*, 28, 3207 (1972).
31. Bugayong, Sequeira, and Chidambaram, *Acta Crystallogr. (B) (Kbh.)*, 28, 3214 (1972).
32. Bhat and Vijayan, unpublished results.
33. Lehmann, Verbist, Hamilton, and Koetzle, *J. Chem. Soc. Perkin Trans. 2*, p. 133 (1973).
34. Saenger and Wagner, *Acta Crystallogr. (B) (Kbh.)*, 28, 2237 (1972).
35. Dow, Jensen, Mazumdar, Srinivasan, and Ramachandran, *Acta Crystallogr. (B) (Kbh.)*, 26, 1662 (1970).
36. Furberg and Solbakk, *Acta Chem. Scand.*, 27, 1226 (1973).
37. Mazumdar, Venkatesan, Mez, and Donohue, *Z. Kristallogr.*, 130, 328 (1969).
38. Madden, McGandy, Seeman, Harding, and Hoy, *Acta Crystallogr. (B) (Kbh.)*, 28, 2382 (1972).
39. Madden, McGandy, and Seeman, *Acta Crystallogr. (B) (Kbh.)*, 28, 2377 (1972).
40. Lehmann, Koetzle, and Hamilton, *Int. J. Pept. Protein Res.* 4, 229 (1972).
41. Edington and Harding, *Acta Crystallogr. (B) (Kbh.)*, 30, 204 (1974).
42. Bennett, Davidson, Harding, and Morelle, *Acta Crystallogr. (B) (Kbh.)*, 26, 1722 (1970).
43. Oda and Koyama, *Acta Crystallogr. (B) (Kbh.)*, 28, 639 (1972).
44. Derissen, Endeman, and Peerdeman, *Acta Crystallogr. (B) (Kbh.)*, 24, 1349 (1968).
45. Rao, *Acta Crystallogr. (B) (Kbh.)*, 29, 1718 (1973).
46. Lehmann, Koetzle and Hamilton, *J. Cryst. Mol. Struct.*, 2, 225 (1972).
47. Sequeira, Rajagopal, and Chidambaram, *Acta Crystallogr. (B) (Kbh.)*, 28, 2514 (1972).
48. Verbist, Lehmann, Koetzle, and Hamilton, *Acta Crystallogr. (B) (Kbh.)*, 28, 3006 (1972).
49. Koetzle, Frey, Lehmann, and Hamilton, *Acta Crystallogr. (B) (Kbh.)*, 29, 2571 (1973).
50. Kayushima and Vainshtein, *Sov. Phys. Crystallogr.*, 10, 698 (1966).
51. Mitsui, Tsuboi, and Iitaka, *Acta Crystallogr. (B) (Kbh.)* 25, 2182 (1969).
52. Koetzle, Lehmann, and Hamilton, *Acta Crystallogr. (B) (Kbh.)*, 29, 231 (1973).
53. Ponnuswamy and Sasisekharan, *Int. J. Protein Res.*, 2, 37 (1970).
54. Ponnuswamy and Sasisekharan, *Int. J. Protein Res.*, 3, 9 (1971).
55. Ramachandran, Kolaskar, Ramakrishnan, and Sasisekharan, *Biochim. Biophys. Acta,* 359, 298 (1974).
56. Kolaskar, Lakshminarayanan, Sarathy, and Sasisekharan, *Biochim. Biophys. Acta,* in press.

PROTEIN STRUCTURES

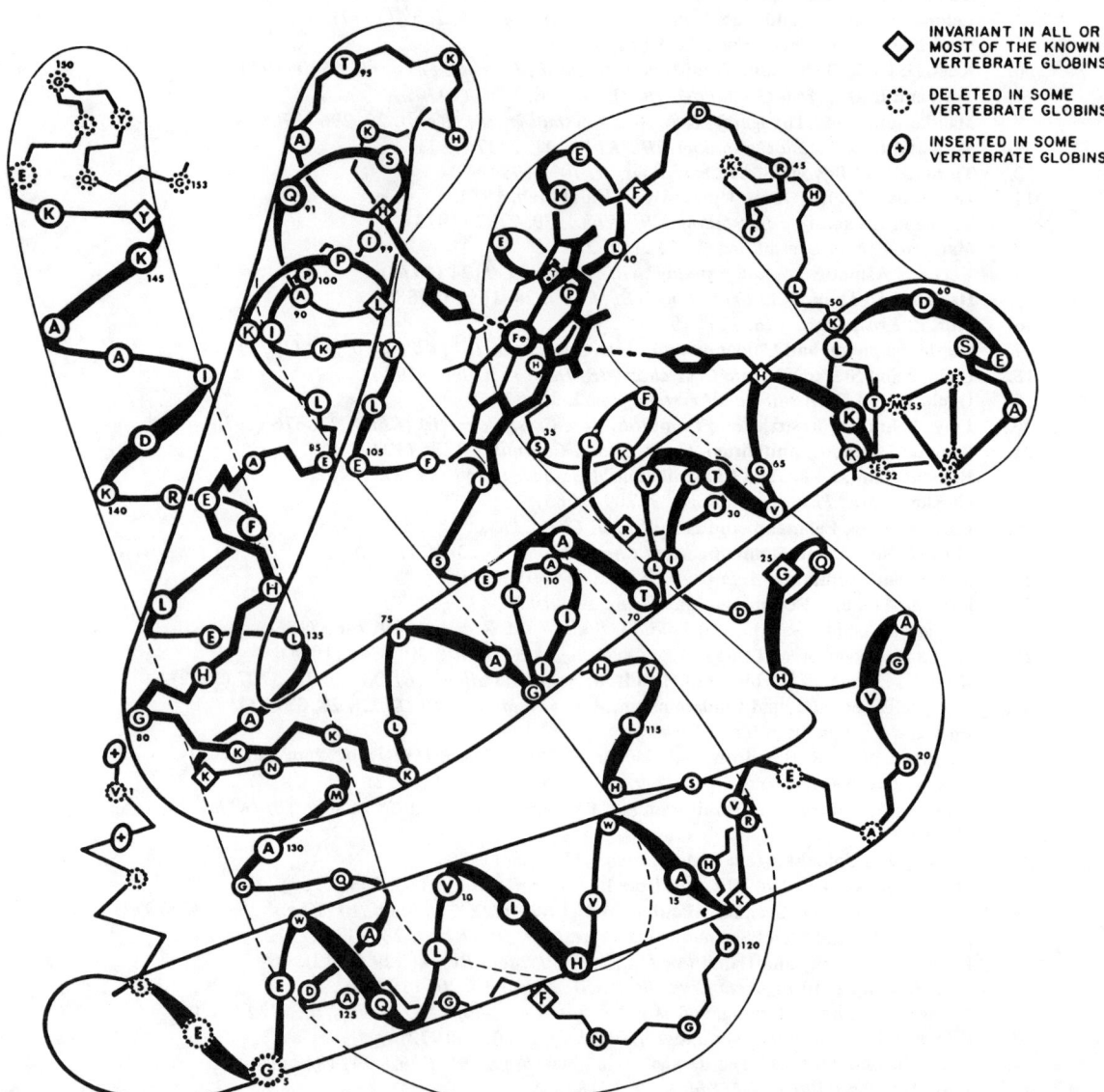

FIGURE 1. The three-dimensional conformation of sperm whale myoglobin determined by Kendrew and collaborators. Letters representing the amino acid side chains according to the single letter code are placed at the positions of the respective α-carbon atoms. Relatively invariant positions as well as insertions and deletions observed in vertebrate hemoglobin chains are indicated by the special symbols shown in the inset. (Adapted from Dickerson, in *The Proteins*, Vol. 2, Neurath, Ed., Academic Press, New York, 1964, 634. Reprinted from Dayhoff, *Atlas of Protein Sequence and Structure*, Vol. 5, National Biomedical Research Foundation, Washington, D.C., 1972, 27. With permission.)

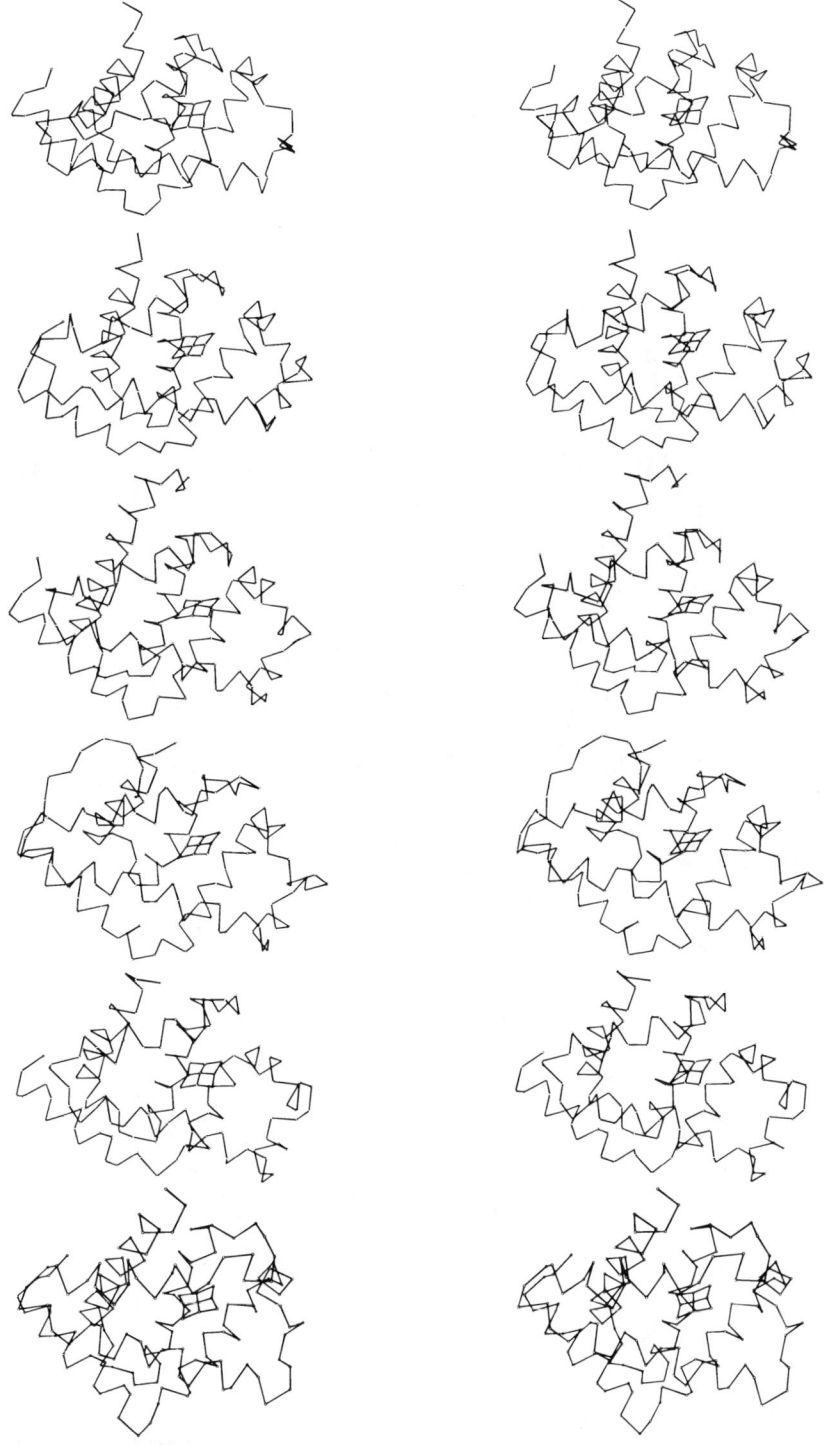

FIGURE 2. Stereo drawings of the backbones of six globins illustrating the way in which the fold of the polypeptide chain has been retained through evolution in spite of substantial changes in the amino acid sequence. From top to bottom: (a) α chain of horse methemoglobin (M. F. Perutz); (b) β chain of horse methemoglobin (M. F. Perutz); (c) sperm whale myoglobin (H. C. Watson and J. C. Kendrew); (d) lamprey hemoglobin (W. E. Love); (e) *Chironomus* hemoglobin (R. Huber); (f) *Glycera* hemoglobin (W. E. Love). (From Love, Klock, Lattmann, Padlan, Ward, Jr., and Hendrickson, *Cold Spring Harbor Symp. Quant. Biol.*, 36, 349 (1971). With permission.) (Stereo viewers may be purchased, for example, from Ward's Natural Science Establishment, Inc., Rochester, New York, Model 25W2951.)

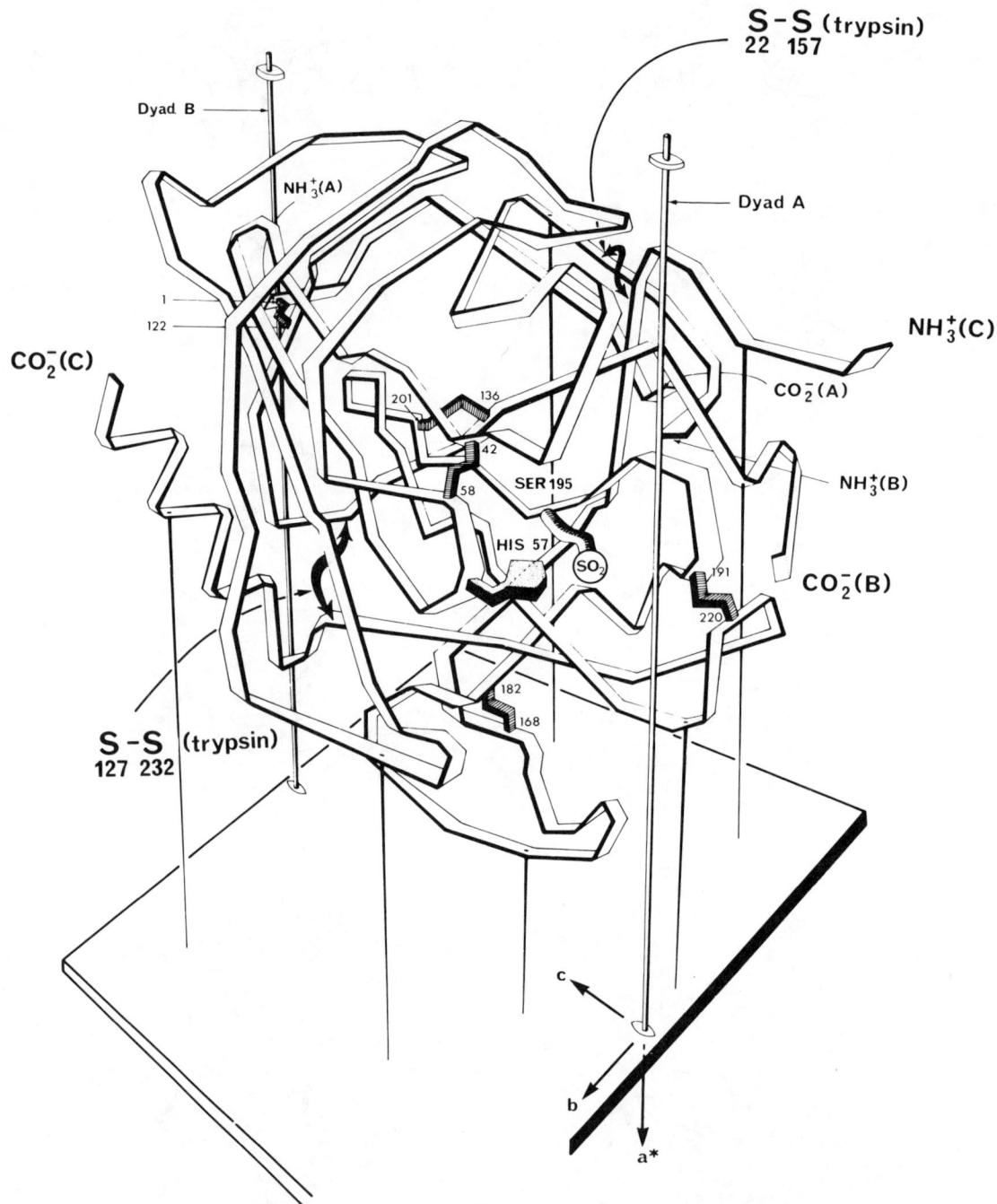

FIGURE 3. Schematic drawing representing the conformation of the polypeptide chains in α-chymotrypsin. The positions of amino acids in positions homologous to the two additional cystines in trypsin are shown suggesting that the polypeptide chain of trypsin is folded in a manner very similar to that of α-chymotrypsin. (From Sigler, Blow, Matthews, and Henderson, *J. Mol. Biol.*, 35, 143 (1968). With permission.)

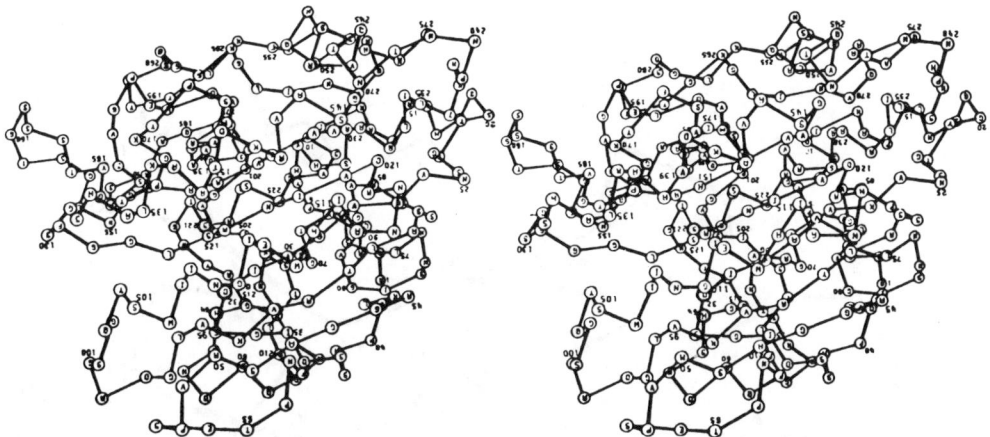

FIGURE 4. Stereo pair illustrating the conformation of subtilisin, an extracellular protease from the bacterium *Bacillus amyloliquefaciens*. The overall three-dimensional structure of subtilisin is totally dissimilar to that of α-chymotrypsin (Figure 3), yet the active sites of these two proteases are strikingly similar. (From Dayhoff, *Atlas of Protein Sequence and Structure,* Vol. 5, National Biomedical Research Foundation, Washington, D.C., 1972, D-115. With permission.)

FIGURE 5. Perspective drawing illustrating the backbone of bovine carboxypeptidase A. The zinc atom essential for catalytic activity is shown as a stippled ball. (From Lipscomb, Hartsuck, Quiocho, and Reeke, Jr., *Proc. Natl. Acad. Sci. U.S.A.,* 64, 28 (1969).)

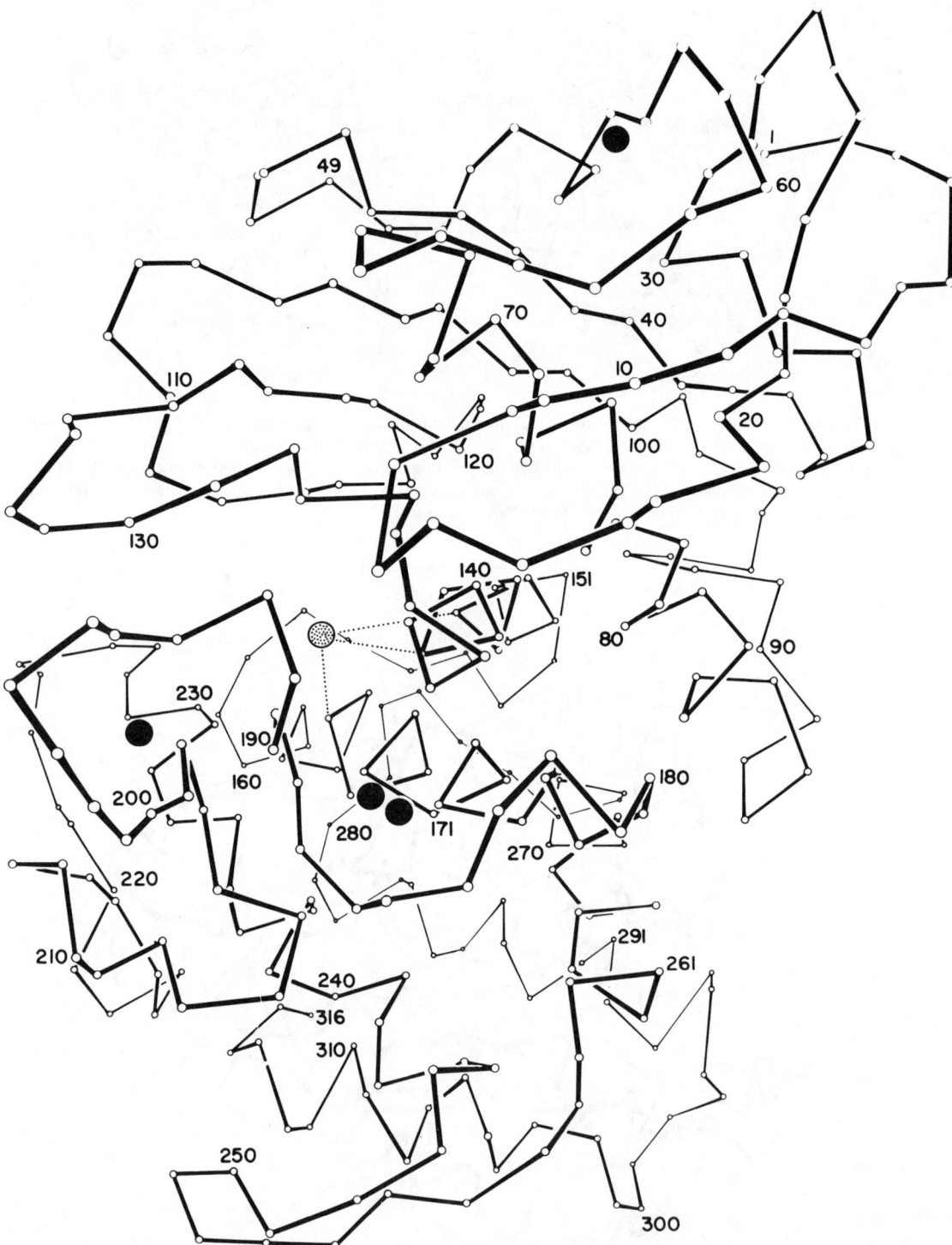

FIGURE 6. Perspective drawing illustrating the conformation of thermolysin, a thermostable extracellular protease from *Bacillus thermoproteolyticus*. The zinc ion at the active site is drawn stippled, while four calcium ions essential for thermostability are drawn as solid circles. As in the case for α-chymotrypsin and subtilisin, the overall conformations of carboxypeptidase A and thermolysin are quite different; in this instance, however, the respective active sites of these two proteases have some elements in common, but also have striking differences. (From Colman, Jansonius, and Matthews, *J. Mol. Biol.*, 70, 701 (1972). With permission.)

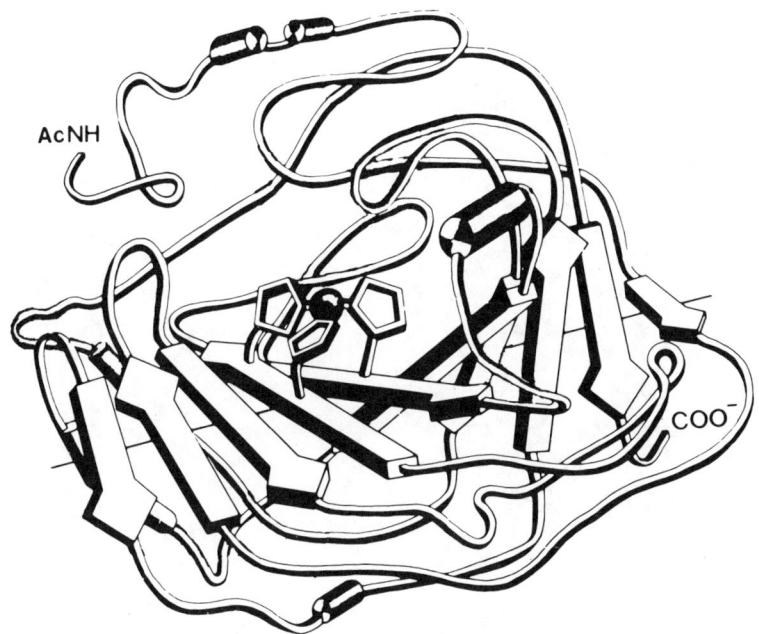

FIGURE 7. Schematic illustration of the three-dimensional structure of human erythrocyte carbonic anhydrase C. The helices are represented by cylinders and regions of β-sheet by arrows. The active site zinc, liganded by three histidines, is also shown. (From Kannan, Notstrand, Fridborg, Lövgren, and Petef, *Proc. Natl. Acad. Sci. U.S.A.*, 72, 51 (1975).)

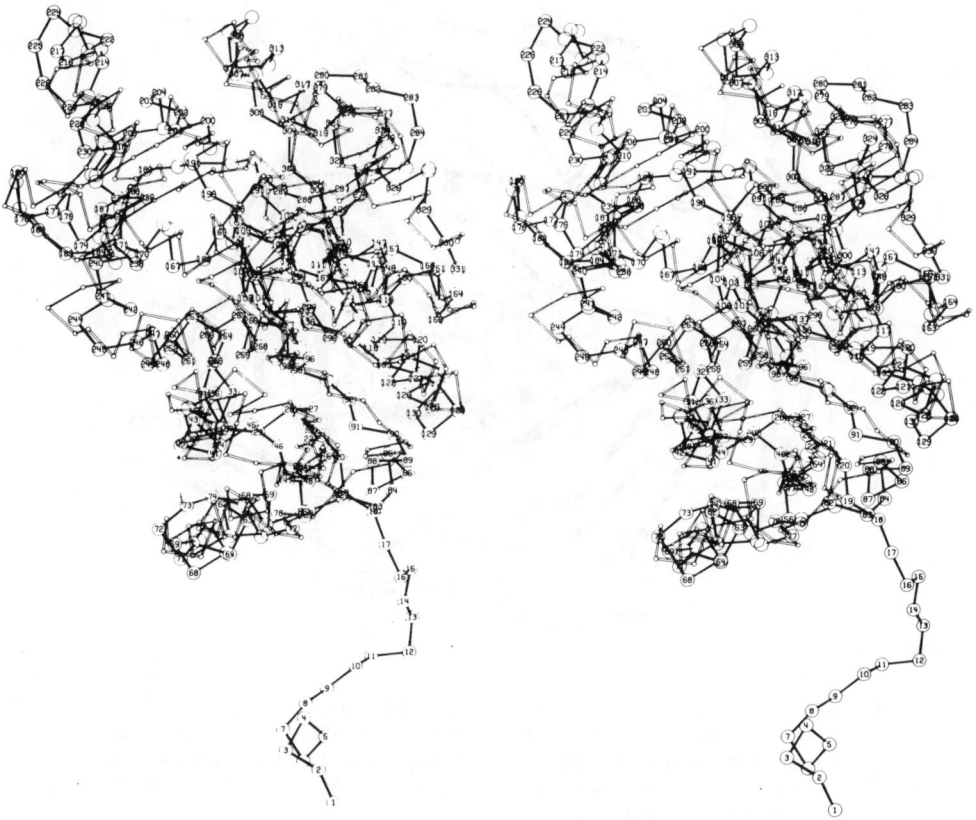

FIGURE 8. Stereo pair illustrating the similarity in the three-dimensional structures of the respective subunits of dogfish lactate dehydrogenase (dark bonds) and soluble porcine malate dehydrogenase. The malate subunits normally associate to form dimers whereas the lactate subunits form tetramers. The difference in aggregation is presumably due in part to the long "tail," present only in the lactate subunits, which interacts extensively with other subunits of the tetramer. (From Rao and Rossmann, *J. Mol. Biol.*, 76, 241 (1973). With permission.)

Compiled by B. W. Matthews.

Index

INDEX

A

Abbreviations
  amino acid derivatives, journal reference to rules, 87
  amino acids, table of, 62
  chemical names, journal reference to rules, 87
  distinguished from symbols, 6
  immunoglobulins, 179
  nucleic acids, journal reference to, 87
  peptide hormones, list, 177–178
  peptides, journal reference to rules, 87
  polynucleotides, journal references to, 87
  polypeptide chains, conformation, journal reference, 87
  polypeptide chains, nomenclature rules, 63–77
  polypeptides, synthetic, journal reference, 87
  use of in biochemical nomenclature, 6–7
Absorbance, UV and visible region, table of values, 383–514
Absorption coefficient, molar, see Molar absorptivity
Acetoacetate decarboxylase
  molar absorptivity and/or absorbance values, 383
Acetylcholinesterase
  molar absorptivity and/or absorbance values, 383
  subunit, constitution of, 344
Acetyl hexosamine, content of, in glyco-proteins, 258–274
β-D-N-Acetylhexose amidase
  placenta, human
    subunit, constitution of, 333
O-Acetylserine sulfhydrase A
  molar absorptivity and/or absorbance values, 383
Achiral molecule, nomenclature for, 36–58
Acid phosphomonoesterase I
  prostate, human
    subunit, constitution of, 333
Actin
  identification of, in nonmuscle cells, table, 307–313
  molar absorptivity and/or absorbance values, 384
  myosins, nonmuscle, interaction, 320
  nonmuscle, interaction of several with myosin, 316
  nonmuscle, physical chemical properties of several, 314–315
  refractive index increments, 372
F-Actin
  molar absorptivity and/or absorbance values, 384
G-Actin
  molar absorptivity and/or absorbance values, 384
  molecular parameters, 306
α-Actinin
  molecular parameters, 306
Activation energy
  myosins, nonmuscle, table of values, 319
Active site titrations
  discussion, 672
  reagents, 681
Actomyosin
  identification of, in nonmuscle cells, table, 307–313
  refractive index increments, 372
Acyl carrier protein
  cyanogen bromide cleavage, 199
  molar absorptivity and/or absorbance values, 384
Acylphosphatase
  molar absorptivity and/or absorbance values, 384
Adenine phosphoribosyltransferase
  erythrocyte, human
    subunit, constitution of, 326
Adenosine triphosphate sulfurylase
  molar absorptivity and/or absorbance values, 385
Adenovirus
  molar absorptivity and/or absorbance values, 385
Adrenodoxin
  molar absorptivity and/or absorbance values, 385
  table of cofactors, molecular weights, stoichiometry and sources, 279
Aequorin
  molar absorptivity and/or absorbance values, 385
Agglutinin
  carbohydrate content, 270
  molar absorptivity and/or absorbance values, 385
  subunit, constitution of, 327, 335
Alanine
  symbols for atoms and bonds in side chains, 73
Albumin
  molar absorptivity and/or absorbance values, 386–387
  molecular parameters, 242
  refractive index increments, 372–373
Albumin, plasma
  ligand binding
    drugs, list of references, 568–574
    endogenous ligands, list of references, 556–557
    fatty acids, list of references, 558–562
    inorganic anions, list of references, 563
    measurement of, list of techniques, 555
    metal ions, list of references, 563
    steroids, list of references, 563–564
    visible dyes, list of references, 564–568
Alcohol dehydrogenase
  molar absorptivity and/or absorbance values, 388
  subunit, constitution of, 328, 330, 337, 338
  table of cofactors, molecular weights, stoichiometry and sources, 284
Aldehyde dehydrogenase
  molar absorptivity and/or absorbance values, 388
  subunit, constitution of, 343
Aldehyde oxidase
  molar absorptivity and/or absorbance values, 389
  table of cofactors, molecular weights, stoichiometry and sources, 278
Aldolase
  cofactors, molecular weights, stoichiometry, sources, 284
  cyanogen bromide cleavage, 199
  enzymatic activity, subcellular fractions, 698, 715
  molar absorptivity and/or absorbance values, 389
  subunit, constitution of, 331, 335, 338, 377
  viscosity, intrinsic, 721

Aldolase C
  subunit, constitution of, 338
Aldolase, L-Rhamnulose l-phosphate
  molar absorptivity and/or absorbance values, 390
Aldolase, 3-deoxy-2-keto-6-phosphogluconate
  molar absorptivity and/or absorbance values, 390
Aldose l-epimerase
  molar absorptivity and/or absorbance values, 390
Alkaline phosphatase
  carbohydrate content, 274
  molecular weights, stoichiometry and sources, 284
  subunit, constitution of, 331, 335, 337
Alliin lyase
  molar absorptivity and/or absorbance values, 391
Amandin
  molar absorptivity and/or absorbance values, 392
  refractive index increments, 373
Amine oxidase
  molar absorptivity and/or absorbance values, 392
  table of cofactors, molecular weights, stoichiometry and sources, 282
Amino acid composition
  proteinase inhibitors
    animal sources, table, 649–653
    microbial sources, 664
    plant sources, table, 605–610
  retinol-binding protein, human, 303
  Tamm-Horsfall glycoprotein, 303
  urokinase, 302
Amino acid group
  dimensions of, 742–758
D-Amino-acid oxidase
  molar absorptivity and/or absorbance values, 392
  subunit, constitution of, 333
L-Amino-acid oxidase
  molar absorptivity and/or absorbance values, 392
  subunit, constitution of, 336
Amino acid sequences
  β_2 microglobulin, 305
  proteinase inhibitors
    animal sources, table, 656–659
    microbial source, 665
    plant sources, 611–617
  rules for one-letter notation, 59–62
Amino acid side chains
  dimensions of, 742–758
Amino acids
  abbreviations, table of, 62
  peptic hydrolysis in human α-and β-chains, 216
  pK values for several, in proteins, tables, 689–696
  polymerized, use of symbols for, 13–14
  residue, notation, 63–64
  sequences, rules for one-letter notation, 59–62
  symbols for, use of, 8–9
ϵ-Amino groups
  pK values in proteins, table, 694
α-Amino groups
  pK values in proteins, table, 693
Aminopeptidase
  enzymatic hydrolysis, conjugated proteins, 211
  molar absorptivity and/or absorbance values, 393
  molecular weights, stoichiometry and sources, 285
  subunit, constitution of, 346, 347

Aminopeptidase (microsomal)
  molar absorptivity and/or absorbance values, 393
Aminopeptidase P
  molar absorptivity and/or absorbance values, 393
Amino protein group, chemical modification of
  reagents used, table of specificities, 203–204
AMP Deaminase
  subunit, constitution of, 344
Amylase
  enzymatic activity, subcellular fractions, 698, 714
  molar absorptivity, and/or absorbance values, 393
  table of cofactors, molecular weights, stoichiometry and sources, 278
α-Amylase
  carbohydrate content, 274
  molar absorptivity and/or absorbance values, 393
  refractive index increments, 374
  subunit, constitution of, 332
β-Amylase
  molar absorptivity and/or absorbance values, 394
  subunit, constitution of, 340
Analgesics, ligands binding to plasma albumin, 570
Anthranilate synthase
  molar absorptivity and/or absorbance values, 394
  subunit, constitution of, 331
Antibiotics
  ligands binding to plasma albumin, 568–569
Anticoagulants
  ligands binding to plasma albumin, 571
Antigen
  molar absorptivity and/or absorbance values, 395
Antihemophilic factor
  physical data and characteristics, 254–255
Antithrombin III
  animal sources, amino acid composition, 653
  carbohydrate composition, 666–667
  mammalian and chicken blood, specificity and properties, 636–637
  molecular parameters, 243
α_1-Antitrypsin
  animal sources, amino acid composition, 653
  carbohydrate composition, 666–667
  mammalian and chicken blood, specificity and properties, 636–637
  molecular parameters, 242
Apolipoprotein Glu-II
  molar absorptivity and/or absorbance values, 395
α-L-Arabinofuranosidase
  molar absorptivity and/or absorbance values, 395
Arachin
  molar absorptivity and/or absorbance values, 395
  refractive index increments, 374
  subunit, constitution of, 346
Arginase
  enzymatic activity, subcellular fractions, 698
  molar absorptivity and/or absorbance values, 395
  subunit, constitution of, 335
Arginine
  symbols for atoms and bonds in side chains, 74
Arginine decarboxylase
  molar absorptivity and/or absorbance values, 395
  subunit, constitution of, 345, 350

Arginine kinase
  molar absorptivity and/or absorbance values, 395
Argininosuccinase
  molar absorptivity and/or absorbance values, 396
  subunit, constitution of, 340
Asparaginase
  molar absorptivity and/or absorbance values, 396
Asparagine
  symbols for atoms and bonds in side chains, 73
Aspartate aminotransferase
  molar absorptivity and/or absorbance values, 397
Aspartate transcarbamoylase
  molar absorptivity and/or absorbance values, 397
Aspartic acid
  symbols for atoms and bonds in side chains, 73
Aspartokinase
  molar absorptivity and/or absorbance values, 398, 450
  subunit, constitution of, 335, 336
Aspergillopeptidase A, B
  molar absorptivity and/or absorbance values, 398
ATPase activity
  enzymatic, subcellular fractions, 713
  myosins, nonmuscle, table of values, 319
Avidin
  carbohydrate content, 268
  molar absorptivity and/or absorbance values, 398
  subunit, constitution of, 329
Axial ratio
  Tamm-Horsfall mucoprotein, 304
Azurin
  cyanogen bromide cleavage, 199
  molar absorptivity and/or absorbance values, 399
  molecular weights, stoichiometry and sources, 282

B

Barbiturates, ligands binding to plasma albumin, 569
Bibliographic references in biochemical journals, CEBJ recommendations, 17–20
Binding sites, number of, for purified lectins, 547–551
Biochemical nomenclature, see Nomenclature, biochemical
Bird egg whites
  proteinase inhibitors from, specificities and properties, 624–629
Blood coagulation factors, nomenclature, 195
Blood coagulation proteins
  table of data and characteristics, 254–255
Blood, mammalian and chicken
  proteinase inhibitors from, specificity and properties, 636–641
Blood plasma
  fibrin stabilizing factor system of, 670
Boat conformation in stereochemistry, 42–43
Bond angles
  amino acid side chains, amino acid group, peptide linkage, 742–758
Bond lengths
  amino acid side chains, amino acid group, peptide linkage, 742–758

Bowman-Birk inhibitor
  legume sources, specificity and properties, 588–589
  plant sources, amino acid composition, 606
Brain
  molar absorptivity and/or absorbance values, 399
Bromelain
  carbohydrate content, 274
  molar absorptivity and/or absorbance values, 399

C

Caffeine, ligands binding to plasma albumins, 573
Calcitonin
  molar absorptivity and/or absorbance values, 399
Carbamoylphosphate synthase
  molar absorptivity and/or absorbance values, 400
  subunit, constitution of, 345
Carbohydrate composition
  proteinase inhibitors, table of selected, 666–667
  Tamm-Horsfall glycoprotein, 303
Carbohydrate content
  glycoproteins
    blood group substances, table, 262–263
    body fluids, 267
    cell membranes, 266
    eggs, table of values, 268–269
    enzymes, table of values, 273–274
    hormones, table of values, 271–272
    milk, table of values, 268–269
    mucins, table, 262–263
    plants, table of values, 270
    plasma, human, table of values, 258
    plasma, nonhuman, table of values, 259–260
    tissues, table of values, 264–265
    viruses, enveloped, 267
  lectins, purified, table of values, 546–551
  proteinase inhibitors
    animal sources, 618–641
    cereal grains, 598–599
Carbohydrate nomenclature, journal references, 87
Carbohydrate-protein linkages, types, 257
Carbohydrates, protein
  physical-chemical data for, index to ultracentrifuge studies, 232
Carbonic anhydrase
  molar absorptivity and/or absorbance values, 400–401
  molecular weights, stoichiometry and sources, 283
  subunit, constitution of, 339
Carbonic anhydrase C, human erythrocyte structure, schematic illustrating, 765
Carboxylesterase
  carbohydrate content, 274
  molar absorptivity and/or absorbance values, 402
  subunit, constitution of, 339
Carboxyl protein group, chemical modification of reagents used, table of specificities, 203–204
S-Carboxymethylboxymethylkerateine 2
  refractive index increments, 374

Carboxypeptidase
  molar absorptivity and/or absorbance values, 402
  molecular weights, stoichiometry and sources, 284
Carboxypeptidase A
  cyanogen bromide cleavage, 199
  enzymatic hydrolysis of proteins, 209
  molar absorptivity and/or absorbance values, 402
  molecular weights, stoichiometry and sources, 283
  structure, drawing of, 763
Cardiac glycosides, ligands binding to plasma albumins, 572
Casein
  carbohydrate content, 268, 269
  molar absorptivity and/or absorbance values, 403
β-Casein
  refractive index increments, 374
$\alpha_{s1,2}$-Caseins
  refractive index increments, 374
Catalase
  enzymatic activity, subcellular fractions, 699, 710
  molar absorptivity and/or absorbance values, 403
  subunit, constitution of, 343
Cellulase
  molar absorptivity and/or absorbance values, 404
Cenuloplasmin
  molecular parameters, 245
Cereal grains, proteinase inhibitors from, specificities and properties, 598–599
Cerebrocuprein
  molar absorptivity and/or absorbance values, 404
Ceruloplasmin
  human
    molar absorptivity and/or absorbance values, 404
    molecular weight, stoichiometry and sources, 281
  pig
    molar absorptivity and/or absorbance values, 405
  plasma, human
    carbohydrate content, molecular weights, sedimentation coefficients, 258
  rabbit
    molar absorptivity and/or absorbance values, 405
  rat
    molar absorptivity and/or absorbance values, 405
  serum, human
    subunit, constitution of, 336
Chain weight
  contractile proteins, table of values, 306
Chair conformation in stereochemistry, 42
Chirality
  definition and nomenclature rules, 34–58
  sequence rule, use of, 65
Chlorocruorin
  refractive index increments, 374
Chloroperoxidase
  carbohydrate content, 273
Cholecystokinin-pancreozymin
  cyanogen bromide cleavage, 199
Cholinesterase
  carbohydrate content, 273
  molar absorptivity and/or absorbance values, 405
  molecular parameters, 245
  subunit, constitution of, 345

Chondroitin sulfate
  table of data and characteristics, 686
Chorionic gonadotropin
  molar absorptivity and/or absorbance values, 406
  subunit, constitution of, 327
Chorismate mutase
  molar absorptivity and/or absorbance values, 406
Christmas factor
  physical data and characteristics, 254–255
Chymotrypsin
  cyanogen bromide cleavage, 199
  enzymatic hydrolysis, conjugated proteins, 211
  enzymatic hydrolysis of proteins, 209
  molar absorptivity and/or absorbance values, 406
  specificity for hydrolysis of peptide bonds, 212
Chymotrypsin A
  molar absorptivity and/or absorbance values, 406
Chymotrypsin C
  molar absorptivity and/or absorbance values, 406
Chymotrypsin II, human
  molar absorptivity and/or absorbance values, 406
α-Chymotrypsin
  conformation, schematic drawing of, 762
  molar absorptivity and/or absorbance values, 406
  refractive index increments, 374
Chymotrypsin inhibitor
  lower animal sources, specificity and properties, 620–621
Chymotrypsin inhibitor I
  subunit, constitution of, 327
Chymotrypsinogen
  molar absorptivity and/or absorbance values, 407
  refractive index increments, 374
  viscosity, intrinsic, 721
Chymotrypsinogen A, B, C
  molar absorptivity and/or absorbance values, 407
α-Chymotrypsinogen
  molar absorptivity and/or absorbance values, 407
Chymotryptic hydrolysis
  human globin, α-chains of, 213
  proteins, condition for, 214
Citation of bibliographic references in biochemical journals, CEBJ recommendations, 17–20
Citrate synthase
  molar absorptivity and/or absorbance values, 407
  subunit, constitution of, 333, 343
Coagulation factors
  plasma, human
    carbohydrate content, molecular weight, sedimentation coefficients, 258
  plasma, nonhuman
    carbohydrate content, 260
Cocoonase
  molar absorptivity and/or absorbance values, 408
Coenzymes
  abbreviations for, use of, 6
  nomenclature, journal references for, 87
Collagen
  cyanogen bromide cleavage, 199, 202
  refractive index increments, 374
  subunit, constitution of, 342

Collagenase
 enzymatic hydrolysis of proteins, 209
 molar absorptivity and/or absorbance values, 408
Conalbumin
 carbohydrate content, 268
 molar absorptivity and/or absorbance values, 408
 molecular weight, stoichiometry, source, function, 276
 refractive index increments, 374
Configuration
 polypeptide chemistry, use of term, 77
Conformation
 polypeptide chemistry, use of term, 77
 proteins, drawings illustrating, 760–766
Conformational map, typical, 71
Conformer molecule in stereochemistry, 42–46
Contractile proteins
 molecular parameters, table, 306
Corrinoids
 nomenclature, journal references for, 87
Corticosteroid-transporting protein
 plasma, nonhuman, carbohydrate content, 259
Creatine kinase
 molar absorptivity and/or absorbance values, 409
 subunit, constitution of, 330
Crotonase
 molar absorptivity and/or absorbance values, 409
 subunit, constitution of, 338, 339
Crotoxin
 molar absorptivity and/or absorbance values, 409
Cryoglobulin
 molar absorptivity and/or absorbance values, 409
 refractive index increments, 374
Crystallin
 molar absorptivity and/or absorbance values, 410, 411
Cuproproteins
 nomenclature rules, 121
Cyanogen bromide cleavage, peptides and proteins, table, 199–202
Cyclic polypeptides, see Polypeptides, cyclic
Cyclitols
 nomenclature, journal references for, 87
Cysteine
 symbols for atoms and bonds in side chains, 73
Cysteine β 93 SH
 pK values in human hemoglobin, 696
Cysteine desulfyhydrase
 molar absorptivity and/or absorbance values, 412
 subunit, constitution of, 342
Cysteine synthase
 molar absorptivity and/or absorbance values, 412
Cysteine synthetase
 subunit, constitution of, 345
Cystine
 destruction of, during acid hydrolysis, 206
 symbols for atoms and bonds in side chains, 73
Cytochrome b
 molar absorptivity and/or absorbance values, 412
Cytochrome b_2
 enzymatic activity, subcellular fractions, 707
 subunit, constitution of, 343

Cytochrome c
 cyanogen bromide cleavage, 199
 enzymatic activity, subcellular fractions, 702
 molar absorptivity and/or absorbance values, 413
Cytochrome c_1
 molar absorptivity and/or absorbance values, 414
Cytochrome c_2
 cyanogen bromide cleavage, 199
 molar absorptivity and/or absorbance values, 414
Cytochrome c_{550}
 molar absorptivity and/or absorbance values, 414
Cytochrome c_{551}
 cyanogen bromide cleavage, 199
 molar absorptivity and/or absorbance values, 415
Cytochrome c_{552}
 molar absorptivity and/or absorbance values, 415
Cytochrome oxidase
 molar absorptivity and/or absorbance values, 417
 subunit, constitution of, 340
 table of cofactors, molecular weights, stoichiometry and sources, 283
Cytochrome c peroxidase
 molar absorptivity and/or absorbance values, 417
Cytochromes
 definition and nomenclature, 116–120
 list of, 119
Cytocuprein
 apocytocuprein, human
  molar absorptivity and/or absorbance values, 417
 molar absorptivity and/or absorbance values, 417

D

Deoxycytidylate deaminase (T-2 bacteriophage induced)
 subunit, constitution of, 336
Deoxycytidylate deaminase
 subunit, constitution of, 328
Deoxyribonuclease
 carbohydrate content, 273
Deoxyribonuclease I
 molar absorptivity and/or absorbance values, 417
Deoxyribonuclease inhibitor
 molar absorptivity and/or absorbance values, 418
Deoxyribonuclease, pancreatic
 refractive index increments, 374
Desoxyribonuclease
 carbohydrate content, 274
Diamine oxidase
 molar absorptivity and/or absorbance values, 418
 table of cofactors, molecular weights, stoichiometry and sources, 282
Diffusion coefficients
 proteins
  carbohydrates, index to studies containing, 232
  fibrous, index to studies containing, 229
  globular, index to studies containing, 222–228
  organelles, index to studies containing, 232
  particles, index to studies containing, 232
  plasma, human, table of values, 242–250
  virus, index to studies containing, 232
 Tamm-Horsfall mucoprotein, 304
 urokinase, 302

Dihydrofolate reductase
  molar absorptivity and/or absorbance values, 418
Dihydrolipoyl transacetylase
  molar absorptivity and/or absorbance values, 419
  subunit, constitution of, 350, 351
Dipeptidase
  enzymatic activity, subcellular fractions, 698
  molar absorptivity and/or absorbance values, 419
  molecular weights, stoichiometry, and sources, 286
Dipeptidase M
  subunit, constitution of, 332
Dissociation constant
  proteinase inhibitors
    animal sources, 618–641
    plant sources, 588–596, 600
Disulfide protein group, chemical modification of
  reagents used, table of specificities, 203–204
Diuretics
  ligands binding to plasma albumin, 571
DNA polymerase
  enzymatic activity, subcellular fractions, 699, 718
DPNase
  carbohydrate content, 273
  molar absorptivity and/or absorbance values, 419
DPNase inhibitor
  carbohydrate content, 274
Drugs, ligands binding to plasma albumin, 568–574
Dyes, visible, ligands binding to plasma albumin, 564–568

E

Edestin
  refractive index increments, 374
  subunit, constitution of, 345
EDTA, inhibition of transamitase activity, 669–673
Egg whites, birds
  proteinase inhibitors from, specificities and properties, 624–629
Eggs
  glycoproteins from, carbohydrate content, 268–269
Elastase
  molar absorptivity and/or absorbance values, 419
Elastase inhibitor
  legume sources, specificity and properties, 588–589
  mammalian organs and secretions, specificities and properties, 632–641
Elastase-like enzyme
  molar absorptivity and/or absorbance values, 419
Electrophoretic mobility
  proteins
    human plasma, table of values, 242–250
    urokinase, 302
Enantiomers, definition and nomenclature, 36–58
Endo-γ-Glutamine: ϵ-lysine transferase
  transamilase activity, discussion and data, 669–683
Endogenous ligands binding to plasma albumin, 556–557
Endopolygalacturonase
  molar absorptivity and/or absorbance values, 420
Enolase
  enzymatic activity, subcellular fractions, 698, 715
  molar absorptivity and/or absorbance values, 420
  subunit, constitution of, 331, 332, 346

Enterotoxin A, B, C, E
  molar absorptivity and/or absorbance values, 420–421
Enzymatic activity
  nonmuscle myosins, data table, 319
  subcellular fractions, tables, 697–719
Enzymatic cross-linking
  discussion and data, 668–683
Enzymatic hydrolysis
  conjugated proteins, table, 211
  proteins and polypeptides, 209
Enzymatic incorporation of amines
  proteins serving as substrates, list of, 680
Enzyme kinetics
  Michaelis constant, equation and symbols, 111
  rate constants, definition, 112
  symbols used in, 111–114
Enzyme, thrombin-like
  molar absorptivity and/or absorbance values, 421
Enzymes
  classification of, 93–109
  electron-transfer proteins, nomenclature, 114–121
  genetically variant, nomenclature, 86
  glycoproteins, carbohydrate content, 273–274
  immobilization of
    absorption method, table of supports, 723
    covalent bond method, table of polymeric supports, 725–734
    entrapment method, table of supports, 724
    fixation on solid supports, multifunctional reagents for, 735
    methods used, description, 722
  immobilized, uses of, 722
  isoenzymes, definition and nomenclature of, 84–86
  isozymes, definition and nomenclature of, 84–86
  list of recommended and systematic names, 121–172
  metalloenzymes, table of cofactors, molecular weights, stoichiometry and sources, 278–286
  multiple forms, list of groups, 85
  names, list of recommended and systematic, 121–172
  nomenclature, multiple forms of, 84–86
  nomenclature, recommendations, 91–172
  proteolytic, digestion cross-linked proteins, 669
  transamidase activity, detection and sources, 670, 671
  urine, normal human, molecules weight of those found in, 301
Enzymic activity
  definition of, 109
  units of, defined, 109–110
Epimerases
  definition and nomenclature, 98
  numbering and classification of, 108
Erabutoxin a
  molar absorptivity and/or absorbance values, 421
Erythrocruorin
  molar absorptivity and/or absorbance values, 421
  subunit, constitution of, 351
Erythrocuprein
  molar absorptivity and/or absorbance values, 421, 422
  subunit, constitution of, 328
Erythropoietin
  carbohydrate content, 272–273
  molar absorptivity and/or absorbance values, 422

Esterase
  enzymatic activity, subcellular fractions, 698, 706
  molar absorptivity and/or absorbance values, 422
  subunit, constitution of, 337
Excelsin
  molar absorptivity and/or absorbance values, 422
  subunit, constitution of, 345
Exogenous ligands binding to plasma albumin, 558–574

F

Factor I-XIII, see also synonyms
  molar absorptivity and/or absorbance values, 422
  nomenclature, synonyms and symbols, 195
  subunit, constitution of, 327
  table of data and characteristics, 254–255
Fatty acid synthetase
  molar absorptivity and/or absorbance values, 422
  subunit, constitution of, 348, 350
Fatty acids, ligands binding to plasma albumin, 558–562
Ferredoxin
  cyanogen bromide cleavage, 199
  molar absorptivity and/or absorbance values, 423–424
  nomenclature recommendations, 89–90, 120
  table of cofactors, molecular weights, stoichiometry and sources, 279
Ferritin
  molar absorptivity and/or absorbance values, 425
  molecular weight, stoichiometry, source, function, 276
Fetuin
  animal sources, amino acid composition, 653
  enzymatic hydrolysis, conjugated proteins, 211
  mammalian and chicken blood, specificity and properties, 638–639
  molar absorptivity and/or absorbance values, 425
  plasma, nonhuman
    carbohydrate content, 259
Fibrin
  molar absorptivity and/or absorbance values, 425
Fibrin stabilizing factor
  disc gel electrophoresis, 678
  measurement of, graph, 677
  molar absorptivity and/or absorbance values, 425
  physical data and characteristics, 254–255
  system in blood plasma, description, 670
Fibrinogen
  molar absorptivity and/or absorbance values, 425–426
  molecular parameters, 248
  physical data and characteristics, 254–255
  plasma, human
    carbohydrate content, molecular weights, sedimentation coefficients, 258
  refractive index increments, 375
Fibrinogen, iodinated
  refractive index increments, 375
Fibrinoligase, transamidase activity, 671–681
Fibroin
  molar absorptivity and/or absorbance values, 426
Fibrous proteins
  physical-chemical data for, index to ultracentrifuge studies, 229

Ficin
  molar absorptivity and/or absorbance values, 426
Filaments of nonmuscle myosins, table of data, 318
Fischer projection formula, description, 46
Fixation of enzymes, multifunctional reagents for, 735
Flavodoxin
  molar absorptivity and/or absorbance values, 426–427
Flavoproteins
  definition and nomenclature, 116
  molar absorptivity and/or absorbance values, 428
Folic acid
  nomenclature, journal references for, 87
Follicle stimulating hormone
  molar absorptivity and/or absorbance values, 428
Formiminoglutamase
  molar absorptivity and/or absorbance values, 428
Formulas for stereochemistry, use of, 46–48
Fraction
  soluble, enzymatic activity, table, 715–719
  subcellular, enzymatic activity, tables, 697–719
Fragment
  definition, 181
Frictional coefficients
  proteins
    carbohydrate, index to studies containing, 232
    globular, index to studies containing, 232
    organelles, index to studies containing, 232
    particles, index to studies containing, 232
    virus, index to studies containing, 232
Frictional ratio
  proteins
    human plasma, table of values, 242–250
  Tamm-Horsfall mucoprotein, 304
Fructose diphosphatase
  subunit, constitution of, 336
Fructose diphosphate aldolase
  subunit, constitution of, 329
Fucose, content of, in glycoproteins, 258–267, 271–274
α-L-Fucosidase
  carbohydrate content, 274
  subunit, constitution of, 341
Fumarase
  enzymatic activity, subcellular fractions, 702
  molar absorptivity and/or absorbance values, 429
  subunit, constitution of, 340

G

α-Galactosidase
  carbohydrate content, 274
  molar absorptivity and/or absorbance values, 429
β-Galactosidase
  cyanogen bromide cleavage, 199
  molar absorptivity and/or absorbance values, 429
  subunit, constitution of, 348
Gastrin
  cyanogen bromide cleavage, 199
  molar absorptivity and/or absorbance values, 429
Gelatin
  refractive index increments, 375
Genetic factors
  immunoglobulins, human, notation for, 186

Globin
  molar absorptivity and/or absorbance values, 430
  refractive index increments, 375, 376
  stereo drawings of several, 761
Globular proteins
  physical-chemical data for, index to ultracentrifuge studies, 222–228
Globulin
  carbohydrate content, 270
  molar absorptivity and/or absorbance values, 430
  plasma, human
    carbohydrate content, molecular weights, sedimentation coefficients, 258
  subunit, constitution of, 349
α_2-Globulin
  refractive index increments, 376
β_1-Globulin
  refractive index increments, 376
γ-Globulin
  refractive index increments, 376
Glucagon
  molar absorptivity and/or absorbance values, 431
Glucose oxidase
  carbohydrate content, 273
  molar absorptivity and/or absorbance values, 431
Glucose 6-phosphate dehydrogeanse, 336
β-Glucuronidase
  carbohydrate content, 274
  molar absorptivity and/or absorbance values, 432
Glutamate decarboxylase
  molar absorptivity and/or absorbance values, 432
  subunit, constitution of, 331, 345
Glutamate dehydrogenase
  molar absorptivity and/or absorbance values, 432
  subunit, constitution of, 344, 345
Glutamic acid
  symbols for atoms and bonds in side chains, 74
Glutamic dehydrogenase
  cofactors, molecular weights, stoichiometry and sources, 284
  molar absorptivity and/or absorbance values, 432
  refractive index increments, 376
Glutamine
  symbols for atoms and bonds in side chains, 74
Glutamine synthetase
  molar absorptivity and/or absorbance values, 433
  subunit, constitution of, 345, 346, 347, 349
γ-Glutamyl transferase
  carbohydrate content, 274
Glutathione peroxidase
  molar absorptivity and/or absorbance values, 433
  molecular weight, stoichiometry and source, 285
Glutathione reductase
  molar absorptivity and/or absorbance values, 433
  subunit, constitution of, 333
Glyceraldehyde-3-phosphate dehydrogenase
  molar absorptivity and/or absorbance values, 434
  subunit, constitution of, 349
  viscosity, intrinsic, 721
D-Glyceraldehyde-3-phosphate dehydrogenase
  refractive index increments, 376
Glycerol-3-phosphate dehydrogenase
  molar absorptivity and/or absorbance values, 434

Glycocyaminekinase
  molar absorptivity and/or absorbance values, 435
Glycoproteins
  blood group substances
    carbohydrate content, 262–263
  body fluids
    carbohydrate content, 267
  cell membranes
    carbohydrate content, 266
  definition of, 257
  eggs
    carbohydrate content, 268–269
  enzymes
    carbohydrate content, 273–274
  hormones
    carbohydrate content, 271–272
  lectins, purified, table of properties, 546–551
  milk
    carbohydrate content, 268–269
  molar absorptivity and/or absorbance values, 435–436
  molecular parameters, 242
  mouse plasma, tumor
    molar absorptivity and/or absorbance values, 435
  mucins
    carbohydrate content, 262–263
  plants
    carbohydrate content, 270
  plasma, human
    carbohydrate content, molecular weights, sedimentation coefficients, 258
  plasma, nonhuman
    carbohydrate content, 259–260
  Tamm-Horsfall, human urine
    molar absorptivity and/or absorbance values, 435
  Tamm-Horsfall, rabbit urine
    molar absorptivity and/or absorbance values, 435
  Tamm-Horsfall, rat intestinal mucosa
    molar absorptivity and/or absorbance values, 435
  tissues
    carbohydrate content, 264–265
  viruses, enveloped
    carbohydrate content, 267
Glycosaminoglycans
  table of data and characteristics, 686–687
Golgi material
  enzymatic activity, table, 714
Gonadotrophin-transporting protein
  plasma, nonhuman
    carbohydrate content, 259
Growth hormone
  bovine
    pituitary gland, constitution of subunit, 327
  cyanogen bromide cleavage, 199
  molar absorptivity and/or absorbance values, 437
Guanidinyl protein group, chemical modification of reagents used, table of specificities, 203–204

H

Hageman factor
  physical data and characteristics, 254–255
  plasma, nonhuman
    carbohydrate content, 260

Haptoglobin
  molar absorptivity and/or absorbance values, 437
  plasma, human
    carbohydrate content, molecular weights, sedimentation coefficients, 258
Haptoglobulin, human
  molar absorptivity and/or absorbance values, 437
Haptoglobin 1-1, human
  molar absorptivity and/or absorbance values, 437
  subunit, constitution of, 331
Harvard system in citation of bibliographic references, 18–19
Helical segments, nomenclature rules in polypeptide chain, 76
Helicorubin
  definition and nomenclature, 118
α-Helix content
  tropomyosin from nonmuscle values, 321
Hemagglutinin
  carbohydrate content, 270
  molar absorptivity and/or absorbance values, 437
  subunit, constitution of, 327
Hemerythrin
  molar absorptivity and/or absorbance values, 437
  molecular weight, stoichiometry, source, function, 276
  refractive index increments, 376
  subunit, constitution of, 327, 334
Hemocyanin
  molar absorptivity and/or absorbance values, 438
  molecular weight, stoichiometry, source, function, 277
  subunit, constitution of, 350
Hemoglobin
  beef
    molar absorptivity and/or absorbance values, 443
  bovine
    refractive index increments, 376
  canine
    refractive index increments, 376
  horse
    molar absorptivity and/or absorbance values, 443
  human
    molar absorptivity and/or absorbance values, 444
    pK values, α-amino groups, table, 693
    pK values, cysteine β 93 SH, 696
    pK values, histidine, 690
    pK values, histidine 146 β, 690
    pK values, tyrosine, 695
    refractive index increments, 376
  human Hb, subunits
    molar absorptivity and/or absorbance values, 447
  molar absorptivity and/or absorbance values, 439–445
  mouse
    molar absorptivity and/or absorbance values, 444
  sheep
    molar absorptivity and/or absorbance values, 443
  subunit, constitution of, 328, 351
  viscosity, intrinsic, 721
Hemoglobin I
  subunit, constitution of, 326
Hemoglobin III
  subunit, constitution of, 329
Hemoglobins, synthetic
  molar absorptivity and/or absorbance values, 446

Hemopexin
  blood, human
    molar absorptivity and/or absorbance values, 447
  blood, rabbit
    molar absorptivity and/or absorbance values, 447
  molecular parameters, 246
  plasma, human
    carbohydrate content, molecular weight, sedimentation coefficient, 258
Hemovanadin, source of, 276
Heparin
  table of data and characteristics, 686
Heptoglobin
  molecular parameters, 245
Hexokinase
  enzymatic activity, subcellular fractions, 698, 702, 715
  molar absorptivity and/or absorbance values, 448
  subunit, constitution of, 333
Hexose, content of, in glycoproteins, 258
High potential, see High potential iron protein
High potential iron protein
  molar absorptivity and/or absorbance values, 448
Hirudin
  animal sources, amino acid composition, 649
  lower animal sources, specificity and properties, 620–621
Histidine
  pK values in various proteins, table, 689–692
  symbols for atoms and bonds in side chains, 74
Histidine 146 β
  pK value in human hemoglobin, 690
Histidine ammonia lyase
  molar absorptivity and/or absorbance values, 448
Histidine decarboxylase
  molar absorptivity and/or absorbance values, 448
  subunit, constitution of, 334, 340
Histidinol dehydrogenase
  subunit, constitution of, 331
  molar absorptivity and/or absorbance values, 449
Histidyl tRNA synthetase
  subunit, constitution of, 331
Histones
  calf thymus, cyanogen bromide cleavage, 199
  characterization, data table, 294
  molar absorptivity and/or absorbance values, 449
  sequences, table for several, 295–299
Hormone
  chorionic gonadotropin, human
    carbohydrate content, 271
    molar absorptivity and/or absorbance values, 450
  chorionic somatomammotropin, human
    molar absorptivity and/or absorbance values, 450
  follicle-stimulating, human
    carbohydrate content, 271
    molar absorptivity and/or absorbance values, 450
    subunit, constitution of, 326
  growth, beef pituitary
    molar absorptivity and/or absorbance values, 450
  growth, human pituitary
    molar absorptivity and/or absorbance values, 450
  growth, sheep pituitary
    molar absorptivity and/or absorbance values, 450

interstitial cell stimulating
  carbohydrate content, 271
  subunit, constitution of, 326
lactogen (MPL-2)
  molar absorptivity and/or absorbance values, 450
luteinizing
  carbohydrate content, 272–273
  molar absorptivity and/or absorbance values, 468
parathyroid
  cyanogen bromide cleavage, 199
  molar absorptivity and/or absorbance values, 450
parathyroid, pig
  molar absorptivity and/or absorbance values, 450
prolactin, sheep
  molar absorptivity and/or absorbance values, 450
Hormones
  glycoproteins from, carbohydrate content, 271–272
  peptide, list of, 177–178
  peptide, nomenclature recommendations, 175–178
Human globin
  chymotryptic hydrolysis of α-chains of, 213
Human hemoglobin
  hydrolysis peptides by leucine, 217
  hydrolysis peptides by papain, 219
Human organs and secretions
  proteinase inhibitors from, specificity and properties, 632–639
Human plasma
  glycoproteins from, carbohydrate content, 258
Human plasma albumin
  ligand binding
    drugs, list of references, 568–574
    endogenous ligands, list of references, 556–557
    fatty acids, list of references, 558–562
    inorganic anions, list of references, 563
    metal ions, list of references, 563–564
    steroids, list of references, 563–564
    visible dyes, list of references, 564–568
Human plasma proteins
  content in normal plasma, 242–250
  molecular parameters, table, 242–250
Human serum
  proteinase inhibitors from, specificity and properties, 636–641
Hyaluronic acid
  table of data and characteristics, 686
Hydrogenase
  molar absorptivity and/or absorbance values, 451
  subunit, constitution of, 328
  table of cofactors, molecular weights, stoichiometry and sources, 280
Hydrolases
  definition and nomenclature, 97
  numbering and classification of, 107
Hydrolysis
  acid, of peptides, relative rate table, 208
  acid, of proteins, data tables, 206–207
  acid, recovery of tryptophan from various proteins, 207
  chymotryptic, of α-chains of human globin, 213
  chymotryptic, proteins, conditions for, 214
  enzymatic, conjugated proteins, table, 211
  enzymatic, of proteins and polypeptides, 209
  peptic, aromatic amino acids in human α- and β-chains, 216
  peptic, leucine in human α- and β-chains, 217
  peptide bonds, specificity of chymotrypsin, 212
  peptide bonds, specificity of papain, 218
  peptide bonds, specificity of pepsin, 215
  peptide bonds, specificity of subtilisin, 221
  peptides from α- and β-chains of human hemoglobin by papain, 219
  peptides in acid solution, table, 208
  thiol esters, kinetic constants, 682
Hydroxyl protein group, chemical modification of reagents used, table of specificities, 203–204
Hydroxyproline
  symbols for atoms and bonds in side chains, 74
Hydroxypyruvate reductase
  molar absorptivity and/or absorbance values, 451
  subunit, constitution of, 329
Hypoglycemic agents, ligands binding to plasma albumin, 572
Hypothalamic factors, nomenclature, 175–177

I

Imidazole protein group, chemical modification of reagents used, table of specificities, 203–204
Immobilization of enzymes, methods and tables of supports, 722–734
Immobilized enzymes, uses of, 722
Immunoglobulin
  plasma, human
    carbohydrate content, molecular weights, sedimentation coefficients, 258
  plasma, nonhuman
    carbohydrate content, 259
Immunoglobulin IgG
  human
    cyanogen bromide cleavage, 199
  rabbit
    cyanogen bromide cleavage, 199
Immunoglobulin molecules
  formulas for, 192
  regions, definition of, 191
Immunoglobulins
  beef
    molar absorptivity and/or absorbance values, 454
  chicken
    molar absorptivity and/or absorbance values, 453
  definition, 179
  dog
    molar absorptivity and/or absorbance values, 454
  goat
    molar absorptivity and/or absorbance values, 453
  guinea pig
    molar absorptivity and/or absorbance values, 454
  horse
    molar absorptivity and/or absorbance values, 454
  human
    genetic factors, notation, 186
    molar absorptivity and/or absorbance values, 452
    nomenclature recommendations for, 173–174, 179–195
    subclasses, notation for, 184

molecular parameters, 249
mouse
  molar absorptivity and/or absorbance values, 454
pig
  molar absorptivity and/or absorbance values, 453
rabbit
  molar absorptivity and/or absorbance values, 454
rat
  molar absorptivity and/or absorbance values, 454
sheep
  molar absorptivity and/or absorbance values, 454
Immunoglobulins, specific
 γ-globulin fraction, chicken
  molar absorptivity and/or absorbance values, 456
 $γ_1$-globulin, human
  molar absorptivity and/or absorbance values, 456
 $γ_1$, guinea pig
  molar absorptivity and/or absorbance values, 456
 $γ_2$, guinea pig
  molar absorptivity and/or absorbance values, 456
 horse
  molar absorptivity and/or absorbance values, 456
 γG, horse
  molar absorptivity and/or absorbance values, 456
 γG, human
  molar absorptivity and/or absorbance values, 456
 γM, human
  molar absorptivity and/or absorbance values, 456
 γM, human, subunit
  molar absorptivity and/or absorbance values, 456
 γG, lemon shark
  molar absorptivity and/or absorbance values, 456
 γA, rabbit
  molar absorptivity and/or absorbance values, 456
 γG, rabbit
  molar absorptivity and/or absorbance values, 456
Cl-Inactivator
 carbohydrate composition, 666–667
 molar absorptivity and/or absorbance values, 399
 molecular parameters, 244
Indole protein group, chemical modification of
 reagents used, table of specificities, 203–204
Infrared spectra, see IR spectra
Inheritance, mode of
 blood coagulation proteins, 255
Inhibition constant
 enzyme kinetics, definition, 112
Inhibitor, protease, see Protease inhibitor
Inhibitors, proteinase, see Proteinase inhibitors
Inorganic anions, ligands binding to plasma albumin, 563
Insulin
 beef
  molar absorptivity and/or absorbance values, 457
 brain, bovine
  subunit, constitution of, 326
  refractive index increments, 376
  viscosity, intrinsic, 721
International Union of Immunological Sciences
 human immunoglobulins, nomenclature recommendations, 173–174

Invertase
 carbohydrate content, 274
 molar absorptivity and/or absorbance values, 458
 subunit, constitution of, 341
Iodoacetamide
 inhibition of transamidase activity, 672
IR spectra
 polysaccharides, fingerprint region, 686–687
Iron protein, see High potential iron protein
Iron proteins
 classification of, 90
Iron sulfur proteins
 nomenclature recommendations, 89–90
 table of cofactors, molecular weights, stoichiometry and sources, 278–280
Isocitrate dehydrogenase
 molar absorptivity and/or absorbance values, 458
Isoelectric point
 proteinase inhibitors
  animal sources, 618–641
  microbial sources, 661–662
  plant sources, 586–600
 proteins, human plasma, 242–250
 retinol-binding protein, 304
 Tamm-Horsfall mucoprotein, 304
Isoenzymes
 definition of, 84
 nomenclature of, 84–86
Isolectin content, purified lectins, 547–551
Isoleucine
 symbols for atoms and bonds in side chains, 73
Isomerases
 definition and nomenclature, 98
 molar absorptivity and/or absorbance values, 458
 numbering and classification of, 108
Isomerism, types of, discussion and nomenclature, 21–58
Isotopic labeling, nomenclature rules for, 16
Isozymes
 definition of, 84
 nomenclature of, 84–86
IUPAC Commission on the Nomenclature of Organic Chemistry
 rules affecting biochemical nomenclature, 3–4
IUPAC tentative rules, nomenclature in fundamental stereochemistry, 21–58
IUPAC-IUB Commission on Biochemical Nomenclature
 amino acid sequences, one-letter notation, tentative rules, 59–62
 enzyme nomenclature, recommendations, 91–172
 enzymes, nomenclature of multiple forms of, 84–86
 iron-sulfur proteins, nomenclature recommendations, 89–90
 journal citations for locating rules of, 5
 list of rules and recommendations, 3–4
 peptide hormones, nomenclature recommendations, 175–178
 peptides, natural, rules for naming synthetic modifications of, 79–83
 polypeptide chains, abbreviations and symbols, tentative rules, 63–77

J

Journal references, use of, CEBJ recommendations, 17–20

K

Kallikrein A, B
  molar absorptivity and/or absorbance values, 458–459
Kallikrein inactivator
  molar absorptivity and/or absorbance values, 459
Kallikrein inhibitor
  mammalian and chicken blood, specificity and properties, 638–639
  mammalian organs and secretions, specificity and properties, 630–631
  plant sources, amino acid composition, 606
  potato tuber sources, specificity and properties, 586–587
Katal, unit of enzymic activity, definition, 109–110
Keratinase
  molar absorptivity and/or absorbance values, 459
α-Keratose
  refractive index increments, 377
Kininogen
  cyanogen bromide cleavage, 199
  molecular parameters, 245
Kunitz inhibitor
  legume sources, specificity and properties, 588–589
  mammalian organs and secretions, specificity and properties, 630–631
  plant sources, amino acid composition, 606

L

Lac repressor
  molar absorptivity and/or absorbance values, 460
  subunit, constitution of, 338
α-Lactalbumin
  cyanogen bromide cleavage, 199
  molar absorptivity and/or absorbance values, 460
  refractive index increments, 377
β-Lactamase I, II
  molar absorptivity and/or absorbance values, 461
Lactate dehydrogenase
  molar absorptivity and/or absorbance values, 461–462
  subunit, constitution of, 337
Lactate dehydrogenase, dogfish
  structure, illustration, 766
Lactic dehydrogenase
  refractive index increments, 377
Lactoferrin
  molar absorptivity and/or absorbance values, 462
  molecular parameters, 250
β-Lactoglobulin
  molar absorptivity and/or absorbance values, 462–463
  refractive index increments, 377
  subunit, constitution of, 327
  viscosity, intrinsic, 721
Lactoperoxidase
  carbohydrate content, 274
  molar absorptivity and/or absorbance values, 463
Lactose synthetase, A protein
  molar absorptivity and/or absorbance values, 463
Lactosiderophilin lactotransferrin, human milk
  molar absorptivity and/or absorbance values, 463
Lactotransferrin
  carbohydrate content, 269
  molecular weight, stoichiometry, source, function, 276
Lectin
  carbohydrate content, 270
  subunit, constitution of, 334, 337
Lectin, purified, from various sources
  binding sites, number of, 547–551
  carbohydrate content, 546–551
  isolectins, number of, 547–551
  metal requirement, 547–551
  mitogenic activity on lymphocytes, 547–551
  molecular weight, 546–551
  specificity, 546–551
  subunits, number of, 546–551
Leghemoglobin
  molar absorptivity and/or absorbance values, 464
Legumes, proteinase inhibitors from, specificities and properties, 588–597
Legumin
  molar absorptivity and/or absorbance values, 466
  refractive index increments, 378
Leucine
  peptic hydrolysis in human α- and β-chains, 217
  symbols for atoms and bonds in side chains, 73
Leucine aminopeptidase
  enzymatic hydrolysis of proteins, 209
  molar absorptivity and/or absorbance values, 466
  molecular weights, stoichiometry and sources, 285
  subunit, constitution of, 344, 345
Ligands binding to plasma albumin, table, 556–574
Ligases
  definition and nomenclature, 98
  numbering and classification of, 108
Lipase
  carbohydrate content, 274
  enzymatic activity, subcellular fractions, 698, 711, 714
  molar absorptivity and/or absorbance values, 466
Lipids
  nomenclature, journal references for, 87
Lipoprotein, high density
  subunit, constitution of, 347
Lipoprotein, HDL_2, human
  molar absorptivity and/or absorbance values, 467
Lipoprotein very low density, human ApoLP-Val
  molar absorptivity and/or absorbance values, 467
Lipoprotein very low density, rat serum
  molar absorptivity and/or absorbance values, 467
α_1-Lipoprotein
  molecular parameters, 242
  refractive index increments, 378
α_2-Lipoprotein
  molecular parameters, 245
β-Lipoprotein
  molecular parameters, 247
  refractive index increments, 378
β_1-Lipoprotein
  refractive index increments, 378

Lipovitellin
  molar absorptivity and/or absorbance values, 467
  refractive index increments, 379
  subunit, constitution of, 347
α-Livetin
  refractive index increments, 379
β-Livetin
  refractive index increments, 379
γ-Livetin
  refractive index increments, 379
Lombricine kinase
  molar absorptivity and/or absorbance values, 467
Luciferase
  subunit, constitution of, 330, 332
Luteinizing hormone
  carbohydrate content, 272–273
  molar absorptivity and/or absorbance values, 468
Lyases
  definition and nomenclature, 98
  numbering and classification of, 108
Lysin
  molar absorptivity and/or absorbance values, 468
Lysine
  symbols for atoms and bonds in side chains, 73
Lysine decarboxylase
  molar absorptivity and/or absorbance values, 468
  subunit, constitution of, 349
β-Lysine mutase
  subunit, constitution of, 339
Lysine tRNA synthetase
  subunit, constitution of, 336
Lysosomes
  enzymatic activity, table, 711
  molecular parameters, 250
Lysozyme
  cyanogen bromide cleavage, 199
  molar absorptivity and/or absorbance values, 468, 469
  refractive index increments, 379
  viscosity, intrinsic, 721
Lysyl tRNA synthetase
  molar absorptivity and/or absorbance values, 469

M

Macroglobulin
  plasma, nonhuman
    carbohydrate content, 259
α_2-Macroglobulin
  carbohydrate composition, 666–667
  human
    carbohydrate content, molecular weight, sedimentation coefficients, 258
    molar absorptivity and/or absorbance values, 470
  molar absorptivity and/or absorbance values, 469, 470
  molecular parameters, 245
  molecular weight, stoichiometry, source, function, 277
Malate dehydrogenase
  molar absorptivity and/or absorbance values, 470
  subunit, constitution of, 328, 329, 330, 338
Malate dehydrogenase, porcine
  structure, illustration, 766
Malate-lactate transhydrogenase
  molar absorptivity and/or absorbance values, 470

Malic enzyme
  molar absorptivity and/or absorbance values, 470–471
  subunit, constitution of, 343, 344
Mammalian organs and secretions
  proteinase inhibitors from, specificity and properties, 630–635
Melilotate hydroxylase
  molar absorptivity and/or absorbance values, 471
Mercaptopyruvate sulfurtransferase
  molar absorptivity and/or absorbance values, 471
  subunit, constitution of, 326
Metal ions, ligands binding to plasma albumin, 563
Metal requirement, purified lectins, 547–551
Metalloenzymes
  cofactors, molecular weights, stoichiometry and sources, 278–286
Metalloproteins
  molecular weight, source, stoichiometry, function, 276–277
  nomenclature rules, 121
Metapyrocatechase
  molar absorptivity and/or absorbance values, 472
  table of cofactors, molecular weights, stoichiometry and sources, 280
Methadone, ligands binding to plasma albumins, 574
Methionine
  symbols for atoms and bonds in side chains, 73
Methionyl tRNA synthetase
  molar absorptivity and/or absorbance values, 472
β-Methylaspartase
  molar absorptivity and/or absorbance values, 472
Michaelis constant
  equation and symbols used, 111
Microbodies
  enzymatic activity, table, 710
β_2-Microglobulin
  amino acid sequences, 305
  molar absorptivity and/or absorbance values, 472
  properties, table, 305
Microsomes
  enzymatic activity, table, 706–709
Milk
  glycoproteins from, carbohydrate content, 268–269
Mitochondria, isolated
  enzymatic activity, table, 702–704
Mitogen
  carbohydrate content, 270
Mitogenic activity on lymphocytes, purified lectins, 547–551
Molar absorption coefficient, see Molar absorptivity
Molar absorptivity
  β_2 microglobulin, 305
  retinol-binding protein, 304
  Tamm-Horsfall mucoprotein, 304
  urokinase, 302
Molar absorptivity, UV and visible region, table of values for proteins, 383–514
Molar activity
  enzymes, defined, 109
Molecular weight
  blood coagulation proteins, table of values, 255
  contractile proteins, table of values, 306
  enzymes found in normal human urine, table, 301

glycoproteins
  plasma, human, table of values, 258
histones, table for several, 294
lectins, purified, table of values, 546–551
β_2 microglobulin, 305
myosins, nonmuscle, table of values, 317
nonmuscle actins, of monomer, 314–315
polysaccharides, table of values, 686–687
proteinase inhibitors
  animal sources, 618–641
  microbial sources, 661–662
  plant sources, 586–600
proteins
  carbohydrate, index to studies containing, 232
  fibrous, index to studies containing, 229
  globular, index to studies containing, 222–228
  human plasma, table of values, 242–250
  organelles, index to studies containing, 232
  particles, index to studies containing, 222–228
  subunits of, table, 326–351
  table of values, 326–351
  virus, index to studies containing, 232
retinol-binding protein, 304
Tamm-Horsfall mucoprotein, 304
transamidases, purified, 681
tropomyosin from nonmuscle cells, of subunits, 321
urokinase, 302
Molybdoproteins
  nomenclature rules, 121
Monellin
  molar absorptivity and/or absorbance values, 472
Monoamine oxidase
  carbohydrate content, 273, 274
  molar absorptivity and/or absorbance values, 473
  subunit, constitution of, 335, 345, 350
  table of cofactors, molecular weights, stoichiometry, and sources, 282
Mucopolysaccharides
  table of data and characteristics, 686–687
Mucoprotein
  Tamm-Horsfall, properties, of, 304
Mutases
  definition and nomenclature, 98
Myeloperoxidase
  molar absorptivity and/or absorbance values, 473
Myoglobin
  conformation for sperm whale, 760
  cyanogen bromide cleavage, 199
  molar absorptivity and/or absorbance values, 476–479
  viscosity, intrinsic, 721
Myosin
  identification of, in nonmuscle cells, table, 307–313
  interaction of nonmuscle actins with, 316
  molar absorptivity and/or absorbance values, 481
  molecular parameters, 306
  nonmuscle
    enzymatic activity of, 319
    filament data for several, 318
    interaction with action, 320
    physical properties for several, 317
  refractive index increments, 379
  viscosity, intrinsic, 721

Myosin A
  cyanogen bromide cleavage, 199

N

Nerve growth factor
  molar absorptivity and/or absorbance values, 482
  subunit, constitution of, 326
Neurotoxin
  molar absorptivity and/or absorbance values, 482
Newman projection formula, description, 46–47
Nitrate reductase
  cofactors, molecular weights, stoichiometry and sources, 285
  molar absorptivity and/or absorbance values, 484
  subunit, constitution of, 349
Nitrogenase
  molar absorptivity and/or absorbance values, 484
  subunit, constitution of, 344, 345
Nomenclature
  amino acid sequences, one-letter symbols, rules, 59–62
  biochemical, list of rules and recommendations affecting, 3–4
  biochemical, location of rules in the literature, 5
  biochemical, use of abbreviations in, 6–7
  blood coagulation factors, 195
  carbohydrate, references for rules, 87
  coenzymes, references for rules, 87
  conformational map, typical, 71
  corrinoids, references for rules, 87
  cyclitols, references for rules, 87
  cytochromes, 116–120
  electron-transfer proteins, 114–121
  enzymes, genetically variant, 86
  enzymes, multiple forms of, 84–86
  enzymes, names, list of recommended and systematic, 121–172
  enzymes, recommendations, 91–172
  folic acid, references for rules, 87
  immunoglobulins, genetic factors, 186
  immunoglobulins, human, recommendations for, 173–174, 179–195
  iron-sulfur proteins, recommendations, 89–90, 120
  isotopically labeled compounds, CEBJ rules for, 16
  lipids, references for rules, 87
  peptide hormones, recommendations, 175–178
  peptides, natural, rules for naming synthetic modifications of, 79–83
  polypeptide chains, rules for abbreviations and symbols, 63–77
  quinones, references for rules, 87
  sequence-rule procedure in fundamental stereochemistry, 50–58
  sequence rule, use of, 65–67
  stereochemistry, fundamental, IUPAC rules for nomenclature, 21–58
  steroids, references for rules, 87
  torsion angle, choice of, 66–67
  vitamins, references for rules, 87

Nonmuscle actins
  interaction of, with myosin, 316
  physical chemical properties of, 314–315
Nonmuscle cells
  proteins in, identification of several, 307–313
  tropomyosin from, properties of, 361
Nonmuscle myosins
  enzymatic activity of, data table, 319
  filaments of, data table, 318
  interaction with action, data table, 320
  physical properties of, 317
Nuclease
  cyanogen bromide cleavage, 199
  molar absorptivity and/or absorbance values, 485
Nuclei
  enzymatic activity, subcellular fraction, 698–700
Nucleic acids
  abbreviations for, use of, 7
Nucleoli
  enzymatic activity for several, table, 701
Nucleoside diphosphokinase
  subunit, constitution of, 333, 336
Nucleotides
  abbreviations for, use of, 6

O

One-letter symbols
  amino-acid sequences, journal references, 87
  amino acid sequences, tentative rules, 59–62
  amino acids, table of, 62
  use of in nomenclature, 8, 15
Organelles, protein
  physical-chemical data for, index to ultracentrifuge studies, 232
Ornithine transcarbamylase
  molar absorptivity and/or absorbance values, 485
  subunit, constitution of, 334
Orosomucoid
  molar absorptivity and/or absorbance values, 485
  molecular parameters, 242
Ovalbumin
  carbohydrate content, 268
  enzymatic hydrolysis, conjugated proteins, 211
  molar absorptivity and/or absorbance values, 485
  refractive index increments, 379
Ovalbumin A_1
  viscosity, intrinsic, 721
Ovoglycoprotein
  carbohydrate content, 268
  molar absorptivity and/or absorbance values, 486
Ovoinhibitor
  bird egg whites, specificities and properties, 628–629
  carbohydrate composition, 666–667
  molar absorptivity and/or absorbance values, 486
Ovomacroglobulin
  molar absorptivity and/or absorbance values, 486
  subunit, constitution of, 349
Ovomucin
  carbohydrate content, 268
  molar absorptivity and/or absorbance values, 486
  viscosity, intrinsic, 721
Ovomucoid
  animal sources, amino acid composition, 650
  bird egg whites, specificities and properties, 624–629
  carbohydrate composition, 666–667
  carbohydrate content, 268
  molar absorptivity and/or absorbance values, 486
  viscosity, intrinsic, 721
Ovorubin
  lower animal sources, specificity and properties, 620–621
  molar absorptivity and/or absorbance values, 486
Ovotransferrin
  molar absorptivity and/or absorbance values, 486
Oxidoreductases
  definition and nomenclature of, 96
  intramolecular, numbering and classification, 108
  numbering and classification of, 104–106

P

Pancreatin
  enzymatic hydrolysis, conjugated proteins, 211
Papain
  enzymatic hydrolysis, conjugated proteins, 211
  enzymatic hydrolysis of proteins, 209
  fragments produced by digestion with, 182
  hydrolysis peptides from α-and β-chains of human hemoglobin, 219
  molar absorptivity and/or absorbance values, 487
  specificity for hydrolysis of peptide bonds, 218
  viscosity, intrinsic, 721
Paracrystal period
  tropomyosin from nonmuscle cells, 321
Paramyosin
  molar absorptivity and/or absorbance values, 488
  molecular parameters, 306
  subunit, constitution of, 342
  viscosity, intrinsic, 721
Parathyroid hormone
  molar absorptivity and/or absorbance values, 488
Partial specific volume
  β_2 microglobulin, 305
  proteins
    carbohydrate, index to studies containing, 232
    fibrous, index to studies containing, 222–228
    globular, index to studies containing, 222–228
    organelles, index to studies containing, 232
    particles, index to studies containing, 232
    plasma, human, table of values, 242–250
    virus, index to studies containing, 232
  retinol-binding protein, 304
  Tamm-Horsfall mucoprotein, 304
  urokinase, 302
Particles, protein
  physical-chemical data for, index to ultracentrifuge studies, 232
Penicillocarboxypeptidase-S
  molar absorptivity and/or absorbance values, 488
Pepsin
  enzymatic hydrolysis, conjugated proteins, 211
  enzymatic hydrolysis of proteins, 209

molar absorptivity and/or absorbance values, 489
specificity for hydrolysis of peptide bonds, 215
refractive index increments, 380
Pepsin inhibitors
  animal sources, amino acid composition, 649
  lower animal sources, specificity and properties, 618–619
Pepsinogen
  carbohydrate content, 274
  molar absorptivity and/or absorbance values, 489
  viscosity, intrinsic, 721
Peptic hydrolysis
  aromatic amino acids in human α-and β-chains, 216
  leucine in human α-and β-chains, 217
Peptidase
  subunit, constitution of, 346
Peptide bonds
  hydrolysis, specificity of chymotrypsin, 212
  hydrolysis, specificity of papain, 218
  hydrolysis, specificity of pepsin, 215
  hydrolysis, specificity of subtilisin, 221
Peptide hormones
  list of, 177–178
  nomenclature recommendations, 175–178
Peptide linkage
  dimensions of, 742–758
Peptides
  cyanogen bromide cleavage of, table, 199–202
  hydrolysis from α-and β-chains of human hemoglobin, 219
  hydrolysis in acid solution, relative rates, 208
  natural
    abbreviations, use of, for synthetic modification, 14–15
    rules for naming synthetic modifications of, 79–83
    semitrivial names for variants, 79
  symbols for, use of, 8–15
Peroxidase
  carbohydrate content, 273
  molar absorptivity and/or absorbance values, 489, 492
  subunit, constitution of, 340
Peroxisomes
  enzymatic activity, table, 710
Phenol protein group, chemical modification of
  reagents used, table of specificities, 203–204
Phenylalanine
  symbols for atoms and bonds in side chains, 74
L-Phenylalanine ammonia lyase
  molar absorptivity and/or absorbance values, 492
Phosphatase
  molar absorptivity and/or absorbance values, 493
  subunit, constitution of, 334
Phospho*enol*pyruvate carboxykinase
  molar absorptivity and/or absorbance values, 494
Phospho*enol*pyruvate carboxylase
  molar absorptivity and/or absorbance values, 494
Phosphofructokinase
  molar absorptivity and/or absorbance values, 494
  subunit, constitution of, 349
Phosphoglucoisomerase
  molar absorptivity and/or absorbance values, 494

Phosphoglucomutase
  molar absorptivity and/or absorbance values, 494
6-Phosphogluconate dehydrogenase
  molar absorptivity and/or absorbance values, 494
  subunit, constitution of, 332, 333
Phosphoglucose isomerase
  molar absorptivity and/or absorbance values, 495
Phospholipase A_2
  molar absorptivity and/or absorbance values, 495
Phosphoribosyl transferase
  viscosity, intrinsic, 721
Phosphorylase
  enzymatic hydrolysis, conjugated proteins, 211
  molar absorptivity and/or absorbance values, 496
  subunit, constitution of, 347
Phosphorylase kinase
  molar absorptivity and/or absorbance values, 497
  subunit, constitution of, 350
Phosphorylase, purine-nucleoside
  molar absorptivity and/or absorbance values, 497
Phosphorylated compounds
  symbols for, use of, 8
Phycocyanin
  molar absorptivity and/or absorbance values, 497
  subunit, constitution of, 327, 346
Phycocyanin, anacystis
  molar absorptivity and/or absorbance values, 498
C-Phycocyanin
  molar absorptivity and/or absorbance values, 498
Phycoerythrin
  molar absorptivity and/or absorbance values, 498
Phytochrome
  molar absorptivity and/or absorbance values, 498
  subunit, constitution of, 344
Pigeon droppings
  molar absorptivity and/or absorbance, values, 498
Pigment
  molar absorptivity and/or absorbance values, 498
Pituitary hormones, nomenclature, 176–178
pK
  α-amino groups in proteins, table of values, 693
  ε-amino groups in proteins, table of values, 694
  carboxyl side chain, values estimated in lysozymes, 696
  cysteine β93 SH in human hemoglobin, 696
  histidine in
    carbonic anhydrase, values for residues, 691
    cytochrome c, specific values, 691
    hemoglobin, human, table of values, 690
    hemoglobin, human, values for His 146β, 690
    myoglobin, values for residues, 690
    nuclease, table of values, 691
    proteins, various, table of values, 692
    ribonuclease A, nonspecific values, table 689
    ribonuclease A, specific values, table, 689
  proteins, techniques for determining values in, 689
  tyrosine, table of values, 694
  tyrosine in
    hemoglobin, table of values, 695
    myoglobin, table of values, 695
Plants
  glycoproteins from, carbohydrate content, 270

Plasma
  human, basic B_2
    molar absorptivity and/or absorbance values, 498
Plasma albumin, see Albumin, plasma
Plasma concentration
  blood coagulation proteins, table of values, 255
Plasma membranes
  enzymatic activity, table, 713
Plasma proteins
  human
    refractive index increments, 380
Plasmin
  human
    molar absorptivity and/or absorbance values, 498
  human, urokinase activated
    molar absorptivity and/or absorbance values, 498
    molecular parameters, 246
Plasminogen
  human
    molar absorptivity and/or absorbance values, 499
  human, A
    molar absorptivity and/or absorbance values, 499
  human, B
    molar absorptivity and/or absorbance values, 499
    molecular parameters, 245
    physical data and characteristics, 254–255
  plasma, human
    carbohydrate content, molecular weights, sedimentation coefficients, 258
Plastocyanin
  molar absorptivity and/or absorbance values, 499
  molecular weights, stoichiometry and sources, 282
Polymerized amino acids, see Amino acids, polymerized
Polynucleotide phosphorylase
  subunit, constitution of, 340
  molar absorptivity and/or absorbance values, 499
Polypeptide chains
  blood coagulation proteins, table of values, 255
  immunoglobulin molecules, notation for, 180
  nomenclature rules, 63–77
  primary structure of a segment, definition, 77
  secondary structure of a segment, definition, 77
Polypeptides
  enzymatic hydrolysis of, 209
  symbols for, use of, 10–11
Polypeptides, cyclic
  symbols for, use of, 11
Polypeptides, synthetic
  symbols for, use of, 13–14
Polysaccharide depolymerase
  molar absorptivity and/or absorbance values, 500
  subunit, constitution of, 347
Polysaccharides
  connective tissue, table of data and characteristics, 686–687
Postalbumin, 4.6S
  molar absorptivity and/or absorbance values, 500
  molecular parameters, 242
Potato tuber sources, proteinase inhibitors from, specificities and properties, 586–587

Prealbumin
  molar absorptivity and/or absorbance values, 500
  molecular parameters, 242
  subunit, constitution of, 328
Prekallikrein
  molecular parameters, 248
  physical data and characteristics, 254–255
Proaccelerin
  physical data and characteristics, 254–255
Procaine, ligands binding to plasma albumins, 574
Procarboxypeptidase A
  molar absorptivity and/or absorbance values, 500
Procarboxypeptidase B
  molar absorptivity and/or absorbance values, 501
Progesterone
  plasma, nonhuman
    carbohydrate content, 260
Progesterone-binding globulin
  molar absorptivity and/or absorbance values, 501
Proinsulin
  molar absorptivity and/or absorbance values, 501
Prolactin
  molar absorptivity and/or absorbance values, 501
Proline
  symbols for atoms and bonds in side chains, 74
Pronase
  enzymatic hydrolysis, conjugated proteins, 211
Properdin
  molecular parameters, 250
Protease
  cofactors, molecular weights, stoichiometry, sources, 284
  molar absorptivity and/or absorbance values, 501
Protease, acid
  molar absorptivity and/or absorbance values, 502
Protease inhibitor
  molar absorptivity and/or absorbance values, 457, 503,
Protein
  primary structure of a segment, definition, 77
  quaternary structure of a molecule, definition, 77
  tertiary structure of a molecule, definition, 77
Proteinase inhibitors
  animal sources
    amino acid composition, table, 649–653
    amino acid sequences, table, 656–659
    carbohydrate content, 618–628
    dissociation constant, 618–641
    isoelectric point, 618–641
    molecular weight, 618–641
    specificity, 618–641
  carbohydrate composition of selected, 666–667
  microbial sources
    amino acid composition, 664
    amino acid sequence, 665
    isoelectric point, 661–662
    molecular weight, 661–662
    specificity, 661–662
  plant sources
    amino acid composition, table, 605–610
    amino acid sequences, 611–617

carbohydrate content, 598–599
dissociation constant, 588–596, 600
isoelectric point, 586–600
molecular weight, 586–600
specificity, 586–600
Proteinase inhibitor I
    subunit, constitution of, 327
Protein composition
    Tamm-Horsfall glycoprotein, 303
Protein, high density
    subunit, constitution of, 341
Protein kinase
    subunit, constitution of, 344
Protein P11
    subunit, constitution of, 329
Mo-Fe Protein
    subunit, constitution of, 340
C-protein
    molecular parameters, 306
M-proteins
    molecular parameters, 306
Protein toxin B
    subunit, constitution of, 335
Protein phosphokinase
    subunit, constitution of, 337
Proteins, see also individual proteins
    absorbance, UV and visible region, table
        of values, 383–514
    blood coagulation, table of data and characteristics,
        254–255
    carbohydrate content tables for glycoproteins
        from various sources, 257–274
    carbohydrate linkages, types of, 257
    carbohydrates
        physical-chemical data for, index to
            ultracentrifuge studies, 232
    chymotryptic hydrolysis, conditions for, 214
    contractile, molecular parameters, 306
    cyanogen bromide cleavage of, table, 199–202
    electron-transfer, nomenclature, 114–121
    enzymatic cross-linking, 669–683
    enzymatic hydrolysis of, 209
    fibrous, physical-chemical data for, index to
        ultracentrifuge studies, 229
    globular, physical-chemical data for, index to
        ultracentrifuge studies, 222–228
    histones, sequence and characterization tables, 294–299
    human plasma, molecular parameters, table, 242–250
    hydrolysis of, tables of data for several, 206–221
    iron, classification of, 90
    iron-sulfur, nomenclature recommendations, 89–90,
        120
    metalloproteins, table of molecular weights, sources,
        stoichiometry and function, 276–277
    modification of, specificities of reagents used, 203–204
    molar absorptivity, UV and visible region, table
        of values, 383–514
    nonmuscle cells, identification of several in, 307–313
    organelles
        physical-chemical data for, index to ultra-
            centrifuge studies, 232
    particles, physical-chemical data for, index to
        ultracentrifuge studies, 232

pK values for several amino acids in, 689–696
refractive index increments, table, 372–380
retinol-binding protein, properties, 304
serum vertebrate
    sedimentation coefficients for, 230–231
structures, drawings of several, 760–766
substrates for enzymatic incorporation of amines, 680
subunit constitution of, table, 326–351
symbols for, use of, 8–15
techniques for determining pK values in, 689
viruses, physical-chemical data for, index to
    ultracentrifuge studies, 232
viscosity, intrinsic, table for several in native
    and denatured states, 721
Prothrombin
    beef
        molar absorptivity and/or absorbance values, 503
    human
        molar absorptivity and/or absorbance values, 503
    physical data and characteristics, 254–255
    plasma, human
        carbohydrate content, molecular weights,
            sedimentation coefficients, 258
    plasma, nonhuman
        carbohydrate content, 260
Putrescine oxidase
    molar absorptivity and/or absorbance values, 504
Pyrocatechase
    molar absorptivity and/or absorbance values, 504
    table of cofactors, molecular weights, stoichiometry
        and sources, 278
Pyruvate carboxylase
    subunit, constitution of, 341, 348, 349
Pyruvate dehydrogenase
    molar absorptivity and/or absorbance values, 504
    subunit, constitution of, 332, 338
Pyruvate dehydrogenase complex
    subunit, constitution of, 351
Pyruvate kinase
    molar absorptivity and/or absorbance values, 504
    subunit, constitution of, 340, 341, 342, 343

Q

Quinones
    nomenclature, journal references for, 87

R

Racemases
    definition and nomenclature, 98
    numbering and classification of, 108
Radio-opaque agents, ligands binding to plasma albumins,
    572
C-Reactive protein
    molar absorptivity and/or absorbance values, 409
    subunit, constitution of, 337
Reagents
    protein modification, table of specificities for those in
        common use, 203–204
Refractive index increments, table for proteins, 372–380

Relaxing protein
  molar absorptivity and/or absorbance values, 505
Retinol-binding protein
  human, amino acid composition, 303
  molar absorptivity and/or absorbance values, 505
  properties of, 304
Rheumatoid factor
  plasma, human
    carbohydrate content, molecular weights, sedimentation coefficients, 258
Rhodanese
  molar absorptivity and/or absorbance values, 505
  subunit, constitution of, 327
Rhodopsin
  molar absorptivity and/or absorbance values, 505
Ribitol dehydrogenase
  molar absorptivity and/or absorbance values, 505
  subunit, constitution of, 334
Ribonuclease
  bovine
    cyanogen bromide cleavage, 199
  bovine pancreatic
    cyanogen bromide cleavage, 199
    carbohydrate content, 273, 174
  cross-linked bovine pancreatic
    cyanogen bromide cleavage, 199
  enzymatic activity, subcellular fractions, 701
  molar absorptivity and/or absorbance values, 505, 506
  refractive index increments, 380
  viscosity, intrinsic, 721
Ribonuclease A
  cyanogen bromide cleavage, 199
Ribonucleotide-diphosphate reductase
  molar absorptivity and/or absorbance values, 506
Ribonucleotide reductase
  subunit, constitution of, 347
L-Ribulokinase
  molar absorptivity and/or absorbance values, 506
  subunit, constitution of, 333
RNA polymerase
  enzymatic activity, subcellular fractions, 700, 701
  molar absorptivity and/or absorbance values, 504
  subunit, constitution of, 347–350
Rubredoxin
  molar absorptivity and/or absorbance values, 506
  table of cofactors, molecular weights, stoichiometry and sources, 278
Rubredoxin reductase
  molar absorptivity and/or absorbance values, 507
Rubredoxins
  nomenclature recommendations, 89–90, 120

S

Saccharin, ligands binding to plasma albumins, 574
Sedimentation coefficient
  glycoproteins
    plasma, human, table of values, 258
  intrinsic, contractile proteins, 306
  β_2 microglobulin, 305
  myosins, nonmuscle, table of values, 317
  proteins
    carbohydrates, index to studies containing, 232
    fibrous, index to studies containing, 229
    globular, index to studies containing, 222–228
    organelles, index to studies containing, 232
    particles, index to studies containing, 232
    plasma, human, table of values, 242–250
    serum, vertebrate, table of values, 230–231
    virus, index to studies containing, 232
  retinol-binding protein, 304
  Tamm-Horsfall mucoprotein, 304
  urokinase, 302
Sequence data
  histones, table for several, 295–299
Sequence rule, use of in nomenclature, 65–67
Sequence-rule procedure in nomenclature of fundamental stereochemistry, 50–58
Serine
  destruction of, during acid hydrolysis, 206
  symbols for atoms and bonds in side chains, 73
Serine dehydratase
  molar absorptivity and/or absorbance values, 507
  subunit, constitution of, 328
D-Serine dehydratase
  molar absorptivity and/or absorbance values, 507
Serum albumin
  viscosity, intrinsic, 721
Serum content
  retinol-binding protein, 304
Serum proteins
  vertebrate, sedimentation coefficients, table, 230–231
Shape, Tamm-Horsfall mucoprotein, 304
Sialic acid, content of, in glycoproteins, 258–269, 271–274
Siderophilin
  molar absorptivity and/or absorbance values, 507
Soybean trypsin inhibitor
  refractive index increments, 380
Specific activity
  enzymes, defined, 109
Specific rotation
  polysaccharides, table of values, 686–687
Specificity
  lectins, purified, table of values, 546–551
  protein modification, table for reagents used in, 203–204
  proteinase inhibitors
    animal sources, 618–641
    microbial sources, 661–662
    plant sources, 586–600
  trypsin toward synthetic substrates, 211
Spectrin, human erythrocytes
  molar absorptivity and/or absorbance values, 508
Sperm whale myoglobin
  conformation, three-dimensional, 760
Stellacyanin
  carbohydrate content, 270
  molar absorptivity and/or absorbance values, 508
Stereochemistry, fundamental, IUPAC rules for nomenclature, 21–58
Stereo drawings of several proteins, 761, 763, 766
Stereoformulas, use of, 46–48

Steroids
  ligands binding to plasma albumin, 563
  nomenclature, journal references for, 87
Stokes radius
  myosins, nonmuscle, table of values, 317
Streptokinase
  cyanogen bromide cleavage, 199
  molar absorptivity and/or absorbance values, 508
Structures
  proteins, drawings of several, 760–766
Stuart factor
  physical data and characteristics, 254–255
Substrate concentration
  enzyme kinetics, definition, 112
Substrate constant
  enzyme kinetics, definition, 112
Substrate specificity
  myosins, nonmuscle, table of values, 319
Subtilisin
  conformation, stereo pair illustrating, 763
  enzymatic hydrolysis, conjugated proteins, 211
  enzymatic hydrolysis of proteins, 209
  molar absorptivity and/or absorbance values, 508
  specificity for hydrolysis of peptide bonds, 221
Subtilisin, Thiol-
  molar absorptivity and/or absorbance values, 508
Subtilisin BPN'
  molar absorptivity and/or absorbance values, 508
Subunit composition
  myosins, nonmuscle, table of values, 317
Subunit content
  lectins, purified, 546–551
Subunit constitution, proteins
  index, alphabetical, to table, 352–360
  table of data, 326–351
Subunit size
  Tamm-Horsfall mucoprotein, 304
Subunit structures
  dogfish lactate dehydrogenase, illustration, 766
  porcine malate dehydrogenase, illustration, 766
  transamidases, purified, 681
Succinyl-CoA synthetase
  molar absorptivity and/or absorbance values, 508
  subunit, constitution of, 337
Sugar, neutral, content of in glycoproteins, 259–274
S-Sulfopepsin
  cyanogen bromide cleavage, 199
Sulfhydryl protein group, chemical modification of
  reagents used, table of specificities, 203–204
Sulfonamides
  ligands binding to plasma albumin, 568
Superoxide dismutase
  subunit, constitution of, 326, 327
Supports, tables of, used in immobilization of enzymes, 723–734
Symbols
  L-amino acids, side chains of, 73–75
  chemical names, journal reference to rules, 87
  distinguished from abbreviations, 7
  enzyme kinetics, description of, 111–114
  nucleic acids, journal reference to, 87
  polynucleotides, journal references to, 87
  polypeptide chains, conformation, journal reference, 87
  polypeptide chains, nomenclature rules, 63–77
  use of, in biochemical nomenclature, 7–15
Synthetases
  definition and nomenclature, 98
  numbering and classification of, 108
Synthetic polypeptides, see Polypeptides, synthetic

T

T-2 Tail sheath
  refractive index increments, 380
Taka-amylase A
  carbohydrate content, 247
Tamm-Horsfall glycoprotein
  amino acid composition, 303
  carbohydrate composition, 303
  carbohydrate content, 267
  protein composition, 303
Tamm-Horsfall mucoprotein, see Mucoprotein, Tamm-Horsfall
Tartronic semialdehyde reductase
  molar absorptivity and/or absorbance values, 509
Taurocyamine kinase
  molar absorptivity and/or absorbance values, 509
  subunit, constitution of, 331
Tautomerases
  definition and nomenclature, 98
Tetanus toxin
  molar absorptivity and/or absorbance values, 509
Thermolysin
  cofactors, molecular weights, stoichiometry, sources, 284
  conformation, drawing illustrating, 764
  molar absorptivity and/or absorbance values, 509
Thio ether protein group, chemical modification of
  reagents used, table of specificities, 203–204
Thiocholine
  enhancement of the steady state formation of, 683
Thiol esters
  hydrolysis, kinetic constants, 682
Thioredoxin
  cyanogen bromide cleavage, 199
  molar absorptivity and/or absorbance values, 509
Thioredoxin reductase
  molar absorptivity and/or absorbance values, 509
  subunit, constitution of, 329
Three-letter symbols
  amino acids, table of, 62
  use of, in nomenclature, 8
Threonine
  destruction of, during acid hydrolysis, 206
  symbols for atoms and bonds in side chains, 73
Threonine deaminase
  molar absorptivity and/or absorbance values, 509
  subunit, constitution of, 338, 340, 341
Thrombin
  beef
    molar absorptivity and/or absorbance values, 510
  human
    molar absorptivity and/or absorbance values, 510
  zymogen conversion during coagulation of plasma, 670
Thromboplastin
  physical data and characteristics, 254–255

Thyrocalcitonin
  cyanogen bromide cleavage, 199
Thyroglobulin
  carbohydrate content, 271–272
  cyanogen bromide cleavage, 199
  molar absorptivity and/or absorbance values, 510
  refractive index increments, 380
  subunit, constitution of, 349
  viscosity, intrinsic, 721
Thyroid stimulating hormone
  beef
    molar absorptivity and/or absorbance values, 510
  human
    molar absorptivity and/or absorbance values, 510
Thyrotrophin
  carbohydrate content, 271–272
Torsion angle
  choice of, in nomenclature, 66–67
  designation of, 75
  peptides of L-amino acids, table of, 69
  principal, use of, 68–75
  regular structures, table, 70
Toxin
  animal sources, amino acid composition, 650
  lower animal sources, specificity and properties, 622–623
Tranquilizers
  ligands binding to plasma albumin, 571
Transamidase activity
  detection of, 670
  inhibition of, 669–674
  sources of, 671
  tissues, list of references, 680
Transamidases
  molecular weight, subunit structure, 681
Trauscortin
  molecular parameters, 242
  plasma, human
    carbohydrate content, molecular weight, sedimentation coefficients, 258
    molecular absorptivity and/or absorbance values, 510
Transferases
  definition and nomenclature of, 96–97
  intramolecular, numbering and classification of, 108
  numbering and classification of, 106
Transferrin
  carbohydrate content, 268
  molecular parameters, 246
  molecular weight, stoichiometry, source, function, 276
  plasma, human
    carbohydrate content, molecular weights, sedimentation coefficients, 258
    molar absorptivity and/or absorbance values, 510
  plasma, nonhuman
    carbohydrate content, 260
Tropomyosin
  molecular parameters, 306
  nonmuscle cells, properties, 321
  refractive index increments, 380
  viscosity, intrinsic, 721

Tropomyosin B
  molar absorptivity and/or absorbance values, 510
  refractive index increments, 380
  subunit, constitution of, 329
Troponin
  molar absorptivity and/or absorbance values, 510
  molecular parameters, 306
  subunit, constitution of, 331
Trypsin
  enzymatic hydrolysis, conjugated proteins, 211
  enzymatic hydrolysis of proteins, 209
  molar absorptivity and/or absorbance values, 510
  refractive index increments, 380
  specificity toward synthetic substrates, 211
Trypsin inhibitor
  animal sources, amino acid composition, 649
  beef pancreas
    molar absorptivity and/or absorbance values, 511
  inter-α, human
    molar absorptivity and/or absorbance values, 511
  legume and grain sources, specificity and properties, 590–600
  lower animal sources, specificity and properties, 618–621
  mammalian organs and secretions, specificities and properties, 632–641
  pig pancreas
    molar absorptivity and/or absorbance values, 511
  plant sources, amino acid composition, 607–609
  soybean
    molar absorptivity and/or absorbance values, 511
Trypsin inhibitors
  carbohydrate composition, 666–667
Trypsin-Trypsin inhibitor complex
  molar absorptivity and/or absorbance values, 511
Trypsinogen
  bovine pancreatic
    cyanogen bromide cleavage, 199
  molar absorptivity and/or absorbance values, 511
  refractive index increments, 380
Tryptophan
  recovery from various proteins after acid hydrolysis, 207
  symbols for atoms and bonds in side chains, 75
Tryptophanase
  molar absorptivity and/or absorbance values, 512
  subunit, constitution of, 341, 342
Tryptophan oxygenase
  molar absorptivity and/or absorbance values, 511
  subunit constitution of, 335, 339
Tryptophan synthetase
  cyanogen bromide cleavage, 199
  molar absorptivity and/or absorbance values, 512
  subunit, constitution of, 337, 338
Tryptophanyl tRNA synthetase
  molar absorptivity and/or absorbance values, 512
  subunit, constitution of, 330, 334
Tyrosinase
  molecular weights, stoichiometry and sources, 282
Tyrosine
  destruction of, during acid hydrolysis, 206
  pK values in various proteins, tables, 694–695
  symbols for atoms and bonds in side chains, 75

U

Umecyanin
  molar absorptivity and/or absorbance values, 512
  molecular weights, stoichiometry and sources, 282
Urease
  molar absorptivity and/or absorbance values, 513
  subunit, constitution of, 348
Uricase
  enzymatic activity, subcellular fractions, 699
  molar absorptivity and/or absorbance values, 513
  molecular weights, stoichiometry and sources, 282
  subunit, constitution of, 336
Urine
  human
    enzymes found in, molecular weights, 301
Urocanase
  molar absorptivity and/or absorbance values, 513
  subunit, constitution of, 334
Urokinase
  amino acid composition, 302
  human urine
    molar absorptivity and/or absorbance values, 513
  human placenta
    molar absorptivity and/or absorbance values, 513
  properties, 302
Urokinase inhibitor
  mammalian organs and secretions, specificities and properties, 632–641

V

Valine
  symbols for atoms and bonds in side chains, 73

Virus, protein
  molar absorptivity and/or absorbance values, 514
  physical-chemical data for, index to ultracentrifuge studies, 232
Viscosity
  intrinsic
    contractile proteins, 306
    polysaccharides, table of values, 686–687
    proteins, human plasma, 242–250
    proteins in native and denatured states, 721
    Tamm-Horsfall mucoprotein, 304
Visual pigment
  molar absorptivity and/or absorbance values, 514
Vitamins
  abbreviations for, use of, 6
  nomenclature, journal references for, 87
Volume, partial specific, see Partial specific volume

W

Warfarin, ligands binding to plasma albumin, 572

X

Xanthine oxidase
  molar absorptivity and/or absorbance values, 514
  table of cofactors, molecular weights, stoichiometry, and sources, 278

Z

Zymogen
  conversion during coagulation of plasma, 670
Zymogen granules
  enzymatic activity, table, 714